NONLOCAL NONLINEAR FRACTIONAL-ORDER BOUNDARY VALUE PROBLEMS

Other World Scientific Titles by the Author

Quantum Calculus: New Concepts, Impulsive IVPs and BVPs, Inequalities
ISBN: 978-981-3141-52-0

Boundary Value Problems for Fractional Differential Equations and Systems
ISBN: 978-981-122-445-4

NONLOCAL NONLINEAR FRACTIONAL-ORDER BOUNDARY VALUE PROBLEMS

Bashir Ahmad
King Abdulaziz University, Saudi Arabia

Sotiris K Ntouyas
University of Ioannina, Greece

World Scientific

NEW JERSEY • LONDON • SINGAPORE • BEIJING • SHANGHAI • HONG KONG • TAIPEI • CHENNAI • TOKYO

Published by

World Scientific Publishing Co. Pte. Ltd.
5 Toh Tuck Link, Singapore 596224
USA office: 27 Warren Street, Suite 401-402, Hackensack, NJ 07601
UK office: 57 Shelton Street, Covent Garden, London WC2H 9HE

Library of Congress Cataloging-in-Publication Data
Names: Ahmad, Bashir (Mathematics professor), author. | Ntouyas, Sotiris, 1950– author.
Title: Nonlocal nonlinear fractional-order boundary value problems / Bashir Ahmad, Sotiris K. Ntouyas.
Description: New Jersey : World Scientific, [2021] | Includes bibliographical references and index.
Identifiers: LCCN 2020048415 | ISBN 9789811230400 (hardcover) |
ISBN 9789811230417 (ebook for institutions) | ISBN 9789811230424 (ebook for individuals)
Subjects: LCSH: Nonlinear boundary value problems. | Fractional differential equations.
Classification: LCC QA379 .A355 2021 | DDC 515/.355--dc23
LC record available at https://lccn.loc.gov/202004841

British Library Cataloguing-in-Publication Data
A catalogue record for this book is available from the British Library.

For any available supplementary material, please visit
https://www.worldscientific.com/worldscibooks/10.1142/12102#t=suppl

Desk Editor: Liu Yumeng

Preface

Nonlinear boundary value problems extensively appear in the mathematical modeling of several real world phenomena [155]. The topic of boundary value problems is an important and interesting area of investigation in view of its extensive applications in diverse disciplines of applied sciences and engineering. The point of central importance in the study of nonlinear boundary value problems is to examine the effect of nonlinearity on the solutions of the associated problem. The literature on the topic is much enriched and it ranges from its theoretical development to the analytic and numerical methods for solving boundary value problems. Classical boundary conditions fail to describe certain changes in the processes happening inside the domain. This led to the birth of nonlocal conditions [71,99], which connect the boundary values of the unknown functions to their values at some interior positions of the domain. Computational fluid dynamics (CFD) technique are directly related to the boundary data [122] and the assumption of circular cross-section in the fluid flow problems cannot be justified in many cases. The concept of integral boundary conditions acts as a natural tool to resolve this issue as such conditions can be applied to arbitrary shaped structures. Integral boundary conditions are also found to be quite useful in the study of thermal conduction and hydrodynamic problems [76,79,100]. In heat conduction problems, the concept of non-uniformity can be relaxed by using the integro multi-point boundary conditions, which can accommodate the non-uniformities in form of points or sub-segments on the heat sources. In fact, the integro multi-point conditions relate the sum of the values of the unknown function (e.g., temperature) at the nonlocal positions (points and sub-segments) and the value of the unknown function over the given domain. Integro multi-point conditions are also useful in the scattering problems when scattering boundary consists of finitely many sub-strips (finitely many edge-scattering problems). For details and applications in engineering problems, see [3,134,142,148].

The subject of fractional calculus, initiated with the speculations of Leibniz (1697) and Euler (1730) about fractional-order derivatives, has developed into an important branch of mathematical analysis. During the last few decades, the tools of fractional calculus have played an important role in improving the mathematical modeling of many issues related to applied and technical sciences. Fractional-order models are considered to be more realistic and informative than their integer-order counterparts. An interesting feature of fractional-order operators is their nonlocal nature that accounts for hereditary characteristics of many materials and processes in contrast to the corresponding integer-order derivative operators. Examples include long-time memory in financial time series [132], transport dynamics in systems governed by anomalous diffusion [97], transport of passive tracers carried by fluid flow in a porous medium in groundwater hydrology [136], two-species fractional reaction-diffusion system [95], fractional dynamics [110], chaotic neuron model [121], self-similar processes like protein dynamics [90], fractional quantum fields [117], visco-elastic behavior [96], etc.

Numerous applications of fractional calculus motivated many mathematicians to develop the theoretical aspect of this interesting branch of mathematical analysis. In particular, boundary value problems of fractional differential equations and inclusions supplemented with classical, local and nonlocal, and integral type boundary conditions have been addressed by many researchers. Differential inclusions (generalization of differential equations and inequalities) serve as an important tool in the investigation of dynamical systems, stochastic processes, optimal control, queuing networks, finance, climate control, etc., for details, see the text [109]. Differential equations containing more than one fractional order derivative operator, such as Bagley-Torvik equation [140] and Basset equation [120], also constitute an important area of investigation in view of their application in the modeling of motion of a rigid plate immersed in a Newtonian fluid. In the study of variational principles, fractional differential equations involving both left and right fractional derivative operators give rise to a special class of Euler–Lagrange equations, for details, see [70] and the references cited therein. Other applications of fractional derivative operators include steady heat-transfer in fractal media [147], electromagnetic waves phenomena in a variety of dielectric media with susceptibility [141], etc.

There has also been shown a great interest in the fractional differential systems in view of the occurrence of such systems in the mathematical models of physical and engineering problems. Examples include anomalous diffusion [138], disease models [77,84,126,129], synchronization of chaotic systems [87,89], ecological models [104], etc. In a recent survey [83], one can find a detailed treatment of the fractional analogue of Bhalekar–Gejji system, Lorenz system, Liu system, Chen system and Rössler system as a characteristic representative of fractional order autonomous dynamical systems.

The main objective for writing this book is to present some recent results on single-valued and multi-valued boundary value problems involving different kinds of fractional derivative and integral operators, and several kinds of nonlocal multi-point, integral, integro-differential boundary conditions. Much of the content of this book contains the recent research published by the authors on the topic. The book is organized as follows.

Chapter 1 contains fundamental background material for multivalued analysis, differential inclusions and fractional calculus. Also, we enlist a number of fixed point theorems due to Banach, Boyd and Wong, Covitz and Nadler, Dhage, Krasnosel'skii (single and multi-valued maps), Amini-Harandi (Endpoint fixed point theorem), Guo-Krasnoselskii, Krasnosel'skii-Zabreiko, Leray-Schauder alternative, Leggett-Williams, Leray-Schauder nonlinear alternative for single and multi-valued maps, O'Regan, Petryshyn, Sadovski, Schauder, Schaefer and Wegrzyk. These tools of fixed point theory play a key role in obtaining the existence results presented in this book.

Chapter 2 is devoted to the study of fractional-order single-valued and multivalued boundary value problems involving different kinds of nonlocal boundary conditions. In Section 2.1, existence results for four-point nonlocal boundary value problems of nonlinear integro-differential equations and inclusions of fractional order $q \in (1, 2]$ are obtained, while Section 2.2 deals with a three-point nonlocal boundary value problem of nonlinear integro-differential equations of arbitrary order. In Section 2.3, we present the existence results for a new class of three-point nonlocal boundary value problems of mixed fractional differential equations and inclusions involving both left Caputo and right Riemann-Liouville fractional derivative operators. Section 2.4 is concerned with the solvability of a differential equation involving right-Caputo and left-Riemann-Liouville fractional derivative operators of different orders and right-left Riemann-Liouville fractional integrals, supplemented with four-point nonlocal boundary conditions. In Section 2.5, we discuss the existence of solutions for hybrid fractional differential equations and inclusions equipped with nonlocal boundary data. Section 2.6 is concerned with the existence and uniqueness results for a nonlinear nonlocal boundary value problem of coupled fractional differential equations, while Section 2.7 contains the sufficient criteria for the existence of solutions for a coupled system of mixed-order fractional differential equations equipped with nonlocal multi-point coupled boundary conditions.

In Chapter 3 we investigate nonlocal boundary value problems involving different kinds of integral boundary conditions. Section 3.1 is concerned with the existence of solutions for a boundary value problem of fractional differential equations and inclusions of arbitrary order equipped with four-point integral boundary conditions. In Section 3.2, we discuss the existence of solutions for fractional differential equations and inclusions subject to four-point nonlocal Riemann-Liouville fractional integral boundary conditions. Section 3.3 deals with a nonlocal boundary value problems of fractional integro-differential equations and inclusions supplemented with nonlocal strip boundary conditions. Section 3.4 studies the solvability of a coupled system of fractional differential equations with nonlocal and integral boundary conditions. The existence of solutions for a nonlinear neutral Liouville-Caputo-type fractional differential equation with Riemann-Liouville integral boundary conditions is discussed in Section 3.5. In Section 3.6, we investigate a coupled system of nonlocal fractional differential equations equipped with coupled and uncoupled slit-strip type boundary conditions. The existence results for fractional differential equations and inclusions subject to nonlocal nonlinear flux type integral boundary conditions are studied in Section 3.7. We present the solvability criteria for a coupled system of fractional differential equations equipped with flux type nonlocal nonlinear integral boundary conditions in Section 3.8. In Section 3.9, we introduce the concept of nonlocal Erdélyi-Kober type integral boundary conditions and solve fractional differential equations and inclusions equipped with these conditions. Existence results for Riemann-Liouville fractional differential inclusions supplemented with multi-strip boundary conditions involving Erdélyi-Kober fractional integrals are obtained in Section 3.10, while the case of perturbed fractional differential inclusions with nonlocal multi-point Erdélyi-Kober fractional integral boundary conditions is studied in Section 3.11.

Chapter 4 deals with the existence of solutions for fractional boundary value problems of differential equations with different kinds of boundary conditions involving fractional derivatives. In Section 4.1, we obtain existence results for Riemann-Liouville fractional integro-differential equations with fractional nonlocal integral boundary conditions. Section 4.2 contains existence results for differential equations and inclusions involving mixed Riemann-Liouville and Caputo fractional derivative operators, and equipped with nonlocal fractional integro-differential boundary conditions. Section 4.3 is concerned with the existence and uniqueness of solutions for a fractional order nonlinear coupled system containing mixed Riemann-Liouville and Caputo fractional derivatives, supplemented with integro-differential boundary conditions. In Section 4.4, we investigate the existence criteria for solutions of systems of Caputo and Riemannn-Liouville coupled fractional differential equations and inclusions of different orders subject to coupled integral fractional boundary conditions. Section 4.5 presents the existence criteria for solutions of a nonlinear fractional integro-differential equation and fractional differential inclusions equipped with three-point nonlocal fractional boundary conditions. In Section 4.6, we obtain some existence results for an integro-differential equation of fractional order with nonlocal three-point boundary conditions containing fractional derivatives of different orders. Section 4.7 is concerned with the existence and uniqueness of solutions for a coupled system of nonlinear fractional differential equations equipped with uncoupled multi-point nonlocal boundary conditions involving fractional derivatives on an unbounded domain.

In Chapter 5, we discuss the existence and uniqueness results for different kinds of nonlocal fractional-order boundary value problems with discrete and integral boundary conditions. In Section 5.1, we prove some existence theorems for a semilinear Caputo fractional differential equation equipped with nonlocal and discrete and integral boundary conditions. Section 5.2 studies nonlinear fractional differential equations and inclusions of arbitrary order complemented with Riemann-Liouville type multi-strip boundary conditions. Section 5.3 is concerned with the existence and uniqueness of solutions for a nonlinear system of coupled fractional differential equations equipped with nonlocal coupled integro-multistrip-multipoint boundary conditions. In Section 5.4, we study a boundary value problem of fractional differential equations involving mixed nonlinearities, and nonlocal multi-point and Riemann-Stieltjes integral-multi-strip boundary conditions. The existence results for a multi-term fractional differential equation equipped with nonlocal multi-point and strip boundary conditions are derived in Section 5.5. Section 5.6 investigates a fully coupled two-parameter system of sequential fractional integro-differential equations supplemented with nonlocal integro-multipoint boundary conditions. Section 5.7 is concerned with the existence of solutions for Caputo type sequential fractional differential equations and inclusions complemented with nonlocal integro-multipoint boundary conditions. In Section 5.8, we discuss the existence and uniqueness of solution for a coupled system of neutral fractional order differential equations containing mixed Caputo and Riemann-Liouville sequential fractional derivative operators, supplemented with nonlocal multi-point and Riemann-Stieltjes integral multi-strip conditions. Section 5.9 is concerned with the study of an abstract nonlinear multi-term Caputo fractional differential equation with nonlinearity depending on the unknown function together with its lower-order Caputo fractional derivative operator subject to Riemann-Stieltjes integro-multipoint boundary conditions.

Chapter 6 is concerned with the existence of positive solutions for nonlocal fractional order boundary value problems. In Section 6.1, we apply a monotone iterative method to obtain positive extremal solutions for a fractional differential equation with the nonlinear term depending on lower-order fractional derivative operator on a half line. Section 6.2 deals with the existence and approximation of positive solutions for a nonlinear Riemann-Liouville fractional differential equation with nonlocal fractional integro-differential boundary conditions on an unbounded domain via a monotone iterative procedure. Section 6.3 studies a new class of boundary value problems for one-dimensional lower-order nonlinear Hadamard fractional differential equations equipped with nonlocal multipoint discrete and Hadamard integral boundary conditions. In Section 6.4, a monotone iterative method is applied to obtain the existence of an extremal solution for a nonlinear system involving the right-handed Riemann–Liouville fractional derivative operators with nonlocal coupled integral boundary conditions. By employing the fixed point theory together with the monotone iterative technique, the existence of a unique solution for a nonlinear fractional integrodifferential equation on a semi-infinite domain in a Banach space is discussed in Section 6.5. We derive extremal solutions for a new fractional-order integro-initial value problem involving a Caputo-type generalized fractional derivative operator and a Steiltjes type fractional integral in Section 6.6.

The purpose of Chapter 7 is to investigate boundary value problems involving generalized fractional integrals and derivatives. Section 7.1 investigates the solvability of fractional differential equations and inclusions subject to nonlocal generalized Riemann-Liouville integral boundary conditions. Section 7.2 studies perturbed fractional differential equations and inclusions with generalized Riemann-Liouville fractional integral boundary conditions. In Section 7.3, we study a boundary value problem consisting of a generalized fractional differential equation supplemented with Stieltjes and generalized fractional integral boundary conditions (the generalized fractional derivative operator and integral are in the sense of Katugampola). Section 7.4 is concerned with the study of generalized fractional differential equations and inclusions equipped with three kinds of nonlocal multi-point and generalized fractional integral boundary conditions. We investigate a boundary value problem of fractional differential equations involving multiple Caputo type fractional derivative operators and generalized nonlocal fractional integro-differential boundary conditions in Section 7.5. A new class of boundary value problems of Caputo-type generalized fractional differential equations and inclusions supplemented with nonlocal generalized fractional integral (Katugampola) and multipoint boundary conditions is discussed in Section 7.6.

We gratefully acknowledge the contribution of our co-authors on the topic of research.

We are especially grateful to the Editor for his continued support and encouragement during the preparation of this monograph.

Bashir Ahmad
Sotiris K. Ntouyas

Contents

Chapter 1

Preliminaries

In this chapter, we fix terminology and describe some basic concepts from real analysis and fractional calculus, which are used throughout this book.

1.1 Definitions and results from multi-valued analysis

In this section, we introduce notation and necessary details from multi-valued analysis.

Let $(X, \|\cdot\|)$ be a normed space and $\mathcal{P}(X)$ denote the family of all non-empty subsets of X. Let us introduces the spaces: $\mathcal{P}_{cl}(X) = \{Y \in \mathcal{P}(X) : Y \text{ is closed}\}$, $\mathcal{P}_b(X) = \{Y \in \mathcal{P}(X) : Y \text{ is bounded}\}$, $\mathcal{P}_{cp}(X) = \{Y \in \mathcal{P}(X) : Y \text{ is compact}\}$, $\mathcal{P}_{cp,c}(X) = \{Y \in \mathcal{P}(X) : Y \text{ is compact and convex}\}$.

Definition 1.1. A multi-valued map $G : X \to \mathcal{P}(X)$ is
(i) *convex (closed) valued* if $G(x)$ is convex (closed) for all $x \in X$;
(ii) *bounded* on bounded sets if $G(\mathbb{B}) = \cup_{x\in\mathbb{B}} G(x)$ is bounded in X for all $\mathbb{B} \in \mathcal{P}_b(X)$ (i.e., $\sup_{x\in\mathbb{B}}\{\sup\{|y| : y \in G(x)\}\} < \infty$);
(iii) *measurable* if the function $t \longmapsto d(y, G(t)) = \inf\{|y - z| : z \in G(t)\}$ is measurable for every $y \in \mathbb{R}$;
(iv) *upper semi-continuous (u.s.c.)* on X if for each $x_0 \in X$, the set $G(x_0)$ is a non-empty closed subset of X, and if for each open set N of X containing $G(x_0)$, there exists an open neighborhood $\mathcal{N}_0$ of x_0 such that $G(\mathcal{N}_0) \subseteq N$;
(v) *lower semi-continuous (l.s.c.)* if the set $\{y \in X : G(y) \cap B \neq \emptyset\}$ is open for any open set B in E;
(vi) *completely continuous* if $G(\mathbb{B})$ is relatively compact for every $\mathbb{B} \in \mathcal{P}_b(X)$.

Definition 1.2. A multi-valued map $G : X \to \mathcal{P}(X)$ *has a fixed point* if there is $x \in X$ such that $x \in G(x)$. The fixed point set of the multi-valued operator G will be denoted by $FixG$.

Definition 1.3. A multi-valued map $F : [a, b] \times \mathbb{R} \to \mathcal{P}(\mathbb{R})$ is said to be Carathéodory if $t \longmapsto F(t, x)$ is measurable for each $x \in \mathbb{R}$ and $x \longmapsto F(t, x)$ is upper semi-continuous (u.s.c.) for almost all $t \in [a, b]$.

Further a Carathéodory function F is called L^1-Carathéodory if there exists $\varphi_\alpha \in L^1([a,b],\mathbb{R}^+)$ such that $\|F(t,x)\| = \sup\{|v| : v \in F(t,x)\} \leq \varphi_\alpha(t)$ for each $\alpha > 0$, for all $x \in \mathbb{R}$ with $\|x\| \leq \alpha$ and for a.e. $t \in [a,b]$. (Here $\|x\| = \sup_{t\in[a,b]} |x(t)|$).

For each $x \in C([a,b],\mathbb{R})$, define the set of selections of F by

$$S_{F,x} := \{v \in L^1([a,b],\mathbb{R}) : v(t) \in F(t,x(t)) \text{ for a.e. } t \in [a,b]\}.$$

We define the graph of G to be the set $Gr(G) = \{(x,y) \in X \times Y, y \in G(x)\}$ and recall two useful results regarding closed graphs and upper semi-continuity.

Lemma 1.1. *[81, Proposition 1.2] If $G : X \to \mathcal{P}_{cl}(Y)$ is u.s.c., then $Gr(G)$ is a closed subset of $X \times Y$; i.e., for every sequence $\{x_n\}_{n\in\mathbb{N}} \subset X$ and $\{y_n\}_{n\in\mathbb{N}} \subset Y$, if when $n \to \infty$, $x_n \to x_*$, $y_n \to y_*$ and $y_n \in G(x_n)$, then $y_* \in G(x_*)$. Conversely, if G is completely continuous and has a closed graph, then it is upper semi-continuous.*

Lemma 1.2. *[114] Let X be a separable Banach space. Let $F : [a,b]\times\mathbb{R} \to \mathcal{P}_{cp,c}(X)$ be an L^1-Carathéodory multi-valued map and let Θ be a linear continuous mapping from $L^1([a,b],X)$ to $C([a,b],X)$. Then the operator*

$$\Theta \circ S_F : C([a,b],X) \to \mathcal{P}_{cp,c}(C([a,b],X)), \quad x \mapsto (\Theta \circ S_F)(x) = \Theta(S_{F,x,y})$$

is a closed graph operator in $C([a,b],X) \times C([a,b],X)$.

Let X be a non-empty closed subset of a Banach space E and $G : X \to \mathcal{P}(E)$ be a multi-valued operator with non-empty closed values. G is lower semi-continuous (l.s.c.) if the set $\{y \in X : G(y) \cap B \neq \emptyset\}$ is open for any open set B in E.

Let A be a subset of $[a,b] \times \mathbb{R}$. A is $\mathcal{L} \otimes \mathcal{B}$ measurable if A belongs to the σ-algebra generated by all sets of the form $\mathcal{J} \times \mathcal{D}$, where $\mathcal{J}$ is Lebesgue measurable in $[a,b]$ and $\mathcal{D}$ is Borel measurable in $\mathbb{R}$. A subset $\mathcal{A}$ of $L^1([a,b],\mathbb{R})$ is decomposable if for all $x, y \in \mathcal{A}$ and measurable $\mathcal{J} \subset [a,b]$, the function $x\chi_{\mathcal{J}} + y\chi_{J-\mathcal{J}} \in \mathcal{A}$, where $\chi_{\mathcal{J}}$ stands for the characteristic function of $\mathcal{J}$.

Definition 1.4. Let Y be a separable metric space and let $N : Y \to \mathcal{P}(L^1([a,b],\mathbb{R}))$ be a multi-valued operator. We say N has a property (BC) if N is lower semi-continuous (l.s.c.) and has non-empty closed and decomposable values.

Let $F : [a,b]\times\mathbb{R} \to \mathcal{P}(\mathbb{R})$ be a multi-valued map with non-empty compact values. Define a multi-valued operator $\mathcal{F} : C([a,b] \times \mathbb{R}) \to \mathcal{P}(L^1([a,b],\mathbb{R}))$ associated with F as

$$\mathcal{F}(x) = \{w \in L^1([a,b],\mathbb{R}) : w(t) \in F(t,x(t)) \text{ for a.e. } t \in [a,b]\},$$

which is called the Nemytskii operator associated with F.

Definition 1.5. Let $F : [a,b] \times \mathbb{R} \to \mathcal{P}(\mathbb{R})$ be a multi-valued function with non-empty compact values. We say F is of lower semi-continuous type (l.s.c. type) if its associated Nemytskii operator $\mathcal{F}$ is lower semi-continuous and has non-empty closed and decomposable values.

Lemma 1.3. *[73] Let Y be a separable metric space and let $N : Y \to \mathcal{P}(L^1([a,b],\mathbb{R}))$ be a multi-valued operator satisfying the property (BC). Then N has a continuous selection, that is, there exists a continuous function (single-valued) $g : Y \to L^1([a,b],\mathbb{R})$ such that $g(x) \in N(x)$ for every $x \in Y$.*

For more details on multi-valued analysis we refer to the books of Deimling [81], Gorniewitz [91] and Hu and Papageorgiou [98].

1.2 Definitions and results from fractional calculus

Definition 1.6. Let $\alpha \geq 0$ and f be a real function defined on $[a,b], -\infty < a \leq t \leq b < +\infty$. The left *Riemann-Liouville fractional integral* of order α for a continuous function $f : [a,b] \to \mathbb{R}$ is defined by $(I^0_{a+}f)(x) = f(x)$ and

$$(I^\alpha_{a+}f)(x) = \frac{1}{\Gamma(\alpha)}\int_a^t (t-s)^{\alpha-1}f(s)ds, \quad \alpha > 0,$$

where $\Gamma(\cdot)$ is the Gamma function. Similarly, we can define the right *Riemann-Liouville fractional integral* of order $\alpha > 0$ as

$$(I^\alpha_{b-}f)(x) = \frac{1}{\Gamma(\alpha)}\int_t^b (s-t)^{\alpha-1}f(s)ds. \tag{1.1}$$

Definition 1.7. For a continuous function $f : [a,b] \to \mathbb{R}$, the left *Riemann-Liouville fractional derivative* of fractional order $\alpha > 0$ is defined by

$$D^\alpha_{a+}f(t) = \frac{d^n}{dt^n}I^{1-\alpha}_{a+} = \frac{1}{\Gamma(n-\alpha)}\left(\frac{d}{dt}\right)^n \int_a^t (t-s)^{n-\alpha-1}f(s)ds, \quad n = [\alpha]+1,$$

where $[\alpha]$ denotes the integer part of the real number α. On the other hand, the right *Riemann-Liouville fractional derivative* of order $\alpha > 0$ can be defined as

$$D^\alpha_{b-}f(t) = \left(\frac{d}{dt}\right)^n I^{1-\alpha}_{b-}f(t) = \frac{1}{\Gamma(n-\alpha)}\left(\frac{d}{dt}\right)^n \int_t^b (s-t)^{n-1-\alpha}f(s)ds.$$

For $\alpha < 0$ and the convenience of reader we use the denotation $D^\alpha y = I^{-\alpha}y$. Moreover, for $\beta \in [0,\alpha)$, it is valid that $D^\beta I^\alpha y = I^{\alpha-\beta}y$.

Definition 1.8. The left *Caputo fractional derivative operator* of order $\alpha \geq 0$ is given by $({}^C_aD^0 f)(t) = f(t)$ and $({}^CD^\alpha_{a+}f)(t) = (I^{m-\alpha}_{a+}D^m f)(t)$ for $\alpha > 0$, where m is the smallest integer greater or equal to α. Besides, it can be formulated by

$${}^CD^\alpha_{a+}f(t) = \frac{1}{\Gamma(n-\alpha)}\int_a^t (t-s)^{n-\alpha-1}f^{(n)}(s)ds, \quad n = [\alpha]+1, \tag{1.2}$$

where $\alpha \notin \mathbb{N}_0$ and $f \in AC^n([a,b],\mathbb{R}) = \{h : [a,b] \to \mathbb{R} : h, h', \ldots, h^{(n-1)} \in C([a,b],\mathbb{R})$ and $h^{(n-1)}$ is absolutely continuous$\}$ (see also [108]). Likewise, the right *Caputo fractional derivative operator* of order $\alpha \geq 0$ can be defined as $({}^CD^\alpha_{b-}f)(t) = (I^{m-\alpha}_{b-}D^m f)(t)$ for $\alpha > 0$.

Analogue to (1.2), we define the right *Caputo fractional derivative operator* of order $\alpha \geq 0$ as

$$ {}^C D_{b-}^{\alpha} f(t) = \frac{1}{\Gamma(n-\alpha)} \int_t^{b-} (s-t)^{n-\alpha-1} f^{(n)}(s) ds, \quad n = [\alpha] + 1, $$

for $f \in AC^n([a,b])$.

Remark 1.1. The Caputo fractional derivative operator ${}^C D_a^{\alpha}$ of order $\alpha \in \mathbb{R}$ $(m-1 < \alpha < m,\ m \in \mathbb{N})$ can be expressed in terms of Riemann-Liouville fractional derivative operator as follows

$$ {}^C D_{a+}^{\alpha} u(t) = D_{a+}^{\alpha} \left[u(t) - u(a) - u'(a) \frac{(t-a)}{1!} - \ldots - u^{(m-1)}(a) \frac{(t-a)^{(m-1)}}{(m-1)!} \right]. $$

Remark 1.2. In the rest of the book, we will call the 'left Riemann-Liouville fractional integral (derivative)' as 'Riemann-Liouville fractional integral (derivative)' and 'left Caputo fractional derivative operator' as the 'Caputo fractional derivative operator'.

From the definitions of Riemann-Liouville fractional derivative operator and integral, we can obtain the following lemmas.

Lemma 1.4. *[108] Let $q > 0$ and $y \in C(0,T) \cap L(0,T)$. Then the fractional differential equation $D^q y(t) = 0$ has a unique solution*

$$ y(t) = c_1 t^{q-1} + c_2 t^{q-2} + \cdots + c_n t^{q-n}, $$

where $c_i \in \mathbb{R}$, $i = 1, 2, \ldots, n$ and $n-1 < q < n$.

Lemma 1.5. *[108] Let $q > 0$. Then for $y \in C(0,T) \cap L(0,T)$ it holds*

$$ I^q D^q y(t) = y(t) + c_1 t^{q-1} + c_2 t^{q-2} + \cdots + c_n t^{q-n}, $$

where $c_i \in \mathbb{R}$, $i = 1, 2, \ldots, n$ and $n-1 < q < n$.

Definition 1.9. The Erdélyi-Kober fractional integral of order $\delta > 0$ with $\eta > 0$ and $\gamma \in \mathbb{R}$ of a continuous function $f : (0,\infty) \to \mathbb{R}$ is defined by

$$ I_{\eta}^{\gamma,\delta} f(t) = \frac{\eta t^{-\eta(\delta+\gamma)}}{\Gamma(\delta)} \int_0^t \frac{s^{\eta\gamma+\eta-1} f(s)}{(t^{\eta} - s^{\eta})^{1-\delta}} ds, $$

provided the right side is pointwise defined on $\mathbb{R}_+$.

Remark 1.3. For $\eta = 1$ the above integral operator reduces to the Kober operator

$$ I_{\gamma}^{\delta} f(t) = \frac{t^{-(\delta+\gamma)}}{\Gamma(\delta)} \int_0^t \frac{s^{\gamma} f(s)}{(t-s)^{1-\delta}} ds, \quad \gamma,\ \delta > 0, $$

that was introduced by Kober in [111]. For $\gamma = 0$, the Kober operator takes the form of the Riemann-Liouville fractional integral with a power weight:

$$ I_0^{\delta} f(t) = \frac{t^{-\delta}}{\Gamma(\delta)} \int_0^t \frac{f(s)}{(t-s)^{1-\delta}} ds, \ \delta > 0. $$

Lemma 1.6. *Let $\delta, \eta > 0$ and $\gamma, q \in \mathbb{R}$. Then we have*

$$I_\eta^{\gamma,\delta} t^q = \frac{t^q \Gamma(\gamma + (q/\eta) + 1)}{\Gamma(\gamma + (q/\eta) + \delta + 1)}. \tag{1.3}$$

Proof. From the Definition 1.9, we have

$$\begin{aligned} I_\eta^{\gamma,\delta} t^q &= \frac{\eta t^{-\eta(\delta+\gamma)}}{\Gamma(\delta)} \int_0^t \frac{s^{\eta\gamma+\eta-1} \cdot s^q}{(t^\eta - s^\eta)^{1-\delta}} ds \\ &= \frac{t^q}{\Gamma(\delta)} \int_0^1 u^{\gamma+\frac{q}{\eta}} (1-u)^{\delta-1} du \\ &= \frac{t^q}{\Gamma(\delta)} B\left(\gamma + \frac{q}{\eta} + 1, \delta\right) \\ &= \frac{t^q \Gamma(\gamma + (q/\eta) + 1)}{\Gamma(\gamma + (q/\eta) + \delta + 1)}, \end{aligned}$$

where we have used the beta function:

$$B(x,y) = \int_0^1 u^{x-1}(1-u)^{y-1} du \text{ with its property } B(x,y) = \frac{\Gamma(x)\Gamma(y)}{\Gamma(x+y)}, \; x, y > 0.$$

The proof is complete. □

Before introducing the generalized fractional integrals and derivative, let us fix our terminology by defining appropriate spaces.

For $c \in \mathbb{R}, 1 \leq p \leq \infty$, let $X_c^p(a,b)$ denote the space of all complex-valued Lebesgue measurable functions ϕ on (a,b) equipped with the norm:

$$\|\phi\|_{X_c^p} = \left(\int_a^b |x^c \phi(x)|^p \frac{dx}{x}\right)^{1/p} < \infty.$$

Denote by $L^1(a,b)$ the space of all Lebesgue measurable functions φ on (a,b) equipped with the norm:

$$\|\varphi\|_{L^1} = \int_a^b |\varphi(x)| dx < \infty.$$

Let $\mathcal{C} = C([a,b], \mathbb{R})$ denotes the Banach space of all continuous functions from $[a,b]$ to $\mathbb{R}$ endowed with the norm defined by $\|y\| = \sup_{t \in [a,b]} |y(t)|$.

For $0 \leq \epsilon < 1$, we define $C_{\epsilon,\rho}([a,b], \mathbb{R}) = \{f : [a,b] \to \mathbb{R} : (t^\rho - a^\rho)^\epsilon f(t) \in C([a,b], \mathbb{R})\}$ endowed with the norm $\|f\|_{C_{\epsilon,\rho}} = \|(t^\rho - a^\rho)^\epsilon f(t)\|_C$. Moreover, we define the class of functions f that have absolutely continuous δ^{n-1}-derivative, denoted by $AC_\delta^n([a,b], \mathbb{R})$, as follows: $AC_\delta^n([a,b], \mathbb{R}) = \Big\{f : [a,b] \to \mathbb{R} : \delta^{n-1} f \in AC([a,b], \mathbb{R}), \delta = t^{1-\rho}\frac{d}{dt}\Big\}$, which is equipped with the norm $\|f\|_{C_\delta^n} = \sum_{k=0}^{n-1} \|\delta^k f\|_C$. More generally, let $C_{\delta,\epsilon}^n([a,b], \mathbb{R}) = \Big\{f : [a,b] \to \mathbb{R} : \delta^{n-1} f \in C([a,b], \mathbb{R}), \delta^n f \in C_{\epsilon,\rho}([a,b], \mathbb{R}), \delta = t^{1-\rho}\frac{d}{dt}\Big\}$ be the space of functions endowed with the norm $\|f\|_{C^n_{\delta,\epsilon}} = \sum_{k=0}^{n-1} \|\delta^k f\|_C + \|\delta^n f\|_{C_{\epsilon,\rho}}$. Here we use the convention $C^n_{\delta,0} = C^n_\delta$.

Definition 1.10. [105] The generalized fractional integral of order $\alpha > 0$ and $\rho > 0$ of $f \in X^p_c(a,b)$ for $-\infty < a < t < b < \infty$, is defined by

$$({}^\rho I^\alpha_{a+} f)(t) = \frac{\rho^{1-\alpha}}{\Gamma(\alpha)} \int_a^t \frac{s^{\rho-1}}{(t^\rho - s^\rho)^{1-\alpha}} f(s)ds. \tag{1.4}$$

Note that the integral in (1.4) is called the left-sided fractional integral. Similarly we can define right-sided fractional integral ${}^\rho I^\alpha_{b-} f$ as

$$({}^\rho I^\alpha_{b-} f)(t) = \frac{\rho^{1-\alpha}}{\Gamma(\alpha)} \int_t^b \frac{s^{\rho-1}}{(s^\rho - t^\rho)^{1-\alpha}} f(s)ds. \tag{1.5}$$

Definition 1.11. [106] The generalized left and right fractional derivative operators, associated with the generalized fractional integrals (1.4) and (1.5), are respectively defined, for $0 \le a < t < b < \infty$, by

$$\begin{aligned}({}^\rho D^\alpha_{a+} f)(t) &= \left(t^{1-\rho}\frac{d}{dt}\right)^n ({}^\rho I^{n-\alpha}_{a+} f)(t)\\ &= \frac{\rho^{\alpha-n+1}}{\Gamma(n-\alpha)} \left(t^{1-\rho}\frac{d}{dt}\right)^n \int_a^t \frac{s^{\rho-1}}{(t^\rho - s^\rho)^{\alpha-n+1}} f(s)ds \end{aligned} \tag{1.6}$$

and

$$\begin{aligned}({}^\rho D^\alpha_{b-} f)(t) &= \left(-t^{1-\rho}\frac{d}{dt}\right)^n ({}^\rho I^{n-\alpha}_{b-} f)(t)\\ &= \frac{\rho^{\alpha-n+1}}{\Gamma(n-\alpha)} \left(-t^{1-\rho}\frac{d}{dt}\right)^n \int_t^b \frac{s^{\rho-1}}{(s^\rho - t^\rho)^{\alpha-n+1}} f(s)ds, \end{aligned} \tag{1.7}$$

if the integrals exist.

Definition 1.12. [103] For $\alpha \ge 0$ and $f \in AC^n_\delta[a,b]$, the Liouville-Caputo-type generalized fractional derivative operator ${}^\rho_c D^\alpha_a$ is defined via the above generalized fractional derivative operator by

$${}^\rho_c D^\alpha_{a+} f(x) = {}^\rho D^\alpha_{a+} \left[f(t) - \sum_{k=0}^{n-1} \frac{\delta^k f(a)}{k!}\left(\frac{t^\rho - a^\rho}{\rho}\right)^k\right](x), \quad \delta = x^{1-\rho}\frac{d}{dx}. \tag{1.8}$$

Similarly we have

$${}^\rho_c D^\alpha_{b-} f(x) = {}^\rho D^\alpha_{b-} \left[f(t) - \sum_{k=0}^{n-1} \frac{(-1)^k \delta^k f(b)}{k!}\left(\frac{b^\rho - t^\rho}{\rho}\right)^k\right](x), \quad \delta = x^{1-\rho}\frac{d}{dx}, \tag{1.9}$$

where $n = [\alpha] + 1$.

Lemma 1.7. *[103] Let $\alpha \geq 0, n = [\alpha]+1$ and $f \in AC^n_\delta[a, b]$, where $0 < a < b < \infty$. Then,*

1. if $\alpha \notin \mathbb{N}$,

$$ {}^\rho_c D^\alpha_{a+} f(t) = \frac{1}{\Gamma(n-\alpha)} \int_a^t \left(\frac{t^\rho - s^\rho}{\rho}\right)^{n-\alpha-1} \frac{(\delta^n f)(s)dt}{s^{1-\rho}} = {}^\rho I^{n-\alpha}_{a+}(\delta^n f)(t), \quad (1.10) $$

$$ {}^\rho_c D^\alpha_{b-} f(t) = \frac{1}{\Gamma(n-\alpha)} \int_t^b \left(\frac{s^\rho - t^\rho}{\rho}\right)^{n-\alpha-1} \frac{(-1)^n(\delta^n f)(s)dt}{s^{1-\rho}} = {}^\rho I^{n-\alpha}_{b-}(\delta^n f)(t). \quad (1.11) $$

2. If $\alpha \in \mathbb{N}$

$$ {}^\rho_c D^\alpha_{a+} f = \delta^n f, \quad {}^\rho_c D^\alpha_{b-} f = (-1)^n \delta^n f. \quad (1.12) $$

Lemma 1.8. *[103] Let $f \in AC^n_\delta[a, b]$ or $C^n_\delta[a, b]$ and $\alpha \in \mathbb{R}$. Then*

$$ {}^\rho I^\alpha_{a+} {}^\rho_c D^\alpha_{a+} f(x) = f(x) - \sum_{k=0}^{n-1} \frac{(\delta^k f)(a)}{k!} \left(\frac{x^\rho - a^\rho}{\rho}\right)^k, $$

$$ {}^\rho I^\alpha_{b-} {}^\rho_c D^\alpha_{b-} f(x) = f(x) - \sum_{k=0}^{n-1} \frac{(-1)^k(\delta^k f)(a)}{k!} \left(\frac{b^\rho - x^\rho}{\rho}\right)^k. $$

In particular, for $0 < \alpha \leq 1$, we have

$$ {}^\rho I^\alpha_{a+} {}^\rho_c D^\alpha_{a+} f(x) = f(x) - f(a), \ {}^\rho I^\alpha_{b-} {}^\rho_c D^\alpha_{b-} f(x) = f(x) - f(b). $$

Definition 1.13. [107] Let $f \in X^p_c(a, b)$, $\alpha > 0$ and $\rho, \eta, \kappa \in \mathbb{R}$. The generalized fractional integral is defined by

$$ ({}^\rho I^{\alpha,\beta}_{\eta,\kappa} f)(t) = \frac{\rho^{1-\beta} t^\kappa}{\Gamma(\alpha)} \int_0^t \frac{s^{\rho(\eta+1)-1}}{(t^\rho - s^\rho)^{1-\alpha}} f(s)ds, \quad 0 \leq t < b \leq \infty, \quad (1.13) $$

if the integral exists.

Remark 1.4. The fractional integral (1.13) contains some well-known fractional integrals as its particular cases (see also [107]):

(i) Setting $\kappa = 0, \eta = 0$ and taking the limit $\rho \to 1$ in (1.13), the integral operator (1.13) reduces to the Riemann-Liouville fractional integral

$$ J^\alpha f(t) = \frac{1}{\Gamma(\alpha)} \int_0^t (t-s)^{\alpha-1} f(s)ds. $$

(ii) Setting $\beta = \alpha, \kappa = 0, \eta = 0$, and taking the limit $\rho \to 0+$ with L'Hospital's rule in (1.13), the integral operator (1.13) reduces to the Hadamard fractional integral

$$ H^\alpha f(t) = \frac{1}{\Gamma(\alpha)} \int_0^t \left(\log \frac{t}{s}\right)^{\alpha-1} \frac{f(s)}{s} ds. $$

(iii) Setting $\beta = 0$ and $\kappa = -\rho(\alpha+\eta)$ in (1.13), the integral operator (1.13) reduces to the Erdélyi-Kober fractional integral

$$I^{\alpha}_{\eta,-\rho(\alpha+\eta)} f(t) = \frac{\rho t^{-\rho(\alpha+\eta)}}{\Gamma(\alpha)} \int_0^t \frac{s^{\rho(\eta+1)-1}}{(t^\rho - s^\rho)^{1-\alpha}} f(s)ds.$$

(iv) Setting $\beta = \alpha, \kappa = 0$ and $\eta = 0$ in (1.13), the integral operator (1.13) reduces to the Katugampola fractional integral

$${}^{\rho}I^{\alpha} f(t) = \frac{\rho^{1-\alpha}}{\Gamma(\alpha)} \int_0^t \frac{s^{\rho-1}}{(t^\rho - s^\rho)^{1-\alpha}} f(s)ds.$$

From Definition 1.13 we easily deduce that

Lemma 1.9. *Let $\alpha, \beta > 0$ and $\eta, \kappa, \rho \in \mathbb{R}$. Then*

$${}^{\rho}I^{\alpha,\beta}_{\eta,\kappa} t^q = \frac{t^{\kappa+\rho(\eta+\alpha)+q}}{\rho^\beta} \frac{\Gamma\left(\eta + \frac{q}{\rho} + 1\right)}{\Gamma\left(\eta + \frac{q}{\rho} + \alpha + 1\right)}.$$

1.3 Fixed point theorems

The tools of the fixed point theory play a major role in developing the existence theory for boundary value problems. We collect here the fixed point theorems used throughout this book.

Theorem 1.1. *(Banach fixed point theorem) [82] Let X be a Banach space, $D \subset X$ closed and $\mathcal{F} : D \to D$ a strict contraction, i.e., $\|\mathcal{F}x - \mathcal{F}y\| \leq k\|x - y\|$ for some $k \in (0,1)$ and all $x, y \in D$. Then $\mathcal{F}$ has a fixed point in D.*

Theorem 1.2. *(Schauder fixed point theorem) [154] Let E be a Banach space, and $K \subset E$ a non-empty, bounded, closed and convex subset of E. Let $\mathcal{T} : K \longrightarrow K$ be a completely continuous operator. Then $\mathcal{T}$ has a fixed point in K.*

Lemma 1.10. *(Schaefer fixed point theorem) [92] Let X be a Banach space. Assume that $\mathcal{T} : X \to X$ is a completely continuous operator and the set $V = \{u \in X \mid u = \mu\mathcal{T}u, 0 < \mu < 1\}$ is bounded. Then $\mathcal{T}$ has a fixed point in X.*

Theorem 1.3. *(Rothe fixed point theorem) [137] Let Ω be an open bounded subset of a Banach space X with $0 \in \Omega$ and the operator $\mathcal{F} : \overline{\Omega} \longrightarrow X$ is completely continuous satisfying $\|\mathcal{F}x\| \leq \|x\|$, $\forall\ x \in \partial\Omega$. Then the operator $\mathcal{F}$ has a fixed point in $\overline{\Omega}$.*

Theorem 1.4. *(Krasnosel'skiĭ's fixed point theorem) [112] Let M be a closed, bounded, convex and non-empty subset of a Banach space X. Let A, B be the operators such that (a) $Ax + By \in M$ whenever $x, y \in M$; (b) A is a compact and continuous; (c) B is a contraction mapping. Then there exists $z \in M$ such that $z=Az+Bz$.*

Theorem 1.5. *(Krasnolsk'ii-Zabreiko's fixed point theorem)[113] Let $(E, \|\cdot\|)$ be a Banach space, and $\mathcal{T} : E \longrightarrow E$ be a completely continuous operator. Assume that $A : E \longrightarrow E$ is a bounded linear operator such that 1 is not an eigenvalue of A and*

$$\lim_{\|u\|\to\infty} \frac{\|\mathcal{T}u - Au\|}{\|u\|} = 0.$$

Then $\mathcal{T}$ has a fixed point in E.

Theorem 1.6. *(Burton and Kirk's fixed point theorem) [74] Let X be a Banach space, and $A, B : X \to X$ be two operators such that A is a contraction and B is completely continuous. Then either*

(a) the operator equation $y = A(y) + B(y)$ has a solution, or
(b) the set $\mathcal{E} = \left\{u \in X : \lambda A\left(\frac{u}{\lambda}\right) + \lambda B(u) = u\right\}$ is unbounded for $\lambda \in (0, 1)$.

Theorem 1.7. *(Leray-Schauder alternative) [92, page 4] Let $\mathcal{F} : E \to E$ be a completely continuous operator (i.e., a map that restricted to any bounded set in E is compact). Let $\mathcal{E}(\mathcal{F}) = \{x \in E : x = \lambda\mathcal{F}(x)$ for some $0 < \lambda < 1\}$. Then either the set $\mathcal{E}(\mathcal{F})$ is unbounded, or $\mathcal{F}$ has at least one fixed point.*

Theorem 1.8. *(Leray-Schauder's nonlinear alternative for single-valued maps) ([92]) Let E be a Banach space, C be a closed, convex subset of E, U be an open subset of C and $0 \in U$. Suppose that $\mathcal{F} : \overline{U} \to C$ is a continuous, compact (that is, $\mathcal{F}(\overline{U})$ is a relatively compact subset of C) map. Then either*

(i) $\mathcal{F}$ has a fixed point in $\overline{U}$, or
(ii) there is a $u \in \partial U$ and $\lambda \in (0, 1)$ with $u = \lambda\mathcal{F}(u)$.

The next fixed point theorem, due to O'Regan, is a nonlinear alternative for Krasnosel'skiĭ's fixed point theorem.

Theorem 1.9. *[127] Denote by O an open set in a closed, convex set K of a Banach space X. Assume $0 \in O$. Also assume that $\mathcal{F}(\bar{O})$ is bounded and that $\mathcal{F} : \bar{O} \to K$ is given by $\mathcal{F} = \mathcal{F}_1 + \mathcal{F}_2$, in which $\mathcal{F}_1 : \bar{O} \to K$ is continuous and completely continuous and $\mathcal{F}_2 : \bar{O} \to K$ is nonlinear contraction (that is, there exists a non-negative non-decreasing function $\phi : [0, \infty) \to [0, \infty)$ satisfying $\phi(z) < z$ for $z > 0$, such that $\|\mathcal{F}_2 x - \mathcal{F}_2 y\| \le \phi(\|x - y\|)$ for all $x, y \in O$.) Then, either*

(C1) $\mathcal{F}$ has a fixed point $u \in \bar{O}$; or
(C2) there exist a point $u \in \partial O$ and $\lambda \in (0, 1)$ with $u = \lambda\mathcal{F}(u)$, where $\bar{O}$ and ∂O, respectively, represent the closure and boundary of O.

The following hybrid fixed point theorem for three operators in a Banach algebra E is due to Dhage [85].

Theorem 1.10. *[85] Let S be a non-empty, closed convex and bounded subset of a Banach algebra E and let $A, C : E \to E$ and $B : S \to E$ be three operators such that*

(a_1) *A and C are Lipschitzian with Lipschitz constants δ and ρ respectively;*
(b_1) *B is compact and continuous;*
(c_1) *$x = AxBy + Cx \Rightarrow x \in S$ for all $y \in S$,*
(d_1) *$\delta M + \rho < 1$, where $M = \|B(S)\|$.*

Then the operator equation $x = AxBx + Cx$ has a solution.

The following hybrid fixed point theorem of Schaefer type for three operators in a Banach algebra X, is due to Dhage [86, Theorem 4.13].

Theorem 1.11. *[86] Let X be a Banach algebra and let $A, C : X \to X$ be two single-valued operators and $B : X \to \mathcal{P}_{cp,cv}(X)$ be a multi-valued operator such that*

(a) A and C are Lipschitzian with the Lipschitz constants q_1 and q_2, respectively;
(b) B is compact and upper semi-continuous;
(c) $Mq_1 + q_2 < 1/2$, where $M = \| \cup B(X)\|_{\mathcal{P}}$.

Then, either (i) the operator inclusion $x \in AxBx + Cx$ has a solution or (ii) the set $\mathcal{E} = \{u \in X : \mu u \in AuBu + Cu,\ \mu > 1\}$ is unbounded.

Definition 1.14. Let E be a Banach space and let $\mathcal{A} : E \to E$ be a mapping. $\mathcal{A}$ is said to be a nonlinear contraction if there exists a continuous non-decreasing function $\Psi : \mathbb{R}^+ \to \mathbb{R}^+$ such that $\Psi(0) = 0$ and $\Psi(\varepsilon) < \varepsilon$ for all $\varepsilon > 0$ with the property:

$$\|\mathcal{A}x - \mathcal{A}y\| \leq \Psi(\|x - y\|), \qquad \forall x, y \in E.$$

Theorem 1.12. *(Boyd and Wong fixed point theorem) [72] Let E be a Banach space and let $\mathcal{A} : E \to E$ be a nonlinear contraction. Then $\mathcal{A}$ has a unique fixed point in E.*

Definition 1.15. [92] Let $\Phi : D(\Phi) \subseteq X \to X$ be a bounded and continuous operator on Banach space X. Then Φ is called a condensing map if $\alpha(\Phi(B)) < \alpha(B)$ for all bounded sets $B \subset D(\Phi)$, where α denotes the Kuratowski measure of noncompactness.

Theorem 1.13. *[154] The map $K_1 + K_2$ is a k-set contraction with $0 \leq k < 1$, and thus also condensing if the following conditions hold:*

(i) $K_1, K_2 : D \subseteq X \to X$ are operators on the Banach space X;
(ii) K_1 is k-contractive, that is, $\|K_1x - K_1y\| \leq k\|x - y\|$ for all $x, y \in D$ and fixed $k \in [0, 1)$;
(iii) K_2 is compact.

Theorem 1.14. *(Sadovski fixed point theorem) [135] Let B be a convex bounded and closed subset of a Banach space X and $\Phi : B \to B$ be a condensing map. Then Φ has a fixed point.*

The next fixed point theorem is based in degree theory.

Theorem 1.15. *[92] Suppose that $A : \bar{\Omega} \to E$ is completely continuous operator. If*

$$Ax \neq \lambda x, \qquad \forall x \in \partial\Omega, \lambda \geq 1.$$

Then $\deg(I - A, \Omega, \theta) = 1$ *and A has at least one fixed point in $\bar{\Omega}$.*

The next fixed point theorem is the well-known nonlinear alternative of Leray-Schauder for multi-valued maps.

Theorem 1.16. *(Leray-Schauder nonlinear alternative for Kakutani maps) [92] Let E be a Banach space, C a closed convex subset of E, U an open subset of C and $0 \in U$. Suppose that $\mathcal{F} : \overline{U} \to \mathcal{P}_{cp,c}(C)$ is a upper semi-continuous compact map. Then either*

(i) $\mathcal{F}$ has a fixed point in $\overline{U}$, or
(ii) there is a $u \in \partial U$ and $\lambda \in (0,1)$ with $u \in \lambda\mathcal{F}(u)$.

Our next theorem is Krasnosel'skiǐ's fixed point theorem for multi-valued maps.

Theorem 1.17. *(Krasnosel'skiǐ's multi-valued fixed point theorem) [130] Let X be a Banach space, $Y \in \mathcal{P}_{b,cl,c}(X)$ and $A, B : Y \to \mathcal{P}_{cp,c}(X)$ two multi-valued operators. If the following conditions are satisfied*

(i) $Ay + By \subset Y$ for all $y \in Y$;
(ii) A is contraction;
(iii) B is u.s.c and compact,

then, there exists $y \in Y$ such that $y \in Ay + By$.

The next fixed point theorem is known as Nonlinear alternative for contractive multi-valued maps. [131, Corollary 3.8].

Theorem 1.18. *(Nonlinear alternative for contractive maps) Let X be a Banach space, and D a bounded neighbourhood of $0 \in X$. Let $Z_1 : X \to \mathcal{P}_{cp,c}(X)$ and $Z_2 : \bar{D} \to \mathcal{P}_{cp,c}(X)$ two multi-valued operators satisfying*

(a) Z_1 is contraction, and
(b) Z_2 is u.s.c and compact.

Then, if $G = Z_1 + Z_2$, either

(i) G has a fixed point in $\bar{D}$ or
(ii) there is a point $u \in \partial D$ and $\lambda \in (0,1)$ with $u \in \lambda G(u)$.

Before stating the next fixed point theorem we need some preliminaries. Let (X, d) be a metric space induced from the normed space $(X; \|\cdot\|)$. Consider the Pompeiu-Hausdoff metric $H_d : \mathcal{P}(X) \times \mathcal{P}(X) \to \mathbb{R} \cup \{\infty\}$ given by

$$H_d(A, B) = \max\{\sup_{a \in A} d(a, B), \sup_{b \in B} d(A, b)\},$$

where $d(A,b) = \inf_{a\in A} d(a;b)$ and $d(a,B) = \inf_{b\in B} d(a;b)$. Then $(\mathcal{P}_{b,cl}(X), H_d)$ is a metric space and $(\mathcal{P}_{cl}(X), H_d)$ is a generalized metric space (see [109]).

Definition 1.16. A multi-valued operator $\mathcal{H} : X \to \mathcal{P}_{cl}(X)$ is called:

(a) κ-Lipschitz if and only if there exists $\kappa > 0$ such that

$$H_d(\mathcal{H}(x), \mathcal{H}(y)) \le \kappa d(x,y) \quad \text{for each} \quad x, y \in X;$$

(b) a contraction if and only if it if κ-Lipschitz with $\kappa < 1$.

The next fixed point theorem is the well known Covitz and Nadler fixed point theorem for multi-valued contractions.

Theorem 1.19. *(Covitz and Nadler fixed point theorem) [80] Let (X,d) be a complete metric space. If $N : X \to \mathcal{P}_{cl}(X)$ is a contraction, then $FixN \neq \emptyset$.*

An element $x \in X$ is called *an endpoint of a multifunction* $F : X \to \mathcal{P}(X)$ whenever $Fx = \{x\}$ ([69]). Also, we say that F *has an approximate endpoint property* whenever $\inf_{x\in X} \sup_{y\in Fx} d(x,y) = 0$ ([69]).

A function $f : \mathbb{R} \to \mathbb{R}$ is called upper semi-continuous whenever $\limsup_{n\to\infty} f(\lambda_n) \le f(\lambda)$ for all sequence $\{\lambda_n\}_{n\ge 1}$ with $\lambda_n \to \lambda$.

The following endpoint fixed point theorem is due to Amini-Harandi [69].

Theorem 1.20. *(Endpoint fixed point theorem) [69] Let $\psi : [0,\infty) \to [0,\infty)$ be an upper semi-continuous function such that*

$$\psi(t) < t \quad \text{and} \quad \liminf_{t\to\infty}(t - \psi(t)) > 0 \quad \text{for all} \quad t > 0,$$

(X,d) a complete metric space and $S : X \to \mathcal{P}_{cl,bd}(X)$ ($\mathcal{P}_{cl,bd}(X)$ is the collection of all non-empty closed and bounded subsets of X) a multifunction such that

$$H_d(Sx, Sy) \le \psi(d(x,y)) \quad \text{for all} \quad x, y \in X.$$

Then S has a unique endpoint if and only if S has approximate endpoint property.

Definition 1.17. A function $\delta : \mathbb{R}_+ \to \mathbb{R}_+$ is said to be a strict comparison function (see [92]) if it is continuous, strictly increasing and $\sum_{n=1}^{\infty} \delta^n(t) < \infty$, for each $t > 0$.

Definition 1.18. A multi-valued operator $N : X \to \mathcal{P}_{cl}(X)$ is called a generalized contraction if and only if there is a strict comparison function $\delta : \mathbb{R}_+ \to \mathbb{R}_+$ such that $H_d(N(x), N(y)) \le \delta(d(x,y))$ for each $x, y \in X$.

Theorem 1.21. *(Wegrzyk's fixed point theorem) [146] Let (X,d) be a complete metric space. If $\mathcal{Q} : X \to \mathcal{P}_{cl}(X)$ is a generalized contraction, then $Fix\mathcal{Q} \neq \emptyset$.*

The following fixed point theorem is based on index theory.

Lemma 1.11. *[82] Let P be a cone in a Banach space E. For $\rho > 0$, define $\Omega_\rho = \{x \in P \mid \|x\| < \rho\}$. Assume that $\mathcal{T} : \overline{\Omega}_\rho \to P$ is a compact map such that $x \neq \mathcal{T}x$ for $x \in \partial\Omega_\rho$.*

(i) If $\|x\| < \|\mathcal{T}x\|$ for $x \in \partial\Omega_\rho$, then $i(\mathcal{T}, \Omega_\rho, P) = 0$.

(ii) If $\|x\| > \|\mathcal{T}x\|$ for $x \in \partial\Omega_\rho$, then $i(\mathcal{T}, \Omega_\rho, P) = 1$.

The next Guo-Krasnosel'skiĭ fixed point theorem is used to prove the existence of at least one positive solution.

Theorem 1.22. *(Guo-Krasnosel'skiĭ fixed point theorem) [93] Let E be a Banach space, and let $P \subset E$ be a cone. Assume that Ω_1, Ω_2 are open subsets of E with $0 \in \Omega_1$, $\overline{\Omega}_1 \subset \Omega_2$ and let $\mathcal{T} : P \cap (\overline{\Omega}_2 \setminus \Omega_1) \to P$ be a completely continuous operator such that:*

(i) $\|\mathcal{T}u\| \geq \|u\|$, $u \in P \cap \partial\Omega_1$, and $\|\mathcal{T}u\| \leq \|u\|$, $u \in P \cap \partial\Omega_2$; or
(ii) $\|\mathcal{T}u\| \leq \|u\|$, $u \in P \cap \partial\Omega_1$, and $\|\mathcal{T}u\| \geq \|u\|$, $u \in P \cap \partial\Omega_2$.

Then $\mathcal{T}$ has a fixed point in $P \cap (\overline{\Omega}_2 \setminus \Omega_1)$.

The following Leggett-Williams fixed point theorem is used to prove the existence of at least three positive solutions.

Definition 1.19. A continuous mapping $\theta : P \to [1, \infty)$ is said to be a non-negative continuous concave functional on the cone P of a real Banach space E provided that $\theta(\lambda u + (1-\lambda)v) \geq \lambda\theta(u) + (1-\lambda)\theta(v)$ for all $u, v \in P$ and $\lambda \in [0, 1]$.

Let $a, b, d > 0$ be constants. We defined $P_d = \{u \in P : \|u\| < d\}$, $\overline{P}_d = \{u \in P : \|u\| \leq d\}$ and $P(\theta, a, b) = \{u \in P : \theta(u) \geq a, \|u\| \leq b\}$.

Theorem 1.23. *(Leggett-Williams fixed point theorem) [115] Let P be a cone in the real Banach space E and $c > 0$ be a constant. Assume that there exists a concave non-negative continuous functional θ on P with $\theta(u) \leq \|u\|$ for all $u \in \overline{P}_c$. Let $\mathcal{T} : \overline{P}_c \to \overline{P}_c$ be a completely continuous operator. Suppose that there exist constants $0 < a < b < d \leq c$ such that the following conditions hold:*

(i) $\{u \in P(\theta, b, d) : \theta(u) > b\} \neq \emptyset$ and $\theta(\mathcal{T}u) > b$ for $u \in P(\theta, b, d)$;
(ii) $\|\mathcal{T}u\| < a$ for $u \leq a$;
(iii) $\theta(\mathcal{T}u) > b$ for $u \in P(\theta, b, c)$ with $\|\mathcal{T}u\| > d$.

Then $\mathcal{T}$ has at least three fixed points u_1, u_2 and u_3 in $\overline{P}_c$. Furthermore, $\|u_1\| < a$, $b < \theta(u_2)$, $a < \|u_3\|$ with $\theta(u_3) < b$.

Chapter 2

Nonlocal boundary value problems with different kinds of boundary conditions

We study boundary value problems for fractional differential equations and inclusions equipped with different kinds of boundary conditions. Nonlocal boundary value problems investigated in this chapter include fractional-order nonlinear integro-differential equations, inclusions, mixed fractional differential equations and inclusions involving both right Caputo and left Riemann-Liouville fractional derivative operators, right and left Riemann-Liouville fractional integrals, hybrid fractional differential equations and inclusions, and coupled systems of fractional differential equations.

2.1 Four-point nonlocal boundary value problems of nonlinear integro-differential equations and inclusions of fractional order

In this section, we study the existence and uniqueness of solutions for four-point nonlocal boundary value problems of nonlinear integro-differential equations and inclusions of fractional order $q \in (1, 2]$. Our results are based on some standard fixed point theorems.

2.1.1 *Nonlinear fractional order integro-differential equations*

We consider the following four-point nonlocal boundary value problem for a nonlinear fractional integro-differential equation:

$$\begin{cases} {}^{C}D^{q}x(t) = f(t, x(t), (\phi x)(t), (\psi x)(t)), \quad 0 < t < 1, \quad 1 < q \leq 2, \\ x'(0) + ax(\eta_1) = 0, \qquad bx'(1) + x(\eta_2) = 0, \quad 0 < \eta_1 \leq \eta_2 < 1, \end{cases} \tag{2.1}$$

where ${}^{C}D$ is the Caputo's fractional derivative operator, $f : [0, 1] \times X \times X \times X \to X$ is continuous, for $\gamma, \delta : [0, 1] \times [0, 1] \to [0, \infty)$,

$$(\phi x)(t) = \int_0^t \gamma(t, s)x(s)ds, \quad (\psi x)(t) = \int_0^t \delta(t, s)x(s)ds,$$

and $a, b \in (0, 1)$. Here, $(X, \|\cdot\|)$ is a Banach space and $\mathrm{C} = \mathrm{C}([0, 1], X)$ denotes the Banach space of all continuous functions from $[0, 1] \to X$ endowed with a topology of uniform convergence with the norm denoted by $\|\cdot\|$.

We remark that the boundary conditions in (2.1) arise in the study of heat flow problems involving a rod of unit length with two controllers at the end positions $t = 0$ and $t = 1$ adding or removing heat according to the temperatures detected by two sensors at $t = \eta_1$ and $t = \eta_2$.

Lemma 2.1. *For a given $\sigma \in C[0,1]$, the unique solution of the boundary value problem:*

$$\begin{cases} {}^C D^q x(t) = \sigma(t), \quad 0 < t < 1, \quad 1 < q \leq 2, \\ x'(0) + ax(\eta_1) = 0, \qquad bx'(1) + x(\eta_2) = 0, \quad 0 < \eta_1 \leq \eta_2 < 1, \end{cases} \tag{2.2}$$

is given by

$$\begin{aligned} x(t) = & \int_0^t \frac{(t-s)^{q-1}}{\Gamma(q)}\sigma(s)ds + \frac{a(b+\eta_2 - t)}{1+a(\eta_1-\eta_2-b)}\int_0^{\eta_1}\frac{(\eta_1-s)^{q-1}}{\Gamma(q)}\sigma(s)ds \\ & + \frac{at-(1+a\eta_1)}{1+a(\eta_1-\eta_2-b)}\Big[b\int_0^1\frac{(1-s)^{q-2}}{\Gamma(q-1)}\sigma(s)ds + \int_0^{\eta_2}\frac{(\eta_2-s)^{q-1}}{\Gamma(q)}\sigma(s)ds\Big]. \end{aligned} \tag{2.3}$$

Proof. For arbitrary constants c_1 and c_2, the solution of linear fractional differential equation in (2.2) can be written as

$$x(t) = I^q\sigma(t) - c_1 t - c_2 = \int_0^t \frac{(t-s)^{q-1}}{\Gamma(q)}\sigma(s)ds - c_1 t - c_2. \tag{2.4}$$

Applying the boundary conditions of (2.2) in (2.4), we find that

$$\begin{aligned} c_1 = & \frac{a}{1+a(\eta_1-\eta_2-b)}\Big[\int_0^{\eta_1}\frac{(\eta_1-s)^{q-1}}{\Gamma(q)}\sigma(s)ds - \int_0^{\eta_2}\frac{(\eta_2-s)^{q-1}}{\Gamma(q)}\sigma(s)ds \\ & - b\int_0^1\frac{(1-s)^{q-2}}{\Gamma(q-1)}\sigma(s)ds\Big], \\ c_2 = & \frac{1}{1+a(\eta_1-\eta_2-b)}\Big[-a(\eta_2+b)\int_0^{\eta_1}\frac{(\eta_1-s)^{q-1}}{\Gamma(q)}\sigma(s)ds \\ & + (1+a\eta_1)\Big\{\int_0^{\eta_2}\frac{(\eta_2-s)^{q-1}}{\Gamma(q)}\sigma(s)ds + b\int_0^1\frac{(1-s)^{q-2}}{\Gamma(q-1)}\sigma(s)ds\Big\}\Big]. \end{aligned}$$

Substituting the values of c_1 and c_2 in (2.4), we obtain the solution given by (2.3). The converse follows by direct computation. This completes the proof. □

In the sequel, we need the assumption:

(2.1.1) There exist positive functions $L_1(t),\ L_2(t),\ L_3(t)$ such that

$$\begin{aligned} & |f(t,x(t),(\phi x)(t),(\psi x)(t)) - f(t,y(t),(\phi y)(t),(\psi y)(t))| \\ & \leq L_1(t)|x-y| + L_2(t)|\phi x - \phi y| + L_3(t)|\psi x - \psi y|, \forall t \in [0,1],\ x,y \in X. \end{aligned}$$

For the sake of computational brevity, we set the notation:

$$\lambda_1 = \sup_{t\in[0,1]}\left|\frac{a(b+\eta_2-t)}{1+a(\eta_1-\eta_2-b)}\right|, \quad \lambda_2 = \sup_{t\in[0,1]}\left|\frac{at-(1+a\eta_1)}{1+a(\eta_1-\eta_2-b)}\right|,$$

$$\gamma_0 = \sup_{t\in[0,1]} \Big|\int_0^t \gamma(t,s)ds\Big|, \quad \delta_0 = \sup_{t\in[0,1]} \Big|\int_0^t \delta(t,s)ds\Big|,$$
$$I_L^q = \max\Big\{ \sup_{t\in[0,1]} |I^q L_1(t)|, \ \sup_{t\in[0,1]} |I^q L_2(t)|, \ \sup_{t\in[0,1]} |I^q L_3(t)|\Big\},$$
$$I^{q-1}L(1) = \max\{|I^{q-1}L_1(1)|, \ |I^{q-1}L_2(1)|, \ |I^{q-1}L_3(1)|\},$$
$$I^q L(\eta_i) = \max\{|I^q L_1(\eta_i)|, \ |I^q L_2(\eta_i)|, \ |I^q L_3(\eta_i)|\}, \quad i = 1, 2. \tag{2.5}$$

Theorem 2.1. *Assume that $f : [0,1] \times X \times X \times X \to X$ is a jointly continuous function and satisfies the assumption (2.1.1). Then the boundary value problem (2.1) has a unique solution provided that*

$$\Delta := (1 + \gamma_0 + \delta_0)\Big\{ I_L^q + \lambda_1 I^q L(\eta_1) + \lambda_2 (b I^{q-1} L(1) + I^q L(\eta_2))\Big\} < 1, \tag{2.6}$$

where the quantities appearing in (2.6) are defined in (2.5).

Proof. Define $\mathbf{F} : \mathrm{C} \to \mathrm{C}$ by

$$\begin{aligned}(\mathbf{F}x)(t) = &\int_0^t \frac{(t-s)^{q-1}}{\Gamma(q)} f(s, x(s), (\phi x)(s), (\psi x)(s))ds \\ &+ \frac{a(b+\eta_2 - t)}{1 + a(\eta_1 - \eta_2 - b)} \int_0^{\eta_1} \frac{(\eta_1 - s)^{q-1}}{\Gamma(q)} f(s, x(s), (\phi x)(s), (\psi x)(s))ds \\ &+ \frac{at - (1 + a\eta_1)}{1 + a(\eta_1 - \eta_2 - b)} \Big[b\int_0^1 \frac{(1-s)^{q-2}}{\Gamma(q-1)} f(s, x(s), (\phi x)(s), (\psi x)(s))ds \\ &+ \int_0^{\eta_2} \frac{(\eta_2 - s)^{q-1}}{\Gamma(q)} f(s, x(s), (\phi x)(s), (\psi x)(s))ds\Big], t \in [0,1].\end{aligned}$$

Let us set $\sup_{t\in[0,1]} |f(t,0,0,0)| = M$, and choose

$$r \geq \frac{M}{(1-\Delta)\Gamma(q+1)} \Big\{ 1 + \lambda_1 \eta_1^q + \lambda_2 (bq + \eta_2^q)\Big\},$$

where Δ is given by (2.6). Now we show that $\mathbf{F}B_r \subset B_r$, where $B_r = \{x \in \mathrm{C} : \|x\| \leq r\}$. For $x \in B_r$, we have

$$\begin{aligned}|\mathbf{F}x| = &\sup_{t\in[0,1]} |(\mathbf{F}x)(t)| \\ \leq &\sup_{t\in[0,1]} \Big\{ \int_0^t \frac{(t-s)^{q-1}}{\Gamma(q)} |f(s, x(s), (\phi x)(s), (\psi x)(s))|ds \\ &+ \Big|\frac{a(b+\eta_2 - t)}{1 + a(\eta_1 - \eta_2 - b)}\Big| \int_0^{\eta_1} \frac{(\eta_1 - s)^{q-1}}{\Gamma(q)} |f(s, x(s), (\phi x)(s), (\psi x)(s))|ds \\ &+ \Big|\frac{at - (1 + a\eta_1)}{1 + a(\eta_1 - \eta_2 - b)}\Big| \Big[b\int_0^1 \frac{(1-s)^{q-2}}{\Gamma(q-1)} |f(s, x(s), (\phi x)(s), (\psi x)(s))|ds \\ &+ \int_0^{\eta_2} \frac{(\eta_2 - s)^{q-1}}{\Gamma(q)} |f(s, x(s), (\phi x)(s), (\psi x)(s))|ds\Big]\Big\} \\ \leq &\sup_{t\in[0,1]} \Big\{ \int_0^t \frac{(t-s)^{q-1}}{\Gamma(q)} (|f(s, x(s), (\phi x)(s), (\psi x)(s)) - f(s,0,0,0)| + |f(s,0,0,0)|)ds\end{aligned}$$

$$+\Big|\frac{a(b+\eta_2-t)}{1+a(\eta_1-\eta_2-b)}\Big|\int_0^{\eta_1}\frac{(\eta_1-s)^{q-1}}{\Gamma(q)}(|f(s,x(s),(\phi x)(s),(\psi x)(s))-f(s,0,0,0)|$$
$$+|f(s,0,0,0)|)ds$$
$$+\Big|\frac{at-(1+a\eta_1)}{1+a(\eta_1-\eta_2-b)}\Big|\Big[b\int_0^1\frac{(1-s)^{q-2}}{\Gamma(q-1)}(|f(s,x(s),(\phi x)(s),(\psi x)(s))-f(s,0,0,0)|$$
$$+|f(s,0,0,0)|)ds$$
$$+\int_0^{\eta_2}\frac{(\eta_2-s)^{q-1}}{\Gamma(q)}(|f(s,x(s),(\phi x)(s),(\psi x)(s))-f(s,0,0,0)|+|f(s,0,0,0)|)ds\Big]\Big\}$$

$$\leq \sup_{t\in[0,1]}\Big\{\int_0^t\frac{(t-s)^{q-1}}{\Gamma(q)}(L_1(s)|x(s)|+\gamma_0L_2(s)|x(s)|+\delta_0L_3(s)|x(s)|+M)ds$$
$$+\Big|\frac{a(b+\eta_2-t)}{1+a(\eta_1-\eta_2-b)}\Big|\int_0^{\eta_1}\frac{(\eta_1-s)^{q-1}}{\Gamma(q)}(L_1(s)|x(s)|+\gamma_0L_2(s)|x(s)|$$
$$+\delta_0L_3(s)|x(s)|+M)ds$$
$$+\Big|\frac{at-(1+a\eta_1)}{1+a(\eta_1-\eta_2-b)}\Big|\Big[b\int_0^1\frac{(1-s)^{q-2}}{\Gamma(q-1)}(L_1(s)|x(s)|+\gamma_0L_2(s)|x(s)|$$
$$+\delta_0L_3(s)|x(s)|+M)ds$$
$$+\int_0^{\eta_2}\frac{(\eta_2-s)^{q-1}}{\Gamma(q)}(L_1(s)|x(s)|+\gamma_0L_2(s)|x(s)|+\delta_0L_3(s)|x(s)|+M)ds\Big]\Big\}$$
$$\leq \sup_{t\in[0,1]}\Big\{(I^qL_1(t)+\gamma_0I^qL_2(t)+\delta_0I^qL_3(t))r+\frac{Mt^q}{\Gamma(q+1)}$$
$$+\Big|\frac{a(b+\eta_2-t)}{1+a(\eta_1-\eta_2-b)}\Big|\Big((I^qL_1(\eta_1)+\gamma_0I^qL_2(\eta_1)+\delta_0I^qL_3(\eta_1))r+\frac{M\eta_1^q}{\Gamma(q+1)}\Big)$$
$$+\Big|\frac{at-(1+a\eta_1)}{1+a(\eta_1-\eta_2-b)}\Big|\Big[b\Big((I^{q-1}L_1(1)+\gamma_0I^{q-1}L_2(1)+\delta_0I^{q-1}L_3(1))r+\frac{M}{\Gamma(q)}\Big)$$
$$+(I^qL_1(\eta_2)+\gamma_0I^qL_2(\eta_2)+\delta_0I^qL_3(\eta_2))r+\frac{M\eta_2^q}{\Gamma(q+1)}\Big]\Big\}$$
$$\leq \Delta r+\frac{M}{\Gamma(q+1)}\Big\{1+\lambda_1\eta_1^q+\lambda_2(bq+\eta_2^q)\Big\}\leq r.$$

Now, for $x,y\in \mathrm{C}$ and for each $t\in[0,1]$, we obtain

$$\|\mathbf{F}x-\mathbf{F}y\| = \sup_{t\in[0,1]}\{|(\mathbf{F}x)(t)-(\mathbf{F}y)(t)|\}$$
$$\leq \sup_{t\in[0,1]}\Big\{\int_0^t\frac{(t-s)^{q-1}}{\Gamma(q)}|f(s,x(s),(\phi x)(s),(\psi x)(s)-f(s,y(s),(\phi y)(s),(\psi y)(s))|ds$$
$$+\Big|\frac{a(b+\eta_2-t)}{1+a(\eta_1-\eta_2-b)}\Big|\int_0^{\eta_1}\frac{(\eta_1-s)^{q-1}}{\Gamma(q)}|f(s,x(s),(\phi x)(s),(\psi x)(s)$$
$$-f(s,y(s),(\phi y)(s),(\psi y)(s))|ds$$
$$+\Big|\frac{at-(1+a\eta_1)}{1+a(\eta_1-\eta_2-b)}\Big|\Big[b\int_0^1\frac{(1-s)^{q-2}}{\Gamma(q-1)}|f(s,x(s),(\phi x)(s),(\psi x)(s)$$
$$-f(s,y(s),(\phi y)(s),(\psi y)(s))|ds$$
$$+\int_0^{\eta_2}\frac{(\eta_2-s)^{q-1}}{\Gamma(q)}|f(s,x(s),(\phi x)(s),(\psi x)(s)-f(s,y(s),(\phi y)(s),(\psi y)(s))|ds\Big]\Big\}$$

$$
\begin{aligned}
&\le \sup_{t\in[0,1]}\Big\{\int_0^t \frac{(t-s)^{q-1}}{\Gamma(q)}\Big(L_1(s)\|x-y\| + L_2(s)|\phi x-\phi y| + L_3(s)|\psi x-\psi y|\Big)ds \\
&\quad+\Big|\frac{a(b+\eta_2-t)}{1+a(\eta_1-\eta_2-b)}\Big|\int_0^{\eta_1}\frac{(\eta_1-s)^{q-1}}{\Gamma(q)}\Big(L_1(s)|x-y| + L_2(s)|\phi x-\phi y| \\
&\quad+L_3(s)|\psi x-\psi y|\Big)ds \\
&\quad+\Big|\frac{at-(1+a\eta_1)}{1+a(\eta_1-\eta_2-b)}\Big|\Big[b\int_0^1\frac{(1-s)^{q-2}}{\Gamma(q-1)}\Big(L_1(s)|x-y| + L_2(s)|\phi x-\phi y| \\
&\quad+L_3(s)|\psi x-\psi y|\Big)ds \\
&\quad+\int_0^{\eta_2}\frac{(\eta_2-s)^{q-1}}{\Gamma(q)}\Big(L_1(s)|x-y| + L_2(s)|\phi x-\phi y| + L_3(s)|\psi x-\psi y|\Big)ds\Big]\Big\} \\
&\le \sup_{t\in[0,1]}\Big\{\Big(I^qL_1(t)+\gamma_0 I^qL_2(t)+\delta_0 I^qL_3(t)\Big) \\
&\quad+\Big|\frac{a(b+\eta_2-t)}{1+a(\eta_1-\eta_2-b)}\Big|\Big(I^qL_1(\eta_1)+\gamma_0 I^qL_2(\eta_1)+\delta_0 I^qL_3(\eta_1)\Big) \\
&\quad+\Big|\frac{at-(1+a\eta_1)}{1+a(\eta_1-\eta_2-b)}\Big|\Big[b\Big((I^{q-1}L_1(1)+\gamma_0 I^{q-1}L_2(1)+\delta_0 I^{q-1}L_3(1))\Big) \\
&\quad+(I^qL_1(\eta_2)+\gamma_0 I^qL_2(\eta_2)+\delta_0 I^qL_3(\eta_2))\Big]\Big\}\|x-y\| \\
&\le (1+\gamma_0+\delta_0)\Big\{I_L^q+\lambda_1 I^qL(\eta_1)+\lambda_2(bI^{q-1}L(1)+I^qL(\eta_2))\Big\}\|x-y\| \\
&= \Delta\|x-y\|.
\end{aligned}
$$

Since $\Delta < 1$ by (2.6), therefore $\mathbf{F}$ is a contraction. Thus, the conclusion of the theorem follows by the contraction mapping principle. □

Example 2.1. Consider the following boundary value problem

$$
\begin{cases}
{}^cD^{\frac{3}{2}}x(t) = \dfrac{t}{8}\dfrac{|x|}{1+|x|} + \dfrac{1}{5}\displaystyle\int_0^t \frac{e^{-(s-t)}}{5}x(s)ds + \frac{1}{5}\int_0^t \frac{e^{-(s-t)/2}}{5}x(s)ds, \quad t\in[0,1], \\
x'(0) + \dfrac{1}{2}x\Big(\dfrac{1}{3}\Big) = 0, \qquad \dfrac{1}{4}x'(1) + x\Big(\dfrac{2}{3}\Big) = 0,
\end{cases}
\tag{2.7}
$$

Here, $q = 3/2$, $\gamma(t,s) = e^{-(s-t)}/5$, $\delta = e^{-(s-t)/2}/5$, $a = 1/2$, $\eta_1 = \frac{1}{3}$, $\eta_2 = 2/3$ and $b = 1/4$. With

$$
\gamma_0 = \frac{e-1}{5},\ \delta_0 = \frac{2(\sqrt{e}-1)}{5},\ \sup_{t\in[o,1]} I^qL(t) = \frac{4}{15\sqrt{\pi}},\ I^{q-1}L(1) = \frac{2}{5\sqrt{\pi}},
$$

$$
I^qL(1/3) = \frac{4}{45\sqrt{3\pi}},\ I^qL(2/3) = \frac{8\sqrt{2}}{45\sqrt{3\pi}},
$$

we find that

$$
\Delta = \frac{(e+2(\sqrt{e}+1))}{5\sqrt{\pi}}\Big\{\frac{22}{51} + \frac{4(11+56\sqrt{2})}{765\sqrt{3}}\Big\} \approx 0.6364432527 < 1.
$$

Thus, by Theorem 2.1, the boundary value problem (2.7) has a unique solution on $[0,1]$.

In the next theorem, we use Krasnosel'skiĭ's fixed point theorem.

Theorem 2.2. *Suppose that $f:[0,1]\times X\times X\times X\to X$ is jointly continuous and maps bounded subsets of $[0,1]\times X\times X\times X$ into relatively compact subsets of X. In addition, the condition (2.1.1) and the following assumption hold:*

(2.2.1) $|f(t,x(t),(\phi x)(t),(\psi x)(t))|\le\mu(t),\ \ \forall(t,x,\phi x,\psi x)\in[0,1]\times X\times X\times X,\ \ \mu\in C([0,1],\mathbb{R}^+)$.

If

$$\Delta_1:=(1+\gamma_0+\delta_0)\Big\{\lambda_1 I^q L(\eta_1)+\lambda_2(bI^{q-1}L(1)+I^qL(\eta_2))\Big\}<1,$$

where the quantities appearing in the above inequality are defined in (2.5), then there exists at least one solution of the boundary value problem (2.1) on $[0,1]$.

Proof. Let us fix

$$r\ge\frac{\|\mu\|}{\Gamma(q+1)}\Big\{1+\lambda_1\eta_1^q+\lambda_2(bq+\eta_2^q)\Big\},$$

and consider $B_r=\{x\in\mathcal{C}:\|x\|\le r\}$. We define the operators Θ_1 and Θ_2 on B_r as

$$(\Theta_1x)(t)=\int_0^t\frac{(t-s)^{q-1}}{\Gamma(q)}f(s,x(s),(\phi x)(s),(\psi x)(s))ds,$$

$$\begin{aligned}(\Theta_2x)(t)=&\ \frac{a(b+\eta_2-t)}{1+a(\eta_1-\eta_2-b)}\int_0^{\eta_1}\frac{(\eta_1-s)^{q-1}}{\Gamma(q)}f(s,x(s),(\phi x)(s),(\psi x)(s))ds\\&+\frac{at-(1+a\eta_1)}{1+a(\eta_1-\eta_2-b)}\Big[b\int_0^1\frac{(1-s)^{q-2}}{\Gamma(q-1)}f(s,x(s),(\phi x)(s),(\psi x)(s))ds\\&+\int_0^{\eta_2}\frac{(\eta_2-s)^{q-1}}{\Gamma(q)}f(s,x(s),(\phi x)(s),(\psi x)(s))ds\Big].\end{aligned}$$

For $x,y\in B_r$, we find that

$$\|\Theta_1x+\Theta_2y\|\ \le\frac{\|\mu\|}{\Gamma(q+1)}\Big\{1+\lambda_1\eta_1^q+\lambda_2(bq+\eta_2^q)\Big\}\le r.$$

Thus, $\Theta_1x+\Theta_2y\in B_r$. It follows from the assumption (2.1.1) that Θ_2 is a contraction mapping for $\Delta_1<1$. Continuity of f implies that the operator Θ_1 is continuous. Also, Θ_1 is uniformly bounded on B_r as

$$\|\Theta_1x\|\le\frac{\|\mu\|}{\Gamma(q+1)}.$$

To show that the operator Θ_1 is compact, we use the classical Arzela-Ascoli's Theorem. Let $\mathcal{A}$ be a bounded subset of $\mathcal{C}$. We have to show that $\Theta_1(\mathcal{A})$ is equicontinuous and for each t, the set $\Theta_1(\mathcal{A})(t)$ is relatively compact in X. In view of

(2.1.1) and (2.2.1) we define $\sup_{(t,x,\phi x,\psi x)\in[0,1]\times B_r\times B_r\times B_r}\|f(t,x,\phi x,\psi x)\| = f_{\max}$, and consequently we have

$$\begin{aligned}
&|(\Theta_1 x)(t_1) - (\Theta_1 x)(t_2)| \\
&= \Big|\frac{1}{\Gamma(q)}\int_0^{t_1}\Big((t_2-s)^{q-1}-(t_1-s)^{q-1}\Big)f(s,x(s),\phi x(s),\psi x(s))ds \\
&\quad + \int_{t_1}^{t_2}(t_2-s)^{q-1}f(s,x(s),\phi x(s),\psi x(s))ds\Big| \\
&\le \frac{f_{\max}}{\Gamma(q+1)}|2(t_2-t_1)^q + t_1^q - t_2^q|,
\end{aligned}$$

which is independent of x. Thus, Θ_1 is equi-continuous. Using the fact that f maps bounded subsets into relatively compact subsets, we have that $\Theta_1(\mathcal{A})(t)$ is relatively compact in X for every t. Therefore, Θ_1 is relatively compact on B_r. Hence, By Arzela-Ascoli's Theorem, Θ_1 is compact on B_r. Thus all the assumptions of Theorem 1.4 are satisfied and the conclusion of Theorem 3.2 implies that the boundary value problem (2.1) has at least one solution on $[0,1]$. □

2.1.2 *Fractional differential inclusions*

Now we study a nonlocal four-point boundary value problem for fractional differential inclusions given by

$$\begin{cases} {}^C D^q x(t) \in F(t,x(t)), & t\in[0,1], \quad 1<q\le 2, \\ x'(0)+ax(\eta_1)=0, & bx'(1)+x(\eta_2)=0, \qquad a,b\in(0,1) \end{cases} \tag{2.8}$$

where $0<\eta_1\le\eta_2<1$, ${}^C D^q$ denotes the Caputo fractional derivative of order q, and $F:[0,1]\times\mathbb{R}\to\mathcal{P}(\mathbb{R})$ is a multi-valued map, $\mathcal{P}(\mathbb{R})$ is the family of all non-empty subsets of $\mathbb{R}$.

Definition 2.1. A function $x\in C^2([0,1])$ is a solution of the problem (2.8) if there exists a function $f\in L^1([0,1],\mathbb{R})$ such that $f(t)\in F(t,x(t))$ a.e. on $[0,1]$ and

$$\begin{aligned}
x(t) &= \int_0^t \frac{(t-s)^{q-1}}{\Gamma(q)}f(s)ds + a\mu_3(\mu_1-t)\int_0^{\eta_1}\frac{(\eta_1-s)^{q-1}}{\Gamma(q)}f(s)ds \\
&\quad +\mu_3(at-\mu_2)\Big[b\int_0^1\frac{(1-s)^{q-2}}{\Gamma(q-1)}f(s)ds + \int_0^{\eta_2}\frac{(\eta_2-s)^{q-1}}{\Gamma(q)}f(s)ds\Big].
\end{aligned}\tag{2.9}$$

Theorem 2.3. *Assume that*

(2.3.1) $F:[0,1]\times\mathbb{R}\to\mathcal{P}(\mathbb{R})$ *is Carathéodory and has non-empty compact convex values;*

(2.3.2) there exists a continuous non-decreasing function $\psi:[0,\infty)\to(0,\infty)$ *and a function* $p\in L^1([0,1],\mathbb{R}_+)$ *such that*

$$\|F(t,x)\|_{\mathcal{P}} := \sup\{|y| : y\in F(t,x)\} \le p(t)\psi(\|x\|) \text{ for each } (t,x)\in[0,1]\times\mathbb{R};$$

(2.3.3) there exists a number $M > 0$ such that

$$\frac{\Gamma(q)M}{\Big[1+\lambda_1\eta_1^{q-1}+\lambda_2\Big(b(q-1)+\eta_1^{q-1}\Big)\Big]\psi(M)\|p\|_{L^1}} > 1,$$

where

$$\lambda_1 = \sup_{t\in[0,1]} |a\mu_3(\mu_1 - t)|, \quad \lambda_2 = \sup_{t\in[0,1]} |\mu_3(at-\mu_2)|. \tag{2.10}$$

Then the boundary value problem (2.8) has at least one solution on $[0,1]$.

Proof. Define an operator $\Omega : C([0,1],\mathbb{R}) \to \mathcal{P}(C([0,1],\mathbb{R}))$ by

$$\Omega(x) = \left\{ \begin{array}{l} h \in C([0,1],\mathbb{R}) : \\ h(t) = \left\{ \begin{array}{l} \displaystyle\int_0^t \frac{(t-s)^{q-1}}{\Gamma(q)} f(s)ds \\ \displaystyle +a\mu_3(\mu_1 - t)\int_0^{\eta_1} \frac{(\eta_1-s)^{q-1}}{\Gamma(q)} f(s)ds \\ \displaystyle +\mu_3(at-\mu_2)\Big[b\int_0^1 \frac{(1-s)^{q-2}}{\Gamma(q-1)} f(s)ds + \int_0^{\eta_2} \frac{(\eta_2-s)^{q-1}}{\Gamma(q)} f(s)ds\Big], \end{array} \right. \end{array} \right\}$$

for $f \in S_{F,x}$. We will show that Ω satisfies the assumptions of the nonlinear alternative of Leray-Schauder type. The proof consists of several steps. As a first step, we show that $\Omega(x)$ *is convex for each* $x \in C([0,1],\mathbb{R})$. For that, let $h_1, h_2 \in \Omega(x)$. Then there exist $f_1, f_2 \in S_{F,x}$ such that for each $t \in [0,1]$, we have

$$\begin{aligned} h_i(t) = & \int_0^t \frac{(t-s)^{q-1}}{\Gamma(q)} f_i(s)ds + a\mu_3(\mu_1-t)\int_0^{\eta_1} \frac{(\eta_1-s)^{q-1}}{\Gamma(q)} f_i(s)ds \\ & +\mu_3(at-\mu_2)\Big[b\int_0^1 \frac{(1-s)^{q-2}}{\Gamma(q-1)} f_i(s)ds + \int_0^{\eta_2} \frac{(\eta_2-s)^{q-1}}{\Gamma(q)} f_i(s)ds\Big], \ i=1,2. \end{aligned}$$

Let $0 \le \omega \le 1$. Then, for each $t \in [0,1]$, we have

$$\begin{aligned} [\omega h_1 + (1-\omega)h_2](t) = & \int_0^t \frac{(t-s)^{q-1}}{\Gamma(q)} [\omega f_1(s) + (1-\omega) f_2(s)](s)ds \\ & +a\mu_3(\mu_1-t)\int_0^{\eta_1} \frac{(\eta_1-s)^{q-1}}{\Gamma(q)} [\omega f_1(s) + (1-\omega) f_2(s)](s)ds \\ & +\mu_3(at-\mu_2)\Big[b\int_0^1 \frac{(1-s)^{q-2}}{\Gamma(q-1)} [\omega f_1(s) + (1-\omega) f_2(s)](s)ds \\ & +\int_0^{\eta_2} \frac{(\eta_2-s)^{q-1}}{\Gamma(q)} [\omega f_1(s) + (1-\omega) f_2(s)](s)ds\Big]. \end{aligned}$$

Since $S_{F,x}$ is convex (F has convex values), therefore it follows that $\omega h_1+(1-\omega)h_2 \in \Omega(x)$.

Next, we show that $\Omega(x)$ *maps bounded sets into bounded sets in* $C([0,1],\mathbb{R})$. For a positive number r, let $B_r = \{x \in C([0,1],\mathbb{R}) : \|x\| \le r\}$ be a bounded set in $C([0,1],\mathbb{R})$. Then, for each $h \in \Omega(x), x \in B_r$, there exists $f \in S_{F,x}$ such that

$$h(t) = \int_0^t \frac{(t-s)^{q-1}}{\Gamma(q)} f(s)ds + a\mu_3(\mu_1-t)\int_0^{\eta_1} \frac{(\eta_1-s)^{q-1}}{\Gamma(q)} f(s)ds$$

$$+\mu_3(at-\mu_2)\Big[b\int_0^1 \frac{(1-s)^{q-2}}{\Gamma(q-1)}f(s)ds+\int_0^{\eta_2}\frac{(\eta_2-s)^{q-1}}{\Gamma(q)}f(s)ds\Big]$$

and

$$\begin{aligned}|h(t)| &\le \int_0^t \frac{|t-s|^{q-1}}{\Gamma(q)}|f(s)|ds+|a\mu_3(\mu_1-t)|\int_0^{\eta_1}\frac{|\eta_1-s|^{q-1}}{\Gamma(q)}|f(s)|ds\\ &\quad +|\mu_3(at-\mu_2)|\Big[b\int_0^1\frac{|1-s|^{q-2}}{\Gamma(q-1)}|f(s)|ds+\int_0^{\eta_2}\frac{|\eta_2-s|^{q-1}}{\Gamma(q)}|f(s)|ds\Big]\\ &\le \frac{1}{\Gamma(q)}\Big[1+\lambda_1\eta_1^{q-1}+\lambda_2\Big(b(q-1)+\eta_2^{q-1}\Big)\Big]\psi(\|x\|)\int_0^1 p(s)ds,\end{aligned}$$

where λ_1 and λ_2 are given by (2.10). Thus,

$$\|h\|\le \frac{1}{\Gamma(q)}\Big[1+\lambda_1\eta_1^{q-1}+\lambda_2\Big(b(q-1)+\eta_2^{q-1}\Big)\Big]\psi(\|x\|)\int_0^1 p(s)ds.$$

Now we show that Ω *maps bounded sets into equi-continuous sets of* $C([0,1],\mathbb{R})$. Let $t',t''\in[0,1]$ with $t'<t''$ and $x\in B_r$, where B_r is a bounded set of $C([0,1],\mathbb{R})$. For each $h\in\Omega(x)$, we obtain

$$\begin{aligned}&|h(t'')-h(t')|\\ &=\Big|\int_0^{t''}\frac{(t''-s)^{q-1}}{\Gamma(q)}f(s)ds-\int_0^{t'}\frac{(t'-s)^{q-1}}{\Gamma(q)}f(s)ds\\ &\quad +a\mu_3(t''-t')\Big[-\int_0^{\eta_1}\frac{(\eta_1-s)^{q-1}}{\Gamma(q)}f(s)ds\\ &\quad +b\int_0^1\frac{(1-s)^{q-2}}{\Gamma(q-1)}f(s)ds+\int_0^{\eta_2}\frac{(\eta_2-s)^{q-1}}{\Gamma(q)}f(s)ds\Big]\Big|\\ &\le \psi(r)\Big|\int_0^{t'}\frac{[(t''-s)^{q-1}-(t'-s)^{q-1}]}{\Gamma(q)}p(s)ds\Big|+\Big|\int_{t'}^{t''}\frac{(t''-s)^{q-1}}{\Gamma(q)}f(s)ds\Big|\\ &\quad +\psi(r)\Big|a\mu_3(t''-t')\Big[\int_0^{\eta_1}\frac{(\eta_1-s)^{q-1}}{\Gamma(q)}p(s)ds\\ &\quad +b\int_0^1\frac{(1-s)^{q-2}}{\Gamma(q-1)}p(s)ds+\int_0^{\eta_2}\frac{(\eta_2-s)^{q-1}}{\Gamma(q)}p(s)ds\Big]\Big|.\end{aligned}$$

Obviously the right hand side of the above inequality tends to zero independently of $x\in B_{r'}$ as $t''-t'\to 0$. As Ω satisfies the above three assumptions, therefore it follows by Ascoli-Arzelá theorem that $\Omega: C([0,1],\mathbb{R})\to\mathcal{P}(C([0,1],\mathbb{R}))$ is completely continuous.

In our next step, we show that Ω *has a closed graph.* Let $x_n\to x_*, h_n\in\Omega(x_n)$ and $h_n\to h_*$. Then we need to show that $h_*\in\Omega(x_*)$. Associated with $h_n\in\Omega(x_n)$, there exists $f_n\in S_{F,x_n}$ such that for each $t\in[0,1]$,

$$\begin{aligned}h_n(t)&=\int_0^t\frac{(t-s)^{q-1}}{\Gamma(q)}f_n(s)ds+a\mu_3(\mu_1-t)\int_0^{\eta_1}\frac{(\eta_1-s)^{q-1}}{\Gamma(q)}f_n(s)ds\\ &\quad +\mu_3(at-\mu_2)\Big[b\int_0^1\frac{(1-s)^{q-2}}{\Gamma(q-1)}f_n(s)ds+\int_0^{\eta_2}\frac{(\eta_2-s)^{q-1}}{\Gamma(q)}f_n(s)ds\Big].\end{aligned}$$

Thus we have to show that there exists $f_* \in S_{F,x_*}$ such that for each $t \in [0,1]$,

$$h_*(t) = \int_0^t \frac{(t-s)^{q-1}}{\Gamma(q)} f_*(s)ds + a\mu_3(\mu_1 - t)\int_0^{\eta_1} \frac{(\eta_1 - s)^{q-1}}{\Gamma(q)} f_*(s)ds$$
$$+\mu_3(at-\mu_2)\Big[b\int_0^1 \frac{(1-s)^{q-2}}{\Gamma(q-1)} f_*(s)ds + \int_0^{\eta_2} \frac{(\eta_2-s)^{q-1}}{\Gamma(q)} f_*(s)ds\Big].$$

Let us consider the continuous linear operator $\Theta : L^1([0,1],\mathbb{R}) \to C([0,1],\mathbb{R})$ so that

$$f \mapsto \Theta(f) = \int_0^t \frac{(t-s)^{q-1}}{\Gamma(q)} f(s)ds + a\mu_3(\mu_1 - t)\int_0^{\eta_1} \frac{(\eta_1 - s)^{q-1}}{\Gamma(q)} f(s)ds$$
$$+\mu_3(at-\mu_2)\Big[b\int_0^1 \frac{(1-s)^{q-2}}{\Gamma(q-1)} f(s)ds + \int_0^{\eta_2} \frac{(\eta_2-s)^{q-1}}{\Gamma(q)} f(s)ds\Big].$$

Observe that

$$\|h_n(t) - h_*(t)\| = \Big\| \int_0^t \frac{(t-s)^{q-1}}{\Gamma(q)}(f_n(s) - f_*(s))ds$$
$$+a\mu_3(\mu_1 - t)\int_0^{\eta_1} \frac{(\eta_1 - s)^{q-1}}{\Gamma(q)}(f_n(s) - f_*(s))ds$$
$$+\mu_3(at-\mu_2)\Big[b\int_0^1 \frac{(1-s)^{q-2}}{\Gamma(q-1)}(f_n(s) - f_*(s))ds$$
$$+\int_0^{\eta_2} \frac{(\eta_2-s)^{q-1}}{\Gamma(q)}(f_n(s) - f_*(s))ds\Big]\Big\| \to 0 \quad \text{as} \quad n \to \infty.$$

Thus, it follows by Lemma 1.2 that $\Theta \circ S_F$ is a closed graph operator. Further, we have $h_n(t) \in \Theta(S_{F,x_n})$. Since $x_n \to x_*$, it follows that

$$h_*(t) = \int_0^t \frac{(t-s)^{q-1}}{\Gamma(q)} f_*(s)ds + a\mu_3(\mu_1 - t)\int_0^{\eta_1} \frac{(\eta_1 - s)^{q-1}}{\Gamma(q)} f_*(s)ds$$
$$+\mu_3(at-\mu_2)\Big[b\int_0^1 \frac{(1-s)^{q-2}}{\Gamma(q-1)} f_*(s)ds + \int_0^{\eta_2} \frac{(\eta_2-s)^{q-1}}{\Gamma(q)} f_*(s)ds\Big],$$

for some $f_* \in S_{F,x_*}$.

Finally, *we discuss a priori bounds on solutions.* Let x be a solution of (2.8). Then there exists $f \in L^1([0,1],\mathbb{R})$ with $f \in S_{F,x}$ such that, for $t \in [0,1]$, we have

$$x(t) = \int_0^t \frac{(t-s)^{q-1}}{\Gamma(q)} f(s)ds + a\mu_3(\mu_1 - t)\int_0^{\eta_1} \frac{(\eta_1 - s)^{q-1}}{\Gamma(q)} f(s)ds$$
$$+\mu_3(at-\mu_2)\Big[b\int_0^1 \frac{(1-s)^{q-2}}{\Gamma(q-1)} f(s)ds + \int_0^{\eta_2} \frac{(\eta_2-s)^{q-1}}{\Gamma(q)} f(s)ds\Big].$$

In view of (2.3.2), for each $t \in [0,1]$, we obtain

$$|x(t)| \leq \frac{1}{\Gamma(q)}\Big[1 + \lambda_1\eta_1^{q-1} + \lambda_2\Big(b(q-1) + \eta_2^{q-1}\Big)\Big]\psi(\|x\|)\int_0^1 p(s)ds,$$

where λ_1 and λ_2 are given by (2.10). Consequently, we have

$$\frac{\Gamma(q)\|x\|}{\Big[1 + \lambda_1\eta_1^{q-1} + \lambda_2\Big(b(q-1) + \eta_1^{q-1}\Big)\Big]\psi(\|x\|)\|p\|_{L^1}} \leq 1,$$

In view of (2.3.3), there exists M such that $\|x\| \neq M$. Let us set

$$U = \{x \in C([0,1],\mathbb{R}) : \|x\| < M+1\}.$$

Note that the operator $\Omega : \overline{U} \to \mathcal{P}(C([0,1],\mathbb{R}))$ is upper semi-continuous and completely continuous. From the choice of U, there is no $x \in \partial U$ such that $x \in \mu\Omega(x)$ for some $\mu \in (0,1)$. Consequently, by the nonlinear alternative of Leray-Schauder type (Theorem 1.8), we deduce that Ω has a fixed point $x \in \overline{U}$ which is a solution of the problem (2.8). This completes the proof. □

As a next result, we study the case when F is not necessarily convex valued. Our strategy to deal with this problems is based on the nonlinear alternative of Leray-Schauder type together with the selection theorem of Bressan and Colombo (Lemma 1.3) for lower semi-continuous maps with decomposable values.

Theorem 2.4. *Assume that (2.3.2), (2.3.3) and the following conditions hold:*

(2.4.1) $F : [0,1]\times\mathbb{R} \to \mathcal{P}(\mathbb{R})$ *is a non-empty compact-valued multi-valued map such that*

(a) $(t,x) \longmapsto F(t,x)$ *is* $\mathcal{L}\otimes\mathcal{B}$ *measurable,*

(b) $x \longmapsto F(t,x)$ *is lower semi-continuous for each* $t \in [0,1]$.

Then the boundary value problem (2.8) has at least one solution on $[0,1]$.

Proof. It follows from (2.3.2) and (2.4.1) that F is of l.s.c. type. Then from Lemma 1.3, there exists a continuous function $f : C([0,1],\mathbb{R}) \to L^1([0,1],\mathbb{R})$ such that $f(x) \in \mathcal{F}(x)$ for all $x \in C([0,1],\mathbb{R})$.

Consider the problem

$$\begin{cases} {}^C D^q x(t) = f(x(t)), & t \in [0,1], \quad 1 < q \le 2, \\ x'(0) + a x(\eta_1) = 0, & b x'(1) + x(\eta_2) = 0, \quad 0 < \eta_1 \le \eta_2 < 1, \quad a,b \in (0,1). \end{cases} \tag{2.11}$$

$0 < \eta_1 \le \eta_2 < 1$. Observe that if $x \in C^2([0,1])$ is a solution of (2.11), then x is a solution to the problem (2.8). In order to transform the problem (2.11) into a fixed point problem, we define the operator $\overline{\Omega}$ as

$$\begin{aligned} \overline{\Omega}x(t) &= \int_0^t \frac{(t-s)^{q-1}}{\Gamma(q)} f(x(s))ds + a\mu_3(\mu_1 - t)\int_0^{\eta_1} \frac{(\eta_1-s)^{q-1}}{\Gamma(q)} f(x(s))ds \\ &\quad + \mu_3(at-\mu_2)\Big[b\int_0^1 \frac{(1-s)^{q-2}}{\Gamma(q-1)} f(x(s))ds + \int_0^{\eta_2} \frac{(\eta_2-s)^{q-1}}{\Gamma(q)} f(x(s))ds\Big]. \end{aligned}$$

It can easily be shown that $\overline{\Omega}$ is continuous and completely continuous. The remaining part of the proof is similar to that of Theorem 2.3. So we omit it. This completes the proof. □

Now we prove the existence of solutions for the problem (2.8) with a non-convex valued right hand side by applying a fixed point theorem for multi-valued map due to Covitz and Nadler (Theorem 1.19).

Theorem 2.5. *Assume that the following conditions hold:*

(2.5.1) $F : [0,1] \times \mathbb{R} \to \mathcal{P}_{cp}(\mathbb{R})$ *is such that* $F(\cdot, x) : [0,1] \to \mathcal{P}_{cp}(\mathbb{R})$ *is measurable for each* $x \in \mathbb{R}$.

(2.5.2) $H_d(F(t,x), F(t,\bar{x})) \le m(t)|x - \bar{x}|$ *for almost all* $t \in [0,1]$ *and* $x, \bar{x} \in \mathbb{R}$ *with* $m \in L^1([0,1], \mathbb{R})$ *and* $d(0, F(t,0)) \le m(t)$ *for almost all* $t \in [0,1]$.

Then the boundary value problem (2.8) has at least one solution on $[0,1]$ *if*

$$\frac{\|m\|_{L^1}}{\Gamma(q)} \left[1 + \lambda_1 \eta_1^{q-1} + \lambda_2 \left(b(q-1) + \eta_2^{q-1}\right)\right] < 1.$$

Proof. Observe that the set $S_{F,x}$ is non-empty for each $x \in C([0,1], \mathbb{R})$ by the assumption (2.5.1), so F has a measurable selection (see Theorem III.6 [78]). Now we show that the operator Ω satisfies the assumptions of Theorem 1.19. To show that $\Omega(x) \in \mathcal{P}_{cl}((C[0,1], \mathbb{R}))$ for each $x \in C([0,1], \mathbb{R})$, let $\{u_n\}_{n \ge 0} \in \Omega(x)$ be such that $u_n \to u$ $(n \to \infty)$ in $C([0,1], \mathbb{R})$. Then $u \in C([0,1], \mathbb{R})$ and there exists $v_n \in S_{F,x}$ such that, for each $t \in [0,1]$,

$$\begin{aligned} u_n(t) = & \int_0^t \frac{(t-s)^{q-1}}{\Gamma(q)} v_n(s) ds + a\mu_3(\mu_1 - t) \int_0^{\eta_1} \frac{(\eta_1 - s)^{q-1}}{\Gamma(q)} v_n(s) ds \\ & + \mu_3(at - \mu_2) \left[b \int_0^1 \frac{(1-s)^{q-2}}{\Gamma(q-1)} v_n(s) ds + \int_0^{\eta_2} \frac{(\eta_2 - s)^{q-1}}{\Gamma(q)} v_n(s) ds \right]. \end{aligned}$$

As F has compact values, we pass onto a subsequence to obtain that v_n converges to v in $L^1([0,1], \mathbb{R})$. Thus, $v \in S_{F,x}$ and for each $t \in [0,1]$,

$$\begin{aligned} u_n(t) \to u(t) = & \int_0^t \frac{(t-s)^{q-1}}{\Gamma(q)} v(s) ds + a\mu_3(\mu_1 - t) \int_0^{\eta_1} \frac{(\eta_1 - s)^{q-1}}{\Gamma(q)} v(s) ds \\ & + \mu_3(at - \mu_2) \left[b \int_0^1 \frac{(1-s)^{q-2}}{\Gamma(q-1)} v(s) ds + \int_0^{\eta_2} \frac{(\eta_2 - s)^{q-1}}{\Gamma(q)} v(s) ds \right]. \end{aligned}$$

Hence $u \in \Omega(x)$.

Next we show that there exists $\gamma < 1$ such that

$$H_d(\Omega(x), \Omega(\bar{x})) \le \gamma \|x - \bar{x}\| \quad \text{for each} \quad x, \bar{x} \in C([0,1], \mathbb{R}).$$

Let $x, \bar{x} \in C([0,1], \mathbb{R})$ and $h_1 \in \Omega(x)$. Then there exists $v_1(t) \in F(t, x(t))$ such that, for each $t \in [0,1]$,

$$\begin{aligned} h_1(t) = & \int_0^t \frac{(t-s)^{q-1}}{\Gamma(q)} v_1(s) ds + a\mu_3(\mu_1 - t) \int_0^{\eta_1} \frac{(\eta_1 - s)^{q-1}}{\Gamma(q)} v_1(s) ds \\ & + \mu_3(at - \mu_2) \left[b \int_0^1 \frac{(1-s)^{q-2}}{\Gamma(q-1)} v_1(s) ds + \int_0^{\eta_2} \frac{(\eta_2 - s)^{q-1}}{\Gamma(q)} v_1(s) ds \right]. \end{aligned}$$

By (2.5.2), we have

$$H_d(F(t,x), F(t,\bar{x})) \le m(t)|x(t) - \bar{x}(t)|.$$

So, there exists $w \in F(t, \bar{x}(t))$ such that

$$|v_1(t) - w| \le m(t)|x(t) - \bar{x}(t)|, \quad t \in [0,1].$$

Define $U : [0,1] \to \mathcal{P}(\mathbb{R})$ by

$$U(t) = \{w \in \mathbb{R} : |v_1(t) - w| \leq m(t)|x(t) - \bar{x}(t)|\}.$$

Since the multi-valued operator $U(t) \cap F(t, \bar{x}(t))$ is measurable (Proposition III.4 [78]), there exists a function $v_2(t)$ which is a measurable selection for V. So $v_2(t) \in F(t, \bar{x}(t))$ and for each $t \in [0,1]$, we have $|v_1(t) - v_2(t)| \leq m(t)|x(t) - \bar{x}(t)|$.

For each $t \in [0,1]$, let us define

$$\begin{aligned} h_2(t) = &\int_0^t \frac{(t-s)^{q-1}}{\Gamma(q)} v_2(s)ds + a\mu_3(\mu_1 - t)\int_0^{\eta_1} \frac{(\eta_1 - s)^{q-1}}{\Gamma(q)} v_2(s)ds \\ &+ \mu_3(at - \mu_2)\Big[b\int_0^1 \frac{(1-s)^{q-2}}{\Gamma(q-1)} v_2(s)ds + \int_0^{\eta_2} \frac{(\eta_2 - s)^{q-1}}{\Gamma(q)} v_2(s)ds\Big]. \end{aligned}$$

Thus

$$\begin{aligned} |h_1(t) - h_2(t)| \leq &\int_0^t \frac{|t-s|^{q-1}}{\Gamma(q)} |v_1(s) - v_2(s)|ds \\ &+ |a\mu_3(\mu_1 - t)| \int_0^{\eta_1} \frac{|\eta_1 - s|^{q-1}}{\Gamma(q)} |v_1(s) - v_2(s)|ds \\ &+ |\mu_3(at - \mu_2)| \Big[b\int_0^1 \frac{|1-s|^{q-2}}{\Gamma(q-1)} |v_1(s) - v_2(s)|ds \\ &+ \int_0^{\eta_2} \frac{|\eta_2 - s|^{q-1}}{\Gamma(q)} |v_1(s) - v_2(s)|ds\Big] \\ \leq &\frac{1}{\Gamma(q)}\Big[1 + \lambda_1 \eta_1^{q-1} + \lambda_2\Big(b(q-1) + \eta_2^{q-1}\Big)\Big] \int_0^1 m(s)\|x - \bar{x}\|ds. \end{aligned}$$

Hence

$$\|h_1 - h_2\| \leq \frac{\|m\|_{L^1}}{\Gamma(q)}\Big[1 + \lambda_1 \eta_1^{q-1} + \lambda_2\Big(b(q-1) + \eta_2^{q-1}\Big)\Big]\|x - \bar{x}\|.$$

Analogously, interchanging the roles of x and $\overline{x}$, we obtain

$$\begin{aligned} H_d(\Omega(x), \Omega(\bar{x})) &\leq \gamma\|x - \bar{x}\| \\ &\leq \frac{\|m\|_{L^1}}{\Gamma(q)}\Big[1 + \lambda_1 \eta_1^{q-1} + \lambda_2\Big(b(q-1) + \eta_2^{q-1}\Big)\Big]\|x - \bar{x}\|. \end{aligned}$$

Since Ω is a contraction, it follows by Theorem 1.19 that Ω has a fixed point x which is a solution of (2.8). This completes the proof. □

2.2 On nonlocal boundary value problems for nonlinear integro-differential equations of arbitrary fractional order

In this section, we discuss the existence of solutions of a nonlocal boundary value problem for nonlinear integro-differential equations of fractional order q:

$$\begin{cases} {}^C D^q x(t) = f(t, x(t), (\phi x)(t), (\psi x)(t)), \ q \in (m-1, m], \ m \in \mathbb{N}, \ m \geq 2, 0 < t < 1, \\ x(0) = \beta x(\eta), \ x'(0) = 0, \ x''(0) = 0, \ldots, x^{(m-2)}(0) = 0, \ x(1) = \alpha x(\eta), \\ 0 < \eta < 1, \ (\alpha - \beta)\eta^{m-1} \neq 1 - \beta, \quad \beta, \ \alpha \in \mathbb{R}, \end{cases} \tag{2.12}$$

where ${}^C D$ is the Caputo fractional derivative operator, $f : [0,1] \times X \times X \times X \to X$ is continuous, and for $\gamma, \delta : [0,1] \times [0,1] \to [0,\infty)$,

$$(\phi x)(t) = \int_0^t \gamma(t,s)x(s)ds, \quad (\psi x)(t) = \int_0^t \delta(t,s)x(s)ds.$$

Here, $(X, \|.\|)$ is a Banach space and $\mathcal{C} = \mathrm{C}([0,1], X)$ denotes the Banach space of all continuous functions from $[0,1] \to X$ endowed with a topology of uniform convergence with the norm denoted by $\|.\|$. One can notice that the boundary conditions considered in the problem (2.12) give rise to different special cases for different values of the parameters β and α.

By a solution of (2.12), we mean a function $x \in \mathcal{C}$ of class $C^m[0,1]$ which satisfies the nonlocal fractional boundary value problem (2.12).

To study the nonlinear problem (2.12), we first consider the associated linear problem and obtain its solution.

Lemma 2.2. *For a given $\sigma \in C[0,1]$, the unique solution of the boundary value problem*

$$\begin{cases} {}^C D^q x(t) = \sigma(t), \ \ 0 < t < 1, \quad q \in (m-1, m], \ m \in \mathbb{N}, \ m \geq 2, \\ x(0) = \beta x(\eta), \ x'(0) = 0, \ x''(0) = 0, \ldots, x^{(m-2)}(0) = 0, \ x(1) = \alpha x(\eta), \\ 0 < \eta < 1, \ (\alpha - \beta)\eta^{m-1} \neq 1 - \beta, \quad \beta, \ \alpha \in \mathbb{R}, \end{cases} \tag{2.13}$$

is given by

$$\begin{aligned} x(t) = &\int_0^t \frac{(t-s)^{q-1}}{\Gamma(q)}\sigma(s)ds + \Big(\frac{\beta\eta^{m-1} + (1-\beta)t^{m-1}}{\beta - 1 + (\alpha-\beta)\eta^{m-1}}\Big)\int_0^1 \frac{(1-s)^{q-1}}{\Gamma(q)}\sigma(s)ds \\ &-\Big(\frac{\beta + (\alpha-\beta)t^{m-1}}{\beta - 1 + (\alpha-\beta)\eta^{m-1}}\Big)\int_0^\eta \frac{(\eta-s)^{q-1}}{\Gamma(q)}\sigma(s)ds. \end{aligned} \tag{2.14}$$

Proof. The general solution of (2.13) can be written as

$$x(t) = \int_0^t \frac{(t-s)^{q-1}}{\Gamma(q)}\sigma(s)ds - c_0 - c_1 t - c_2 t^2 - \ldots - c_{m-1}t^{m-1}, \tag{2.15}$$

where $c_0, c_1, c_2, \ldots, c_{m-1} \in \mathbb{R}$ are arbitrary constants. Applying the boundary conditions of (2.13) in (2.15), we find that

$$\begin{aligned} c_0 = &\Big(\frac{\beta}{\beta - 1 + (\alpha-\beta)\eta^{m-1}}\Big)\int_0^\eta \frac{(\eta-s)^{q-1}}{\Gamma(q)}\sigma(s)ds \\ &- \Big(\frac{\beta\eta^{m-1}}{\beta - 1 + (\alpha-\beta)}\eta^{m-1}\Big)\int_0^1 \frac{(1-s)^{q-1}}{\Gamma(q)}\sigma(s)ds, \end{aligned}$$

$c_1 = 0, \ldots, c_{m-2} = 0$, and

$$\begin{aligned} c_{m-1} = &\Big(\frac{\alpha - \beta}{\beta - 1 + (\alpha-\beta)\eta^{m-1}}\Big)\int_0^\eta \frac{(\eta-s)^{q-1}}{\Gamma(q)}\sigma(s)ds \\ &+ \Big(\frac{\beta - 1}{\beta - 1 + (\alpha-\beta)\eta^{m-1}}\Big)\int_0^1 \frac{(1-s)^{q-1}}{\Gamma(q)}\sigma(s)ds. \end{aligned}$$

Substituting the values of $c_0, \ c_1, \ldots, c_{m-1}$ in (2.15), we obtain (2.14). The converse can be proven by direct computation. This completes the proof. □

2.2.1 *Existence and uniqueness results*

Theorem 2.6. *Assume that:*

(2.6.1) There exist positive functions $L_1(t),\ L_2(t),\ L_3(t)$ *such that*

$$|f(t,x(t),(\phi x)(t),(\psi x)(t)) - f(t,y(t),(\phi y)(t),(\psi y)(t))| \le L_1(t)|x-y| + L_2(t)|\phi x - \phi y| + L_3(t)|\psi x - \psi y|,\ \forall t\in[0,1],\ x,y\in X.$$

(2.6.2)

$$\Lambda = (1+\gamma_0+\delta_0)\Big\{I_L^q + \frac{(|\beta\eta^{m-1}| + |1-\beta|)I^qL(1) + (|\beta| + |\alpha-\beta|)I^qL(\eta)}{|\beta - 1 + (\alpha-\beta)\eta^{m-1}|}\Big\} < 1,$$

where

$$\begin{aligned}
&\gamma_0 = \sup_{t\in[0,1]}\Big|\int_0^t \gamma(t,s)ds\Big|,\quad \delta_0 = \sup_{t\in[0,1]}\Big|\int_0^t \delta(t,s)ds\Big|,\\
&I_L^q = \max\Big\{\sup_{t\in[0,1]}|I^qL_1(t)|,\ \sup_{t\in[0,1]}|I^qL_2(t)|,\ \sup_{t\in[0,1]}|I^qL_3(t)|\Big\},\\
&I^qL(1) = \max\{|I^qL_1(1)|,\ |I^qL_2(1)|,\ |I^qL_3(1)|\},\\
&I^qL(\eta) = \max\{|I^qL_1(\eta)|,\ |I^qL_2(\eta)|,\ |I^qL_3(\eta)|\}.
\end{aligned}\tag{2.16}$$

Then the boundary value problem (2.12) has a unique solution on $[0,1]$.

Proof. Define $\mathcal{G}:\mathcal{C}\to\mathcal{C}$ by

$$\begin{aligned}
(\mathcal{G}x)(t) = &\int_0^t \frac{(t-s)^{q-1}}{\Gamma(q)} f(s,x(s),(\phi x)(s),(\psi x)(s))ds\\
&+\Big(\frac{\beta\eta^{m-1} + (1-\beta)t^{m-1}}{\beta - 1 + (\alpha-\beta)\eta^{m-1}}\Big)\int_0^1 \frac{(1-s)^{q-1}}{\Gamma(q)} f(s,x(s),(\phi x)(s),(\psi x)(s))ds\\
&-\Big(\frac{\beta + (\alpha-\beta)t^{m-1}}{\beta - 1 + (\alpha-\beta)\eta^{m-1}}\Big)\int_0^\eta \frac{(\eta-s)^{q-1}}{\Gamma(q)} f(s,x(s),(\phi x)(s),(\psi x)(s))ds, t\in[0,1].
\end{aligned}$$

Let us set $\sup_{t\in[0,1]}|f(t,0,0,0)| = M$, and choose

$$r \ge \frac{M}{(1-\Lambda)\Gamma(q+1)}\Big\{1 + \frac{(|\beta\eta^{m-1}| + |1-\beta|) + (|\beta| + |\alpha-\beta|)\eta^q}{|\beta - 1 + (\alpha-\beta)\eta^{m-1}|}\Big\},$$

where Λ is given in (2.16). Now we show that $\mathcal{G}B_r \subset B_r$, where $B_r = \{x\in\mathcal{C} : \|x\| \le r\}$. For $x\in B_r$, we have

$$\begin{aligned}
\|\mathcal{G}x\| = &\sup_{t\in[0,1]}|(\mathcal{G}x)(t)|\\
\le &\sup_{t\in[0,1]}\Big\{\int_0^t \frac{(t-s)^{q-1}}{\Gamma(q)}|f(s,x(s),(\phi x)(s),(\psi x)(s))|ds\\
&+\Big|\frac{\beta\eta^{m-1} + (1-\beta)t^{m-1}}{\beta - 1 + (\alpha-\beta)\eta^{m-1}}\Big|\int_0^1 \frac{(1-s)^{q-1}}{\Gamma(q)}|f(s,x(s),(\phi x)(s),(\psi x)(s))|ds\\
&+\Big|\frac{\beta + (\alpha-\beta)t^{m-1}}{\beta - 1 + (\alpha-\beta)\eta^{m-1}}\Big|\int_0^\eta \frac{(\eta-s)^{q-1}}{\Gamma(q)}|f(s,x(s),(\phi x)(s),(\psi x)(s))|ds\Big\}
\end{aligned}$$

$$
\begin{aligned}
&\leq \sup_{t\in[0,1]}\Big\{\int_0^t \frac{(t-s)^{q-1}}{\Gamma(q)}(|f(s,x(s),(\phi x)(s),(\psi x)(s))-f(s,0,0,0)|+|f(s,0,0,0)|)ds\\
&\quad+\Big|\frac{\beta\eta^{m-1}+(1-\beta)t^{m-1}}{\beta-1+(\alpha-\beta)\eta^{m-1}}\Big|\int_0^1 \frac{(1-s)^{q-1}}{\Gamma(q)}(|f(s,x(s),(\phi x)(s),(\psi x)(s))\\
&\quad-f(s,0,0,0)|+|f(s,0,0,0)|)ds\\
&\quad+\Big|\frac{\beta+(\alpha-\beta)t^{m-1}}{\beta-1+(\alpha-\beta)\eta^{m-1}}\Big|\int_0^\eta \frac{(\eta-s)^{q-1}}{\Gamma(q)}(|f(s,x(s),(\phi x)(s),(\psi x)(s))\\
&\quad-f(s,0,0,0)|+|f(s,0,0,0)|)ds\Big\}\\
&\leq\Big\{\int_0^t \frac{(t-s)^{q-1}}{\Gamma(q)}(L_1(s)\|x(s)\|+\gamma_0L_2(s)\|x(s)\|+\delta_0L_3(s)\|x(s)\|+M)ds\\
&\quad+\Big|\frac{\beta\eta^{m-1}+(1-\beta)t^{m-1}}{\beta-1+(\alpha-\beta)\eta^{m-1}}\Big|\int_0^1 \frac{(1-s)^{q-1}}{\Gamma(q)}(L_1(s)\|x(s)\|+\gamma_0L_2(s)\|x(s)\|\\
&\quad+\delta_0L_3(s)\|x(s)\|+M)ds\\
&\quad+\Big|\frac{\beta+(\alpha-\beta)t^{m-1}}{\beta-1+(\alpha-\beta)\eta^{m-1}}\Big|\int_0^\eta \frac{(\eta-s)^{q-1}}{\Gamma(q)}(L_1(s)\|x(s)\|+\gamma_0L_2(s)\|x(s)\|\\
&\quad+\delta_0L_3(s)\|x(s)\|+M)ds\\
&\leq(1+\gamma_0+\delta_0)\Big\{I_L^q+\frac{(|\beta\eta^{m-1}|+|1-\beta|)}{|\beta-1+(\alpha-\beta)\eta^{m-1}|}I^qL(1)\\
&\quad+\frac{(|\beta|+|\alpha-\beta|)}{|\beta-1+(\alpha-\beta)\eta^{m-1}|}I^qL(\eta)\Big\}r\\
&\leq\Lambda r+\frac{M}{\Gamma(q+1)}\Big\{1+\frac{(|\beta\eta^{m-1}|+|1-\beta|)+(|\beta|+|\alpha-\beta|)\eta^q}{|\beta-1+(\alpha-\beta)\eta^{m-1}|}\Big\}\leq r.
\end{aligned}
$$

Now, for $x,y\in\mathcal{C}$ and for each $t\in[0,1]$, we obtain

$$
\begin{aligned}
&|\mathcal{G}x-\mathcal{G}y|=\sup_{t\in[0,1]}|(\mathcal{G}x)(t)-(\mathcal{G}y)(t)|\\
&\leq\sup_{t\in[0,1]}\Big\{\int_0^t \frac{(t-s)^{q-1}}{\Gamma(q)}|f(s,x(s),(\phi x)(s),(\psi x)(s)-f(s,y(s),(\phi y)(s),(\psi y)(s))|ds\\
&\quad+\Big|\frac{\beta\eta^{m-1}+(1-\beta)t^{m-1}}{\beta-1+(\alpha-\beta)\eta^{m-1}}\Big|\int_0^1 \frac{(1-s)^{q-1}}{\Gamma(q)}|f(s,x(s),(\phi x)(s),(\psi x)(s)\\
&\quad-f(s,y(s),(\phi y)(s),(\psi y)(s))|ds\\
&\quad+\Big|\frac{\beta+(\alpha-\beta)t^{m-1}}{\beta-1+(\alpha-\beta)\eta^{m-1}}\Big|\int_0^\eta \frac{(\eta-s)^{q-1}}{\Gamma(q)}|f(s,x(s),(\phi x)(s),(\psi x)(s)\\
&\quad-f(s,y(s),(\phi y)(s),(\psi y)(s))|ds\Big\}\\
&\leq\sup_{t\in[0,1]}\Big\{\Big(I^qL_1(t)+\gamma_0I^qL_2(t)+\delta_0I^qL_3(t)\Big)\\
&\quad+\Big|\frac{\beta\eta^{m-1}+(1-\beta)t^{m-1}}{\beta-1+(\alpha-\beta)\eta^{m-1}}\Big|\Big(I^qL_1(1)+\gamma_0I^qL_2(1)+\delta_0I^qL_3(1)\Big)\\
&\quad+\Big|\frac{\beta+(\alpha-\beta)t^{m-1}}{\beta-1+(\alpha-\beta)\eta^{m-1}}\Big|\Big(I^qL_1(\eta)+\gamma_0I^qL_2(\eta)+\delta_0I^qL_3(\eta)\Big)\Big\}\|x-y\|
\end{aligned}
$$

$$\leq (1+\gamma_0+\delta_0)\Big\{I_L^q + \frac{(|\beta\eta^{m-1}|+|1-\beta|)I^qL(1)+(|\beta|+|\alpha-\beta|)I^qL(\eta)}{|\beta-1+(\alpha-\beta)\eta^{m-1}|}\Big\}\|x-y\|$$
$$\leq \Lambda\|x-y\|,$$

where we have used the assumption (2.6.2) in the last inequality. Clearly $\mathcal{G}$ is a contraction. Thus, by the contraction mapping principle, we obtain the conclusion of the theorem. This completes the proof. □

Example 2.2. Consider the following boundary value problem

$$\begin{cases} {}^cD^{\frac{5}{2}}x(t) = \dfrac{t}{2}\dfrac{|x|}{1+|x|} + \dfrac{1}{5}\displaystyle\int_0^t \dfrac{e^{-(s-t)}}{5}x(s)ds + \dfrac{1}{5}\displaystyle\int_0^t \dfrac{e^{-(s-t)/2}}{5}x(s)ds, \quad t\in[0,1], \\ x(0) = \frac{1}{2}x(\frac{1}{3}), \quad x'(0)=0, \quad x(1)=x(\frac{1}{3}). \end{cases} \tag{2.17}$$

Here, $m=3$, $q=\frac{5}{2}$, $\gamma(t,s)=\frac{e^{-(s-t)}}{5}$, $\delta=\frac{e^{-(s-t)/2}}{5}$ $\beta=\frac{1}{2}$, $\alpha=1$, $\eta=\frac{1}{3}$. With $\gamma_0=\frac{e-1}{5}$, $\delta_0=\frac{2(\sqrt{e}-1)}{5}$, we find that

$$\Lambda = \frac{12(e+2(\sqrt{e}+1))}{125\sqrt{\pi}} < 1.$$

Thus, by Theorem 2.6, the boundary value problem (2.17) has a unique solution on $[0,1]$.

Theorem 2.7. *Suppose that $f:[0,1]\times X\times X\times X\to X$ is jointly continuous and maps bounded subsets of $[0,1]\times X\times X\times X$ into relatively compact subsets of X and the assumptions (2.6.1) and (2.6.2) hold with*

$$\Lambda_1 = (1+\gamma_0+\delta_0)\Big\{\frac{(|\beta\eta^{m-1}|+|1-\beta|)I^qL(1)+(|\beta|+|\alpha-\beta|)I^qL(\eta)}{|\beta-1+(\alpha-\beta)\eta^{m-1}|}\Big\} < 1.$$

In addition we assume that:

(2.7.1) $|f(t,x(t),(\phi x)(t),(\psi x)(t))| \leq \mu(t)$, $\forall (t,x,\phi x,\psi x)\in[0,1]\times X\times X\times X$, $\mu\in C([0,1],R^+)$.

Then there exists at least one solution of the boundary value problem (2.12) on $[0,1]$.

Proof. Let us fix

$$r \geq \frac{\|\mu\|}{\Gamma(q+1)}\Big\{1+\frac{(|\beta\eta^{m-1}|+|1-\beta|)+(|\beta|+|\alpha-\beta|)\eta^q)}{|\beta-1+(\alpha-\beta)\eta^{m-1}|}\Big\},$$

and consider $B_r=\{x\in\mathcal{C}:\|x\|\leq r\}$. We define the operators Θ_1 and Θ_2 on B_r by

$$(\Theta_1x)(t) = \int_0^t \frac{(t-s)^{q-1}}{\Gamma(q)}f(s,x(s),(\phi x)(s),(\psi x)(s))ds,$$

$$(\Theta_2x)(t) = \Big(\frac{\beta\eta^{m-1}+(1-\beta)t^{m-1}}{\beta-1+(\alpha-\beta)\eta^{m-1}}\Big)\int_0^1 \frac{(1-s)^{q-1}}{\Gamma(q)}f(s,x(s),(\phi x)(s),(\psi x)(s))ds$$
$$-\Big(\frac{\beta+(\alpha-\beta)t^{m-1}}{\beta-1+(\alpha-\beta)\eta^{m-1}}\Big)\int_0^\eta \frac{(\eta-s)^{q-1}}{\Gamma(q)}f(s,x(s),(\phi x)(s),(\psi x)(s))ds.$$

For $x, y \in B_r$, we find that

$$\|\Theta_1 x + \Theta_2 y\| \leq \frac{\|\mu\|}{\Gamma(q+1)}\Big\{1 + \frac{(|\beta\eta^{m-1}| + |1-\beta|) + (|\beta| + |\alpha-\beta|\eta^q)}{|\beta - 1 + (\alpha-\beta)\eta^{m-1}|}\Big\} \leq r.$$

Thus, $\Theta_1 x + \Theta_2 y \in B_r$. It follows from the assumption (2.6.1) that Θ_2 is a contraction mapping for $\Lambda_1 < 1$. Continuity of f implies that the operator Θ_1 is continuous. Also, Θ_1 is uniformly bounded on B_r as

$$\|\Theta_1 x\| \leq \frac{\|\mu\|}{\Gamma(q+1)}.$$

To show that the operator Θ_1 is compact, we use the classical Arzela-Ascoli's Theorem. Let $\mathcal{A}$ be a bounded subset of $\mathcal{C}$. We have to show that $\Phi(\mathcal{A})$ is equi-continuous and for each t, the set $\Phi(\mathcal{A})(t)$ is relatively compact in X. In view of (2.6.1), (2.7.1), we define $\sup_{(t,x,\phi x,\psi x)\in[0,1]\times B_r\times B_r\times B_r} \|f(t,x,\phi x,\psi x)\| = f_{\max}$, and consequently we have

$$\begin{aligned}
&|(\Theta_1 x)(t_1) - (\Theta_1 x)(t_2)| \\
&= \Big|\frac{1}{\Gamma(q)}\int_0^{t_1}\Big((t_2-s)^{q-1} - (t_1-s)^{q-1}\Big) f(s,x(s),\phi x(s),\psi x(s))ds \\
&\quad + \int_{t_1}^{t_2}(t_2-s)^{q-1} f(s,x(s),\phi x(s),\psi x(s))ds\Big| \\
&\leq \frac{f_{\max}}{\Gamma(q+1)}|2(t_2-t_1)^q + t_1^q - t_2^q|,
\end{aligned}$$

which is independent of x. Thus, Θ_1 is equi-continuous. Using the fact that f maps bounded subsets into relatively compact subsets, we have that $\Theta_1(\mathcal{A})(t)$ is relatively compact in X for every t. Therefore, Θ_1 is relatively compact on B_r. Hence, by Arzela-Ascoli's Theorem, Θ_1 is compact on B_r. Thus all the assumptions of Theorem 1.4 are satisfied and so the boundary value problem (2.12) has at least one solution on $[0,1]$. □

2.3 Existence theory for nonlocal boundary value problems involving mixed fractional derivative operators

In this section, we introduce a new class of nonlocal boundary value problems (BVP for short) of mixed fractional differential equations and inclusions involving both left Caputo and right Riemann-Liouville fractional derivative operators and obtain some existence and uniqueness results for the problems at hand. In precise terms, we investigate the problems:

$$\begin{cases} {}^C D_{1-}^{\alpha}\, {}^{RL}D_{0+}^{\beta} y(t) = f(t,y(t)), & t \in J := [0,1], \\ y(0) = y'(0) = 0, \quad y(1) = \delta y(\eta), & 0 < \eta < 1, \end{cases} \tag{2.18}$$

and

$$\begin{cases} {}^C D_{1-}^{\alpha}\, {}^{RL}D_{0+}^{\beta} y(t) \in F(t,y(t)), & t \in J := [0,1], \\ y(0) = y'(0) = 0, \quad y(1) = \delta y(\eta), & 0 < \eta < 1, \end{cases} \tag{2.19}$$

where ${}^{C}D^{\alpha}_{1-}$ and ${}^{RL}D^{\beta}_{0+}$ denote the right Caputo fractional derivative operator operator of order $\alpha \in (1,2]$ and the left Riemann-Liouville fractional derivative operator operator of order $\beta \in (0,1]$ respectively, $f : J \times \mathbb{R} \to \mathbb{R}$ is a given function, $F : [0,1] \times \mathbb{R} \to \mathcal{P}(\mathbb{R})$ is a multi-valued map, $\mathcal{P}(\mathbb{R})$ is the family of all non-empty subsets of $\mathbb{R}$ and $\delta \in \mathbb{R}$ is an appropriate constant.

The following lemma, dealing with a linear variant of the problem (2.18), plays an important role in the forthcoming analysis.

Lemma 2.3. *Let $h \in C(J, \mathbb{R})$ and $\delta \neq \eta^{-(\beta+1)}$. The function y is a solution of the problem*

$$\begin{cases} {}^{C}D^{\alpha}_{1-}\,{}^{RL}D^{\beta}_{0+}y(t) = h(t), \quad t \in J := [0,1], \\ y(0) = y'(0) = 0, \quad y(1) - \delta y(\eta), \quad 0 < \eta < 1, \end{cases} \tag{2.20}$$

if and only if

$$\begin{aligned} y(t) = {}& \frac{1}{\Gamma(\beta)} \int_0^t (t-s)^{\beta-1} I^{\alpha}_{1-}h(s)ds \\ & + \frac{t^{\beta+1}}{(1-\delta\eta^{\beta+1})\Gamma(\beta)} \left[\delta \int_0^{\eta} (\eta - s)^{\beta-1} I^{\alpha}_{1-}h(s)ds \right. \\ & \left. - \int_0^1 (1-s)^{\beta-1} I^{\alpha}_{1-}h(s)ds \right], \end{aligned} \tag{2.21}$$

where $I^{\alpha}_{1-}y(s)$ is defined by (1.1).

Proof. Applying the right fractional integral I^{α}_{1-} to the mixed fractional differential equation in (2.20) and then the left fractional integral I^{β}_{0+} to the resulting equation, and using the properties of Caputo and Riemann-Liouville fractional derivative operators, we get

$$\begin{aligned} y(t) &= I^{\beta}_{0+}(I^{\alpha}_{1-}h(t) + c_0 + c_1 t) + c_2 t^{\beta-1} \\ &= I^{\beta}_{0+}I^{\alpha}_{1-}h(t) + c_0 \frac{t^{\beta}}{\Gamma(\beta+1)} + c_1 \frac{t^{\beta+1}}{\Gamma(\beta+2)} + c_2 t^{\beta-1}. \end{aligned} \tag{2.22}$$

Inserting the conditions $y(0) = 0$ and $y'(0) = 0$ in (2.22) yields $c_2 = 0$ and $c_0 = 0$ respectively and consequently, (2.22) reduces to

$$y(t) = I^{\beta}_{0+}I^{\alpha}_{1-}h(t) + c_1 \frac{t^{\beta+1}}{\Gamma(\beta+2)}. \tag{2.23}$$

Making use of the condition $y(1) = \delta y(\eta)$ in the equation (2.23) yields

$$c_1 = \frac{\Gamma(\beta+2)}{1-\delta\eta^{\beta+1}} \left[\delta I^{\beta}_{0+}I^{\alpha}_{1-}h(t)|_{t=\eta} - I^{\beta}_{0+}I^{\alpha}_{1-}h(t)|_{t=1}\right],$$

which, on substituting in (2.23), completes the solution (2.21). The converse follows by direct computation. The proof is completed. □

2.3.1 *Existence and uniqueness results for problem (2.18)*

Let $\mathcal{X} = C([0,1],\mathbb{R})$ denote the Banach space of all continuous functions from $[0,1] \to \mathbb{R}$ equipped with the norm $\|y\| = \sup\{|y(t)| : t \in [0,1]\}$.

By Lemma 2.3, the problem (2.18) can be transformed into a fixed point problem as

$$y = \mathcal{G}y, \tag{2.24}$$

where the operator $\mathcal{G} : \mathcal{X} \to \mathcal{X}$ is defined by

$$\begin{aligned}\mathcal{G}y(t) &= \frac{1}{\Gamma(\beta)}\int_0^t (t-s)^{\beta-1} I_{1-}^{\alpha} f(s,y(s))ds \\ &\quad + \frac{t^{\beta+1}}{(1-\delta\eta^{\beta+1})\Gamma(\beta)}\Big[\delta\int_0^{\eta}(\eta-s)^{\beta-1} I_{1-}^{\alpha} f(s,y(s))ds \\ &\quad - \int_0^1 (1-s)^{\beta-1} I_{1-}^{\alpha} f(s,y(s))ds\Big].\end{aligned} \tag{2.25}$$

Our first result deals with the existence and uniqueness of solutions for the problem (2.18).

Theorem 2.8. *Let $f : [0,1] \times \mathbb{R} \to \mathbb{R}$ be a continuous function such that*

(2.8.1) $|f(t,x) - f(t,y)| \le L|x-y|$, for all $t \in [0,1]$, $x, y \in \mathbb{R}$, $L > 0$.

Then the problem (2.18) has a unique solution on $[0,1]$ if

$$L\Omega_1 < 1, \tag{2.26}$$

where

$$\Omega_1 = \frac{1}{\Gamma(\alpha+1)\Gamma(\beta+1)}\Big[1 + \frac{|\delta|\eta^{\beta}+1}{|1-\delta\eta^{\beta+1}|}\Big]. \tag{2.27}$$

Proof. Let us define $\sup_{t\in[0,1]}|f(t,0)| = M$ and select $r \ge \dfrac{M\Omega_1}{1-L\Omega_1}$ to establish that $\mathcal{G}\mathcal{B}_r \subset \mathcal{B}_r$, where $\mathcal{B}_r = \{y \in \mathcal{X} : \|y\| \le r\}$ and $\mathcal{G}$ is defined by (2.25). Using the condition (2.8.1), we have

$$\begin{aligned}|f(t,y)| &= |f(t,y) - f(t,0) + f(t,0)| \le |f(t,y) - f(t,0)| + |f(t,0)| \\ &\le L\|y\| + M \le Lr + M.\end{aligned} \tag{2.28}$$

Then, for $y \in \mathcal{B}_r$, we obtain

$$\begin{aligned}\|\mathcal{G}y\| &\le \sup_{t\in[0,1]}\Big\{\int_0^t \frac{(t-s)^{\beta-1}}{\Gamma(\beta)}\int_s^1 \frac{(u-s)^{\alpha-1}}{\Gamma(\alpha)}|f(u,y(u))|duds \\ &\quad + \frac{t^{\beta+1}}{|1-\delta\eta^{\beta+1}|}\Big[|\delta|\int_0^{\eta}\frac{(\eta-s)^{\beta-1}}{\Gamma(\beta)}\int_s^1\frac{(u-s)^{\alpha-1}}{\Gamma(\alpha)}|f(u,y(u))|duds \\ &\quad + \int_0^1\frac{(1-s)^{\beta-1}}{\Gamma(\beta)}\int_s^1\frac{(u-s)^{\alpha-1}}{\Gamma(\alpha)}|f(u,y(u))|duds\Big]\Big\} \\ &\le (Lr+M)\Omega_1 < r.\end{aligned}$$

This show that $\mathcal{G}y \in \mathcal{B}_r$, $y \in \mathcal{B}_r$. Thus $\mathcal{G}\mathcal{B}_r \subset \mathcal{B}_r$. Next we show that $\mathcal{G}$ is a contraction. For that, let $y, z \in \mathcal{X}$. Then, for each $t \in [0,1]$, we have

$$\begin{aligned}
&\|(\mathcal{G}y) - (\mathcal{G}z)\| \\
&\leq \sup_{t\in[0,1]} \left\{ \int_0^t \frac{(t-s)^{\beta-1}}{\Gamma(\beta)} \int_s^1 \frac{(u-s)^{\alpha-1}}{\Gamma(\alpha)} |f(u,y(u)) - f(u,z(u))| duds \right. \\
&\quad + \frac{t^{\beta+1}}{|1-\delta\eta^{\beta+1}|} \left[|\delta| \int_0^\eta \frac{(\eta-s)^{\beta-1}}{\Gamma(\beta)} \int_s^1 \frac{(u-s)^{\alpha-1}}{\Gamma(\alpha)} |f(u,y(u)) - f(u,z(u))| duds \right. \\
&\quad \left.\left. + \int_0^1 \frac{(1-s)^{\beta-1}}{\Gamma(\beta)} \int_s^1 \frac{(u-s)^{\alpha-1}}{\Gamma(\alpha)} |f(u,y(u)) - f(u,z(u))| duds \right] \right\} \\
&\leq L\Omega_1 \|y-z\|,
\end{aligned}$$

which, in view of the given condition $L\Omega_1 < 1$, implies that $\mathcal{G}$ is a contraction. In consequence, it follows by the contraction mapping principle that there exists a unique solution for the problem (2.18) on $[0,1]$. This completes the proof. □

Our next existence result for the problem (2.18), is based on Krasnosel'skiǐ's fixed point theorem (Theorem 1.4).

Theorem 2.9. *Let $f : [0,1] \times \mathbb{R} \to \mathbb{R}$ be a continuous function satisfying the condition (2.8.1). In addition we assume that:*

(2.9.1) $|f(t,y)| \leq m(t)$, for all $(t,y) \in [0,1] \times \mathbb{R}$ and $m \in C([0,1], \mathbb{R}^+)$.

Then there exists at least one solution for the problem (2.18) on $[0,1]$ if

$$\frac{L}{\Gamma(\alpha+1)\Gamma(\beta+1)} < 1. \tag{2.29}$$

Proof. Setting $\sup_{t\in[0,1]} |m(t)| = \|m\|$, we fix

$$\varrho \geq \|m\|\Omega_1, \tag{2.30}$$

where Ω_1 is defined by (2.27), and consider $B_\varrho = \{y \in \mathcal{X} : \|y\| \leq \varrho\}$. Introduce the operators $\mathcal{G}_1$ and $\mathcal{G}_2$ on B_ϱ as follows:

$$\mathcal{G}_1 y(t) = \int_0^t \frac{(t-s)^{\beta-1}}{\Gamma(\beta)} \int_s^1 \frac{(u-s)^{\alpha-1}}{\Gamma(\alpha)} f(u,y(u)) duds,$$

and

$$\begin{aligned}
\mathcal{G}_2 y(t) = \frac{t^{\beta+1}}{(1-\delta\eta^{\beta+1})} &\left[\delta \int_0^\eta \frac{(\eta-s)^{\beta-1}}{\Gamma(\beta)} \int_s^1 \frac{(u-s)^{\alpha-1}}{\Gamma(\alpha)} f(u,y(u)) duds \right. \\
&\left. - \int_0^1 \frac{(1-s)^{\beta-1}}{\Gamma(\beta)} \int_s^1 \frac{(u-s)^{\alpha-1}}{\Gamma(\alpha)} f(u,y(u)) duds \right].
\end{aligned}$$

Observe that $\mathcal{G} = \mathcal{G}_1 + \mathcal{G}_2$. Now we verify the hypotheses of Krasnosel'skiǐ's fixed point theorem in the following steps.

(i) For $y, z \in B_\varrho$, we have

$$\begin{aligned}
\|\mathcal{G}_1 y + \mathcal{G}_2 z\| &= \sup_{t\in[0,1]} |(\mathcal{G}_1 y)(t) + (\mathcal{G}_2 z)(t)| \\
&\le \sup_{t\in[0,1]} \left\{ \int_0^t \frac{(t-s)^{\beta-1}}{\Gamma(\beta)} \int_s^1 \frac{(u-s)^{\alpha-1}}{\Gamma(\alpha)} |f(u, y(u))| duds \right. \\
&\quad + \frac{t^{\beta+1}}{|1-\delta\eta^{\beta+1}|} \left[|\delta| \int_0^\eta \frac{(\eta-s)^{\beta-1}}{\Gamma(\beta)} \int_s^1 \frac{(u-s)^{\alpha-1}}{\Gamma(\alpha)} |f(u, z(u))| duds \right. \\
&\quad \left.\left. + \int_0^1 \frac{(1-s)^{\beta-1}}{\Gamma(\beta)} \int_s^1 \frac{(u-s)^{\alpha-1}}{\Gamma(\alpha)} |f(u, z(u))| duds \right] \right\} \\
&\le \|m\|\Omega_1 \le \varrho,
\end{aligned}$$

where we have used (2.30). Thus $\mathcal{G}_1 y + \mathcal{G}_2 z \in B_\varrho$.

(ii) It is easy to show that $\mathcal{G}_1$ is a contraction, by using the assumption (2.8.1) together with (2.29).

(iii) Using the continuity of f, it is easy to show that the operator $\mathcal{G}_2$ is continuous. Further, $\mathcal{G}_2$ is uniformly bounded on B_ϱ as

$$\|\mathcal{G}_2 x\| = \sup_{t\in[0,1]} |(\mathcal{G}_2 y)(t)| \le \frac{\|m\|(|\delta|\eta^\beta + 1)}{|1-\delta\eta^{\beta+1}|\Gamma(\alpha+1)\Gamma(\beta+1)}.$$

In order to establish that $\mathcal{G}_2$ is compact, we define $\sup_{(t,y)\in[0,1]\times B_\varrho} |f(t,y)| = \overline{f}$. Thus, for $0 < t_1 < t_2 < 1$, we have

$$\begin{aligned}
|(\mathcal{G}_2 y)(t_2) - (\mathcal{G}_1 2y)(t_1)| &\le \frac{\overline{f}|t_2^{\beta+1} - t_1^{\beta+1}|}{|1-\delta\eta^{\beta+1}|} \left[|\delta| \int_0^\eta \frac{(\eta-s)^{\beta-1}}{\Gamma(\beta)} \int_s^1 \frac{(u-s)^{\alpha-1}}{\Gamma(\alpha)} duds \right. \\
&\quad \left. + \int_0^1 \frac{(1-s)^{\beta-1}}{\Gamma(\beta)} \int_s^1 \frac{(u-s)^{\alpha-1}}{\Gamma(\alpha)} duds \right] \\
&\le \frac{\overline{f}|t_2^{\beta+1} - t_1^{\beta+1}|}{|1-\delta\eta^{\beta+1}|} \left[\frac{|\delta|\eta^\beta + 1}{\Gamma(\alpha+1)\Gamma(\beta+1)} \right] \to 0 \text{ as } t_1 \to t_2,
\end{aligned}$$

independent of y. This shows that $\mathcal{G}_2$ is relatively compact on B_ϱ. As all the conditions of the Arzelá-Ascoli theorem are satisfied, so $\mathcal{G}_2$ is compact on B_ϱ. In view of steps *(i)-(iii)*, the conclusion of Krasnosel'skiĭ's fixed point theorem applies and hence there exists at least one solution for the problem (2.18) on $[0, 1]$. The proof is finished. □

Remark 2.1. Interchanging the role of the operators $\mathcal{G}_1$ and $\mathcal{G}_2$ in the foregoing result, we can obtain a second result by requiring the condition:

$$L \frac{|\delta|\eta^\beta + 1}{\Gamma(\alpha+1)\Gamma(\beta+1)} < 1,$$

instead of (2.29).

The following existence result is based on Leray-Schauder nonlinear alternative (Theorem 1.8).

Theorem 2.10. *Let $f : [0,1] \times \mathbb{R} \to \mathbb{R}$ be a continuous function. Assume that*

(2.10.1) There exist a function $g \in C([0,1], \mathbb{R}^+)$, and a non-decreasing function $\psi : \mathbb{R}^+ \to \mathbb{R}^+$ such that $|f(t,y)| \le g(t)\psi(\|y\|)$, $\forall (t,y) \in [0,1] \times \mathbb{R}$;

(2.10.2) There exists a constant $K > 0$ such that

$$\frac{K}{\|g\|\psi(K)\Omega_1} > 1.$$

Then the problem (2.18) has at least one solution on $[0,1]$.

Proof. Consider the operator $\mathcal{G} : \mathcal{X} \to \mathcal{X}$ defined by (2.25). We show that $\mathcal{G}$ maps bounded sets into bounded sets in $\mathcal{X} = C([0,1], \mathbb{R})$. For a positive number r, let $\mathcal{B}_r = \{y \in C([0,1], \mathbb{R}) : \|y\| \le r\}$ be a bounded set in $\mathcal{X}$. Then, by using the fact that $(p-s)^{\alpha-1} \le 1$ $(1 < \alpha \le 2)$, we have

$$\begin{aligned}|\mathcal{G}y(t)| &\le \|g\|\psi(r)\Bigg\{\int_0^t \frac{(t-s)^{\beta-1}}{\Gamma(\beta)}\int_s^1 \frac{(u-s)^{\alpha-1}}{\Gamma(\alpha)}duds \\ &\quad + \frac{t^{\beta+1}}{|1-\delta\eta^{\beta+1}|}\Bigg[|\delta|\int_0^\eta \frac{(\eta-s)^{\beta-1}}{\Gamma(\beta)}\int_s^1 \frac{(u-s)^{\alpha-1}}{\Gamma(\alpha)}duds \\ &\quad + \int_0^1 \frac{(1-s)^{\beta-1}}{\Gamma(\beta)}\int_s^1 \frac{(u-s)^{\alpha-1}}{\Gamma(\alpha)}duds\Bigg]\Bigg\} \\ &\le \|g\|\psi(r)\Omega_1,\end{aligned}$$

which, on taking the norm for $t \in [0,1]$, yields

$$\|\mathcal{G}y\| \le \|g\|\psi(r)\Omega_1,$$

where Ω_1 is defined by (2.27). Next we show that $\mathcal{G}$ maps bounded sets into equicontinuous sets of $\mathcal{X}$. Let $t_1, t_2 \in [0,1]$ with $t_1 < t_2$ and $y \in \mathcal{B}_r$, where $\mathcal{B}_r$ is a bounded set of $\mathcal{X}$. Then, using the fact that $(p-s)^{\alpha-1} \le 1$ $(1 < \alpha \le 2)$, we obtain

$$\begin{aligned}&|\mathcal{G}y(t_2) - \mathcal{G}y(t_1)| \\ &\le \|g\|\psi(r)\Bigg\{\Bigg|\int_0^{t_1} \frac{[(t_2-s)^{\beta-1} - (t_1-s)^{\beta-1}]}{\Gamma(\beta)\Gamma(\alpha+1)}ds + \int_{t_1}^{t_2} \frac{(t_2-s)^{\beta-1}}{\Gamma(\beta)\Gamma(\alpha+1)}ds\Bigg| \\ &\quad + \frac{|t_2^{\beta+1} - t_1^{\beta+1}|}{|1-\delta\eta^{\beta+1}|}\Bigg[|\delta|\int_0^\eta \frac{(\eta-s)^{\beta-1}}{\Gamma(\beta)}\int_s^1 \frac{(u-s)^{\alpha-1}}{\Gamma(\alpha)}duds \\ &\quad + \int_0^1 \frac{(1-s)^{\beta-1}}{\Gamma(\beta)}\int_s^1 \frac{(u-s)^{\alpha-1}}{\Gamma(\alpha)}duds\Bigg]\Bigg\} \\ &\le \|g\|\psi(r)\Bigg\{\frac{2(t_2-t_1)^\beta + t_2^\beta - t_1^\beta}{\Gamma(\beta+1)\Gamma(\alpha+1)} + \frac{|t_2^{\beta+1} - t_1^{\beta+1}|}{|1-\delta\eta^{\beta+1}|}\Bigg[\frac{|\delta|\eta^\beta + 1}{\Gamma(\alpha+1)\Gamma(\beta+1)}\Bigg]\Bigg\},\end{aligned}$$

which tends to zero independently of $y \in \mathcal{B}_r$ as $t_2 - t_1 \to 0$. As $\mathcal{G}$ satisfies the above assumptions, therefore it follows by the Arzelá-Ascoli theorem that $\mathcal{G} : \mathcal{X} \to \mathcal{X}$ is completely continuous.

The result will follow from the Leray-Schauder nonlinear alternative (Theorem 1.8) once it is shown that the set of all solutions to the equation $y = \lambda\mathcal{G}y$ is bounded for $\lambda \in [0,1]$. For that, let y be a solution of $y = \lambda \mathcal{G} y$ for $\lambda \in [0,1]$. Then, for $t \in [0,1]$, we have

$$\begin{aligned}|y(t)| = |\lambda \mathcal{G} y(t)| &\leq g(t)\psi(\|y\|)\Bigg\{ \int_0^t \frac{(t-s)^{\beta-1}}{\Gamma(\beta)} \int_s^1 \frac{(u-s)^{\alpha-1}}{\Gamma(\alpha)} duds \\ &\quad + \frac{t^{\beta+1}}{(1-\delta\eta^{\beta+1})}\Big[\delta \int_0^\eta \frac{(\eta-s)^{\beta-1}}{\Gamma(\beta)} \int_s^1 \frac{(u-s)^{\alpha-1}}{\Gamma(\alpha)} duds \\ &\quad + \int_0^1 \frac{(1-s)^{\beta-1}}{\Gamma(\beta)} \int_s^1 \frac{(u-s)^{\alpha-1}}{\Gamma(\alpha)} duds\Big]\Bigg\} \\ &\leq \|g\|\psi(\|y\|)\Omega_1,\end{aligned}$$

which implies that

$$\frac{\|y\|}{\|g\|\psi(\|y\|)\Omega_1} \leq 1.$$

In view of (2.10.2), there is no solution y such that $\|y\| \neq K$. Let us set

$$U = \{y \in \mathcal{X} : \|y\| < K\}.$$

The operator $\mathcal{G} : \overline{U} \to \mathcal{X}$ is continuous and completely continuous. From the choice of U, there is no $u \in \partial U$ such that $u = \lambda \mathcal{G}(u)$ for some $\lambda \in (0,1)$. Consequently, by the nonlinear alternative of Leray-Schauder type (Theorem 1.8), we deduce that $\mathcal{G}$ has a fixed point $u \in \overline{U}$ which is a solution of the problem (2.18). This completes the proof. □

2.3.2 *Examples*

In this subsection we construct examples for the illustration of the results obtained in the last section. For that, we consider the following problem:

$$\begin{cases} {}^C D_{1-}^{3/2}\, {}^{RL} D_{0+}^{1/2} y(t) = f(t, y(t)), \ t \in J := [0,1], \\ y(0) = y'(0) = 0, \quad y(1) = 2y(3/4). \end{cases} \tag{2.31}$$

Here $\alpha = 3/2, \beta = 1/2, \eta = 3/4, \delta = 2$ ($\delta \neq 1/\eta^{\beta+1}$), and

$$f(t,y) = \frac{e^t}{51}\tan^{-1} y + \frac{|y|}{(t^2+51)(1+|y|)} + \frac{1}{\sqrt{t^2+1}}. \tag{2.32}$$

With the given values of the parameters, it is found that

$$\Omega_1 = \frac{8}{3\pi}\Big[1 + \frac{\sqrt{3}+1}{|1 - \frac{3\sqrt{3}}{4}|}\Big] \approx 8.84,$$

$|f(t,y_1) - f(t,y_2)| \leq L|y_1 - y_2|$, $L = (e+1)/51$ and $L\Omega_1 \approx 0.643 < 1$. Clearly all the assumptions of Theorem 2.8 hold true and consequently its conclusion applies to the problem (2.31).

In order to illustrate Theorem 2.9, we notice that (2.29) is satisfied as

$$\frac{L}{\Gamma(\alpha+1)\Gamma(\beta+1)} = \frac{8(e+1)}{153\pi} \approx 0.061 < 1,$$

and

$$|f(t,y)| \leq m(t) = \frac{\pi e^t}{102} + \frac{1}{(t^2+9)} + \frac{1}{\sqrt{t^2+1}}.$$

As the hypothesis of Theorem 2.9 is satisfied, we deduce from the conclusion of Theorem 2.9 that the problem (2.31) has at least one solution on $[0,1]$.

Now we demonstrate the application of Theorem 2.10 by considering the nonlinear function

$$f(t,y) = \frac{1}{8\sqrt{t^2+4}}\Big(\sin y + \cos y + 3\Big). \tag{2.33}$$

Clearly $|f(t,y)| \leq g(t)\psi(\|y\|)$, where $g(t) = \frac{1}{8\sqrt{t^2+4}}$, $\psi(\|y\| = (4 + \|y\|)$. By the condition (2.10.2), we find that $K > 5.072625$. Thus all the conditions of Theorem 2.10 hold true and consequently, the problem (2.31) with $f(t,y)$ given by (2.33) has at least one solution on $[0,1]$.

2.3.3 *Existence results for problem (2.19)*

Definition 2.2. A function $y \in C^3([0,1],\mathbb{R})$ is said to be a solution of the boundary value problem (2.19) if $y(0) = y'(0) = 0$, $y(1) = \delta y(\eta)$, and there exists a function $v \in S_{F,y}$ such that $v(t) \in F(t,y(t))$ and

$$\begin{aligned} y(t) = &\int_0^t \frac{(t-s)^{\beta-1}}{\Gamma(\beta)} \int_s^1 \frac{(u-s)^{\alpha-1}}{\Gamma(\alpha)} v(u)duds \\ &+ \frac{t^{\beta+1}}{(1-\delta\eta^{\beta+1})}\Big[\delta \int_0^\eta \frac{(\eta-s)^{\beta-1}}{\Gamma(\beta)} \int_s^1 \frac{(u-s)^{\alpha-1}}{\Gamma(\alpha)} v(u)duds \\ &- \int_0^1 \frac{(1-s)^{\beta-1}}{\Gamma(\beta)} \int_s^1 \frac{(u-s)^{\alpha-1}}{\Gamma(\alpha)} v(u)duds\Big]. \end{aligned}$$

In the case when F has convex values we prove an existence result based on nonlinear alternative of Leray-Schauder type (Theorem 1.16).

Theorem 2.11. *Assume that:*

(2.11.1) $F : [0,1] \times \mathbb{R} \to \mathcal{P}(\mathbb{R})$ *is* L^1*-Carathéodory and has non-empty compact and convex values;*

(2.11.2) there exist a function $\phi \in C([0,1],\mathbb{R}^+)$*, and a non-decreasing function* $\Omega : \mathbb{R}^+ \to \mathbb{R}^+$ *such that*

$$\|F(t,y)\|_{\mathcal{P}} := \sup\{|w| : w \in F(t,y)\} \leq \phi(t)\Omega(|y|)$$

for each $(t,y) \in [0,1] \times \mathbb{R}$;

(2.11.3) there exists a constant $M > 0$ *such that*

$$\frac{M}{\|\phi\|\Omega_1\Omega(M)} > 1,$$

where Ω_1 *is defined by (2.27).*

Then the boundary value problem (2.19) has at least one solution on $[0,1]$.

Proof. Define an operator $\Omega_F : \mathcal{X} \to \mathcal{P}(\mathcal{X})$ by

$$\Omega_F(y) = \left\{ \begin{array}{l} h \in \mathcal{X} : \\ h(t) = \left\{ \begin{array}{l} \displaystyle\int_0^t \frac{(t-s)^{\beta-1}}{\Gamma(\beta)} \int_s^1 \frac{(u-s)^{\alpha-1}}{\Gamma(\alpha)} v(u)duds \\ \displaystyle + \frac{t^{\beta+1}}{(1-\delta\eta^{\beta+1})}\Big[\delta \int_0^\eta \frac{(\eta-s)^{\beta-1}}{\Gamma(\beta)} \int_s^1 \frac{(u-s)^{\alpha-1}}{\Gamma(\alpha)} v(u)duds \\ \displaystyle - \int_0^1 \frac{(1-s)^{\beta-1}}{\Gamma(\beta)} \int_s^1 \frac{(u-s)^{\alpha-1}}{\Gamma(\alpha)} v(u)duds\Big], \end{array} \right. \end{array} \right\}$$

for $v \in S_{F,y}$. We will show that Ω_F satisfies the assumptions of the nonlinear alternative of Leray-Schauder type. The proof consists of several steps. As a first step, we show that Ω_F *is convex for each* $y \in C([0,1],\mathbb{R})$. This step is obvious since $S_{F,y}$ is convex (F has convex values), and therefore we omit the proof.

In the second step, we show that Ω_F *maps bounded sets (balls) into bounded sets in* $\mathcal{X}$. For a positive number ρ, let $B_\rho = \{y \in \mathcal{X} : \|y\| \le \rho\}$ be a bounded ball in $\mathcal{X}$. Then, for each $h \in \Omega_F(y), y \in B_\rho$, there exists $v \in S_{F,y}$ such that

$$\begin{aligned} h(t) &= \int_0^t \frac{(t-s)^{\beta-1}}{\Gamma(\beta)} \int_s^1 \frac{(u-s)^{\alpha-1}}{\Gamma(\alpha)} v(u)duds \\ &\quad + \frac{t^{\beta+1}}{(1-\delta\eta^{\beta+1})}\Big[\delta \int_0^\eta \frac{(\eta-s)^{\beta-1}}{\Gamma(\beta)} \int_s^1 \frac{(u-s)^{\alpha-1}}{\Gamma(\alpha)} v(u)duds \\ &\quad - \int_0^1 \frac{(1-s)^{\beta-1}}{\Gamma(\beta)} \int_s^1 \frac{(u-s)^{\alpha-1}}{\Gamma(\alpha)} v(u)duds\Big]. \end{aligned}$$

Then, by using the fact that $(p-s)^{\alpha-1} \le 1 \ (1 < \alpha \le 2)$, we have

$$\begin{aligned} |h(t)| &\le \|\phi\|\Omega(\|y\|)\Bigg\{ \int_0^t \frac{(t-s)^{\beta-1}}{\Gamma(\beta)} \int_s^1 \frac{(u-s)^{\alpha-1}}{\Gamma(\alpha)} duds \\ &\quad + \frac{t^{\beta+1}}{|1-\delta\eta^{\beta+1}|}\Big[|\delta| \int_0^\eta \frac{(\eta-s)^{\beta-1}}{\Gamma(\beta)} \int_s^1 \frac{(u-s)^{\alpha-1}}{\Gamma(\alpha)} duds \\ &\quad + \int_0^1 \frac{(1-s)^{\beta-1}}{\Gamma(\beta)} \int_s^1 \frac{(u-s)^{\alpha-1}}{\Gamma(\alpha)} duds\Big]\Bigg\} \\ &\le \|\phi\|\Omega(\|y\|)\Omega_1, \end{aligned}$$

which, on taking the norm for $t \in [0,1]$, yields

$$\|h\| \le \|\phi\|\Omega(r)\Omega_1.$$

Now we show that Ω_F *maps bounded sets into equi-continuous sets of* $\mathcal{X}$. Let $t_1, t_2 \in [0,1]$ with $t_1 < t_2$ and $y \in B_\rho$. For each $h \in \Omega_F(y)$, using the fact that $(p-s)^{\alpha-1} \le 1 \ (1 < \alpha \le 2)$, we obtain

$$|h(t_2) - h(t_1)|$$

$$\leq \|g\|\Omega(r)\bigg\{\bigg|\int_0^{t_1}\frac{[(t_2-s)^{\beta-1}-(t_1-s)^{\beta-1}]}{\Gamma(\beta)\Gamma(\alpha+1)}ds+\int_{t_1}^{t_2}\frac{(t_2-s)^{\beta-1}}{\Gamma(\beta)\Gamma(\alpha+1)}ds\bigg|$$
$$+\frac{|t_2^{\beta+1}-t_1^{\beta+1}|}{|1-\delta\eta^{\beta+1}|}\bigg[|\delta|\int_0^{\eta}\frac{(\eta-s)^{\beta-1}}{\Gamma(\beta)}\int_s^1\frac{(u-s)^{\alpha-1}}{\Gamma(\alpha)}duds$$
$$+\int_0^1\frac{(1-s)^{\beta-1}}{\Gamma(\beta)}\int_s^1\frac{(u-s)^{\alpha-1}}{\Gamma(\alpha)}duds\bigg]\bigg\}$$
$$\leq \|g\|\Omega(r)\bigg\{\frac{2(t_2-t_1)^{\beta}+t_2^{\beta}-t_1^{\beta}}{\Gamma(\beta+1)\Gamma(\alpha+1)}+\frac{|t_2^{\beta+1}-t_1^{\beta+1}|}{|1-\delta\eta^{\beta+1}|}\bigg[\frac{|\delta|\eta^{\beta}+1}{\Gamma(\alpha+1)\Gamma(\beta+1)}\bigg]\bigg\},$$

which tends to zero independently of $y \in \mathcal{B}_\rho$ as $t_2-t_1 \to 0$. As Ω_F satisfies the above assumptions, therefore it follows by the Arzelá-Ascoli theorem that $\Omega_F : \mathcal{X} \to \mathcal{P}(\mathcal{X})$ is completely continuous.

In our next step, we show that Ω_F *is upper semi-continuous.* To this end it is sufficient to show that Ω_F *has a closed graph* by Lemma 1.1. Let $y_n \to y_*, h_n \in \Omega_F(y_n)$ and $h_n \to h_*$. Then we need to show that $h_* \in \Omega_F(y_*)$. Associated with $h_n \in \Omega_F(x_n)$, there exists $v_n \in S_{F,x_n}$ such that for each $t \in [0,1]$,

$$h_n(t)=\int_0^{t}\frac{(t-s)^{\beta-1}}{\Gamma(\beta)}\int_s^1\frac{(u-s)^{\alpha-1}}{\Gamma(\alpha)}v_n(u)duds$$
$$+\frac{t^{\beta+1}}{(1-\delta\eta^{\beta+1})}\bigg[\delta\int_0^{\eta}\frac{(\eta-s)^{\beta-1}}{\Gamma(\beta)}\int_s^1\frac{(u-s)^{\alpha-1}}{\Gamma(\alpha)}v_n(u)duds$$
$$-\int_0^1\frac{(1-s)^{\beta-1}}{\Gamma(\beta)}\int_s^1\frac{(u-s)^{\alpha-1}}{\Gamma(\alpha)}v_n(u)duds\bigg].$$

Thus it suffices to show that there exists $v_* \in S_{F,y_*}$ such that for each $t \in [0,1]$,

$$h_*(t)=\int_0^{t}\frac{(t-s)^{\beta-1}}{\Gamma(\beta)}\int_s^1\frac{(u-s)^{\alpha-1}}{\Gamma(\alpha)}v_*(u)duds$$
$$+\frac{t^{\beta+1}}{(1-\delta\eta^{\beta+1})}\bigg[\delta\int_0^{\eta}\frac{(\eta-s)^{\beta-1}}{\Gamma(\beta)}\int_s^1\frac{(u-s)^{\alpha-1}}{\Gamma(\alpha)}v_*(u)duds$$
$$-\int_0^1\frac{(1-s)^{\beta-1}}{\Gamma(\beta)}\int_s^1\frac{(u-s)^{\alpha-1}}{\Gamma(\alpha)}v_*(u)duds\bigg].$$

Let us consider the linear operator $\Theta : L^1([0,1],\mathbb{R}) \to \mathcal{X}$ given by

$$v \mapsto \Theta(v)(t)=\int_0^{t}\frac{(t-s)^{\beta-1}}{\Gamma(\beta)}\int_s^1\frac{(u-s)^{\alpha-1}}{\Gamma(\alpha)}v(u)duds$$
$$+\frac{t^{\beta+1}}{(1-\delta\eta^{\beta+1})}\bigg[\delta\int_0^{\eta}\frac{(\eta-s)^{\beta-1}}{\Gamma(\beta)}\int_s^1\frac{(u-s)^{\alpha-1}}{\Gamma(\alpha)}v(u)duds$$
$$-\int_0^1\frac{(1-s)^{\beta-1}}{\Gamma(\beta)}\int_s^1\frac{(u-s)^{\alpha-1}}{\Gamma(\alpha)}v(u)duds\bigg].$$

Observe that

$$\|h_n(t)-h_*(t)\|$$

$$= \Big\| \int_0^t \frac{(t-s)^{\beta-1}}{\Gamma(\beta)} \int_s^1 \frac{(u-s)^{\alpha-1}}{\Gamma(\alpha)}(v_n - v_*)(u)duds$$
$$+ \frac{t^{\beta+1}}{(1-\delta\eta^{\beta+1})}\Big[\delta \int_0^\eta \frac{(\eta-s)^{\beta-1}}{\Gamma(\beta)} \int_s^1 \frac{(u-s)^{\alpha-1}}{\Gamma(\alpha)}(v_n - v_*)(u)duds$$
$$- \int_0^1 \frac{(1-s)^{\beta-1}}{\Gamma(\beta)} \int_s^1 \frac{(u-s)^{\alpha-1}}{\Gamma(\alpha)}(v_n - v_*)(u)duds\Big]\Big\| \to 0 \text{ as } n \to \infty.$$

Thus, it follows by Lemma 1.2 that $\Theta \circ S_F$ is a closed graph operator. Further, we have $h_n(t) \in \Theta(S_{F,y_n})$. Since $y_n \to y_*$, therefore, we have

$$h_*(t) = \int_0^t \frac{(t-s)^{\beta-1}}{\Gamma(\beta)} \int_s^1 \frac{(u-s)^{\alpha-1}}{\Gamma(\alpha)} v_*(u)duds$$
$$+ \frac{t^{\beta+1}}{(1-\delta\eta^{\beta+1})}\Big[\delta \int_0^\eta \frac{(\eta-s)^{\beta-1}}{\Gamma(\beta)} \int_s^1 \frac{(u-s)^{\alpha-1}}{\Gamma(\alpha)} v_*(u)duds$$
$$- \int_0^1 \frac{(1-s)^{\beta-1}}{\Gamma(\beta)} \int_s^1 \frac{(u-s)^{\alpha-1}}{\Gamma(\alpha)} v_*(u)duds\Big], \text{ for some } v_* \in S_{F,y_*}.$$

Finally, we show there exists an open set $U \subseteq C([0,1],\mathbb{R})$ with $y \notin \theta\Omega_F(y)$ for any $\theta \in (0,1)$ and all $y \in \partial U$. Let $\theta \in (0,1)$ and $y \in \theta\Omega_F(y)$. Then there exists $v \in L^1([0,1],\mathbb{R})$ with $v \in S_{F,y}$ such that, for $t \in [0,1]$, we obtain

$$|y(t)| = |\theta\Omega_F(y)(t)| \le \int_0^t \frac{(t-s)^{\beta-1}}{\Gamma(\beta)} \int_s^1 \frac{(u-s)^{\alpha-1}}{\Gamma(\alpha)}|v(u)|duds$$
$$+ \frac{t^{\beta+1}}{|1-\delta\eta^{\beta+1}|}\Big[|\delta| \int_0^\eta \frac{(\eta-s)^{\beta-1}}{\Gamma(\beta)} \int_s^1 \frac{(u-s)^{\alpha-1}}{\Gamma(\alpha)}|v(u)|duds$$
$$+ \int_0^1 \frac{(1-s)^{\beta-1}}{\Gamma(\beta)} \int_s^1 \frac{(u-s)^{\alpha-1}}{\Gamma(\alpha)}|v(u)|duds\Big]$$
$$\le \|\phi\|\Omega(\|y\|)\Omega_1,$$

which implies that

$$\frac{\|y\|}{\|\phi\|\Omega(\|y\|)\Omega_1} \le 1.$$

In view of (2.11.3), there exists M such that $\|y\| \neq M$. Let us set

$$U = \{y \in \mathcal{X} : \|y\| < M\}.$$

Note that the operator $\Omega_F : \overline{U} \to \mathcal{P}(\mathcal{X})$ is upper semi-continuous and completely continuous. From the choice of U, there is no $y \in \partial U$ such that $y \in \theta\Omega_F(y)$ for some $\theta \in (0,1)$. Consequently, by the nonlinear alternative of Leray-Schauder type (Theorem 1.16), we deduce that Ω_F has a fixed point $y \in \overline{U}$ which is a solution of the problem (2.19). This completes the proof. □

Next we prove an existence result for the problem (2.19) with a non-convex valued right-hand side by applying a fixed point theorem for multi-valued maps due to Covitz and Nadler (Theorem 1.19).

Theorem 2.12. *Assume that:*

(2.12.1) $F : [0,1] \times \mathbb{R} \to \mathcal{P}_{cp}(\mathbb{R})$ *is such that* $F(\cdot, y(t)) : [0,1] \to \mathcal{P}_{cp}(\mathbb{R})$ *is measurable for each* $y \in \mathbb{R}$;

(2.12.2) $H_d(F(t,y), F(t,\bar{y}) \leq q(t)|y - \bar{y}|$ *for almost all* $t \in [0,1]$ *and* $y, \bar{y} \in \mathbb{R}$ *with* $q \in C([0,1], \mathbb{R}^+)$ *and* $d(0, F(t,0)) \leq q(t)$ *for almost all* $t \in [0,1]$.

Then the problem (2.19) has at least one solution on $[0,1]$ *if*

$$\|q\|\Omega_1 < 1, \tag{2.34}$$

where Ω_1 *is defined by (2.27).*

Proof. Consider the operator $\Omega_F : \mathcal{X} \to \mathcal{P}(\mathcal{X})$ defined in the beginning of the proof of Theorem 2.11. Observe that the set $S_{F,y}$ is non-empty for each $y \in \mathcal{X}$ by the assumption (2.12.1). So F has a measurable selection (see Theorem III.6 [78]). Now we show that the operator Ω_F satisfies the assumptions of Theorem 1.19. To show that $\Omega_F(y) \in \mathcal{P}_{cl}(\mathcal{X})$ for each $y \in \mathcal{X}$, let $\{u_n\}_{n\geq 0} \in \Omega_F(y)$ be such that $u_n \to u$ as $n \to \infty$ in $\mathcal{X}$. Then $u \in \mathcal{X}$ and there exists $v_n \in S_{F,x}$ such that, for each $t \in [0,1]$,

$$\begin{aligned} u_n(t) = &\int_0^t \frac{(t-s)^{\beta-1}}{\Gamma(\beta)} \int_s^1 \frac{(u-s)^{\alpha-1}}{\Gamma(\alpha)} v_n(u)duds \\ &+ \frac{t^{\beta+1}}{(1-\delta\eta^{\beta+1})} \Big[\delta \int_0^\eta \frac{(\eta-s)^{\beta-1}}{\Gamma(\beta)} \int_s^1 \frac{(u-s)^{\alpha-1}}{\Gamma(\alpha)} v_n(u)duds \\ &- \int_0^1 \frac{(1-s)^{\beta-1}}{\Gamma(\beta)} \int_s^1 \frac{(u-s)^{\alpha-1}}{\Gamma(\alpha)} v_n(u)duds\Big]. \end{aligned}$$

As F has compact values, we pass onto a subsequence (if necessary) to obtain that v_n converges to v in $L^1([0,1], \mathbb{R})$. Thus, $v \in S_{F,y}$ and for each $t \in [0,1]$, we have

$$\begin{aligned} u_n(t) \to u(t) = &\int_0^t \frac{(t-s)^{\beta-1}}{\Gamma(\beta)} \int_s^1 \frac{(u-s)^{\alpha-1}}{\Gamma(\alpha)} v(u)duds \\ &+ \frac{t^{\beta+1}}{(1-\delta\eta^{\beta+1})} \Big[\delta \int_0^\eta \frac{(\eta-s)^{\beta-1}}{\Gamma(\beta)} \int_s^1 \frac{(u-s)^{\alpha-1}}{\Gamma(\alpha)} v(u)duds \\ &- \int_0^1 \frac{(1-s)^{\beta-1}}{\Gamma(\beta)} \int_s^1 \frac{(u-s)^{\alpha-1}}{\Gamma(\alpha)} v(u)duds\Big]. \end{aligned}$$

Hence, $u \in \Omega_F(y)$.

Next we show that there exists $\hat{\theta} := \|q\|\Omega_1 < 1$ such that

$$H_d(\Omega_F(y), \Omega_F(\bar{y})) \leq \hat{\theta}\|y - \bar{y}\| \quad \text{for each} \quad y, \bar{y} \in \mathcal{X}.$$

Let $y, \bar{y} \in \mathcal{X}$ and $h_1 \in \Omega_F(y)$. Then there exists $v_1(t) \in F(t, y(t))$ such that, for each $t \in [0,1]$,

$$h_1(t) = \int_0^t \frac{(t-s)^{\beta-1}}{\Gamma(\beta)} \int_s^1 \frac{(u-s)^{\alpha-1}}{\Gamma(\alpha)} v_1(u)duds$$

$$+\frac{t^{\beta+1}}{(1-\delta\eta^{\beta+1})}\Big[\delta\int_0^\eta\frac{(\eta-s)^{\beta-1}}{\Gamma(\beta)}\int_s^1\frac{(u-s)^{\alpha-1}}{\Gamma(\alpha)}v_1(u)duds$$
$$-\int_0^1\frac{(1-s)^{\beta-1}}{\Gamma(\beta)}\int_s^1\frac{(u-s)^{\alpha-1}}{\Gamma(\alpha)}v_1(u)duds\Big].$$

By (2.12.2), we have

$$H_d(F(t,y),F(t,\bar{y}))\le q(t)|y-\bar{y}|.$$

So, there exists $w\in F(t,\bar{y})$ such that

$$|v_1(t)-w|\le q(t)|y(t)-\bar{y}(t)|,\quad t\in[0,1].$$

Define $U:[0,1]\to\mathcal{P}(\mathbb{R})$ by

$$U(t)=\{w\in\mathbb{R}:|v_1(t)-w|\le q(t)|y(t)-\bar{y}(t)|\}.$$

Since the multi-valued operator $U(t)\cap F(t,\bar{y})$ is measurable (Proposition III.4 [78]), there exists a function $v_2(t)$ which is a measurable selection for $U(t)\cap F(t,\bar{y})$. So $v_2(t)\in F(t,\bar{y})$ and for each $t\in[0,1]$, we have $|v_1(t)-v_2(t)|\le q(t)|y(t)-\bar{y}(t)|$. For each $t\in[0,1]$, let us define

$$\begin{aligned}h_2(t)=&\int_0^t\frac{(t-s)^{\beta-1}}{\Gamma(\beta)}\int_s^1\frac{(u-s)^{\alpha-1}}{\Gamma(\alpha)}v_2(u)duds\\&+\frac{t^{\beta+1}}{(1-\delta\eta^{\beta+1})}\Big[\delta\int_0^\eta\frac{(\eta-s)^{\beta-1}}{\Gamma(\beta)}\int_s^1\frac{(u-s)^{\alpha-1}}{\Gamma(\alpha)}v_2(u)duds\\&-\int_0^1\frac{(1-s)^{\beta-1}}{\Gamma(\beta)}\int_s^1\frac{(u-s)^{\alpha-1}}{\Gamma(\alpha)}v_2(u)duds\Big].\end{aligned}$$

Thus

$$\begin{aligned}&|h_1(t)-h_2(t)|\\&\le\int_0^t\frac{(t-s)^{\beta-1}}{\Gamma(\beta)}\int_s^1\frac{(u-s)^{\alpha-1}}{\Gamma(\alpha)}|v_1-v_2|(u)duds\\&\quad+\frac{t^{\beta+1}}{|1-\delta\eta^{\beta+1}|}\Big[|\delta|\int_0^\eta\frac{(\eta-s)^{\beta-1}}{\Gamma(\beta)}\int_s^1\frac{(u-s)^{\alpha-1}}{\Gamma(\alpha)}|v_1-v_2|(u)duds\\&\quad+\int_0^1\frac{(1-s)^{\beta-1}}{\Gamma(\beta)}\int_s^1\frac{(u-s)^{\alpha-1}}{\Gamma(\alpha)}|v_1-v_2|(u)duds\Big]\\&\le\|q\|\Omega_1\|y-\bar{y}\|,\end{aligned}$$

which yields $\|h_1-h_2\|\le\|q\|\Omega_1\|y-\bar{y}\|$.

Analogously, interchanging the roles of y and $\bar{y}$, we can obtain

$$H_d(\Omega_F(y),\Omega_F(\bar{y}))\le\|q\|\Omega_1\|y-\bar{y}\|.$$

By the condition (2.34), it follows that Ω_F is a contraction and hence it has a fixed point y by Theorem 1.19, which is a solution of the problem (2.19). This completes the proof. □

2.4 Nonlinear integro-differential equations involving mixed right and left fractional derivatives and integrals with nonlocal boundary data

We introduce a new kind of integro-differential equation involving right-Caputo and left-Riemann-Liouville fractional derivative operators of different orders and right-left Riemann-Liouville fractional integrals and solve it subject to nonlocal boundary conditions. In precise terms, we prove existence and uniqueness of solutions for the problem given by

$$ {}^{C}D^{\alpha}_{1-}\,{}^{RL}D^{\beta}_{0+}y(t) + \lambda I^{p}_{1-}I^{q}_{0+}h(t,y(t)) = f(t,y(t)), \quad t \in J := [0,1], \tag{2.35}$$

$$ y(0) = y(\zeta) = 0, \quad y(1) = \delta y(\mu), \quad 0 < \xi < \mu < 1, \tag{2.36}$$

where ${}^{C}D^{\alpha}_{1-}$ and ${}^{RL}D^{\beta}_{0+}$ denote the right Caputo fractional derivative operator operator of order $\alpha \in (1,2]$ and the left Riemann-Liouville fractional derivative operator operator of order $\beta \in (0,1]$, I^{p}_{1-} and I^{q}_{0+} denote the right and left Riemann-Liouville fractional integrals of orders $p, q > 0$ respectively, $f, h : [0,1] \times \mathbb{R} \to \mathbb{R}$ are given continuous functions and $\delta, \lambda \in \mathbb{R}$. It is imperative to notice that the integro-differential equation (2.35) contains mixed type (integral and non-integral) nonlinearities.

In the following lemma, we solve a linear variant of the problem (2.35)–(2.36).

Lemma 2.4. *Let $H, F \in C[0,1] \cap L(0,1)$. Then the linear problem*

$$\begin{cases} {}^{C}D^{\alpha}_{1-}\,{}^{RL}D^{\beta}_{0+}y(t) + \lambda I^{p}_{1-}I^{q}_{0+}H(t) = F(t), & t \in J := [0,1], \\ y(0) = y(\xi) = 0, \quad y(1) = \delta y(\mu), \end{cases} \tag{2.37}$$

is equivalent to the fractional integral equation:

$$\begin{aligned} y(t) &= \int_0^t \frac{(t-s)^{\beta-1}}{\Gamma(\beta)} \Big[I^{\alpha}_{1-}F(s) - \lambda I^{\alpha+p}_{1-}I^{q}_{0+}H(s)\Big] ds \\ &\quad + a_1(t)\Big\{\delta \int_0^{\mu} \frac{(\mu-s)^{\beta-1}}{\Gamma(\beta)} \Big[I^{\alpha}_{1-}F(s) - \lambda I^{\alpha+p}_{1-}I^{q}_{0+}H(s)\Big] ds \\ &\quad - \int_0^1 \frac{(1-s)^{\beta-1}}{\Gamma(\beta)} \Big[I^{\alpha}_{1-}F(s) - \lambda I^{\alpha+p}_{1-}I^{q}_{0+}H(s)\Big] ds\Big\} \\ &\quad + a_2(t) \int_0^{\xi} \frac{(\xi-s)^{\beta-1}}{\Gamma(\beta)} \Big[I^{\alpha}_{1-}F(s) - \lambda I^{\alpha+p}_{1-}I^{q}_{0+}H(s)\Big] ds, \end{aligned} \tag{2.38}$$

where

$$\begin{aligned} a_1(t) &= \tfrac{1}{\Lambda}\Big[\xi^{\beta+1}t^{\beta} - \xi^{\beta}t^{\beta+1}\Big], \\ a_2(t) &= \tfrac{1}{\Lambda}\Big[t^{\beta}(1-\delta\mu^{\beta+1}) - t^{\beta+1}(1-\delta\mu^{\beta})\Big], \end{aligned} \tag{2.39}$$

and it is assumed that

$$ \Lambda = \xi^{\beta+1}(1-\delta\mu^{\beta}) - \xi^{\beta}(1-\delta\mu^{\beta+1}) \neq 0. \tag{2.40}$$

Proof. Applying the left and right fractional integrals I_{1-}^{α} and I_{0+}^{β} successively to the integro-differential equation in (2.37), we get

$$y(t) = I_{0+}^{\beta} I_{1-}^{\alpha} F(t) - \lambda I_{0+}^{\beta} I_{1-}^{\alpha+p} I_{0+}^{q} H(t) \\ + c_0 \frac{t^{\beta}}{\Gamma(\beta+1)} + c_1 \frac{t^{\beta+1}}{\Gamma(\beta+2)} + c_2 t^{\beta-1}, \tag{2.41}$$

where c_0, c_1 and c_2 are unknown arbitrary constants.

In view of the condition $y(0) = 0$, it follows from (2.41) that $c_2 = 0$. Inserting $c_2 = 0$ in (2.41) and then using the nonlocal boundary conditions $y(\xi) = 0, y(1) = \delta y(\mu)$ in the resulting equation, we obtain a system of equations in c_0 and c_1 given by

$$\begin{aligned} c_0\Big(\frac{1-\delta\mu^{\beta}}{\Gamma(\beta+1)}\Big) + c_1\Big(\frac{1-\delta\mu^{\beta+1}}{\Gamma(\beta+2)}\Big) &= \delta A_1 - A_2, \\ c_0\Big(\frac{\xi^{\beta}}{\Gamma(\beta+1)}\Big) + c_1\Big(\frac{\xi^{\beta+1}}{\Gamma(\beta+2)}\Big) &= -A_3, \end{aligned} \tag{2.42}$$

where

$$A_1 = I_{0+}^{\beta} I_{1-}^{\alpha} F(\mu) - \lambda I_{0+}^{\beta} I_{1-}^{\alpha+p} I_{0+}^{q} H(\mu), \quad A_2 = I_{0+}^{\beta} I_{1-}^{\alpha} F(1) - \lambda I_{0+}^{\beta} I_{1-}^{\alpha+p} I_{0+}^{q} H(1),$$
$$A_3 = I_{0+}^{\beta} I_{1-}^{\alpha} F(\xi) - \lambda I_{0+}^{\beta} I_{1-}^{\alpha+p} I_{0+}^{q} H(\xi).$$

Solving the system (2.42), we find that

$$c_0 = \frac{\Gamma(\beta+1)}{\Lambda}\Big[\xi^{\beta+1}(\delta A_1 - A_2) + (1-\delta\mu^{\beta+1})A_3\Big],$$
$$c_1 = \frac{-\Gamma(\beta+2)}{\Lambda}\Big[\xi^{\beta}(\delta A_1 - A_2) + (1-\delta\mu^{\beta})A_3\Big],$$

where Λ is defined by (2.40). Substituting the values of c_0 and c_1 together with the notation (2.39) in (2.41), we obtain the solution (2.38). The converse follows by direct computation. This completes the proof. □

Let $\mathcal{X} = C([0,1], \mathbb{R})$ denote the Banach space of all continuous functions from $[0,1] \to \mathbb{R}$ equipped with the norm $\|y\| = \sup\{|y(t)| : t \in [0,1]\}$. By Lemma 2.4, we define an operator $\mathcal{G} : \mathcal{X} \to \mathcal{X}$ associated with the problem (2.35)–(2.36) as

$$\begin{aligned} \mathcal{G}y(t) = &\int_0^t \frac{(t-s)^{\beta-1}}{\Gamma(\beta)}\Big[I_{1-}^{\alpha} f(s,y(s)) - \lambda I_{1-}^{\alpha+p} I_{0+}^{q} h(s,y(s))\Big] ds \\ &+ a_1(t)\Bigg[\delta \int_0^{\mu} \frac{(\mu-s)^{\beta-1}}{\Gamma(\beta)}\Big[I_{1-}^{\alpha} f(s,y(s)) - \lambda I_{1-}^{\alpha+p} I_{0+}^{q} h(s,y(s))\Big] ds \\ &- \int_0^1 \frac{(1-s)^{\beta-1}}{\Gamma(\beta)}\Big[I_{1-}^{\alpha} f(s,y(s)) - \lambda I_{1-}^{\alpha+p} I_{0+}^{q} h(s,y(s))\Big] ds\Bigg] \\ &+ a_2(t)\int_0^{\xi} \frac{(\xi-s)^{\beta-1}}{\Gamma(\beta)}\Big[I_{1-}^{\alpha} f(s,y(s)) - \lambda I_{1-}^{\alpha+p} I_{0+}^{q} h(s,y(s))\Big] ds. \end{aligned}$$

Notice that the fixed points of the operator $\mathcal{G}$ are solutions of the problem (2.35)–(2.36).

In the forthcoming analysis, we use the following estimates:

$$\begin{aligned}
&\int_0^t \frac{(t-s)^{\beta-1}}{\Gamma(\beta)} I_{1-}^{\alpha+p} I_{0+}^{q} ds \\
&\quad = \int_0^t \frac{(t-s)^{\beta-1}}{\Gamma(\beta)} \int_s^1 \frac{(u-s)^{\alpha+p-1}}{\Gamma(\alpha+p)} \int_0^u \frac{(u-r)^{q-1}}{\Gamma(q)} dr\, du\, ds \\
&\quad \leq \frac{t^\beta}{\Gamma(\beta+1)\Gamma(\alpha+p+1)\Gamma(q+1)},
\end{aligned}$$

$$\begin{aligned}
\int_0^t \frac{(t-s)^{\beta-1}}{\Gamma(\beta)} I_{1-}^{\alpha} ds &= \int_0^t \frac{(t-s)^{\beta-1}}{\Gamma(\beta)} \int_s^1 \frac{(u-s)^{\alpha-1}}{\Gamma(\alpha)} du\, ds \\
&\leq \frac{t^\beta}{\Gamma(\alpha+1)\Gamma(\beta+1)},
\end{aligned}$$

where we have used $u^q \leq 1$, $(1-s)^{\alpha+p} < 1$; $p, q > 0, 1 < \alpha \leq 2$.

In the sequel, we set

$$\Omega_1 = \frac{\Delta}{\Gamma(\alpha+1)}, \quad \Omega_2 = \frac{|\lambda|\Delta}{\Gamma(\alpha+p+1)\Gamma(q+1)}, \tag{2.43}$$

where

$$\Delta = \frac{1}{\Gamma(\beta+1)}\Big[1 + \overline{a}_1(|\delta|\mu^\beta + 1) + \overline{a}_2\xi^\beta\Big], \ \overline{a}_1 = \max_{t\in[0,1]} |a_1(t)|, \ \overline{a}_2 = \max_{t\in[0,1]} |a_2(t)|.$$

2.4.1 *Existence results*

In the following, we prove our first existence result for the problem (2.35)–(2.36), which relies on Krasnosel'skiĭ's fixed point theorem (Theorem 1.4).

Theorem 2.13. *Assumed that:*

(2.13.1) There exist $L > 0$ such that $|f(t,x) - f(t,y)| \leq L|x-y|$, $\forall t \in [0,1]$, $x, y \in \mathbb{R}$;

(2.13.2) There exist $K > 0$ such that $|h(t,x) - h(t,y)| \leq K|x-y|$, $\forall t \in [0,1]$, $x, y \in \mathbb{R}$;

(2.13.3) $|f(t,y)| \leq \sigma(t)$ and $|h(t,y)| \leq \rho(t)$, where $\sigma, \rho \in C([0,1], \mathbb{R}^+)$.

Then the problem (2.35)–(2.36) has at least one solution on $[0,1]$ if $L\gamma_1 + K\gamma_2 < 1$, where

$$\gamma_1 = \frac{1}{\Gamma(\beta+1)\Gamma(\alpha+1)}, \quad \gamma_2 = \frac{|\lambda|}{\Gamma(\beta+1)\Gamma(\alpha+p+1)\Gamma(q+1)}. \tag{2.44}$$

Proof. Introduce the ball $B_\theta = \{y \in \mathcal{X} : \|y\| \leq \theta\}$, where $\|\sigma\| = \sup_{t\in[0,1]} |\sigma(t)|$, $\|\rho\| = \sup_{t\in[0,1]} |\rho(t)|$ and

$$\theta \geq \|\sigma\|\Omega_1 + \|\rho\|\Omega_2. \tag{2.45}$$

Let us split the operator $\mathcal{G}:\mathcal{X}\to\mathcal{X}$ on B_θ as $\mathcal{G}=\mathcal{G}_1+\mathcal{G}_2$, where

$$\mathcal{G}_1y(t)=\int_0^t\frac{(t-s)^{\beta-1}}{\Gamma(\beta)}I_{1-}^{\alpha}f(s,y(s))ds-\lambda\int_0^t\frac{(t-s)^{\beta-1}}{\Gamma(\beta)}I_{1-}^{\alpha+p}I_{0+}^{q}h(s,y(s))ds,$$

$$\begin{aligned}&\mathcal{G}_2y(t)\\&=a_1(t)\Bigg[\delta\Bigg(\int_0^\mu\frac{(\mu-s)^{\beta-1}}{\Gamma(\beta)}I_{1-}^{\alpha}f(s,y(s))ds-\lambda\int_0^\mu\frac{(\mu-s)^{\beta-1}}{\Gamma(\beta)}I_{1-}^{\alpha+p}I_{0+}^{q}h(s,y(s))ds\Bigg)\\&\quad-\Bigg(\int_0^1\frac{(1-s)^{\beta-1}}{\Gamma(\beta)}I_{1-}^{\alpha}f(s,y(s))ds-\lambda\int_0^1\frac{(1-s)^{\beta-1}}{\Gamma(\beta)}I_{1-}^{\alpha+p}I_{0+}^{q}h(s,y(s))ds\Bigg)\Bigg]\\&\quad+a_2(t)\Bigg[\int_0^\xi\frac{(\xi-s)^{\beta-1}}{\Gamma(\beta)}I_{1-}^{\alpha}f(s,y(s))ds-\lambda\int_0^\xi\frac{(\xi-s)^{\beta-1}}{\Gamma(\beta)}I_{1-}^{\alpha+p}I_{0+}^{q}h(s,y(s))ds\Bigg].\end{aligned}$$

Now, we verify that the operators $\mathcal{G}_1$ and $\mathcal{G}_2$ satisfy the hypothesis of Krasnosel'skiĭ's theorem (Theorem 1.4) in three steps.

(a) For $y,x\in B_\theta$, we have

$$\begin{aligned}&\|\mathcal{G}_1y+\mathcal{G}_2x\|\\&\le\sup_{t\in[0,1]}\Bigg\{\int_0^t\frac{(t-s)^{\beta-1}}{\Gamma(\beta)}I_{1-}^{\alpha}|f(s,y(s))|ds+|\lambda|\int_0^t\frac{(t-s)^{\beta-1}}{\Gamma(\beta)}I_{1-}^{\alpha+p}I_{0+}^{q}|h(s,y(s))|ds\\&\quad+|a_1(t)|\Bigg\{|\delta|\Bigg(\int_0^\mu\frac{(\mu-s)^{\beta-1}}{\Gamma(\beta)}I_{1-}^{\alpha}|f(s,x(s))|ds\\&\quad+|\lambda|\int_0^\mu\frac{(\mu-s)^{\beta-1}}{\Gamma(\beta)}I_{1-}^{\alpha+p}I_{0+}^{q}|h(s,x(s))|ds\Bigg)\\&\quad+\Bigg(\int_0^1\frac{(1-s)^{\beta-1}}{\Gamma(\beta)}I_{1-}^{\alpha}|f(s,x(s))|ds+|\lambda|\int_0^1\frac{(1-s)^{\beta-1}}{\Gamma(\beta)}I_{1-}^{\alpha+p}I_{0+}^{q}|h(s,x(s))|ds\Bigg)\Bigg\}\\&\quad+|a_2(t)|\Bigg\{\int_0^\xi\frac{(\xi-s)^{\beta-1}}{\Gamma(\beta)}I_{1-}^{\alpha}|f(s,x(s))|ds\\&\quad+|\lambda|\int_0^\xi\frac{(\xi-s)^{\beta-1}}{\Gamma(\beta)}I_{1-}^{\alpha+p}I_{0+}^{q}|h(s,x(s))|ds\Bigg\}\Bigg\}\\&\le\|\sigma\|\sup_{t\in[0,1]}\Bigg\{\int_0^t\frac{(t-s)^{\beta-1}}{\Gamma(\beta)}I_{1-}^{\alpha}ds+|a_1(t)|\Bigg[|\delta|\int_0^\mu\frac{(\mu-s)^{\beta-1}}{\Gamma(\beta)}I_{1-}^{\alpha}ds\\&\quad+\int_0^1\frac{(1-s)^{\beta-1}}{\Gamma(\beta)}I_{1-}^{\alpha}ds\Bigg]+|a_2(t)|\int_0^\xi\frac{(\xi-s)^{\beta-1}}{\Gamma(\beta)}I_{1-}^{\alpha}ds\Bigg\}\\&\quad+\|\rho\||\lambda|\sup_{t\in[0,1]}\Bigg\{\int_0^t\frac{(t-s)^{\beta-1}}{\Gamma(\beta)}I_{1-}^{\alpha+p}I_{0+}^{q}ds+|a_1(t)|\Bigg[|\delta|\int_0^\mu\frac{(\mu-s)^{\beta-1}}{\Gamma(\beta)}I_{1-}^{\alpha+p}I_{0+}^{q}ds\\&\quad+\int_0^1\frac{(1-s)^{\beta-1}}{\Gamma(\beta)}I_{1-}^{\alpha+p}I_{0+}^{q}ds\Bigg]+|a_2(t)|\int_0^\xi\frac{(\xi-s)^{\beta-1}}{\Gamma(\beta)}I_{1-}^{\alpha+p}I_{0+}^{q}ds\Bigg\}\\&\le\Bigg\{\frac{\|\sigma\|}{\Gamma(\alpha+1)}+\frac{\|\rho\||\lambda|}{\Gamma(\alpha+p+1)\Gamma(q+1)}\Bigg\}\Delta\\&=\|\sigma\|\Omega_1+\|\rho\|\Omega_2<\theta,\end{aligned}$$

where we used (2.45). Thus $\mathcal{G}_1y+\mathcal{G}_2x\in B_\theta$.

(b) Using (2.13.1) and (2.13.2), it is easy to show that

$$\begin{aligned}
\|\mathcal{G}_1 y - \mathcal{G}_1 x\| \le & \sup_{t\in[0,1]} \left\{ \int_0^t \frac{(t-s)^{\beta-1}}{\Gamma(\beta)} I_{1-}^{\alpha} |f(s,y(s)) - f(s,x(s))| ds \right. \\
& \left. + |\lambda| \int_0^t \frac{(t-s)^{\beta-1}}{\Gamma(\beta)} I_{1-}^{\alpha+p} I_{0+}^{q} |h(s,y(s)) - h(s,x(s))| ds \right\} \\
\le & (L\gamma_1 + K\gamma_2)\|y - x\|,
\end{aligned}$$

which, in view of the condition: $L\gamma_1 + K\gamma_2 < 1$, implies that the operator $\mathcal{G}_1$ is a contraction.

(c) Continuity of the functions f, h implies that the operator $\mathcal{G}_2$ is continuous. In addition, $\mathcal{G}_2$ is uniformly bounded on B_θ as

$$\begin{aligned}
& \|\mathcal{G}_2 y\| \\
& \le \sup_{t\in[0,1]} \left\{ |a_1(t)| \left\{ |\delta| \int_0^\mu \frac{(\mu-s)^{\beta-1}}{\Gamma(\beta)} \left[I_{1-}^{\alpha} |f(s,y(s))| + |\lambda| I_{1-}^{\alpha+p} I_{0+}^{q} |h(s,y(s))| \right] ds \right. \right. \\
& \quad \left. + \int_0^1 \frac{(1-s)^{\beta-1}}{\Gamma(\beta)} \left[I_{1-}^{\alpha} |f(s,y(s))| + |\lambda| I_{1-}^{\alpha+p} I_{0+}^{q} |h(s,y(s))| \right] ds \right\} \\
& \quad \left. + |a_2(t)| \int_0^\xi \frac{(\xi-s)^{\beta-1}}{\Gamma(\beta)} \left[I_{1-}^{\alpha} |f(s,y(s))| + |\lambda| I_{1-}^{\alpha+p} I_{0+}^{q} |h(s,y(s))| \right] ds \right\} \\
& \le \|\sigma\| \sup_{t\in[0,1]} \left\{ |a_1(t)| \left[|\delta| \int_0^\mu \frac{(\mu-s)^{\beta-1}}{\Gamma(\beta)} I_{1-}^{\alpha} ds + \int_0^1 \frac{(1-s)^{\beta-1}}{\Gamma(\beta)} I_{1-}^{\alpha} ds \right] \right. \\
& \quad \left. + |a_2(t)| \int_0^\xi \frac{(\xi-s)^{\beta-1}}{\Gamma(\beta)} I_{1-}^{\alpha} ds \right\} \\
& \quad + \|\rho\| |\lambda| \sup_{t\in[0,1]} \left\{ |a_1(t)| \left[|\delta| \int_0^\mu \frac{(\mu-s)^{\beta-1}}{\Gamma(\beta)} I_{1-}^{\alpha+p} I_{0+}^{q} ds \right. \right. \\
& \quad \left. \left. + \int_0^1 \frac{(1-s)^{\beta-1}}{\Gamma(\beta)} I_{1-}^{\alpha+p} I_{0+}^{q} ds \right] + |a_2(t)| \int_0^\xi \frac{(\xi-s)^{\beta-1}}{\Gamma(\beta)} I_{1-}^{\alpha+p} I_{0+}^{q} ds \right\} \\
& \le \|\sigma\|(\Omega_1 - \gamma_1) + \|\rho\|(\Omega_2 - \gamma_2),
\end{aligned}$$

where Ω_i, and γ_i $(i = 1, 2)$ are defined by (2.43) and (2.44) respectively.

To show the compactness of $\mathcal{G}_2$, we fix

$$\sup_{(t,y)\in[0,1]\times B_\theta} |f(t,y)| = \overline{f}, \qquad \sup_{(t,y)\in[0,1]\times B_\theta} |h(t,y)| = \overline{h}.$$

Then, for $0 < t_1 < t_2 < 1$, we have

$$\begin{aligned}
& |(\mathcal{G}_2 y)(t_2) - (\mathcal{G}_2 y)(t_1)| \\
& \le |a_1(t_2) - a_1(t_1)| \left\{ |\delta| \int_0^\mu \frac{(\mu-s)^{\beta-1}}{\Gamma(\beta)} \left[I_{1-}^{\alpha} |f(s,y(s))| + |\lambda| I_{1-}^{\alpha+p} I_{0+}^{q} |h(s,y(s))| \right] ds \right. \\
& \quad \left. + \int_0^1 \frac{(1-s)^{\beta-1}}{\Gamma(\beta)} \left[I_{1-}^{\alpha} |f(s,y(s))| + |\lambda| I_{1-}^{\alpha+p} I_{0+}^{q} |h(s,y(s))| \right] ds \right\}
\end{aligned}$$

$$+|a_2(t_2)-a_2(t_1)|\left\{\int_0^{\xi}\frac{(\xi-s)^{\beta-1}}{\Gamma(\beta)}\left[I_{1-}^{\alpha}|f(s,y(s))|+|\lambda|I_{1-}^{\alpha+p}I_{0+}^{q}|h(s,y(s))|\right]ds\right\}$$
$$\leq(\gamma_1\overline{f}+\gamma_2\overline{h})\left\{(\xi^{\beta+1}|t_2^{\beta}-t_1^{\beta}|+\xi^{\beta}|t_1^{\beta+1}-t_2^{\beta+1}|)\frac{(|\delta|\mu^{\beta}+1)}{|\Lambda|}\right.$$
$$\left.+(|1-\delta\mu^{\beta+1}||t_2^{\beta}-t_1^{\beta}|+|1-\delta\mu^{\beta}||t_1^{\beta+1}-t_2^{\beta+1}|)\frac{\xi^{\beta}}{|\Lambda|}\right\},$$

which tends to zero independent of y as $t_2\to t_1$. This shows that $\mathcal{G}_2$ is equi-continuous. It is clear from the foregoing arguments that the operator $\mathcal{G}_2$ is relatively compact on B_θ. Hence, by the Arzelá-Ascoli theorem, $\mathcal{G}_2$ is compact on B_θ.

In view of the arguments (a)-(c), the hypotheses of the Krasnosel'skiĭ's fixed point theorem (Theorem 1.4) holds true. Thus, the operator $\mathcal{G}_1+\mathcal{G}_2=\mathcal{G}$ has a fixed point, which implies that the problem (2.35)–(2.36) has at least one solution on $[0,1]$. The proof is finished. □

Remark 2.2. If we interchange the roles of the operators $\mathcal{G}_1$ and $\mathcal{G}_2$ in the previous result, the condition $L\gamma_1+K\gamma_2<1$, is replaced with the following one:

$$L(\Omega_1-\gamma_1)+K(\Omega_2-\gamma_2)<1,$$

where Ω_1,Ω_2 and γ_1,γ_2 are defined by (2.43) and (2.44) respectively.

The following existence result relies on Leray-Schauder nonlinear alternative (Theorem 1.8).

Theorem 2.14. *Suppose that the following conditions hold:*

(2.14.1) There exist continuous non-decreasing functions $\phi,\psi:[0,\infty)\to(0,\infty)$ such that $\forall(t,y)\in[0,1]\times\mathbb{R}$, $|f(t,y)|\leq\omega_1(t)\phi(\|y\|)$ and $|h(t,y)|\leq\omega_2(t)\psi(\|y\|)$, where $\omega_1,\omega_2\in C([0,T],\mathbb{R}^+)$.

(2.14.2) There exist a constant $M>0$ such that

$$\frac{M}{\|\omega_1\|\phi(M)\Omega_1+\|\omega_2\|\psi(M)\Omega_2}>1,$$

Then the problem (2.35)–(2.36) has at least one solution on $[0,1]$.

Proof. First we show that the operator $\mathcal{G}$ is completely continuous. This will be established in several steps.

(i) $\mathcal{G}$ maps bounded sets into bounded sets in $\mathcal{X}$.

Let $y\in\mathcal{B}_r=\{y\in\mathcal{X}:\|y\|\leq r\}$, where r is a fixed number. Then, using the strategy employed in the proof of Theorem 2.13, we obtain

$$\|\mathcal{G}y(t)\|\leq\left\{\frac{\|\omega_1\|\phi(r)}{\Gamma(\alpha+1)}+\frac{\|\omega_2\|\psi(r)|\lambda|}{\Gamma(\alpha+p+1)\Gamma(q+1)}\right\}\Delta$$
$$=\|\omega_1\|\phi(r)\Omega_1+\|\omega_2\|\psi(r)\Omega_2<\infty.$$

(ii) $\mathcal{G}$ maps bounded sets into equi-continuous sets.

Let $0 < t_1 < t_2 < 1$ and $y \in \mathcal{B}_r$, where $\mathcal{B}_r$ is bounded set in $\mathcal{X}$. Then we obtain

$$
\begin{aligned}
&|\mathcal{G}y(t_2) - \mathcal{G}y(t_1)| \\
\leq &\left| \int_0^{t_1} \frac{(t_2 - s)^{\beta-1} - (t_1 - s)^{\beta-1}}{\Gamma(\beta)} \left[I_{1-}^{\alpha} |f(s, y(s))| + |\lambda| I_{1-}^{\alpha+p} I_{0+}^{q} |h(s, x(s))| \right] ds \right| \\
&+ \left| \int_{t_1}^{t_2} \frac{(t_2 - s)^{\beta-1}}{\Gamma(\beta)} \left[I_{1-}^{\alpha} |f(s, y(s))| + |\lambda| I_{1-}^{\alpha+p} I_{0+}^{q} |h(s, x(s))| \right] ds \right| \\
&+ \left| a_1(t_2) - a_1(t_1) \right| \Bigg\{ |\delta| \Bigg(\int_0^{\mu} \frac{(\mu - s)^{\beta-1}}{\Gamma(\beta)} \Big[I_{1-}^{\alpha} |f(s, y(s))| \\
&+ |\lambda| I_{1-}^{\alpha+p} I_{0+}^{q} |h(s, x(s))| \Big] ds \Bigg) \\
&+ \left(\int_0^1 \frac{(1 - s)^{\beta-1}}{\Gamma(\beta)} \left[I_{1-}^{\alpha} |f(s, y(s))| + |\lambda| I_{1-}^{\alpha+p} I_{0+}^{q} |h(s, x(s))| \right] ds \right) \Bigg\} \\
&+ \left| a_2(t_2) - a_2(t_1) \right| \left\{ \int_0^{\xi} \frac{(\xi - s)^{\beta-1}}{\Gamma(\beta)} \left[I_{1-}^{\alpha} |f(s, y(s))| + |\lambda| I_{1-}^{\alpha+p} I_{0+}^{q} |h(s, x(s))| \right] ds \right\} \\
\leq &\left[\frac{\|\omega_1\| \phi(r)}{\Gamma(\beta + 1)\Gamma(\alpha + 1)} + \frac{\|\omega_2\| \psi(r) |\lambda|}{\Gamma(\beta + 1)\Gamma(\alpha + p + 1)\Gamma(q + 1)} \right] \\
&\times \Bigg\{ 2(t_2 - t_1)^{\beta} + |t_2{}^{\beta} - t_1{}^{\beta}| + \frac{(|\delta| \mu^{\beta} + 1)}{|\Lambda|} \left(\xi^{\beta+1} |t_2^{\beta} - t_1^{\beta}| + \xi^{\beta} |t_2^{\beta+1} - t_1^{\beta+1}| \right) \\
&+ \frac{\xi^{\beta}}{|\Lambda|} \left(|1 - \delta \mu^{\beta+1}| |t_2^{\beta} - t_1^{\beta}| + |1 - \delta \mu^{\beta}| |t_2^{\beta+1} - t_1^{\beta+1}| \right) \Bigg\}.
\end{aligned}
$$

Notice that the right-hand side of the above inequality tends to 0 as $t_2 \to t_1$, independent of $y \in \mathcal{B}_r$. In view of the foregoing arguments, it follows by the Arzelá–Ascoli theorem that $\mathcal{G} : \mathcal{X} \to \mathcal{X}$ is completely continuous.

The conclusion of the Leray–Schauder nonlinear alternative (Theorem 1.8) will be applicable once it is shown that there exists an open set $U \subset C([0, 1], \mathbb{R})$ with $y \neq \nu \mathcal{G}y$ for $\nu \in (0, 1)$ and $y \in \partial U$. Let $y \in C([0, 1], \mathbb{R})$ such that $y = \nu \mathcal{G}y$ for $\nu \in (0, 1)$. As argued in proving that the operator $\mathcal{G}$ is bounded, one can obtain that

$$|y(t)| = |\nu \mathcal{G}y(t)| \leq |\omega_1(t)| \phi(\|y\|) \Omega_1 + |\omega_2(t)| \psi(\|y\|) \Omega_2,$$

which can be written as

$$\frac{\|y\|}{\|\omega_1\| \phi(\|y\|) \Omega_1 + \|\omega_2\| \psi(\|y\|) \Omega_2} \leq 1.$$

On the other hand, we can find a positive number M such that $\|y\| \neq M$ by assumption (2.14.2). Let us set

$$U = \{y \in \mathcal{X} : \|y\| < M\}.$$

Clearly, ∂U contains a solution only when $\|y\| = M$. In other words, there is no solution $y \in \partial U$ such that $y = \nu \mathcal{G}y$ for some $\nu \in (0, 1)$. Therefore, $\mathcal{G}$ has a fixed point $y \in \overline{U}$ which is a solution of the problem (2.35)–(2.36). The proof is finished. □

2.4.2 *Uniqueness result*

Here we prove a uniqueness result for the problem (2.35)–(2.36) with the aid of Banach contraction mapping principle.

Theorem 2.15. *If the conditions (2.13.1) and (2.13.2) hold, then the problem (2.35)–(2.36) has a unique solution on* $[0,1]$ *if*

$$L\Omega_1 + K\Omega_2 < 1, \tag{2.46}$$

where Ω_1 *and* Ω_2 *are defined by (2.43).*

Proof. In the first step, we show that $\mathcal{G}\mathcal{B}_r \subset \mathcal{B}_r$, where $\mathcal{B}_r = \{y \in \mathcal{X} : \|y\| \le r\}$ with

$$r \ge \frac{f_0\Omega_1 + h_0\Omega_2}{1-(L\Omega_1+K\Omega_2)}, \; f_0 = \sup_{t\in[0,1]} |f(t,0)|, \; h_0 = \sup_{t\in[0,1]} |h(t,0)|.$$

For $y \in \mathcal{B}_r$ and using the condition (2.13.1), we have

$$\begin{aligned} |f(t,y)| &= |f(t,y) - f(t,0) + f(t,0)| \le |f(t,y) - f(t,0)| + |f(t,0)| \\ &\le L\|y\| + f_0 \le Lr + f_0. \end{aligned} \tag{2.47}$$

Similarly, using (2.13.2), we get

$$|h(t,y)| \le Kr + h_0. \tag{2.48}$$

In view of (2.47) and (2.48), we obtain

$$\begin{aligned} \|\mathcal{G}y\| &= \sup_{t\in[0,1]} |\mathcal{G}y(t)| \\ &\le \sup_{t\in[0,1]} \Bigg\{ \int_0^t \frac{(t-s)^{\beta-1}}{\Gamma(\beta)} \Big[I_{1-}^{\alpha} |f(s,y(s))| + |\lambda| I_{1-}^{\alpha+p} I_{0+}^{q} |h(s,y(s))| \Big] ds \\ &\quad + |a_1(t)| \Bigg\{ |\delta| \int_0^{\mu} \frac{(\mu-s)^{\beta-1}}{\Gamma(\beta)} \Big[I_{1-}^{\alpha} |f(s,y(s))| + |\lambda| I_{1-}^{\alpha+p} I_{0+}^{q} |h(s,y(s))| \Big] ds \\ &\quad + \int_0^1 \frac{(1-s)^{\beta-1}}{\Gamma(\beta)} \Big[I_{1-}^{\alpha} |f(s,y(s))| + |\lambda| I_{1-}^{\alpha+p} I_{0+}^{q} |h(s,y(s))| \Big] ds \Bigg\} \\ &\quad + |a_2(t)| \int_0^{\xi} \frac{(\xi-s)^{\beta-1}}{\Gamma(\beta)} \Big[I_{1-}^{\alpha} |f(s,y(s))| + |\lambda| I_{1-}^{\alpha+p} I_{0+}^{q} |h(s,y(s))| \Big] ds \Bigg\} \\ &\le (Lr + f_0) \sup_{t\in[0,1]} \Bigg\{ \int_0^t \frac{(t-s)^{\beta-1}}{\Gamma(\beta)} I_{1-}^{\alpha} ds + |a_1(t)| \Bigg[|\delta| \int_0^{\mu} \frac{(\mu-s)^{\beta-1}}{\Gamma(\beta)} I_{1-}^{\alpha} ds \\ &\quad + \int_0^1 \frac{(1-s)^{\beta-1}}{\Gamma(\beta)} I_{1-}^{\alpha} ds \Bigg] + |a_2(t)| \int_0^{\xi} \frac{(\xi-s)^{\beta-1}}{\Gamma(\beta)} I_{1-}^{\alpha} ds \Bigg\} \\ &\quad + (Kr + h_0)|\lambda| \sup_{t\in[0,1]} \Bigg\{ \int_0^t \frac{(t-s)^{\beta-1}}{\Gamma(\beta)} I_{1-}^{\alpha+p} I_{0+}^{q} ds \\ &\quad + |a_1(t)| \Bigg[|\delta| \int_0^{\mu} \frac{(\mu-s)^{\beta-1}}{\Gamma(\beta)} I_{1-}^{\alpha+p} I_{0+}^{q} ds + \int_0^1 \frac{(1-s)^{\beta-1}}{\Gamma(\beta)} I_{1-}^{\alpha+p} I_{0+}^{q} ds \Bigg] \end{aligned}$$

$$
\begin{aligned}
&+\big|a_2(t)\big|\int_0^{\xi}\frac{(\xi-s)^{\beta-1}}{\Gamma(\beta)}I_{1-}^{\alpha+p}I_{0+}^{q}ds\Big\} \\
&\le\left\{\frac{(Lr+f_0)}{\Gamma(\alpha+1)}+\frac{(Kr+h_0)|\lambda|}{\Gamma(\alpha+p+1)\Gamma(q+1)}\right\}\Delta \\
&=(Lr+f_0)\Omega_1+(Kr+h_0)\Omega_2<r,
\end{aligned}
$$

which implies that $\mathcal{G}y\in\mathcal{B}_r$, for any $y\in\mathcal{B}_r$. Therefore, $\mathcal{G}\mathcal{B}_r\subset\mathcal{B}_r$. Next, we prove that $\mathcal{G}$ is a contraction. For that, let $x,y\in\mathcal{X}$ and $t\in[0,1]$. Then, by the conditions (B_1) and (B_2), we obtain

$$
\begin{aligned}
\|\mathcal{G}y-\mathcal{G}x\| =& \sup_{t\in[0,1]}\big|(\mathcal{G}y)(t)-(\mathcal{G}x)(t)\big| \\
\le& \sup_{t\in[0,1]}\left\{\int_0^t\frac{(t-s)^{\beta-1}}{\Gamma(\beta)}I_{1-}^{\alpha}|f(s,y(s))-f(s,x(s))|ds\right. \\
&+|\lambda|\int_0^t\frac{(t-s)^{\beta-1}}{\Gamma(\beta)}I_{1-}^{\alpha+p}I_{0+}^{q}|h(s,y(s))-h(s,x(s))|ds \\
&+\big|a_1(t)\big|\left[|\delta|\left(\int_0^{\mu}\frac{(\mu-s)^{\beta-1}}{\Gamma(\beta)}I_{1-}^{\alpha}|f(s,y(s))-f(s,x(s))|ds\right.\right. \\
&\left.+|\lambda|\int_0^{\mu}\frac{(\mu-s)^{\beta-1}}{\Gamma(\beta)}I_{1-}^{\alpha+p}I_{0+}^{q}|h(s,y(s))-h(s,x(s))|ds\right) \\
&+\left(\int_0^1\frac{(1-s)^{\beta-1}}{\Gamma(\beta)}I_{1-}^{\alpha}|f(s,y(s))-f(s,x(s))|ds\right. \\
&\left.\left.+|\lambda|\int_0^1\frac{(1-s)^{\beta-1}}{\Gamma(\beta)}I_{1-}^{\alpha+p}I_{0+}^{q}|h(s,y(s))-h(s,x(s))|ds\right)\right] \\
&+\big|a_2(t)\big|\left[\int_0^{\xi}\frac{(\xi-s)^{\beta-1}}{\Gamma(\beta)}I_{1-}^{\alpha}|f(s,y(s))-f(s,x(s))|ds\right. \\
&\left.\left.+|\lambda|\int_0^{\xi}\frac{(\xi-s)^{\beta-1}}{\Gamma(\beta)}I_{1-}^{\alpha+p}I_{0+}^{q}|h(s,y(s))-h(s,x(s))|dr\,du\,ds\right]\right\} \\
\le& L\|y-x\|\sup_{t\in[0,1]}\left\{\int_0^t\frac{(t-s)^{\beta-1}}{\Gamma(\beta)}I_{1-}^{\alpha}ds\right. \\
&+\big|a_1(t)\big|\left[|\delta|\int_0^{\mu}\frac{(\mu-s)^{\beta-1}}{\Gamma(\beta)}I_{1-}^{\alpha}ds+\int_0^1\frac{(1-s)^{\beta-1}}{\Gamma(\beta)}I_{1-}^{\alpha}ds\right] \\
&\left.+\big|a_2(t)\big|\int_0^{\xi}\frac{(\xi-s)^{\beta-1}}{\Gamma(\beta)}I_{1-}^{\alpha}ds\right\} \\
&+K\|y-x\||\lambda|\sup_{t\in[0,1]}\left\{\int_0^t\frac{(t-s)^{\beta-1}}{\Gamma(\beta)}I_{1-}^{\alpha+p}I_{0+}^{q}ds\right. \\
&+\big|a_1(t)\big|\left[|\delta|\int_0^{\mu}\frac{(\mu-s)^{\beta-1}}{\Gamma(\beta)}I_{1-}^{\alpha+p}I_{0+}^{q}ds+\int_0^1\frac{(1-s)^{\beta-1}}{\Gamma(\beta)}I_{1-}^{\alpha+p}I_{0+}^{q}ds\right] \\
&\left.+\big|a_2(t)\big|\int_0^{\xi}\frac{(\xi-s)^{\beta-1}}{\Gamma(\beta)}I_{1-}^{\alpha+p}I_{0+}^{q}ds\right\}
\end{aligned}
$$

$$\leq \left\{\frac{L\Delta}{\Gamma(\alpha+1)} + \frac{K|\lambda|\Delta}{\Gamma(\alpha+p+1)\Gamma(q+1)}\right\}\|y-x\|$$
$$= (L\Omega_1 + K\Omega_2)\|y-x\|.$$

From the above inequality, it follows by the assumption $(L\Omega_1 + K\Omega_2) < 1$ that $\mathcal{G}$ is a contraction. Therefore, we deduce by Banach contraction mapping principle that there exists a unique fixed point for the operator $\mathcal{G}$, which corresponds to a unique solution for the problem (2.35)–(2.36) on $[0,1]$. The proof is completed. □

2.4.3 *Examples*

In this subsection, we construct examples to illustrate the existence and uniqueness results obtained in the last two subsections. Let us consider the following problem:

$$\begin{cases} {}^{C}D_{1-}^{3/2}\,{}^{RL}D_{0+}^{1/2}y(t) + 2I_{1-}^{4/3}I_{0+}^{5/4}h(t,y(t)) = f(t,y(t)),\ t\in J:=[0,1], \\ y(0) = y(2/3) = 0,\quad y(1) = \frac{1}{2}y(3/4). \end{cases} \tag{2.49}$$

Here $\alpha = 3/2, \beta = 1/2, \lambda = 2, p = 4/3, q = 5/4, \mu = 3/4, \delta = 1/2, \xi = 2/3$, and

$$f(t,y) = \frac{1}{(t^2+8)}\left(\tan^{-1}y + e^{-t}\right), \qquad h(t,y) = \frac{1}{2\sqrt{t^2+9}}\left(\frac{|y|}{1+|y|} + e^{-t}\right).$$

Using the given data, it is found that

$$\overline{a}_1 = \max_{t\in[0,1]}|a_1(t)| = |a_1(t)|_{t=1} \approx 1.121394517474712,$$
$$\overline{a}_2 = \max_{t\in[0,1]}|a_2(t)| = |a_2(t)|_{t=t_{a_2}} \approx 1.168623082364286,$$

where

$$t_{a_2} = \frac{\beta(1-\delta\mu^{\beta+1})}{(1-\delta\mu^{\beta})(\beta+1)} \approx 0.396975661732535.$$

In consequence, we get

$$\Omega_1 \approx 3.022797441671726, \Omega_2 \approx 1.451691300771574, |\Lambda| \approx 0.242702744426469,$$

where Ω_1,Ω_2 are defined by (2.43) and Λ is given by (2.40).

(i) For illustrating Theorem 2.13, we have

$$|f(t,y)| \leq \sigma(t) = \frac{e^{-t} + (\pi/2)}{t^2+8}, \qquad |h(t,y)| \leq \rho(t) = \frac{e^{-t}+1}{2\sqrt{t^2+9}},$$

and that

$$L\gamma_1 + K\gamma_2 \approx 0.174044436618777 < 1,$$

where $\gamma_1 \approx 0.848826363156775$, $\gamma_2 \approx 0.407646847345084$. Clearly, the hypothesis of Theorem 2.13 is satisfied and consequently its conclusion applies to the problem (2.49).

(ii) In order to apply Theorem 2.14, we take

$$f(t,y) = \frac{1}{\sqrt{t^2+25}}\left(y\cos y + \pi/2\right), \qquad h(t,y) = \frac{1}{5\sqrt{t^2+4}}\left(\sin y + 1/4\right), \tag{2.50}$$

and note that $\omega_1(t) = \frac{1}{\sqrt{t^2+25}}$, $\|\omega_1\| = 1/5$, $\omega_2(t) = \frac{1}{5\sqrt{t^2+4}}$, $\|\omega_2\| = 1/10$, $\phi(\|y\|) = \|y\| + \pi/2$ and $\psi(\|y\|) = \|y\| + 1/4$. By the condition (B_5), we find that $M > 3.939452045479877$. Thus, all the conditions of Theorem 2.14 are satisfied and, hence the problem (2.49) with $f(t, y)$ and $h(t, y)$ given by (2.50) has at least one solution on $[0, 1]$.

(iii) It is easy to show that $f(t, y)$ and $h(t, y)$ satisfy the conditions (2.13.1) and (2.13.2) respectively with $L = 1/8$ and $K = 1/6$ and that $L\Omega_1 + K\Omega_2 \approx 0.61979821 < 1$. Thus, all the assumptions of Theorem 2.15 hold true and hence the problem (2.49) has a unique solution on $[0, 1]$.

2.5 Nonlocal hybrid boundary value problem for fractional integro-differential equations and inclusions

In this section, we introduce a new class of nonlocal fractional boundary value problems involving hybrid fractional differential equations and inclusions. The class of hybrid fractional differential equations involves the fractional derivative operator of an unknown function hybrid with the nonlinearity depending on it. Existence results are obtained by means of hybrid fixed point theorems for three operators in a Banach algebra due to Dhage (Theorems 1.10 and 1.11).

2.5.1 *Nonlocal hybrid single-valued boundary value problem*

We consider a nonlocal boundary value problem of hybrid fractional integro-differential equations given by

$$\begin{cases} {}^CD^{\alpha}\left[\dfrac{x(t) - \sum\limits_{i=1}^{m} I^{\beta_i} h_i(t, x(t))}{f(t, x(t))}\right] = g(t, x(t)), \quad t \in J := [0, 1], \\ x(0) = \mu(x), \quad x(1) = A, \end{cases} \tag{2.51}$$

where ${}^CD^{\alpha}$ denotes the Caputo fractional derivative operator of order α, $1 < \alpha \leq 2$, I^{ϕ} is the Riemann-Liouville fractional integral of order $\phi > 0$, $\phi \in \{\beta_1, \beta_2, \ldots, \beta_m\}$, $f \in C(J \times \mathbb{R}, \mathbb{R}\setminus\{0\})$, $g \in C(J \times \mathbb{R}, \mathbb{R})$, $h_i \in C(J \times \mathbb{R}, \mathbb{R})$, $1 < \beta_i \leq 2$, $i = 1, 2, \ldots, m$, $\mu : C([0, 1], \mathbb{R}) \to \mathbb{R}$ and $A \in \mathbb{R}$.

The main tool of our study includes a hybrid fixed point theorem for three operators in a Banach algebra due to Dhage (Theorem 1.10).

We denote by $E = C(J, \mathbb{R})$ the space of continuous real-valued functions defined on $J = [0, 1]$. Define a norm $\|\cdot\|$ and a multiplication in E by

$$\|x\| = \sup_{t \in J} |x(t)| \quad \text{and} \quad (xy)(t) = x(t)y(t), \ \forall t \in J.$$

Clearly E is a Banach algebra with respect to above supremum norm and the multiplication in it.

Lemma 2.5. *Let $y \in AC([0, 1], \mathbb{R})$. Then x is a solution of the hybrid fractional integro-differential problem*

$$\begin{cases} {}^C D^\alpha \left[\dfrac{x(t) - \sum_{i=1}^m I^{\beta_i} h_i(t, x(t))}{f(t, x(t))} \right] = y(t), \quad t \in J := [0,1], \\ x(0) = \mu(x), \quad x(1) = A, \end{cases} \tag{2.52}$$

if and only if

$$\begin{aligned} x(t) = \sum_{i=1}^m I^{\beta_i} h_i(t, x(t)) + f(t, x(t)) \Bigg\{ \int_0^t \frac{(t-s)^{\alpha-1}}{\Gamma(\alpha)} y(s) ds \\ + \left(\frac{A - \sum_{i=1}^m \int_0^1 \frac{(1-s)^{\beta_i - 1}}{\Gamma(\beta_i)} h_i(s, x(s)) ds}{f(1, A)} - \int_0^1 \frac{(1-s)^{\alpha-1}}{\Gamma(\alpha)} y(s) ds \right) t \\ + (1-t) \frac{\mu(x)}{f(0, \mu(x))} \Bigg\}, \quad t \in J. \end{aligned} \tag{2.53}$$

Proof. Applying the Riemann-Liouville fractional integral of order α to both sides of (2.52), we have

$$\frac{x(t) - \sum_{i=1}^m I^{\beta_i} h_i(t, x(t))}{f(t, x(t))} = \int_0^t \frac{(t-s)^{\alpha-1}}{\Gamma(\alpha)} y(s) ds + c_1 t + c_2, \tag{2.54}$$

where $c_1, c_2 \in \mathbb{R}$. Using the given boundary conditions, we find that

$$c_2 = \frac{\mu(x)}{f(0, \mu(x))}$$

and

$$c_1 = \frac{A - \displaystyle\sum_{i=1}^m \int_0^1 \frac{(1-s)^{\beta_i - 1}}{\Gamma(\beta_i)} h_i(s, x(s)) ds}{f(1, A)} - \int_0^1 \frac{(1-s)^{\alpha-1}}{\Gamma(\alpha)} y(s) ds - \frac{\mu(x)}{f(0, \mu(x))}.$$

Substituting the values of c_1, c_2 in (2.54), we get the solution (2.53). The converse follows by direct computation. This completes the proof. □

Theorem 2.16. *Assume that:*

(2.16.1) The functions $f : J \times \mathbb{R} \to \mathbb{R} \setminus \{0\}$ and $h_i : J \times \mathbb{R} \to \mathbb{R}$, $i = 1, 2, \ldots, m$, are continuous and there exist positive functions ϕ, ψ_i, $i = 1, 2, \ldots, m$, with bounds $\|\phi\|$ and $\|\psi_i\|$, $i = 1, 2, \ldots, m$, respectively, such that

$$|f(t, x) - f(t, y)| \le \phi(t) |x - y| \tag{2.55}$$

and

$$|h_i(t, x) - h_i(t, y)| \le \psi_i(t) |x - y|, \; i = 1, 2, \ldots, m, \tag{2.56}$$

for $t \in J$ and $x, y \in \mathbb{R}$;

(2.16.2) there exists a function $p \in C(J, \mathbb{R}^+)$ and a continuous non-decreasing function $\Psi : [0, \infty) \to (0, \infty)$ such that

$$|g(t, x(t))| \leq p(t)\Psi(|x|), \qquad (t, x) \in J \times \mathbb{R}; \tag{2.57}$$

(2.16.3) there exists a constant $M_0 > 0$ such that

$$|h_i(t, x)| \leq M_0, \quad (t, x) \in J \times \mathbb{R}, \quad i = 1, 2, \ldots, m;$$

(2.16.4) there exists a constant $M_1 > 0$ such that

$$\left| \frac{\mu(x)}{f(0, \mu(x))} \right| \leq M_1, \ \forall x \in C(J, \mathbb{R});$$

(2.16.5) there exists a number $r > 0$ such that

$$r \geq \frac{F_0 \left[\dfrac{2\|p\|\Psi(r)}{\Gamma(\alpha+1)} + \dfrac{|A| + M_0 \sum\limits_{i=1}^{m} \frac{1}{\Gamma(\beta_i+1)}}{|f(1, A)|} + M_1 \right] + \sum\limits_{i=1}^{m} \dfrac{K_i}{\Gamma(\beta_i+1)}}{1 - \|\phi\| \left[\dfrac{2\|p\|\Psi(r)}{\Gamma(\alpha+1)} + \dfrac{|A| + M_0 \sum\limits_{i=1}^{m} \frac{1}{\Gamma(\beta_i+1)}}{|f(1, A)|} + M_1 \right] - \sum\limits_{i=1}^{m} \dfrac{\|\psi_i\|}{\Gamma(\beta_i+1)}} \tag{2.58}$$

where $F_0 = \sup_{t \in J} |f(t, 0)|$, $K_i = \sup_{t \in J} |h_i(t, 0)|$, $i = 1, 2, \ldots, m$, and

$$\|\phi\| \left[\frac{2\|p\|\Psi(r)}{\Gamma(\alpha+1)} + \frac{|A| + M_0 \sum\limits_{i=1}^{m} \frac{1}{\Gamma(\beta_i+1)}}{|f(1, A)|} + M_1 \right] + \sum_{i=1}^{m} \frac{\|\psi_i\|}{\Gamma(\beta_i+1)} < 1. \tag{2.59}$$

Then there exists at least one solution for the nonlocal problem (2.51) on J.

Proof. We consider a subset S of E given by

$$S = \{x \in E : \|x\| \leq r\},$$

where r satisfies inequality (2.58). Notice that S is closed, convex and bounded subset of the Banach space E. In view of Lemma 2.5, we define an operator

$$\begin{aligned} Tx(t) = \sum_{i=1}^{m} I^{\beta_i} h_i(t, x(t)) + f(t, x(t)) \Bigg\{ &\int_0^t \frac{(t-s)^{\alpha-1}}{\Gamma(\alpha)} g(s, x(s)) ds \\ &+ \left(\frac{A - \sum\limits_{i=1}^{m} \int_0^1 \frac{(1-s)^{\beta_i - 1}}{\Gamma(\beta_i)} h_i(s, x(s)) ds}{f(1, A)} \right. \\ &\left. - \int_0^1 \frac{(1-s)^{\alpha-1}}{\Gamma(\alpha)} g(s, x(s)) ds \right) t + (1-t) \frac{\mu(x)}{f(0, \mu(x))} \Bigg\}, \quad t \in J. \end{aligned} \tag{2.60}$$

Notice that the fixed point problem $Tx = x$ is equivalent to the problem (2.51). Next we introduce three operators $\mathcal{A} : E \to E$, $\mathcal{B} : S \to E$ and $\mathcal{C} : E \to E$ as follows:

$$\mathcal{A}x(t) = f(t, x(t)), \ t \in J, \tag{2.61}$$

$$
\begin{aligned}
\mathcal{B}x(t) &= \int_0^t \frac{(t-s)^{\alpha-1}}{\Gamma(\alpha)} g(s,x(s))ds \\
&+ \left(\frac{A - \sum_{i=1}^{m} \int_0^1 \frac{(1-s)^{\beta_i-1}}{\Gamma(\beta_i)} h_i(s,x(s))ds}{f(1,A)} \right. \\
&\left. - \int_0^1 \frac{(1-s)^{\alpha-1}}{\Gamma(\alpha)} g(s,x(s))ds \right) t + (1-t)\frac{\mu(x)}{f(0,\mu(x))}, \quad t \in J, \qquad (2.62)
\end{aligned}
$$

and

$$
\begin{aligned}
\mathcal{C}x(t) &= \sum_{i=1}^{m} I^{\beta_i} h_i(t,x(t)) \\
&= \sum_{i=1}^{m} \int_0^t \frac{(t-s)^{\beta_i-1}}{\Gamma(\beta_i)} h_i(s,x(s))ds, \quad t \in J. \qquad (2.63)
\end{aligned}
$$

Observe that $Tx = \mathcal{A}x\mathcal{B}x + \mathcal{C}x$. Now, we show that the operators $\mathcal{A}$, $\mathcal{B}$ and $\mathcal{C}$ satisfy all the conditions of Theorem 1.10 in a series of steps.

Step 1. *We first show that $\mathcal{A}$ and $\mathcal{C}$ are Lipschitzian on E.*

Let $x, y \in E$. Then by (2.16.1), for $t \in J$, we have

$$|\mathcal{A}x(t) - \mathcal{A}y(t)| = |f(t,x(t)) - f(t,y(t))| \le \phi(t)|x(t) - y(t)| \le \|\phi\|\|x-y\|,$$

which implies $\|\mathcal{A}x - \mathcal{A}y\| \le \|\phi\|\|x-y\|$ for all $x, y \in E$. Therefore, $\mathcal{A}$ is a Lipschitzian on E with Lipschitz constant $\|\phi\|$.

Analogously, for any $x, y \in E$, we have

$$
\begin{aligned}
|\mathcal{C}x(t) - \mathcal{C}y(t)| &= \left| \sum_{i=1}^{m} I^{\beta_i} h_i(t,x(t)) - \sum_{i=1}^{m} I^{\beta_i} h_i(t,y(t)) \right| \\
&\le \sum_{i=1}^{m} \int_0^t \frac{(t-s)^{\beta_i-1}}{\Gamma(\beta_i)} \psi_i(s)|x(s) - y(s)|ds \le \|x-y\| \sum_{i=1}^{m} \frac{\|\psi_i\|}{\Gamma(\beta_i+1)},
\end{aligned}
$$

which implies that

$$\|\mathcal{C}x - \mathcal{C}y\| \le \sum_{i=1}^{m} \frac{\|\psi_i\|}{\Gamma(\beta_i+1)} \|x-y\|.$$

Thus, $\mathcal{C}$ is a Lipschitzian on E with Lipschitz constant $\sum_{i=1}^{m} \frac{\|\psi_i\|}{\Gamma(\beta_i+1)}$.

Step 2. *The operator $\mathcal{B}$ is completely continuous on S.*

We first show that the operator $\mathcal{B}$ is continuous on E. Let $\{x_n\}$ be a sequence in S converging to a point $x \in S$. Then by Lebesgue dominated convergence theorem, for all $t \in J$, we obtain

$$\lim_{n\to\infty} \mathcal{B}x_n(t)$$

$$
\begin{aligned}
&= \lim_{n\to\infty}\Bigg\{\int_0^t \frac{(t-s)^{\alpha-1}}{\Gamma(\alpha)} g(s,x_n(s))ds \\
&\quad + \left(\frac{A-\sum\limits_{i=1}^{m}\int_0^1 \frac{(1-s)^{\beta_i-1}}{\Gamma(\beta_i)} h_i(s,x_n(s))ds}{f(1,A)} - \int_0^1 \frac{(1-s)^{\alpha-1}}{\Gamma(\alpha)} g(s,x_n(s))ds\right)t \\
&\quad +(1-t)\frac{\mu(x_n)-h(0,\mu(x_n))}{f(0,\mu(x_n))}\Bigg\} \\
&= \int_0^t \frac{(t-s)^{\alpha-1}}{\Gamma(\alpha)} \lim_{n\to\infty} g(s,x_n(s))ds + \left(\frac{A-\sum\limits_{i=1}^{m}\int_0^1 \frac{(1-s)^{\beta_i-1}}{\Gamma(\beta_i)} \lim\limits_{n\to\infty} h_i(s,x_n(s))ds}{f(1,A)}\right. \\
&\quad \left.- \int_0^1 \frac{(1-s)^{\alpha-1}}{\Gamma(\alpha)} \lim_{n\to\infty} g(s,x_n(s))ds\right)t + (1-t)\lim_{n\to\infty}\frac{\mu(x_n)-h(0,\mu(x_n))}{f(0,\mu(x_n))} \\
&= \int_0^t \frac{(t-s)^{\alpha-1}}{\Gamma(\alpha)} g(s,x(s))ds \\
&\quad + \left(\frac{A-\sum\limits_{i=1}^{m}\int_0^1 \frac{(1-s)^{\beta_i-1}}{\Gamma(\beta_i)} h_i(s,x(s))ds}{f(1,A)} - \int_0^1 \frac{(1-s)^{\alpha-1}}{\Gamma(\alpha)} g(s,x(s))ds\right)t \\
&\quad +(1-t)\frac{\mu(x)}{f(0,\mu(x))} \\
&= \mathcal{B}x(t).
\end{aligned}
$$

This implies that $Bx_n \to Bx$ point-wise on J. Further it can be shown that $\{Bx_n\}$ is an equi-continuous sequence of functions. So $Bx_n \to Bx$ uniformly and the operator B is continuous on S.

Next we will prove that the set $\mathcal{B}(S)$ is uniformly bounded in S. For any $x \in S$, we have

$$
\begin{aligned}
|\mathcal{B}x(t)| &\le \int_0^t \frac{(t-s)^{\alpha-1}}{\Gamma(\alpha)}|g(s,x(s))|ds + \frac{|A| + M_0\sum\limits_{i=1}^{m}\frac{1}{\Gamma(\beta_i+1)}}{|f(1,A)|} \\
&\quad + \int_0^1 \frac{(1-s)^{\alpha-1}}{\Gamma(\alpha)}|g(s,x(s))|ds + M_1 \\
&\le 2\int_0^1 \frac{(t-s)^{\alpha-1}}{\Gamma(\alpha)} p(s)\Psi(r)ds + \frac{|A| + M_0\sum\limits_{i=1}^{m}\frac{1}{\Gamma(\beta_i+1)}}{|f(1,A)|} + M_1 \\
&\le \frac{2\|p\|\Psi(r)}{\Gamma(\alpha+1)} + \frac{|A| + M_0\sum\limits_{i=1}^{m}\frac{1}{\Gamma(\beta_i+1)}}{|f(1,A)|} + M_1 \\
&:= K_1,
\end{aligned}
$$

for all $t \in J$. Therefore, $\|\mathcal{B}\| \le K_1$ which shows that $\mathcal{B}$ is uniformly bounded on S.

Now, we will show that $\mathcal{B}(S)$ is an equi-continuous set in E. Let $\tau_1, \tau_2 \in J$ with $\tau_1 < \tau_2$ and $x \in S$. Then we have

$$\begin{aligned}
&|\mathcal{B}x(\tau_2) - \mathcal{B}x(\tau_1)| \\
&\leq \left| \int_0^{\tau_2} \frac{(\tau_2 - s)^{\alpha-1}}{\Gamma(\alpha)} g(s, x(s))ds - \int_0^{\tau_1} \frac{(\tau_1 - s)^{\alpha-1}}{\Gamma(\alpha)} g(s, x(s))ds \right| \\
&\quad + |\tau_2 - \tau_1| \left(\frac{|A| + M_0 \sum_{i=1}^{m} \frac{1}{\Gamma(\beta_i+1)}}{|f(1, A)|} + \int_0^1 \frac{(1-s)^{\alpha-1}}{\Gamma(\alpha)} g(s, x(s))ds \right) + |\tau_2 - \tau_1| M_1 \\
&\leq \int_0^{\tau_1} \left| \frac{(\tau_2 - s)^{\alpha-1} - (\tau_1 - s)^{\alpha-1}}{\Gamma(\alpha)} \right| |g(s, x(s))|ds + \int_{\tau_1}^{\tau_2} \frac{(\tau_2 - s)^{\alpha-1}}{\Gamma(\alpha)} |g(s, x(s))|ds \\
&\quad + |\tau_2 - \tau_1| \left(\frac{|A| + M_0 \sum_{i=1}^{m} \frac{1}{\Gamma(\beta_i+1)}}{|f(1, A)|} + \int_0^1 \frac{(1-s)^{\alpha-1}}{\Gamma(\alpha)} |g(s, x(s))|ds \right) + |\tau_2 - \tau_1| M_1 \\
&\leq \int_0^{\tau_1} \left| \frac{(\tau_2 - s)^{\alpha-1} - (\tau_1 - s)^{\alpha-1}}{\Gamma(\alpha)} \right| \|p\|\Psi(r)ds + \int_{\tau_1}^{\tau_2} \frac{(\tau_2 - s)^{\alpha-1}}{\Gamma(\alpha)} \|p\|\Psi(r)ds \\
&\quad + |\tau_2 - \tau_1| \left(\frac{|A| + M_0 \sum_{i=1}^{m} \frac{1}{\Gamma(\beta_i+1)}}{|f(1, A)|} + \int_0^1 \frac{(1-s)^{\alpha-1}}{\Gamma(\alpha)} \|p\|\Psi(r)ds \right) + |\tau_2 - \tau_1| M_1
\end{aligned}$$

which is independent of $x \in S$. As $\tau_1 \to \tau_2$, the right-hand side of the above inequality tends to zero. Therefore, it follows from the Arzelá-Ascoli theorem that $\mathcal{B}$ is a completely continuous operator on S.

Step 3. *The hypothesis* (c_1) *of Theorem 1.10 is satisfied.*

Let $x \in E$ and $y \in S$ be arbitrary elements such that $x = \mathcal{A}x\mathcal{B}y + \mathcal{C}x$. Then we have

$$\begin{aligned}
&|x(t)| \\
&\leq |\mathcal{A}x(t)||\mathcal{B}y(t)| + |\mathcal{C}x(t)| \\
&\leq |f(t, x(t))| \left\{ \int_0^t \frac{(t-s)^{\alpha-1}}{\Gamma(\alpha)} |g(s, x(s))|ds + \frac{|A| + M_0 \sum_{i=1}^{m} \frac{1}{\Gamma(\beta_i+1)}}{|f(1, A)|} \right. \\
&\quad \left. + \int_0^1 \frac{(1-s)^{\alpha-1}}{\Gamma(\alpha)} |g(s, x(s))|ds + M_1 \right\} + \sum_{i=1}^{m} \int_0^t \frac{(t-s)^{\beta_i - 1}}{\Gamma(\beta_i)} |h_i(s, x(s))|ds \\
&\leq (|f(t, x(t)) - f(t, 0))| + |f(t, 0)|) \left\{ \int_0^t \frac{(t-s)^{\alpha-1}}{\Gamma(\alpha)} \|p\|\Psi(r)ds \right. \\
&\quad \left. + \frac{|A| + M_0 \sum_{i=1}^{m} \frac{1}{\Gamma(\beta_i+1)}}{|f(1, A)|} + \int_0^1 \frac{(1-s)^{\alpha-1}}{\Gamma(\alpha)} \|p\|\Psi(r)ds + M_1 \right\}
\end{aligned}$$

$$+\sum_{i=1}^{m}\int_0^t \frac{(t-s)^{\beta_i-1}}{\Gamma(\beta_i)}\left(|h_i(s,x(s))-h_i(s,0)|+|h_i(s,0)|\right)ds$$

$$\leq (r\|\phi\|+F_0)\left(\frac{2\|p\|\Psi(r)}{\Gamma(\alpha+1)}+\frac{|A|+M_0\sum_{i=1}^{m}\frac{1}{\Gamma(\beta_i+1)}}{|f(1,A)|}+M_1\right)+\sum_{i=1}^{m}\frac{r\|\psi_i\|+K_0}{\Gamma(\beta_i+1)},$$

which leads to

$$\|x\| \leq (r\|\phi\|+F_0)\left(\frac{2\|p\|\Psi(r)}{\Gamma(\alpha+1)}+\frac{|A|+M_0\sum_{i=1}^{m}\frac{1}{\Gamma(\beta_i+1)}}{|f(1,A)|}+M_1\right)+\sum_{i=1}^{m}\frac{r\|\psi_i\|+K_0}{\Gamma(\beta_i+1)}$$
$$\leq r.$$

Therefore, $x \in S$.

Step 4. *Finally we show that $\delta M+\rho<1$, that is, (d_1) of Lemma 1.10 holds.*

Since

$$M=\|\mathcal{B}(S)\|=\sup_{x\in S}\left\{\sup_{t\in J}|\mathcal{B}x(t)|\right\}\leq \frac{2\|p\|\Psi(r)}{\Gamma(\alpha+1)}+\frac{|A|+M_0\sum_{i=1}^{m}\frac{1}{\Gamma(\beta_i+1)}}{|f(1,A)|}+M_1,$$

therefore, by (2.16.4), we have

$$\|\phi\|M+\sum_{i=1}^{m}\frac{\|\psi_i\|}{\Gamma(\beta_i+1)}<1,$$

where $\delta=\|\phi\|$ and $\rho=\sum_{i=1}^{m}\frac{\|\psi_i\|}{\Gamma(\beta_i+1)}$.

Thus all the conditions of Theorem 1.10 are satisfied and hence the operator equation $x=\mathcal{A}x\mathcal{B}x+\mathcal{C}x$ has a solution in S. In consequence, the problem (2.51) has a solution on J. This completes the proof. □

Example 2.3. Consider the following nonlocal hybrid boundary value problem

$$\begin{cases} {}^{C}D^{\frac{3}{2}}\left[\dfrac{x(t)-\sum_{i=1}^{5}I^{\frac{2i+1}{2}}h_i(t,x(t))}{f(t,x(t))}\right]=\dfrac{(x(t)+1)e^{-|x(t)|}\sin^2 t}{5+2t+t^2}, & t\in[0,1],\\ x(0)=\gamma\cos^2\left(\sum_{j=1}^{n-2}x(\xi_i)\right), \qquad x(1)=\dfrac{2}{3}, \end{cases} \tag{2.64}$$

where $0<\gamma<1$, $\xi_j\in(0,1)$, $\xi_j<\xi_{j+1}$, $j=1,2,\ldots,n-2$, $n\geq 3$,

$$h_i(t,x(t))=\frac{|x(t)|}{(24+i+t)(1+|x(t)|)},\quad i=1,2,\ldots,5,$$

and

$$f(t,x(t))=\frac{t}{2(50+t)}\left(|x(t)|+\tan^{-1}x(t)\right)+\frac{t+1}{3}.$$

Here $\alpha = 3/2$, $m = 5$, $\beta_1 = 3/2$, $\beta_2 = 5/2$, $\beta_3 = 7/2$, $\beta_4 = 9/2$, $\beta_5 = 11/2$, $A = 2/3$ and $\mu(x) = \gamma\cos^2(\sum_{j=1}^{n-2} x(\xi_j))$. Clearly

$$|f(t,x) - f(t,y)| \le \left(\frac{t}{50+t}\right)|x-y|,$$

$$|h_i(t,x(t)) - h_i(t,y(t))| \le \left(\frac{1}{24+i+t}\right)|x-y|, \quad \text{for } x,y \in \mathbb{R}, i = 1,2,\ldots,5.$$

Setting $\phi(t) = t/(50+t)$ and $\psi(t) = 1/(24+i+t)$, we get $\|\phi\| = 1/51$ and $\|\psi_i\| = 1/(24+i)$, $i = 1,2,\ldots,5$. It is easy to see that

$$|g(t,x)| = \left|\frac{(x+1)e^{-|x|}\sin^2 t}{5+2t+t^2}\right| \le \left(\frac{(|x|+1)}{5+2t+t^2}\right), \quad \left|\frac{\mu(x)}{f(0,\mu(x))}\right| \le 3, \quad |h_i(t,x)| \le \frac{1}{25}.$$

Setting $p(t) = 1/(5+2t+t^2)$, $\Psi(|x|) = |x|+1$, $M_0 = 1/25$ and $M_1 = 3$, we get $\|p\| = 1/5$, $F_0 = \sup_{t\in J}|f(t,0)| = 2/3$ and $K_0 = \sup_{t\in J}|h_i(t,0)| = 0$. Using these values, it follows by (2.58)-(2.59) that the constant r satisfies the inequality $4.531120800 < r < 108.7320755$. Thus all the conditions of Theorem 2.16 are satisfied. Hence, the problem (2.64) has at least one solution on $[0,1]$.

2.5.2 *Nonlocal hybrid multi-valued boundary value problem*

Here we study the existence of solutions for a nonlocal boundary value problem of hybrid fractional integro-differential inclusions given by

$$\begin{cases} {}^C D^\alpha \left[\dfrac{x(t) - \sum\limits_{i=1}^{m} I^{\beta_i} h_i(t,x(t))}{f(t,x(t))}\right] \in G(t,x(t)), \quad t \in J := [0,1], \\ x(0) = \mu(x), \quad x(1) = A, \end{cases} \tag{2.65}$$

where D^α denotes the Caputo fractional derivative operator of order α, $1 < \alpha \le 2$, I^ϕ, f, h_i $(i = 1,2), \mu, A$ are as in problem (2.51) and $G : J \times \mathbb{R} \to \mathcal{P}(\mathbb{R})$ is a multi-valued map, $\mathcal{P}(\mathbb{R})$ is the family of all subsets of $\mathbb{R}$.

A hybrid fixed point theorem of Schaefer type for three operators in a Banach algebra X (Theorem 1.11) due to Dhage [86, Theorem 4.13] plays a key role in proving the existence result for the nonlocal boundary value problem (2.65).

Theorem 2.17. *Assume that (2.16.1), (2.16.3), (2.16.4) hold. In addition we suppose that:*

(2.17.1) $G : J \times \mathbb{R} \to \mathcal{P}_{cp,cv}(\mathbb{R}^+)$ *is* L^1*-Carathéodory multi-valued map;*
(2.17.2) There exists a continuous function $p \in C(J,\mathbb{R}^+)$ *such that*

$$\|G(t,x)\|_{\mathcal{P}} := \sup\{|y| : y \in G(t,x)\} \le p(t) \quad \text{for each} \quad (t,x) \in J \times \mathbb{R}.$$

Then the nonlocal boundary value problem (2.65) *has at least one solution on* J *provided that*

$$\gamma := \|\phi\|\left(\frac{2\|p\|}{\Gamma(\alpha+1)} + \frac{|A| + M_0 \sum\limits_{i=1}^{m} \frac{1}{\Gamma(\beta_i+1)}}{|f(1,A)|} + M_1\right) + \sum_{i=1}^{m} \frac{\|\psi_i\|}{\Gamma(\beta_i+1)} < \frac{1}{2}. \tag{2.66}$$

Proof. Let $X = C(J, \mathbb{R})$ be the space of continuous real-valued functions defined on $J = [0,1]$. Define a norm $\|\cdot\|$ and a multiplication in X by

$$\|x\| = \sup_{t\in J} |x(t)| \quad \text{and} \quad (xy)(t) = x(t)y(t), \ \forall t \in J.$$

Clearly X is a Banach algebra with respect to above norm and the multiplication in it.

Define three operators $\mathcal{A}$, $\mathcal{B}$ and $\mathcal{C}$ on X by

$$\mathcal{A}x(t) = f(t, x(t)), \ \ t \in J, \tag{2.67}$$

$$\mathcal{B}x(t) = \left\{ u(t) : u(t) = \left\{ \begin{array}{l} \displaystyle\int_0^t \frac{(t-s)^{\alpha-1}}{\Gamma(\alpha)} v(s)ds + (1-t)\frac{\mu(x)}{f(0,\mu(x))} \\ \displaystyle + \left(\frac{A - \sum_{i=1}^m \int_0^1 \frac{(1-s)^{\beta_i-1}}{\Gamma(\beta_i)} h_i(s,x(s))ds}{f(1,A)} \right. \\ \displaystyle \left. - \int_0^1 \frac{(1-s)^{\alpha-1}}{\Gamma(\alpha)} v(s)ds \right) t, \ v \in S_{G,x}, \ t \in J, \end{array} \right. \right\} \tag{2.68}$$

and

$$\mathcal{C}x(t) = \sum_{i=1}^m I^{\beta_i} h_i(t, x(t)) = \sum_{i=1}^m \int_0^t \frac{(t-s)^{\beta_i-1}}{\Gamma(\beta_i)} h_i(s, x(s))ds, \ \ t \in J. \tag{2.69}$$

We will show that the operators $\mathcal{A}$, $\mathcal{B}$ and $\mathcal{C}$ satisfy the hypothesis of Theorem 1.11 in a series of steps.

Step 1. *We first show that the operators $\mathcal{A}$ and $\mathcal{C}$ define single-valued operators $\mathcal{A}, \mathcal{C} : X \to X$ and $\mathcal{B} : X \to \mathcal{P}_{cp,cv}(X)$.*

The claim concerning $\mathcal{A}$ and $\mathcal{C}$ is obvious, because the functions f and h_i are continuous on $J \times \mathbb{R}$. We only prove the claim for the multi-valued operator $\mathcal{B}$ on X. First, we show that $\mathcal{B}$ has compact values on X. Observe that the operator $\mathcal{B}$ is equivalent to the composition $K \circ S_{G,x}$ of two operators on $L^1(J, \mathbb{R})$, where $K : L^1(J, \mathbb{R}) \to X$ is the continuous operator defined by

$$\begin{aligned} Kv(t) = &\int_0^t \frac{(t-s)^{\alpha-1}}{\Gamma(\alpha)} v(s)ds + (1-t)\frac{\mu(x)}{f(0,\mu(x))} \\ &+ \left(\frac{A - \sum_{i=1}^m \int_0^1 \frac{(1-s)^{\beta_i-1}}{\Gamma(\beta_i)} h_i(s,z(s))ds}{f(1,A)} - \int_0^1 \frac{(1-s)^{\alpha-1}}{\Gamma(\alpha)} v(s)ds \right) t \end{aligned}$$

and $S_{G,x} := \{v \in L^1(J, \mathbb{R}) : v(t) \in G(t, x(t)) \text{ on } J\}$.

To show that $\mathcal{B}$ has compact values, it then suffices to prove that the composition operator $K \circ S_{G,x}$ has compact values on X. Let $x \in X$ be arbitrary and let $\{v_n\}$ be a sequence in $S_{G,x}$. Then, by the definition of $S_{G,x}$, $v_n(t) \in G(t, x(t))$ a.e. for $t \in J$. Since $G(t, x(t))$ is compact, there is a convergent subsequence of $v_n(t)$ (for simplicity call it $v_n(t)$ itself) that converges in measure to some $v(t)$, where $v(t) \in G(t, x(t))$ a.e. for $t \in J$. From the continuity of K, it follows that $Kv_n(t) \to Kv(t)$ pointwise on J as $n \to \infty$. In order to show that the convergence is uniform, we need to show that $\{Kv_n\}$ is an equi-continuous sequence. Let $\tau_1, \tau_2 \in J$ with $\tau_1 < \tau_2$. Then, we have

$$\begin{aligned}
&|Kv_n(\tau_2) - Kv_n(\tau_1)| \\
&\le \left| \int_0^{\tau_2} \frac{(\tau_2 - s)^{\alpha-1}}{\Gamma(\alpha)} v_n(s)ds - \int_0^{\tau_1} \frac{(\tau_1 - s)^{\alpha-1}}{\Gamma(\alpha)} v_n(s)ds \right| \\
&\quad + |\tau_2 - \tau_1| \left(\frac{|A| + M_0 \sum_{i=1}^{m} \frac{1}{\Gamma(\beta_i+1)}}{|f(1,A)|} + \int_0^1 \frac{(1-s)^{\alpha-1}}{\Gamma(\alpha)} |v_n(s)| ds \right) + |\tau_2 - \tau_1| M_1 \\
&\le \int_0^{\tau_1} \left| \frac{(\tau_2 - s)^{\alpha-1} - (\tau_1 - s)^{\alpha-1}}{\Gamma(\alpha)} \right| |v_n(s)| ds + \int_{\tau_1}^{\tau_2} \frac{(\tau_2 - s)^{\alpha-1}}{\Gamma(\alpha)} |v_n(s)| ds \\
&\quad + |\tau_2 - \tau_1| \left(\frac{|A| + M_0 \sum_{i=1}^{m} \frac{1}{\Gamma(\beta_i+1)}}{|f(1,A)|} + \int_0^1 \frac{(1-s)^{\alpha-1}}{\Gamma(\alpha)} |v_n(s)| ds \right) + |\tau_2 - \tau_1| M_1 \\
&\le \int_0^{\tau_1} \left| \frac{(\tau_2 - s)^{\alpha-1} - (\tau_1 - s)^{\alpha-1}}{\Gamma(\alpha)} \right| \|p\| ds + \int_{\tau_1}^{\tau_2} \frac{(\tau_2 - s)^{\alpha-1}}{\Gamma(\alpha)} \|p\| ds \\
&\quad + |\tau_2 - \tau_1| \left(\frac{|A| + M_0 \sum_{i=1}^{m} \frac{1}{\Gamma(\beta_i+1)}}{|f(1,A)|} + \int_0^1 \frac{(1-s)^{\alpha-1}}{\Gamma(\alpha)} \|p\| ds \right) + |\tau_2 - \tau_1| M_1 \\
&\le \frac{\|p\|}{\Gamma(\alpha+1)} (\tau_2^\alpha - \tau_1^\alpha) + |\tau_2 - \tau_1| \left(\frac{|A| + M_0 \sum_{i=1}^{m} \frac{1}{\Gamma(\beta_i+1)}}{|f(1,A)|} + \frac{\|p\|}{\Gamma(\alpha+1)} + M_1 \right).
\end{aligned}$$

As $\tau_1 \to \tau_2$, the right-hand side of the above inequality tends to zero. Therefore, it follows by the Arzelá-Ascoli theorem that $\{Kv_n\}$ is an equi-continuous sequence.

Next, we show that $\mathcal{B}$ is convex valued. Let $u_1, u_2 \in \mathcal{B}(x)$. Then there exist $v_1, v_2 \in S_{G,x}$ such that

$$\begin{aligned}
u_i(t) = &\int_0^t \frac{(t-s)^{\alpha-1}}{\Gamma(\alpha)} v_i(s)ds + (1-t)\frac{\mu(x)}{f(0,\mu(x))} \\
&+ \left(\frac{A - \sum_{i=1}^{m} \int_0^1 \frac{(1-s)^{\beta_i-1}}{\Gamma(\beta_i)} h_i(s, z(s))ds}{f(1,A)} - \int_0^1 \frac{(1-s)^{\alpha-1}}{\Gamma(\alpha)} v_i(s)ds \right) t,
\end{aligned}$$

for $i = 1, 2$. Now for any $\lambda \in [0, 1]$, we have

$$\begin{aligned}
&\lambda u_1(t) + (1-\lambda) u_2(t) \\
&= \int_0^t \frac{(t-s)^{\alpha-1}}{\Gamma(\alpha)} \lambda v_1(s)ds + \lambda(1-t)\frac{\mu(x)}{f(0,\mu(x))}
\end{aligned}$$

$$+\left(\frac{\lambda A-\sum_{i=1}^{m}\int_0^1\frac{(1-s)^{\beta_i-1}}{\Gamma(\beta_i)}\lambda h_i(s,x(s))ds}{f(1,A)}-\int_0^1\frac{(1-s)^{\alpha-1}}{\Gamma(\alpha)}\lambda v_1(s)ds\right)t$$
$$+\int_0^t\frac{(t-s)^{\alpha-1}}{\Gamma(\alpha)}(1-\lambda)v_2(s)ds+(1-\lambda)(1-t)\frac{\mu(x)}{f(0,\mu(x))}$$
$$+\left(\frac{(1-\lambda)A-\sum_{i=1}^{m}\int_0^1\frac{(1-s)^{\beta_i-1}}{\Gamma(\beta_i)}(1-\lambda)h_i(s,x(s))ds}{f(1,A)}\right.$$
$$\left.-\int_0^1\frac{(1-s)^{\alpha-1}}{\Gamma(\alpha)}(1-\lambda)v_2(s)ds\right)t$$
$$=\int_0^t\frac{(t-s)^{\alpha-1}}{\Gamma(\alpha)}[\lambda v_1(s)+(1-\lambda v_2(s)]ds$$
$$+(1-t)\frac{\mu(x)}{f(0,\mu(x))}+\left(\frac{A-\sum_{i=1}^{m}\int_0^1\frac{(1-s)^{\beta_i-1}}{\Gamma(\beta_i)}h_1(s,x(s))ds}{f(1,A)}\right.$$
$$\left.-\int_0^1\frac{(1-s)^{\alpha-1}}{\Gamma(\alpha)}\lambda[\lambda v_1(s)+(1-\lambda v_2(s)]ds\right)t.$$

Since $G(t,x(t))$ is convex, $\lambda v_1(t)+(1-\lambda)v_2(t)\in G(t,x(t))$ for all $t\in J$ and so $\lambda u_1+(1-\lambda)u_2\in S_{G,x}$. Thus $\lambda u_1+(1-\lambda)u_2\in\mathcal{B}(x)$, which proves that $\mathcal{B}(x)$ is a convex subset of X. As a result $\mathcal{B}$ defines a multi-valued operator $\mathcal{B}:X\to\mathcal{P}_{cp,cv}(X)$.

Step 2. *We show now that $\mathcal{A}$ and $\mathcal{C}$ are single-valued Lipschitz operators on X.*

Let $x,y\in X$. Then by (2.16.1), for $t\in J$, we have

$$|\mathcal{A}x(t)-\mathcal{A}y(t)|=|f(t,x(t))-f(t,y(t))|\le\phi(t)|x(t)-y(t)|\le\|\phi\|\|x-y\|,$$

which implies $\|\mathcal{A}x-\mathcal{A}y\|\le\|\phi\|\|x-y\|$ for all $x,y\in X$. Therefore, $\mathcal{A}$ is a Lipschitzian on X with Lipschitz constant $\|\phi\|$.

Analogously, for any $x,y\in X$, we have

$$\begin{aligned}|\mathcal{C}x(t)-\mathcal{C}y(t)|&=\left|\sum_{i=1}^{m}I^{\beta_i}[h_i(t,x(t))-h_i(t,y(t))]\right|\\&\le\sum_{i=1}^{m}\int_0^t\frac{(t-s)^{\beta_i-1}}{\Gamma(\beta_i)}\psi_i(s)|x(s)-y(s)|ds\\&\le\|x-y\|\sum_{i=1}^{m}\frac{\|\psi_i\|}{\Gamma(\beta_i+1)},\end{aligned}$$

which implies that

$$\|\mathcal{C}x - \mathcal{C}y\| \leq \sum_{i=1}^{m} \frac{\|\psi_i\|}{\Gamma(\beta_i + 1)} \|x - y\|.$$

Thus, $\mathcal{C}$ is a Lipschitzian on X with Lipschitz constant $\sum_{i=1}^{m} \frac{\|\psi_i\|}{\Gamma(\beta_i + 1)}$.

Step 3. *The operator $\mathcal{B}$ is completely continuous and upper semi-continuous on X.*

Let S be a bounded subset of X. Then, there is a constant $r > 0$, such that $\|x\| \leq r$, for all $x \in S$. First, we prove that $\mathcal{B}$ is compact operator on S. To do this, it is enough to prove that $\mathcal{B}(S)$ is a uniformly bounded and equi-continuous set in X. Let $x \in \mathcal{B}(S)$ be arbitrary. Then, there is a $v \in S_{G,x}$, such that

$$x(t) = \int_0^t \frac{(t-s)^{\alpha-1}}{\Gamma(\alpha)} v(s)ds + (1-t)\frac{\mu(x)}{f(0, \mu(x))}$$
$$+ \left(\frac{A - \sum_{i=1}^{m} \int_0^1 \frac{(1-s)^{\beta_i - 1}}{\Gamma(\beta_i)} h_i(s, x(s))ds}{f(1, A)} - \int_0^1 \frac{(1-s)^{\alpha-1}}{\Gamma(\alpha)} v(s)ds \right) t,$$

for any $x \in S$. Then, we have

$$\begin{aligned}
|x(t)| &\leq \int_0^t \frac{(t-s)^{\alpha-1}}{\Gamma(\alpha)} |v(s)|ds + M_1 + \frac{|A| + M_0 \sum_{i=1}^{m} \frac{1}{\Gamma(\beta_i+1)}}{|f(1, A)|} \\
&\quad + \int_0^1 \frac{(1-s)^{\alpha-1}}{\Gamma(\alpha)} |v(s)|ds \\
&\leq 2 \int_0^1 \frac{(1-s)^{\alpha-1}}{\Gamma(\alpha)} p(s)ds + M_1 + \frac{|A| + M_0 \sum_{i=1}^{m} \frac{1}{\Gamma(\beta_i+1)}}{|f(1, A)|} \\
&\leq \frac{2\|p\|}{\Gamma(\alpha + 1)} + M_1 + \frac{|A| + M_0 \sum_{i=1}^{m} \frac{1}{\Gamma(\beta_i+1)}}{|f(1, A)|} \\
&:= K_1,
\end{aligned}$$

for all $t \in J$. Therefore, $\|x\| \leq K_1$, which shows that $\mathcal{B}(S)$ is uniformly bounded set in X.

Again, proceeding with the arguments as in Step 1, we can see that $\mathcal{B}(S)$ is an equi-continuous set in X.

Next, we show that $\mathcal{B}$ is an upper semi-continuous multi-valued mapping on X. It is known, by Lemma 1.1, that $\mathcal{B}$ will be upper semi-continuous if, we establish that it has a closed graph, since it is already shown to be completely continuous. Thus, we will prove that $\mathcal{B}$ *has a closed graph.*

Let $\{x_n\}$ be a sequence in X such that $x_n \to x^*$. Let $\{y_n\}$ be a sequence such that $y_n \in \mathcal{B}x_n$ and $y_n \to y^*$. We shall show that $y^* \in \mathcal{B}x^*$. Since $y_n \in \mathcal{B}x_n$, there exists a $v_n \in S_{G,x_n})$ such that

$$y_n(t) = \int_0^t \frac{(t-s)^{\alpha-1}}{\Gamma(\alpha)} v_n(s)ds + (1-t)\frac{\mu(x)}{f(0,\mu(x))} + \left(\frac{A - \sum_{i=1}^{m} \int_0^1 \frac{(1-s)^{\beta_i - 1}}{\Gamma(\beta_i)} h_i(s,x(s))ds}{f(1,A)} - \int_0^1 \frac{(1-s)^{\alpha-1}}{\Gamma(\alpha)} v_n(s)ds \right) t, \ t \in J.$$

We must prove that there is a $v^* \in S_{G,x^*}$ such that

$$y^*(t) = \int_0^t \frac{(t-s)^{\alpha-1}}{\Gamma(\alpha)} v^*(s)ds + (1-t)\frac{\mu(x)}{f(0,\mu(x))} + \left(\frac{A - \sum_{i=1}^{m} \int_0^1 \frac{(1-s)^{\beta_i - 1}}{\Gamma(\beta_i)} h_i(s,x(s))ds}{f(1,A)} - \int_0^1 \frac{(1-s)^{\alpha-1}}{\Gamma(\alpha)} v^*(s)ds \right) t, \ t \in J.$$

Consider the continuous linear operator $L : L^1(J,\mathbb{R}) \to C(J,\mathbb{R})$ defined by

$$Lv(t) = \int_0^t \frac{(t-s)^{\alpha-1}}{\Gamma(\alpha)} v(s)ds + (1-t)\frac{\mu(x)}{f(0,\mu(x))} + \left(\frac{A - \sum_{i=1}^{m} \int_0^1 \frac{(1-s)^{\beta_i - 1}}{\Gamma(\beta_i)} h_i(s,x(s))ds}{f(1,A)} - \int_0^1 \frac{(1-s)^{\alpha-1}}{\Gamma(\alpha)} v(s)ds \right) t, \ t \in J.$$

Observe that

$$\|y_n(t) - y^*(t)\| = \left\| \int_0^t \frac{(t-s)^{\alpha-1}}{\Gamma(\alpha)} [v_n(s) - v^*(s)]ds - \int_0^1 \frac{(1-s)^{\alpha-1}}{\Gamma(\alpha)} [v_n(s) - v^*(s)]ds \right\| \to 0,$$

as $n \to \infty$. From Lemma 1.2, it follows that $L \circ S_{G,x}$ is a closed graph operator. Further, we have $y_n(t) \in L(S_{G,x})$. Since $y_n \to y^*$, therefore, we have

$$y^*(t) = \int_0^t \frac{(t-s)^{\alpha-1}}{\Gamma(\alpha)} v^*(s)ds + (1-t)\frac{\mu(x)}{f(0,\mu(x))} + \left(\frac{A - \sum_{i=1}^{m} \int_0^1 \frac{(1-s)^{\beta_i - 1}}{\Gamma(\beta_i)} h_i(s,x(s))ds}{f(1,A)} - \int_0^1 \frac{(1-s)^{\alpha-1}}{\Gamma(\alpha)} v^*(s)ds \right) t, \ t \in J.$$

As a result, we have that $\mathcal{B}$ is a compact and upper semi-continuous operator on X.

Step 4. *We show that the condition (c) of Theorem 1.11 holds, that is,* $Mq_1 + q_2 < 1/2$.

This is obvious by (2.66).

Step 5. *Finally, we show that the conclusion (ii) of Theorem 1.11 does not hold.*

Let x be any solution of the boundary value problem (2.65) such that $\mu x \in \mathcal{A}x\mathcal{B}x + \mathcal{C}x$ for some $\mu > 1$. Then there is a $v \in S_{G,x}$ such that

$$x(t) = \lambda \sum_{i=1}^{m} I^{\beta_i} h_i(t, x(t)) + \lambda f(t, x(t)) \Bigg\{ \int_0^t \frac{(t-s)^{\alpha-1}}{\Gamma(\alpha)} v(s)ds + \left(\frac{A - \sum_{i=1}^{m} \int_0^1 \frac{(1-s)^{\beta_i-1}}{\Gamma(\beta_i)} h_i(s, x(s))ds}{f(1,A)} - \int_0^1 \frac{(1-s)^{\alpha-1}}{\Gamma(\alpha)} v(s)ds \right) t + (1-t)\frac{\mu(x)}{f(0,\mu(x))} \Bigg\}, \quad t \in J,$$

where $\lambda = \frac{1}{\mu} < 1$. Then, we have

$$\begin{aligned}
&|x(t)| \\
&\le |f(t,x(t))| \Bigg\{ \int_0^t \frac{(t-s)^{\alpha-1}}{\Gamma(\alpha)} |v(s)|ds + \frac{|A| + M_0 \sum_{i=1}^{m} \frac{1}{\Gamma(\beta_i+1)}}{|f(1,A)|} \\
&\quad + \int_0^1 \frac{(1-s)^{\alpha-1}}{\Gamma(\alpha)} |v(s)|ds + M_1 \Bigg\} + \sum_{i=1}^{m} \int_0^t \frac{(t-s)^{\beta_i-1}}{\Gamma(\beta_i)} |h_i(s, x(s))|ds \\
&\le (|f(t,x(t) - f(t,0))| + |f(t,0)|) \Bigg\{ \int_0^t \frac{(t-s)^{\alpha-1}}{\Gamma(\alpha)} \|p\| ds \\
&\quad + \frac{|A| + M_0 \sum_{i=1}^{m} \frac{1}{\Gamma(\beta_i+1)}}{|f(1,A)|} + \int_0^1 \frac{(1-s)^{\alpha-1}}{\Gamma(\alpha)} \|p\| ds + M_1 \Bigg\} \\
&\quad + \sum_{i=1}^{m} \int_0^t \frac{(t-s)^{\beta_i-1}}{\Gamma(\beta_i)} (|h_i(s, x(s)) - h_i(s,0)| + |h_i(s,0)|)\, ds \\
&\le (\|\phi\|\|x\| + F_0) \left(\frac{2\|p\|}{\Gamma(\alpha+1)} + \frac{|A| + M_0 \sum_{i=1}^{m} \frac{1}{\Gamma(\beta_i+1)}}{|f(1,A)|} + M_1 \right) + \sum_{i=1}^{m} \frac{\|\psi_i\|\|x\| + K_i}{\Gamma(\beta_i+1)},
\end{aligned}$$

where $F_0 = \sup_{t \in J} |f(t,0)|$, $K_i = \sup_{t \in J} |h_i(t,0)|$, $i = 1, 2, \ldots, m$. Taking the supremum for $t \in [0,1]$ of the above inequality, we obtain a constant $M > 0$ such that

$$\|x\| \le M := \frac{F_0 \left(\frac{2\|p\|}{\Gamma(\alpha+1)} + \frac{|A| + M_0 \sum_{i=1}^{m} \frac{1}{\Gamma(\beta_i+1)}}{|f(1,A)|} + M_1 \right) + \sum_{i=1}^{m} \frac{K_i}{\Gamma(\beta_i+1)}}{1 - \|\phi\| \left(\frac{2\|p\|}{\Gamma(\alpha+1)} + \frac{|A| + M_0 \sum_{i=1}^{m} \frac{1}{\Gamma(\beta_i+1)}}{|f(1,A)|} + M_1 \right) - \sum_{i=1}^{m} \frac{\|\psi_i\|}{\Gamma(\beta_i+1)}},$$

which means that the set $\mathcal{E} = \{x \in X : \mu x \in \mathcal{A}x\mathcal{B}x + \mathcal{C}x,\ \mu > 1\}$ is bounded, since by (2.66), $\gamma < 1/2$. As a result, the conclusion (ii) of Theorem 1.11 does not hold. Hence, the conclusion (i) holds and consequently the boundary value problem (2.65) has at least one solution on J. This completes the proof. □

Example 2.4. Consider the following nonlocal hybrid boundary value problem

$$\begin{cases} {}^{C}D^{5/3}\left[\dfrac{x(t) - \sum_{i=1}^{6} I^{\frac{2i-1}{2}} h_i(t, x(t))}{f(t, x(t))}\right] \in G(t, x(t)), \quad t \in [0, 1], \\ x(0) = \gamma e^{\left(-\sum_{j=1}^{n-2} x^2(\xi_i)\right)}, \qquad x(1) = \dfrac{3}{4}, \end{cases} \tag{2.70}$$

where $0 < \gamma < 1$, $\xi_j \in (0, 1)$, $\xi_j < \xi_{j+1}$, $j = 1, 2, \ldots, n-2$, $n \geq 3$,

$$h_i(t, x(t)) = \frac{|x(t)|}{(20 + i + t)(1 + |x(t)|)}, \quad i = 1, 2, \ldots, 5,$$

$G : [0, 1] \times \mathbb{R} \to \mathcal{P}(\mathbb{R})$ given by

$$G(t, x(t)) = \left[\frac{\sqrt{t} + 1}{3} \cos^2 x(t),\ \frac{t^2 + 3}{2} + e^{-x^2(t)} + \frac{2}{3} \sin |x(t)|\right],$$

and

$$f(t, x(t)) = \frac{1}{2(35 + t)}\left(\frac{x^2(t) + |x(t)|}{1 + |x(t)|}\right) + t^{1/3} + 1.$$

Here $\alpha = 5/3$, $m = 6$, $\beta_i = (2i - 1)/2$, $i = 1, \ldots, 6$, $A = 3/4$ and $\mu(x) = \gamma \exp(-\sum_{j=1}^{n-2} x^2(\xi_j))$. Clearly

$$|f(t, x) - f(t, y)| \leq \left(\frac{1}{35 + t}\right)|x - y|,$$

$$|h_i(t, x) - h_i(t, y)| \leq \left(\frac{1}{20 + i + t}\right)|x - y|, \quad \text{for } x, y \in \mathbb{R}, i = 1, 2, \ldots, 6.$$

Setting $\phi(t) = 1/(35 + t)$ and $\psi(t) = 1/(20 + i + t)$, we get $\|\phi\| = 1/35$ and $\|\psi_i\| = 1/(20 + i)$, $i = 1, 2, \ldots, 6$. It is easy to see that

$$\left|\frac{\mu(x)}{f(0, \mu(x))}\right| \leq \frac{35}{36}, \ |h_i(t, x)| \leq \frac{1}{21}, \ f(1, A) = \frac{193}{96}.$$

Further, we have

$$\|G(t, x)\| = \sup\{|y| : y \in G(t, x)\} \leq \frac{t^2}{2} + \frac{19}{6} := p(t), \ x \in \mathbb{R}.$$

Setting constants $M_0 = 1/21$, $M_1 = 35/36$, $\|p\| = 11/3$ and using these values, we find that $\gamma < 1/2$. Thus all the conditions of Theorem 2.17 are satisfied. Hence, the problem (2.70) has at least one solution on $[0, 1]$.

2.6 Coupled systems of nonlinear fractional differential equations with nonlocal boundary conditions

In this section, we consider a nonlinear nonlocal boundary value problem for a coupled system of fractional differential equations given by

$$\begin{cases} {}^CD^{\alpha}u(t) = f(t, v(t), {}^CD^{p}v(t)), & t \in (0,1), \\ {}^CD^{\beta}v(t) = g(t, u(t), {}^CD^{q}u(t)), & t \in (0,1), \\ u(0) = h_1(u), \ u(1) = u_1, \ \ v(0) = h_2(v), \ v(1) = v_1, \end{cases} \tag{2.71}$$

where $f,\ g \in C([0,1] \times \mathbb{R} \times \mathbb{R},\ \mathbb{R})$, $h_i \in C(C([0,1],\mathbb{R}),\mathbb{R})$, $1 < \alpha,\ \beta \leq 2$, $p,\ q > 0$, $\alpha - q \geq 1$, $\beta - p \geq 1$, $u_1,\ v_1 \in \mathbb{R}$, and ${}^CD$ is the Caputo fractional derivative operator. It is important to note that the nonlinear terms in this coupled system involve the fractional derivative operators of the unknown functions. Moreover, the nonlocal term $h_1(u)$ in (2.71) may be understood as $h_1(u) = \sum_{j=1}^{p} \kappa_j u(t_j)$ where $\kappa_j, j = 1, \ldots, p$, are given constants and $0 < t_1 < \ldots < t_p \leq 1$ and similar is the case with the other nonlocal term $h_2(v)$.

Let $C(J)$ Denote the space of all continuous functions defined on $J = [0,1]$. Let $X = \{u(t) \mid u \in C(J) \text{ and } {}^CD^{q}u \in C(J)\}$ be a Banach space endowed with the norm $\|u\|_X = \max_{t\in J}|u(t)| + \max_{t\in J}|{}^CD^{q}u(t)|$ and $Y = \{v(t) \mid v \in C(J) \text{ and } {}^CD^{p}v \in C(J)\}$ be a Banach space equipped with the norm $\|v\|_Y = \max_{t\in J}|v(t)| + \max_{t\in J}|{}^CD^{p}v(t)|$. Thus, $(X \times Y, \|\cdot\|_{X\times Y})$ is a Banach with the norm defined by $\|(u,v)\|_{X\times Y} = \max\{\|u\|_X, \|v\|_Y\}$ for $(u,v) \in X \times Y$.

Lemma 2.6. *Let $\zeta \in C(J)$ be a given function and $1 < \alpha \leq 2$. Then a function u is a solution of the fractional integral equation*

$$u(t) = \frac{1}{\Gamma(\alpha)}\int_0^t (t-s)^{\alpha-1}\zeta(s)ds - \frac{t}{\Gamma(\alpha)}\int_0^1 (1-s)^{\alpha-1}\zeta(s)ds + (1-t)h_1(u) + tu_1$$

if and only if u is a solution of the boundary value problem

$$\begin{cases} {}^CD^{\alpha}u(t) = \zeta(t), & t \in (0,1), \\ u(0) = h_1(u), \quad u(1) = u_1. \end{cases} \tag{2.72}$$

Proof. A direct computation proves the lemma. □

Observe that the solution of (2.72) can be expressed in terms of a Green's function as

$$u(t) = \int_0^1 K_1(t,s)\zeta(s)ds + (1-t)h_1(u) + tu_1,$$

where $K_1(t,s)$ is the Green's function and is given by

$$K_1(t,s) = \frac{1}{\Gamma(\alpha)}\begin{cases} (t-s)^{\alpha-1} - t(1-s)^{\alpha-1}, & 0 \leq s \leq t, \\ -t(1-s)^{\alpha-1}, & t \leq s \leq 1. \end{cases}$$

Similarly, the general solution of ${}^CD^{\beta}v(t) = \zeta(t)$, $v(0) = h_2(v)$, $v(1) = v_1$ is

$$v(t) = \int_0^1 K_2(t,s)\zeta(s)ds + (1-t)h_2(v) + tv_1,$$

where $K_2(t,s)$ can be obtained from $K_1(t,s)$ by replacing α with β. Let (K_1, K_2) denote the Green's function for the boundary value problem (2.71).

Consider the coupled system of integral equations

$$\begin{cases} u(t) = \displaystyle\int_0^1 K_1(t,s)f(s,v(s),{}^CD^pv(s))ds + (1-t)h_1(u) + tu_1, \\ v(t) = \displaystyle\int_0^1 K_2(t,s)g(s,u(s),{}^CD^qu(s))ds + (1-t)h_2(v) + tv_1. \end{cases} \tag{2.73}$$

Lemma 2.7. *Assume that $f, g : J \times \mathbb{R} \times \mathbb{R} \to \mathbb{R}$ are continuous functions. Then $(u,v) \in X \times Y$ is a solution of (2.71) if and only if $(u,v) \in X \times Y$ is a solution of (2.73).*

Proof. It is easy to show that $(u,v) \in X \times Y$ is a solution of (2.73) if $(u,v) \in X \times Y$ is a solution of (2.71). In order to prove the converse of the foregoing statement, let $w(t)$ denote the right hand side of the first equation in (2.73). Then,

$$w(t) = I^{\alpha}f(t,v(t),{}^CD^pv(t)) - tI^{\alpha}f(1,v(1),{}^CD^pv(1)) + (1-t)h_1(u) + tu_1.$$

Using the fact that ${}^CD^{\alpha}t^{\alpha-m} = 0$, $m = 1, 2, \ldots, N$ (N is the smallest integer such that $N \geq \alpha$), we obtain

$$\begin{aligned} {}^CD^{\alpha}w(t) &= {}^CD^{\alpha}[I^{\alpha}f(t,v(t),{}^CD^pv(t)) - tI^{\alpha}f(1,v(1),{}^CD^pv(1))] \\ &= f(t,v(t),{}^CD^pv(t)), \end{aligned}$$

that is, ${}^CD^{\alpha}u(t) = f(t,v(t),{}^CD^pv(t))$. It can easily be verified that $u(0) = h_1(u)$ and $u(1) = u_1$. In a similar manner, it can be shown that ${}^CD^{\beta}v(t) = g(t,u(t),{}^CD^qu(t))$, $v(0) = h_2(v)$, and $v(1) = v_1$. Thus, $(u,v) \in X \times Y$ is a solution of (2.71), which completes the proof. □

Let us define an operator $\mathcal{F} : X \times Y \to X \times Y$ by

$$\mathcal{F}(u,v)(t) = (\mathcal{F}_1v(t), \mathcal{F}_2u(t)), \tag{2.74}$$

where

$$\mathcal{F}_1v(t) = \int_0^1 K_1(t,s)f(s,v(s),{}^CD^pv(s))ds + (1-t)h_1(u) + tu_1,$$

$$\mathcal{F}_2u(t) = \int_0^1 K_2(t,s)g(s,u(s),{}^CD^qu(s))ds + (1-t)h_2(v) + tv_1.$$

In view of the continuity of K_1, K_2, f, g, h_1, and h_2, it follows that $\mathcal{F}$ is continuous. Moreover, by Lemma 2.7, the fixed point of the operator $\mathcal{F}$ coincides with the solution of (2.71).

For convenience, we define the following quantities where $a(t)$ and $b(t)$ are non-negative functions defined on $(0,1)$:

$$\Lambda_1 = \frac{\{\Gamma(\alpha-q+1)+\Gamma(\alpha+1)\}\Gamma(2-q)+\alpha\Gamma(\alpha-q)}{\Gamma(\alpha+1)\Gamma(\alpha-q+1)\Gamma(2-q)},$$

$$\Lambda_2 = \frac{\{\Gamma(\beta-p+1)+\Gamma(\beta+1)\}\Gamma(2-p)+\beta\Gamma(\beta-p)}{\Gamma(\beta+1)\Gamma(\beta-p+1)\Gamma(2-p)},$$

$$\mu = \max_{t\in J}\int_0^1 |K_1(t,s)a(s)|ds + \frac{\Gamma(\alpha-q)+\Gamma(\alpha)\Gamma(2-q)}{\Gamma(\alpha)\Gamma(\alpha-q)\Gamma(2-q)}\int_0^1 (1-s)^{\alpha-q-1}a(s)ds,$$

$$\nu = \max_{t\in J}\int_0^1 |K_2(t,s)b(s)|ds + \frac{\Gamma(\beta-p)+\Gamma(\beta)\Gamma(2-p)}{\Gamma(\beta)\Gamma(\beta-p)\Gamma(2-p)}\int_0^1 (1-s)^{\beta-p-1}b(s)ds,$$

$$\eta_1 = \frac{1+\Gamma(2-q)}{\Gamma(2-q)},\ \eta_2 = \frac{1+\Gamma(2-p)}{\Gamma(2-p)}.$$

Now we are in a position to present the main result.

2.6.1 *Main result*

The following theorem is the main result in this section.

Theorem 2.18. *Let $f,\ g: J\times\mathbb{R}\times\mathbb{R}\to\mathbb{R}$ be continuous functions. Assume that there exist non-negative functions $a(t),\ b(t)\in L(0,1)$ and constants $\epsilon_i,\ \delta_i\geq 0$, $0<\rho_i,\ \sigma_i<1$, and $\gamma_1,\ \gamma_2\geq 0$, $i=1,2$, such that*

$$|f(t,x,y)|\leq a(t)+\epsilon_1|x|^{\rho_1}+\epsilon_2|y|^{\rho_2},\quad |g(t,x,y)|\leq b(t)+\delta_1|x|^{\sigma_1}+\delta_2|y|^{\sigma_2},$$

$$|h_1(u)|+|u_1|\leq\gamma_1,\quad \text{and}\quad |h_2(v)|+|v_1|\leq\gamma_2.$$

Then there exists a solution to the nonlocal boundary value problem (2.71).

Proof. Define a ball W in the Banach space $X\times Y$ by

$$W=\{(u(t),v(t))\mid (u(t),v(t))\in X\times Y,\ \|(u(t),v(t))\|_{X\times Y}\leq R,\ t\in J\},$$

where

$$R\geq\max\{(4\Lambda_1\epsilon_1)^{\frac{1}{1-\rho_1}},(4\Lambda_1\epsilon_2)^{\frac{1}{1-\rho_2}},(4\Lambda_2\delta_1)^{\frac{1}{1-\sigma_1}},(4\Lambda_2\delta_2)^{\frac{1}{1-\sigma_2}},4\mu,4\nu,4\gamma_1\eta_1,4\gamma_2\eta_2\}.$$

First, we prove that $\mathcal{F}: W\to W$. Using the fact that ${}^CD^q t^{\alpha-1}=\Gamma(\alpha)t^{\alpha-q-1}/\Gamma(\alpha-q)$, we obtain

$$\begin{aligned}|\mathcal{F}_1 v(t)| &= \left|\int_0^1 K_1(t,s)f(s,v(s),{}^CD^p v(s))ds+(1-t)h_1(u)+tu_1\right|\\ &\leq \int_0^1 |K_1(t,s)a(s)|ds+(\epsilon_1|R|^{\rho_1}+\epsilon_2|R|^{\rho_2})\int_0^1 |K_1(t,s)|ds\\ &\quad +(1-t)|h_1(u)|+t|u_1|\\ &= \int_0^1 |K_1(t,s)a(s)|ds+(\epsilon_1|R|^{\rho_1}+\epsilon_2|R|^{\rho_2})\Big[-\int_0^t\frac{(t-s)^{\alpha-1}}{\Gamma(\alpha)}ds\end{aligned}$$

$$+ t\int_0^1 \frac{(1-s)^{\alpha-1}}{\Gamma(\alpha)}ds\Bigg] + (1-t)|h_1(u)| + t|u_1|$$
$$= \int_0^1 |K_1(t,s)a(s)|ds + \frac{(\epsilon_1|R|^{\rho_1} + \epsilon_2|R|^{\rho_2})}{\Gamma(\alpha+1)}(t - t^\alpha) + (1-t)|h_1(u)| + t|u_1|$$
$$\leq \int_0^1 |K_1(t,s)a(s)|ds + \frac{(\epsilon_1|R|^{\rho_1} + \epsilon_2|R|^{\rho_2})}{\Gamma(\alpha+1)} + |h_1(u)| + |u_1|,$$

and

$$\begin{aligned}
&|{}^C D^q \digamma_1 v(t)| \\
&= |{}^C D^q I^\alpha f(t, v(t), {}^C D^p v(t)) \\
&\quad - I^\alpha f(1, v(1), {}^C D^p v(1)) {}^C D^q t + {}^C D^q((1-t)h_1(u) + tu_1)| \\
&= \Bigg| I^{\alpha-q} f(t, v(t), {}^C D^p v(t)) - I^\alpha f(1, v(1), {}^C D^p v(1)) \frac{1}{\Gamma(2-q)} t^{1-q} \\
&\quad - h_1(u)\frac{1}{\Gamma(2-q)}(t-1)^{1-q} + u_1 \frac{1}{\Gamma(2-q)} t^{1-q}\Bigg| \\
&\leq \frac{1}{\Gamma(\alpha-q)}\left[\int_0^t (t-s)^{\alpha-q-1}a(s)ds + (\epsilon_1|R|^{\rho_1} + \epsilon_2|R|^{\rho_2})\int_0^t (t-s)^{\alpha-q-1}ds\right] \\
&\quad + \frac{1}{\Gamma(2-q)\Gamma(\alpha)}\left[\int_0^1 (1-s)^{\alpha-1}a(s)ds + (\epsilon_1|R|^{\rho_1} + \epsilon_2|R|^{\rho_2})\int_0^1 (1-s)^{\alpha-1}ds\right] \\
&\quad + \frac{1}{\Gamma(2-q)}(|h_1(u)| + |u_1|) \\
&\leq \frac{\Gamma(2-q)\Gamma(\alpha) + \Gamma(\alpha-q)}{\Gamma(\alpha-q)\Gamma(2-q)\Gamma(\alpha)}\Bigg[\int_0^1 (1-s)^{\alpha-q-1}a(s)ds + (\epsilon_1|R|^{\rho_1} \\
&\quad + \epsilon_2|R|^{\rho_2})\int_0^1 (1-s)^{\alpha-q-1}ds\Bigg] + \frac{1}{\Gamma(2-q)}\Big(|h_1(u)| + |u_1|\Big) \\
&= \frac{\Gamma(2-q)\Gamma(\alpha) + \Gamma(\alpha-q)}{\Gamma(\alpha-q)\Gamma(2-q)\Gamma(\alpha)}\Bigg[\int_0^1 (1-s)^{\alpha-q-1}a(s)ds \\
&\quad + \frac{(\epsilon_1|R|^{\rho_1} + \epsilon_2|R|^{\rho_2})}{(\alpha-q)}\Bigg] + \frac{1}{\Gamma(2-q)}\Big(|h_1(u)| + |u_1|\Big).
\end{aligned}$$

Thus,

$$\begin{aligned}
\|\digamma_1 v(t)\|_X = \max_{t\in J}|\digamma_1 v(t)| + \max_{t\in J}|{}^C D^q \digamma_1 v(t)| &\leq \mu + (\epsilon_1|R|^{\rho_1} + \epsilon_2|R|^{\rho_2})\Lambda_1 + \gamma_1\eta_1 \\
&\leq \frac{R}{4} + \frac{R}{4} + \frac{R}{4} + \frac{R}{4} = R.
\end{aligned}$$

Similarly, it can be shown that

$$\|\digamma_2 u(t)\|_Y \leq \nu + (\delta_1|R|^{\sigma_1} + \delta_2|R|^{\sigma_2})\Lambda_2 + \gamma_2\eta_2 \leq \frac{R}{4} + \frac{R}{4} + \frac{R}{4} + \frac{R}{4} = R.$$

Hence, we conclude that $\|\digamma(u,v)\|_{X\times Y} \leq R$. Since $\digamma_1 v(t)$, $\digamma_2 u(t)$, ${}^C D^q \digamma_1 v(t)$, and ${}^C D^p \digamma_2 u(t)$ are continuous on J, we have $\digamma : W \to W$.

Now we show that $\mathcal{F}$ is a completely continuous operator. To do this, we first set

$$M = \max_{t\in J} |f(t, v(t), {}^C D^p v(t))| \text{ and } N = \max_{t\in J} |g(t, u(t), {}^C D^q u(t))|.$$

For $(u, v) \in W$ and $t, \tau \in J$ with $t < \tau$, we have

$$\begin{aligned}
&|\mathcal{F}_1 v(t) - \mathcal{F}_1 v(\tau)| \\
&= \left| \int_0^1 (K_1(t,s) - K_1(\tau,s)) f(s, v(s), {}^C D^p v(s)) ds + (\tau - t) h_1(u) + (t-\tau) u_1 \right| \\
&\le M \left[\int_0^t |K_1(t,s) - K_1(\tau,s)| ds + \int_t^\tau |K_1(t,s) - K_1(\tau,s)| ds \right. \\
&\quad \left. + \int_\tau^1 |K_1(t,s) - K_1(\tau,s)| ds \right] + (\tau - t)(|h_1(u)| + |u_1|) \\
&\le \frac{M}{\Gamma(\alpha)} \left[\int_0^t \{(\tau - s)^{\alpha-1} - (t-s)^{\alpha-1} + (\tau - t)(1-s)^{\alpha-1}\} ds \right. \\
&\quad \left. + \int_t^\tau \{(\tau - t)(1-s)^{\alpha-1} + (\tau - s)^{\alpha-1}\} ds + \int_\tau^1 (\tau - t)(1-s)^{\alpha-1} ds \right] \\
&\quad + (\tau - t)(|h_1(u)| + |u_1|) \\
&= \frac{M}{\Gamma(\alpha+1)} [\tau^\alpha - t^\alpha + \tau - t] + (\tau - t)(|h_1(u)| + |u_1|)
\end{aligned}$$

and

$$\begin{aligned}
&|{}^C D^q \mathcal{F}_1 v(t) - {}^C D^q \mathcal{F}_1 v(\tau)| \\
&= \left| I^{\alpha-q} f(t, v(t), {}^C D^p v(t)) - I^{\alpha-q} f(\tau, v(\tau), {}^C D^p v(\tau)) \right. \\
&\quad + \left(\frac{\tau^{1-q} - t^{1-q}}{\Gamma(2-q)} \right) I^\alpha f(1, v(1), {}^C D^p v(1)) \\
&\quad \left. + \left(\frac{(\tau-1)^{1-q} - (t-1)^{1-q}}{\Gamma(2-q)} \right) h_1(u) - \left(\frac{\tau^{1-q} - t^{1-q}}{\Gamma(2-q)} \right) u_1 \right| \\
&\le \frac{1}{\Gamma(\alpha - q)} \left[\left| \int_0^t (t-s)^{\alpha-q-1} f(s, v(s), {}^C D^p v(s)) ds \right. \right. \\
&\quad \left. - \int_0^\tau (t-s)^{\alpha-q-1} f(s, v(s), {}^C D^p v(s)) ds \right| \\
&\quad + \left| \int_0^\tau (t-s)^{\alpha-q-1} f(s, v(s), {}^C D^p v(s)) ds \right. \\
&\quad \left. \left. - \int_0^\tau (\tau-s)^{\alpha-q-1} f(s, v(s), {}^C D^p v(s)) ds \right| \right] \\
&\quad + \left| \left(\frac{\tau^{1-q} - t^{1-q}}{\Gamma(2-q)} \right) I^\alpha f(1, v(1), {}^C D^p v(1)) \right. \\
&\quad \left. + \left(\frac{(\tau-1)^{1-q} - (t-1)^{1-q}}{\Gamma(2-q)} \right) h_1(u) - \left(\frac{\tau^{1-q} - t^{1-q}}{\Gamma(2-q)} \right) u_1 \right|
\end{aligned}$$

$$\leq \frac{M}{\Gamma(\alpha-q)}\left[\int_0^\tau ((\tau-s)^{\alpha-q-1}-(t-s)^{\alpha-q-1})ds+\int_t^\tau (t-s)^{\alpha-q-1}ds\right]$$
$$+M\left(\frac{\tau^{1-q}-t^{1-q}}{\Gamma(2-q)\Gamma(\alpha+1)}\right)+\left(\frac{(\tau-1)^{1-q}-(t-1)^{1-q}}{\Gamma(2-q)}\right)|h_1(u)|$$
$$+\left(\frac{\tau^{1-q}-t^{1-q}}{\Gamma(2-q)}\right)|u_1|$$
$$=\frac{M}{\Gamma(\alpha-q+1)}(\tau^{\alpha-q}-t^{\alpha-q})+M\left(\frac{\tau^{1-q}-t^{1-q}}{\Gamma(2-q)\Gamma(\alpha+1)}\right)$$
$$+\left(\frac{(\tau-1)^{1-q}-(t-1)^{1-q}}{\Gamma(2-q)}\right)|h_1(u)|+\left(\frac{\tau^{1-q}-t^{1-q}}{\Gamma(2-q)}\right)|u_1|.$$

Analogously, it can be proved that

$$|\mathcal{F}_2u(t)-\mathcal{F}_2u(\tau)|\leq \frac{N}{\Gamma(\beta+1)}(\tau^\beta-t^\beta+\tau-t)+(\tau-t)(|h_2(v)|+|v_1|)$$

and

$$|{}^CD^p\mathcal{F}_2u(t)-{}^CD^q\mathcal{F}_2u(\tau)|\leq \frac{N}{\Gamma(\beta-p+1)}(\tau^{\beta-p}-t^{\beta-p})+N\left(\frac{\tau^{1-p}-t^{1-p}}{\Gamma(2-p)\Gamma(\beta+1)}\right)$$
$$+\left(\frac{(\tau-1)^{1-p}-(t-1)^{1-p}}{\Gamma(2-p)}\right)|h_2(v)|+\left(\frac{\tau^{1-p}-t^{1-p}}{\Gamma(2-p)}\right)|v_1|.$$

Since the functions $t^\alpha, t^{\alpha-1}, t^\beta, t^{\beta-1}, t^{\alpha-q}, t^{\alpha-q-1}, t^{\beta-p}, t^{\beta-p-1}$ are uniformly continuous on J, it follows from the above estimates that $\mathcal{F}W$ is an equi-continuous set. Also, it is uniformly bounded as $\mathcal{F}W\subset W$. Thus, we conclude that $\mathcal{F}$ is a completely continuous operator. Hence, by Schauder's fixed point theorem, there exists a solution of (2.71), and this completes the proof of the theorem. □

Remark 2.3. Letting $a(t)\equiv 0$ and $b(t)\equiv 0$ in the assumptions of Theorem 2.18, we obtain a simpler version of the theorem in which f and g satisfy the conditions

$$|f(t,x,y)|\leq \epsilon_1|x|^{\rho_1}+\epsilon_2|y|^{\rho_2} \text{ and } |g(t,x,y)|\leq \delta_1|x|^{\sigma_1}+\delta_2|y|^{\sigma_2}, \tag{2.75}$$

and R is chosen to satisfy

$$R\geq \max\{(3\Lambda_1\epsilon_1)^{\frac{1}{1-\rho_1}},(3\Lambda_1\epsilon_2)^{\frac{1}{1-\rho_2}},(3\Lambda_2\delta_1)^{\frac{1}{1-\sigma_1}},(3\Lambda_2\delta_2)^{\frac{1}{1-\sigma_2}},3\gamma_1\eta_1,3\gamma_2\eta_2\}.$$

In this case, we obtain the following estimates:

$$\|\mathcal{F}_1v(t)\|_X\leq(\epsilon_1|R|^{\rho_1}+\epsilon_2|R|^{\rho_2})\Lambda_1+\gamma_1\eta_1\leq \frac{R}{3}+\frac{R}{3}+\frac{R}{3}=R,$$
$$\|\mathcal{F}_2u(t)\|_Y\leq(\delta_1|R|^{\sigma_1}+\delta_2|R|^{\sigma_2})\Lambda_2+\gamma_2\eta_2\leq \frac{R}{3}+\frac{R}{3}+\frac{R}{3}=R.$$

It is also observed that a proper choice of the values of ϵ_i and δ_i makes the conclusion of Theorem 3.1 valid for the case where ρ_i, $\sigma_i\geq 1$ $(i=1,2)$. For instance, we may assume ρ_i, $\sigma_i\geq 1$ $(i=1,2)$ in (3.1) with $\epsilon_i<\frac{1}{2\Lambda_1}$ for $\rho_i=1$ and $\delta_i<\frac{1}{2\Lambda_2}$ for $\sigma_i=1$.

Example 2.5. Consider the boundary value problem

$$\begin{cases} {}^CD^{\frac{9}{7}}u(t) = (t-\frac{3}{5})^4[(v(t))^{\rho_1} + ({}^CD^{\frac{2}{5}}v(t))^{\rho_2}], & t \in (0,1), \\ {}^CD^{\frac{7}{5}}v(t) = (t-\frac{3}{5})^4[(u(t))^{\sigma_1} + ({}^CD^{\frac{1}{7}}u(t))^{\sigma_2}], & t \in (0,1), \\ u(0) = h_1(u),\ u(1) = u_1,\ \ v(0) = h_2(v),\ v(1) = v_1, \end{cases} \tag{2.76}$$

where $0 < \rho_i,\ \sigma_i \leq 1$ or $\rho_i,\ \sigma_i \geq 1$ $i = 1, 2$. Here, $a(t) = 0 = b(t)$ and $\epsilon_i = \frac{81}{625} = 0.1296 = \delta_i$. Let us take $\rho_i,\ \sigma_i \geq 1,\ i = 1, 2$. In a straightforward manner, it is found that $\frac{1}{2\Lambda_1} \approx 0.176935619$ and $\frac{1}{2\Lambda_2} \approx 0.1630565354$; clearly $\epsilon_i < \frac{1}{2\Lambda_1}$ and $\delta_i < \frac{1}{2\Lambda_2}$. Thus, the conclusion of Theorem 2.18 applies to (2.76), and hence there exists a solution of (2.76).

2.7 On fully coupled nonlocal multi-point boundary value problems of nonlinear mixed-order fractional differential equations on an arbitrary domain

We introduce and study a new class of coupled systems of mixed-order fractional differential equations equipped with nonlocal multi-point coupled boundary conditions. In precise terms, we consider the following fully coupled system:

$$\begin{cases} {}^CD^{\xi}x(t) = \varphi(t, x(t), y(t)), & t \in [a,b], \quad 0 < \xi < 1, \\ {}^CD^{\zeta}y(t) = \psi(t, x(t), y(t)), & t \in [a,b], \quad 1 < \zeta < 2, \\ px(a) + qy(b) = x_0,\ y(a) = 0,\ \ y'(b) = \sum_{i=1}^{m} \delta_i x(\sigma_i),\ a < \sigma_i < b, \end{cases} \tag{2.77}$$

where ${}^CD^{\chi}$ is Caputo fractional derivative operator of order $\chi \in \{\xi, \zeta\}$, $\varphi, \psi : [a,b] \times \mathbb{R} \times \mathbb{R} \to \mathbb{R}$ are given functions $p, q, \delta_i \in \mathbb{R}, i = 1, 2, \ldots, m$.

Here we emphasize that the novelty of the present work lies in the fact that we consider a coupled system of fractional differential equations of different order on an arbitrary domain equipped with coupled nonlocal multi-point boundary conditions. Moreover, several new results appear as special cases of the obtained work. It is imperative to notice that much of the work related to the coupled systems of fractional differential equations deals with the fixed domain. Thus our results are more general and contribute significantly to the existing literature on the topic.

Now we present an important result to analyze the problem (2.77).

Lemma 2.8. *Let* $\lambda := p + q(b-a)\sum_{i=1}^{m} \delta_i \neq 0$ *and* $\bar{\varphi}, \bar{\psi} \in C[a,b]$. *Then the unique solution of the system*

$$\begin{cases} {}^CD^{\xi}x(t) = \bar{\varphi}(t), & t \in [a,b], \quad 0 < \xi < 1, \\ {}^CD^{\zeta}y(t) = \bar{\psi}(t), & t \in [a,b], \quad 1 < \zeta < 2, \\ px(a) + qy(b) = x_0,\ y(a) = 0,\ \ y'(b) = \sum_{i=1}^{m} \delta_i x(\sigma_i), \end{cases} \tag{2.78}$$

is given by the following pair of integral equations

$$x(t) = I^{\xi}\bar{\varphi}(t) + \frac{1}{\lambda}\Big[x_0 - qI^{\varsigma}\bar{\psi}(b) + q(b-a)(I^{\varsigma-1}\bar{\psi}(b) - \sum_{i=1}^{m}\delta_i I^{\xi}\bar{\varphi}(\sigma_i))\Big], \tag{2.79}$$

$$y(t) = I^{\varsigma}\bar{\psi}(t) - \frac{t-a}{\lambda}\Big[p(I^{\varsigma-1}\bar{\psi}(b) - \sum_{i=1}^{m}\delta_i I^{\xi}\bar{\varphi}(\sigma_i)) - \sum_{i=1}^{m}\delta_i(x_0 - qI^{\varsigma}\bar{\psi}(b)\Big]. \tag{2.80}$$

Proof. The solutions of fractional differential equations in (2.78) can be written as

$$x(t) = I^{\xi}\bar{\varphi}(t) + c_1 \quad \text{and} \quad y(t) = I^{\varsigma}\bar{\psi}(t) + c_2 + c_3(t-a), \tag{2.81}$$

where $c_i \in \mathbb{R}\,(i = 1,2,3)$ are arbitrary constants. Using the condition $y(a) = 0$ in (2.81), we get $c_2 = 0$, while making use of the conditions $px(a) + qy(b) = x_0$ and $y'(b) = \sum_{i=1}^{m}\delta_i x(\sigma_i)$ in (2.81) yields the following system

$$\begin{aligned} &pc_1 + q(b-a)c_3 = x_0 - qI^{\varsigma}\bar{\psi}(b), \\ &\sum_{i=1}^{m}\delta_i\, c_1 - c_3 = I^{\varsigma-1}\bar{\psi}(b) - \sum_{i=1}^{m}\delta_i I^{\xi}\bar{\varphi}(\sigma_i). \end{aligned}$$

Solving the above system for c_1 and c_3, we find that

$$c_1 = \frac{1}{\lambda}\Big[x_0 - qI^{\varsigma}\bar{\psi}(b) + q(b-a)(I^{\varsigma-1}\bar{\psi}(b) - \sum_{i=1}^{m}\delta_i I^{\xi}\bar{\varphi}(\sigma_i))\Big]$$

and

$$c_3 = -\frac{1}{\lambda}\Big[p(I^{\varsigma-1}\bar{\psi}(b) - \sum_{i=1}^{m}\delta_i I^{\xi}\bar{\varphi}(\sigma_i)) - \sum_{i=1}^{m}\delta_i(x_0 - qI^{\varsigma}\bar{\psi}(b)\Big],$$

which, on substituting in (2.81), completes the solutions (2.79) and (2.80). The converse follows by direct computation. □

In the following, for brevity, we use the notations:

$$L_1 = \frac{(b-a)^{\xi}}{\Gamma(\xi+1)} + \frac{q(b-a)}{|\lambda|\Gamma(\xi+1)}\sum_{i=1}^{m}|\delta_i|(\sigma_i-a)^{\xi}, \; M_1 = \frac{q}{|\lambda|}\frac{(b-a)^{\varsigma}(\varsigma+1)}{\Gamma(\varsigma+1)}, \tag{2.82}$$

$$L_2 = \frac{p(b-a)}{|\lambda|\Gamma(\xi+1)}\sum_{i=1}^{m}|\delta_i|(\sigma_i-a)^{\xi}, \; M_2 = \frac{(b-a)^{\varsigma}}{\Gamma(\varsigma+1)}\left(1 + \frac{p\varsigma}{|\lambda|} + q(b-a)\sum_{i=1}^{m}|\delta_i|\right). \tag{2.83}$$

2.7.1 *Main results*

In view of Lemma 2.8, we define an operator $T : X \times X \to X \times X$ by

$$T(x,y)(t) = \begin{pmatrix} T_1(x,y)(t) \\ T_2(x,y)(t) \end{pmatrix},$$

where $(X \times X, \|(x,y)\|)$ is a Banach space equipped with norm $\|(x,y)\| = \|x\| + \|y\|, x, y \in X$ ($X = \{x(t)|x(t) \in C([a,b],\mathbb{R})\}$ is a Banach space associated with the norm $\|x\| = \sup\{|x(t)|, t \in [a,b]\}$),

$$T_1(x,y)(t) = I^{\xi}\bar{\varphi}(t) + \frac{1}{\lambda}\Big[x_0 - qI^{\zeta}\bar{\psi}(b) + q(b-a)(I^{\zeta-1}\bar{\psi}(b) - \sum_{i=1}^{m}\delta_i I^{\xi}\bar{\varphi}(\sigma_i))\Big],$$

$$T_2(x,y)(t) = I^{\zeta}\bar{\psi}(t) - \frac{t-a}{\lambda}\Big[p(I^{\zeta-1}\bar{\psi}(b) - \sum_{i=1}^{m}\delta_i I^{\xi}\bar{\varphi}(\sigma_i)) - \sum_{i=1}^{m}\delta_i(x_0 - qI^{\zeta}\bar{\psi}(b)\Big],$$

$\bar{\varphi}(t) = \varphi(t, x(t), y(t))$ and $\bar{\psi}(t) = \psi(t, x(t), y(t))$.

In our first result, we establish the uniqueness of solutions for the system (2.77) by applying contraction mapping principle due to Banach.

Theorem 2.19. *Assume that*

(2.19.1) $\varphi, \psi : [a,b] \times \mathbb{R} \times \mathbb{R} \to \mathbb{R}$ are continuous functions and there exist positive constants ℓ_1 and ℓ_2 such that for all $t \in [a,b]$ and $x_i, y_i \in \mathbb{R}$, $i = 1,2$, we have

$$|\varphi(t,x_1,x_2) - \varphi(t,y_1,y_2)| \le \ell_1(|x_1 - y_1| + |x_2 - y_2|),$$
$$|\psi(t,x_1,x_2) - \psi(t,y_1,y_2)| \le \ell_2(|x_1 - y_1| + |x_2 - y_2|).$$

If

$$(L_1 + L_2)\ell_1 + (M_1 + M_2)\ell_2 < 1, \tag{2.84}$$

where L_1, M_1 and L_2, M_2 are respectively given by (2.82) and (2.83), then the system (2.77) has a unique solution on $[a,b]$.

Proof. Define $\sup_{t\in[a,b]} \varphi(t,0,0) = N_1 < \infty$ and $\sup_{t\in[a,b]} \psi(t,0,0) = N_2 < \infty$ and $r > 0$ such that

$$r > \frac{\frac{|x_0|}{|\lambda|}\Big(1 + (b-a)\sum_{i=1}^{m}|\delta_i|\Big) + (L_1 + L_2)N_1 + (M_1 + M_2)N_2}{1 - (L_1 + L_2)\ell_1 - (M_1 + M_2)\ell_2}.$$

In the first step, we show that $TB_r \subset B_r$, where $B_r = \{(x,y) \in X \times X : \|(x,y)\| \le r\}$. By the assumption (2.19.1), for $(x,y) \in B_r$, $t \in [a,b]$, we have

$$\begin{aligned}|\varphi(t,x(t),y(t))| &\le |\varphi(t,x(t),y(t)) - \varphi(t,0,0)| + |\varphi(t,0,0)| \\ &\le \ell_1(|x(t)| + |y(t)|) + N_1 \\ &\le \ell_1(\|x\| + \|y\|) + N_1 \le \ell_1 r + N_1.\end{aligned} \tag{2.85}$$

Similarly, we can get

$$|\psi(t,x(t),y(t))| \le \ell_2(\|x\| + \|y\|) + N_2 \le \ell_2 r + N_2. \tag{2.86}$$

Using (2.85) and (2.86), we obtain

$$|T_1(x,y)(t)| \le \frac{(b-a)^{\xi}}{\Gamma(\xi+1)}\|\bar{\varphi}\| + \frac{1}{|\lambda|}\Bigg[|x_0| + q\frac{(b-a)^{\zeta}}{\Gamma(\zeta+1)}\|\bar{\psi}\|$$

$$+q(b-a)\Big(\frac{(b-a)^{\zeta-1}}{\Gamma(\zeta)}\|\bar{\psi}\|+\sum_{i=1}^{m}|\delta_i|\frac{(\sigma_1-a)^{\xi}}{\Gamma(\xi+1)}\|\bar{\varphi}\|\Big)\Big]$$
$$\le \frac{|x_0|}{|\lambda|}+\left[\frac{(b-a)^{\xi}}{\Gamma(\xi+1)}+\frac{q(b-a)}{|\lambda|\Gamma(\xi+1)}\sum_{i=1}^{m}|\delta_i|(\sigma_i-a)^{\xi}\right](\ell_1 r+N_1)$$
$$+\frac{q}{|\lambda|}\frac{(b-a)^{\zeta}(\zeta+1)}{\Gamma(\zeta+1)}(\ell_2 r+N_2)$$
$$=\frac{|x_0|}{|\lambda|}+(L_1\ell_1+M_1\ell_2)r+L_1N_1+M_1N_2,$$

which implies that

$$\|T_1(x,y)\|\le\frac{|x_0|}{|\lambda|}+(L_1\ell_1+M_1\ell_2)r+L_1N_1+M_1N_2,$$

where we have taken the norm for $t\in[a,b]$. Likewise, we can find that

$$\|T_2(x,y)\|\le\frac{(b-a)|x_0|}{|\lambda|}\sum_{i=1}^{m}|\delta_i|+(L_2\ell_1+M_2\ell_2)r+L_2N_1+M_2N_2.$$

Consequently,

$$\|T(x,y)\|\le\frac{|x_0|}{|\lambda|}+\frac{(b-a)|x_0|}{|\lambda|}\sum_{i=1}^{m}|\delta_i|$$
$$+[(L_1+L_2)\ell_1+(M_1+M_2)\ell_2]r+(L_1+L_2)N_1+(M_1+M_2)N_2\le r.$$

Now, for $(x_1,y_1),(x_2,y_2)\in X\times X$ and for any $t\in[a,b]$, we get

$$|T_1(x_2,y_2)(t)-T_1(x_1,y_1)(t)|$$
$$\le\left[\frac{(b-a)^{\xi}}{\Gamma(\xi+1)}+\frac{q(b-a)}{|\lambda|\Gamma(\xi+1)}\sum_{i=1}^{m}|\delta_i|(\sigma_i-a)^{\xi}\right]\ell_1(\|x_2-x_1\|+\|y_2-y_1\|)$$
$$+\frac{q}{|\lambda|}\frac{(b-a)^{\zeta}(\zeta+1)}{\Gamma(\zeta+1)}\ell_2(\|x_2-x_1\|+\|y_2-y_1\|)$$
$$=(L_1\ell_1+M_1\ell_2)(\|x_2-x_1\|+\|y_2-y_1\|),$$

which implies that

$$\|T_1(x_2,y_2)-T_1(x_1,y_1)\|\le(L_1\ell_1+M_1\ell_2)(\|x_2-x_1\|+\|y_2-y_1\|). \tag{2.87}$$

Similarly, we find that

$$\|T_2(x_2,y_2)(t)-T_2(x_1,y_1)\|\le(L_2\ell_1+M_2\ell_2)(\|x_2-x_1\|+\|y_2-y_1\|). \tag{2.88}$$

It follows from (2.87) and (2.88) that

$$\|T(x_2,y_2)-T(x_1,y_1)\|\le[(L_1+L_2)\ell_1+(M_1+M_2)\ell_2](\|x_2-x_1\|+\|y_2-y_1\|).$$

From the above inequality, we deduce that T is a contraction in view of the condition (2.84). Hence it follows by by Banach's fixed point theorem that there exists a unique fixed point for the operator T, which corresponds to a unique solution of problem (2.77) on $[a,b]$. □

In the following result, we apply Leray-Schauder alternative (Theorem 1.7) to prove the existence of solutions for the problem (2.77).

Theorem 2.20. *Assume that*

(2.20.1) $\varphi, \psi : [a,b] \times \mathbb{R} \times \mathbb{R} \to \mathbb{R}$ *are continuous functions and there exist real constants* $k_i, \gamma_i \geq 0,\ (i = 1,2)$ *and* $k_0 > 0, \gamma_0 > 0$ *such that* $\forall x, y \in \mathbb{R}$,

$$|\varphi(t,x,y)| \leq k_0 + k_1|x| + k_2|y|,$$
$$|\psi(t,x,y)| \leq \gamma_0 + \gamma_1|x| + \gamma_2|y|.$$

Then the system (2.77) has at least one solution on $[a,b]$ *provided that*

$$(L_1 + L_2)k_1 + (M_1 + M_2)\gamma_1 < 1 \quad \text{and} \quad (L_1 + L_2)k_2 + (M_1 + M_2)\gamma_2 < 1, \tag{2.89}$$

where L_1, M_1 *and* L_2, M_2 *are respectively given by (2.82) and (2.83).*

Proof. Observe that continuity of the operator $T : X \times X \to X \times X$ follows from that of functions φ and ψ. Next, let $\Omega \subset X \times X$ be bounded such that

$$|\varphi(t,x(t),y(t))| \leq K_1, \quad |\psi(t,x(t),y(t))| \leq K_2, \quad \forall (x,y) \in \Omega,$$

for positive constants K_1 and K_2. Then for any $(x,y) \in \Omega$, we have

$$\begin{aligned}
|T_1(x,y)(t)| &\leq \frac{|x_0|}{|\lambda|} + \left[\frac{(b-a)^{\xi}}{\Gamma(\xi+1)} + \frac{q(b-a)}{|\lambda|\Gamma(\xi+1)}\sum_{i=1}^{m}|\delta_i|(\sigma_i - a)^{\xi}\right]K_1 \\
&\quad + \frac{q}{|\lambda|}\frac{(b-a)^{\zeta}(\zeta+1)}{\Gamma(\zeta+1)}K_2 \\
&= \frac{|x_0|}{|\lambda|} + L_1K_1 + M_1K_2,
\end{aligned}$$

which implies that

$$\|T_1(x,y)\| \leq \frac{|x_0|}{|\lambda|} + L_1K_1 + M_1K_2.$$

Similarly, it can be shown that

$$\|T_2(x,y)\| \leq \frac{(b-a)|x_0|}{|\lambda|}\sum_{i=1}^{m}|\delta_i| + L_2K_1 + M_2K_2.$$

From the foregoing arguments, we deduce that the operator T is uniformly bounded, that is,

$$\|T(x,y)\| \leq \frac{|x_0|}{|\lambda|} + \frac{(b-a)|x_0|}{|\lambda|}\sum_{i=1}^{m}|\delta_i| + (L_1 + L_2)K_1 + (M_1 + M_2)K_2.$$

Next, we show that T is equi-continuous. Let $t_1, t_2 \in [a,b]$ with $t_1 < t_2$. Then we have

$$|T_1(x(t_2),y(t_2)) - T_1(x(t_1),y(t_1))|$$

$$\leq K_1 \left| \frac{1}{\Gamma(\xi)} \int_a^{t_2} (t_2 - s)^{\xi-1} ds - \frac{1}{\Gamma(\xi)} \int_a^{t_1} (t_1 - s)^{\xi-1} ds \right|$$
$$\leq K_1 \left\{ \frac{1}{\Gamma(\xi)} \int_a^{t_1} [(t_2 - s)^{\xi-1} - (t_1 - s)^{\xi-1}] ds + \frac{1}{\Gamma(q)} \int_{t_1}^{t_2} (t_2 - s)^{\xi-1} ds \right\}$$
$$\leq \frac{K_1}{\Gamma(\xi+1)} [2(t_2 - t_1)^{\xi} + |t_2^{\xi} - t_1^{\xi}|].$$

Analogously, we can obtain

$$\begin{aligned} &|T_2(x(t_2), y(t_2)) - T_2(x(t_1), y(t_1))| \\ \leq &\frac{K_2}{\Gamma(\zeta+1)} [2(t_2 - t_1)^{\zeta} + |t_2^{\zeta} - t_1^{\zeta}|] \\ &+ \frac{|t_2 - t_1|}{|\lambda|} \left[p\Big(\frac{(b-a)^{\zeta-1}}{\Gamma(\zeta)} K_2 + \sum_{i=1}^{m} |\delta_i| \frac{(\sigma_1 - a)^{\xi}}{\Gamma(\xi+1)} K_1 \Big) + \sum_{i=1}^{m} |\delta_i| q \frac{(b-a)^{\zeta}}{\Gamma(\zeta+1)} K_2 \right]. \end{aligned}$$

This shows that the operator $T(x, y)$ is equi-continuous. In consequence, we deduce that the operator $T(x, y)$ is completely continuous.

Finally, we consider the set $\mathcal{P} = \{(x, y) \in X \times X | (x, y) = \lambda T(x, y),\ 0 \leq \lambda \leq 1\}$ and show that it is bounded. Let $(x, y) \in \mathcal{P}$ with $(x, y) = \lambda T(x, y)$. For any $t \in [a, b]$, we have $x(t) = \lambda T_1(x, y)(t)$, $y(t) = \lambda T_2(x, y)(t)$. Then

$$\begin{aligned} |x(t)| &\leq \frac{|x_0|}{|\lambda|} + L_1(k_0 + k_1|x| + k_2|y|) + M_1(\gamma_0 + \gamma_1|x| + \gamma_2|y|) \\ &= \frac{|x_0|}{|\lambda|} + L_1k_0 + M_1\gamma_0 + (L_1k_1 + M_1\gamma_1)|x| + (L_1k_2 + M_1\gamma_2)|y|, \end{aligned}$$

and

$$\begin{aligned} |y(t)| &\leq \frac{(b-a)|x_0|}{|\lambda|} \sum_{i=1}^{m} |\delta_i| + L_2(k_0 + k_1|x| + k_2|y|) + M_2(\gamma_0 + \gamma_1|x| + \gamma_2|y|) \\ &= \frac{(b-a)|x_0|}{|\lambda|} \sum_{i=1}^{m} |\delta_i| + L_2k_0 + M_2\gamma_0 + (L_2k_1 + M_2\gamma_1)|x| + (L_2k_2 + M_2\gamma_2)|y|. \end{aligned}$$

In consequence, it follows that

$$\|x\| \leq \frac{|x_0|}{|\lambda|} + L_1k_0 + M_1\gamma_0 + (L_1k_1 + M_1\gamma_1)\|x\| + (L_1k_2 + M_1\gamma_2)\|y\|$$

and

$$\|y\| \leq \frac{(b-a)|x_0|}{|\lambda|} \sum_{i=1}^{m} |\delta_i| + L_2k_0 + M_2\gamma_0 + (L_2k_1 + M_2\gamma_1)\|x\| + (L_2k_2 + M_2\gamma_2)\|y\|,$$

which lead to

$$\begin{aligned} \|x\| + \|y\| \leq &\frac{|x_0|}{|\lambda|} + \frac{(b-a)|x_0|}{|\lambda|} \sum_{i=1}^{m} |\delta_i| + (L_1 + L_2)k_0 + (M_1 + M_2)\gamma_0 \\ &+ [(L_1 + L_2)k_1 + (M_1 + M_2)\gamma_1]\|x\| + [(L_1 + L_2)k_2 + (M_1 + M_2)\gamma_2)]\|y\|. \end{aligned}$$

Thus

$$\|(x,y)\| \le \frac{1}{M_0}\Big[\frac{|x_0|}{|\lambda|}\Big(1+(b-a)\sum_{i=1}^{m}|\delta_i|\Big)+(L_1+L_2)k_0+(M_1+M_2)\gamma_0\Big],$$

where $M_0 = \min\{1-[(L_1+L_2)k_1+(M_1+M_2)\gamma_1], 1-[(L_1+L_2)k_2+(M_1+M_2)\gamma_2)]\}$. Hence the set $\mathcal{P}$ is bounded. Thus, by Leray-Schauder alternative (Theorem 1.7), we deduce that the operator T has at least one fixed point, which corresponds to the fact that the problem (2.77) has at least one solution on $[a,b]$. □

2.7.2 *Examples*

Let us consider the following mixed-type coupled fractional differential systems

$$\begin{cases} {}^{C}D^{3/4}x(t) = \varphi(t,x(t),y(t)), \quad {}^{C}D^{3/2}y(t) = \psi(t,x(t),y(t)), \quad t\in[1,2], \\ x(a)+\dfrac{1}{2}y(b) = 1, \ y(1)=0, \quad y'(2) = \displaystyle\sum_{i=1}^{3}\delta_i x(\sigma_i), \end{cases} \tag{2.90}$$

where $\xi = 3/4, \zeta = 3/2, p = 1, q = 1/2, x_0 = 1, \delta_1 = 1/4, \delta_2 = 1/2, \delta_3 = 3/4, \sigma_1 = 5/4, \sigma_2 = 3/2, \sigma_3 = 7/4$, With the given data, it is found that $L_1 \simeq 1.395874, L_2 \simeq 0.615619, M_1 \simeq 0.537323, M_2 \simeq 1.961230$.

(a) In order to illustrate Theorem 2.19, we take

$$\varphi(t,x,y) = \frac{e^{-t}}{\sqrt{3+t^2}}\Big(\frac{|x|}{1+|x|}+\tan^{-1}y\Big)+\cos t, \ \psi(t,x,y) = \frac{1}{(5+t^4)}\Big(\sin x+|y|\Big)+e^{-t}, \tag{2.91}$$

which clearly satisfy the condition (2.19.1) with $\ell_1 = 1/2e$ and $\ell_2 = 1/6$. Moreover $(L_1+L_2)\ell_1+(M_1+M_2)\ell_2 < 0.786419$. Thus the hypothesis of Theorem 2.19 holds true and consequently there exists a unique solution of the problem (2.90) with $\varphi(t,x,y)$ and $\psi(t,x,y)$ given by (2.91) on $[1,2]$.

(b) In order to illustrate Theorem 2.20, we take

$$\begin{aligned} \varphi(t,x,y) &= e^{-2t}+\frac{1}{8}x\cos y+\frac{e^{-t}}{3}\sin y, \\ \psi(t,x,y) &= t\sqrt{t^2+3}+\frac{e^{-t}}{3\pi}x\tan^{-1}y+\frac{1}{\sqrt{48+t^2}}y. \end{aligned} \tag{2.92}$$

It is easy to check that the conditions (2.20.1) is satisfied with $k_0 = 1/2e, k_1 = 1/8, k_2 = 1/3e, \gamma_0 = 2\sqrt{7}, \gamma_1 = 1/6e, \gamma_2 = 1/7$. Furthermore, $(L_1+L_2)k_1+(M_1+M_2)\gamma_1 \simeq 0.404631 < 1$ and $(L_1+L_2)k_2+(M_1+M_2)\gamma_2 \simeq 0.603598 < 1$. Clearly the hypotheses of Theorem 2.20 are satisfied and hence the conclusion of Theorem 2.20 applies to problem (2.90) with $\varphi(t,x,y)$ and $\psi(t,x,y)$ given by (2.92).

2.8 Notes and remarks

The content of this chapter is based on the papers [61], [21], [4], [41], [16], [57], [58], [17] and [42].

Chapter 3

Nonlocal fractional boundary value problems with integral boundary conditions

In this chapter, we investigate nonlocal boundary value problems for fractional differential equations, integro-differential equations, inclusions, and coupled systems of fractional differential equations supplemented with different kinds of nonlocal integral boundary conditions. Our study includes Riemann-Liouville fractional integral boundary conditions, nonlocal strip boundary conditions, combinations of nonlocal and integral boundary conditions, coupled and uncoupled slit-strip type boundary conditions, flux type integral boundary conditions, nonlocal Erdélyi-Kober type integral boundary conditions and nonlocal multi-point Erdélyi-Kober fractional integral boundary conditions.

3.1 A four-point nonlocal integral boundary value problem for fractional differential equations and inclusions of arbitrary order

In this section, we study a four-point nonlocal integral boundary value problem for fractional differential equations and inclusions of arbitrary order.

3.1.1 *Nonlinear fractional differential equations*

In this subsection, we consider a boundary value problem of nonlinear fractional differential equations of an arbitrary order with four-point integral boundary conditions given by

$$\begin{cases} {}^cD^q x(t) = f(t, x(t)), \quad 0 < t < 1, \quad m-1 < q \leq m, \ m \geq 2, \\ x(0) = \alpha \displaystyle\int_0^\xi x(s)ds, \ x'(0) = 0, \ x''(0) = 0, \ldots, x^{(m-2)}(0) = 0, \\ x(1) = \beta \displaystyle\int_0^\eta x(s)ds, \quad 0 < \xi, \eta < 1, \end{cases} \tag{3.1}$$

where ${}^cD^q$ denotes the Caputo fractional derivative operator of order q, $f : [0,1] \times X \to X$ is continuous and $\alpha, \beta \in \mathbb{R}$. Here, $(X, \|\cdot\|)$ is a Banach space and $\mathcal{C} = C([0,1], X)$ denotes the Banach space of all continuous functions from $[0,1] \to X$ endowed with a topology of uniform convergence with the norm denoted by $\|\cdot\|$.

It is imperative to note that four-point integral boundary conditions in the problem (3.1) reduce to three-point ones when we choose either $\alpha = 0$ or $\beta = 0$.

Lemma 3.1. *For a given $\sigma \in C[0,1]$, the unique solution of the boundary value problem*

$$\begin{cases} {}^cD^q x(t) = \sigma(t), \quad 0 < t < 1, \quad m-1 < q \le m, \\ x(0) = \alpha \displaystyle\int_0^\xi x(s)ds, \ x'(0) = 0, \ x''(0) = 0, \dots, x^{(m-2)}(0) = 0, \\ x(1) = \beta \displaystyle\int_0^\eta x(s)ds, \quad 0 < \xi, \eta < 1, \end{cases} \tag{3.2}$$

is given by

$$\begin{aligned} x(t) = &\int_0^t \frac{(t-s)^{q-1}}{\Gamma(q)}\sigma(s)ds + \frac{\alpha}{m\Delta}\Big[(\beta\eta^m - m)\int_0^\xi \Big(\int_0^s \frac{(s-k)^{q-1}}{\Gamma(q)}\sigma(k)dk\Big)ds \\ &- \beta\xi^m \int_0^\eta \Big(\int_0^s \frac{(s-k)^{q-1}}{\Gamma(q)}\sigma(k)dk\Big)ds + \xi^m \int_0^1 \frac{(1-s)^{q-1}}{\Gamma(q)}\sigma(s)ds\Big] \\ &- \frac{t^{m-1}}{\Delta}\Big[\alpha(\beta\eta - 1)\int_0^\xi \Big(\int_0^s \frac{(s-k)^{q-1}}{\Gamma(q)}\sigma(k)dk\Big)ds \\ &- \beta(\alpha\xi - 1)\int_0^\eta \Big(\int_0^s \frac{(s-k)^{q-1}}{\Gamma(q)}\sigma(k)dk\Big)ds + (\alpha\xi - 1)\int_0^1 \frac{(1-s)^{q-1}}{\Gamma(q)}\sigma(s)ds\Big], \end{aligned}$$

where it is assumed that

$$\Delta = \frac{\alpha\xi^m(\beta\eta - 1) - (\alpha\xi - 1)(\beta\eta^m - m)}{m} \ne 0. \tag{3.3}$$

Proof. It is well known that the general solution of the fractional differential equation in (3.2) can be written as

$$x(t) = \int_0^t \frac{(t-s)^{q-1}}{\Gamma(q)}\sigma(s)ds - c_0 - c_1 t - c_2 t^2 - \dots - c_{m-1}t^{m-1}, \tag{3.4}$$

where $c_0, c_1, c_2, \dots, c_{m-1}$ are arbitrary constants. Applying the boundary conditions for the problem (3.2), it follows from (3.4) that $c_1 = 0, \dots, c_{m-2} = 0$,

$$\begin{aligned} c_0 = &-\frac{\alpha}{m\Delta}\Big[(\beta\eta^m - m)\int_0^\xi \Big(\int_0^s \frac{(s-k)^{q-1}}{\Gamma(q)}\sigma(k)dk\Big)ds \\ &- \beta\xi^m \int_0^\eta \Big(\int_0^s \frac{(s-k)^{q-1}}{\Gamma(q)}\sigma(k)dk\Big)ds + \xi^m \int_0^1 \frac{(1-s)^{q-1}}{\Gamma(q)}\sigma(s)ds\Big] \end{aligned}$$

and

$$\begin{aligned} c_{m-1} = &\frac{1}{\Delta}\Big[\alpha(\beta\eta - 1)\int_0^\xi \Big(\int_0^s \frac{(s-k)^{q-1}}{\Gamma(q)}\sigma(k)dk\Big)ds \\ &- \beta(\alpha\xi - 1)\int_0^\eta \Big(\int_0^s \frac{(s-k)^{q-1}}{\Gamma(q)}\sigma(k)dk\Big)ds + (\alpha\xi - 1)\int_0^1 \frac{(1-s)^{q-1}}{\Gamma(q)}\sigma(s)ds\Big], \end{aligned}$$

where Δ is given by (3.3). Substituting the values of $c_0,\ c_1, \ldots, c_{m-1}$ in (3.4), we obtain

$$x(t) = \int_0^t \frac{(t-s)^{q-1}}{\Gamma(q)}\sigma(s)ds + \frac{\alpha}{m\Delta}\Big[(\beta\eta^m - m)\int_0^\xi \Big(\int_0^s \frac{(s-k)^{q-1}}{\Gamma(q)}\sigma(k)dk\Big)ds$$
$$- \beta\xi^m \int_0^\eta \Big(\int_0^s \frac{(s-k)^{q-1}}{\Gamma(q)}\sigma(k)dk\Big)ds + \xi^m \int_0^1 \frac{(1-s)^{q-1}}{\Gamma(q)}\sigma(s)ds\Big]$$
$$- \frac{t^{m-1}}{\Delta}\Big[\alpha(\beta\eta - 1)\int_0^\xi \Big(\int_0^s \frac{(s-k)^{q-1}}{\Gamma(q)}\sigma(k)dk\Big)ds$$
$$- \beta(\alpha\xi - 1)\int_0^\eta \Big(\int_0^s \frac{(s-k)^{q-1}}{\Gamma(q)}\sigma(k)dk\Big)ds + (\alpha\xi - 1)\int_0^1 \frac{(1-s)^{q-1}}{\Gamma(q)}\sigma(s)ds\Big].$$

The converse follows by direct computation. This completes the proof. □

In view of Lemma 3.1, we define an operator $\mathbf{F} : \mathcal{C} \to \mathcal{C}$ by

$$(\mathbf{F}x)(t) = \int_0^t \frac{(t-s)^{q-1}}{\Gamma(q)} f(s, x(s))ds$$
$$+ \frac{\alpha}{m\Delta}\Big[(\beta\eta^m - m)\int_0^\xi \Big(\int_0^s \frac{(s-k)^{q-1}}{\Gamma(q)} f(k, x(k))dk\Big)ds$$
$$- \beta\xi^m \int_0^\eta \Big(\int_0^s \frac{(s-k)^{q-1}}{\Gamma(q)} f(k, x(k))dk\Big)ds$$
$$+ \xi^m \int_0^1 \frac{(1-s)^{q-1}}{\Gamma(q)} f(s, x(s))ds\Big]$$
$$- \frac{t^{m-1}}{\Delta}\Big[\alpha(\beta\eta - 1)\int_0^\xi \Big(\int_0^s \frac{(s-k)^{q-1}}{\Gamma(q)} f(k, x(k))dk\Big)ds$$
$$- \beta(\alpha\xi - 1)\int_0^\eta \Big(\int_0^s \frac{(s-k)^{q-1}}{\Gamma(q)} f(k, x(k))dk\Big)ds$$
$$+ (\alpha\xi - 1)\int_0^1 \frac{(1-s)^{q-1}}{\Gamma(q)} f(s, x(s))ds\Big], \quad t \in [0, 1]. \tag{3.5}$$

For convenience, let us set

$$\vartheta = \max_{t\in[0,1]}\Big\{\frac{t^q}{\Gamma(q+1)} + \frac{|\alpha|}{m|\Delta|}\Big[\frac{|\beta\eta^m - m|\xi^{q+1} + (|\beta|\eta^{q+1} + (q+1))\xi^m}{\Gamma(q+2)}\Big]$$
$$+ \frac{t^{m-1}}{|\Delta|}\Big[\frac{|\alpha(\beta\eta - 1)|\xi^{q+1} + |\alpha\xi - 1|(|\beta|\eta^{q+1} + (q+1))}{\Gamma(q+2)}\Big]\Big\}$$
$$= \frac{1}{\Gamma(q+1)}\Big(1 + \frac{\vartheta_1 + \vartheta_2}{m|\Delta|(q+1)}\Big), \tag{3.6}$$

where

$$\vartheta_1 = |\alpha|(|\beta\eta^m - m| + m|\beta\eta - 1|)\xi^{q+1},$$

$$\vartheta_2 = (|\alpha|\xi^m + m|\alpha\xi - 1|)(|\beta|\eta^{q+1} + (q+1)).$$

Observe that the problem (3.1) has solutions if the operator equation $\mathbf{F}x = x$ has fixed points.

Lemma 3.2. *The operator $\mathbf{F} : \mathcal{C} \to \mathcal{C}$ is completely continuous.*

Proof. Clearly, continuity of the operator $\mathbf{F}$ follows from that of f. Let $\Omega \subset \mathcal{C}$ be bounded. Then, $\forall x \in \Omega$, by the assumption (H), there exists $L_1 > 0$ such that $|f(t,x)| \leq L_1$. Thus, we have

$$\begin{aligned}
|(\mathbf{F})(t)| \leq& \int_0^t \frac{(t-s)^{q-1}}{\Gamma(q)}|f(s,x(s))|ds \\
&+\frac{|\alpha|}{m|\Delta|}\Big[|\beta\eta^m - m|\int_0^\xi \Big(\int_0^s \frac{(s-k)^{q-1}}{\Gamma(q)}|f(k,x(k))|dk\Big)ds \\
&+|\beta|\xi^m \int_0^\eta \Big(\int_0^s \frac{(s-k)^{q-1}}{\Gamma(q)}|f(k,x(k))|dk\Big)ds \\
&+\xi^m \int_0^1 \frac{(1-s)^{q-1}}{\Gamma(q)}|f(s,x(s))|ds\Big] \\
&+\frac{|t^{m-1}|}{|\Delta|}\Big[|\alpha(\beta\eta-1)|\int_0^\xi \Big(\int_0^s \frac{(s-k)^{q-1}}{\Gamma(q)}|f(k,x(k))|dk\Big)ds \\
&+|\beta(\alpha\xi-1)|\int_0^\eta \Big(\int_0^s \frac{(s-k)^{q-1}}{\Gamma(q)}|f(k,x(k))|dk\Big)ds \\
&+|\alpha\xi-1|\int_0^1 \frac{(1-s)^{q-1}}{\Gamma(q)}|f(s,x(s))|ds\Big] \\
\leq& L_1\Big\{\int_0^t \frac{(t-s)^{q-1}}{\Gamma(q)}ds + \frac{|\alpha|}{m|\Delta|}\Big[|\beta\eta^m - m|\int_0^\xi \Big(\int_0^s \frac{(s-k)^{q-1}}{\Gamma(q)}dk\Big)ds \\
&+|\beta|\xi^m \int_0^\eta \Big(\int_0^s \frac{(s-k)^{q-1}}{\Gamma(q)}dk\Big)ds + \xi^m\int_0^1 \frac{(1-s)^{q-1}}{\Gamma(q)}ds\Big] \\
&+\frac{|t^{m-1}|}{|\Delta|}\Big[|\alpha(\beta\eta-1)|\int_0^\xi \Big(\int_0^s \frac{(s-k)^{q-1}}{\Gamma(q)}dk\Big)ds \\
&+|\beta(\alpha\xi-1)|\int_0^\eta \Big(\int_0^s \frac{(s-k)^{q-1}}{\Gamma(q)}dk\Big)ds + |\alpha\xi-1|\int_0^1 \frac{(1-s)^{q-1}}{\Gamma(q)}ds\Big]\Big\} \\
\leq& \frac{L_1}{\Gamma(q+1)}\Big(1+\frac{\vartheta_1+\vartheta_2}{m|\Delta|(q+1)}\Big) = L_2, \qquad (3.7)
\end{aligned}$$

which implies that $\|(\mathbf{F}x)\| \leq L_2$. Furthermore,

$$\begin{aligned}
|(\mathbf{F}x)'(t)| =& \int_0^t \frac{(t-s)^{q-2}}{\Gamma(q-1)}|f(s,x(s))|ds \\
&+\frac{|(m-1)t^{m-2}|}{|\Delta|}\Big[|\alpha(\beta\eta-1)|\int_0^\xi \Big(\int_0^s \frac{(s-k)^{q-1}}{\Gamma(q)}|f(k,x(k))|dk\Big)ds \\
&+|\beta(\alpha\xi-1)|\int_0^\eta \Big(\int_0^s \frac{(s-k)^{q-1}}{\Gamma(q)}|f(k,x(k))|dk\Big)ds \\
&+|\alpha\xi-1|\int_0^1 \frac{(1-s)^{q-1}}{\Gamma(q)}|f(s,x(s))|ds\Big]
\end{aligned}$$

$$\leq L_1\Big\{\int_0^t \frac{(t-s)^{q-2}}{\Gamma(q-1)}ds$$
$$+\frac{|(m-1)t^{m-2}|}{|\Delta|}\Big[|\alpha(\beta\eta-1)|\int_0^\xi\Big(\int_0^s \frac{(s-k)^{q-1}}{\Gamma(q)}dk\Big)ds$$
$$+|\beta(\alpha\xi-1)|\int_0^\eta\Big(\int_0^s \frac{(s-k)^{q-1}}{\Gamma(q)}dk\Big)ds+|\alpha\xi-1|\int_0^1 \frac{(1-s)^{q-1}}{\Gamma(q)}ds\Big]\Big\}$$
$$\leq L_1\Big\{\frac{1}{\Gamma(q)}+\frac{|m-1|}{|\Delta|}\Big[\frac{|\alpha(\beta\eta-1)|\xi^{q+1}+|\beta(\alpha\xi-1)|\eta^{q+1}}{\Gamma(q+2)}+\frac{|(\alpha\xi-1)|}{\Gamma(q+1)}\Big]\Big\}$$
$$= L_3. \tag{3.8}$$

Hence, for $t_1, t_2 \in [0,1]$, we have

$$|(\mathbf{F}x)(t_2)-(\mathbf{F}x)(t_1)| \leq \int_{t_1}^{t_2} |(\mathbf{F}x)'(s)|ds \leq L_3(t_2-t_1).$$

This implies that $\mathbf{F}$ is equi-continuous on $[0,1]$. Thus, by the Arzela-Ascoli theorem, we have that $\mathbf{F}(\Omega)(t)$ is relatively compact in X for every t, and so the operator $\mathbf{F}:\mathcal{C}\to\mathcal{C}$ is completely continuous. □

3.1.2 *Existence and uniqueness results in the Banach space*

Theorem 3.1. *Assume that:*

(3.1.1) $f:[0,1]\times X\to X$ *is a jointly continuous function and maps bounded subsets of* $[0,1]\times X$ *into relatively compact subsets of* X.

If $\lim_{x\to 0}\frac{f(t,x)}{x}=0$, *where the limit is uniform with respect to* t. *Then the problem (3.1) has at least one solution on* $[0,1]$.

Proof. Since $\lim_{x\to 0}\frac{f(t,x)}{x}=0$, therefore there exists a constant $r>0$ such that $|f(t,x)|\leq\delta|x|$ for $0<|x|<r$, where $\delta>0$ is such that

$$\vartheta\delta\leq 1, \tag{3.9}$$

where ϑ is given by (3.6). Define $\Omega_1=\{x\in\mathcal{C}\mid \|x\|<r\}$ and take $x\in\mathcal{C}$ such that $\|x\|=r$, that is, $x\in\partial\Omega_1$. By Lemma 3.2, $\mathbf{F}$ is completely continuous and that

$$|\mathbf{F}x(t)|\leq \max_{t\in[0,1]}\Big\{\frac{t^q}{\Gamma(q+1)}+\frac{|\alpha|}{m|\Delta|}\Big[\frac{|\beta\eta^m-m|\xi^{q+1}+(|\beta|\eta^{q+1}+(q+1))\xi^m}{\Gamma(q+2)}\Big]$$
$$+\frac{t^{m-1}}{|\Delta|}\Big[\frac{|\alpha(\beta\eta-1)|\xi^{q+1}+|\alpha\xi-1|(|\beta|\eta^{q+1}+(q+1))}{\Gamma(q+2)}\Big]\Big\}\delta\|x\|$$
$$=\vartheta\delta\|x\|, \tag{3.10}$$

which, in view of (3.9), yields $\|\mathbf{F}x\|\leq\|x\|$, $x\in\partial\Omega_1$. Therefore, by Theorem 1.3, the operator $\mathbf{F}$ has at least one fixed point, which in turn implies that the problem (3.1) has at least one solution on $[0,1]$. □

Theorem 3.2. *Assume that* $f : [0,1] \times X \to X$ *is a jointly continuous function and satisfies the condition*

$$|f(t,x) - f(t,y)| \leq L|x-y|, \forall t \in [0,1],\ L > 0,\ x, y \in X,$$

with $L < 1/\vartheta$*, where* ϑ *is given by (3.6). Then the boundary value problem (3.1) has a unique solution on* $[0,1]$.

Proof. Setting $\sup_{t\in[0,1]} |f(t,0)| = M$ and choosing $r \geq \dfrac{\vartheta M}{1-L\vartheta}$, we show that $\mathbf{F}B_r \subset B_r$, where $B_r = \{x \in \mathcal{C} : \|x\| \leq r\}$. For $x \in B_r$, and any $t \in [0,1]$, we note that $|f(t,x(t))| = |f(t,x(t)) - f(t,0) + f(t,0)| \leq |f(t,x(t)) - f(t,0)| + |f(t,0)| \leq L|x(t)| + M \leq L\|x\| + M \leq Lr + M$. Therefore, we have

$$\begin{aligned}
|(\mathbf{F}x)(t)| \leq & \sup_{t\in[0,1]} \Big\{ \int_0^t \frac{(t-s)^{q-1}}{\Gamma(q)} |f(s,x(s))| ds \\
& + \frac{|\alpha|}{m|\Delta|} \Big[|\beta\eta^m - m| \int_0^\xi \Big(\int_0^s \frac{(s-k)^{q-1}}{\Gamma(q)} |f(k,x(k))| dk \Big) ds \\
& + |\beta|\xi^m \int_0^\eta \Big(\int_0^s \frac{(s-k)^{q-1}}{\Gamma(q)} |f(k,x(k))| dk \Big) ds \\
& + \xi^m \int_0^1 \frac{(1-s)^{q-1}}{\Gamma(q)} |f(s,x(s))| ds \Big] \\
& + \frac{|t^{m-1}|}{|\Delta|} \Big[|\alpha(\beta\eta - 1)| \int_0^\xi \Big(\int_0^s \frac{(s-k)^{q-1}}{\Gamma(q)} |f(k,x(k))| dk \Big) ds \\
& + |\beta(\alpha\xi - 1)| \int_0^\eta \Big(\int_0^s \frac{(s-k)^{q-1}}{\Gamma(q)} |f(k,x(k))| dk \Big) ds \\
& + |\alpha\xi - 1| \int_0^1 \frac{(1-s)^{q-1}}{\Gamma(q)} |f(s,x(s))| ds \Big] \Big\} \\
\leq & (Lr + M) \sup_{t\in[0,1]} \Big[\int_0^t \frac{(t-s)^{q-1}}{\Gamma(q)} ds \\
& + \frac{|\alpha|}{m|\Delta|} \Big\{ |\beta\eta^m - m| \int_0^\xi \Big(\int_0^s \frac{(s-k)^{q-1}}{\Gamma(q)} dk \Big) ds \\
& + |\beta|\xi^m \int_0^\eta \Big(\int_0^s \frac{(s-k)^{q-1}}{\Gamma(q)} dk \Big) ds + \xi^m \int_0^1 \frac{(1-s)^{q-1}}{\Gamma(q)} ds \Big\} \\
& + \frac{|t^{m-1}|}{|\Delta|} \Big\{ |\alpha(\beta\eta - 1)| \int_0^\xi \Big(\int_0^s \frac{(s-k)^{q-1}}{\Gamma(q)} dk \Big) ds \\
& + |\beta(\alpha\xi - 1)| \int_0^\eta \Big(\int_0^s \frac{(s-k)^{q-1}}{\Gamma(q)} dk \Big) ds + |\alpha\xi - 1| \int_0^1 \frac{(1-s)^{q-1}}{\Gamma(q)} ds \Big\} \Big] \\
\leq & \frac{(Lr+M)}{\Gamma(q+1)} \Big(1 + \frac{\vartheta_1 + \vartheta_2}{m|\Delta|(q+1)} \Big) = (Lr + M)\vartheta \leq r.
\end{aligned}$$

Now, for $x, y \in \mathcal{C}$ and for each $t \in [0,1]$, we obtain

$$
\begin{aligned}
&|(\mathbf{F}x)(t) - (\mathbf{F}y)(t)| \\
&\leq \sup_{t\in[0,1]} \Big[\int_0^t \frac{(t-s)^{q-1}}{\Gamma(q)}|f(s,x(s)) - f(s,y(s))|ds \\
&\quad + \frac{|\alpha|}{m|\Delta|}\Big\{|\beta\eta^m - m|\int_0^\xi \Big(\int_0^s \frac{(s-k)^{q-1}}{\Gamma(q)}|f(k,x(k)) - f(k,y(k))|dk\Big)ds \\
&\quad + |\beta|\xi^m \int_0^\eta \Big(\int_0^s \frac{(s-k)^{q-1}}{\Gamma(q)}|f(k,x(k)) - f(k,y(k))|dk\Big)ds \\
&\quad + \xi^m \int_0^1 \frac{(1-s)^{q-1}}{\Gamma(q)}|f(s,x(s)) - f(s,y(s))|ds\Big\} \\
&\quad + \frac{|t^{m-1}|}{|\Delta|}\Big\{|\alpha(\beta\eta - 1)|\int_0^\xi \Big(\int_0^s \frac{(s-k)^{q-1}}{\Gamma(q)}|f(k,x(k)) - f(k,y(k))|dk\Big)ds \\
&\quad + |\beta(\alpha\xi - 1)|\int_0^\eta \Big(\int_0^s \frac{(s-k)^{q-1}}{\Gamma(q)}|f(k,x(k)) - f(k,y(k))|dk\Big)ds \\
&\quad + |\alpha\xi - 1|\int_0^1 \frac{(1-s)^{q-1}}{\Gamma(q)}|f(s,x(s)) - f(s,y(s))|ds\Big\}\Big] \\
&\leq L\|x-y\| \sup_{t\in[0,1]} \Big[\int_0^t \frac{(t-s)^{q-1}}{\Gamma(q)}ds + \frac{|\alpha|}{m|\Delta|}\Big\{|\beta\eta^m - m|\int_0^\xi \Big(\int_0^s \frac{(s-k)^{q-1}}{\Gamma(q)}dk\Big)ds \\
&\quad + |\beta|\xi^m \int_0^\eta \Big(\int_0^s \frac{(s-k)^{q-1}}{\Gamma(q)}dk\Big)ds + \xi^m \int_0^1 \frac{(1-s)^{q-1}}{\Gamma(q)}ds\Big\} \\
&\quad + \frac{|t^{m-1}|}{|\Delta|}\Big\{|\alpha(\beta\eta - 1)|\int_0^\xi \Big(\int_0^s \frac{(s-k)^{q-1}}{\Gamma(q)}dk\Big)ds \\
&\quad + |\beta(\alpha\xi - 1)|\int_0^\eta \Big(\int_0^s \frac{(s-k)^{q-1}}{\Gamma(q)}dk\Big)ds + |(\alpha\xi - 1)|\int_0^1 \frac{(1-s)^{q-1}}{\Gamma(q)}ds\Big\}\Big] \\
&\leq \frac{L}{\Gamma(q+1)}\left(1 + \frac{\vartheta_1 + \vartheta_2}{m|\Delta|(q+1)}\right)\|x-y\| \\
&= L\vartheta\|x-y\|,
\end{aligned}
$$

where ϑ is given by (3.6). As $L < 1/\vartheta$, therefore $\mathbf{F}$ is a contraction. Thus, the conclusion of the theorem follows by the contraction mapping principle (Banach fixed point theorem). □

Our next existence result is based on Leray-Schauder degree theory (Theorem 1.15).

Theorem 3.3. *Suppose that (3.1.1) holds. Furthermore, it is assumed that there exist constants* $0 \leq \kappa < \frac{1}{\theta}$, *where* θ *is given by (3.6) and* $M > 0$ *such that* $|f(t,x)| \leq \kappa|x| + M$ *for all* $t \in [0,1], x \in X$. *Then the boundary value problem (3.1) has at least one solution.*

Proof. Consider a fixed point problem

$$x = \mathbf{F}x, \tag{3.11}$$

where $\mathbf{F}$ is defined by (3.5). In view of the fixed point problem (3.11), we just need to prove the existence of at least one solution $x \in \mathcal{C}$ satisfying (3.11). Define a suitable ball B_R with radius $R > 0$ as

$$B_R = \{x \in \mathcal{C} : \|x\| < R\},$$

where R will be fixed later. Then, it is sufficient to show that $\mathbf{F} : \overline{B}_R \to \mathcal{C}$ satisfies

$$x \neq \lambda \mathbf{F}x, \quad \forall\, x \in \partial B_R \text{ and } \forall\, \lambda \in [0,1]. \tag{3.12}$$

Let us set

$$H(\lambda, x) = \lambda \mathbf{F}x, \quad x \in X, \quad \lambda \in [0,1].$$

Then, by the Arzelá-Ascoli Theorem, $h_\lambda(x) = x - H(\lambda, x) = x - \lambda \mathbf{F}x$ is completely continuous. If (3.12) is true, then the following Leray-Schauder degrees are well defined and by the homotopy invariance of topological degree, it follows that

$$\begin{aligned}\deg(h_\lambda, B_R, 0) &= \deg(I - \lambda \mathbf{F}, B_R, 0) = \deg(h_1, B_R, 0)\\ &= \deg(h_0, B_R, 0) = \deg(I, B_R, 0) = 1 \neq 0, \quad 0 \in B_r,\end{aligned}$$

where I denotes the identity operator. By the nonzero property of Leray-Schauder degree, $h_1(t) = x - \lambda \mathbf{F}x = 0$ for at least one $x \in B_R$. In order to prove (3.12), we assume that $x = \lambda \mathbf{F}x$ for some $\lambda \in [0,1]$. Then, for all $t \in [0,1]$, we can obtain (as before) that

$$\|x\| = \sup_{t\in[0,1]} |\lambda \mathbf{F}x(t)| \leq \frac{(\kappa\|x\| + M)}{\Gamma(q+1)}\Big(1 + \frac{\vartheta_1 + \vartheta_2}{m|\Delta|(q+1)}\Big) = (\kappa\|x\| + M)\,\vartheta,$$

which can alternatively be written as

$$\|x\| \leq \frac{M\vartheta}{1 - \kappa\vartheta}.$$

Letting $R = \dfrac{M\vartheta}{1-\kappa\vartheta} + 1$, (3.12) holds. This completes the proof. □

3.1.2.1 *Examples*

Let us define

$$\mathcal{E} = \{x = (x_1, x_2, \ldots, x_p, \ldots) : x_p \to 0\}$$

with the norm $\|x\| = \sup_p |x_p|$.

Example 3.1. Consider the problem

$$\begin{cases} {}^C D^q x_p(t) = (121 + x_p^3(t))^{\frac{1}{2}} + 2(t+1)(x_p - \sin x_p(t)) - 11, \quad 0 < t < 1,\\ x_p(0) = \alpha \displaystyle\int_0^\xi x_p(s)ds, \ x_p'(0) = 0, \ x_p''(0) = 0, \ldots, x_p^{(m-2)}(0) = 0,\\ x_p(1) = \beta \displaystyle\int_0^\eta x_p(s)ds, \quad 0 < \xi, \eta < 1, \end{cases} \tag{3.13}$$

where $m - 1 < q \leq m, \ m \geq 2$.

It can easily be verified that all the assumptions of Theorem 3.1 hold. Consequently, the conclusion of Theorem 3.1 implies that the problem (3.13) has at least one solution.

Example 3.2. Consider the following four-point integral fractional boundary value problem

$$\begin{cases} {}^cD^{13/2}x_p(t) = \dfrac{512}{(t+2)^2}\dfrac{|x_p|}{1+|x_p|}, \quad t \in [0,1], \\ x_p(0) = \dfrac{1}{2}\displaystyle\int_0^{1/4} x_p(s)ds, \; x_p'(0) = 0, \ldots, x_p^{(v)}(0) = 0, \; x_p(1) = \displaystyle\int_0^{3/4} x_p(s)ds. \end{cases} \tag{3.14}$$

Here, $q = 13/2$, $m = 7$, $\alpha = 1/2$, $\beta = 1$, $\xi = 1/4$, $\eta = 3/4$, and $f_p(t,x) = \dfrac{512}{(t+2)^2}\dfrac{|x_p|}{1+|x_p|}$, $f = (f_1, f_2, \ldots, f_p, \ldots)$. As $|f_p(t,x) - f_p(t,y)| \leq 128|x_p - y_p|$, therefore, $|f(t,x) - f(t,y)| \leq 128\|x - y\|$ with $L = 128$. Further,

$$L\vartheta = \frac{L}{\Gamma(q+1)}\left(1 + \frac{\vartheta_1 + \vartheta_2}{m|\Delta|(q+1)}\right) = 128 \times (1.087591 \times 10^{-03}) \approx 0.139212 < 1.$$

Thus, by the conclusion of Theorem 3.2, the boundary value problem (3.14) has a unique solution on $[0,1]$.

Example 3.3. Consider the following boundary value problem

$$\begin{cases} {}^cD^{13/2}x_p(t) = \dfrac{1}{(4\pi)}\sin(2\pi x_p) + \dfrac{|x_p|}{1+|x_p|}, \quad t \in [0,1], \\ x_p(0) = 2\displaystyle\int_0^{1/4} x_p(s)ds, \; x_p'(0) = 0, \ldots, x_p^{(m-2)}(0) = 0, \; x_p(1) = 3\displaystyle\int_0^{3/4} x_p(s)ds. \end{cases} \tag{3.15}$$

Here, $q = 13/2$, $m = 7$, $\alpha = 2$, $\beta = 3, \xi = 1/3$, $\eta = 2/3$, and

$$\Big|f_p(t,x)\Big| = \left|\frac{1}{(4\pi)}\sin(2\pi x_p) + \frac{|x_p|}{1+|x_p|}\right| \leq \frac{1}{2}|x_p| + 1.$$

So $|f(t,x)| \leq \frac{1}{2}\|x\| + 1$. Clearly $M = 1$, $\kappa = 1/2$, $\vartheta = 1.081553 \times 10^{-03}$, and $\kappa < 1/\vartheta$. Thus, all the conditions of Theorem 3.3 are satisfied and consequently the problem (3.15) has at least one solution.

3.1.3 *Nonlinear fractional differential inclusions*

Here we are concerned with the existence of solutions for a boundary value problem for fractional differential inclusions of arbitrary order supplemented with four-point integral boundary conditions given by

$$\begin{cases} {}^cD^q x(t) \in F(t, x(t)), \quad 0 < t < 1, \quad 1 \;\; m-1 < q \leq m, \\ x(0) = \alpha\displaystyle\int_0^{\xi} x(s)ds, \; x'(0) = 0, \; x''(0) = 0, \ldots, x^{(m-2)}(0) = 0, \\ x(1) = \beta\displaystyle\int_0^{\eta} x(s)ds, \quad 0 < \xi, \eta < 1, \end{cases} \tag{3.16}$$

where ${}^cD^q$ denotes the Caputo fractional derivative operator of order q, $F : [0,1] \times \mathbb{R} \to \mathcal{P}(\mathbb{R})$ is a multi-valued map, $\mathcal{P}(\mathbb{R})$ is the family of all subsets of $\mathbb{R}$ and $\alpha, \beta \in \mathbb{R}$.

Definition 3.1. A function $x \in C^m([0,1])$ is a solution of the problem (3.16) if there exists a function $f \in L^1([0,1],\mathbb{R})$ such that $f(t) \in F(t, x(t))$ a.e. on $[0,1]$ and

$$\begin{aligned} x(t) = &\int_0^t \frac{(t-s)^{q-1}}{\Gamma(q)} f(s)ds + \frac{\alpha}{m\Delta}\Big[(\beta\eta^m - m)\int_0^\xi \Big(\int_0^s \frac{(s-m)^{q-1}}{\Gamma(q)} f(m)dm\Big)ds \\ &- \beta\xi^m \int_0^\eta \Big(\int_0^s \frac{(s-m)^{q-1}}{\Gamma(q)} f(m)dm\Big)ds + \xi^m \int_0^1 \frac{(1-s)^{q-1}}{\Gamma(q)} f(s)ds\Big] \\ &- \frac{t^{m-1}}{\Delta}\Big[\alpha(\beta\eta - 1)\int_0^\xi \Big(\int_0^s \frac{(s-m)^{q-1}}{\Gamma(q)} f(m)dm\Big)ds \\ &- \beta(\alpha\xi - 1)\int_0^\eta \Big(\int_0^s \frac{(s-m)^{q-1}}{\Gamma(q)} f(m)dm\Big)ds + (\alpha\xi - 1)\int_0^1 \frac{(1-s)^{q-1}}{\Gamma(q)} f(s)ds\Big]. \end{aligned}$$

3.1.3.1 *The Carathéodory case*

Theorem 3.4. *Assume that:*

(3.4.1) $F : [0,1] \times \mathbb{R} \to \mathcal{P}(\mathbb{R})$ *is* L^1*-Carathéodory and has convex values;*

(3.4.2) there exists a continuous non-decreasing function $\psi : [0,\infty) \to (0,\infty)$ *and a function* $p \in C([0,1],\mathbb{R}^+)$ *such that*

$\|F(t,x)\|_{\mathcal{P}} := \sup\{|y| : y \in F(t,x)\} \le p(t)\psi(\|x\|)$ *for each* $(t,x) \in [0,1]\times\mathbb{R}$;

(3.4.3) there exists a number $M > 0$ *such that*

$$\frac{M}{\frac{1}{\Gamma(q+1)}\left(1 + \frac{\vartheta_1 + \vartheta_2}{m|\Delta|(q+1)}\right)\psi(M)\|p\|} > 1, \tag{3.17}$$

where ϑ_1, ϑ_2 *are defined in (3.6).*

Then the boundary value problem (3.16) has at least one solution on $[0,1]$.

Proof. Define an operator

$$\Omega(x) = \left\{ \begin{array}{l} h \in C([0,1],\mathbb{R}) : \\ h(t) = \left\{ \begin{array}{l} \frac{1}{\Gamma(q)}\int_0^t (t-s)^{q-1} f(s)ds \\ + \frac{\alpha}{m\Delta}\Big[(\beta\eta^m - m)\int_0^\xi \Big(\int_0^s \frac{(s-m)^{q-1}}{\Gamma(q)} f(m)dm\Big)ds \\ -\beta\xi^m \int_0^\eta \Big(\int_0^s \frac{(s-m)^{q-1}}{\Gamma(q)} f(m)dm\Big)ds + \xi^m \int_0^1 \frac{(1-s)^{q-1}}{\Gamma(q)} f(s)ds\Big] \\ -\frac{t^{m-1}}{\Delta}\Big[\alpha(\beta\eta - 1)\int_0^\xi \Big(\int_0^s \frac{(s-m)^{q-1}}{\Gamma(q)} f(m)dm\Big)ds \\ -\beta(\alpha\xi - 1)\int_0^\eta \Big(\int_0^s \frac{(s-m)^{q-1}}{\Gamma(q)} f(m)dm\Big)ds \\ +(\alpha\xi - 1)\int_0^1 \frac{(1-s)^{q-1}}{\Gamma(q)} f(s)ds\Big], \quad t \in [0,1] \end{array} \right. \end{array} \right\}$$

for $f \in S_{F,x}$. We will show that Ω satisfies the assumptions of the nonlinear alternative of Leray- Schauder type (Theorem 1.8). The proof consists of several steps. As a first step, we show that $\Omega(x)$ is convex for each $x \in C([0,1],\mathbb{R})$. For that, let $h_1, h_2 \in \Omega(x)$. Then there exist $f_1, f_2 \in S_{F,x}$ such that for each $t \in [0,1]$, we have

$$\begin{aligned} h_i(t) = & \int_0^t \frac{(t-s)^{q-1}}{\Gamma(q)} f_i(s)ds + \frac{\alpha}{m\Delta}\Big[(\beta\eta^m - m)\int_0^\xi \Big(\int_0^s \frac{(s-m)^{q-1}}{\Gamma(q)} f_i(m)dm\Big)ds \\ & -\beta\xi^m \int_0^\eta \Big(\int_0^s \frac{(s-m)^{q-1}}{\Gamma(q)} f_i(m)dm\Big)ds + \xi^m \int_0^1 \frac{(1-s)^{q-1}}{\Gamma(q)} f_i(s)ds\Big] \\ & -\frac{t^{m-1}}{\Delta}\Big[\alpha(\beta\eta - 1)\int_0^\xi \Big(\int_0^s \frac{(s-m)^{q-1}}{\Gamma(q)} f_i(m)dm\Big)ds \\ & -\beta(\alpha\xi - 1)\int_0^\eta \Big(\int_0^s \frac{(s-m)^{q-1}}{\Gamma(q)} f_i(m)dm\Big)ds \\ & +(\alpha\xi - 1)\int_0^1 \frac{(1-s)^{q-1}}{\Gamma(q)} f_i(s)ds\Big], \quad i = 1,2. \end{aligned}$$

Let $0 \le \omega \le 1$. Then, for each $t \in [0,1]$, we have

$$\begin{aligned} & [\omega h_1 + (1-\omega)h_2](t) \\ = & \int_0^t \frac{(t-s)^{q-1}}{\Gamma(q)} [\omega f_1(s) + (1-\omega)f_2(s)]ds \\ & + \frac{\alpha}{m\Delta}\Big[(\beta\eta^m - m)\int_0^\xi \Big(\int_0^s \frac{(s-m)^{q-1}}{\Gamma(q)} [\omega f_1(m) + (1-\omega)f_2(m)]dm\Big)ds \\ & -\beta\xi^m \int_0^\eta \Big(\int_0^s \frac{(s-m)^{q-1}}{\Gamma(q)} [\omega f_1(m) + (1-\omega)f_2(m)]dm\Big)ds \\ & +\xi^m \int_0^1 \frac{(1-s)^{q-1}}{\Gamma(q)} [\omega f_1(s) + (1-\omega)f_2(s)]ds\Big] \\ & -\frac{t^{m-1}}{\Delta}\Big[\alpha(\beta\eta - 1)\int_0^\xi \Big(\int_0^s \frac{(s-m)^{q-1}}{\Gamma(q)} [\omega f_1(m) + (1-\omega)f_2(m)]dm\Big)ds \\ & -\beta(\alpha\xi - 1)\int_0^\eta \Big(\int_0^s \frac{(s-m)^{q-1}}{\Gamma(q)} [\omega f_1(m) + (1-\omega)f_2(m)]dm\Big)ds \\ & +(\alpha\xi - 1)\int_0^1 \frac{(1-s)^{q-1}}{\Gamma(q)} [\omega f_1(s) + (1-\omega)f_2(s)]ds\Big]. \end{aligned}$$

Since $S_{F,x}$ is convex (F has convex values), therefore it follows that $\omega h_1 + (1-\omega)h_2 \in \Omega(x)$.

Next, we show that $\Omega(x)$ maps bounded sets into bounded sets in $C([0,1],\mathbb{R})$. For a positive number r, let $B_r = \{x \in C([0,1],\mathbb{R}) : \|x\| \le r\}$ be a bounded set in $C([0,1],\mathbb{R})$. Then, for each $h \in \Omega(x), x \in B_r$, there exists $f \in S_{F,x}$ such that

$$\begin{aligned} & h(t) \\ = & \int_0^t \frac{(t-s)^{q-1}}{\Gamma(q)} f(s)ds + \frac{\alpha}{m\Delta}\Big[(\beta\eta^m - m)\int_0^\xi \Big(\int_0^s \frac{(s-m)^{q-1}}{\Gamma(q)} f(m)dm\Big)ds \end{aligned}$$

$$-\beta\xi^m\int_0^\eta\Big(\int_0^s\frac{(s-m)^{q-1}}{\Gamma(q)}f(m)dm\Big)ds+\xi^m\int_0^1\frac{(1-s)^{q-1}}{\Gamma(q)}f(s)ds\Big]$$
$$-\frac{t^{m-1}}{\Delta}\Big[\alpha(\beta\eta-1)\int_0^\xi\Big(\int_0^s\frac{(s-m)^{q-1}}{\Gamma(q)}f(m)dm\Big)ds$$
$$-\beta(\alpha\xi-1)\int_0^\eta\Big(\int_0^s\frac{(s-m)^{q-1}}{\Gamma(q)}f(m)dm\Big)ds+(\alpha\xi-1)\int_0^1\frac{(1-s)^{q-1}}{\Gamma(q)}f(s)ds\Big].$$

Then

$$\begin{aligned}
&|h(t)|\\
&\le\int_0^t\frac{(t-s)^{q-1}}{\Gamma(q)}|f(s)|ds+\frac{|\alpha|}{m|\Delta|}\Big[|\beta\eta^m-m|\int_0^\xi\Big(\int_0^s\frac{(s-m)^{q-1}}{\Gamma(q)}|f(m)|dm\Big)ds\\
&\quad+|\beta|\xi^m\int_0^\eta\Big(\int_0^s\frac{(s-m)^{q-1}}{\Gamma(q)}|f(m)|dm\Big)ds+\xi^m\int_0^1\frac{(1-s)^{q-1}}{\Gamma(q)}|f(s)|ds\Big]\\
&\quad+\frac{t^{m-1}}{|\Delta|}\Big[|\alpha(\beta\eta-1)|\int_0^\xi\Big(\int_0^s\frac{(s-m)^{q-1}}{\Gamma(q)}|f(m)|dm\Big)ds\\
&\quad+|\beta(\alpha\xi-1)|\int_0^\eta\Big(\int_0^s\frac{(s-m)^{q-1}}{\Gamma(q)}|f(m)|dm\Big)ds\\
&\quad+|\alpha\xi-1|\int_0^1\frac{(1-s)^{q-1}}{\Gamma(q)}|f(s)|ds\Big]\\
&\le\Big(\frac{1}{\Gamma(q+1)}+\frac{|\alpha||\beta\eta^m-m|\xi^{q+1}+|\alpha||\beta|\xi^m\eta^{q+1}+|\alpha|\xi^m}{m|\Delta|\Gamma(q)}\\
&\quad+\frac{|\alpha||\beta\eta-1|\xi^{q+1}+|\beta||\alpha\xi-1|\eta^{q+1}+|\alpha\xi-1|}{|\Delta|\Gamma(q)}\Big)\|p\|\psi(\|x\|)\\
&=\frac{1}{\Gamma(q+1)}\Big(1+\frac{\vartheta_1+\vartheta_2}{m|\Delta|(q+1)}\Big)\|p\|\psi(\|x\|).
\end{aligned}$$

Thus,

$$\|h\|\le\frac{1}{\Gamma(q+1)}\Big(1+\frac{\vartheta_1+\vartheta_2}{m|\Delta|(q+1)}\Big)\|p\|\psi(r).$$

Now we show that Ω maps bounded sets into equi-continuous sets of $C([0,1],\mathbb{R})$. Let $t',t''\in[0,1]$ with $t'<t''$ and $x\in B_r$, where B_r is a bounded set of $C([0,1],\mathbb{R})$. For each $h\in\Omega(x)$, we obtain

$$\begin{aligned}
&|h(t'')-h(t')|\\
&=\Big|\frac{1}{\Gamma(q)}\int_0^{t''}(t''-s)^{q-1}f(s)ds-\frac{1}{\Gamma(q)}\int_0^{t'}(t'-s)^{q-1}f(s)ds\\
&\quad-\frac{(t'')^{m-1}-(t')^{m-1}}{\Delta}\Big[\alpha(\beta\eta-1)\int_0^\xi\Big(\int_0^s\frac{(s-m)^{q-1}}{\Gamma(q)}f(m)dm\Big)ds\\
&\quad-\beta(\alpha\xi-1)\int_0^\eta\Big(\int_0^s\frac{(s-m)^{q-1}}{\Gamma(q)}f(m)dm\Big)ds
\end{aligned}$$

$$
\begin{aligned}
&+(\alpha\xi-1)\int_0^1\frac{(1-s)^{q-1}}{\Gamma(q)}f(s)ds\Big]\Big| \\
&\le \left|\frac{1}{\Gamma(q)}\int_0^{t'}[(t''-s)^{q-1}-(t'-s)^{q-1}]f(s)ds\right| \\
&+\left|\frac{1}{\Gamma(q)}\int_{t'}^{t''}(t''-s)^{q-1}f(s)ds\right| \\
&+\left|\frac{(t'')^{m-1}-(t')^{m-1}}{\Delta}\Big[\alpha(\beta\eta-1)\int_0^{\xi}\Big(\int_0^s\frac{(s-m)^{q-1}}{\Gamma(q)}f(m)dm\Big)ds\Big]\right| \\
&+\left|\frac{(t'')^{m-1}-(t')^{m-1}}{\Delta}\Big[\beta(\alpha\xi-1)\int_0^{\eta}\Big(\int_0^s\frac{(s-m)^{q-1}}{\Gamma(q)}f(m)dm\Big)ds\Big]\right| \\
&+\left|\frac{(t'')^{m-1}-(t')^{m-1}}{\Delta}\Big[(\alpha\xi-1)\int_0^1\frac{(1-s)^{q-1}}{\Gamma(q)}f(s)ds\Big]\right|.
\end{aligned}
$$

Obviously the right hand side of the above inequality tends to zero independently of $x\in B_{r'}$ as $t''-t'\to 0$. As Ω satisfies the above three assumptions, therefore it follows by the Arzelá-Ascoli theorem that $\Omega: C([0,1],\mathbb{R})\to\mathcal{P}(C([0,1],\mathbb{R}))$ is completely continuous.

In our next step, we show that Ω has a closed graph. Let $x_n\to x_*, h_n\in\Omega(x_n)$ and $h_n\to h_*$. Then we need to show that $h_*\in\Omega(x_*)$. Associated with $h_n\in\Omega(x_n)$, there exists $f_n\in S_{F,x_n}$ such that for each $t\in[0,1]$,

$$
\begin{aligned}
h_n(t)&=\int_0^t\frac{(t-s)^{q-1}}{\Gamma(q)}f_n(s)ds+\frac{\alpha}{m\Delta}\Big[(\beta\eta^m-m)\int_0^{\xi}\Big(\int_0^s\frac{(s-m)^{q-1}}{\Gamma(q)}f_n(m)dm\Big)ds \\
&-\beta\xi^m\int_0^{\eta}\Big(\int_0^s\frac{(s-m)^{q-1}}{\Gamma(q)}f_n(m)dm\Big)ds+\xi^m\int_0^1\frac{(1-s)^{q-1}}{\Gamma(q)}f_n(s)ds\Big] \\
&-\frac{t^{m-1}}{\Delta}\Big[\alpha(\beta\eta-1)\int_0^{\xi}\Big(\int_0^s\frac{(s-m)^{q-1}}{\Gamma(q)}f_n(m)dm\Big)ds \\
&-\beta(\alpha\xi-1)\int_0^{\eta}\Big(\int_0^s\frac{(s-m)^{q-1}}{\Gamma(q)}f_n(m)dm\Big)ds \\
&+(\alpha\xi-1)\int_0^1\frac{(1-s)^{q-1}}{\Gamma(q)}f_n(s)ds\Big].
\end{aligned}
$$

Thus we have to show that there exists $f_*\in S_{F,x_*}$ such that for each $t\in[0,1]$,

$$
\begin{aligned}
h_*(t)&=\int_0^t\frac{(t-s)^{q-1}}{\Gamma(q)}f_*(s)ds+\frac{\alpha}{m\Delta}\Big[(\beta\eta^m-m)\int_0^{\xi}\Big(\int_0^s\frac{(s-m)^{q-1}}{\Gamma(q)}f_*(m)dm\Big)ds \\
&-\beta\xi^m\int_0^{\eta}\Big(\int_0^s\frac{(s-m)^{q-1}}{\Gamma(q)}f_*(m)dm\Big)ds+\xi^m\int_0^1\frac{(1-s)^{q-1}}{\Gamma(q)}f_*(s)ds\Big] \\
&-\frac{t^{m-1}}{\Delta}\Big[\alpha(\beta\eta-1)\int_0^{\xi}\Big(\int_0^s\frac{(s-m)^{q-1}}{\Gamma(q)}f_*(m)dm\Big)ds \\
&-\beta(\alpha\xi-1)\int_0^{\eta}\Big(\int_0^s\frac{(s-m)^{q-1}}{\Gamma(q)}f_*(m)dm\Big)ds
\end{aligned}
$$

$$+(\alpha\xi-1)\int_0^1\frac{(1-s)^{q-1}}{\Gamma(q)}f_*(s)ds\Big].$$

Let us consider the continuous linear operator $\Theta : L^1([0,1],\mathbb{R}) \to C([0,1],\mathbb{R})$ given by

$$\begin{aligned}f\mapsto\Theta(f)(t)=&\int_0^t\frac{(t-s)^{q-1}}{\Gamma(q)}f(s)ds\\&+\frac{\alpha}{m\Delta}\Big[(\beta\eta^m-m)\int_0^\xi\Big(\int_0^s\frac{(s-m)^{q-1}}{\Gamma(q)}f(m)dm\Big)ds\\&-\beta\xi^m\int_0^\eta\Big(\int_0^s\frac{(s-m)^{q-1}}{\Gamma(q)}f(m)dm\Big)ds+\xi^m\int_0^1\frac{(1-s)^{q-1}}{\Gamma(q)}f(s)ds\Big]\\&-\frac{t^{m-1}}{\Delta}\Big[\alpha(\beta\eta-1)\int_0^\xi\Big(\int_0^s\frac{(s-m)^{q-1}}{\Gamma(q)}f(m)dm\Big)ds\\&-\beta(\alpha\xi-1)\int_0^\eta\Big(\int_0^s\frac{(s-m)^{q-1}}{\Gamma(q)}f(m)dm\Big)ds\\&+(\alpha\xi-1)\int_0^1\frac{(1-s)^{q-1}}{\Gamma(q)}f(s)ds\Big].\end{aligned}$$

Observe that

$$\begin{aligned}&\|h_n(t)-h_*(t)\|\\&=\int_0^t\frac{(t-s)^{q-1}}{\Gamma(q)}(f_n(s)-f_*(s))ds\\&+\frac{\alpha}{m\Delta}\Big[(\beta\eta^m-m)\int_0^\xi\Big(\int_0^s\frac{(s-m)^{q-1}}{\Gamma(q)}(f_n(m)-f_*(m))dm\Big)ds\\&-\beta\xi^m\int_0^\eta\Big(\int_0^s\frac{(s-m)^{q-1}}{\Gamma(q)}(f_n(m)-f_*(m))dm\Big)ds+\xi^m\int_0^1\frac{(1-s)^{q-1}}{\Gamma(q)}f(s)ds\Big]\\&-\frac{t^{m-1}}{\Delta}\Big[\alpha(\beta\eta-1)\int_0^\xi\Big(\int_0^s\frac{(s-m)^{q-1}}{\Gamma(q)}(f_n(m)-f_*(m))dm\Big)ds\\&-\beta(\alpha\xi-1)\int_0^\eta\Big(\int_0^s\frac{(s-m)^{q-1}}{\Gamma(q)}(f_n(m)-f_*(m))dm\Big)ds\\&+(\alpha\xi-1)\int_0^1\frac{(1-s)^{q-1}}{\Gamma(q)}(f_n(s)-f_*(s))ds\Big]\Big\|\to 0.\end{aligned}$$

as $n\to\infty$. Thus, it follows by Lemma 1.2 that $\Theta\circ S_F$ is a closed graph operator. Further, we have $h_n(t)\in\Theta(S_{F,x_n})$. Since $x_n\to x_*$, therefore, we have

$$\begin{aligned}h_*(t)=&\int_0^t\frac{(t-s)^{q-1}}{\Gamma(q)}f_*(s)ds+\frac{\alpha}{m\Delta}\Big[(\beta\eta^m-m)\int_0^\xi\Big(\int_0^s\frac{(s-m)^{q-1}}{\Gamma(q)}f_*(m)dm\Big)ds\\&-\beta\xi^m\int_0^\eta\Big(\int_0^s\frac{(s-m)^{q-1}}{\Gamma(q)}f(m)dm\Big)ds+\xi^m\int_0^1\frac{(1-s)^{q-1}}{\Gamma(q)}f_*(s)ds\Big]\\&-\frac{t^{m-1}}{\Delta}\Big[\alpha(\beta\eta-1)\int_0^\xi\Big(\int_0^s\frac{(s-m)^{q-1}}{\Gamma(q)}f_*(m)dm\Big)ds\end{aligned}$$

$$-\beta(\alpha\xi-1)\int_0^\eta\Big(\int_0^s \frac{(s-m)^{q-1}}{\Gamma(q)}f_*(m)dm\Big)ds$$
$$+(\alpha\xi-1)\int_0^1 \frac{(1-s)^{q-1}}{\Gamma(q)}f_*(s)ds\Big]$$

for some $f_* \in S_{F,x_*}$.

Finally, we discuss *a priori* bounds on solutions. Let x be a solution of (3.16). Then there exists $f \in L^1([0,1],\mathbb{R})$ with $f \in S_{F,x}$ such that, for $t \in [0,1]$, we have

$$\begin{aligned} x(t) = &\int_0^t \frac{(t-s)^{q-1}}{\Gamma(q)}f(s)ds + \frac{\alpha}{m\Delta}\Big[(\beta\eta^m - m)\int_0^\xi\Big(\int_0^s \frac{(s-m)^{q-1}}{\Gamma(q)}f(m)dm\Big)ds \\ &- \beta\xi^m\int_0^\eta\Big(\int_0^s \frac{(s-m)^{q-1}}{\Gamma(q)}f(m)dm\Big)ds + \xi^m\int_0^1 \frac{(1-s)^{q-1}}{\Gamma(q)}f(s)ds\Big] \\ &- \frac{t^{m-1}}{\Delta}\Big[\alpha(\beta\eta-1)\int_0^\xi\Big(\int_0^s \frac{(s-m)^{q-1}}{\Gamma(q)}f(m)dm\Big)ds \\ &- \beta(\alpha\xi-1)\int_0^\eta\Big(\int_0^s \frac{(s-m)^{q-1}}{\Gamma(q)}f(m)dm\Big)ds + (\alpha\xi-1)\int_0^1 \frac{(1-s)^{q-1}}{\Gamma(q)}f(s)ds\Big]. \end{aligned}$$

In view of (3.4.2), and using the computations in second step above, for each $t \in [0,1]$, we obtain

$$|x(t)| \le \frac{1}{\Gamma(q+1)}\left(1+\frac{\vartheta_1+\vartheta_2}{m|\Delta|(q+1)}\right)\|p\|\psi(\|x\|).$$

Consequently, we have

$$\frac{\|x\|}{\dfrac{1}{\Gamma(q+1)}\left(1+\dfrac{\vartheta_1+\vartheta_2}{m|\Delta|(q+1)}\right)\|p\|\psi(\|x\|)} \le 1.$$

In view of (3.4.3), there exists M such that $\|x\| \neq M$. Let us set

$$U = \{x \in C([0,1],\mathbb{R}) : \|x\| < M+1\}.$$

Note that the operator $\Omega : \overline{U} \to \mathcal{P}(C([0,1],\mathbb{R}))$ is upper semi-continuous and completely continuous. From the choice of U, there is no $x \in \partial U$ such that $x \in \mu\Omega(x)$ for some $\mu \in (0,1)$. Consequently, by the nonlinear alternative of Leray-Schauder type (Theorem 1.8), we deduce that Ω has a fixed point $x \in \overline{U}$ which is a solution of the problem (3.16). This completes the proof. □

Example 3.4. Consider the nonlocal fractional inclusion boundary value problem

$$\begin{cases} {}^cD^{3/2}x(t) \in F(t,x(t)), \quad t \in [0,1], \\ x(0) = \displaystyle\int_0^{1/3} x(s)ds, \quad x(1) = \int_0^{2/3} x(s)ds, \end{cases} \tag{3.18}$$

where $m=2, q=3/2,\ \alpha=1,\ \beta=1,\ \xi=1/3,\ \eta=2/3$ and $F:[0,1]\times\mathbb{R}\to\mathcal{P}(\mathbb{R})$ is a multi-valued map given by

$$x \to F(t,x) = \left[\frac{|x|^3}{|x|^3+3}+t^3+5,\ \frac{|x|}{|x|+1}+t+1\right].$$

For $f \in F$, we have

$$|f| \leq \max\left(\frac{|x|^3}{|x|^3+3}+t^3+5, \frac{|x|}{|x|+1}+t+1\right) \leq 7, \quad x \in \mathbb{R}.$$

Thus,

$$\|F(t,x)\|_{\mathcal{P}} := \sup\{|y| : y \in F(t,x)\} \leq 7 = p(t)\psi(\|x\|), \quad x \in \mathbb{R},$$

with $p(t) = 1$, $\psi(\|x\|) = 7$. Further, using the condition (3.17), it is found that $M > 13.6548017875$. Clearly, all the conditions of Theorem 3.4 are satisfied. So there exists at least one solution of the problem (3.18) on $[0,1]$.

3.1.3.2 *The lower semi-continuous case*

As a next result, we study the case when F is not necessarily convex valued. Our strategy to deal with this problems is based on the nonlinear alternative of Leray-Schauder type together with the selection theorem of Bressan and Colombo (Lemma 1.3) for lower semi-continuous maps with decomposable values.

Theorem 3.5. *Assume that (3.4.2), (3.4.3) and the following conditions hold:*

(3.5.1) $F : [0,1]\times\mathbb{R} \to \mathcal{P}(\mathbb{R})$ *is a non-empty compact-valued multi-valued map such that:*

- *(a)* $(t,x) \longmapsto F(t,x)$ *is* $\mathcal{L}\otimes\mathcal{B}$ *measurable,*
- *(b)* $x \longmapsto F(t,x)$ *is lower semi-continuous for each* $t \in [0,1]$.

Then the boundary value problem (3.16) has at least one solution on $[0,1]$.

Proof. It follows from (3.5.1) and (3.5.2) that F is of l.s.c. type. Then from Lemma 1.3, there exists a continuous function $f : C([0,1],\mathbb{R}) \to L^1([0,1],\mathbb{R})$ such that $f(x) \in \mathcal{F}(x)$ for all $x \in C([0,1],\mathbb{R})$. Consider the problem

$$\begin{cases} {}^cD^q x(t) = f(x(t)), \quad 0 < t < 1, \quad 1 \;\; m-1 < q \leq m, \\ x(0) = \alpha \displaystyle\int_0^{\xi} x(s)ds, \;\; x'(0) = 0, \;\; x''(0) = 0, \ldots, x^{(m-2)}(0) = 0, \\ x(1) = \beta \displaystyle\int_0^{\eta} x(s)ds, \quad 0 < \xi, \eta < 1. \end{cases} \tag{3.19}$$

Observe that $x \in C^2([0,1])$ is a solution of (3.2), so x is a solution to the problem (3.16). In order to transform the problem (3.2) into a fixed point problem, we define the operator $\overline{\Omega}$ as

$$\begin{aligned} \overline{\Omega}x(t) = & \int_0^t \frac{(t-s)^{q-1}}{\Gamma(q)} f(x(s))ds \\ & + \frac{\alpha}{m\Delta}\Big[(\beta\eta^m - m)\int_0^{\xi}\Big(\int_0^s \frac{(s-m)^{q-1}}{\Gamma(q)} f(x(m))dm\Big)ds \\ & - \beta\xi^m \int_0^{\eta}\Big(\int_0^s \frac{(s-m)^{q-1}}{\Gamma(q)} f(x(m))dm\Big)ds + \xi^m \int_0^1 \frac{(1-s)^{q-1}}{\Gamma(q)} f(x(s))ds\Big] \end{aligned}$$

$$-\frac{t^{m-1}}{\Delta}\Big[\alpha(\beta\eta-1)\int_0^{\xi}\Big(\int_0^{s}\frac{(s-m)^{q-1}}{\Gamma(q)}f(x(m))dm\Big)ds$$
$$-\beta(\alpha\xi-1)\int_0^{\eta}\Big(\int_0^{s}\frac{(s-m)^{q-1}}{\Gamma(q)}f(x(m))dm\Big)ds$$
$$+(\alpha\xi-1)\int_0^{1}\frac{(1-s)^{q-1}}{\Gamma(q)}f(x(s))ds\Big].$$

It can easily be shown that $\overline{\Omega}$ is continuous and completely continuous. The remaining part of the proof is similar to that of Theorem 3.4. So we omit it. This completes the proof. □

3.1.3.3 *The Lipschitz case*

Now we prove the existence of solutions for the problem (3.16) with a non-convex valued right hand side by applying a fixed point theorem for multi-valued maps due to Covitz and Nadler (Theorem 1.19).

Theorem 3.6. *Assume that the following conditions hold:*

(3.6.1) $F:[0,1]\times\mathbb{R}\to\mathcal{P}_{cp}(\mathbb{R})$ *is such that* $F(.,x):[0,1]\to\mathcal{P}_{cp}(\mathbb{R})$ *is measurable for each* $x\in\mathbb{R}$.

(3.6.2) $H_d(F(t,x),F(t,\bar{x}))\le\lambda(t)|x-\bar{x}|$ *for almost all* $t\in[0,1]$ *and* $x,\bar{x}\in\mathbb{R}$ *with* $\lambda\in C([0,1],\mathbb{R}^+)$ *and* $d(0,F(t,0))\le\lambda(t)$ *for almost all* $t\in[0,1]$.

Then the boundary value problem (3.16) has at least one solution on $[0,1]$ *if*

$$\frac{1}{\Gamma(q+1)}\left(1+\frac{\vartheta_1+\vartheta_2}{m|\Delta|(q+1)}\right)\|\lambda\|<1.$$

Proof. Observe that the set $S_{F,x}$ is non-empty for each $x\in C([0,1],\mathbb{R})$ by the assumption (3.6.1), so F has a measurable selection (see Theorem III.6 [78]). Now we show that the operator Ω satisfies the assumptions of Theorem 1.19. To show that $\Omega(x)\in\mathcal{P}_{cl}((C[0,1],\mathbb{R}))$ for each $x\in C([0,1],\mathbb{R})$, let $\{u_n\}_{n\ge0}\in\Omega(x)$ be such that $u_n\to u$ $(n\to\infty)$ in $C([0,1],\mathbb{R})$. Then $u\in C([0,1],\mathbb{R})$ and there exists $v_n\in S_{F,x}$ such that, for each $t\in[0,1]$,

$$u_n(t)=\int_0^t\frac{(t-s)^{q-1}}{\Gamma(q)}v_n(s)ds+\frac{\alpha}{m\Delta}\Big[(\beta\eta^m-m)\int_0^{\xi}\Big(\int_0^{s}\frac{(s-m)^{q-1}}{\Gamma(q)}v_n(m)dm\Big)ds$$
$$-\beta\xi^m\int_0^{\eta}\Big(\int_0^{s}\frac{(s-m)^{q-1}}{\Gamma(q)}v_n(m)dm\Big)ds+\xi^m\int_0^{1}\frac{(1-s)^{q-1}}{\Gamma(q)}v_n(s)ds\Big]$$
$$-\frac{t^{m-1}}{\Delta}\Big[\alpha(\beta\eta-1)\int_0^{\xi}\Big(\int_0^{s}\frac{(s-m)^{q-1}}{\Gamma(q)}v_n(m)dm\Big)ds$$
$$-\beta(\alpha\xi-1)\int_0^{\eta}\Big(\int_0^{s}\frac{(s-m)^{q-1}}{\Gamma(q)}v_n(m)dm\Big)ds$$
$$+(\alpha\xi-1)\int_0^{1}\frac{(1-s)^{q-1}}{\Gamma(q)}v_n(s)ds\Big].$$

As F has compact values, we pass onto a subsequence to obtain that v_n converges to v in $L^1([0,1],\mathbb{R})$. Thus, $v \in S_{F,x}$ and for each $t \in [0,1]$,

$$\begin{aligned} u_n(t) \to u(t) &= \int_0^t \frac{(t-s)^{q-1}}{\Gamma(q)} v(s)ds \\ &+ \frac{\alpha}{m\Delta}\Big[(\beta\eta^m - m)\int_0^\xi \Big(\int_0^s \frac{(s-m)^{q-1}}{\Gamma(q)} v(m)dm\Big)ds \\ &-\beta\xi^m \int_0^\eta \Big(\int_0^s \frac{(s-m)^{q-1}}{\Gamma(q)} v(m)dm\Big)ds + \xi^m \int_0^1 \frac{(1-s)^{q-1}}{\Gamma(q)} v(s)ds\Big] \\ &-\frac{t^{m-1}}{\Delta}\Big[\alpha(\beta\eta - 1)\int_0^\xi \Big(\int_0^s \frac{(s-m)^{q-1}}{\Gamma(q)} v(m)dm\Big)ds \\ &-\beta(\alpha\xi - 1)\int_0^\eta \Big(\int_0^s \frac{(s-m)^{q-1}}{\Gamma(q)} v(m)dm\Big)ds \\ &+(\alpha\xi - 1)\int_0^1 \frac{(1-s)^{q-1}}{\Gamma(q)} f(s)ds\Big]. \end{aligned}$$

Hence, $u \in \Omega(x)$.

Next we show that there exists $\gamma < 1$ such that

$$H_d(\Omega(x), \Omega(\bar{x})) \le \gamma\|x - \bar{x}\| \quad \text{for each} \quad x, \bar{x} \in C([0,1],\mathbb{R}).$$

Let $x, \bar{x} \in C([0,1],\mathbb{R})$ and $h_1 \in \Omega(x)$. Then there exists $v_1(t) \in F(t, x(t))$ such that, for each $t \in [0,1]$,

$$\begin{aligned} h_1(t) &= \int_0^t \frac{(t-s)^{q-1}}{\Gamma(q)} v_1(s)ds + \frac{\alpha}{m\Delta}\Big[(\beta\eta^m - m)\int_0^\xi \Big(\int_0^s \frac{(s-m)^{q-1}}{\Gamma(q)} v_1(m)dm\Big)ds \\ &-\beta\xi^m \int_0^\eta \Big(\int_0^s \frac{(s-m)^{q-1}}{\Gamma(q)} v_1(m)dm\Big)ds + \xi^m \int_0^1 \frac{(1-s)^{q-1}}{\Gamma(q)} v_1(s)ds\Big] \\ &-\frac{t^{m-1}}{\Delta}\Big[\alpha(\beta\eta - 1)\int_0^\xi \Big(\int_0^s \frac{(s-m)^{q-1}}{\Gamma(q)} v_1(m)dm\Big)ds \\ &-\beta(\alpha\xi - 1)\int_0^\eta \Big(\int_0^s \frac{(s-m)^{q-1}}{\Gamma(q)} v_1(m)dm\Big)ds \\ &+(\alpha\xi - 1)\int_0^1 \frac{(1-s)^{q-1}}{\Gamma(q)} v_1(s)ds\Big]. \end{aligned}$$

By (3.6.2), we have

$$H_d(F(t,x), F(t,\bar{x})) \le \lambda(t)|x(t) - \bar{x}(t)|.$$

So, there exists $w \in F(t, \bar{x}(t))$ such that

$$|v_1(t) - w| \le \lambda(t)|x(t) - \bar{x}(t)|, \quad t \in [0,1].$$

Define $U : [0,1] \to \mathcal{P}(\mathbb{R})$ by

$$U(t) = \{w \in \mathbb{R} : |v_1(t) - w| \le \lambda(t)|x(t) - \bar{x}(t)|\}.$$

Since the multi-valued operator $V(t) \cap F(t, \bar{x}(t))$ is measurable (Proposition III.4 [78]), there exists a function $v_2(t)$ which is a measurable selection for V. So $v_2(t) \in F(t, \bar{x}(t))$ and for each $t \in [0,1]$, we have $|v_1(t) - v_2(t)| \leq \lambda(t)|x(t) - \bar{x}(t)|$.

For each $t \in [0,1]$, let us define

$$\begin{aligned} h_2(t) = &\int_0^t \frac{(t-s)^{q-1}}{\Gamma(q)} v_2(s)ds + \frac{\alpha}{m\Delta}\Big[(\beta\eta^m - m)\int_0^\xi \Big(\int_0^s \frac{(s-m)^{q-1}}{\Gamma(q)} v_2(m)dm\Big)ds \\ &-\beta\xi^m \int_0^\eta \Big(\int_0^s \frac{(s-m)^{q-1}}{\Gamma(q)} v_2(m)dm\Big)ds + \xi^m \int_0^1 \frac{(1-s)^{q-1}}{\Gamma(q)} v_2(s)ds\Big] \\ &-\frac{t^{m-1}}{\Delta}\Big[\alpha(\beta\eta - 1)\int_0^\xi \Big(\int_0^s \frac{(s-m)^{q-1}}{\Gamma(q)} v_2(m)dm\Big)ds \\ &-\beta(\alpha\xi - 1)\int_0^\eta \Big(\int_0^s \frac{(s-m)^{q-1}}{\Gamma(q)} v_2(m)dm\Big)ds \\ &+(\alpha\xi - 1)\int_0^1 \frac{(1-s)^{q-1}}{\Gamma(q)} f(s)ds\Big]. \end{aligned}$$

Thus,

$$\begin{aligned} |h_1(t) - h_2(t)| = &\int_0^t \frac{(t-s)^{q-1}}{\Gamma(q)} |v_1(s) - v_2(s)|ds \\ &+\frac{|\alpha|}{m|\Delta|}\Big[|\beta\eta^m - m|\int_0^\xi \Big(\int_0^s \frac{(s-m)^{q-1}}{\Gamma(q)} |v_1(m) - v_2(m)|dm\Big)ds \\ &+|\beta|\xi^m \int_0^\eta \Big(\int_0^s \frac{(s-m)^{q-1}}{\Gamma(q)} |v_1(m) - v_2(m)|dm\Big)ds \\ &+\xi^m \int_0^1 \frac{(1-s)^{q-1}}{\Gamma(q)} |v_1(s) - v_2(s)|ds\Big] \\ &+\frac{1}{|\Delta|}\Big[|\alpha(\beta\eta - 1)|\int_0^\xi \Big(\int_0^s \frac{(s-m)^{q-1}}{\Gamma(q)} |v_1(m) - v_2(m)|dm\Big)ds \\ &+|\beta(\alpha\xi - 1)|\int_0^\eta \Big(\int_0^s \frac{(s-m)^{q-1}}{\Gamma(q)} |v_1(m) - v_2(m)|dm\Big)ds \\ &+|(\alpha\xi - 1)|\int_0^1 \frac{(1-s)^{q-1}}{\Gamma(q)} |v_1(s) - v_2(s)|ds\Big]. \end{aligned}$$

Hence,

$$\|h_1 - h_2\| \leq \frac{1}{\Gamma(q+1)}\left(1 + \frac{\vartheta_1 + \vartheta_2}{m|\Delta|(q+1)}\right)\|\lambda\|\|x - \overline{x}\|.$$

Analogously, interchanging the roles of x and $\overline{x}$, we obtain

$$\begin{aligned} H_d(\Omega(x), \Omega(\bar{x})) &\leq \gamma\|x - \bar{x}\| \\ &\leq \frac{1}{\Gamma(q+1)}\left(1 + \frac{\vartheta_1 + \vartheta_2}{m|\Delta|(q+1)}\right)\|\lambda\|\|x - \overline{x}\|. \end{aligned}$$

Since Ω is a contraction, it follows by Theorem 1.19 that Ω has a fixed point x which is a solution of (3.16). This completes the proof. □

3.2 Caputo type fractional differential equations and inclusions with nonlocal Riemann-Liouville integral boundary conditions

In this section, we study nonlocal fractional boundary value problems for fractional differential equations and inclusions with nonlocal Riemann-Liouville integral boundary conditions.

3.2.1 *Caputo type fractional differential equations with nonlocal Riemann-Liouville integral boundary conditions*

In this section, we discuss the existence of solutions for a fractional differential equations of order $q \in (1, 2]$ with four-point nonlocal Riemann-Liouville fractional integral boundary conditions of different orders given by

$$\begin{cases} {}^cD^q x(t) = f(t, x(t)), \quad 1 < q \leq 2, \quad t \in [0, 1], \\ x(0) = aI^\beta x(\eta) = a\displaystyle\int_0^\eta \frac{(\eta - s)^{\beta-1}}{\Gamma(\beta)} x(s)ds, \quad \beta > 0, \\ x(1) = bI^\alpha x(\sigma) = b\displaystyle\int_0^\sigma \frac{(\sigma - s)^{\alpha-1}}{\Gamma(\alpha)} x(s)ds, \quad \alpha > 0, \end{cases} \tag{3.20}$$

where ${}^cD^q$ denotes the Caputo fractional derivative operator of order q, f is a given continuous function, and a, b, η, σ are real constants with $0 < \eta, \sigma < 1$.

Lemma 3.3. *Let $\nu \neq 0$. For any $y \in C[0, 1]$, the unique solution of the linear fractional boundary value problem*

$$\begin{cases} {}^cD^q x(t) = y(t), \quad 1 < q \leq 2, \; t \in [0, 1], \\ x(0) = a\displaystyle\int_0^\eta \frac{(\eta - s)^{\beta-1}}{\Gamma(\beta)} x(s)ds, \\ x(1) = b\displaystyle\int_0^\sigma \frac{(\sigma - s)^{\alpha-1}}{\Gamma(\alpha)} x(s)ds, \quad \beta, \alpha > 0, \end{cases} \tag{3.21}$$

is

$$x(t) = I^q y(t) + (\nu_1 - \nu_4 t)I^{q+\beta} y(\eta) + (\nu_2 + \nu_3 t)(bI^{q+\alpha} y(\sigma) - I^q y(1)), \tag{3.22}$$

where

$$\begin{aligned} \nu_1 &= \frac{a}{\nu}\Big(1 - \frac{b\sigma^{\alpha+1}}{\Gamma(\alpha + 2)}\Big), \quad \nu_2 = \frac{a\eta^{\beta+1}}{\nu\Gamma(\beta + 2)}, \\ \nu_3 &= \frac{1}{\nu}\Big(1 - \frac{a\eta^\beta}{\Gamma(\beta + 1)}\Big), \quad \nu_4 = \frac{a}{\nu}\Big(1 - \frac{b\sigma^\alpha}{\Gamma(\alpha + 1)}\Big), \\ \nu &= \Big(1 - \frac{a\eta^\beta}{\Gamma(\beta + 1)}\Big)\Big(1 - \frac{b\sigma^{\alpha+1}}{\Gamma(\alpha + 2)}\Big) + \frac{a\eta^{\beta+1}}{\Gamma(\beta + 2)}\Big(1 - \frac{b\sigma^\alpha}{\Gamma(\alpha + 1)}\Big). \end{aligned} \tag{3.23}$$

Proof. For some constants $c_0, c_1 \in \mathbb{R}$ and $1 < q \leq 2$, the general solution of ${}^cD^q x(t) = y(t)$ can be written as

$$x(t) = I^q y(t) + c_0 + c_1 t. \tag{3.24}$$

Using the boundary conditions of (3.21) in (3.24), we find a system of equations:

$$\begin{aligned} &\Big(1-\frac{a\eta^{\beta}}{\Gamma(\beta+1)}\Big)c_0-\frac{a\eta\beta+1}{\Gamma(\beta+2)}c_1=aI^{q+\beta}y(\eta)\\ &\Big(1-\frac{b\sigma^{\alpha}}{\Gamma(\alpha+1)}\Big)c_0+\Big(1-\frac{b\sigma^{\alpha+1}}{\Gamma(\alpha+2)}\Big)c_1=bI^{q+\alpha}y(\sigma)-I^{q}y(1), \end{aligned} \tag{3.25}$$

which, on solving, yields

$$c_0=\frac{1}{\nu}\bigg\{a\Big(1-\frac{b\sigma^{\alpha+1}}{\Gamma(\alpha+2)}\Big)I^{q+\beta}y(\eta)+\frac{a\eta^{\beta+1}}{\Gamma(\beta+2)}\Big(bI^{q+\alpha}y(\sigma)-I^{q}y(1)\Big)\bigg\}$$

and

$$c_1=\frac{1}{\nu}\bigg\{a\Big(\frac{b\sigma^{\alpha}}{\Gamma(\alpha+1)}-1\Big)I^{q+\beta}y(\eta)+\Big(1-\frac{a\eta^{\beta}}{\Gamma(\beta+1)}\Big)\Big(bI^{q+\alpha}y(\sigma)-I^{q}y(1)\Big)\bigg\}.$$

Substituting the values of c_0, c_1 in (3.24), we get (3.22).

We note that the solution $x(t)$ given by (3.22) satisfies the problem (3.21). It is easy to verify that the solution $x(t)$ satisfies the fractional differential equation of the problem (3.21). However, we will show that the solution (3.22) satisfies the boundary conditions of (3.21). Precisely, from the condition

$$x(0)=a\int_0^{\eta}\frac{(\eta-s)^{\beta-1}}{\Gamma(\beta)}x(s)ds,$$

we get

$$\begin{aligned} \nu_1I^{q+\beta}y(\eta)+\nu_2\big(bI^{q+\alpha}y(\sigma)-I^{q}y(1)\big)&=a\Big(1+\frac{\nu_1\eta^{\beta}}{\Gamma(\beta+1)}-\frac{\nu_4\eta^{\beta+1}}{\Gamma(\beta+2)}\Big)I^{q+\beta}y(\eta)\\ &+a\Big(\frac{\nu_2\eta^{\beta}}{\Gamma(\beta+1)}+\frac{\nu_3\eta^{\beta+1}}{\Gamma(\beta+2)}\Big)(bI^{q+\alpha}y(\sigma)-I^{q}y(1)), \end{aligned}$$

which is true as

$$\nu_1=a\Big(1+\frac{\nu_1\eta^{\beta}}{\Gamma(\beta+1)}-\frac{\nu_4\eta^{\beta+1}}{\Gamma(\beta+2)}\Big),\quad \nu_2=a\Big(\frac{\nu_2\eta^{\beta}}{\Gamma(\beta+1)}+\frac{\nu_3\eta^{\beta+1}}{\Gamma(\beta+2)}\Big).$$

Similarly, by the condition

$$x(1)=b\int_0^{\sigma}\frac{(\sigma-s)^{\alpha-1}}{\Gamma(\alpha)}x(s)ds,$$

we have

$$\begin{aligned} &(\nu_1-\nu_4)I^{q+\beta}y(\eta)+b(\nu_2+\nu_3)I^{q+\alpha}y(\sigma)+(1-\nu_2-\nu_3)I^{q}y(1)\\ =\;&b\Big(\frac{\nu_1\sigma^{\alpha}}{\Gamma(\alpha+1)}-\frac{\nu_4\sigma^{\alpha+1}}{\Gamma(\alpha+2)}\Big)I^{q+\beta}y(\eta)+b\Big(1+\frac{b\nu_2\sigma^{\alpha}}{\Gamma(\alpha+1)}+\frac{b\nu_3\sigma^{\alpha+1}}{\Gamma(\alpha+2)}\Big)I^{q+\alpha}y(\sigma)\\ &-b\Big(\frac{\nu_2\sigma^{\alpha}}{\Gamma(\alpha+1)}+\frac{\nu_3\sigma^{\alpha+1}}{\Gamma(\alpha+2)}\Big)I^{q}y(1), \end{aligned}$$

which is valid in view of the relations

$$\nu_1-\nu_4=b\Big(\frac{\nu_1\sigma^{\alpha}}{\Gamma(\alpha+1)}-\frac{\nu_4\sigma^{\alpha+1}}{\Gamma(\alpha+2)}\Big);$$

$$b(\nu_2+\nu_3) = b\Big(1+\frac{b\nu_2\sigma^\alpha}{\Gamma(\alpha+1)}+\frac{b\nu_3\sigma^{\alpha+1}}{\Gamma(\alpha+2)}\Big);$$
$$1-\nu_2-\nu_3 = -b\Big(\frac{\nu_2\sigma^\alpha}{\Gamma(\alpha+1)}+\frac{\nu_3\sigma^{\alpha+1}}{\Gamma(\alpha+2)}\Big).$$

This shows the equivalence of the problem (3.21) and the integral equation (3.22). This completes the proof. □

3.2.2 *Existence and uniqueness results*

To define a fixed point problem equivalent to problem (3.20), let us define an operator $\mathfrak{F} : C([0,1],\mathbb{R}) \to C([0,1],\mathbb{R})$ as

$$\begin{aligned}(\mathfrak{F}x)(t) &= \int_0^t \frac{(t-s)^{q-1}}{\Gamma(q)} f(s,x(s))ds + (\nu_1 - t\nu_4)\int_0^\eta \frac{(\eta-s)^{q+\beta-1}}{\Gamma(q+\beta)} f(s,x(s))ds \\ &+(\nu_2+\nu_3 t)\Bigg\{ b\int_0^\sigma \frac{(\sigma-s)^{q+\alpha-1}}{\Gamma(q+\alpha)} f(s,x(s))ds - \int_0^1 \frac{(1-s)^{q-1}}{\Gamma(q)} f(s,x(s))ds\Bigg\},\end{aligned} \tag{3.26}$$

where $\nu_1,\ \nu_2,\ \nu_3$ and ν_4 are given by (3.23).

In view of Lemma 3.3 and (3.26), we obtain a fixed point problem $x = \mathfrak{F}x$. Thus it follows that problem (3.20) has a solution if and only if we show that the operator $\mathfrak{F}$ has a fixed point.

For computational convenience, we set

$$\Lambda = \frac{1+|\nu_2|+|\nu_3|}{\Gamma(q+1)} + \frac{\eta^{q+\beta}(|\nu_1|+|\nu_4|)}{\Gamma(q+\beta+1)} + \frac{b\sigma^{q+\alpha}(|\nu_2|+|\nu_3|)}{\Gamma(q+\alpha+1)}. \tag{3.27}$$

Now we are in a position to present the main results.

Theorem 3.7. *Let $f : [0,1]\times\mathbb{R} \to \mathbb{R}$ be a continuous function satisfying the assumption*

(3.7.1) $|f(t,x)-f(t,y)| \le L|x-y|, \quad \forall t\in[0,1], \quad x,y\in\mathbb{R}.$

Then problem (3.20) has a unique solution provided that $L\Lambda < 1$, where Λ is given by (3.27).

Proof. Let us set $\sup_{t\in[0,1]}|f(t,0)| = M$ and define $B_r = \{x\in C([0,1],\mathbb{R}) : |x| \le r\}$, where

$$r \ge M\Lambda/(1-L\Lambda).$$

As a first step, we show that $\mathfrak{F}B_r \subset B_r$. For $x\in B_r$, we have

$$\begin{aligned}|(\mathfrak{F}x)(t)| \le \sup_{t\in[0,1]}\Bigg\{ &\int_0^t \frac{(t-s)^{q-1}}{\Gamma(q)}(|f(s,x(s))-f(s,0)|+|f(s,0)|)ds \\ &+|\nu_1-\nu_4 t|\int_0^\eta \frac{(\eta-s)^{q+\beta-1}}{\Gamma(q+\beta)}(|f(s,x(s))-f(s,0)|+|f(s,0)|)ds\end{aligned}$$

$$
\begin{aligned}
&+|\nu_2+\nu_3 t|\Big(b\int_0^\sigma \frac{(\sigma-s)^{q+\alpha-1}}{\Gamma(q+\alpha)}(|f(s,x(s))-f(s,0)|+|f(s,0)|)ds\\
&+\int_0^1 \frac{(1-s)^{q-1}}{\Gamma(q)}(|f(s,x(s))-f(s,0)|+|f(s,0)|)ds\Big)\Big\}\\
&\leq (Lr+M)\sup_{t\in[0,1]}\Big\{\frac{t^q}{\Gamma(q+1)}+(|\nu_1|+|\nu_4|)\frac{\eta^{q+\beta}}{\Gamma(q+\beta+1)}\\
&+(|\nu_2|+|\nu_3|)\Big(\frac{b\sigma^{q+\alpha}}{\Gamma(q+\alpha+1)}+\frac{1}{\Gamma(q+1)}\Big)\Big\}\\
&\leq (Lr+M)\Big\{\frac{1+|\nu_2|+|\nu_3|}{\Gamma(q+1)}+\frac{\eta^{q+\beta}(|\nu_1|+|\nu_4|)}{\Gamma(q+\beta+1)}+\frac{b\sigma^{q+\alpha}(|\nu_2|+|\nu_3|)}{\Gamma(q+\alpha+1)}\Big\}\\
&\leq (Lr+M)\Lambda\\
&\leq r.
\end{aligned}
$$

Thus, $\mathfrak{F}B_r \subset B_r$. Now for $x, y \in C([0,1],\mathbb{R})$, we obtain

$$
\begin{aligned}
\|(\mathfrak{F}x)-(\mathfrak{F}y)\| \leq & \sup_{t\in[0,1]}\Big\{\int_0^t \frac{(t-s)^{q-1}}{\Gamma(q)}|f(s,x(s))-f(s,y(s))|ds\\
&+|\nu_1-\nu_4 t|\int_0^\eta \frac{(\eta-s)^{q+\beta-1}}{\Gamma(q+\beta)}|f(s,x(s))-f(s,y(s))|ds\\
&+|\nu_2+\nu_3 t|\Big(b\int_0^\sigma \frac{(\sigma-s)^{q+\alpha-1}}{\Gamma(q+\alpha)}|f(s,x(s))-f(s,y(s))|ds\\
&+\int_0^1 \frac{(1-s)^{q-1}}{\Gamma(q)}|f(s,x(s))-f(s,y(s))|ds\Big)\Big\}\\
\leq & L\sup_{t\in[0,1]}\Big\{\frac{t^q}{\Gamma(q+1)}+(|\nu_1|+|\nu_4|)\frac{\eta^{q+\beta}}{\Gamma(q+\beta+1)}\\
&+(|\nu_2|+|\nu_3|)\left(\frac{b\sigma^{q+\alpha}}{\Gamma(q+\alpha+1)}+\frac{1}{\Gamma(q+1)}\right)\Big\}\|x-y\|\\
\leq & L\Lambda\|x-y\|.
\end{aligned}
$$

By the assumption $L < 1/\Lambda$, it follows that the operator $\mathfrak{F}$ is a contraction. Therefore, by Banach's contraction mapping principle, problem (3.20) has a unique solution. □

Example 3.5. Consider the following fractional boundary value problem

$$
\begin{cases}
{}^cD^{\frac{3}{2}}x(t) = \dfrac{1}{(t^2+2)^2}\Big(\dfrac{|x|}{1+|x|}\Big)+\sin^2 t, \quad t\in[0,1],\\
x(0) = I^{1/2}x(1/4), \quad x(1) = I^{1/3}x(1/2).
\end{cases}
\tag{3.28}
$$

Here, $q = 3/2$, $a = b = 1$, $\beta = 1/2$, $\alpha = 1/3$, $\eta = 1/4$, $\sigma = 1/2$ and $f(t,x) = \dfrac{1}{(t^2+2)^2}\dfrac{|x|}{1+|x|} + \sin^2 t$. Clearly, $L = 1/4$ as $|f(t,x) - f(t,y)| \leq (1/4)|x-y|$.

Furthermore,

$$L\Lambda = L\left\{ \frac{(1+|\nu_2|+|\nu_3|)}{\Gamma(q+1)} + \frac{\eta^{q+\beta}(|\nu_1|+|\nu_4|)}{\Gamma(q+\beta+1)} + \frac{b\sigma^{q+\alpha}(|\nu_2|+|\nu_3|)}{\Gamma(q+\alpha+1)} \right\} \simeq 0.4229 < 1.$$

Thus all the assumptions of Theorem 3.7 are satisfied. So, by the conclusion of Theorem 3.7, problem (3.28) has a unique solution on $[0,1]$.

Theorem 3.8. *Let $f : [0,1] \times \mathbb{R} \to \mathbb{R}$ be a continuous function satisfying the assumption (3.7.1). In addition, it is assumed that*

(3.8.1) $|f(t,x)| \le \mu(t)$, $\forall (t,x) \in [0,1] \times \mathbb{R}$ *where* $\mu \in C([0,1], \mathbb{R}^+)$;
(3.8.2)

$$L\left\{ \frac{(|\nu_2|+|\nu_3|)}{\Gamma(q+1)} + \frac{\eta^{q+\beta}(|\nu_1|+|\nu_4|)}{\Gamma(q+\beta+1)} + \frac{b\sigma^{q+\alpha}(|\nu_2|+|\nu_3|)}{\Gamma(q+\alpha+1)} \right\} < 1.$$

Then problem (3.20) has at least one solution on $[0,1]$.

Proof. Denoting $\sup_{t\in[0,T]} |\mu(t)| = \|\mu\|$, we consider the set $B_r = \{x \in C([0,1],\mathbb{R}) : \|x\| \le r\}$, where $r \ge \|\mu\|\Lambda$, Λ is given by (3.27). For $t \in [0,1]$, let us introduce the operators Φ and Ψ on B_r as

$$\begin{aligned}
(\Phi x)(t) &= \int_0^t \frac{(t-s)^{q-1}}{\Gamma(q)} f(s,x(s))ds, \\
(\Psi x)(t) &= (\nu_1 - t\nu_4) \int_0^\eta \frac{(\eta-s)^{q+\beta-1}}{\Gamma(q+\beta)} f(s,x(s))ds \\
&\quad + (\nu_2+\nu_3 t)\left\{ b\int_0^\sigma \frac{(\sigma-s)^{q+\alpha-1}}{\Gamma(q+\alpha)} f(s,x(s))ds - \int_0^1 \frac{(1-s)^{q-1}}{\Gamma(q)} f(s,x(s))ds \right\}.
\end{aligned}$$

For $x,\ y \in B_r$, we find that $\|\Phi x + \Psi y\| \le \|\mu\|\Lambda \le r$ which implies that $\Phi x + \Psi y \in B_r$.

In view of (3.8.2), Ψ is a contraction. Continuity of f implies that the operator Φ is continuous. Also, Φ is uniformly bounded on B_r as

$$\|\Phi x\| \le \frac{\|\mu\|}{\Gamma(q+1)}.$$

Now, we prove the compactness of the operator Φ. By the condition (3.7.1), we define $\sup\{|f(t,x)| : (t,x) \in [0,1] \times B_r\} = f_M$. Then, for $t_1, t_2 \in [0,1]$ with $t_2 > t_1$, we have

$$|(\Phi x)(t_2) - (\Phi y)(t_1)| \le \frac{f_M}{\Gamma(q+1)}[2(t_2-t_1)^q + |t_2^q - t_1^q|],$$

which is independent of x and tends to zero as $t_2 \to t_1$. So Φ is relatively compact on B_r. Hence, by the Arzelá-Ascoli theorem, Φ is compact on B_r. Thus all the assumptions of Theorem 1.4 are satisfied. Therefore, problem (3.20) has at least one solution on $[0,1]$. This completes the proof. □

Theorem 3.9. *Assume that there exists a positive constant L_1 such that $|f(t,x)| \leq L_1$ for $t \in [0,1]$, $x \in \mathbb{R}$. Then problem (3.20) has at least one solution on $[0,1]$.*

Proof. In the first step, it will be shown that the operator $\mathfrak{F}$ defined by (3.26) is completely continuous. Observe that continuity of $\mathfrak{F}$ follows from that of f. Let $\Omega \subset C([0,1],\mathbb{R})$ be bounded. As before, for $x \in \Omega$, it can be shown that $\|(\mathfrak{F}x)\| \leq L_1\Lambda = L_2$. Furthermore, we have

$$\begin{aligned}
|(\mathfrak{F}x)'(t)| &\leq \Bigg| \int_0^t \frac{(t-s)^{q-2}}{\Gamma(q-1)} f(s,x(s))ds - \nu_4 \int_0^\eta \frac{(\eta-s)^{q+\beta-1}}{\Gamma(q+\beta)} f(s,x(s))ds \\
&\quad + \nu_3\Big(b\int_0^\sigma \frac{(\sigma-s)^{q+\alpha-1}}{\Gamma(q+\alpha)} f(s,x(s))ds - \int_0^1 \frac{(1-s)^{q-1}}{\Gamma(q)} f(s,x(s))ds \Big) \Bigg| \\
&\leq L_1 \left\{ \frac{t^{q-1}}{\Gamma(q)} + \frac{|\nu_4|\eta^{q+\beta}}{\Gamma(q+\beta+1)} + |\nu_3|\Big(\frac{b\sigma^{q+\alpha}}{\Gamma(q+\alpha+1)} + \frac{1}{\Gamma(q+1)} \Big) \right\} \\
&\leq L_1 \left\{ \frac{1}{\Gamma(q)} + \frac{|\nu_4|\eta^{q+\beta}}{\Gamma(q+\beta+1)} + |\nu_3|\Big(\frac{b\sigma^{q+\alpha}}{\Gamma(q+\alpha+1)} + \frac{1}{\Gamma(q+1)} \Big) \right\} \\
&= L_3.
\end{aligned}$$

Now, for $t_1, t_2 \in [0,1]$ with $t_1 < t_2$, we get

$$|(\mathfrak{F}x)(t_1) - (\mathfrak{F}x)(t_2)| \leq \int_{t_1}^{t_2} |(\mathfrak{F}x)'(s)|ds \leq L_3(t_2 - t_1).$$

Therefore, $\mathfrak{F}$ is equi-continuous. Thus, by Arzelá-Ascoli theorem, the operator $\mathfrak{F}$ is completely continuous.

Next, we consider the set $V = \{x \in C([0,1],\mathbb{R}) : x = \mu\mathfrak{F}x,\ 0 < \mu < 1\}$. In order to show that V is bounded, let $x \in V$ and $t \in [0,1]$. Then

$$\begin{aligned}
x(t) &= \int_0^t \frac{(t-s)^{q-1}}{\Gamma(q)} f(s,x(s))ds + (\nu_1 - \nu_4 t) \int_0^\eta \frac{(\eta-s)^{q+\beta-1}}{\Gamma(q+\beta)} f(s,x(s))ds \\
&\quad + (\nu_2 + \nu_3 t)\left\{ b\int_0^\sigma \frac{(\sigma-s)^{q+\alpha-1}}{\Gamma(q+\alpha)} f(s,x(s))ds - \int_0^1 \frac{(1-s)^{q-1}}{\Gamma(q)} f(s,x(s))ds \right\}
\end{aligned}$$

with $\|x\| = \sup_{t\in[0,1]} |\mu(\mathfrak{F}x)(t)| \leq L_1\Lambda = M_1$. Therefore, V is bounded. Hence, by Theorem 1.4, problem (3.20) has at least one solution. □

Example 3.6. Consider the fractional boundary value problem

$$\begin{cases} {}^cD^{3/2}x(t) = \dfrac{e^{-\cos^2 x(t)}\Big[1 + 3\cos 2t + 2\ln(2 + 3\sin^2 x(t))\Big]}{2 + |\sin x(t)|}, & t \in [0,1], \\ x(0) = I^{1/2}x(1/4), \quad x(1) = I^{1/3}x(1/2), \end{cases} \tag{3.29}$$

where, $q = 3/2$, $a = b = 1$, $\beta = 1/2$, $\alpha = 1/3$, $\eta = 1/4$, $\sigma = 1/2$. Clearly $|f(t,x)| \leq 2 + \ln 5 = L_1$. So the hypothesis of Theorem 3.9 holds. Hence, the conclusion of Theorem 3.9 applies to problem (3.29).

Our last existence result is based on Leray-Schauder Nonlinear Alternative (Theorem 1.8).

Theorem 3.10. *Suppose that $f : [0,1] \times \mathbb{R} \to \mathbb{R}$ is a continuous function and the following conditions hold:*

(3.10.1) there exist a function $p \in C([0,1], \mathbb{R}^+)$, and a non-decreasing function $\psi : \mathbb{R}^+ \to \mathbb{R}^+$ such that $|f(t,x)| \le p(t)\psi(\|x\|)$, $\forall (t,x) \in [0,1] \times \mathbb{R}$;

(3.10.2) there exists a constant $M > 0$ such that

$$M\left[\psi(M)\|p\|\left(\frac{1+|\nu_2|+|\nu_3|}{\Gamma(q+1)} + \frac{\eta^{q+\beta}(|\nu_1|+|\nu_4|)}{\Gamma(q+\beta+1)} + \frac{b\sigma^{q+\alpha}(|\nu_2|+|\nu_3|)}{\Gamma(q+\alpha+1)}\right)\right]^{-1} > 1.$$

Then problem (3.20) has at least one solution on $[0,1]$.

Proof. The proof consists of several steps.
First we show that the operator $\mathfrak{F} : C([0,1], \mathbb{R}) \to C([0,1], \mathbb{R})$, defined by (3.26), maps bounded sets into bounded sets in $C([0,1], \mathbb{R}^+)$. For a positive number r, let $B_r = \{x \in C([0,1], \mathbb{R}) : \|x\| \le r\}$ be a bounded set in $C([0,1], \mathbb{R})$. Then, for $x \in B_r$, we have

$$\begin{aligned}
\|(\mathfrak{F}x)\| \le & \sup_{t\in[0,1]} \left\{ \int_0^t \frac{(t-s)^{q-1}}{\Gamma(q)} p(s)\psi(\|x\|)ds \right. \\
& +(|\nu_1|+|\nu_4|)\left| \int_0^\eta \frac{(\eta-s)^{q+\beta-1}}{\Gamma(q+\beta)} p(s)\psi(\|x\|)ds \right. \\
& +(|\nu_2|+|\nu_3|)\Big(b\int_0^\sigma \frac{(\sigma-s)^{q+\alpha-1}}{\Gamma(q+\alpha)} p(s)\psi(\|x\|)ds \\
& \left. + \int_0^1 \frac{(1-s)^{q-1}}{\Gamma(q)} p(s)\psi(\|x\|)ds \Big) \right\} \\
& \le \|p\|\psi(\|x\|) \sup_{t\in[0,1]} \left\{ \frac{t^q}{\Gamma(q+1)} + (|\nu_1|+|\nu_4|)\frac{\eta^{q+\beta}}{\Gamma(q+\beta+1)} \right. \\
& \left. +(|\nu_2|+|\nu_3|)\left(\frac{b\sigma^{q+\alpha}}{\Gamma(q+\alpha+1)} + \frac{1}{\Gamma(q+1)}\right)\right\} \\
& \le \|p\|\psi(\|x\|)\left\{ \frac{1+|\nu_2|+|\nu_3|}{\Gamma(q+1)} + \frac{\eta^{q+\beta}(|\nu_1|+|\nu_4|)}{\Gamma(q+\beta+1)} + \frac{b\sigma^{q+\alpha}(|\nu_2|+|\nu_3|)}{\Gamma(q+\alpha+1)} \right\} \\
& \le \|p\|\psi(r)\Lambda,
\end{aligned}$$

where Λ is given by (3.27).

Next, we show that the operator $\mathfrak{F}$ maps bounded sets into equi-continuous sets of $C([0,1], \mathbb{R})$. Let $t_1, t_2 \in [0,1]$ with $t_1 < t_2$ and $x \in B_r$. Then we have

$$\begin{aligned}
& |(\mathfrak{F}x)(t_2) - (\mathfrak{F}x)(t_1)| \\
\le & \left| \int_0^{t_2} \frac{(t_2-s)^{q-1}}{\Gamma(q)} p(s)\psi(r)ds - \int_0^{t_1} \frac{(t_1-s)^{q-1}}{\Gamma(q)} p(s)\psi(r)ds \right.
\end{aligned}$$

$$-\nu_4(t_2-t_1)\int_0^{\eta}\frac{(\eta-s)^{q+\beta-1}}{\Gamma(q+\beta)}p(s)\psi(r)ds$$
$$-\nu_3(t_2-t_1)\Big(b\int_0^{\sigma}\frac{(\sigma-s)^{q+\alpha-1}}{\Gamma(q+\alpha)}p(s)\psi(r)ds-\int_0^1\frac{(1-s)^{q-1}}{\Gamma(q)}p(s)\psi(r)ds\Big)\Bigg|$$
$$\le \psi(r)\|p\|\Bigg[\int_{t_1}^{t_2}\left|\frac{(t_2-s)^{q-1}}{\Gamma(q)}\right|ds+\int_0^{t_1}\left|\frac{(t_1-s)^{q-1}-(t_2-s)^{q-1}}{\Gamma(q)}\right|ds$$
$$+|\nu_4|(t_2-t_1)\int_0^{\eta}\left|\frac{(\eta-s)^{q+\beta-1}}{\Gamma(q+\beta)}\right|ds$$
$$+|\nu_3|(t_2-t_1)\Big(b\int_0^{\sigma}\left|\frac{(\sigma-s)^{q+\alpha-1}}{\Gamma(q+\alpha)}\right|ds+\int_0^1\left|\frac{(1-s)^{q-1}}{\Gamma(q)}ds\right|\Big)\Bigg].$$

Clearly, the right hand side of the above inequality tends to zero independently of $x\in B_r$ as $t_2\to t_1$. Thus, it follows by the Arzelá-Ascoli theorem that the operator $\mathfrak{F}$ is completely continuous.

Let x be a solution of problem (3.20). Then, for $t\in[0,1]$ and $\lambda\in(0,1)$, it can be shown that

$$\|x\|\le\psi(\|x\|)\|p\|\left\{\frac{1+|\nu_2|+|\nu_3|}{\Gamma(q+1)}+\frac{\eta^{q+\beta}(|\nu_1|+|\nu_4|)}{\Gamma(q+\beta+1)}+\frac{b\sigma^{q+\alpha}(|\nu_2|+|\nu_3|)}{\Gamma(q+\alpha+1)}\right\},$$

which can alternatively be written as

$$\|x\|\left[\psi(\|x\|)\|p\|\left(\frac{1+|\nu_2|+|\nu_3|}{\Gamma(q+1)}+\frac{\eta^{q+\beta}(|\nu_1|+|\nu_4|)}{\Gamma(q+\beta+1)}+\frac{b\sigma^{q+\alpha}(|\nu_2|+|\nu_3|)}{\Gamma(q+\alpha+1)}\right)\right]^{-1}\le 1.$$

In view of (3.10.2), there exists M such that $\|x\|\ne M$. Let us set

$$U=\{x\in\mathcal{C}([0,1],\mathbb{R}):\|x\|<M+1\}.$$

Note that the operator $\mathfrak{F}:\overline{U}\to\mathcal{C}([0,1],\mathbb{R})$ is continuous and completely continuous. From the choice of U, there is no $x\in\partial U$ such that $x=\lambda\mathfrak{F}(x)$ for some $\lambda\in(0,1)$. In consequence, by Theorem 1.8, we deduce that $\mathfrak{F}$ has a fixed point $x\in\overline{U}$ which is a solution of problem (3.20). This completes the proof. □

3.2.3 *Fractional differential inclusions with four-point nonlocal Riemann-Liouville type integral boundary conditions*

In this subsection, we study a boundary value problem of fractional differential inclusions with nonlocal fractional integral boundary conditions given by

$$\begin{cases} {}^cD^qx(t)\in F(t,x(t)), \quad 0<t<1, \quad 1<q\le 2, \\ x(0)=aI^{\beta}x(\eta)=a\displaystyle\int_0^{\eta}\frac{(\eta-s)^{\beta-1}}{\Gamma(\beta)}x(s)ds, \quad \beta>0, \\ x(1)=bI^{\alpha}x(\sigma)=b\displaystyle\int_0^{\sigma}\frac{(\sigma-s)^{\alpha-1}}{\Gamma(\alpha)}x(s)ds, \quad \alpha>0, \end{cases} \tag{3.30}$$

where ${}^cD^q$ is the Capuo fractional derivative operator of order q, $F : [0,1] \times \mathbb{R} \to \mathcal{P}(\mathbb{R})$ is a multi-valued map, $\mathcal{P}(\mathbb{R})$ is the family of all subsets of $\mathbb{R}$, a, b are real constants and $\eta, \sigma \in (0,1)$.

Definition 3.2. A function $x \in C^2([0,1], \mathbb{R})$ is a solution of the problem (3.30) if $x(0) = aI^\beta x(\eta)$, $x(1) = bI^\alpha x(\sigma)$, and there exists a function $f \in L^1([0,1], \mathbb{R})$ such that $f(t) \in F(t, x(t))$ a.e. on $[0,1]$ and

$$\begin{aligned} x(t) = &\int_0^t \frac{(t-s)^{q-1}}{\Gamma(q)} f(s)ds + (\nu_1 - t\nu_4) \int_0^\eta \frac{(\eta - s)^{q+\beta-1}}{\Gamma(q+\beta)} f(s)ds \\ &+ (\nu_2 + \nu_3 t)\left\{ b \int_0^\sigma \frac{(\sigma - s)^{q+\alpha-1}}{\Gamma(q+\alpha)} f(s)ds - \int_0^1 \frac{(1-s)^{q-1}}{\Gamma(q)} f(s)ds \right\}, \quad t \in [0,1], \end{aligned}$$

where ν_1, ν_2, ν_3 and ν_4 are given by (3.23).

3.2.3.1 *The Carathéodory case*

Theorem 3.11. *Assume that:*

(3.11.1) $F : [0,1] \times \mathbb{R} \to \mathcal{P}(\mathbb{R})$ *is Carathéodory and has non-empty compact and convex values;*

(3.11.2) there exists a continuous non-decreasing function $\psi : [0, \infty) \to (0, \infty)$ *and a function* $p \in L^1([0,1], \mathbb{R}^+)$ *such that*

$$\|F(t,x)\|_{\mathcal{P}} := \sup\{|y| : y \in F(t,x)\} \le p(t)\psi(\|x\|) \text{ for each } (t,x) \in [0,1] \times \mathbb{R};$$

(3.11.3) there exists a constant $M > 0$ *such that*

$$\frac{M}{\psi(M)\left\{\frac{1 + |\nu_2| + |\nu_3|}{\Gamma(q)} + \frac{\eta^{q+\beta-1}(|\nu_1| + |\nu_4|)}{\Gamma(q+\beta)} + \frac{b\sigma^{q+\alpha-1}(|\nu_2| + |\nu_3|)}{\Gamma(q+\alpha)}\right\}\|p\|_{L^1}} > 1.$$

Then the boundary value problem (3.30) has at least one solution on $[0,1]$.

Proof. Define an operator $\Omega_F : C([0,1], \mathbb{R}) \to \mathcal{P}(C([0,1], \mathbb{R}))$ by

$$\Omega_F(x) = \left\{ \begin{array}{l} h \in C([0,1], \mathbb{R}) : \\ h(t) = \left\{ \begin{array}{l} \int_0^t \frac{(t-s)^{q-1}}{\Gamma(q)} f(s)ds + (\nu_1 - t\nu_4) \int_0^\eta \frac{(\eta - s)^{q+\beta-1}}{\Gamma(q+\beta)} f(s)ds \\ (\nu_2 + \nu_3 t)\left\{ b \int_0^\sigma \frac{(\sigma - s)^{q+\alpha-1}}{\Gamma(q+\alpha)} f(s)ds - \int_0^1 \frac{(1-s)^{q-1}}{\Gamma(q)} f(s)ds \right\}, \end{array} \right. \end{array} \right\}$$

for $f \in S_{F,x}$. We will show that Ω_F satisfies the assumptions of the nonlinear alternative of Leray-Schauder type (Theorem 1.16). The proof consists of several steps. As a first step, we show that Ω_F *is convex for each* $x \in C([0,1], \mathbb{R})$. This step is obvious since $S_{F,x}$ is convex (F has convex values), and therefore we omit the proof.

In the second step, we show that Ω_F *maps bounded sets (balls) into bounded sets in* $C([0,1],\mathbb{R})$. For a positive number ρ, let $B_\rho = \{x \in C([0,1],\mathbb{R}) : \|x\| \leq \rho\}$ be a bounded ball in $C([0,1],\mathbb{R})$. Then, for each $h \in \Omega_F(x), x \in B_\rho$, there exists $f \in S_{F,x}$ such that

$$h(t) = \int_0^t \frac{(t-s)^{q-1}}{\Gamma(q)} f(s)ds + (\nu_1 - t\nu_4)\int_0^\eta \frac{(\eta-s)^{q+\beta-1}}{\Gamma(q+\beta)} f(s)ds$$
$$+(\nu_2+\nu_3 t)\Big\{b\int_0^\sigma \frac{(\sigma-s)^{q+\alpha-1}}{\Gamma(q+\alpha)} f(s)ds - \int_0^1 \frac{(1-s)^{q-1}}{\Gamma(q)} f(s)ds\Big\}, \quad t\in[0,1].$$

Then

$$|h(t)| \leq \int_0^t \frac{(t-s)^{q-1}}{\Gamma(q)} |f(s)|ds + (|\nu_1|+|\nu_4|)\int_0^\eta \frac{(\eta-s)^{q+\beta-1}}{\Gamma(q+\beta)} |f(s)|ds$$
$$+(|\nu_2|+|\nu_3|)\Big\{b\int_0^\sigma \frac{(\sigma-s)^{q+\alpha-1}}{\Gamma(q+\alpha)} |f(s)|ds + \int_0^1 \frac{(1-s)^{q-1}}{\Gamma(q)} |f(s)|ds\Big\},$$

which, in view of(3.11.1), yields

$$\|h\| \leq \psi(\rho)\Big\{\frac{1+|\nu_2|+|\nu_3|}{\Gamma(q)} + \frac{\eta^{q+\beta-1}(|\nu_1|+|\nu_4|)}{\Gamma(q+\beta)} + \frac{b\sigma^{q+\alpha-1}(|\nu_2|+|\nu_3|)}{\Gamma(q+\alpha)}\Big\}\|p\|_{L^1}.$$

Now we show that Ω_F *maps bounded sets into equi-continuous sets of* $C([0,1],\mathbb{R})$. Let $t', t'' \in [0,1]$ with $t' < t''$ and $x \in B_\rho$. For each $h \in \Omega_F(x)$, we obtain

$$|h(t'') - h(t')|$$
$$\leq \left|\psi(\rho)\int_0^{t'}\left[\frac{(t''-s)^{q-1}-(t'-s)^{q-1}}{\Gamma(q)}\right]p(s)ds + \psi(\rho)\int_{t'}^{t''}\frac{(t''-s)^{q-1}}{\Gamma(q)}p(s)ds\right|$$
$$+|\nu_4(t''-t')|\psi(\rho)\int_0^\eta \frac{(\eta-s)^{q+\beta-1}}{\Gamma(q+\beta)}p(s)ds$$
$$+|\nu_3(t''-t')|\psi(\rho)\Big\{b\int_0^\sigma \frac{(\sigma-s)^{q+\alpha-1}}{\Gamma(q+\alpha)}p(s)ds + \int_0^1 \frac{(1-s)^{q-1}}{\Gamma(q)}p(s)ds\Big\}.$$

Obviously the right hand side of the above inequality tends to zero independently of $x \in B_\rho$ as $t'' - t' \to 0$. As Ω_F satisfies the above three assumptions, therefore it follows by the Ascoli-Arzelá theorem that $\Omega_F : C([0,1],\mathbb{R}) \to \mathcal{P}(C([0,1],\mathbb{R}))$ is completely continuous.

In our next step, we show that Ω_F *has a closed graph.* Let $x_n \to x_*, h_n \in \Omega_F(x_n)$ and $h_n \to h_*$. Then we need to show that $h_* \in \Omega_F(x_*)$. Associated with $h_n \in \Omega_F(x_n)$, there exists $f_n \in S_{F,x_n}$ such that for each $t \in [0,1]$,

$$h_n(t) = \int_0^t \frac{(t-s)^{q-1}}{\Gamma(q)} f_n(s)ds + (\nu_1 - t\nu_4)\int_0^\eta \frac{(\eta-s)^{q+\beta-1}}{\Gamma(q+\beta)} f_n(s)ds$$
$$+(\nu_2+\nu_3 t)\Big\{b\int_0^\sigma \frac{(\sigma-s)^{q+\alpha-1}}{\Gamma(q+\alpha)} f_n(s)ds - \int_0^1 \frac{(1-s)^{q-1}}{\Gamma(q)} f_n(s)ds\Big\}.$$

Thus it suffices to show that there exists $f_* \in S_{F,x_*}$ such that for each $t \in [0,1]$,

$$h_*(t) = \int_0^t \frac{(t-s)^{q-1}}{\Gamma(q)} f_*(s)ds + (\nu_1 - t\nu_4)\int_0^\eta \frac{(\eta-s)^{q+\beta-1}}{\Gamma(q+\beta)} f_*(s)ds$$

$$+(\nu_2+\nu_3 t)\Big\{b\int_0^\sigma \frac{(\sigma-s)^{q+\alpha-1}}{\Gamma(q+\alpha)}f_*(s)ds-\int_0^1\frac{(1-s)^{q-1}}{\Gamma(q)}f_*(s)ds\Big\}.$$

Let us consider the linear operator $\Theta : L^1([0,1],\mathbb{R}) \to C([0,1],\mathbb{R})$ given by

$$f\mapsto \Theta(f)(t)=\int_0^t \frac{(t-s)^{q-1}}{\Gamma(q)}f(s)ds+(\nu_1-t\nu_4)\int_0^\eta\frac{(\eta-s)^{q+\beta-1}}{\Gamma(q+\beta)}f(s)ds$$
$$+(\nu_2+\nu_3 t)\Big\{b\int_0^\sigma \frac{(\sigma-s)^{q+\alpha-1}}{\Gamma(q+\alpha)}f(s)ds-\int_0^1\frac{(1-s)^{q-1}}{\Gamma(q)}f(s)ds\Big\}.$$

Observe that

$$\|h_n-h_*\|$$
$$=\Big\|\int_0^t \frac{(t-s)^{q-1}}{\Gamma(q)}(f_n(s)-f_*(s))ds+(\nu_1-t\nu_4)\int_0^\eta\frac{(\eta-s)^{q+\beta-1}}{\Gamma(q+\beta)}(f_n(s)-f_*(s))ds$$
$$+(\nu_2+\nu_3 t)\Big\{b\int_0^\sigma \frac{(\sigma-s)^{q+\alpha-1}}{\Gamma(q+\alpha)}(f_n(s)-f_*(s))ds-\int_0^1\frac{(1-s)^{q-1}}{\Gamma(q)}(f_n(s)-f_*(s))ds\Big\}\Big\|$$
$$\to 0,\quad \text{as}\quad n\to\infty.$$

Thus, it follows by Lemma 1.2 that $\Theta\circ S_F$ is a closed graph operator. Further, we have $h_n(t)\in\Theta(S_{F,x_n})$. Since $x_n\to x_*$, therefore, we have

$$h_*(t)=\int_0^t \frac{(t-s)^{q-1}}{\Gamma(q)}f_*(s)ds+(\nu_1-t\nu_4)\int_0^\eta\frac{(\eta-s)^{q+\beta-1}}{\Gamma(q+\beta)}f_*(s)ds$$
$$+(\nu_2+\nu_3 t)\Big\{b\int_0^\sigma \frac{(\sigma-s)^{q+\alpha-1}}{\Gamma(q+\alpha)}f_*(s)ds-\int_0^1\frac{(1-s)^{q-1}}{\Gamma(q)}f_*(s)ds\Big\},\ t\in[0,1],$$

for some $f_*\in S_{F,x_*}$.

Finally, we show that there exists an open set $U\subseteq C([0,1],\mathbb{R})$ with $x\notin\Omega_F(x)$ for any $\lambda\in(0,1)$ and all $x\in\partial U$. Let $\lambda\in(0,1)$ and $x\in\lambda\Omega_F(x)$. Then there exists $f\in L^1([0,1],\mathbb{R})$ with $f\in S_{F,x}$ such that, for $t\in[0,1]$, we have

$$|x(t)|$$
$$=\Big|\int_0^t \frac{(t-s)^{q-1}}{\Gamma(q)}f(s)ds+(\nu_1-t\nu_4)\int_0^\eta\frac{(\eta-s)^{q+\beta-1}}{\Gamma(q+\beta)}f(s)ds$$
$$+(\nu_2+\nu_3 t)\Big\{b\int_0^\sigma \frac{(\sigma-s)^{q+\alpha-1}}{\Gamma(q+\alpha)}f(s)ds-\int_0^1\frac{(1-s)^{q-1}}{\Gamma(q)}f(s)ds\Big\}\Big|$$
$$\le\psi(\|x\|)\Big\{\frac{1+|\nu_2|+|\nu_3|}{\Gamma(q)}+\frac{\eta^{q+\beta-1}(|\nu_1|+|\nu_4|)}{\Gamma(q+\beta)}+\frac{b\sigma^{q+\alpha-1}(|\nu_2|+|\nu_3|)}{\Gamma(q+\alpha)}\Big\}\|p\|_{L^1}.$$

Consequently, we have

$$\frac{\|x\|}{\psi(\|x\|)\Big\{\frac{1+|\nu_2|+|\nu_3|}{\Gamma(q)}+\frac{\eta^{q+\beta-1}(|\nu_1|+|\nu_4|)}{\Gamma(q+\beta)}+\frac{b\sigma^{q+\alpha-1}(|\nu_2|+|\nu_3|)}{\Gamma(q+\alpha)}\Big\}\|p\|_{L^1}}\le 1.$$

In view of (3.11.3), there exists M such that $\|x\|\neq M$. Let us set

$$U=\{x\in C([0,1],\mathbb{R}):\|x\|<M\}.$$

Note that the operator $\Omega_F : \overline{U} \to \mathcal{P}(C([0,1],\mathbb{R}))$ is upper semi-continuous and completely continuous. From the choice of U, there is no $x \in \partial U$ such that $x \in \lambda\Omega_F(x)$ for some $\lambda \in (0,1)$. Consequently, by the nonlinear alternative of Leray-Schauder type (Theorem 1.16), we deduce that Ω_F has a fixed point $x \in \overline{U}$ which is a solution of the problem (3.30). This completes the proof. □

Example 3.7. Consider the following fractional boundary value problem

$$\begin{cases} {}^cD^{\frac{3}{2}}x(t) \in F(t,x(t)), & 0<t<1, \\ x(0) = I^{1/3}x(1/2), & x(1) = 2I^{2/3}x(3/4). \end{cases} \tag{3.31}$$

Here, $q = 3/2,\ a = 1,\ b = 2,\ \ \beta = 1/3,\ \alpha = 2/3,\ \eta = 1/2,\ \sigma = 3/4$ and $F : [0,1]\times\mathbb{R} \to \mathcal{P}(\mathbb{R})$ is a multi-valued map given by

$$x \to F(t,x) = \left[\frac{|x|^3}{|x|^3+3} + \frac{3}{4}t^3 + \frac{7}{15}, \frac{|x|}{|x|+1} + t + 1\right].$$

For $f \in F$, we have

$$|f| \le \max\left(\frac{|x|^3}{|x|^3+3} + \frac{3}{4}t^3 + \frac{7}{15}, \frac{|x|}{|x|+1} + t + 1\right) \le 3, \quad x \in \mathbb{R}.$$

Thus,

$$\|F(t,x)\|_{\mathcal{P}} := \sup\{|y| : y \in F(t,x)\} \le 3 = p(t)\psi(\|x\|), \quad x \in \mathbb{R},$$

with $p(t) = 1,\ \psi(\|x\|) = 3$. Furthermore, by the condition

$$\frac{M}{\psi(M)\Big\{\frac{1+|\nu_2|+|\nu_3|}{\Gamma(q)} + \frac{\eta^{q+\beta-1}(|\nu_1|+|\nu_4|)}{\Gamma(q+\beta)} + \frac{b\sigma^{q+\alpha-1}(|\nu_2|+|\nu_3|)}{\Gamma(q+\alpha)}\Big\}\|p\|_{L^1}} > 1,$$

we find that $M > M_1$ with $M_1 \simeq 21.138414$. Clearly, all the conditions of Theorem 3.11 are satisfied. So there exists at least one solution of the problem (3.31) on $[0,1]$.

3.2.3.2 *The lower semi-continuous case*

As a next result, we study the case when F is not necessarily convex valued. Our strategy to deal with this problem is based on the nonlinear alternative of Leray-Schauder type together with the selection theorem of Bressan and Colombo (Lemma 1.3) for lower semi-continuous maps with decomposable values.

Theorem 3.12. *Assume that (3.11.2), (3.11.3) and the following condition holds:*

(3.11.1) $F : [0,1]\times\mathbb{R} \to \mathcal{P}(\mathbb{R})$ is a non-empty compact-valued multi-valued map such that

(a) $(t,x) \longmapsto F(t,x)$ is $\mathcal{L}\otimes\mathcal{B}$ measurable,
(b) $x \longmapsto F(t,x)$ is lower semi-continuous for each $t \in [0,1]$.

Then the boundary value problem (3.30) has at least one solution on $[0,1]$.

Proof. It follows from (3.11.2) and (3.11.1) that F is of l.s.c. type. Then, from Lemma 1.3, there exists a continuous function $f : C([0,1],\mathbb{R}) \to L^1([0,1],\mathbb{R})$ such that $f(x) \in \mathcal{F}(x)$ for all $x \in C([0,1],\mathbb{R})$.

Consider the problem

$$\begin{cases} {}^cD^q x(t) = f(x(t)), \quad t \in [0,1], \\ x(0) = aI^\beta x(\eta), \quad x(1) = bI^\alpha x(\sigma), \quad 0 < \beta, \alpha \leq 1. \end{cases} \tag{3.32}$$

Observe that if $x \in C^2([0,1],\mathbb{R})$ is a solution of (3.32), then x is a solution to the problem (3.30). In order to transform the problem (3.32) into a fixed point problem, we define an operator $\overline{\Omega_F}$ as

$$\begin{aligned} &\overline{\Omega_F} x(t) \\ &= \int_0^t \frac{(t-s)^{q-1}}{\Gamma(q)} f(s)ds + (\nu_1 - t\nu_4) \int_0^\eta \frac{(\eta-s)^{q+\beta-1}}{\Gamma(q+\beta)} f(x(s))ds \\ &+(\nu_2+\nu_3 t)\Big\{ b\int_0^\sigma \frac{(\sigma-s)^{q+\alpha-1}}{\Gamma(q+\alpha)} f(x(s))ds - \int_0^1 \frac{(1-s)^{q-1}}{\Gamma(q)} f(x(s))ds \Big\}, \ t \in [0,1]. \end{aligned}$$

It can easily be shown that $\overline{\Omega_F}$ is continuous and completely continuous. The remaining part of the proof is similar to that of Theorem 3.11. So we omit it. This completes the proof. □

3.2.3.3 *The Lipschitz case*

Now we prove the existence of solutions for the problem (3.30) with a non-convex valued right hand side by applying a fixed point theorem for multi-valued maps due to Covitz and Nadler (Theorem 1.19).

Theorem 3.13. *Assume that the following conditions hold:*

(3.13.1) $F : [0,1] \times \mathbb{R} \to \mathcal{P}_{cp}(\mathbb{R})$ *is such that* $F(\cdot, x) : [0,1] \to \mathcal{P}_{cp}(\mathbb{R})$ *is measurable for each* $x \in \mathbb{R}$.

(3.13.2) $H_d(F(t,x), F(t,\bar{x})) \leq m(t)|x - \bar{x}|$ *for almost all* $t \in [0,1]$ *and* $x, \bar{x} \in \mathbb{R}$ *with* $m \in L^1([0,1],\mathbb{R}^+)$ *and* $d(0, F(t,0)) \leq m(t)$ *for almost all* $t \in [0,1]$.

Then the boundary value problem (3.30) has at least one solution on $[0,1]$ *if*

$$\Big\{ \frac{1+|\nu_2|+|\nu_3|}{\Gamma(q)} + \frac{\eta^{q+\beta-1}(|\nu_1|+|\nu_4|)}{\Gamma(q+\beta)} + \frac{b\sigma^{q+\alpha-1}(|\nu_2|+|\nu_3|)}{\Gamma(q+\alpha)} \Big\} \|m\|_{L^1} < 1.$$

Proof. Observe that the set $S_{F,x}$ is non-empty for each $x \in C([0,1],\mathbb{R})$ by the assumption (3.13.1), so F has a measurable selection (see Theorem III.6 [78]). Now we show that the operator Ω_F, defined in the beginning of proof of Theorem 3.11, satisfies the assumptions of Theorem 1.19. To show that $\Omega_F(x) \in \mathcal{P}_{cl}((C[0,1],\mathbb{R}))$ for each $x \in C([0,1],\mathbb{R})$, let $\{u_n\}_{n\geq 0} \in \Omega_F(x)$ be such that $u_n \to u$ $(n \to \infty)$ in $C([0,1],\mathbb{R})$. Then $u \in C([0,1],\mathbb{R})$ and there exists $v_n \in S_{F,x_n}$ such that, for each $t \in [0,1]$,

$$u_n(t) = \int_0^t \frac{(t-s)^{q-1}}{\Gamma(q)} v_n(s)ds + (\nu_1 - t\nu_4) \int_0^\eta \frac{(\eta-s)^{q+\beta-1}}{\Gamma(q+\beta)} v_n(s)ds$$

$$+ (\nu_2 + \nu_3 t)\Big\{b \int_0^\sigma \frac{(\sigma - s)^{q+\alpha-1}}{\Gamma(q+\alpha)} v_n(s)ds - \int_0^1 \frac{(1-s)^{q-1}}{\Gamma(q)} v_n(s)ds\Big\},\ t \in [0,1].$$

As F has compact values, we pass onto a subsequence to obtain that v_n converges to v in $L^1([0,1],\mathbb{R})$. Thus, $v \in S_{F,x}$ and for each $t \in [0,1]$,

$$\begin{aligned} & u_n(t) \to u(t) \\ &= \int_0^t \frac{(t-s)^{q-1}}{\Gamma(q)} v(s)ds + (\nu_1 - t\nu_4) \int_0^\eta \frac{(\eta - s)^{q+\beta-1}}{\Gamma(q+\beta)} v(s)ds \\ &\quad + (\nu_2 + \nu_3 t)\Big\{b \int_0^\sigma \frac{(\sigma - s)^{q+\alpha-1}}{\Gamma(q+\alpha)} v(s)ds - \int_0^1 \frac{(1-s)^{q-1}}{\Gamma(q)} v(s)ds\Big\}. \end{aligned}$$

Hence, $u \in \Omega(x)$.

Next we show that there exists $\gamma < 1$ such that

$$H_d(\Omega_F(x), \Omega_F(\bar{x})) \le \gamma \|x - \bar{x}\| \ \text{ for each } \ x, \bar{x} \in C([0,1],\mathbb{R}).$$

Let $x, \bar{x} \in C([0,1],\mathbb{R})$ and $h_1 \in \Omega(x)$. Then there exists $v_1(t) \in F(t, x(t))$ such that, for each $t \in [0,1]$,

$$\begin{aligned} h_1(t) &= \int_0^t \frac{(t-s)^{q-1}}{\Gamma(q)} v_1(s)ds + (\nu_1 - t\nu_4) \int_0^\eta \frac{(\eta - s)^{q+\beta-1}}{\Gamma(q+\beta)} v_1(s)ds \\ &\quad + (\nu_2 + \nu_3 t)\Big\{b \int_0^\sigma \frac{(\sigma - s)^{q+\alpha-1}}{\Gamma(q+\alpha)} v_1(s)ds - \int_0^1 \frac{(1-s)^{q-1}}{\Gamma(q)} v_1(s)ds\Big\}. \end{aligned}$$

By (3.13.2), we have

$$H_d(F(t,x), F(t,\bar{x})) \le m(t)|x(t) - \bar{x}(t)|.$$

So, there exists $w \in F(t, \bar{x}(t))$ such that

$$|v_1(t) - w| \le m(t)|x(t) - \bar{x}(t)|, \quad t \in [0,1].$$

Define $U : [0,1] \to \mathcal{P}(\mathbb{R})$ by

$$U(t) = \{w \in \mathbb{R} : |v_1(t) - w| \le m(t)|x(t) - \bar{x}(t)|\}.$$

Since the multi-valued operator $U(t) \cap F(t, \bar{x}(t))$ is measurable (Proposition III.4 [78]), there exists a function $v_2(t)$ which is a measurable selection for U. So $v_2(t) \in F(t, \bar{x}(t))$ and for each $t \in [0,1]$, we have $|v_1(t) - v_2(t)| \le m(t)|x(t) - \bar{x}(t)|$.

For each $t \in [0,1]$, let us define

$$\begin{aligned} h_2(t) &= \int_0^t \frac{(t-s)^{q-1}}{\Gamma(q)} v_2(s)ds + (\nu_1 - t\nu_4) \int_0^\eta \frac{(\eta - s)^{q+\beta-1}}{\Gamma(q+\beta)} v_2(s)ds \\ &\quad + (\nu_2 + \nu_3 t)\Big\{b \int_0^\sigma \frac{(\sigma - s)^{q+\alpha-1}}{\Gamma(q+\alpha)} v_2(s)ds - \int_0^1 \frac{(1-s)^{q-1}}{\Gamma(q)} v_2(s)ds\Big\}. \end{aligned}$$

Thus, for $t \in [0,1]$, we have

$$\begin{aligned} & |h_1(t) - h_2(t)| \\ &= \Big| \int_0^t \frac{(t-s)^{q-1}}{\Gamma(q)} [v_1(s) - v_2(s)]ds + (\nu_1 - t\nu_4) \int_0^\eta \frac{(\eta - s)^{q+\beta-1}}{\Gamma(q+\beta)} [v_1(s) - v_2(s)]ds \end{aligned}$$

$$+ (\nu_2+\nu_3 t)\Big\{b\int_0^\sigma \frac{(\sigma-s)^{q+\alpha-1}}{\Gamma(q+\alpha)}[v_1(s)-v_2(s)]ds - \int_0^1 \frac{(1-s)^{q-1}}{\Gamma(q)}[v_1(s)-v_2(s)]ds\Big\}\Big|.$$

Hence,

$$\|h_1-h_2\|$$
$$\leq \Big\{\frac{1+|\nu_2|+|\nu_3|}{\Gamma(q)} + \frac{\eta^{q+\beta-1}(|\nu_1|+|\nu_4|)}{\Gamma(q+\beta)} + \frac{b\sigma^{q+\alpha-1}(|\nu_2|+|\nu_3|)}{\Gamma(q+\alpha)}\Big\}\|m\|_{L^1}\|x-\bar{x}\|.$$

Analogously, interchanging the roles of x and $\overline{x}$, we obtain

$$H_d(\Omega_F(x),\Omega_F(\bar{x})) \leq \gamma\|x-\bar{x}\|$$
$$\leq \Big\{\frac{1+|\nu_2|+|\nu_3|}{\Gamma(q)} + \frac{\eta^{q+\beta-1}(|\nu_1|+|\nu_4|)}{\Gamma(q+\beta)} + \frac{b\sigma^{q+\alpha-1}(|\nu_2|+|\nu_3|)}{\Gamma(q+\alpha)}\Big\}\|m\|_{L^1}\|x-\bar{x}\|.$$

Since Ω_F is a contraction, it follows by Theorem 1.19 that Ω_F has a fixed point x which is a solution of (3.30). This completes the proof. □

3.3 Nonlocal integral boundary value problems of fractional integro-differential equations and inclusions

In this section, we study nonlocal integral boundary value problems of fractional integro-differential equations and inclusions.

3.3.1 *Nonlocal fractional integro-differential equations*

We consider a fractional differential equation with nonlocal integral boundary conditions given by

$$\begin{cases} {}^cD^q x(t) = Af(t,x(t)) + BI^r g(t,x(t)), \quad t\in[0,1], \\ x(0)=\beta x(\theta), \quad x(\xi)=\alpha\displaystyle\int_\eta^1 x(s)ds, \quad 0<\theta<\xi<\eta<1, \end{cases} \tag{3.33}$$

where ${}^cD^q$ denotes the Caputo fractional derivative operator of order q, $f:[0,1]\times\mathbb{R}\to\mathbb{R}$ is a given continuous function, $1<q\leq 2, 0<r<1$, and α,β,A,B are real constants.

Here we remark that the boundary conditions introduced in the problem (3.33) are of nonlocal strip type and describe the situation when the receptors at the end points of the boundary are influenced by the nonlocal contributions due to interior points and strips of the domain for the problem. The problem (3.33) can also be termed as a five-point nonlocal fractional boundary value problem.

Lemma 3.4. *For any $y\in C([0,1],\mathbb{R})$, the unique solution of the linear fractional boundary value problem*

$$\begin{cases} {}^cD^q x(t) = y(t), \quad t\in[0,1], \quad 1<q\leq 2, \\ x(0)=\beta x(\theta), \quad x(\xi)=\alpha\displaystyle\int_\eta^1 x(s)ds, \quad 0<\theta<\xi<\eta<1, \end{cases} \tag{3.34}$$

is

$$\begin{aligned} x(t) &= \int_0^t \frac{(t-s)^{q-1}}{\Gamma(q)} y(s)ds + \frac{\beta\Big(\xi - \frac{\alpha}{2}(1-\eta^2)\Big)}{Q} \int_0^\theta \frac{(\theta-s)^{q-1}}{\Gamma(q)} y(s)ds \\ &+ \frac{\alpha\beta\theta}{Q}\left(\int_0^1 \frac{(1-s)^q}{\Gamma(q+1)} y(s)ds - \int_0^\eta \frac{(\eta-s)^q}{\Gamma(q+1)} y(s)ds \right) - \frac{\beta\theta}{Q}\int_0^\xi \frac{(\xi-s)^q}{\Gamma(q)} y(s)ds \\ &+ \frac{t}{Q}\left[\alpha(1-\beta)\left(\int_0^1 \frac{(1-s)^q}{\Gamma(q+1)} y(s)ds - \int_0^\eta \frac{(\eta-s)^q}{\Gamma(q+1)} y(s)ds \right) \right. \\ &\left. -(1-\beta)\int_0^\xi \frac{(\xi-s)^q}{\Gamma(q)} y(s)ds - \beta(1-\alpha(1-\eta)) \int_0^\theta \frac{(\theta-s)^{q-1}}{\Gamma(q)} y(s)ds \right] \end{aligned} \tag{3.35}$$

where it is assumed that

$$Q = (1-\beta)\Big(\xi - \frac{\alpha}{2}(1-\eta^2)\Big) + \beta\theta\Big(1-\alpha(1-\eta)\Big) \neq 0. \tag{3.36}$$

Proof. It is well known that the general solution of the fractional differential equation in (3.34) can be written as

$$x(t) = c_0 + c_1 t + \int_0^t \frac{(t-s)^{q-1}}{\Gamma(q)} y(s)ds, \tag{3.37}$$

where $c_0, c_1 \in \mathbb{R}$ are arbitrary constants.

Applying the given boundary conditions to (3.37), we obtain the following system

$$(1-\beta)c_0 - \beta\theta c_1 = \beta \int_0^\theta \frac{(\theta-s)^{q-1}}{\Gamma(q)} y(s)ds$$

$$\begin{aligned} \Big(1-\alpha(1-\eta)\Big)c_0 + \Big(\xi - \frac{\alpha}{2}(1-\eta^2)\Big)c_1 &= \alpha\left(\int_0^1 \frac{(1-s)^q}{\Gamma(q+1)} y(s)ds - \int_0^\eta \frac{(\eta-s)^q}{\Gamma(q+1)} y(s)ds \right) \\ &- \int_0^\xi \frac{(\xi-s)^q}{\Gamma(q)} y(s)ds, \end{aligned}$$

which, on solving for c_0 and c_1, leads to

$$\begin{aligned} c_0 &= \frac{\beta\Big(\xi - \frac{\alpha}{2}(1-\eta^2)\Big)}{Q} \int_0^\theta \frac{(\theta-s)^{q-1}}{\Gamma(q)} y(s)ds \\ &+ \frac{\alpha\beta\theta}{Q}\left(\int_0^1 \frac{(1-s)^q}{\Gamma(q+1)} y(s)ds - \int_0^\eta \frac{(\eta-s)^q}{\Gamma(q+1)} y(s)ds \right) - \frac{\beta\theta}{Q}\int_0^\xi \frac{(\xi-s)^q}{\Gamma(q)} y(s)ds \end{aligned}$$

and

$$\begin{aligned} c_1 &= \frac{1}{Q}\left[\alpha(1-\beta)\left(\int_0^1 \frac{(1-s)^q}{\Gamma(q+1)} y(s)ds - \int_0^\eta \frac{(\eta-s)^q}{\Gamma(q+1)} y(s)ds \right) \right. \\ &\left. -(1-\beta)\int_0^\xi \frac{(\xi-s)^q}{\Gamma(q)} y(s)ds - \beta(1-\alpha(1-\eta)) \int_0^\theta \frac{(\theta-s)^{q-1}}{\Gamma(q)} y(s)ds \right]. \end{aligned}$$

Substituting the values of c_0, c_1 in (3.37), we get (3.35). We can prove the converse by direct computation. This completes the proof. □

We denote by $\mathcal{C} = C([0,1],\mathbb{R})$ the Banach space of all continuous functions from $[0,1] \to \mathbb{R}$ endowed with the norm defined by $\|x\| = \sup\{|x(t)| : t \in [0,1]\}$. Also by $L^1([0,1],\mathbb{R})$ we denote the Banach space of measurable functions $x : [0,1] \to \mathbb{R}$ which are Lebesgue integrable and normed by $\|x\|_{L^1} = \int_0^1 |x(t)|dt$.

In the following we will give two existence results for the problem (3.33), based on Sadovskii's fixed point theorem (Theorem 1.14) and a fixed point theorem due to O'Regan (Theorem 1.9).

3.3.1.1 *Existence result via Sadovskii's fixed point theorem*

In view of Lemma 3.4, we define an operator $\mathcal{P} : \mathcal{C} \to \mathcal{C}$ by

$$(\mathcal{P}x)(t) = (\mathcal{P}_1x)(t) + (\mathcal{P}_2x)(t), \quad t \in [0,1], \tag{3.38}$$

where

$$\begin{aligned}
&(\mathcal{P}_1x)(t)\\
&= A\int_0^t \frac{(t-s)^{q-1}}{\Gamma(q)} f(s,x(s))ds + A\frac{\beta\left(\xi - \frac{\alpha}{2}(1-\eta^2)\right)}{Q}\int_0^\theta \frac{(\theta-s)^{q-1}}{\Gamma(q)} f(s,x(s))ds\\
&\quad +A\frac{\alpha\beta\theta}{Q}\left(\int_0^1 \frac{(1-s)^q}{\Gamma(q+1)} f(s,x(s))ds - \int_0^\eta \frac{(\eta-s)^q}{\Gamma(q+1)} f(s,x(s))ds\right)\\
&\quad -A\frac{\beta\theta}{Q}\int_0^\xi \frac{(\xi-s)^q}{\Gamma(q)} f(s,x(s))ds + \frac{At}{Q}\left[\alpha(1-\beta)\left(\int_0^1 \frac{(1-s)^q}{\Gamma(q+1)} f(s,x(s))ds\right.\right.\\
&\quad \left. - \int_0^\eta \frac{(\eta-s)^q}{\Gamma(q+1)} f(s,x(s))ds\right) - (1-\beta)\int_0^\xi \frac{(\xi-s)^q}{\Gamma(q)} f(s,x(s))ds\\
&\quad \left. -\beta(1-\alpha(1-\eta))\int_0^\theta \frac{(\theta-s)^{q-1}}{\Gamma(q)} f(s,x(s))ds\right]
\end{aligned} \tag{3.39}$$

and

$$\begin{aligned}
&(\mathcal{P}_2x)(t)\\
&= B\int_0^t \frac{(t-s)^{q-1}}{\Gamma(q)} \int_0^s \frac{(s-u)^{r-1}}{\Gamma(r)} g(u,x(u))duds\\
&\quad +B\frac{\beta\left(\xi - \frac{\alpha}{2}(1-\eta^2)\right)}{Q}\int_0^\theta \frac{(\theta-s)^{q-1}}{\Gamma(q)}\int_0^s \frac{(s-u)^{r-1}}{\Gamma(r)} g(u,x(u))duds\\
&\quad +B\frac{\alpha\beta\theta}{Q}\left(\int_0^1 \frac{(1-s)^q}{\Gamma(q+1)}\int_0^s \frac{(s-u)^{r-1}}{\Gamma(r)} g(u,x(u))duds\right.\\
&\quad \left. - \int_0^\eta \frac{(\eta-s)^q}{\Gamma(q+1)}\int_0^s \frac{(s-u)^{r-1}}{\Gamma(r)} g(u,x(u))duds\right)\\
&\quad -B\frac{\beta\theta}{Q}\int_0^\xi \frac{(\xi-s)^q}{\Gamma(q)}\int_0^s \frac{(s-u)^{r-1}}{\Gamma(r)} g(u,x(u))duds
\end{aligned}$$

$$
\begin{aligned}
&+\frac{Bt}{Q}\Bigg[\alpha(1-\beta)\Bigg(\int_0^1 \frac{(1-s)^q}{\Gamma(q+1)}\int_0^s \frac{(s-u)^{r-1}}{\Gamma(r)}g(u,x(u))duds \\
&-\int_0^\eta \frac{(\eta-s)^q}{\Gamma(q+1)}\int_0^s \frac{(s-u)^{r-1}}{\Gamma(r)}g(u,x(u))duds\Bigg) \\
&-(1-\beta)\int_0^\xi \frac{(\xi-s)^q}{\Gamma(q)}\int_0^s \frac{(s-u)^{r-1}}{\Gamma(r)}g(u,x(u))duds \\
&-\beta(1-\alpha(1-\eta))\int_0^\theta \frac{(\theta-s)^{q-1}}{\Gamma(q)}\int_0^s \frac{(s-u)^{r-1}}{\Gamma(r)}g(u,x(u))duds\Bigg]. \qquad (3.40)
\end{aligned}
$$

Theorem 3.14. *Let $f, g : [0,1]\times\mathbb{R}\to\mathbb{R}$ be continuous functions satisfying the conditions:*

(3.14.1) f satisfies the Lipschitz condition:

$$|f(t,x)-f(t,y)|\le L|x-y|, \quad L>0, \quad \textit{for all} \quad (t,x),(t,y)\in[0,1]\times\mathbb{R},$$

(3.14.2) there exist a function $m\in C([0,1],\mathbb{R}^+)$, and a non-decreasing function $\psi:\mathbb{R}^+\to\mathbb{R}^+$ such that

$$|g(t,x)|\le m(t)\psi(\|x\|), \quad \forall (t,x)\in[0,1]\times\mathbb{R}.$$

Then the boundary value problem (3.33) has at least one solution provided that

$$
\begin{aligned}
\gamma := |A|L\Bigg\{&\frac{1}{\Gamma(q+1)}+\frac{\left|\beta\left(\xi-\frac{\alpha}{2}(1-\eta^2)\right)\right|\theta^q}{|Q|\Gamma(q+1)}+\frac{|\alpha\beta|\theta}{|Q|}\left(\frac{1}{\Gamma(q+2)}+\frac{\eta^{q+1}}{\Gamma(q+2)}\right) \\
&+\frac{|\beta|\theta\xi^q}{|Q|\Gamma(q+1)}+\frac{1}{|Q|}\Bigg[|\alpha(1-\beta)|\left(\frac{1}{\Gamma(q+2)}+\frac{\eta^{q+1}}{\Gamma(q+2)}\right) \\
&+\frac{|1-\beta|\xi^q}{\Gamma(q+1)}+\frac{|\beta(1-\alpha(1-\eta))|\theta^q}{\Gamma(q+1)}\Bigg]\Bigg\}<1. \qquad (3.41)
\end{aligned}
$$

Proof. Let $B_\nu=\{x\in\mathcal{C} : \|x\|\le\nu\}$ be a closed bounded and convex subset of $\mathcal{C}:=C([0,1],\mathbb{R})$, where ν will be fixed later. We define a map $\mathcal{P}: B_\nu\to\mathcal{C}$ as

$$(\mathcal{P}x)(t)=(\mathcal{P}_1x)(t)+(\mathcal{P}_2x)(t), \quad t\in[0,1],$$

where $\mathcal{P}_1$ and $\mathcal{P}_2$ are defined by (3.39) and (3.40) respectively. Notice that the problem (3.33) is equivalent to a fixed point problem $\mathcal{P}x=x$.

Step 1. $(\mathcal{P}x)(B_\nu)\subset B_\nu$.

Let us set $M=\sup_{t\in[0,1]}|f(t,0)|$ and select $\nu\ge\dfrac{\omega}{1-\gamma}$, where

$$\omega=|A|M\Bigg[\frac{1}{\Gamma(q+1)}+\frac{\left|\beta\left(\xi-\frac{\alpha}{2}(1-\eta^2)\right)\right|\theta^q}{|Q|\Gamma(q+1)}+\frac{|\alpha\beta|\theta}{|Q|}\left(\frac{1}{\Gamma(q+2)}+\frac{\eta^{q+1}}{\Gamma(q+2)}\right)$$

$$+\frac{|\beta|\theta\xi^q}{|Q|\Gamma(q+1)}+\frac{1}{|Q|}\left[|\alpha(1-\beta)|\left(\frac{1}{\Gamma(q+2)}+\frac{\eta^{q+1}}{\Gamma(q+2)}\right)\right.$$

$$\left.+\frac{|1-\beta|\xi^q}{\Gamma(q+1)}+\frac{|\beta(1-\alpha(1-\eta))|\theta^q}{\Gamma(q+1)}\right]$$

$$+|B|\|m\|\psi(r)\left[\frac{1}{\Gamma(q+r+1)}+\frac{\left|\beta\left(\xi-\frac{\alpha}{2}(1-\eta^2)\right)\right|\theta^{q+r}}{|Q|\Gamma(q+r+1)}\right.$$

$$+\frac{|\alpha\beta|\theta}{|Q|}\left(\frac{1}{\Gamma(q+r+2)}+\frac{\eta^{q+r+1}}{\Gamma(q+r+2)}\right)+\frac{|\beta|\theta\xi^{q+r}}{|Q|\Gamma(q+r+1)}$$

$$+\frac{1}{|Q|}\left\{|\alpha(1-\beta)|\left(\frac{1}{\Gamma(q+r+2)}+\frac{\eta^{q+r+1}}{\Gamma(q+r+2)}\right)\right.$$

$$\left.\left.+\frac{|1-\beta|\xi^{q+r}}{\Gamma(q+r+1)}+\frac{|\beta(1-\alpha(1-\eta))|\theta^{q+r}}{\Gamma(q+r+1)}\right\}\right].$$

Using $|f(t,x(t))\leq|f(t,x(t))-f(t,0)|+|f(t,0)|\leq L\nu+M$, for $x\in B_\nu,\ t\in[0,1]$, we get

$$|(\mathcal{P}_1x)(t)|\leq|A|(L\nu+M)\left\{\frac{1}{\Gamma(q+1)}+\frac{\left|\beta\left(\xi-\frac{\alpha}{2}(1-\eta^2)\right)\right|\theta^q}{|Q|\Gamma(q+1)}\right.$$

$$+\frac{|\alpha\beta|\theta}{|Q|}\left(\frac{1}{\Gamma(q+2)}+\frac{\eta^{q+1}}{\Gamma(q+2)}\right)+\frac{|\beta|\theta\xi^q}{|Q|\Gamma(q+1)}$$

$$+\frac{1}{|Q|}\left[|\alpha(1-\beta)|\left(\frac{1}{\Gamma(q+2)}+\frac{\eta^{q+1}}{\Gamma(q+2)}\right)\right.$$

$$\left.\left.+\frac{|1-\beta|\xi^q}{\Gamma(q+1)}+\frac{|\beta(1-\alpha(1-\eta))|\theta^q}{\Gamma(q+1)}\right]\right\},$$

$$|(\mathcal{P}_2x)(t)|\leq|B|\|m\|\psi(\nu)\left\{\frac{1}{\Gamma(q+r+1)}+\frac{\left|\beta\left(\xi-\frac{\alpha}{2}(1-\eta^2)\right)\right|\theta^{q+r}}{|Q|\Gamma(q+r+1)}\right.$$

$$+\frac{|\alpha\beta|\theta}{|Q|}\left(\frac{1}{\Gamma(q+r+2)}+\frac{\eta^{q+r+1}}{\Gamma(q+r+2)}\right)+\frac{|\beta|\theta\xi^{q+r}}{|Q|\Gamma(q+r+1)}$$

$$+\frac{1}{|Q|}\left[|\alpha(1-\beta)|\left(\frac{1}{\Gamma(q+r+2)}+\frac{\eta^{q+r+1}}{\Gamma(q+r+2)}\right)\right.$$

$$\left.\left.+\frac{|1-\beta|\xi^{q+r}}{\Gamma(q+r+1)}+\frac{|\beta(1-\alpha(1-\eta))|\theta^{q+r}}{\Gamma(q+r+1)}\right]\right\},$$

where we have used the following relations:

$$\int_0^t\frac{(t-s)^{q-1}}{\Gamma(q)}\int_0^s\frac{(s-u)^{r-1}}{\Gamma(r)}duds=\int_0^t\frac{(t-s)^{q-1}}{\Gamma(q)}\frac{(s)^r}{\Gamma(r+1)}ds$$

$$= \frac{t^{q+r}}{\Gamma(q)\Gamma(r+1)} \int_0^1 (1-v)^{q-1} v^r dv = \frac{t^{q+r}}{\Gamma(q+r+1)},$$
$$\int_0^1 (1-u)^{\alpha-1} u^\beta du = \frac{\Gamma(\alpha)\Gamma(\beta+1)}{\Gamma(\alpha+\beta+1)}.$$

Consequently

$$\begin{aligned}
|(\mathcal{P}x)(t)| &\le |(\mathcal{P}_1 x)(t)| + |(\mathcal{P}_2 x)(t)| \\
&\le |A|(L\nu + M)\Bigg\{ \frac{1}{\Gamma(q+1)} + \frac{\left|\beta\left(\xi - \frac{\alpha}{2}(1-\eta^2)\right)\right|\theta^q}{|Q|\Gamma(q+1)} \\
&\quad + \frac{|\alpha\beta|\theta}{|Q|}\left(\frac{1}{\Gamma(q+2)} + \frac{\eta^{q+1}}{\Gamma(q+2)}\right) + \frac{|\beta|\theta\xi^q}{|Q|\Gamma(q+1)} \\
&\quad + \frac{1}{|Q|}\Bigg[|\alpha(1-\beta)|\left(\frac{1}{\Gamma(q+2)} + \frac{\eta^{q+1}}{\Gamma(q+2)}\right) \\
&\quad + \frac{|1-\beta|\xi^q}{\Gamma(q+1)} + \frac{|\beta(1-\alpha(1-\eta))|\theta^q}{\Gamma(q+1)} \Bigg] \Bigg\} \\
&\quad + |B|\|m\|\psi(\nu)\Bigg\{ \frac{1}{\Gamma(q+r+1)} + \frac{\left|\beta\left(\xi - \frac{\alpha}{2}(1-\eta^2)\right)\right|\theta^{q+r}}{|Q|\Gamma(q+r+1)} \\
&\quad + \frac{|\alpha\beta|\theta}{|Q|}\left(\frac{1}{\Gamma(q+r+2)} + \frac{\eta^{q+r+1}}{\Gamma(q+r+2)}\right) + \frac{|\beta|\theta\xi^{q+r+r}}{|Q|\Gamma(q+r+1)} \\
&\quad + \frac{1}{|Q|}\Bigg[|\alpha(1-\beta)|\left(\frac{1}{\Gamma(q+r+2)} + \frac{\eta^{q+r+1}}{\Gamma(q+r+2)}\right) \\
&\quad + \frac{|1-\beta|\xi^{q+r}}{\Gamma(q+r+1)} + \frac{|\beta(1-\alpha(1-\eta))|\theta^{q+r}}{\Gamma(q+r+1)} \Bigg] \Bigg\} \\
&\le \gamma\nu + \omega \le \nu,
\end{aligned}$$

which implies that $(\mathcal{P}x)(B_\nu) \subset B_\nu$.

Step 2. $\mathcal{P}_1$ is continuous and γ-contractive.

To show the continuity of $\mathcal{P}_1$ for $t \in [0,1]$, let us consider a sequence x_n converging to x. Then, by the assumption (3.14.1), we have

$$\begin{aligned}
\|(\mathcal{P}_1 x_n) - (\mathcal{P}_1 x)\| &\le |A|L\Bigg\{ \frac{1}{\Gamma(q+1)} + \frac{\left|\beta\left(\xi - \frac{\alpha}{2}(1-\eta^2)\right)\right|\theta^q}{|Q|\Gamma(q+1)} \\
&\quad + \frac{|\alpha\beta|\theta}{|Q|}\left(\frac{1}{\Gamma(q+2)} + \frac{\eta^{q+1}}{\Gamma(q+2)}\right) + \frac{|\beta|\theta\xi^q}{|Q|\Gamma(q+1)} \\
&\quad + \frac{1}{|Q|}\Bigg[|\alpha(1-\beta)|\left(\frac{1}{\Gamma(q+2)} + \frac{\eta^{q+1}}{\Gamma(q+2)}\right)
\end{aligned}$$

$$+\frac{|1-\beta|\xi^q}{\Gamma(q+1)}+\frac{|\beta(1-\alpha(1-\eta))|\theta^q}{\Gamma(q+1)}\Bigg]\Bigg\}\|x_n-x\|.$$

Next, we show that $\mathcal{P}_1$ is γ-contractive. For $x, y \in B_\nu$, we get

$$\begin{aligned}|(\mathcal{P}_1x)(t)-(\mathcal{P}_1y)(t)| \le |A|L\Bigg\{&\frac{1}{\Gamma(q+1)}+\frac{\left|\beta\left(\xi-\frac{\alpha}{2}(1-\eta^2)\right)\right|\theta^q}{|Q|\Gamma(q+1)}\\ &+\frac{|\alpha\beta|\theta}{|Q|}\left(\frac{1}{\Gamma(q+2)}+\frac{\eta^{q+1}}{\Gamma(q+2)}\right)+\frac{|\beta|\theta\xi^q}{|Q|\Gamma(q+1)}\\ &+\frac{1}{|Q|}\Bigg[|\alpha(1-\beta)|\left(\frac{1}{\Gamma(q+2)}+\frac{\eta^{q+1}}{\Gamma(q+2)}\right)\\ &+\frac{|1-\beta|\xi^q}{\Gamma(q+1)}+\frac{|\beta(1-\alpha(1-\eta))|\theta^q}{\Gamma(q+1)}\Bigg]\Bigg\}\|x-y\|,\end{aligned}$$

which, on taking the norm for $t \in [0,1]$ and using (3.41), implies that $\mathcal{P}_1$ is γ-contractive.

Step 3. $\mathcal{P}_2$ is compact.

In Step 1, it has been shown that $\mathcal{P}_2$ is uniformly bounded. Now we show that $\mathcal{P}_2$ maps bounded sets into equi-continuous sets of $C([0,1],\mathbb{R})$. Let $t_1, t_2 \in [0,1]$ with $t_1 < t_2$ and $x \in B_\nu$. Then we obtain

$$\begin{aligned}&|(\mathcal{P}_2x)(t_2)-(\mathcal{P}_2x)(t_1)|\\ \le\ &\frac{|B|\psi(\nu)}{\Gamma(q+r+1)}\int_0^{t_1}[(t_2-s)^{q-1}-(t_1-s)^{q-1}]m(s)ds\\ &+\frac{|B|\psi(\nu)}{\Gamma(q+r+1)}\int_{t_1}^{t_2}(t_2-s)^{q-1}m(s)ds\\ &+\frac{|B|\psi(\nu)\|m\||t_2-t_1|}{|Q|}\Bigg[|\alpha(1-\beta)|\left(\frac{1}{\Gamma(q+r+2)}+\frac{\eta^{q+r+1}}{\Gamma(q+r+2)}\right)\\ &+\frac{|1-\beta|\xi^{q+r}}{\Gamma(q+r+1)}+\frac{|\beta(1-\alpha(1-\eta))|\theta^{q+r}}{\Gamma(q+r+1)}\Bigg].\end{aligned}$$

Obviously the right hand side of the above inequality tends to zero independently of $x \in B_\nu$ as $t_2 - t_1 \to 0$. Therefore it follows by the Arzelá-Ascoli theorem that $\mathcal{P}_2 : C([0,1],\mathbb{R}) \to C([0,1],\mathbb{R})$ is completely continuous. Thus $\mathcal{P}_2$ is compact on $[0,1]$.

Step 4. $\mathcal{P}$ is condensing.

Since $\mathcal{P}_1$ is continuous, γ-contraction and $\mathcal{P}_2$ is compact, therefore, by Lemma 1.13, $\mathcal{P} : B_\nu \to B_\nu$ with $\mathcal{P} = \mathcal{P}_1 + \mathcal{P}_2$ is a condensing map on B_ν.

Consequently, by Theorem 1.14, the map $\mathcal{P}$ has a fixed point which, in turn, implies that the problem (3.33) has a solution. □

Example 3.8. Consider a fractional integro-differential equation with nonlocal integral boundary conditions given by

$$\begin{cases} D^{3/2}x(t) = f(t,x(t)) + I^{3/4}g(t,x(t)), \quad t \in [0,1], \\ x(0) = \dfrac{1}{2}x(1/4)), \quad x(1/3) = \displaystyle\int_{2/3}^{1} x(s)ds, \end{cases} \tag{3.42}$$

where $q = 3/2,\ A = B = 1,\ r = 3/4,\ \theta = 1/4,\ \xi = 1/3,\ \eta = 2/3,\ \alpha = 1,\ \beta = 1/2,\ f(t,x) = \dfrac{1}{(2+t)^3}\tan^{-1}x + t + 1,\ g(t,x) = \dfrac{t^2}{1+t^3}\Big(1 + \dfrac{|x|}{1+|x|}\Big).$

Clearly $L = 1/8$ as $|f(t,x) - f(t,y)| \le \frac{1}{8}|x-y|$, and $|g(t,x)| \le m(t)\psi(\|x\|)$ with $m(t) = t^2/(1+t^3)$ and $\psi(\|x\|) = 2$. Furthermore $|Q| = 1/9$, and the condition (3.41) yields $\gamma \simeq 0.522371 < 1$. Thus all the conditions of Theorem 3.14 are satisfied and consequently the problem (3.42) has a solution.

3.3.1.2 *Existence result via O'Regan's fixed point theorem*

Our next existence result relies on a fixed point theorem due to O'Regan (Theorem 1.9).

For convenience we set:

$$\begin{aligned} p_0 = |A|\Bigg\{ & \frac{1}{\Gamma(q+1)} + \frac{\Big|\beta\Big(\xi - \frac{\alpha}{2}(1-\eta^2)\Big)\Big|\theta^q}{|Q|\Gamma(q+1)} + \frac{|\alpha\beta|\theta}{|Q|}\left(\frac{1}{\Gamma(q+2)} + \frac{\eta^{q+1}}{\Gamma(q+2)}\right) \\ & + \frac{|\beta|\theta\xi^q}{|Q|\Gamma(q+1)} + \frac{1}{|Q|}\Bigg[|\alpha(1-\beta)|\left(\frac{1}{\Gamma(q+2)} + \frac{\eta^{q+1}}{\Gamma(q+2)}\right) \\ & + \frac{|1-\beta|\xi^q}{\Gamma(q+1)} + \frac{|\beta(1-\alpha(1-\eta))|\theta^q}{\Gamma(q+1)}\Bigg]\Bigg\}, \end{aligned}$$

and

$$\begin{aligned} k_0 = |B|\Bigg\{ & \frac{1}{\Gamma(q+r+1)} + \frac{\Big|\beta\Big(\xi - \frac{\alpha}{2}(1-\eta^2)\Big)\Big|\theta^{q+r}}{|Q|\Gamma(q+r+1)} \\ & + \frac{|\alpha\beta|\theta}{|Q|}\left(\frac{1}{\Gamma(q+r+2)} + \frac{\eta^{q+r+1}}{\Gamma(q+r+2)}\right) + \frac{|\beta|\theta\xi^{q+r}}{|Q|\Gamma(q+r+1)} \\ & + \frac{1}{|Q|}\Bigg[|\alpha(1-\beta)|\left(\frac{1}{\Gamma(q+r+2)} + \frac{\eta^{q+r+1}}{\Gamma(q+r+2)}\right) \\ & + \frac{|1-\beta|\xi^{q+r}}{\Gamma(q+r+1)} + \frac{|\beta(1-\alpha(1-\eta))|\theta^{q+r}}{\Gamma(q+r+1)}\Bigg]\Bigg\}. \end{aligned}$$

Let $\Omega_\sigma = \{x \in C([0,1],\mathbb{R}) : \|x\| < \sigma\}$, and define

$$M_\sigma = \max\{|f(t,x)| : (t,x) \in [0,1] \times [-\sigma,\sigma]\}.$$

Theorem 3.15. *Let $f, g : [0,1] \times \mathbb{R} \to \mathbb{R}$ be continuous functions. Assume that*

(3.14.1) there exists a nonnegative function $p \in C([0,1], \mathbb{R})$ and a non-decreasing function $\zeta : [0,\infty) \to (0,\infty)$ such that

$$|f(t,u)| \le p(t)\zeta(\|u\|) \ \text{for any} \ (t,u) \in [0,1] \times \mathbb{R};$$

(3.14.2) there exist a positive constant $\ell < k_0^{-1}$ and a continuous function $\phi : [0,\infty) \to (0,\infty)$ such that $\phi(z) \le \ell z$ and $|g(t,u) - g(t,v)| \le \phi(\|u - v\|)$ for all $t \in [0,1]$ and $u, v \in \mathbb{R}$;

(3.14.3) $\sup\limits_{r \in (0,\infty)} \dfrac{r}{k_0 K + p_0 \zeta(r)\|p\|} > \dfrac{1}{1 - k_0 \ell}$, *where* $K = \sup_{t \in [0,1]} |g(t,0)|$.

Then the boundary value problem (3.33) has at least one solution on $[0,1]$.

Proof. By the assumption (3.14.3), there exists a number $r_0 > 0$ such that

$$\frac{r_0}{k_0 K + p_0 \zeta(r_0)\|p\|} > \frac{1}{1 - k_0 \ell}. \tag{3.43}$$

We shall show that the operators $\mathcal{P}_1$ and $\mathcal{P}_2$ defined by (3.39) and (3.40) respectively, satisfy all the conditions of Lemma 1.9. The proof consists of a series of steps.
Step 1. *The operator $\mathcal{P}_1$ is continuous and completely continuous.* We first show that $\mathcal{P}_1(\bar{\Omega}_{r_0})$ is bounded. For any $x \in \bar{\Omega}_{r_0}$, we have

$$\begin{aligned}
\|(\mathcal{P}_1 x)\| \le |A| M_\sigma \Bigg\{ & \frac{1}{\Gamma(q+1)} + \frac{\left|\beta\left(\xi - \frac{\alpha}{2}(1-\eta^2)\right)\right|\theta^q}{|Q|\Gamma(q+1)} \\
& + \frac{|\alpha\beta|\theta}{|Q|}\left(\frac{1}{\Gamma(q+2)} + \frac{\eta^{q+1}}{\Gamma(q+2)}\right) + \frac{|\beta|\theta\xi^q}{|Q|\Gamma(q+1)} \\
& + \frac{1}{|Q|}\Bigg[|\alpha(1-\beta)|\left(\frac{1}{\Gamma(q+2)} + \frac{\eta^{q+1}}{\Gamma(q+2)}\right) \\
& + \frac{|1-\beta|\xi^q}{\Gamma(q+1)} + \frac{|\beta(1-\alpha(1-\eta))|\theta^q}{\Gamma(q+1)}\Bigg]\Bigg\}.
\end{aligned}$$

Thus the operator $\mathcal{P}_1(\bar{\Omega}_{r_0})$ is uniformly bounded. For any $t_1, t_2 \in [0,1], t_1 < t_2$, we have

$$\begin{aligned}
& |(\mathcal{P}_1 x)(t_2) - (\mathcal{P}_1 x)(t_1)| \\
\le\ & |A| M_r \left[\int_0^{t_1} [(t_2 - s)^{q-1} - (t_1 - s)^{q-1}] ds + \int_{t_1}^{t_2} (t_2 - s)^{q-1} ds\right] \\
& + \frac{|A| M_r |t_2 - t_1|}{|Q|}\Bigg[|\alpha(1-\beta)|\left(\frac{1}{\Gamma(q+2)} + \frac{\eta^{q+1}}{\Gamma(q+2)}\right) \\
& + \frac{|1-\beta|\xi^q}{\Gamma(q+1)} + \frac{|\beta(1-\alpha(1-\eta))|\theta^q}{\Gamma(q+1)}\Bigg],
\end{aligned}$$

which is independent of x and tends to zero as $t_2 - t_1 \to 0$. Thus, $\mathcal{P}_1$ is equicontinuous. Hence, by the Arzelá-Ascoli Theorem, $\mathcal{P}_1(\bar{\Omega}_{r_0})$ is a relatively compact set. Now, let $x_n \in \bar{\Omega}_{r_0}$ with $\|x_n - x\| \to 0$. Then the limit $\|x_n(t) - x(t)\| \to 0$ is uniformly valid on $[0,1]$. From the uniform continuity of $f(t,x)$ on the compact set $[0,1] \times [-r_0, r_0]$, it follows that $\|f(t,x_n(t)) - f(t,x(t))\| \to 0$ is uniformly valid on $[0,1]$. Hence $\|\mathcal{P}_1 x_n - \mathcal{P}_1 x\| \to 0$ as $n \to \infty$. This shows the continuity of $\mathcal{P}_1$.

Step 2. *The operator $\mathcal{P}_2 : \bar{\Omega}_{r_0} \to C([0,1],\mathbb{R})$ is contractive.*

$$\begin{aligned}
&|(\mathcal{P}_2 x)(t) - (\mathcal{P}_2 y)(t)| \\
&\le |B| \Bigg\{ \frac{1}{\Gamma(q+r+1)} + \frac{\left|\beta\left(\xi - \frac{\alpha}{2}(1-\eta^2)\right)\right| \theta^{q+r}}{|Q|\Gamma(q+r+1)} \\
&\quad + \frac{|\alpha\beta|\theta}{|Q|}\left(\frac{1}{\Gamma(q+r+2)} + \frac{\eta^{q+r+1}}{\Gamma(q+r+2)}\right) + \frac{|\beta|\theta\xi^{q+r}}{|Q|\Gamma(q+r+1)} \\
&\quad + \frac{1}{|Q|}\Bigg[|\alpha(1-\beta)|\left(\frac{1}{\Gamma(q+r+2)} + \frac{\eta^{q+r+1}}{\Gamma(q+r+2)}\right) \\
&\quad + \frac{|1-\beta|\xi^{q+r}}{\Gamma(q+r+1)} + \frac{|\beta(1-\alpha(1-\eta))|\theta^{q+r}}{\Gamma(q+r+1)}\Bigg]\Bigg\} \phi(\|x-y\|).
\end{aligned}$$

This, together with (3.14.2), implies that

$$\|(\mathcal{P}_2 x) - (\mathcal{P}_2 y)\| \le \phi(\|x-y\|),$$

so $\mathcal{P}_2 : \bar{\Omega}_{r_0} \to C([0,1],\mathbb{R})$ is a nonlinear contraction.

Step 3. *The set $\mathcal{P}(\bar{\Omega}_{r_0})$ is bounded.* Using the inequality

$$|g(t,x)| \le |g(t,x) - g(t,0)| + |g(t,0)| \le \phi(\|x\|) + K \le \ell r_0 + K,$$

we have

$$\begin{aligned}
\|\mathcal{P}_2(x)\| \le |B|(\ell r_0 + K) &\Bigg\{ \frac{1}{\Gamma(q+r+1)} + \frac{\left|\beta\left(\xi - \frac{\alpha}{2}(1-\eta^2)\right)\right| \theta^{q+r}}{|Q|\Gamma(q+r+1)} \\
&+ \frac{|\alpha\beta|\theta}{|Q|}\left(\frac{1}{\Gamma(q+r+2)} + \frac{\eta^{q+r+1}}{\Gamma(q+r+2)}\right) + \frac{|\beta|\theta\xi^{q+r}}{|Q|\Gamma(q+r+1)} \\
&+ \frac{1}{|Q|}\Bigg[|\alpha(1-\beta)|\left(\frac{1}{\Gamma(q+r+2)} + \frac{\eta^{q+r+1}}{\Gamma(q+r+2)}\right) \\
&+ \frac{|1-\beta|\xi^{q+r}}{\Gamma(q+r+1)} + \frac{|\beta(1-\alpha(1-\eta))|\theta^{q+r}}{\Gamma(q+r+1)}\Bigg]\Bigg\},
\end{aligned}$$

for any $x \in \bar{\Omega}_{r_0}$. This, with the boundedness of the set $\mathcal{P}_1(\bar{\Omega}_{r_0})$, implies that the set $\mathcal{P}(\bar{\Omega}_{r_0})$ is bounded.

Step 4. *Finally, it will be shown that the case (C2) in Theorem 1.9 does not hold.* On the contrary, we suppose that (C2) holds. Then, we have that there exist $\lambda \in (0,1)$ and $x \in \partial\Omega_{r_0}$ such that $x = \lambda \mathcal{P} x$. So, we have $\|x\| = r_0$ and

$$
\begin{aligned}
&x(t)\\
&= \lambda A\int_0^t \frac{(t-s)^{q-1}}{\Gamma(q)}f(s,x(s))ds + \lambda A\frac{\beta\left(\xi-\frac{\alpha}{2}(1-\eta^2)\right)}{Q}\int_0^\theta \frac{(\theta-s)^{q-1}}{\Gamma(q)}f(s,x(s))ds\\
&+\lambda A\frac{\alpha\beta\theta}{Q}\left(\int_0^1 \frac{(1-s)^q}{\Gamma(q+1)}f(s,x(s))ds - \int_0^\eta \frac{(\eta-s)^q}{\Gamma(q+1)}f(s,x(s))ds\right)\\
&-\lambda A\frac{\beta\theta}{Q}\int_0^\xi \frac{(\xi-s)^q}{\Gamma(q)}f(s,x(s))ds + \frac{\lambda At}{Q}\Bigg[\alpha(1-\beta)\left(\int_0^1 \frac{(1-s)^q}{\Gamma(q+1)}f(s,x(s))ds\right.\\
&\left.-\int_0^\eta \frac{(\eta-s)^q}{\Gamma(q+1)}f(s,x(s))ds\right) - (1-\beta)\int_0^\xi \frac{(\xi-s)^q}{\Gamma(q)}f(s,x(s))ds\\
&-\beta(1-\alpha(1-\eta))\int_0^\theta \frac{(\theta-s)^{q-1}}{\Gamma(q)}f(s,x(s))ds\Bigg]\\
&+\lambda B\int_0^t \frac{(t-s)^{q-1}}{\Gamma(q)}\int_0^s \frac{(s-u)^{r-1}}{\Gamma(r)}g(u,x(u))duds\\
&+\lambda B\frac{\beta\left(\xi-\frac{\alpha}{2}(1-\eta^2)\right)}{Q}\int_0^\theta \frac{(\theta-s)^{q-1}}{\Gamma(q)}\int_0^s \frac{(s-u)^{r-1}}{\Gamma(r)}g(u,x(u))duds\\
&+\lambda B\frac{\alpha\beta\theta}{Q}\left(\int_0^1 \frac{(1-s)^q}{\Gamma(q+1)}\int_0^s \frac{(s-u)^{r-1}}{\Gamma(r)}g(u,x(u))duds\right.\\
&\left.-\int_0^\eta \frac{(\eta-s)^q}{\Gamma(q+1)}\int_0^s \frac{(s-u)^{r-1}}{\Gamma(r)}g(u,x(u))duds\right)\\
&-\lambda B\frac{\beta\theta}{Q}\int_0^\xi \frac{(\xi-s)^q}{\Gamma(q)}\int_0^s \frac{(s-u)^{r-1}}{\Gamma(r)}g(u,x(u))duds\\
&+\frac{\lambda Bt}{Q}\Bigg[\alpha(1-\beta)\left(\int_0^1 \frac{(1-s)^q}{\Gamma(q+1)}\int_0^s \frac{(s-u)^{r-1}}{\Gamma(r)}g(u,x(u))duds\right.\\
&\left.-\int_0^\eta \frac{(\eta-s)^q}{\Gamma(q+1)}\int_0^s \frac{(s-u)^{r-1}}{\Gamma(r)}g(u,x(u))duds\right)\\
&-(1-\beta)\int_0^\xi \frac{(\xi-s)^q}{\Gamma(q)}\int_0^s \frac{(s-u)^{r-1}}{\Gamma(r)}g(u,x(u))duds\\
&-\beta(1-\alpha(1-\eta))\int_0^\theta \frac{(\theta-s)^{q-1}}{\Gamma(q)}\int_0^s \frac{(s-u)^{r-1}}{\Gamma(r)}g(u,x(u))duds\Bigg].
\end{aligned}
$$

Using the assumptions (3.14.1) and (3.14.2), we get

$$r_0 \le k_0 K + p_0\zeta(r_0)\|p\| + k_0\ell r_0,$$

which leads to a contradiction:

$$\frac{r_0}{k_0K + p_0\zeta(r_0)\|p\|} \le \frac{1}{1-k_0\ell}.$$

Thus the operators $\mathcal{P}_1$ and $\mathcal{P}_2$ satisfy all the conditions of Theorem 1.9. Hence, the operator $\mathcal{P}$ has at least one fixed point $x \in \bar{\Omega}_{r_0}$, which is the solution of the problem (3.33). This completes the proof. □

Example 3.9. Consider the following nonlocal integral boundary value problem for a fractional integro-differential equation

$$\begin{cases} D^{3/2}x(t) = f(t,x(t)) + I^{3/4}g(t,x(t)), & t \in [0,1], \\ x(0) = \dfrac{1}{2}x(1/4)), \quad x(1/3) = \displaystyle\int_{2/3}^{1} x(s)ds, \end{cases} \tag{3.44}$$

where $q = 3/2$, $A = B = 1$, $r = 3/4$, $\theta = 1/4$, $\xi = 1/3$, $\eta = 2/3$, $\alpha = 1$, $\beta = 1/2$, $f(t,x) = \dfrac{1}{27}\Big(2\sqrt{(1+t)} - 1\Big)\sin x$, $g(t,x) = \dfrac{1}{(2+t)^3}\Big(3 + \dfrac{|x|}{1+|x|}\Big)$.

Observe that $|f(t,x)| \leq \frac{1}{27}\Big(2\sqrt{(1+t)} - 1\Big)x$, $|g(t,x) - g(t,y)| \leq \frac{1}{8}|x-y|$, and $\sup_{t\in[0,1]} g(t,0) = 3/8$. Further, we set $p(t) = \dfrac{1}{3}\Big(2\sqrt{(1+t)} - 1\Big)$, $\zeta(x) = x/9$, $\ell = 1/8, K = 3/8$. With the given data, it is found that $k_0 \simeq 1.495606$, $p_0 \simeq 4.178968$, $|Q| = 1/9$, $\|p\| = (2\sqrt{2}-1)/3$, and

$$\sup_{r\in(0,\infty)} \left\{ \frac{r}{k_0 K + p_0\zeta(r)\|p\|} \right\} \simeq 3.533597, \quad \frac{1}{1-k_0\ell} \simeq 1.229953.$$

Clearly, all the conditions of Theorem 3.15 are satisfied and hence there exists a solution for the problem (3.44).

3.3.2 *Nonlocal integral boundary value problem for fractional integro-differential inclusions*

In this subsection, we consider the following boundary value problem of fractional integro-differential inclusions with nonlocal integral boundary conditions:

$$\begin{cases} {}^cD^q x(t) \in AF(t,x(t)) + BI^r G(t,x(t)), & t \in [0,1], \\ x(0) = \beta x(\theta), \quad x(\xi) = \alpha \displaystyle\int_{\eta}^{1} x(s)ds, \quad 0 < \theta < \xi < \eta < 1, \end{cases} \tag{3.45}$$

where ${}^cD^q$ denotes the Caputo fractional derivative operator of order q, $F, G : [0,1] \times \mathbb{R} \to \mathcal{P}(\mathbb{R})$ are multi-valued maps, $\mathcal{P}(\mathbb{R})$ is the family of all non-empty subsets of $\mathbb{R}$, $1 < q \leq 2, 0 < r < 1$, and α, β, A, B are real constants

Definition 3.3. A function $x \in C^2([0,1],\mathbb{R})$ is said to be a solution of the problem (3.45) if $x(0) = \beta x(\theta)$, $x(\xi) = \alpha \displaystyle\int_{\eta}^{1} x(s)ds$, and there exist functions $f \in S_{F,x}, g \in S_{G,x}$ such that

$$\begin{aligned} x(t) &= A\int_0^t \frac{(t-s)^{q-1}}{\Gamma(q)} f(s)ds + A\frac{\beta\Big(\xi - \frac{\alpha}{2}(1-\eta^2)\Big)}{Q}\int_0^\theta \frac{(\theta-s)^{q-1}}{\Gamma(q)} f(s)ds \\ &\quad + A\frac{\alpha\beta\theta}{Q}\left(\int_0^1 \frac{(1-s)^q}{\Gamma(q+1)} f(s)ds - \int_0^\eta \frac{(\eta-s)^q}{\Gamma(q+1)} f(s)ds\right) \\ &\quad - A\frac{\beta\theta}{Q}\int_0^\xi \frac{(\xi-s)^q}{\Gamma(q)} f(s)ds + \frac{At}{Q}\Bigg[\alpha(1-\beta)\left(\int_0^1 \frac{(1-s)^q}{\Gamma(q+1)} f(s)ds\right. \end{aligned}$$

$$
\begin{aligned}
&-\int_0^\eta \frac{(\eta-s)^q}{\Gamma(q+1)}f(s)ds\Bigg) - (1-\beta)\int_0^\xi \frac{(\xi-s)^q}{\Gamma(q)}f(s)ds \\
&-\beta(1-\alpha(1-\eta))\int_0^\theta \frac{(\theta-s)^{q-1}}{\Gamma(q)}f(s)ds\Bigg] \\
&+B\int_0^t \frac{(t-s)^{q-1}}{\Gamma(q)}\int_0^s \frac{(s-u)^{r-1}}{\Gamma(r)}g(u)duds \\
&+B\frac{\beta\left(\xi-\frac{\alpha}{2}(1-\eta^2)\right)}{Q}\int_0^\theta \frac{(\theta-s)^{q-1}}{\Gamma(q)}\int_0^s \frac{(s-u)^{r-1}}{\Gamma(r)}g(u)duds \\
&+B\frac{\alpha\beta\theta}{Q}\Bigg(\int_0^1 \frac{(1-s)^q}{\Gamma(q+1)}\int_0^s \frac{(s-u)^{r-1}}{\Gamma(r)}g(u)duds \\
&-\int_0^\eta \frac{(\eta-s)^q}{\Gamma(q+1)}\int_0^s \frac{(s-u)^{r-1}}{\Gamma(r)}g(u)duds\Bigg) \\
&-B\frac{\beta\theta}{Q}\int_0^\xi \frac{(\xi-s)^q}{\Gamma(q)}\int_0^s \frac{(s-u)^{r-1}}{\Gamma(r)}g(u)duds \\
&+\frac{Bt}{Q}\Bigg[\alpha(1-\beta)\Bigg(\int_0^1 \frac{(1-s)^q}{\Gamma(q+1)}\int_0^s \frac{(s-u)^{r-1}}{\Gamma(r)}g(u)duds \\
&-\int_0^\eta \frac{(\eta-s)^q}{\Gamma(q+1)}\int_0^s \frac{(s-u)^{r-1}}{\Gamma(r)}g(u)duds\Bigg) \\
&-(1-\beta)\int_0^\xi \frac{(\xi-s)^q}{\Gamma(q)}\int_0^s \frac{(s-u)^{r-1}}{\Gamma(r)}g(u)duds \\
&-\beta(1-\alpha(1-\eta))\int_0^\theta \frac{(\theta-s)^{q-1}}{\Gamma(q)}\int_0^s \frac{(s-u)^{r-1}}{\Gamma(r)}g(u)duds\Bigg].
\end{aligned}
$$

In the sequel, we set

$$
\begin{aligned}
Z_1 = |A|\Bigg\{&\frac{1}{\Gamma(q+1)} + \frac{\left|\beta\left(\xi-\frac{\alpha}{2}(1-\eta^2)\right)\right|\theta^q}{|Q|\Gamma(q+1)} + \frac{|\alpha\beta|\theta}{|Q|}\left(\frac{1}{\Gamma(q+2)}+\frac{\eta^{q+1}}{\Gamma(q+2)}\right) \\
&+\frac{|\beta|\theta\xi^q}{|Q|\Gamma(q+1)} + \frac{1}{|Q|}\Bigg[|\alpha(1-\beta)|\left(\frac{1}{\Gamma(q+2)}+\frac{\eta^{q+1}}{\Gamma(q+2)}\right) \\
&+\frac{|1-\beta|\xi^q}{\Gamma(q+1)} + \frac{|\beta(1-\alpha(1-\eta))|\theta^q}{\Gamma(q+1)}\Bigg]\Bigg\}, \qquad (3.46)
\end{aligned}
$$

$$
\begin{aligned}
Z_2 = |B|\Bigg\{&\frac{1}{\Gamma(q+r+1)} + \frac{\left|\beta\left(\xi-\frac{\alpha}{2}(1-\eta^2)\right)\right|\theta^{q+r}}{|Q|\Gamma(q+r+1)} \\
&+\frac{|\alpha\beta|\theta}{|Q|}\left(\frac{1}{\Gamma(q+r+2)}+\frac{\eta^{q+r+1}}{\Gamma(q+r+2)}\right) + \frac{|\beta|\theta\xi^{q+r}}{|Q|\Gamma(q+r+1)}
\end{aligned}
$$

$$+\frac{1}{|Q|}\left[|\alpha(1-\beta)|\left(\frac{1}{\Gamma(q+r+2)}+\frac{\eta^{q+r+1}}{\Gamma(q+r+2)}\right)\right.$$
$$\left.+\frac{|1-\beta|\xi^{q+r}}{\Gamma(q+r+1)}+\frac{|\beta(1-\alpha(1-\eta))|\theta^{q+r}}{\Gamma(q+r+1)}\right]\Bigg\}. \tag{3.47}$$

3.3.2.1 *The Carathéodory case*

Theorem 3.16. *Assume that*

(3.16.1) $F:[0,1]\times\mathbb{R}\to\mathcal{P}_{cp,c}(\mathbb{R})$ *is* L^1*-Carathéodory multi-valued map;*

(3.16.2) there exists a function $k\in C([0,1],\mathbb{R}^+)$ *such that*

$$H(F(t,x),F(t,y))\le k(t)\|x-y\| \quad a.e. \quad t\in[0,1],$$

for all $x,y\in C([0,1],\mathbb{R})$ *and* $Z_1\|k\|<1$, *where* Z_1 *is given by (3.46);*

(3.16.3) $G:[0,1]\times\mathbb{R}\to\mathcal{P}_{cp,c}(\mathbb{R})$ *is* L^1*-Carathéodory multi-valued map;*

(3.16.4) there exists a function $m\in C([0,1],\mathbb{R})$ *with* $m(t)>0$ *for a.e.* $t\in[0,1]$ *and a non-decreasing function* $\psi:\mathbb{R}^+\to(0,\infty)$ *such that*

$$\|G(t,x)\|_{\mathcal{P}}:=\sup\{|v|:v\in G(t,x)\}\le m(t)\psi(\|x\|) \ \ a.e. \ t\in[0,1],$$

for all $x\in\mathbb{R}$;

(3.16.5) there exists a number $M>0$ *such that*

$$\frac{(1-Z_1\|k\|)M}{Z_1F_0+Z_2\|m\|\psi(M)}>1, \tag{3.48}$$

where Z_1,Z_2 *are given by (3.46) and (3.47) respectively and* $F_0=\int_0^1|F(t,0)|dt$.

Then the problem (3.45) has a solution on $[0,1]$.

Proof. To transform the problem (3.45) into a fixed point problem, let us define an operator $\mathcal{N}:C([0,1],\mathbb{R})\longrightarrow\mathcal{P}(C([0,1],\mathbb{R}))$ as

$$\mathcal{N}=\mathcal{N}_1+\mathcal{N}_2,$$

with

$$\mathcal{N}_1(x)=\{h\in C([0,1],\mathbb{R}):h(t)=(F_1x)(t)\},$$

where

$$(F_1x)(t)=A\int_0^t\frac{(t-s)^{q-1}}{\Gamma(q)}f(s)ds+A\frac{\beta\Big(\xi-\frac{\alpha}{2}(1-\eta^2)\Big)}{Q}\int_0^\theta\frac{(\theta-s)^{q-1}}{\Gamma(q)}f(s)ds$$
$$+A\frac{\alpha\beta\theta}{Q}\left(\int_0^1\frac{(1-s)^q}{\Gamma(q+1)}f(s)ds-\int_0^\eta\frac{(\eta-s)^q}{\Gamma(q+1)}f(s)ds\right)$$
$$-A\frac{\beta\theta}{Q}\int_0^\xi\frac{(\xi-s)^q}{\Gamma(q)}f(s)ds+\frac{At}{Q}\left[\alpha(1-\beta)\left(\int_0^1\frac{(1-s)^q}{\Gamma(q+1)}f(s)ds\right.\right.$$

$$-\int_0^\eta \frac{(\eta-s)^q}{\Gamma(q+1)}f(s)ds\Bigg) - (1-\beta)\int_0^\xi \frac{(\xi-s)^q}{\Gamma(q)}f(s)ds$$
$$-\beta(1-\alpha(1-\eta))\int_0^\theta \frac{(\theta-s)^{q-1}}{\Gamma(q)}f(s)ds\Bigg],$$

for $f \in S_{F,x}$ and

$$\mathcal{N}_2(x) = \{h \in C([0,1],\mathbb{R}) : h(t) = (F_2x)(t)\},$$

where

$$\begin{aligned}(F_2x)(t) &= B\int_0^t \frac{(t-s)^{q-1}}{\Gamma(q)}\int_0^s \frac{(s-u)^{r-1}}{\Gamma(r)}g(u)duds\\ &+B\frac{\beta\left(\xi-\frac{\alpha}{2}(1-\eta^2)\right)}{Q}\int_0^\theta \frac{(\theta-s)^{q-1}}{\Gamma(q)}\int_0^s \frac{(s-u)^{r-1}}{\Gamma(r)}g(u)duds\\ &+B\frac{\alpha\beta\theta}{Q}\Bigg(\int_0^1 \frac{(1-s)^q}{\Gamma(q+1)}\int_0^s \frac{(s-u)^{r-1}}{\Gamma(r)}g(u)duds\\ &-\int_0^\eta \frac{(\eta-s)^q}{\Gamma(q+1)}\int_0^s \frac{(s-u)^{r-1}}{\Gamma(r)}g(u)duds\Bigg)\\ &-B\frac{\beta\theta}{Q}\int_0^\xi \frac{(\xi-s)^q}{\Gamma(q)}\int_0^s \frac{(s-u)^{r-1}}{\Gamma(r)}g(u)duds\\ &+\frac{Bt}{Q}\Bigg[\alpha(1-\beta)\Bigg(\int_0^1 \frac{(1-s)^q}{\Gamma(q+1)}\int_0^s \frac{(s-u)^{r-1}}{\Gamma(r)}g(u)duds\\ &-\int_0^\eta \frac{(\eta-s)^q}{\Gamma(q+1)}\int_0^s \frac{(s-u)^{r-1}}{\Gamma(r)}g(u)duds\Bigg)\\ &-(1-\beta)\int_0^\xi \frac{(\xi-s)^q}{\Gamma(q)}\int_0^s \frac{(s-u)^{r-1}}{\Gamma(r)}g(u)duds\\ &-\beta(1-\alpha(1-\eta))\int_0^\theta \frac{(\theta-s)^{q-1}}{\Gamma(q)}\int_0^s \frac{(s-u)^{r-1}}{\Gamma(r)}g(u)duds\Bigg],\end{aligned}$$

for $g \in S_{G,x}$.

We shall show that the operators $\mathcal{N}_1$ and $\mathcal{N}_2$ satisfy all the conditions of Theorem 1.18 on $[0,1]$. For the sake of clarity, we split the proof into a sequence of steps and claims.

Step 1. We show that $\mathcal{N}_1$ *is a multi-valued contraction on* $C([0,1],\mathbb{R})$.

Let $x, y \in C([0,1],\mathbb{R})$ and $u_1 \in \mathcal{N}_1(x)$. Then $u_1 \in \mathcal{P}(C([0,1],\mathbb{R}))$ and

$$u_1(t) = A\int_0^t \frac{(t-s)^{q-1}}{\Gamma(q)}v_1(s)ds + A\frac{\beta\left(\xi-\frac{\alpha}{2}(1-\eta^2)\right)}{Q}\int_0^\theta \frac{(\theta-s)^{q-1}}{\Gamma(q)}v_1(s)ds$$

$$+A\frac{\alpha\beta\theta}{Q}\left(\int_0^1\frac{(1-s)^q}{\Gamma(q+1)}v_1(s)ds-\int_0^\eta\frac{(\eta-s)^q}{\Gamma(q+1)}v_1(s)ds\right)$$
$$-A\frac{\beta\theta}{Q}\int_0^\xi\frac{(\xi-s)^q}{\Gamma(q)}v_1(s)ds+\frac{At}{Q}\left[\alpha(1-\beta)\left(\int_0^1\frac{(1-s)^q}{\Gamma(q+1)}v_1(s)ds\right.\right.$$
$$\left.-\int_0^\eta\frac{(\eta-s)^q}{\Gamma(q+1)}v_1ds\right)-(1-\beta)\int_0^\xi\frac{(\xi-s)^q}{\Gamma(q)}v_1(s)ds$$
$$\left.-\beta(1-\alpha(1-\eta))\int_0^\theta\frac{(\theta-s)^{q-1}}{\Gamma(q)}v_1(s)ds\right],$$

for some $v_1\in S_{F,x}$. Since $H(F(t,x),F(t,y)\le k(t)\|x-y\|$, therefore there exists $w\in F(t,y)$ such that $|v_1(t)-w(t)|\le k(t)\|x-y\|$. Thus the multi-valued operator U is defined by $U(t)=S_{F,y}\cap K(t)$, where

$$K(t)=\{w\in\mathbb{R}|\ |v_1(t)-w(t)|\le k(t)\|x-y\|\}$$

has non-empty values and is measurable. Let v_2 be a measurable selection for U (which exists by Kuratowski-Ryll-Nardzewski's selection theorem [91]). Then $v_2\in F(t,y)$ and $|v_1(t)-v_2(t)|\le k(t)\|x-y\|$ a.e. on $[0,1]$.

Define

$$u_2(t)=A\int_0^t\frac{(t-s)^{q-1}}{\Gamma(q)}v_2(s)ds+A\frac{\beta\Big(\xi-\frac{\alpha}{2}(1-\eta^2)\Big)}{Q}\int_0^\theta\frac{(\theta-s)^{q-1}}{\Gamma(q)}v_2(s)ds$$
$$+A\frac{\alpha\beta\theta}{Q}\left(\int_0^1\frac{(1-s)^q}{\Gamma(q+1)}v_2(s)ds-\int_0^\eta\frac{(\eta-s)^q}{\Gamma(q+1)}v_2(s)ds\right)$$
$$-A\frac{\beta\theta}{Q}\int_0^\xi\frac{(\xi-s)^q}{\Gamma(q)}v_2(s)ds+\frac{At}{Q}\left[\alpha(1-\beta)\left(\int_0^1\frac{(1-s)^q}{\Gamma(q+1)}v_2(s)ds\right.\right.$$
$$\left.-\int_0^\eta\frac{(\eta-s)^q}{\Gamma(q+1)}v_2ds\right)-(1-\beta)\int_0^\xi\frac{(\xi-s)^q}{\Gamma(q)}v_2(s)ds$$
$$\left.-\beta(1-\alpha(1-\eta))\int_0^\theta\frac{(\theta-s)^{q-1}}{\Gamma(q)}v_2(s)ds\right].$$

It follows that $u_2\in\mathcal{N}_1(y)$ and

$$|u_1(t)-u_2(t)|$$
$$\le|A|\int_0^t\frac{(t-s)^{q-1}}{\Gamma(q)}|v_1(s)-v_2(s)|ds$$
$$+|A|\frac{\beta\Big(\xi-\frac{\alpha}{2}(1-\eta^2)\Big)}{Q}\int_0^\theta\frac{(\theta-s)^{q-1}}{\Gamma(q)}|v_1(s)-v_2(s)|ds$$
$$+|A|\frac{\alpha\beta\theta}{Q}\left(\int_0^1\frac{(1-s)^q}{\Gamma(q+1)}|v_1(s)-v_2(s)|ds+\int_0^\eta\frac{(\eta-s)^q}{\Gamma(q+1)}|v_1(s)-v_2(s)|ds\right)$$

$$+|A|\frac{\beta\theta}{Q}\int_0^{\xi}\frac{(\xi-s)^q}{\Gamma(q)}|v_1(s)-v_2(s)|ds$$
$$+\frac{|A|t}{Q}\Bigg[\alpha(1-\beta)\Bigg(\int_0^1\frac{(1-s)^q}{\Gamma(q+1)}|v_1(s)-v_2(s)|ds$$
$$+\int_0^{\eta}\frac{(\eta-s)^q}{\Gamma(q+1)}v_1ds\Bigg)-(1-\beta)\int_0^{\xi}\frac{(\xi-s)^q}{\Gamma(q)}|v_1(s)-v_2(s)|ds$$
$$-\beta(1-\alpha(1-\eta))\int_0^{\theta}\frac{(\theta-s)^{q-1}}{\Gamma(q)}|v_1(s)-v_2(s)|ds\Bigg]$$
$$\leq|A|\Bigg\{\frac{1}{\Gamma(q+1)}+\frac{\Big|\beta\Big(\xi-\frac{\alpha}{2}(1-\eta^2)\Big)\Big|\theta^q}{|Q|\Gamma(q+1)}+\frac{|\alpha\beta|\theta}{|Q|}\Bigg(\frac{1}{\Gamma(q+2)}+\frac{\eta^{q+1}}{\Gamma(q+2)}\Bigg)$$
$$+\frac{|\beta|\theta\xi^q}{|Q|\Gamma(q+1)}+\frac{1}{|Q|}\Bigg[|\alpha(1-\beta)|\Bigg(\frac{1}{\Gamma(q+2)}+\frac{\eta^{q+1}}{\Gamma(q+2)}\Bigg)$$
$$+\frac{|1-\beta|\xi^q}{\Gamma(q+1)}+\frac{|\beta(1-\alpha(1-\eta))|\theta^q}{\Gamma(q+1)}\Bigg]\Bigg\}\|k\|\|x-y\|.$$

Taking the supremum over the interval $[0,1]$, we obtain

$$\|u_1-u_2\|$$
$$\leq|A|\Bigg\{\frac{1}{\Gamma(q+1)}+\frac{\Big|\beta\Big(\xi-\frac{\alpha}{2}(1-\eta^2)\Big)\Big|\theta^q}{|Q|\Gamma(q+1)}+\frac{|\alpha\beta|\theta}{|Q|}\Bigg(\frac{1}{\Gamma(q+2)}+\frac{\eta^{q+1}}{\Gamma(q+2)}\Bigg)$$
$$+\frac{|\beta|\theta\xi^q}{|Q|\Gamma(q+1)}+\frac{1}{|Q|}\Bigg[|\alpha(1-\beta)|\Bigg(\frac{1}{\Gamma(q+2)}+\frac{\eta^{q+1}}{\Gamma(q+2)}\Bigg)$$
$$+\frac{|1-\beta|\xi^q}{\Gamma(q+1)}+\frac{|\beta(1-\alpha(1-\eta))|\theta^q}{\Gamma(q+1)}\Bigg]\Bigg\}\|k\|\|x-y\|.$$

Combining the previous inequality with the corresponding one obtained by interchanging the roles of x and y, we get that

$$H(\mathcal{N}_1(x),\mathcal{N}_1(y))$$
$$\leq|A|\Bigg\{\frac{1}{\Gamma(q+1)}+\frac{\Big|\beta\Big(\xi-\frac{\alpha}{2}(1-\eta^2)\Big)\Big|\theta^q}{|Q|\Gamma(q+1)}+\frac{|\alpha\beta|\theta}{|Q|}\Bigg(\frac{1}{\Gamma(q+2)}+\frac{\eta^{q+1}}{\Gamma(q+2)}\Bigg)$$
$$+\frac{|\beta|\theta\xi^q}{|Q|\Gamma(q+1)}+\frac{1}{|Q|}\Bigg[|\alpha(1-\beta)|\Bigg(\frac{1}{\Gamma(q+2)}+\frac{\eta^{q+1}}{\Gamma(q+2)}\Bigg)$$
$$+\frac{|1-\beta|\xi^q}{\Gamma(q+1)}+\frac{|\beta(1-\alpha(1-\eta))|\theta^q}{\Gamma(q+1)}\Bigg]\Bigg\}\|k\|\|x-y\|,$$

for all $x,y\in C([0,1],\mathbb{R})$. This shows that $\mathcal{N}_1$ is a multi-valued contraction as

$$Z_1\|k\|=|A|\Bigg\{\frac{1}{\Gamma(q+1)}+\frac{\Big|\beta\Big(\xi-\frac{\alpha}{2}(1-\eta^2)\Big)\Big|\theta^q}{|Q|\Gamma(q+1)}+\frac{|\alpha\beta|\theta}{|Q|}\Bigg(\frac{1}{\Gamma(q+2)}+\frac{\eta^{q+1}}{\Gamma(q+2)}\Bigg)$$

$$+\frac{|\beta|\theta\xi^q}{|Q|\Gamma(q+1)}+\frac{1}{|Q|}\Bigg[|\alpha(1-\beta)|\left(\frac{1}{\Gamma(q+2)}+\frac{\eta^{q+1}}{\Gamma(q+2)}\right)$$
$$+\frac{|1-\beta|\xi^q}{\Gamma(q+1)}+\frac{|\beta(1-\alpha(1-\eta))|\theta^q}{\Gamma(q+1)}\Bigg]\Bigg\}\|k\|<1.$$

Step 2. We shall show that the operator $\mathcal{N}_2$ *is u.s.c. and compact.* It is well known (Lemma 1.1) that if an operator is completely continuous and has a closed graph, then it is u.s.c. Therefore we will prove that $\mathcal{N}_2$ is completely continuous and has a closed graph. This step involves several claims.

Claim I. *$\mathcal{N}_2$ maps bounded sets into bounded sets in $C([0,1],\mathbb{R})$.*

Let $B_r=\{x\in C([0,1],\mathbb{R}):\|x\|\le r\}$ be a bounded set in $C([0,1],\mathbb{R})$ and $u\in\mathcal{N}_2(x)$ for some $x\in B_r$. In calculating the upper bounds for the operator $\mathcal{P}_2$, the beta-function will be used. For instance,

$$\begin{aligned}\int_0^t\frac{(t-s)^{q-1}}{\Gamma(q)}\int_0^s\frac{(s-u)^{r-1}}{\Gamma(r)}duds&=\int_0^t\frac{(t-s)^{q-1}}{\Gamma(q)}\frac{(s)^r}{\Gamma(r+1)}ds\\&=\frac{t^{q+r}}{\Gamma(q)\Gamma(r+1)}\int_0^1(1-v)^{q-1}v^r dv\\&=\frac{t^{q+r}}{\Gamma(q+r+1)}\end{aligned}$$

since $\int_0^1(1-u)^{\alpha-1}u^\beta du=\frac{\Gamma(\alpha)\Gamma(\beta+1)}{\Gamma(\alpha+\beta+1)}$. Thus we have

$$\begin{aligned}|(\mathcal{P}_2x)(t)|\le|B|\|m\|\psi(r)\Bigg\{&\frac{1}{\Gamma(q+r+1)}+\frac{\left|\beta\left(\xi-\frac{\alpha}{2}(1-\eta^2)\right)\right|\theta^{q+r}}{|Q|\Gamma(q+r+1)}\\&+\frac{|\alpha\beta|\theta}{|Q|}\left(\frac{1}{\Gamma(q+r+2)}+\frac{\eta^{q+r+1}}{\Gamma(q+r+2)}\right)+\frac{|\beta|\theta\xi^{q+r}}{|Q|\Gamma(q+r+1)}\\&+\frac{1}{|Q|}\Bigg[|\alpha(1-\beta)|\left(\frac{1}{\Gamma(q+r+2)}+\frac{\eta^{q+r+1}}{\Gamma(q+r+2)}\right)\\&+\frac{|1-\beta|\xi^{q+r}}{\Gamma(q+r+1)}+\frac{|\beta(1-\alpha(1-\eta))|\theta^{q+r}}{\Gamma(q+r+1)}\Bigg]\Bigg\}.\end{aligned}$$

Hence $\mathcal{N}_2$ is bounded.

Claim II. *$\mathcal{N}_2$ maps bounded sets into equi-continuous sets.*

As in the proof of Claim I, let B_r be a bounded set and $u\in\mathcal{N}_2(x)$ for some $x\in B_r$. Let $t_1,t_2\in[0,1]$ with $t_1<t_2$ and $x\in B_r$. Then we obtain

$$|(\mathcal{P}_2x)(t_2)-(\mathcal{P}_2x)(t_1)|$$

$$
\begin{aligned}
&\leq \frac{|B|\psi(r)}{\Gamma(q+r+1)}\int_0^{t_1}[(t_2-s)^{q-1}-(t_1-s)^{q-1}]m(s)ds \\
&+\frac{|B|\psi(r)}{\Gamma(q+r+1)}\int_{t_1}^{t_2}(t_2-s)^{q-1}m(s)ds \\
&+\frac{|B|\psi(r)\|m\||t_2-t_1|}{|Q|}\Bigg[|\alpha(1-\beta)|\left(\frac{1}{\Gamma(q+r+2)}+\frac{\eta^{q+r+1}}{\Gamma(q+r+2)}\right) \\
&+\frac{|1-\beta|\xi^{q+r}}{\Gamma(q+r+1)}+\frac{|\beta(1-\alpha(1-\eta))|\theta^{q+r}}{\Gamma(q+r+1)}\Bigg].
\end{aligned}
$$

Obviously the right hand side of the above inequality tends to zero independently of $x \in B_r$ as $t_1 - t_2 \to 0$. Therefore it follows by the Arzelá-Ascoli theorem that $\mathcal{N}_2 : C([0,1],\mathbb{R}) \to \mathcal{P}(C([0,1],\mathbb{R}))$ is completely continuous.

Claim III. *Next we prove that $\mathcal{N}_2$ has a closed graph.*

Let $x_n \to x_*, h_n \in \mathcal{N}_2(x_n)$ and $h_n \to h_*$. Then we need to show that $h_* \in \mathcal{B}(x_*)$. Associated with $h_n \in \mathcal{B}(x_n)$, there exists $v_n \in S_{G,x_n}$ such that for each $t \in [0,1]$,

$$
\begin{aligned}
h_n(t) = B&\int_0^t \frac{(t-s)^{q-1}}{\Gamma(q)}\int_0^s \frac{(s-u)^{r-1}}{\Gamma(r)}v_n(u)duds \\
&+B\frac{\beta\left(\xi-\frac{\alpha}{2}(1-\eta^2)\right)}{Q}\int_0^\theta \frac{(\theta-s)^{q-1}}{\Gamma(q)}\int_0^s \frac{(s-u)^{r-1}}{\Gamma(r)}v_n(u)duds \\
&+B\frac{\alpha\beta\theta}{Q}\left(\int_0^1 \frac{(1-s)^q}{\Gamma(q+1)}\int_0^s \frac{(s-u)^{r-1}}{\Gamma(r)}v_n(u)duds\right. \\
&\left.-\int_0^\eta \frac{(\eta-s)^q}{\Gamma(q+1)}\int_0^s \frac{(s-u)^{r-1}}{\Gamma(r)}v_n(u)duds\right) \\
&-B\frac{\beta\theta}{Q}\int_0^\xi \frac{(\xi-s)^q}{\Gamma(q)}\int_0^s \frac{(s-u)^{r-1}}{\Gamma(r)}v_n(u)duds \\
&+\frac{Bt}{Q}\Bigg[\alpha(1-\beta)\left(\int_0^1 \frac{(1-s)^q}{\Gamma(q+1)}\int_0^s \frac{(s-u)^{r-1}}{\Gamma(r)}v_n(u)duds\right. \\
&\left.-\int_0^\eta \frac{(\eta-s)^q}{\Gamma(q+1)}\int_0^s \frac{(s-u)^{r-1}}{\Gamma(r)}v_n(u)duds\right) \\
&-(1-\beta)\int_0^\xi \frac{(\xi-s)^q}{\Gamma(q)}\int_0^s \frac{(s-u)^{r-1}}{\Gamma(r)}v_n(u)duds \\
&-\beta(1-\alpha(1-\eta))\int_0^\theta \frac{(\theta-s)^{q-1}}{\Gamma(q)}\int_0^s \frac{(s-u)^{r-1}}{\Gamma(r)}v_n(u)duds\Bigg].
\end{aligned}
$$

Thus it suffices to show that there exists $v_* \in S_{G,x_*}$ such that for each $t \in [0,1]$,

$$
h_*(t) = B\int_0^t \frac{(t-s)^{q-1}}{\Gamma(q)}\int_0^s \frac{(s-u)^{r-1}}{\Gamma(r)}v_*(u)duds
$$

$$
\begin{aligned}
&+B\frac{\beta\left(\xi-\frac{\alpha}{2}(1-\eta^2)\right)}{Q}\int_0^\theta\frac{(\theta-s)^{q-1}}{\Gamma(q)}\int_0^s\frac{(s-u)^{r-1}}{\Gamma(r)}v_*(u)duds\\
&+B\frac{\alpha\beta\theta}{Q}\left(\int_0^1\frac{(1-s)^q}{\Gamma(q+1)}\int_0^s\frac{(s-u)^{r-1}}{\Gamma(r)}v_*(u)duds\right.\\
&\left.-\int_0^\eta\frac{(\eta-s)^q}{\Gamma(q+1)}\int_0^s\frac{(s-u)^{r-1}}{\Gamma(r)}v_*(u)duds\right)\\
&-B\frac{\beta\theta}{Q}\int_0^\xi\frac{(\xi-s)^q}{\Gamma(q)}\int_0^s\frac{(s-u)^{r-1}}{\Gamma(r)}v_*(u)duds\\
&+\frac{Bt}{Q}\left[\alpha(1-\beta)\left(\int_0^1\frac{(1-s)^q}{\Gamma(q+1)}\int_0^s\frac{(s-u)^{r-1}}{\Gamma(r)}v_*(u)duds\right.\right.\\
&\left.-\int_0^\eta\frac{(\eta-s)^q}{\Gamma(q+1)}\int_0^s\frac{(s-u)^{r-1}}{\Gamma(r)}v_*(u)duds\right)\\
&-(1-\beta)\int_0^\xi\frac{(\xi-s)^q}{\Gamma(q)}\int_0^s\frac{(s-u)^{r-1}}{\Gamma(r)}v_*(u)duds\\
&\left.-\beta(1-\alpha(1-\eta))\int_0^\theta\frac{(\theta-s)^{q-1}}{\Gamma(q)}\int_0^s\frac{(s-u)^{r-1}}{\Gamma(r)}v_*(u)duds\right].
\end{aligned}
$$

Let us consider the linear operator $\Theta : L^1([0,1],\mathbb{R}) \to C([0,1],\mathbb{R})$ given by

$$
\begin{aligned}
v\mapsto\Theta(v)(t)=B&\int_0^t\frac{(t-s)^{q-1}}{\Gamma(q)}\int_0^s\frac{(s-u)^{r-1}}{\Gamma(r)}v(u)duds\\
&+B\frac{\beta\left(\xi-\frac{\alpha}{2}(1-\eta^2)\right)}{Q}\int_0^\theta\frac{(\theta-s)^{q-1}}{\Gamma(q)}\int_0^s\frac{(s-u)^{r-1}}{\Gamma(r)}v(u)duds\\
&+B\frac{\alpha\beta\theta}{Q}\left(\int_0^1\frac{(1-s)^q}{\Gamma(q+1)}\int_0^s\frac{(s-u)^{r-1}}{\Gamma(r)}v(u)duds\right.\\
&\left.-\int_0^\eta\frac{(\eta-s)^q}{\Gamma(q+1)}\int_0^s\frac{(s-u)^{r-1}}{\Gamma(r)}v(u)duds\right)\\
&-B\frac{\beta\theta}{Q}\int_0^\xi\frac{(\xi-s)^q}{\Gamma(q)}\int_0^s\frac{(s-u)^{r-1}}{\Gamma(r)}v(u)duds\\
&+\frac{Bt}{Q}\left[\alpha(1-\beta)\left(\int_0^1\frac{(1-s)^q}{\Gamma(q+1)}\int_0^s\frac{(s-u)^{r-1}}{\Gamma(r)}v(u)duds\right.\right.\\
&\left.-\int_0^\eta\frac{(\eta-s)^q}{\Gamma(q+1)}\int_0^s\frac{(s-u)^{r-1}}{\Gamma(r)}v(u)duds\right)\\
&-(1-\beta)\int_0^\xi\frac{(\xi-s)^q}{\Gamma(q)}\int_0^s\frac{(s-u)^{r-1}}{\Gamma(r)}v(u)duds
\end{aligned}
$$

$$-\beta(1-\alpha(1-\eta))\int_0^\theta \frac{(\theta-s)^{q-1}}{\Gamma(q)}\int_0^s \frac{(s-u)^{r-1}}{\Gamma(r)}v(u)duds\Bigg].$$

Observe that

$$\begin{aligned}
&\|h_n-h_*\|\\
&=\Bigg\|B\int_0^t \frac{(t-s)^{q-1}}{\Gamma(q)}\int_0^s \frac{(s-u)^{r-1}}{\Gamma(r)}(v_n(u)-v_*(u))duds\\
&+B\frac{\beta\Big(\xi-\frac{\alpha}{2}(1-\eta^2)\Big)}{Q}\int_0^\theta \frac{(\theta-s)^{q-1}}{\Gamma(q)}\int_0^s \frac{(s-u)^{r-1}}{\Gamma(r)}(v_n(u)-v_*(u))duds\\
&+B\frac{\alpha\beta\theta}{Q}\Bigg(\int_0^1 \frac{(1-s)^q}{\Gamma(q+1)}\int_0^s \frac{(s-u)^{r-1}}{\Gamma(r)}(v_n(u)-v_*(u))duds\\
&-\int_0^\eta \frac{(\eta-s)^q}{\Gamma(q+1)}\int_0^s \frac{(s-u)^{r-1}}{\Gamma(r)}(v_n(u)-v_*(u))duds\Bigg)\\
&-B\frac{\beta\theta}{Q}\int_0^\xi \frac{(\xi-s)^q}{\Gamma(q)}\int_0^s \frac{(s-u)^{r-1}}{\Gamma(r)}(v_n(u)-v_*(u))duds\\
&+\frac{Bt}{Q}\Bigg[\alpha(1-\beta)\Bigg(\int_0^1 \frac{(1-s)^q}{\Gamma(q+1)}\int_0^s \frac{(s-u)^{r-1}}{\Gamma(r)}(v_n(u)-v_*(u))duds\\
&-\int_0^\eta \frac{(\eta-s)^q}{\Gamma(q+1)}\int_0^s \frac{(s-u)^{r-1}}{\Gamma(r)}(v_n(u)-v_*(u))duds\Bigg)\\
&-(1-\beta)\int_0^\xi \frac{(\xi-s)^q}{\Gamma(q)}\int_0^s \frac{(s-u)^{r-1}}{\Gamma(r)}(v_n(u)-v_*(u))duds\\
&-\beta(1-\alpha(1-\eta))\int_0^\theta \frac{(\theta-s)^{q-1}}{\Gamma(q)}\int_0^s \frac{(s-u)^{r-1}}{\Gamma(r)}(v_n(u)-v_*(u))duds\Bigg]\Bigg\|\to 0,
\end{aligned}$$

as $n\to\infty$. Thus, it follows by Lemma 1.2 that $\Theta\circ S_G$ is a closed graph operator. Further, we have $h_n(t)\in\Theta(S_{G,x_n})$. Since $x_n\to x_*$, therefore, we have

$$\begin{aligned}
h_*(t)=&\,B\int_0^t \frac{(t-s)^{q-1}}{\Gamma(q)}\int_0^s \frac{(s-u)^{r-1}}{\Gamma(r)}v_*(u)duds\\
&+B\frac{\beta\Big(\xi-\frac{\alpha}{2}(1-\eta^2)\Big)}{Q}\int_0^\theta \frac{(\theta-s)^{q-1}}{\Gamma(q)}\int_0^s \frac{(s-u)^{r-1}}{\Gamma(r)}v_*(u)duds\\
&+B\frac{\alpha\beta\theta}{Q}\Bigg(\int_0^1 \frac{(1-s)^q}{\Gamma(q+1)}\int_0^s \frac{(s-u)^{r-1}}{\Gamma(r)}v_*(u)duds\\
&-\int_0^\eta \frac{(\eta-s)^q}{\Gamma(q+1)}\int_0^s \frac{(s-u)^{r-1}}{\Gamma(r)}v_*(u)duds\Bigg)\\
&-B\frac{\beta\theta}{Q}\int_0^\xi \frac{(\xi-s)^q}{\Gamma(q)}\int_0^s \frac{(s-u)^{r-1}}{\Gamma(r)}v_*(u)duds
\end{aligned}$$

$$
\begin{aligned}
&+\frac{Bt}{Q}\Bigg[\alpha(1-\beta)\Bigg(\int_0^1 \frac{(1-s)^q}{\Gamma(q+1)}\int_0^s \frac{(s-u)^{r-1}}{\Gamma(r)}v_*(u)duds \\
&-\int_0^\eta \frac{(\eta-s)^q}{\Gamma(q+1)}\int_0^s \frac{(s-u)^{r-1}}{\Gamma(r)}v_*(u)duds\Bigg) \\
&-(1-\beta)\int_0^\xi \frac{(\xi-s)^q}{\Gamma(q)}\int_0^s \frac{(s-u)^{r-1}}{\Gamma(r)}v_*(u)duds \\
&-\beta(1-\alpha(1-\eta))\int_0^\theta \frac{(\theta-s)^{q-1}}{\Gamma(q)}\int_0^s \frac{(s-u)^{r-1}}{\Gamma(r)}v_*(u)duds\Bigg],
\end{aligned}
$$

for some $v_* \in S_{G,x_*}$. Hence $\mathcal{N}_2$ has a closed graph (and therefore has closed values). In consequence, $\mathcal{N}_2$ is compact valued.

Therefore the operators $\mathcal{N}_1$ and $\mathcal{N}_2$ satisfy all the conditions of Theorem 1.18. So the conclusion of Theorem 1.18 applies and either condition (i) or condition (ii) holds. We show that the conclusion (ii) is not possible. If $x \in \lambda\mathcal{N}_1(x) + \lambda\mathcal{N}_2(x)$ for $\lambda \in (0,1)$, then there exist $f \in S_{F,x}$ and $g \in S_{G,x}$ such that

$$
\begin{aligned}
x(t) = {} & \lambda A\int_0^t \frac{(t-s)^{q-1}}{\Gamma(q)}f(s)ds + \lambda A\frac{\beta\Big(\xi-\frac{\alpha}{2}(1-\eta^2)\Big)}{Q}\int_0^\theta \frac{(\theta-s)^{q-1}}{\Gamma(q)}f(s)ds \\
&+\lambda A\frac{\alpha\beta\theta}{Q}\Bigg(\int_0^1 \frac{(1-s)^q}{\Gamma(q+1)}f(s)ds - \int_0^\eta \frac{(\eta-s)^q}{\Gamma(q+1)}f(s)ds\Bigg) \\
&-\lambda A\frac{\beta\theta}{Q}\int_0^\xi \frac{(\xi-s)^q}{\Gamma(q)}f(s)ds + \lambda\frac{At}{Q}\Bigg[\alpha(1-\beta)\Bigg(\int_0^1 \frac{(1-s)^q}{\Gamma(q+1)}f(s)ds \\
&-\int_0^\eta \frac{(\eta-s)^q}{\Gamma(q+1)}f(s)ds\Bigg) - (1-\beta)\int_0^\xi \frac{(\xi-s)^q}{\Gamma(q)}f(s)ds \\
&-\beta(1-\alpha(1-\eta))\int_0^\theta \frac{(\theta-s)^{q-1}}{\Gamma(q)}f(s)ds\Bigg] \\
&+\lambda B\int_0^t \frac{(t-s)^{q-1}}{\Gamma(q)}\int_0^s \frac{(s-u)^{r-1}}{\Gamma(r)}g(u)duds \\
&+\lambda B\frac{\beta\Big(\xi-\frac{\alpha}{2}(1-\eta^2)\Big)}{Q}\int_0^\theta \frac{(\theta-s)^{q-1}}{\Gamma(q)}\int_0^s \frac{(s-u)^{r-1}}{\Gamma(r)}g(u)duds \\
&+\lambda B\frac{\alpha\beta\theta}{Q}\Bigg(\int_0^1 \frac{(1-s)^q}{\Gamma(q+1)}\int_0^s \frac{(s-u)^{r-1}}{\Gamma(r)}g(u)duds \\
&-\int_0^\eta \frac{(\eta-s)^q}{\Gamma(q+1)}\int_0^s \frac{(s-u)^{r-1}}{\Gamma(r)}g(u)duds\Bigg) \\
&-\lambda B\frac{\beta\theta}{Q}\int_0^\xi \frac{(\xi-s)^q}{\Gamma(q)}\int_0^s \frac{(s-u)^{r-1}}{\Gamma(r)}g(u)duds
\end{aligned}
$$

$$+\lambda\frac{Bt}{Q}\Bigg[\alpha(1-\beta)\Bigg(\int_0^1\frac{(1-s)^q}{\Gamma(q+1)}\int_0^s\frac{(s-u)^{r-1}}{\Gamma(r)}g(u)duds$$
$$-\int_0^\eta\frac{(\eta-s)^q}{\Gamma(q+1)}\int_0^s\frac{(s-u)^{r-1}}{\Gamma(r)}g(u)duds\Bigg)$$
$$-(1-\beta)\int_0^\xi\frac{(\xi-s)^q}{\Gamma(q)}\int_0^s\frac{(s-u)^{r-1}}{\Gamma(r)}g(u)duds$$
$$-\beta(1-\alpha(1-\eta))\int_0^\theta\frac{(\theta-s)^{q-1}}{\Gamma(q)}\int_0^s\frac{(s-u)^{r-1}}{\Gamma(r)}g(u)duds\Bigg].$$

By the hypothesis (3.16.2), for all $t\in[0,1]$, we have

$$|f(t,x)| = H(F(t,x),0)\le H(F(t,x),F(t,0))+H(F(t,0),0)$$
$$\le H(F(t,x),F(t,0))+\|F(t,0)\|.$$

Hence for any $a\in F(t,x)$,

$$|a|\le|f(t,x)|\le H(F(t,x),F(t,0))+\|F(t,0)\|$$
$$\le k(t)\|x\|+\|F(t,0)\|,$$

for all $t\in[0,1]$. Then, we have

$$|x(t)|$$
$$\le|A|\Bigg\{\frac{1}{\Gamma(q+1)}+\frac{\left|\beta\left(\xi-\frac{\alpha}{2}(1-\eta^2)\right)\right|\theta^q}{|Q|\Gamma(q+1)}+\frac{|\alpha\beta|\theta}{|Q|}\left(\frac{1}{\Gamma(q+2)}+\frac{\eta^{q+1}}{\Gamma(q+2)}\right)$$
$$+\frac{|\beta|\theta\xi^q}{|Q|\Gamma(q+1)}+\frac{1}{|Q|}\Bigg[|\alpha(1-\beta)|\left(\frac{1}{\Gamma(q+2)}+\frac{\eta^{q+1}}{\Gamma(q+2)}\right)$$
$$+\frac{|1-\beta|\xi^q}{\Gamma(q+1)}+\frac{|\beta(1-\alpha(1-\eta))|\theta^q}{\Gamma(q+1)}\Bigg]\Bigg\}(\|k\|\|x\|+F_0)$$
$$+|B|\|m\|\psi(\|x\|)\Bigg\{\frac{1}{\Gamma(q+r+1)}+\frac{\left|\beta\left(\xi-\frac{\alpha}{2}(1-\eta^2)\right)\right|\theta^{q+r}}{|Q|\Gamma(q+r+1)}$$
$$+\frac{|\alpha\beta|\theta}{|Q|}\left(\frac{1}{\Gamma(q+r+2)}+\frac{\eta^{q+r+1}}{\Gamma(q+r+2)}\right)+\frac{|\beta|\theta\xi^{q+r}}{|Q|\Gamma(q+r+1)}$$
$$+\frac{1}{|Q|}\Bigg[|\alpha(1-\beta)|\left(\frac{1}{\Gamma(q+r+2)}+\frac{\eta^{q+r+1}}{\Gamma(q+r+2)}\right)$$
$$+\frac{|1-\beta|\xi^{q+r}}{\Gamma(q+r+1)}+\frac{|\beta(1-\alpha(1-\eta))|\theta^{q+r}}{\Gamma(q+r+1)}\Bigg]\Bigg\}$$
$$=Z_1(\|k\|\|x\|+F_0)+Z_2\|m\|\psi(\|x\|).$$

Thus,

$$\|x\|\le Z_1(\|k\|\|x\|+F_0)+Z_2\|m\|\psi(\|x\|). \tag{3.49}$$

Now, if condition (ii) of Theorem 1.18 holds, then there exists $\lambda \in (0,1)$ and $x \in \partial B_M$ such that $x = \lambda\mathcal{N}(x)$. Then, x is a solution with $\|x\| = M$ and consequently, the inequality (3.49) yields

$$\frac{(1 - Z_1\|k\|)M}{Z_1F_0 + Z_2\|q\|_{L^1}\psi(M)} \leq 1,$$

which contradicts (3.48). Hence, $\mathcal{N}$ has a fixed point in $[0,1]$ by Theorem 1.18, which in fact is a solution of the problem (3.45). This completes the proof. □

3.3.2.2 *The lower semi-continuous case*

This section is devoted to study the case when the maps in (3.45) are not necessarily convex valued. We establish the existence result for the problem at hand by applying the nonlinear alternative of Leray-Schauder type and a selection theorem due to Bressan and Colombo (Theorem 1.3) for lower semi-continuous maps with decomposable values.

Theorem 3.17. *Assume that (3.16.2), (3.16.4), (3.16.5) and the following condition holds:*

(3.17.1) $F, G : [0,1] \times \mathbb{R} \to \mathcal{P}(\mathbb{R})$ *are non-empty compact-valued multi-valued maps such that:*

(a) $(t,x) \longmapsto F(t,x)$, $(t,x) \longmapsto G(t,x)$ *are* $\mathcal{L} \otimes \mathcal{B}$ *measurable,*

(b) $x \longmapsto F(t,x)$, $x \longmapsto F(t,x)$ *are lower semi-continuous for each* $t \in [0,1]$.

Then the boundary value problem (3.45) has at least one solution on $[0,1]$.

Proof. It follows from (3.16.2), (3.16.4) and (3.17.1) that F and G are of l.s.c. type. Then from Theorem 1.3, there exist continuous functions $f, g : C([0,1],\mathbb{R}) \to L^1([0,1],\mathbb{R})$ such that $f(x) \in \mathcal{F}(x)$, $g(x) \in \mathcal{G}(x)$ for all $x \in C([0,1],\mathbb{R})$.

Consider the problem

$$\begin{cases} {}^cD^q x(t) = Af(x(t)) + BI^r g(x(t)), \quad t \in [0,1], \\ x(0) = \beta x(\theta), \quad x(\xi) = \alpha \displaystyle\int_\eta^1 x(s)ds, \quad 0 < \theta < \xi < \eta < 1. \end{cases} \tag{3.50}$$

Observe that if $x \in AC^2([0,1])$ is a solution of (3.50), then x is a solution to the problem (3.45). Now, we define two multi-valued operators $\widetilde{\mathcal{N}}_1, \widetilde{\mathcal{N}}_2 : C([0,1],\mathbb{R}) \longrightarrow \mathcal{P}(C([0,1],\mathbb{R}))$ by

$$\begin{aligned} \widetilde{\mathcal{N}}_1 x(t) = A\int_0^t \frac{(t-s)^{q-1}}{\Gamma(q)} f(x(s))ds + A\frac{\beta\left(\xi - \frac{\alpha}{2}(1-\eta^2)\right)}{Q}\int_0^\theta \frac{(\theta-s)^{q-1}}{\Gamma(q)} f(x(s))ds \\ + A\frac{\alpha\beta\theta}{Q}\left(\int_0^1 \frac{(1-s)^q}{\Gamma(q+1)} f(x(s))ds - \int_0^\eta \frac{(\eta-s)^q}{\Gamma(q+1)} f(x(s))ds\right) \end{aligned}$$

$$-A\frac{\beta\theta}{Q}\int_0^{\xi}\frac{(\xi-s)^q}{\Gamma(q)}f(x(s))ds+\frac{At}{Q}\Bigg[\alpha(1-\beta)\bigg(\int_0^1\frac{(1-s)^q}{\Gamma(q+1)}f(x(s))ds$$
$$-\int_0^{\eta}\frac{(\eta-s)^q}{\Gamma(q+1)}f(x(s))ds\bigg)-(1-\beta)\int_0^{\xi}\frac{(\xi-s)^q}{\Gamma(q)}f(x(s))ds$$
$$-\beta(1-\alpha(1-\eta))\int_0^{\theta}\frac{(\theta-s)^{q-1}}{\Gamma(q)}f(x(s))ds\Bigg],$$

and

$$\widetilde{\mathcal{N}}_2x(t)=B\int_0^t\frac{(t-s)^{q-1}}{\Gamma(q)}\int_0^s\frac{(s-u)^{r-1}}{\Gamma(r)}g(x(u))duds$$
$$+B\frac{\beta\Big(\xi-\frac{\alpha}{2}(1-\eta^2)\Big)}{Q}\int_0^{\theta}\frac{(\theta-s)^{q-1}}{\Gamma(q)}\int_0^s\frac{(s-u)^{r-1}}{\Gamma(r)}g(x(u))duds$$
$$+B\frac{\alpha\beta\theta}{Q}\bigg(\int_0^1\frac{(1-s)^q}{\Gamma(q+1)}\int_0^s\frac{(s-u)^{r-1}}{\Gamma(r)}g(x(u))duds$$
$$-\int_0^{\eta}\frac{(\eta-s)^q}{\Gamma(q+1)}\int_0^s\frac{(s-u)^{r-1}}{\Gamma(r)}g(x(u))duds\bigg)$$
$$-B\frac{\beta\theta}{Q}\int_0^{\xi}\frac{(\xi-s)^q}{\Gamma(q)}\int_0^s\frac{(s-u)^{r-1}}{\Gamma(r)}g(x(u))duds$$
$$+\frac{Bt}{Q}\Bigg[\alpha(1-\beta)\bigg(\int_0^1\frac{(1-s)^q}{\Gamma(q+1)}\int_0^s\frac{(s-u)^{r-1}}{\Gamma(r)}g(x(u))duds$$
$$-\int_0^{\eta}\frac{(\eta-s)^q}{\Gamma(q+1)}\int_0^s\frac{(s-u)^{r-1}}{\Gamma(r)}g(x(u))duds\bigg)$$
$$-(1-\beta)\int_0^{\xi}\frac{(\xi-s)^q}{\Gamma(q)}\int_0^s\frac{(s-u)^{r-1}}{\Gamma(r)}g(x(u))duds$$
$$-\beta(1-\alpha(1-\eta))\int_0^{\theta}\frac{(\theta-s)^{q-1}}{\Gamma(q)}\int_0^s\frac{(s-u)^{r-1}}{\Gamma(r)}g(x(u))duds\Bigg].$$

Clearly $\widetilde{\mathcal{N}}_1, \widetilde{\mathcal{N}}_2 : C([0,1],\mathbb{R}) \to C([0,1],\mathbb{R})$ are continuous. Also the argument in Theorem 3.16 guarantees that $\widetilde{\mathcal{N}}_1$ and $\widetilde{\mathcal{N}}_2$ satisfy all the conditions of the nonlinear alternative for contractive maps in the single-valued setting (Theorem 1.8) and hence the problem (3.50) has a solution. □

Example 3.10. Consider a nonlocal integral boundary value problem for a fractional integro-differential equation given by

$$\begin{cases} D^{3/2}x(t)\in F(t,x(t))+I^{3/4}G(t,x(t)), & t\in[0,1],\\ x(0)=\dfrac{1}{2}x(1/4)), \quad x(1/3)=\displaystyle\int_{2/3}^1 x(s)ds,, \end{cases} \tag{3.51}$$

where $q=3/2,\ A=B=1,\ r=3/4,\ \theta=1/4,\ \xi=1/3,\ \eta=2/3,\ \alpha=1,\ \beta=1/2.$

Let

$$F(t,x)=\left[-\frac{18}{(3+t)^2}-\frac{\sin x}{(4+t)^2}-2,-\frac{1}{10}\right],\quad G(t,x)=\left[\frac{|x|^3}{8(|x|^3+3)},\frac{|x|}{9(|x|+1)}\right].$$

Then we have

$$\sup\{|u|:u\in F(t,x)\}\leq 2+\frac{1}{(4+t)^2}+\frac{18}{(3+t)^2},\quad H(F(t,x),F(t,\bar{x}))\leq k(t)|x-\bar{x}|$$

with $k(t)=\dfrac{1}{(4+t)^2}$.

With the given data, it is found that $|Q|=1/9,\ Z_1\simeq 4.178968,\ Z_2\simeq 1.495606,$ $\|k\|_{L^1}=1/20,\ \|m\|=1/8,\ \psi(M)=1,\ F_0=3.5.$

Clearly $Z_1\|k\|<1$, and by the condition:

$$\frac{(1-Z_1\|k\|)M}{Z_1F_0+Z_2\|m\|\psi(M)}>1,$$

it is found that $M>M_1$, where $M_1\simeq 18.726133$. Thus, all the assumptions of Theorem 3.16 are satisfied. Hence, the conclusion of Theorem 3.16 applies to the problem (3.51).

3.4 On a coupled system of fractional differential equations with coupled nonlocal and integral boundary conditions

In this section, we study a coupled system of nonlinear fractional differential equations:

$$\begin{cases} {}^cD^{\alpha}x(t)=f(t,x(t),y(t),{}^cD^{\gamma}y(t)), & t\in[0,T],\quad 1<\alpha\leq 2,\ 0<\gamma<1 \\ {}^cD^{\beta}y(t)=g(t,x(t),{}^cD^{\delta}x(t),y(t)), & t\in[0,T],\quad 1<\beta\leq 2,\ 0<\delta<1, \end{cases}\tag{3.52}$$

supplemented with coupled nonlocal and integral boundary conditions of the form:

$$\begin{cases} x(0)=h(y), & \displaystyle\int_0^T y(s)ds=\mu_1x(\eta), \\ y(0)=\phi(x), & \displaystyle\int_0^T x(s)ds=\mu_2y(\xi),\quad \eta,\xi\in(0,T), \end{cases}\tag{3.53}$$

where ${}^cD^i$ denotes the Caputo fractional derivative operator of order i, $i=\alpha,\beta,\gamma,\delta$, $f,g:[0,T]\times\mathbb{R}\times\mathbb{R}\times\mathbb{R}\to\mathbb{R}$, $h,\phi:C([0,T],\mathbb{R})\to\mathbb{R}$ are given continuous functions, and μ_1,μ_2 are real constants.

Now we prove an auxiliary result which is pivotal to define the solution for the problem (3.52)-(3.53).

Lemma 3.5. *(Auxiliary Lemma) Let $\Delta\neq 0$, $\omega,z\in C[0,1]$. Then the unique solution of the problem*

$$\begin{cases} {}^cD^{\alpha}x(t)=\omega(t), & t\in[0,T],\quad 1<\alpha\leq 2, \\ {}^cD^{\beta}y(t)=z(t), & t\in[0,T],\quad 1<\beta\leq 2, \\ x(0)=h(y), & \displaystyle\int_0^T y(s)ds=\mu_1x(\eta), \\ y(0)=\phi(x), & \displaystyle\int_0^T x(s)ds=\mu_2y(\xi), \end{cases}\tag{3.54}$$

is

$$x(t) = \int_0^t \frac{(t-s)^{\alpha-1}}{\Gamma(\alpha)}\omega(s)ds + (\sigma_1 t + 1)h(y) + \sigma_2 t\phi(x)$$
$$+\frac{t}{\Delta}\Big[\mu_2\xi\Big(\mu_1\int_0^\eta \frac{(\eta-s)^{\alpha-1}}{\Gamma(\alpha)}\omega(s)ds - \int_0^T \frac{(T-s)^\beta}{\Gamma(\beta+1)}z(s)ds\Big)$$
$$+\frac{T^2}{2}\Big(\mu_2\int_0^\xi \frac{(\xi-s)^{\beta-1}}{\Gamma(\beta)}z(s)ds - \int_0^T \frac{(T-s)^\alpha}{\Gamma(\alpha+1)}\omega(s)ds\Big)\Big], \quad (3.55)$$

and

$$y(t) = \int_0^t \frac{(t-s)^{\beta-1}}{\Gamma(\beta)}z(s)ds + (\sigma_3 t + 1)\phi(x) + t\sigma_4 h(y)$$
$$+\frac{t}{\Delta}\Big[\mu_1\eta\Big(\mu_2\int_0^\xi \frac{(\xi-s)^{\beta-1}}{\Gamma(\beta)}z(s)ds - \int_0^T \frac{(T-s)^\alpha}{\Gamma(\alpha+1)}\omega(s)ds\Big)$$
$$+\frac{T^2}{2}\Big(\mu_1\int_0^\eta \frac{(\eta-s)^{\alpha-1}}{\Gamma(\alpha)}\omega(s)ds - \int_0^T \frac{(T-s)^\beta}{\Gamma(\beta+1)}z(s)ds\Big)\Big], \quad (3.56)$$

where

$$\Delta = \frac{T^4 - 4\mu_1\mu_2\eta\xi}{4}, \ \sigma_1 = \frac{2\mu_1\mu_2\xi - T^3}{2\Delta}, \ \sigma_2 = \frac{T\mu_2(T-2\xi)}{2\Delta},$$
$$\sigma_3 = \frac{2\mu_1\mu_2\eta - T^3}{2\Delta}, \sigma_4 = \frac{T\mu_1(T-2\eta)}{2\Delta}.$$

Proof. It is well known that the general solutions of the fractional differential equations in (3.54) can be written as

$$x(t) = c_1 t + c_2 + \int_0^t \frac{(t-s)^{\alpha-1}}{\Gamma(\alpha)}\omega(s)ds, \quad (3.57)$$

$$y(t) = d_1 t + d_2 + \int_0^t \frac{(t-s)^{\beta-1}}{\Gamma(\beta)}z(s)ds, \quad (3.58)$$

where $c_i, d_i \in \mathbb{R}, i = 1, 2$ are arbitrary constants.

Applying the boundary conditions of the problem (3.54) to (3.57) and (3.58), it is found that $c_2 = h(y),\ d_2 = \phi(x)$,

$$c_1 = \frac{1}{\Delta}\left[\frac{T^2}{2}\left(\mu_1\int_0^\eta \frac{(\eta-s)^{\alpha-1}}{\Gamma(\alpha)}\omega(s)ds + \mu_1 h(y) - \int_0^T \frac{(T-s)^\beta}{\Gamma(\beta+1)}z(s)ds - T\phi(x)\right)\right.$$
$$\left.+\mu_1\eta\left(\mu_2\int_0^\xi \frac{(\xi-s)^{\beta-1}}{\Gamma(\beta)}z(s)ds + \mu_2\phi(x) - \int_0^T \frac{(T-s)^\alpha}{\Gamma(\alpha+1)}\omega(s)ds - Th(y)\right)\right]$$

and

$$d_1 = \frac{1}{\Delta}\left[\frac{T^2}{2}\left(\mu_2\int_0^\xi \frac{(\xi-s)^{\beta-1}}{\Gamma(\beta)}z(s)ds + \mu_2\phi(x) - \int_0^T \frac{(T-s)^\alpha}{\Gamma(\alpha+1)}\omega(s)ds - Th(y)\right)\right.$$
$$\left.+\mu_2\xi\left(\mu_1\int_0^\eta \frac{(\eta-s)^{\alpha-1}}{\Gamma(\alpha)}\omega(s)ds + \mu_1 h(y) - \int_0^T \frac{(T-s)^\beta}{\Gamma(\beta+1)}z(s)ds - T\phi(x)\right)\right].$$

Substituting the values of c_1, c_2, d_1, d_2 in (3.57) and (3.58), we get (3.55) and (3.56). The converse of the lemma can be proved by direct computation. This completes the proof. □

3.4.1 *Existence results*

We define the space

$$X = \{x | x \in C([0,T],\mathbb{R}) \text{ and } {}^cD^\delta x \in C([0,T],\mathbb{R})\},$$

equipped with the norm

$$\|x\|_X = \|x\| + \|{}^cD^\delta x\| = \sup_{t\in[0,T]} |x(t)| + \sup_{t\in[0,T]} |{}^cD^\delta x(t)|.$$

Then it is well known that $(X, \|\cdot\|_X)$ is a Banach space ([139]).

Also the space

$$Y = \{y | y \in C([0,T],\mathbb{R}) \text{ and } {}^cD^\gamma y \in C([0,T],\mathbb{R})\},$$

endowed with the norm

$$\|y\|_Y = \|y\| + \|{}^cD^\gamma y\| = \sup_{t\in[0,T]} |y(t)| + \sup_{t\in[0,T]} |{}^cD^\gamma y(t)|,$$

represented as $(Y, \|\cdot\|_Y)$ is a Banach space.

Obviously the product space $(X \times Y, \|\cdot\|_{X\times Y})$ is a Banach space with norm $\|(x,y)\|_{X\times Y} = \|x\|_X + \|y\|_Y$ for $(x,y) \in X \times Y$. In view of Lemma 3.5, we define an operator $\mathcal{F} : X \times Y \to X \times Y$ by

$$\mathcal{F}(x,y)(t) := (\mathcal{F}_1(x,y)(t), \mathcal{F}_2(x,y)(t)),$$

where

$$\begin{aligned}
\mathcal{F}_1(x,y)(t) = & \int_0^t \frac{(t-s)^{\alpha-1}}{\Gamma(\alpha)} f(s,x(s),y(s),{}^cD^\gamma y(s))ds + (\sigma_1 t + 1)h(y) + \sigma_2 t\phi(x) \\
& + \frac{t}{\Delta}\Big[\mu_2\xi\Big(\mu_1 \int_0^\eta \frac{(\eta-s)^{\alpha-1}}{\Gamma(\alpha)} f(s,x(s),y(s),{}^cD^\gamma y(s))ds \\
& - \int_0^T \frac{(T-s)^\beta}{\Gamma(\beta+1)} g(s,x(s),{}^cD^\delta x(s),y(s))ds\Big) \\
& + \frac{T^2}{2}\Big(\mu_2 \int_0^\xi \frac{(\xi-s)^{\beta-1}}{\Gamma(\beta)} g(s,x(s),{}^cD^\delta x(s),y(s))ds \\
& - \int_0^T \frac{(T-s)^\alpha}{\Gamma(\alpha+1)} f(s,x(s),y(s),{}^cD^\gamma y(s))ds\Big)\Big],
\end{aligned} \tag{3.59}$$

and

$$\begin{aligned}
\mathcal{F}_2(x,y)(t) = & \int_0^t \frac{(t-s)^{\beta-1}}{\Gamma(\beta)} g(s,x(s),{}^cD^\delta x(s),y(s))ds + (\sigma_3 t + 1)\phi(x) + t\sigma_4 h(y) \\
& + \frac{t}{\Delta}\Big[\mu_1\eta\Big(\mu_2 \int_0^\xi \frac{(\xi-s)^{\beta-1}}{\Gamma(\beta)} g(s,x(s),{}^cD^\delta x(s),y(s))ds \\
& - \int_0^T \frac{(T-s)^\alpha}{\Gamma(\alpha+1)} f(s,x(s),y(s),{}^cD^\gamma y(s))ds\Big) \\
& + \frac{T^2}{2}\Big(\mu_1 \int_0^\eta \frac{(\eta-s)^{\alpha-1}}{\Gamma(\alpha)} f(s,x(s),y(s),{}^cD^\gamma y(s))ds
\end{aligned}$$

$$-\int_0^T \frac{(T-s)^\beta}{\Gamma(\beta+1)} g(s, x(s), {}^cD^\delta x(s), y(s))ds\Big)\Big]. \tag{3.60}$$

For computation convenience, we introduce the notation:

$$\Lambda = \vartheta_1\ell + \vartheta_2\ell_1 + (|\sigma_1|T+1)L + |\sigma_2|TL_1, \; M = \vartheta_1 N_1 + \vartheta_2 N_2, \tag{3.61}$$

$$\Lambda_1 = \bar{\vartheta}_1\ell + \frac{\vartheta_2\ell_1}{T} + |\sigma_1|L + |\sigma_2|L_1, \; M_1 = \bar{\vartheta}_1 N_1 + \frac{\vartheta_2 N_2}{T}, \tag{3.62}$$

$$\Lambda' = \vartheta_3\ell_1 + \vartheta_4\ell + (|\sigma_3|T+1)L_1 + |\sigma_4|TL, \; M' = \vartheta_3 N_2 + \vartheta_4 N_1, \tag{3.63}$$

$$\Lambda_1' = \bar{\vartheta}_3\ell_1 + \frac{\vartheta_4\ell}{T} + |\sigma_3|L_1 + |\sigma_4|L, \; M_1' = \bar{\vartheta}_3 N_2 + \frac{\vartheta_4 N_1}{T}, \tag{3.64}$$

where

$$\vartheta_1 = \left[\frac{T^\alpha}{\Gamma(\alpha+1)} + \frac{T}{|\Delta|}\left(|\mu_1\mu_2|\xi\frac{\eta^\alpha}{\Gamma(\alpha+1)} + \frac{T^2}{2}\frac{T^{\alpha+1}}{\Gamma(\alpha+2)}\right)\right],$$

$$\vartheta_2 = \frac{T|\mu_2|}{|\Delta|}\left(\xi\frac{T^{\beta+1}}{\Gamma(\beta+2)} + \frac{T^2}{2}\frac{\xi^\beta}{\Gamma(\beta+1)}\right),$$

$$\bar{\vartheta}_1 = \left[\frac{T^{\alpha-1}}{\Gamma(\alpha)} + \frac{1}{|\Delta|}\left(|\mu_1\mu_2|\xi\frac{\eta^\alpha}{\Gamma(\alpha+1)} + \frac{T^2}{2}\frac{T^{\alpha+1}}{\Gamma(\alpha+2)}\right)\right],$$

$$\vartheta_3 = \left[\frac{T^\beta}{\Gamma(\beta+1)} + \frac{T}{|\Delta|}\left(|\mu_1\mu_2|\eta\frac{\xi^\beta}{\Gamma(\beta+1)} + \frac{T^2}{2}\frac{T^{\beta+1}}{\Gamma(\beta+2)}\right)\right],$$

$$\bar{\vartheta}_3 = \left[\frac{T^{\beta-1}}{\Gamma(\beta)} + \frac{1}{|\Delta|}\left(|\mu_1\mu_2|\eta\frac{\xi^\beta}{\Gamma(\beta+1)} + \frac{T^2}{2}\frac{T^{\beta+1}}{\Gamma(\beta+2)}\right)\right],$$

$$\vartheta_4 = \frac{T|\mu_1|}{|\Delta|}\left(\eta\frac{T^{\alpha+1}}{\Gamma(\alpha+2)} + \frac{T^2}{2}\frac{\eta^\alpha}{\Gamma(\alpha+1)}\right),$$

$$N_1 = \sup_{t\in[0,T]} f(t,0,0,0) < \infty, \qquad N_2 = \sup_{t\in[0,T]} g(t,0,0,0) < \infty.$$

The first result is concerned with the existence and uniqueness of solutions for the problem (3.52)-(3.53) and is based on Banach's contraction mapping principle.

Theorem 3.18. *Assume that:*

(3.18.1) $f, g : [0,T]\times\mathbb{R}^3 \to \mathbb{R}$ *are continuous functions and there exist constants* $\ell > 0, \ell_1 > 0$ *such that for all* $t \in [0,T]$ *and* $u_i, v_i \in \mathbb{R}, i = 1,2,3,$

$$|f(t,u_1,u_2,u_3) - f(t,v_1,v_2,v_3)| \le \ell(|u_1-v_1| + |u_2-v_2| + |u_3-v_3|)$$

and

$$|g(t,u_1,u_2,u_3) - g(t,v_1,v_2,v_3)| \le \ell_1(|u_1-v_1| + |u_2-v_2| + |u_3-v_3|);$$

(3.18.2) $h, \phi : C([0,T],\mathbb{R}) \to \mathbb{R}$ *are continuous functions with* $h(0) = \phi(0) = 0$ *and there exist constants* $L > 0, L_1 > 0$ *such that*

$$|h(x_1)-h(x_2)| \le L\|x_1-x_2\|, \;\; |\phi(x_1)-\phi(x_2)| \le L_1\|x_1-x_2\|, \;\; \forall x_1, x_2 \in C([0,T],\mathbb{R}).$$

If

$$\Lambda+\frac{T^{1-\gamma}}{\Gamma(2-\gamma)}\Lambda_1<\frac{1}{2},\quad \Lambda'+\frac{T^{1-\delta}}{\Gamma(2-\delta)}\Lambda_1'<\frac{1}{2},$$

where $\Lambda,\Lambda_1,\Lambda',\Lambda_1'$ *are given in (3.61)-(3.64), then the boundary value problem (3.52)-(3.53) has a unique solution on* $[0,T]$.

Proof. Let us fix $\sup_{t\in[0,T]} f(t,0,0,0)=N_1<\infty$ and $\sup_{t\in[0,T]} g(t,0,0,0)=N_2<\infty$ and define

$$r\ge \max\left\{\frac{M+\dfrac{T^{1-\gamma}}{\Gamma(2-\gamma)}M_1}{\dfrac{1}{2}-\left(\Lambda+\dfrac{T^{1-\gamma}}{\Gamma(2-\gamma)}\Lambda_1\right)},\frac{M'+\dfrac{T^{1-\delta}}{\Gamma(2-\delta)}M_1'}{\dfrac{1}{2}-\left(\Lambda'+\dfrac{T^{1-\delta}}{\Gamma(2-\delta)}\Lambda_1'\right)}\right\},$$

where $\Lambda,\Lambda_1,\Lambda',\Lambda_1'$ and M,M_1,M',M_1' are given in (3.61)-(3.64).
We first show that $\mathcal{F}B_r\subset B_r$, where $B_r=\{(x,y)\in X\times Y:\|(x,y)\|_{X\times Y}\le r\}$. For $(x,y)\in B_r$, we find the following estimates based on the assumptions (3.18.1) and (3.18.2):

$$\begin{aligned}&|f(t,x(t),y(t),{}^cD^\gamma y(t))|\le |f(t,x(t),y(t),{}^cD^\gamma y(t))-f(t,0,0,0)|+|f(t,0,0,0)|\\&\le \ell(|x(t)|+|y(t)|+|D^\gamma y(t)|)+N_1\le \ell(\|x\|_X+\|y\|_Y)+N_1\le \ell r+N_1,\end{aligned}$$

and

$$|g(t,x(t),{}^cD^\delta x(t),y(t))|\le \ell_1(\|x\|_X+\|y\|_Y)+N_2\le \ell_1 r+N_2,$$

$$|h(y)|\le L\|y\|\le L\|y\|_Y\le Lr,\ |\phi(x)|\le L_1\|x\|_X\le L_1 r.$$

Using these estimates, we get

$$\begin{aligned}&|\mathcal{F}_1(x,y)(t)|\\&\le \int_0^t\frac{(t-s)^{\alpha-1}}{\Gamma(\alpha)}|f(s,x(s),y(s),{}^cD^\gamma y(s))|ds+(|\sigma_1|T+1)|h(y)|+T|\sigma_2||\phi(x)|\\&+\frac{T}{|\Delta|}\Big[|\mu_2|\xi\Big(|\mu_1|\int_0^\eta\frac{(\eta-s)^{\alpha-1}}{\Gamma(\alpha)}|f(s,x(s),y(s),{}^cD^\gamma y(s))|ds\\&+\int_0^T\frac{(T-s)^\beta}{\Gamma(\beta+1)}|g(s,x(s),{}^cD^\delta x(s),y(s))|ds\Big)\\&+\frac{T^2}{2}\Big(|\mu_2|\int_0^\xi\frac{(\xi-s)^{\beta-1}}{\Gamma(\beta)}|g(s,x(s),{}^cD^\delta x(s),y(s)|ds\\&+\int_0^T\frac{(T-s)^\alpha}{\Gamma(\alpha+1)}|f(s,x(s),y(s),{}^cD^\gamma y(s))|ds\Big)\Big]\\&\le \Lambda r+M,\end{aligned}$$

where Λ and M are given by (3.61). Using the notation (3.62), we get

$$\begin{aligned}
&|\mathcal{F}_1(x,y)'(t)| \\
&\leq \int_0^t \frac{(t-s)^{\alpha-2}}{\Gamma(\alpha-1)}|f(s,x(s),y(s),{}^c D^\gamma y(s))|ds + |\sigma_1||h(y)| + |\sigma_2||\phi(x)| \\
&\quad + \frac{1}{|\Delta|}\Big[|\mu_2|\xi\Big(|\mu_1|\int_0^\eta \frac{(\eta-s)^{\alpha-1}}{\Gamma(\alpha)}|f(s,x(s),y(s),{}^c D^\gamma y(s))ds \\
&\quad + \int_0^T \frac{(T-s)^\beta}{\Gamma(\beta+1)}|g(s,x(s),{}^c D^\delta x(s),y(s))|ds\Big) \\
&\quad + \frac{T^2}{2}\Big(|\mu_2|\int_0^\xi \frac{(\xi-s)^{\beta-1}}{\Gamma(\beta)}|g(s,x(s),{}^c D^\delta x(s),y(s))|ds \\
&\quad + \int_0^T \frac{(T-s)^\alpha}{\Gamma(\alpha+1)}|f(s,x(s),y(s),{}^c D^\gamma y(s))|ds\Big)\Big] \\
&\leq \Lambda_1 r + M_1,
\end{aligned}$$

which implies that

$$|{}^c D^\gamma \mathcal{F}_1(x,y)(t)| \leq \int_0^t \frac{(t-s)^{-\gamma}}{\Gamma(1-\gamma)}|\mathcal{F}_1(x,y)'(s)|ds \leq \frac{T^{1-\gamma}}{\Gamma(2-\gamma)}(\Lambda_1 r + M_1).$$

Thus

$$\begin{aligned}
\|\mathcal{F}_1(x,y)\|_X &= \|\mathcal{F}_1(x,y)\| + \|{}^c D^\gamma \mathcal{F}_1(x,y)\| \\
&\leq \left(\Lambda + \frac{T^{1-\gamma}}{\Gamma(2-\gamma)}\Lambda_1\right)r + \left(M + \frac{T^{1-\gamma}}{\Gamma(2-\gamma)}M_1\right) \leq \frac{r}{2}. \quad (3.65)
\end{aligned}$$

In a similar manner, using (3.63) and (3.64), we obtain

$$|\mathcal{F}_2(x,y)(t)| \leq \Lambda' r + M', \quad |\mathcal{F}_2(x,y))'(t)| \leq \Lambda_1' r + M_1',$$

and

$$|{}^c D^\delta \mathcal{F}_2(x,y)(t)| \leq \int_0^t \frac{(t-s)^{-\delta}}{\Gamma(1-\delta)}|\mathcal{F}_2(x,y)'(s)|ds \leq \frac{T^{1-\delta}}{\Gamma(2-\delta)}(\Lambda_1' r + M_1').$$

In consequence, we get

$$\begin{aligned}
\|\mathcal{F}_2(x,y)\|_Y &= \|\mathcal{F}_2(x,y)\| + \|D^\delta \mathcal{F}_2(x,y))\| \\
&\leq \left(\Lambda' + \frac{T^{1-\delta}}{\Gamma(2-\delta)}\Lambda_1'\right)r + \left(M' + \frac{T^{1-\delta}}{\Gamma(2-\delta)}M_1'\right) \leq \frac{r}{2}. \quad (3.66)
\end{aligned}$$

Thus, it follows from (3.65) and (3.66) that $\|\mathcal{F}(x,y)(t)\|_{X\times Y} \leq r$.

For $x_i, y_i \in B_r$, $i = 1,2$ and for each $t \in [0,T]$, we have

$$\begin{aligned}
&|\mathcal{F}_1(x_1,y_1)(t) - \mathcal{F}_1(x_2,y_2)(t)| \\
&\leq \int_0^t \frac{(t-s)^{\alpha-1}}{\Gamma(\alpha)}|f(s,x_1(s),y_1(s),{}^c D^\gamma y_1(s)) - f(s,x_2(s),y_2(s),{}^c D^\gamma y_2(s))|ds \\
&\quad +(|\sigma_1|T+1)|h(y_1)-h(y_2)| + |\sigma_2|T|\phi(x_1)-\phi(x_2)| \\
&\leq \int_0^T \frac{(T-s)^{\alpha-1}}{\Gamma(\alpha)}\ell\Big(\|x_1-x_2\| + \|y_1-y_2\| + \|{}^c D^\gamma y_1 - {}^c D^\gamma y_2\|\Big)ds
\end{aligned}$$

$$+(|\sigma_1|T+1)L\|y_1-y_2\|+|\sigma_2|TL_1\|x_1-x_2\|$$
$$+\frac{T}{|\Delta|}\Big[|\mu_2|\xi\Big(|\mu_1|\int_0^\eta\frac{(\eta-s)^{\alpha-1}}{\Gamma(\alpha)}\ell\Big(\|x_1-x_2\|+\|y_1-y_2\|+\|{}^cD^\gamma y_1-{}^cD^\gamma y_2\|\Big)ds$$
$$+\int_0^T\frac{(T-s)^\beta}{\Gamma(\beta+1)}\ell_1\Big(\|x_1-x_2\|+\|y_1-y_2\|+\|{}^cD^\delta x_1-{}^cD^\delta x_2\|\Big)ds\Big)$$
$$+\frac{T^2}{2}\Big(|\mu_2|\int_0^\xi\frac{(\xi-s)^{\beta-1}}{\Gamma(\beta)}\ell_1\Big(\|x_1-x_2\|+\|y_1-y_2\|+\|{}^cD^\delta x_1-{}^cD^\delta x_2\|\Big)ds$$
$$+\int_0^T\frac{(T-s)^\alpha}{\Gamma(\alpha+1)}\ell\Big(\|x_1-x_2\|+\|y_1-y_2\|+\|{}^cD^\gamma y_1-{}^cD^\gamma y_2\|\Big)ds\Big)\Big]$$
$$\leq\Lambda(\|x_1-x_2\|_X+\|y_1-y_2\|_Y).$$

Also we have

$$|\mathcal{F}_1(x_1,y_1)'(t)-\mathcal{F}_1(x_2,y_2)'(t)|\leq\Lambda_1(\|x_1-x_2\|_X+\|y_1-y_2\|_Y).$$

Thus we obtain

$$\begin{aligned}&|{}^cD^\gamma\mathcal{F}_1(x_1,y_1)(t)-{}^cD^\gamma\mathcal{F}_1(x_2,y_2)(t)|\\&\leq\int_0^t\frac{(t-s)^{-\gamma}}{\Gamma(1-\gamma)}|\mathcal{F}_1(x_1,y_1)'(s)-\mathcal{F}_1(x_2,y_2)'(s)|ds\\&\leq\frac{T^{1-\gamma}}{\Gamma(2-\gamma)}\Lambda_1(\|x_1-x_2\|_X+\|y_1-y_2\|_Y).\end{aligned}$$

From the above inequalities, we get

$$\begin{aligned}&\|\mathcal{F}_1(x_1,y_1)-\mathcal{F}_1(x_2,y_2)\|_X\\&=\|\mathcal{F}_1(x_1,y_1)-\mathcal{F}_1(x_2,y_2)\|+\|{}^cD^\gamma(\mathcal{F}_1(x_1,y_1)-{}^cD^\gamma(\mathcal{F}_1(x_2,y_2)\|\\&\leq\left\{\Lambda+\frac{T^{1-\gamma}}{\Gamma(2-\gamma)}\Lambda_1\right\}(\|x_1-x_2\|_X+\|y_1-y_2\|_Y).\end{aligned}$$

In a similar manner, we can find that

$$\begin{aligned}&\|\mathcal{F}_2(x_1,y_1)-\mathcal{F}_2(x_2,y_2)\|_Y\\&=\|\mathcal{F}_2(x_1,y_1)-\mathcal{F}_2(x_2,y_2)\|+\|{}^cD^\delta\mathcal{F}_2(x_1,y_1)-{}^cD^\delta(\mathcal{F}_2(x_2,y_2)\|\\&\leq\left\{\Lambda'+\frac{T^{1-\delta}}{\Gamma(2-\delta)}\Lambda_1'\right\}(\|x_1-x_2\|_X+\|y_1-y_2\|_Y).\end{aligned}$$

Consequently, we obtain

$$\|\mathcal{F}(x,y)(t)\|_{X\times Y}\leq\left(\Lambda+\frac{T^{1-\gamma}}{\Gamma(2-\gamma)}\Lambda_1+\Lambda'+\frac{T^{1-\delta}}{\Gamma(2-\delta)}\Lambda_1'\right)(\|x_1-x_2\|_X+\|y_1-y_2\|_Y),$$

which shows that $\mathcal{F}$ is a contraction in view of the assumptions: $\Lambda+\frac{T^{1-\gamma}}{\Gamma(2-\gamma)}\Lambda_1+\Lambda'+\frac{T^{1-\delta}}{\Gamma(2-\delta)}\Lambda_1'<1$. Hence, by Banach's fixed point theorem, the operator $\mathcal{F}$ has a unique fixed point which corresponds to the unique solution of problem (3.52)-(3.53). This completes the proof. □

Example 3.11. Consider the following coupled system of fractional differential equations

$$\begin{cases} {}^cD^{5/4}x(t) = \dfrac{1}{12\sqrt{3600+t^2}}\Big(\dfrac{|x(t)|}{1+|x(t)|} + \sin(y(t)) + \tan^{-1}({}^cD^{1/4}y(t))\Big) + \dfrac{e^{-t}}{1+t^4}, \\ {}^cD^{3/2}y(t) = \dfrac{e^{-t}}{12\sqrt{1600+t^4}}\Big(\sin(x(t)) + \dfrac{|{}^cD^{1/2}x(t)|}{1+|D^{1/2}x(t)|} + y(t)\Big) + \cos t + 1, t \in [0,1], \end{cases} \tag{3.67}$$

supplemented with nonlocal coupled integral boundary conditions:

$$x(0) = \frac{1}{180}\tan^{-1}(y(t)),\ \int_0^1 y(s)ds = x(2/3),\ y(0) = \frac{1}{144}x(t),\ \int_0^1 x(s)ds = y(1/3). \tag{3.68}$$

Using the given data, we find that $\ell = 1/240$, $\ell_1 = 1/160$, $L = 1/180$, $L_1 = 1/144, \Delta = 1/36$, $-\sigma_1 = \sigma_2 = \sigma_3 = -\sigma_4 = 6$, $\vartheta_1 = 14.32371725$, $\vartheta_2 = 6.21669342$, $\bar{\vartheta}_1 = 14.5443698$, $\vartheta_3 = 9.64297953$, $\bar{\vartheta}_3 = 10.01910592$, $\vartheta_4 = 18.98484724$, $\Lambda + \frac{1}{\Gamma(7/4)}\Lambda_1 \simeq 0.36891137 < 1/2$, $\Lambda' + \frac{1}{\Gamma(3/2)}\Lambda_1' \simeq 0.46586225 < 1/2$. Obviously all the conditions of Theorem 3.18 are satisfied. Thus, by the conclusion of Theorem 3.18, there exists a unique solution for the problem (3.67)-(3.68).

In the next result, we show the existence of solutions for the problem (3.52)-(3.53) by applying Leray-Schauder alternative (Theorem 1.7).

To facilitate the proof, we introduce the following notation:

$$\begin{aligned} A = {} & \vartheta_1 k_0 + \vartheta_2\lambda_0 + \frac{T^{1-\gamma}}{\Gamma(2-\gamma)}\Big(\bar{\vartheta}_1 k_0 + \frac{\vartheta_2}{T}\lambda_0\Big) \\ & +\vartheta_3\lambda_0 + \vartheta_4 k_0 + \frac{T^{1-\delta}}{\Gamma(2-\delta)}\Big(\bar{\vartheta}_3\lambda_0 + \frac{\vartheta_4}{T}k_0\Big), \end{aligned} \tag{3.69}$$

$$\begin{aligned} B = {} & \vartheta_1 k_1 + \vartheta_2 \max\{\lambda_1,\lambda_2\} + (|\sigma_1|T+1)K \\ & + \frac{T^{1-\gamma}}{\Gamma(2-\gamma)}\Big\{\bar{\vartheta}_1 k_1 + \frac{\vartheta_2}{T}\max\{\lambda_1,\lambda_2\} + |\sigma_1|K\Big\} \\ & +\vartheta_3\max\{\lambda_1,\lambda_2\} + \vartheta_4 k_1 + |\sigma_4|TK \\ & + \frac{T^{1-\delta}}{\Gamma(2-\delta)}\Big\{\bar{\vartheta}_3\max\{\lambda_1,\lambda_2\} + \frac{\vartheta_4}{T}k_1 + |\sigma_4|K\Big\}. \end{aligned} \tag{3.70}$$

$$\begin{aligned} D = {} & \vartheta_1 \max\{k_2,k_3\} + \vartheta_3\lambda_3 + |\sigma_2|TK_1 \\ & + \frac{T^{1-\gamma}}{\Gamma(2-\gamma)}\Big\{\bar{\vartheta}_1 \max\{k_2,k_3\} + \frac{\vartheta_2}{T}\lambda_3 + |\sigma_2|TK_1\Big\} \\ & +\vartheta_3\lambda_3 + \vartheta_3\max\{k_1,k_2\} + (|\sigma_3|T+1)|K_1 \\ & + \frac{T^{1-\delta}}{\Gamma(2-\delta)}\Big\{\bar{\vartheta}_3\lambda_3 + \frac{\vartheta_4}{T}\max\{k_2,k_3\} + |\sigma_4|K_1\Big\}. \end{aligned} \tag{3.71}$$

Theorem 3.19. *Assume that:*

(3.19.1) $f, g : [0,T] \times \mathbb{R}^3 \to \mathbb{R}$ *are continuous and there exist real constants* k_i, $\lambda_i \geq 0$ $(i = 1, 2, 3)$ *and* $k_0 > 0, \lambda_0 > 0$ *such that* $\forall x_i \in \mathbb{R}$ $(i = 1, 2, 3)$,

$$|f(t, x_1, x_2, x_3)| \leq k_0 + k_1|x_1| + k_2|x_2| + k_3|x_3|,$$
$$|g(t, x_1, x_2, x_3)| \leq \lambda_0 + \lambda_1|x_1| + \lambda_2|x_2| + \lambda_3|x_3|.$$

(3.19.2) $h, \phi : C([0,T], \mathbb{R}) \to \mathbb{R}$ *are continuous with* $h(0) = \phi(0) = 0$ *and there exist constants* $K, K_1 > 0$, *such that*

$$|h(y)| \leq K\|y\|, \quad |\phi(x)| \leq K_1\|x\|, \quad \forall y, x \in C([0,T], \mathbb{R}).$$

In addition it is assumed that $\max\{B, D\} < 1$, *where* B, D *are defined by (3.70) and (3.71) respectively. Then there exists at least one solution for problem (3.52)-(3.53) on* $[0,T]$.

Proof. In the first step, we show that the operator $\mathcal{F} : X \times Y \to X \times Y$ is completely continuous. By continuity of the functions f, g, h and ϕ, it follows that the operator $\mathcal{F}$ is continuous.

Let $\Omega \subset X \times Y$ be bounded. Then there exist positive constants M_f and M_g such that $|f(t, x(t), y(t), {}^c D^\gamma y(t)| \leq M_f, |g(t, x(t), {}^c D^\delta x(t), y(t)| \leq M_g, \forall (x, y) \in \Omega$, and constants $M_h, M_\phi > 0$ such that $|h(y)| \leq M_h, |\phi(x)| \leq M_\phi$ for all $y, x \in C([0,T], \mathbb{R})$. Then, for any $(x, y) \in \Omega$, one can find that

$$|\mathcal{F}_1(x, y)(t)| \leq \vartheta_1 M_f + \vartheta_2 M_g + (|\sigma_1|T + 1)M_h + |\sigma_2|TM_\phi,$$
$$|\mathcal{F}_1(x, y)'(t)| \leq \bar{\vartheta}_1 M_f + \frac{\vartheta_2}{T} M_g + |\sigma_1|M_h + |\sigma_2|M_\phi,$$

$$\begin{aligned} |{}^c D^\gamma(\mathcal{F}_1(x, y)(t) &\leq \int_0^t \frac{(t-s)^{-\gamma}}{\Gamma(1-\gamma)} |\mathcal{F}_1(x, y)'(s)ds \\ &\leq \frac{T^{1-\gamma}}{\Gamma(2-\beta)} \left\{ \bar{\vartheta}_1 M_f + \frac{\vartheta_2}{T} M_g + |\sigma_1|M_h + |\sigma_2|M_\phi \right\}. \end{aligned}$$

Thus

$$\begin{aligned} \|\mathcal{F}_1(x, y)\|_X &= \|\mathcal{F}_1(x, y)\| + \|{}^c D^\gamma(\mathcal{F}_1(x, y)\| \\ &\leq \vartheta_1 M_f + \frac{\vartheta_2}{T} M_g + (|\sigma_1|T + 1)M_h + |\sigma_2|TM_\phi \\ &\quad + \frac{T^{1-\gamma}}{\Gamma(2-\beta)} \left\{ \bar{\vartheta}_1 M_f + \frac{\vartheta_2}{T} M_g + |\sigma_1|M_h + |\sigma_2|M_\phi \right\}. \end{aligned}$$

Similarly, one can obtain that

$$\begin{aligned} \|\mathcal{F}_2(x, y)\|_Y &= \|\mathcal{F}_2(x, y)\| + \|{}^c D^\delta(\mathcal{F}_2(x, y))\| \\ &\leq \vartheta_3 M_g + \vartheta_4 M_f + (|\sigma_3|T + 1)M_\phi + |\sigma_4|TM_h \\ &\quad + \frac{T^{1-\delta}}{\Gamma(2-\delta)} \left\{ \bar{\vartheta}_3 M_g + \frac{\vartheta_4}{T} M_f + |\sigma_3|M_\phi + |\sigma_4|M_h \right\}. \end{aligned}$$

From the above inequalities, it follows that the operator T is uniformly bounded.

Next, we show that $\mathcal{F}$ is equi-continuous. Let $t_1, t_2 \in [0, T]$ with $t_1 < t_2$. Then we have

$$\begin{aligned}
&|\mathcal{F}_1(x,y)(t_2)) - \mathcal{F}_1(x,y)(t_1)| \\
\leq &\Big| \int_0^{t_1} \frac{[(t_2-s)^{\alpha-1} - (t_1-s)^{\alpha-1}]}{\Gamma(\alpha)} f(s,x(s),y(s),D^\gamma y(s))ds \Big| \\
&+\Big| \int_{t_1}^{t_2} \frac{(t_2-s)^{\alpha-1}}{\Gamma(\alpha)} f(s,x(s),y(s),D^\gamma y(s))ds \Big| \\
&+|\sigma_1|M_h(t_2-t_1) + |\sigma_2|M_\phi(t_2-t_1) \\
&+\frac{t_2-t_1}{|\Delta|}\Big[|\mu_2|\xi|\mu_1|\frac{\eta^\alpha}{\Gamma(\alpha+1)}M_f + |\mu_2|\xi\frac{T^{\beta+1}}{\Gamma(\beta+2)}M_g \\
&+\frac{T^2}{2}|\mu_2|\frac{\xi^\beta}{\Gamma(\beta+1)}M_g + \frac{T^2}{2}\frac{T^{\alpha+1}}{\Gamma(\alpha+2)}M_f\Big] \\
\leq &\frac{M_f}{\Gamma(\alpha+1)}[(t_2-t_1)^\alpha + |t_2^\alpha - t_1^\alpha|] + |\sigma_1|M_h(t_2-t_1) + |\sigma_2|M_\phi(t_2-t_1) \\
&+\frac{t_2-t_1}{|\Delta|}\Big[|\mu_2|\xi|\mu_1|\frac{\eta^\alpha}{\Gamma(\alpha+1)}M_f + |\mu_2|\xi\frac{T^{\beta+1}}{\Gamma(\beta+2)}M_g \\
&+\frac{T^2}{2}|\mu_2|\frac{\xi^\beta}{\Gamma(\beta+1)}M_g + \frac{T^2}{2}\frac{T^{\alpha+1}}{\Gamma(\alpha+2)}M_f\Big],
\end{aligned}$$

and

$$|\mathcal{F}_1(x,y)'(t_2)) - \mathcal{F}_1(x,y)'(t_1)| \leq \frac{M_f}{\Gamma(\alpha)}[(t_2-t_1)^{\alpha-1} + |t_2^{\alpha-1} - t_1^{\alpha-1}|].$$

Thus

$$\begin{aligned}
|{}^cD^\gamma\mathcal{F}_1(x,y)(t_2) - {}^cD^\gamma\mathcal{F}_1(x,y)(t_1)| &= \int_0^t \frac{(t-s)^{-\gamma}}{\Gamma(1-\gamma)}|\mathcal{F}_1(x,y)'(t_2) - \mathcal{F}_1(x,y)'(t_1)|ds \\
&\leq \frac{T^{1-\gamma}}{\Gamma(2-\gamma)}\frac{M_f}{\Gamma(\alpha)}[(t_2-t_1)^{\alpha-1} + |t_2^{\alpha-1} - t_1^{\alpha-1}|].
\end{aligned}$$

Hence we have that $|\mathcal{F}_1(x,y)(t_2) - \mathcal{F}_1(x,y)(t_1)|_X \to 0$ independent of x and y as $t_2 \to t_1$. Analogously, one can obtain

$$\begin{aligned}
&|\mathcal{F}_2(x,y)(t_2) - \mathcal{F}_2(x,y)(t_1)| \\
\leq &\frac{M_g}{\Gamma(\beta+1)}[(t_2-t_1)^\beta + |t_2^\beta - t_1^\beta|] + |\sigma_3|M_\phi(t_2-t_1) + |\sigma_4|M_h(t_2-t_1) \\
&+\frac{t_2-t_1}{|\Delta|}\Big[|\mu_1|\eta|\mu_2|\frac{\xi^\beta}{\Gamma(\beta+1)}M_g + |\mu_1|\eta\frac{T^{\alpha+1}}{\Gamma(\alpha+2)}M_f \\
&+\frac{T^2}{2}|\mu_1|\frac{\eta^\alpha}{\Gamma(\alpha+1)}M_f + \frac{T^2}{2}\frac{T^{\beta+1}}{\Gamma(\beta+2)}M_g\Big],
\end{aligned}$$

$$|\mathcal{F}_2(x,y)'(t_2)) - \mathcal{F}_2(x,y)'(t_1)| \leq \frac{M_g}{\Gamma(\beta)}[(t_2-t_1)^{\beta-1} + |t_2^{\beta-1} - t_1^{\beta-1}|],$$

and

$$|^cD^\delta \mathcal{F}_2(x,y)(t_2) -^c D^\delta \mathcal{F}_2(x,y)(t_1)| \leq \frac{T^{1-\delta}}{\Gamma(2-\delta)}\frac{M_g}{\Gamma(\beta)}[(t_2-t_1)^{\beta-1} + |t_2^{\beta-1} - t_1^{\beta-1}|],$$

which imply that $|\mathcal{F}_2(x,y)(t_2) - \mathcal{F}_2(x,y)(t_1)|_Y \to 0$ independent of x and y as $t_2 \to t_1$. Therefore, the operator $\mathcal{F}(x,y)$ is equi-continuous. From the foregoing arguments, we infer that the operator $\mathcal{F}(x,y)$ is completely continuous by Arzelá-Ascoli theorem.

Finally, it will be verified that the set $\mathcal{E} = \{(x,y) \in X \times Y | (x,y) = \lambda\mathcal{F}(x,y), 0 \leq \lambda \leq 1\}$ is bounded. Let $(x,y) \in \mathcal{E}$, then $(x,y) = \lambda\mathcal{F}(x,y)$. For any $t \in [0,T]$, we have

$$x(t) = \lambda\mathcal{F}_1(x,y)(t), \quad y(t) = \lambda\mathcal{F}_2(x,y)(t).$$

Then

$$\begin{aligned}
&|x(t)| \\
&\leq \vartheta_1(k_0 + k_1\|x\|_X + \max\{k_2,k_3\}\|y\|_Y) + \vartheta_2(\lambda_0 + \max\{\lambda_1,\lambda_2\}\|x\|_X + \lambda_3\|y\|_Y) \\
&\quad +(|\sigma_1|T+1)K\|x\|_X + |\sigma_2|TK_1\|y\|_Y,
\end{aligned}$$

$$\begin{aligned}
&|x'(t)| \\
&\leq \bar{\vartheta}_1(k_0 + k_1\|x\|_X + \max\{k_2,k_3\}\|y\|_Y) + \frac{\vartheta_2}{T}(\lambda_0 + \max\{\lambda_1,\lambda_2\}\|x\|_X + \lambda_3\|y\|_Y) \\
&\quad +|\sigma_1|K\|x\|_X + |\sigma_2|K_1\|y\|_Y,
\end{aligned}$$

and

$$\begin{aligned}
|^cD^\gamma x(t)| \leq \frac{T^{1-\gamma}}{\Gamma(2-\gamma)}\Bigg\{&\bar{\vartheta}_1(k_0 + k_1\|x\|_X + \max\{k_2,k_3\}\|y\|_Y) \\
&+\frac{\vartheta_2}{T}(\lambda_0 + \max\{\lambda_1,\lambda_2\}\|x\|_X + \lambda_3\|y\|_Y) + |\sigma_1|K\|x\|_X + |\sigma_2|K_1\|y\|_Y\Bigg\}.
\end{aligned}$$

In view of the above estimates, we get

$$\begin{aligned}
&\|x\|_X \\
&\leq \vartheta_1(k_0 + k_1\|x\|_X + \max\{k_2,k_3\}\|y\|_Y) + \vartheta_2(\lambda_0 + \max\{\lambda_1,\lambda_2\}\|x\|_X + \lambda_3\|y\|_Y) \\
&\quad +(|\sigma_1|T+1)K\|x\|_X + |\sigma_2|TK_1\|y\|_Y \\
&\quad +\frac{T^{1-\gamma}}{\Gamma(2-\gamma)}\Bigg\{\bar{\vartheta}_1(k_0 + k_1\|x\|_X + \max\{k_2,k_3\}\|y\|_Y) \\
&\quad +\frac{\vartheta_2}{T}(\lambda_0 + \max\{\lambda_1,\lambda_2\}\|x\|_X + \lambda_3\|y\|_Y) + |\sigma_1|K\|x\|_X + |\sigma_2|K_1\|y\|_Y\Bigg\}.
\end{aligned}$$

In a similar manner, we can have

$$\begin{aligned}
&\|y\|_Y \\
&\leq \vartheta_3(\lambda_0 + \max\{\lambda_1,\lambda_2\}\|x\|_X + \lambda_3\|y\|_Y) + \vartheta_4(k_0 + k_1\|x\|_X + \max\{k_2,k_3\}\|y\|_Y)
\end{aligned}$$

$$+(|\sigma_3|T+1)K_1\|y\|_Y+|\sigma_4|TK\|x\|_X$$
$$+\frac{T^{1-\delta}}{\Gamma(2-\delta)}\Bigg\{\bar{\vartheta}_3(\lambda_0+\max\{\lambda_1,\lambda_2\}\|x\|_X+\lambda_3\|y\|_Y)$$
$$+\frac{\vartheta_4}{T}(k_0+k_1\|x\|_X+\max\{k_2,k_3\}\|y\|_Y)+|\sigma_3|K_1\|y\|_Y+|\sigma_4|K\|x\|_X\Bigg\}.$$

Then, making use of the notation (3.69), (3.70) and (3.71), we find that

$$\|x\|_X+\|y\|_Y\leq A+\max\{B,D\}\|(x,y)\|_{X\times Y},$$

which implies that

$$\|(x,y)\|_{X\times Y}\leq\frac{A}{1-\max\{B,D\}}.$$

This shows that the set $\mathcal{E}$ is bounded. Thus, by Theorem 1.7, the operator $\mathcal{F}$ has at least one fixed point. Hence problem (3.52)-(3.53) has at least one solution on $[0,T]$. This completes the proof. □

Example 3.12. Let us consider the following system

$$\begin{cases} {}^cD^{5/4}x(t)=\dfrac{1}{\sqrt{16+t^2}}\cos t+\dfrac{1}{240}e^{-3t}\sin x(t)+\dfrac{1}{300}\dfrac{y(t)|x(t)|}{1+|x(t)|}+\dfrac{1}{2(190+t)}{}^cD^{1/4}y(t),\\ {}^cD^{3/2}y(t)=\dfrac{e^{-t}}{30+t^2}+\dfrac{x(t)}{120(1+y^2(t))}+\dfrac{1}{2(120+t)}\sin({}^cD^{1/2}x(t))\\ \qquad+\dfrac{1}{2\sqrt{6400+t}}e^{-t}\sin(y(t)),\quad t\in[0,1], \end{cases} \tag{3.72}$$

subject to the conditions:

$$x(0)=\frac{1}{180}y(t),\ \int_0^1 y(s)ds=x(2/3),\ y(0)=\frac{1}{140}x(t),\ \int_0^1 x(s)ds=y(1/3). \tag{3.73}$$

Clearly

$$|f(t,x(t),y(t),{}^cD^{1/4}y(t))|\leq\frac{1}{4}+\frac{1}{240}\|x\|+\frac{1}{300}\|y\|+\frac{1}{380}\|{}^cD^{1/4}y\|,$$
$$|g(t,x(t),{}^cD^{1/2}x(t),y(t))|\leq\frac{1}{30}+\frac{1}{120}\|x\|+\frac{1}{240}\|{}^cD^{1/2}x\|+\frac{1}{160}\|y\|,$$
$$|h(y)|\leq\frac{1}{180}\|y\|,\ |\phi(x)|\leq\frac{1}{140}\|x\|.$$

Thus $k_0=1/4$, $k_1=1/240$, $k_2=1/300$, $k_3=1/380$, $\lambda_0=1/30$, $\lambda_1=1/120$, $\lambda_2=1/240$, $\lambda_3=1/160, K=1/180, K_1=1/140$. Using (3.70) and (3.71), it is found that $B\simeq 0.73394093$, $D\simeq 0.63340189$ and that $\max\{B,D\}=0.73394093<1$. Thus the hypothesis of Theorem 3.19 is satisfied. Therefore, there exists a solution for the problem (3.72)-(3.73) on $[0,1]$.

3.5 On nonlinear neutral Caputo-type fractional differential equations with Riemann-Liouville integral boundary conditions

In this section, we consider a neutral fractional Caputo-type single-valued boundary value problem involving Riemann-Liouville integral boundary conditions given by

$$\begin{cases} {}^cD^\rho[{}^cD^\sigma y(t) - g(t,y(t))] = f(t,y(t)), \ t \in J := [0,T], \\ y(\eta) = 0, \quad I^\gamma y(T) = 0, \quad \eta \in (0,T), \end{cases} \tag{3.74}$$

where ${}^cD^\theta$ is Caputo fractional derivative operator of order $\vartheta \in \{\rho, \sigma,\}$, $0 < \vartheta < 1$, I^γ is the Riemann-Liouville fractional integral of order γ, and $f, g : J \times \mathbb{R} \to \mathbb{R}$ are given functions.

We also investigate the inclusions case of the problem (3.74):

$$\begin{cases} {}^cD^\rho[{}^cD^\sigma y(t) - g(t,y(t))] \in F(t,y(t)), \ t \in J := [0,T], \\ y(\eta) = 0, \quad I^\gamma y(T) = 0, \quad \eta \in (0,T), \end{cases} \tag{3.75}$$

where $F : [0,T] \times \mathbb{R} \to \mathcal{P}(\mathbb{R})$ is a multi-valued function ($\mathcal{P}(\mathbb{R})$ is the family of all non-empty subjects of $\mathbb{R}$).

We prove the existence of solutions for the problem (3.74) with the aid of a fixed point theorem due to Sadovskii, while the existence result for the problem (3.75) is obtained by applying the nonlinear alternative for contractive maps.

The following lemma, concerning with a linear variant of the problem (3.74), plays a crucial role in the forthcoming analysis.

Lemma 3.6. *Let $\bar{f} \in C[0,T]$ and*

$$\Lambda = \frac{T^\gamma \eta^\sigma}{\Gamma(\sigma+1)\Gamma(\gamma+1)} - \frac{T^{\sigma+\gamma}}{\Gamma(\sigma+\gamma+1)} \neq 0. \tag{3.76}$$

The function y is a solution of the problem

$$\begin{cases} {}^cD^\rho[{}^cD^\sigma y(t) - \bar{g}(t)] = \bar{f}(t), \ t \in J := [0,T], \\ y(\eta) = 0, \quad I^\gamma y(T) = 0, \end{cases} \tag{3.77}$$

if and only if

$$\begin{aligned} y(t) &= I^\sigma \bar{g}(s)(t) + I^{\rho+\sigma}\bar{f}(s)(t) \\ &\quad + \frac{t^\sigma}{\Lambda\Gamma(\sigma+1)}\Big[I^{\sigma+\gamma}\bar{g}(s)(T) + I^{\rho+\sigma+\gamma}\bar{f}(s)(T) \\ &\quad - \frac{T^\gamma}{\Gamma(\gamma+1)}\Big(I^\sigma\bar{g}(s)(\eta) + I^{\rho+\sigma}\bar{f}(s)(\eta)\Big)\Big] \\ &\quad + \frac{1}{\Lambda}\Big[\frac{T^{\sigma+\gamma}}{\Gamma(\sigma+\gamma+1)}\Big(I^\sigma\bar{g}(s)(\eta) + I^{\rho+\sigma}\bar{f}(s)(\eta)\Big) \\ &\quad - \frac{\eta^\sigma}{\Gamma(\sigma+1)}\Big(I^{\sigma+\gamma}\bar{g}(s)(T) + I^{\rho+\sigma+\gamma}\bar{f}(s)(T)\Big)\Big]. \end{aligned} \tag{3.78}$$

Proof. The solution of the differential equation in (3.77) can be written as

$$^cD^\sigma y(t) - \bar{g}(s) = I^\rho \bar{f}(s)(t) + c_1,$$

where $c_1 \in \mathbb{R}$ is an arbitrary constant and moreover

$$y(t) = I^\sigma \bar{g}(s)(t) + I^{\rho+\sigma}\bar{f}(s)(t) + \frac{t^\sigma}{\Gamma(\sigma+1)}c_1 + c_2, \tag{3.79}$$

where $c_2 \in \mathbb{R}$ is an arbitrary constant. Applying Riemann-Liouville fractional integral of order γ to (3.79), we have

$$I^\gamma y(t) = I^{\sigma+\gamma}\bar{g}(s)(t) + I^{\rho+\sigma+\gamma}\bar{f}(s)(t) + \frac{t^{\sigma+\gamma}}{\Gamma(\sigma+\gamma+1)}c_1 + \frac{t^\gamma}{\Gamma(\gamma+1)}c_2.$$

Using the above values of $y(t)$ and $I^\gamma y(t)$ in the boundary conditions given by (3.77), we get

$$I^\sigma \bar{g}(s)(\eta) + I^{\rho+\sigma}\bar{f}(s)(\eta) + \frac{\eta^\sigma}{\Gamma(\sigma+1)}c_1 + c_2 = 0,$$

$$I^{\sigma+\gamma}\bar{g}(s)(T) + I^{\rho+\sigma+\gamma}\bar{f}(s)(T) + \frac{T^{\sigma+\gamma}}{\Gamma(\sigma+\gamma+1)}c_1 + \frac{T^\gamma}{\Gamma(\gamma+1)}c_2 = 0.$$

Solving the above system for c_1 and c_2, we find that

$$c_1 = \frac{1}{\Lambda}\Big\{I^{\sigma+\gamma}\bar{g}(s)(T) + I^{\rho+\sigma+\gamma}\bar{f}(s)(T) - \frac{T^\gamma}{\Gamma(\gamma+1)}\Big(I^\sigma \bar{g}(s)(\eta) + I^{\rho+\sigma}\bar{f}(s)(\eta)\Big)\Big\}$$

and

$$c_2 = \frac{1}{\Lambda}\Big\{\frac{T^{\sigma+\gamma}}{\Gamma(\sigma+\gamma+1)}\Big(I^\sigma \bar{g}(s)(\eta) + I^{\rho+\sigma}\bar{f}(s)(\eta)\Big) - \frac{\eta^\sigma}{\Gamma(\sigma+1)}\Big(I^{\sigma+\gamma}\bar{g}(s)(T) + I^{\rho+\sigma+\gamma}\bar{f}(s)(T)\Big)\Big\}.$$

Inserting the values of c_1 and c_2 in (3.79), we get the solution (3.78). The converse follows by direct computation. This completes the proof. □

In the sequel, we denote by $\mathcal{C} := C([0,T],\mathbb{R})$ the Banach space of all continuous functions $x : [0,T] \to \mathbb{R}$ equipped with the norm $\|x\| = \sup\{|x(t)|,\ t \in [0,T]\}$, and by $L^1([0,T],\mathbb{R})$ the space of functions $y : [0,T] \to \mathbb{R}$ with the norm $\|y\|_{L^1} = \int_0^T |y(t)|dt$.

Furthermore, we set the notation:

$$\Lambda_1 = \frac{T^\sigma}{\Gamma(\sigma+1)} + \frac{T^\sigma}{|\Lambda|\Gamma(\sigma+1)}\left(\frac{T^{\sigma+\gamma}}{\Gamma(\sigma+\gamma+1)} + \frac{T^\gamma}{\Gamma(\gamma+1)}\frac{\eta^\sigma}{\Gamma(\sigma+1)}\right) + \frac{1}{|\Lambda|}\frac{T^{\sigma+\gamma}}{\Gamma(\sigma+\gamma+1)}\frac{2\eta^\sigma}{\Gamma(\sigma+1)}, \tag{3.80}$$

and

$$\Lambda_2 = \frac{T^{\rho+\sigma}}{\Gamma(\rho+\sigma+1)}$$

$$+\frac{T^{\sigma}}{|\Lambda|\Gamma(\sigma+1)}\left(\frac{T^{\rho+\sigma+\gamma}}{\Gamma(\rho+\sigma+\gamma+1)}+\frac{T^{\gamma}}{\Gamma(\gamma+1)}\frac{\eta^{\rho+\sigma}}{\Gamma(\rho+\sigma+1)}\right)$$
$$+\frac{1}{|\Lambda|}\left\{\frac{T^{\sigma+\gamma}}{\Gamma(\sigma+\gamma+1)}\frac{\eta^{\rho+\sigma}}{\Gamma(\rho+\sigma+1)}+\frac{T^{\rho+\sigma+\gamma}}{\Gamma(\rho+\sigma+\gamma+1)}\frac{\eta^{\sigma}}{\Gamma(\sigma+1)}\right\}. \tag{3.81}$$

3.5.1 *Existence result for problem (3.74)*

Here we derive an existence result for the problem (3.74) by applying Sadovskii's fixed point (Theorem 1.14).

Theorem 3.20. *Let $f, g : J \times \mathbb{R} \to \mathbb{R}$ be continuous functions. Assume that*

(3.20.1) there exists a nonnegative constant $k < \Lambda_1^{-1}$ such that

$$|g(t,x_1)-g(t,x_2)| \le k|x_1-x_2|, \quad \textit{for } t\in J \textit{ and every } x_1,x_2\in\mathbb{R};$$

(3.20.2) $|g(t,x)| \le q(t)$, $|f(t,x)| \le z(t)$, $\forall (t,x)\in J\times\mathbb{R}$, and $q,z \in C(J,\mathbb{R}^+)$.

Then the problem (3.74) has at least one solution on J.

Proof. Let $B_r = \{y \in \mathcal{C} : \|y\| \le r\}$ be a closed bounded and convex subset of $\mathcal{C}$, where r is a fixed constant. In view of Lemma 3.6, we introduce an operator $\mathcal{Q} : \mathcal{C} \to \mathcal{C}$ by replacing $\bar{g}(t) = g(t, y(t))$, $\bar{f}(t) = f(t, y(t))$ in (3.78) as follows

$$\begin{aligned}
\mathcal{Q}y(t) &= I^{\sigma}g(s,y(s))(t)+I^{\rho+\sigma}f(s,y(s))(t)\\
&\quad+\frac{t^{\sigma}}{\Lambda\Gamma(\sigma+1)}\Big[I^{\sigma+\gamma}g(s,y(s))(T)+I^{\rho+\sigma+\gamma}f(s,y(s))(T)\\
&\quad-\frac{T^{\gamma}}{\Gamma(\gamma+1)}\Big(I^{\sigma}g(s,y(s))(\eta)+I^{\rho+\sigma}f(s,y(s))(\eta)\Big)\Big]\\
&\quad+\frac{1}{\Lambda}\Big[\frac{T^{\sigma+\gamma}}{\Gamma(\sigma+\gamma+1)}\Big(I^{\sigma}g(s,y(s))(\eta)+I^{\rho+\sigma}f(s,y(s))(\eta)\Big)\\
&\quad-\frac{\eta^{\sigma}}{\Gamma(\sigma+1)}\Big(I^{\sigma+\gamma}g(s,y(s))(T)+I^{\rho+\sigma+\gamma}f(s,y(s))(T)\Big)\Big], \quad t\in J.
\end{aligned} \tag{3.82}$$

Let us define the operators $\mathcal{Q}_1, \mathcal{Q}_2 : B_r \to \mathcal{C}$ by

$$\begin{aligned}
(\mathcal{Q}_1y)(t) &= I^{\sigma}g(s,y(s))(t)\\
&\quad+\frac{t^{\sigma}}{\Lambda\Gamma(\sigma+1)}\Big[I^{\sigma+\gamma}g(s,y(s))(T)-\frac{T^{\gamma}}{\Gamma(\gamma+1)}I^{\sigma}g(s,y(s))(\eta)\Big]\\
&\quad+\frac{1}{\Lambda}\Big[\frac{T^{\sigma+\gamma}}{\Gamma(\sigma+\gamma+1)}I^{\sigma}g(s,y(s))(\eta)-\frac{\eta^{\sigma}}{\Gamma(\sigma+1)}I^{\sigma+\gamma}g(s,y(s))(T)\Big],
\end{aligned}$$

and

$$\begin{aligned}
&(\mathcal{Q}_2y)(t)\\
&= I^{\rho+\sigma}f(s,y(s))(t)\\
&\quad+\frac{t^{\sigma}}{\Lambda\Gamma(\sigma+1)}\Big[I^{\rho+\sigma+\gamma}f(s,y(s))(T)-\frac{T^{\gamma}}{\Gamma(\gamma+1)}I^{\rho+\sigma}f(s,y(s))(\eta)\Big]
\end{aligned}$$

$$+\frac{1}{\Lambda}\Big[\frac{T^{\sigma+\gamma}}{\Gamma(\sigma+\gamma+1)}I^{\rho+\sigma}f(s,y(s))(\eta)-\frac{\eta^{\sigma}}{\Gamma(\sigma+1)}I^{\rho+\sigma+\gamma}f(s,y(s))(T)\Big],$$

and note that

$$(\mathcal{Q}y)(t)=(\mathcal{Q}_1y)(t)+(\mathcal{Q}_2y)(t),\quad t\in J. \tag{3.83}$$

Obviously the existence of a fixed point for the sum of operators $\mathcal{Q}_1+\mathcal{Q}_2$ will imply that operator $\mathcal{Q}$ has a fixed point. In order to show that $\mathcal{Q}_1+\mathcal{Q}_2$ has a fixed point, it will be shown that the operators $\mathcal{Q}_1$ and $\mathcal{Q}_2$ satisfy the hypothesis of Theorem 1.14. This will be established in several steps.

Step 1. $\mathcal{Q}B_r\subset B_r$.

Let us select $r\geq\|q\|\Lambda_1+\|z\|\Lambda_2$, where Λ_1,Λ_2 are defined by (3.80) and (3.81) respectively. For any $x\in B_r$, we have

$$\begin{aligned}
\|\mathcal{Q}y\| &\leq I^{\sigma}|g(s,x(s))|(T)+I^{\rho+\sigma}|f(s,y(s))|(T)\\
&\quad+\frac{T^{\sigma}}{|\Lambda|\Gamma(\sigma+1)}\Big[I^{\sigma+\gamma}|g(s,y(s))|(T)+I^{\rho+\sigma+\gamma}|f(s,y(s))|(T)\\
&\quad+\frac{T^{\gamma}}{\Gamma(\gamma+1)}\Big(I^{\sigma}|g(s,y(s))|(\eta)+I^{\rho+\sigma}|f(s,y(s))|(\eta)\Big)\Big]\\
&\quad+\frac{1}{|\Lambda|}\Big[\frac{T^{\sigma+\gamma}}{\Gamma(\sigma+\gamma+1)}\Big(I^{\sigma}|g(s,y(s))|(\eta)+I^{\rho+\sigma}|f(s,y(s))|(\eta)\Big)\\
&\quad+\frac{\eta^{\sigma}}{\Gamma(\sigma+1)}\Big(I^{\sigma+\gamma}|g(s,y(s))|(T)+I^{\rho+\sigma+\gamma}|f(s,y(s))|(T)\Big)\Big]\\
&\leq\|q\|\Bigg[\frac{T^{\sigma}}{\Gamma(\sigma+1)}+\frac{T^{\sigma}}{|\Lambda|\Gamma(\sigma+1)}\Big(\frac{T^{\sigma+\gamma}}{\Gamma(\sigma+\gamma+1)}+\frac{T^{\gamma}}{\Gamma(\gamma+1)}\frac{\eta^{\sigma}}{\Gamma(\sigma+1)}\Big)\\
&\quad+\frac{1}{|\Lambda|}\frac{T^{\sigma+\gamma}}{\Gamma(\sigma+\gamma+1)}\frac{2\eta^{\sigma}}{\Gamma(\sigma+1)}\Bigg]+\|z\|\Bigg[\frac{T^{\rho+\sigma}}{\Gamma(\rho+\sigma+1)}\\
&\quad+\frac{T^{\sigma}}{|\Lambda|\Gamma(\sigma+1)}\Big(\frac{T^{\rho+\sigma+\gamma}}{\Gamma(\rho+\sigma+\gamma+1)}+\frac{T^{\gamma}}{\Gamma(\gamma+1)}\frac{\eta^{\rho+\sigma}}{\Gamma(\rho+\sigma+1)}\Big)\\
&\quad+\frac{1}{|\Lambda|}\Big\{\frac{T^{\sigma+\gamma}}{\Gamma(\sigma+\gamma+1)}\frac{\eta^{\rho+\sigma}}{\Gamma(\rho+\sigma+1)}+\frac{T^{\rho+\sigma+\gamma}}{\Gamma(\rho+\sigma+\gamma+1)}\frac{\eta^{\sigma}}{\Gamma(\sigma+1)}\Big\}\Bigg]\\
&=\|q\|\Lambda_1+\|z\|\Lambda_2\\
&\leq r,
\end{aligned}$$

which implies that $\mathcal{Q}B_r\subset B_r$.

Step 2. $\mathcal{Q}_2$ is compact.

Observe that the operator $\mathcal{Q}_2$ is uniformly bounded in view of Step 1. Let $t_1,t_2\in J$ with $t_1<t_2$ and $x\in B_r$. Then we obtain

$$|(\mathcal{Q}_2y)(t_2)-(\mathcal{Q}_2y)(t_1)|\leq\frac{\|z\|}{\Gamma(\rho+\sigma+1)}[t_2^{\rho+\sigma}-t_1^{\rho+\sigma}+2(t_2-t_1)^{\rho+\sigma}]$$

$$+\frac{|t_2^\sigma - t_1^\sigma|\|z\|}{|\Lambda|\Gamma(\sigma+1)}\left\{\frac{T^{\rho+\sigma+\gamma}}{\Gamma(\rho+\sigma+\gamma+1)}+\frac{\eta^{\rho+\sigma}}{\Gamma(\rho+\sigma+1)}\right\},$$

which is independent of y and tends to zero as $t_2 - t_1 \to 0$. Thus, $\mathcal{Q}_2$ is equicontinuous. Hence, by the Arzelá-Ascoli Theorem, $\mathcal{Q}_2(B_r)$ is a relatively compact set.

Step 3. $\mathcal{Q}_1$ is γ-contractive.

Let $y, z \in B_r$. Then, we have

$$\begin{aligned}
&\|\mathcal{Q}_1 y - \mathcal{Q}_1 z\| \\
&\le I^\sigma |g(s,y(s)) - g(s,z(s))|(T) \\
&\quad + \frac{t^\sigma}{|\Lambda|\Gamma(\sigma+1)}\Big[I^{\sigma+\gamma}|g(s,y(s)) - g(s,z(s))|(T) \\
&\quad + \frac{T^\gamma}{\Gamma(\gamma+1)}\Big(I^\sigma|g(s,y(s)) - g(s,z(s))|(\eta)\Big)\Big] \\
&\quad + \frac{1}{|\Lambda|}\Big[\frac{T^{\sigma+\gamma}}{\Gamma(\sigma+\gamma+1)}\Big(I^\sigma|g(s,y(s)) - g(s,z(s))|(\eta)\Big) \\
&\quad + \frac{\eta^\sigma}{\Gamma(\sigma+1)}\Big(I^{\sigma+\gamma}|g(s,y(s)) - g(s,z(s))|(T)\Big)\Big] \\
&\le \left\{\frac{T^\sigma}{\Gamma(\sigma+1)} + \frac{T^\sigma}{|\Lambda|\Gamma(\sigma+1)}\left(\frac{T^{\sigma+\gamma}}{\Gamma(\sigma+\gamma+1)} + \frac{T^\gamma}{\Gamma(\gamma+1)}\frac{\eta^\sigma}{\Gamma(\sigma+1)}\right)\right. \\
&\quad \left. + \frac{1}{|\Lambda|}\frac{T^{\sigma+\gamma}}{\Gamma(\sigma+\gamma+1)}\frac{2\eta^\sigma}{\Gamma(\sigma+1)}\right\}k\|y-z\|,
\end{aligned}$$

which is γ-contractive, since $\gamma = k\Lambda_1 < 1$.

Step 4. $\mathcal{Q}$ is condensing.

In view of the fact that $\mathcal{Q}_1$ is continuous, γ-contraction and $\mathcal{Q}_2$ is compact, it follows by Lemma 1.13 that $\mathcal{Q} : B_r \to B_r$ with $\mathcal{Q} = \mathcal{Q}_1 + \mathcal{Q}_2$ is a condensing map on B_r.

From the foregoing arguments, we deduce by Theorem 1.14 that the operator $\mathcal{Q}$ has a fixed point. In consequence, the problem (3.74) has a solution on $[0, T]$. □

3.5.2 *Existence result for problem (3.75)*

Definition 3.4. A function $y \in C^2(J, \mathbb{R})$ is a solution of the problem (3.75) if $y(\eta) = 0$, $I^\gamma y(T) = 0$, and there exists function $v \in L^1(J, \mathbb{R})$ such that $v(t) \in F(t, y(t))$ a.e. on J and

$$\begin{aligned}
y(t) &= I^\sigma g(s,y(s))(t) + I^{\rho+\sigma}v(s)(t) \\
&\quad + \frac{t^\sigma}{\Lambda\Gamma(\sigma+1)}\Big\{I^{\sigma+\gamma}g(s,y(s))(T) + I^{\rho+\sigma+\gamma}v(s)(T)
\end{aligned}$$

$$
\begin{aligned}
&- \frac{T^\gamma}{\Gamma(\gamma+1)}\Big(I^\sigma g(s,y(s))(\eta) + I^{\rho+\sigma}v(s)(\eta)\Big)\Big]\\
&+ \frac{1}{\Lambda}\Big[\frac{T^{\sigma+\gamma}}{\Gamma(\sigma+\gamma+1)}\Big(I^\sigma g(s,y(s))(\eta) + I^{\rho+\sigma}v(s)(\eta)\Big)\\
&- \frac{\eta^\sigma}{\Gamma(\sigma+1)}\Big(I^{\sigma+\gamma}g(s,y(s))(T) + I^{\rho+\sigma+\gamma}v(s)(T)\Big)\Big], \quad t \in J,
\end{aligned}
$$

where $\Lambda \neq 0$ is defined by (3.76).

Now the platform is set to present an existence result for the problem (3.75).

Theorem 3.21. *Assume that (H_1) holds. In addition we assume that*

(3.21.1) $F : J \times \mathbb{R} \to \mathcal{P}_{cp,c}(\mathbb{R})$ *is L^1-Carathéodory;*

(3.21.2) there exists a continuous non-decreasing function $\Phi : [0,\infty) \to (0,\infty)$ and a function $p \in L^1(J,\mathbb{R}^+)$ such that

$$\|F(t,y)\|_{\mathcal{P}} := \sup\{|x| : x \in F(t,y)\} \leq p(t)\Phi(\|y\|) \quad \textit{for each} \quad (t,y) \in J \times \mathbb{R};$$

(3.21.3) there exists a constant $\varpi > 0$ such that

$$\frac{(1-k\Lambda_1)\varpi}{\Phi(\varpi)\Psi_1 + \Lambda_1 g_0} > 1, \tag{3.84}$$

where $g_0 = \sup_{t\in[0.T]}|g(t,0)|$ and

$$
\begin{aligned}
\Psi_1 = I^{\rho+\sigma}p(s)(T) &+ \frac{T^\sigma}{|\Lambda|\Gamma(\sigma+1)}\Big[I^{\rho+\sigma+\gamma}p(s)(T) + \frac{T^\gamma}{\Gamma(\gamma+1)}I^{\rho+\sigma}p(s)(\eta)\Big]\\
&+ \frac{1}{|\Lambda|}\Big[\frac{T^{\sigma+\gamma}}{\Gamma(\sigma+\gamma+1)}I^{\rho+\sigma}p(s)(\eta) + \frac{\eta^\sigma}{\Gamma(\sigma+1)}I^{\rho+\sigma+\gamma}p(s)(T)\Big].
\end{aligned}
$$

Then the boundary value problem (3.75) has at least one solution on J.

Proof. In order to transform the problem (3.75) into a fixed point problem, define an operator $\mathcal{N} : \mathcal{C} \longrightarrow \mathcal{P}(\mathcal{C})$ by

$$
\mathcal{N}(y) = \left\{
\begin{aligned}
&h \in \mathcal{C} :\\
&h(t) = \left\{
\begin{aligned}
&I^\sigma g(s,y(s))(t) + I^{\rho+\sigma}v(s)(t)\\
&+\frac{t^\sigma}{\Lambda\Gamma(\sigma+1)}\Big[I^{\sigma+\gamma}g(s,y(s))(T) + I^{\rho+\sigma+\gamma}v(s)(T)\\
&-\frac{T^\gamma}{\Gamma(\gamma+1)}\Big(I^\sigma g(s,y(s))(\eta) + I^{\rho+\sigma}v(s)(\eta)\Big)\Big]\\
&+\frac{1}{\Lambda}\Big[\frac{T^{\sigma+\gamma}}{\Gamma(\sigma+\gamma+1)}\Big(I^\sigma g(s,y(s))(\eta) + I^{\rho+\sigma}v(s)(\eta)\Big)\\
&-\frac{\eta^\sigma}{\Gamma(\sigma+1)}\Big(I^{\sigma+\gamma}g(s,y(s))(T) + I^{\rho+\sigma+\gamma}v(s)(T)\Big)\Big],
\end{aligned}
\right.
\end{aligned}
\right\}
$$

for $v \in S_{F,y}$.

Next we introduce an operator $\mathcal{A}_1 : \mathcal{C} \longrightarrow \mathcal{C}$ by

$$(\mathcal{A}_1 y)(t) = I^\sigma g(s,y(s))(t)$$

$$+\frac{t^{\sigma}}{\Lambda\Gamma(\sigma+1)}\Big[I^{\sigma+\gamma}g(s,y(s))(T)-\frac{T^{\gamma}}{\Gamma(\gamma+1)}I^{\sigma}g(s,y(s))(\eta)\Big]$$
$$+\frac{1}{\Lambda}\Big[\frac{T^{\sigma+\gamma}}{\Gamma(\sigma+\gamma+1)}I^{\sigma}g(s,y(s))(\eta)-\frac{\eta^{\sigma}}{\Gamma(\sigma+1)}I^{\sigma+\gamma}g(s,y(s))(T)\Big],$$

and a multi-valued operator $\mathcal{A}_2:\mathcal{C}\longrightarrow\mathcal{P}(\mathcal{C})$ by

$$\mathcal{A}_2(y)(t)=\left\{\begin{array}{l} h\in\mathcal{C}: \\ h(t)=\left\{\begin{array}{l} I^{\rho+\sigma}v(s)(t) \\ +\frac{t^{\sigma}}{\Lambda\Gamma(\sigma+1)}\Big[I^{\rho+\sigma+\gamma}v(s)(T)-\frac{T^{\gamma}}{\Gamma(\gamma+1)}I^{\rho+\sigma}v(s)(\eta)\Big] \\ +\frac{1}{\Lambda}\Big[\frac{T^{\sigma+\gamma}}{\Gamma(\sigma+\gamma+1)}I^{\rho+\sigma}v(s)(\eta) \\ -\frac{\eta^{\sigma}}{\Gamma(\sigma+1)}I^{\rho+\sigma+\gamma}v(s)(T)\Big], \end{array}\right. \end{array}\right\}$$

for $v\in S_{F,x}$. Notice that $\mathcal{N}=\mathcal{A}_1+\mathcal{A}_2$. Let us now show that the operators $\mathcal{A}_1$ and $\mathcal{A}_2$ satisfy the conditions of Theorem 1.18 on J. First, we show that the operators $\mathcal{A}_1$ and $\mathcal{A}_2$ define the operators $\mathcal{A}_1,\mathcal{A}_2:B_r\to\mathcal{P}_{cp,c}(\mathcal{C})$ where $B_r=\{y\in\mathcal{C}:\|y\|\le r\}$ is a bounded set in $\mathcal{C}$. Observe that the operator $\mathcal{A}_2$ is equivalent to the composition $\mathcal{L}\circ S_F$, where $\mathcal{L}$ is the continuous linear operator on $L^1(J,\mathbb{R})$ into $\mathcal{C}$, given by

$$\mathcal{L}(v)(t)=I^{\rho+\sigma}v(s)(t)+\frac{t^{\sigma}}{\Lambda\Gamma(\sigma+1)}\Big[I^{\rho+\sigma+\gamma}v(s)(T)-\frac{T^{\gamma}}{\Gamma(\gamma+1)}I^{\rho+\sigma}v(s)(\eta)\Big]$$
$$+\frac{1}{\Lambda}\Big[\frac{T^{\sigma+\gamma}}{\Gamma(\sigma+\gamma+1)}I^{\rho+\sigma}v(s)(\eta)-\frac{\eta^{\sigma}}{\Gamma(\sigma+1)}I^{\rho+\sigma+\gamma}v(s)(T)\Big].$$

Suppose that $y\in B_r$ is arbitrary and let $\{v_n\}$ be a sequence in $S_{F,y}$. Then, by definition of $S_{F,y}$, we have $v_n(t)\in F(t,y(t))$ for almost all $t\in J$. As $F(t,y(t))$ is compact for all $t\in J$, there is indeed a convergent subsequence of $\{v_n(t)\}$ (denote it by $\{v_n(t)\}$ again) that converges in measure to some $v(t)\in S_{F,y}$ for almost all $t\in J$. On the other hand, $\mathcal{L}$ is continuous, so $\mathcal{L}(v_n)(t)\to\mathcal{L}(v)(t)$ pointwise on J.

To establish that the convergence is uniform, we show that $\{\mathcal{L}(v_n)\}$ is an equi-continuous sequence. Let $t_1,t_2\in J$ with $t_1<t_2$. Then, we have

$$\begin{aligned}&|\mathcal{L}(v_n)(t_2)-\mathcal{L}(v_n)(t_1)|\\&\le\Big|\frac{1}{\Gamma(\rho+\sigma)}\Big[\int_0^{t_2}(t_2-s)^{\rho+\sigma-1}v_n(s)ds-\int_0^{t_1}(t_1-s)^{\rho+\sigma-1}v_n(s)ds\Big]\Big|\\&\quad+\frac{|t_2^{\sigma}-t_1^{\sigma}|}{|\Lambda|\Gamma(\sigma+1)}\Big\{I^{\rho+\sigma+\gamma}|v_n(s)|(T)+I^{\rho+\sigma}|v_n(s)|(\eta)\Big\}\\&\le\frac{\Phi(r)}{\Gamma(\rho+\sigma)}\Big|\int_0^{t_1}[t_2-s)^{\rho+\sigma-1}-(t_1-s)^{\rho+\sigma-1}]p(s)ds+\int_{t_1}^{t_2}(t_1-s)^{\rho-1}p(s)ds\Big|\\&\quad+\Phi(r)\frac{|t_2^{\sigma}-t_1^{\sigma}|}{|\Lambda|\Gamma(\sigma+1)}\left\{\frac{T^{\rho+\sigma+\gamma}}{\Gamma(\rho+\sigma+\gamma+1)}+\frac{\eta^{\rho+\sigma}}{\Gamma(\rho+\sigma+1)}\right\}\to 0\,\text{as}\,t_2\to t_1.\end{aligned}$$

This implies that the sequence $\{\mathcal{L}(v_n)\}$ is equi-continuous and hence it is a uniformly convergent subsequence by the Arzelá-Ascoli theorem. So, there exists a subsequence of $\{v_n\}$ (denote it again by $\{v_n\}$) such that $\mathcal{L}(v_n) \to \mathcal{L}(v)$. Note that $\mathcal{L}(v) \in \mathcal{L}(S_{F,y})$. Hence, $\mathcal{A}_2(y) = \mathcal{L}(S_{F,y})$ is compact for all $x \in B_r$. So $\mathcal{A}_2(y)$ is compact.

Now, we show that $\mathcal{A}_2(y)$ is convex for all $y \in \mathcal{C}$. Let $h_1, h_2 \in \mathcal{A}_2(y)$. We select $v_1, v_2 \in S_{F,y}$ such that

$$h_j(t) = I^{\rho+\sigma}v_j(s)(t) + \frac{t^\sigma}{\Lambda\Gamma(\sigma+1)}\Big[I^{\rho+\sigma+\gamma}v_j(s)(T) - \frac{T^\gamma}{\Gamma(\gamma+1)}I^{\rho+\sigma}v_j(s)(\eta)\Big]$$
$$+\frac{1}{\Lambda}\Big[\frac{T^{\sigma+\gamma}}{\Gamma(\sigma+\gamma+1)}I^{\rho+\sigma}v_j(s)(\eta) - \frac{\eta^\sigma}{\Gamma(\sigma+1)}I^{\rho+\sigma+\gamma}v_j(s)(T)\Big],$$

for almost all $t \in J$. Let $0 \le \theta \le 1$. Then, we have

$$\begin{aligned}[\theta h_1 + (1-\theta)h_2](t) &= I^{\rho+\sigma}[\theta v_1(s) + (1-\theta)v_2(s)](t)\\
&\quad + \frac{t^\sigma}{\Lambda\Gamma(\sigma+1)}\Big[I^{\rho+\sigma+\gamma}[\theta v_1(s)+(1-\theta)v_2(s)](T)\\
&\quad - \frac{T^\gamma}{\Gamma(\gamma+1)}\Big(I^{\rho+\sigma}[\theta v_1(s)+(1-\theta)v_2(s)](\eta)\Big)\Big]\\
&\quad + \frac{1}{\Lambda}\Big[\frac{T^{\sigma+\gamma}}{\Gamma(\sigma+\gamma+1)}\Big(I^{\rho+\sigma}[\theta v_1(s)+(1-\theta)v_2(s)](\eta)\Big)\\
&\quad - \frac{\eta^\sigma}{\Gamma(\sigma+1)}\Big(I^{\rho+\sigma+\gamma}[\theta v_1(s)+(1-\theta)v_2(s)](T)\Big)\Big].\end{aligned}$$

Since F has convex values, so $S_{F,y}$ is convex and $\theta v_1(s) + (1-\theta)v_2(s) \in S_{F,y}$. Thus

$$\theta h_1 + (1-\theta)h_2 \in \mathcal{A}_2(y).$$

Consequently, $\mathcal{A}_2$ is convex-valued. Obviously $\mathcal{A}_1$ is compact and convex-valued.

The rest of the proof consists of several steps and claims.

Step 1. We show that $\mathcal{A}_1$ *is a contraction on* $\mathcal{C}$.

This was proved in Step 3 of Theorem 3.20.

Step 2. $\mathcal{A}_2$ *is compact and upper semi-continuous.* This will be established in several claims.

CLAIM I. $\mathcal{A}_2$ *maps bounded sets into bounded sets in* $\mathcal{C}$. Let $B_r = \{y \in \mathcal{C} : \|y\| \le r\}$ be a bounded set in $\mathcal{C}$. Then, for each $h \in \mathcal{A}_2(y), y \in B_r$, there exists $v \in S_{F,y}$ such that

$$h(t) = I^{\rho+\sigma}v(s)(t) + \frac{t^\sigma}{\Lambda\Gamma(\sigma+1)}\Big[I^{\rho+\sigma+\gamma}v(s)(T) - \frac{T^\gamma}{\Gamma(\gamma+1)}I^{\rho+\sigma}v(s)(\eta)\Big]$$
$$+\frac{1}{\Lambda}\Big[\frac{T^{\sigma+\gamma}}{\Gamma(\sigma+\gamma+1)}I^{\rho+\sigma}v(s)(\eta) - \frac{\eta^\sigma}{\Gamma(\sigma+1)}I^{\rho+\sigma+\gamma}v(s)(T)\Big].$$

Then, for $t \in J$, we have

$$|h(t)| \leq \Phi(r)\Big\{I^{\rho+\sigma}p(s)(t) + \frac{T^\sigma}{|\Lambda|\Gamma(\sigma+1)}\Big[I^{\rho+\sigma+\gamma}p(s)(T) + \frac{T^\gamma}{\Gamma(\gamma+1)}I^{\rho+\sigma}p(s)(\eta)\Big] + \frac{1}{|\Lambda|}\Big[\frac{T^{\sigma+\gamma}}{\Gamma(\sigma+\gamma+1)}I^{\rho+\sigma}p(s)(\eta) + \frac{\eta^\sigma}{\Gamma(\sigma+1)}I^{\rho+\sigma+\gamma}p(s)(T)\Big]\Big\}.$$

Thus,

$$\|h\| \leq \Phi(r)\Big\{I^{\rho+\sigma}p(s)(t) + \frac{T^\sigma}{|\Lambda|\Gamma(\sigma+1)}\Big[I^{\rho+\sigma+\gamma}p(s)(T) + \frac{T^\gamma}{\Gamma(\gamma+1)}I^{\rho+\sigma}p(s)(\eta)\Big] + \frac{1}{|\Lambda|}\Big[\frac{T^{\sigma+\gamma}}{\Gamma(\sigma+\gamma+1)}I^{\rho+\sigma}p(s)(\eta) + \frac{\eta^\sigma}{\Gamma(\sigma+1)}I^{\rho+\sigma+\gamma}p(s)(T)\Big]\Big\}.$$

CLAIM II. *$\mathcal{A}_2$ maps bounded sets into equi continuous sets.* Let $t_1, t_2 \in J$ with $t_1 < t_2$ and $y \in B_r$. Then, for each $h \in \mathcal{A}_2(x)$, we obtain

$$\begin{aligned}|h(t_2) - h(t_1)| &\leq \frac{1}{\Gamma(\rho+\sigma)}\Big|\int_0^{t_1}(t_2-s)^{\rho+\sigma-1} - (t_1-s)^{\rho+\sigma-1}\Big]v(s)ds \\ &\quad + \frac{1}{\Gamma(\rho+\sigma)}\int_{t_1}^{t_2}(t_2-s)^{\rho+\sigma-1}v(s)ds\Big| \\ &\quad + \frac{|t_2^\sigma - t_1^\sigma|}{|\Lambda|\Gamma(\sigma+1)}\Big\{I^{\rho+\sigma+\gamma}|v_n(s)|(T) + I^{\rho+\sigma}|v_n(s)|(\eta)\Big\} \\ &\leq \frac{\Phi(r)}{\Gamma(\rho+\sigma)}\Big|\int_0^{t_1}(t_2-s)^{\rho+\sigma-1} - (t_1-s)^{\rho+\sigma-1}\Big]p(s)ds \\ &\quad + \frac{1}{\Gamma(\rho+\sigma)}\int_{t_1}^{t_2}(t_2-s)^{\rho+\sigma-1}p(s)ds\Big| \\ &\quad + \Phi(r)\frac{|t_2^\sigma - t_1^\sigma|}{|\Lambda|\Gamma(\sigma+1)}\Big\{\frac{T^{\rho+\sigma+\gamma}}{\Gamma(\rho+\sigma+\gamma+1)} + \frac{\eta^{\rho+\sigma}}{\Gamma(\rho+\sigma+1)}\Big\}.\end{aligned}$$

Obviously the right hand side of the above inequality tends to zero independently of $y \in B_r$ as $t_2 - t_1 \to 0$. Therefore it follows by the Arzelá-Ascoli theorem that $\mathcal{A}_2 : \mathcal{C} \to \mathcal{P}(\mathcal{C})$ is completely continuous.

Next we show that $\mathcal{A}_2$ is an upper semi-continuous multi-valued mapping. By Lemma 1.1, we know that $\mathcal{A}_2$ will be upper semi-continuous if it has a closed graph; as it is already shown that $\mathcal{A}_2$ completely continuous, we just need to establish the following claim.

CLAIM III. *$\mathcal{A}_2$ has a closed graph.* Let $y_n \to y_*, h_n \in \mathcal{A}_2(y_n)$ and $h_n \to h_*$. Then we need to show that $h_* \in \mathcal{A}_2(x_*)$. Associated with $h_n \in \mathcal{A}_2(y_n)$, there exists $v_n \in S_{F,y_n}$ such that for each $t \in J$,

$$h_n(t) = I^{\rho+\sigma}v_n(s)(t) + \frac{t^\sigma}{\Lambda\Gamma(\sigma+1)}\Big[I^{\rho+\sigma+\gamma}v_n(s)(T) - \frac{T^\gamma}{\Gamma(\gamma+1)}I^{\rho+\sigma}v_n(s)(\eta)\Big] + \frac{1}{\Lambda}\Big[\frac{T^{\sigma+\gamma}}{\Gamma(\sigma+\gamma+1)}I^{\rho+\sigma}v_n(s)(\eta) - \frac{\eta^\sigma}{\Gamma(\sigma+1)}I^{\rho+\sigma+\gamma}v_n(s)(T)\Big].$$

Thus it suffices to show that there exists $v_* \in S_{F,y_*}$ such that for each $t \in J$,

$$h_*(t) = I^{\rho+\sigma}v_*(s)(t) + \frac{t^\sigma}{\Lambda\Gamma(\sigma+1)}\Big[I^{\rho+\sigma+\gamma}v_*(s)(T) - \frac{T^\gamma}{\Gamma(\gamma+1)}I^{\rho+\sigma}v_*(s)(\eta)\Big] + \frac{1}{\Lambda}\Big[\frac{T^{\sigma+\gamma}}{\Gamma(\sigma+\gamma+1)}I^{\rho+\sigma}v_*(s)(\eta) - \frac{\eta^\sigma}{\Gamma(\sigma+1)}I^{\rho+\sigma+\gamma}v_*(s)(T)\Big].$$

Let us consider the linear operator $\Theta : L^1(J,\mathbb{R}) \to \mathcal{C}$ given by

$$\begin{aligned} v \mapsto \Theta(v)(t) &= I^{\rho+\sigma}v(s)(t) + \frac{t^\sigma}{\Lambda\Gamma(\sigma+1)}\Big[I^{\rho+\sigma+\gamma}v(s)(T) - \frac{T^\gamma}{\Gamma(\gamma+1)}I^{\rho+\sigma}v(s)(\eta)\Big] \\ &+ \frac{1}{\Lambda}\Big[\frac{T^{\sigma+\gamma}}{\Gamma(\sigma+\gamma+1)}I^{\rho+\sigma}v(s)(\eta) - \frac{\eta^\sigma}{\Gamma(\sigma+1)}I^{\rho+\sigma+\gamma}v(s)(T)\Big]. \end{aligned}$$

Observe that

$$\begin{aligned} &\|h_n - h_*\| \\ &= \Big\|I^{\rho+\sigma}(v_n(t)-v_*(t))(t) \\ &+ \frac{t^\sigma}{\Lambda\Gamma(\sigma+1)}\Big[I^{\rho+\sigma+\gamma}v(s)(T) - \frac{T^\gamma}{\Gamma(\gamma+1)}I^{\rho+\sigma}(v_n(t)-v_*(t))(\eta)\Big] \\ &+ \frac{1}{\Lambda}\Big[\frac{T^{\sigma+\gamma}}{\Gamma(\sigma+\gamma+1)}I^{\rho+\sigma}(v_n(t)-v_*(t))(\eta) \\ &- \frac{\eta^\sigma}{\Gamma(\sigma+1)}I^{\rho+\sigma+\gamma}(v_n(t)-v_*(t))(T)\Big]\Big\| \to 0, \quad \text{as } n \to \infty. \end{aligned}$$

Thus, it follows by Lemma 1.2 that $\Theta \circ S_F$ is a closed graph operator. Further, we have $h_n(t) \in \Theta(S_{F,y_n})$. Since $y_n \to y_*$, we have that

$$h_*(t) = I^{\rho+\sigma}v_*(s)(t) + \frac{t^\sigma}{\Lambda\Gamma(\sigma+1)}\Big[I^{\rho+\sigma+\gamma}v_*(s)(T) - \frac{T^\gamma}{\Gamma(\gamma+1)}I^{\rho+\sigma}v_*(s)(\eta)\Big] + \frac{1}{\Lambda}\Big[\frac{T^{\sigma+\gamma}}{\Gamma(\sigma+\gamma+1)}I^{\rho+\sigma}v_*(s)(\eta) - \frac{\eta^\sigma}{\Gamma(\sigma+1)}I^{\rho+\sigma+\gamma}v_*(s)(T)\Big],$$

for some $v_* \in S_{F,y_*}$. Hence $\mathcal{A}_2$ has a closed graph (and therefore has closed values). In consequence, the operator $\mathcal{A}_2$ is compact valued and upper semi-continuous.

Thus the operators $\mathcal{A}_1$ and $\mathcal{A}_2$ satisfy the hypotheses of Theorem 1.18 and hence its conclusion will imply either condition (i) or condition (ii) holds. We show that the conclusion (ii) is not possible. If $y \in \lambda\mathcal{A}_1(y) + \lambda\mathcal{A}_2(y)$ for $\lambda \in (0,1)$, then there exist $v \in S_{F,y}$ such that

$$\begin{aligned} y(t) &= \lambda I^\sigma g(s,y(s))(t) + \lambda I^{\rho+\sigma}v(s)(t) \\ &+ \lambda\frac{t^\sigma}{\Lambda\Gamma(\sigma+1)}\Big[I^{\sigma+\gamma}g(s,y(s))(T) + I^{\rho+\sigma+\gamma}v(s)(T) \\ &- \frac{T^\gamma}{\Gamma(\gamma+1)}\Big(I^\sigma g(s,y(s))(\eta) + I^{\rho+\sigma}v(s)(\eta)\Big)\Big] \end{aligned}$$

$$+\lambda\frac{1}{\Lambda}\Big[\frac{T^{\sigma+\gamma}}{\Gamma(\sigma+\gamma+1)}\Big(I^{\sigma}g(s,y(s))(\eta)+I^{\rho+\sigma}v(s)(\eta)\Big)$$
$$-\frac{\eta^{\sigma}}{\Gamma(\sigma+1)}\Big(I^{\sigma+\gamma}g(s,y(s))(T)+I^{\rho+\sigma+\gamma}v(s)(T)\Big)\Big],\quad t\in J.$$

By our assumptions, and using the estimate $|g(t,y)|=|g(t,y)-g(t,0)+g(t,0)|\leq k\|y\|+g_0$, we obtain

$$\begin{aligned}
&|y(t)|\\
&\leq (k\|y\|+g_0)\left[\frac{T^{\sigma}}{\Gamma(\sigma+1)}+\frac{T^{\sigma}}{|\Lambda|\Gamma(\sigma+1)}\left(\frac{T^{\sigma+\gamma}}{\Gamma(\sigma+\gamma+1)}+\frac{T^{\gamma}}{\Gamma(\gamma+1)}\frac{\eta^{\sigma}}{\Gamma(\sigma+1)}\right)\right.\\
&\left.+\frac{1}{|\Lambda|}\frac{T^{\sigma+\gamma}}{\Gamma(\sigma+\gamma+1)}\frac{2\eta^{\sigma}}{\Gamma(\sigma+1)}\right]+\Phi(\|y\|)\Big\{I^{\rho+\sigma}p(s)(T)\\
&+\frac{T^{\sigma}}{|\Lambda|\Gamma(\sigma+1)}\Big[I^{\rho+\sigma+\gamma}p(s)(T)+\frac{T^{\gamma}}{\Gamma(\gamma+1)}\Big(I^{\rho+\sigma}p(s)(\eta)\Big)\Big]\\
&+\frac{1}{|\Lambda|}\Big[\frac{T^{\sigma+\gamma}}{\Gamma(\sigma+\gamma+1)}\Big(I^{\rho+\sigma}p(s)(\eta)\Big)+\frac{\eta^{\sigma}}{\Gamma(\sigma+1)}\Big(I^{\rho+\sigma+\gamma}p(s)(T)\Big)\Big]\Big\}\\
&=\Lambda_1[k\|y\|+g_0]+\Phi(\|x\|)\Psi_1.
\end{aligned}$$

Thus

$$(1-k\Lambda_1)\|y\|\leq\Phi(\|y\|)\Psi_1+\Lambda_1 g_0. \tag{3.85}$$

If condition (ii) of Theorem 1.18 holds, then there exists $\lambda\in(0,1)$ and $y\in\partial B_{\varpi}$ with $y=\lambda\mathcal{N}(y)$. Then, y is a solution of (3.75) with $\|y\|=\varpi$. Now, by the inequality (3.85), we get

$$\frac{(1-k\Lambda_1)\varpi}{\Phi(\varpi)\Psi_1+\Lambda_1 g_0}\leq 1,$$

which contradicts (3.84). Hence the operator $\mathcal{N}$ has a fixed point in J by Theorem 1.18. In consequence, it follows that there exists a solution for the problem (3.75). This completes the proof. □

3.5.3 *Examples*

Here we illustrate the results obtained in the previous sections.

(a) Consider the problem

$$\begin{cases} D^{1/4}[D^{1/2}y(t)-g(t,y(t))]=f(t,y(t)),\ t\in J:=[0,2],\\ y(1)=0,\quad I^{1/4}y(2)=0.\end{cases} \tag{3.86}$$

Here $\rho=1/4,\sigma=1/2,\gamma=1/4,\eta=1,T=2$, and

$$g(t,y(t))=\frac{\sin y(t)+\cos y(t)}{2\sqrt{1600+t^2}},\quad f(t,y)=t+2+\frac{|y|^3}{1+|y|^3}.$$

With the given data, we find that $k=1/40$, $|\Lambda|\simeq 0.3495,\Lambda_1\simeq 28.4363,\Lambda_2\simeq 29.6369$ (Λ_1 and Λ_2 are respectively given by (3.80) and (3.81)). Further, $|g(t,y)|\leq$

$(1600 + t^2)^{-1/2} = q(t)$ and by the condition (3.20.1), it is found that $k = 1/40$. Clearly $k < \Lambda^{-1}$. Thus all the conditions of Theorem 3.20 are satisfied. Hence it follows by the conclusion of Theorem 3.20 that the problem (3.86) has at least one solution on $[0, 2]$.

(b) Consider the fractional inclusions problem:

$$\begin{cases} D^{1/4}[D^{1/2}y(t) - g(t, y(t))] \in F(t, y(t)), \quad t \in [0, 2], \\ y(1) = 0, \quad I^{1/4}y(2) = 0, \end{cases} \tag{3.87}$$

where $g(t, x(t))$ is the same as considered in the problem (3.86) and

$$F(t, y(t)) = \left[\frac{t}{200}(\sin y + 1), \frac{t^2 e^{-t}}{40}\left(\frac{1}{64}\tan^{-1} y + \frac{\pi}{16}(y + \frac{1}{2})\right) \right]. \tag{3.88}$$

Obviously $|F(t, y)| \leq p(t)\Phi(\|y\|)$, where $p(t) = t, \Phi(\|y\|) = (1 + |y|)/200$. With the given values, it is found that $g_0 = 1/80$ and $\Psi_1 \simeq 24.66275$. By the condition (3.21.1), that is, $\dfrac{(1 - k\Lambda_1)M}{\Phi(M)\Psi_1 + \Lambda_1 g_0} > 1$, we find that $M > 2.88795$. Since all the conditions of Theorem 3.21 are satisfied, therefore we deduce that there exists at least one solution for the problem (3.87) on $[0, 2]$.

3.5.4 *Nonlinear nonlocal boundary conditions case*

Here we briefly discuss a boundary value problem of neutral fractional differential equations supplemented with nonlinear nonlocal boundary conditions involving fractional integral. Precisely, we consider the following problem:

$$\begin{cases} {}^cD^{\rho}[{}^cD^{\sigma}y(t) - g(t, y(t))] = f(t, y(t)), \ t \in J := [0, T], \\ y(\eta) = \phi(y), \quad I^{\gamma}y(T) = h(y), \end{cases} \tag{3.89}$$

where $\phi, h : C([0, T], \mathbb{R}) \to \mathbb{R}$ are given appropriate functions.

Like problem (3.74), we can obtain an integral operator $\mathcal{P} : \mathcal{C} \to \mathcal{C}$ associated with the problem (3.89) which is given by

$$\begin{aligned} \mathcal{P}y(t) &= I^{\sigma}g(s, y(s))(t) + I^{\rho+\sigma}f(s, y(s))(t) \\ &+ \frac{t^{\sigma}}{\Lambda\Gamma(\sigma + 1)}\left[\frac{T^{\gamma}}{\Gamma(\gamma + 1)}\Big(\phi(y) - I^{\sigma}g(s, y(s))(\eta) - I^{\rho+\sigma}f(s, y(s))(\eta)\Big)\right. \\ &- \left.\Big(h(y) - I^{\sigma+\gamma}g(s, y(s))(T) - I^{\rho+\sigma+\gamma}f(s, y(s))(T)\Big)\right] \\ &+ \frac{1}{\Lambda}\left[\frac{\eta^{\sigma}}{\Gamma(\sigma + 1)}\Big(h(y) - I^{\sigma+\gamma}g(s, y(s))(T) - I^{\rho+\sigma+\gamma}f(s, y(s))(T)\Big)\right. \\ &- \left.\frac{T^{\sigma+\gamma}}{\Gamma(\sigma + \gamma + 1)}\Big(\phi(y) - I^{\sigma}g(s, y(s))(\eta) - I^{\rho+\sigma}f(s, y(s))(\eta)\Big)\right], \ t \in [0, T]. \end{aligned} \tag{3.90}$$

Similar to the problem (3.74), we can investigate the existence of solutions for the problem (3.89) with the aid of the operator (3.90) and imposing appropriate conditions (like Lipschitz condition) on ϕ and h.

3.6 A coupled system of nonlocal fractional differential equations with coupled and uncoupled slit-strips integral boundary conditions

This section is devoted to the study of a coupled system of nonlinear fractional differential equations:

$$\begin{cases} {}^cD^q x(t) = f(t, x(t), y(t)), & t \in [0,1], \quad 1 < q \le 2, \\ {}^cD^p y(t) = g(t, x(t), y(t)), & t \in [0,1], \quad 1 < p \le 2, \end{cases} \tag{3.91}$$

supplemented with coupled and uncoupled slit-strips type integral boundary conditions respectively given by

$$\begin{cases} x(0) = 0, \quad x(\zeta) = a\displaystyle\int_0^\eta y(s)ds + b\int_\xi^1 y(s)ds, \quad 0 < \eta < \zeta < \xi < 1, \\ y(0) = 0, \quad y(\zeta) = a\displaystyle\int_0^\eta x(s)ds + b\int_\xi^1 x(s)ds, \quad 0 < \eta < \zeta < \xi < 1, \end{cases} \tag{3.92}$$

and

$$\begin{cases} x(0) = 0, \quad x(\zeta) = a\displaystyle\int_0^\eta x(s)ds + b\int_\xi^1 x(s)ds, \quad 0 < \eta < \zeta < \xi < 1, \\ y(0) = 0, \quad y(\zeta) = a\displaystyle\int_0^\eta y(s)ds + b\int_\xi^1 y(s)ds, \quad 0 < \eta < \zeta < \xi < 1, \end{cases} \tag{3.93}$$

where ${}^cD^q, {}^cD^p$ denote the Caputo fractional derivative operators of order q and p respectively, $f, g : [0,1] \times \mathbb{R} \times \mathbb{R} \to \mathbb{R}$ are given continuous functions, and a, b are real constants.

Here we remark that the concept of coupled and uncoupled integral slit-strip type boundary conditions is new. We can interpret these conditions physically as the contribution due to finite strips of arbitrary lengths on the given interval is related to the value of the unknown function at an arbitrary (nonlocal) position in the region off these strips.

Now we prove an auxiliary result which is pivotal to define the solution for the problem (3.91)-(3.92).

Lemma 3.7. *(Auxiliary Lemma). Given $\phi, \psi \in AC([0,1], \mathbb{R})$, the following system*

$$\begin{cases} {}^cD^q x(t) = \phi(t), \quad t \in [0,1], \quad 1 < q \le 2, \\ {}^cD^p y(t) = \psi(t), \quad t \in [0,1], \quad 1 < p \le 2, \\ x(0) = 0, \quad x(\zeta) = a\displaystyle\int_0^\eta y(s)ds + b\int_\xi^1 y(s)ds, \quad 0 < \eta < \zeta < \xi < 1, \\ y(0) = 0, \quad y(\zeta) = a\displaystyle\int_0^\eta x(s)ds + b\int_\xi^1 x(s)ds, \quad 0 < \eta < \zeta < \xi < 1, \end{cases} \tag{3.94}$$

can equivalently be written as a system of integral equations

$$x(t) = \frac{t}{\zeta^2 - \Delta^2}\Big[\zeta\Big\{a\int_0^\eta\int_0^s \frac{(s-\tau)^{p-1}}{\Gamma(p)}\psi(\tau)d\tau ds + b\int_\xi^1\int_0^s \frac{(s-\tau)^{p-1}}{\Gamma(p)}\psi(\tau)d\tau ds$$

$$- \int_0^{\zeta} \frac{(\zeta - s)^{q-1}}{\Gamma(q)}\phi(s)ds\Big\} + \Delta\Big\{a \int_0^{\eta}\int_0^{s} \frac{(s-\tau)^{q-1}}{\Gamma(q)}\phi(\tau)d\tau ds$$
$$+b\int_{\xi}^{1}\int_0^{s} \frac{(s-\tau)^{q-1}}{\Gamma(q)}\phi(\tau)d\tau ds - \int_0^{\zeta} \frac{(\zeta - s)^{p-1}}{\Gamma(p)}\psi(s)ds\Big\}\Big]$$
$$+\int_0^{t} \frac{(t-s)^{q-1}}{\Gamma(q)}\phi(s)ds, \tag{3.95}$$

$$y(t) = \frac{t}{\zeta^2 - \Delta^2}\Big[\Delta\Big\{a \int_0^{\eta}\int_0^{s} \frac{(s-\tau)^{p-1}}{\Gamma(p)}\psi(\tau)d\tau ds + b\int_{\xi}^{1}\int_0^{s} \frac{(s-\tau)^{p-1}}{\Gamma(p)}\psi(\tau)d\tau ds$$
$$- \int_0^{\zeta} \frac{(\zeta - s)^{q-1}}{\Gamma(q)}\phi(s)ds\Big\} + \zeta\Big\{a \int_0^{\eta}\int_0^{s} \frac{(s-\tau)^{q-1}}{\Gamma(q)}\phi(\tau)d\tau ds$$
$$+b\int_{\xi}^{1}\int_0^{s} \frac{(s-\tau)^{q-1}}{\Gamma(q)}\phi(\tau)d\tau ds - \int_0^{\zeta} \frac{(\zeta - s)^{p-1}}{\Gamma(p)}\psi(s)ds\Big\}\Big]$$
$$+\int_0^{t} \frac{(t-s)^{p-1}}{\Gamma(p)}\psi(s)ds, \tag{3.96}$$

where

$$\Delta = [a\eta^2 + b(1-\xi^2)]/2 \neq 0. \tag{3.97}$$

Proof. It is well known that the general solution of the fractional differential equations in (3.94) can be written as

$$x(t) = c_0 + c_1 t + \int_0^{t} \frac{(t-s)^{q-1}}{\Gamma(q)}\phi(s)ds, \tag{3.98}$$

$$y(t) = c_2 + c_3 t + \int_0^{t} \frac{(t-s)^{p-1}}{\Gamma(p)}\psi(s)ds, \tag{3.99}$$

where $c_0, c_1 \in \mathbb{R}$ are arbitrary constants.

Applying the conditions $x(0) = 0, y(0) = 0$, to (3.98) and (3.98), it is found that $c_0 = 0,\ c_2 = 0$. In view of the nonlocal conditions

$$x(\zeta) = a\int_0^{\eta} y(s)ds + b\int_{\xi}^{1} y(s)ds,\ y(\zeta) = a\int_0^{\eta} x(s)ds + b\int_{\xi}^{1} x(s)ds,$$

we obtain the following system of equations from (3.98) and (3.98) after inserting $c_0 = 0$ and $c_2 = 0$:

$$\zeta c_1 - \Delta c_3 = a\int_0^{\eta}\int_0^{s} \frac{(s-\tau)^{p-1}}{\Gamma(p)}\psi(\tau)d\tau ds + b\int_{\xi}^{1}\int_0^{s} \frac{(s-\tau)^{p-1}}{\Gamma(p)}\psi(\tau)d\tau ds$$
$$- \int_0^{\zeta} \frac{(\zeta - s)^{q-1}}{\Gamma(q)}\phi(s)ds, \tag{3.100}$$

$$-\Delta c_1 + \zeta c_3 = a\int_0^{\eta}\int_0^{s} \frac{(s-\tau)^{q-1}}{\Gamma(q)}\phi(\tau)d\tau ds + b\int_{\xi}^{1}\int_0^{s} \frac{(s-\tau)^{q-1}}{\Gamma(q)}\phi(\tau)d\tau ds$$

$$-\int_0^{\zeta}\frac{(\zeta-s)^{p-1}}{\Gamma(p)}\psi(s)ds, \tag{3.101}$$

where

$$\Delta=[a\eta^2+b(1-\xi^2)]/2. \tag{3.102}$$

Solving the system (3.100)-(3.101), we obtain

$$c_1=\frac{1}{\zeta^2-\Delta^2}\Big[\zeta\Big\{a\int_0^{\eta}\int_0^{s}\frac{(s-\tau)^{p-1}}{\Gamma(p)}\psi(\tau)d\tau ds+b\int_{\xi}^{1}\int_0^{s}\frac{(s-\tau)^{p-1}}{\Gamma(p)}\psi(\tau)d\tau ds$$
$$-\int_0^{\zeta}\frac{(\zeta-s)^{q-1}}{\Gamma(q)}\phi(s)ds\Big\}+\Delta\Big\{a\int_0^{\eta}\int_0^{s}\frac{(s-\tau)^{q-1}}{\Gamma(q)}\phi(\tau)d\tau ds$$
$$+b\int_{\xi}^{1}\int_0^{s}\frac{(s-\tau)^{q-1}}{\Gamma(q)}\phi(\tau)d\tau ds-\int_0^{\zeta}\frac{(\zeta-s)^{p-1}}{\Gamma(p)}\psi(s)ds\Big\}\Big],$$

$$c_3=\frac{1}{\zeta^2-\Delta^2}\Big[\Delta\Big\{a\int_0^{\eta}\int_0^{s}\frac{(s-\tau)^{p-1}}{\Gamma(p)}\psi(\tau)d\tau ds+b\int_{\xi}^{1}\int_0^{s}\frac{(s-\tau)^{p-1}}{\Gamma(p)}\psi(\tau)d\tau ds$$
$$-\int_0^{\zeta}\frac{(\zeta-s)^{q-1}}{\Gamma(q)}\phi(s)ds\Big\}+\zeta\Big\{a\int_0^{\eta}\int_0^{s}\frac{(s-\tau)^{q-1}}{\Gamma(q)}\phi(\tau)d\tau ds$$
$$+b\int_{\xi}^{1}\int_0^{s}\frac{(s-\tau)^{q-1}}{\Gamma(q)}\phi(\tau)d\tau ds-\int_0^{\zeta}\frac{(\zeta-s)^{p-1}}{\Gamma(p)}\psi(s)ds\Big\}\Big].$$

Substituting the values of c_0, c_1, c_2, c_3 in (3.98) and (3.99), we get (3.95) and (3.96). The converse follows by direct computation. This completes the proof. □

3.6.1 *Existence results*

Let us introduce the space $X=\{x(t)|x(t)\in C([0,1])\}$ endowed with the norm $\|x\|=\max\{|x(t)|, t\in[0,1]\}$. Obviously $(X,\|\cdot\|)$ is a Banach space. Also let $Y=\{y(t)|y(t)\in C([0,1])\}$ be endowed with the norm $\|y\|=\max\{|y(t)|, t\in[0,1]\}$. Obviously the product space $(X\times Y,\|(x,y)\|)$ is a Banach space with norm $\|(x,y)\|=\|x\|+\|y\|$.

In view of Lemma 3.7, we define an operator $T: X\times Y\to X\times Y$ by

$$T(x,y)(t)=\begin{pmatrix}T_1(x,y)(t)\\ T_2(x,y)(t)\end{pmatrix},$$

where

$$T_1(x,y)(t)$$
$$=\frac{t}{\zeta^2-\Delta^2}\Big[\zeta\Big\{a\int_0^{\eta}\int_0^{s}\frac{(s-\tau)^{p-1}}{\Gamma(p)}g(\tau,x(\tau),y(\tau))d\tau ds$$
$$+b\int_{\xi}^{1}\int_0^{s}\frac{(s-\tau)^{p-1}}{\Gamma(p)}g(\tau,x(\tau),y(\tau))d\tau ds$$

$$-\int_0^\zeta \frac{(\zeta-s)^{q-1}}{\Gamma(q)} f(s,x(s),y(s))ds\Bigg\} + \Delta\Bigg\{a\int_0^\eta\int_0^s \frac{(s-\tau)^{q-1}}{\Gamma(q)} f(\tau,x(\tau),y(\tau))d\tau ds$$

$$+b\int_\xi^1\int_0^s \frac{(s-\tau)^{q-1}}{\Gamma(q)} f(\tau,x(\tau),y(\tau))d\tau ds - \int_0^\zeta \frac{(\zeta-s)^{p-1}}{\Gamma(p)} g(s,x(s),y(s))ds\Bigg\}\Bigg]$$

$$+\int_0^t \frac{(t-s)^{q-1}}{\Gamma(q)} f(s,x(s),y(s))ds,$$

$$T_2(x,y)(t)$$

$$= \frac{t}{\zeta^2-\Delta^2}\Bigg[\Delta\Bigg\{a\int_0^\eta\int_0^s \frac{(s-\tau)^{p-1}}{\Gamma(p)} g(\tau,x(\tau),y(\tau))d\tau ds$$

$$+b\int_\xi^1\int_0^s \frac{(s-\tau)^{p-1}}{\Gamma(p)} g(\tau,x(\tau),y(\tau))d\tau ds$$

$$-\int_0^\zeta \frac{(\zeta-s)^{q-1}}{\Gamma(q)} f(s,x(s),y(s))ds\Bigg\} + \zeta\Bigg\{a\int_0^\eta\int_0^s \frac{(s-\tau)^{q-1}}{\Gamma(q)} f(\tau,x(\tau),y(\tau))d\tau ds$$

$$+b\int_\xi^1\int_0^s \frac{(s-\tau)^{q-1}}{\Gamma(q)} f(\tau,x(\tau),y(\tau))d\tau ds - \int_0^\zeta \frac{(\zeta-s)^{p-1}}{\Gamma(p)} g(s,x(s),y(s))ds\Bigg\}\Bigg]$$

$$+\int_0^t \frac{(t-s)^{p-1}}{\Gamma(p)} g(s,x(s),y(s))ds.$$

For the sake of convenience, we set

$$M_1 = \frac{1}{\Gamma(p+1)} + \frac{1}{|\zeta^2-\Delta^2|}\left[\frac{\zeta^{q+1}}{\Gamma(q+1)} + |\Delta||a|\frac{\eta^{q+1}}{\Gamma(q+2)} + |\Delta||b|\frac{1-\xi^{q+1}}{\Gamma(q+2)}\right], \tag{3.103}$$

$$M_2 = \frac{1}{|\zeta^2-\Delta^2|}\left[\zeta|a|\frac{\eta^{p+1}}{\Gamma(p+2)} + \zeta|b|\frac{1-\xi^{p+1}}{\Gamma(p+2)} + |\Delta|\frac{\zeta^p}{\Gamma(p+1)}\right], \tag{3.104}$$

$$M_3 = \frac{1}{|\zeta^2-\Delta^2|}\left[\zeta|a|\frac{\eta^{q+1}}{\Gamma(q+2)} + \zeta|b|\frac{1-\xi^{q+1}}{\Gamma(q+2)} + |\Delta|\frac{\zeta^q}{\Gamma(q+1)}\right], \tag{3.105}$$

$$M_4 = \frac{1}{\Gamma(p+1)} + \frac{1}{|\zeta^2-\Delta^2|}\left[\frac{\zeta^{p+1}}{\Gamma(p+1)} + |\Delta||a|\frac{\eta^{p+1}}{\Gamma(p+2)} + |\Delta||b|\frac{1-\xi^{p+1}}{\Gamma(p+2)}\right], \tag{3.106}$$

and

$$M_0 = \min\{1-(M_1+M_3)k_1-(M_2+M_4)\lambda_1,\ 1-(M_1+M_3)k_2-(M_2+M_4)\lambda_2\}, \tag{3.107}$$

$k_i,\ \lambda_i \geq 0\ (i=1,2)$.

The first result is concerned with the existence and uniqueness of solutions for the problem (3.91)-(3.92) and is based on Banach's contraction mapping principle.

Theorem 3.22. *Assume that $f, g : [0,1] \times \mathbb{R}^2 \to \mathbb{R}$ are continuous functions and there exist constants $m_i, n_i, i = 1,2$ such that for all $t \in [0,1]$ and $u_i, v_i \in \mathbb{R}, i = 1,2,$*

$$|f(t,u_1,u_2) - f(t,v_1,v_2)| \le m_1|u_1 - v_1| + m_2|u_2 - v_2|$$

and

$$|g(t,u_1,u_2) - g(t,v_1,v_2)| \le n_1|u_1 - v_1| + n_2|u_2 - v_2|$$

In addition, assume that

$$(M_1 + M_3)(m_1 + m_2) + (M_2 + M_4)(n_1 + n_2) < 1,$$

where $M_i, i = 1,2,3,4$ are given by (3.103)-(3.106). Then the boundary value problem (3.91)-(3.92) has a unique solution.

Proof. Define $\sup_{t\in[0,1]} f(t,0,0) = N_1 < \infty$ and $\sup_{t\in[0,1]} g(t,0,0) = N_2 < \infty$ such that

$$r \ge \frac{(M_1 + M_3)N_1 + (M_2 + M_4)N_2}{1 - [(M_1 + M_3)(m_1 + m_2) + (M_2 + M_4)(n_1 + n_2)]}.$$

We show that $TB_r \subset B_r$, where $B_r = \{(x,y) \in X \times Y : \|(x,y)\| \le r\}$.

For $(x,y) \in B_r$, we have

$$\begin{aligned}
&|T_1(x,y)(t)| \\
&= \max_{t\in[0,1]} \Bigg[\frac{t}{|\zeta^2 - \Delta^2|} \Bigg[\zeta \Bigg\{ |a| \int_0^\eta \int_0^s \frac{(s-\tau)^{p-1}}{\Gamma(p)} (|g(\tau,x(\tau),y(\tau)) - g(\tau,0,0)| \\
&\quad + |g(\tau,0,0)|) d\tau ds + |b| \int_\xi^1 \int_0^s \frac{(s-\tau)^{p-1}}{\Gamma(p)} (|g(\tau,x(\tau),y(\tau)) - g(\tau,0,0)| \\
&\quad + |g(\tau,0,0)|) d\tau ds + \int_0^\zeta \frac{(\zeta - s)^{q-1}}{\Gamma(q)} \left(|f(s,x(s),y(s)) - f(s,0,0)| + |f(s,0,0)| \right) ds \Bigg\} \\
&\quad + |\Delta| \Bigg\{ |a| \int_0^\eta \int_0^s \frac{(s-\tau)^{q-1}}{\Gamma(q)} \left(|f(\tau,x(\tau),y(\tau)) - f(\tau,0,0)| + |f(\tau,0,0)| \right) d\tau ds \\
&\quad + |b| \int_\xi^1 \int_0^s \frac{(s-\tau)^{q-1}}{\Gamma(q)} \left(|f(\tau,x(\tau),y(\tau)) - f(\tau,0,0)| + |f(\tau,0,0)| \right) d\tau ds \\
&\quad + \int_0^\zeta \frac{(\zeta - s)^{p-1}}{\Gamma(p)} \left(|g(s,x(s),y(s)) - g(s,0,0)| + |g(s,0,0)| \right) ds \Bigg\} \Bigg] \\
&\quad + \int_0^t \frac{(t-s)^{q-1}}{\Gamma(q)} \left(|f(s,x(s),y(s)) - f(s,0,0)| + |f(s,0,0)| \right) ds \Bigg] \\
&\le \frac{1}{|\zeta^2 - \Delta^2|} \Bigg[\zeta \Bigg\{ |a| \int_0^\eta \int_0^s \frac{(s-\tau)^{p-1}}{\Gamma(p)} (n_1\|x\| + n_2\|y\| + N_2) d\tau ds \\
&\quad + |b| \int_\xi^1 \int_0^s \frac{(s-\tau)^{p-1}}{\Gamma(p)} (n_1\|x\| + n_2\|y\| + N_2) d\tau ds
\end{aligned}$$

$$
\begin{aligned}
&+\int_0^{\zeta}\frac{(\zeta-s)^{q-1}}{\Gamma(q)}(m_1\|x\|+m_2\|y\|+N_1)ds\Bigg\}\\
&+|\Delta|\Bigg\{|a|\int_0^{\eta}\int_0^{s}\frac{(s-\tau)^{q-1}}{\Gamma(q)}(m_1\|x\|+m_2\|y\|+N_1)d\tau ds\\
&+|b|\int_{\xi}^{1}\int_0^{s}\frac{(s-\tau)^{q-1}}{\Gamma(q)}(m_1\|x\|+m_2\|y\|+N_1)d\tau ds\\
&+\int_0^{\zeta}\frac{(\zeta-s)^{p-1}}{\Gamma(p)}(n_1\|x\|+n_2\|y\|+N_2)ds\Bigg\}\Bigg]\\
&+\int_0^{t}\frac{(t-s)^{q-1}}{\Gamma(q)}(m_1\|x\|+m_2\|y\|+N_1)ds\\
\le{}& \frac{1}{|\zeta^2-\Delta^2|}\left[\zeta|a|\frac{\eta^{p+1}}{\Gamma(p+2)}+\zeta|b|\frac{1-\xi^{p+1}}{\Gamma(p+2)}+|\Delta|\frac{\zeta^{p}}{\Gamma(p+1)}\right](n_1\|x\|+n_2\|y\|+N_2)\\
&+\Bigg\{\frac{1}{|\zeta^2-\Delta^2|}\left[\frac{\zeta^{q+1}}{\Gamma(q+1)}+|\Delta||a|\frac{\eta^{q+1}}{\Gamma(q+2)}+|\Delta||b|\frac{1-\xi^{q+1}}{\Gamma(q+2)}\right]\\
&+\frac{1}{\Gamma(q+1)}\Bigg\}(m_1\|x\|+m_2\|y\|+N_1)\\
={}&(M_2n_1+M_1m_1)\|x\|+(M_2n_2+M_1m_2)\|y\|+M_2N_2+M_1N_1\\
\le{}&(M_2n_1+M_1m_1+M_2n_2+M_1m_2)r+M_2N_2+M_1N_1.
\end{aligned}
$$

In the same way, one can obtain that

$$|T_2(x,y)(t)|\le(M_4n_1+M_3m_1+M_4n_2+M_3m_2)r+M_4N_2+M_3N_1.$$

Consequently, $\|T(x,y)(t)\|\le r$.
Now for $(x_2,y_2),(x_1,y_1)\in X\times Y$, and for any $t\in[0,1]$, we get

$$
\begin{aligned}
&|T_1(x_2,y_2)(t)-T_1(x_1,y_1)(t)|\\
\le{}&\frac{t}{|\zeta^2-\Delta^2|}\Bigg[\zeta\Bigg\{|a|\int_0^{\eta}\int_0^{s}\frac{(s-\tau)^{p-1}}{\Gamma(p)}|g(\tau,x_2(\tau),y_2(\tau))-g(\tau,x_1(\tau),y_1(\tau))|d\tau ds\\
&+|b|\int_{\xi}^{1}\int_0^{s}\frac{(s-\tau)^{p-1}}{\Gamma(p)}|g(\tau,x_2(\tau),y_2(\tau))-g(\tau,x_1(\tau),y_1(\tau))|d\tau ds\\
&+\int_0^{\zeta}\frac{(\zeta-s)^{q-1}}{\Gamma(q)}|f(s,x_2(s),y_2(s))-f(s,x_1(s),y_1(s))|ds\Bigg\}\\
&+|\Delta|\Bigg\{|a|\int_0^{\eta}\int_0^{s}\frac{(s-\tau)^{q-1}}{\Gamma(q)}|f(\tau,x_2(\tau),y_2(\tau))-f(\tau,x_1(\tau),y_1(\tau))|d\tau ds\\
&+|b|\int_{\xi}^{1}\int_0^{s}\frac{(s-\tau)^{q-1}}{\Gamma(q)}|f(\tau,x_2(\tau),y_2(\tau))-f(\tau,x_1(\tau),y_1(\tau))|d\tau ds\\
&+\int_0^{\zeta}\frac{(\zeta-s)^{p-1}}{\Gamma(p)}|g(s,x_2(s),y_2(s))-g(s,x_1(s),y_1(s))|ds\Bigg\}\Bigg]
\end{aligned}
$$

$$
\begin{aligned}
&+\int_0^t \frac{(t-s)^{q-1}}{\Gamma(q)}|g(s,x_2(s),y_2(s))-g(s,x_1(s),y_1(s))|ds \\
&\le \frac{1}{|\zeta^2-\Delta^2|}\left[\zeta|a|\frac{\eta^{p+1}}{\Gamma(p+2)}+\zeta|b|\frac{1-\xi^{p+1}}{\Gamma(p+2)}+|\Delta|\frac{\zeta^p}{\Gamma(p+1)}\right](n_1\|x_2-x_1\|+n_2\|y_2-y_1\|) \\
&+\left\{\frac{1}{|\zeta^2-\Delta^2|}\left[\frac{\zeta^{q+1}}{\Gamma(q+1)}+|\Delta||a|\frac{\eta^{q+1}}{\Gamma(q+2)}+|\Delta||b|\frac{1-\xi^{q+1}}{\Gamma(q+2)}\right]\right. \\
&\left.+\frac{1}{\Gamma(q+1)}\right\}(m_1\|x_2-x_1\|+m_2\|y_2-y_1\|) \\
&\le (M_2n_1+M_1m_1)\|x_2-x_1\|+(M_2n_2+M_1m_2)\|y_2-y_1\|,
\end{aligned}
$$

and consequently we obtain

$$
\|T_1(x_2,y_2)(t)-T_1(x_1,y_1)\| \le (M_2n_1+M_1m_1+M_2n_2+M_1m_2)[\|x_2-x_1\|+\|y_2-y_1\|]. \tag{3.108}
$$

Similarly, we have

$$
\|T_2(x_2,y_2)(t)-T_2(x_1,y_1)\| \le (M_4n_1+M_3m_1+M_4n_2+M_3m_2)[\|x_2-x_1\|+\|y_2-y_1\|]. \tag{3.109}
$$

It follows from (3.108) and (3.109) that

$$
\begin{aligned}
&\|T(x_2,y_2)(t)-T(x_1,y_1)(t)\| \\
&\le [(M_1+M_3)(m_1+m_2)+(M_2+M_4)(n_1+n_2)](\|u_2-u_1\|+\|v_2-v_1\|).
\end{aligned}
$$

Since $(M_1+M_3)(m_1+m_2)+(M_2+M_4)(n_1+n_2)<1$, therefore, T is a contraction operator. So, by Banach's fixed point theorem, the operator T has a unique fixed point, which is the unique solution of problem (3.91)-(3.92). This completes the proof. □

Example 3.13. Consider the following system of coupled fractional differential equations with slit-strips integral boundary conditions

$$
\begin{cases}
{}^cD^{5/4}x(t) = \dfrac{2}{55}x(t)+\dfrac{3}{61}\dfrac{|y(t)|}{(1+|y(t)|)}+\dfrac{3}{2}, \quad t\in[0,1], \\
{}^cD^{3/2}y(t) = \dfrac{1}{27}\dfrac{|\cos x(t)|}{(1+|\cos x(t)|)}+\dfrac{2}{41}\sin y(t)+3, \quad t\in[0,1], \\
x(0)=0, \quad x(1/2)=\displaystyle\int_0^{1/3} y(s)ds+\int_{2/3}^1 y(s)ds, \\
y(0)=0, \quad y(1/2)=\displaystyle\int_0^{1/3} x(s)ds+\int_{2/3}^1 x(s)ds.
\end{cases} \tag{3.110}
$$

Here $q=5/4$, $p=3/2, a=1$, $b=1$, $\zeta=1/2$, $\eta=1/3, \xi=2/3$. With the given values, it is found that $\Delta=1/3, m_1=2/55$, $m_2=3/61$, $n_1=1/27$, $n_2=2/41$, $M_1\simeq 2.731029$, $M_2\simeq 1.397944$, $M_3\simeq 1.854888$, $M_4\simeq 2.216142$, and

$$
(M_1+M_3)(m_1+m_2)+(M_2+M_4)(n_1+n_2)\simeq 0.702454<1.
$$

Thus all the conditions of Theorem 3.22 are satisfied. Therefore, by the conclusion of Theorem 3.22, the problem (3.110) has a unique solution on $[0,1]$.

In the next result, we prove the existence of solutions for the problem (3.91)-(3.92) by applying Leray-Schauder alternative (Theorem 1.7).

Theorem 3.23. *Assume that there exist real constants* $k_i,\ \lambda_i \geq 0$ $(i=1,2)$ *and* $k_0>0, \lambda_0>0$ *such that* $\forall x_i \in \mathbb{R},\ (i=1,2)$ *we have*

$$|f(t,x_1,x_2)| \leq k_0 + k_1|x_1| + k_2|x_2|,$$

$$|g(t,x_1,x_2)| \leq \lambda_0 + \lambda_1|x_1| + \lambda_2|x_2|.$$

In addition it is assumed that

$$(M_1+M_3)k_1 + (M_2+M_4)\lambda_1 < 1 \quad \textit{and} \quad (M_1+M_3)k_2 + (M_2+M_4)\lambda_2] < 1,$$

where $M_i, i=1,2,3,4$ *are given by (3.103)-(3.106). Then there exists at least one solution for the boundary value problem (3.91)-(3.92).*

Proof. First we show that the operator $T: X\times Y \to X \times Y$ is completely continuous. By continuity of functions f and g, the operator T is continuous.

Let $\Omega \subset X\times Y$ be bounded. Then there exist positive constants L_1 and L_2 such that

$$|f(t,x(t),y(t)| \leq L_1, \quad |g(t,x(t),y(t)| \leq L_2, \quad \forall (x,y)\in\Omega.$$

In consequence, for any $(x,y)\in\Omega$, we can obtain that

$$\begin{aligned}\|T_1(x,y)\| \leq{}& \frac{1}{|\zeta^2-\Delta^2|}\left[\zeta|a|\frac{\eta^{p+1}}{\Gamma(p+2)} + \zeta|b|\frac{1-\xi^{p+1}}{\Gamma(p+2)} + |\Delta|\frac{\zeta^p}{\Gamma(p+1)}\right]L_2 \\ &+\left\{\frac{1}{|\zeta^2-\Delta^2|}\left[\frac{\zeta^{q+1}}{\Gamma(q+1)} + |\Delta||a|\frac{\eta^{q+1}}{\Gamma(q+2)} + |\Delta||b|\frac{1-\xi^{q+1}}{\Gamma(q+2)}\right]\right. \\ &\left.+\frac{1}{\Gamma(q+1)}\right\}L_1 \\ ={}& M_2L_2 + M_1L_1,\end{aligned}$$

and

$$\begin{aligned}\|T_2(x,y)\| \leq{}& \left\{\frac{1}{|\zeta^2-\Delta^2|}\left[|\Delta||a|\frac{\eta^{p+1}}{\Gamma(p+2)} + |\Delta||b|\frac{1-\xi^{p+1}}{\Gamma(p+2)} + \zeta\frac{\zeta^{p+1}}{\Gamma(p+1)}\right]\right. \\ &\left.+\frac{1}{\Gamma(q+1)}\right\}L_2 \\ &+\frac{1}{|\zeta^2-\Delta^2|}\left[|\Delta|\frac{\zeta^q}{\Gamma(q+1)} + \zeta|a|\frac{\eta^{q+1}}{\Gamma(q+2)} + \zeta|b|\frac{1-\xi^{q+1}}{\Gamma(q+2)}\right]L_1 \\ ={}& M_4L_2 + M_3L_1.\end{aligned}$$

Thus, it follows from the above inequalities that the operator T is uniformly bounded.

Next, we show that T is equi-continuous. Let $t_1, t_2 \in [0,1]$ with $t_1 < t_2$. Then, by a simple computation, we find that

$$\begin{aligned}
&|T_1(x(t_2),y(t_2)) - T_1(x(t_1),y(t_1))| \\
&\le \frac{L_1}{\Gamma(q)}\int_0^{t_1}[(t_2-s)^{q-1}-(t_1-s)^{q-1}]ds + \frac{L_1}{\Gamma(q)}\int_{t_1}^{t_2}(t_2-s)^{q-1}ds \\
&+\frac{t_2-t_1}{|\zeta^2-\Delta^2|}\left\{\left[\zeta|a|\frac{\eta^{p+1}}{\Gamma(p+2)}+\zeta|b|\frac{1-\xi^{p+1}}{\Gamma(p+2)}+|\Delta|\frac{\zeta^p}{\Gamma(p+1)}\right]L_2\right. \\
&\left.+\left[\frac{\zeta^{q+1}}{\Gamma(q+1)}+|\Delta||a|\frac{\eta^{q+1}}{\Gamma(q+2)}+|\Delta||b|\frac{1-\xi^{q+1}}{\Gamma(q+2)}\right]L_1\right\},
\end{aligned}$$

and

$$\begin{aligned}
&|T_2(x(t_2),y(t_2)) - T_2(x(t_1),y(t_1))| \\
&\le \frac{L_2}{\Gamma(p)}\int_0^{t_1}[(t_2-s)^{p-1}-(t_1-s)^{p-1}]ds + \frac{L_2}{\Gamma(p)}\int_{t_1}^{t_2}(t_2-s)^{p-1}ds \\
&+\frac{t_2-t_1}{|\zeta^2-\Delta^2|}\left\{\left[|\Delta||a|\frac{\eta^{p+1}}{\Gamma(p+2)}+|\Delta||b|\frac{1-\xi^{p+1}}{\Gamma(p+2)}+\zeta\frac{\zeta^{p+1}}{\Gamma(p+1)}\right]L_2\right. \\
&\left.+\left[|\Delta|\frac{\zeta^{q}}{\Gamma(q+1)}+\zeta|a|\frac{\eta^{q+1}}{\Gamma(q+2)}+\zeta|b|\frac{1-\xi^{q+1}}{\Gamma(q+2)}\right]L_1\right\}.
\end{aligned}$$

Therefore, the operator $T(x,y)$ is equi-continuous, and thus the operator $T(x,y)$ is completely continuous.

Finally, it will be verified that the set $\mathcal{E} = \{(x,y) \in X \times Y | (x,y) = \lambda T(x,y), 0 \le \lambda \le 1\}$ is bounded. Let $(x,y) \in \mathcal{E}$, then $(x,y) = \lambda T(x,y)$. For any $t \in [0,1]$, we have

$$x(t) = \lambda T_1(x,y)(t), \quad y(t) = \lambda T_2(x,y)(t).$$

Then

$$\begin{aligned}
&|x(t)| \\
&\le \frac{1}{|\zeta^2-\Delta^2|}\left[\zeta|a|\frac{\eta^{p+1}}{\Gamma(p+2)}+\zeta|b|\frac{1-\xi^{p+1}}{\Gamma(p+2)}+|\Delta|\frac{\zeta^p}{\Gamma(p+1)}\right](\lambda_0+\lambda_1\|x\|+\lambda_2\|y\|) \\
&+\left\{\frac{1}{|\zeta^2-\Delta^2|}\left[\frac{\zeta^{q+1}}{\Gamma(q+1)}+|\Delta||a|\frac{\eta^{q+1}}{\Gamma(q+2)}+|\Delta||b|\frac{1-\xi^{q+1}}{\Gamma(q+2)}\right]\right. \\
&\left.+\frac{1}{\Gamma(q+1)}\right\}(k_0+k_1\|x\|+k_2\|y\|),
\end{aligned}$$

$$
\begin{aligned}
&|y(t)| \\
&\leq \left\{ \frac{1}{|\zeta^2 - \Delta^2|} \left[|\Delta||a| \frac{\eta^{p+1}}{\Gamma(p+2)} + |\Delta||b| \frac{1-\xi^{p+1}}{\Gamma(p+2)} + \zeta \frac{\zeta^{p+1}}{\Gamma(p+1)} \right] \right. \\
&\left. + \frac{1}{\Gamma(q+1)} \right\} (\lambda_0 + \lambda_1\|x\| + \lambda_2\|y\|) \\
&+ \frac{1}{|\zeta^2 - \Delta^2|} \left[|\Delta| \frac{\zeta^q}{\Gamma(q+1)} + \zeta|a| \frac{\eta^{q+1}}{\Gamma(q+2)} + \zeta|b| \frac{1-\xi^{q+1}}{\Gamma(q+2)} \right] (k_0 + k_1\|x\| + k_2\|y\|).
\end{aligned}
$$

In consequence, we get

$$\|x\| \leq M_1(k_0 + k_1\|x\| + k_2\|y\|) + M_2(\lambda_0 + \lambda_1\|x\| + \lambda_2\|y\|)$$

and

$$\|y\| \leq M_3(k_0 + k_1\|x\| + k_2\|y\|) + M_4(\lambda_0 + \lambda_1\|x\| + \lambda_2\|y\|),$$

which imply that

$$
\begin{aligned}
\|x\| + \|y\| \leq\ & (M_1 + M_3)k_0 + (M_2 + M_4)\lambda_0 + [(M_1 + M_3)k_1 + (M_2 + M_4)\lambda_1]\|x\| \\
& + [(M_1 + M_3)k_2 + (M_2 + M_4)\lambda_2]\|y\|.
\end{aligned}
$$

Thus we obtain

$$\|(x, y)\| \leq \frac{(M_1 + M_3)k_0 + (M_2 + M_4)\lambda_0}{M_0},$$

for any $t \in [0, 1]$, where M_0 is defined by (3.107), which shows that $\mathcal{E}$ is bounded. Thus, by Lemma 1.7, the operator T has at least one fixed point. Hence the boundary value problem (3.91)-(3.92) has at least one solution. The proof is complete. □

Example 3.14. Let us consider the problem (3.110) with the following values

$$
\begin{aligned}
f(t, x(t), y(t)) &= \frac{1}{2} + \frac{2}{41} \sin x(t) + \frac{2}{43\pi} y(t) \tan^{-1} x(t), \\
g(t, x(t), y(t)) &= \frac{2}{3} + \frac{1}{11} x(t) + \frac{1}{17} \sin y(t).
\end{aligned}
$$

Clearly $|f(t, x, y)| \leq k_0 + k_1|x| + k_2|y|$, $|g(t, x, y)| = \lambda_0 + \lambda_1|x| + \lambda_2|y|$, where $k_0 = 1/2$, $k_1 = 2/41$, $k_2 = 1/43$, $\lambda_0 = 2/3$, $\lambda_1 = 1/11$ $\lambda_2 = 1/17$. Furthermore, $(M_1 + M_3)k_1 + (M_2 + M_4)\lambda_1 \simeq 0.552257 < 1$, $(M_1 + M_3)k_2 + (M_2 + M_4)\lambda_2 \simeq 0.319243 < 1$. Thus all the conditions for Theorem 3.23 hold true and consequently the conclusion of Theorem 3.23 applies to the problem (3.110) with the given values of $f(t, x, y)$ and $g(t, x, y)$.

3.6.2 *Uncoupled slit-strips integral boundary conditions case*

In relation to the linear variant of the problem (3.91) and (3.93), we consider the following lemma.

Lemma 3.8. *(Auxiliary Lemma) For $\chi \in C([0, 1], \mathbb{R})$, the unique solution of the problem*

$$
\begin{cases}
{}^cD^q x(t) = \chi(t), \quad 1 < q \leq 2, \quad t \in [0, 1] \\
x(0) = 0, \quad x(\zeta) = a \displaystyle\int_0^\eta x(s)ds + b \int_\xi^1 x(s)ds, \quad 0 < \eta < \zeta < \xi < 1,
\end{cases}
\tag{3.111}
$$

is given by

$$x(t) = \int_0^t \frac{(t-s)^{q-1}}{\Gamma(q)}\chi(s)ds + \frac{t}{A}\Bigg\{ a\int_0^\eta \int_0^s \frac{(s-\tau)^{q-1}}{\Gamma(q)}\chi(\tau)d\tau ds + b\int_\xi^1 \int_0^s \frac{(s-\tau)^{q-1}}{\Gamma(q)}\chi(\tau)d\tau ds - \int_0^\zeta \frac{(\zeta-s)^{q-1}}{\Gamma(q)}\chi(s)d\ ds\Bigg\}, \tag{3.112}$$

where

$$A = \zeta - \frac{a\eta^2}{2} - \frac{b(1-\xi^2)}{2} \neq 0. \tag{3.113}$$

Proof. We just provide the outline of the proof. The general solution of the fractional differential equation in (3.111) can be written as

$$x(t) = e_0 + e_1 t + \int_0^t \frac{(t-s)^{q-1}}{\Gamma(q)}y(s)ds, \tag{3.114}$$

where $e_0, e_1 \in \mathbb{R}$ are arbitrary constants. Applying the given boundary conditions, we find that $e_0 = 0$, and

$$e_1 = \frac{1}{A}\Bigg\{ a\int_0^\eta \int_0^s \frac{(s-\tau)^{q-1}}{\Gamma(q)}y(\tau)d\tau ds + b\int_\xi^1 \int_0^s \frac{(s-\tau)^{q-1}}{\Gamma(q)}y(\tau)d\tau ds - \int_0^\zeta \frac{(\zeta-s)^{q-1}}{\Gamma(q)}y(s)d\ ds\Bigg\}.$$

Substituting the values of e_0, e_1 in (3.114), we get (3.112). This completes the proof. □

3.6.3 *Existence results for uncoupled case*

In view of Lemma 3.8, we define an operator $\mathfrak{T} : X \times Y \to X \times Y$ associated with the problem (3.91) and (3.93) by

$$\mathfrak{T}(u,v)(t) = \begin{pmatrix} \mathfrak{T}_1(u,v)(t) \\ \mathfrak{T}_2(u,v)(t) \end{pmatrix}$$

where

$$\begin{aligned}&\mathfrak{T}_1(u,v)(t)\\ &= \int_0^t \frac{(t-s)^{q-1}}{\Gamma(q)}f(s,u(s),v(s))ds + \frac{t}{A}\Bigg\{ a\int_0^\eta \int_0^s \frac{(s-\tau)^{q-1}}{\Gamma(q)}f(\tau,u(\tau),v(\tau))d\tau ds \\ &\quad + b\int_\xi^1 \int_0^s \frac{(s-\tau)^{q-1}}{\Gamma(q)}f(\tau,u(\tau),v(\tau))d\tau ds - \int_0^\zeta \frac{(\zeta-s)^{q-1}}{\Gamma(q)}f(s,u(s),v(s))ds\Bigg\},\end{aligned}$$

and

$$\mathfrak{T}_2(u,v)(t)$$

$$= \int_0^t \frac{(t-s)^{p-1}}{\Gamma(p)} h(s,u(s),v(s))ds + \frac{t}{A}\Bigg\{ a\int_0^\eta \int_0^s \frac{(s-\tau)^{p-1}}{\Gamma(p)} h(\tau,u(\tau),v(\tau))d\tau ds$$

$$+b\int_\xi^1 \int_0^s \frac{(s-\tau)^{p-1}}{\Gamma(p)} h(\tau,u(\tau),v(\tau))d\tau ds - \int_0^\zeta \frac{(\zeta-s)^{p-1}}{\Gamma(p)} f(s,u(s),v(s))ds\Bigg\}.$$

In the sequel, we set

$$\mu_1 = \frac{1}{\Gamma(q+1)} + \frac{1}{|A|}\left\{ |a|\frac{\eta^{q+1}}{\Gamma(q+2)} + |b|\frac{1-\xi^{q+1}}{\Gamma(q+2)} + \frac{\zeta^q}{\Gamma(q+1)} \right\}, \tag{3.115}$$

$$\mu_2 = \frac{1}{\Gamma(p+1)} + \frac{1}{|A|}\left\{ |a|\frac{\eta^{p+1}}{\Gamma(p+2)} + |b|\frac{1-\xi^{p+1}}{\Gamma(p+2)} + \frac{\zeta^p}{\Gamma(p+1)} \right\}. \tag{3.116}$$

Now we present the existence and uniqueness result for the problem (3.91) and (3.93). We do not provide the proof of this result as it is similar to that of Theorem 3.22.

Theorem 3.24. *Assume that $f, g : [0,1] \times \mathbb{R}^2 \to \mathbb{R}$ are continuous functions and there exist constants $\bar{m}_i, \bar{n}_i, i = 1,2$ such that for all $t \in [0,1]$ and $u_i, v_i \in \mathbb{R}, i = 1,2,$*

$$|f(t,u_1,u_2) - g(t,v_1,v_2)| \le \bar{m}_1|u_1 - v_1| + \bar{m}_2|u_2 - v_2|$$

and

$$|g(t,u_1,u_2) - h(t,v_1,v_2)| \le \bar{n}_1|u_1 - v_1| + \bar{n}_2|u_2 - v_2|.$$

In addition, assume that

$$\mu_1(\bar{m}_1 + \bar{m}_2) + \mu_2(\bar{n}_1 + \bar{n}_2) < 1,$$

where μ_1 and μ_2 are given by (3.115) and (3.116) respectively. Then the boundary value problem (3.91) and (3.93) has a unique solution.

Example 3.15. Consider the following system of coupled fractional differential equations with uncoupled slit-strips integral boundary conditions

$$\begin{cases} {}^cD^{5/4}x(t) = \dfrac{|x(t)|}{24(1+|x(t)|)} + \dfrac{1}{20}\tan^{-1} y + 1, \quad t \in [0,1], \\ {}^cD^{3/2}y(t) = \dfrac{1}{35}\sin x(t) + \dfrac{1}{25}y(t) + 4, \quad t \in [0,1], \\ x(0) = 0, \quad x(1/2) = \displaystyle\int_0^{1/3} x(s)ds + \int_{2/3}^1 x(s)ds, \\ y(0) = 0, \quad y(1/2) = \displaystyle\int_0^{1/3} y(s)ds + \int_{2/3}^1 y(s)ds. \end{cases} \tag{3.117}$$

Here $q = 5/4$, $p = 3/2$, $a = 1$, $b = 1$, $\zeta = 1/2$, $\eta = 1/3, \xi = 2/3$. With the given values, it is found that $A = 1/6, \bar{m}_1 = 1/24$, $\bar{m}_2 = 1/20$, $\bar{n}_1 = 1/35$, $\bar{n}_2 = 1/25$, $\mu_1 \simeq 4.716276$, $\mu_2 \simeq 3.614087$. In consequence, $\mu_1(\bar{m}_1+\bar{m}_2)+\mu_2(\bar{n}_1+\bar{n}_2) \simeq$

$0.680148 < 1$. Thus all the conditions of Theorem 3.24 are satisfied. Therefore, there exists a unique solution for the problem (3.117) on $[0, 1]$.

The second result dealing with the existence of solutions for the problem (3.91) and (3.93) is analogous to Theorem 3.23 and is given below.

Theorem 3.25. *Assume that there exist real constants* ρ_i, $\nu_i \geq 0$ $(i = 1, 2)$ *and* $\rho_0 > 0, \nu_0 > 0$ *such that* $\forall x_i \in \mathbb{R}$, $(i = 1, 2)$ *we have*

$$|f(t, x_1, x_2)| \leq \rho_0 + \rho_1|x_1| + \rho_2|x_2|,$$

$$|g(t, x_1, x_2)| \leq \nu_0 + \nu_1|x_1| + \nu_2|x_2|.$$

In addition it is assumed that

$$\mu_1\rho_1 + \mu_2\nu_1 < 1 \quad and \quad \mu_1\rho_2 + \mu_2\nu_2 < 1,$$

where μ_1 *and* μ_2 *are given by (3.115) and (3.116) respectively. Then the boundary value problem (3.91)-(3.93) has at least one solution.*

Proof. Setting

$$\mu_0 = \min\{1 - (\mu_1\rho_1 + \mu_2\nu_1),\ 1 - (\mu_1\rho_2 + \mu_2\nu_2)\}, \quad \rho_i,\ \nu_i \geq 0\ (i = 1, 2), \tag{3.118}$$

with μ_1 and μ_2 given by (3.115) and (3.116) respectively, the proof is similar to that of Theorem 3.23. So we omit it. □

3.7 Fractional-order one-dimensional semi-linear problems under nonlocal integral boundary conditions

In this section, we investigate a new class of boundary value problems of nonlinear fractional differential equations and inclusions with nonlocal nonlinear flux type nonlocal integral boundary conditions.

3.7.1 *Nonlinear fractional differential equations*

In this section, we investigate a new class of boundary value problems of nonlinear fractional differential equations and nonlocal nonlinear flux type nonlocal integral boundary conditions. Precisely, we consider the following fractional differential equation:

$${}^cD^q x(t) = f(t, x(t)), \qquad t \in [0, 1], \quad 1 < q \leq 2, \tag{3.119}$$

supplemented with the nonlocal nonlinear flux type integral boundary conditions

$$x'(0) = \alpha \int_0^\xi x'(s)ds, \quad x(1) = \beta g(x'(\eta)), \quad 0 \leq \xi, \eta \leq 1 \tag{3.120}$$

and

$$x(0) = \beta g(x'(\eta)), \ \ x'(1) = \alpha \int_0^\xi x'(s)ds, \quad 0 < \xi, \eta < 1, \tag{3.121}$$

where ${}^cD^q$ denotes the Caputo fractional derivative operator of order q, $f:[0,1]\times\mathbb{R}\to\mathbb{R}$, $g:\mathbb{R}\to\mathbb{R}$ are continuous functions and α,β are appropriate real constants with $\alpha\xi\neq 1$.

To define the solution for the problem (3.119)-(3.120), we need the following lemma.

Lemma 3.9. *For $y\in C([0,1],\mathbb{R})$, the integral solution of the linear fractional differential equation*

$$ {}^cD^qx(t)=y(t),\quad t\in[0,1],\quad 1<q\leq 2, \tag{3.122}$$

supplemented with the nonlocal conditions (3.120) *is given by*

$$\begin{aligned} x(t)=&\int_0^t\frac{(t-s)^{q-1}}{\Gamma(q)}y(s)ds\\ &+\frac{\alpha(t-1)}{1-\alpha\xi}\int_0^\xi\int_0^s\frac{(s-\tau)^{q-2}}{\Gamma(q-1)}y(\tau)d\tau ds-\int_0^1\frac{(1-s)^{q-1}}{\Gamma(q)}y(s)ds\\ &+\beta g\left(\int_0^\eta\frac{(\eta-s)^{q-2}}{\Gamma(q-1)}y(s)ds+\frac{\alpha}{1-\alpha\xi}\int_0^\xi\int_0^s\frac{(s-\tau)^{q-2}}{\Gamma(q-1)}y(\tau)d\tau ds\right). \end{aligned}\tag{3.123}$$

Proof. It is well known that the general solution of the fractional differential equation in (3.122) can be written as

$$x(t)=c_0+c_1t+\int_0^t\frac{(t-s)^{q-1}}{\Gamma(q)}y(s)ds, \tag{3.124}$$

where $c_0,c_1\in\mathbb{R}$ are arbitrary constants.

Applying the given boundary conditions, we find that

$$\begin{aligned} c_0=&\beta g\left(\int_0^\eta\frac{(\eta-s)^{q-2}}{\Gamma(q-1)}y(s)ds+\frac{\alpha}{1-\alpha\xi}\int_0^\xi\int_0^s\frac{(s-\tau)^{q-2}}{\Gamma(q-1)}y(\tau)d\tau ds\right)\\ &-\int_0^1\frac{(1-s)^{q-1}}{\Gamma(q)}y(s)ds-\frac{\alpha}{1-\alpha\xi}\int_0^\xi\int_0^s\frac{(s-\tau)^{q-2}}{\Gamma(q-1)}y(\tau)d\tau ds \end{aligned}$$

and

$$c_1=\frac{\alpha}{1-\alpha\xi}\int_0^\xi\int_0^s\frac{(s-\tau)^{q-2}}{\Gamma(q-1)}y(\tau)d\tau ds.$$

Substituting the values of c_0,c_1 in (3.124), we get (3.123). The converse follows by direct computation. This completes the proof. □

3.7.2 *Existence results for problem (3.119)-(3.120)*

In this section, we discuss the existence of solutions for problem (3.119)-(3.120) by applying some standard tools of fixed point theory. Lets us denote by $\mathcal{C}=C([0,1],\mathbb{R})$ the Banach space of all continuous functions from $[0,1]\to\mathbb{R}$ endowed with a topology of uniform convergence with the norm defined by $\|x\|=\sup\{|x(t)|:t\in[0,1]\}$.

Using Lemma 3.9, we define an operator $F : \mathcal{C} \to \mathcal{C}$ related to problem (3.119)-(3.120) by

$$\begin{aligned}(Fx)(t) = &\int_0^t \frac{(t-s)^{q-1}}{\Gamma(q)} f(s, x(s))ds \\ &+ \frac{\alpha(t-1)}{1-\alpha\xi} \int_0^\xi \int_0^s \frac{(s-\tau)^{q-2}}{\Gamma(q-1)} f(\tau, x(\tau))d\tau ds \\ &- \int_0^1 \frac{(1-s)^{q-1}}{\Gamma(q)} f(s, x(s))ds + \beta g\Bigg(\int_0^\eta \frac{(\eta-s)^{q-2}}{\Gamma(q-1)} f(s, x(s))ds \\ &+ \frac{\alpha}{1-\alpha\xi} \int_0^\xi \int_0^s \frac{(s-\tau)^{q-2}}{\Gamma(q-1)} f(\tau, x(\tau))d\tau ds \Bigg), \quad t \in [0,1]. \qquad (3.125)\end{aligned}$$

Observe that the fixed point problem $Fx = x$ is equivalent to problem (3.119)-(3.120).

In the sequel, let us set

$$\Lambda = \frac{1}{\Gamma(q+1)} \left(2 + |\beta| q \eta^{q-1} + \frac{|\alpha| \xi^q}{|1-\alpha\xi|}(1+|\beta|) \right). \qquad (3.126)$$

Our first existence result is based on Leray-Schauder nonlinear alternative (Theorem 1.8).

Theorem 3.26. *Let $f : [0,1] \times \mathbb{R} \to \mathbb{R}$, $g : \mathbb{R} \to \mathbb{R}$ be continuous functions. Assume that the following conditions hold:*

(3.26.1) there exist a function $p \in C([0,1], \mathbb{R}^+)$, and a non-decreasing function $\psi : \mathbb{R}^+ \to \mathbb{R}^+$ such that $|f(t,x)| \leq p(t)\psi(\|x\|)$ for each $(t,x) \in [0,1] \times \mathbb{R}$;

(3.26.2) $|g(v)| \leq |v|, \quad \forall v \in \mathbb{R}$;

(3.26.3) there exists a number $M > 0$ such that

$$\frac{M}{\|p\|\psi(M)\Lambda} > 1,$$

where Λ is defined in (3.126).

Then the boundary value problem (3.119)-(3.120) has at least one solution on $[0,1]$.

Proof. Obviously the operator $F : \mathcal{C} \to \mathcal{C}$ defined by (3.125) is continuous. Next, we show that F maps bounded sets into bounded sets in $C([0,1], \mathbb{R})$. For a positive number ρ, let $B_\rho = \{x \in C([0,1], \mathbb{R}) : \|x\| \leq \rho\}$ be a bounded set in $C([0,1], \mathbb{R})$. Then, for each $x \in B_\rho$, we have

$$\begin{aligned}&|(Fx)(t)| \\ &\leq \int_0^t \frac{(t-s)^{q-1}}{\Gamma(q)} |f(s, x(s))|ds + \frac{|\alpha(t-1)|}{|1-\alpha\xi|} \int_0^\xi \int_0^s \frac{(s-\tau)^{q-2}}{\Gamma(q-1)} |f(\tau, x(\tau))|d\tau ds \\ &\quad + \int_0^1 \frac{(1-s)^{q-1}}{\Gamma(q)} |f(s, x(s))|ds + |\beta| g\Bigg(\int_0^\eta \frac{(\eta-s)^{q-2}}{\Gamma(q-1)} |f(s, x(s))|ds\end{aligned}$$

$$
\begin{aligned}
&+\frac{|\alpha|}{|1-\alpha\xi|}\int_0^\xi\int_0^s\frac{(s-\tau)^{q-2}}{\Gamma(q-1)}|f(\tau,x(\tau))|d\tau ds\Bigg)\\
&\le \|p\|\psi(\|x\|)\Bigg\{\frac{1}{\Gamma(q+1)}+\frac{|\alpha|}{|1-\alpha\xi|}\frac{\xi^q}{\Gamma(q+1)}+\frac{1}{\Gamma(q+1)}\\
&+|\beta|\left(\frac{\eta^{q-1}}{\Gamma(q)}+\frac{|\alpha|}{|1-\alpha\xi|}\frac{\xi^q}{\Gamma(q+1)}\right)\Bigg\}\\
&=\|p\|\psi(\|x\|)\Lambda.
\end{aligned}
$$

Thus,

$$\|Fx\|\le\|p\|\psi(\rho)\Lambda.$$

Now we show that F maps bounded sets into equi-continuous sets of $C([0,1],\mathbb{R})$. Let $t_1,t_2\in[0,1]$ with $t_1<t_2$ and $x\in B_\rho$. Then we have

$$
\begin{aligned}
&|h(t_2)-h(t_1)|\\
&\le\left|\frac{1}{\Gamma(q)}\int_0^{t_2}(t_2-s)^{q-1}f(s,x(s))ds-\frac{1}{\Gamma(q)}\int_0^{t_1}(t_1-s)^{q-1}f(s,x(s))ds\right|\\
&\quad+\frac{|\alpha||t_2-t_1|}{1-\alpha\xi}\int_0^\xi\int_0^s\frac{(s-\tau)^{q-2}}{\Gamma(q-1)}|f(\tau,x(\tau))|d\tau ds\\
&\le\frac{1}{\Gamma(q)}\int_0^{t_1}[(t_2-s)^{q-1}-(t_1-s)^{q-1}]p(s)\psi(\rho)ds+\frac{1}{\Gamma(q)}\int_{t_1}^{t_2}(t_2-s)^{q-1}p(s)\psi(\rho)ds\\
&\quad+\frac{|\alpha||t_2-t_1|}{|1-\alpha\xi|}\int_0^\xi\int_0^s\frac{(s-\tau)^{q-2}}{\Gamma(q-1)}p(s)\psi(\rho)d\tau ds\\
&\le\frac{\|p\|\psi(\rho)}{\Gamma(q)}\int_0^{t_1}[(t_2-s)^{q-1}-(t_1-s)^{q-1}]ds+\frac{\|p\|\psi(\rho)}{\Gamma(q)}\int_{t_1}^{t_2}(t_2-s)^{q-1}ds\\
&\quad+\frac{\|p\|\psi(\rho)|t_2-t_1|}{|1-\alpha\xi|}\frac{|\alpha|\xi^q}{\Gamma(q+1)}.
\end{aligned}
$$

Notice that the right hand side of the above inequality tends to zero independently of $x\in B_\rho$ as $t_2-t_1\to 0$. Therefore it follows by the Arzelá-Ascoli theorem that $F:C([0,1],\mathbb{R})\to C([0,1],\mathbb{R})$ is completely continuous.

The result will follow from the Leray-Schauder nonlinear alternative (Theorem 1.8) once we have proved the boundedness of the set of all solutions to equations $x=\lambda Fx$ for $\lambda\in[0,1]$.

Let x be a solution. Then, for $t\in[0,1]$, and using the computations in proving that F is bounded, we have

$$
\begin{aligned}
|(Fx)(t)|\le\|p\|\psi(\|x\|)\Bigg\{&\frac{1}{\Gamma(q+1)}+\frac{|\alpha|}{|1-\alpha\xi|}\frac{\xi^q}{\Gamma(q+1)}+\frac{1}{\Gamma(q+1)}\\
&+|\beta|\left(\frac{\eta^{q-1}}{\Gamma(q)}+\frac{|\alpha|}{|1-\alpha\xi|}\frac{\xi^q}{\Gamma(q+1)}\right)\Bigg\}
\end{aligned}
$$

$$= \|p\|\psi(\|x\|)\Lambda.$$

Consequently, we have

$$\frac{\|x\|}{\|p\|\psi(\|x\|)\Lambda} \leq 1.$$

In view of (3.26.3), there exists M such that $\|x\| \neq M$. Let us set

$$U = \{x \in C([0,1], \mathbb{R}) : \|x\| < M + 1\}.$$

Note that the operator $F : U \to C([0,1], \mathbb{R})$ is continuous and completely continuous. From the choice of U, there is no $x \in \partial U$ such that $x = \lambda F x$ for some $\lambda \in (0,1)$. Consequently, by the Leray-Schauder alternative (Theorem 1.8), we deduce that F has a fixed point $x \in U$ which is a solution of the problem (3.119)-(3.120). □

Example 3.16. Consider the problem

$$\begin{cases} {}^cD^{3/2}x(t) = \dfrac{1}{64}\left(\dfrac{x^2}{|x|+1} + \dfrac{\sqrt{|x|}}{1+\sqrt{|x|}}\right) + \dfrac{1}{2}, \quad 0 \leq t \leq 1, \\ x'(0) = \dfrac{1}{2}\displaystyle\int_0^{1/3} x'(s)ds, \quad x(1) = \dfrac{1}{3}g(x'(3/4)). \end{cases} \tag{3.127}$$

Here, $q = 3/2,\ \alpha = 1/2,\ \xi = 1/3,\ \beta = 1/3,\ \eta = 3/4$ and $g(v) = \begin{cases} \sqrt{v}, & |v| \geq 1, \\ v^2, & |v| < 1. \end{cases}$

With the given values, we find that $\Lambda \approx 1.9465507$.

Clearly,

$$|f(t,x)| = \left|\frac{1}{8}(1+t^2)\left(\frac{x^2}{|x|+1} + \frac{\sqrt{|x|}}{1+\sqrt{|x|}}\right) + \frac{1}{2}\right| \leq \frac{1}{8}(1+t^2)(|x|+1).$$

Choosing $p(t) = (1/8)(1+t^2)$ and $\psi(|x|) = |x| + 1$, $\dfrac{M}{\psi(M)\|p\|\Lambda} > 1$, implies that $M > 0.9479419$. Hence, by Theorem 3.26, the boundary value problem (3.127) has at least one solution on $[0,1]$.

Now we establish an existence and uniqueness result for the (3.119)-(3.120) by applying a celebrated fixed point theorem due to Banach.

Theorem 3.27. *Let $f : [0,1] \times \mathbb{R} \to \mathbb{R}$, $g : \mathbb{R} \to \mathbb{R}$ be continuous functions. Assume that (3.26.2) holds. In addition we suppose that:*

(3.27.1) $|f(t,x) - f(t,y)| \leq L|x-y|, \forall t \in [0,1],\ L > 0,\ x, y \in \mathbb{R}.$

Then the boundary value problem (3.119)-(3.120) has a unique solution if

$$L\Lambda < 1,$$

where Λ is given by (3.126).

Proof. Let us fix $\sup_{t\in[0,1]}|f(t,0)| = M$, and choose $r \geq \dfrac{M\Lambda}{1-L\Lambda}$. Then we show that $FB_r \subset B_r$, where $F : \mathcal{C} \to \mathcal{C}$ is given by (3.125) and $B_r = \{x \in \mathcal{C} : \|x\| \leq r\}$. For $x \in B_r$, we have

$$
\begin{aligned}
|(Fx)(t)| \leq & \sup_{t\in[0,1]}\Bigg\{\int_0^t \frac{(t-s)^{q-1}}{\Gamma(q)}|f(s,x(s))|ds \\
& +\frac{|\alpha(t-1)|}{|1-\alpha\xi|}\int_0^\xi\int_0^s \frac{(s-\tau)^{q-2}}{\Gamma(q-1)}|f(\tau,x(\tau))|d\tau ds \\
& +\int_0^1 \frac{(1-s)^{q-1}}{\Gamma(q)}|f(s,x(s))|ds + |\beta|g\Bigg(\int_0^\eta \frac{(\eta-s)^{q-2}}{\Gamma(q-1)}|f(s,x(s))|ds \\
& +\frac{|\alpha|}{|1-\alpha\xi|}\int_0^\xi\int_0^s \frac{(s-\tau)^{q-2}}{\Gamma(q-1)}|f(\tau,x(\tau))|d\tau ds\Bigg)\Bigg\} \\
\leq & \sup_{t\in[0,1]}\Bigg\{\int_0^t \frac{(t-s)^{q-1}}{\Gamma(q)}(|f(s,x(s))-f(s,0)|+|f(s,0)|)ds \\
& +\frac{|\alpha|}{1-\alpha\xi}\int_0^\xi\int_0^s \frac{(s-\tau)^{q-2}}{\Gamma(q-1)}(|f(\tau,x(\tau))-f(s,0)|+|f(s,0)|)d\tau ds \\
& +\int_0^1 \frac{(1-s)^{q-1}}{\Gamma(q)}(|f(s,x(s))-f(s,0)|+|f(s,0)|)ds \\
& +|\beta|g\Bigg(\int_0^\eta \frac{(\eta-s)^{q-2}}{\Gamma(q-1)}(|f(s,x(s))-f(s,0)|+|f(s,0)|)ds \\
& +\frac{|\alpha|}{|1-\alpha\xi|}\int_0^\xi\int_0^s \frac{(s-\tau)^{q-2}}{\Gamma(q-1)}(|f(\tau,x(\tau))-f(s,0)|+|f(s,0)|)d\tau ds\Bigg)\Bigg\} \\
\leq & (Lr+M)\Bigg\{\frac{1}{\Gamma(q+1)}+\frac{|\alpha|}{|1-\alpha\xi|}\frac{\xi^q}{\Gamma(q+1)}+\frac{1}{\Gamma(q+1)} \\
& +|\beta|\Bigg(\frac{\eta^{q-1}}{\Gamma(q)}+\frac{|\alpha|}{|1-\alpha\xi|}\frac{\xi^q}{\Gamma(q+1)}\Bigg)\Bigg\} \\
= & (L\Lambda+M)\Lambda \leq r.
\end{aligned}
$$

Now, for $x, y \in \mathcal{C}$ and for each $t \in [0,1]$, we obtain

$$
\begin{aligned}
& |(Fx)(t)-(Fy)(t)| \\
\leq & \sup_{t\in[0,1]}\Bigg\{\int_0^t \frac{(t-s)^{q-1}}{\Gamma(q)}|f(s,x(s))-f(s,y(s))|ds \\
& +\frac{|\alpha(t-1)|}{|1-\alpha\xi|}\int_0^\xi\int_0^s \frac{(s-\tau)^{q-2}}{\Gamma(q-1)}|f(\tau,x(\tau))-f(\tau,y(\tau)|d\tau ds \\
& +\int_0^1 \frac{(1-s)^{q-1}}{\Gamma(q)}|f(s,x(s))-f(s,y(s))|ds
\end{aligned}
$$

$$+|\beta|g\left(\int_0^\eta \frac{(\eta-s)^{q-2}}{\Gamma(q-1)}|f(s,x(s))-f(s,y(s))|ds\right.$$

$$\left.\left.+\frac{|\alpha|}{|1-\alpha\xi|}\int_0^\xi\int_0^s \frac{(s-\tau)^{q-2}}{\Gamma(q-1)}|f(\tau,x(\tau))-f(\tau,y(\tau)|d\tau ds\right)\right\}$$

$$\leq L\Lambda\|x-y\|,$$

and consequently $\|(Fx)-(Fy)\| \leq L\Lambda\|x-y\|$. As $L\Lambda < 1$, therefore F is a contraction. Thus, the conclusion of the theorem follows by the contraction mapping principle (Banach fixed point theorem). □

Example 3.17. Consider the problem

$$\begin{cases} {}^cD^{3/2}x(t) = \dfrac{\sin^2(\pi t)}{(e^t+3)^2}\cdot\dfrac{|x(t)|}{|x(t)|+1}+\dfrac{\sqrt{3}}{2}, & 0\leq t\leq 1,\\ x'(0) = \dfrac{1}{2}\displaystyle\int_0^{1/3} x'(s)ds, \quad x(1) = \dfrac{1}{3}g(x'(3/4)). \end{cases} \tag{3.128}$$

Here the boundary conditions are the same as in the problem (3.127) and $f(t,x) = (\sin^2(\pi t)/(e^t+3)^2)(|x|/(1+|x|))+(\sqrt{3}/2)$. Since $|f(t,x)-f(t,y)| \leq (1/16)|x-y|$, (3.27.1) is satisfied with $L = 1/16$. Thus $L\Lambda \approx 0.4866376 < 1$. Hence, by Theorem 3.27, the boundary value problem (3.128) has a unique solution on $[0,1]$.

Our next existence result is based on Krasnosel'skii's fixed point theorem (Theorem 1.4).

Theorem 3.28. *Let $f : [0,1]\times\mathbb{R}\to\mathbb{R}$, $g:\mathbb{R}\to\mathbb{R}$ be continuous functions. Assume that the assumptions (3.26.2) and (3.27.1) hold. In addition, we suppose that:*

(3.28.1) $|f(t,x)| \leq \nu(t), \quad \forall(t,x)\in[0,1]\times\mathbb{R}$, *and* $\nu\in C([0,1],\mathbb{R}^+)$.

If

$$\frac{1}{\Gamma(q+1)}\left(1+|\beta|q\eta^{q-1}+\frac{|\alpha|\xi^q}{|1-\alpha\xi|}(1+|\beta|)\right) < 1, \tag{3.129}$$

then the boundary value problem (3.119)-(3.120) has at least one solution on $[0,1]$.

Proof. Letting $\sup_{t\in[0,1]}|\nu(t)| = \|\nu\|$, we fix $\overline{r}\geq\Lambda\|\nu\|$ and consider $B_{\overline{r}} = \{x\in\mathcal{C} : \|x\|\leq\overline{r}\}$. We define the operators $\mathcal{P}$ and $\mathcal{Q}$ on $B_{\overline{r}}$ as

$$(\mathcal{P}x)(t) = \int_0^t \frac{(t-s)^{q-1}}{\Gamma(q)}f(s,x(s))ds,$$

$$(\mathcal{Q}x)(t) = \frac{\alpha(t-1)}{1-\alpha\xi}\int_0^\xi\int_0^s \frac{(s-\tau)^{q-2}}{\Gamma(q-1)}f(\tau,x(\tau))d\tau ds$$

$$-\int_0^1 \frac{(1-s)^{q-1}}{\Gamma(q)}f(s,x(s))ds+\beta g\left(\int_0^\eta \frac{(\eta-s)^{q-2}}{\Gamma(q-1)}f(s,x(s))ds\right.$$

$$+\frac{\alpha}{1-\alpha\xi}\int_0^{\xi}\int_0^{s}\frac{(s-\tau)^{q-2}}{\Gamma(q-1)}f(\tau,x(\tau))d\tau ds\Bigg),\quad t\in[0,1].$$

For $x, y \in B_{\bar{r}}$, we find that

$$\|\mathcal{P}x+\mathcal{Q}y\|\le \Lambda\|\nu\|\le \bar{r}.$$

Thus, $\mathcal{P}x+\mathcal{Q}y\in B_{\bar{r}}$. It follows from the assumption (3.27.1) together with (3.129) that $\mathcal{Q}$ is a contraction mapping. Continuity of f implies that the operator $\mathcal{P}$ is continuous. Also, $\mathcal{P}$ is uniformly bounded on $B_{\bar{r}}$ as

$$\|\mathcal{P}x\|\le \frac{1}{\Gamma(q+1)}\|\mu\|.$$

Now we prove the compactness of the operator $\mathcal{P}$.

We define $\sup_{(t,x)\in[0,1]\times B_{\bar{r}}}|f(t,x)|=f_s<\infty$, and consequently, for $t_1, t_2\in[0,1]$ with $t_2<t_1$, we have

$$|(\mathcal{P}x)(t_2)-(\mathcal{P}x)(t_1)|\le \frac{f_s}{\Gamma(q)}\left|\int_0^{t_1}[(t_2-s)^{q-1}-(t_1-s)^{q-1}]ds+\int_{t_1}^{t_2}(t_2-s)^{q-1}ds\right|,$$

which is independent of x. Thus, $\mathcal{P}$ is equi-continuous. So $\mathcal{P}$ is relatively compact on $B_{\bar{r}}$. Hence, by the Arzelá-Ascoli Theorem, $\mathcal{P}$ is compact on $B_{\bar{r}}$. Thus all the assumptions of Theorem 1.4 are satisfied. So the conclusion of Theorem 1.4 implies that the boundary value problem (3.119)-(3.120) has at least one solution on $[0,1]$. □

Example 3.18. Consider the problem

$$\begin{cases} {}^cD^{3/2}x(t)=\dfrac{e^{-t^2}\sin^2(2t)}{(t+6)^2}\cdot\dfrac{|x(t)|}{|x(t)|+1}+\dfrac{t-1}{t+1}, & 0\le t\le 1,\\[2ex] x'(0)=\dfrac{1}{4}\displaystyle\int_0^{1/3}x'(s)ds,\quad x(1)=\dfrac{1}{9}g(x'(3/4)). \end{cases}\tag{3.130}$$

Here, $q=3/2$, $\alpha=1/4$, $\xi=1/3$, $\beta=1/9$, $\eta=3/4$ and $g(v)=\begin{cases}\sqrt{v}, & |v|\ge 1,\\ v^2, & |v|<1.\end{cases}$ Since $|f(t,x)-f(t,y)|\le(1/36)|x-y|$, (A_4) is satisfied with $L=1/36$. Also

$$\frac{1}{\Gamma(q+1)}\left(1+|\beta|q\eta^{q-1}+\frac{|\alpha|\xi^q}{|1-\alpha\xi|}(1+|\beta|)\right)\approx 0.9258047<1.$$

Clearly,

$$|f(t,x)|=\left|\frac{e^{-t^2}\sin^2(2t)}{(t+6)^2}\cdot\frac{|x(t)|}{|x(t)|+1}+\frac{t-1}{t+1}\right|\le\frac{e^{-t^2}}{36}+\frac{t-1}{t+1}.$$

Hence, by Theorem 3.28, the boundary value problem (3.130) has at least one solution on $[0,1]$.

Theorem 3.29. *Let $f:[0,1]\times\mathbb{R}\to\mathbb{R}$, $g:\mathbb{R}\to\mathbb{R}$ be continuous functions. Assume that the assumption (3.26.2) holds. In addition we suppose that:*

(3.29.1) $|f(t,x)-f(t,y)| \le h(t)\frac{|x-y|}{H^*+|x-y|}$, *for* $t\in[0,1], x,y\ge 0$, *where* $h:[0,1]\to\mathbb{R}^+$ *is continuous and* H^* *is the constant defined by*

$$\begin{aligned} H^* := &\int_0^1 \frac{(t-s)^{q-1}}{\Gamma(q)}h(s)ds \\ &+\frac{|\alpha|}{|1-\alpha\xi|}\int_0^\xi\int_0^s \frac{(s-\tau)^{q-2}}{\Gamma(q-1)}h(\tau)d\tau ds+\int_0^1\frac{(1-s)^{q-1}}{\Gamma(q)}h(s)ds \\ &+|\beta|g\left(\int_0^\eta \frac{(\eta-s)^{q-2}}{\Gamma(q-1)}h(s)ds+\frac{|\alpha|}{|1-\alpha\xi|}\int_0^\xi\int_0^s\frac{(s-\tau)^{q-2}}{\Gamma(q-1)}h(\tau)d\tau ds\right). \end{aligned}$$

Then the boundary value problem (3.119)-(3.120) has a unique solution on $[0,1]$.

Proof. Consider the operator $F:\mathcal{C}\to\mathcal{C}$ defined by (3.125) and define a continuous non-decreasing function $\Psi:\mathbb{R}^+\to\mathbb{R}^+$ as

$$\Psi(\varepsilon)=\frac{H^*\varepsilon}{H^*+\varepsilon}, \qquad \forall\varepsilon\ge 0.$$

Note that the function Ψ satisfies $\Psi(0)=0$ and $\Psi(\varepsilon)<\varepsilon$ for all $\varepsilon>0$.

For any $x,y\in\mathcal{C}$ and for each $t\in[0,1]$, we have

$$\begin{aligned} &|(Fx)(t)-(Fy)(t)| \\ \le &\int_0^t \frac{(t-s)^{q-1}}{\Gamma(q)}|f(s,x(s))-f(s,y(s))|ds \\ &+\frac{|\alpha|}{|1-\alpha\xi|}\int_0^\xi\int_0^s\frac{(s-\tau)^{q-2}}{\Gamma(q-1)}|f(\tau,x(\tau))-f(\tau,y(\tau)|d\tau ds \\ &+\int_0^1\frac{(1-s)^{q-1}}{\Gamma(q)}|f(s,x(s))-f(s,y(s))|ds \\ &+|\beta|g\left(\int_0^\eta\frac{(\eta-s)^{q-2}}{\Gamma(q-1)}|f(s,x(s))-f(s,y(s))|ds\right. \\ &\left.+\frac{|\alpha|}{|1-\alpha\xi|}\int_0^\xi\int_0^s\frac{(s-\tau)^{q-2}}{\Gamma(q-1)}|f(\tau,x(\tau))-f(\tau,y(\tau)|d\tau ds\right) \\ \le &\frac{\Psi(\|x-y\|)}{H^*}\left\{\int_0^1\frac{(t-s)^{q-1}}{\Gamma(q)}h(s)ds\right. \\ &+\frac{|\alpha|}{|1-\alpha\xi|}\int_0^\xi\int_0^s\frac{(s-\tau)^{q-2}}{\Gamma(q-1)}h(\tau)d\tau ds+\int_0^1\frac{(1-s)^{q-1}}{\Gamma(q)}h(s)ds \\ &+|\beta|g\left(\int_0^\eta\frac{(\eta-s)^{q-2}}{\Gamma(q-1)}h(s)ds\right. \\ &\left.\left.+\frac{|\alpha|}{|1-\alpha\xi|}\int_0^\xi\int_0^s\frac{(s-\tau)^{q-2}}{\Gamma(q-1)}h(\tau)d\tau ds\right)\right\} \\ \le &\Psi(\|x-y\|). \end{aligned}$$

This implies that $\|Fx-Fy\| \leq \Psi(\|x-y\|)$. Therefore F is a nonlinear contraction. Hence, by Theorem 1.12, the operator F has a unique fixed point which is the unique solution of the boundary value problem (3.119) – (3.120). This completes the proof. □

Theorem 3.30. *Let $f : [0,1]\times\mathbb{R} \to \mathbb{R}$, $g : \mathbb{R} \to \mathbb{R}$ be continuous functions. Assume that the assumption (3.26.2) holds. In addition we suppose that:*

(3.30.1) there exist constants $0 \leq \kappa < \Lambda^{-1}$, and $K > 0$ such that

$$|f(t,x) \leq \kappa|x| + K \quad \textit{for all} \quad (t,x) \in [0,1] \times \mathbb{R},$$

where Λ is defined by (3.126).

Then the boundary value problem (3.119)-(3.120) has at least one solution on $[0,1]$.

Proof. Let us consider the fixed point problem

$$x = Fx, \tag{3.131}$$

where $F : \mathcal{C} \to \mathcal{C}$ is given by (3.125). We shall prove the existence of at least one solution $x \in C[0,1]$ satisfying (3.131). Set a ball $B_R \subset C[0,1]$ as

$$B_R = \{x \in \mathcal{C} : \max_{t\in C[0,1]} |x(t)| < R\},$$

with a constant radius $R > 0$. Hence, we shall show that $F : \overline{B}_R \to C[0,1]$ satisfies a condition

$$x \neq \theta Fx, \qquad \forall x \in \partial B_R, \qquad \forall \theta \in [0,1]. \tag{3.132}$$

We set

$$H(\theta,x) = \theta Fx, \qquad x \in \mathcal{C}, \qquad \theta \in [0,1].$$

As shown in Theorem 3.26 we have that the operator F that is continuous, uniformly bounded and equi-continuous. Then, by the Arzelá-Ascoli theorem, a continuous map h_θ defined by $h_\theta(x) = x - H(\theta,x) = x - \theta Fx$ is completely continuous. If (3.132) holds, then the following Leray-Schauder degrees are well defined and by the homotopy invariance of topological degree, it follow that

$$\begin{aligned}\deg(h_\theta, B_R, 0) &= \deg(I-\theta F, B_R, 0) = \deg(h, B_R, 0)\\ &= \deg(h_0, B_R, 0) = \deg(I, B_R, 0) = 1 \neq 0, \qquad 0 \in B_R,\end{aligned}$$

where I denotes the unit operator. By the nonzero property of Leray-Schauder degree, $h_1(x) = x - Fx = 0$ for at least one $x \in B_R$. Let us assume that $x = \theta Fx$ for some $\theta \in [0,1]$ and for all $t \in [0,1]$ so that

$$\begin{aligned}&|x(t)| = |\theta(Fx)(t)|\\ &\leq \int_0^t \frac{(t-s)^{q-1}}{\Gamma(q)}|f(s,x(s))|ds + \frac{|\alpha(t-1)|}{|1-\alpha\xi|}\int_0^\xi\int_0^s \frac{(s-\tau)^{q-2}}{\Gamma(q-1)}|f(\tau,x(\tau))|d\tau ds\end{aligned}$$

$$
\begin{aligned}
&+\int_0^1 \frac{(1-s)^{q-1}}{\Gamma(q)}|f(s,x(s))|ds + |\beta| g\bigg(\int_0^\eta \frac{(\eta-s)^{q-2}}{\Gamma(q-1)}|f(s,x(s))|ds \\
&+\frac{|\alpha|}{|1-\alpha\xi|}\int_0^\xi\int_0^s \frac{(s-\tau)^{q-2}}{\Gamma(q-1)}|f(\tau,x(\tau))|d\tau ds\bigg) \\
&\le (\kappa|x|+K)\bigg\{\frac{1}{\Gamma(q+1)}+\frac{|\alpha|}{|1-\alpha\xi|}\frac{\xi^q}{\Gamma(q+1)}+\frac{1}{\Gamma(q+1)} \\
&+|\beta|\bigg(\frac{\eta^{q-1}}{\Gamma(q)}+\frac{|\alpha|}{|1-\alpha\xi|}\frac{\xi^q}{\Gamma(q+1)}\bigg)\bigg\} \\
&= (\kappa|x|+K)\Lambda,
\end{aligned}
$$

which on taking norm $\sup_{t\in[0,1]}|x(t)| = \|x\|$ and solving for $\|x\|$, yields

$$\|x\| \le \frac{K\Lambda}{1-\kappa\Lambda}.$$

If $R = \dfrac{K\Lambda}{1-\kappa\Lambda}+1$, inequality (3.132) holds. This completes the proof. □

Example 3.19. Consider the problem

$$
\begin{cases}
{}^cD^{3/2}x(t) = \dfrac{1}{2\pi}\sin\dfrac{\pi}{2}x + \dfrac{|x|}{|x|+1}+1, \quad 0\le t\le 1, \\
x'(0) = \dfrac{1}{2}\displaystyle\int_0^{1/3} x'(s)ds, \quad x(1) = \dfrac{1}{3}g(x'(3/4)).
\end{cases}
\tag{3.133}
$$

Here, $q=3/2,\ \alpha=1/2,\ \xi=1/3,\ \beta=1/3,\ \eta=3/4$ and $g(v) = \begin{cases}\sqrt{v}, & |v|\ge 1,\\ v^2, & |v|<1.\end{cases}$

With the given values, we find that $\Lambda \approx 1.9465507$. Since

$$|f(t,x)| = \left|\frac{1}{2\pi}\sin\frac{\pi}{2}x + \frac{|x|}{|x|+1}+1\right| \le \frac{1}{4}|x|+1,$$

(3.30.1) is satisfied with $\kappa = 1/4$ and $K=1$ such that

$$\kappa = \frac{1}{4} < \frac{1}{\Lambda} \approx 0.9465507.$$

Hence, by Theorem 3.30, the boundary value problem (3.133) has at least one solution on $[0,1]$.

3.7.3 *Existence results for the boundary value problem (3.119) and (3.121)*

This section is devoted to the study of problem (3.119) and (3.121). First of all, we consider the following lemma to define the solution for problem (3.119) and (3.121).

Lemma 3.10. *For $y\in C([0,1],\mathbb{R})$, the integral solution of the linear boundary value problem*

$$
\begin{cases}
{}^cD^q x(t) = y(t), \quad t\in[0,1], \quad 1<q\le 2, \\
x(0) = \beta g(x'(\eta)) \quad x'(1) = \alpha\displaystyle\int_0^\xi x'(s)ds, \quad 0\le\xi,\eta\le 1,
\end{cases}
\tag{3.134}
$$

is given by

$$
\begin{aligned}
x(t) = & \int_0^t \frac{(t-s)^{q-1}}{\Gamma(q)} y(s)ds \\
& + \frac{t}{1-\alpha\xi}\left(\alpha \int_0^{\xi}\int_0^{s} \frac{(s-\tau)^{q-2}}{\Gamma(q-1)} y(\tau)d\tau ds - \int_0^1 \frac{(1-s)^{q-1}}{\Gamma(q)} y(s)ds\right) \\
& + \beta g\left(\int_0^{\eta} \frac{(\eta-s)^{q-2}}{\Gamma(q-1)} y(s)ds + \frac{\alpha}{1-\alpha\xi}\int_0^{\xi}\int_0^{s} \frac{(s-\tau)^{q-2}}{\Gamma(q-1)} y(\tau)d\tau ds \right. \\
& \left. - \frac{1}{1-\alpha\xi}\int_0^1 \frac{(1-s)^{q-1}}{\Gamma(q)} y(s)ds\right). \qquad (3.135)
\end{aligned}
$$

Proof. We omit the proof as it is similar to that of Lemma 3.9. □

In relation to problem (3.119) and (3.121), we define an operator $G : \mathcal{C} \to \mathcal{C}$ as

$$
\begin{aligned}
& (Gx)(t) \\
& = \int_0^t \frac{(t-s)^{q-1}}{\Gamma(q)} f(s, x(s))ds \\
& + \frac{t}{1-\alpha\xi}\left(\alpha \int_0^{\xi}\int_0^{s} \frac{(s-\tau)^{q-2}}{\Gamma(q-1)} f(\tau, x(\tau))d\tau ds - \int_0^1 \frac{(1-s)^{q-1}}{\Gamma(q)} f(s, x(s))ds\right) \\
& + \beta g\left(\int_0^{\eta} \frac{(\eta-s)^{q-2}}{\Gamma(q-1)} f(s, x(s))ds + \frac{\alpha}{1-\alpha\xi}\int_0^{\xi}\int_0^{s} \frac{(s-\tau)^{q-2}}{\Gamma(q-1)} f(\tau, x(\tau))d\tau ds \right. \\
& \left. - \frac{1}{1-\alpha\xi}\int_0^1 \frac{(1-s)^{q-1}}{\Gamma(q)} f(s, x(s))ds\right)
\end{aligned}
$$

and set

$$
\bar{\Lambda} = \frac{1}{\Gamma(q+1)}\left(1 + |\beta| q\eta^{q-1} + \frac{(1+|\alpha|\xi^q)(1+|\beta|)}{|1-\alpha\xi|}\right). \qquad (3.136)
$$

With the above operator and the estimate (3.136), we can reproduce all the existence results obtained in the last section for the boundary value problem (3.119) and (3.121).

3.7.4 *Nonlinear fractional differential inclusions*

In this subsection, we investigate the existence of solutions for a fractional differential inclusion:

$$
{}^cD^q x(t) \in F(t, x(t)), \qquad t \in [0,1], \quad 1 < q \leq 2, \qquad (3.137)
$$

supplemented with nonlocal nonlinear flux type integral boundary conditions

$$x(0) = \beta g(x'(\eta)), \quad x'(1) = \alpha \int_0^{\xi} x'(s)ds, \quad 0 < \xi, \eta < 1, \tag{3.138}$$

and

$$x'(0) = \alpha \int_0^{\xi} x'(s)ds, \quad x(1) = \beta g(x'(\eta)) \quad 0 < \xi, \eta < 1, \tag{3.139}$$

where ${}^cD^q$ denotes the Caputo fractional derivative operator of order q, $F : [0,1] \times \mathbb{R} \to \mathcal{P}(\mathbb{R})$ is a multi-valued map, $\mathcal{P}(\mathbb{R})$ is the family of all non-empty subsets of $\mathbb{R}$, $g : \mathbb{R} \to \mathbb{R}$ is a continuous function and α, β are appropriate real constants with $\alpha\xi \neq 1$.

Definition 3.5. A function $x \in C^2([0,1], \mathbb{R})$ is said to be a solution of the boundary value problem (3.137)-(3.138) if $x(0) = \beta g(x'(\eta))$, $x'(1) = \alpha \int_0^{\xi} x'(s)ds$, and there exists a function $v \in S_{F,x}$ such that

$$\begin{aligned} x(t) &= \int_0^t \frac{(t-s)^{q-1}}{\Gamma(q)} v(s)ds \\ &\quad + \frac{t}{1-\alpha\xi}\left(\alpha \int_0^{\xi}\int_0^s \frac{(s-\tau)^{q-2}}{\Gamma(q-1)} v(\tau)d\tau ds - \int_0^1 \frac{(1-s)^{q-1}}{\Gamma(q)} v(s)ds\right) \\ &\quad + \beta g\Bigg(\int_0^{\eta} \frac{(\eta-s)^{q-2}}{\Gamma(q-1)} v(s)ds + \frac{\alpha}{1-\alpha\xi}\int_0^{\xi}\int_0^s \frac{(s-\tau)^{q-2}}{\Gamma(q-1)} v(\tau)d\tau ds \\ &\quad - \frac{1}{1-\alpha\xi}\int_0^1 \frac{(1-s)^{q-1}}{\Gamma(q)} v(s)ds\Bigg), \quad t \in [0,1]. \end{aligned} \tag{3.140}$$

Now we are in a position to present our first existence result for the problem (3.137)-(3.138) which deals with a nonconvex valued right of (3.137).

Theorem 3.31. *(The Lipschitz case). Assume that*

(3.31.1) $F : [0,1] \times \mathbb{R} \to \mathcal{P}_{cp}(\mathbb{R})$ *is such that* $F(\cdot, x) : [0,1] \to \mathcal{P}_{cp}(\mathbb{R})$ *is measurable for each* $x \in \mathbb{R}$;

(3.31.2) $H_d(F(t,x), F(t,\bar{x})) \leq m(t)|x - \bar{x}|$ *for almost all* $t \in [0,1]$ *and* $x, \bar{x} \in \mathbb{R}$ *with* $m \in C([0,1], \mathbb{R}^+)$ *and* $d(0, F(t,0)) \leq m(t)$ *for almost all* $t \in [0,1]$;

(3.31.3) $g : \mathbb{R} \to \mathbb{R}$ *is continuous and* $|g(v)| \leq |v|, \forall v \in \mathbb{R}$.

Then the boundary value problem (3.137)-(3.138) has at least one solution on $[0,1]$ *if* $\|m\|\Lambda < 1$, *where* Λ *is given by* (3.126).

Proof. Define an operator $\Omega_F : C([0,1],\mathbb{R}) \to \mathcal{P}(C([0,1],\mathbb{R}))$ by

$$\Omega_F(x) = \left\{ \begin{array}{l} h \in C([0,1],\mathbb{R}) : \\ h(t) = \left\{ \begin{array}{l} \displaystyle\int_0^t \frac{(t-s)^{q-1}}{\Gamma(q)} v(s)ds \\ \displaystyle + \frac{t}{1-\alpha\xi}\left(\alpha\int_0^\xi\int_0^s \frac{(s-\tau)^{q-2}}{\Gamma(q-1)} v(\tau)d\tau ds - \int_0^1 \frac{(1-s)^{q-1}}{\Gamma(q)} v(s)ds\right) \\ \displaystyle + \beta g\Bigg(\int_0^\eta \frac{(\eta-s)^{q-2}}{\Gamma(q-1)} v(s)ds + \frac{\alpha}{1-\alpha\xi}\int_0^\xi\int_0^s \frac{(s-\tau)^{q-2}}{\Gamma(q-1)} v(\tau)d\tau ds \\ \displaystyle - \frac{1}{1-\alpha\xi}\int_0^1 \frac{(1-s)^{q-1}}{\Gamma(q)} v(s)ds\Bigg) \end{array} \right. \end{array} \right\}$$

for $v \in S_{F,x}$.

Observe that the set $S_{F,x}$ is non-empty for each $x \in C([0,1],\mathbb{R})$ by the assumption (3.31.1), so F has a measurable selection (see Theorem III.6 [78]). Now we show that the operator Ω_F satisfies the assumptions of Theorem 1.19. To show that $\Omega_F(x) \in \mathcal{P}_{cl}((C[0,1],\mathbb{R}))$ for each $x \in C([0,1],\mathbb{R})$, let $\{u_n\}_{n\geq 0} \in \Omega_F(x)$ be such that $u_n \to u$ $(n \to \infty)$ in $C([0,1],\mathbb{R})$. Then $u \in C([0,1],\mathbb{R})$ and there exists $v_n \in S_{F,x_n}$ such that, for each $t \in [0,1]$,

$$\begin{aligned} u_n(t) = {} & \int_0^t \frac{(t-s)^{q-1}}{\Gamma(q)} v_n(s)ds \\ & + \frac{t}{1-\alpha\xi}\left(\alpha\int_0^\xi\int_0^s \frac{(s-\tau)^{q-2}}{\Gamma(q-1)} v_n(\tau)d\tau ds - \int_0^1 \frac{(1-s)^{q-1}}{\Gamma(q)} v_n(s)ds\right) \\ & + \beta g\Bigg(\int_0^\eta \frac{(\eta-s)^{q-2}}{\Gamma(q-1)} v_n(s)ds + \frac{\alpha}{1-\alpha\xi}\int_0^\xi\int_0^s \frac{(s-\tau)^{q-2}}{\Gamma(q-1)} v_n(\tau)d\tau ds \\ & - \frac{1}{1-\alpha\xi}\int_0^1 \frac{(1-s)^{q-1}}{\Gamma(q)} v_n(s)ds\Bigg). \end{aligned}$$

As F has compact values, we pass onto a subsequence (if necessary) to obtain that v_n converges to v in $L^1([0,1],\mathbb{R})$. Thus, $v \in S_{F,x}$ and for each $t \in [0,1]$, we have

$$\begin{aligned} v_n(t) \to v(t) = {} & \int_0^t \frac{(t-s)^{q-1}}{\Gamma(q)} v(s)ds \\ & + \frac{t}{1-\alpha\xi}\left(\alpha\int_0^\xi\int_0^s \frac{(s-\tau)^{q-2}}{\Gamma(q-1)} v(\tau)d\tau ds - \int_0^1 \frac{(1-s)^{q-1}}{\Gamma(q)} v(s)ds\right) \\ & + \beta g\Bigg(\int_0^\eta \frac{(\eta-s)^{q-2}}{\Gamma(q-1)} v(s)ds + \frac{\alpha}{1-\alpha\xi}\int_0^\xi\int_0^s \frac{(s-\tau)^{q-2}}{\Gamma(q-1)} v(\tau)d\tau ds \\ & - \frac{1}{1-\alpha\xi}\int_0^1 \frac{(1-s)^{q-1}}{\Gamma(q)} v(s)ds\Bigg). \end{aligned}$$

Hence, $u \in \Omega_F(x)$.

Next we show that there exists $\delta < 1$ such that

$$H_d(\Omega_F(x), \Omega_F(\bar{x})) \leq \delta \|x - \bar{x}\| \quad \text{for each} \quad x, \bar{x} \in C^2([0,1], \mathbb{R}).$$

Let $x, \bar{x} \in C^2([0,1], \mathbb{R})$ and $h_1 \in \Omega_F(x)$. Then there exists $v_1(t) \in F(t, x(t))$ such that, for each $t \in [0,1]$,

$$\begin{aligned} h_1(t) = & \int_0^t \frac{(t-s)^{q-1}}{\Gamma(q)} v_1(s)ds \\ & + \frac{t}{1-\alpha\xi}\left(\alpha \int_0^\xi \int_0^s \frac{(s-\tau)^{q-2}}{\Gamma(q-1)} v_1(\tau)d\tau ds - \int_0^1 \frac{(1-s)^{q-1}}{\Gamma(q)} v_1(s)ds\right) \\ & + \beta g\Bigg(\int_0^\eta \frac{(\eta-s)^{q-2}}{\Gamma(q-1)} v_1(s)ds + \frac{\alpha}{1-\alpha\xi}\int_0^\xi \int_0^s \frac{(s-\tau)^{q-2}}{\Gamma(q-1)} v_1(\tau)d\tau ds \\ & - \frac{1}{1-\alpha\xi}\int_0^1 \frac{(1-s)^{q-1}}{\Gamma(q)} v_1(s)ds\Bigg). \end{aligned}$$

By (3.31.2), we have

$$H_d(F(t,x), F(t,\bar{x})) \leq m(t)|x(t) - \bar{x}(t)|.$$

So, there exists $w \in F(t, \bar{x}(t))$ such that

$$|v_1(t) - w| \leq m(t)|x(t) - \bar{x}(t)|, \quad t \in [0,1].$$

Define $U : [0,1] \to \mathcal{P}(\mathbb{R})$ by

$$U(t) = \{w \in \mathbb{R} : |v_1(t) - w| \leq m(t)|x(t) - \bar{x}(t)|\}.$$

Since the multi-valued operator $U(t) \cap F(t, \bar{x}(t))$ is measurable (Proposition III.4 [78]), there exists a function $v_2(t)$ which is a measurable selection for U. So $v_2(t) \in F(t, \bar{x}(t))$ and for each $t \in [0,1]$, we have $|v_1(t) - v_2(t)| \leq m(t)|x(t) - \bar{x}(t)|$.

For each $t \in [0,1]$, let us define

$$\begin{aligned} h_2(t) = & \int_0^t \frac{(t-s)^{q-1}}{\Gamma(q)} v_2(s)ds \\ & + \frac{t}{1-\alpha\xi}\left(\alpha \int_0^\xi \int_0^s \frac{(s-\tau)^{q-2}}{\Gamma(q-1)} v_2(\tau)d\tau ds - \int_0^1 \frac{(1-s)^{q-1}}{\Gamma(q)} v_2(s)ds\right) \\ & + \beta g\Bigg(\int_0^\eta \frac{(\eta-s)^{q-2}}{\Gamma(q-1)} v_2(s)ds + \frac{\alpha}{1-\alpha\xi}\int_0^\xi \int_0^s \frac{(s-\tau)^{q-2}}{\Gamma(q-1)} v_2(\tau)d\tau ds \\ & - \frac{1}{1-\alpha\xi}\int_0^1 \frac{(1-s)^{q-1}}{\Gamma(q)} v_2(s)ds\Bigg). \end{aligned}$$

Thus,

$$|h_1(t) - h_2(t)| \leq \int_0^t \frac{(t-s)^{q-1}}{\Gamma(q)} |v_1(s) - v_2(s)|ds$$

$$
\begin{aligned}
&+\frac{t}{|1-\alpha\xi|}\left(|\alpha|\int_0^\xi\int_0^s \frac{(s-\tau)^{q-2}}{\Gamma(q-1)}|v_1(\tau)-v_2(\tau)|d\tau ds\right.\\
&\left.+\int_0^1 \frac{(1-s)^{q-1}}{\Gamma(q)}|v_1(s)-v_2(s)|ds\right)\\
&+|\beta|g\left(\int_0^\eta \frac{(\eta-s)^{q-2}}{\Gamma(q-1)}|v_1(s)-v_2(s)|(s)ds\right.\\
&+\frac{|\alpha|}{|1-\alpha\xi|}\int_0^\xi\int_0^s \frac{(s-\tau)^{q-2}}{\Gamma(q-1)}|v_1(\tau)-v_2(\tau)|d\tau ds\\
&\left.+\frac{1}{|1-\alpha\xi|}\int_0^1 \frac{(1-s)^{q-1}}{\Gamma(q)}|v_1(s)-v_2(s)|(s)ds\right)\\
\leq\ &\|m\|\left\{\frac{1}{\Gamma(q+1)}+\frac{1}{|1-\alpha\xi|}\left(\frac{|\alpha|\xi^q}{\Gamma(q+1)}+\frac{1}{\Gamma(q+1)}\right)\right.\\
&\left.+|\beta|\left(\frac{\eta^{q-1}}{\Gamma(q)}+\frac{|\alpha|}{|1-\alpha\xi|}\frac{\xi^q}{\Gamma(q+1)}+\frac{1}{|1-\alpha\xi|}\frac{1}{\Gamma(q+1)}\right)\right\}\|x-\bar{x}\|\\
=\ &\|m\|\Lambda\|x-\bar{x}\|.
\end{aligned}
$$

Hence

$$\|h_1-h_2\|\leq\|m\|\Lambda\|x-\bar{x}\|.$$

Analogously, interchanging the roles of x and $\overline{x}$, we obtain

$$H_d(\Omega_F(x),\Omega_F(\bar{x}))\leq\|m\|\Lambda\|x-\bar{x}\|.$$

Since Ω_F is a contraction, it follows by Theorem 1.19 that Ω_F has a fixed point x which is a solution of (3.137)-(3.138). This completes the proof. □

Our next result is concerned with the existence of the solutions for the problem (3.137)-(3.138) with a convex valued multi-valued map.

Theorem 3.32. *(The upper semi-continuous case). Assume that (3.31.3) holds. In addition we suppose that*

(3.32.1) $F:[0,1]\times\mathbb{R}\to\mathcal{P}(\mathbb{R})$ *is Carathéodory and has non-empty compact and convex values;*

(3.32.2) there exists a continuous non-decreasing function $\psi:[0,\infty)\to(0,\infty)$ *and a function* $\phi\in C([0,1],\mathbb{R}^+)$ *such that*
$\|F(t,x)\|_{\mathcal{P}}:=\sup\{|y|:y\in F(t,x)\}\leq\phi(t)\psi(\|x\|)$ *for each* $(t,x)\in[0,1]\times\mathbb{R}$;

(3.32.3) there exists a constant $M>0$ *such that*

$$\frac{M}{\psi(M)\|\phi\|\Lambda}>1,$$

where Λ *is defined by (3.126).*

Then the boundary value problem (3.137)-(3.138) has at least one solution on $[0,1]$.

Proof. Consider the operator $\Omega_F : C([0,1],\mathbb{R}) \to \mathcal{P}(C([0,1],\mathbb{R}))$ defined in the beginning of the proof of Theorem 3.31. We will show that Ω_F satisfies the assumptions of the nonlinear alternative of Leray-Schauder type. The proof consists of several steps. As a first step, we show that Ω_F *is convex for each* $x \in C([0,1],\mathbb{R})$. This step is obvious since $S_{F,x}$ is convex (F has convex values), and therefore we omit the proof.

In the second step, we show that Ω_F *maps bounded sets (balls) into bounded sets in* $C([0,1],\mathbb{R})$. For a positive number ρ, let $B_\rho = \{x \in C([0,1],\mathbb{R}) : \|x\| \le \rho\}$ be a bounded ball in $C([0,1],\mathbb{R})$. Then, for each $h \in \Omega_F(x), x \in B_\rho$, there exists $v \in S_{F,x}$ such that

$$\begin{aligned} h(t) = &\int_0^t \frac{(t-s)^{q-1}}{\Gamma(q)} v(s)ds \\ &+\frac{t}{1-\alpha\xi}\left(\alpha \int_0^\xi \int_0^s \frac{(s-\tau)^{q-2}}{\Gamma(q-1)} v(\tau)d\tau ds - \int_0^1 \frac{(1-s)^{q-1}}{\Gamma(q)} v(s)ds\right) \\ &+\beta g\Bigg(\int_0^\eta \frac{(\eta-s)^{q-2}}{\Gamma(q-1)} v(s)ds + \frac{\alpha}{1-\alpha\xi}\int_0^\xi \int_0^s \frac{(s-\tau)^{q-2}}{\Gamma(q-1)} v(\tau)d\tau ds \\ &-\frac{1}{1-\alpha\xi}\int_0^1 \frac{(1-s)^{q-1}}{\Gamma(q)} v(s)ds\Bigg), \quad t \in [0,1]. \end{aligned}$$

Then, for $t \in [0,1]$, we have

$$\begin{aligned} |h(t)| \le &\int_0^t \frac{(t-s)^{q-1}}{\Gamma(q)} |v(s)|ds \\ &+\frac{t}{|1-\alpha\xi|}\left(|\alpha| \int_0^\xi \int_0^s \frac{(s-\tau)^{q-2}}{\Gamma(q-1)} |v(\tau)|d\tau ds + \int_0^1 \frac{(1-s)^{q-1}}{\Gamma(q)} |v(s)|ds\right) \\ &+|\beta| g\Bigg(\int_0^\eta \frac{(\eta-s)^{q-2}}{\Gamma(q-1)} |v(s)|ds + \frac{|\alpha|}{|1-\alpha\xi|}\int_0^\xi \int_0^s \frac{(s-\tau)^{q-2}}{\Gamma(q-1)} |v(\tau)|d\tau ds \\ &+\frac{1}{|1-\alpha\xi|}\int_0^1 \frac{(1-s)^{q-1}}{\Gamma(q)} |v(s)|ds\Bigg) \\ \le\ &\psi(\|x\|)\|\phi\|\Bigg\{\frac{1}{\Gamma(q+1)} + \frac{1}{|1-\alpha\xi|}\left(\frac{|\alpha|\xi^q}{\Gamma(q+1)} + \frac{1}{\Gamma(q+1)}\right) \\ &+|\beta|\left(\frac{\eta^{q-1}}{\Gamma(q)} + \frac{|\alpha|}{|1-\alpha\xi|}\frac{\xi^q}{\Gamma(q+1)} + \frac{1}{|1-\alpha\xi|}\frac{1}{\Gamma(q+1)}\right)\Bigg\} \\ =\ &\psi(\|x\|)\|\phi\|\Lambda. \end{aligned}$$

Consequently

$$\|h\| \leq \psi(\rho)\|\phi\|\Lambda.$$

Now we show that Ω_F *maps bounded sets into equi-continuous sets of* $C([0,1],\mathbb{R})$. Let $t_1, t_2 \in [0,1]$ with $t_1 < t_2$ and $x \in B_\rho$. For each $h \in \Omega_F(x)$, we obtain

$$\begin{aligned}
&|h(t_2) - h(t_1)| \\
&\leq \left| \frac{1}{\Gamma(\alpha)} \int_0^{t_2} (t_2 - s)^{q-1} v(s) ds - \frac{1}{\Gamma(\alpha)} \int_0^{t_1} (t_1 - s)^{q-1} v(s) ds \right| \\
&\quad + \frac{|t_2 - t_1|}{|1 - \alpha\xi|} \left(|\alpha| \int_0^{\xi} \int_0^{s} \frac{(s-\tau)^{q-2}}{\Gamma(q-1)} |v(\tau)| d\tau ds + \int_0^1 \frac{(1-s)^{q-1}}{\Gamma(q)} |v(s)| ds \right) \\
&\leq \frac{1}{\Gamma(\alpha)} \int_0^{t_1} [(t_2 - s)^{q-1} - (t_1 - s)^{q-1}] \phi(s)\psi(r) ds + \frac{1}{\Gamma(\alpha)} \int_{t_1}^{t_2} (t_2 - s)^{q-1} \phi(s)\psi(r) ds \\
&\quad + \frac{|t_2 - t_1|}{|1 - \alpha\xi|} \left(|\alpha| \int_0^{\xi} \int_0^{s} \frac{(s-\tau)^{q-2}}{\Gamma(q-1)} \phi(s)\psi(r) d\tau ds + \int_0^1 \frac{(1-s)^{q-1}}{\Gamma(q)} \phi(s)\psi(r) ds \right) \\
&\leq \frac{\|\phi\|\psi(\rho)}{\Gamma(\alpha)} \int_0^{t_1} [(t_2 - s)^{q-1} - (t_1 - s)^{q-1}] ds + \frac{\|\phi\|\psi(\rho)}{\Gamma(\alpha)} \int_{t_1}^{t_2} (t_2 - s)^{q-1} ds \\
&\quad + \frac{\|\phi\|\psi(\rho)|t_2 - t_1|}{|1 - \alpha\xi|} \left(\frac{|\alpha|\xi^q}{\Gamma(q+1)} + \frac{1}{\Gamma(q+1)} \right).
\end{aligned}$$

Obviously the right hand side of the above inequality tends to zero independently of $x \in B_\rho$ as $t_2 - t_1 \to 0$. As Ω_F satisfies the above three assumptions, therefore it follows by the Ascoli-Arzelá theorem that $\Omega_F : C([0,1],\mathbb{R}) \to \mathcal{P}(C([0,1],\mathbb{R}))$ is completely continuous.

In our next step, we show that Ω_F *is upper semi-continuous.* To this end it is sufficient to show that Ω_F *has a closed graph,* by Lemma 1.1. Let $x_n \to x_*, h_n \in \Omega_F(x_n)$ and $h_n \to h_*$. Then we need to show that $h_* \in \Omega_F(x_*)$. Associated with $h_n \in \Omega_F(x_n)$, there exists $v_n \in S_{F,x_n}$ such that for each $t \in [0,1]$,

$$\begin{aligned}
h_n(t) = &\int_0^t \frac{(t-s)^{q-1}}{\Gamma(q)} v_n(s) ds \\
&+ \frac{t}{1 - \alpha\xi} \left(\alpha \int_0^{\xi} \int_0^{s} \frac{(s-\tau)^{q-2}}{\Gamma(q-1)} v_n(\tau) d\tau ds - \int_0^1 \frac{(1-s)^{q-1}}{\Gamma(q)} v_n(s) ds \right) \\
&+ \beta g \left(\int_0^{\eta} \frac{(\eta - s)^{q-2}}{\Gamma(q-1)} v_n(s) ds + \frac{\alpha}{1 - \alpha\xi} \int_0^{\xi} \int_0^{s} \frac{(s-\tau)^{q-2}}{\Gamma(q-1)} v_n(\tau) d\tau ds \right. \\
&\left. - \frac{1}{1 - \alpha\xi} \int_0^1 \frac{(1-s)^{q-1}}{\Gamma(q)} v_n(s) ds \right).
\end{aligned}$$

Thus it suffices to show that there exists $v_* \in S_{F,x_*}$ such that for each $t \in [0,1]$,

$$\begin{aligned}
h_*(t) = & \int_0^t \frac{(t-s)^{q-1}}{\Gamma(q)} v_*(s)ds \\
& + \frac{t}{1-\alpha\xi}\left(\alpha \int_0^\xi \int_0^s \frac{(s-\tau)^{q-2}}{\Gamma(q-1)} v_*(\tau)d\tau ds - \int_0^1 \frac{(1-s)^{q-1}}{\Gamma(q)} v_*(s)ds\right) \\
& + \beta g\left(\int_0^\eta \frac{(\eta-s)^{q-2}}{\Gamma(q-1)} v_*(s)ds + \frac{\alpha}{1-\alpha\xi}\int_0^\xi \int_0^s \frac{(s-\tau)^{q-2}}{\Gamma(q-1)} v_*(\tau)d\tau ds \right. \\
& \left. - \frac{1}{1-\alpha\xi}\int_0^1 \frac{(1-s)^{q-1}}{\Gamma(q)} v_*(s)ds\right).
\end{aligned}$$

Let us consider the linear operator $\Theta : L^1([0,1],\mathbb{R}) \to C([0,1],\mathbb{R})$ given by

$$\begin{aligned}
v \mapsto \Theta(v)(t) = & \int_0^t \frac{(t-s)^{q-1}}{\Gamma(q)} v(s)ds \\
& + \frac{t}{1-\alpha\xi}\left(\alpha \int_0^\xi \int_0^s \frac{(s-\tau)^{q-2}}{\Gamma(q-1)} v(\tau)d\tau ds - \int_0^1 \frac{(1-s)^{q-1}}{\Gamma(q)} v(s)ds\right) \\
& + \beta g\left(\int_0^\eta \frac{(\eta-s)^{q-2}}{\Gamma(q-1)} v(s)ds + \frac{\alpha}{1-\alpha\xi}\int_0^\xi \int_0^s \frac{(s-\tau)^{q-2}}{\Gamma(q-1)} v(\tau)d\tau ds \right. \\
& \left. - \frac{1}{1-\alpha\xi}\int_0^1 \frac{(1-s)^{q-1}}{\Gamma(q)} v(s)ds\right).
\end{aligned}$$

Observe that

$$\begin{aligned}
& \|h_n(t) - h_*(t)\| \\
= & \left\| \int_0^t \frac{(t-s)^{q-1}}{\Gamma(q)} (v_n(s) - v_*(s))ds \right. \\
& + \frac{t}{1-\alpha\xi}\left(\alpha \int_0^\xi \int_0^s \frac{(s-\tau)^{q-2}}{\Gamma(q-1)} (v_n(\tau) - v_*(\tau))d\tau ds - \int_0^1 \frac{(1-s)^{q-1}}{\Gamma(q)} v(s)ds\right) \\
& + \beta g\left(\int_0^\eta \frac{(\eta-s)^{q-2}}{\Gamma(q-1)} (v_n(s) - v_*(s))ds \right. \\
& + \frac{\alpha}{1-\alpha\xi}\int_0^\xi \int_0^s \frac{(s-\tau)^{q-2}}{\Gamma(q-1)} (v_n(\tau) - v_*(\tau))d\tau ds \\
& \left.\left. - \frac{1}{1-\alpha\xi}\int_0^1 \frac{(1-s)^{q-1}}{\Gamma(q)} (v_n(s) - v_*(s))ds\right) \right\| \to 0,
\end{aligned}$$

as $n \to \infty$.

Thus, it follows by Lemma 1.2 that $\Theta \circ S_F$ is a closed graph operator. Further, we have $h_n(t) \in \Theta(S_{F,x_n})$. Since $x_n \to x_*$, therefore, we have

$$\begin{aligned}
h_*(t) = & \int_0^t \frac{(t-s)^{q-1}}{\Gamma(q)} v_*(s)ds \\
& + \frac{t}{1-\alpha\xi}\left(\alpha \int_0^\xi \int_0^s \frac{(s-\tau)^{q-2}}{\Gamma(q-1)} v_*(\tau)d\tau ds - \int_0^1 \frac{(1-s)^{q-1}}{\Gamma(q)} v_*(s)ds\right)
\end{aligned}$$

$$+\beta g\bigg(\int_0^\eta \frac{(\eta-s)^{q-2}}{\Gamma(q-1)}v_*(s)ds+\frac{\alpha}{1-\alpha\xi}\int_0^\xi\int_0^s \frac{(s-\tau)^{q-2}}{\Gamma(q-1)}v_*(\tau)d\tau ds$$
$$-\frac{1}{1-\alpha\xi}\int_0^1 \frac{(1-s)^{q-1}}{\Gamma(q)}v_*(s)ds\bigg),$$

for some $v_* \in S_{F,x_*}$.

Finally, we show there exists an open set $U \subseteq C([0,1],\mathbb{R})$ with $x \notin \Omega_F(x)$ for any $\lambda \in (0,1)$ and all $x \in \partial U$. Let $\lambda \in (0,1)$ and $x \in \lambda\Omega_F(x)$. Then there exists $v \in L^1([0,1],\mathbb{R})$ with $v \in S_{F,x}$ such that, for $t \in [0,1]$, we have

$$x(t)=\lambda\int_0^t \frac{(t-s)^{q-1}}{\Gamma(q)}v(s)ds$$
$$+\lambda\frac{t}{1-\alpha\xi}\bigg(\alpha\int_0^\xi\int_0^s \frac{(s-\tau)^{q-2}}{\Gamma(q-1)}v(\tau)d\tau ds-\int_0^1 \frac{(1-s)^{q-1}}{\Gamma(q)}v(s)ds\bigg)$$
$$+\lambda\beta g\bigg(\int_0^\eta \frac{(\eta-s)^{q-2}}{\Gamma(q-1)}v(s)ds+\frac{\alpha}{1-\alpha\xi}\int_0^\xi\int_0^s \frac{(s-\tau)^{q-2}}{\Gamma(q-1)}v(\tau)d\tau ds$$
$$-\frac{1}{1-\alpha\xi}\int_0^1 \frac{(1-s)^{q-1}}{\Gamma(q)}v(s)ds\bigg).$$

Using the computations of the second step above we have

$$|x(t)| \le \psi(\|x\|)\|\phi\|\bigg\{\frac{1}{\Gamma(q+1)}+\frac{1}{|1-\alpha\xi|}\bigg(\frac{|\alpha|\xi^q}{\Gamma(q+1)}+\frac{1}{\Gamma(q+1)}\bigg)$$
$$+|\beta|\bigg(\frac{\eta^{q-1}}{\Gamma(q)}+\frac{|\alpha|}{|1-\alpha\xi|}\frac{\xi^q}{\Gamma(q+1)}+\frac{1}{|1-\alpha\xi|}\frac{1}{\Gamma(q+1)}\bigg)\bigg\}$$
$$=\psi(\|x\|)\|\phi\|\Lambda.$$

Consequently, we have

$$\frac{\|x\|}{\psi(\|x\|)\|\phi\|\Lambda}\le 1.$$

In view of (3.32.3), there exists M such that $\|x\| \neq M$. Let us set

$$U=\{x \in C([0,1],\mathbb{R}) : \|x\| < M\}.$$

Note that the operator $\Omega_F : \overline{U} \to \mathcal{P}(C([0,1],\mathbb{R}))$ is upper semi-continuous and completely continuous. From the choice of U, there is no $x \in \partial U$ such that $x \in \lambda\Omega_F(x)$ for some $\lambda \in (0,1)$. Consequently, by the nonlinear alternative of Leray-Schauder type (Theorem 1.16), we deduce that Ω_F has a fixed point $x \in \overline{U}$ which is a solution of the problem (3.137)-(3.138). This completes the proof. □

Finally, we assume that the multi-valued map F in (3.137) is not necessarily convex valued and formulate (and prove) the existence result as follows.

Theorem 3.33. *(The lower semi-continuous case). Assume that (3.31.1), (3.32.2), (3.32.3) and the following condition hold:*

(3.33.1) $F : [0,1] \times \mathbb{R} \to \mathcal{P}(\mathbb{R})$ *is a non-empty compact-valued multi-valued map such that:*

(a) $(t, x) \longmapsto F(t, x)$ *is* $\mathcal{L} \otimes \mathcal{B}$ *measurable,*

(b) $x \longmapsto F(t, x)$ *is lower semi-continuous for each* $t \in [0, 1]$.

Then the boundary value problem (3.137)-(3.138) has at least one solution on $[0,1]$.

Proof. It follows from (3.32.2) and (3.33.1) that F is of l.s.c. type. Then from Lemma 1.3, there exists a continuous function $f : C^2([0,1],\mathbb{R}) \to L^1([0,1],\mathbb{R})$ such that $f(x) \in \mathcal{F}(x)$ for all $x \in C([0,1],\mathbb{R})$.

Consider the problem

$$\begin{cases} {}^cD^q x(t) = f(x(t)), \quad t \in [0,1], \ 1 < q \le 2, \\ x(0) = \beta g(x'(\eta)), \ x'(1) = \alpha \displaystyle\int_0^{\xi} x'(s)ds, \ 0 < \xi, \eta < 1. \end{cases} \tag{3.141}$$

Observe that if $x \in C^2([0,1],\mathbb{R})$ is a solution of (3.141), then x is a solution to the problem (3.137)-(3.138). In order to transform the problem (3.141) into a fixed point problem, we define the operator $\overline{\Omega_F}$ as

$$\begin{aligned} \overline{\Omega_F} x(t) &= \int_0^t \frac{(t-s)^{q-1}}{\Gamma(q)} f(x(s))ds \\ &+ \frac{t}{1-\alpha\xi}\left(\alpha \int_0^{\xi}\int_0^s \frac{(s-\tau)^{q-2}}{\Gamma(q-1)} f(x(\tau))d\tau ds - \int_0^1 \frac{(1-s)^{q-1}}{\Gamma(q)} f(x(s))ds\right) \\ &+ \beta g\Bigg(\int_0^{\eta} \frac{(\eta-s)^{q-2}}{\Gamma(q-1)} f(x(s))ds + \frac{\alpha}{1-\alpha\xi}\int_0^{\xi}\int_0^s \frac{(s-\tau)^{q-2}}{\Gamma(q-1)} f(x(\tau))d\tau ds \\ &- \frac{1}{1-\alpha\xi}\int_0^1 \frac{(1-s)^{q-1}}{\Gamma(q)} f(x(s))ds\Bigg) \quad t \in [0,1]. \end{aligned}$$

It can easily be shown that $\overline{\Omega_F}$ is continuous and completely continuous. The remaining part of the proof is similar to that of Theorem 3.32. So we omit it. This completes the proof. □

3.7.5 *Examples*

Consider the problem

$$\begin{cases} {}^cD^{3/2} x(t) \in F(t, x(t)), \ 0 \le t \le 1, \\ x(0) = \dfrac{1}{3} g(x'(3/4)), \ x'(1) = \dfrac{1}{2}\displaystyle\int_0^{1/3} x'(s)ds. \end{cases} \tag{3.142}$$

Here, $q = 3/2$, $\alpha = 1/2$, $\xi = 1/3$, $\beta = 1/3$, $\eta = 3/4$ and $g(v) = \begin{cases} \sqrt{v}, \ |v| \ge 1, \\ v^2, \ |v| < 1. \end{cases}$

With the given values, we find that $\Lambda \approx 2.3980168$.

(i) Consider the multi-valued map $F : [0,1] \times \mathbb{R} \to \mathcal{P}(\mathbb{R})$ given by

$$F(t,x) = \left[0, \frac{1}{6}(t+1)\sin x + \frac{2}{3}\right].$$

Then we have

$$\sup\{|v| : v \in F(t,x)\} \le \frac{1}{6}(t+1) + \frac{2}{3},$$

and

$$H_d(F(t,x), F(t,\bar{x}) \le \frac{1}{6}(t+1)|x - \bar{x}|.$$

Let $m(t) = \frac{1}{6}(t+1)$. Then $\|m\| = \frac{1}{3}$ and $\|m\|\Lambda \approx 0.7993389 < 1$. Hence by Theorem 3.31 the problem (3.142) has a solution.

(ii) Let $F : [0,1] \times \mathbb{R} \to \mathcal{P}(\mathbb{R})$ is a multi-valued map given by

$$F(t,x) = \left[e^{-t}\left(\frac{|x|^3}{|x|^3+5} + 9\right), 2e^{-t}\left(\frac{|x|^3}{|x|^3+3} + 1\right)\right].$$

For $v \in F$, we have

$$|v(t)| \le \max\left(e^{-t}\left(\frac{|x|^3}{|x|^3+5} + 9\right), 2e^{-t}\left(\frac{|x|^3}{|x|^3+3} + 1\right)\right) \le 10e^{-t}, x \in \mathbb{R}.$$

Thus

$$\|F(t,x)\|_{\mathcal{P}} := \sup\{|y| : y \in F(t,x)\} \le 10e^{-t} = \phi(t)\psi(\|x\|),$$

with $\phi(t) = e^{-t}$, $\psi(\|x\|) = 10$.
By the assumption (3.32.3), we find that $M > 23.980168$. It follows by Theorem 3.32 that problem (3.142) has a solution.

3.7.6 *Existence results for the problem (3.137) and (3.139)*

In relation to problem (3.137) and (3.139), we define an operator $G_F : C([0,1],\mathbb{R}) \to \mathcal{P}(C([0,1],\mathbb{R}))$ by

$$G_F(x) = \left\{ \begin{array}{l} h \in C([0,1],\mathbb{R}) : \\ h(t) = \left\{ \begin{array}{l} \displaystyle\int_0^t \frac{(t-s)^{q-1}}{\Gamma(q)} v(s)ds \\ \displaystyle +(t-1)\frac{\alpha}{1-\alpha\xi}\int_0^\xi \int_0^s \frac{(s-\tau)^{q-2}}{\Gamma(q-1)} v(\tau)d\tau ds \\ \displaystyle -\int_0^1 \frac{(1-s)^{q-1}}{\Gamma(q)} v(s)ds \\ \displaystyle +\beta g\left(\int_0^\eta \frac{(\eta-s)^{q-2}}{\Gamma(q-1)} v(s)ds \right. \\ \displaystyle \left. +\frac{\alpha}{1-\alpha\xi}\int_0^\xi \int_0^s \frac{(s-\tau)^{q-2}}{\Gamma(q-1)} v(\tau)d\tau ds \right) \end{array} \right. \end{array} \right\}$$

for $v \in S_{F,x}$, and set

$$\bar{\Lambda} = \frac{1}{\Gamma(q+1)}\left(2 + q|\beta|\eta^{q-1} + \frac{|\alpha|(1+|\beta|)\xi^q}{|1-\alpha\xi|}\right). \tag{3.143}$$

With the above operator and the estimate (3.143), we can reproduce all the existence results obtained in the last section for the boundary value problem (3.137) and (3.139).

3.8 On solvability of a coupled system of fractional differential equations supplemented with a new kind of flux type integral boundary conditions

In this section, we consider a more generalized version of flux type integral boundary conditions and develop the existence criteria for a coupled system of Caputo type fractional differential equations equipped with these new conditions. Precisely, we investigate the following coupled system of Caputo type fractional differential equations

$$\begin{cases} {}^cD^q x(t) = f(t, x(t), y(t)), & t \in [0,1], \quad 1 < q \leq 2, \\ {}^cD^p y(t) = h(t, x(t), y(t)), & t \in [0,1], \quad 1 < p \leq 2, \end{cases} \tag{3.144}$$

supplemented with the nonlocal nonlinear flux type integral boundary conditions:

$$\begin{cases} x'(0) = \alpha \displaystyle\int_0^\xi x'(s)ds, \quad x(1) = \beta \int_0^1 g(x'(s))ds, \quad 0 \leq \xi \leq 1, \\ y'(0) = \alpha_1 \displaystyle\int_0^\theta y'(s)ds, \quad y(1) = \beta_1 \int_0^1 g(y'(s))ds, \quad 0 \leq \theta \leq 1, \end{cases} \tag{3.145}$$

where ${}^cD^q, {}^cD^p$ denote the Caputo fractional derivative operators of order q and p respectively, $f, h : [0,1] \times \mathbb{R} \times \mathbb{R} \to \mathbb{R}$, $g : \mathbb{R} \to \mathbb{R}$ are appropriately chosen functions, and $\alpha, \beta, \alpha_1, \beta_1$ are real constants.

To define the solution for the problem (3.144)-(3.145), we use the following lemma.

Lemma 3.11. *Let $\alpha\xi \neq 1$. For $\phi \in C([0,1], \mathbb{R})$, the linear problem consisting by the equation*

$$ {}^cD^q x(t) = \phi(t), \qquad t \in [0,1], \quad 1 < q \leq 2, \tag{3.146}$$

supplemented with the boundary conditions

$$x'(0) = \alpha \int_0^\xi x'(s)ds, \quad x(1) = \beta \int_0^1 g(x'(s))ds, \quad 0 \leq \xi \leq 1, \tag{3.147}$$

is equivalent to the integral equation

$$x(t) = \int_0^t \frac{(t-s)^{q-1}}{\Gamma(q)}\phi(s)ds + \frac{\alpha(t-1)}{1-\alpha\xi}\int_0^\xi \int_0^s \frac{(s-\tau)^{q-2}}{\Gamma(q-1)}\phi(\tau)d\tau ds$$

$$
\begin{aligned}
&-\int_0^1 \frac{(1-s)^{q-1}}{\Gamma(q)}\phi(s)ds \\
&+\beta\int_0^1 g\left(\int_0^s \frac{(s-\tau)^{q-2}}{\Gamma(q-1)}\phi(\tau)d\tau + \frac{\alpha}{1-\alpha\xi}\int_0^\xi\int_0^s \frac{(s-\tau)^{q-2}}{\Gamma(q-1)}\phi(\tau)d\tau ds\right)ds.
\end{aligned} \tag{3.148}
$$

Proof. It is well known that the general solution of the fractional differential equation (3.146) can be written as

$$
x(t) = c_0 + c_1 t + \int_0^t \frac{(t-s)^{q-1}}{\Gamma(q)}\phi(s)ds, \tag{3.149}
$$

where $c_0, c_1 \in \mathbb{R}$ are arbitrary constants. Using the boundary conditions (3.147) in (3.149), we find that

$$
\begin{aligned}
c_0 = {}&\beta\int_0^1 g\left(\int_0^s \frac{(s-\tau)^{q-2}}{\Gamma(q-1)}\phi(\tau)d\tau + \frac{\alpha}{1-\alpha\xi}\int_0^\xi\int_0^s \frac{(s-\tau)^{q-2}}{\Gamma(q-1)}\phi(\tau)d\tau ds\right)ds \\
&-\int_0^1 \frac{(1-s)^{q-2}}{\Gamma(q-1)}\phi(s)ds - \frac{\alpha}{1-\alpha\xi}\int_0^\xi\int_0^s \frac{(s-\tau)^{q-2}}{\Gamma(q-1)}\phi(\tau)d\tau ds
\end{aligned}
$$

and

$$
c_1 = \frac{\alpha}{1-\alpha\xi}\int_0^\xi\int_0^s \frac{(s-\tau)^{q-2}}{\Gamma(q-1)}\phi(\tau)d\tau ds.
$$

Substituting the values of c_0, c_1 in (3.149), we get (3.148). The converse follows by direct computation. This completes the proof. □

3.8.1 *Main results*

Let us introduce the space $X = \{u(t)|u(t) \in C([0,1],\mathbb{R})\}$ endowed with the norm $\|u\| = \sup\{|u(t)|, t \in [0,1]\}$. Obviously $(X, \|\cdot\|)$ is a Banach space. Also $Y = \{v(t)|v(t) \in C([0,1],\mathbb{R})\}$ endowed with the norm $\|v\| = \sup\{|v(t)|, t \in [0,1]\}$ is a Banach space. Then the product space $(X \times Y, \|(u,v)\|)$ is also a Banach space equipped with norm $\|(u,v)\| = \|u\| + \|v\|$.

In view of Lemma 3.11, we define the operator $T : X \times Y \to X \times Y$ by

$$
T(u,v)(t) = \begin{pmatrix} T_1(u,v)(t) \\ T_2(u,v)(t) \end{pmatrix},
$$

where

$$
\begin{aligned}
&T_1(u,v)(t) \\
&= \int_0^t \frac{(t-s)^{q-1}}{\Gamma(q)} f(s,u(s),v(s))ds + \frac{\alpha(t-1)}{1-\alpha\xi}\int_0^\xi\int_0^s \frac{(s-\tau)^{q-2}}{\Gamma(q-1)} f(\tau,u(\tau),v(\tau))d\tau ds \\
&\quad -\int_0^1 \frac{(1-s)^{q-1}}{\Gamma(q)} f(s,u(s),v(s))ds + \beta\int_0^1 g\left(\int_0^s \frac{(s-\tau)^{q-2}}{\Gamma(q-1)} f(\tau,u(\tau),v(\tau))d\tau\right.
\end{aligned}
$$

$$+\frac{\alpha}{1-\alpha\xi}\int_0^\xi\int_0^s\frac{(s-\tau)^{q-2}}{\Gamma(q-1)}f(\tau,u(\tau),v(\tau))d\tau ds\Bigg)ds,$$

and

$$\begin{aligned}&T_2(u,v)(t)\\&=\int_0^t\frac{(t-s)^{p-1}}{\Gamma(p)}h(s,u(s),v(s))ds+\frac{\alpha_1(t-1)}{1-\alpha_1\xi}\int_0^\xi\int_0^s\frac{(s-\tau)^{p-2}}{\Gamma(p-1)}h(\tau,u(\tau),v(\tau))d\tau ds\\&-\int_0^1\frac{(1-s)^{p-1}}{\Gamma(p)}h(s,u(s),v(s))ds+\beta_1\int_0^1 g\Bigg(\int_0^s\frac{(s-\tau)^{p-2}}{\Gamma(p-1)}h(\tau,u(\tau),v(\tau))d\tau\\&+\frac{\alpha_1}{1-\alpha_1\xi}\int_0^\xi\int_0^s\frac{(s-\tau)^{p-2}}{\Gamma(p-1)}h(\tau,u(\tau),v(\tau))d\tau ds\Bigg)ds.\end{aligned}$$

For the sake of convenience, we set

$$M_1=\frac{1}{\Gamma(q+1)}\left(2+\frac{|\alpha|\xi^q}{|1-\alpha\xi|}\right)+\frac{|\beta|}{\Gamma(q+2)}\left(q+1+\frac{|\alpha|}{|1-\alpha\xi|}\right),\tag{3.150}$$

$$M_2=\frac{1}{\Gamma(p+1)}\left(2+\frac{|\alpha_1|\theta^p}{|1-\alpha_1\theta|}\right)+\frac{|\beta_1|}{\Gamma(p+2)}\left(p+1+\frac{|\alpha_1|}{|1-\alpha_1\theta|}\right),\tag{3.151}$$

and

$$M_0=\min\{1-(M_1k_1+M_2\lambda_1),\ 1-(M_1k_2+M_2\lambda_2)\},\ \ k_i,\ \lambda_i\geq 0\ (i=1,2).\tag{3.152}$$

The first result is based on Leray-Schauder alternative (Theorem 1.7).

Theorem 3.34. *Assume that:*

(3.34.1) there exist real constants $k_i,\ \lambda_i\geq 0\ (i=1,2)$ *and* $k_0>0, \lambda_0>0$ *such that* $\forall x_i\in\mathbb{R},\ (i=1,2)$,

$$|f(t,x_1,x_2)|\leq k_0+k_1|x_1|+k_2|x_2|,\ |h(t,x_1,x_2)|\leq \lambda_0+\lambda_1|x_1|+\lambda_2|x_2|;$$

(3.34.2) $|g(v)|\leq|v|,\ \ \forall v\in\mathbb{R}$.

In addition it is assumed that

$$M_1k_1+M_2\lambda_1<1\ \ \textit{and}\ \ M_1k_2+M_2\lambda_2<1,$$

where M_1 *and* M_2 *are given by (3.150) and (3.151) respectively. Then the system (3.144)-(3.145) has at least one solution.*

Proof. First we show that the operator $T: X\times Y\to X\times Y$ is completely continuous. By continuity of functions f,h and g, the operator T is continuous.

Let $\Omega\subset X\times Y$ be bounded. Then there exist positive constants L_1 and L_2 such that

$$|f(t,u(t),v(t)|\leq L_1,\quad |h(t,u(t),v(t)|\leq L_2,\ \ \forall(u,v)\in\Omega.$$

Then, for any $(u,v) \in \Omega$, one can obtain that

$$\|T_1(u,v)\| \le L_1\left\{\frac{1}{\Gamma(q+1)}\left(2+\frac{|\alpha|\xi^q}{|1-\alpha\xi|}\right)+\frac{|\beta|}{\Gamma(q+2)}\left(q+1+\frac{|\alpha|}{|1-\alpha\xi|}\right)\right\} = L_1 M_1,$$

$$\|T_2(u,v)\| \le L_2\left\{\frac{1}{\Gamma(p+1)}\left(2+\frac{|\alpha_1|\theta^p}{|1-\alpha_1\theta|}\right)+\frac{|\beta_1|}{\Gamma(p+2)}\left(p+1+\frac{|\alpha_1|}{|1-\alpha_1\theta|}\right)\right\} = L_2 M_2.$$

Thus, it follows from the above inequalities that the operator T is uniformly bounded.

Next, we show that T is equi-continuous. Let $t_1, t_2 \in [0,1]$ with $t_1 < t_2$. Then we have

$$\begin{aligned}
&|T_1(u(t_2),v(t_2)) - T_1(u(t_1),v(t_1))|\\
&\le L_1\left\{\left|\frac{1}{\Gamma(q)}\int_0^{t_2}(t_2-s)^{q-1}ds - \frac{1}{\Gamma(q)}\int_0^{t_1}(t_1-s)^{q-1}vds\right|\right.\\
&\quad\left.+\frac{|\alpha||t_2-t_1|}{|1-\alpha\xi|}\int_0^{\xi}\int_0^{s}\frac{(s-\tau)^{q-2}}{\Gamma(q-1)}d\tau ds\right\}\\
&\le L_1\left\{\frac{1}{\Gamma(q)}\int_0^{t_1}[(t_2-s)^{q-1}-(t_1-s)^{q-1}]ds + \frac{1}{\Gamma(q)}\int_{t_1}^{t_2}(t_2-s)^{q-1}ds\right.\\
&\quad\left.+\frac{|\alpha||t_2-t_1|}{|1-\alpha\xi|}\frac{\xi^q}{\Gamma(q+1)}\right\}\\
&\le \frac{L_1}{\Gamma(q+1)}[2(t_2-t_1)^q + |t_2^q - t_1^q|] + L_1\frac{|\alpha||t_2-t_1|}{|1-\alpha\xi|}\frac{\xi^q}{\Gamma(q+1)}.
\end{aligned}$$

Analogously, we can obtain

$$\begin{aligned}
&|T_2(u(t_2),v(t_2)) - T_2(u(t_1),v(t_1))|\\
&\le \frac{L_2}{\Gamma(p+1)}[2(t_2-t_1)^p + |t_2^p - t_1^p|] + L_2\frac{|\alpha_1||t_2-t_1|}{|1-\alpha_1\theta|}\frac{\theta^p}{\Gamma(p+1)}.
\end{aligned}$$

Therefore, the operator $T(u,v)$ is equi-continuous, and thus the operator $T(u,v)$ is completely continuous.

Finally, it will be verified that the set $\mathcal{E} = \{(u,v) \in X \times Y | (u,v) = \lambda T(u,v), 0 \le \lambda \le 1\}$ is bounded. Let $(u,v) \in \mathcal{E}$, with $(u,v) = \lambda T(u,v)$. For any $t \in [0,1]$, we have

$$u(t) = \lambda T_1(u,v)(t), \quad v(t) = \lambda T_2(u,v)(t).$$

Then

$$\begin{aligned}
&|u(t)|\\
&\le \left\{\frac{1}{\Gamma(q+1)}\left(2+\frac{|\alpha|\xi^q}{|1-\alpha\xi|}\right)+\frac{|\beta|}{\Gamma(q+2)}\left(q+1+\frac{|\alpha|}{|1-\alpha\xi|}\right)\right\}(k_0 + k_1\|u\| + k_2\|v\|),
\end{aligned}$$

and

$$|v(t)|$$
$$\leq \left\{ \frac{1}{\Gamma(p+1)} \left(2 + \frac{|\alpha_1|\theta^p}{|1-\alpha_1\theta|} \right) + \frac{|\beta_1|}{\Gamma(p+2)} \left(p+1+\frac{|\alpha_1|}{|1-\alpha_1\theta|} \right) \right\} (\lambda_0 + \lambda_1\|u\| + \lambda_2\|v\|).$$

Hence we have

$$\|u\| \leq M_1(k_0 + k_1\|u\| + k_2\|v\|),\ \|v\| \leq M_2(\lambda_0 + \lambda_1\|u\| + \lambda_2\|v\|),$$

which imply that

$$\|u\| + \|v\| = (M_1k_0 + M_2\lambda_0) + (M_1k_1 + M_2\lambda_1)\|u\| + (M_1k_2 + M_2\lambda_2)\|v\|.$$

Consequently,

$$\|(u,v)\| \leq \frac{M_1k_0 + M_2\lambda_0}{M_0},$$

for any $t \in [0,1]$, where M_0 is defined by (3.152), which proves that $\mathcal{E}$ is bounded. Thus, by Lemma 1.7, the operator T has at least one fixed point. Hence the boundary value problem (3.144)-(3.145) has at least one solution. The proof is complete. □

In the second result, we prove existence and uniqueness of solutions of the boundary value problem (3.144)-(3.145) via Banach's contraction mapping principle.

Theorem 3.35. *Assume that (3.35.2) holds and*

(3.35.1) $f,h : [0,1] \times \mathbb{R}^2 \to \mathbb{R}$ *are continuous functions and there exist constants* $m_i, n_i, i = 1,2$ *such that for all* $t \in [0,1]$ *and* $u_i, v_i \in \mathbb{R}, i = 1,2,$

$$|f(t,u_1,u_2) - f(t,v_1,v_2)| \leq m_1|u_1 - v_1| + m_2|u_2 - v_2|$$

and

$$|h(t,u_1,u_2) - h(t,v_1,v_2)| \leq n_1|u_1 - v_1| + n_2|u_2 - v_2|.$$

In addition, assume that

$$M_1(m_1 + m_2) + M_2(n_1 + n_2) < 1,$$

where M_1 *and* M_2 *are given by (3.150) and (3.151) respectively. Then the system (3.144)-(3.145) has a unique solution on [0,1].*

Proof. Define $\sup_{t\in[0,1]} f(t,0,0) = N_1 < \infty$ and $\sup_{t\in[0,1]} g(t,0,0) = N_2 < \infty$ such that

$$r \geq \frac{N_1M_1 + N_2M_2}{1 - M_1(m_1+m_2) - M_2(n_1+n_2)}.$$

We show that $TB_r \subset B_r$, where $B_r = \{(u,v) \in X \times Y : \|(u,v)\| \leq r\}$. For $(u,v) \in B_r$, we have

$$
\begin{aligned}
&|T_1(u,v)(t)| \\
&\le \int_0^t \frac{(t-s)^{q-1}}{\Gamma(q)}(|f(s,u(s),v(s))-f(t,0,0)|+|f(t,0,0)|)ds \\
&\quad+\frac{|\alpha|}{|1-\alpha\xi|}\int_0^\xi\int_0^s \frac{(s-\tau)^{q-2}}{\Gamma(q-1)}(|f(\tau,u(\tau),v(\tau))-f(t,0,0)|+|f(t,0,0)|)d\tau ds \\
&\quad+\int_0^1 \frac{(1-s)^{q-1}}{\Gamma(q)}(|f(s,u(s),v(s))-f(t,0,0)|+|f(t,0,0)|)ds \\
&\quad+|\beta|\int_0^1 g\Bigg(\int_0^s \frac{(s-\tau)^{q-2}}{\Gamma(q-1)}(|f(\tau,u(\tau),v(\tau))-f(t,0,0)|+|f(t,0,0)|)d\tau \\
&\quad+\frac{|\alpha|}{|1-\alpha\xi|}\int_0^\xi\int_0^s \frac{(s-\tau)^{q-2}}{\Gamma(q-1)}(|f(\tau,u(\tau),v(\tau))-f(t,0,0)|+|f(t,0,0)|)d\tau ds\Bigg)ds \\
&\le \Bigg\{\frac{1}{\Gamma(q+1)}+\frac{|\alpha|\xi^q}{|1-\alpha\xi|\Gamma(q+1)}+\frac{1}{\Gamma(q+1)} \\
&\quad+|\beta|\left(\frac{1}{\Gamma(q+1)}+\frac{|\alpha|}{|1-\alpha\xi|\Gamma(q+2)}\right)\Bigg\}(m_1\|u\|+m_2\|v\|+N_1) \\
&= M_1[(m_1+m_2)r+N_1].
\end{aligned}
$$

Hence

$$\|T_1(u,v)(t)\| \le M_1[(m_1+m_2)r+N_1].$$

In the same way, we can obtain that

$$\|T_2(u,v)(t)\| \le M_2[(n_1+n_2)r+N_2].$$

Consequently, $\|T(u,v)(t)\| \le r$.

Now for $(u_2,v_2),(u_1,v_1)\in X\times Y$, and for any $t\in[0,e]$, we get

$$
\begin{aligned}
&|T_1(u_2,v_2)(t)-T_1(u_1,v_1)(t)| \\
&\le \int_0^t \frac{(t-s)^{q-1}}{\Gamma(q)}|f(s,u_2(s),v_2(s))-f(s,u_1(s),v_1(s))|ds \\
&\quad+\frac{|\alpha|}{|1-\alpha\xi|}\int_0^\xi\int_0^s \frac{(s-\tau)^{q-2}}{\Gamma(q-1)}|f(\tau,u_2(\tau),v_2(\tau))-|f(\tau,u_1(\tau),v_1(\tau))|d\tau ds \\
&\quad+\int_0^1 \frac{(1-s)^{q-1}}{\Gamma(q)}|f(s,u_2(s),v_2(s))-f(s,u_1(s),v_1(s))|ds \\
&\quad+|\beta|\int_0^1 g\Bigg(\int_0^s \frac{(s-\tau)^{q-2}}{\Gamma(q-1)}|f(s,u_2(s),v_2(s))-f(s,u_1(s),v_1(s))|d\tau \\
&\quad+\frac{|\alpha|}{|1-\alpha\xi|}\int_0^\xi\int_0^s \frac{(s-\tau)^{q-2}}{\Gamma(q-1)}|f(s,u_2(s),v_2(s))-f(s,u_1(s),v_1(s))|d\tau ds\Bigg)ds \\
&\le \Bigg\{\frac{1}{\Gamma(q+1)}+\frac{|\alpha|\xi^q}{|1-\alpha\xi|\Gamma(q+1)}+\frac{1}{\Gamma(q+1)}
\end{aligned}
$$

$$+|\beta|\left(\frac{1}{\Gamma(q+1)}+\frac{|\alpha|}{|1-\alpha\xi|\Gamma(q+2)}\right)\Bigg\}(m_1\|u_2-u_1\|+m_2\|v_2-v_1\|)$$
$$\leq M_1(m_1+m_2)(\|u_2-u_1\|+\|v_2-v_1\|),$$

and consequently, we obtain

$$\|T_1(u_2,v_2)(t)-T_1(u_1,v_1)\| \leq M_1(m_1+m_2)(\|u_2-u_1\|+\|v_2-v_1\|). \quad (3.153)$$

Similarly,

$$\|T_2(u_2,v_2)(t)-T_2(u_1,v_1)\| \leq M_2(n_1+n_2)(\|u_2-u_1\|+\|v_2-v_1\|). \quad (3.154)$$

It follows from (3.153) and (3.154) that

$$\|T(u_2,v_2)(t)-T(u_1,v_1)(t)\| \leq [M_1(m_1+m_2)+M_2(n_1+n_2)](\|u_2-u_1\|+\|v_2-v_1\|).$$

Since $M_1(m_1+m_2)+M_2(n_1+n_2)<1$, therefore, T is a contraction operator. So, by Banach's fixed point theorem, the operator T has a unique fixed point, which is the unique solution of problem (3.144)-(3.145). This completes the proof. □

Example 3.20. Consider the following system of fractional boundary value problem

$$\begin{cases} {}^cD^{3/2}x(t)=\dfrac{1}{4(t+2)^2}\dfrac{|x(t)|}{1+|x(t)|}+1+\dfrac{1}{32}\sin^2 y(t), & t\in[0,1],\\ {}^cD^{3/2}y(t)=\dfrac{1}{32\pi}\sin(2\pi x(t))+\dfrac{|y(t)|}{16(1+|y(t)|)}+\dfrac{1}{2}, & t\in[0,1],\\ x'(0)=\dfrac{1}{2}\displaystyle\int_0^{1/3}x'(s)ds, \quad x(1)=\dfrac{1}{3}\int_0^1 g(x'(s))ds, \\ y'(0)=\dfrac{4}{5}\displaystyle\int_0^{1/4}y'(s)ds, \quad y(1)=\dfrac{3}{4}\int_0^1 g(y'(s))ds. \end{cases} \quad (3.155)$$

Here $q=p=3/2$, $\alpha=1/2$, $\alpha_1=4/5$, $\xi=1/3$, $\theta=1/4$, $\beta=1/3$, $\beta_1=3/4$, $g(v)=\begin{cases}\sqrt{v}, & |v|\geq 1,\\ v^2, & |v|<1.\end{cases}$ $f(t,u,v)=\dfrac{1}{4(t+2)^2}\dfrac{|u|}{1+|u|}+1+\dfrac{1}{32}\sin^2 v$, and $h(t,u,v)=\dfrac{1}{32\pi}\sin(2\pi u)+\dfrac{|v|}{16(1+|v|)}+\dfrac{1}{2}$. With the given data, we find that $M_1\approx 1.9027815$, $M_2\approx 1.6365646$. Note that $|f(t,u_1,u_2)-f(t,v_1,v_2)|\leq \frac{1}{16}|u_1-u_2|+\frac{1}{16}|v_1-v_2|$, $|g(t,u_1,u_2)-g(t,v_1,v_2)|\leq \frac{1}{16}|u_1-u_2|+\frac{1}{16}|v_1-v_2|$, and $M_1(m_1+m_2)+M_2(n_1+n_2)\approx 0.4424181<1$. Thus all the conditions of Theorem 3.35 are satisfied and consequently, its conclusion applies to the problem (3.155).

3.9 A study of fractional differential equations and inclusions with nonlocal Erdélyi-Kober type integral boundary conditions

In this section, we introduce a new class of boundary value problems of fractional differential equations and inclusions supplemented with Erdélyi-Kober fractional integral conditions. In precise terms, we consider the following nonlocal problems:

$$\begin{cases} {}^cD^q x(t) = f(t, x(t)), \quad t \in [0,T], \\ x(0) = g(x(t)), \quad x(T) = \alpha I_\eta^{\gamma,\delta} x(\xi), \quad 0 < \xi < T, \end{cases} \tag{3.156}$$

and

$$\begin{cases} {}^cD^q x(t) \in F(t, x(t)), \quad t \in [0,T], \\ x(0) = g(x), \quad x(T) = \alpha I_\eta^{\gamma,\delta} x(\xi), \quad 0 < \xi < T, \end{cases} \tag{3.157}$$

where ${}^cD^q$ is the Caputo fractional derivative operator of order $1 < q \leq 2$, $f : [0,T] \times \mathbb{R} \to \mathbb{R}$ is a continuous function, $g : C([0,T], \mathbb{R}) \to \mathbb{R}$ is a continuous function, $I_\eta^{\gamma,\delta}$ is the Erdélyi-Kober fractional integral of order $\delta > 0$ with $\eta > 0$, $\gamma \in \mathbb{R}$ and $\alpha \in \mathbb{R}$, $F : [0,T] \times \mathbb{R} \to \mathcal{P}(\mathbb{R})$ is a multi-valued function, ($\mathcal{P}(\mathbb{R})$ is the family of all non-empty subsets of $\mathbb{R}$).

The following lemma plays a key role in the forthcoming analysis.

Lemma 3.12. *For any $y \in C([0,T], \mathbb{R})$, the boundary value problem*

$$\begin{cases} {}^cD^q x(t) = y(t), \quad 1 < q \leq 2, \ t \in [0,T], \\ x(0) = g(x), \quad x(T) = \alpha I_\eta^{\gamma,\delta} x(\xi), \quad 0 < \xi < T, \end{cases} \tag{3.158}$$

is equivalent to the following integral equation

$$\begin{aligned} x(t) = {} & J^q y(t) + \frac{t}{\Lambda}\Big\{\alpha I_\eta^{\gamma,\delta} J^q y(\xi) - J^q y(T)\Big\} \\ & + \Big[1 + \frac{t}{\Lambda} \frac{\alpha\Gamma(\gamma+1) - \Gamma(\gamma+\delta+1)}{\Gamma(\gamma+\delta+1)}\Big] g(x), \end{aligned} \tag{3.159}$$

where it is assumed that

$$\Lambda = T - \alpha\xi \frac{\Gamma(\gamma + 1/\eta + 1)}{\Gamma(\gamma + 1/\eta + \delta + 1)} \neq 0. \tag{3.160}$$

Proof. It is well known that the general solution of the fractional differential equation in (3.158) can be written as

$$x(t) = c_0 + c_1 t + J^q y(t), \tag{3.161}$$

where $c_0, c_1 \in \mathbb{R}$ are arbitrary constants. Using the first condition ($x(0) = g(x)$) given by (3.158) in (3.161), we get $c_0 = g(x)$. Applying the Erdélyi-Kober fractional integral operator of order $\delta > 0$ with $\eta > 0$ and $\gamma \in \mathbb{R}$ on (3.161) and using Lemma 1.6, we obtain

$$I_\eta^{\gamma,\delta} x(t) = I_\eta^{\gamma,\delta} J^q y(t) + c_1 t \frac{\Gamma(\gamma + 1/\eta + 1)}{\Gamma(\gamma + 1/\eta + \delta + 1)} + c_0 \frac{\Gamma(\gamma+1)}{\Gamma(\gamma+\delta+1)}. \tag{3.162}$$

Using (3.162) in the second condition of (3.158), we get

$$J^q y(T) + c_1 T + c_0 = \alpha I_\eta^{\gamma,\delta} J^q y(\xi) + c_1 \alpha\xi \frac{\Gamma(\gamma + 1/\eta + 1)}{\Gamma(\gamma + 1/\eta + \delta + 1)} + c_0 \frac{\alpha\Gamma(\gamma+1)}{\Gamma(\gamma+\delta+1)},$$

which, together with $c_0 = g(x)$, yields

$$c_1 = \frac{1}{\Lambda}\bigg\{\alpha I_\eta^{\gamma,\delta} J^q y(t) - J^q y(T) + g(x) \frac{\alpha\Gamma(\gamma+1) - \Gamma(\gamma+\delta+1)}{\Gamma(\gamma+\delta+1)}\bigg\}.$$

Substituting the values of c_0, c_1 in (3.161), we get (3.159). The converse follows by direct computation. This completes the proof. □

In the sequel, we use the following expressions:

$$J^q f(s,x(s))(z) = \frac{1}{\Gamma(q)} \int_0^z (z-s)^{q-1} f(s,x(s))ds, \quad z \in \{t,T\},$$

for $t \in [0,T]$ and

$$I_\eta^{\gamma,\delta} J^q f(s,x(s))(\xi) = \frac{\eta \xi^{-\eta(\delta+\gamma)}}{\Gamma(q)\Gamma(\delta)} \int_0^\xi \int_0^r \frac{r^{\eta\gamma+\eta-1}(r-s)^{q-1}}{(\xi^\eta - r^\eta)^{1-\delta}} f(s,x(s))ds\, dr,$$

where $\xi \in (0,T)$.

3.9.1 *Existence and uniqueness results for problem (3.156)*

We denote by $\mathcal{C} = C([0,T],\mathbb{R})$ the Banach space of all continuous functions from $[0,T] \to \mathbb{R}$ endowed with a topology of uniform convergence with the norm defined by $\|x\| = \sup\{|x(t)| : t \in [0,T]\}$. Also by $L^1([0,T],\mathbb{R})$ we denote the Banach space of measurable functions $x : [0,T] \to \mathbb{R}$ which are Lebesgue integrable and normed by $\|x\|_{L^1} = \int_0^T |x(t)|dt$.

In view of Lemma 3.12, we define an operator $\mathcal{P} : \mathcal{C} \to \mathcal{C}$ by

$$\begin{aligned}(\mathcal{P}x)(t) &= J^q f(s,x(s))(t) + \frac{t}{\Lambda}\Big\{\alpha I_\eta^{\gamma,\delta} J^q f(s,x(s))(\xi) - J^q f(s,x(s))(T)\Big\} \\ &\quad + \left[1 + \frac{t}{\Lambda}\frac{\alpha\Gamma(\gamma+1) - \Gamma(\gamma+\delta+1)}{\Gamma(\gamma+\delta+1)}\right] g(x), \quad t \in [0,T]. \end{aligned} \tag{3.163}$$

For the sake of convenience, we set

$$p_0 := \frac{T^q}{\Gamma(q+1)}\left\{1 + \frac{T}{|\Lambda|}\left(1 + \frac{\alpha \xi^q T^{-q} \Gamma(\gamma + q/\eta + 1)}{\Gamma(\gamma + q/\eta + \delta + 1)}\right)\right\}, \tag{3.164}$$

and

$$k_0 := 1 + \frac{T}{|\Lambda|}\frac{|\alpha|\Gamma(\gamma+1) + \Gamma(\gamma+\delta+1)}{\Gamma(\gamma+\delta+1)}. \tag{3.165}$$

Theorem 3.36. *Let $f : [0,T] \times \mathbb{R} \to \mathbb{R}$ be a continuous function. Assume that*

(3.36.1) $|f(t,x) - f(t,y)| \le L\|x-y\|, \forall t \in [0,T],\ L > 0,\ x,y \in \mathbb{R}$;
(3.36.2) $g : C([0,T],\mathbb{R}) \to \mathbb{R}$ *is a continuous function satisfying the condition:*

$$|g(u) - g(v)| \le \ell\|u-v\|, \ \ell < k_0^{-1}, \ \ \forall\, u,v \in C([0,T],\mathbb{R}),\ \ell > 0;$$

(3.36.3) $\gamma := Lp_0 + k_0\ell < 1$.

Then the boundary value problem (3.156) has a unique solution on $[0,T]$.

Proof. For $x,y \in \mathcal{C}$ and for each $t \in [0,T]$, from the definition of $\mathcal{P}$ and assumptions (3.36.1) and (3.36.2), we obtain

$$\begin{aligned}&|(\mathcal{P}x)(t) - (\mathcal{P}y)(t)| \\ &\le \sup_{t\in[0,T]}\Bigg\{ J^q|f(s,x(s)) - f(s,y(s))|(t) + \frac{t}{|\Lambda|} J^q|f(s,x(s)) - f(s,y(s))|(T)\end{aligned}$$

$$
\begin{aligned}
&+\frac{|\alpha|t}{|\Lambda|}I_\eta^{\gamma,\delta}J^q|f(s,x(s))-f(s,y(s))|(\xi)\\
&+\left|1+\frac{t}{\Lambda}\frac{|\alpha|\Gamma(\gamma+1)-\Gamma(\gamma+\delta+1)}{\Gamma(\gamma+\delta+1)}\right||g(x)-g(y)|\Bigg\}\\
\le\ & L\|x-y\|J^q(1)(T)+L\|x-y\|\frac{T}{|\Lambda|}J^q(1)(T)+L\|x-y\|\frac{|\alpha|T}{|\Lambda|}I_\eta^{\gamma,\delta}J^q(1)(\xi)\\
&+\left(1+\frac{T}{|\Lambda|}\frac{|\alpha|\Gamma(\gamma+1)+\Gamma(\gamma+\delta+1)}{\Gamma(\gamma+\delta+1)}\right)|g(x)-g(y)|\\
\le\ & (Lp_0+k_0\ell)\|x-y\|.
\end{aligned}
$$

Hence

$$\|(\mathcal{P}x)-(\mathcal{P}y)\|\le\gamma\|x-y\|.$$

As $\gamma<1$ by (3.36.3), the operator $\mathcal{P}$ is a contraction map from the Banach space $\mathcal{C}$ into itself. Hence the conclusion of the theorem follows by the contraction mapping principle (Banach fixed point theorem). □

Example 3.21. Consider the following fractional boundary value problem

$$
\begin{cases}
{}^cD^{3/2}x(t)=\dfrac{\sin^2(\pi t)}{2(e^t+9)}\left(\dfrac{|x(t)|}{|x(t)|+1}+1\right)|x(t)|+\dfrac{\sqrt{3}}{4}, & t\in[0,1],\\[2ex]
x(0)=\dfrac{1}{2}+\dfrac{1}{12}\tan^{-1}(x(1/8)),\quad x(1)=\dfrac{1}{2}I_{5/2}^{1/2,3/2}x(3/4).
\end{cases}
\tag{3.166}
$$

Here, $q=3/2,\ T=1,\ \alpha=1/2,\ \xi=3/4,\ \eta=5/2,\ \gamma=1/2,\ \delta=3/2,$

$$f(t,x)=\frac{\sin^2(\pi t)}{2(e^t+9)}\Big(\frac{|x|}{|x|+1}+1\Big)|x|+\frac{\sqrt{3}}{4},\quad g(x)=(1/12)\tan^{-1}(x(4/5)).$$

Since $|f(t,x)-f(t,y)|\le\frac{1}{10}\|x-y\|, |g(x)-g(y)|\le\frac{1}{12}\|x-y\|$, therefore, (3.36.1) and (3.36.2) are respectively satisfied with $L=1/10$ and $\ell=1/12$. Using the given values, it is found that $\Lambda\approx0.879021,\ p_0\approx1.686284,\ k_0=2.389679$. Clearly $\gamma=Lp_0+k_0\ell\approx0.367768<1$. Thus, the conclusion of Theorem 3.36 applies and the boundary value problem (3.166) has a solution on $[0,1]$.

3.9.2 *Existence results for problem (3.157)*

To prove our main result in this section, we use the nonlinear alternative for contractive maps.

Definition 3.6. A function $x\in C^2([0,T],\mathbb{R})$ is a solution of the problem (3.157) if $x(0)=g(x),\ x(T)=\alpha I_\eta^{\gamma,\delta}x(\xi)$, and there exists a function $f\in L^1([0,T],\mathbb{R})$ such that $f(t)\in F(t,x(t))$ for a.e. $t\in[0,T]$ and

$$
\begin{aligned}
x(t)=J^qf(s)(t)&+\frac{t}{\Lambda}\Big\{\alpha I_\eta^{\gamma,\delta}J^qf(s)(\xi)-J^qf(s)(T)\Big\}\\
&+\Big[1+\frac{t}{\Lambda}\frac{\alpha\Gamma(\gamma+1)-\Gamma(\gamma+\delta+1)}{\Gamma(\gamma+\delta+1)}\Big]g(x).
\end{aligned}
$$

Theorem 3.37. *Assume that (3.36.2) holds. In addition we suppose that:*

(3.37.1) $F : [0, T] \times \mathbb{R} \to \mathcal{P}_{cp,c}(\mathbb{R})$ *is* L^1-*Carathéodory multi-valued map;*

(3.37.2) there exists a continuous non-decreasing function $\psi : [0, \infty) \to (0, \infty)$ *and a function* $p \in C([0, T], \mathbb{R}^+)$ *such that*

$$\|F(t, x)\|_{\mathcal{P}} := \sup\{|y| : y \in F(t, x)\} \le p(t)\psi(\|x\|) \text{ for each } (t, x) \in [0, T]\times\mathbb{R};$$

(3.37.3) there exists a number $M > 0$ *such that*

$$\frac{(1 - k_0\ell)M}{\psi(M)p_0\|p\|} > 1, \tag{3.167}$$

where p_0, k_0 *are defined in (3.164) and (3.165) respectively.*

Then the boundary value problem (3.157) has at least one solution on $[0, T]$.

Proof. To transform the problem (3.157) into a fixed point problem, we define an operator $\mathcal{N} : \mathcal{C} \longrightarrow \mathcal{P}(\mathcal{C})$ by

$$\mathcal{N}(x) = \left\{ \begin{array}{l} h \in \mathcal{C} : \\ h(t) = \left\{ \begin{array}{l} J^q f(s)(t) + \dfrac{t}{\Lambda}\Big\{\alpha I_\eta^{\gamma,\delta} J^q f(s)(\xi) - J^q f(s)(T)\Big\} \\ +\Big[1 + \dfrac{t}{\Lambda}\dfrac{\alpha\Gamma(\gamma + 1) - \Gamma(\gamma + \delta + 1)}{\Gamma(\gamma + \delta + 1)}\Big]g(x), \end{array} \right. \end{array} \right\}$$

for $f \in S_{F,x}$.

Next we introduce an operators $\mathcal{A} : \mathcal{C} \longrightarrow \mathcal{C}$ by

$$\mathcal{A}x(t) = \left[1 + \frac{t}{\Lambda}\frac{\alpha\Gamma(\gamma + 1) - \Gamma(\gamma + \delta + 1)}{\Gamma(\gamma + \delta + 1)}\right]g(x), \tag{3.168}$$

and a multi-valued operator $\mathcal{B} : \mathcal{C} \longrightarrow \mathcal{P}(\mathcal{C})$ by

$$\mathcal{B}(x) = \left\{ \begin{array}{l} h \in \mathcal{C} : \\ h(t) = J^q f(s)(t) + \dfrac{t}{\Lambda}\Big\{\alpha I_\eta^{\gamma,\delta} J^q f(s)(\xi) - J^q f(s)(T)\Big\}. \end{array} \right\} \tag{3.169}$$

Observe that $\mathcal{N} = \mathcal{A} + \mathcal{B}$. We shall show that the operators $\mathcal{A}$ and $\mathcal{B}$ satisfy all the conditions of Theorem 1.18 on $[0, T]$. The proof consists of several steps and claims.

Step 1. We show that $\mathcal{A}$ *is a contraction on* $\mathcal{C}$. For $x, y \in \mathcal{C}$, we have

$$\begin{aligned} |\mathcal{A}x(t) - \mathcal{A}y(t)| &= \left|1 + \frac{t}{\Lambda}\frac{\alpha\Gamma(\gamma + 1) - \Gamma(\gamma + \delta + 1)}{\Gamma(\gamma + \delta + 1)}\right||g(x) - g(y)| \\ &\le \left(1 + \frac{T}{|\Lambda|}\frac{|\alpha|\Gamma(\gamma + 1) + \Gamma(\gamma + \delta + 1)}{\Gamma(\gamma + \delta + 1)}\right)|g(x) - g(y)|, \\ &\le k_0\ell\|x - y\|, \end{aligned}$$

which, on taking supremum over $t \in [0, T]$, yields

$$\|\mathcal{A}x - \mathcal{A}y\| \le L_0\|x - y\|, \quad L_0 = k_0\ell < 1.$$

This shows that $\mathcal{A}$ is a contraction as $L_0 < 1$.

Step 2. *$\mathcal{B}$ is compact and convex valued and it is completely continuous.* This will be established in several claims.

CLAIM I. *$\mathcal{B}$ maps bounded sets into bounded sets in $\mathcal{C}$.* Let $B_r = \{x \in \mathcal{C} : \|x\| \le r\}$ be a bounded set in $\mathcal{C}$. Then, for each $h \in \mathcal{B}(x), x \in B_r$, there exists $f \in S_{F,x}$ such that

$$h(t) = J^q f(s)(t) + \frac{t}{\Lambda}\Big\{\alpha I_\eta^{\gamma,\delta} J^q f(s)(\xi) - J^q f(s)(T)\Big\}.$$

Then, for $t \in [0,T]$, we have

$$\begin{aligned}
|h(t)| &\le J^q|f(s)|(T) + \frac{T}{|\Lambda|}\{|\alpha| I_\eta^{\gamma,\delta} J^q |f(s)|(\xi) - J^q|f(s)|(T)\}\\
&\le \psi(\|x\|) J^q p(s)(T) + \psi(\|x\|)\frac{|\alpha| T}{|\Lambda|} I_\eta^{\gamma,\delta} J^q p(s)(\xi) + \psi(\|x\|)\frac{T}{|\Lambda|} J^q p(s)(T)\\
&\le \|p\|\psi(\|x\|)\bigg\{\frac{T^q}{\Gamma(q+1)} + \frac{T}{|\Lambda|}\frac{T^q}{\Gamma(q+1)} + \frac{T}{|\Lambda|}\frac{|\alpha|\xi^q T^{-q}\Gamma(\gamma+q/\eta+1)}{\Gamma(q+1)\Gamma(\gamma+q/\eta+\delta+1)}\bigg\}.
\end{aligned}$$

Thus,

$$\|h\| \le \frac{T^q\|p\|\psi(r)}{\Gamma(q+1)}\bigg\{1 + \frac{T}{|\Lambda|}\bigg(1 + \frac{|\alpha|\xi^q T^{-q}\Gamma(\gamma+q/\eta+1)}{\Gamma(\gamma+q/\eta+\delta+1)}\bigg)\bigg\}.$$

CLAIM II. *$\mathcal{B}$ maps bounded sets into equi-continuous sets.* Let $\tau_1, \tau_2 \in [0,T]$ with $\tau_1 < \tau_2$ and $x \in B_r$. Then, for each $h \in \mathcal{B}(x)$, we obtain

$$\begin{aligned}
&|h(\tau_2) - h(\tau_1)|\\
&\le |J^q f(s)(\tau_2) - J^q f(s)(\tau_1)| + \frac{|\tau_2-\tau_1|}{|\Lambda|} J^q|f(s)|(T)\\
&\quad + \frac{|\alpha||\tau_2-\tau_1|}{|\Lambda|}|I_\eta^{\gamma,\delta} J^q|f(s)|(\xi)\\
&\le \frac{\psi(r)}{\Gamma(q)}\bigg|\int_0^{\tau_1}[(\tau_2-s)^{q-1} - (\tau_1-s)^{q-1}]p(s)ds + \int_{\tau_1}^{\tau_2}(\tau_2-s)^{q-1}p(s)ds\bigg|\\
&\quad + \frac{|\tau_2-\tau_1|}{|\Lambda|}\bigg(J^q|f(s)|(T) + |\alpha||I_\eta^{\gamma,\delta} J^q|f(s)|(\xi)\bigg)\\
&\le \frac{\psi(r)}{\Gamma(q+1)}[2(\tau_2-\tau_1)^q + |\tau_2^q - \tau_1^q|]\\
&\quad + \frac{|\tau_2-\tau_1|}{|\Lambda|}\bigg(J^q|f(s)|(T) + |\alpha||I_\eta^{\gamma,\delta} J^q|f(s)|(\xi)\bigg).
\end{aligned}$$

Obviously the right hand side of the above inequality tends to zero independently of $x \in B_r$ as $\tau_2 - \tau_1 \to 0$. Therefore it follows by the Ascoli-Arzelá theorem that $\mathcal{B} : \mathcal{C} \to \mathcal{P}(\mathcal{C})$ is completely continuous.

In our next step, we show that $\mathcal{B}$ is upper semi-continuous. By Lemma 1.1, $\mathcal{B}$ will be upper semi-continuous if we establish that it has a closed graph, since $\mathcal{B}$ is already shown to be completely continuous. We prove it in the following claim.
CLAIM III. *$\mathcal{B}$ has a closed graph.* Let $x_n \to x_*, h_n \in \mathcal{B}(x_n)$ and $h_n \to h_*$. Then we need to show that $h_* \in \mathcal{B}(x_*)$. Associated with $h_n \in \mathcal{B}(x_n)$, there exists $f_n \in S_{F,x_n}$ such that, for each $t \in [0,T]$,

$$h(t) = J^q f_n(s)(t) + \frac{t}{\Lambda}\Big\{\alpha I_\eta^{\gamma,\delta} J^q f_n(s)(\xi) - J^q f_n(s)(T)\Big\}.$$

Thus it suffices to show that there exists $f_* \in S_{F,x_*}$ such that for each $t \in [0,T]$,

$$h_*(t) = J^q f_n(s)(t) + \frac{t}{\Lambda}\Big\{\alpha I_\eta^{\gamma,\delta} J^q f_*(s)(\xi) - J^q f_*(s)(T)\Big\}.$$

Let us consider the linear operator $\Theta : L^1([0,T],\mathbb{R}) \to \mathcal{C}$ given by

$$f \mapsto \Theta(f)(t) = J^q f_n(s)(t) + \frac{t}{\Lambda}\Big\{\alpha I_\eta^{\gamma,\delta} J^q f_n(s)(\xi) - J^q f_n(s)(T)\Big\}.$$

Observe that

$$\begin{aligned}&\|h_n(t) - h_*(t)\|\\ &= \Big\| J^q(f_n(s) - f_*(s))(t)\\ &\quad + \frac{t}{\Lambda}\Big\{\alpha I_\eta^{\gamma,\delta} J^q(f_n(s) - f_*(s))(\xi) - J^q(f_n(s) - f_*(s))(T)\Big\}\Big\| \to 0,\end{aligned}$$

as $n \to \infty$.

Thus, it follows by Lemma 1.2 that $\Theta \circ S_F$ is a closed graph operator. Further, we have that $h_n(t) \in \Theta(S_{F,x_n})$. Since $x_n \to x_*$, therefore

$$h_*(t) = J^q f_n(s)(t) + \frac{t}{\Lambda}\Big\{\alpha I_\eta^{\gamma,\delta} J^q f_*(s)(\xi) - J^q f_*(s)(T)\Big\},$$

for some $f_* \in S_{F,x_*}$. Hence $\mathcal{B}$ has a closed graph (and therefore has closed values). In consequence, the operator $\mathcal{B}$ is upper semi-continuous and compact.

Thus the operators $\mathcal{A}$ and $\mathcal{B}$ satisfy all the conditions of Theorem 1.18 and hence its conclusion implies either condition (i) or condition (ii) holds. We show that the conclusion (ii) is not possible. If $x \in \lambda\mathcal{A}(x) + \lambda\mathcal{B}(x)$ for $\lambda \in (0,1)$, then there exists $f \in S_{F,x}$ such that

$$\begin{aligned}x(t) &= J^q f_n(s)(t) + \frac{t}{\Lambda}\Big\{\alpha I_\eta^{\gamma,\delta} J^q f(s)(\xi) - J^q f(s)(T)\Big\}\\ &\quad + \left[1 + \frac{t}{\Lambda}\frac{\alpha\Gamma(\gamma+1) - \Gamma(\gamma+\delta+1)}{\Gamma(\gamma+\delta+1)}\right] g(x), \quad t \in [0,T].\end{aligned}$$

Following the method for proof of Claim I, we get

$$\begin{aligned}|x(t)| &\le \frac{T^q\|p\|\psi(\|x\|)}{\Gamma(q+1)}\left\{1 + \frac{T}{|\Lambda|}\left(1 + \frac{|\alpha|\xi^q T^{-q}\Gamma(\gamma+q/\eta+1)}{\Gamma(\gamma+q/\eta+\delta+1)}\right)\right\}\\ &\quad + \left(1 + \frac{T}{|\Lambda|}\frac{|\alpha|\Gamma(\gamma+1) + \Gamma(\gamma+\delta+1)}{\Gamma(\gamma+\delta+1)}\right)\ell\|x\|.\end{aligned}$$

Thus

$$\|x\| \leq \psi(\|x\|)p_0\|p\| + k_0\ell\|x\|. \tag{3.170}$$

If condition (ii) of Theorem 1.18 holds, then there exists $\lambda \in (0,1)$ and $x \in \partial B_r$ with $x = \lambda\mathcal{N}(x)$. Then, x is a solution of (3.157) with $\|x\| = M$. Now, by the inequality (3.170), we get

$$\frac{(1-k_0\ell)M}{\psi(M)p_0\|p\|} \leq 1,$$

which contradicts (3.167). Hence, $\mathcal{N}$ has a fixed point in $[0,T]$ by Theorem 1.18, and consequently the problem (3.157) has a solution. This completes the proof. □

Example 3.22. Consider the following boundary value problem of fractional differential inclusions

$$\begin{cases} D^{3/2}x(t) \in F(t,x(t)), \ t \in [0,1], \ 0 < t < 1, \\ x(0) = \dfrac{1}{8}x(4/5), \quad x(1) = \dfrac{1}{2}I_{5/2}^{1/2,3/2}x(3/4), \end{cases} \tag{3.171}$$

where

$$F(t,x(t)) = \Big[\frac{2}{\sqrt{t^2+64}}\Big(\frac{|x(t)|}{2}\Big(\frac{|x(t)|}{|x(t)|+1}+1\Big)+\frac{1}{5}\Big), \frac{e^{-t}}{(10+t)}\Big(\sin x(t)+\frac{1}{15}\Big)\Big].$$

Clearly $|F(t,x)| \leq p(t)\psi(|x|)$, where $p(t) = 2/\sqrt{t^2+64}$, $\psi_1(|x|) = |x| + 1/5$ and $\ell = 1/8$. Using the values: $\Lambda \approx 0.879021$, $p_0 \approx 1.686284$, $k_0 = 2.389679$ (see Example 3.21) and the condition (3.37.3), we find that $M > M_1 \simeq 0.301425$. Since the hypothesis of Theorem 3.37 is satisfied, the problem (3.171) has a solution on $[0,1]$.

3.10 Existence results for fractional differential inclusions with Erdélyi-Kober fractional integral conditions

In order to enrich the work on fractional-order boundary value problems involving integral boundary conditions further, we consider a problem consisting of fractional differential inclusions and Erdélyi-Kober fractional integral boundary conditions. Precisely, we investigate the following problem:

$$\begin{cases} D^q x(t) \in F(t,x(t)), \quad 0 < t < T, \quad 1 < q \leq 2, \\ x(0) = 0, \\ \alpha x(T) = \displaystyle\sum_{i=1}^m \beta_i \frac{\eta t^{-\eta_i(\delta_i+\gamma_i)}}{\Gamma(\delta_i)} \int_0^{\xi_i} \frac{s^{\eta_i\gamma_i+\eta_i-1}x(s)}{(t^{\eta_i}-s^{\eta_i})^{1-\delta_i}}ds := \sum_{i=1}^m \beta_i I_{\eta_i}^{\gamma_i,\delta_i}x(\xi_i), \end{cases} \tag{3.172}$$

where $1 < q \leq 2$, D^q is the standard Riemann-Liouville fractional derivative operator of order q, $I_{\eta_i}^{\gamma_i,\delta_i}$ is the Erdélyi-Kober fractional integral of order $\delta_i > 0$

with $\eta_i > 0$ and $\gamma_i \in \mathbb{R}$, $i = 1, 2, \dots, m$, $F : [0, T] \times \mathbb{R} \to \mathcal{P}(\mathbb{R})$ is a multi-valued map, $\mathcal{P}(\mathbb{R})$ is the family of all non-empty subsets of $\mathbb{R}$, and $\alpha, \beta_i \in \mathbb{R}$, $\xi_i \in (0, T)$, $i = 1, 2, \dots, m$ are given constants.

Lemma 3.13. *Let $1 < q \le 2$, $\delta_i, \eta_i > 0$, $\alpha, \gamma_i, \beta_i \in \mathbb{R}$, $\xi_i \in (0, T)$, $i = 1, 2, \dots, m$ and $h \in C([0, T], \mathbb{R})$. Then the linear Riemann-Liouville fractional differential equation subject to the Erdélyi-Kober fractional integral boundary conditions:*

$$\begin{cases} D^q x(t) = h(t), \quad t \in (0, T), \\ x(0) = 0, \qquad \alpha x(T) = \displaystyle\sum_{i=1}^{m} \beta_i I_{\eta_i}^{\gamma_i, \delta_i} x(\xi_i), \end{cases} \tag{3.173}$$

is equivalent to the following integral equation

$$x(t) = J^q h(t) - \frac{t^{q-1}}{\Lambda} \left(\alpha J^q h(T) - \sum_{i=1}^{m} \beta_i I_{\eta_i}^{\gamma_i, \delta_i} J^q h(\xi_i) \right), \tag{3.174}$$

where it is assumed that

$$\Lambda := \alpha T^{q-1} - \sum_{i=1}^{m} \frac{\beta_i \xi_i{}^{q-1} \Gamma(\gamma_i + (q-1)/\eta_i + 1)}{\Gamma(\gamma_i + (q-1)/\eta_i + \delta_i + 1)} \neq 0. \tag{3.175}$$

Proof. Solving the linear fractional differential equation in (3.173), we get an integral equation

$$x(t) = J^q h(t) - c_1 t^{q-1} - c_2 t^{q-2}, \tag{3.176}$$

where $c_1, c_2 \in \mathbb{R}$. The first condition ($x(0) = 0$) of (3.173) implies that $c_2 = 0$. Applying the Erdélyi-Kober fractional integral operator of order $\delta_i > 0$ with $\eta_i > 0$ and $\gamma_i \in \mathbb{R}$ on (3.176) and using Lemma 1.6, we obtain

$$I_{\eta_i}^{\gamma_i, \delta_i} x(t) = I_{\eta_i}^{\gamma_i, \delta_i} J^q h(t) - c_1 \frac{t^{q-1} \Gamma(\gamma + (q-1)/\eta + 1)}{\Gamma(\gamma + (q-1)/\eta + \delta + 1)},$$

which, together with the second condition of (3.173), yields

$$\alpha J^q h(T) - c_1 \alpha T^{q-1} = \sum_{i=1}^{m} \beta_i I_{\eta_i}^{\gamma_i, \delta_i} J^q h(\xi_i) - c_1 \sum_{i=1}^{m} \frac{\beta_i \xi_i^{q-1} \Gamma(\gamma_i + (q-1)/\eta_i + 1)}{\Gamma(\gamma_i + (q-1)/\eta_i + \delta_i + 1)}.$$

Thus

$$c_1 = \frac{1}{\Lambda} \left(\alpha J^q h(T) - \sum_{i=1}^{m} \beta_i I_{\eta_i}^{\gamma_i, \delta_i} J^q h(\xi_i) \right),$$

where Λ is given by (3.175). Substituting the values of c_1 and c_2 in (3.176), we obtain the desired solution (3.174). The converse follows by direct computation. □

3.10.1 *Existence result via endpoint theory*

In this subsection we prove the existence of solutions for the problem (3.172) via endpoint theory. Based on Lemma 3.13, letting $\mathcal{U} = C([0,T],\mathbb{R})$, we introduce an operator $\mathcal{G} : \mathcal{U} \to \mathcal{P}(\mathcal{U})$ as follows:

$$\mathcal{G}(u) = \left\{ \begin{array}{l} h \in \mathcal{U} : \\ h(t) = \left\{ \begin{array}{l} J^q v(s)(t) \\ -\dfrac{t^{q-1}}{\Lambda}\left(\alpha J^q v(s)(T) - \displaystyle\sum_{i=1}^{m} \beta_i I_{\eta_i}^{\gamma_i,\delta_i} J^q v(s)(\xi_i)\right), \end{array} \right. \end{array} \right\} \tag{3.177}$$

for $v \in S_{F,u}$, where $S_{F,u}$ denotes the set of selections of F defined by

$$S_{F,u} := \{v \in L^1(J,\mathbb{R}) : v(t) \in F(t,u(t)) \text{ for a.e. } t \in [0,T]\}.$$

Now we are in a position to present our main existence result.

Theorem 3.38. *Suppose that* $\psi : [0,\infty) \to [0,\infty)$ *is a non-decreasing upper semi-continuous mapping such that* $\liminf_{t\to\infty}(t - \psi(t)) > 0$ *and* $\psi(t) < t$ *for all* $t > 0$*. Also, let* $F : [0,T] \times \mathbb{R} \to \mathcal{P}_{cp}(\mathbb{R})$ *be an integrable bounded multifunction such that* $F(\cdot,u) : [0,T] \to \mathcal{P}_{cp}(\mathbb{R})$ *is measurable for all* $u \in \mathbb{R}$*, where* $\mathcal{P}_{cp}(\mathbb{R})$ *is the collection of all non-empty compact subsets of* $\mathbb{R}$*. Assume that there exists a function* $\ell \in C([0,T],[0,\infty))$ *such that*

$$H_d(F(t,u(t)) - F(t,v(t))) \leq \left(\|\ell\|\theta\right)^{-1}\ell(t)\psi(|u(t) - v(t)|),$$

where

$$\theta = \frac{T^q}{\Gamma(q+1)} + \frac{|\alpha|T^{2q-1}}{|\Lambda|\Gamma(q+1)} + \frac{T^{q-1}}{|\Lambda|\Gamma(q+1)}\sum_{i=1}^{m}\frac{|\beta_i|\xi_i^q\Gamma(\gamma_i + (q/\eta_i)+1)}{\Gamma(\gamma_i + (q/\eta_i) + \delta_i + 1)}. \tag{3.178}$$

If the multifunction $\mathcal{G}$ *has the approximate endpoint property, then the inclusion problem (3.172) has a solution.*

Proof. Our proof will be complete when we establish that the multifunction $\mathcal{G} : \mathcal{U} \to \mathcal{P}(\mathcal{U})$ defined by (3.177) has an endpoint. To do this, we show that the operator $\mathcal{G}(u)$ is a closed subset of $\mathcal{P}(\mathcal{U})$ for all $u \in \mathcal{U}$. Since the multi-valued map $t \mapsto F(t,u(t))$ is measurable and has closed values for all $u \in \mathcal{U}$, so it has measurable selection. This implies that $S_{F,u}$ is non-empty for all $u \in \mathcal{U}$. Let $\{z_n\}_{n\geq 1}$ be a sequence in $\mathcal{G}(u)$ with $z_n \to z$ for $u \in \mathcal{U}$. For every $n \in \mathbb{N}$, choose $v_n \in S_{F,u_n}$ such that

$$z_n(t) = J^q v_n(s)(t) - \frac{t^{q-1}}{\Lambda}\left(\alpha J^q v_n(s)(T) - \sum_{i=1}^{m}\beta_i I_{\eta_i}^{\gamma_i,\delta_i} J^q v_n(s)(\xi_i)\right).$$

By compactness of F, the sequence $\{v_n\}_{n\geq 1}$ has a subsequence which converges to some $v \in L^1([0,T])$. We denote this subsequence again by $\{v_n\}_{n\geq 1}$. It is clear that $v \in S_{F,u}$ and for all $t \in [0,T]$,

$$z_n(t) \to z(t) = J^q v(s)(t) - \frac{t^{q-1}}{\Lambda}\left(\alpha J^q v(s)(T) - \sum_{i=1}^{m}\beta_i I_{\eta_i}^{\gamma_i,\delta_i} J^q v(s)(\xi_i)\right).$$

This shows that $z \in \mathcal{G}(u)$ and so $\mathcal{G}$ is closed-valued. On the other hand, $\mathcal{G}(u)$ is a bounded set for all $u \in \mathcal{U}$, because F is a compact multi-valued map.

Finally, we show that $H_d(\mathcal{G}(u), \mathcal{G}(w)) \leq \psi(\|u-w\|)$. Let $u, w \in \mathcal{U}$ and $h_1 \in \mathcal{G}(w)$. Choose $v_1 \in S_{F,w}$ such that

$$h_1(t) = J^q v_1(s)(t) - \frac{t^{q-1}}{\Lambda}\left(\alpha J^q v_1(s)(T) - \sum_{i=1}^{m} \beta_i I_{\eta_i}^{\gamma_i,\delta_i} J^q v_1(s)(\xi_i)\right),$$

for almost all $t \in J$. Since

$$H_d(F(t,u(t)) - F(t,w(t))) \leq \left(\|\ell\|\theta\right)^{-1}\ell(t)\psi(|u(t)-w(t)|), \text{ for all } t \in [0,T],$$

there exists $z \in F(t,u(t))$ provided that

$$|v_1(t) - z| \leq \left(\|\ell\|\theta\right)^{-1}\ell(t)\psi(|u(t)-w(t)|), \text{ for all } t \in [0,T].$$

Now, we consider the multi-valued map $\mathcal{V} : [0,T] \to \mathcal{P}(\mathbb{R})$ given by

$$\mathcal{V}(t) = \left\{ z \in \mathbb{R} : \ |v_1(t) - z| \leq \left(\|\ell\|\theta\right)^{-1}\ell(t)\psi(|u(t)-w(t)|) \right\}.$$

Since v_1 and $\varphi = \left(\|\ell\|\theta\right)^{-1}\ell\psi(|u-w|)$ are measurable, the multifunction $\mathcal{V}(\cdot) \cap F(\cdot, u(\cdot))$ is measurable. Choose $v_2(t) \in F(t,u(t))$ such that

$$|v_1(t) - v_2(t)| \leq \left(\|\ell\|\theta\right)^{-1}\ell(t)\psi(|u(t)-w(t)|), \text{ for all } t \in [0,T].$$

We define the element $h_2 \in \mathcal{G}(u)$ as follows:

$$h_2(t) = J^q v_2(s)(t) - \frac{t^{q-1}}{\Lambda}\left(\alpha J^q v_2(s)(T) - \sum_{i=1}^{m} \beta_i I_{\eta_i}^{\gamma_i,\delta_i} J^q v_2(s)(\xi_i)\right), \text{ for all } t \in [0,T].$$

Let $\sup_{t\in[0,T]} |\ell(t)| = \|\ell\|$. Then, one can get

$$\begin{aligned}
&|h_1(t) - h_2(t)| \\
&\leq J^q|v_1(s) - v_2(s)|(t) \\
&\quad + \frac{t^{q-1}}{|\Lambda|}\left(\alpha J^q|v_1(s) - v_2(s)|(T) + \sum_{i=1}^{m} \beta_i I_{\eta_i}^{\gamma_i,\delta_i} J^q|v_1(s) - v_2(s)|(\xi_i)\right) \\
&\leq \left(\|\ell\|\theta\right)^{-1}\left(\|\ell\|\theta\right)\psi(\|u-w\|) \\
&= \psi(\|u-w\|).
\end{aligned}$$

Hence, $\|h_1 - h_2\| \leq \psi(\|u-w\|)$. Therefore $H_d(\mathcal{G}(u), \mathcal{G}(w)) \leq \psi(\|u-w\|)$ for all $u, w \in \mathcal{U}$. By hypothesis, since the multifunction $\mathcal{G}$ has approximate endpoint property, by Lemma 1.20, there exists $u^* \in \mathcal{U}$ such that $\mathcal{G}(u^*) = \{u^*\}$. Consequently, the problem (3.172) has the solution u^* on $[0,T]$. This completes the proof. □

Example 3.23. Consider the following boundary value problem for fractional differential inclusion with Erdélyi-Kober fractional integral conditions

$$\begin{cases} D^{3/2}x(t) \in F(t,x(t)), & t \in [0,5], \\ x(0) = 0, \quad \dfrac{2}{3}x(5) = \dfrac{e}{2} I_{\frac{\sqrt{3}}{5}}^{\frac{5}{3},\frac{3}{7}} x\left(\dfrac{4}{3}\right) + \dfrac{\pi}{3} I_{\frac{\sqrt{2}}{5}}^{\frac{2}{9},\frac{\sqrt{3}}{8}} x\left(\dfrac{3}{2}\right) + \dfrac{\sqrt{\pi}}{6} I_{\frac{e}{3}}^{\frac{\sqrt{e}}{2},\frac{e^2}{4}} x\left(\dfrac{2}{7}\right). \end{cases} \tag{3.179}$$

Here $q = 3/2$, $m = 3$, $T = 5$, $\alpha = 2/3$, $\beta_1 = e/2$, $\beta_2 = \pi/3$, $\beta_3 = \sqrt{\pi}/6$, $\eta_1 = \sqrt{3}/5$, $\eta_2 = \sqrt{2}/5$, $\eta_3 = e/3$, $\gamma_1 = 5/3$, $\gamma_2 = 2/9$, $\gamma_3 = \sqrt{e}/2$, $\delta_1 = 3/7$, $\delta_2 = \sqrt{3}/8$, $\delta_3 = e^2/4$, $\xi_1 = 4/3$, $\xi_2 = 3/2$, $\xi_3 = 2/7$. Using the given data, we find that

$$\theta = \frac{T^q}{\Gamma(q+1)} + \frac{|\alpha|T^{2q-1}}{|\Lambda|\Gamma(q+1)} + \frac{T^{q-1}}{|\Lambda|\Gamma(q+1)}\sum_{i=1}^{m}\frac{|\beta_i|\xi_i^q\Gamma(\gamma_i + (q/\eta_i) + 1)}{\Gamma(\gamma_i + (q/\eta_i) + \delta_i + 1)} \approx 44.000.$$

Let $F : [0,5] \times \mathbb{R} \to \mathcal{P}(\mathbb{R})$ be a multi-valued map given by

$$x \to F(t,x) = \left[0, \frac{3t}{10}\frac{|x|}{1+|x|} + \frac{2}{3}\right]. \tag{3.180}$$

Setting $\ell(t) = (3t)/10$, $t \in [0,5]$, we have $\|\ell\| = 3/2$. Choosing $\psi(y) = y/2$, it is clear the function ψ is non-decreasing upper semi-continuous on $[0,5]$ such that $\liminf_{y\to\infty}(y - \psi(y)) > 0$ and $\psi(y) < y$ for all $y > 0$. Also

$$H_d(F(t,x) - F(t,\bar{x})) \leq \frac{3t}{10}|x - \bar{x}| < \Big(\|\ell\|\theta\Big)^{-1}\frac{3t}{10}\psi(|x - \bar{x}|),$$

for x, $\bar{x} \in \mathbb{R}$. Let $\mathcal{U} = C([0,5],\mathbb{R})$. We define an operator $\mathcal{G} : \mathcal{U} \to \mathcal{P}(\mathcal{U})$ by

$$\mathcal{G}(u) = \{z \in \mathcal{U} : \text{there exists } v \in S_{F,u} \text{ such that } z(t) = w(t) \text{ for all } t \in [0,5]\},$$

where

$$w(t) = J^q v(s)(t) - \frac{t^{q-1}}{\Lambda}\left(\alpha J^q v(s)(T) - \sum_{i=1}^{m}\beta_i I_{\eta_i}^{\gamma_i,\delta_i} J^q v(s)(\xi_i)\right), \quad t \in [0,5].$$

Since $\sup_{u\in\mathcal{G}(0)}\|u\| = 0$, thus $\inf_{u\in\mathcal{U}}\sup_{s\in\mathcal{G}(u)}\|u - s\| = 0$. Consequently, the operator $\mathcal{G}$ has the approximate endpoint property. Thus all the conditions of Theorem 3.38 are satisfied. Therefore, by the conclusion of Theorem 3.38, the problem (3.179) with $F(t,x)$ given by (3.180) has at least one solution on $[0,5]$.

3.10.2 *Existence result via Krasnosel'skiĭ's multi-valued fixed point theorem*

In this subsection, we prove the existence of solutions for the problem (3.172) with a convex valued right hand side, by applying multi-valued version of Krasnosel'skiĭ's fixed point theorem (Theorem 1.17) and assuming F to be L^1-Carathéodory.

Theorem 3.39. *Assume that:*

(3.39.1) $F : [0,T] \times \mathbb{R} \to \mathcal{P}_{cp,c}(\mathbb{R})$ *is* L^1*-Carathéodory multi-valued map;*
(3.39.2) there exists a continuous function $p \in C([0,T],\mathbb{R}^+)$ *such that*

$$\|F(t,x)\|_{\mathcal{P}} := \sup\{|y| : y \in F(t,x)\} \leq p(t), \quad \textit{for each } (t,x) \in [0,T] \times \mathbb{R};$$

(3.39.3) there exists a function $k \in \mathcal{U}$ *such that*

$$H_d(F(t,x), F(t,y)) \leq \|k\|\|x - y\|,$$

for a.e $t \in [0,T]$ *and all* $x, y \in \mathcal{U}$, *with*

$$\|k\|\frac{T^{q-1}}{|\Lambda|\Gamma(q+1)}\sum_{i=1}^{m}\frac{|\beta_i|\xi_i^q\Gamma(\gamma_i + (q/\eta_i) + 1)}{\Gamma(\gamma_i + (q/\eta_i) + \delta_i + 1)} < 1.$$

Then the boundary value problem (3.172) has at least one solution on $[0, T]$.

Proof. Let us introduce the multi-valued operators $\mathcal{A} : \mathcal{U} \longrightarrow \mathcal{P}(\mathcal{U})$ by

$$\mathcal{A}(x) = \left\{ h \in \mathcal{U} : h(t) = \frac{t^{q-1}}{\Lambda} \sum_{i=1}^{m} \beta_i I_{\eta_i}^{\gamma_i,\delta_i} J^q f(s)(\xi_i) \right\}, \tag{3.181}$$

and $\mathcal{B} : \mathcal{U} \longrightarrow \mathcal{P}(\mathcal{U})$ by

$$\mathcal{B}(x) = \left\{ h \in \mathcal{U} : h(t) = J^\alpha f(s)(t) - \alpha \frac{t^{q-1}}{\Lambda} J^q f(s)(T) \right\}. \tag{3.182}$$

Observe that $\mathcal{G} = \mathcal{A} + \mathcal{B}$, where the operator $\mathcal{G} : \mathcal{U} \longrightarrow \mathcal{P}(\mathcal{U})$ is given by (3.177). We shall show that the operators $\mathcal{A}$ and $\mathcal{B}$ satisfy the conditions of Theorem 1.17 on $[0, T]$ in several steps. We begin by showing that the operators $\mathcal{A}$ and $\mathcal{B}$ define the multi-valued operators $\mathcal{A}, \mathcal{B} : B_r \to \mathcal{P}_{cp,c}(\mathcal{U})$, where $B_r = \{x \in \mathcal{U} : \|x\| \leq r\}$ is a bounded set in $\mathcal{U}$. We will only prove that the operator $\mathcal{B}$ is compact-valued on B_r and convex for all $x \in \mathcal{U}$ as the similar steps work for the operator $\mathcal{A}$. Note that the operator $\mathcal{B}$ is equivalent to the composition $\mathcal{L} \circ S_F$, where $\mathcal{L}$ is the continuous linear operator on $L^1([0, T], \mathbb{R})$ into $\mathcal{U}$, defined by

$$\mathcal{L}(v)(t) = J^\alpha v(s)(t) - \alpha \frac{t^{q-1}}{\Lambda} J^q v(s)(T).$$

Suppose that $x \in B_r$ is arbitrary and let $\{v_n\}$ be a sequence in $S_{F,x}$. Then, by definition of $S_{F,x}$, we have $v_n(t) \in F(t, x(t))$ for almost all $t \in [0, T]$. Since $F(t, x(t))$ is compact for all $t \in J$, there is a convergent subsequence of $\{v_n(t)\}$ (we denote it by $\{v_n(t)\}$ again) that converges in measure to some $v(t) \in S_{F,x}$ for almost all $t \in J$. On the other hand, $\mathcal{L}$ is continuous, so $\mathcal{L}(v_n)(t) \to \mathcal{L}(v)(t)$ pointwise on $[0, T]$.

In order to show that the convergence is uniform, we have to show that $\{\mathcal{L}(v_n)\}$ is an equi-continuous sequence. Let $t_1, t_2 \in [0, T]$ with $t_1 < t_2$. Then, we have

$$\begin{aligned}
&|\mathcal{L}(v_n)(t_2) - \mathcal{L}(v_n)(t_1)| \\
&\leq |J^\alpha v_n(s)(t_2) - J^\alpha v_n(s)(t_1)| + |\alpha| \frac{|t_2^{q-1} - t_1^{q-1}|}{|\Lambda|} J^q |v_n(s)|(T) \\
&\leq \frac{1}{\Gamma(q)} \left| \int_0^{t_1} [(t_2 - s)^{q-1} - (t_1 - s)^{q-1}] p(s) ds + \int_{t_1}^{t_2} (t_2 - s)^{q-1} p(s) ds \right| \\
&\quad + |\alpha| \frac{|t_2^{q-1} - t_1^{q-1}|}{|\Lambda|} J^q |v_n(s)|(T) \\
&\leq \frac{\|p\|}{\Gamma(q+1)} [t_2^q - t_1^q] + |\alpha| \|p\| \frac{|t_2^{q-1} - t_1^{q-1}|}{|\Lambda|} \frac{T^q}{\Gamma(q+1)}.
\end{aligned}$$

We see that the right hand of the above inequality tends to zero as $t_2 \to t_1$. Thus, the sequence $\{\mathcal{L}(v_n)\}$ is equi-continuous and hence it follows by the Arzelá-Ascoli theorem that there exists a uniformly convergent subsequence $\{v_n\}$ (we denote it again by $\{v_n\}$) such that $\mathcal{L}(v_n) \to \mathcal{L}(v)$. Note that $\mathcal{L}(v) \in \mathcal{L}(S_{F,x})$. Hence, $\mathcal{B}(x) = \mathcal{L}(S_{F,x})$ is compact for all $x \in B_r$. So $\mathcal{B}(x)$ is compact.

To establish that $\mathcal{B}(x)$ is convex for all $x \in \mathcal{U}$, let $z_1, z_2 \in \mathcal{B}(x)$. We select $f_1, f_2 \in S_{F,x}$ such that

$$z_i(t) = J^{\alpha} f_i(s)(t) - \alpha \frac{t^{q-1}}{\Lambda} J^q f_i(s)(T), \quad i = 1, 2,$$

for almost all $t \in [0, T]$. Let $0 \le \lambda \le 1$. Then, we have

$$\begin{aligned}[\lambda z_1 + (1-\lambda) z_2](t) &= J^{\alpha}[\lambda f_1(s) + (1-\lambda) f_2(s)](t) \\ &\quad -\alpha \frac{t^{q-1}}{\Lambda} J^q [\lambda f_1(s) + (1-\lambda) f_2(s)](T).\end{aligned}$$

Since F has convex values, so $S_{F,u}$ is convex and $\lambda f_1(s) + (1-\lambda) f_2(s) \in S_{F,x}$. Thus $\lambda z_1 + (1-\lambda) z_2 \in \mathcal{B}(x)$. Consequently, $\mathcal{B}$ is convex-valued.

The rest of the proof consists of the following steps and claims.

Step 1. We show that $\mathcal{A}$ is a multi-valued contraction on $\mathcal{U}$. Let $x, y \in \mathcal{U}$ and $h_1 \in \mathcal{A}x$. Then there exists $v_1(t) \in F(t, x(t))$ such that, for each $t \in [0, T]$,

$$h_1(t) = \frac{t^{q-1}}{\Lambda} \sum_{i=1}^{m} \beta_i I_{\eta_i}^{\gamma_i, \delta_i} J^q v_1(s)(\xi_i).$$

Since $H_d(F(t,x), F(t,y)) \le k(t)\|x - y\|$, there exists $w \in F(t, y)$ such that

$$|v_1(t) - w| \le k(t)\|x - y\|.$$

Thus the multi-valued operator U defined by $U(t) = S_{F,y} \cap K(t)$, where

$$K(t) = \{w \in \mathbb{R} : |v_1(t) - w| \le k(t)\|x - y\|\},$$

is non-empty and measurable. Let v_2 be a measurable selection for U (which exists by Kuratowski-Ryll-Nardzewski's selection theorem [98]). Then $v_2(t) \in F(t, y(t))$ and for each $t \in [0, T]$, $|v_1(t) - v_2(t)| \le k(t)\|x - y\|$ a.e. on $[0, T]$.

For each $t \in [0, T]$, let us define

$$h_2(t) = \frac{t^{q-1}}{\Lambda} \sum_{i=1}^{m} \beta_i I_{\eta_i}^{\gamma_i, \delta_i} J^q v_2(s)(\xi_i),$$

such that $h_2 \in \mathcal{A}y$ and

$$\begin{aligned}|h_1(t) - h_2(t)| &\le \frac{t^{q-1}}{|\Lambda|} \sum_{i=1}^{m} \beta_i I_{\eta_i}^{\gamma_i, \delta_i} J^q |v_1(s) - v_2(s)|(\xi_i) \\ &\le \|k\| \frac{T^{q-1}}{|\Lambda| \Gamma(q+1)} \sum_{i=1}^{m} \frac{|\beta_i| \xi_i^q \Gamma(\gamma_i + (q/\eta_i) + 1)}{\Gamma(\gamma_i + (q/\eta_i) + \delta_i + 1)} \|x - y\|.\end{aligned}$$

Taking supremum for $t \in [0, T]$, we obtain

$$\|h_1 - h_2\| \le \|k\| \frac{T^{q-1}}{|\Lambda| \Gamma(q+1)} \sum_{i=1}^{m} \frac{|\beta_i| \xi_i^q \Gamma(\gamma_i + (q/\eta_i) + 1)}{\Gamma(\gamma_i + (q/\eta_i) + \delta_i + 1)} \|x - y\|,$$

which together with the analogous inequality obtained by interchanging the roles of x and y yields

$$H_d(\mathcal{A}x, \mathcal{A}y) \leq \|k\| \frac{T^{q-1}}{|\Lambda|\Gamma(q+1)} \sum_{i=1}^{m} \frac{|\beta_i|\xi_i^q \Gamma(\gamma_i + (q/\eta_i) + 1)}{\Gamma(\gamma_i + (q/\eta_i) + \delta_i + 1)} \|x - y\|,$$

for each $x, y \in \mathcal{U}$. This shows that $\mathcal{A}$ is a multi-valued contraction, since

$$\|k\| \frac{T^{q-1}}{|\Lambda|\Gamma(q+1)} \sum_{i=1}^{m} \frac{|\beta_i|\xi_i^q \Gamma(\gamma_i + (q/\eta_i) + 1)}{\Gamma(\gamma_i + (q/\eta_i) + \delta_i + 1)} < 1.$$

Step 2. *$\mathcal{B}$ is compact and upper semi-continuous.* This will be established in several claims.

Claim I. *$\mathcal{B}$ maps bounded sets into bounded sets in $\mathcal{U}$.*

For a positive number ρ, let $B_\rho = \{x \in \mathcal{U} : \|x\| \leq \rho\}$ be a bounded ball in $\mathcal{U}$. Then, for each $h \in \mathcal{G}(x), x \in B_\rho$, there exists $v \in S_{F,x}$ such that

$$h(t) = J^q v(s)(t) - \frac{t^{q-1}}{\Lambda} \alpha J^q v(s)(T).$$

Then we have

$$\begin{aligned} |h(x)| &\leq J^q |v(s)|(T) + \frac{|\alpha|T^{q-1}}{|\Lambda|} J^q |v(s)|(T) \\ &\leq J^q p(s)(T) + \frac{|\alpha|T^{q-1}}{|\Lambda|} J^q p(s)(T) \\ &\leq \|p\| \left(\frac{T^q}{\Gamma(q+1)} + \frac{|\alpha|T^{2q-1}}{|\Lambda|\Gamma(q+1)} \right) \end{aligned}$$

and consequently,

$$\|h\| \leq \|p\| \left(\frac{T^q}{\Gamma(q+1)} + \frac{|\alpha|T^{2q-1}}{|\Lambda|\Gamma(q+1)} \right).$$

Claim II. *$\mathcal{B}$ maps bounded sets into equi-continuous sets.*

Let $\tau_1, \tau_2 \in [0, T]$ with $\tau_1 < \tau_2$ and $x \in B_\rho$. For each $h \in \mathcal{G}(x)$, we obtain

$$\begin{aligned} &|h(\tau_2) - h(\tau_1)| \\ &\leq |J^q v(s)(\tau_2) - J^q v(s)(\tau_1)| + \frac{|\alpha||\tau_2^{q-1} - \tau_1^{q-1}|}{|\Lambda|} J^q |v(s)|(T) \\ &\leq \frac{1}{\Gamma(q)} \left| \int_0^{\tau_1} [(\tau_2 - s)^{q-1} - (\tau_1 - s)^{q-1}] p(s) ds + \int_{\tau_1}^{\tau_2} (\tau_2 - s)^{q-1} p(s) ds \right| \\ &\quad + \frac{|\tau_2^{q-1} - \tau_1^{q-1}|}{|\Lambda|} |\alpha| J^q p(s)(T) \\ &\leq \frac{\|p\|}{\Gamma(q+1)} [t_2^q - t_1^q] + |\alpha| \|p\| \frac{|t_2^{q-1} - t_1^{q-1}|}{|\Lambda|} \frac{T^q}{\Gamma(q+1)}. \end{aligned}$$

Obviously the right hand side of the above inequality tends to zero independently of $x \in B_r$ as $\tau_2 - \tau_1 \to 0$. Therefore we deduce by the Ascoli-Arzelá theorem that $\mathcal{B} : \mathcal{U} \to \mathcal{P}(\mathcal{U})$ is completely continuous.

By Claims I and II, $\mathcal{B}$ is completely continuous. By Lemma 1.1, $\mathcal{B}$ will be upper semi-continuous (since it is completely continuous) if we show that it has a closed graph. In the next claim, we prove it.

Claim III. *$\mathcal{B}$ has a closed graph.*

Let $x_n \to x_*, h_n \in \mathcal{B}(x_n)$ and $h_n \to h_*$. Then we need to show that $h_* \in \mathcal{B}(x_*)$. Associated with $h_n \in \mathcal{B}(x_n)$, there exists $f_n \in S_{F,x_n}$ such that for each $t \in [0,T]$,

$$h(t) = J^{\alpha} f_n(s)(t) - \alpha \frac{t^{q-1}}{\Lambda} J^q f_n(s)(T).$$

Thus it suffices to show that there exists $f_* \in S_{F,x_*}$ such that for each $t \in [0,T]$,

$$h_*(t) = J^{\alpha} f_*(s)(t) - \alpha \frac{t^{q-1}}{\Lambda} J^{\alpha} f_*(s)(T).$$

Let us consider the linear operator $\Theta : L^1([0,T],\mathbb{R}) \to \mathcal{U}$ given by

$$f \mapsto \Theta(f)(t) = J^{\alpha} f(s)(t) - \alpha \frac{t^{q-1}}{\Lambda} J^{\alpha} f(s)(T).$$

Observe that

$$\|h_n(t) - h_*(t)\| = \Big\| J^{\alpha}(f_n(s) - f_*(s))(t) \\ -\alpha \frac{t^{q-1}}{\Lambda} f_*(s))(\xi) - J^{\alpha}(f_n(s) - f_*(s))(T) \Big\| \to 0,$$

as $n \to \infty$.
Thus, it follows by Lemma 1.2 that $\Theta \circ S_F$ is a closed graph operator. Further, we have $h_n(t) \in \Theta(S_{F,x_n})$. Since $x_n \to x_*$, we have that

$$h_*(t) = J^{\alpha} f_*(s)(t) - \alpha \frac{t^{q-1}}{\Lambda} J^{\alpha} f_*(s)(T),$$

for some $f_* \in S_{F,x_*}$. Hence $\mathcal{B}$ has a closed graph (and therefore has closed values). In consequence, the operator $\mathcal{B}$ is compact and upper semi-continuous.

Step 3. Here, we show that $\mathcal{A}(x) + \mathcal{B}(x) \subset B_r$ for all $x \in B_r$. Suppose $x \in B_r$, with $r > \|p\|\theta$ (θ defined by (3.178)) and $h \in \mathcal{B}$ are arbitrary elements. Choose $f \in S_{F,x}$ such that

$$h(t) = J^{\alpha} f(s)(t) - \frac{t^{q-1}}{\Lambda} \left(\alpha J^q f(s)(T) - \sum_{i=1}^{m} \beta_i I_{\eta_i}^{\gamma_i,\delta_i} J^q f(s)(\xi_i) \right), \quad t \in [0,T].$$

Then we have

$$|h(x)| \le J^q |f(s)|(T) + \frac{|\alpha| T^{q-1}}{|\Lambda|} J^q |f(s)|(T) + \frac{T^{q-1}}{|\Lambda|} \sum_{i=1}^{m} |\beta_i| I_{\eta_i}^{\gamma_i,\delta_i} J^q |f(s)|(\xi_i)$$

$$\leq J^q p(s)(T) + \frac{|\alpha| T^{q-1}}{|\Lambda|} J^q p(s)(T) + \frac{T^{q-1}}{|\Lambda|} \sum_{i=1}^{m} |\beta_i| I_{\eta_i}^{\gamma_i,\delta_i} J^q p(s)(\xi_i)$$

$$\leq \|p\| \left(\frac{T^q}{\Gamma(q+1)} + \frac{|\alpha| T^{2q-1}}{|\Lambda|\Gamma(q+1)} + \frac{T^{q-1}}{|\Lambda|\Gamma(q+1)} \sum_{i=1}^{m} \frac{|\beta_i| \xi_i^q \Gamma(\gamma_i + (q/\eta_i) + 1)}{\Gamma(\gamma_i + (q/\eta_i) + \delta_i + 1)} \right).$$

Thus

$$\|h\| \leq \|p\|\theta < r. \tag{3.183}$$

Hence $\|h\| \leq r$, which means that $\mathcal{A}(x) + \mathcal{B}(x) \subset B_r$ for all $x \in B_r$.

Thus, the operators $\mathcal{A}$ and $\mathcal{B}$ satisfy all the conditions of Theorem 1.17 and hence we conclude that $x \in \mathcal{A}(x) + \mathcal{B}(x)$ has a solution in B_r. Therefore problem (3.172) has a solution in B_r and the proof is completed. □

Example 3.24. Let us consider problem (3.179) with the multi-valued map $F : [0,5] \times \mathbb{R} \to \mathcal{P}(\mathbb{R})$ given by

$$x \to F(t,x) = \left[\frac{\cos t}{\sqrt{t+25}}, \frac{1}{10\sqrt{(1+t^2)}} \tan^{-1} x + \frac{\cos t}{\sqrt{t+25}} \right]. \tag{3.184}$$

Clearly

$$\|F(t,x)\|_{\mathcal{P}} := \sup\{|y| : y \in F(t,x)\} \leq \frac{\pi}{20\sqrt{(1+t^2)}} + \frac{\cos t}{\sqrt{t+25}} = p(t),$$

and $H_d(F(t,x), F(t,y)) \leq \frac{1}{10}|x-y|$, for $x, y \in \mathbb{R}$. With $\|k\| = 1/10$, we find that

$$\|k\| \frac{T^{q-1}}{|\Lambda|\Gamma(q+1)} \sum_{i=1}^{m} \frac{\Gamma(\gamma_i + (q/\eta_i) + 1)}{\Gamma(\gamma_i + (q/\eta_i) + \delta_i + 1)} \approx 0.818538 < 1.$$

Thus the hypotheses of Theorem 3.39 hold and consequently, problem (3.179) with $F(t,x)$ given by 3.184 has a solution by Theorem 3.39.

3.10.3 *Existence result via Wegrzyk's fixed point theorem*

In this subsection, we prove the existence of solutions for problem (3.172) with a not necessary nonconvex valued right hand side, by applying Wegrzyk's fixed point theorem (Theorem 1.21).

Theorem 3.40. *Assume that:*

(3.40.1) $F : [0,T] \times \mathbb{R} \to \mathcal{P}_{cp}(\mathbb{R})$ *is such that* $F(\cdot,x) : [0,T] \to \mathcal{P}_{cp}(\mathbb{R})$ *is measurable for each* $x \in \mathbb{R}$;

(3.40.2) $H_d(F(t,x), F(t,\bar{x})) \leq m(t)\delta(|x-\bar{x}|)$ *for almost all* $t \in [0,T]$ *and* $x, \bar{x} \in \mathbb{R}$ *with* $m \in C([0,T], \mathbb{R}^+)$ *and* $d(0, F(t,0)) \leq m(t)$ *for almost all* $t \in [0,T]$, *where* $\delta : \mathbb{R}_+ \to \mathbb{R}_+$ *is strictly increasing.*

Then the boundary value problem (3.172) has at least one solution on $[0,T]$ *if* $\varepsilon\delta:\mathbb{R}_+\to\mathbb{R}_+$ *is a strict comparison function, where* $\varepsilon=\|m\|\theta$, *i.e.,*

$$\varepsilon=\|m\|\left\{\frac{T^q}{\Gamma(q+1)}+\frac{|\alpha|T^{2q-1}}{|\Lambda|\Gamma(q+1)}+\frac{T^{q-1}}{|\Lambda|\Gamma(q+1)}\sum_{i=1}^{m}\frac{|\beta_i|\xi_i^q\Gamma(\gamma_i+(q/\eta_i)+1)}{\Gamma(\gamma_i+(q/\eta_i)+\delta_i+1)}\right\}.$$

Proof. Let $\varepsilon\delta:\mathbb{R}_+\to\mathbb{R}_+$ be a strict comparison function. It follows by the assumptions (3.40.1) and (3.40.1) that $F(\cdot,x(\cdot))$ is measurable and has a measurable selection $v(\cdot)$ (see Theorem III.6 [78]). Also $m\in C([0,1],\mathbb{R})$ and

$$\begin{aligned}|v(t)|&\le d(0,F(t,0))+H_d(F(t,0),F(t,x(t)))\\&\le m(t)+m(t)\delta(|x(t)|)\\&\le(1+\delta(||x||)m(t).\end{aligned}$$

Thus the set $S_{F,x}$ is non-empty for each $x\in\mathcal{U}$. Now we show that the operator $\mathcal{G}$ defined by (3.177) satisfies the assumptions of Lemma 1.21. We show that $\mathcal{G}(x)\in\mathcal{P}_{cl}((\mathcal{U})$ for each $x\in\mathcal{U}$. Let $\{u_n\}_{n\ge0}\in\mathcal{G}(x)$ be such that $u_n\to u$ in $\mathcal{U}$ as $n\to\infty$. Then $u\in\mathcal{U}$ and there exists $v_n\in S_{F,x_n}$ such that, for each $t\in[0,T]$,

$$u_n(t)=J^qv_n(s)(t)-\frac{t^{q-1}}{\Lambda}\left(\alpha J^qv_n(s)(T)-\sum_{i=1}^{m}\beta_iI_{\eta_i}^{\gamma_i,\delta_i}J^qv_n(s)(\xi_i)\right).$$

As $\mathcal{G}$ has compact values, we pass onto a subsequence (if necessary) to obtain that v_n converges to v in $L^1([0,T],\mathbb{R})$. Thus, $v\in S_{F,x}$ and for each $t\in[0,T]$, we have

$$u_n(t)\to v(t)=J^qv(s)(t)-\frac{t^{q-1}}{\Lambda}\left(\alpha J^qv(s)(T)-\sum_{i=1}^{m}\beta_iI_{\eta_i}^{\gamma_i,\delta_i}J^qv(s)(\xi_i)\right).$$

Hence, $u\in\mathcal{G}(x)$.

Next we show that

$$H_d(\mathcal{G}(x),\mathcal{G}(\bar{x}))\le\varepsilon\delta(\|x-\bar{x}\|)\quad\text{for each}\quad x,\bar{x}\in\mathcal{U}.$$

Let $x,\bar{x}\in\mathcal{U}$ and $h_1\in\mathcal{G}(x)$. Then there exists $v_1(t)\in F(t,x(t))$ such that, for each $t\in[0,T]$,

$$h_1(t)=J^qv_1(s)(t)-\frac{t^{q-1}}{\Lambda}\left(\alpha J^qv_1(s)(T)-\sum_{i=1}^{m}\beta_iI_{\eta_i}^{\gamma_i,\delta_i}J^qv_1(s)(\xi_i)\right).$$

By (3.40.2), we have

$$H_d(F(t,x),F(t,\bar{x}))\le m(t)\delta(|x(t)-\bar{x}(t)|).$$

So, there exists $w\in F(t,\bar{x}(t))$ such that

$$|v_1(t)-w(t)|\le m(t)\delta(|x(t)-\bar{x}(t)|),\quad t\in[0,T].$$

Define $U:[0,T]\to\mathcal{P}(\mathbb{R})$ by

$$U(t)=\{w\in\mathbb{R}:|v_1(t)-w|\le m(t)\delta(|x(t)-\bar{x}(t)|)\}.$$

Since the multi-valued operator $U(t) \cap F(t, \bar{x}(t))$ is measurable (Proposition III.4 [78]), there exists a function $v_2(t)$ which is a measurable selection for U. So $v_2(t) \in F(t, \bar{x}(t))$ and for each $t \in [0,T]$, we have $|v_1(t) - v_2(t)| \leq m(t)\delta(|x(t) - \bar{x}(t)|)$.

For each $t \in [0,T]$, let us define

$$h_2(t) = J^q v_2(s)(t) - \frac{t^{q-1}}{\Lambda}\left(\alpha J^q v_2(s)(T) - \sum_{i=1}^{m} \beta_i I_{\eta_i}^{\gamma_i,\delta_i} J^q v_2(s)(\xi_i)\right).$$

Thus we have

$$\begin{aligned}
&|h_1(t) - h_2(t)| \\
&\leq J^q|v_1(s) - v_2(s)|(t) \\
&\quad + \frac{t^{q-1}}{|\Lambda|}\left(\alpha J^q|v_1(s) - v_2(s)|(T) + \sum_{i=1}^{m} \beta_i I_{\eta_i}^{\gamma_i,\delta_i} J^q|v_1(s) - v_2(s)|(\xi_i)\right) \\
&\leq \|m\|\left(\frac{T^q}{\Gamma(q+1)} + \frac{|\alpha|T^{2q-1}}{|\Lambda|\Gamma(q+1)}\right. \\
&\quad \left. + \frac{T^{q-1}}{|\Lambda|\Gamma(q+1)} \sum_{i=1}^{m} \frac{|\beta_i|\xi_i^q \Gamma(\gamma_i + (q/\eta_i) + 1)}{\Gamma(\gamma_i + (q/\eta_i) + \delta_i + 1)}\right)\delta(\|x - \bar{x}\|),
\end{aligned}$$

which yields

$$\begin{aligned}
\|h_1 - h_2\| \leq \|m\|&\left(\frac{T^q}{\Gamma(q+1)} + \frac{|\alpha|T^{2q-1}}{|\Lambda|\Gamma(q+1)}\right. \\
&\left. + \frac{T^{q-1}}{|\Lambda|\Gamma(q+1)} \sum_{i=1}^{m} \frac{|\beta_i|\xi_i^q \Gamma(\gamma_i + (q/\eta_i) + 1)}{\Gamma(\gamma_i + (q/\eta_i) + \delta_i + 1)}\right)\delta(\|x - \bar{x}\|).
\end{aligned}$$

Analogously, interchanging the roles of x and $\bar{x}$, we obtain

$$\begin{aligned}
&H_d(\mathcal{G}(x), \mathcal{G}(\bar{x})) \leq \varepsilon\delta(\|x - \bar{x}\|) \\
&= \|m\|\left(\frac{T^q}{\Gamma(q+1)} + \frac{|\alpha|T^{2q-1}}{|\Lambda|\Gamma(q+1)} + \frac{T^{q-1}}{|\Lambda|\Gamma(q+1)} \sum_{i=1}^{m} \frac{|\beta_i|\xi_i^q \Gamma(\gamma_i + (q/\eta_i) + 1)}{\Gamma(\gamma_i + (q/\eta_i) + \delta_i + 1)}\right),
\end{aligned}$$

for each $x, \bar{x} \in \mathcal{U}$. Therefore, $\mathcal{G}$ is a generalized contraction. Thus it follows by Lemma 1.21 that $\mathcal{G}$ has a fixed point x which is a solution of (3.172). This completes the proof. □

Remark 3.1. Theorem 3.40 holds for several values of the function δ, for example, $\delta(t) = \ln(1+t)^3$, $\delta(t) = t$, etc.

3.11 On perturbed fractional differential inclusions with nonlocal multi-point Erdélyi-Kober fractional integral boundary conditions

In this section, we investigate perturbed fractional differential inclusions equipped with nonlocal multi-point Erdélyi-Kober fractional integral boundary conditions given by

$$\begin{cases} D^q x(t) \in F(t, x(t)) + G(t, x(t)), & 0 < t < T, \quad 1 < q \le 2, \\ x(0) = 0, \qquad \alpha x(T) = \sum_{i=1}^{m} \beta_i I_{\eta_i}^{\gamma_i, \delta_i} x(\xi_i), \end{cases} \tag{3.185}$$

where D^q is the standard Riemann-Liouville fractional derivative operator of order q, $I_{\eta_i}^{\gamma_i,\delta_i}$ is the Erdélyi-Kober fractional integral of order $\delta_i > 0$ with $\eta_i > 0$ and $\gamma_i \in \mathbb{R}$, $i = 1, 2, \ldots, m$, $F, G : [0, T] \times \mathbb{R} \to \mathcal{P}(\mathbb{R})$ are multi-valued maps, $\mathcal{P}(\mathbb{R})$ is the family of all non-empty subsets of $\mathbb{R}$, and $\alpha, \beta_i \in \mathbb{R}$, $\xi_i \in (0, T)$, $i = 1, 2, \ldots, m$ are given constants.

3.11.1 *Existence results*

Throughout this subsection, for convenience, we use the following expressions

$$J^q f(z) = \frac{1}{\Gamma(q)} \int_0^z (z - s)^{q-1} f(s) ds, \quad z \in \{t, T\},$$

for $t \in [0, T]$ and

$$I_{\eta_i}^{\gamma_i,\delta_i} J^q f(\xi_i) = \frac{\eta_i \xi_i^{-\eta_i(\delta_i+\gamma_i)}}{\Gamma(q)\Gamma(\delta_i)} \int_0^{\xi_i} \int_0^r \frac{r^{\eta_i\gamma_i+\eta_i-1}(r - s)^{q-1}}{(\xi_i^{\eta_i} - r^{\eta_i})^{1-\delta_i}} f(s) ds\, dr,$$

where $\xi_i \in (0, T)$ for $i = 1, 2, \ldots, m$.

Let us list the following assumptions:

(H_1) $F : [0, T] \times \mathbb{R} \to \mathcal{P}_{cp,c}(\mathbb{R})$ is L^1-Carathéodory;

(H_2) there exists a continuous non-decreasing function $\Phi : [0, \infty) \to (0, \infty)$ and a function $p \in L^1([0, T], \mathbb{R}^+)$ such that

$$\|F(t, x)\|_{\mathcal{P}} := \sup\{|y| : y \in F(t, x)\} \le p(t)\Phi(\|x\|) \text{ for each } (t, x) \in [0, T] \times \mathbb{R};$$

(H_3) the multi-valued map $t \to G(t, x)$ is measurable for each $x \in \mathbb{R}$ and integrably bounded, i.e., there exists a function $M \in L^1([0, T], \mathbb{R}^+)$ such that

$$\|G(t, x)\|_{\mathcal{P}} := \sup\{|g| : g(t) \in G(t, x)\} \le M(t), \text{ for a.e. } t \in [0, T] \text{ and } x \in \mathbb{R};$$

(H_4) $G : [0, T] \times \mathbb{R} \to \mathcal{P}_{cp,c}(\mathbb{R})$ and there exists a function $\ell \in L^1([0, T], \mathbb{R})$ such that

$$H_d(G(t, x), G(t, y)) \le \ell(t)|x - y|, \ t \in [0, T]$$

for all $x, y \in \mathbb{R}$ with

$$J^q \ell(T) + \frac{T^{q-1}}{|\Lambda|} \left(\alpha J^q \ell(T) + \sum_{i=1}^{m} \beta_i I_{\eta_i}^{\gamma_i,\delta_i} J^q \ell(\xi_i) \right) < 1; \tag{3.186}$$

(H_5) there exists a constant $r > 0$ such that

$$\frac{r}{\Phi(r)\Psi_1 + \Psi_2} > 1, \tag{3.187}$$

where

$$\Psi_1 = J^q p(T) + \frac{|\alpha|T^{q-1}}{|\Lambda|} J^q p(T) + \frac{T^{q-1}}{|\Lambda|} \sum_{i=1}^{m} |\beta_i| I_{\eta_i}^{\gamma_i,\delta_i} J^q p(\xi_i)$$

and

$$\Psi_2 = J^q M(T) + \frac{|\alpha|T^{q-1}}{|\Lambda|} J^q M(T) + \frac{T^{q-1}}{|\Lambda|} \sum_{i=1}^{m} |\beta_i| I_{\eta_i}^{\gamma_i,\delta_i} J^q M(\xi_i).$$

We transform the problem (3.185) into a fixed point problem by introducing the operator

$$\mathcal{N}(x) = \left\{ \begin{array}{l} h \in C([0,T],\mathbb{R}) : \\ h(t) = \left\{ \begin{array}{l} J^q[f(t) + g(t)] \\ -\dfrac{t^{q-1}}{\Lambda}\left(\alpha J^q[f(T) + g(T)] \right. \\ \left. -\displaystyle\sum_{i=1}^{m} \beta_i I_{\eta_i}^{\gamma_i,\delta_i} J^q[f(\xi_i) + g(\xi_i)] \right) \end{array} \right. \end{array} \right\} \tag{3.188}$$

for $f \in S_{F,x}$ and $g \in S_{G,x}$.

Define the operators $\mathcal{F}, \mathcal{G} : C([0,T],\mathbb{R}) \to \mathcal{P}(C([0,T],\mathbb{R}))$ by

$$\mathcal{F}(x) = \left\{ \begin{array}{l} h \in C([0,T],\mathbb{R}) : \\ h(t) = \left\{ \begin{array}{l} J^q f(t) \\ -\dfrac{t^{q-1}}{\Lambda}\left(\alpha J^q f(T) - \displaystyle\sum_{i=1}^{m} \beta_i I_{\eta_i}^{\gamma_i,\delta_i} J^q f(\xi_i) \right) \end{array} \right. \end{array} \right\} \tag{3.189}$$

for $f \in S_{F,x}$, and

$$\mathcal{G}(x) = \left\{ \begin{array}{l} z \in C([0,T],\mathbb{R}) : \\ z(t) = \left\{ \begin{array}{l} J^q g(t) \\ -\dfrac{t^{q-1}}{\Lambda}\left(\alpha J^q g(T) - \displaystyle\sum_{i=1}^{m} \beta_i I_{\eta_i}^{\gamma_i,\delta_i} J^q g(\xi_i) \right) \end{array} \right. \end{array} \right\} \tag{3.190}$$

for $g \in S_{G,x}$. Observe that $\mathcal{N} = \mathcal{F} + \mathcal{G}$. We shall show that the operators $\mathcal{F}$ and $\mathcal{G}$ satisfy all the conditions of Theorem 1.18 on $[0,T]$.

Lemma 3.14. *The operators $\mathcal{F}$ and $\mathcal{G}$, defined by (3.189) and (3.190) respectively, are compact and convex-valued.*

Proof. First, we show that the operators $\mathcal{F}$ and $\mathcal{G}$ define the multi-valued operators $\mathcal{F}, \mathcal{G} : B_r \to \mathcal{P}_{cp,c}(C([0,T],\mathbb{R}))$, where $B_r = \{x \in C([0,T],\mathbb{R}) : \|x\| \leq r\}$ is a bounded set in $C([0,T],\mathbb{R})$. We shall prove that $\mathcal{F}$ is compact-valued on B_r. Note that the operator $\mathcal{F}$ is equivalent to the composition $\mathcal{L} \circ S_F$, where $\mathcal{L}$ is the continuous linear operator on $L^1([0,T],\mathbb{R})$ into $C([0,T],\mathbb{R})$, defined by

$$\mathcal{L}(v)(t) = J^q v(t) + \frac{t^{q-1}}{\Lambda}\Big\{\alpha I_\eta^{\gamma,\delta} J^q v(\xi) - J^q v(T)\Big\}.$$

Suppose that $x \in B_r$ is arbitrary and let $\{v_n\}$ be a sequence in $S_{F,x}$. Then, by definition of $S_{F,x}$, we have $v_n(t) \in F(t, x(t))$ for almost all $t \in [0,T]$. Since $F(t, x(t))$ is compact for all $t \in [0,T]$, there is a convergent subsequence of $\{v_n(t)\}$ (we denote it by $\{v_n(t)\}$ again) that converges in measure to some $v(t) \in S_{F,x}$ for almost all $t \in [0,T]$. On the other hand, $\mathcal{L}$ is continuous, so $\mathcal{L}(v_n)(t) \to \mathcal{L}(v)(t)$ pointwise on $[0,T]$.

In order to show that the convergence is uniform, we have to show that $\{\mathcal{L}(v_n)\}$ is an equi-continuous sequence. Let $t_1, t_2 \in [0,T]$ with $t_1 < t_2$. Then, we have

$$\begin{aligned}
&|\mathcal{L}(v_n)(t_2) - \mathcal{L}(v_n)(t_1)| \\
&\leq |J^q v_n(t_2) - J^q v_n(t_1)| + \frac{|t_2^{q-1} - t_1^{q-1}|}{|\Lambda|} J^q |v_n(T)| + \frac{|\alpha||t_2^{q-1} - t_1^{q-1}|}{|\Lambda|} |I_\eta^{\gamma,\delta} J^q |v_n(\xi)| \\
&\leq \frac{\psi(r)}{\Gamma(q)} \left| \int_0^{t_1} [(t_2 - s)^{q-1} - (t_1 - s)^{q-1}] p(s) ds + \int_{t_1}^{t_2} (t_2 - s)^{q-1} p(s) ds \right| \\
&\quad + \frac{|t_2^{q-1} - t_1^{q-1}|}{|\Lambda|} \left(J^q |v_n(T)| + |\alpha||I_\eta^{\gamma,\delta} J^q |v_n(\xi)| \right).
\end{aligned}$$

Notice that the right hand of the above inequality tends to zero as $t_2 \to t_1$. Thus, the sequence $\{\mathcal{L}(v_n)\}$ is equi-continuous and by using the Arzelá-Ascoli theorem, we deduce that there exists a uniformly convergent subsequence. So, there is a subsequence of $\{v_n\}$ (we denote it again by $\{v_n\}$) such that $\mathcal{L}(v_n) \to \mathcal{L}(v)$. Note that, $\mathcal{L}(v) \in \mathcal{L}(S_{F,x})$. Hence, $\mathcal{F}(x) = \mathcal{L}(S_{F,x})$ is compact for all $x \in B_r$. So $\mathcal{F}(x)$ is compact.

Now we show that $\mathcal{F}(x)$ is convex for all $x \in C([0,T],\mathbb{R})$. Let $h_1, h_2 \in \mathcal{F}(x)$. Then we select $f_1, f_2 \in S_{F,x}$ such that

$$h_i(t) = J^q f_i(t) + \frac{t^{q-1}}{\Lambda}\Big\{\alpha I_\eta^{\gamma,\delta} J^q f_i(\xi) - J^q f_i(T)\Big\}, \quad i = 1, 2,$$

for almost all $t \in [0,T]$. Let $0 \leq \lambda \leq 1$. Then, we have

$$\begin{aligned}
&[\lambda h_1 + (1-\lambda)h_2](t) = J^q[\lambda f_1(t) + (1-\lambda) f_2(t)] \\
&+ \frac{t^{q-1}}{\Lambda}\Big\{\alpha I_\eta^{\gamma,\delta} J^q [\lambda f_1(\xi) + (1-\lambda) f_2(\xi)] - J^q[\lambda f_1(T) + (1-\lambda) f_2(T)]\Big\}.
\end{aligned}$$

Since F has convex values, so $S_{F,u}$ is convex and $\lambda f_1(s) + (1-\lambda) f_2(s) \in S_{F,x}$. Thus $\lambda h_1 + (1-\lambda)h_2 \in \mathcal{F}(x)$. Consequently, $\mathcal{F}$ is convex-valued. Similarly, $\mathcal{G}$ is compact and convex-valued. □

Lemma 3.15. *Assume that* (H_3) *and* (H_4) *hold. Then the operator* $\mathcal{G}$ *defined by (3.190) is a contraction.*

Proof. Let $x, \bar{x} \in C^2([0,T],\mathbb{R})$ and $h_1 \in \mathcal{G}(x)$. Then there exists $g_1(t) \in G(t, x(t))$ such that, for each $t \in [0,T]$,

$$h_1(t) = J^q g_1(t) - \frac{t^{q-1}}{\Lambda}\left(\alpha J^q g_1(T) - \sum_{i=1}^{m} \beta_i I_{\eta_i}^{\gamma_i,\delta_i} J^q g_1(\xi_i)\right).$$

By (H_4), we have

$$H_d(F(t,x), F(t,\bar{x})) \leq \ell(t)|x(t) - \bar{x}(t)|.$$

So, there exists $w \in G(t, \bar{x}(t))$ such that

$$|v_1(t) - w(t)| \leq \ell(t)|x(t) \quad \bar{x}(t)|, \quad t \subset [0,T].$$

Define $U : [0,T] \to \mathcal{P}(\mathbb{R})$ by

$$U(t) = \{w \in \mathbb{R} : |v_1(t) - w| \leq \ell(t)|x(t) - \bar{x}(t)|\}.$$

Since the multi-valued operator $U(t) \cap G(t, \bar{x}(t))$ is measurable (Proposition III.4 [78]), there exists a function $v_2(t)$ which is a measurable selection for U. So $v_2(t) \in G(t, \bar{x}(t))$ and for each $t \in [0,T]$, we have $|v_1(t) - v_2(t)| \leq \ell(t)|x(t) - \bar{x}(t)|$.
For each $t \in [0,T]$, let us define

$$h_2(t) = J^q g_2(t) - \frac{t^{q-1}}{\Lambda}\left(\alpha J^q g_2(T) - \sum_{i=1}^{m} \beta_i I_{\eta_i}^{\gamma_i,\delta_i} J^q g_2(\xi_i)\right).$$

Thus we get

$$\begin{aligned}
&|h_1(t) - h_2(t)| \\
&\leq J^q|g_1(t) - g_2(t)| \\
&\quad + \frac{t^{q-1}}{|\Lambda|}\left(\alpha J^q|g_1(T) - g_2(T)| + \sum_{i=1}^{m} \beta_i I_{\eta_i}^{\gamma_i,\delta_i} J^q|g_1(\xi_i) - g_2(\xi_i)|\right) \\
&\leq \left\{J^q\ell(T) + \frac{T^{q-1}}{|\Lambda|}\left(\alpha J^q\ell(T) + \sum_{i=1}^{m} \beta_i I_{\eta_i}^{\gamma_i,\delta_i} J^q\ell(\xi_i)\right)\right\}\|x - \bar{x}\|.
\end{aligned}$$

which leads to

$$\|h_1 - h_2\| \leq \left\{J^q\ell(T) + \frac{T^{q-1}}{|\Lambda|}\left(\alpha J^q\ell(T) + \sum_{i=1}^{m} \beta_i I_{\eta_i}^{\gamma_i,\delta_i} J^q\ell(\xi_i)\right)\right\}\|x - \bar{x}\|.$$

Analogously, interchanging the roles of x and $\overline{x}$, we obtain

$$H_d(\mathcal{F}(x), \mathcal{F}(\bar{x})) \leq \delta\|x - \bar{x}\|,$$

where

$$\delta = J^q\ell(T) + \frac{T^{q-1}}{|\Lambda|}\left(\alpha J^q\ell(T) + \sum_{i=1}^{m} \beta_i I_{\eta_i}^{\gamma_i,\delta_i} J^q\ell(\xi_i)\right).$$

Clearly $\mathcal{G}$ is a contraction, since $\delta < 1$ by (3.186). This completes the proof. □

Lemma 3.16. *Assume that* (H_1) *and* (H_2) *hold. Then the operator* $\mathcal{F}$ *defined by (3.189) is upper semi-continuous.*

Proof. For the sake of convenience, we break the proof into several steps.

Step 1. *$\mathcal{F}$ maps bounded sets (balls) into bounded sets in* $C([0,T],\mathbb{R})$.

For a positive number ρ, let $B_\rho = \{x \in C([0,T],\mathbb{R}) : \|x\| \le \rho\}$ be a bounded ball in $C([0,T],\mathbb{R})$. Then, for each $h \in \mathcal{F}(x), x \in B_\rho$, there exists $f \in S_{F,x}$ such that

$$h(t) = J^q f(t) - \frac{t^{q-1}}{\Lambda}\left(\alpha J^q f(T) - \sum_{i=1}^{m} \beta_i I_{\eta_i}^{\gamma_i,\delta_i} J^q f(\xi_i)\right).$$

Then we have

$$\begin{aligned}
|h(t)| &\le J^q|f(T)| + \frac{|\alpha|T^{q-1}}{|\Lambda|}J^q|f(T)| + \frac{T^{q-1}}{|\Lambda|}\sum_{i=1}^{m}|\beta_i|I_{\eta_i}^{\gamma_i,\delta_i}J^q|f(\xi)| \\
&\le \Phi(\|x\|)J^q p(T) + \Phi(\|x\|)\frac{|\alpha|T^{q-1}}{|\Lambda|}J^q p(T) \\
&\quad +\Phi(\|x\|)\frac{T^{q-1}}{|\Lambda|}\sum_{i=1}^{m}|\beta_i|I_{\eta_i}^{\gamma_i,\delta_i}J^q p(\xi_i),
\end{aligned}$$

and consequently,

$$\|h\| \le \Phi(r)\left\{J^q p(T) + \frac{|\alpha|T^{q-1}}{|\Lambda|}J^q p(T) + \frac{T^{q-1}}{|\Lambda|}\sum_{i=1}^{m}|\beta_i|I_{\eta_i}^{\gamma_i,\delta_i}J^q p(\xi_i)\right\}.$$

Step 2. *$\mathcal{F}$ maps bounded sets into equi-continuous sets of* $C([0,T],\mathbb{R})$.

Let $\tau_1, \tau_2 \in [0,T]$ with $\tau_1 < \tau_2$ and $x \in B_\rho$. For each $h \in \mathcal{F}(x)$, we obtain

$$\begin{aligned}
&|h(\tau_2) - h(\tau_1)| \\
&\le |J^q v(\tau_2) - J^q v(\tau_1)| + \frac{|\alpha||\tau_2^{q-1} - \tau_1^{q-1}|}{|\Lambda|}J^q|v(T)| \\
&\quad + \frac{|\tau_2^{q-1} - \tau_1^{q-1}|}{|\Lambda|}\sum_{i=1}^{m}|\beta_i|I_{\eta_i}^{\gamma_i,\delta_i}J^q|v(\xi)| \\
&\le \frac{\Phi(r)}{\Gamma(q)}\left|\int_0^{\tau_1}[(\tau_2 - s)^{q-1} - (\tau_1 - s)^{q-1}]p(s)ds + \int_{\tau_1}^{\tau_2}(\tau_2 - s)^{q-1}p(s)ds\right| \\
&\quad + \frac{|\tau_2^{q-1} - \tau_1^{q-1}|\Phi(r)}{|\Lambda|}\left(|\alpha|J^q p(T) + \sum_{i=1}^{m}|\beta_i|I_{\eta_i}^{\gamma_i,\delta_i}J^q(\xi_i)\right).
\end{aligned}$$

Obviously the right hand side of the above inequality tends to zero independently of $x \in B_\rho$ as $\tau_2 - \tau_1 \to 0$. As $\mathcal{F}$ satisfies the above three assumptions, therefore it follows by the Arzelá-Ascoli theorem that $\mathcal{F} : C([0,T],\mathbb{R}) \to \mathcal{P}(C([0,T],\mathbb{R}))$ is completely continuous.

Since $\mathcal{F}$ is completely continuous, in order to prove that it is u.s.c., it is enough to prove that it has a closed graph, which is shown in the next step.

Step 3. *$\mathcal{F}$ has a closed graph.*

Let $x_n \to x_*, h_n \in \mathcal{F}(x_n)$ and $h_n \to h_*$. Then we need to show that $h_* \in \mathcal{F}(x_*)$. Associated with $h_n \in \mathcal{F}(x_n)$, there exists $v_n \in S_{F,x_n}$ such that for each $t \in [0,T]$,

$$h_n(t) = J^q v_n(t) - \frac{t^{q-1}}{\Lambda}\left(\alpha J^q v_n(T) - \sum_{i=1}^{m} \beta_i I_{\eta_i}^{\gamma_i,\delta_i} J^q v_n(\xi_i)\right).$$

Thus it suffices to show that there exists $v_* \in S_{F,x_*}$ such that for each $t \in [0,T]$,

$$h_*(t) = J^q v_*(t) - \frac{t^{q-1}}{\Lambda}\left(\alpha J^q v_*(T) - \sum_{i=1}^{m} \beta_i I_{\eta_i}^{\gamma_i,\delta_i} J^q v_*(\xi_i)\right).$$

Let us consider the linear operator $\Theta : L^1([0,T],\mathbb{R}) \to C([0,T],\mathbb{R})$ given by

$$f \mapsto \Theta(v)(t) = J^q v(t) - \frac{t^{q-1}}{\Lambda}\left(\alpha J^q v(T) - \sum_{i=1}^{m} \beta_i I_{\eta_i}^{\gamma_i,\delta_i} J^q v(\xi_i)\right).$$

Observe that

$$\begin{aligned}\|h_n(t) - h_*(t)\| = \Bigg\| J^q(v_n(t) - v_*(t)) - \frac{t^{q-1}}{\Lambda}\Bigg(\alpha J^q(v_n(T) - v_*(T)) \\ - \sum_{i=1}^{m} \beta_i I_{\eta_i}^{\gamma_i,\delta_i} J^q(v_n(\xi_i) - v_*(\xi_i))\Bigg)\Bigg\| \to 0,\end{aligned}$$

as $n \to \infty$.

Thus it follows by Lemma 1.2 that $\Theta \circ S_F$ is a closed graph operator. Further, we have that $h_n(t) \in \Theta(S_{F,x_n})$. Since $x_n \to x_*$, therefore, we have

$$h_*(t) = J^q v_*(t) - \frac{t^{q-1}}{\Lambda}\left(\alpha J^q v_*(T) - \sum_{i=1}^{m} \beta_i I_{\eta_i}^{\gamma_i,\delta_i} J^q v_*(\xi_i)\right),$$

for some $v_* \in S_{F,x_*}$. Thus the operator $\mathcal{F}$ is upper semi-continuous. □

Theorem 3.41. *Assume that $(H_1) - (H_5)$ are satisfied. Then the boundary value problem (3.185) has at least one solution on $[0,T]$.*

Proof. Define an open ball $B_r = \{x \in C([0,T],\mathbb{R}) : \|x\| \leq r\}$, where r satisfies the inequality (3.191) given in condition (H_5). As a consequence of Lemmas 3.14, 3.15, 3.16 the operators $\mathcal{F}$ and $\mathcal{G}$ satisfy all the conditions of Theorem 1.18 and hence its conclusion implies either condition (i) or condition (ii) holds. We show that the conclusion (ii) is not possible. If $x \in \lambda\mathcal{F}(x) + \lambda\mathcal{G}(x)$ for $\lambda \in (0,1)$, then there exist $f \in S_{F,x}$ and $g \in S_{G,x}$ such that

$$x(t) = \lambda J^q f(t) - \lambda\frac{t^{q-1}}{\Lambda}\left(\alpha J^q f(T) - \sum_{i=1}^{m} \beta_i I_{\eta_i}^{\gamma_i,\delta_i} J^q f(\xi_i)\right)$$

$$+\lambda J^q g(t) - \lambda \frac{t^{q-1}}{\Lambda}\left(\alpha J^q g(T) - \sum_{i=1}^{m} \beta_i I_{\eta_i}^{\gamma_i,\delta_i} J^q g(\xi_i)\right).$$

In view of (H_2) and (H_3), we obtain

$$|x(t)| \leq \Phi(\|x\|)\left\{ J^q p(T) + \frac{|\alpha| T^{q-1}}{|\Lambda|} J^q p(T) + \frac{T^{q-1}}{|\Lambda|}\sum_{i=1}^{m} |\beta_i| I_{\eta_i}^{\gamma_i,\delta_i} J^q p(\xi_i)\right\}$$
$$+J^q M(T) + \frac{|\alpha| T^{q-1}}{|\Lambda|} J^q M(T) + \frac{T^{q-1}}{|\Lambda|}\sum_{i=1}^{m} |\beta_i| I_{\eta_i}^{\gamma_i,\delta_i} J^q M(\xi_i),$$

or

$$\|x\| \leq \Phi(\|x\|)\left\{ J^q p(T) + \frac{|\alpha| T^{q-1}}{|\Lambda|} J^q p(T) + \frac{T^{q-1}}{|\Lambda|}\sum_{i=1}^{m} |\beta_i| I_{\eta_i}^{\gamma_i,\delta_i} J^q p(\xi_i)\right\}$$
$$+J^q M(T) + \frac{|\alpha| T^{q-1}}{|\Lambda|} J^q M(T) + \frac{T^{q-1}}{|\Lambda|}\sum_{i=1}^{m} |\beta_i| I_{\eta_i}^{\gamma_i,\delta_i} J^q M(\xi_i). \quad (3.191)$$

If condition (ii) of Theorem 1.18 holds, then there exists $\lambda \in (0,1)$ and $x \in \partial B_r$ with $x = \lambda\mathcal{F}(x) + \lambda\mathcal{G}(x)$. Then, x is a solution of (3.185) with $\|x\| = r$. Now, by the inequality (3.191), we get

$$\frac{r}{\Phi(r)\Psi_1 + \Psi_2} \leq 1,$$

which contradicts (3.187). Hence, $\mathcal{N}$ has a fixed point in $[0,T]$ by Theorem 1.18, and consequently the problem (3.185) has a solution. This completes the proof. □

3.11.2 *Example*

Here, we will illustrate our main result with the help of an example. Let us consider the following boundary value problem for Riemann-Liouville fractional differential inclusions with Erdélyi-Kober fractional integral boundary conditions

$$\begin{cases} D^{3/2}x(t) \in F(t,x(t)) + G(t,x(t)), \quad t \in (0,1), \\ x(0) = 0, \\ \alpha x(1) = \dfrac{3}{2} I_{\frac{1}{5}}^{\frac{1}{3},\frac{1}{7}} x\left(\dfrac{1}{4}\right) + I_{\frac{2}{5}}^{\frac{2}{3},\frac{4}{7}} x\left(\dfrac{1}{2}\right) + 2 I_{\frac{3}{5}}^{\frac{3}{4},\frac{6}{7}} x\left(\dfrac{3}{4}\right). \end{cases} \quad (3.192)$$

Here $q = 3/2$, $m = 3$, $T = 1$, $\beta_1 = 3/2$, $\beta_2 = 1$, $\beta_3 = 2$, $\eta_1 = 1/5$, $\eta_2 = 2/5$, $\eta_3 = 3/5$, $\gamma_1 = 1/3$, $\gamma_2 = 2/3$, $\gamma_3 = 3/4$, $\delta_1 = 1/7$, $\delta_2 = 4/7$, $\delta_3 = 6/7$, $\xi_1 = 1/4$, $\xi_2 = 1/2$, $\xi_3 = 3/4$, the multi-valued maps $F, G : [0,1] \times \mathbb{R} \to \mathcal{P}(\mathbb{R})$ are given by

$$x \to F(t,x) = \left[\frac{t\cos^2 x}{(16+t^2)(1+\cos^2 x)}, \frac{x^2 \sin t}{(1+x^2)(1+t^4)} + \frac{1}{2}\right], \quad (3.193)$$

$$x \to G(t,x) = \left[\frac{2}{\pi\sqrt{t+9}}\tan^{-1} x, \frac{t|x|}{1+|x|} + 1\right] \quad (3.194)$$

and α will be fixed later. For $f \in F$, we have

$$|f| \leq \max\left(\frac{t\cos^2 x}{(16+t^2)(1+\cos^2 x)}, \frac{x^2 \sin t}{(1+x^2)(1+t^4)} + \frac{1}{2}\right) \leq \frac{3}{2}, \ x \in \mathbb{R}.$$

According to the condition (H_2), let us fix $p(t) = 3/2, \Phi(\|x\|) = 1$. In view of the assumptions (H_3) and (H_4), we have $M(t) = 1 + t, \ell(t) = t$. Using the given data, we find that the condition (3.186) holds for any $\alpha > 3.479637$. In our analysis, we take $\alpha = 4$. With this choice, $\Lambda \approx 2.185595$, $\Psi_1 \approx 3.532777$, $\Psi_2 \approx 3.260379$. Hence by the condition (3.187) given by (H_5), there exists $r > 6.793156$. Thus all the conditions of Theorem 3.41 are satisfied. In consequence, there exists a solution for problem (3.192) on $[0, 1]$.

Remark 3.2. The existence result obtained in this Section corresponds to the Dirichlet boundary value problem of perturbed fractional differential inclusions if we take $\alpha = 1$ and $\beta_i = 0$, $i = 1, \ldots, m$ in the statement of the problem (3.185).

3.12 Notes and remarks

The contents of this chapter are adapted from the papers [43], [24], [56], [25], [67], [54], [44], [45], [30], [31], [53], [46], [60], [32] and [33].

Chapter 4

Nonlinear fractional-order boundary value problems with boundary conditions involving fractional derivative operators

In this chapter, we investigate boundary value problems for fractional differential equations and inclusions with boundary conditions involving fractional derivative operators. Nonlocal boundary value problems investigated in this chapter include fractional integro-differential equations with fractional nonlocal integral boundary conditions, differential equations, inclusions and coupled systems for mixed Caputo and Riemann-Liouville fractional derivative operators with nonlocal fractional integro-differential boundary conditions, fractional integro-differential equations and inclusions with three-point boundary conditions, and coupled systems of fractional differential equations with multi-point fractional boundary conditions on an undounded domain.

4.1 Riemann-Liouville fractional integro-differential equations with fractional nonlocal integral boundary conditions

In this section, we study the existence and uniqueness of solutions for the following nonlinear fractional integro-differential equation:

$$D^{\alpha}u(t) = f(t, u(t), (\phi u)(t), (\psi u)(t)), \quad t \in [0, T], \quad \alpha \in (1, 2], \tag{4.1}$$

subject to the boundary conditions of fractional order given by

$$D^{\alpha-2}u(0^+) = 0, \tag{4.2}$$

$$D^{\alpha-1}u(0^+) = \nu I^{\alpha-1}u(\eta), \ 0 < \eta < T, \quad \nu \text{ is a constant}, \tag{4.3}$$

where D^{α} denotes the Riemann-Liouville fractional derivative operator of order α, $f : [0, T] \times \mathbb{R} \times \mathbb{R} \times \mathbb{R} \to \mathbb{R}$ is continuous, and

$$(\phi x)(t) = \int_0^t \gamma(t, s)x(s)\mathrm{d}s, \quad (\psi x)(t) = \int_0^t \delta(t, s)x(s)\mathrm{d}s,$$

with γ and δ being continuous functions on $[0, T] \times [0, T]$.

To define the solution for the nonlinear problem (4.1) and (4.2)–(4.3), we consider the following linear equation

$$D^{\alpha}u(t) = \sigma(t), \ \alpha \in (1, 2], \ \ t \in [0, T], \ T > 0, \tag{4.4}$$

where $\sigma \in C[0,T]$.

We define

$$A = \nu \int_0^\eta \frac{s^{\alpha-1}(\eta - s)^{\alpha-2}}{\Gamma(\alpha-1)} \mathrm{d}s = \frac{\nu\Gamma(\alpha)\eta^{2\alpha-2}}{\Gamma(2\alpha-1)}, \tag{4.5}$$

such that $A \neq \Gamma(\alpha)$.

The general solution of (4.4) is given by

$$u(t) = c_1 t^{\alpha-1} + c_0 t^{\alpha-2} + I^\alpha \sigma(t), \tag{4.6}$$

with I^α the usual Riemann-Liouville fractional integral of order α.

From (4.6), we have

$$D^{\alpha-1}u(t) = c_1\Gamma(\alpha) + I^1\sigma(t), \tag{4.7}$$

$$D^{\alpha-2}u(t) = c_1\Gamma(\alpha)t + c_0\Gamma(\alpha-1) + I^2\sigma(t). \tag{4.8}$$

Using the condition (4.2) and (4.3) in (4.7) and (4.8), we find that $c_0 = 0$ and

$$c_1 = \frac{\nu}{[\Gamma(\alpha) - A]} \int_0^\eta \frac{(\eta - s)^{\alpha-2}}{\Gamma(\alpha-1)} \left(\int_0^s \frac{(s-x)^{\alpha-1}}{\Gamma(\alpha)} \sigma(x)\mathrm{d}x \right) \mathrm{d}s,$$

where A is defined by (4.5).
Substituting the values of c_0 and c_1 in (4.6), the unique solution of (4.4) subject to the boundary condition (4.2)-(4.3) is given by

$$\begin{aligned} u(t) &= \int_0^t \frac{(t-s)^{\alpha-1}}{\Gamma(\alpha)} \sigma(s)\mathrm{d}s \\ &\quad + \frac{\nu t^{\alpha-1}}{[\Gamma(\alpha) - A]} \int_0^\eta \frac{(\eta - s)^{\alpha-2}}{\Gamma(\alpha-1)} \left(\int_0^s \frac{(s-x)^{\alpha-1}}{\Gamma(\alpha)} \sigma(x)\mathrm{d}x \right) \mathrm{d}s \\ &= \int_0^t \frac{(t-s)^{\alpha-1}}{\Gamma(\alpha)} \sigma(s)\mathrm{d}s + \frac{\nu t^{\alpha-1}}{[\Gamma(\alpha) - A]} I^{2\alpha-1}\sigma(\eta). \end{aligned} \tag{4.9}$$

4.1.1 *Main results*

Let $\mathcal{C} = C([0,T],\mathbb{R})$ denotes the Banach space of all continuous functions from $[0,T] \to \mathbb{R}$ endowed with the norm defined by $\|u\| = \sup\{|u(t)|, t \in [0,T]\}$.

If u is a solution of (4.1) and (4.2)–(4.3), then

$$\begin{aligned} u(t) &= \int_0^t \frac{(t-s)^{\alpha-1}}{\Gamma(\alpha)} f(s, u(s), (\phi u)(s), (\psi u)(s))\mathrm{d}s \\ &\quad + \nu_1 t^{\alpha-1} \int_0^\eta \frac{(\eta - s)^{2\alpha-2}}{\Gamma(2\alpha-1)} f(s, u(s), (\phi u)(s), (\psi u)(s))\mathrm{d}s, \end{aligned}$$

where

$$\nu_1 = \frac{\nu}{[\Gamma(\alpha) - A]}.$$

Define an operator $\mathcal{P} : \mathcal{C} \to \mathcal{C}$ as

$$(\mathcal{P}u)(t) = \int_0^t \frac{(t-s)^{\alpha-1}}{\Gamma(\alpha)} f(s, u(s), (\phi u)(s), (\psi u)(s)) \mathrm{d}s$$
$$+\nu_1 t^{\alpha-1} \int_0^\eta \frac{(\eta-s)^{2\alpha-2}}{\Gamma(2\alpha-1)} f(s, u(s), (\phi u)(s), (\psi u)(s)) \mathrm{d}s, \quad t \in [0, T].$$

Observe that the problem (4.1) and (4.2)–(4.3) has solutions if and only if the operator equation $\mathcal{P}u = u$ has fixed points.

Lemma 4.1. *The operator $\mathcal{P}$ is compact.*

Proof. (i) Let $\mathbf{B}$ be a bounded set in $C[0, T]$. Then, there exists a constant M such that $|f(t, u(t), (\phi u)(t), (\psi u)(t))| \leq M, \ \forall u \in \mathbf{B}, \ t \in [0, T]$. Thus

$$|(\mathcal{P}u)(t)| \leq M \int_0^t \frac{(t-s)^{\alpha-1}}{\Gamma(\alpha)} \mathrm{d}s + M|\nu_1| t^{\alpha-1} \int_0^\eta \frac{(\eta-s)^{2\alpha-2}}{\Gamma(2\alpha-1)} \mathrm{d}s$$
$$\leq MT^{\alpha-1} \left(\frac{T}{\Gamma(\alpha+1)} + \frac{|\nu_1| \eta^{2\alpha-1}}{\Gamma(2\alpha)} \right),$$

which implies that

$$\|(\mathcal{P}u)\| \leq MT^{\alpha-1} \left(\frac{T}{\Gamma(\alpha+1)} + \frac{|\nu_1| \eta^{2\alpha-1}}{\Gamma(2\alpha)} \right) < \infty.$$

Hence, $\mathcal{P}(\mathbf{B})$ is uniformly bounded.

(ii) For any $t_1, \ t_2 \in [0, T], u \in \mathbf{B}$, we have

$$|(\mathcal{P}u)(t_1) - (\mathcal{P}u)(t_2)|$$
$$= \left| \int_0^{t_1} \frac{(t_1-s)^{\alpha-1}}{\Gamma(\alpha)} f(s, u(s), (\phi u)(s), (\psi u)(s)) \mathrm{d}s \right.$$
$$- \int_0^{t_2} \frac{(t_2-s)^{\alpha-1}}{\Gamma(\alpha)} f(s, u(s), (\phi u)(s), (\psi u)(s)) \mathrm{d}s$$
$$\left. +\nu_1 (t_1^{\alpha-1} - t_2^{\alpha-1}) \int_0^\eta \frac{(\eta-s)^{2\alpha-2}}{\Gamma(2\alpha-1)} f(s, u(s), (\phi u)(s), (\psi u)(s)) \mathrm{d}s \right|$$
$$\leq M \Bigg(\left| \frac{1}{\Gamma(\alpha)} \int_0^{t_1} \left[(t_1-s)^{\alpha-1} - (t_2-s)^{\alpha-1} \right] \mathrm{d}s - \frac{1}{\Gamma(\alpha)} \int_{t_1}^{t_2} (t_2-s)^{\alpha-1} \mathrm{d}s \right|$$
$$+ \left| \nu_1 (t_1^{\alpha-1} - t_2^{\alpha-1}) \int_0^\eta \frac{(\eta-s)^{2\alpha-2}}{\Gamma(2\alpha-1)} \mathrm{d}s \right| \Bigg) \to 0 \quad \text{as} \quad t_1 \to t_2.$$

Thus, $\mathcal{P}(\mathbf{B})$ is equi-continuous. Consequently, the operator $\mathcal{P}$ is compact. This completes the proof. □

Theorem 4.1. *Assume that there exists a constant $M > 0$ such that*

$$|f(t, u(t), (\phi u)(t), (\psi u)(t))| \leq M, \ \forall \ t \in [0, T], \ u \in \mathbb{R}.$$

Then, the problem (4.1) and (4.2)–(4.3) has at least one solution on $[0, T]$.

Proof. We consider the set

$$V = \{u \in \mathbb{R} \mid u = \mu \mathcal{P} u, \ 0 < \mu < 1\},$$

and show that the set V is bounded. Let $u \in V$, then $u = \mu \mathcal{P} u, \ 0 < \mu < 1$. For any $t \in [0, T]$, we have

$$|u(t)| \leq \mu \left[\int_0^t \frac{(t-s)^{\alpha-1}}{\Gamma(\alpha)} |f(s, u(s), (\phi u)(s), (\psi u)(s))| \mathrm{d}s \right.$$
$$\left. + |\nu_1| t^{\alpha-1} \int_0^\eta \frac{(\eta-s)^{2\alpha-2}}{\Gamma(2\alpha-1)} |f(s, u(s), (\phi u)(s), (\psi u)(s))| \mathrm{d}s \right].$$

As in part (i) of Lemma 4.1, we have

$$\|(\mathcal{P} u)\| \leq M T^{\alpha-1} \left(\frac{T}{\Gamma(\alpha+1)} + \frac{|\nu_1| \eta^{2\alpha-1}}{\Gamma(2\alpha)} \right) < \infty.$$

This implies that the set V is bounded independently of $\mu \in (0, 1)$. Using Lemma 4.1 and Theorem 1.10, we obtain that the operator $\mathcal{P}$ has at least a fixed point, which implies that the problem (4.1) and (4.2)–(4.3) has at least one solution. This completes the proof. □

Theorem 4.2. *Assume that*

(4.2.1) there exist positive functions $L_1(t), \ L_2(t), \ L_3(t)$ such that

$$|f(t, u(t), (\phi u)(t), (\psi u)(t)) - f(t, v(t), (\phi v)(t), (\psi v)(t))|$$
$$\leq L_1(t)|u - v| + L_2(t)|\phi u - \phi v| + L_3(t)|\psi u - \psi v|, \ \forall t \in [0, 1], \ u, v \in \mathbb{R}.$$

(4.2.2) $\Lambda = (\xi_1 + |\nu_1| T^{\alpha-1} \xi_2)(1 + \gamma_0 + \delta_0) < 1$, where

$$\gamma_0 = \sup_{t \in [0,1]} \left| \int_0^t \gamma(t, s) \mathrm{d}s \right|, \quad \delta_0 = \sup_{t \in [0,1]} \left| \int_0^t \delta(t, s) \mathrm{d}s \right|,$$

$$\xi_1 = \sup_{t \in [0,T]} \{|I^q L_1(t)|, \ |I^q L_2(t)|, \ |I^q L_3(t)|\},$$

$$\xi_2 = \max\{|I^{2\alpha-1} L_1(\eta)|, \ |I^{2\alpha-1} L_2(\eta)|, \ |I^{2\alpha-1} L_3(\eta)|\},$$

Then the problem (4.1) and (4.2)–(4.3) has a unique solution on $C[0, T]$.

Proof. Let us set $\sup_{t\in[0,T]}|f(t,0,0,0)| = M$, and choose $r \geq \frac{\varepsilon M}{1-\Lambda}$. Then we show that $\mathcal{P}B_r \subset B_r$, where $B_r = \{x \in \mathcal{C} : \|u\| \leq r\}$. For $x \in B_r$, we have

$$
\begin{aligned}
&|(\mathcal{P}u)(t)| \\
&= \sup_{t\in[0,T]}\bigg|\int_0^t \frac{(t-s)^{\alpha-1}}{\Gamma(\alpha)} f(s,u(s),(\phi u)(s),(\psi u)(s))\mathrm{d}s \\
&\quad +\nu_1 t^{\alpha-1}\int_0^\eta \frac{(\eta-s)^{2\alpha-2}}{\Gamma(2\alpha-1)} f(s,u(s),(\phi u)(s),(\psi u)(s))\mathrm{d}s\bigg| \\
&\leq \sup_{t\in[0,T]}\bigg(\int_0^t \frac{(t-s)^{\alpha-1}}{\Gamma(\alpha)}(|f(s,x(s),(\phi x)(s),(\psi x)(s)) - f(s,0,0,0)| \\
&\quad +|f(s,0,0,0)|)\mathrm{d}s \\
&\quad +|\nu_1|t^{\alpha-1}\int_0^\eta \frac{(\eta-s)^{2\alpha-2}}{\Gamma(2\alpha-1)}(|f(s,x(s),(\phi x)(s),(\psi x)(s)) - f(s,0,0,0)| \\
&\quad +|f(s,0,0,0)|)\mathrm{d}s\bigg) \\
&\leq \sup_{t\in[0,T]}\bigg(\int_0^t \frac{(t-s)^{\alpha-1}}{\Gamma(\alpha)}(L_1(s)|x(s)| + L_2(s)|(\phi x)(s)| + L_3(s)|(\psi x)(s)| + M)\mathrm{d}s \\
&\quad +|\nu_1|t^{\alpha-1}\int_0^\eta \frac{(\eta-s)^{2\alpha-2}}{\Gamma(2\alpha-1)}(L_1(s)|x(s)| + L_2(s)|(\phi x)(s)| + L_3(s)|(\psi x)(s)| + M)\mathrm{d}s\bigg) \\
&\leq \sup_{t\in[0,T]}\bigg(\int_0^t \frac{(t-s)^{\alpha-1}}{\Gamma(\alpha)}(L_1(s)|x(s)| + \gamma_0 L_2(s)|x(s)| \\
&\quad +\delta_0 L_3(s)|x(s)| + M)\mathrm{d}s \\
&\quad +|\nu_1|t^{\alpha-1}\int_0^\eta \frac{(\eta-s)^{2\alpha-2}}{\Gamma(2\alpha-1)}(L_1(s)|x(s)| + \gamma_0 L_2(s)|x(s)| \\
&\quad +\delta_0 L_3(s)|x(s)| + M)\mathrm{d}s\bigg) \\
&\leq \sup_{t\in[0,T]}\bigg((I^\alpha L_1(t) + \gamma_0 I^\alpha L_2(t) + \delta_0 I^\alpha L_3(t))r + \frac{Mt^q}{\Gamma(q+1)} \\
&\quad +|\nu_1|t^{\alpha-1}(I^{(2\alpha-1)}L_1(\eta) + \gamma_0 I^{(2\alpha-1)}L_2(\eta) + \delta_0 I^{(2\alpha-1)}L_3(\eta))r + \frac{M\eta^{2\alpha-1}}{\Gamma(2\alpha)}\bigg) \\
&\leq (\xi_1 + |\nu_1|T^{\alpha-1}\xi_2)(1+\gamma_0+\delta_0)r + MT^{\alpha-1}\left(\frac{T}{\Gamma(\alpha+1)} + \frac{|\nu_1|\eta^{\alpha-1}}{\Gamma(2\alpha)}\right) \\
&= \Lambda r + M\varepsilon \leq r
\end{aligned}
$$

In view of (4.2.1), for every $t \in [0,T]$, we have

$$
\begin{aligned}
&|(\mathcal{P}u)(t) - (\mathcal{P}v)(t)| \\
&\le \sup_{t\in[0,T]} \Big(\int_0^t \frac{(t-s)^{\alpha-1}}{\Gamma(\alpha)} |f(s,u(s),(\phi u)(s),(\psi u)(s) - f(s,v(s),(\phi v)(s),(\psi v)(s))|ds \\
&\quad + |\nu_1| t^{\alpha-1} \int_0^\eta \frac{(\eta-s)^{2\alpha-2}}{\Gamma(2\alpha-1)} |f(s,u(s),(\phi u)(s),(\psi u)(s) - f(s,v(s),(\phi v)(s),(\psi v)(s))|\mathrm{d}s \Big) \\
&\le \sup_{t\in[0,T]} \Big(\int_0^t \frac{(t-s)^{\alpha-1}}{\Gamma(\alpha)} \Big(L_1(s)|u-v| + L_2(s)|\phi u - \phi v| + L_3(s)|\psi u - \psi v| \Big) \mathrm{d}s \\
&\quad + |\nu_1| t^{\alpha-1} \int_0^\eta \frac{(\eta-s)^{2\alpha-2}}{\Gamma(2\alpha-1)} | \Big(L_1(s)|u-v| + L_2(s)|\phi u - \phi v| + L_3(s)|\psi u - \psi v| \Big) \|\mathrm{d}s \Big) \\
&\le \sup_{t\in[0,T]} \Big((I^\alpha L_1(t) + \gamma_0 I^\alpha L_2(t) + \delta_0 I^\alpha L_3(t) \Big) \|u-v\| \\
&\quad + |\nu_1| T^{\alpha-1} \Big(I^{(2\alpha-1)} L_1(\eta) + \gamma_0 I^{(2\alpha-1)} L_2(\eta) + \delta_0 I^{(2\alpha-1)} L_3(\eta) \Big) \|u-v\| \\
&\le (\xi_1 + |\nu_1| T^{\alpha-1} \xi_2)(1+\gamma_0+\delta_0)\|u-v\| = \Lambda \|u-v\|.
\end{aligned}
$$

By assumption (4.2.2), $\Lambda < 1$, therefore, the operator $\mathcal{P}$ is a contraction. Hence, by Banach fixed point theorem, we deduce that $\mathcal{P}$ has a unique fixed point which in fact is a unique solution of problem (4.1) and (4.2)–(4.3). This completes the proof. □

The next theorem relies on Krasnosel'skiĭ's fixed point theorem.

Theorem 4.3. *Assume that $f : [0,T] \times \mathbb{R} \times \mathbb{R} \times \mathbb{R} \to \mathbb{R}$ is a continuous function satisfying (4.2.1). In addition we suppose that:*

(4.3.1) $|f(t,u)| \le \mu(t)$, $\forall (t,u) \in [0,T] \times \mathbb{R}$, and $\mu \in C([0,T],\mathbb{R}^+)$.

If

$$\frac{|\nu_1| T^{\alpha-1} \eta^{2\alpha-1}}{\Gamma(2\alpha)} < 1, \tag{4.10}$$

then the boundary value problem (4.1) and (4.2)–(4.3) has at least one solution on $[0,T]$.

Proof. Letting $\sup_{t\in[0,T]} |\mu(t)| = \|\mu\|$, we fix

$$\overline{r} \ge \|\mu\| T^{\alpha-1} \left(\frac{T}{\Gamma(\alpha+1)} + \frac{|\nu_1| \eta^{2\alpha-1}}{\Gamma(2\alpha)} \right),$$

and consider $B_{\overline{r}} = \{u \in \mathcal{C} : \|u\| \le \overline{r}\}$. We define the operators $\mathcal{P}_1$ and $\mathcal{P}_2$ on $B_{\overline{r}}$ as

$$(\mathcal{P}_1 u)(t) = \int_0^t \frac{(t-s)^{\alpha-1}}{\Gamma(\alpha)} f(s,u(s),(\phi u)(s),(\psi u)(s))\mathrm{d}s,$$

$$(\mathcal{P}_2 u)(t) = \nu_1 t^{\alpha-1} \int_0^{\eta} \frac{(\eta - s)^{2\alpha-2}}{\Gamma(2\alpha-1)} f(s, u(s), (\phi u)(s), (\psi u)(s))ds.$$

For $u, v \in B_{\overline{r}}$, we find that

$$\|\mathcal{P}_1 u + \mathcal{P}_2 v\| \leq \|\mu\| T^{\alpha-1}\left(\frac{T}{\Gamma(\alpha+1)} + \frac{|\nu_1|\eta^{2\alpha-1}}{\Gamma(2\alpha)}\right) \leq \overline{r}.$$

Thus, $\mathcal{P}_1 u + \mathcal{P}_2 v \in B_{\overline{r}}$. It follows from the assumption (4.2.1) together with (4.10) that $\mathcal{P}_2$ is a contraction mapping. Continuity of f implies that the operator $\mathcal{P}_1$ is continuous. Also, $\mathcal{P}_1$ is uniformly bounded on $B_{\overline{r}}$ as

$$\|\mathcal{P}_1 u\| \leq \frac{\|\mu\| T^{\alpha}}{\Gamma(\alpha+1)}.$$

Now we prove the compactness of the operator $\mathcal{P}_1$.

In view of (4.2.1), we define $\sup_{(t,x,\phi x,\psi x)\in[0,T]\times B_r\times B_r\times B_r} |f(t, x, \phi x, \psi x)| = \overline{f}$, and consequently we have

$$\begin{aligned}
&|(\mathcal{P}_1 u)(t_1) - (\mathcal{P}_2 u)(t_2)| \\
&= \left|\frac{1}{\Gamma(\alpha)} \int_0^{t_1} \left[(t_1 - s)^{\alpha-1} - (t_2 - s)^{\alpha-1}\right](s, u(s), (\phi u)(s), (\psi u)(s))ds \right. \\
&\quad \left. - \frac{1}{\Gamma(\alpha)} \int_{t_1}^{t_2} (t_2 - s)^{\alpha-1} f(s, u(s), (\phi u)(s), (\psi u)(s))ds\right| \\
&\leq \frac{\overline{f}}{\Gamma(\alpha+1)} |2(t_2 - t_1)^{\alpha} + t_1^{\alpha} - t_2^{\alpha}|,
\end{aligned}$$

which is independent of u and tends to zero as $t_2 \to t_1$. So, $\mathcal{P}_1$ is relatively compact on $B_{\overline{r}}$. Hence, by the Arzelá–Ascoli Theorem, $\mathcal{P}_1$ is compact on $B_{\overline{r}}$. Thus, all the assumptions of Theorem 1.4 are satisfied. So the conclusion of Theorem 1.4 implies that the boundary value problem (4.1) and (4.2)–(4.3) has at least one solution on $[0, T]$. This completes the proof. □

4.2 Existence theorems for mixed Riemann-Liouville and Caputo fractional differential equations and inclusions with nonlocal fractional integro-differential boundary conditions

We introduce and investigate a new boundary value problem involving both Riemann-Liouville and Caputo fractional derivative operators, and nonlocal fractional integro-differential boundary conditions given by

$$\begin{cases} {}^{RL}D^q({}^C D^r x(t)) = f(t, x(t)), \quad 0 < t < T, \\ x'(\xi) = \lambda\, {}^C D^{\nu} x(\eta), \quad x(T) = \mu\, I^p x(\zeta), \quad \xi, \eta, \zeta \in (0, T), \end{cases} \tag{4.11}$$

where ${}^{RL}D^q$ denote the Riemann-Liouville fractional derivative operator of order q, $0 < q < 1$, ${}^C D^r$, ${}^C D^{\nu}$ denote the Caputo fractional derivative operators of orders r

and ν respectively, $0 < r < 1, 0 < \nu < q + r$, I^p is the Riemann-Liouville fractional integral operator of order $p > 0$, $f : J \times \mathbb{R} \to \mathbb{R}$ and $\lambda, \mu \in \mathbb{R}$.

The main results for the problem (4.11) are based on Banach contraction mapping principle, Krasnosel'skiĭ fixed point theorem and nonlinear alternative of Leray-Schauder type. Then we extend our study to the multi-valued analogue of the problem (4.11) given by

$$\begin{cases} {}^{RL}D^q({}^C D^r x(t)) \in F(t, x(t)), \quad 0 < t < T, \\ x'(\xi) = \lambda \, {}^C D^\nu x(\eta), \quad x(T) = \mu \, I^p x(\zeta), \quad \xi, \eta, \zeta \in (0, T), \end{cases} \tag{4.12}$$

where $F : J \times \mathbb{R} \to \mathcal{P}(\mathbb{R})$ is a multi-valued map ($\mathcal{P}(\mathbb{R})$ is the family of all non-empty subsets of $\mathbb{R}$) and all other constants are the same as defined in problem (4.11). We derive the existence results for the inclusion boundary value problem (4.12) with the aid of standard fixed point theorems for multi-valued maps. In case of convex valued right-hand side of the inclusion, we use Leray-Schauder nonlinear alternative for multi-valued maps. In the case of non-convex valued right hand side of the inclusion, we apply a fixed point theorem for multi-valued contractions due to Covitz and Nadler (Theorem 1.19).

Lemma 4.2. *Let*

$$\begin{aligned} \Lambda_1 &= \frac{\Gamma(q)}{\Gamma(q+r-1)}\xi^{q+r-2} - \lambda\frac{\Gamma(q)}{\Gamma(q+r-\nu)}\eta^{q+r-\nu-1} \neq 0, \\ \Lambda_2 &= \mu\frac{\Gamma(q)}{\Gamma(q+r+p)}\zeta^{q+r+p-1} - \frac{\Gamma(q)}{\Gamma(q+r)}T^{q+r-1}, \; \Lambda_3 = 1 - \mu\frac{1}{\Gamma(1+p)}\zeta^p \neq 0, \end{aligned} \tag{4.13}$$

and $y \in C([0,T], \mathbb{R})$. Then the unique solution of the linear problem

$$\begin{cases} {}^{RL}D^q({}^C D^r x(t)) = y(t), \quad 0 < t < T, \\ x'(\xi) = \lambda \, {}^C D^\nu x(\eta), \quad x(T) = \mu \, I^p x(\zeta), \quad \xi, \eta, \zeta \in (0, T), \end{cases} \tag{4.14}$$

is given by

$$\begin{aligned} x(t) &= I^{q+r}y(t) + \frac{1}{\Lambda_1}\Big[\lambda I^{q+r-\nu}y(\eta) - I^{q+r-1}y(\xi)\Big]\Big[\frac{\Gamma(q)}{\Gamma(q+r)}t^{q+r-1} - \frac{\Lambda_2}{\Lambda_3}\Big] \\ &\quad + \frac{1}{\Lambda_3}\Big[\mu I^{q+r+p}y(\zeta) - I^{q+r}y(T)\Big]. \end{aligned} \tag{4.15}$$

Proof. Applying the Riemann-Liouville fractional integral of order q to both sides of equation in (4.14), we get

$${}^C D^r x(t) = I^q y(t) + c_1 t^{q-1}, \tag{4.16}$$

where $c_1 \in \mathbb{R}$. Next, applying Riemann-Liouville fractional integral of order r to both sides (4.16), we get

$$x(t) = I^{q+r}y(t) + c_1\frac{\Gamma(q)}{\Gamma(q+r)}t^{q+r-1} + c_2, \tag{4.17}$$

where $c_2 \in \mathbb{R}$. From (4.17), we have

$$x'(t) = I^{q+r-1}y(t) + c_1\frac{\Gamma(q)}{\Gamma(q+r-1)}t^{q+r-2}, \tag{4.18}$$

$$^{C}D^{\nu}x(t) = I^{q+r-\nu}y(t) + c_1\frac{\Gamma(q)}{\Gamma(q+r-\nu)}t^{q+r-\nu-1}, \tag{4.19}$$

$$I^{p}x(t) = I^{q+r+p}y(t) + c_1\frac{\Gamma(q)}{\Gamma(q+r+p)}t^{q+r+p-1} + c_2\frac{1}{\Gamma(1+p)}t^{p}. \tag{4.20}$$

Using (4.17)-(4.20) in the boundary conditions of (4.14), we obtain

$$c_1 = \frac{1}{\Lambda_1}[\lambda I^{q+r-\nu}y(\eta) - I^{q+r-1}y(\xi)],$$

$$c_2 = \frac{1}{\Lambda_3}\Big[\mu I^{q+r+p}y(\zeta) - I^{q+r}y(T) - \frac{\Lambda_2}{\Lambda_1}\{\lambda I^{q+r-\nu}y(\eta) - I^{q+r-1}y(\xi)\}\Big],$$

which, on substituting in (4.17), yields the solution (4.15). The converse follows by direct computation. The proof is completed. □

4.2.1 *Main results for single-valued problem (4.11)*

Let $\mathcal{C} = C([0,T],\mathbb{R})$ denote the Banach space of all continuous functions from $[0,T]$ to $\mathbb{R}$ endowed with the norm defined by $\|x\| = \sup_{t\in[0,T]}|x(t)|$. By Lemma 4.2, we define an operator $\mathcal{A}:\mathcal{C}\to\mathcal{C}$ by

$$\begin{aligned}
&(\mathcal{A}x)(t)\\
&= I^{q+r}f(s,x(s))(t)\\
&\quad+\frac{1}{\Lambda_1}\Big[\lambda I^{q+r-\nu}f(s,x(s))(\eta) - I^{q+r-1}f(s,x(s))(\xi)\Big]\Big[\frac{\Gamma(q)}{\Gamma(q+r)}t^{q+r-1} - \frac{\Lambda_2}{\Lambda_3}\Big]\\
&\quad+\frac{1}{\Lambda_3}\Big[\mu I^{q+r+p}f(s,x(s))(\zeta) - I^{q+r}f(s,x(s))(T)\Big],
\end{aligned} \tag{4.21}$$

with $\Lambda_1,\Lambda_3\neq 0$. It should be noticed that problem (4.11) has solutions if and only if the operator $\mathcal{A}$ has fixed points.

For the sake of convenience, we put

$$\Phi_0 = \frac{\Gamma(q)}{\Gamma(q+r)}T^{q+r-1} + \frac{|\Lambda_2|}{|\Lambda_3|}, \tag{4.22}$$

$$\begin{aligned}
\Phi &= \frac{T^{q+r}}{\Gamma(q+r+1)} + \frac{\Phi_0}{|\Lambda_1|}\left[|\lambda|\frac{\eta^{q+r-\nu}}{\Gamma(q+r-\nu+1)} + \frac{\xi^{q+r-1}}{\Gamma(q+r)}\right]\\
&\quad+\frac{1}{|\Lambda_3|}\left[|\mu|\frac{\zeta^{q+r+p}}{\Gamma(q+r+p+1)} + \frac{T^{q+r}}{\Gamma(q+r+1)}\right].
\end{aligned} \tag{4.23}$$

Our first result, dealing with the existence of a unique solution, is based on the Banach contraction mapping principle.

Theorem 4.4. *Let $f:[0,T]\times\mathbb{R}\to\mathbb{R}$ be a continuous function satisfying the Lipschitz condition:*

(4.4.1) there exists a positive constant L such that

$$|f(t,x) - f(t,y)| \leq L|x-y|,\ t \in [0,T],\ \ x,y \in \mathbb{R}.$$

If

$$L\Phi < 1, \tag{4.24}$$

then the boundary value problem (4.11) has a unique solution on $[0,T]$, where Φ is given by (4.23).

Proof. We transform the problem (4.11) into a fixed point problem, $x = \mathcal{A}x$, where the operator $\mathcal{A}$ is defined by (4.21). By using the Banach's contraction mapping principle, we shall show that $\mathcal{A}$ has a fixed point which is the unique solution of problem (4.11).

We set $\sup_{t\in[0,T]} |f(t,0)| = M < \infty$ and choose

$$r \geq \frac{M\Phi}{1-L\Phi}.$$

Now, we show that $\mathcal{A}B_r \subset B_r$, where $B_r = \{x \in \mathcal{C} : \|x\| \leq r\}$. For any $x \in B_r$, we obtain by the assumption (4.4.1) that

$$\begin{aligned}
\|\mathcal{A}x\| \leq & \sup_{t\in[0,T]} \Bigg\{ I^{q+r}|f(s,x(s))|(t) \\
& + \frac{\Phi_0}{|\Lambda_1|}\Big[|\lambda| I^{q+r-\nu}|f(s,x(s))|(\eta) + I^{q+r-1}|f(s,x(s))|(\xi)\Big] \\
& + \frac{1}{|\Lambda_3|}\Big[|\mu| I^{q+r+p}|f(s,x(s))|(\zeta) + I^{q+r}|f(s,x(s))|(T)\Big]\Bigg\} \\
\leq & \sup_{t\in[0,T]} \Bigg\{ I^{q+r}(|f(s,x(s)) - f(s,0)| + |f(s,0)|)(t) \\
& + \frac{\Phi_0}{|\Lambda_1|}\Big[|\lambda| I^{q+r-\nu}(|f(s,x(s)) - f(s,0)| + |f(s,0)|)(\eta) \\
& + I^{q+r-1}(|f(s,x(s)) - f(s,0)| + |f(s,0)|)(\xi)\Big] \\
& + \frac{1}{|\Lambda_3|}\Big[|\mu| I^{q+r+p}(|f(s,x(s)) - f(s,0)| + |f(s,0)|)(\zeta) \\
& + I^{q+r}(|f(s,x(s)) - f(s,0)| + |f(s,0)|)(T)\Big]\Bigg\} \\
\leq & (Lr+M)\Bigg\{\frac{T^{q+r}}{\Gamma(q+r+1)} + \frac{\Phi_0}{|\Lambda_1|}\Bigg[|\lambda|\frac{\eta^{q+r-\nu}}{\Gamma(q+r-\nu+1)} + \frac{\xi^{q+r-1}}{\Gamma(q+r)}\Bigg] \\
& + \frac{1}{|\Lambda_3|}\Bigg[|\mu|\frac{\zeta^{q+r+p}}{\Gamma(q+r+p+1)} + \frac{T^{q+r}}{\Gamma(q+r+1)}\Bigg]\Bigg\} \\
= & (Lr+M)\Phi \\
\leq & r,
\end{aligned}$$

which implies that $\mathcal{A}B_r \subset B_r$. For $t \in [0,T]$ and $x, y \in \mathcal{C}$, we have

$$\begin{aligned}
&|\mathcal{A}x(t) - \mathcal{A}y(t)| \\
&\leq I^{q+r}|f(s,x(s)) - f(s,y(s))|(t) \\
&\quad + \frac{\Phi_0}{|\Lambda_1|}\Big[|\lambda| I^{q+r-\nu}|f(s,x(s))|(\eta) + I^{q+r-1}|f(s,x(s)) - f(s,y(s))|(\xi)\Big] \\
&\quad + \frac{1}{|\Lambda_3|}\Big[|\mu| I^{q+r+p}|f(s,x(s)) - f(s,y(s))|(\zeta) + I^{q+r}|f(s,x(s)) - f(s,y(s))|(T)\Big] \\
&\leq L\|x-y\|\left\{\frac{T^{q+r}}{\Gamma(q+r+1)} + \frac{\Phi_0}{|\Lambda_1|}\left[|\lambda|\frac{\eta^{q+r-\nu}}{\Gamma(q+r-\nu+1)} + \frac{\xi^{q+r-1}}{\Gamma(q+r)}\right]\right. \\
&\quad \left. + \frac{1}{|\Lambda_3|}\left[|\mu|\frac{\zeta^{q+r+p}}{\Gamma(q+r+p+1)} + \frac{T^{q+r}}{\Gamma(q+r+1)}\right]\right\} \\
&= L\Phi\|x-y\|,
\end{aligned}$$

which, on taking the norm for $t \in [0,T]$, yields $\|\mathcal{A}x - \mathcal{A}y\| \leq L\Phi\|x-y\|$. As $L\Phi < 1$, therefore $\mathcal{A}$ is a contraction. Hence, by the conclusion of Banach contraction mapping principle, the operator $\mathcal{A}$ has a unique fixed point which corresponds to a unique solution of the problem (4.11). The proof is completed. □

Next, we prove an existence result for the given problem by using Krasnosel'skiĭ fixed point theorem.

Theorem 4.5. *Assume that $f : [0,T] \times \mathbb{R} \to \mathbb{R}$ is a continuous function satisfying the assumption (4.4.1). In addition we suppose that:*

(4.5.1) $|f(t,x)| \leq \delta(t)$, $\forall (t,x) \in [0,T] \times \mathbb{R}$ and $\delta \in C([0,T],\mathbb{R}^+)$.

Then the boundary value problem (4.11) has at least one solution on $[0,T]$ if

$$L\Big(\Phi - \frac{T^{q+r}}{\Gamma(q+r+1)}\Big) < 1, \tag{4.25}$$

where Φ is given by (4.23).

Proof. We define $\sup_{t\in[0,T]} |\delta(t)| = \|\delta\|$ and choose a suitable constant $\overline{r}$ as

$$\overline{r} \geq \|\delta\|\Phi.$$

Furthermore, we define operators $\mathcal{P}$ and $\mathcal{Q}$ on $B_{\overline{r}} = \{x \in \mathcal{C} : \|x\| \leq \overline{r}\}$ by

$$\begin{aligned}
(\mathcal{P}x)(t) &= I^{q+r}f(s,x(s))(t), \quad t \in [0,T], \\
(\mathcal{Q}x)(t) &= \frac{1}{\Lambda_1}\Big[\lambda I^{q+r-\nu}f(s,x(s))(\eta) - I^{q+r-1}f(s,x(s))(\xi)\Big]\Big[\frac{\Gamma(q)}{\Gamma(q+r)}t^{q+r-1} - \frac{\Lambda_2}{\Lambda_3}\Big] \\
&\quad + \frac{1}{\Lambda_3}\Big[\mu I^{q+r+p}f(s,x(s))(\zeta) - I^{q+r}f(s,x(s))(T)\Big], \quad t \in [0,T].
\end{aligned}$$

Observe that $\mathcal{A}x = \mathcal{P}x + \mathcal{Q}x$. For $x, y \in B_{\overline{r}}$, we have

$$\begin{aligned}\|\mathcal{P}x + \mathcal{Q}y\| &\le \|\delta\|\Bigg\{\frac{T^{q+r}}{\Gamma(q+r+1)} + \frac{\Phi_0}{|\Lambda_1|}\left[|\lambda|\frac{\eta^{q+r-\nu}}{\Gamma(q+r-\nu+1)} + \frac{\xi^{q+r-1}}{\Gamma(q+r)}\right] \\ &\quad + \frac{1}{|\Lambda_3|}\left[|\mu|\frac{\zeta^{q+r+p}}{\Gamma(q+r+p+1)} + \frac{T^{q+r}}{\Gamma(q+r+1)}\right]\Bigg\} \\ &= \|\delta\|\Phi \\ &\le \overline{r}.\end{aligned}$$

This shows that $\mathcal{P}x + \mathcal{Q}y \in B_{\overline{r}}$. It follows from the assumption (4.4.1) together with (4.25) that $\mathcal{Q}$ is a contraction. Since the function f is continuous, we have that the operator $\mathcal{P}$ is continuous. It is easy to verify that

$$\|\mathcal{P}x\| \le \|\delta\|\frac{T^{q+r}}{\Gamma(q+r+1)}.$$

Therefore, $\mathcal{P}$ is uniformly bounded on $B_{\overline{r}}$. Next, we prove the compactness of the operator $\mathcal{P}$. Let us set $\sup_{(t,x)\in[0,T]\times B_{\overline{r}}} |f(t,x)| = \overline{f} < \infty$. Let $t_1, t_2 \in [0,T]$ with $t_1 < t_2$. Then we have

$$\begin{aligned}&|(\mathcal{P}x)(t_2) - (\mathcal{P}x)(t_1)| \\ &\le \frac{\overline{f}}{\Gamma(q+r)}\left|\int_0^{t_1}[(t_2-s)^{q+r-1} - (t_1-s)^{q+r-1}ds\right| + \frac{\overline{f}}{\Gamma(q+r)}\left|\int_{t_1}^{t_2}(t_2-s)^{q+r-1}ds\right| \\ &\le \frac{\overline{f}}{\Gamma(q+r+1)}[|t_2^{q+r} - t_1^{q+r}| + 2(t_2-t_1)^{q+r}],\end{aligned}$$

which is independent of x and tends to zero as $t_2 \to t_1$. Thus, $\mathcal{P}$ is equi-continuous. So $\mathcal{P}$ is relatively compact on $B_{\overline{r}}$. Hence, by the Arzelá-Ascoli theorem, $\mathcal{P}$ is compact on $B_{\overline{r}}$. Thus all the assumptions of Theorem 1.4 are satisfied. So, the boundary value problem (4.11) has at least one solution on $[0,T]$. The proof is completed. □

Remark 4.1. In the above theorem we can interchange the roles of the operators $\mathcal{P}$ and $\mathcal{Q}$ to obtain a second result replacing (4.25) by the following condition:

$$L\frac{T^{q+r}}{\Gamma(q+r+1)} < 1.$$

Now we prove our next existence result by means of Leray-Schauder nonlinear alternative.

Theorem 4.6. *Assume that $f : [0,T]\times\mathbb{R} \to \mathbb{R}$ is a continuous function. In addition we suppose that:*

(4.6.1) there exist a continuous non-decreasing function $\psi : [0,\infty) \to (0,\infty)$ and a function $\phi \in C([0,T],\mathbb{R}^+)$ such that

$$|f(t,x)| \le \phi(t)\psi(\|x\|) \quad \textit{for each } (t,x) \in [0,T]\times\mathbb{R};$$

(4.6.2) there exists a constant $N > 0$ such that

$$\frac{N}{\Phi\|\phi\|\psi(N)} > 1,$$

where Φ is given by (4.23).

Then the boundary value problem (4.11) has at least one solution on $[0, T]$.

Proof. Firstly, we shall show that the operator $\mathcal{A}$, defined by (4.21), *maps bounded sets (balls) into bounded sets in $\mathcal{C}$*. For a positive number R, let $B_R = \{x \in \mathcal{C} : \|x\| \le R\}$ be a bounded ball in $\mathcal{C}$. Then, for $t \in [0, T]$, we have

$$\begin{aligned}
|\mathcal{A}x(t)| &\le I^{q+r}|f(s, x(s))|(t) \\
&\quad + \frac{\Phi_0}{|\Lambda_1|}\Big[|\lambda| I^{q+r-\nu}|f(s, x(s))|(\eta) + I^{q+r-1}|f(s, x(s))|(\xi)\Big] \\
&\quad + \frac{1}{|\Lambda_3|}\Big[|\mu| I^{q+r+p}|f(s, x(s))|(\zeta) + I^{q+r}|f(s, x(s))|(T)\Big] \\
&\le \phi(t)\psi(\|x\|)\Bigg\{\frac{T^{q+r}}{\Gamma(q+r+1)} + \frac{\Phi_0}{|\Lambda_1|}\left[|\lambda|\frac{\eta^{q+r-\nu}}{\Gamma(q+r-\nu+1)} + \frac{\xi^{q+r-1}}{\Gamma(q+r)}\right] \\
&\quad + \frac{1}{|\Lambda_3|}\left[|\mu|\frac{\zeta^{q+r+p}}{\Gamma(q+r+p+1)} + \frac{T^{q+r}}{\Gamma(q+r+1)}\right]\Bigg\},
\end{aligned}$$

which leads to $\|\mathcal{A}x\| \le \Phi\|\phi\|\psi(R)$.

Secondly, we show that $\mathcal{A}$ *maps bounded sets into equi-continuous sets of $\mathcal{C}$*. Let $\nu_1, \nu_2 \in [0, T]$ with $\nu_1 < \nu_2$ and $x \in B_R$. Then, as argued in the proof of Theorem 4.5, we have

$$\begin{aligned}
&|(\mathcal{A}x)(\nu_2) - (\mathcal{A}x)(\nu_1)| \\
&\le I^{q+r}(|f(s, x(s))(\nu_2) - f(s, x(s))(\nu_1)|) \\
&\quad + \frac{|\nu_2^{q+r-1} - \nu_1^{q+r-1}|}{|\Lambda_1|}\frac{\Gamma(q)}{\Gamma(q+r)}\left[|\lambda|\frac{\eta^{q+r-\nu}}{\Gamma(q+r-\nu+1)} + \frac{\xi^{q+r-1}}{\Gamma(q+r)}\right] \\
&\le \frac{\|\phi\|\psi(R)}{\Gamma(r+1)}[|t_2^r - t_2^r| + 2(t_2 - t_1)^r] \\
&\quad + \frac{|\nu_2^{q+r-1} - \nu_1^{q+r-1}|}{|\Lambda_1|}\frac{\Gamma(q)}{\Gamma(q+r)}\left[|\lambda|\frac{\eta^{q+r-\nu}}{\Gamma(q+r-\nu+1)} + \frac{\xi^{q+r-1}}{\Gamma(q+r)}\right].
\end{aligned}$$

Obviously the right hand side of the above inequality tends to zero independently of $x \in B_R$ as $\nu_2 \to \nu_1$. Therefore it follows by the Arzelá-Ascoli theorem that $\mathcal{F} : \mathcal{C} \to \mathcal{C}$ is completely continuous.

The result will follow from the Leray-Schauder nonlinear alternative once it is shown that the set of all solutions to the equation $x = \theta\mathcal{A}x$ is bounded for $\theta \in [0, 1]$. For that, let x be a solution of $x = \theta\mathcal{A}x$ for $\theta \in [0, 1]$. Then, for $t \in [0, T]$, and following the similar computations as in the first step, we have

$$\|x\| \le \Phi\|\phi\|\psi(\|x\|).$$

Consequently, we have

$$\frac{\|x\|}{\Phi\|\phi\|\psi(\|x\|)} \leq 1.$$

In view of (4.6.2), there exists N such that $\|x\| \neq N$. Let us set

$$U = \{x \in \mathcal{C} : \|x\| < N\}. \tag{4.26}$$

Note that the operator $\mathcal{A} : \overline{U} \to \mathcal{C}$ is continuous and completely continuous. From the choice of U, there is no $x \in \partial U$ such that $x = \theta \mathcal{A} x$ for some $\theta \in (0,1)$. Consequently, by nonlinear alternative of Leray-Schauder type (Theorem 1.8) we deduce that $\mathcal{A}$ has a fixed point in $\overline{U}$, which is a solution of the boundary value problem (4.11). This completes the proof. □

4.2.2 *Existence results for multi-valued problem (4.12)*

Definition 4.1. A function $x \in C^2(J,\mathbb{R})$ is said to be a solution of the problem (4.12) if there exists a function $v \in L^1(J,\mathbb{R})$ with $v(t) \in F(t,x)$ a.e. on J such that

$$x(t) = I^{q+r}v(s)(t) + \frac{1}{\Lambda_1}\Big[\lambda I^{q+r-\nu}v(s)(\eta) - I^{q+r-1}v(s)(\xi)\Big]\Big[\frac{\Gamma(q)}{\Gamma(q+r)}t^{q+r-1} - \frac{\Lambda_2}{\Lambda_3}\Big]$$
$$+\frac{1}{\Lambda_3}\Big[\mu I^{q+r+p}v(s)(\zeta) - I^{q+r}v(s)(T)\Big],$$

and $x'(\xi) = \lambda^C D^{\nu} x(\eta),\ x(T) = \mu I^p x(\zeta)$.

4.2.2.1 *The upper semi-continuous case*

Our first result, dealing with the convex-valued F, is based on Leray-Schauder nonlinear alternative for multi-valued maps.

Theorem 4.7. *Assume that:*

(4.7.1) $F : J \times \mathbb{R} \to \mathcal{P}_{cp,c}(\mathbb{R})$ *is* L^1*-Carathéodory;*
(4.7.2) there exists a continuous non-decreasing function $Q : [0,\infty) \to (0,\infty)$ *and a function* $P \in C(J,\mathbb{R}^+)$ *such that*
$\|F(t,x)\|_{\mathcal{P}} := \sup\{|y| : y \in F(t,x)\} \leq P(t)Q(\|x\|)$ *for each* $(t,x) \in J \times \mathbb{R}$;
(4.7.3) there exists a constant $M > 0$ *such that*

$$\frac{M}{\Phi\|P\|Q(M)} > 1,$$

where Φ *is given by (4.23).*

Then the boundary value problem (4.12) has at least one solution on J.

Proof. Firstly, we transform the problem (4.12) into a fixed point problem. Consider the multi-valued map: $N : C(J,\mathbb{R}) \to \mathcal{P}(C(J,\mathbb{R}))$ defined by

$$N(x) = \left\{ \begin{array}{l} h \in C(J,\mathbb{R}) : \\ h(t) = \left\{ \begin{array}{l} I^{q+r}v(s)(t) \\ +\dfrac{1}{\Lambda_1}\Big[\lambda I^{q+r-\nu}v(s)(\eta) - I^{q+r-1}v(s)(\xi)\Big]\Big[\dfrac{\Gamma(q)}{\Gamma(q+r)}t^{q+r-1} - \dfrac{\Lambda_2}{\Lambda_3}\Big] \\ +\dfrac{1}{\Lambda_3}\Big[\mu I^{q+r+p}v(s)(\zeta) - I^{q+r}v(s)(T)\Big], \end{array} \right. \end{array} \right\}$$

for $v \in S_{F,x}$. Clearly the fixed points of N are solutions of problem (4.12). Now we proceed to show that the operator N satisfies all condition of Theorem 1.16. This is done in several steps.

Step 1. *$N(x)$ is convex for each $x \in C(J,\mathbb{R})$.*

Indeed, if h_1, h_2 belongs to $N(x)$, then there exist $v_1, v_2 \in S_{F,x}$ such that for each $t \in J$, we have

$$\begin{aligned} h_i(t) &= I^{q+r}v_i(s)(t) \\ &\quad +\frac{1}{\Lambda_1}\Big[\lambda I^{q+r-\nu}v_i(s)(\eta) - I^{q+r-1}v_i(s)(\xi)\Big]\Big[\frac{\Gamma(q)}{\Gamma(q+r)}t^{q+r-1} - \frac{\Lambda_2}{\Lambda_3}\Big] \\ &\quad +\frac{1}{\Lambda_3}\Big[\mu I^{q+r+p}v_i(s)(\zeta) - I^{q+r}v_i(s)(T)\Big], \quad i = 1, 2. \end{aligned}$$

Let $0 \le \theta \le 1$. Then, for each $t \in J$, we have

$$\begin{aligned} &[\theta h_1 + (1-\theta)h_2](t) \\ &= I^{q+r}[\theta v_1(s) + (1-\theta)v_2(s)])(s)(t) \\ &+\frac{1}{\Lambda_1}\Big[\lambda I^{q+r-\nu}[\theta v_1(s) + (1-\theta)v_2(s)])(s)(\eta) \\ &-I^{q+r-1}[\theta v_1(s) + (1-\theta)v_2(s)])(s)(\xi)\Big]\Big[\frac{\Gamma(q)}{\Gamma(q+r)}t^{q+r-1} - \frac{\Lambda_2}{\Lambda_3}\Big] \\ &+\frac{1}{\Lambda_3}\Big[\mu I^{q+r+p}[\theta v_1(s) + (1-\theta)v_2(s)])(s)(\zeta) - I^{q+r}[\theta v_1(s) + (1-\theta)v_2(s)])(s)(T)\Big]. \end{aligned}$$

Since F has convex values ($S_{F,x}$ is convex), therefore, $\theta h_1 + (1-\theta)h_2 \in N(x)$.

Step 2. *$N(x)$ maps bounded sets (balls) into bounded sets in $C(J,\mathbb{R})$.*

For a positive number r, let $B_r = \{x \in C(J,\mathbb{R}) : \|x\| \le r\}$ be a bounded ball in $C(J,\mathbb{R})$. Then, for each $h \in N(x), x \in B_r$, there exists $v \in S_{F,x}$ such that

$$\begin{aligned} h(t) &= I^{q+r}v(s)(t) \\ &\quad +\frac{1}{\Lambda_1}\Big[\lambda I^{q+r-\nu}v(s)(\eta) - I^{q+r-1}v(s)(\xi)\Big]\Big[\frac{\Gamma(q)}{\Gamma(q+r)}t^{q+r-1} - \frac{\Lambda_2}{\Lambda_3}\Big] \\ &\quad +\frac{1}{\Lambda_3}\Big[\mu I^{q+r+p}v(s)(\zeta) - I^{q+r}v(s)(T)\Big]. \end{aligned}$$

In view of (4.7.2), for each $t \in J$, we have

$$\begin{aligned}|h(t)| &\leq I^{q+r}|v(s)|(t) \\ &+\frac{\Phi_0}{|\Lambda_1|}\Big[|\lambda| I^{q+r-\nu}|v(s)|(\eta) + I^{q+r-1}|v(s)|(\xi)\Big] \\ &+\frac{1}{|\Lambda_3|}\Big[|\mu| I^{q+r+p}v(s)(\zeta) + I^{q+r}v(s)(T)\Big] \\ &\leq P(t)Q(\|x\|)\Bigg\{\frac{T^{q+r}}{\Gamma(q+r+1)} + \frac{\Phi_0}{|\Lambda_1|}\left[|\lambda|\frac{\eta^{q+r-\nu}}{\Gamma(q+r-\nu+1)} + \frac{\xi^{q+r-1}}{\Gamma(q+r)}\right] \\ &+\frac{1}{|\Lambda_3|}\left[|\mu|\frac{\zeta^{q+r+p}}{\Gamma(q+r+p+1)} + \frac{T^{q+r}}{\Gamma(q+r+1)}\right]\Bigg\},\end{aligned}$$

which yields

$$\|h\| \leq \Phi\|P\|Q(r).$$

Step 3. *$N(x)$ maps bounded sets into equi-continuous sets of $C(J,\mathbb{R})$.*

Let x be any element in B_r and $h \in N(x)$. Then there exists a function $v \in S_{F,x}$ such that for each $t \in J$, we have

$$\begin{aligned}h(t) &= I^{q+r}v(s)(t) \\ &+\frac{1}{\Lambda_1}\Big[\lambda I^{q+r-\nu}v(s)(\eta) - I^{q+r-1}v(s)(\xi)\Big]\left[\frac{\Gamma(q)}{\Gamma(q+r)}t^{q+r-1} - \frac{\Lambda_2}{\Lambda_3}\right] \\ &+\frac{1}{\Lambda_3}\Big[\mu I^{q+r+p}v(s)(\zeta) - I^{q+r}v(s)(T)\Big].\end{aligned}$$

Let $\tau_1,\ \tau_2 \in J,\ \tau_1 < \tau_2$. Then

$$\begin{aligned}|h(\tau_2) - h(\tau_1)| &\leq I^{q+r}(|v(s)(\nu_2) - v(s)(\nu_1)|) \\ &+\frac{|\nu_2^{q+r-1} - \nu_1^{q+r-1}|}{|\Lambda_1|}\frac{\Gamma(q)}{\Gamma(q+r)}\left[|\lambda|\frac{\eta^{q+r-\nu}}{\Gamma(q+r-\nu+1)} + \frac{\xi^{q+r-1}}{\Gamma(q+r)}\right] \\ &\leq \frac{\|P\|Q(r)}{\Gamma(r+1)}[|t_2^r - t_2^r| + 2(t_2 - t_1)^r] \\ &+\frac{|\nu_2^{q+r-1} - \nu_1^{q+r-1}|}{|\Lambda_1|}\frac{\Gamma(q)}{\Gamma(q+r)}\left[|\lambda|\frac{\eta^{q+r-\nu}}{\Gamma(q+r-\nu+1)} + \frac{\xi^{q+r-1}}{\Gamma(q+r)}\right].\end{aligned}$$

The right hand of the above inequality tends to zero independently of $x \in B_r$ as $\tau_1 \to \tau_2$. As a consequence of Steps 1-3 together with Arzelá-Ascoli theorem, we conclude that $N : C(J,\mathbb{R}) \to \mathcal{P}(C(J,\mathbb{R}))$ is completely continuous.

Since N is completely continuous, it is enough to show that it has a closed graph in view of Lemma 1.1, which will imply that N is u.s.c. This is done in the following step.

Step 4. *N has a closed graph.*

Let $x_n \to x_*$, $h_n \in N(x_n)$ and $h_n \to h_*$. Then we need to show that $h_* \in N(x_*)$. Observe that $h_n \in N(x_n)$ implies that there exists $v_n \in S_{F,x_n}$ such that for each $t \in J$,

$$\begin{aligned} h_n(t) &= I^{q+r}v_n(s)(t) \\ &+\frac{1}{\Lambda_1}\Big[\lambda I^{q+r-\nu}v_n(s)(\eta) - I^{q+r-1}v_n(s)(\xi)\Big]\Big[\frac{\Gamma(q)}{\Gamma(q+r)}t^{q+r-1} - \frac{\Lambda_2}{\Lambda_3}\Big] \\ &+\frac{1}{\Lambda_3}\Big[\mu I^{q+r+p}v_n(s)(\zeta) - I^{q+r}v_n(s)(T)\Big]. \end{aligned}$$

Therefore, we must show that there exists $v_* \in S_{F,x_*}$ such that for each $t \in J$,

$$\begin{aligned} h_*(t) &- I^{q+r}v_*(s)(t) \\ &+\frac{1}{\Lambda_1}\Big[\lambda I^{q+r-\nu}v_*(s)(\eta) - I^{q+r-1}v_*(s)(\xi)\Big]\Big[\frac{\Gamma(q)}{\Gamma(q+r)}t^{q+r-1} - \frac{\Lambda_2}{\Lambda_3}\Big] \\ &+\frac{1}{\Lambda_3}\Big[\mu I^{q+r+p}v_*(s)(\zeta) - I^{q+r}v_*(s)(T)\Big]. \end{aligned}$$

Consider the continuous linear operator $\Theta : L^1(J,\mathbb{R}) \to C(J,\mathbb{R})$ defined by

$$\begin{aligned} v \to \Theta(v)(t) &= I^{q+r}v(s)(t) \\ &+\frac{1}{\Lambda_1}\Big[\lambda I^{q+r-\nu}v(s)(\eta) - I^{q+r-1}v(s)(\xi)\Big]\Big[\frac{\Gamma(q)}{\Gamma(q+r)}t^{q+r-1} - \frac{\Lambda_2}{\Lambda_3}\Big] \\ &+\frac{1}{\Lambda_3}\Big[\mu I^{q+r+p}v(s)(\zeta) - I^{q+r}v(s)(T)\Big]. \end{aligned}$$

Obviously $\|h_n - h_*\| \to 0$ as $n \to \infty$. So it follows from Lemma 1.2, that $\Theta \circ S_{F,x}$ is a closed graph operator. Moreover, we have

$$h_n \in \Theta(S_{F,x_n}).$$

Since $x_n \to x_*$, Lemma 1.2 implies that

$$\begin{aligned} h_*(t) &= I^{q+r}v_*(s)(t) \\ &+\frac{1}{\Lambda_1}\Big[\lambda I^{q+r-\nu}v_*(s)(\eta) - I^{q+r-1}v_*(s)(\xi)\Big]\Big[\frac{\Gamma(q)}{\Gamma(q+r)}t^{q+r-1} - \frac{\Lambda_2}{\Lambda_3}\Big] \\ &+\frac{1}{\Lambda_3}\Big[\mu I^{q+r+p}v_*(s)(\zeta) - I^{q+r}v_*(s)(T)\Big], \end{aligned}$$

for some $v_* \in S_{F,x_*}$.

Step 5. *We show there exists an open set $U \subseteq C(J,\mathbb{R})$ with $x \notin \theta N(x)$ for any $\theta \in (0,1)$ and all $x \in \partial U$.*

Let $\theta \in (0,1)$ and $x \in \theta N(x)$. Then there exists $v \in L^1(J,\mathbb{R})$ with $v \in S_{F,x}$ such that, for $t \in J$, we have

$$\begin{aligned} x(t) &= \theta I^{q+r}v(s)(t) \\ &+\theta\frac{1}{\Lambda_1}\Big[\lambda I^{q+r-\nu}v(s)(\eta) - I^{q+r-1}v(s)(\xi)\Big]\Big[\frac{\Gamma(q)}{\Gamma(q+r)}t^{q+r-1} - \frac{\Lambda_2}{\Lambda_3}\Big] \end{aligned}$$

$$+\theta\frac{1}{\Lambda_3}\Big[\mu I^{q+r+p}v(s)(\zeta)-I^{q+r}v(s)(T)\Big].$$

Using the method of computation employed in Step 2, for each $t\in J$, we get

$$|x(t)|\leq\Phi\|P\|Q(\|x\|),$$

which can alternatively be written as

$$\frac{\|x\|}{\Phi\|P\|Q(\|x\|)}\leq 1.$$

In view of (4.7.3), there exists M such that $\|x\|\neq M$. Let us set

$$U=\{x\in C(J,\mathbb{R}):\|x\|<M\}.$$

The operator $N:\overline{U}\to\mathcal{P}(C(J,\,\mathbb{R}))$ is a compact multi-valued map, u.s.c. with convex closed values. From the choice of U, there is no $x\in\partial U$ such that $x\in\theta N(x)$ for some $\theta\in(0,1)$. Consequently, by the nonlinear alternative of Leray-Schauder type (Theorem 1.16), we deduce that N has a fixed point $x\in\overline{U}$ which is a solution of the boundary value problem (4.12). This completes the proof. □

4.2.2.2 *The Lipschitz case*

Now we prove the existence of solutions for the boundary value problem (4.12) with a non-convex valued right hand side by applying a fixed point theorem for multi-valued maps due to Covitz and Nadler (Theorem 1.19).

Theorem 4.8. *Assume that the following conditions hold:*

(4.8.1) $F:J\times\mathbb{R}\to\mathcal{P}_{cp}(\mathbb{R})$ *is such that* $F(\cdot,x):J\to\mathcal{P}_{cp}(\mathbb{R})$ *is measurable for each* $x\in\mathbb{R}$;

(4.8.2) $H_d(F(t,x),F(t,\bar{x}))\leq m(t)|x-\bar{x}|$ *for almost all* $t\in J$ *and* $x,\bar{x}\in\mathbb{R}$ *with* $m\in C(J,\mathbb{R}^+)$ *and* $d(0,F(t,0))\leq m(t)$ *for almost all* $t\in J$.

Then the boundary value problem (4.12) has at least one solution on J *if*

$$\Phi\|m\|<1,$$

where Φ *is given by (4.23).*

Proof. We transform the boundary value problem (4.12) into a fixed point problem by considering the operator $N:C(J,\mathbb{R})\to\mathcal{P}(C(J,\mathbb{R}))$ defined in the beginning of the proof of Theorem 4.7. Actually we show that the operator N satisfies the assumptions of Theorem 1.19. The proof will be given in two steps.

Step 1. $N(x)$ is non-empty and closed for every $v\in S_{F,x}$.

Since the set-valued map $F(\cdot,x(\cdot))$ is measurable, by the measurable selection theorem (e.g., [78, Theorem III.6]), it admits a measurable selection $v:J\to\mathbb{R}$. Moreover, by the assumption (4.8.1), we have

$$|v(t)|\leq m(t)+m(t)|x(t)|,$$

i.e., $v \in L^1(J,\mathbb{R})$ and hence F is integrably bounded. Therefore, $S_{F,x} \neq \emptyset$.

Next we show that $N(x)$ is closed for each $x \in C(J,\mathbb{R})$. Let $\{u_n\}_{n\geq 0} \in N(x)$ be such that $u_n \to u$ $(n \to \infty)$ in $C(J,\mathbb{R})$. Then $u \in C(J,\mathbb{R})$ and there exists $v_n \in S_{F,x_n}$ such that, for each $t \in J$,

$$\begin{aligned} u_n(t) &= I^{q+r}v_n(s)(t) \\ &+\frac{1}{\Lambda_1}\Big[\lambda I^{q+r-\nu}v_n(s)(\eta) - I^{q+r-1}v_n(s)(\xi)\Big]\Big[\frac{\Gamma(q)}{\Gamma(q+r)}t^{q+r-1} - \frac{\Lambda_2}{\Lambda_3}\Big] \\ &+\frac{1}{\Lambda_3}\Big[\mu I^{q+r+p}v_n(s)(\zeta) - I^{q+r}v_n(s)(T)\Big]. \end{aligned}$$

As F has compact values, we pass onto a subsequence (if necessary) to obtain that v_n converges to v in $L^1(J,\mathbb{R})$. Thus $v \in S_{F,x}$ and for each $t \in J$, we have

$$\begin{aligned} u_n(t) \to v(t) &= I^{q+r}v(s)(t) \\ &+\frac{1}{\Lambda_1}\Big[\lambda I^{q+r-\nu}v(s)(\eta) - I^{q+r-1}v(s)(\xi)\Big]\Big[\frac{\Gamma(q)}{\Gamma(q+r)}t^{q+r-1} - \frac{\Lambda_2}{\Lambda_3}\Big] \\ &+\frac{1}{\Lambda_3}\Big[\mu I^{q+r+p}v(s)(\zeta) - I^{q+r}v(s)(T)\Big]. \end{aligned}$$

Hence $u \in N(x)$.

Step 2. Next we show that there exists $0 < \hat{\theta} < 1$ $(\hat{\theta} = \Phi\|m\|)$ such that

$$H_d(N(x), N(\bar{x})) \leq \hat{\theta}\|x - \bar{x}\| \quad \text{for each} \quad x, \bar{x} \in C(J,\mathbb{R}).$$

Let $x, \bar{x} \in C(J,\mathbb{R})$ and $h_1 \in N(x)$. Then there exists $v_1(t) \in F(t, x(t))$ such that, for each $t \in J$,

$$\begin{aligned} h_1(t) &= I^{q+r}v_1(s)(t) \\ &+\frac{1}{\Lambda_1}\Big[\lambda I^{q+r-\nu}v_1(s)(\eta) - I^{q+r-1}v_1(s)(\xi)\Big]\Big[\frac{\Gamma(q)}{\Gamma(q+r)}t^{q+r-1} - \frac{\Lambda_2}{\Lambda_3}\Big] \\ &+\frac{1}{\Lambda_3}\Big[\mu I^{q+r+p}v_1(s)(\zeta) - I^{q+r}v_1(s)(T)\Big]. \end{aligned}$$

By (4.8.2), we have

$$H_d(F(t,x), F(t,\bar{x})) \leq m(t)|x(t) - \bar{x}(t)|.$$

So, there exists $w(t) \in F(t, \bar{x}(t))$ such that

$$|v_1(t) - w| \leq m(t)|x(t) - \bar{x}(t)|, \quad t \in J.$$

Define $U : J \to \mathcal{P}(\mathbb{R})$ by

$$U(t) = \{w \in \mathbb{R} : |v_1(t) - w| \leq m(t)|x(t) - \bar{x}(t)|\}.$$

Since the multi-valued operator $U(t) \cap F(t, \bar{x}(t))$ is measurable (Proposition III.4 [78]), there exists a function $v_2(t)$ which is a measurable selection for U. So $v_2(t) \in F(t, \bar{x}(t))$ and for each $t \in J$, we have $|v_1(t) - v_2(t)| \leq m(t)|x(t) - \bar{x}(t)|$.

For each $t \in J$, let us define

$$h_2(t) = I^{q+r}v_2(s)(t) + \frac{1}{\Lambda_1}\Big[\lambda I^{q+r-\nu}v_2(s)(\eta) - I^{q+r-1}v_2(s)(\xi)\Big]\Big[\frac{\Gamma(q)}{\Gamma(q+r)}t^{q+r-1} - \frac{\Lambda_2}{\Lambda_3}\Big] + \frac{1}{\Lambda_3}\Big[\mu I^{q+r+p}v_2(s)(\zeta) - I^{q+r}v_2(s)(T)\Big].$$

In consequence, we get

$$\begin{aligned}
&|h_1(t) - h_2(t)| \\
&\le I^{q+r}|v_2(s) - v_1(s)|(t) \\
&\quad + \frac{\Phi_0}{|\Lambda_1|}\Big[|\lambda| I^{q+r-\nu}|v_2(s) - v_1(s)|(\eta) + I^{q+r-1}|v_2(s) - v_1(s)|(\xi)\Big] \\
&\quad + \frac{1}{|\Lambda_3|}\Big[|\mu| I^{q+r+p}|v_2(s) - v_1(s)|(\zeta) + I^{q+r}|v_2(s) - v_1(s)|(T)\Big] \\
&\le \Bigg\{\frac{T^{q+r}}{\Gamma(q+r+1)} + \frac{\Phi_0}{|\Lambda_1|}\Bigg[|\lambda|\frac{\eta^{q+r-\nu}}{\Gamma(q+r-\nu+1)} + \frac{\xi^{q+r-1}}{\Gamma(q+r)}\Bigg] \\
&\quad + \frac{1}{|\Lambda_3|}\Bigg[|\mu|\frac{\zeta^{q+r+p}}{\Gamma(q+r+p+1)} + \frac{T^{q+r}}{\Gamma(q+r+1)}\Bigg]\Bigg\}\|m\|\|x - \overline{x}\|.
\end{aligned}$$

Hence

$$\|h_1 - h_2\| \le \Phi\|m\|\|x - \overline{x}\|.$$

Analogously, interchanging the roles of x and $\overline{x}$, we obtain

$$H_d(N(x), N(\overline{x})) \le \Phi\|m\|\|x - \overline{x}\|.$$

Since N is a contraction, it follows by Lemma 1.19 that N has a fixed point x which is a solution of (4.12). This completes the proof. □

4.2.3 *Examples*

Consider the following nonlinear Riemann-Liouville and Caputo type fractional boundary value problem

$$\begin{cases} {}^{RL}D^{4/5}\left({}^{C}D^{1/2}x(t)\right)(t) = f(t, x(t)) \ \text{ or } \ \in F(t, x(t)), \quad 0 < t < 3, \\ x'(1/2) = 1/12\ {}^{C}D^{1/2}x(1/3), \quad x(3) = 2\ I^{2/3}x(1/2). \end{cases} \tag{4.27}$$

Here $q = 4/5$, $r = 1/2$, $\xi = 1/2$, $\lambda = 1/12$, $\nu = 1/2$, $\eta = 1/3$, $T = 3$, $\mu = 2$, $p = 1/2$, $\zeta = 1/2$. With these data we find $\Lambda_1 \approx 0.13659$, $\Lambda_2 \approx 2.371158$, $\Lambda_3 \approx -1.256322$, $\Phi_0 \approx 6.525634$, $\Phi \approx 49.056696$.

Single-valued case.

(i) Let

$$f(t, x) = \frac{\cos^2(\pi t)}{t^2 + 250}\Big(\frac{|x|}{|x| + 1}\Big) + \cos(2\pi t). \tag{4.28}$$

Note that $|f(t, x) - f(t, y)| \le (1/(250))|x - y|$ and thus (4.4.1) is satisfied with $L = 1/(250)$. Since $L\Phi \approx 0.736559 < 1$ by Theorem 4.4, the boundary value problem (4.27), with f given by (4.28), has a unique solution on $[0, 3]$.

(ii) With the function f given by (4.28), we remark that $f(t,x)| \leq \frac{1}{t^2+250}+1$ and $L\Big(\Phi - \dfrac{T^{q+r}}{\Gamma(q+r+1)}\Big) \approx 0.181925 < 1$.
Hence, by Theorem 4.5, the boundary value problem (4.27) with f given by (4.28) has at least one solution on $[0,3]$.

(iii) Next consider

$$f(t,x) = \frac{1}{t+350}(x(t)\cos x(t) + 5). \tag{4.29}$$

It is easy to find that $|f(t,x)| \leq \frac{1}{t+50}(\|x\|+5)$. Then by condition (4.6.2) with $\psi(\|x\|) = \|x\|+5$ snf $\|\phi\| = 1/(350)$, we find that $N > N_1 \approx 0.815048$. Hence, by Theorem 4.6, the boundary value problem (4.27), with f given by (4.29), has at least one solution on $[0,3]$.

Multi-valued case.

(i) Consider the multi-valued map $F : [0,3] \times \mathbb{R} \to \mathcal{P}(\mathbb{R})$ given by

$$x \to F(t,x) = \left[\frac{1}{t^2+200}\Big(\frac{x}{3}\Big(\frac{|x|}{|x|+1}+2\Big)+1\Big), \frac{e^{-t}}{19+t}\Big(\sin x + \frac{1}{100}\Big)\right]. \tag{4.30}$$

Clearly the multi-valued map F satisfies condition (4.7.1) and that

$$\|F(t,x)\|_{\mathcal{P}} \leq \frac{1}{t^2+200}(\|x\|+1) := P(t)Q(|x|),$$

which yields $\|P\| = 1/200$ and $Q(\|x\|) = \|x\|+1$. Therefore the condition (4.7.2) is fulfilled. By direct computation, there exists a constant $M > M_1 \approx 0.3250$ satisfying condition (4.7.3). Hence all assumptions of Theorem 4.7 hold and hence the problem (4.27), with F given by (4.30), has at least one solution on $[0,3]$.

(ii) Let the multi-valued map $F : [0,3] \times \mathbb{R} \to \mathcal{P}(\mathbb{R})$ be defined by

$$x \to F(t,x) = \left[0, \frac{1}{200+12t}\left(\frac{x^2+2|x|}{1+|x|}\right) + \frac{1}{400}\right]. \tag{4.31}$$

Choosing $m(t) = 1/(200+12t)$, we can show that $H_d(F(t,x),F(t,y)) \leq m(t)|x-y|$ and $d(0,F(t,0)) = 1/400 \leq m(t)$ for almost all $t \in [0,3]$. In addition we get $\|m\| = 1/200$ which leads to $\Phi\|m\| \approx 0.122641 < 1$. By the conclusion of Theorem 4.8, the problem (4.27), with F given by (4.31), has at least one solution on $[0,3]$.

Remark 4.2. Interchanging the position of Riemann-Liouville and Caputo fractional derivative operators in problem (4.11), we get the following boundary value problem:

$$\begin{cases} {}^{C}D^{r}({}^{RL}D^{q}x(t)) = f(t,x(t)), & 0 < t < T, \\ x(0) = 0, \quad x(T) = \mu\, I^{p}x(\zeta), & \zeta \in (0,T). \end{cases} \tag{4.32}$$

In this case the condition $x(0) = 0$ is necessary for the well-possedness of the problem.

The solution for the problem (4.32) is given by the integral equation

$$x(t) = I^{r+q} f(s, x(s))(t) + \frac{t^q}{\Lambda\Gamma(1+q)}[\mu I^{r+p+q} f(s, x(s))(\zeta) - I^{r+q} f(s, x(s))(T)], \tag{4.33}$$

where it is assumed that

$$\Lambda = \frac{T^q}{\gamma(1+q)} - \frac{\zeta^{q+p}}{\Gamma(q+p+1)} \neq 0.$$

Another more general boundary value problem consisting of Riemann-Liouville and Caputo fractional derivative operators of neutral type is

$$\begin{cases} {}^{RL}D^q[{}^C D^r x(t) - g(t, x(t))] = f(t, x(t)), \quad 0 < t < T, \\ x'(\xi) = \lambda\ {}^C D^\nu x(\eta), \quad x(T) = \mu\ I^p x(\zeta), \quad \xi, \eta, \zeta \in (0, T), \end{cases} \tag{4.34}$$

where $g : J \times \mathbb{R} \to \mathbb{R}$ a continuous function while all other quantities are the same as defined in (4.11). Note that the problem (4.11) is a special case of the problem (4.34) when $g = 0$.

The solution for the problem (4.34) is given by

$$\begin{aligned} x(t) &= I^r g(s, x(s))(t) + I^{q+r} f(s, x(s))(t) \\ &+ \frac{1}{\Lambda}\left[t^{q+r-1}\frac{\Gamma(q)}{\Gamma(q+r)} + \frac{D}{\Omega}\right]\left[\lambda I^{r-\nu} g(s, x(s))(\eta) + \lambda I^{q+r-\nu} f(s, x(s))(\eta)\right. \\ &\left. - I^{r-1} g(s, x(s))(\xi) - I^{q+r-1} f(s, x(s))(\xi)\right] + \frac{1}{\Omega}\left[\mu I^{r+p} g(s, x(s))(\zeta)\right. \\ &\left. + \mu I^{q+r+p} f(s, x(s))(\zeta) - I^r g(s, x(s))(T) - I^{q+r} f(s, x(s))(T)\right]. \end{aligned}$$

where

$$\Omega = 1 - \frac{\mu\zeta^p}{\Gamma(1+p)} \neq 0,$$

$$D = \mu\frac{\Gamma(q)}{\Gamma(q+r+p)}\zeta^{q+r+p-1} - \frac{\Gamma(q)}{\Gamma(q+r)}T^{q+r-1},$$

$$\Lambda = \frac{\Gamma(q)}{\Gamma(q+r-1)}\xi^{q+r-2} - \lambda\frac{\Gamma(q)}{\Gamma(q+r-\nu)}\eta^{q+r-\nu-1} \neq 0.$$

We can obtain the existence results for the problem (4.34) by following the procedure used in the previous sections.

4.3 Fractional order systems for mixed Riemann-Liouville and Caputo fractional derivative operators with integro-differential boundary conditions

In this section, we study fractional order nonlinear coupled systems involving mixed Riemann-Liouville and Caputo fractional derivatives. Precisely, we are concerned with the existence and uniqueness of solutions for the following fractional order nonlinear coupled system containing mixed Riemann-Liouville and Caputo fractional derivative operators, subject to integro-differential boundary conditions:

$$\begin{cases} {}^{RL}D^q({}^CD^r x(t)) = f(t,x(t),y(t)), \quad 0<t<T, \\ {}^CD^r({}^{RL}D^q y(t)) = g(t,x(t),y(t)), \quad 0<t<T, \\ x'(\xi) = \lambda\ {}^CD^\nu y(\eta), \quad x(T) = \mu\ I^p y(\zeta), \quad \xi,\eta,\zeta \in (0,T), \\ y(0)=0, \quad y(T) = \mu_1 I^{p_1} x(\zeta_1), \quad \zeta_1 \in (0,T), \end{cases} \tag{4.35}$$

where ${}^{RL}D^q$ is the standard Riemann-Liouville fractional derivative operator of order $q \in (0,1)$, ${}^CD^r, {}^CD^\nu$ are the Caputo fractional derivative operators of order $r \in (0,1)$ and $\nu \in (0,1)$ respectively with $q+r>1$, I^p, I^{p_1} are the Riemann-Lioville fractional integral operators of order $p>0, p_1>0$, respectively, $f,g : J\times\mathbb{R}\times\mathbb{R} \to \mathbb{R}$ are continuous functions and $\lambda, \mu, \mu_1 \in \mathbb{R}$.

In the following lemma, we solve the linear variant of the coupled system (4.35), which plays a fundamental role in the forthcoming analysis.

Lemma 4.3. *Let* $\Lambda := -A\Gamma_2 + B(A_1\Gamma_2 - A_2\Gamma_1) + A_2\Gamma_0 \neq 0$ *and* $f_1, g_1 \in C([0,T],\mathbb{R})$. *Then the linear system*

$$\begin{cases} {}^{RL}D^q({}^CD^r x(t)) = f_1(t), \quad 0<t<T, \\ {}^CD^r({}^{RL}D^q y(t)) = g_1(t), \quad 0<t<T, \\ x'(\xi) = \lambda\ {}^CD^\nu y(\eta), \quad x(T) = \mu\ I^p y(\zeta), \quad \xi,\eta,\zeta \in (0,T), \\ y(0)=0, \quad y(T) = \mu_1 I^{p_1} x(\zeta_1), \quad \zeta_1 \in (0,T), \end{cases} \tag{4.36}$$

is equivalent to a system of integral equations:

$$\begin{aligned} x(t) &= I^{q+r} f_1(t) + \frac{\Gamma(q)}{\Gamma(q+r)} \frac{t^{q+r-1}}{\Lambda} \Big[-\Gamma_2\Big(I^{q+r}g_1(T) - \mu_1 I^{p_1+q+r} f_1(\zeta_1)\Big) \\ &\quad + B\Gamma_2\Big(\mu I^{q+r+p} g_1(\xi) - I^{q+r} f_1(T)\Big) \\ &\quad + (\Gamma_0 - B\Gamma_1)\Big(\lambda I^{q+r-\nu} g_1(\eta) - I^{q+r-1} f_1(\xi)\Big)\Big] \\ &\quad + \frac{1}{\Lambda}\Big[(A_1\Gamma_2 - A_2\Gamma_1)\Big(I^{q+r}g_1(T) - \mu_1 I^{p_1+q+r} f_1(\zeta_1)\Big) \\ &\quad + (A_2\Gamma_0 - A\Gamma_2)\Big(\mu I^{q+r+p} g_1(\zeta) - I^{q+r} f_1(T)\Big) \\ &\quad + (A\Gamma_1 - A_1\Gamma_0)\Big(\lambda I^{q+r-\nu} g_1(\eta) - I^{q+r-1} f_1(\xi)\Big)\Big], \end{aligned} \tag{4.37}$$

$$\begin{aligned} y(t) &= I^{q+r} g_1(t) + \frac{\Gamma(q)}{\Gamma(q+r)} \frac{t^q}{\Lambda} \Big[-A_2\Big(I^{q+r}g_1(T) - \mu_1 I^{p_1+q+r} f_1(\zeta_1)\Big) \\ &\quad + A_2 B\Big(\mu I^{q+r+p} g_1(\zeta) - I^{q+r} f_1(T)\Big) \\ &\quad + (A - A_1 B)\Big(\lambda I^{q+r-\nu} g_1(\eta) - I^{q+r-1} f_1(\xi)\Big)\Big], \end{aligned} \tag{4.38}$$

where

$$A = \mu_1 \frac{\Gamma(q)}{\Gamma(q+r)} \zeta_1^{p_1+q+r-1}, \quad A_1 = \frac{\Gamma(q)}{\Gamma(q+r)} T^{q+r-1}, \quad A_2 = \frac{\Gamma(q)}{\Gamma(q+r-1)} \xi^{q+r-2}$$

$$B = \frac{\mu_1}{\Gamma(1+p_1)} \zeta_1^{p_1}, \ \Gamma_0 = \frac{1}{\Gamma(1+q)} T^q, \ \Gamma_1 = \frac{\mu}{\Gamma(q+r+1)}, \ \Gamma_2 = \lambda \frac{1}{\Gamma(q+1-\nu)} \eta^{q-\nu}.$$

Proof. Applying the Riemann-Liouville fractional integral operator of order q to both sides of the first equation in (4.36) we get

$$^{C}D^{r}x(t) = I^{q}f_1(t) + c_1 t^{q-1}, \tag{4.39}$$

where $c_1 \in \mathbb{R}$ is an unknown arbitrary constant. Operating the Riemann-Liouville fractional integral of order r on both sides (4.39) yields

$$x(t) = I^{q+r}f_1(t) + c_1\frac{\Gamma(q)}{\Gamma(q+r)}t^{q+r-1} + c_2, \tag{4.40}$$

where $c_2 \in \mathbb{R}$ is an unknown arbitrary constant.

Now applying firstly the Riemann-Liouville fractional integral operator of order r to both sides of the second equation in (4.36), and then Riemann-Liouville fractional integral operator of order q to the resulting equation, we obtain

$$y(t) = I^{q+r}g_1(t) + d_1\frac{1}{\Gamma(1+q)}t^{q} + d_2 t^{q-1}, \tag{4.41}$$

where $d_1, d_2 \in \mathbb{R}$ are unknown arbitrary constants. Using $y(0) = 0$ in (4.41) implies that $d_2 = 0$. Thus, from (4.40) and (4.41), we have

$$x'(t) = I^{q+r-1}f_1(t) + c_1\frac{\Gamma(q)}{\Gamma(q+r-1)}t^{q+r-2},$$

$$^{C}D^{\nu}y(t) = I^{q+r-\nu}g_1(t) + d_1\frac{\Gamma(q)}{\Gamma(q+1-\nu)}t^{q-\nu},$$

$$I^{p}y(t) = I^{q+r+p}g_1(t) + d_1\frac{\Gamma(q)}{\Gamma(q+p+1)}t^{q+p},$$

$$I^{p_1}x(t) = I^{q+r+p_1}f_1(t) + c_1\frac{\Gamma(q)}{\Gamma(q+r+p_1)}t^{q+r+p_1-1} + c_2\frac{1}{\Gamma(1+p_1)}t^{p_1}.$$

Making use of the remaining boundary conditions given by (4.36) into (4.40) and (4.41) together with above expressions, and setting

$$\begin{aligned} P &= I^{q+r}g_1(T) - \mu_1 I^{p_1+q+r}f_1(\zeta_1), \\ Q &= \mu I^{q+r+p}g_1(\xi) - I^{q+r}f_1(T), \\ R &= \lambda I^{q+r-\nu}g_1(\eta) - I^{q+r-1}f_1(\xi), \end{aligned}$$

we obtain the following system:

$$\begin{aligned} Ac_1 + Bc_2 - \Gamma_0 d_1 &= P, \\ A_1c_1 + c_2 - \Gamma_1 d_1 &= Q, \\ A_2c_1 - \Gamma_2 d_1 &= R. \end{aligned}$$

Solving the above system for the unknown constants c_1, c_2, and d_1 yields

$$\begin{aligned} c_1 &= \frac{1}{\Lambda}\Big[-\Gamma_2 P + B\Gamma_2 Q + (\Gamma_0 - B\Gamma_1)R\Big], \\ c_2 &= \frac{1}{\Lambda}\Big[(A_1\Gamma_2 - A_2\Gamma_1)P + (A_2\Gamma_0 - A\Gamma_2)Q + (A\Gamma_1 - A_1\Gamma_0)R\Big], \\ d_1 &= \frac{1}{\Lambda}\Big[-A_2P + A_2BQ + (A - A_1B)R\Big]. \end{aligned}$$

Inserting the values of c_1, c_2 and d_1 in (4.40) and (4.41), we find the required solution given by (4.37) and (4.38). The converse follows by direct computation. The proof is completed. □

4.3.1 *Main results*

Let us introduce the space $X = \{x(t)|x(t) \in C([0,T],\mathbb{R})\}$ endowed with the norm $\|x\| = \sup\{|x(t)|, t \in [0,T]\}$. Obviously $(X, \|\cdot\|)$ is a Banach space. Then the product space $(X \times X, \|(x,y)\|)$ is also a Banach space equipped with norm $\|(x,y)\| = \|x\| + \|y\|$.

In view of Lemma 4.3, we define an operator $T : X \times X \to X \times X$ by

$$T(x,y)(t) = \begin{pmatrix} T_1(x,y)(t) \\ T_2(x,y)(t) \end{pmatrix}, \tag{4.42}$$

where

$$\begin{aligned} T_1(x,y)(t) &= I^{q+r}\bar{f}(t) + \frac{\Gamma(q)}{\Gamma(q+r)}\frac{t^{q+r-1}}{\Lambda}\Big[-\Gamma_2\Big(I^{q+r}\bar{g}(T) - \mu_1 I^{p_1+q+r}\bar{f}(\zeta_1)\Big) \\ &+B\Gamma_2\Big(\mu I^{q+r+p}\bar{g}(\xi) - I^{q+r}\bar{f}(T)\Big) \\ &+(\Gamma_0 - B\Gamma_1)\Big(\lambda I^{q+r-\nu}\bar{g}(\eta) - I^{q+r-1}\bar{f}(\xi)\Big)\Big] \\ &+\frac{1}{\Lambda}\Big[(A_1\Gamma_2 - A_2\Gamma_1)\Big(I^{q+r}\bar{g}(T) - \mu_1 I^{p_1+q+r}\bar{f}(\zeta_1)\Big) \\ &+(A_2\Gamma_0 - A\Gamma_2)\Big(\mu I^{q+r+p}\bar{g}(\zeta) - I^{q+r}\bar{f}(T)\Big) \\ &+(A\Gamma_1 - A_1\Gamma_0)\Big(\lambda I^{q+r-\nu}\bar{g}(\eta) - I^{q+r-1}\bar{f}(\xi)\Big)\Big] \end{aligned} \tag{4.43}$$

and

$$\begin{aligned} T_2(x,y)(t) &= I^{q+r}g_1(t) + \frac{\Gamma(q)}{\Gamma(q+r)}\frac{t^q}{\Lambda}\Big[-A_2\Big(I^{q+r}\bar{g}(T) - \mu_1 I^{p_1+q+r}\bar{f}(\zeta_1)\Big) \\ &+A_2B\Big(\mu I^{q+r+p}\bar{g}(\zeta) - I^{q+r}\bar{f}(T)\Big) \\ &+(A - A_1B)\Big(\lambda I^{q+r-\nu}\bar{g}(\eta) - I^{q+r-1}\bar{f}(\xi)\Big)\Big], \end{aligned} \tag{4.44}$$

where

$$\bar{f}(t) = f(t,x(t),y(t)), \quad \bar{g}(t) = g(t,x(t),y(t)).$$

For computational convenience, we introduce the notations:

$$Q_0 = \frac{\Gamma(q)}{\Gamma(q+r)}\frac{T^{q+r-1}}{|\Lambda|}, \quad \bar{Q}_0 = \frac{\Gamma(q)}{\Gamma(q+r)}\frac{T^q}{|\Lambda|}, \tag{4.45}$$

$$\begin{aligned} Q_1 &= \frac{T^{q+r}}{\Gamma(q+r+1)} + Q_0\Big[|\mu_1|\frac{\zeta_1^{p_1+q+r}}{\Gamma(p_1+q+r+1)} + |B\Gamma_2|\frac{T^{q+r}}{\Gamma(q+r+1)} \\ &+|\Gamma_0 - B\Gamma_1|\frac{\xi^{q+r-1}}{\Gamma(q+r)}\Big] + \frac{1}{|\Lambda|}\Big[|A_1\Gamma_2 - A_2\Gamma_1||\mu_1|\frac{\zeta_1^{p_1+q+r}}{\Gamma(p_1+q+r+1)} \\ &+|A_2\Gamma_0 - A\Gamma_2|\frac{T^{q+r}}{\Gamma(q+r+1)} + |A\Gamma_1 - A_1\Gamma|\frac{\xi^{q+r-1}}{\Gamma(q+r)}\Big], \end{aligned} \tag{4.46}$$

$$Q_2 = Q_0\Big[|\Gamma_2|\frac{T^{q+r}}{\Gamma(q+r+1)} + |B\Gamma_2||\mu|\frac{\xi^{q+r+p}}{\Gamma(q+r+p+1)} + |\Gamma_0 - B\Gamma_1||\lambda|\frac{\eta^{q+r-\nu}}{\Gamma(q+r-\nu+1)}\Big] + \frac{1}{|\Lambda|}\Big[|A_1\Gamma_2 - A_2\Gamma_1|\frac{T^{q+r}}{\Gamma(q+r+1)} + |A_2\Gamma_0 - A\Gamma_2||\mu|\frac{\zeta^{q+r+p}}{\Gamma(q+r+p+1)} + |A\Gamma_1 - A_1\Gamma||\lambda|\frac{\eta^{q+r-\nu}}{\Gamma(q+r-\nu+1)}\Big], \tag{4.47}$$

$$Q_3 = \bar{Q}_0\Big[|A_2||\mu_1|\frac{\zeta_1^{p_1+q+r}}{\Gamma(p_1+q+r+1)} + |A_2B|\frac{T^{q+r}}{\Gamma(q+r+1)} + |A - A_1B|\frac{\xi^{q+r-1}}{\Gamma(q+r)}\Big], \tag{4.48}$$

$$Q_4 = \frac{T^{q+r}}{\Gamma(q+r+1)} + \bar{Q}_0\Big[|A_2|\frac{T^{q+r}}{\Gamma(q+r+1)} + |A_2B||\mu|\frac{\zeta^{q+r+p}}{\Gamma(q+r+p+1)} + |A_2B||\lambda|\frac{\eta^{q+r-\nu}}{\Gamma(q+r-\nu)}\Big]. \tag{4.49}$$

Now the platform is set to present our main results. In the following theorem, we prove the existence and uniqueness of solutions to the system (4.35) via contraction mapping principle due to Banach.

Theorem 4.9. *Assume that:*

(4.9.1) $f, g : [0,T]\times\mathbb{R}\times\mathbb{R} \to \mathbb{R}$ *are continuous functions and that there exist positive constants* ℓ_1 *and* ℓ_2 *such that for all* $t \in [0,T]$ *and* $x_i, y_i \in \mathbb{R},\ i = 1,2,$ *we have*

$$|f(t,x_1,x_2) - f(t,y_1,y_2)| \le \ell_1(|x_1 - y_1| + |x_2 - y_2|),$$
$$|g(t,x_1,x_2) - g(t,y_1,y_2)| \le \ell_2(|x_1 - y_1| + |x_2 - y_2|).$$

Then there exists a unique solution for the the system (4.35) on $[0,T]$, *provided that*

$$(Q_1 + Q_3)\ell_1 + (Q_2 + Q_4)\ell_2 < 1,$$

where $Q_i, i = 0,1,2,3,4$ *are given by (4.45)-(4.49).*

Proof. Define $\sup_{t\in[0,T]} f(t,0,0) = N_1 < \infty$, $\sup_{t\in[0,T]} g(t,0,0) = N_2 < \infty$ and a positive number r such that

$$r > \frac{(Q_1+Q_3)N_1 + (Q_2+Q_4)N_2}{1-(Q_1+Q_3)\ell_1 - (Q_2+Q_4)\ell_2}.$$

Then we show that $TB_r \subset B_r$, where $B_r = \{(x,y) \in X \times X : \|(x,y)\| \le r\}$.

By the assumption (4.9.1), for $(u,v) \in B_r,\ t \in [0,T]$, we have

$$\begin{aligned} |f(t,x(t),y(t))| &\le |f(t,x(t),y(t)) - f(t,0,0)| + |f(t,0,0)| \\ &\le \ell_1(|x(t)| + |y(t)|) + N_1 \\ &\le \ell_1(\|x\| + \|y\|) + N_1 \le \ell_1 r + N_1 \end{aligned}$$

and

$$|g(t,x(t),y(t))| \le \ell_2(\|x\|+\|y\|)+N_2 \le \ell_2 r + N_2.$$

In consequence, we obtain

$$\begin{aligned}
&|T_1(x,y)(t)|\\
&\le I^{q+r}|\bar f(t)| + \frac{\Gamma(q)}{\Gamma(q+r)}\frac{T^{q+r-1}}{|\Lambda|}\Big[|\Gamma_2|\Big(I^{q+r}|\bar g(T)| + |\mu_1|I^{p_1+q+r}|\bar f(\zeta_1)|\Big)\\
&\quad +|B\Gamma_2|\Big(|\mu|I^{q+r+p}|\bar g(\xi)| + I^{q+r}|\bar f(T)|\Big)\\
&\quad +|\Gamma_0 - B\Gamma_1|\Big(|\lambda|I^{q+r-\nu}|\bar g(\eta)| + I^{q+r-1}|\bar f(\xi)|\Big)\Big]\\
&\quad +\frac{1}{|\Lambda|}\Big[|A_1\Gamma_2 - A_2\Gamma_1|\Big(I^{q+r}|\bar g(T)| + |\mu_1|I^{p_1+q+r}|\bar f(\zeta_1)|\Big)\\
&\quad +|A_2\Gamma_0 - A\Gamma_2|\Big(|\mu|I^{q+r+p}|\bar g(\zeta)| + I^{q+r}|\bar f(T)|\Big)\\
&\quad +|A\Gamma_1 - A_1\Gamma_0|\Big(|\lambda|I^{q+r-\nu}|\bar g(\eta)| + I^{q+r-1}|\bar f(\xi)|\Big)\Big]\\
&\le \frac{T^{q+r}}{\Gamma(q+r+1)}(\ell_1 r + N_1) + \frac{\Gamma(q)}{\Gamma(q+r)}\frac{T^{q+r-1}}{|\Lambda|}\Big[|\Gamma_2|\frac{T^{q+r}}{\Gamma(q+r+1)}(\ell_2 r + N_2)\\
&\quad +|\mu_1|\frac{\zeta_1^{p_1+q+r}}{\Gamma(p_1+q+r+1)}(\ell_1 r + N_1)\Big)\\
&\quad +|B\Gamma_2|\Big(|\mu|\frac{\xi^{q+r+p}}{\Gamma(q+r+p+1)}(\ell_2 r + N_2) + \frac{T^{q+r}}{\Gamma(q+r+1)}(\ell_1 r + N_1)\Big)\\
&\quad +|\Gamma_0 - B\Gamma_1|\Big(|\lambda|\frac{\eta^{q+r-\nu}}{\Gamma(q+r-\nu+1)}(\ell_2 r + N_2) + \frac{\xi^{q+r-1}}{\Gamma(q+r)}(\ell_1 r + N_1)\Big)\Big]\\
&\quad +\frac{1}{|\Lambda|}\Big[|A_1\Gamma_2 - A_2\Gamma_1|\Big(\frac{T^{q+r}}{\Gamma(q+r+1)}(\ell_2 r + N_2)\\
&\quad +|\mu_1|\frac{\zeta_1^{p_1+q+r}}{\Gamma(p_1+q+r+1)}(\ell_1 r + N_1)\Big)\\
&\quad +|A_2\Gamma_0 - A\Gamma_2|\Big(|\mu|\frac{\zeta^{q+r+p}}{\Gamma(q+r+p+1)}(\ell_2 r + N_2) + \frac{T^{q+r}}{\Gamma(q+r+1)}(\ell_1 r + N_1)\Big)\\
&\quad +|A\Gamma_1 - A_1\Gamma_0|\Big(|\lambda|\frac{\eta^{q+r-\nu}}{\Gamma(q_r-\nu+1)}(\ell_2 r + N_2) + \frac{\xi^{q+r-1}}{\Gamma(q+r)}(\ell_1 r + N_1)\Big)\Big]\\
&= (Q_1\ell_1 + Q_2\ell_2)r + Q_1N_1 + Q_2N_2,
\end{aligned}$$

which implies that

$$\|T_1(x,y)\| \le (Q_1\ell_1 + Q_2\ell_2)r + Q_1N_1 + Q_2N_2.$$

In the same way, we can find that

$$\|T_2(x,y)\| \le (Q_3\ell_1 + Q_4\ell_2)r + Q_3N_1 + Q_4N_2.$$

From the above inequalities, it follows that

$$\|T(x,y)\| \le [(Q_1+Q_3)\ell_1 + (Q_2+Q_4)\ell_2]r + (Q_1+Q_3)N_1 + (Q_2+Q_4)N_2 \le r.$$

Next, for $(x_2, y_2), (x_1, y_1) \in X \times X$, and for any $t \in [0, T]$, we get

$$
\begin{aligned}
&|T_1(x_2, y_2)(t) - T_1(x_1, y_1)(t)| \\
&\leq \frac{T^{q+r}}{\Gamma(q+r+1)}\ell_1(\|x_2 - x_1\| + \|y_2 - y_1\|) \\
&+\frac{\Gamma(q)}{\Gamma(q+r)}\frac{T^{q+r-1}}{|\Lambda|}\Big[|\Gamma_2|\Big(\frac{T^{q+r}}{\Gamma(q+r+1)}\ell_2(\|x_2 - x_1\| + \|y_2 - y_1\|) \\
&+|\mu_1|\frac{\zeta_1^{p_1+q+r}}{\Gamma(p_1+q+r+1)}\ell_1(\|x_2 - x_1\| + \|y_2 - y_1\|)\Big) \\
&+|B\Gamma_2|\Big(|\mu|\frac{\xi^{q+r+p}}{\Gamma(q+r+p+1)}\ell_2(\|x_2 - x_1\| + \|y_2 - y_1\|) \\
&+\frac{T^{q+r}}{\Gamma(q+r+1)}\ell_1(\|x_2 - x_1\| + \|y_2 - y_1\|)\Big) \\
&+|\Gamma_0 - B\Gamma_1|\Big(|\lambda|\frac{\eta^{q+r-\nu}}{\Gamma(q+r-\nu+1)}\ell_2(\|x_2 - x_1\| + \|y_2 - y_1\|) \\
&+\frac{\xi^{q+r-1}}{\Gamma(q+r)}\ell_1(\|x_2 - x_1\| + \|y_2 - y_1\|)\Big)\Big] \\
&+\frac{1}{|\Lambda|}\Big[|A_1\Gamma_2 - A_2\Gamma_1|\Big(\frac{T^{q+r}}{\Gamma(q+r+1)}\ell_2(\|x_2 - x_1\| + \|y_2 - y_1\|) \\
&+|\mu_1|\frac{\zeta_1^{p_1+q+r}}{\Gamma(p_1+q+r+1)}\ell_1(\|x_2 - x_1\| + \|y_2 - y_1\|)\Big) \\
&+|A_2\Gamma_0 - A\Gamma_2|\Big(|\mu|\frac{\zeta^{q+r+p}}{\Gamma(q+r+p+1)}\ell_2(\|x_2 - x_1\| + \|y_2 - y_1\|) \\
&+\frac{T^{q+r}}{\Gamma(q+r+1)}\ell_1(\|x_2 - x_1\| + \|y_2 - y_1\|)\Big) \\
&+|A\Gamma_1 - A_1\Gamma_0|\Big(|\lambda|\frac{\eta^{q+r-\nu}}{\Gamma(q_r-\nu+1)}\ell_2(\|x_2 - x_1\| + \|y_2 - y_1\|) \\
&+\frac{\xi^{q+r-1}}{\Gamma(q+r)}\ell_1(\|x_2 - x_1\| + \|y_2 - y_1\|)\Big)\Big] \\
&\leq (Q_1\ell_1 + Q_2\ell_2)(\|x_2 - x_1\| + \|y_2 - y_1\|),
\end{aligned}
$$

which leads to

$$\|T_1(x_2, y_2) - T_1(x_1, y_1)\| \leq (Q_1\ell_1 + Q_2\ell_2)(\|x_2 - x_1\| + \|y_2 - y_1\|). \tag{4.50}$$

Similarly, one can obtain

$$\|T_2(x_2, y_2)(t) - T_2(x_1, y_1)\| \leq (Q_3\ell_1 + Q_4\ell_2)(\|x_2 - x_1\| + \|y_2 - y_1\|). \tag{4.51}$$

From (4.50) and (4.51), we deduce that

$$\|T(x_2, y_2) - T(x_1, y_1)\| \leq [(Q_1 + Q_3)\ell_1 + (Q_2 + Q_4)\ell_2](\|x_2 - x_1\| + \|y_2 - y_1\|).$$

Since $(Q_1 + Q_3)\ell_1 + (Q_2 + Q_4)\ell_2 < 1$, therefore, T is a contraction. So, by Banach fixed point theorem, the operator T has a unique fixed point, which corresponds to a unique solution of problem (4.35). This completes the proof. □

The second result is based on Leray-Schauder alternative (Theorem 1.7).

Theorem 4.10. *Assume that:*

(4.10.1) $f, g : [0,T] \times \mathbb{R} \times \mathbb{R} \to \mathbb{R}$ *are continuous functions and that there exist real constants* $k_i, \gamma_i \geq 0$, $(i = 0, 1, 2)$ *with* $k_0 > 0, \gamma_0 > 0$ *such that,* $\forall x_i \in \mathbb{R}$, $(i = 1, 2)$,

$$|f(t, x_1, x_2)| \leq k_0 + k_1|x_1| + k_2|x_2|,$$
$$|g(t, x_1, x_2)| \leq \gamma_0 + \gamma_1|x_1| + \gamma_2|x_2|.$$

If

$$(Q_1 + Q_3)k_1 + (Q_2 + Q_4)\gamma_1 < 1 \quad \text{and} \quad (Q_1 + Q_3)k_2 + (Q_2 + Q_4)\gamma_2 < 1, \tag{4.52}$$

where $Q_i, i = 0, 1, 2, 3, 4$, *are given by (4.45)-(4.49), then the system (4.35) has at least one solution on* $[0, T]$.

Proof. Firstly we show that the operator $T : X \times X \to X \times X$ defined by (4.42) is completely continuous. Notice that continuity of the operator T follows from that of the functions f and g.

Let $\Omega \subset X \times X$ be bounded. Then there exist positive constants L_1 and L_2 such that $|f(t, x(t), y(t))| \leq L_1, |g(t, x(t), y(t))| \leq L_2, \quad \forall (x, y) \in \Omega$. Then, for any $(x, y) \in \Omega$, we have

$$\begin{aligned}
|T_1(x,y)(t)| \leq & \frac{T^{q+r}}{\Gamma(q+r+1)}L_1 + \frac{\Gamma(q)}{\Gamma(q+r)}\frac{T^{q+r-1}}{|\Lambda|}\Big[|\Gamma_2|\Big(\frac{T^{q+r}}{\Gamma(q+r+1)}L_2 \\
& + |\mu_1|\frac{\zeta_1^{p_1+q+r}}{\Gamma(p_1+q+r+1)}L_1\Big) \\
& + |B\Gamma_2|\Big(|\mu|\frac{\xi^{q+r+p}}{\Gamma(q+r+p+1)}L_2 + \frac{T^{q+r}}{\Gamma(q+r+1)}L_1\Big) \\
& + |\Gamma_0 - B\Gamma_1|\Big(|\lambda|\frac{\eta^{q+r-\nu}}{\Gamma(q+r-\nu+1)}L_2 + \frac{\xi^{q+r-1}}{\Gamma(q+r)}L_1\Big)\Big] \\
& + \frac{1}{|\Lambda|}\Big[|A_1\Gamma_2 - A_2\Gamma_1|\Big(\frac{T^{q+r}}{\Gamma(q+r+1)}L_2 + |\mu_1|\frac{\zeta_1^{p_1+q+r}}{\Gamma(p_1+q+r+1)}L_1\Big) \\
& + |A_2\Gamma_0 - A\Gamma_2|\Big(|\mu|\frac{\zeta^{q+r+p}}{\Gamma(q+r+p+1)}L_2 + \frac{T^{q+r}}{\Gamma(q+r+1)}L_1\Big) \\
& + |A\Gamma_1 - A_1\Gamma_0|\Big(|\lambda|\frac{\eta^{q+r-\nu}}{\Gamma(q_r-\nu+1)}L_2 + \frac{\xi^{q+r-1}}{\Gamma(q+r)}L_1\Big)\Big] \\
= & \, Q_1L_1 + Q_2L_2,
\end{aligned}$$

which implies that

$$\|T_1(x,y)\| \leq Q_1L_1 + Q_2L_2.$$

In a similar manner, one can find that

$$\|T_2(x,y)\| \leq Q_3L_1 + Q_4L_2.$$

Thus, it follows from the above inequalities that the operator T is uniformly bounded, since $\|T(x,y)\| \le (Q_1+Q_3)L_1 + (Q_2+Q_4)L_2$.

Next, we show that T is equi-continuous. Let $t_1, t_2 \in [0,T]$ with $t_1 < t_2$. Then we have

$$
\begin{aligned}
&|T_1(x(t_2),y(t_2)) - T_1(x(t_1),y(t_1))| \\
\le\ & L_1\Bigg\{\frac{1}{\Gamma(q+r)}\int_0^{t_1}[(t_2-s)^{q+r-1}-(t_1-s)^{q+r-1}]ds \\
&+\frac{1}{\Gamma(q+r)}\int_{t_1}^{t_2}(t_2-s)^{q+r-1}ds\Bigg\} \\
&+\frac{\Gamma(q)}{\Gamma(q+r)}\frac{|t_2^{q+r-1}-t_1^{q+r-1}|}{|\Lambda|}\Bigg[|\Gamma_2|\Big(\frac{T^{q+r}}{\Gamma(q+r+1)}L_2+|\mu_1|\frac{\zeta_1^{p_1+q+r}}{\Gamma(p_1+q+r+1)}L_1\Big) \\
&+|B\Gamma_2|\Big(|\mu|\frac{\xi^{q+r+p}}{\Gamma(q+r+p+1)}L_2+\frac{T^{q+r}}{\Gamma(q+r+1)}L_1\Big) \\
&+|\Gamma_0-B\Gamma_1|\Big(|\lambda|\frac{\eta^{q+r-\nu}}{\Gamma(q+r-\nu+1)}L_2+\frac{\xi^{q+r-1}}{\Gamma(q+r)}L_1\Big)\Bigg] \\
\le\ & \frac{L_1}{\Gamma(q+r+1)}[2(t_2-t_1)^{q+r}+|t_2^{q+r}-t_1^{q+r}|] \\
&+\frac{\Gamma(q)}{\Gamma(q+r)}\frac{|t_2^{q+r-1}-t_1^{q+r-1}|}{|\Lambda|}\Bigg[|\Gamma_2|\Big(\frac{T^{q+r}}{\Gamma(q+r+1)}L_2+|\mu_1|\frac{\zeta_1^{p_1+q+r}}{\Gamma(p_1+q+r+1)}L_1\Big) \\
&+|B\Gamma_2|\Big(|\mu|\frac{\xi^{q+r+p}}{\Gamma(q+r+p+1)}L_2+\frac{T^{q+r}}{\Gamma(q+r+1)}L_1\Big) \\
&+|\Gamma_0-B\Gamma_1|\Big(|\lambda|\frac{\eta^{q+r-\nu}}{\Gamma(q+r-\nu+1)}L_2+\frac{\xi^{q+r-1}}{\Gamma(q+r)}L_1\Big)\Bigg].
\end{aligned}
$$

Analogously, we can obtain

$$
\begin{aligned}
&|T_2(x(t_2),y(t_2)) - T_2(x(t_1),y(t_1))| \\
\le\ & \frac{L_2}{\Gamma(q+r+1)}[2(t_2-t_1)^{q+r}+|t_2^{q+r}-t_1^{q+r}|] \\
&+\frac{\Gamma(q)}{\Gamma(q+r)}\frac{|t_2^{q}-t_1^{q}|}{|\Lambda|}\Bigg[|A_2|\Big(\frac{T^{q+r}}{\Gamma(q+r+1)}L_2+|\mu_1|\frac{\zeta_1^{p_1+q+r}}{\Gamma(p_1+q+r+1)}L_1\Big) \\
&+|A_2B|\Big(|\mu|\frac{\zeta^{q+r+p}}{\Gamma(q+r+p+1)}L_2+\frac{T^{q+r}}{\Gamma(q+r+1)}L_1\Big) \\
&+|A-A_1B|\Big(|\lambda|\frac{\eta^{q+r-\nu}}{\Gamma(q+r-\nu)}L_2+\frac{\xi^{q+r-1}}{\Gamma(q+r)}L_1\Big)\Bigg].
\end{aligned}
$$

Thus the operator $T(x,y)$ is equi-continuous. In view of the foregoing arguments, we deduce that the operator $T(x,y)$ is completely continuous.

Finally, it will be verified that the set $\mathcal{E} = \{(x,y) \in X \times X | (x,y) = \lambda T(x,y), 0 \le \lambda \le 1\}$ is bounded. Let $(x,y) \in \mathcal{E}$, with $(x,y) = \lambda T(x,y)$. For any $t \in [0,T]$, we have

$$x(t) = \lambda T_1(x,y)(t), \quad y(t) = \lambda T_2(x,y)(t).$$

Then

$$\begin{aligned}|x(t)| &\le Q_1(k_0 + k_1|x| + k_2|y|) + Q_2(\gamma_0 + \gamma_1|x| + \gamma_2|y|)\\ &= Q_1k_0 + Q_2\gamma_0 + (Q_1k_1 + Q_2\gamma_1)|x| + (Q_1k_2 + Q_2\gamma_2)|y|.\end{aligned}$$

and

$$\begin{aligned}|y(t)| &\le Q_3(k_0 + k_1|x| + k_2|y|) + Q_4(\gamma_0 + \gamma_1|x| + \gamma_2|y|)\\ &= Q_3k_0 + Q_4\gamma_0 + (Q_3k_1 + Q_4\gamma_1)|x| + (Q_3k_2 + Q_4\gamma_2)|y|.\end{aligned}$$

In consequence, we have

$$\|x\| \le Q_1k_0 + Q_2\gamma_0 + (Q_1k_1 + Q_2\gamma_1)\|x\| + (Q_1k_2 + Q_2\gamma_2)\|y\|$$

and

$$\|y\| \le Q_3k_0 + Q_4\gamma_0 + (Q_3k_1 + Q_4\gamma_1)\|x\| + (Q_3k_2 + Q_4\gamma_2)\|y\|,$$

which imply that

$$\begin{aligned}\|x\| + \|y\| \le\ & (Q_1 + Q_3)k_0 + (Q_2 + Q_4)\gamma_0 + [(Q_1 + Q_3)k_1 + (Q_2 + Q_4)\gamma_1]\|x\|\\ & + [(Q_1 + Q_3)k_2 + (Q_2 + Q_4)\gamma_2)]\|y\|.\end{aligned}$$

Thus we have

$$\|(x, y)\| \le \frac{(Q_1 + Q_3)k_0 + (Q_2 + Q_4)\gamma_0}{M_0},$$

where $M_0 = \min\{1-[(Q_1+Q_3)k_1+(Q_2+Q_4)\gamma_1], 1-[(Q_1+Q_3)k_2+(Q_2+Q_4)\gamma_2)]\}$, which establishes that the set $\mathcal{E}$ is bounded. Thus, by Theorem 1.7, the operator T has at least one fixed point. Hence the system (4.35) has at least one solution. The proof is complete. □

Example 4.1. Consider the following boundary value problem of coupled nonlinear fractional differential equations

$$\begin{cases} {}^{RL}D^{4/5}\left({}^{C}D^{1/2}x(t)\right)(t) = \dfrac{1}{8(t+2)^2}\dfrac{|x|}{1+|x|} + 1 + \dfrac{1}{64}\sin^2 y, \quad 0 < t < 3,\\ {}^{C}D^{1/2}({}^{RL}D^{4/5}y(t)) = \dfrac{1}{64\pi}\sin(2\pi x) + \dfrac{|y|}{32(1+|y|)} + \dfrac{1}{2}, \quad 0 < t < 3,\\ x'(1/2) = (1/12)\ {}^{C}D^{1/2}y(1/3), \quad x(3) = 2\ I^{2/3}y(1/2),\\ y(0) = 0, \quad y(3) = (1/2)I^{1/2}x(1/4). \end{cases} \tag{4.53}$$

Here $q = 4/5$, $r = 1/2$, $\xi = 1/2$, $\lambda = 1/12$, $\nu = 1/2$, $\eta = 1/3$, $T = 3$, $\mu = 2$, $p = 2/3$, $\zeta = 1/2$, $\mu_1 = 1/2$, $p_1 = 1/2$, $\zeta_1 = 1/4$. With these data we find $A \approx 0.069498$, $A_1 \approx 1.546083$, $A_2 \approx 0.684503$, $B \approx 0.282040$, $\Gamma_0 \approx 2.585643$, $\Gamma_1 \approx 1.714220$, $\Gamma_2 \approx 0.092930$, $\Lambda \approx 1.473003$, $Q_0 \approx 1.049991$, $\bar{Q}_0 \approx 1.817985$, $Q_1 \approx 6.24681$, $Q_2 \approx 4.728334$, $Q_3 \approx 1.889017$, $Q_4 \approx 8.732422$. From $f(t, x, y) = \dfrac{1}{8(t+2)^2}\dfrac{|x|}{1+|x|}+$

$1+\frac{1}{64}\sin^2 y$, and $g(t,x,y)=\frac{1}{64\pi}\sin(2\pi x)+\frac{|y|}{32(1+|y|)}+\frac{1}{2}$, we have $\ell_1=\ell_2=\frac{1}{32}$ as $|f(t,x_1,x_2)-f(t,y_1,y_2)|\leq \frac{1}{32}(|x_1-x_2|+|y_1-y_2|)$, $|g(t,x_1,x_2)-g(t,y_1,y_2)|\leq \frac{1}{32}(|x_1-x_2|+|y_1-y_2|)$. Then $(Q_1+Q_3)\ell_1+(Q_2+Q_4)\ell_2\approx 0.674892<1$. Thus all the conditions of Theorem 4.9 are satisfied and consequently, its conclusion applies to the problem (4.53).

Remark 4.3. By fixing the parameters involved in the given problem, we can obtain some new results as special cases of the present work. For instance, if we take $\lambda=0=\mu,\mu_1=0$, our results correspond to the uncoupled three-point nonlocal boundary conditions of the form: $x'(\xi)=0, x(T)=0; y(0)=0, y(T)=0, 0<\xi<T$. Letting $\mu_1=0$ in the results of this paper, we obtain the ones for a coupled system of differential equations involving both Riemann-Liouville and Caputo fractional derivative operators, subject to the boundary conditions: $x'(\xi)=\lambda\ {}^CD^{\nu}y(\eta),\quad x(T)=\mu\ I^py(\zeta); y(0)=0, y(T)=0, \xi,\eta,\zeta\in(0,T)$.

Moreover, if we interchange Riemann-Liouville and Caputo fractional derivative operators in the coupled system (4.35), we obtain the following problem:

$$\begin{cases} {}^CD^r({}^{RL}D^q(x(t))=f(t,x(t),y(t)), & 0<t<T,\\ {}^{RL}D^q({}^CD^r(y(t))=g(t,x(t),y(t)), & 0<t<T,\\ x(0)=0,\quad x(T)=\mu\ I^py(\zeta), & \zeta\in(0,T),\\ y'(\xi)=\lambda\ {}^CD^{\nu}y(\eta),\quad y(T)=\mu_1I^{p_1}x(\zeta_1), & \xi,\eta,\zeta_1\in(0,T).\end{cases}$$

We can investigate this problem like the system (4.35).

4.4 A study of coupled systems of mixed order fractional differential equations and inclusions with coupled integral fractional boundary conditions

In this section, we consider the following systems of Caputo and Riemannn-Liouville type mixed order coupled fractional differential equations and inclusions:

$$\begin{cases} {}^CD^{\alpha}x(t)=f(t,x(t),y(t)), & t\in[0,T],\ 0<\alpha\leq 1,\\ {}^{RL}D^{\beta}y(t)=g(t,x(t),y(t)), & t\in[0,T],\quad 1<\beta\leq 2,\end{cases}\tag{4.54}$$

$$\begin{cases} {}^CD^{\alpha}x(t)\in F(t,x(t),y(t)), & t\in[0,T],\ 0<\alpha\leq 1,\\ {}^{RL}D^{\beta}y(t)\in G(t,x(t),y(t)), & t\in[0,T],\quad 1<\beta\leq 2,\end{cases}\tag{4.55}$$

subject to the following coupled fractional boundary conditions:

$$\begin{cases} x(0)=\lambda^cD^py(\eta), & 0<p<1,\\ y(0)=0,\quad y(T)=\gamma I^qx(\xi),\end{cases}\tag{4.56}$$

where ${}^CD^{\alpha},{}^CD^p$, are the Caputo fractional derivatives operators of order α and p respectively, ${}^{RL}D^{\beta}$ is the Riemann-Liouville fractional derivative operator of order β, I^q is the Riemann-Liouville fractional integral operator of order q, $f,g:[0,T]\times$

$\mathbb{R} \times \mathbb{R} \to \mathbb{R}$, $F, G : [0,T] \times \mathbb{R} \times \mathbb{R} \to \mathcal{P}(\mathbb{R})$ are given continuous functions, $\mathcal{P}(\mathbb{R})$ is the family of all non-empty subsets of $\mathbb{R}$, $\eta, \xi \in (0,T)$ and $\lambda, \gamma \in \mathbb{R}$.

Here we emphasize that the proposed single and multi-valued problems include:

- *fractional derivative operators of different orders $\alpha \in (0,1]$ and $\beta \in (1,2]$;*
- *the first and second equations in the given systems are respectively Caputo and Riemann-Liouville types;*
- *the boundary conditions are of nonlocal type and contain both fractional derivative operators and integrals.*

The following auxiliary lemma which concerns the linear variant of system (4.54), plays a key role in the sequel.

Lemma 4.4. *Let $\phi, h \subset C([0,T], \mathbb{R})$. Then the solution of the linear fractional differential system*

$$\begin{cases} {}^{C}D^{\alpha}x(t) = \phi(t), \quad t \in [0,T], \quad 1 < \alpha \leq 2 \\ {}^{RL}D^{\beta}y(t) = h(t), \quad t \in [0,T], \quad 1 < \beta \leq 2, \\ x(0) = \lambda D^{p}y(\eta), \\ y(0) = 0, \quad y(T) = \gamma I^{q}x(\xi), \end{cases} \tag{4.57}$$

is equivalent to the system of integral equations

$$x(t) = I^{\alpha}\phi(t) + \frac{\lambda}{\Lambda}\Big[- T^{\beta-1}I^{\beta-p}h(\eta) + \frac{\Gamma(\beta)}{\Gamma(\beta-p)}\eta^{\beta-p-1}\Big(\gamma I^{q+\alpha}\phi(\xi) - I^{\beta}h(T)\Big)\Big], \tag{4.58}$$

and

$$y(t) = I^{\beta}h(t) + \frac{t^{\beta-1}}{\Lambda}\Big[I^{\beta}h(T) - \gamma I^{q+\alpha}\phi(\xi) - \lambda\gamma\frac{\xi^{q}}{\Gamma(1+q)}I^{\beta-p}h(\eta)\Big], \tag{4.59}$$

where it is assumed that

$$\Lambda := T^{\beta-1} + \lambda\gamma\frac{\Gamma(\beta)\xi^{q}\eta^{\beta-p-1}}{\Gamma(1+q)\Gamma(\beta-p)} \neq 0. \tag{4.60}$$

Proof. Applying the Riemann-Liouville operators I^{α} and I^{β} to the Caputo and Riemann-Liouville fractional differential equations respectively in (4.57) and using the composition laws of fractional order integral and differential operators, we obtain

$$x(t) = I^{\alpha}\phi(t) + c_0 \quad \text{and} \quad y(t) = I^{\beta}h(t) + c_1 t^{\beta-1} + c_2 t^{\beta-2}, \tag{4.61}$$

where c_0, c_1, c_2 are arbitrary constants.

Using the boundary conditions of (4.57) in (4.61), we get $c_2 = 0$ and a system of algebraic equations in the unknown constants c_0 and c_1:

$$\begin{aligned} c_0 - \lambda\frac{\Gamma(\beta)}{\Gamma(\beta-p)}\eta^{\beta-p-1}c_1 &= \lambda I^{\beta-p}h(\eta), \\ \gamma\frac{\xi^{q}}{\Gamma(1+q)}c_0 - T^{\beta-1}c_1 &= I^{\beta}h(T) - \gamma I^{q+\alpha}\phi(\xi). \end{aligned}$$

Solving the above system, we get

$$c_0 = \frac{\lambda}{\Lambda}\Big[-T^{\beta-1}I^{\beta-p}h(\eta) + \frac{\Gamma(\beta)}{\Gamma(\beta-p)}\eta^{\beta-p-1}\Big(\gamma I^{q+\alpha}\phi(\xi) - I^{\beta}h(T)\Big)\Big],$$

and

$$c_1 = \frac{1}{\Lambda}\Big[I^{\beta}h(T) - \gamma I^{q+\alpha}\phi(\xi) - \lambda\gamma\frac{\xi^q}{\Gamma(1+q)}I^{\beta-p}h(\eta)\Big].$$

Substituting the values of c_0, c_1, c_2 in (4.61), we get the solution (4.58) and (4.59). We can prove the converse of the lemma by direct computation. The proof is completed. □

4.4.1 *The single-valued system (4.54)-(4.56)*

Let $X = \{x(t)|x(t) \in C([0,T],\mathbb{R})\}$ be the Banach space endowed with the norm $\|x\| = \sup\{|x(t)|, t \in [0,T]\}$. Also let the product space $(X \times X, \|(x,y)\|)$ be the Banach space equipped with norm $\|(x,y)\| = \|x\| + \|y\|$.

We define an operator $\mathcal{H} : X \times X \to X \times X$ by

$$\mathcal{H}(x,y)(t) = \begin{pmatrix} \mathcal{H}_1(x,y)(t) \\ \mathcal{H}_2(x,y)(t) \end{pmatrix}, \tag{4.62}$$

where

$$\mathcal{H}_1(x,y)(t) = I^{\alpha}\widehat{f}(t) + \frac{\lambda}{\Lambda}\Big[-T^{\beta-1}I^{\beta-p}\widehat{g}(\eta) + \frac{\Gamma(\beta)}{\Gamma(\beta-p)}\eta^{\beta-p-1}\Big(\gamma I^{q+\alpha}\widehat{f}(\xi) - I^{\beta}\widehat{g}(T)\Big)\Big],$$

and

$$\mathcal{H}_2(x,y)(t) = I^{\beta}\widehat{g}(t) + \frac{t^{\beta-1}}{\Lambda}\Big[I^{\beta}\widehat{g}(T) - \gamma I^{q+\alpha}\widehat{f}(\xi) - \lambda\gamma\frac{\xi^q}{\Gamma(1+q)}I^{\beta-p}\widehat{g}(\eta)\Big],$$

and $\widehat{f}(t) = f(t,x(t),y(t))$, $\widehat{g}(t) = g(t,x(t),y(t))$.

For convenience we set the notation:

$$M_1 = \frac{T^{\alpha}}{\Gamma(1+\alpha)} + \frac{1}{|\Lambda|}|\lambda||\gamma|\frac{\Gamma(\beta)}{\Gamma(\beta-p)}\frac{\eta^{\beta-p-1}\xi^{q+\alpha}}{\Gamma(q+\alpha+1)}, \tag{4.63}$$

$$M_2 = \frac{T^{\beta-1}\eta^{\beta-p-1}|\lambda|}{|\Lambda|}\Big[\frac{T\Gamma(\beta)}{\Gamma(\beta-p)\Gamma(\beta+1)} + \frac{\eta}{\Gamma(1+\beta)}\Big], \tag{4.64}$$

$$M_3 = \frac{T^{\beta-1}|\gamma|\xi^{q+\alpha}}{|\Lambda|\Gamma(q+\alpha+1)}, \tag{4.65}$$

$$M_4 = \frac{T^{\beta}}{\Gamma(1+\beta)}\Big(1 + \frac{T^{\beta-1}}{|\Lambda|}\Big) + \frac{T^{\beta-1}}{|\Lambda|}|\lambda||\gamma|\frac{\xi^q\eta^{\beta-p}}{\Gamma(1+q)\Gamma(\beta-p+1)}. \tag{4.66}$$

Our first existence result is based on Leray-Schauder alternative (Theorem 1.7).

Theorem 4.11. *Assume that:*

(4.11.1) $f, g : [0,T] \times \mathbb{R} \times \mathbb{R} \to \mathbb{R}$ *are continuous functions and that there exist real constants* $k_i, \gamma_i \geq 0,\ (i = 0,1,2)$ *with* $k_0 > 0, \gamma_0 > 0$ *such that,* $\forall x_i \in \mathbb{R},\ (i = 1,2)$,

$$|f(t,x_1,x_2)| \leq k_0 + k_1|x_1| + k_2|x_2|,$$
$$|g(t,x_1,x_2)| \leq \gamma_0 + \gamma_1|x_1| + \gamma_2|x_2|.$$

If

$$(M_1 + M_3)k_1 + (M_2 + M_4)\gamma_1 < 1 \quad \text{and} \quad (M_1 + M_3)k_2 + (M_2 + M_4)\gamma_2 < 1, \tag{4.67}$$

where $M_i, i = 1,2,3,4$, *are given by (4.63)-(4.66), then the system (4.54)-(4.56) has at least one solution on* $[0,T]$.

Proof. Firstly we show that the operator $\mathcal{H} : X \times X \to X \times X$ defined by (4.62) is completely continuous. Notice that continuity of the operator $\mathcal{H}$ follows from that of the functions f and g.

Let $\Omega \subset X \times X$ be bounded. Then, there exist positive constants L_1 and L_2 such that $|f(t,x(t),y(t))| \leq L_1, |g(t,x(t),y(t))| \leq L_2,\ \ \forall (x,y) \in \Omega$. In consequence, for any $(x,y) \in \Omega$, we have

$$\begin{aligned} |\mathcal{H}_1(x,y)(t)| &\leq \frac{T^\alpha}{\Gamma(1+\alpha)}L_1 + \frac{|\lambda|}{|\Lambda|}\Big[T^{\beta-1}\frac{\eta^{\beta-p}}{\Gamma(\beta-p+1)}L_2 \\ &\quad + |\lambda|\frac{\Gamma(\beta)}{\Gamma(\beta-p)}\eta^{\beta-p-1}\Big(\frac{T^\beta}{\Gamma(1+\beta)}L_2 + |\gamma|\frac{\xi^{q+\alpha}}{\Gamma(q+\alpha+1)}L_1\Big)\Big] \\ &= M_1L_1 + M_2L_2, \end{aligned}$$

which implies that

$$\|\mathcal{H}_1(x,y)\| \leq M_1L_1 + M_2L_2.$$

In a similar way, we can find that

$$\|\mathcal{H}_2(x,y)\| \leq M_3L_1 + M_4L_2.$$

From the above inequalities we conclude that the operator $\mathcal{H}$ is uniformly bounded, since $\|\mathcal{H}(x,y)\| \leq (M_1 + M_3)L_1 + (M_2 + M_4)L_2$.

Next, we show that $\mathcal{H}$ is equi-continuous. Let $t_1, t_2 \in [0,T]$ with $t_1 < t_2$. Then we have

$$\begin{aligned} &|\mathcal{H}_1(x(t_2),y(t_2)) - \mathcal{H}_1(x(t_1),y(t_1))| \\ &\leq L_1\left|\frac{1}{\Gamma(\alpha)}\int_0^{t_2}(t_2-s)^{\alpha-1}ds - \frac{1}{\Gamma(\alpha)}\int_0^{t_1}(t_1-s)^{\alpha-1}ds\right| \\ &\leq \frac{L_1}{\Gamma(\alpha)}\left\{\int_0^{t_1}[(t_2-s)^{\alpha-1} - (t_1-s)^{\alpha-1}]ds + \int_{t_1}^{t_2}(t_2-s)^{\alpha-1}ds\right\} \\ &\leq \frac{L_1}{\Gamma(\alpha+1)}[2(t_2-t_1)^\alpha + |t_2^\alpha - t_1^\alpha|]. \end{aligned}$$

Analogously, we can obtain

$$
\begin{aligned}
&|\mathcal{H}_2(x(t_2), y(t_2)) - \mathcal{H}_2(x(t_1), y(t_1))| \\
&\leq L_2\Big[\frac{T^\beta}{\Gamma(1+\beta)} + |\lambda||\gamma|\frac{\xi^q\eta^{\beta-p}}{\Gamma(1+q)\Gamma(\beta-p+1)}\Big]\frac{t_2^{\beta-1} - t_1^{\beta-1}}{|\Lambda|} \\
&+L_1\frac{|\gamma|\xi^{q+\alpha}}{\Gamma(q+\alpha+1)}\frac{t_2^{\beta-1} - t_1^{\beta-1}}{|\Lambda|} + \frac{L_2}{\Gamma(\beta+1)}[2(t_2-t_1)^\beta + |t_2^\beta - t_1^\beta|].
\end{aligned}
$$

Thus the operator $\mathcal{H}(x, y)$ is equi-continuous. In view of the foregoing arguments, we deduce that the operator $\mathcal{H}(x, y)$ is completely continuous.

Finally, it will be verified that the set $\mathcal{Z} = \{(x, y) \in X \times X | (x, y) = \theta\mathcal{H}(x, y), 0 \leq \theta \leq 1\}$ is bounded. Let $(x, y) \in \mathcal{Z}$, with $(x, y) = \theta\mathcal{H}(x, y)$. For any $t \in [0, T]$, we have

$$
x(t) = \theta\mathcal{H}_1(x, y)(t), \quad y(t) = \theta\mathcal{H}_2(x, y)(t).
$$

Then

$$
\begin{aligned}
|x(t)| &\leq M_1(k_0 + k_1|x| + k_2|y|) + M_2(\gamma_0 + \gamma_1|x| + \gamma_2|y|) \\
&= M_1k_0 + M_2\gamma_0 + (M_1k_1 + M_2\gamma_1)|x| + (M_1k_2 + M_2\gamma_2)|y|,
\end{aligned}
$$

and

$$
\begin{aligned}
|y(t)| &\leq M_3(k_0 + k_1|x| + k_2|y|) + M_4(\gamma_0 + \gamma_1|x| + \gamma_2|y|) \\
&= M_3k_0 + M_4\gamma_0 + (M_3k_1 + M_4\gamma_1)|x| + (M_3k_2 + M_4\gamma_2)|y|.
\end{aligned}
$$

In consequence, we have

$$
\|x\| \leq M_1k_0 + M_2\gamma_0 + (M_1k_1 + M_2\gamma_1)\|x\| + (M_1k_2 + M_2\gamma_2)\|y\|
$$

and

$$
\|y\| \leq M_3k_0 + M_4\gamma_0 + (M_3k_1 + M_4\gamma_1)\|x\| + (M_3k_2 + M_4\gamma_2)\|y\|,
$$

which imply that

$$
\begin{aligned}
\|x\| + \|y\| \leq (M_1 + M_3)k_0 + (M_2 + M_4)\gamma_0 + [(M_1 + M_3)k_1 + (M_2 + M_4)\gamma_1]\|x\| \\
+[(M_1 + M_3)k_2 + (M_2 + M_4)\gamma_2)]\|y\|.
\end{aligned}
$$

Thus we have

$$
\|(x, y)\| \leq \frac{(M_1 + M_3)k_0 + (M_2 + M_4)\gamma_0}{M_0},
$$

where $M_0 = \min\{1-[(M_1+M_3)k_1+(M_2+M_4)\gamma_1], 1-[(M_1+M_3)k_2+(M_2+M_4)\gamma_2)]\}$, which establishes that the set $\mathcal{Z}$ is bounded. Thus, by Leray-Schauder alternative (Theorem 1.7), the operator $\mathcal{H}$ has at least one fixed point. Hence the system (4.54)-(4.56) has at least one solution. The proof is complete. □

The uniqueness of solutions for the problem (4.54)-(4.56) is proved in the next theorem via Banach's contraction mapping principle.

Theorem 4.12. *Assume that:*

(4.12.1) $f, g : [0,T] \times \mathbb{R} \times \mathbb{R} \to \mathbb{R}$ *are continuous functions and that there exist positive constants* ℓ_1 *and* ℓ_2 *such that for all* $t \in [0,T]$ *and* $x_i, y_i \in \mathbb{R}$, $i = 1, 2$,

$$|f(t,x_1,x_2) - f(t,y_1,y_2)| \leq \ell_1(|x_1 - y_1| + |x_2 - y_2|),$$
$$|g(t,x_1,x_2) - g(t,y_1,y_2)| \leq \ell_2(|x_1 - y_1| + |x_2 - y_2|).$$

Then the system (4.54)-(4.56) has a unique solution on $[0,T]$, *provided that*

$$(M_1 + M_3)\ell_1 + (M_2 + M_4)\ell_2 < 1,$$

where $M_i, i = 1,2,3,4$ *are given by (4.63)-(4.66).*

Proof. Put $\sup_{t\in[0,T]} f(t,0,0) = N_1 < \infty$, $\sup_{t\in[0,T]} g(t,0,0) = N_2 < \infty$ and choose a positive number r such that

$$r > \frac{(M_1 + M_3)N_1 + (M_2 + M_4)N_2}{1 - (M_1 + M_3)\ell_1 - (M_2 + M_4)\ell_2}.$$

Then we show that $\mathcal{H}B_r \subset B_r$, where $B_r = \{(x,y) \in X \times X : \|(x,y)\| \leq r\}$ and $\mathcal{H}$ is defined by (4.62).

By the assumption (4.12.1), for $(u,v) \in B_r$, $t \in [0,T]$, we have

$$\begin{aligned}|f(t,x(t),y(t))| &\leq |f(t,x(t),y(t)) - f(t,0,0)| + |f(t,0,0)| \\ &\leq \ell_1(|x(t)| + |y(t)|) + N_1 \\ &\leq \ell_1(\|x\| + \|y\|) + N_1 \leq \ell_1 r + N_1\end{aligned}$$

and

$$|g(t,x(t),y(t))| \leq \ell_2(\|x\| + \|y\|) + N_2 \leq \ell_2 r + N_2.$$

In consequence, we obtain

$$\begin{aligned}&|\mathcal{H}_1(x,y)(t)| \\ &\leq \frac{T^\alpha}{\Gamma(1+\alpha)}(\ell_1 r + N_1) + \frac{|\lambda|}{|\Lambda|}\left[T^{\beta-1}\frac{\eta^{\beta-p}}{\Gamma(\beta-p+1)}(\ell_2 r + N_2)\right. \\ &\quad \left. + |\lambda|\frac{\Gamma(\beta)}{\Gamma(\beta-p)}\eta^{\beta-p-1}\left(\frac{T^\beta}{\Gamma(1+\beta)}(\ell_2 r + N_2) + |\gamma|\frac{\xi^{q+\alpha}}{\Gamma(q+\alpha+1)}(\ell_1 r + N_1)\right)\right] \\ &= (\ell_1 r + N_1)M_1 + (\ell_2 r + N_2)M_2 \\ &= (M_1\ell_1 + M_2\ell_2)r + M_1 N_1 + M_2 N_2,\end{aligned}$$

which implies that

$$\|\mathcal{H}_1(x,y)\| \leq (M_1\ell_1 + M_2\ell_2)r + M_1 N_1 + M_2 N_2.$$

In the same way, we can find that

$$\|\mathcal{H}_2(x,y)\| \leq (M_3\ell_1 + M_4\ell_2)r + M_3 N_1 + M_4 N_2.$$

From the above inequalities, it follows that

$$\|\mathcal{H}(x,y)\| \leq [(M_1 + M_3)\ell_1 + (M_2 + M_4)\ell_2]r + (M_1 + M_3)N_1 + (M_2 + M_4)N_2 \leq r.$$

Next, for $(x_2, y_2), (x_1, y_1) \in X \times X$, and for any $t \in [0, T]$, we get

$$\begin{aligned}
&|\mathcal{H}_1(x_2, y_2)(t) - \mathcal{H}_1(x_1, y_1)(t)| \\
&\le \frac{T^\alpha}{\Gamma(1+\alpha)}\ell_1(\|x_2 - x_1\| + \|y_2 - y_1\|) \\
&+ \frac{|\lambda|}{|\Lambda|}\Big[T^{\beta-1}\frac{\eta^{\beta-p}}{\Gamma(\beta-p+1)}\ell_2(\|x_2 - x_1\| + \|y_2 - y_1\|)) \\
&+ |\lambda|\frac{\Gamma(\beta)}{\Gamma(\beta-p)}\eta^{\beta-p-1}\Big(\frac{T^\beta}{\Gamma(1+\beta)}\ell_2(\|x_2 - x_1\| + \|y_2 - y_1\|) \\
&+ |\gamma|\frac{\xi^{q+\alpha}}{\Gamma(q+\alpha+1)}\ell_1(\|x_2 - x_1\| + \|y_2 - y_1\|)\Big)\Big] \\
&\le (M_1\ell_1 + M_2\ell_2)(\|x_2 - x_1\| + \|y_2 - y_1\|),
\end{aligned}$$

which leads to

$$\|\mathcal{H}_1(x_2, y_2) - \mathcal{H}_1(x_1, y_1)\| \le (M_1\ell_1 + M_2\ell_2)(\|x_2 - x_1\| + \|y_2 - y_1\|). \tag{4.68}$$

Similarly, one can obtain

$$\|\mathcal{H}_2(x_2, y_2)(t) - \mathcal{H}_2(x_1, y_1)\| \le (M_3\ell_1 + M_4\ell_2)(\|x_2 - x_1\| + \|y_2 - y_1\|). \tag{4.69}$$

From (4.68) and (4.69), we deduce that

$$\|\mathcal{H}(x_2, y_2) - \mathcal{H}(x_1, y_1)\| \le [(M_1 + M_3)\ell_1 + (M_2 + M_4)\ell_2](\|x_2 - x_1\| + \|y_2 - y_1\|).$$

Since $(M_1 + M_3)\ell_1 + (M_2 + M_4)\ell_2 < 1$, therefore, $\mathcal{H}$ is a contraction. So, by Banach's contraction mapping principle, the operator $\mathcal{H}$ has a unique fixed point, which corresponds to a unique solution of problem (4.54)-(4.56). This completes the proof. □

Example 4.2. Consider the following fractional boundary value problem

$$\begin{cases}
{}^cD^{1/2}x(t) = \dfrac{1}{4(t+2)^2}\dfrac{|x(t)|}{1+|x(t)|} + 1 + \dfrac{1}{t^3+32}\sin^2 y(t), \quad t \in [0,1], \\[2ex]
{}^{RL}D^{3/2}x(t) = \dfrac{1}{32\pi}\sin(2\pi x(t)) + \dfrac{|y(t)|}{16(1+|y(t)|)} + \dfrac{1}{2}, \quad t \in [0,1], \\[2ex]
u(0) = \sqrt{3}D^{1/2}y\left(\dfrac{1}{3}\right), \\[2ex]
y(0) = 0, \quad y(1) = \sqrt{2}I^{1/2}x\left(\dfrac{1}{2}\right).
\end{cases} \tag{4.70}$$

Here, $\alpha = 1/2$, $\lambda = \sqrt{3}$, $p = 1/2$, $\eta = 1/3$, $\beta = 3/2$, $\gamma = \sqrt{2}$, $q = 1/2$, $\xi = 1/2$, and $f(t, x, y) = \dfrac{1}{4(t+2)^2}\dfrac{|x|}{1+|x|} + 1 + \dfrac{1}{t^3+32}\sin^2 y$ and $g(t, x, y) = \frac{1}{32\pi}\sin(2\pi x) + \frac{|y|}{16(1+|y|)} + \frac{1}{2}$. Note that $|f(t, x_1, y_1) - f(t, x_2, y_2)| \le \dfrac{1}{16}|x_1 - x_2| + \dfrac{1}{16}|y_1 - y_2|$, $|g(t, x_1, y_1) - g(t, x_2, y_2)| \le \dfrac{1}{16}|x_1 - x_2| + \dfrac{1}{16}|y_1 - y_2|$. Using the given data in (4.63)-(4.66), it is found that $M_1 \approx 1.5256638$, $M_2 \approx 0.58161945$, $M_3 \approx 0.258819045$, $M_4 \approx$

1.26605098. Clearly $\ell_1 = 1/16, \ell_2 = 1/16$ and consequently $(M_1 + M_3)\ell_1 + (M_2 + M_4)\ell_2 \approx 0.22700958 < 1$.

Thus all the conditions of Theorem 4.12 are satisfied and consequently, its conclusion applies to the problem (4.70).

4.4.2 *The multi-valued system (4.55)-(4.56)*

Definition 4.2. A function $(x, y) \in C^1([0,T],\mathbb{R}) \times C^2([0,T],\mathbb{R})$ satisfying the coupled boundary conditions $x(0) = \lambda^c D^p y(\eta)$, $y(0) = 0$, $y(T) = \gamma I^q x(\xi)$ and for which exist functions $f, g \in L^1([0,T],\mathbb{R})$ such that $f(t) \in F(t, x(t), y(t))$, $g(t) \in G(t, x(t), y(t))$ a.e. on $t \in [0,T]$ and

$$x(t) = I^\alpha f(t) + \frac{1}{\Lambda}\Big[-\lambda T^{\beta-1} I^{\beta-p} g(\eta) + \lambda \frac{\Gamma(\beta)}{\Gamma(\beta-p)} \eta^{\beta-p-1}\Big(\gamma I^{q+\alpha} f(\xi) - I^\beta g(T)\Big)\Big], \tag{4.71}$$

and

$$y(t) = I^\beta g(t) + \frac{t^{\beta-1}}{\Lambda}\Big[I^\beta g(T) - \gamma I^{q+\alpha} f(\xi) - \lambda\gamma \frac{\xi^q}{\Gamma(1+q)} I^{\beta-p} g(\eta)\Big], \tag{4.72}$$

is called a solution of the coupled system (4.55)-(4.56).

For each $(x, y) \in X \times X$, the sets of selections of F, G are defined by

$$S_{F,(x,y)} = \{f \in L^1([0,T],\mathbb{R}) : f(t) \in F(t, x(t), y(t)) \text{ for a.e. } t \in [0,T]\},$$

and

$$S_{G,(x,y)} = \{g \in L^1([0,T],\mathbb{R}) : g(t) \in G(t, x(t), y(t)) \text{ for a.e. } t \in [0,T]\}.$$

In view of Lemma 4.4, we define the operators $\mathcal{K}_1, \mathcal{K}_2 : X \times X \to \mathcal{P}(X \times X)$ as

$$\begin{aligned}\mathcal{K}_1(x,y)(t) = \{h_1 \in X \times X : \text{there exist } f \in S_{F,(x,y)}, g \in S_{G,(x,y)} \text{ such that}\\ h_1(x,y)(t) = Q_1(x,y)(t), \forall t \in [0,T]\},\end{aligned} \tag{4.73}$$

and

$$\begin{aligned}\mathcal{K}_2(x,y)(t) = \{h_2 \in X \times X : \text{there exists } f \in S_{F,(x,y)}, g \in S_{G,(x,y)} \text{ such that}\\ h_2(x,y)(t) = Q_2(x,y)(t), \forall t \in [0,T]\},\end{aligned} \tag{4.74}$$

where

$$\begin{aligned} &Q_1(x,y)(t)\\ &= I^\alpha f(t) + \frac{1}{\Lambda}\Big[-\lambda T^{\beta-1} I^{\beta-p} g(\eta) + \lambda \frac{\Gamma(\beta)}{\Gamma(\beta-p)} \eta^{\beta-p-1}\Big(\gamma I^{q+\alpha} f(\xi) - I^\beta g(T)\Big)\Big],\end{aligned}$$

and

$$Q_2(x,y)(t) = I^\beta g(t) + \frac{t^{\beta-1}}{\Lambda}\Big[I^\beta g(T) - \gamma I^{q+\alpha} f(\xi) - \lambda\gamma \frac{\xi^q}{\Gamma(1+q)} I^{\beta-p} g(\eta)\Big].$$

Then we define an operator $\mathcal{K} : X \times X \to \mathcal{P}(X \times X)$ by

$$\mathcal{K}(x,y)(t) = \begin{pmatrix} \mathcal{K}_1(x,y)(t) \\ \mathcal{K}_2(x,y)(t) \end{pmatrix},$$

where $\mathcal{K}_1$ and $\mathcal{K}_2$ are defined by (4.73) and (4.74) respectively.

4.4.2.1 *The Carathéodory case*

Our first result dealing with convex values F and G, is proved via Leray-Schauder nonlinear alternative for multi-valued maps (Theorem 1.16).

Theorem 4.13. *Suppose that the following conditions are satisfied:*

(4.13.1) $F, G : [0,T] \times \mathbb{R}^2 \to \mathcal{P}(\mathbb{R})$ *are* L^1*-Carathéodory and have convex values;*
(4.13.2) there exist continuous non-decreasing functions $\psi_1, \psi_2, \phi_1, \phi_2 : [0,\infty) \to (0,\infty)$ *and functions* $p_1, p_2 \in C([0,T], \mathbb{R}_+)$ *such that*

$$\|F(t,x,y)\|_{\mathcal{P}} := \sup\{|f| : f \in F(t,x,y)\} \leq p_1(t)[\psi_1(\|x\|) + \phi_1(\|y\|)],$$

and

$$\|G(t,x,y)\|_{\mathcal{P}} := \sup\{|g| : g \in G(t,x,y)\} \leq p_2(t)[\psi_2(\|x\|) + \phi_2(\|y\|)],$$

for each $(t,x,y) \in [0,T] \times \mathbb{R}^2$;
(4.13.3) there exists a number $N > 0$ *such that*

$$\frac{N}{(M_1+M_3)\|p_1\|(\psi_1(N)+\phi_1(N)) + (M_2+M_4)\|p_2\|(\psi_2(N)+\phi_2(N))} > 1,$$

where M_i $(i = 1,2,3,4)$ *are given by* (4.63)-(4.66). *Then the coupled system (4.55)-(4.56) has at least one solution on* $[0,T]$.

Proof. Consider the operators $\mathcal{K}_1, \mathcal{K}_2 : X \times X \to \mathcal{P}(X \times X)$ defined by (4.73) and (4.74). From (4.13.1), it follows that the sets $S_{F,(x,y)}$ and $S_{G,(x,y)}$ are non-empty for each $(x,y) \in X \times X$. Then, for $f \in S_{F,(x,y)}$, $g \in S_{G,(x,y)}$ for $(x,y) \in X \times X$, we have

$$\begin{aligned} h_1(x,y)(t) = I^\alpha f(t) + \frac{1}{\Lambda}\Big[&-\lambda T^{\beta-1} I^{\beta-p} g(\eta) \\ &+\lambda \frac{\Gamma(\beta)}{\Gamma(\beta-p)} \eta^{\beta-p-1}\Big(\gamma I^{q+\alpha} f(\xi) - I^\beta g(T)\Big)\Big], \end{aligned}$$

and

$$h_2(x,y)(t) = I^\beta g(t) + \frac{t^{\beta-1}}{\Lambda}\Big[I^\beta g(T) - \gamma I^{q+\alpha} f(\xi) - \lambda\gamma \frac{\xi^q}{\Gamma(1+q)} I^{\beta-p} g(\eta)\Big],$$

where $h_1 \in \mathcal{K}_1(x,y), h_2 \in \mathcal{K}_2(x,y)$ and so $(h_1,h_2) \in \mathcal{K}(x,y)$.

It will be established in several steps that the operator $\mathcal{K}$ satisfies the hypotheses of Leray-Schauder nonlinear alternative. First we show that $\mathcal{K}(x,y)$ is convex-valued. Let $(h_i, \bar{h}_i) \in (\mathcal{K}_1, \mathcal{K}_2), i = 1,2$. Then there exist $f_i \in S_{F,(x,y)}, g_i \in S_{G,(x,y)}, i = 1,2$ such that for each $t \in [0,T]$, we have

$$h_i(t) = I^\alpha f_i(t) + \frac{1}{\Lambda}\Big[-\lambda T^{\beta-1} I^{\beta-p} g_i(\eta) + \lambda \frac{\Gamma(\beta)}{\Gamma(\beta-p)} \eta^{\beta-p-1}\Big(\gamma I^{q+\alpha} f_i(\xi) - I^\beta g_i(T)\Big)\Big],$$

and

$$\bar{h}_i(t) = I^\beta g_i(t) + \frac{t^{\beta-1}}{\Lambda}\Big[I^\beta g_i(T) - \gamma I^{q+\alpha} f_i(\xi) - \lambda\gamma \frac{\xi^q}{\Gamma(1+q)} I^{\beta-p} g_i(\eta)\Big].$$

Let $0 \le \omega \le 1$. Then, for each $t \in [0,T]$, we have

$$\begin{aligned}
&[\omega h_1 + (1-\omega)h_2](t)\\
&= I^{\alpha}[\omega f_1(s) + (1-\omega) f_2(s)](t) + \frac{1}{\Lambda}\Big[-\lambda T^{\beta-1} I^{\beta-p}[\omega g_1(s) + (1-\omega) g_2(s)](\eta)\\
&\quad +\lambda \frac{\Gamma(\beta)}{\Gamma(\beta-p)}\eta^{\beta-p-1}\Big(I^{\beta}[\omega g_1(s) + (1-\omega) g_2(s)](T)\\
&\quad -\gamma I^{q+\alpha}[\omega f_1(s) + (1-\omega) f_2(s)](\xi)\Big)\Big],
\end{aligned}$$

and

$$\begin{aligned}
&[\omega \bar{h}_1 + (1-\omega)\bar{h}_2](t)\\
&= I^{\beta}[\omega g_1(s) + (1-\omega) g_2(s)](t) + \frac{t^{\beta-1}}{\Lambda}\Big[I^{\beta}[\omega g_1(s) + (1-\omega) g_2(s)](T)\\
&\quad -\gamma I^{q+\alpha}[\omega f_1(s) + (1-\omega) f_2(s)](\xi) - \lambda\gamma \frac{\xi^q}{\Gamma(1+q)} I^{\beta-p}[\omega g_1(s) + (1-\omega) g_2(s)](\eta)\Big].
\end{aligned}$$

We deduce that $S_{F,(x,y)}$, $S_{G,(x,y)}$ are convex-valued, since F, G are convex-valued. Obviously $\omega h_1 + (1-\omega)h_2 \in \mathcal{K}_1$, $\omega \bar{h}_1 + (1-\omega)\bar{h}_2 \in \mathcal{K}_2$ and hence $\omega(h_1, \bar{h}_1) + (1-\omega)(h_2, \bar{h}_2) \in \mathcal{K}$.

Now we show that $\mathcal{K}$ maps bounded sets into bounded sets in $X \times X$. For a positive number r, let $B_r = \{(x,y) \in X \times X : \|(x,y)\| \le r\}$ be a bounded set in $X \times X$. Then, there exist $f \in S_{F,(x,y)}, g \in S_{G,(x,y)}$ such that

$$\begin{aligned}
h_1(x,y)(t) = I^{\alpha} f(t) &+ \frac{1}{\Lambda}\Big[-\lambda T^{\beta-1} I^{\beta-p} g(\eta)\\
&+\lambda \frac{\Gamma(\beta)}{\Gamma(\beta-p)}\eta^{\beta-p-1}\Big(\gamma I^{q+\alpha} f(\xi) - I^{\beta} g(T)\Big)\Big],
\end{aligned}$$

and

$$h_2(x,y)(t) = I^{\beta} g(t) + \frac{t^{\beta-1}}{\Lambda}\Big[I^{\beta} g(T) - \gamma I^{q+\alpha} f(\xi) - \lambda\gamma \frac{\xi^q}{\Gamma(1+q)} I^{\beta-p} g(\eta)\Big].$$

Then we have

$$\begin{aligned}
&|h_1(x,y)(t)|\\
&\le I^{\alpha}|f(t)| + \frac{1}{|\Lambda|}\Big[|\lambda| T^{\beta-1} I^{\beta-p}|g(\eta)| + |\lambda|\frac{\Gamma(\beta)}{\Gamma(\beta-p)}\eta^{\beta-p-1}\Big(I^{\beta}|g(T)| + |\gamma| I^{q+\alpha}|f(\xi)|\Big)\Big]\\
&\le \frac{T^{\alpha}}{\Gamma(1+\alpha)}\|p_1\|(\psi_1(r) + \phi_1(r)) + \frac{1}{|\Lambda|}\Big[|\lambda| T^{\beta-1}\frac{\eta^{\beta-p}}{\Gamma(\beta-p+1)}\|p_2\|(\psi_2(r) + \phi_2(r))\\
&\quad +|\lambda|\frac{\Gamma(\beta)}{\Gamma(\beta-p)}\eta^{\beta-p-1}\Big(\frac{T^{\beta}}{\Gamma(1+\beta)}\|p_2\|(\psi_2(r) + \phi_2(r))\\
&\quad +|\gamma|\frac{\xi^{q+\alpha}}{\Gamma(q+\alpha+1)}\|p_1\|(\psi_1(r) + \phi_1(r))\Big)\Big]\\
&= M_1\|p_1\|(\psi_1(r) + \phi_1(r)) + M_2\|p_2\|(\psi_2(r) + \phi_2(r)),
\end{aligned}$$

and

$$|h_2(x,y)(t)| \le M_3\|p_1\|(\psi_1(r) + \phi_1(r)) + M_4\|p_2\|(\psi_2(r) + \phi_2(r)).$$

Thus,

$$\|h_1(x,y)\| \le M_1\|p_1\|(\psi_1(r)+\phi_1(r)) + M_2\|p_2\|(\psi_2(r)+\phi_2(r)),$$

and

$$\|h_2(x,y)\| \le M_3\|p_1\|(\psi_1(r)+\phi_1(r)) + M_4\|p_2\|(\psi_2(r)+\phi_2(r)).$$

Hence we obtain

$$\begin{aligned}\|(h_1,h_2)\| &= \|h_1(x,y)\| + \|h_2(x,y)\|\\ &\le (M_1+M_3)\|p_1\|(\psi_1(r)+\phi_1(r)) + (M_2+M_4)\|p_2\|(\psi_2(r)+\phi_2(r)).\end{aligned}$$

Next, we show that $\mathcal{K}$ is equi-continuous. Let $t_1, t_2 \in [0,T]$ with $t_1 < t_2$. Then, there exist $f \in S_{F,(x,y)}, g \in S_{G,(x,y)}$ such that

$$\begin{aligned}h_1(x,y)(t) = I^\alpha f(t) + \frac{1}{\Lambda}\Big[&-\lambda T^{\beta-1} I^{\beta-p} g(\eta)\\ &+\lambda \frac{\Gamma(\beta)}{\Gamma(\beta-p)}\eta^{\beta-p-1}\Big(\gamma I^{q+\alpha} f(\xi) - I^\beta g(T)\Big)\Big],\end{aligned}$$

and

$$h_2(x,y)(t) = I^\beta g(t) + \frac{t^{\beta-1}}{\Lambda}\Big[I^\beta g(T) - \gamma I^{q+\alpha} f(\xi) - \lambda\gamma \frac{\xi^q}{\Gamma(1+q)} I^{\beta-p} g(\eta)\Big].$$

Then we have

$$\begin{aligned}&|h_1(x,y)(t_2) - h_1(x,y)(t_1)|\\ &\le \|p_1\|(\psi_1(r)+\phi_1(r))\left|\frac{1}{\Gamma(\alpha)}\int_0^{t_2}(t_2-s)^{\alpha-1}ds - \frac{1}{\Gamma(\alpha)}\int_0^{t_1}(t_1-s)^{\alpha-1}ds\right|\\ &\le \frac{\|p_1\|(\psi_1(r)+\phi_1(r))}{\Gamma(\alpha)}\left\{\int_0^{t_1}[(t_2-s)^{\alpha-1}-(t_1-s)^{\alpha-1}]ds + \int_{t_1}^{t_2}(t_2-s)^{\alpha-1}ds\right\}\\ &\le \frac{\|p_1\|(\psi_1(r)+\phi_1(r))}{\Gamma(\alpha+1)}[2(t_2-t_1)^\alpha + |t_2^\alpha - t_1^\alpha|].\end{aligned}$$

Analogously, we can obtain

$$\begin{aligned}&|h_2(x,y)(t_2) - h_2(x,y)(t_1)|\\ &\le \|p_2\|(\psi_2(r)+\phi_2(r))\Big[\frac{T^\beta}{\Gamma(1+\beta)} + |\lambda||\gamma|\frac{\xi^q\eta^{\beta-p}}{\Gamma(1+q)\Gamma(\beta-p+1)}\Big]\frac{t_2^{\beta-1}-t_1^{\beta-1}}{|\Lambda|}\\ &\quad+\|p_1\|(\psi_1(r)+\phi_1(r))\frac{|\gamma|\xi^{q+\alpha}}{\Gamma(q+\alpha+1)}\frac{t_2^{\beta-1}-t_1^{\beta-1}}{|\Lambda|}\\ &\quad+\frac{\|p_2\|(\psi_2(r)+\phi_2(r))}{\Gamma(\beta+1)}[2(t_2-t_1)^\beta + |t_2^\beta - t_1^\beta|].\end{aligned}$$

Therefore, the operator $\mathcal{K}(x,y)$ is equi-continuous, and thus, by the Ascoli-Arzelá theorem, the operator $\mathcal{K}(x,y)$ is completely continuous. We know from Lemma 1.1 that a completely continuous operator is upper semi-continuous if it has a closed graph. Thus we need to prove that $\mathcal{K}$ has a closed graph. Let $(x_n,y_n) \to (x_*,y_*)$,

$(h_n, \bar{h}_n) \in \mathcal{K}(x_n, y_n)$ and $(h_n, \bar{h}_n) \to (h_*, \bar{h}_*)$, then we need to show $(h_*, \bar{h}_*) \in \mathcal{K}(x_*, y_*)$. Observe that $(h_n, \bar{h}_n) \in \mathcal{K}(x_n, y_n)$ implies that there exist $f_n \in S_{F,(x_n,y_n)}$ and $g_n \in S_{G,(x_n,y_n)}$ such that

$$\begin{aligned} &h_n(x_n, y_n)(t) \\ &= I^\alpha f_n(t) + \frac{1}{\Lambda}\Big[-\lambda T^{\beta-1} I^{\beta-p} g_n(\eta) + \lambda \frac{\Gamma(\beta)}{\Gamma(\beta-p)} \eta^{\beta-p-1}\Big(\gamma I^{q+\alpha} f_n(\xi) - I^\beta g_n(T)\Big)\Big], \end{aligned}$$

and

$$\bar{h}_n(x_n, y_n)(t) = I^\beta g_n(t) + \frac{t^{\beta-1}}{\Lambda}\Big[I^\beta g_n(T) - \gamma I^{q+\alpha} f_n(\xi) - \lambda\gamma \frac{\xi^q}{\Gamma(1+q)} I^{\beta-p} g_n(\eta)\Big].$$

Let us consider the continuous linear operators $\Phi_1, \Phi_2 : L^1([0,T], X \times X) \to C([0,T], X \times X)$ given by

$$\begin{aligned} \Phi_1(x,y)(t) = I^\alpha f(t) + \frac{1}{\Lambda}\Big[&-\lambda T^{\beta-1} I^{\beta-p} g(\eta) \\ &+\lambda \frac{\Gamma(\beta)}{\Gamma(\beta-p)} \eta^{\beta-p-1}\Big(\gamma I^{q+\alpha} f(\xi) - I^\beta g(T)\Big)\Big], \end{aligned}$$

and

$$\Phi_2(x,y)(t) = I^\beta g(t) + \frac{t^{\beta-1}}{\Lambda}\Big[I^\beta g(T) - \gamma I^{q+\alpha} f(\xi) - \lambda\gamma \frac{\xi^q}{\Gamma(1+q)} I^{\beta-p} g(\eta)\Big].$$

From [114] we know that $(\Phi_1, \Phi_2) \circ (S_F, S_G)$ is a closed graph operator. Further, we have $(h_n, \bar{h}_n) \in (\Phi_1, \Phi_2) \circ (S_{F,(x_n,y_n)}, S_{G,(x_n,y_n)})$ for all n. Since $(x_n, y_n) \to (x_*, y_*)$, $(h_n, \bar{h}_n) \to (h_*, \bar{h}_*)$ it follows that $f_* \in S_{F,(x,y)}$ and $g_* \in S_{G,(x,y)}$ such that

$$\begin{aligned} &h_*(x_*, y_*)(t) \\ &= I^\alpha f_*(t) + \frac{1}{\Lambda}\Big[-\lambda T^{\beta-1} I^{\beta-p} g_*(\eta) + \lambda \frac{\Gamma(\beta)}{\Gamma(\beta-p)} \eta^{\beta-p-1}\Big(\gamma I^{q+\alpha} f_*(\xi) - I^\beta g_*(T)\Big)\Big], \end{aligned}$$

and

$$\bar{h}_*(x_*, y_*)(t) + I^\beta g_*(t) + \frac{t^{\beta-1}}{\Lambda}\Big[I^\beta g_*(T) - \gamma I^{q+\alpha} f_*(\xi) - \lambda\gamma \frac{\xi^q}{\Gamma(1+q)} I^{\beta-p} g_*(\eta)\Big],$$

that is, $(h_n, \bar{h}_n) \in \mathcal{K}(x_*, y_*)$.

Finally, we establish the a priori bounds on solutions. Let $(x, y) \in \nu\mathcal{K}(x, y)$. Then there exist $f \in S_{F,(x,y)}$ and $g \in S_{G,(x,y)}$ such that

$$\begin{aligned} x(t) = \nu I^\alpha f(t) + \nu\frac{1}{\Lambda}\Big[&-\lambda T^{\beta-1} I^{\beta-p} g(\eta) \\ &+\lambda \frac{\Gamma(\beta)}{\Gamma(\beta-p)} \eta^{\beta-p-1}\Big(\gamma I^{q+\alpha} f(\xi) - I^\beta g(T)\Big)\Big], \end{aligned}$$

and

$$y(t) = \nu I^\beta g(t) + \nu\frac{t^{\beta-1}}{\Lambda}\Big[I^\beta g(T) - \gamma I^{q+\alpha} f(\xi) - \lambda\gamma \frac{\xi^q}{\Gamma(1+q)} I^{\beta-p} g(\eta)\Big].$$

For each $t \in [0,T]$, we obtain

$$\|x\| \le M_1 \|p_1\|(\psi_1(\|x\|) + \phi_1(\|y\|)) + M_2\|p_2\|(\psi_2(\|x\|) + \phi_2(\|y\|)),$$

and

$$\|y\| \leq M_3\|p_1\|(\psi_1(\|x\|) + \phi_1(\|y\|)) + M_4\|p_2\|(\psi_2(\|x\|) + \phi_2(\|y\|)),$$

following the same arguments as in second step.

Thus

$$\begin{aligned}\|(x,y)\| &= \|x\| + \|y\| \\ &\leq (M_1 + M_3)\|p_1\|(\psi_1(\|x\|) + \phi_1(\|y\|)) \\ &\quad +(M_2 + M_4)\|p_2\|(\psi_2(\|x\|) + \phi_2(\|y\|)),\end{aligned}$$

which implies that

$$\frac{\|(x,y)\|}{(M_1 + M_3)\|p_1\|(\psi_1(\|x\|) + \phi_1(\|y\|)) + (M_2 + M_4)\|p_2\|(\psi_2(\|x\|) + \phi_2(\|y\|))} \leq 1.$$

In view of (4.13.3), there exists N such that $\|(x,y)\| \neq N$. Let us set

$$U = \{(x,y) \in X \times X : \|(x,y)\| < N\}.$$

Note that the operator $\mathcal{K} : \overline{U} \to \mathcal{P}_{cp,cv}(X) \times \mathcal{P}_{cp,cv}(X)$ is completely continuous and upper semi-continuous. There is no $(x,y) \in \partial U$ such that $(x,y) \in \nu\mathcal{K}(x,y)$ for some $\nu \in (0,1)$ by the choice of U. Hence, by the nonlinear alternative of Leray-Schauder type (Theorem 1.16), we deduce that $\mathcal{K}$ has a fixed point $(x,y) \in \overline{U}$ which is a solution of the coupled system (4.55)-(4.56). This completes the proof. □

4.4.2.2 *The Lipschitz case*

This subsection is concerned with the case when the multi-valued maps in the system (4.55) have non-convex values.

In the forthcoming result, we make use of the fixed point theorem for multi-valued maps due to Covitz and Nadler (Theorem 1.19).

Theorem 4.14. *If*

(4.14.1) $F, G : [0,T] \times \mathbb{R}^2 \to \mathcal{P}_{cp}(\mathbb{R})$ *are such that* $F(\cdot, x, y) : [0,T] \to \mathcal{P}_{cp}(\mathbb{R})$ *and* $G(\cdot, x, y) : [0,T] \to \mathcal{P}_{cp}(\mathbb{R})$ *are measurable for each* $x, y \in \mathbb{R}$;

(4.14.2)

$$H_d(F(t,x,y), F(t,\bar{x},\bar{y}) \leq m_1(t)(|x - \bar{x}| + |y - \bar{y}|)$$

and

$$H_d(G(t,x,y), G(t,\bar{x},\bar{y}) \leq m_2(t)(|x - \bar{x}| + |y - \bar{y}|)$$

for almost all $t \in [0,T]$ *and* $x, y, \bar{x}, \bar{y} \in \mathbb{R}$ *with* $m_1, m_2 \in C([0,T], \mathbb{R}^+)$ *and* $d(0, F(t,0,0)) \leq m_1(t)$, $d(0, G(t,0,0)) \leq m_2(t)$ *for almost all* $t \in [0,T]$

hold, then the coupled system (4.55)-(4.56) has at least one solution on $[0,T]$ *provided that*

$$(M_1 + M_3)\|m_1\| + (M_2 + M_4)\|m_2\| < 1. \tag{4.75}$$

Proof. The sets $S_{F,(x,y)}$ and $S_{G,(x,y)}$ are non-empty for each $(x,y) \in X \times Y$ by the assumption (4.14.1), so F and G have measurable selections (see Theorem III.6 [78]). Now we show that the operator $\mathcal{K}$ satisfies the assumptions of Covitz and Nadler's fixed point theorem (Theorem 1.19).

First we show that $\mathcal{K}(x,y) \in \mathcal{P}_{cl}(X) \times \mathcal{P}_{cl}(X)$ for each $(x,y) \in X \times X$. Let $(h_n, \bar{h}_n) \in \mathcal{K}(x_n, y_n)$ such that $(h_n, \bar{h}_n) \to (h, \bar{h})$ in $X \times X$. Then $(h, \bar{h}) \in X \times X$ and there exist $f_n \in S_{F,(x_n,y_n)}$ and $g_n \in S_{G,(x_n,y_n)}$ such that

$$\begin{aligned} &h_n(x_n, y_n)(t) \\ &= I^\alpha f_n(t) + \frac{1}{\Lambda}\Big[- \lambda T^{\beta-1} I^{\beta-p} g_n(\eta) + \lambda \frac{\Gamma(\beta)}{\Gamma(\beta-p)} \eta^{\beta-p-1}\Big(I^\beta g_n(T) - \gamma I^{q+\alpha} f_n(\xi)\Big)\Big], \end{aligned}$$

and

$$\bar{h}_n(x_n, y_n)(t) = I^\beta g_n(t) + \frac{t^{\beta-1}}{\Lambda}\Big[I^\beta g_n(T) - \gamma I^{q+\alpha} f_n(\xi) - \lambda\gamma \frac{\xi^q}{\Gamma(1+q)} I^{\beta-p} g_n(\eta)\Big].$$

Since F and G have compact values, we pass onto a subsequences (denoted as sequences) to get that f_n and g_n converge to f and g in $L^1([0,T],\mathbb{R})$ respectively. Thus $f \in S_{F,(x,y)}$ and $g \in S_{G,(x,y)}$ for each $t \in [0,T]$ and that

$$\begin{aligned} &h_n(x_n, y_n)(t) \to h(x,y)(t) \\ &= I^\alpha f(t) + \frac{1}{\Lambda}\Big[- \lambda T^{\beta-1} I^{\beta-p} g(\eta) + \lambda \frac{\Gamma(\beta)}{\Gamma(\beta-p)} \eta^{\beta-p-1}\Big(\gamma I^{q+\alpha} f(\xi) - I^\beta g(T)\Big)\Big], \end{aligned}$$

and

$$\begin{aligned} &\bar{h}_n(x_n, y_n)(t) \to \bar{h}(x,y)(t) \\ &= I^\beta g(t) + \frac{t^{\beta-1}}{\Lambda}\Big[I^\beta g(T) - \gamma I^{q+\alpha} f(\xi) - \lambda\gamma \frac{\xi^q}{\Gamma(1+q)} I^{\beta-p} g(\eta)\Big]. \end{aligned}$$

Hence $(h, \bar{h}) \in \mathcal{K}$, which implies that $\mathcal{K}$ is closed.

Next we show that there exists $\widehat{\theta} < 1$ (defined by (4.75) such that

$$H_d(\mathcal{K}(x,y), \mathcal{K}(\bar{x},\bar{y})) \le \widehat{\theta}(\|x-\bar{x}\| + \|y-\bar{y}\|) \quad \text{for each} \quad x, \bar{x}, y, \bar{y} \in X.$$

Let $(x,\bar{x}), (y,\bar{y}) \in X \times X$ and $(h_1, \bar{h_1}) \in \mathcal{K}(x,y)$. Then there exist $f_1 \in S_{F,(x,y)}$ and $g_1 \in S_{G,(x,y)}$ such that, for each $t \in [0,T]$, we have

$$\begin{aligned} &h_1(x,y)(t) \\ &= I^\alpha f_1(t) + \frac{1}{\Lambda}\Big[- \lambda T^{\beta-1} I^{\beta-p} g_1(\eta) + \lambda \frac{\Gamma(\beta)}{\Gamma(\beta-p)} \eta^{\beta-p-1}\Big(I^\beta g_1(T) - \gamma I^{q+\alpha} f_1(\xi)\Big)\Big], \end{aligned}$$

and

$$\bar{h}_1(x,y)(t) = I^\beta g_1(t) + \frac{t^{\beta-1}}{\Lambda}\Big[I^\beta g_1(T) - \gamma I^{q+\alpha} f_1(\xi) - \lambda\gamma \frac{\xi^q}{\Gamma(1+q)} I^{\beta-p} g_1(\eta)\Big].$$

By (4.14.2), we have

$$H_d(F(t,x,y), F(t,\bar{x},\bar{y}) \le m_1(t)(|x(t)-\bar{x}(t)| + |y(t)-\bar{y}(t)|),$$

and

$$H_d(G(t,x,y),G(t,\bar{x},\bar{y}) \leq m_2(t)(|x(t)-\bar{x}(t)|+|y(t)-\bar{y}(t)|).$$

So there exist $f \in F(t,x(t),y(t))$ and $g \in G(t,x(t),y(t))$ such that

$$|f_1(t)-w| \leq m_1(t)(|x(t)-\bar{x}(t)|+|y(t)-\bar{y}(t)|),$$

and

$$|g_1(t)-z| \leq m_2(t)(|x(t)-\bar{x}(t)|+|y(t)-\bar{y}(t)|).$$

Define $V_1, V_2 : [0,T] \to \mathcal{P}(\mathbb{R})$ by

$$V_1(t) = \{f \in L^1([0,T],\mathbb{R}) : |f_1(t)-w| \leq m_1(t)(|x(t)-\bar{x}(t)|+|y(t)-\bar{y}(t)|)\},$$

and

$$V_2(t) = \{g \in L^1([0,T],\mathbb{R}) : |g_1(t)-z| \leq m_2(t)(|x(t)-\bar{x}(t)|+|y(t)-\bar{y}(t)|)\}.$$

Since the multi-valued operators $V_1(t) \cap F(t,x(t),y(t))$ and $V_2(t) \cap G(t,x(t),y(t))$ are measurable (Proposition III.4 [78]), there exist functions $f_2(t), g_2(t)$ which are measurable selection for V_1, V_2 and $f_2(t) \in F(t,x(t),y(t))$, $g_2(t) \in G(t,x(t),y(t))$ such that for a.e. $t \in [0,T]$, we have

$$|f_1(t)-f_2(t)| \leq m_1(t)(|x(t)-\bar{x}(t)|+|y(t)-\bar{y}(t)|),$$

and

$$|g_1(t)-g_2(t)| \leq m_g(t)(|x(t)-\bar{x}(t)|+|y(t)-\bar{y}(t)|).$$

Let

$$\begin{aligned} &h_2(x,y)(t) \\ &= I^\alpha f_2(t) + \frac{1}{\Lambda}\Big[-\lambda T^{\beta-1}I^{\beta-p}g_2(\eta) + \lambda\frac{\Gamma(\beta)}{\Gamma(\beta-p)}\eta^{\beta-p-1}\Big(I^\beta g_2(T) - \gamma I^{q+\alpha}f_2(\xi)\Big)\Big], \end{aligned}$$

and

$$\bar{h}_2(x,y)(t) = I^\beta g_2(t) + \frac{t^{\beta-1}}{\Lambda}\Big[I^\beta g_2(T) - \gamma I^{q+\alpha}f_2(\xi) - \lambda\gamma\frac{\xi^q}{\Gamma(1+q)}I^{\beta-p}g_2(\eta)\Big].$$

Thus,

$$\begin{aligned} &|h_1(x,y)(t) - h_2(x,y)(t)| \\ &\leq I^\alpha|f_1(s)-f_2(s)|(t) + \frac{1}{|\Lambda|}\Big[|\lambda|T^{\beta-1}I^{\beta-p}|g_1(s)-g_2(s)|(\eta) \\ &\quad +|\lambda|\frac{\Gamma(\beta)}{\Gamma(\beta-p)}\eta^{\beta-p-1}\Big(I^\beta|g_1(s)-g_2(s)|(T) + |\gamma|I^{q+\alpha}|f_1(s)-f_2(s)|(\xi)\Big)\Big] \\ &\leq I^\alpha m_1(s)(|x(s)-\bar{x}(s)|+|y(s)-\bar{y}(s)|)(t) \\ &\quad +\frac{1}{|\Lambda|}\Big[|\lambda|T^{\beta-1}I^{\beta-p}m_2(s)(|x(s)-\bar{x}(s)|+|y(s)-\bar{y}(s)|)(\eta) \\ &\quad +|\lambda|\frac{\Gamma(\beta)}{\Gamma(\beta-p)}\eta^{\beta-p-1}\Big(I^\beta m_2(s)(|x(s)-\bar{x}(s)|+|y(s)-\bar{y}(s)|)(T) \end{aligned}$$

$$+|\gamma|I^{q+\alpha}m_1(s)(|x(s)-\bar{x}(s)|+|y(s)-\bar{y}(s)|)(\xi)\Big)\Big]$$
$$\leq M_1\|m_1\|(\|x-\bar{x}\|+\|y-\bar{y}\|)+M_2\|m_2\|(\|x-\bar{x}\|+\|y-\bar{y}\|).$$

Hence

$$\|h_1(x,y)-h_2(x,y)\|\leq (M_1\|m_1\|+M_2\|m_2\|)(\|x-\bar{x}\|+\|y-\bar{y}\|).$$

In a similar manner, we can establish that

$$\|\bar{h}_1(x,y)-\bar{h}_2(x,y)\|\leq (M_3\|m_1\|+M_4\|m_2\|)(\|x-\bar{x}\|+\|y-\bar{y}\|).$$

Thus

$$\|(h_1,\bar{h}_1),(h_2,\bar{h}_2)\|\leq [(M_1+M_3)\|m_1\|+(M_2+M_4)\|m_2\|](\|x-\bar{x}\|+\|y-\bar{y}\|).$$

Analogously, interchanging the roles of (x,y) and (x,y), we can obtain

$$H_d(T(x,y),T(\bar{x},\bar{y}))\leq [(M_1+M_3)\|m_1\|+(M_2+M_4)\|m_2\|](\|x-\bar{x}\|+\|y-\bar{y}\|).$$

Therefore $\mathcal{K}$ is a contraction in view of the assumption (4.75). Hence it follows by Covitz and Nadler's fixed point theorem (Theorem 1.19) that $\mathcal{K}$ has a fixed point (x,y), which is a solution of the problem (4.55)-(4.56). This completes the proof. □

4.5 Existence of solutions for nonlinear fractional integro-differential equations and inclusions with three-point nonlocal fractional boundary conditions

In this section, we study existence and uniqueness of solutions for nonlinear fractional integro-differential equations and inclusions with three-point nonlocal fractional boundary conditions.

4.5.1 *Fractional differential equations*

In this subsection, we study the following nonlinear fractional integro-differential equations with three-point nonlocal fractional boundary conditions

$$\begin{cases} D^q x(t)+f(t,x(t),(\phi x)(t),(\psi x)(t))=0, & 0<t<1,\ \ 1<q\leq 2,\\ D^{(q-1)/2}x(0)=0, & aD^{(q-1)/2}x(1)+x(\eta)=0,\ \ 0<\eta<1,\end{cases}\tag{4.76}$$

where D^q is the standard Riemann-Liouville fractional derivative operator of order q, $f:[0,1]\times X\times X\times X\to X$ is continuous, for $\gamma,\delta:[0,1]\times[0,1]\to[0,\infty)$,

$$(\phi x)(t)=\int_0^t\gamma(t,s)x(s)ds,\quad (\psi x)(t)=\int_0^t\delta(t,s)x(s)ds,$$

and $a\in\mathbb{R}$ satisfies the condition $a\Gamma(q)+\eta^{(q-1)}\Gamma(\frac{q+1}{2})\neq 0$. Here, $(X,\|\cdot\|)$ is a Banach space and $\mathrm{C}=\mathrm{C}([0,1],X)$ denotes the Banach space of all continuous functions from $[0,1]\to X$ endowed with a topology of uniform convergence with the norm denoted by $\|\cdot\|$.

We remark that fractional boundary conditions result in existence of both electric and magnetic surface currents on the strip and are similar to the impedance boundary conditions with pure imaginary impedance and in the physical optics approximation, the ratio of the surface currents is the same as for the impedance strip.

Lemma 4.5. *For a given $\sigma \in C[0,1] \cap L(0,1)$, the unique solution of the boundary value problem*

$$\begin{cases} D^q x(t) + \sigma(t) = 0, & 0 < t < 1, \quad 1 < q \leq 2, \\ D^{(q-1)/2} x(0) = 0, & aD^{(q-1)/2} x(1) + x(\eta) = 0, \quad 0 < \eta < 1, \end{cases} \tag{4.77}$$

is given by

$$x(t) = -\int_0^t \frac{(t-s)^{q-1}}{\Gamma(q)} \sigma(s) ds + \frac{\Gamma(\frac{q+1}{2}) t^{q-1}}{[a\Gamma(q) + \eta^{(q-1)} \Gamma(\frac{q+1}{2})]} \left\{ \int_0^\eta \frac{(\eta - s)^{q-1}}{\Gamma(q)} \sigma(s) ds \right.$$
$$\left. + a \int_0^1 \frac{(1-s)^{(q-1)/2}}{\Gamma(\frac{q+1}{2})} \sigma(s) ds \right\}. \tag{4.78}$$

Proof. The fractional differential equation in (4.77) is equivalent to the integral equation

$$x(t) = -I^q \sigma(t) + b_1 t^{q-1} + b_2 t^{q-2} = -\int_0^t \frac{(t-s)^{q-1}}{\Gamma(q)} \sigma(s) ds + b_1 t^{q-1} + b_2 t^{q-2}, \tag{4.79}$$

where $b_1, b_2 \in \mathbb{R}$ are arbitrary constants. Applying the boundary conditions of (4.77) in (4.79), we find that $b_2 = 0$ and

$$b_1 = \frac{\Gamma(\frac{q+1}{2})}{[a\Gamma(q) + \eta^{(q-1)} \Gamma(\frac{q+1}{2})]} \left(\int_0^\eta \frac{(\eta - s)^{q-1}}{\Gamma(q)} \sigma(s) ds \right.$$
$$\left. + a \int_0^1 \frac{(1-s)^{(q-1)/2}}{\Gamma(\frac{q+1}{2})} \sigma(s) ds \right).$$

Substituting the values of b_1 and b_2 in (4.79), we obtain (4.78). The converse follows by direct computation. This completes the proof. □

4.5.1.1 *Existence and uniqueness result via Banach's fixed point theorem*

In the following we use the notation:

$$\gamma_0 = \sup_{t \in [0,1]} \left| \int_0^t \gamma(t,s) ds \right|, \quad \delta_0 = \sup_{t \in [0,1]} \left| \int_0^t \delta(t,s) ds \right|,$$

and

$$I_L^q = \sup_{t \in [0,1]} \{ |I^q L_1(t)|, \ |I^q L_2(t)|, \ |I^q L_3(t)| \},$$
$$I^{(q+1)/2} L(1) = \max\{ |I^{(q+1)/2} L_1(1)|, \ |I^{(q+1)/2} L_2(1)|, \ |I^{(q+1)/2} L_3(1)| \},$$
$$I^q L(\eta) = \max\{ |I^q L_1(\eta)|, \ |I^q L_2(\eta)|, \ |I^q L_3(\eta)| \}.$$

Theorem 4.15. *Assume that $f : [0,1] \times X \times X \times X \to X$ is a continuous function and satisfies the assumptions:*

(4.15.1) There exist positive functions $L_1(t),\ L_2(t),\ L_3(t)$ such that

$$\begin{aligned}&|f(t,x(t),(\phi x)(t),(\psi x)(t))-f(t,y(t),(\phi y)(t),(\psi y)(t))|\\ &\leq L_1(t)\|x-y\|+L_2(t)\|\phi x-\phi y\|+L_3(t)\|\psi x-\psi y\|,\forall t\in[0,1],\ x,y\in X.\end{aligned}$$

(4.15.2) There exists a number κ such that $\Lambda\leq\kappa<1$, where

$$\Lambda=(1+\gamma_0+\delta_0)\Big\{I_L^q+\lambda_1\Big(I^qL(\eta)+|a|I^{(q+1)/2}L(1)\Big)\Big\},$$
$$\lambda_1=\frac{\Gamma(\frac{q+1}{2})}{[a\Gamma(q)+\eta^{(q-1)}\Gamma(\frac{q+1}{2})]}.$$

Then the boundary value problem (4.76) has a unique solution provided that $\Lambda<1$.

Proof. Define $\mathcal{F}:\mathrm{C}\to\mathrm{C}$ by

$$\begin{aligned}(\mathcal{F}x)(t)&=-\int_0^t\frac{(t-s)^{q-1}}{\Gamma(q)}f(s,x(s),(\phi x)(s),(\psi x)(s))ds\\ &+\frac{\Gamma(\frac{q+1}{2})t^{q-1}}{[a\Gamma(q)+\eta^{(q-1)}\Gamma(\frac{q+1}{2})]}\left\{\int_0^\eta\frac{(\eta-s)^{q-1}}{\Gamma(q)}f(s,x(s),(\phi x)(s),(\psi x)(s))ds\right.\\ &\left.+a\int_0^1\frac{(1-s)^{(q-1)/2}}{\Gamma(\frac{q+1}{2})}f(s,x(s),(\phi x)(s),(\psi x)(s))ds\right\},t\in[0,1].\end{aligned}$$

Let us set $\sup_{t\in[0,1]}|f(t,0,0,0)|=M$, and choose

$$r\geq\frac{M}{(1-\lambda)}\left\{\frac{1+\lambda_1\eta^q}{\Gamma(q+1)}+\frac{\lambda_1|a|}{\Gamma(\frac{q+3}{2})}\right\},$$

where λ is such that $\Lambda\leq\lambda<1$. Now we show that $\mathcal{F}B_r\subset B_r$, where $B_r=\{x\in\mathrm{C}:\|x\|\leq r\}$. For $x\in B_r$, we have

$$\begin{aligned}&|(\mathcal{F}x)(t)|\\ &\leq\int_0^t\frac{(t-s)^{q-1}}{\Gamma(q)}|f(s,x(s),(\phi x)(s),(\psi x)(s))|ds\\ &+\left|\frac{\Gamma(\frac{q+1}{2})t^{q-1}}{[a\Gamma(q)+\eta^{(q-1)}\Gamma(\frac{q+1}{2})]}\right|\left\{\int_0^\eta\frac{(\eta-s)^{q-1}}{\Gamma(q)}|f(s,x(s),(\phi x)(s),(\psi x)(s))|ds\right.\\ &\left.+|a|\int_0^1\frac{(1-s)^{(q-1)/2}}{\Gamma(\frac{q+1}{2})}|f(s,x(s),(\phi x)(s),(\psi x)(s))|ds\right\}\\ &\leq\int_0^t\frac{(t-s)^{q-1}}{\Gamma(q)}(|f(s,x(s),(\phi x)(s),(\psi x)(s))-f(s,0,0,0)|+|f(s,0,0,0)|)ds\\ &+\left|\frac{\Gamma(\frac{q+1}{2})t^{q-1}}{[a\Gamma(q)+\eta^{(q-1)}\Gamma(\frac{q+1}{2})]}\right|\left\{\int_0^\eta\frac{(\eta-s)^{q-1}}{\Gamma(q)}(|f(s,x(s),(\phi x)(s),(\psi x)(s))\right.\\ &-f(s,0,0,0)|+|f(s,0,0,0)\|)ds+|a|\int_0^1\frac{(1-s)^{(q-1)/2}}{\Gamma(\frac{q+1}{2})}(|f(s,x(s),(\phi x)(s),(\psi x)(s))\\ &\left.-f(s,0,0,0)|+|f(s,0,0,0)|)ds\right\}\end{aligned}$$

$$
\begin{aligned}
&\leq \int_0^t \frac{(t-s)^{q-1}}{\Gamma(q)}(L_1(s)|x(s)| + L_2(s)|(\phi x)(s)| + L_3(s)|(\psi x)(s)| + M)ds \\
&+\Big|\frac{\Gamma(\frac{q+1}{2})t^{q-1}}{[a\Gamma(q) + \eta^{(q-1)}\Gamma(\frac{q+1}{2})]}\Big|\Big\{ \int_0^\eta \frac{(\eta-s)^{q-1}}{\Gamma(q)}(L_1(s)|x(s)| + L_2(s)|(\phi x)(s)| \\
&+L_3(s)|(\psi x)(s)| + M)ds \\
&+|a|\int_0^1 \frac{(1-s)^{(q-1)/2}}{\Gamma(\frac{q+1}{2})}(L_1(s)|x(s)| + L_2(s)|(\phi x)(s)| + L_3(s)|(\psi x)(s)| + M)ds\Big\} \\
&\leq \int_0^t \frac{(t-s)^{q-1}}{\Gamma(q)}(L_1(s)\|x\| + \gamma_0 L_2(s)\|x\| + \delta_0 L_3(s)\|x\| + M)ds \\
&+\Big|\frac{\Gamma(\frac{q+1}{2})t^{q-1}}{[a\Gamma(q) + \eta^{(q-1)}\Gamma(\frac{q+1}{2})]}\Big|\Big\{ \int_0^\eta \frac{(\eta-s)^{q-1}}{\Gamma(q)}(L_1(s)\|x\| + \gamma_0 L_2(s)\|x\| \\
&+\delta_0 L_3(s)\|x\| + M)ds \\
&+|a|\int_0^1 \frac{(1-s)^{(q-1)/2}}{\Gamma(\frac{q+1}{2})}(L_1(s)\|x\| + \gamma_0 L_2(s)\|x\| + \delta_0 L_3(s)\|x\| + M)ds\Big\} \\
&\leq (I^q L_1(t) + \gamma_0 I^q L_2(t) + \delta_0 I^q L_3(t))r + \frac{Mt^q}{\Gamma(q+1)} \\
&+\Big|\frac{\Gamma(\frac{q+1}{2})t^{q-1}}{[a\Gamma(q) + \eta^{(q-1)}\Gamma(\frac{q+1}{2})]}\Big|\Big\{\Big((I^q L_1(\eta) + \gamma_0 I^q L_2(\eta) + \delta_0 I^q L_3(\eta))r + \frac{M\eta^q}{\Gamma(q+1)}\Big) \\
&+|a|\Big((I^{(q+1)/2}L_1(1) + \gamma_0 I^{(q+1)/2}L_2(1) + \delta_0 I^{(q+1)/2}L_3(1))r + \frac{M}{\Gamma(\frac{q+3}{2})}\Big)\Big\} \\
&\leq (1+\gamma_0+\delta_0)\Big\{I_L^q + \lambda_1\Big(I^q L(\eta) + |a|I^{(q+1)/2}L(1)\Big)\Big\}r + M\Big\{\frac{1+\lambda_1\eta^q}{\Gamma(q+1)} + \frac{\lambda_1|a|}{\Gamma(\frac{q+3}{2})}\Big\} \\
&\leq (\Lambda + 1 - \lambda)r \leq r.
\end{aligned}
$$

Now, for $x, y \in \mathrm{C}$ and for each $t \in [0,1]$, we obtain

$$
\begin{aligned}
&|(\mathcal{F}x)(t) - (\mathcal{F}y)(t)| \\
&\leq \int_0^t \frac{(t-s)^{q-1}}{\Gamma(q)}|f(s, x(s), (\phi x)(s), (\psi x)(s) - f(s, y(s), (\phi y)(s), (\psi y)(s))|ds \\
&+\Big|\frac{\Gamma(\frac{q+1}{2})t^{q-1}}{[a\Gamma(q) + \eta^{(q-1)}\Gamma(\frac{q+1}{2})]}\Big|\Big\{ \int_0^\eta \frac{(\eta-s)^{q-1}}{\Gamma(q)}|f(s, x(s), (\phi x)(s), (\psi x)(s)) \\
&-f(s, y(s), (\phi y)(s), (\psi y)(s))|ds \\
&+|a|\int_0^1 \frac{(1-s)^{(q-1)/2}}{\Gamma(\frac{q+1}{2})}|f(s, x(s), (\phi x)(s), (\psi x)(s)) \\
&-f(s, y(s), (\phi y)(s), (\psi y)(s))|ds\Big\} \\
&\leq \int_0^t \frac{(t-s)^{q-1}}{\Gamma(q)}\Big(L_1(s)\|x-y\| + L_2(s)\|\phi x - \phi y\| + L_3(s)\|\psi x - \psi y\|\Big)ds
\end{aligned}
$$

$$+\left|\frac{\Gamma(\frac{q+1}{2})t^{q-1}}{[a\Gamma(q)+\eta^{(q-1)}\Gamma(\frac{q+1}{2})]}\right|\Bigg\{\int_0^\eta \frac{(\eta-s)^{q-1}}{\Gamma(q)}\Big(L_1(s)\|x-y\|+L_2(s)\|\phi x-\phi y\|$$
$$+L_3(s)\|\psi x-\psi y\|\Big)ds$$
$$+|a|\int_0^1 \frac{(1-s)^{(q-1)/2}}{\Gamma(\frac{q+1}{2})}\Big(L_1(s)\|x-y\|+L_2(s)\|\phi x-\phi y\|+L_3(s)\|\psi x-\psi y\|\Big)ds\Bigg\}$$
$$\leq \Big(I^qL_1(t)+\gamma_0 I^qL_2(t)+\delta_0 I^qL_3(t)\Big)\|x-y\|$$
$$+\left|\frac{\Gamma(\frac{q+1}{2})t^{q-1}}{[a\Gamma(q)+\eta^{(q-1)}\Gamma(\frac{q+1}{2})]}\right|\Bigg\{\Big(I^qL_1(\eta)+\gamma_0 I^qL_2(\eta)+\delta_0 I^qL_3(\eta)\Big)$$
$$+|a|\Big(I^{(q+1)/2}L_1(1)+\gamma_0 I^{(q+1)/2}L_2(1)+\delta_0 I^{(q+1)/2}L_3(1)\Big)\Bigg\}\|x-y\|$$
$$\leq (1+\gamma_0+\delta_0)\Big\{I_L^q+\lambda_1\Big(I^qL(\eta)+|a|I^{(q+1)/2}L(1)\Big)\Big\}\|x-y\|$$
$$=\Lambda\|x-y\|,$$

where we have used the assumption (4.15.2). As $\Lambda < 1$, therefore $\mathcal{F}$ is a contraction. Thus, the conclusion of the theorem follows by the contraction mapping principle. □

4.5.1.2 *Existence result via Krasnosel'skiĭ's fixed point theorem*

The following theorem is based on Krasnosel'skiĭ's fixed point theorem.

Theorem 4.16. *Let $f:[0,1]\times X\times X\times X\to X$ be continuous, which maps bounded sets into bounded sets and satisfies the assumption (4.15.1). In addition we suppose that*

(4.16.1) $|f(t,x(t),(\phi x)(t),(\psi x)(t))| \leq \mu(t), \quad \forall (t,x,\phi x,\psi x) \in [0,1]\times X\times X\times X, \quad \mu\in C([0,1],R^+).$

If

$$\Lambda_1=\lambda_1(1+\gamma_0+\delta_0)\Big(I^qL(\eta)+|a|I^{(q+1)/2}L(1)\Big)<1,$$

where Λ_1 is given in (4.15.2), then there exists at least one solution of the boundary value problem (4.76) on $[0,1]$.

Proof. Let us fix

$$r\geq \|\mu\|\left\{\frac{1+\lambda_1\eta^q}{\Gamma(q+1)}+\frac{\lambda_1|a|}{\Gamma(\frac{q+3}{2})}\right\},$$

and consider $B_r=\{x\in \mathrm{C}:\|x\|\leq r\}$. We define the operators Θ_1 and Θ_2 on B_r as

$$(\Theta_1 x)(t)=-\int_0^t \frac{(t-s)^{q-1}}{\Gamma(q)}f(s,x(s),(\phi x)(s),(\psi x)(s))ds,$$

$$(\Theta_2 x)(t) = \frac{\Gamma(\frac{q+1}{2})t^{q-1}}{[a\Gamma(q)+\eta^{(q-1)}\Gamma(\frac{q+1}{2})]}\left\{\int_0^{\eta}\frac{(\eta-s)^{q-1}}{\Gamma(q)}f(s,x(s),(\phi x)(s),(\psi x)(s))ds\right.$$
$$\left.+a\int_0^1\frac{(1-s)^{(q-1)/2}}{\Gamma(\frac{q+1}{2})}f(s,x(s),(\phi x)(s),(\psi x)(s))ds\right\}.$$

For $x, y \in B_r$, we find that

$$\|\Theta_1 x+\Theta_2 y\| \le \|\mu\|\left\{\frac{1+\lambda_1\eta^q}{\Gamma(q+1)}+\frac{\lambda_1|a|}{\Gamma(\frac{q+3}{2})}\right\}\le r.$$

Thus, $\Theta_1 x+\Theta_2 y \in B_r$. It follows from the assumption (4.15.1) that Θ_2 is a contraction mapping for $\Lambda_1 < 1$.

Continuity of f implies that the operator $(\Theta_1 x)(t)$ is continuous. Also, $(\Theta_1 x)$ is uniformly bounded on B_r as

$$\|\Theta_1 x\| \le \frac{\|\mu\|}{\Gamma(q+1)}.$$

Now, we show that $(\Theta_1 x)$ is equi-continuous. Since f is bounded on the compact set $[0,1]\times B_r\times B_r\times B_r$, therefore, we define

$$\sup_{(t,x,\phi x,\psi x)\in[0,1]\times B_r\times B_r\times B_r}|f(t,x,\phi x,\psi x)| = f_{\max}.$$

Consequently, for $t_1, t_2 \in [0,1]$, we have

$$\begin{aligned}
&|(\Theta_1 x)(t_1)-(\Theta_1 x)(t_2)|\\
&=\left|\frac{1}{\Gamma(q)}\int_0^{t_1}\left((t_2-s)^{q-1}-(t_1-s)^{q-1}\right)f(s,x(s),\phi x(s),\psi x(s))ds\right.\\
&\quad\left.+\int_{t_1}^{t_2}(t_2-s)^{q-1}f(s,x(s),\phi x(s),\psi x(s))ds\right|\\
&\le\frac{f_{\max}}{\Gamma(q+1)}|2(t_2-t_1)^q+t_1^q-t_2^q|,
\end{aligned}$$

which is independent of x. So, Θ_1 is relatively compact on B_r. Hence, By Arzela-Ascoli's Theorem, and the assumption that f maps bounded sets into bounded sets, Θ_1 is compact on B_r. Thus all the assumptions of Theorem 1.4 are satisfied and the conclusion of Theorem 1.4 implies that the boundary value problem (4.76) has at least one solution on $[0,1]$. □

Example 4.3. Consider the following boundary value problem

$$\begin{cases}{}^cD^{\frac{3}{2}}x(t)+\frac{t}{8}\frac{|x|}{1+|x|}+\frac{1}{5}\int_0^t\frac{e^{-(s-t)}}{5}x(s)ds\\
+\frac{1}{5}\int_0^t\frac{e^{-(s-t)/2}}{5}x(s)ds=0,\quad t\in[0,1],\\
D^{1/4}x(0)=0,\qquad aD^{1/4}x(1)+x\left(\frac{1}{3}\right)=0.\end{cases}\tag{4.80}$$

Here, $q = \frac{3}{2}$, $\gamma(t,s) = \frac{e^{-(s-t)}}{5}$, $\delta(t,s) = \frac{e^{-(s-t)/2}}{5}$ $a = 1$, $\eta = \frac{1}{3}$. With $\gamma_0 = \frac{e-1}{5}$, $\delta_0 = \frac{2(\sqrt{e}-1)}{5}$, we find that

$$\Lambda = \frac{8(e+2(\sqrt{e}+1))(9\sqrt{3\pi}+\Gamma(1/4))}{225\sqrt{\pi}(2\sqrt{3\pi}+\Gamma(1/4))} < 1.$$

Thus, by Theorem 4.15, the boundary value problem (4.80) has a unique solution on $[0,1]$.

Remark 4.4. In case $L_1(t)$, $L_2(t)$, and $L_3(t)$ are constant functions, that is $L_1(t) = L_1$, $L_2(t) = L_2$, and $L_3(t) = L_3$ (L_1, L_2, L_3 are positive real numbers), then Lipschitz generalized variant reduces to the classical Lipschitz condition and Λ in the assumption (4.15.2) takes the form

$$\Lambda = (L_1 + \gamma_0 L_2 + \delta_0 L_3)\left\{\frac{1+\lambda_1\eta^q}{\Gamma(q+1)} + \frac{\lambda_1|a|}{\Gamma(\frac{q+3}{2})}\right\}.$$

In the limit $q \to 2$, our results correspond to a second order integro-differential equation with fractional boundary conditions:

$$\begin{aligned} &D^2x(t) + f(t,x(t),(\phi x)(t),(\psi x)(t)) = 0, \quad 0<t<1,\\ &D^{1/2}x(0) = 0, \qquad aD^{1/2}x(1) + x(\eta) = 0, \quad 0<\eta<1. \end{aligned}$$

4.5.2 *Fractional differential inclusions*

In this subsection, we consider the following fractional differential inclusions with three-point nonlocal fractional boundary conditions

$$\begin{cases} D^q x(t) \in F(t,x(t)), \quad t\in[0,1], \quad 1<q\le 2,\\ D^{(q-1)/2}x(0) = 0, \qquad aD^{(q-1)/2}x(1) + x(\eta) = 0, \quad 0<\eta<1, a\in\mathbb{R}^+, \end{cases} \tag{4.81}$$

where D^q is the standard Riemann-Liouville fractional derivative operator of order q, and $F : [0,1]\times\mathbb{R} \to \mathcal{P}(\mathbb{R})$ is a multi-valued map, $\mathcal{P}(\mathbb{R})$ is the family of all non-empty subsets of $\mathbb{R}$.

Definition 4.3. A function $x \in C^2([0,1],\mathbb{R})$ is a solution of the problem (4.81) if there exists a function $f \in L^1([0,1],\mathbb{R})$ such that $f(t) \in F(t,x(t))$ a.e. on $[0,1]$ and

$$\begin{aligned} x(t) = \int_0^t \frac{(t-s)^{q-1}}{\Gamma(q)} f(s)ds - \frac{\Gamma(\frac{q+1}{2})t^{q-1}}{[a\Gamma(q) + \eta^{(q-1)}\Gamma(\frac{q+1}{2})]}\\ \times\left\{\int_0^\eta \frac{(\eta-s)^{q-1}}{\Gamma(q)} f(s)ds + a\int_0^1 \frac{(1-s)^{(q-1)/2}}{\Gamma(\frac{q+1}{2})} f(s)ds\right\}. \end{aligned}$$

4.5.2.1 *The Carathéodory case*

In this subsection we consider the case when F has convex values and prove an existence result based on nonlinear alternative of Leray-Schauder type (Theorem 1.16), assuming that F is Carathéodory.

Theorem 4.17. *Assume that:*

(4.17.1) $F : [0,1] \times \mathbb{R} \to \mathcal{P}(\mathbb{R})$ *is Carathéodory and has non-empty compact convex values;*

(4.17.2) there exists a continuous non-decreasing function $\psi : [0,\infty) \to (0,\infty)$ *and a function* $p \in L^1([0,1],\mathbb{R}_+)$ *such that*

$$\|F(t,x)\|_{\mathcal{P}} := \sup\{|y| : y \in F(t,x)\} \leq p(t)\psi(\|x\|) \text{ for each } (t,x) \in [0,1]\times\mathbb{R};$$

(4.17.3) there exists a number $M > 0$ *such that*

$$\frac{\Gamma(q)M}{2\psi(M)\|p\|_{L^1}} > 1. \tag{3.1}$$

Then the boundary value problem (4.81) has at least one solution on $[0,1]$.

Proof. Define an operator

$$\Omega(x) = \left\{ \begin{array}{l} h \in C([0,1],\mathbb{R}) : \\ h(t) = \left\{ \begin{array}{l} h(t) = \displaystyle\int_0^t \frac{(t-s)^{q-1}}{\Gamma(q)} f(s)ds \\ -\dfrac{\Gamma(\frac{q+1}{2})t^{q-1}}{[a\Gamma(q) + \eta^{(q-1)}\Gamma(\frac{q+1}{2})]} \left(\displaystyle\int_0^\eta \frac{(\eta-s)^{q-1}}{\Gamma(q)} f(s)ds \right. \\ \left. + a \displaystyle\int_0^1 \frac{(1-s)^{(q-1)/2}}{\Gamma(\frac{q+1}{2})} f(s)ds \right), \qquad 0 \leq t \leq 1, \end{array} \right. \end{array} \right\}$$

for $f \in S_{F,x}$. It will be shown that the operator Ω satisfies the assumptions of the nonlinear alternative of Leray-Schauder type. The proof consists of several steps. **(i)** $\Omega(x)$ *is convex for each* $x \in C([0,1],\mathbb{R})$. For that, let $h_1, h_2 \in \Omega(x)$. Then there exist $f_1, f_2 \in S_{F,x}$ such that for each $t \in [0,1]$, we have

$$h_i(t) = \int_0^t \frac{(t-s)^{q-1}}{\Gamma(q)} f_i(s)ds - \frac{\Gamma(\frac{q+1}{2})t^{q-1}}{[a\Gamma(q) + \eta^{(q-1)}\Gamma(\frac{q+1}{2})]}$$
$$\times \left(\int_0^\eta \frac{(\eta-s)^{q-1}}{\Gamma(q)} f_i(s)ds + a\int_0^1 \frac{(1-s)^{(q-1)/2}}{\Gamma(\frac{q+1}{2})} f_i(s)ds \right), \quad i = 1,2.$$

Let $0 \leq \lambda \leq 1$. Then, for each $t \in [0,1]$, we have

$$[\lambda h_1 + (1-\lambda)h_2](t) = \int_0^t \frac{(t-s)^{q-1}}{\Gamma(q)} [\lambda f_1(s) + (1-\lambda)f_2(s)]ds$$
$$- \frac{\Gamma(\frac{q+1}{2})t^{q-1}}{[a\Gamma(q) + \eta^{(q-1)}\Gamma(\frac{q+1}{2})]} \left(\int_0^\eta \frac{(\eta-s)^{q-1}}{\Gamma(q)} [\lambda f_1(s) + (1-\lambda)f_2(s)]ds \right.$$

$$+ a\int_0^1 \frac{(1-s)^{(q-1)/2}}{\Gamma(\frac{q+1}{2})}[\lambda f_1(s)+(1-\lambda)f_2(s)]ds\Bigg).$$

Since $S_{F,x}$ is convex (F has convex values), therefore it follows that $\lambda h_1+(1-\lambda)h_2 \in \Omega(x)$.

(ii) *$\Omega(x)$ maps bounded sets into bounded sets in* $C([0,1],\mathbb{R})$. For a positive number r, let $B_r = \{x \in C([0,1],\mathbb{R}) : \|x\| \le r\}$ be a bounded set in $C([0,1],\mathbb{R})$. Then, for each $h \in \Omega(x), x \in B_r$, there exists $f \in S_{F,x}$ such that

$$h(t) = \int_0^t \frac{(t-s)^{q-1}}{\Gamma(q)}f(s)ds - \frac{\Gamma(\frac{q+1}{2})t^{q-1}}{[a\Gamma(q)+\eta^{(q-1)}\Gamma(\frac{q+1}{2})]}$$
$$\times\left(\int_0^\eta \frac{(\eta-s)^{q-1}}{\Gamma(q)}f(s)ds + a\int_0^1 \frac{(1-s)^{(q-1)/2}}{\Gamma(\frac{q+1}{2})}f(s)ds\right).$$

Using the relations $0 \le t-s \le 1$, $0 \le \eta - s \le \eta$ and $0 \le 1-s \le 1$, we have

$$|h(t)| \le \int_0^t \frac{|t-s|^{q-1}}{\Gamma(q)}|f(s)|ds + \left|\frac{\Gamma(\frac{q+1}{2})t^{q-1}}{[a\Gamma(q)+\eta^{(q-1)}\Gamma(\frac{q+1}{2})]}\right|$$
$$\times\left(\int_0^\eta \frac{|\eta-s|^{q-1}}{\Gamma(q)}|f(s)|ds + |a|\int_0^1 \frac{(1-s)^{(q-1)/2}}{\Gamma(\frac{q+1}{2})}|f(s)|ds\right)$$
$$\le \frac{2}{\Gamma(q)}\psi(\|x\|)\int_0^1 p(s)ds.$$

Thus,

$$\|h\| \le \frac{2}{\Gamma(q)}\psi(r)\|p\|_{L^1}.$$

(iii) *Ω maps bounded sets into equi-continuous sets of* $C([0,1],\mathbb{R})$. Let $t', t'' \in [0,1]$ with $t' < t''$ and $x \in B_r$, where B_r is a bounded set of $C([0,1],\mathbb{R})$. For each $h \in \Omega(x)$, we obtain

$$|h(t'') - h(t')|$$
$$= \Bigg|\int_0^{t''} \frac{(t''-s)^{q-1}}{\Gamma(q)}f(s)ds - \int_0^{t'} \frac{(t'-s)^{q-1}}{\Gamma(q)}f(s)ds$$
$$-\frac{\Gamma(\frac{q+1}{2})[(t'')^{q-1}-(t')^{q-1}]}{[a\Gamma(q)+\eta^{(q-1)}\Gamma(\frac{q+1}{2})]}\Bigg(\int_0^\eta \frac{(\eta-s)^{q-1}}{\Gamma(q)}f(s)ds$$
$$+a\int_0^1 \frac{(1-s)^{(q-1)/2}}{\Gamma(\frac{q+1}{2})}f(s)ds\Bigg)\Bigg|$$
$$\le \left|\int_0^{t'} \frac{[(t''-s)^{q-1}-(t'-s)^{q-1}]}{\Gamma(q)}f(s)ds\right| + \left|\int_{t'}^{t''} \frac{(t''-s)^{q-1}}{\Gamma(q)}f(s)ds\right|$$
$$+\left|\frac{\Gamma(\frac{q+1}{2})[(t'')^{q-1}-(t')^{q-1}]}{[a\Gamma(q)+\eta^{(q-1)}\Gamma(\frac{q+1}{2})]}\right|\Bigg(\int_0^\eta \frac{(\eta-s)^{q-1}}{\Gamma(q)}f(s)ds$$

$$+a\int_0^1 \frac{(1-s)^{(q-1)/2}}{\Gamma(\frac{q+1}{2})}f(s)ds\Bigg)\Bigg|.$$

Obviously the right hand side of the above inequality tends to zero independently of $x \in B_r$ as $t''-t' \to 0$. As Ω satisfies the above three assumptions, therefore it follows by Ascoli-Arzelá theorem that $\Omega : C([0,1],\mathbb{R}) \to \mathcal{P}(C([0,1],\mathbb{R}))$ is completely continuous.

(iv) Ω *has a closed graph.* Let $x_n \to x_*, h_n \in \Omega(x_n)$ and $h_n \to h_*$. Then we need to show that $h_* \in \Omega(x_*)$. Associated with $h_n \in \Omega(x_n)$, there exists $f_n \in S_{F,x_n}$ such that for each $t \in [0,1]$,

$$h_n(t) = \int_0^t \frac{(t-s)^{q-1}}{\Gamma(q)}f_n(s)ds - \frac{\Gamma(\frac{q+1}{2})t^{q-1}}{[a\Gamma(q)+\eta^{(q-1)}\Gamma(\frac{q+1}{2})]}$$
$$\times\left(\int_0^\eta \frac{(\eta-s)^{q-1}}{\Gamma(q)}f_n(s)ds + a\int_0^1 \frac{(1-s)^{(q-1)/2}}{\Gamma(\frac{q+1}{2})}f_n(s)ds\right).$$

Thus we have to show that there exists $f_* \in S_{F,x_*}$ such that, for each $t \in [0,1]$,

$$h_*(t) = \int_0^t \frac{(t-s)^{q-1}}{\Gamma(q)}f_*(s)ds - \frac{\Gamma(\frac{q+1}{2})t^{q-1}}{[a\Gamma(q)+\eta^{(q-1)}\Gamma(\frac{q+1}{2})]}$$
$$\times\left(\int_0^\eta \frac{(\eta-s)^{q-1}}{\Gamma(q)}f_*(s)ds + a\int_0^1 \frac{(1-s)^{(q-1)/2}}{\Gamma(\frac{q+1}{2})}f_*(s)ds\right).$$

Let us consider the continuous linear operator $\Theta : L^1([0,1],\mathbb{R}) \to C([0,1],\mathbb{R})$ so that

$$f \mapsto \Theta(f)(t) = \int_0^t \frac{(t-s)^{q-1}}{\Gamma(q)}f(s)ds - \frac{\Gamma(\frac{q+1}{2})t^{q-1}}{[a\Gamma(q)+\eta^{(q-1)}\Gamma(\frac{q+1}{2})]}$$
$$\times\left(\int_0^\eta \frac{(\eta-s)^{q-1}}{\Gamma(q)}f(s)ds + a\int_0^1 \frac{(1-s)^{(q-1)/2}}{\Gamma(\frac{q+1}{2})}f(s)ds\right).$$

Observe that

$$\|h_n(t)-h_*(t)\| = \Bigg\|\int_0^t \frac{(t-s)^{q-1}}{\Gamma(q)}(f_n(s)-f_*(s))ds$$
$$-\frac{\Gamma(\frac{q+1}{2})t^{q-1}}{[a\Gamma(q)+\eta^{(q-1)}\Gamma(\frac{q+1}{2})]}\Bigg(\int_0^\eta \frac{(\eta-s)^{q-1}}{\Gamma(q)}(f_n(s)-f_*(s))ds$$
$$+a\int_0^1 \frac{(1-s)^{(q-1)/2}}{\Gamma(\frac{q+1}{2})}(f_n(s)-f_*(s))ds\Bigg)\Bigg\| \to 0 \quad \text{as} \quad n \to \infty.$$

Thus, it follows by Lemma 1.2 that $\Theta \circ S_F$ is a closed graph operator. Further, we have $h_n(t) \in \Theta(S_{F,x_n})$. Since $x_n \to x_*$, therefore, we have

$$h_*(t) = \int_0^t \frac{(t-s)^{q-1}}{\Gamma(q)}f_*(s)ds - \frac{\Gamma(\frac{q+1}{2})t^{q-1}}{[a\Gamma(q)+\eta^{(q-1)}\Gamma(\frac{q+1}{2})]}$$

$$\times \left(\int_0^\eta \frac{(\eta - s)^{q-1}}{\Gamma(q)} f_*(s)ds + a \int_0^1 \frac{(1-s)^{(q-1)/2}}{\Gamma(\frac{q+1}{2})} f_*(s)ds \right),$$

for some $f_* \in S_{F,x_*}$.

(v) Finally, *we discuss a priori bounds on solutions.* Let x be a solution of (4.81). Then there exists $f \in L^1([0,1],\mathbb{R})$ with $f \in S_{F,x}$ such that, for $t \in [0,1]$, we have

$$x(t) = \int_0^t \frac{(t-s)^{q-1}}{\Gamma(q)} f(s)ds - \frac{\Gamma(\frac{q+1}{2})t^{q-1}}{[a\Gamma(q) + \eta^{(q-1)}\Gamma(\frac{q+1}{2})]}$$
$$\times \left(\int_0^\eta \frac{(\eta - s)^{q-1}}{\Gamma(q)} f(s)ds + a \int_0^1 \frac{(1-s)^{(q-1)/2}}{\Gamma(\frac{q+1}{2})} f(s)ds \right).$$

In view of (4.17.2), for each $t \in [0,1]$, we obtain

$$|x(t)| \leq \frac{2}{\Gamma(q)} \psi(\|x\|) \int_0^1 p(s)ds.$$

Consequently, we have

$$\frac{\Gamma(q)\|x\|}{2\psi(\|x\|)\|p\|_{L^1}} \leq 1,$$

In view of (4.17.3), there exists M such that $\|x\| \neq M$. Let us set

$$U = \{x \in C([0,1],\mathbb{R}) : \|x\| < M + 1\}.$$

Note that the operator $\Omega : \overline{U} \to \mathcal{P}(C([0,1],\mathbb{R}))$ is upper semi-continuous and completely continuous. From the choice of U, there is no $x \in \partial U$ such that $x \in \mu\Omega(x)$ for some $\mu \in (0,1)$. Consequently, by the nonlinear alternative of Leray-Schauder type (Theorem 1.8), we deduce that Ω has a fixed point $x \in \overline{U}$ which is a solution of the problem (4.81). This completes the proof. □

4.5.2.2 *The lower semi-continuous case*

As a next result, we study the case when F is not necessarily convex valued. Our strategy to deal with this problems is based on the nonlinear alternative of Leray Schauder type together with the selection theorem of Bressan and Colombo (Theorem 1.3) for lower semi-continuous maps with decomposable values.

Theorem 4.18. *Assume that (4.17.2), (4.17.3) and the following condition holds:*

(4.18.1) $F : [0,1] \times \mathbb{R} \to \mathcal{P}(\mathbb{R})$ *is a non-empty compact-valued multi-valued map such that*

- *(a)* $(t,x) \longmapsto F(t,x)$ *is* $\mathcal{L} \otimes \mathcal{B}$ *measurable,*
- *(b)* $x \longmapsto F(t,x)$ *is lower semi-continuous for each* $t \in [0,1]$.

Then the boundary value problem (4.81) has at least one solution on $[0,1]$.

Proof. It follows from (4.17.2) and (4.18.1) that F is of l.s.c. type ([88]). Then, from Theorem 1.3, there exists a continuous function $f : C([0,1],\mathbb{R}) \to L^1([0,1],\mathbb{R})$ such that $f(x) \in \mathcal{F}(x)$ for all $x \in C([0,1],\mathbb{R})$.

Consider the problem

$$\begin{cases} D^q x(t) = f(x(t)), \ \ 0 < t < 1, \ \ 1 < q \le 2, \\ D^{(q-1)/2}x(0) = 0, \quad aD^{(q-1)/2}x(1) + x(\eta) = 0, \ \ 0 < \eta < 1, \ a \in \mathbb{R}^+. \end{cases} \tag{4.82}$$

Observe that if $x \in C^1([0,1],\mathbb{R})$ is a solution of (4.82), then x is a solution to the problem (4.81). In order to transform the problem (4.82) into a fixed point problem, we define an operator $\overline{\Omega}$ as

$$\overline{\Omega}x(t) = \int_0^t \frac{(t-s)^{q-1}}{\Gamma(q)} f(x(s))ds - \frac{\Gamma(\frac{q+1}{2})t^{q-1}}{[a\Gamma(q) + \eta^{(q-1)}\Gamma(\frac{q+1}{2})]}$$
$$\times \left(\int_0^\eta \frac{(\eta-s)^{q-1}}{\Gamma(q)} f(x(s))ds + a\int_0^1 \frac{(1-s)^{(q-1)/2}}{\Gamma(\frac{q+1}{2})} f(x(s))ds \right).$$

It can easily be shown that $\overline{\Omega}$ is continuous and completely continuous. The remaining part of the proof is similar to that of Theorem 4.17. So we omit it. This completes the proof. □

4.5.2.3 *The Lipschitz case*

Now we prove the existence of solutions for the problem (4.81) with a non-convex valued right hand side by applying a fixed point theorem for multi-valued maps due to Covitz and Nadler (Theorem 1.19).

Theorem 4.19. *Assume that:*

(4.19.1) $F : [0,1] \times \mathbb{R} \to \mathcal{P}_{cp}(\mathbb{R})$ *is such that* $F(\cdot, x) : [0,1] \to \mathcal{P}_{cp}(\mathbb{R})$ *is measurable for each* $x \in \mathbb{R}$*;*

(4.19.2) $H_d(F(t,x), F(t,\bar{x})) \le m(t)|x - \bar{x}|$ *for almost all* $t \in [0,1]$ *and* $x, \bar{x} \in \mathbb{R}$ *with* $m \in L^1([0,1],\mathbb{R})$ *and* $d(0, F(t,0)) \le m(t)$ *for almost all* $t \in [0,1]$.

Then the boundary value problem (4.81) has at least one solution on $[0,1]$ *if*

$$\frac{2\|m\|_{L^1}}{\Gamma(q)} < 1.$$

Proof. Observe that the set $S_{F,x}$ is non-empty for each $x \in C([0,1],\mathbb{R})$ by the assumption (4.19.1), so F has a measurable selection (Theorem III.6 [78]). Now we show that the operator $\Omega : C([0,1],\mathbb{R}) \to \mathcal{P}(C([0,1],\mathbb{R}))$ defined at the beginning of the proof of Theorem 4.17, satisfies the assumptions of Theorem 1.19. To show that $\Omega(x) \in \mathcal{P}_{cl}((C[0,1],\mathbb{R}))$ for each $x \in C([0,1],\mathbb{R})$, let $\{u_n\}_{n\ge 0} \in \Omega(x)$ be such that $u_n \to u$ $(n \to \infty)$ in $C([0,1],\mathbb{R})$. Then $u \in C([0,1],\mathbb{R})$ and there exists $v_n \in S_{F,x}$ such that, for each $t \in [0,1]$,

$$u_n(t) = \int_0^t \frac{(t-s)^{q-1}}{\Gamma(q)} v_n(s)ds - \frac{\Gamma(\frac{q+1}{2})t^{q-1}}{[a\Gamma(q) + \eta^{(q-1)}\Gamma(\frac{q+1}{2})]}$$

$$\times\left(\int_0^\eta \frac{(\eta-s)^{q-1}}{\Gamma(q)}v_n(s)ds + a\int_0^1 \frac{(1-s)^{(q-1)/2}}{\Gamma(\frac{q+1}{2})}v_n(s)ds\right).$$

As F has compact values, we pass onto a subsequence to obtain that v_n converges to v in $L^1([0,1],\mathbb{R})$. Thus, $v \in S_{F,x}$ and for each $t \in [0,1]$,

$$u_n(t) \to u(t) = \int_0^t \frac{(t-s)^{q-1}}{\Gamma(q)}v(s)ds - \frac{\Gamma(\frac{q+1}{2})t^{q-1}}{[a\Gamma(q)+\eta^{(q-1)}\Gamma(\frac{q+1}{2})]}$$
$$\times\left(\int_0^\eta \frac{(\eta-s)^{q-1}}{\Gamma(q)}v(s)ds + a\int_0^1 \frac{(1-s)^{(q-1)/2}}{\Gamma(\frac{q+1}{2})}v(s)ds\right).$$

Hence $u \in \Omega(x)$.

Next we show that there exists $\gamma < 1$ such that

$$H_d(\Omega(x),\Omega(\bar{x})) \le \gamma\|x-\bar{x}\| \quad \text{for each} \quad x,\bar{x} \in C([0,1],\mathbb{R}).$$

Let $x,\bar{x} \in C([0,1],\mathbb{R})$ and $h_1 \in \Omega(x)$. Then there exists $v_1(t) \in F(t,x(t))$ such that, for each $t \in [0,1]$,

$$h_1(t) = \int_0^t \frac{(t-s)^{q-1}}{\Gamma(q)}v_1(s)ds - \frac{\Gamma(\frac{q+1}{2})t^{q-1}}{[a\Gamma(q)+\eta^{(q-1)}\Gamma(\frac{q+1}{2})]}$$
$$\times\left(\int_0^\eta \frac{(\eta-s)^{q-1}}{\Gamma(q)}v_1(s)ds + a\int_0^1 \frac{(1-s)^{(q-1)/2}}{\Gamma(\frac{q+1}{2})}v_1(s)ds\right).$$

By (4.19.2), we have

$$H_d(F(t,x),F(t,\bar{x})) \le m(t)|x(t)-\bar{x}(t)|.$$

So, there exists $w(t) \in F(t,\bar{x}(t))$ such that

$$|v_1(t)-w(t)| \le m(t)|x(t)-\bar{x}(t)|, \quad t \in [0,1].$$

Define $U : [0,1] \to \mathcal{P}(\mathbb{R})$ by

$$U(t) = \{w \in \mathbb{R} : |v_1(t)-w(t)| \le m(t)|x(t)-\bar{x}(t)|\}.$$

Since the multi-valued operator $V(t) \cap F(t,\bar{x}(t))$ is measurable (see Theorem III.6 [78]), there exists a function $v_2(t)$ which is a measurable selection for V. So $v_2(t) \in F(t,\bar{x}(t))$ and for each $t \in [0,1]$, we have $|v_1(t)-v_2(t)| \le m(t)|x(t)-\bar{x}(t)|$.

For each $t \in [0,1]$, let us define

$$h_2(t) = \int_0^t \frac{(t-s)^{q-1}}{\Gamma(q)}v_2(s)ds - \frac{\Gamma(\frac{q+1}{2})t^{q-1}}{[a\Gamma(q)+\eta^{(q-1)}\Gamma(\frac{q+1}{2})]}$$
$$\times\left(\int_0^\eta \frac{(\eta-s)^{q-1}}{\Gamma(q)}v_2(s)ds + a\int_0^1 \frac{(1-s)^{(q-1)/2}}{\Gamma(\frac{q+1}{2})}v_2(s)ds\right).$$

Thus

$$|h_1(t)-h_2(t)| \le \int_0^t \frac{|t-s|^{q-1}}{\Gamma(q)}|v_1(s)-v_2(s)|ds$$

$$+\left|\frac{\Gamma(\frac{q+1}{2})}{[a\Gamma(q)+\eta^{(q-1)}\Gamma(\frac{q+1}{2})]}\right|\left(\int_0^\eta \frac{|\eta-s|^{q-1}}{\Gamma(q)}|v_1(s)-v_2(s)|ds\right.$$
$$\left.+|a|\int_0^1 \frac{|1-s|^{(q-1)/2}}{\Gamma(\frac{q+1}{2})}|v_1(s)-v_2(s)|ds\right)$$
$$\leq \frac{2}{\Gamma(q)}\int_0^1 m(s)\|x-\overline{x}\|ds.$$

Hence

$$\|h_1(t)-h_2(t)\| \leq \frac{2\|m\|_{L^1}}{\Gamma(q)}\|x-\overline{x}\|.$$

Analogously, interchanging the roles of x and $\overline{x}$, we obtain

$$H_d(\Omega(x),\Omega(\bar{x})) \leq \gamma\|x-\bar{x}\| \leq \frac{2\|m\|_{L^1}}{\Gamma(q)}\|x-\overline{x}\|.$$

Since Ω is a contraction, it follows by Theorem 1.19 that Ω has a fixed point x which is a solution of (4.81). This completes the proof. □

Example 4.4. Consider the following fractional inclusion boundary value problem

$$\begin{cases} D^{7/4}x(t) \in F(t,x(t)), & t\in[0,1], \quad 1<q\leq 2, \\ D^{3/8}x(0)=0, & D^{3/8}x(1)+x(1/4)=0, \end{cases} \tag{4.83}$$

where $a=1$, $q=7/4$, $\eta=1/4$, and $F:[0,1]\times\mathbb{R}\to\mathcal{P}(\mathbb{R})$ is a multi-valued map given by

$$F(t,x)=\left[\frac{\sqrt{t}|x|^3}{(|x|^3+3)},\frac{\sqrt{t}|x|}{2(|x|+1)}\right].$$

For $f\in F$, we have

$$|f|\leq \max\left\{\frac{\sqrt{t}|x|^3}{(|x|^3+3)},\frac{\sqrt{t}|x|}{2(|x|+1)}\right\}\leq\sqrt{t},\quad x\in\mathbb{R},\ t\in[0,1].$$

Thus,

$$\|F(t,x)\|_{\mathcal{P}} := \sup\{|y| : y\in F(t,x)\}\leq\sqrt{t}=p(t)\psi(\|x\|),\quad x\in\mathbb{R},$$

with $p(t)=\sqrt{t}$, $\psi(\|x\|)=1$. Further, using the condition (4.17.3), it is found that $M>4/3\Gamma(7/4)$. Clearly, all the conditions of Theorem 4.17 are satisfied. So there exists at least one solution of the problem (4.83) on $[0,1]$.

Example 4.5. Consider the following fractional inclusion boundary value problem

$$\begin{cases} D^{7/2}x(t) \in F(t,x(t)), & t\in[0,1], \quad 1<q\leq 2, \\ D^{3/8}x(0)=0, & D^{3/8}x(1)+x(1/4)=0, \end{cases} \tag{4.84}$$

where $a = 1$, $q = 7/2$, $\eta = 1/4$, and $F : [0,1] \times \mathbb{R} \to \mathcal{P}(\mathbb{R})$ is a multi-valued map given by

$$F(t,x) = \left[0, \frac{\sin x}{(4+t)^2} + 3\right].$$

Then we have

$$\sup\{|u| : u \in F(t,x)\} \leq 3 + \frac{1}{(4+t)^2},$$

$$H_d(F(t,x), F(t,\bar{x})) \leq \frac{1}{(4+t)^2}|x - \bar{x}|.$$

Let $m(t) = \dfrac{1}{(4+t)^2}$. Then

$$H_d(F(t,x,y), F(t,\bar{x},\bar{y})) \leq m(t)|x - \bar{x}|,$$

and

$$\frac{2\|m\|_{L^1}}{\Gamma(q)} = \frac{4}{75\sqrt{\pi}} < 1.$$

By Theorem 4.19 the problem (4.84) has at least one solution on $[0,1]$.

4.6 Fractional order integro-differential equations with nonlocal three-point fractional boundary conditions

In this section, we discuss the existence of solutions for integro-differential equations of fractional order with nonlocal three-point fractional boundary conditions

$$\begin{cases} -D^{\alpha}x(t) = Af(t,x(t)) + BI^{\beta}g(t,x(t)), \quad 2 < \alpha \leq 3, \quad t \in [0,1], \\ D^{\delta}x(0) = 0, \quad D^{\delta+1}x(0) = 0, \quad D^{\delta}x(1) - D^{\delta}x(\eta) = a, \end{cases} \tag{4.85}$$

where $0 < \delta \leq 1$, $\alpha - \delta > 2$, $0 < \beta < 1$, $0 < \eta < 1$, $D^{(.)}$ denotes the Riemann-Liouville fractional derivative operator of order $(.)$, f, g are given continuous functions, and A, B, a are real constants.

By the substitution $x(t) = I^{\delta}y(t) = D^{-\delta}y(t)$, where $y(t)$ is a suitable continuous function, the problem (4.85) takes the form

$$\begin{cases} -D^{\alpha-\delta}y(t) = Af(t, I^{\delta}y(t)) + BI^{\beta}g(t, I^{\delta}y(t)), \quad t \in [0,1], \\ y(0) = 0, \; y'(0) = 0, \quad y(1) - y(\eta) = a. \end{cases} \tag{4.86}$$

Lemma 4.6. *For any $h \in C(0,1) \cap L(0,1)$, the unique solution of the linear fractional boundary value problem*

$$\begin{cases} -D^{\alpha-\delta}y(t) = h(t), \quad t \in [0,1], \\ y(0) = 0, \; y'(0) = 0, \quad y(1) - y(\eta) = a, \end{cases} \tag{4.87}$$

is

$$y(t) = -I^{\alpha-\delta}h(t) + \frac{t^{\alpha-\delta-1}}{1 - \eta^{\alpha-\delta-1}}\Big(a + I^{\alpha-\delta}h(1) - I^{\alpha-\delta}h(\eta)\Big),$$

where $I^{(.)}(.)$ denotes Riemann-Liouville integral.

Proof. It is well known that the solutions of fractional differential equation in (4.87) can be written as

$$y(t) = -I^{\alpha-\delta}h(t) + c_1 t^{\alpha-\delta-1} + c_2 t^{\alpha-\delta-2} + c_3 t^{\alpha-\delta-3}, \tag{4.88}$$

where $c_1, c_2, c_3 \in \mathbb{R}$ are arbitrary constants. Using the given boundary conditions, we find that $c_2 = 0,\ c_3 = 0$ and

$$c_1 = \frac{1}{1-\eta^{\alpha-\delta-1}}\Big(a + I^{\alpha-\delta}h(1) - I^{\alpha-\delta}h(\eta)\Big).$$

Substituting these values in (4.88), we get

$$y(t) = -I^{\alpha-\delta}h(t) + \frac{t^{\alpha-\delta-1}}{1-\eta^{\alpha-\delta-1}}\Big(a + I^{\alpha-\delta}h(1) - I^{\alpha-\delta}h(\eta)\Big).$$

This completes the proof. □

Thus, the solution of the linear variant of the problem (4.85) can be written as

$$\begin{aligned}
x(t) &= I^{\delta}y(t) \\
&= I^{\delta}\Big[-I^{\alpha-\delta}h(t) + \frac{t^{\alpha-\delta-1}}{1-\eta^{\alpha-\delta-1}}\Big(a + I^{\alpha-\delta}h(1) - I^{\alpha-\delta}h(\eta)\Big)\Big] \\
&= -I^{\alpha}h(t) + \frac{1}{1-\eta^{\alpha-\delta-1}}\Big(a + I^{\alpha-\delta}h(1) - I^{\alpha-\delta}h(\eta)\Big)\int_0^t \frac{(t-s)^{\delta-1}}{\Gamma(\delta)}s^{\alpha-\delta-1}ds \\
&= -I^{\alpha}h(t) + \frac{1}{1-\eta^{\alpha-\delta-1}}\Big(a + I^{\alpha-\delta}h(1) - I^{\alpha-\delta}h(\eta)\Big) \times \\
&\quad \times\Big\{\frac{t^{\alpha-1}}{\Gamma(\delta)}\int_0^1 (1-\nu)^{\delta-1}\nu^{\alpha-\delta-1}d\nu\Big\},
\end{aligned}$$

where we have used the substitution $s = \nu t$ in the integral of the last term. Using the relation for Beta function $B(\cdot,\cdot)$:

$$B(\beta+1,\alpha) = \int_0^1 (1-u)^{\alpha-1}u^{\beta}du = \frac{\Gamma(\alpha)\Gamma(\beta+1)}{\Gamma(\alpha+\beta+1)},$$

we get

$$x(t) = -I^{\alpha}h(t) + \frac{\Gamma(\alpha-\delta)t^{\alpha-1}}{(1-\eta^{\alpha-\delta-1})\Gamma(\alpha)}\Big(a + I^{\alpha-\delta}h(1) - I^{\alpha-\delta}h(\eta)\Big). \tag{4.89}$$

The solution of the original nonlinear problem (4.85) can be obtained by replacing h with the right hand side of the fractional differential equation of (4.85) in (4.89).

Let $\mathcal{C} = C([0,1],\mathbb{R})$ denote the Banach space of all continuous functions from $[0,1] \to \mathbb{R}$ endowed with the norm defined by $\|x\| = \sup\{|x(t)|, t \in [0,1]\}$.

4.6.1 *Existence result via Sadovski's fixed point theorem*

In the following we denote by $L_{1/p}([0,1],\mathbb{R}^+)$ the space of $\frac{1}{p}$-Lebesgue measurable functions from $[0,1]$ to $\mathbb{R}^+$ with norm $\|\mu\| = \left(\int_0^1 |\mu(s)|^{\frac{1}{p}} ds\right)^p$.

Theorem 4.20. *Let $f, g : [0,1] \times \mathbb{R} \to \mathbb{R}$ be continuous functions satisfying the conditions:*

(4.20.1) f is bounded and Lipschitz, i.e., $|f(t,x)| \le m_1(t)$ and

$$|f(t,x) - f(t,y)| \le L_1(t)|x-y|, \quad \text{for all } (t,x),(t,y) \in [0,1]\times\mathbb{R},$$

where $m_1, L_1 \in L_{1/p}([0,1],\mathbb{R}^+)$, $p = \min\{q_1, q_2\}$, $q_1 \in (0,1)$, $q_2 \in (0, \alpha-\delta)$;

(4.20.2) there exist a function $m_2 \subset L_{1/p}([0,1],\mathbb{R}^+)$, and a non-decreasing function $\psi : \mathbb{R}^+ \to \mathbb{R}^+$ such that

$$|g(t,x)| \le m_2(t)\psi(\|x\|), \quad \forall (t,x) \in [0,1]\times\mathbb{R}.$$

Then the boundary value problem (4.85) has at least one solution, provided

$$\gamma := \frac{|A|\|L_1\|}{\Gamma(\alpha)}\left(\frac{1-p}{\alpha-p}\right)^{1-p} + \frac{|A||Q|\|L_1\|}{\Gamma(\alpha-\delta)}\left(\frac{1-p}{\alpha-\delta-p}\right)^{1-p}(1+\eta^{\alpha-\delta-p}) < 1.$$

Proof. Let B_r be a closed bounded and convex subset of $\mathcal{C} := C([0,1],\mathbb{R})$, defined by $B_r = \{x \in \mathcal{C} : \|x\| \le r\}$. Using (4.89), we define a map $\Phi : B_r \to \mathcal{C}$ as

$$\begin{aligned}
(\Phi x)(t) &= -A\int_0^t \frac{(t-s)^{\alpha-1}}{\Gamma(\alpha)} f(s,x(s))ds - B\int_0^t \frac{(t-s)^{\alpha+\beta-1}}{\Gamma(\alpha+\beta)} g(s,x(s))ds \\
&\quad + Qt^{\alpha-1}\left[a + A\int_0^1 \frac{(1-s)^{\alpha-\delta-1}}{\Gamma(\alpha-\delta)} f(s,x(s))ds\right. \\
&\quad + B\int_0^1 \frac{(1-s)^{\alpha-\delta+\beta-1}}{\Gamma(\alpha-\delta+\beta)} g(s,x(s))ds \\
&\quad \left. - A\int_0^\eta \frac{(\eta-s)^{\alpha-\delta-1}}{\Gamma(\alpha-\delta)} f(s,x(s))ds - B\int_0^\eta \frac{(\eta-s)^{\alpha-\delta+\beta-1}}{\Gamma(\alpha-\delta+\beta)} g(s,x(s))ds\right],
\end{aligned}$$

where

$$Q = \frac{\Gamma(\alpha-\delta)}{(1-\eta^{\alpha-\delta-1})\Gamma(\alpha)}, \quad \eta^{\alpha-\delta-1} \neq 1.$$

Let us rewrite Φ as $\Phi = \Phi_1 + \Phi_2$, where

$$\begin{aligned}
(\Phi_1 x)(t) &= -A\int_0^t \frac{(t-s)^{\alpha-1}}{\Gamma(\alpha)} f(s,x(s))ds + aQt^{\alpha-1} \\
&\quad + AQt^{\alpha-1}\left[\int_0^1 \frac{(1-s)^{\alpha-\delta-1}}{\Gamma(\alpha-\delta)} f(s,x(s))ds\right. \\
&\quad \left. - \int_0^\eta \frac{(\eta-s)^{\alpha-\delta-1}}{\Gamma(\alpha-\delta)} f(s,x(s))ds\right], \quad t \in [0,1],
\end{aligned}$$

and

$$
\begin{aligned}
(\Phi_2 x)(t) = & -B\int_0^t \frac{(t-s)^{\alpha+\beta-1}}{\Gamma(\alpha+\beta)} g(s,x(s))ds \\
& + BQt^{\alpha-1}\Bigg[\int_0^1 \frac{(1-s)^{\alpha-\delta+\beta-1}}{\Gamma(\alpha-\delta+\beta)} g(s,x(s))ds \\
& - \int_0^\eta \frac{(\eta-s)^{\alpha-\delta+\beta-1}}{\Gamma(\alpha-\delta+\beta)} g(s,x(s))ds\Bigg], \quad t\in[0,1].
\end{aligned}
$$

Step 1. $\Phi(B_r)\subset B_r$.

We select

$$
\begin{aligned}
r \geq & \frac{|A|\|m_1\|}{\Gamma(\alpha)}\left(\frac{1-p}{\alpha-p}\right)^{1-p} + |a||A| + \frac{|A||Q|\|m_1\|}{\Gamma(\alpha-\delta)}\left(\frac{1-p}{\alpha-\delta-p}\right)^{1-p}\left(1+\eta^{\alpha-\delta-p}\right) \\
& + \frac{|B|\|m_2\|\psi(r)}{\Gamma(\alpha+\beta)}\left(\frac{1-p}{\alpha+\beta-p}\right)^{1-p} \\
& + \frac{|B||Q|\|m_2\|\psi(r)}{\Gamma(\alpha-\delta+\beta)}\left(\frac{1-p}{\alpha-\delta+\beta-p}\right)^{1-p}\left(1+\eta^{\alpha-\delta+\beta-p}\right).
\end{aligned}
$$

For $x\in B_r$ and $t\in[0,1]$, we have

$$
\begin{aligned}
& |(\Phi_1 x)(t)| \\
\leq & |A|\int_0^t \frac{(t-s)^{\alpha-1}}{\Gamma(\alpha)} m_1(s)ds + |a||A| + |A||Q|\int_0^1 \frac{(1-s)^{\alpha-\delta-1}}{\Gamma(\alpha-\delta)} m(s)ds \\
& + |A||Q|\int_0^\eta \frac{(\eta-s)^{\alpha-\delta-1}}{\Gamma(\alpha-\delta)} m(s)ds \\
\leq & \frac{|A|}{\Gamma(\alpha)}\left(\int_0^t (t-s)^{\frac{\alpha-1}{1-p}} ds\right)^{1-p}\left(\int_0^1 m_1^{\frac{1}{p}}(s)ds\right)^p + |a||A| \\
& + \frac{|A||Q|}{\Gamma(\alpha-\delta)}\left(\int_0^1 (1-s)^{\frac{\alpha-\delta-1}{1-p}} ds\right)^{1-p}\left(\int_0^1 m_1^{\frac{1}{p}}(s)ds\right)^p \\
& + \frac{|A||Q|\eta^{\alpha-\delta-p}}{\Gamma(\alpha-\delta)}\left(\int_0^\eta (\eta-s)^{\frac{\alpha-\delta-1}{1-p}} ds\right)^{1-p}\left(\int_0^1 m_1^{\frac{1}{p}}(s)ds\right)^p \\
\leq & \frac{|A|}{\Gamma(\alpha)}\left(\frac{1-p}{\alpha-p}\right)^{1-p}\|m_1\| + |a||A| + \frac{|A||Q|}{\Gamma(\alpha-\delta)}\left(\frac{1-p}{\alpha-\delta-p}\right)^{1-p}\|m_1\| \\
& + \frac{|A||Q|\eta^{\alpha-\delta-p}}{\Gamma(\alpha-\delta)}\left(\frac{1-p}{\alpha-\delta-p}\right)^{1-p}\|m_1\|.
\end{aligned}
$$

By a similar way, we have

$$
\begin{aligned}
|(\Phi_2 x)(t)| \leq & |B|\int_0^t \frac{(t-s)^{\alpha+\beta-1}}{\Gamma(\alpha+\beta)} m_2(s)\psi(r)\, ds \\
& + |B||Q|\int_0^1 \frac{(1-s)^{\alpha-\delta+\beta-1}}{\Gamma(\alpha-\delta+\beta)} m_2(s)\psi(r)\, ds
\end{aligned}
$$

$$+|B||Q|\int_0^{\eta}\frac{(\eta-s)^{\alpha-\delta+\beta-1}}{\Gamma(\alpha-\delta+\beta)}m_2(s)\psi(r)\,ds$$
$$\le \frac{|B|\psi(r)}{\Gamma(\alpha+\beta)}\left(\frac{1-p}{\alpha+\beta-p}\right)^{1-p}\|m_2\|$$
$$+\frac{|B||Q|\psi(r)}{\Gamma(\alpha-\delta+\beta)}\left(\frac{1-p}{\alpha-\delta+\beta-p}\right)^{1-p}\|m_2\|$$
$$+\frac{|B||Q|\psi(r)\,\eta^{\alpha-\delta+\beta-p}}{\Gamma(\alpha-\delta+\beta)}\left(\frac{1-p}{\alpha-\delta+\beta-p}\right)^{1-p}\|m_2\|.$$

Therefore

$$\begin{aligned}
&|(\Phi x)(t)|\\
&\le |(\Phi_1 x)(t)|+|(\Phi_2 x)(t)|\\
&\le \frac{|A|\|m_1\|}{\Gamma(\alpha)}\left(\frac{1-p}{\alpha-p}\right)^{1-p}+|a||A|+\frac{|A||Q|\|m_1\|}{\Gamma(\alpha-\delta)}\left(\frac{1-p}{\alpha-\delta-p}\right)^{1-p}\left(1+\eta^{\alpha-\delta-p}\right)\\
&\quad+\frac{|B|\|m_2\|\psi(r)}{\Gamma(\alpha+\beta)}\left(\frac{1-p}{\alpha+\beta-p}\right)^{1-p}\\
&\quad+\frac{|B||Q|\|m_2\|\psi(r)}{\Gamma(\alpha-\delta+\beta)}\left(\frac{1-p}{\alpha-\delta+\beta-p}\right)^{1-p}\left(1+\eta^{\alpha-\delta+\beta-p}\right)\\
&\le r,
\end{aligned}$$

which implies that $\Phi(B_r)\subset B_r$.

Step 2. Φ_1 is continuous and γ-contraction.

To show the continuity of Φ_1 for $t\in[0,1]$, let us consider a sequence x_n converging to x. Then, by the assumption (4.20.1), we have

$$\begin{aligned}
&\|(\Phi_1 x_n)-(\Phi_1 x)\|\\
&\le \left[\frac{|A|\|L_1\|}{\Gamma(\alpha)}\left(\frac{1-p}{\alpha-p}\right)^{1-p}+\frac{|A||Q|\|L_1\|}{\Gamma(\alpha-\delta)}\left(\frac{1-p}{\alpha-\delta-p}\right)^{1-p}\left(1+\eta^{\alpha-\delta-p}\right)\right]\|x_n-x\|.
\end{aligned}$$

In order to show that Φ_1 is γ_1-contraction, let us consider $x,y\in B_r$. The we have

$$\begin{aligned}
|(\Phi_1 x)(t)-(\Phi_1 y)(t)| \le \Bigg[&\frac{|A|\|L_1\|}{\Gamma(\alpha)}\left(\frac{1-p}{\alpha-p}\right)^{1-p}\\
&+\frac{|A||Q|\|L_1\|}{\Gamma(\alpha-\delta)}\left(\frac{1-p}{\alpha-\delta-p}\right)^{1-p}\left(1+\eta^{\alpha-\delta-p}\right)\Bigg]\|x-y\|.
\end{aligned}$$

By the given assumption, it follows that

$$\gamma_1=\frac{|A|\|L_1\|}{\Gamma(\alpha)}\left(\frac{1-p}{\alpha-p}\right)^{1-p}+\frac{|A||Q|\|L_1\|}{\Gamma(\alpha-\delta)}\left(\frac{1-p}{\alpha-\delta-p}\right)^{1-p}\left(1+\eta^{\alpha-\delta-p}\right)<1,$$

so Φ_1 is γ_1-contraction.

Step 3. Φ_2 is compact.

In Step 1 it is proved that Φ_2 is uniformly bounded. Now we show that Φ_2 maps bounded sets into equi-continuous sets of $C([0,1],\mathbb{R})$. Let $t_1, t_2 \in [0,1]$ with $t_1 < t_2$ and $x \in B_r$. Then we obtain

$$\begin{aligned}
&|(\Phi_2 x)(t_2) - (\Phi_2 x)(t_1)| \\
\leq\ & \frac{|B|}{\Gamma(\alpha+\beta)} \int_0^{t_1} [(t_2-s)^{\alpha+\beta-1} - (t_1-s)^{\alpha+\beta-1}]|g(s,x(s))|ds \\
&+ \frac{|B|}{\Gamma(\alpha+\beta)} \int_{t_1}^{t_2} (t_2-s)^{\alpha+\beta-1}|g(s,x(s))|ds \\
&+ |B||Q||t_2^{\alpha-1} - t_1^{\alpha-1}| \Bigg[\int_0^1 \frac{(1-s)^{\alpha-\delta+\beta-1}}{\Gamma(\alpha-\delta+\beta)} |g(s,x(s))|ds \\
&+ \int_0^\eta \frac{(\eta-s)^{\alpha-\delta+\beta-1}}{\Gamma(\alpha-\delta+\beta)} |g(s,x(s))|ds \Bigg] \\
\leq\ & \frac{|B|}{\Gamma(\alpha+\beta)} \int_0^{t_1} [(t_2-s)^{\alpha+\beta-1} - (t_1-s)^{\alpha+\beta-1}] m_2(s)ds \\
&+ \frac{|B|\psi(r)}{\Gamma(\alpha+\beta)} \int_{t_1}^{t_2} (t_2-s)^{\alpha+\beta-1} m_2(s)ds \\
&+ |B||Q|\psi(r)|t_2^{\alpha-1} - t_1^{\alpha-1}| \Bigg[\int_0^1 \frac{(1-s)^{\alpha-\delta+\beta-1}}{\Gamma(\alpha-\delta+\beta)} m_2(s)ds \\
&+ \int_0^\eta \frac{(\eta-s)^{\alpha-\delta+\beta-1}}{\Gamma(\alpha-\delta+\beta)} m_2(s)ds \Bigg].
\end{aligned}$$

Obviously the right hand side of the above inequality tends to zero independently of $x \in B_r$ as $t_2 - t_1 \to 0$. Therefore it follows by the Arzelá-Ascoli theorem that $\Phi_2 : C([0,1],\mathbb{R}) \to C([0,1],\mathbb{R})$ is completely continuous. Thus Φ_2 is compact on $[0,1]$.

Step 4. Φ is condensing.

As $\Phi = \Phi_1 + \Phi_2$, $\Phi : B_r \to B_r$, Φ_1 is continuous, γ-contraction and Φ_2 is compact, therefore, by Theorem 1.14, Φ is a condensing map on B_r.

Consequently, by Theorem 1.14, the problem (4.85) has a fixed point which implies that the problem (4.85) has a solution. □

In the special case when $L_1(t) = L$, L a constant, we have the following result.

Corollary 4.1. *Let $f, g : [0,1] \times \mathbb{R} \to \mathbb{R}$ be continuous functions. Assume that g satisfies (4.20.2) and f satisfies the following condition:*

(4.1.1) $|f(t,x) - f(t,y)| \le L|x-y|$, *for all* $(t,x),(t,y) \in [0,1] \times \mathbb{R}$, $L > 0$ *is a constant.*

If

$$\gamma' := \frac{|A|L}{\Gamma(\alpha+1)} + \frac{|A||Q|L}{\Gamma(\alpha-\delta+1)}\left(1+\eta^{\alpha-\delta}\right) < 1,$$

then the boundary value problem (4.85) has at least one solution.

Example 4.6. Consider the following integro-differential equations of fractional order with nonlocal fractional boundary conditions:

$$\begin{cases} -D^{5/2}x(t) = Af(t,x(t)) + BI^{\beta}g(t,x(t)), \quad t \in [0,1], \\ D^{1/4}x(0) = 0, \ D^{5/4}x(0) = 0, \ D^{1/4}x(1) - D^{1/4}x(\eta) = a, \end{cases} \tag{4.90}$$

where $A = B = 1$, $\beta = 3/4, \eta = 2/3, f(t,x) = \dfrac{e^{-t}|x|}{(1+e^t)(1+|x|)}$, $g(t,x) = \dfrac{t^2|x|^3}{1+|x|^3}$.

Then we have

$$\begin{aligned} |f(t,x) - f(t,y)| &= \frac{e^{-t}}{1+e^t}\left|\frac{|x|}{1+|x|} - \frac{|y|}{1+|y|}\right| \\ &\le \frac{e^{-t}|x-y|}{1+e^t} \\ &\le L(t)|x-y|, \quad \text{where } L(t) = \frac{1}{2}e^{-t}. \end{aligned}$$

Also, for all $x \in \mathcal{C}$ and each $t \in [0,1]$, we have

$$|f(t,x)| = \frac{e^{-t}}{1+e^t}\left|\frac{|x|}{1+|x|}\right| \le \frac{e^{-t}}{1+e^t} < m_1(t), \quad \text{where } m_1(t) = \frac{1}{2}e^{-t},$$

and $|g(t,x)| \le m_2(t)\psi(\|x\|)$ with $m_2(t) = t^2$ ($\|m_2\| = (3/11)^{3/4}$) and $\psi(\|x\|) = 1$. Selecting $p = 3/4$ and using the given data, we find that

$$\begin{aligned} \gamma :&= \frac{|A|\|L_1\|}{\Gamma(\alpha)}\left(\frac{1-p}{\alpha-p}\right)^{1-p} + \frac{|A||Q|\|L_1\|}{\Gamma(\alpha-\delta)}\left(\frac{1-p}{\alpha-\delta-p}\right)^{1-p}\left(1+\eta^{\alpha-\delta-p}\right) \\ &\simeq 0.825717. \end{aligned}$$

As $\gamma < 1$, therefore, by the conclusion of Theorem 4.20, the problem (4.90) has a solution on $[0,1]$.

4.7 A coupled system of nonlinear fractional differential equations with multi-point fractional boundary conditions on an unbounded domain

In this section, we discuss the existence and uniqueness of solutions of a coupled system of nonlinear fractional differential equations with m-point boundary conditions on an unbounded domain. Precisely, we consider the following problem

$$\begin{cases} D^p u(t) + f(t, v(t)) = 0, \quad 2 < p < 3, \\ D^q v(t) + g(t, u(t)) = 0, \quad 2 < q < 3, \\ u(0) = u'(0) = 0, \quad D^{p-1}u(+\infty) = \sum\limits_{i=1}^{m-2} \beta_i u(\xi_i), \\ v(0) = v'(0) = 0, \quad D^{q-1}v(+\infty) = \sum\limits_{i=1}^{m-2} \gamma_i v(\xi_i), \end{cases} \tag{4.91}$$

where $t \in [0, +\infty)$, $0 < \xi_1 < \xi_2 < \cdots < \xi_{m-2} < +\infty$, D^p, D^q denote Riemann-Liouville fractional derivative operators of order p and q respectively and $\beta_i > 0$, $\gamma_i > 0$ are such that $0 < \sum\limits_{i=1}^{m-2} \beta_i \xi_i^{p-1} < \Gamma(p)$ and $0 < \sum\limits_{i=1}^{m-2} \gamma_i \xi_i^{q-1} < \Gamma(q)$.

For the forthcoming analysis, we define the spaces

$$X = \{u \in C[0, +\infty) : \sup_{0<t<+\infty} \frac{|u(t)|}{1 + t^{p-1}} < +\infty\},$$

$$Y = \{v \in C[0, +\infty) : \sup_{0<t<+\infty} \frac{|v(t)|}{1 + t^{q-1}} < +\infty\},$$

equipped with the norms

$$\|u\|_X = \sup_{0<t<+\infty} \frac{|u(t)|}{1 + t^{p-1}},$$

$$\|v\|_Y = \sup_{0<t<+\infty} \frac{|v(t)|}{1 + t^{q-1}}.$$

Obviously X and Y are Banach spaces.

Lemma 4.7. *Let $h \in C([0, +\infty))$. For $2 < \alpha < 3$, the fractional boundary value problem*

$$\begin{cases} D^\alpha u(t) + \quad h(t) = 0, \\ u(0) = u'(0) = 0, \quad D^{\alpha-1}u(+\infty) = \sum\limits_{i=1}^{m-2} \beta_i u(\xi_i), \end{cases} \tag{4.92}$$

has a unique solution

$$u(t) = \int_0^{+\infty} G(t, s) h(s) ds,$$

where

$$G(t, s) = G^*(t, s) + G^{**}(t, s),$$

with

$$G^*(t, s) = \frac{1}{\Gamma(\alpha)} \begin{cases} t^{\alpha-1} - (t - s)^{\alpha-1}, & 0 \le s \le t \le +\infty, \\ t^{\alpha-1}, & 0 \le t \le s \le +\infty. \end{cases} \tag{4.93}$$

$$G^{**}(t, s) = \frac{\sum\limits_{i=1}^{m-2} \beta_i t^{\alpha-1}}{\Gamma(\alpha) - \sum\limits_{i=1}^{m-2} \beta_i \xi_i^{\alpha-1}} G^*(\xi_i, s). \tag{4.94}$$

Lemma 4.8. *For* $(s,t) \in [0,+\infty) \times [0,+\infty)$, $\dfrac{G(t,s)}{1+t^{\alpha-1}} \le L_1$, *where*

$$L_1 = \frac{1}{\Gamma(\alpha)} + \frac{\sum\limits_{i=1}^{m-2} \beta_i \xi_{m-2}^{\alpha-1}}{\Gamma(\alpha)\Big(\Gamma(\alpha) - \sum\limits_{i=1}^{m-2} \beta_i \xi_i^{\alpha-1}\Big)}. \tag{4.95}$$

4.7.1 *Main results*

This section is devoted to some existence and uniqueness results for problem (4.91).

Define the space

$$Z = \{(u,v) | u \in X,\ v \in Y\}$$

equipped with the norm

$$\|(u,v)\|_Z = \max\{\|u\|_X,\ \|v\|_Y\}.$$

Clearly Z is a Banach space.

Let an operator $Q : Z \to Z$ be defined by

$$\begin{aligned} Q(u,v) &= (Q_1(v),\ Q_2(u)) \\ &= \Big(\int_0^{+\infty} G_1(t,s) f(t,v(s))ds,\ \int_0^{+\infty} G_2(t,s) g(t,u(s))ds\Big), \end{aligned} \tag{4.96}$$

where $G_1(t,s) = G_{11}(t,s) + G_{12}(t,s)$, $G_2(t,s) = G_{21}(t,s) + G_{22}(t,s)$, with

$$G_{11}(t,s) = \frac{1}{\Gamma(p)} \begin{cases} t^{p-1} - (t-s)^{p-1}, & 0 \le s \le t \le +\infty, \\ t^{p-1}, & 0 \le t \le s \le +\infty. \end{cases}$$

$$G_{12}(t,s) = \frac{\sum\limits_{i=1}^{m-2} \beta_i t^{p-1}}{\Gamma(p) - \sum\limits_{i=1}^{m-2} \beta_i \xi_i^{p-1}} G_{11}(\xi_i, s).$$

$$G_{21}(t,s) = \frac{1}{\Gamma(q)} \begin{cases} t^{q-1} - (t-s)^{q-1}, & 0 \le s \le t \le +\infty, \\ t^{q-1}, & 0 \le t \le s \le +\infty. \end{cases}$$

$$G_{22}(t,s) = \frac{\sum\limits_{i=1}^{m-2} \gamma_i t^{q-1}}{\Gamma(q) - \sum\limits_{i=1}^{m-2} \gamma_i \xi_i^{q-1}} G_{21}(\xi_i, s).$$

Observe that the problem (4.91) has a solution if and only if the operator Q defined by (4.96) has a fixed point.

Lemma 4.9. *For* $(s,t) \in [0,+\infty) \times [0,+\infty)$, *we have*

$$\frac{G_1(t,s)}{1+t^{p-1}} \le L \text{ and } \frac{G_2(t,s)}{1+t^{q-1}} \le L,$$

where

$$L = \max\left\{\frac{1}{\Gamma(p)} + \frac{\sum\limits_{i=1}^{m-2} \beta_i \xi_{m-2}^{p-1}}{\Gamma(p)\Big(\Gamma(p) - \sum\limits_{i=1}^{m-2} \beta_i \xi_i^{p-1}\Big)}, \frac{1}{\Gamma(q)} + \frac{\sum\limits_{i=1}^{m-2} \gamma_i \xi_{m-2}^{q-1}}{\Gamma(q)\Big(\Gamma(q) - \sum\limits_{i=1}^{m-2} \gamma_i \xi_i^{q-1}\Big)}\right\}. \quad (4.97)$$

Theorem 4.21. *Assume that*

(4.21.1) there exist non-negative functions $a(t)$, $b(t) \in C[0, +\infty)$ such that

$$|f(t, x)| \leq a(t)|x| + b(t), \ t \in [0, +\infty),$$

and

$$\int_0^{+\infty} (1 + t^{q-1})a(t)dt < \frac{1}{L}, \quad \int_0^{+\infty} b(t)dt < +\infty;$$

(4.21.2) there exist non-negative functions $c(t)$, $d(t) \in C[0, +\infty)$ such that

$$|g(t, y)| \leq c(t)|y| + d(t), \ t \in [0, +\infty),$$

and

$$\int_0^{+\infty} (1 + t^{p-1})c(t)dt < \frac{1}{L}, \quad \int_0^{+\infty} d(t)dt < +\infty.$$

Then the system (4.91) *has a solution.*

Proof. Let us take

$$R > \max\left\{\frac{L\displaystyle\int_0^{+\infty} b(s)ds}{1 - L\displaystyle\int_0^{+\infty} (1 + s^{q-1})a(s)ds}, \frac{L\displaystyle\int_0^{+\infty} d(s)ds}{1 - L\displaystyle\int_0^{+\infty} (1 + s^{p-1})c(s)ds}\right\}$$

and define

$$B_R = \{(u, v) \in Z \mid \|(u, v)\|_Z \leq R\}.$$

Obviously, B_R is a bounded closed and convex set of Z.

As a first step, we show that the operator Q is $B_R \to B_R$. For any $(u, v) \in B_R$, we have

$$\begin{aligned} \|Q_1 v\|_X &= \sup_{0<t<+\infty} \frac{1}{1 + t^{p-1}}\left|\int_0^{+\infty} G_1(t, s)f(s, v(s)ds\right| \\ &\leq \sup_{0<t<\infty} \frac{1}{1 + t^{p-1}} \int_0^{+\infty} G_1(t, s)(a(s)|v(s)| + b(s))ds \\ &\leq L\int_0^{+\infty} (1 + s^{q-1})a(s)ds\|v\|_Y + L\int_0^{+\infty} b(s)ds \\ &< \frac{L\int_0^{+\infty} b(s)ds}{1 - L\int_0^{+\infty}(1 + t^{q-1})a(s)ds} \\ &< R. \end{aligned} \quad (4.98)$$

Similarly, we can get

$$\begin{aligned}\|Q_2u\|_Y &= \sup_{0<t<+\infty}\frac{1}{1+t^{q-1}}\left|\int_0^{+\infty}G_2(t,s)g(s,u(s))ds\right| \\ &\leq L\int_0^{+\infty}(1+s^{p-1})c(s)ds\|u\|_X + L\int_0^{+\infty}d(s)ds \\ &< \frac{L\int_0^{+\infty}d(s)ds}{1-L\int_0^{+\infty}(1+s^{p-1})c(s)ds} \\ &< R. \end{aligned} \tag{4.99}$$

That is, $\|Q(u,v)\|_Z < R$. Thus, $QB_R \subset B_R$.

Next, we show that Q is completely continuous. By continuity of f, g, G_1 and G_2, it follows that Q is continuous. On the other hand, by a similar process used in [116], we can easily prove that the operators Q_1, Q_2 are equi-continuous. Therefore it follows that QB_R is an equi-continuous set. Also, it is uniformly bounded as $QB_R \subset B_R$. Thus, we conclude that Q is a completely continuous operator. Hence, by Schauder fixed point theorem, there exists a solution of (4.91). This completes the proof. □

Example 4.7. Consider the following multi-point boundary value problem on an unbounded domain:

$$\begin{cases} D^{\frac{9}{4}}u(t) + \dfrac{\sin t\ln(1+|v(t)|)}{(1+t^{\frac{7}{4}})(2+t)^2} + (1+\cos 2t)e^{-t} = 0, \\ D^{\frac{11}{4}}v(t) + \dfrac{e^{-5t}\sin|u(t)|}{3(1+t^{\frac{5}{4}})(1+t)^2} + \dfrac{4}{(t+4)^2} = 0, \\ u(0) = u'(0) = 0,\ D^{\frac{5}{4}}u(+\infty) = \frac{2}{5}u\left(\frac{1}{4}\right) + \frac{1}{10}u(1), \\ v(0) = v'(0) = 0,\ D^{\frac{7}{4}}v(+\infty) = \frac{3}{10}u\left(\frac{1}{4}\right) + \frac{1}{5}u(1). \end{cases} \tag{4.100}$$

Here $t \in [0, +\infty)$, $p = \frac{9}{4}$, $q = \frac{11}{4}$, $\xi_1 = \frac{1}{4}$, $\xi_2 = 1$, $\beta_1 = \frac{2}{5}$, $\beta_2 = \frac{1}{10}$, $\gamma_1 = \frac{3}{10}$, $\gamma_2 = \frac{1}{5}$,

$$f(t,v(t)) = \frac{\sin t\ln(1+|v(t)|)}{(1+t^{\frac{7}{4}})(2+t)^2} + (1+\cos 2t)e^{-t}, \quad g(t,u(t)) = \frac{e^{-5t}\sin|u(t)|}{3(1+t^{\frac{5}{4}})(1+t)^2} + \frac{4}{(t+4)^2}.$$

For $a(t) = \dfrac{1}{(1+t^{\frac{7}{4}})(2+t)^2}$, $b(t) = 2e^{-t}$, $c(t) = \dfrac{1}{3(1+t^{\frac{5}{4}})(1+t)^2}$, $d(t) = \dfrac{4}{(t+4)^2}$, by direct calculation, we find that

$$L = \max\left\{\frac{1}{\Gamma(p)} + \frac{\sum_{i=1}^{m-2}\beta_i\xi_{m-2}^{p-1}}{\Gamma(p)\Big(\Gamma(p) - \sum_{i=1}^{m-2}\beta_i\xi_i^{p-1}\Big)}, \frac{1}{\Gamma(q)} + \frac{\sum_{i=1}^{m-2}\gamma_i\xi_{m-2}^{q-1}}{\Gamma(q)\Big(\Gamma(q) - \sum_{i=1}^{m-2}\gamma_i\xi_i^{q-1}\Big)}\right\}$$

$$= \max\left\{\frac{1}{\Gamma(\frac{9}{4})} + \frac{\frac{2}{5}+\frac{1}{10}}{\Gamma(\frac{9}{4})\Big(\Gamma(\frac{9}{4}) - \frac{2}{5}(\frac{1}{4})^{\frac{5}{4}} - \frac{1}{10}\Big)}, \frac{1}{\Gamma(\frac{11}{4})} + \frac{\frac{3}{10}+\frac{1}{5}}{\Gamma(\frac{11}{4})\Big(\Gamma(\frac{11}{4}) - \frac{3}{10}(\frac{1}{4})^{\frac{7}{4}} - \frac{1}{5}\Big)}\right\}$$

$$\approx 1.341213,$$

$$|f(t,x)| \le a(t)|x| + b(t), \qquad |g(t,y)| \le c(t)|y| + d(t),\ t \in [0,+\infty),$$

and

$$\int_0^{+\infty}(1+t^{q-1})a(t)dt = \frac{1}{2} < \frac{1}{L} \approx 0.745594, \quad \int_0^{+\infty} b(t)dt = 2 < +\infty,$$

$$\int_0^{+\infty}(1+t^{p-1})c(t)dt = \frac{1}{3} < \frac{1}{L} \approx 0.745594, \quad \int_0^{+\infty} d(t)dt = 1 < +\infty.$$

Thus all conditions of Theorem 4.21 are satisfied. Therefore, by Theorem 4.21, the couple system of nonlinear fractional differential equations (4.100) has at least one solution.

Theorem 4.22. *Assume that*

(4.22.1) there exist $0 < \rho_1 < 1$ *and non-negative functions* $a_1(t),\ b_1(t) \in C[0,+\infty)$ *such that*

$$|f(t,x)| \le a_1(t)|x|^{\rho_1} + b_1(t),\ t \in [0,+\infty),$$

and

$$\int_0^{+\infty}(1+t^{q-1})a_1(t)dt < +\infty, \quad \int_0^{+\infty} b_1(t)dt < +\infty.$$

(4.22.2) there exist $0 < \rho_2 < 1$ *and non-negative functions* $c_1(t),\ d_1(t) \in C[0,+\infty)$ *such that*

$$|g(t,y)| \le c_1(t)|y|^{\rho_2} + d_1(t),\ t \in [0,+\infty),$$

and

$$\int_0^{+\infty}(1+t^{p-1})c_1(t)dt < +\infty, \quad \int_0^{+\infty} d_1(t)dt < +\infty.$$

Then the system (4.91) *has a solution.*

Proof. In this case, we take

$$R > \max\Big\{2L\int_0^{+\infty} b_1(s)ds, \Big(2L\int_0^{+\infty}(1+s^{q-1})a_1(s)ds\Big)^{\frac{1}{1-\rho_1}}, 2L\int_0^{+\infty} d_1(s)ds, \Big(2L\int_0^{+\infty}(1+s^{p-1})c_1(s)ds\Big)^{\frac{1}{1-\rho_2}}\Big\}. \tag{4.101}$$

The rest of the proof is similar to that of Theorem 4.21. So we omit it. □

Remark 4.5. By taking $\rho_1, \rho_2 > 1$ (instead of $0 < \rho_1 < 1$, $0 < \rho_2 < 1$) in (4.22.1) and (4.22.2), one can show that (4.91) has a solution.

Theorem 4.23. *Assume that*

(4.23.1) the functions f and g satisfy Lipschitz condition, that is, there exist non-negative functions $K_1(t), K_2(t)$ such that

$$|f(t,x) - f(t,y)| \leq K_1(t)|x-y|, \ t \in [0,+\infty),$$

and

$$|g(t,x) - g(t,y)| \leq K_2(t)|x-y|, \ t \in [0,+\infty).$$

Then the problem (4.91) has a unique solution if

$$\mu = L\int_0^{+\infty} K_1(s)(1+s^{q-1})ds < 1, \ \tau = L\int_0^{+\infty} K_2(s)(1+s^{p-1})ds < 1.$$

Proof. For any $(u_1, v_1), (u_2, v_2) \in Z$, we have

$$\begin{aligned}\|Q_1v_2 - Q_1v_1\|_X &= \sup_{0<t<+\infty} \frac{1}{1+t^{p-1}}\left|\int_0^{+\infty} G_1(t,s)[f(s,v_2(s)) - f(s,v_1(s))]ds\right| \\ &\leq \sup_{0<t<+\infty}\int_0^{+\infty} \frac{G_1(t,s)}{1+t^{p-1}}K_1(s)|(v_2-v_1)(s)|ds \\ &\leq L\int_0^{+\infty} K_1(s)(1+s^{q-1})ds\|v_2-v_1\|_Y \\ &= \mu\|v_2-v_1\|_Y. \end{aligned} \tag{4.102}$$

Similarly, it can be shown that

$$\begin{aligned}\|Q_2u_2 - Q_2u_1\|_Y &= \sup_{0<t<+\infty} \frac{1}{1+t^{q-1}}\left|\int_0^{+\infty} G_1(t,s)(g(s,u_2(s)) - f(s,u_2(s)))ds\right| \\ &\leq L\int_0^{+\infty} K_2(s)(1+s^{p-1})ds\|u_2-u_1\|_X \\ &= \tau\|u_2-u_1\|_X. \end{aligned} \tag{4.103}$$

Thus, we get

$$\|Q(u_2,v_2) - Q(u_1,v_1)\|_Z \leq \max\{\mu,\tau\}\|(u_2,v_2)-(u_1,v_1)\|_Z.$$

Obviously, Q is a contraction. Thus, the conclusion of the theorem follows by the contraction mapping principle. This completes the proof. □

Example 4.8. Consider the following problem on an unbounded domain

$$\begin{cases} D^pu(t) + M_1(t)\sin v(t) + N_1(t) = 0, \\ D^qv(t) + \dfrac{M_2(t)}{1+u^2(t)} + N_2(t) = 0, \\ u(0) = u'(0) = 0, \ D^{p-1}u(+\infty) = \frac{2}{5}u\left(\frac{1}{4}\right) + \frac{1}{10}u(1), \\ v(0) = v'(0) = 0, \ D^{q-1}v(+\infty) = \frac{3}{10}u\left(\frac{1}{4}\right) + \frac{1}{5}u(1). \end{cases} \tag{4.104}$$

Here $t \in [0,+\infty)$, $2 < p, q < 3$, $\xi_1 = \frac{1}{4}$, $\xi_2 = 1$, $\beta_1 = \frac{2}{5}$, $\beta_2 = \frac{1}{10}$, $\gamma_1 = \frac{3}{10}$, $\gamma_2 = \frac{1}{5}$, $M_1(t), M_2(t), N_1(t), N_2(t) \in C([0,+\infty),\mathbb{R})$.

With

$$f(t, v(t)) = M_1(t)\sin v(t) + N_1(t), \qquad g(t, u(t)) = \frac{M_2(t)}{1+u^2(t)} + N_2(t),$$

we have

$$|f(t,x) - f(t,y)| = |M_1(t)||\sin x - \sin y| \leq |M_1(t)||x-y|, \ t \in [0, +\infty)$$

and

$$|g(t,x) - g(t,y)| = |M_2(t)||\frac{1}{1+x^2} - \frac{1}{1+y^2}| \leq |M_2(t)||x-y|, \ t \in [0, +\infty),$$

where $K_1(t) = |M_1(t)|$, $K_2(t) = |M_2(t)|$. So, the condition (4.23.1) holds. Let us assume that

$$\mu = L\int_0^{+\infty} |M_1(s)|(1+s^{q-1})ds < 1, \ \tau = L\int_0^{+\infty} |M_2(s)|(1+s^{p-1})ds < 1. \quad (4.105)$$

For example, condition (4.105) holds if we take

$$p = \frac{9}{4}, \ q = \frac{11}{4}, \ M_1(t) = \frac{1}{(1+t^{\frac{7}{4}})(2+t)^2}, \ M_2(t) = \frac{1}{3(1+t^{\frac{5}{4}})(1+t)^2}.$$

Thus all the conditions of Theorem 4.23 are satisfied. Therefore, by the conclusion of Theorem 4.23, the coupled system (4.104) has a unique solution.

4.8 Notes and remarks

The results of this chapter are adapted from the papers [20], [123], [12], [124], [2], [62], [35] and [144].

Chapter 5

Nonlocal fractional-order boundary value problems with discrete and integral boundary conditions

In this chapter, we study a wide class of nonlocal fractional-order boundary value problems equipped with discrete and integral boundary conditions. Precisely, we investigate the existence of solutions for Caputo fractional differential equations, nonlinear systems of coupled fractional differential equations, multi-term fractional differential equations, coupled system of neutral fractional order differential equations containing mixed Caputo and Riemann-Liouville sequential fractional derivative operators, and fractional differential inclusions, complemented with a variety of boundary conditions such as discrete and integral, Riemann-Liouville type multi-strip, coupled integro-multi-strip-multi-point, Riemann-Stieltjes integral-multi-strip conditions.

5.1 Existence theorems for semilinear Caputo fractional differential equations with nonlocal discrete and integral boundary conditions

In this section, we consider a Caputo type fractional differential equation

$$^{C}D^{q}x(t) = f(t, x(t)),\ n-1 < q \le n,\quad t \in [0,1], \tag{5.1}$$

supplemented with the boundary conditions

$$\begin{cases} x(0)=0,\ x'(0)=0,\ x''(0)=0,\ \ldots,\ x^{(n-2)}(0)=0, \\ x(1) = a\displaystyle\int_0^{\xi} x(s)ds + b\sum_{i=1}^{m-2}\alpha_i x(\eta_i),\ \ 0<\xi<\eta_1<\eta_2<\ldots<\eta_{m-2}<1, \end{cases} \tag{5.2}$$

where $^{C}D^{q}$ denotes the Caputo fractional derivative operator of order q, f is a given continuous function, a and b are real constants and α_i's are positive real constants. Here we emphasize that the multi-point strip boundary condition in (5.2) can be interpreted as the value of the unknown function at the right end point $t = 1$ of the given interval is proportional to the sum of the values of the unknown function on the strip $(0, \xi)$ and discrete values of the unknown function at $\eta_i, i = 1, 2, \ldots, m-2$ $(0 < \xi < \eta_i < 1)$.

The following lemma is useful in defining the solution for the problem (5.1)-(5.2).

Lemma 5.1. *For $F \in C([0,1],\mathbb{R})$, the solution of the linear fractional differential equation*

$$^{C}D^{q}x(t) = F(t), \ n-1 < q \le n, \ \ t \in [0,1], \tag{5.3}$$

supplemented with the boundary conditions (5.2) is equivalent to the integral equation

$$x(t) = \int_0^t \frac{(t-s)^{q-1}}{\Gamma(q)}F(s)ds + \frac{t^{n-1}}{\Delta}\Big(a\int_0^{\xi} \frac{(\xi-s)^{q}}{\Gamma(q+1)}F(s)ds$$
$$+b\sum_{i=1}^{m-2}\alpha_i\int_0^{\eta_i}\frac{(\eta_i-s)^{q-1}}{\Gamma(q)}F(s)ds - \int_0^1 \frac{(1-s)^{q-1}}{\Gamma(q)}F(s)ds\Big), \tag{5.4}$$

where it is assumed that

$$\Delta = 1 - \frac{a\xi^n}{n} - b\sum_{i=1}^{m-2}\alpha_i\eta_i^{n-1} \neq 0. \tag{5.5}$$

Proof. Solving the fractional differential equation (5.3) in a standard manner, we get

$$x(t) = \int_0^t \frac{(t-s)^{q-1}}{\Gamma(q)}F(s)ds + c_0 + c_1t + c_2t^2 + \ldots + c_{n-1}t^{n-1}, \tag{5.6}$$

where $c_i \in \mathbb{R}, \ i = 0,1,\ldots,n-1$, are arbitrary constants.

Using the boundary conditions: $x(0) = 0,\ldots,x^{(n-2)}(0) = 0$ in (5.6), we find that $c_i = 0, \ i = 0,1,\ldots,n-2$. Thus (5.6) becomes

$$x(t) = \int_0^t \frac{(t-s)^{q-1}}{\Gamma(q)}F(s)ds + c_{n-1}t^{n-1}. \tag{5.7}$$

Using the last condition of (5.2) in (5.7), we obtain

$$c_{n-1} = \frac{1}{\Delta}\Big(a\int_0^{\xi}\frac{(\xi-s)^{q}}{\Gamma(q+1)}F(s)ds + b\sum_{i=1}^{m-2}\alpha_i\int_0^{\eta_i}\frac{(\eta_i-s)^{q-1}}{\Gamma(q)}F(s)ds$$
$$-\int_0^1\frac{(1-s)^{q-1}}{\Gamma(q)}F(s)ds\Big),$$

where Δ is given by (5.5). Substituting the value of c_{n-1} in (5.7), we get the solution (5.4). Conversely, by direct computation, it can be shown that (5.4) satisfies the problem (5.3)-(5.2). This completes the proof. □

Let $\mathcal{G} = C([0,1],\mathbb{R})$ denote the Banach space of all continuous functions from $[0,1]$ to $\mathbb{R}$ endowed with the norm: $\|x\| = \sup\{|x(t)|, t \in [0,1]\}$.

Relative to the problem (5.1)-(5.2), in view of Lemma 5.1, we define an operator $\mathcal{S}: \mathcal{G} \to \mathcal{G}$ as

$$(\mathcal{S}x)(t) = \int_0^t \frac{(t-s)^{q-1}}{\Gamma(q)}f(s,x(s))ds + \frac{t^{n-1}}{\Delta}\Big(a\int_0^{\xi}\frac{(\xi-s)^{q}}{\Gamma(q+1)}f(s,x(s))ds$$

$$+b\sum_{i=1}^{m-2}\alpha_i\int_0^{\eta_i}\frac{(\eta_i-s)^{q-1}}{\Gamma(q)}f(s,x(s))ds-\int_0^1\frac{(1-s)^{q-1}}{\Gamma(q)}f(s,x(s))ds\Big). \tag{5.8}$$

Notice that the problem (5.1)-(5.2) has solutions if and only if the operator $\mathcal{S}$ has fixed points.

For the sake of computational convenience, we set

$$\sigma=\frac{1}{\Gamma(q+1)}\Big[1+\frac{1}{|\Delta|}\Big(1+|a|\frac{\xi^{q+1}}{(q+1)}+|b|\sum_{i=1}^{m-2}\alpha_i{\eta_i}^q\Big)\Big]. \tag{5.9}$$

5.1.1 *Existence results*

Theorem 5.1. *Let $f:[0,1]\times\mathbb{R}\to\mathbb{R}$ be a continuous function such that the following condition holds:*

(5.1.1) $|f(t,x)-f(t,y)|\le\ell|x-y|,\quad\forall t\in[0,1],\ x,y\in\mathbb{R},\ \ell>0.$

Then the problem (5.1)-(5.2) has a unique solution if $\ell\sigma<1$, where σ is given by (5.9).

Proof. Let us set $\sup_{t\in[0,1]}|f(t,0)|=\omega$, $r>\sigma\omega(1-\sigma\ell)^{-1}$, and show that $\mathcal{S}B_r\subset B_r$, where $B_r=\{x\in\mathcal{G}:\|x\|\le r\}$. For $x\in B_r$, $t\in[0,1]$, using

$$\begin{aligned}|f(t,x(t))|&=|f(t,x(t))-f(t,0)+f(t,0)|\\&\le|f(t,x(t))-f(t,0)|+|f(t,0)|\le\ell r+\omega,\end{aligned}$$

and (5.9), we get

$$\begin{aligned}&\|(\mathcal{S}x)\|\\
&\le\sup_{t\in[0,1]}\Big\{\int_0^t\frac{(t-s)^{q-1}}{\Gamma(q)}|f(s,x(s))|ds+\frac{t^{n-1}}{|\Delta|}\Big[|a|\int_0^\xi\frac{(\xi-s)^q}{\Gamma(q+1)}|f(s,x(s))|ds\\
&\quad+|b|\sum_{i=1}^{m-2}\alpha_i\int_0^{\eta_i}\frac{(\eta_i-s)^{q-1}}{\Gamma(q)}|f(s,x(s))|ds+\int_0^1\frac{(1-s)^{q-1}}{\Gamma(q)}|f(s,x(s))|ds\Big]\Big\}\\
&\le(\ell r+\omega)\sup_{t\in[0,1]}\Big\{\int_0^t\frac{(t-s)^{q-1}}{\Gamma(q)}ds+\frac{t^{n-1}}{|\Delta|}\Big[|a|\int_0^\xi\frac{(\xi-s)^q}{\Gamma(q+1)}ds\\
&\quad+|b|\sum_{i=1}^{m-2}\alpha_i\int_0^{\eta_i}\frac{(\eta_i-s)^{q-1}}{\Gamma(q)}ds+\int_0^1\frac{(1-s)^{q-1}}{\Gamma(q)}ds\Big]\Big\}\\
&\le(\ell r+\omega)\sup_{t\in[0,1]}\Big\{\frac{t^q}{\Gamma(q+1)}+\frac{t^{n-1}}{|\Delta|}\Big[|a|\frac{\xi^{q+1}}{\Gamma(q+2)}\\
&\quad+|b|\sum_{i=1}^{m-2}\alpha_i\Big(\frac{{\eta_i}^q}{\Gamma(q+1)}\Big)+\frac{1}{\Gamma(q+1)}\Big]\Big\}\\
&\le(\ell r+\omega)\sigma\le r.\end{aligned}$$

This shows that $\mathcal{S}B_r \subset B_r$.

Now, for $x, y \in \mathbb{R}$, and for each $t \in [0,1]$, we obtain

$$\begin{aligned}
\|\mathcal{S}x - \mathcal{S}y\| \leq & \sup_{t\in[0,1]} \left\{ \int_0^t \frac{(t-s)^{q-1}}{\Gamma(q)} |f(s,x(s)) - f(s,y(s))| ds \right. \\
& + \frac{t^{n-1}}{|\Delta|} \Big(|a| \int_0^\xi \frac{(\xi-s)^q}{\Gamma(q+1)} |f(s,x(s)) - f(s,y(s))| ds \\
& + |b| \sum_{i=1}^{m-2} \alpha_i \int_0^{\eta_i} \frac{(\eta_i - s)^{q-1}}{\Gamma(q)} |f(s,x(s)) - f(s,y(s))| ds \\
& \left. + \int_0^1 \frac{(1-s)^{q-1}}{\Gamma(q)} |f(s,x(s)) - f(s,y(s))| ds \Big) \right\} \\
& \leq \ell \|x-y\| \sup_{t\in[0,1]} \left\{ \int_0^t \frac{(t-s)^{q-1}}{\Gamma(q)} ds + \frac{t^{n-1}}{\Delta} \Big(|a| \int_0^\xi \frac{(\xi-s)^q}{\Gamma(q+1)} ds \right. \\
& \left. + |b| \sum_{i=1}^{m-2} \alpha_i \int_0^{\eta_i} \frac{(\eta_i - s)^{q-1}}{\Gamma(q)} ds + \int_0^1 \frac{(1-s)^{q-1}}{\Gamma(q)} ds \Big) \right\} \\
& \leq \ell\sigma \|x-y\|,
\end{aligned}$$

where σ is given by (5.9). Since $\ell\sigma < 1$, the operator $\mathcal{S}$ is a contraction. So, by Banach's fixed point theorem, the problem (5.1)-(5.2) has a unique solution on $[0,1]$. This completes the proof. □

In our next result, we make use of a fixed point theorem due to Krasnosel'skiĭ (Theorem 1.4).

Theorem 5.2. *Let $f : [0,1] \times \mathbb{R} \to \mathbb{R}$ be a continuous function satisfying (5.1.1) and*

(5.2.1) $|f(t,x)| \leq \mu(t), \ \ \forall (t,x) \in [0,1] \times \mathbb{R}$, and $\mu \in C([0,1], \mathbb{R}^+)$.

Then there exists at least one solution for the problem (5.1)-(5.2) on $[0,1]$ provided

$$\frac{\ell}{\Gamma(q+1)|\Delta|} \left[1 + |a| \frac{\xi^{q+1}}{(q+1)} + |b| \sum_{i=1}^{m-2} \alpha_i {\eta_i}^q \right] < 1. \tag{5.10}$$

Proof. Letting $\rho \geq \|\mu\|\sigma$, $(\|\mu\| = \sup_{t\in[0,1]} |\mu(t)|)$, we consider $B_\rho = \{x \in \mathcal{G} : \|x\| \leq \rho\}$ and define the operators $\mathcal{S}_1$ and $\mathcal{S}_2$ on B_ρ as

$$(\mathcal{S}_1 x)(t) = \int_0^t \frac{(t-s)^{q-1}}{\Gamma(q)} f(s,x(s)) ds,$$

$$(\mathcal{S}_2 x)(t) = \frac{t^{n-1}}{\Delta} \Big(a \int_0^\xi \frac{(\xi-s)^q}{\Gamma(q+1)} f(s,x(s)) ds$$

$$+b\sum_{i=1}^{m-2}\alpha_i\int_0^{\eta_i}\frac{(\eta_i-s)^{q-1}}{\Gamma(q)}f(s,x(s))ds-\int_0^1\frac{(1-s)^{q-1}}{\Gamma(q)}f(s,x(s))ds\Big).$$

For $u,\ v\in B_\rho$, we can easily show that $\|(\mathcal{S}_1u)+(\mathcal{S}_2v)\|\leq\|\mu\|\sigma\leq\rho$, where σ is given by (5.9). This implies that $\mathcal{S}_1u+\mathcal{S}_2v\in B_\rho$.

Now, we show that the operator $\mathcal{S}_2$ is a contraction. For $x,y\in\mathbb{R},\ \ t\in[0,1]$, we have

$$\begin{aligned}\|\mathcal{S}_2x-\mathcal{S}_2y\| &\leq \sup_{t\in[0,1]}\Big\{\frac{t^{n-1}}{|\Delta|}\Big(|a|\int_0^\xi\frac{(\xi-s)^q}{\Gamma(q+1)}|f(s,x(s))-f(s,y(s))|ds\\
&\quad+|b|\sum_{i=1}^{m-2}\alpha_i\int_0^{\eta_i}\frac{(\eta_i-s)^{q-1}}{\Gamma(q)}|f(s,x(s))-f(s,y(s))|ds\\
&\quad-\int_0^1\frac{(1-s)^{q-1}}{\Gamma(q)}|f(s,x(s))-f(s,y(s))|ds\Big)\Big\}\\
&\leq \ell\|x-y\|\sup_{t\in[0,1]}\Big\{\frac{t^{n-1}}{|\Delta|}\Big(|a|\int_0^\xi\frac{(\xi-s)^q}{\Gamma(q+1)}ds\\
&\quad+|b|\sum_{i=1}^{m-2}\alpha_i\int_0^{\eta_i}\frac{(\eta_i-s)^{q-1}}{\Gamma(q)}ds-\int_0^1\frac{(1-s)^{q-1}}{\Gamma(q)}ds\Big)\Big\}\\
&\leq \|x-y\|\frac{\ell}{\Gamma(q+1)|\Delta|}\Big[1+|a|\frac{\xi^{q+1}}{(q+1)}+|b|\sum_{i=1}^{m-2}\alpha_i{\eta_i}^q\Big],\end{aligned}$$

which, in view of (5.10), implies that $\mathcal{S}_2$ is a contraction.

Next, it will be shown that $\mathcal{S}_1$ is compact and continuous. Continuity of f implies that the operator $\mathcal{S}_1$ is continuous. Also, $\mathcal{S}_1$ is uniformly bounded on B_ρ as $\|\mathcal{S}_1x\|\leq\dfrac{\|\mu\|}{\Gamma(q+1)}$. Furthermore, for $t_1,\ t_2\in[0,1]$ with $t_1<t_2$, we have

$$\begin{aligned}&|(\mathcal{S}_1x)(t_2)-(\mathcal{S}_1x)(t_1)|\\
&=\Big|\int_0^{t_2}\frac{(t_2-s)^{q-1}}{\Gamma(q)}f(s,x(s))ds-\int_0^{t_1}\frac{(t_1-s)^{q-1}}{\Gamma(q)}f(s,x(s))ds\Big|\\
&\leq\frac{\|\mu\|}{\Gamma(q+1)}\Big(|t_2^q-t_1^q|+2(t_2-t_1)^q\Big),\end{aligned}$$

which tends to zero independent of x as $(t_2-t_1)\to 0$. Thus, $\mathcal{S}_1$ is relatively compact on B_ρ. Hence, by the Arzelá-Ascoli theorem, the operator $\mathcal{S}_1$ is compact on B_ρ. Thus, all the conditions of Krasnosel'skiĭ's fixed point theorem are satisfied and consequently there exists a solution for the problem (5.1)-(5.2) on $[0,1]$. This completes the proof. □

Our next result is based on the Schaefer fixed point theorem (Theorem 1.10).

Theorem 5.3. *Assume that there exists a positive constant* L_1 *such that* $|f(t,x)|\leq L_1$ *for all* $t\in[0,1],x\in\mathcal{G}$. *Then there exists at least one solution for problem (5.1)-(5.2) on* $[0,1]$.

Proof. First of all, it will be shown that the operator $\mathcal{S}$ is completely continuous. Obviously, continuity of f implies the continuity of $\mathcal{S}$. Let $A \subset \mathcal{G}$ be bounded. Then, $\forall x \in A$, one can easily establish that $|(\mathcal{S}x)(t)| \le L_1\sigma = L_2$, where σ is given by (5.9). Moreover, for $t_1,\ t_2 \in [0,1]$ with $t_1 < t_2$, it is easy to find that

$$|(\mathcal{S}x)(t_2) - (\mathcal{S}x)(t_1)| \le L_1\Big[\frac{|t_2^q - t_1^q| + 2(t_2-t_1)^q}{\Gamma(q+1)} + \frac{|t_2^{n-1} - t_1^{n-1}|}{\Gamma(q+1)|\Delta|}\Big(1 + |a|\frac{\xi^{q+1}}{(q+1)} + |b|\sum_{i=1}^{m-2}\alpha_i{\eta_i}^q\Big)\Big],$$

which tends to zero independent of x as $(t_2 - t_1) \to 0$. Therefore, $\mathcal{S}$ is equi-continuous on $[0,1]$. So, by Arzelá-Ascoli Theorem, the operator $\mathcal{S}$ is completely continuous.

Next, we define a set $K = \{x \in \mathcal{G} : x = \zeta\mathcal{S}x,\ 0 < \zeta < 1\}$ and show that it is bounded. For $x \in K,\ t \in [0,1]$, we have

$$x(t) = \int_0^t \frac{(t-s)^{q-1}}{\Gamma(q)} f(s,x(s))ds + \frac{t^{n-1}}{\Delta}\Big(a\int_0^\xi \frac{(\xi-s)^q}{\Gamma(q+1)} f(s,x(s))ds + b\sum_{i=1}^{m-2}\alpha_i\int_0^{\eta_i}\frac{(\eta_i-s)^{q-1}}{\Gamma(q)}f(s,x(s))ds - \int_0^1 \frac{(1-s)^{q-1}}{\Gamma(q)}f(s,x(s))ds\Big).$$

As before, we can obtain $|x(t)| = \zeta|(\mathcal{S}x)(t)| \le L_1\sigma = L_2$, which implies that $\|x\| \le L_2,\ \forall x \in K,\ t \in [0,1]$. So K is bounded. In consequence, by the conclusion of Theorem 1.10, problem (5.1)-(5.2) has at least one solution on $[0,1]$. This completes the proof. □

In the final result, we make use of Leray-Schauder nonlinear alternative (Theorem 1.8).

Theorem 5.4. *Let $f : [0,1] \times \mathbb{R} \to \mathbb{R}$ be a continuous function. Further, the following conditions hold:*

(5.4.1) there exist a function $z \in C([0,1],\mathbb{R}^+)$ and a non-decreasing function $\phi : \mathbb{R}^+ \to \mathbb{R}^+$ such that $|f(t,x)| \le z(t)\phi(\|x\|),\ \forall (t,x) \in [0,1] \times \mathbb{R}$;

(5.4.2) there exists a constant $N > 0$ such that

$$N\Big[\frac{\phi(N)\|z\|}{\Gamma(q+1)}\Big(1 + \frac{1}{|\Delta|}\Big[1 + |a|\frac{\xi^{q+1}}{(q+1)} + |b|\sum_{i=1}^{m-2}\alpha_i{\eta_i}^q\Big]\Big)\Big]^{-1} > 1.$$

Then problem (5.1)-(5.2) has at least one solution on $[0,1]$.

Proof. In the first step, it will be shown that the operator $\mathcal{S} : \mathcal{G} \to \mathcal{G}$ defined by (5.8) maps bounded sets into bounded sets in $\mathcal{G}$. For $\varrho > 0$, let $B_\varrho = \{x \in \mathcal{G} : \|x\| \le \varrho\}$ be a bounded set in $\mathcal{G}$. Then, for $x \in B_\varrho$, in view of (5.4.1), we obtain

$$|(\mathcal{S}x)(t)| \le \int_0^t \frac{(t-s)^{q-1}}{\Gamma(q)} z(s)\phi(\|x\|)ds + \frac{t^{n-1}}{|\Delta|}\Big(|a|\int_0^\xi \frac{(\xi-s)^q}{\Gamma(q+1)} z(s)\phi(\|x\|)ds$$

$$+|b|\sum_{i=1}^{m-2}\alpha_i\int_0^{\eta_i}\frac{(\eta_i-s)^{q-1}}{\Gamma(q)}z(s)\phi(\|x\|)ds+\int_0^1\frac{(1-s)^{q-1}}{\Gamma(q)}z(s)\phi(\|x\|)ds\Big),$$

which implies that

$$\|(\mathcal{S}x)\|\leq\frac{\phi(\varrho)\|z\|}{\Gamma(q+1)}\Big(1+\frac{1}{|\Delta|}\Big[1+|a|\frac{\xi^{q+1}}{(q+1)}+|b|\sum_{i=1}^{m-2}\alpha_i{\eta_i}^q\Big]\Big).$$

Next, we show that $\mathcal{S}$ maps bounded sets into equi-continuous sets of $\mathcal{G}$. Let $t_1,\ t_2\in[0,1]$ with $t_1<t_2$ and $x\in B_\varrho$. Then

$$\begin{aligned}|(\mathcal{S}x)(t_2)-(\mathcal{S}x)(t_1)|\leq\phi(\varrho)\|z\|\Big[&\frac{|t_2^q-t_1^q|+2(t_2-t_1)^q}{\Gamma(q+1)}\\&+\frac{|t_2^{n-1}-t_1^{n-1}|}{\Gamma(q+1)|\Delta|}\Big(1+|a|\frac{\xi^{q+1}}{(q+1)}+|b|\sum_{i=1}^{m-2}\alpha_i{\eta_i}^q\Big)\Big],\end{aligned}$$

which tends to zero independent of $x\in B_\varrho$ as $(t_2-t_1)\to0$. Thus, by the Arzelá-Ascoli theorem, the operator $\mathcal{S}$ is completely continuous.

Let x be a solution of the given problem. Then, for $\gamma\in(0,1)$, using the method of computation employed to show the boundedness of the operator $\mathcal{S}$, we obtain

$$|x(t)|=|\gamma(\mathcal{S}x)(t)|\leq\frac{\phi(\|x\|)\|z\|}{\Gamma(q+1)}\Big\{1+\frac{1}{|\Delta|}\Big(1+|a|\frac{\xi^{q+1}}{(q+1)}+|b|\sum_{i=1}^{m-2}\alpha_i{\eta_i}^q\Big)\Big\},$$

which yields

$$\|x\|\Big[\frac{\phi(\|x\|)\|z\|}{\Gamma(q+1)}\Big\{1+\frac{1}{|\Delta|}\Big(1+|a|\frac{\xi^{q+1}}{(q+1)}+|b|\sum_{i=1}^{m-2}\alpha_i{\eta_i}^q\Big)\Big\}\Big]^{-1}\leq1.$$

By the condition (5.4.2), there exists $N>0$ such that $\|x\|\neq N$. Choose $\mathcal{M}=\{x\in\mathcal{G}:\|x\|<N+1\}$. Observe that the operator $\mathcal{S}:\overline{\mathcal{M}}\to\mathcal{G}$ is continuous and completely continuous. From the choice of $\mathcal{M}$, there is no $x\in\partial\mathcal{M}$ such that $x=\gamma\mathcal{S}(x)$ for some $\gamma\in(0,1)$. Thus, by Theorem 1.8, we have that the operator $\mathcal{S}$ has a fixed point $x\in\overline{\mathcal{M}}$ which is a solution of problem (5.1)-(5.2). This completes the proof. □

Example 5.1. Consider a fractional boundary value problem with integral-multi-point boundary conditions given by

$$\begin{cases}{}^CD^{\frac{3}{2}}x(t)=\lambda\Big(\dfrac{1}{\sqrt{t+4}}x+\dfrac{1}{\sqrt{t^2+9}}\tan^{-1}(x)\Big)+\dfrac{1}{9}, & t\in[0,1]\\ x(0)=0,\ x(1)=a\displaystyle\int_0^{1/4}x(s)ds+b\sum_{i=1}^{4}\alpha_ix(\eta_i).\end{cases}\tag{5.11}$$

Here, $q=3/2,\ a=1/3,\ b=1/2,\ \xi=1/4,\ \eta_1=1/3,\ \eta_2=1/2,\ \eta_3=2/3,\ \eta_4=5/6,\ \alpha_1=1/5,\ \alpha_2=2/5,\ \alpha_3=3/5,\ \alpha_4=4/5$, and

$f(t,x) = \lambda\Big(\frac{1}{\sqrt{t+4}}x + \frac{1}{\sqrt{t^2+9}}\tan^{-1}(x)\Big) + \frac{1}{9}$, where $\lambda > 0$ is to be chosen appropriately. Clearly, $|f(t,x) - f(t,y)| \le \frac{5\lambda}{6}\|x-y\|$. So $\ell = \frac{5\lambda}{6}$. Further

$$|\Delta| = \Big|1 - \frac{a\xi^2}{2} - b\sum_{i=1}^{m-2}\alpha_i\eta_i^{n-1}\Big| = \frac{31}{96} \approx 0.32292,$$

$$\sigma = \frac{1}{\Gamma(q+1)}\Big\{1 + \frac{1}{|\Delta|}\Big(1 + |a|\frac{\xi^{q+1}}{q+1} + |b|\sum_{i=1}^{m-2}\alpha_i{\eta_i}^q\Big)\Big\} \approx 4.3903.$$

The assumption $\ell\sigma < 1$ is clearly satisfied for $\lambda < 0.27333$. Thus all the conditions of Theorem 5.1 are satisfied and consequently, problem (5.11) has a unique solution on $[0,1]$.

Example 5.2. Consider the problem (5.11) with

$$f(t,x) = \frac{(t+2)}{15}\Big(\frac{(x+1)^2}{2+(x+1)^2} + \sin x\Big). \tag{5.12}$$

Evidently, $|f(t,x)| \le \frac{t+2}{15}(1+\|x\|)$. Let us choose $z(t) = \frac{t+2}{15}$, $\phi(\|x\|) = 1 + \|x\|$. By the assumption:

$$N\Big[\frac{\phi(N)\|z\|}{\Gamma(q+1)}\Big\{1 + \frac{1}{|\Delta|}\Big(1 + |a|\frac{\xi^{q+1}}{(q+1)} + |b|\sum_{i=1}^{m-2}\alpha_i{\eta_i}^q\Big)\Big\}\Big]^{-1} > 1,$$

we find that $N > N_1$, where $N_1 \simeq 7.2008$. Thus, by Theorem 5.4, there exists at least one solution for problem (5.11) with $f(t,x)$ given by (5.12).

5.1.2 *Riemann-Liouville integral-multi-point conditions*

Replacing $\int_0^\xi x(s)ds$ by $\int_0^\xi \frac{(\xi-s)^{\beta-1}}{\Gamma(\beta)}x(s)ds$, $\beta > 0$ in (5.2), the resulting problem becomes a fractional boundary value problem with Riemann-Liouville type integral multi-point boundary conditions. The integral operator analogous to (5.8) for this new problem is

$$\begin{aligned}(\mathcal{S}_{RL}x)(t) = &\int_0^t \frac{(t-s)^{q-1}}{\Gamma(q)}f(s,x(s))ds \\ &+ \frac{t^{n-1}}{\Lambda}\Big\{a\int_0^\xi \frac{(\xi-s)^{\beta-1}}{\Gamma(\beta)}\Big(\int_0^s \frac{(s-u)^{q-1}}{\Gamma(q)}f(u,x(u))du\Big)ds \\ &+ b\sum_{i=1}^{m-2}\alpha_i\int_0^{\eta_i}\frac{(\eta_i-s)^{q-1}}{\Gamma(q)}f(s,x(s))ds - \int_0^1 \frac{(1-s)^{q-1}}{\Gamma(q)}f(s,x(s))ds\Big\},\end{aligned} \tag{5.13}$$

where it is assumed that

$$\Lambda = 1 - \frac{a\xi^{\beta+n-1}\Gamma(n)}{\Gamma(\beta+n)} - b\sum_{i=1}^{m-2}\alpha_i\eta_i^{n-1} \neq 0. \tag{5.14}$$

The corresponding constant σ_{RL} (analogue to (5.9)) is

$$\sigma_{RL} = \frac{1}{\Gamma(q+1)}\Big\{1 + \frac{1}{|\Lambda|}\Big(1 + |a|\frac{\xi^{q+\beta}}{\Gamma(\beta+1)} + |b|\sum_{i=1}^{m-2}\alpha_i{\eta_i}^q\Big)\Big\}. \tag{5.15}$$

With (5.13) and (5.14), we can obtain the existence results for this new problem similar to the ones established for problem (5.1)-(5.2) in Section 5.1.1. For instance, the conclusion of Theorem 5.1 holds true for problem (5.1)-(5.2) with Riemann-Liouville integral multi-point boundary conditions if $\ell\sigma_{RL} < 1$.

Example 5.3. Consider the following integral multi-point fractional boundary value problem:

$$\begin{cases} {}^{C}D^{\frac{5}{2}}x(t) = \delta\Big(\frac{x}{\sqrt{t^2+4}} + \frac{|x|}{1+|x|}\Big) + \frac{1}{9}, \quad t \in [0,1] \\ x(0) = 0,\ x(1) = a\displaystyle\int_0^{1/8}\frac{(1/8-s)^{1/2}}{\Gamma(3/2)}x(s)ds + b\sum_{i=1}^{4}\alpha_i x(\eta_i), \end{cases} \tag{5.16}$$

where $q = 5/2$, $a = 1/2$, $b = 1/4$, $\xi = 1/4$, $\beta = 3/2$, $\eta_1 = 1/6$, $\eta_2 = 1/3$, $\eta_3 = 1/2$, $\eta_4 = 2/3$, $\alpha_1 = 1/7$, $\alpha_2 = 2/7$, $\alpha_3 = 3/7$, $\alpha_4 = 4/7$, and $f(t,x) = \delta\Big(\frac{x}{\sqrt{t^2+4}} + \frac{|x|}{1+|x|}\Big) + \frac{1}{9}$, $\delta > 0$ is a suitably chosen parameter. Clearly $\ell = 3\delta/2$ as $|f(t,x) - f(t,y)| \leq (3\delta/2)|x-y|$. With the given data, we find that

$$|\Lambda| = \Big|1 - \frac{a\xi^{\beta+n-1}\Gamma(n)}{\Gamma(\beta+n)} - b\sum_{i=1}^{m-2}\alpha_i\eta_i^{n-1}\Big| \approx 0.90073,$$

$$\sigma_{RL} = \frac{1}{\Gamma(q+1)}\Big\{1 + \frac{1}{|\Lambda|}\Big(1 + |a|\frac{\xi^{q+\beta}}{\Gamma(\beta+1)} + |b|\sum_{i=1}^{m-2}\alpha_i{\eta_i}^q\Big)\Big\} \approx 0.66031,$$

and the condition $\ell\sigma_{RL} < 1$ holds true for $\delta < 1.0096$. Thus, all the conditions analogue to Theorem 5.1 are satisfied and consequently, the problem (5.16) has a unique solution on $[0,1]$.

5.1.3 *Stieltjes multi-point strip condition case*

In this section, we replace the condition $x(1) = a\int_0^{\xi}x(s)ds + b\sum_{i=1}^{m-2}\alpha_i x(\eta_i)$ with $x(1) = a\int_0^{\xi}x(s)d\varphi(s) + b\sum_{i=1}^{m-2}\alpha_i x(\eta_i)$, where $\varphi(s)$ is a function of bounded variation, in problem (5.2), which leads to the problem:

$$\begin{cases} {}^{C}D^{q}x(t) = f(t,x(t)),\ n-1 < q \leq n,\ \ t \in [0,1], \\ x(0) = 0,\ x'(0) = 0,\ x''(0) = 0,\ \ldots,\ x^{(n-2)}(0) = 0, \\ x(1) = a\displaystyle\int_0^{\xi}x(s)d\varphi(s) + b\sum_{i=1}^{m-2}\alpha_i x(\eta_i),\quad 0 < \xi < \eta_1 < \eta_2 < \ldots < \eta_{m-2} < 1. \end{cases} \tag{5.17}$$

In relation to the problem (5.17), we obtain an operator $\mathcal{S}_{ms} : \mathcal{G} \to \mathcal{G}$ given by

$$(\mathcal{S}_{ms}x)(t) = \int_0^t \frac{(t-s)^{q-1}}{\Gamma(q)} f(s, x(s))ds$$
$$+\frac{t^{n-1}}{\Delta_{ms}}\Big(a\int_0^{\xi}\Big(\int_0^s \frac{(s-u)^{q-1}}{\Gamma(q)} f(u, x(u))du\Big)d\varphi(s)$$
$$+b\sum_{i=1}^{m-2}\alpha_i\int_0^{\eta_i}\frac{(\eta_i-s)^{q-1}}{\Gamma(q)} f(s, x(s))ds - \int_0^1 \frac{(1-s)^{q-1}}{\Gamma(q)} f(s, x(s))ds\Big), \tag{5.18}$$

where

$$\Delta_{ms} = 1 - \int_0^{\xi} s^{n-1}d\varphi(s) - b\sum_{i=1}^{m-2}\alpha_i\eta_i^{n-1} \neq 0. \tag{5.19}$$

Further we set

$$\sigma_{ms} = \frac{1}{\Gamma(q+1)}\Big[1 + \frac{1}{|\Delta_{ms}|}\Big(1 + \Big|a\int_0^{\xi} s^q d\varphi(s)\Big| + |b|\sum_{i=1}^{m-2}\alpha_i{\eta_i}^q\Big)\Big]. \tag{5.20}$$

The existence results for problem (5.17) similar to the ones for problem (5.1)-(5.2) can be obtained by using the operator $\mathcal{S}_{ms}$ and the constant σ_{ms} respectively defined by (5.18) and (5.20). For instance, the existence and uniqueness result for problem (5.17) can be formulated as follows.

Theorem 5.5. *With $\sigma = \sigma_{ms}$ in the statement of Theorem 5.1, problem (5.17) has a unique solution on $[0, 1]$.*

We do not provide the proof as it is similar to that of Theorem 5.1.

Example 5.4. Consider the following condition instead of the second condition in problem (5.11):

$$x(1) = a\int_0^{1/4} x(s)d\varphi(s) + b\sum_{i=1}^{4}\alpha_i x(\eta_i), \tag{5.21}$$

with $\varphi(s) = \dfrac{s^2}{2} + s + 1$.

From example (5.1), we have that $\ell = 5\lambda/6$. Using the given values, we find that

$$\Big|\Delta_{ms}\Big| = \Big|1 - \int_0^{\xi} s^{n-1}d\varphi(s) - b\sum_{i=1}^{m-2}\alpha_i\eta_i^{n-1}\Big| = \frac{19}{64} \approx 0.29688,$$

$$\sigma_{ms} = \frac{1}{\Gamma(q+1)}\Big[1 + \frac{1}{|\Delta_{ms}|}\Big(1 + \Big|a\int_0^{\xi} s^q d\varphi(s)\Big| + |b|\sum_{i=1}^{m-2}\alpha_i{\eta_i}^q\Big)\Big] \approx 4.7363.$$

Obviously $\ell\sigma_{ms} < 1$ for $\lambda < 0.25336$. Thus, all the conditions of Theorem 5.5 are satisfied and hence the problem (5.11) with the condition (5.21) has a unique solution on $[0, 1]$.

5.2 A study of nonlinear fractional differential equations and inclusions of arbitrary order with Riemann-Liouville type multi-strip boundary conditions

In this section, we consider fractional differential equations and inclusions of arbitrary order $q \in (n-1,\ n], n \geq 2$ with finite many multi-strip Riemann-Liouville type integral boundary conditions.

5.2.1 *Fractional differential equations*

In this subsection we consider the following boundary value problem

$$\begin{cases} {}^C D^q x(t) = f(t, x(t)), \quad t \in [0, T], \\ x(0) = 0,\ x'(0) = 0, \ldots,\ x^{(n-2)}(0) = 0, \\ x(T) = \sum_{i=1}^{m} \gamma_i [I^{\beta_i} x(\eta_i) - I^{\beta_i} x(\zeta_i)], \end{cases} \tag{5.22}$$

where ${}^C D^q$ denotes the Caputo fractional derivative operator of order q, I^{β_i} is the Riemann-Liouville fractional integral of order $\beta_i > 0$, $i = 1, 2, \ldots, m$, $0 < \zeta_1 < \eta_1 < \zeta_2 < \eta_2 < \ldots < \zeta_m < \eta_m < T$, and $\gamma_i \in \mathbb{R}$ are suitable chosen constants.

Lemma 5.2. *For $h \in C[0, T]$, the fractional boundary value problem*

$$\begin{cases} {}^C D^q x(t) = h(t), \quad t \in [0, T],\ q \in (n-1,\ n] \\ x(0) = 0,\ x'(0) = 0, \ldots,\ x^{(n-2)}(0) = 0, \\ x(T) = \sum_{i=1}^{m} \gamma_i [I^{\beta_i} x(\eta_i) - I^{\beta_i} x(\zeta_i)], \end{cases} \tag{5.23}$$

has a unique solution $x(t) \in C^n[0, T]$ given by

$$\begin{aligned} x(t) &= \frac{1}{\Gamma(q)} \int_0^t (t-s)^{q-1} h(s) ds - \frac{t^{n-1}}{\lambda \Gamma(q)} \int_0^T (T-s)^{q-1} h(s) ds \\ &+ \frac{t^{n-1}}{\lambda \Gamma(q)} \sum_{i=1}^{m} \frac{\gamma_i}{\Gamma(\beta_i)} \Big[\int_0^{\eta_i} \int_0^s (\eta_i - s)^{\beta_i - 1} (s-u)^{q-1} h(u) du ds \\ &- \int_0^{\zeta_i} \int_0^s (\zeta_i - s)^{\beta_i - 1} (s-u)^{q-1} h(u) du ds \Big], \end{aligned} \tag{5.24}$$

where it is assumed that

$$\lambda = \left(T^{n-1} - \sum_{i=1}^{m} \gamma_i \frac{(\eta_i^{\beta_i+n-1} - \zeta_i^{\beta_i+n-1}) \Gamma(n)}{\Gamma(\beta_i + n)} \right) \neq 0. \tag{5.25}$$

Proof. The general solution of fractional differential equations in (5.23) can be written as

$$x(t) = \frac{1}{\Gamma(q)} \int_0^t (t-s)^{q-1} h(s) ds - c_0 - c_1 t - \ldots - c_{n-1} t^{n-1}. \tag{5.26}$$

Using the given boundary conditions in (5.26), it is found that $c_i = 0, i = 1, 2, \ldots, n-2 = 0$. Inserting these values in (5.26) and then applying the Riemann-Liouville integral operator I^{β_i} to the resulting equation, we get

$$I^{\beta_i} x(t) = \frac{1}{\Gamma(\beta_i)} \int_0^t (t-s)^{\beta_i - 1} \Big(\frac{1}{\Gamma(q)} \int_0^s (s-u)^{q-1} h(u) du - c_{n-1} s^{n-1} \Big) ds$$

$$= \frac{1}{\Gamma(\beta_i)\Gamma(q)} \int_0^t \int_0^s (t-s)^{\beta_i - 1}(s-u)^{q-1} h(u)duds$$
$$-c_{n-1}\frac{1}{\Gamma(\beta_i)} \int_0^t (t-s)^{\beta_i-1} s^{n-1} ds.$$

Using the condition $x(T) = \sum_{i=1}^m \gamma_i [I^{\beta_i} x(\eta_i) - I^{\beta_i} x(\zeta_i)]$, together with the fact that

$$\frac{1}{\Gamma(\beta_i)} \int_0^t (t-s)^{\beta_i-1} s^{n-1} ds = \frac{t^{\beta_i+n-1}\Gamma(n)}{\Gamma(\beta_i+n)},$$

we obtain

$$\frac{1}{\Gamma(q)} \int_0^T (T-s)^{q-1} h(s)ds - c_{n-1}T^{n-1}$$
$$= \sum_{i=1}^m \frac{\gamma_i}{\Gamma(q)\Gamma(\beta_i)} \Big[\int_0^{\eta_i} \int_0^s (\eta_i - s)^{\beta_i-1}(s-u)^{q-1} h(u)duds$$
$$- \int_0^{\zeta_i} \int_0^s (\zeta_i - s)^{\beta_i-1}(s-u)^{q-1} h(u)duds \Big] - c_{n-1} \sum_{i=1}^m \gamma_i \frac{(\eta_i^{\beta_i+n-1} - \zeta_i^{\beta_i+n-1})\Gamma(n)}{\Gamma(\beta_i+n)},$$

which yields

$$c_{n-1} = \frac{1}{\lambda\Gamma(q)} \int_0^T (T-s)^{q-1} h(s)ds$$
$$-\frac{1}{\lambda\Gamma(q)} \sum_{i=1}^m \frac{\gamma_i}{\Gamma(\beta_i)} \Big[\int_0^{\eta_i} \int_0^s (\eta_i - s)^{\beta_i-1}(s-u)^{q-1} h(u)duds$$
$$- \int_0^{\zeta_i} \int_0^s (\zeta_i - s)^{\beta_i-1}(s-u)^{q-1} h(u)duds \Big],$$

where λ is given by (5.25). Substituting the values of $c_0, c_1, \dots, c_{n-2}, c_{n-1}$ in (5.26), we obtain (5.24). We can prove the converse by direct computation. This completes the proof. □

Let $\mathcal{C} := C([0,T], \mathbb{R})$ denotes the Banach space of all continuous functions defined on $[0,T] \times \mathbb{R}$ endowed with a topology of uniform convergence with the norm $\|x\| = \sup_{t\in[0,T]} |x(t)|$.

By Lemma 5.2, we define an operator $\mathcal{P} : \mathcal{C} \to \mathcal{C}$ associated with the problem (5.22) as as

$$(\mathcal{P}x)(t) = \frac{1}{\Gamma(q)} \int_0^t (t-s)^{q-1} f(s, x(s))ds - \frac{t^{n-1}}{\lambda\Gamma(q)} \int_0^T (T-s)^{q-1} f(s,x(s))ds$$
$$+\frac{t^{n-1}}{\lambda\Gamma(q)} \sum_{i=1}^m \frac{\gamma_i}{\Gamma(\beta_i)} \Big[\int_0^{\eta_i} \int_0^s (\eta_i - s)^{\beta_i-1}(s-u)^{q-1} f(u, x(u))duds$$
$$- \int_0^{\zeta_i} \int_0^s (\zeta_i - s)^{\beta_i-1}(s-u)^{q-1} f(u,x(u))duds \Big], \qquad t \in [0,T]. \tag{5.27}$$

Observe that the problem (5.22) has a solution if and only if the associated fixed point problem $\mathcal{P}x = x$ has a fixed point.

In the first result we prove an existence and uniqueness result by means of Banach's contraction mapping principle.

For the sake of convenience, we set

$$\Lambda = \frac{T^q}{\Gamma(q+1)} + \frac{T^{q+n-1}}{|\lambda|\Gamma(q+1)} + \frac{T^{n-1}}{|\lambda|}\sum_{i=1}^{m}\gamma_i\frac{\eta_i^{q+\beta_i} - \zeta_i^{q+\beta_i}}{\Gamma(q+\beta_i+1)}. \tag{5.28}$$

Theorem 5.6. *Suppose that $f : [0,T]\times\mathbb{R} \to \mathbb{R}$ is a continuous function and satisfies the following assumption:*

(5.6.1) $|f(t,x) - f(t,y)| \le L\|x - y\|, \forall t \in [0,1],\ L > 0,\ x, y \in \mathbb{R}.$

Then the boundary value problem (5.22) has a unique solution provided that

$$L < 1/\Lambda, \tag{5.29}$$

where Λ is given by (5.28).

Proof. With $r \ge M\Lambda/(1 - L\Lambda)$, we define $B_r = \{x \in \mathcal{C} : \|x\| \le r\}$, where $M = \sup_{t\in[0,T]}|f(t,0)| < \infty$ and Λ is given by (5.28). Then we show that $\mathcal{P}B_r \subset B_r$. For $x \in B_r$, by means of the inequality $|f(s,x(s))| \le |f(s,x(s)) - f(s,0)| + |f(s,0)| \le L\|x\| + M \le Lr + M$, it can easily be shown that

$$\|\mathcal{P}x\| = (Lr + M)\,\Lambda \le r.$$

Now, for $x, y \in \mathcal{C}$ and for each $t \in [0,T]$, we obtain

$$\begin{aligned}
&\|(\mathcal{P}x) - (\mathcal{P}y)\| \\
&\le \sup_{t\in[0,T]}\Bigg\{\frac{1}{\Gamma(q)}\int_0^t (t-s)^{q-1}|f(s,x(s)) - f(s,y(s))|ds \\
&+\frac{t^{n-1}}{\lambda\Gamma(q)}\int_0^T (T-s)^{q-1}|f(s,x(s)) - f(s,y(s))|ds \\
&+\frac{t^{n-1}}{\lambda\Gamma(q)}\sum_{i=1}^{m}\frac{\gamma_i}{\Gamma(\beta_i)}\Big[\int_0^{\eta_i}\int_0^s (\eta_i - s)^{\beta_i-1}(s-u)^{q-1}du|f(u,x(u)) - f(u,y(u))|ds \\
&-\int_0^{\zeta_i}\int_0^s (\zeta_i - s)^{\beta_i-1}(s-u)^{q-1}|f(u,x(u)) - f(u,y(u))|duds\Big]\Bigg\} \\
&\le L\Lambda\|x - y\|.
\end{aligned}$$

Note that Λ depends only on the parameters involved in the problem. As $L\Lambda < 1$, therefore $\mathcal{P}$ is a contraction. Hence, by Banach's contraction mapping principle, the problem (5.22) has a unique solution on $[0,T]$. □

Example 5.5. Let us consider the following 4-strip nonlocal boundary value problem:

$$\begin{cases} {}^CD^{9/2}x(t) = f(t,x(t)), \quad t \in [0,2], \\ x(0) = 0,\ x'(0) = 0,\ x''(0) = 0,\ x'''(0) = 0, \\ x(T) = \sum_{i=1}^{4}\gamma_i[I^{\beta_i}x(\eta_i) - I^{\beta_i}x(\zeta_i)], \end{cases} \tag{5.30}$$

where $q = 9/2,\ n = 5,\ \zeta_1 = 1/4,\ \eta_1 = 1/2,\ \zeta_2 = 2/3,\ \eta_2 = 1,\ \zeta_3 = 5/4,\ \eta_3 = 4/3,\ \zeta_4 = 3/2,\ \eta_4 = 7/4,\ \gamma_1 = 5,\ \gamma_2 = 10,\ \gamma_3 = 15,\ \gamma_4 = 25,\ \beta_1 = 5/4,\ \beta_2 = 7/4,\ \beta_3 = 9/4,\ \beta_4 = 11/4$.

With the given values of the parameters, we find that $\lambda \simeq 9.334784$, and $\Lambda \simeq 1.406972$. Let us choose

$$f(t,x(t)) = \frac{1}{\sqrt[3]{(t+8)}}\Big(\tan^{-1}x\Big) + \sqrt{4+3\sin 2t}. \tag{5.31}$$

Clearly $L = 1/2$ as $|f(t,x) - f(t,y)| \le \frac{1}{2}\|x-y\|$ and $L < \frac{1}{\Lambda}$, where $\Lambda \simeq 1.406972$. Therefore all the conditions of Theorem 5.6 hold and consequently there exists a unique solution for the problem (5.30) with $f(t,x(t))$ given by (5.31).

In case of the following unbounded nonlinear function

$$f(t,x(t)) = \frac{x}{7} + \frac{1}{\sqrt[3]{(t+8)}}\Big(\tan^{-1}x\Big) + \sqrt{4+3\sin 2t}, \tag{5.32}$$

we have $L = 9/14$ and $L < 1/\Lambda$ ($\Lambda \simeq 1.406972$). As before, the problem (5.30) with $f(t,x(t))$ given by (5.32) has a unique solution.

In the proof of the following result, we apply the Leray-Schauder alternative (Theorem 1.7).

Theorem 5.7. *Assume that there exists a positive constant L_1 such that $|f(t,x)| \le L_1$ for $t \in [0,T]$, $x \in \mathbb{R}$. Then the problem (5.22) has at least one solution on $[0,T]$.*

Proof. First of all, we show that the operator $\mathcal{P}$ is completely continuous. Note that the operator $\mathcal{P}$ is continuous in view of the continuity of f. Let $\mathcal{B} \subset \mathcal{C}$ be a bounded set. By the assumption that $|f(t,x)| \le L_1$, for $x \in \mathcal{B}$, we have

$$\begin{aligned}
|(\mathcal{P}x)(t)| \le{}& \frac{1}{\Gamma(q)}\int_0^t (t-s)^{q-1}|f(s,x(s))|ds + \frac{t^{n-1}}{|\lambda|\Gamma(q)}\int_0^T (T-s)^{q-1}|f(s,x(s))|ds \\
&+\frac{t^{n-1}}{|\lambda|\Gamma(q)}\sum_{i=1}^m \frac{\gamma_i}{\Gamma(\beta_i)}\Big[\int_0^{\eta_i}\int_0^s (\eta_i - s)^{\beta_i-1}(s-u)^{q-1}|f(u,x(u))|duds \\
&+\int_0^{\zeta_i}\int_0^s (\zeta_i - s)^{\beta_i-1}(s-u)^{q-1}|f(u,x(u))|duds\Big] \\
\le{}& L_1\Bigg[\frac{1}{\Gamma(q)}\int_0^t (t-s)^{q-1}ds + \frac{t^{n-1}}{|\lambda|\Gamma(q)}\int_0^T (T-s)^{q-1}ds \\
&+\frac{t^{n-1}}{|\lambda|\Gamma(q)}\sum_{i=1}^m \frac{\gamma_i}{\Gamma(\beta_i)}\Big[\int_0^{\eta_i}\int_0^s (\eta_i - s)^{\beta_i-1}(s-u)^{q-1}duds \\
&+\int_0^{\zeta_i}\int_0^s (\zeta_i - s)^{\beta_i-1}(s-u)^{q-1}duds\Big] \\
\le{}& L_1\left\{\frac{T^q}{\Gamma(q+1)} + \frac{T^{q+n-1}}{|\lambda|\Gamma(q+1)} + \frac{T^{n-1}}{|\lambda|}\sum_{i=1}^m \gamma_i \frac{\eta_i^{q+\beta_i} - \zeta_i^{q+\beta_i}}{\Gamma(q+\beta_i+1)}\right\} = L_2,
\end{aligned}$$

which implies that $\|(\mathcal{P}x)\| \le L_2$. Further, we find that

$$\begin{aligned}
&|(\mathcal{P}x)'(t)| \\
&= \frac{1}{\Gamma(q-1)}\int_0^t (t-s)^{q-2}|f(s,x(s))|ds + \frac{(n-1)t^{n-2}}{|\lambda|\Gamma(q)}\int_0^T (T-s)^{q-1}|f(s,x(s))|ds \\
&\quad + \frac{(n-1)t^{n-2}}{|\lambda|\Gamma(q)}\sum_{i=1}^m \frac{\gamma_i}{\Gamma(\beta_i)}\Big[\int_0^{\eta_i}\int_0^s (\eta_i-s)^{\beta_i-1}(s-u)^{q-1}|f(u,x(u))|duds \\
&\quad + \int_0^{\zeta_i}\int_0^s (\zeta_i-s)^{\beta_i-1}(s-u)^{q-1}|f(u,x(u))|duds\Big] \\
&\le L_1\Bigg[\frac{1}{\Gamma(q-1)}\int_0^t (t-s)^{q-2}ds + \frac{(n-1)t^{n-2}}{|\lambda|\Gamma(q)}\int_0^T (T-s)^{q-1}ds \\
&\quad + \frac{(n-1)t^{n-2}}{|\lambda|\Gamma(q)}\sum_{i=1}^m \frac{\gamma_i}{\Gamma(\beta_i)}\Big(\int_0^{\eta_i}\int_0^s (\eta_i-s)^{\beta_i-1}(s-u)^{q-1}|f(u,x(u))|duds \\
&\quad + \int_0^{\zeta_i}\int_0^s (\zeta_i-s)^{\beta_i-1}(s-u)^{q-1}|f(u,x(u))|duds\Big)\Bigg] \\
&\le L_1\left\{\frac{T^{q-1}}{\Gamma(q)} + \frac{(n-1)T^{q+n-2}}{|\lambda|\Gamma(q+1)} + \frac{(n-1)T^{n-2}}{|\lambda|}\sum_{i=1}^m \gamma_j \frac{\eta_i^{q+\beta_i} - \zeta_i^{q+\beta_i}}{\Gamma(q+\beta_i+1}\right\} = L_3.
\end{aligned}$$

Hence, for $t_1, t_2 \in [0,T]$, we have

$$|(\mathcal{P}x)(t_2) - (\mathcal{P}x)(t_1)| \le \int_{t_1}^{t_2} |(\mathcal{P}x)'(s)|ds \le L_3(t_2-t_1).$$

This implies that $\mathcal{P}$ is equi-continuous on $[0,T]$. Thus, by the Arzelá-Ascoli theorem, the operator $\mathcal{P}: \mathcal{C} \to \mathcal{C}$ is completely continuous.

Next, we consider the set

$$V = \{x \in \mathcal{C} \mid x = \mu\mathcal{P}x,\ 0 < \mu < 1\},$$

and show that the set V is bounded. Let $x \in V$, then $x = \mu\mathcal{P}x,\ 0 < \mu < 1$. For any $t \in [0,T]$, it is easy to find that

$$\begin{aligned}
|x(t)| &= \mu|(\mathcal{P}x)(t)| \\
&\le L_1\left\{\frac{T^q}{\Gamma(q+1)} + \frac{T^{q+n-1}}{|\lambda|\Gamma(q+1)} + \frac{T^{n-1}}{|\lambda|}\sum_{i=1}^m \gamma_i \frac{\eta_i^{q+\beta_i} - \zeta_i^{q+\beta_i}}{\Gamma(q+\beta_i+1}\right\} = M_1.
\end{aligned}$$

Thus $\|x\| \le M_1$ for any $t \in [0,T]$. So, the set V is bounded. Thus, by the conclusion of Theorem 1.7, the operator $\mathcal{P}$ has at least one fixed point, which implies that (5.22) has at least one solution $[0,T]$. □

Example 5.6. Consider the boundary value problem of Example 5.5 with

$$f(t,x(t)) = \frac{3e^{\sqrt{(2-|x(t)|)^3}}[\cos 4t + 2\ln(1+4\sin^2 x(t))]}{\sqrt{(10+\cos x(t))}}. \tag{5.33}$$

Observe that $|f(t,x)| \le L_1$ with $L_1 = e^{2\sqrt{2}}(1+\ln 25)$. Thus the hypothesis of Theorem 5.7 is satisfied. Hence, by the conclusion of Theorem 5.7, the problem (5.30) with $f(t,x(t))$ given by (5.33) has at least one solution.

Now we prove one more existence result for problem (5.22) by applying Rothe's fixed point theorem (Theorem 1.3).

Theorem 5.8. *Let there exist a small positive number* τ *such that* $|f(t,x)| \leq \nu|x|$ *for* $0 < |x| < \tau$, *with* $0 < \nu \leq 1/\Lambda$, *where* Λ *is given by (5.28). Then the problem* (5.22) *has at least one solution* $[0,T]$.

Proof. Let us define $\mathcal{B}_\tau = \{x \in \mathcal{C} \mid \|x\| < \tau\}$ and take $x \in \mathcal{C}$ such that $\|x\| = \tau$, that is, $x \in \partial\mathcal{B}_\tau$. As before, it can be shown that $\mathcal{P}$ is completely continuous and

$$\begin{aligned}
&\|\mathcal{P}x\| \\
&\leq \sup_{t\in[0,t]} \left\{ \frac{1}{\Gamma(q)} \int_0^t (t-s)^{q-1}|f(s,x(s))|ds + \frac{t^{n-1}}{|\lambda|\Gamma(q)} \int_0^T (T-s)^{q-1}|f(s,x(s))|ds \right. \\
&\quad + \frac{t^{n-1}}{|\lambda|\Gamma(q)} \sum_{i=1}^m \frac{\gamma_i}{\Gamma(\beta_i)} \left[\int_0^{\eta_i} \int_0^s (\eta_i - s)^{\beta_i - 1}(s-u)^{q-1}|f(u,x(u))|duds \right. \\
&\quad \left. - \int_0^{\zeta_i} \int_0^s (\zeta_i - s)^{\beta_i - 1}(s-u)^{q-1}|f(u,x(u))|duds \right] \\
&\leq \Lambda\nu\|x\|,
\end{aligned}$$

which in view of the given condition ($\nu\Lambda \leq 1$), gives $\|\mathcal{P}x\| \leq \|x\|$, $x \in \partial\mathcal{B}_\tau$. Therefore, by Theorem 1.3, the operator $\mathcal{P}$ has at least one fixed point, which in turn implies that the problem (5.22) has at least one solution. □

Example 5.7. Consider the boundary value problem of Example 5.5 with

$$f(t,x(t)) = x(b^5 + x^4(t))^{\frac{1}{5}} + 2(1+\cos(t^4+3))^5(1-\cos x(t)), \ x \neq 0, \ b > 0. \quad (5.34)$$

For sufficiently small x (ignoring x^2 and higher powers of x), we have

$$|x(b^5 + x^4(t))^{\frac{1}{5}} + 2(1+\cos(t^4+3))^5(1-\cos x(t))| \leq b|x|.$$

Choosing $b \leq 1/\Lambda$, all the assumptions of Theorem 5.8 hold. Therefore, the conclusion of Theorem 5.8 implies that the problem (5.30) with $f(t,x(t))$ given by (5.34) has at least one solution.

Our final existence result is based on Leray-Schauder nonlinear alternative (Theorem 1.8).

Theorem 5.9. *Assume that*

(5.9.1) there exist a function $\sigma \in C([0,1],\mathbb{R}^+)$, *and a non-decreasing function* $\psi : \mathbb{R}^+ \to \mathbb{R}^+$ *such that* $|f(t,x)| \leq \sigma(t)\psi(\|x\|)$, $\forall(t,x) \in [0,T] \times \mathbb{R}$;

(5.9.2) there exists a constant $M > 0$ *such that*

$$\frac{M}{\psi(M)\Lambda\|\sigma\|} > 1.$$

Then the boundary value problem (5.22) has at least one solution on $[0,T]$.

Proof. Consider the operator $\mathcal{P} : \mathcal{C} \to \mathcal{C}$ defined by (5.27). We show that $\mathcal{P}$ *maps bounded sets into bounded sets in* $C([0,T],\mathbb{R})$. For a positive number r, let $B_r = \{x \in C([0,T],\mathbb{R}) : \|x\| \le r\}$ be a bounded set in $C([0,T],\mathbb{R})$. Then

$$\begin{aligned}
&\|\mathcal{P}x\| \\
&\le \sup_{t\in[0,T]} \Bigg\{ \frac{1}{\Gamma(q)} \int_0^t (t-s)^{q-1}|f(s,x(s))|ds + \frac{t^{n-1}}{|\lambda|\Gamma(q)} \int_0^T (T-s)^{q-1}|f(s,x(s))|ds \\
&\quad + \frac{t^{n-1}}{|\lambda|\Gamma(q)} \sum_{i=1}^m \frac{\gamma_i}{\Gamma(\beta_i)} \Big[\int_0^{\eta_i} \int_0^s (\eta_i - s)^{\beta_i-1}(s-u)^{q-1}|f(u,x(u))|duds \\
&\quad - \int_0^{\zeta_i} \int_0^s (\zeta_i - s)^{\beta_i-1}(s-u)^{q-1}|f(u,x(u))|duds \Big] \Bigg\} \\
&\le \psi(r) \Bigg\{ \frac{T^q}{\Gamma(q+1)} + \frac{T^{q+n-1}}{|\lambda|\Gamma(q+1)} + \frac{T^{n-1}}{|\lambda|} \sum_{i=1}^m \gamma_i \frac{\eta_i^{q+\beta_i} - \zeta_i^{q+\beta_i}}{\Gamma(q+\beta_i+1} \Bigg\} \|\sigma\|.
\end{aligned}$$

Next we show that F *maps bounded sets into equi-continuous sets of* $C([0,1],\mathbb{R})$. Let $t', t'' \in [0,1]$ with $t' < t''$ and $x \in B_r$, where B_r is a bounded set of $C([0,1],\mathbb{R})$. Then we obtain

$$\begin{aligned}
&|(\mathcal{P}x)(t'') - (\mathcal{P}x)(t')| \\
&\le \frac{1}{\Gamma(q)} \int_0^{t'} |(t''-s)^{q-1} - (t'-s)^{q-1}|\psi(r)\sigma(s)ds + \frac{1}{\Gamma(q)} \int_{t'}^{t''} |t''-s|^{q-1}\psi(r)\sigma(s)ds \\
&\quad + \frac{|(t'')^{n-1} - (t')^{n-1}|}{|\lambda|\Gamma(q)} \int_0^T |T-s|^{q-1}\psi(r)\sigma(s)ds \\
&\quad + \frac{|(t'')^{n-1} - (t')^{n-1}|}{|\lambda|\Gamma(q)} \sum_{i=1}^m \frac{\gamma_i}{\Gamma(\beta_i)} \Big[\int_0^{\eta_i} \int_0^s (\eta_i - s)^{\beta_i-1}(s-u)^{q-1}du\psi(r)\sigma(s)ds \\
&\quad + \int_0^{\zeta_i} \int_0^s (\zeta_i - s)^{\beta_i-1}(s-u)^{q-1}\psi(r)\sigma(s)duds \Big].
\end{aligned}$$

Obviously the right hand side of the above inequality tends to zero independently of $x \in B_r$ as $t'' - t' \to 0$. As $\mathcal{P} : C([0,T],\mathbb{R}) \to C([0,T],\mathbb{R})$ satisfies the foregoing assumptions, therefore it follows by the Arzelá-Ascoli theorem that $\mathcal{P}$ is completely continuous.

Let x be a solution of problem (5.22). Then, for $t \in [0,T]$, and following the similar computations as before, we find that

$$\begin{aligned}
|x(t)| &= |\mu(\mathcal{P}x)(t)| \\
&\le \psi(r) \Bigg\{ \frac{T^q}{\Gamma(q+1)} + \frac{T^{q+n-1}}{|\lambda|\Gamma(q+1)} + \frac{T^{n-1}}{|\lambda|} \sum_{i=1}^m \gamma_i \frac{\eta_i^{q+\beta_i} - \zeta_i^{q+\beta_i}}{\Gamma(q+\beta_i+1} \Bigg\} \|\sigma\|.
\end{aligned}$$

In consequence, we have

$$\frac{\|x\|}{\psi(\|x\|)\Lambda\|\sigma\|} \le 1.$$

Thus, by (5.9.2), there exists M such that $\|x\| \neq M$. Let us set

$$V = \{x \in C([0,T],\mathbb{R}) : \|x\| < M+1\}.$$

Note that the operator $\mathcal{P} : \overline{V} \to C([0,T],\mathbb{R})$ is continuous and completely continuous. From the choice of V, there is no $x \in \partial V$ such that $x = \mu\mathcal{P}(x)$ for some $\mu \in (0,1)$. Consequently, by the nonlinear alternative of Leray-Schauder type (Theorem 1.8), we deduce that $\mathcal{P}$ has a fixed point $x \in \overline{V}$ which is a solution of the problem (5.22). This completes the proof. □

Example 5.8. Consider the boundary value problem of Example 5.5 with

$$f(t,x(t)) = \frac{1}{\sqrt{t+4}}\Big(1 + \frac{|x|}{1+|x|}\Big) \leq \sigma(t)\psi(\|x\|). \tag{5.35}$$

Then $\sigma(t) = 1/\sqrt{t+4}$ and $\psi(\|x\|) = 2$. Using $\|\sigma\| = 1/2$, $\Lambda \simeq 1.406972$, we find by the condition (A_2) that $M > \Lambda$. Thus all the assumptions of Theorem 5.9 are satisfied. Hence, it follows by Theorem 5.9 that the problem (5.30) with $f(t,x(t))$ defined by (5.35) has at least one solution.

Let us choose an unbounded nonlinearity

$$f(t,x(t)) = \frac{1}{\sqrt{t+4}}\Big(1 + \frac{|x|}{1+|x|} + \frac{|x|}{2}\Big). \tag{5.36}$$

Clearly $f(t,x(t)) \leq \sigma(t)\psi(\|x\|)$ with $\sigma(t) = 1/\sqrt{t+4}$ and $\psi(\|x\|) = 2 + \|x\|/2$. Using the earlier arguments, with $\|\sigma\| = 1/2$, $\Lambda \simeq 1.406972$, we find that $M > M_1$, $M_1 \approx 2.170392$. Hence the problem (5.30) with $f(t,x(t))$ given by (5.36) has at least one solution on $[0,2]$.

5.2.2 *Fractional differential inclusions*

Here we study the multi-valued case of the problem (5.22) given by

$$\begin{cases} {}^C D^q x(t) \in F(t,x(t)), \quad t \in [0,T], \\ x(0) = 0,\ x'(0) = 0, \ldots,\ x^{(n-2)}(0) = 0, \\ x(T) = \sum_{i=1}^m \gamma_i [I^{\beta_i} x(\eta_i) - I^{\beta_i} x(\zeta_i)], \end{cases} \tag{5.37}$$

where ${}^C D^q$ denotes the Caputo fractional derivative operator of order q, $F : [0,T] \times \mathbb{R} \to \mathcal{P}(\mathbb{R})$ is a multi-valued map, $\mathcal{P}(\mathbb{R})$ is the family of all subsets of $\mathbb{R}$.

5.2.2.1 *The Carathéodory case*

In this subsection, we are concerned with the existence of solutions for the problem (5.37) when F is a compact and convex valued multi-valued map.

Theorem 5.10. *Suppose that*

(5.10.1) the map $F : [0,T] \times \mathbb{R} \to \mathcal{P}(\mathbb{R})$ is Carathéodory and has non-empty compact and convex values;

(5.10.2) *there exist a continuous non-decreasing function* $\psi : [0,\infty) \longrightarrow (0,\infty)$ *and function* $p \in L^1([0,T],\mathbb{R}^+)$ *such that*

$$\|F(t,x)\|_{\mathcal{P}} := \sup\{|v| : v \in F(t,x)\} \le p(t)\psi(\|x\|)$$

for each $(t,u) \in [0,T] \times \mathbb{R}$*;*

(5.10.3) *there exists a number* $M > 0$ *such that*

$$\frac{M}{\psi(M)\|p\|_{L^1}\Lambda} > 1,$$

where Λ *is given by (5.28).*

Then the BVP (5.37) has at least one solution on $[0,T]$.

Proof. Let us introduce the operator $N : C([0,T],\mathbb{R}) \to \mathcal{P}(C([0,T],\mathbb{R}))$ as

$$N(x) = \left\{ \begin{array}{l} h \in C([0,T],\mathbb{R}) : \\ h(t) = \left\{ \begin{array}{l} \dfrac{1}{\Gamma(q)}\displaystyle\int_0^t (t-s)^{q-1}v(s)ds - \dfrac{t^{n-1}}{\lambda\Gamma(q)}\displaystyle\int_0^T (T-s)^{q-1}v(s)ds \\ +\dfrac{t^{n-1}}{\lambda\Gamma(q)}\displaystyle\sum_{i=1}^m \dfrac{\gamma_i}{\Gamma(\beta_i)}\left[\displaystyle\int_0^{\eta_i}\int_0^s (\eta_i - s)^{\beta_i-1}(s-u)^{q-1}v(u)duds \right. \\ \left. -\displaystyle\int_0^{\zeta_i}\int_0^s (\zeta_i - s)^{\beta_i-1}(s-u)^{q-1}v(u)duds \right], \end{array} \right. \end{array} \right\}$$

for $v \in S_{F,x}$. We will show that the operator N satisfies the assumptions of the nonlinear alternative of Leray-Schauder type (Theorem 1.8). The proof consists of several steps. As a first step, we show that $N(x)$ *is convex for each* $x \in C([0,T],\mathbb{R})$. For that, let $h_1, h_2 \in N(x)$. Then there exist $v_1, v_2 \in S_{F,x}$ such that for each $t \in [0,T]$, we have

$$\begin{aligned} h_i(t) &= \frac{1}{\Gamma(q)}\int_0^t (t-s)^{q-1}v_i(s)ds - \frac{t^{n-1}}{\lambda\Gamma(q)}\int_0^T (T-s)^{q-1}v_i(s)ds \\ &\quad +\frac{t^{n-1}}{\lambda\Gamma(q)}\sum_{i=1}^m \frac{\gamma_i}{\Gamma(\beta_i)}\left[\int_0^{\eta_i}\int_0^s (\eta_i - s)^{\beta_i-1}(s-u)^{q-1}v_i(u)duds \right. \\ &\quad \left. -\int_0^{\zeta_i}\int_0^s (\zeta_i - s)^{\beta_i-1}(s-u)^{q-1}v_i(u)duds \right], \quad i = 1,2. \end{aligned}$$

Let $0 \le \omega \le 1$. Then, for each $t \in [0,T]$, we have

$$\begin{aligned} &[\omega h_1 + (1-\omega)h_2](t) \\ &= \frac{1}{\Gamma(q)}\int_0^t (t-s)^{q-1}v(s)ds - \frac{t^{n-1}}{\lambda\Gamma(q)}\int_0^T (T-s)^{q-1}[\omega v_1(r) + (1-\omega)v_2(r)]ds \\ &\quad +\frac{t^{n-1}}{\lambda\Gamma(q)}\sum_{i=1}^m \frac{\gamma_i}{\Gamma(\beta_i)}\left[\int_0^{\eta_i}\int_0^s (\eta_i - s)^{\beta_i-1}(s-u)^{q-1}[\omega v_1(r) + (1-\omega)v_2(r)]duds \right. \\ &\quad \left. -\int_0^{\zeta_i}\int_0^s (\zeta_i - s)^{\beta_i-1}(s-u)^{q-1}[\omega v_1(r) + (1-\omega)v_2(r)]duds \right]. \end{aligned}$$

Since $S_{F,x}$ is convex (F has convex values), therefore it follows that $\omega h_1+(1-\omega)h_2 \in N(x)$.

Next, we show that $N(x)$ *maps bounded sets into bounded sets in* $C([0,T],\mathbb{R})$. For a positive number ρ, let $B_\rho = \{x \in C([0,T],\mathbb{R}) : \|x\| \leq \rho\}$ be a bounded set in $C([0,T],\mathbb{R})$. Then, for each $h \in N(x), x \in B_\rho$, there exists $v \in S_{F,x}$ such that

$$\begin{aligned}h(t) &= \frac{1}{\Gamma(q)}\int_0^t (t-s)^{q-1}v(s)ds - \frac{t^{n-1}}{\lambda\Gamma(q)}\int_0^T (T-s)^{q-1}v(s)ds\\&\quad+\frac{t^{n-1}}{\lambda\Gamma(q)}\sum_{i=1}^m \frac{\gamma_i}{\Gamma(\beta_i)}\bigg[\int_0^{\eta_i}\int_0^s (\eta_i-s)^{\beta_i-1}(s-u)^{q-1}v(u)duds\\&\quad-\int_0^{\zeta_i}\int_0^s (\zeta_i-s)^{\beta_i-1}(s-u)^{q-1}v(u)duds\bigg].\end{aligned}$$

It is easy to verify that

$$\|h\| \leq \psi(\rho)\|p\|_{L^1}\Lambda.$$

Now we show that N *maps bounded sets into equi-continuous sets of* $C([0,T],\mathbb{R})$. Let $t', t'' \in [0,T]$ with $t' < t''$ and $x \in B_\rho$, where B_ρ, as above, is a bounded set of $C([0,T],\mathbb{R})$. For each $h \in N(x)$, we obtain

$$\begin{aligned}&|h(t'')-h(t')|\\&\leq \frac{\psi(r)}{\Gamma(q)}\bigg[\int_0^{t'} |(t''-s)^{q-1}-(t'-s)^{q-1}|\sigma(s)ds + \int_{t'}^{t''} |t''-s|^{q-1}p(s)ds\\&\quad+\frac{|(t'')^{n-1}-(t')^{n-1}|}{|\lambda|}\int_0^T |T-s|^{q-1}p(s)ds\\&\quad+\frac{|(t'')^{n-1}-(t')^{n-1}|}{|\lambda|}\sum_{i=1}^m \frac{\gamma_i}{\Gamma(\beta_i)}\bigg(\int_0^{\eta_i}\int_0^s (\eta_i-s)^{\beta_i-1}(s-u)^{q-1}dup(s)ds\\&\quad+\int_0^{\zeta_i}\int_0^s (\zeta_i-s)^{\beta_i-1}(s-u)^{q-1}p(s)duds\bigg)\bigg].\end{aligned}$$

Obviously the right hand side of the above inequality tends to zero independently of $x \in B_\rho$ as $t''-t' \to 0$. As N satisfies the above three assumptions, therefore it follows by Ascoli-Arzelá theorem that $N : C([0,T],\mathbb{R}) \to \mathcal{P}(C([0,T],\mathbb{R}))$ is completely continuous.

In our next step, we show that N *has a closed graph.* Let $x_n \to x_*, h_n \in N(x_n)$ and $h_n \to h_*$. Then we need to show that $h_* \in N(x_*)$. Associated with $h_n \in N(x_n)$, there exists $v_n \in S_{F,x_n}$ such that for each $t \in [0,T]$,

$$\begin{aligned}h_n(t) &= \frac{1}{\Gamma(q)}\int_0^t (t-s)^{q-1}v_n(s)ds - \frac{t^{n-1}}{\lambda\Gamma(q)}\int_0^T (T-s)^{q-1}v_n(s)ds\\&\quad+\frac{t^{n-1}}{\lambda\Gamma(q)}\sum_{i=1}^m \frac{\gamma_i}{\Gamma(\beta_i)}\bigg[\int_0^{\eta_i}\int_0^s (\eta_i-s)^{\beta_i-1}(s-u)^{q-1}v_n(u)duds\end{aligned}$$

$$-\int_0^{\zeta_i}\int_0^s(\zeta_i-s)^{\beta_i-1}(s-u)^{q-1}v_n(u)duds\Bigg].$$

Thus we have to show that there exists $v_*\in S_{F,x_*}$ such that for each $t\in[0,T]$,

$$\begin{aligned}h_*(t)=&\frac{1}{\Gamma(q)}\int_0^t(t-s)^{q-1}v_*(s)ds-\frac{t^{n-1}}{\lambda\Gamma(q)}\int_0^T(T-s)^{q-1}v_*(s)ds\\&+\frac{t^{n-1}}{\lambda\Gamma(q)}\sum_{i=1}^m\frac{\gamma_i}{\Gamma(\beta_i)}\Bigg[\int_0^{\eta_i}\int_0^s(\eta_i-s)^{\beta_i-1}(s-u)^{q-1}v_*(u)duds\\&-\int_0^{\zeta_i}\int_0^s(\zeta_i-s)^{\beta_i-1}(s-u)^{q-1}v_*(u)duds\Bigg].\end{aligned}$$

Let us consider the continuous linear operator $\Theta: L^1([0,T],\mathbb{R})\to C([0,T],\mathbb{R})$ so that

$$\begin{aligned}v\mapsto\Theta(v)=&\frac{1}{\Gamma(q)}\int_0^t(t-s)^{q-1}v(s)ds-\frac{t^{n-1}}{\lambda\Gamma(q)}\int_0^T(T-s)^{q-1}v(s)ds\\&+\frac{t^{n-1}}{\lambda\Gamma(q)}\sum_{i=1}^m\frac{\gamma_i}{\Gamma(\beta_i)}\Bigg[\int_0^{\eta_i}\int_0^s(\eta_i-s)^{\beta_i-1}(s-u)^{q-1}v(u)duds\\&-\int_0^{\zeta_i}\int_0^s(\zeta_i-s)^{\beta_i-1}(s-u)^{q-1}v(u)duds\Bigg].\end{aligned}$$

Observe that

$$\begin{aligned}&\|h_n(t)-h_*(t)\|\\&=\frac{1}{\Gamma(q)}\int_0^t(t-s)^{q-1}(v_n(s)-v_*(s))ds-\frac{t^{n-1}}{\lambda\Gamma(q)}\int_0^T(T-s)^{q-1}(v_n(s)-v_*(s))ds\\&+\frac{t^{n-1}}{\lambda\Gamma(q)}\sum_{i=1}^m\frac{\gamma_i}{\Gamma(\beta_i)}\Bigg[\int_0^{\eta_i}\int_0^s(\eta_i-s)^{\beta_i-1}(s-u)^{q-1}(v_n(u)-v_*(u))duds\\&-\int_0^{\zeta_i}\int_0^s(\zeta_i-s)^{\beta_i-1}(s-u)^{q-1}(v_n(u)-v_*(u))duds\Bigg],\end{aligned}$$

which tends to zero as $n\to\infty$.

Thus, it follows by Lemma 1.2 that $\Theta\circ S_F$ is a closed graph operator. Further, we have $h_n(t)\in\Theta(S_{F,x_n})$. Since $x_n\to x_*$, it follows that

$$\begin{aligned}h_*(t)=&\frac{1}{\Gamma(q)}\int_0^t(t-s)^{q-1}v_*(s)ds-\frac{t^{n-1}}{\lambda\Gamma(q)}\int_0^T(T-s)^{q-1}v_*(s)ds\\&+\frac{t^{n-1}}{\lambda\Gamma(q)}\sum_{i=1}^m\frac{\gamma_i}{\Gamma(\beta_i)}\Bigg[\int_0^{\eta_i}\int_0^s(\eta_i-s)^{\beta_i-1}(s-u)^{q-1}v_*(u)duds\\&-\int_0^{\zeta_i}\int_0^s(\zeta_i-s)^{\beta_i-1}(s-u)^{q-1}v_*(u)duds\Bigg],\end{aligned}$$

for some $v_*\in S_{F,x_*}$.

Finally, *we discuss a priori bounds on solutions.* Let x be a solution of (5.37). Then, using the computations proving that $N(x)$ maps bounded sets into bounded sets, we have

$$\|x\| \leq \psi(\|x\|)\|p\|_{L^1}\left\{\frac{T^q}{\Gamma(q+1)} + \frac{T^{q+n-1}}{|\lambda|\Gamma(q+1)} + \frac{T^{n-1}}{|\lambda|}\sum_{i=1}^{m}\gamma_i\frac{\eta_i^{q+\beta_i} - \zeta_i^{q+\beta_i}}{\Gamma(q+\beta_i+1}\right\}.$$

Consequently, in view of (5.10.3), we get

$$\frac{\|x\|}{\psi(\|x\|)\|p\|_{L^1}\Lambda} \leq 1.$$

In view of (5.10.3), there exists M such that $\|x\| \neq M$. Let us set

$$U = \{x \in C([0,T],\mathbb{R}) : \|x\| < M+1\}.$$

Note that the operator $N : \overline{U} \to \mathcal{P}(C([0,T],\mathbb{R}))$ is upper semi-continuous and completely continuous. From the choice of U, there is no $x \in \partial U$ such that $x \in \mu N(x)$ for some $\mu \in (0,1)$. Consequently, by the nonlinear alternative of Leray-Schauder type (Theorem 1.8), we deduce that N has a fixed point $x \in \overline{U}$ which is a solution of the problem (5.37). This completes the proof. □

Example 5.9. Let us consider the following 4-strip nonlocal boundary value problem:

$$\begin{cases} {}^CD^{9/2}x(t) \in F(t,x(t)), \quad t \in [0,2], \\ x(0) = 0,\ x'(0) = 0,\ x''(0) = 0,\ x'''(0) = 0, \\ x(2) = \sum_{i=1}^{4}\gamma_i[I^{\beta_i}x(\eta_i) - I^{\beta_i}x(\zeta_i)], \end{cases} \tag{5.38}$$

where $q = 9/2$, $n = 5$, $T = 2$, $\zeta_1 = 1/4$, $\eta_1 = 1/2$, $\zeta_2 = 2/3$, $\eta_2 = 1$, $\zeta_3 = 5/4$, $\eta_3 = 4/3$, $\zeta_4 = 3/2$, $\eta_4 = 7/4$, $\gamma_1 = 5$, $\gamma_2 = 10$, $\gamma_3 = 15$, $\gamma_4 = 25$, $\beta_1 = 5/4$, $\beta_2 = 7/4$, $\beta_3 = 9/4$, $\beta_4 = 11/4$, and $F : [0,2]\times\mathbb{R} \to \mathcal{P}(\mathbb{R})$ is a multi-valued map given by

$$x \to F(t,x) = \left[\frac{\sqrt[3]{t}|x|^5}{(|x|^5+3)}, \frac{\sqrt[3]{t}|x|}{2(|x|+1)}\right].$$

For $f \in F$, we have

$$|f| \leq \max\left(\frac{\sqrt[3]{t}|x|^5}{(|x|^5+3)}, \frac{\sqrt[3]{t}|x|}{2(|x|+1)}\right) \leq \sqrt[3]{t}, \quad x \in \mathbb{R}$$

with $p(t) = \sqrt[3]{t}$, $\psi(\|x\|) = 1$.

With the given values of the parameters, we find that $\lambda \simeq 9.334784$ and $\Lambda \simeq 1.406972$.
Using the above values in the condition (5.10.3), that is,

$$\frac{M}{\psi(M)\|p\|_{L^1}\Omega} > 1,$$

we find that $M > M_1 \simeq 1.055229$. Clearly, all the conditions of Theorem 5.10 are satisfied. Hence the conclusion of Theorem 5.10 applies to the problem (5.38).

5.2.2.2 *The Lipschitz case*

Now we prove the existence of solutions for the problem (5.37) with a non-convex valued right hand side by applying a fixed point theorem for multi-valued maps due to Covitz and Nadler (Theorem 1.19).

Theorem 5.11. *Assume that the following conditions hold:*

(5.11.1) $F : [0,T]\times\mathbb{R} \to \mathcal{P}_{cp}(\mathbb{R})$ *is such that* $F(\cdot,x) : [0,T] \to \mathcal{P}_{cp}(\mathbb{R})$ *is measurable for each* $x \in \mathbb{R}$;

(5.11.2) $H_d(F(t,x), F(t,\bar{x})) \leq m(t)|x-\bar{x}|$ *for almost all* $t \in [0,T]$ *and* $x, \bar{x} \in \mathbb{R}$ *with* $m \in C([0,T],\mathbb{R}^+)$ *and* $d(0, F(t,0)) \leq m(t)$ *for almost all* $t \in [0,T]$.

Then the boundary value problem (5.37) has at least one solution on $[0,T]$ *if*

$$\Lambda\|m\| < 1.$$

Proof. We transform the boundary value problem (5.37) into a fixed point problem. Consider the operator $N : C([0,T],\mathbb{R}) \to \mathcal{P}(C([0,T],\mathbb{R}))$ defined at the begining of the proof of Theorem 5.10. We show that the operator N satisfies the assumptions of Lemma 1.19. The proof will be given in two steps.

Step 1. $N(x)$ is non-empty and closed for every $v \in S_{F,x}$. Note that the set-valued map $F(\cdot, x(\cdot))$ is measurable by the measurable selection theorem (e.g., [78, Theorem III.6]) and hence it admits a measurable selection $v : I \to \mathbb{R}$. Moreover, by the assumption (5.11.2), we have

$$|v(t)| \leq m(t) + m(t)|x(t)|,$$

i.e., $v \in L^1([0,T],\mathbb{R})$ and hence F is integrably bounded. Therefore, $S_{F,y} \neq \emptyset$.

To show that $N(x) \in \mathcal{P}_{cl}((C[0,T],\mathbb{R}))$ for each $x \in C([0,T],\mathbb{R})$, let $\{u_n\}_{n\geq 0} \in N(x)$ be such that $u_n \to u$ $(n \to \infty)$ in $C([0,T],\mathbb{R})$. Then $u \in C([0,T],\mathbb{R})$ and there exists $v_n \in S_{F,x}$ such that, for each $t \in [0,T]$,

$$\begin{aligned} u_n(t) &= \frac{1}{\Gamma(q)}\int_0^t (t-s)^{q-1}v_n(s)ds - \frac{t^{n-1}}{\lambda\Gamma(q)}\int_0^T (T-s)^{q-1}v_n(s)ds \\ &\quad + \frac{t^{n-1}}{\lambda\Gamma(q)}\sum_{i=1}^m \frac{\gamma_i}{\Gamma(\beta_i)}\left[\int_0^{\eta_i}\int_0^s (\eta_i-s)^{\beta_i-1}(s-u)^{q-1}v_n(u)duds \right. \\ &\quad \left. - \int_0^{\zeta_i}\int_0^s (\zeta_i-s)^{\beta_i-1}(s-u)^{q-1}v_n(u)duds\right]. \end{aligned}$$

As F has compact values, we pass onto a subsequence to obtain that v_n converges to v in $L^1([0,T],\mathbb{R})$. Thus, $v \in S_{F,x}$ and for each $t \in [0,T]$,

$$\begin{aligned} u_n(t) \to u(t) &= \frac{1}{\Gamma(q)}\int_0^t (t-s)^{q-1}v(s)ds - \frac{t^{n-1}}{\lambda\Gamma(q)}\int_0^T (T-s)^{q-1}v(s)ds \\ &\quad + \frac{t^{n-1}}{\lambda\Gamma(q)}\sum_{i=1}^m \frac{\gamma_i}{\Gamma(\beta_i)}\left[\int_0^{\eta_i}\int_0^s (\eta_i-s)^{\beta_i-1}(s-u)^{q-1}v(u)duds \right. \end{aligned}$$

$$-\int_0^{\zeta_i}\int_0^{s}(\zeta_i-s)^{\beta_i-1}(s-u)^{q-1}v(u)duds\Bigg].$$

Hence, $u \in N(x)$.

Step 2. Next we show that there exists $\gamma < 1$ such that

$$H_d(N(x), N(\bar{x})) \leq \gamma\|x-\bar{x}\| \quad \text{for each} \quad x, \bar{x} \in C([0,T], \mathbb{R}).$$

Let $x, \bar{x} \in C([0,T], \mathbb{R})$ and $h_1 \in N(x)$. Then there exists $v_1(t) \in F(t, x(t))$ such that, for each $t \in [0,T]$,

$$\begin{aligned} h_1(t) &= \frac{1}{\Gamma(q)}\int_0^t (t-s)^{q-1}v_1(s)ds - \frac{t^{n-1}}{\lambda\Gamma(q)}\int_0^T (T-s)^{q-1}v_1(s)ds \\ &+\frac{t^{n-1}}{\lambda\Gamma(q)}\sum_{i=1}^m \frac{\gamma_i}{\Gamma(\beta_i)}\Bigg[\int_0^{\eta_i}\int_0^s(\eta_i-s)^{\beta_i-1}(s-u)^{q-1}v_1(u)duds \\ &-\int_0^{\zeta_i}\int_0^s(\zeta_i-s)^{\beta_i-1}(s-u)^{q-1}v_1(u)duds\Bigg]. \end{aligned}$$

By (5.11.2), we have

$$H_d(F(t,x), F(t,\bar{x})) \leq m(t)|x(t)-\bar{x}(t)|.$$

So, there exists $w(t) \in F(t, \bar{x}(t))$ such that

$$|v_1(t)-w(t)| \leq m(t)|x(t)-\bar{x}(t)|, \quad t \in [0,T].$$

Define $U : [0,T] \to \mathcal{P}(\mathbb{R})$ by

$$U(t) = \{w \in \mathbb{R} : |v_1(t)-w| \leq m(t)|x(t)-\bar{x}(t)|\}.$$

Since the multi-valued operator $U(t) \cap F(t, \bar{x}(t))$ is measurable (Proposition III.4 [78]), there exists a function $v_2(t)$ which is a measurable selection for V. So $v_2(t) \in F(t, \bar{x}(t))$ and for each $t \in [0,T]$, we have $|v_1(t)-v_2(t)| \leq m(t)|x(t)-\bar{x}(t)|$.

For each $t \in [0,T]$, let us define

$$\begin{aligned} h_2(t) &= \frac{1}{\Gamma(q)}\int_0^t (t-s)^{q-1}v_2(s)ds - \frac{t^{n-1}}{\lambda\Gamma(q)}\int_0^T (T-s)^{q-1}v_2(s)ds \\ &+\frac{t^{n-1}}{\lambda\Gamma(q)}\sum_{i=1}^m \frac{\gamma_i}{\Gamma(\beta_i)}\Bigg[\int_0^{\eta_i}\int_0^s(\eta_i-s)^{\beta_i-1}(s-u)^{q-1}v_2(u)duds \\ &-\int_0^{\zeta_i}\int_0^s(\zeta_i-s)^{\beta_i-1}(s-u)^{q-1}v_2(u)duds\Bigg]. \end{aligned}$$

Then

$$\begin{aligned} &|h_1(t)-h_2(t)| \\ &\leq \frac{1}{\Gamma(q)}\int_0^t (t-s)^{q-1}|v_1(s)-v_2(s)|ds - \frac{t^{n-1}}{\lambda\Gamma(q)}\int_0^T (T-s)^{q-1}|v_1(s)-v_2(s)|ds \end{aligned}$$

$$+\frac{t^{n-1}}{\lambda\Gamma(q)}\sum_{i=1}^{m}\frac{\gamma_i}{\Gamma(\beta_i)}\left[\int_0^{\eta_i}\int_0^s(\eta_i-s)^{\beta_i-1}(s-u)^{q-1}|v_1(u)-v_2(u)|duds\right.$$

$$\left.+\int_0^{\zeta_i}\int_0^s(\zeta_i-s)^{\beta_i-1}(s-u)^{q-1}|v_1(u)-v_2(u)|duds\right],$$

which leads to

$$\|h_1-h_2\|\leq\left\{\frac{T^q}{\Gamma(q+1)}+\frac{T^{q+n-1}}{|\lambda|\Gamma(q+1)}+\frac{T^{n-1}}{|\lambda|}\sum_{i=1}^{m}\gamma_i\frac{\eta_i^{q+\beta_i}-\zeta_i^{q+\beta_i}}{\Gamma(q+\beta_i+1}\right\}\|m\|\|x-\overline{x}\|.$$

Analogously, interchanging the roles of x and $\overline{x}$, we obtain

$$H_d(N(x),N(\bar{x}))\leq\gamma\|x-\bar{x}\|\leq\Lambda\|m\|\|x-\overline{x}\|.$$

Since N is a contraction, it follows by Lemma 1.19 that N has a fixed point x which is a solution of (5.37). This completes the proof. □

5.3 Fractional order systems with coupled integro-multi-strip-multi-point boundary conditions

In this section, we are concerned with existence of solutions for a system consisting of coupled nonlinear fractional differential equations:

$$\begin{aligned}&D^{\sigma}x(t)=f(t,x(t),y(t)),\quad n-1<\sigma<n,\ n\geq 3,\quad t\in J:=[0,1],\\&D^{\phi}x(t)=g(t,x(t),y(t)),\quad m-1<\phi<m,\ m\geq 3,\quad t\in J:=[0,1],\end{aligned}\tag{5.39}$$

subject to coupled integro-multi-strip-multi-point boundary conditions:

$$\begin{aligned}&x^{(i)}(0)=0,\quad i=0,1,2,\ldots,n-2,\\&\int_0^1 x(s)ds=\sum_{i=2}^{p}\beta_{i-1}\int_{\eta_{i-1}}^{\eta_i}y(s)ds+\sum_{j=1}^{q}\gamma_j\,y(\rho_j),\\&y^{(j)}(0)=0,\quad j=0,1,2,\ldots,m-2,\\&\int_0^1 y(s)ds=\sum_{i=2}^{\mu}\beta'_{i-1}\int_{\theta_{i-1}}^{\theta_i}x(s)ds+\sum_{j=1}^{\lambda}\gamma'_j\,x(\zeta_j),\end{aligned}\tag{5.40}$$

where D^{σ},D^{ϕ} are the standard Riemann-Liouville fractional derivative operators of order σ and ϕ respectively, $f,g:J\times\mathbb{R}\times\mathbb{R}\to\mathbb{R}$ are continuous functions, $0<\eta_1<\eta_2<\ldots<\eta_p<\rho_1<\rho_2<\ldots<\rho_q<1$, $0<\theta_1<\theta_2<\ldots<\theta_\mu<\zeta_1<\zeta_2<\ldots<\zeta_\lambda<1$, and $\beta_{i-1},\gamma_j>0, i=2,\ldots,p,\ j=1,2,\ldots,q$, $\beta'_{i-1},\gamma'_j>0, i=2,3,\ldots,\mu,\ j=1,2,\ldots,\lambda$, with $p,q,\mu,\lambda\in\mathbb{N}$.

The following lemma is of great importance in the proof of our main results.

Lemma 5.3. *Let $h,k\in C(J,\mathbb{R})$. Then the solution of the linear fractional differential system*

$$\begin{cases} D^\sigma x(t) = h(t), \quad n-1<\sigma<n, \quad t \in J := [0,1], \\ D^\phi y(t) = k(t), \quad m-1<\phi<m, \quad t \in J := [0,1] \\ x^{(i)}(0) = 0, \quad i = 0,1,2,\ldots,n-2, \\ \qquad \int_0^1 x(s)ds = \sum_{i=2}^{p} \beta_{i-1} \int_{\eta_{i-1}}^{\eta_i} y(s)ds + \sum_{j=1}^{q} \gamma_j y(\rho_j), \\ y^{(j)}(0) = 0, \quad j = 0,1,2,\ldots,m-2, \\ \qquad \int_0^1 y(s)ds = \sum_{i=2}^{\mu} \beta'_{i-1} \int_{\theta_{i-1}}^{\theta_i} x(s)ds + \sum_{j=1}^{\lambda} \gamma'_j x(\zeta_j), \end{cases} \tag{5.41}$$

is equivalent to the system of integral equations

$$\begin{aligned} x(t) = {} & \frac{1}{\Gamma(\sigma)} \int_0^t (t-s)^{\sigma-1} h(s)ds + \frac{t^{\sigma-1}}{\Omega} \Bigg\{ \frac{1}{\phi} \Big(\sum_{i=2}^{p} \beta_{i-1} \frac{1}{\Gamma(\phi)} \int_{\eta_{i-1}}^{\eta_i} \int_0^s (s-\tau)^{\phi-1} k(s) d\tau ds \\ & + \sum_{j=1}^{q} \gamma_j \frac{1}{\Gamma(\phi)} \int_0^{\rho_j} (\rho_j - s)^{\phi-1} k(s)ds - \frac{1}{\Gamma(\sigma)} \int_0^1 \int_0^t (t-s)^{\sigma-1} h(s)dsdt \Big) \\ & + \Lambda_1 \Big(\sum_{i=2}^{\mu} \beta'_{i-1} \frac{1}{\Gamma(\sigma)} \int_{\theta_{i-1}}^{\theta_i} \int_0^s (s-\tau)^{\sigma-1} h(\tau) d\tau ds \\ & + \sum_{j=1}^{\lambda} \gamma'_j \frac{1}{\Gamma(\sigma)} \int_0^{\zeta_j} (\zeta_j - s)^{\sigma-1} h(s)ds - \frac{1}{\Gamma(\phi)} \int_0^1 \int_0^t (t-s)^{\phi-1} k(s)dsdt \Big) \Bigg\}, \end{aligned} \tag{5.42}$$

and

$$\begin{aligned} y(t) = {} & \frac{1}{\Gamma(\phi)} \int_0^t (t-s)^{\phi-1} k(s)ds + \frac{t^{\phi-1}}{\Omega} \Bigg\{ \frac{1}{\sigma} \Big(\sum_{i=2}^{\mu} \beta'_{i-1} \frac{1}{\Gamma(\sigma)} \int_{\theta_{i-1}}^{\theta_i} \int_0^s (s-\tau)^{\sigma-1} h(s) d\tau ds \\ & + \sum_{j=1}^{\lambda} \gamma'_j \frac{1}{\Gamma(\sigma)} \int_0^{\zeta_j} (\zeta_j - s)^{\sigma-1} h(s)ds - \frac{1}{\Gamma(\phi)} \int_0^1 \int_0^t (t-s)^{\phi-1} h(s)dsdt \Big) \\ & + \Lambda_2 \Big(\sum_{i=2}^{p} \beta_{i-1} \frac{1}{\Gamma(\phi)} \int_{\eta_{i-1}}^{\eta_i} \int_0^s (s-\tau)^{\phi-1} k(\tau) d\tau ds \\ & + \sum_{j=1}^{q} \gamma_j \frac{1}{\Gamma(\phi)} \int_0^{\rho_j} (\rho_j - s)^{\phi-1} k(s)ds - \frac{1}{\Gamma(\sigma)} \int_0^1 \int_0^t (t-s)^{\sigma-1} h(s)dsdt \Big) \Bigg\}, \end{aligned} \tag{5.43}$$

where it is assumed that $\Omega = \dfrac{1}{\sigma\phi} - \Lambda_1\Lambda_2 \neq 0$, with

$$\Lambda_1 = \frac{1}{\phi} \sum_{i=2}^{p} \beta_{i-1} (\eta_i^\phi - \eta_{i-1}^\phi) + \sum_{j=1}^{q} \gamma_j \rho_j^{\phi-1},$$

$$\Lambda_2 = \frac{1}{\sigma} \sum_{i=2}^{\mu} \beta'_{i-1} (\theta_i^\sigma - \theta_{i-1}^\sigma) + \sum_{j=1}^{\lambda} \gamma'_j \zeta_j^{\sigma-1}.$$

Proof. The general solution of the equations $D^{\sigma}x(t) = h(t)$, $n-1 < \sigma < n$ and $D^{\phi}y(t) = k(t)$, $m-1 < \phi < m$ can be written as

$$x(t) = b_1 t^{\sigma-1} + b_2 t^{\sigma-2} + \cdots + b_n t^{\sigma-n} + \frac{1}{\Gamma(\sigma)} \int_0^t (t-s)^{\sigma-1} h(s) ds, \tag{5.44}$$

$$y(t) = d_1 t^{\phi-1} + d_2 t^{\phi-2} + \cdots + d_m t^{\phi-m} + \frac{1}{\Gamma(\phi)} \int_0^t (t-s)^{\phi-1} k(s) ds, \tag{5.45}$$

where b_i $(i = 1, 2, \ldots, n)$, d_j $(j = 1, \ldots, m)$ are arbitrary constants. Using the conditions $x^{(i)}(0) = 0$, $i = 0, 1, 2, \ldots, n-2$, and $y^{(j)}(0) = 0$, $j = 0, 1, 2, \ldots, m-2$, we find that $b_2 = b_3 = \cdots = b_n = 0$ and $d_1 = d_2 = \ldots = d_m = 0$. Thus we have

$$x(t) = b_1 t^{\sigma-1} + \frac{1}{\Gamma(\sigma)} \int_0^t (t-s)^{\sigma-1} g(s) ds, \tag{5.46}$$

$$y(t) = d_1 t^{\phi-1} + \frac{1}{\Gamma(\phi)} \int_0^t (t-s)^{\phi-1} g(s) ds, \tag{5.47}$$

which, together with the integro-multi-strip-multi-point boundary conditions given by (5.41), yields the system

$$\begin{aligned}
\frac{1}{\sigma} b_1 - \Lambda_1 d_1 &= \sum_{i=2}^{p} \beta_{i-1} \frac{1}{\Gamma(\phi)} \int_{\eta_{i-1}}^{\eta_i} \int_0^s (s-\tau)^{\phi-1} k(\tau) d\tau ds \\
&\quad + \sum_{j=1}^{q} \gamma_j \frac{1}{\Gamma(\phi)} \int_0^{\rho_j} (\rho_j - s)^{\phi-1} k(s) ds - \frac{1}{\Gamma(\sigma)} \int_0^1 \int_0^t (t-s)^{\sigma-1} h(s) ds dt, \\
-\Lambda_2 b_1 + \frac{1}{\phi} d_1 &= \sum_{i=2}^{\mu} \beta'_{i-1} \frac{1}{\Gamma(\sigma)} \int_{\theta_{i-1}}^{\theta_i} \int_0^s (s-\tau)^{\sigma-1} h(\tau) d\tau ds \\
&\quad + \sum_{j=1}^{\lambda} \gamma'_j \frac{1}{\Gamma(\sigma)} \int_0^{\zeta_j} (\zeta_j - s)^{\sigma-1} h(s) ds - \frac{1}{\Gamma(\phi)} \int_0^1 \int_0^t (t-s)^{\phi-1} k(s) ds dt.
\end{aligned}$$

Solving the above system, we get

$$\begin{aligned}
b_1 = \frac{1}{\Omega} \Big(\frac{1}{\phi} \Big[&\sum_{i=2}^{p} \beta_{i-1} \frac{1}{\Gamma(\phi)} \int_{\eta_{i-1}}^{\eta_i} \int_0^s (s-\tau)^{\phi-1} k(\tau) d\tau ds \\
&+ \sum_{j=1}^{q} \gamma_j \frac{1}{\Gamma(\phi)} \int_0^{\rho_j} (\rho_j - s)^{\phi-1} k(s) ds - \frac{1}{\Gamma(\sigma)} \int_0^1 \int_0^t (t-s)^{\sigma-1} h(s) ds dt \Big] \\
&+ \Lambda_1 \Big[\sum_{i=2}^{\mu} \beta'_{i-1} \frac{1}{\Gamma(\sigma)} \int_{\theta_{i-1}}^{\theta_i} \int_0^s (s-\tau)^{\sigma-1} h(\tau) d\tau ds \\
&+ \sum_{j=1}^{\lambda} \gamma'_j \frac{1}{\Gamma(\sigma)} \int_0^{\zeta_j} (\zeta_j - s)^{\sigma-1} h(s) ds - \frac{1}{\Gamma(\phi)} \int_0^1 \int_0^t (t-s)^{\phi-1} k(s) ds dt \Big] \Big),
\end{aligned}$$

and

$$d_1 = \frac{1}{\Omega} \Big(\frac{1}{\sigma} \Big[\sum_{i=2}^{\mu} \beta'_{i-1} \frac{1}{\Gamma(\sigma)} \int_{\theta_{i-1}}^{\theta_i} \int_0^s (s-\tau)^{\sigma-1} h(\tau) d\tau ds$$

$$+\sum_{j=1}^{\lambda}\gamma_j'\frac{1}{\Gamma(\sigma)}\int_0^{\zeta_j}(\zeta_j-s)^{\sigma-1}h(s)ds-\frac{1}{\Gamma(\phi)}\int_0^1\int_0^t(t-s)^{\phi-1}k(s)dsdt\Big]$$

$$+\Lambda_2\Big[\sum_{i=2}^{p}\beta_{i-1}\frac{1}{\Gamma(\phi)}\int_{\eta_{i-1}}^{\eta_i}\int_0^s(s-\tau)^{\phi-1}k(\tau)d\tau ds$$

$$+\sum_{j=1}^{q}\gamma_j\frac{1}{\Gamma(\phi)}\int_0^{\rho_j}(\rho_j-s)^{\phi-1}k(s)ds-\frac{1}{\Gamma(\sigma)}\int_0^1\int_0^t(t-s)^{\sigma-1}h(s)dsdt\Big]\Big).$$

Inserting the above values in (5.46) and (5.47) leads to the solution (5.42) and (5.43). The converse follows by direct computation. The proof is completed. □

5.3.1 *Main results*

Let us introduce the space $X=\{x(t)|x(t)\in C([0,1],\mathbb{R})\}$ endowed with the norm $\|x\|=\sup\{|x(t)|,t\in[0,1]\}$. Obviously $(X,\|\cdot\|)$ is a Banach space. Then the product space $(X\times X,\|(x,y)\|)$ is also a Banach space equipped with norm $\|(x,y)\|=\|x\|+\|y\|$.

In view of Lemma 5.3, we define an operator $T:X\times X\to X\times X$ by

$$T(x,y)(t)=\begin{pmatrix}T_1(x,y)(t)\\T_2(x,y)(t)\end{pmatrix},$$

where

$$\begin{aligned}T_1(x,y)(t)&=\frac{1}{\Gamma(\sigma)}\int_0^t(t-s)^{\sigma-1}f(s,x(s),y(s))ds\\&+\frac{t^{\sigma-1}}{\Omega}\Big\{\frac{1}{\phi}\Big(\sum_{i=2}^{p}\beta_{i-1}\frac{1}{\Gamma(\phi)}\int_{\eta_{i-1}}^{\eta_i}\int_0^s(s-\tau)^{\phi-1}g(s,x(s),y(s))d\tau ds\\&+\sum_{j=1}^{q}\gamma_j\frac{1}{\Gamma(\phi)}\int_0^{\rho_j}(\rho_j-s)^{\phi-1}g(s,x(s),y(s))ds\\&-\frac{1}{\Gamma(\sigma)}\int_0^1\int_0^t(t-s)^{\sigma-1}f(s,x(s),y(s))dsdt\Big)\\&+\Lambda_1\Big(\sum_{i=2}^{\mu}\beta_{i-1}'\frac{1}{\Gamma(\sigma)}\int_{\theta_{i-1}}^{\theta_i}\int_0^s(s-\tau)^{\sigma-1}f(\tau,x(\tau),y(\tau))d\tau ds\\&+\sum_{j=1}^{\lambda}\gamma_j'\frac{1}{\Gamma(\sigma)}\int_0^{\zeta_j}(\zeta_j-s)^{\sigma-1}f(s,x(s),y(s))ds\\&-\frac{1}{\Gamma(\phi)}\int_0^1\int_0^t(t-s)^{\phi-1}g(s,x(s),y(s))dsdt\Big)\Big\},\end{aligned}\tag{5.48}$$

and

$$T_2(x,y)(t)=\frac{1}{\Gamma(\phi)}\int_0^t(t-s)^{\phi-1}g(s,x(s),y(s))ds$$

$$+\frac{t^{\phi-1}}{\Omega}\bigg\{\frac{1}{\sigma}\Big(\sum_{i=2}^{\mu}\beta'_{i-1}\frac{1}{\Gamma(\sigma)}\int_{\theta_{i-1}}^{\theta_i}\int_0^s(s-\tau)^{\sigma-1}f(\tau,x(\tau),y(\tau))d\tau ds$$
$$+\sum_{j=1}^{\lambda}\gamma'_j\frac{1}{\Gamma(\sigma)}\int_0^{\zeta_j}(\zeta_j-s)^{\sigma-1}f(s,x(s),y(s))ds$$
$$-\frac{1}{\Gamma(\phi)}\int_0^1\int_0^t(t-s)^{\phi-1}g(s,x(s),y(s))dsdt\Big)$$
$$+\Lambda_2\Big(\sum_{i=2}^{p}\beta_{i-1}\frac{1}{\Gamma(\phi)}\int_{\eta_{i-1}}^{\eta_i}\int_0^s(s-\tau)^{\phi-1}g(\tau,x(\tau),y(\tau))d\tau ds$$
$$+\sum_{j=1}^{q}\gamma_j\frac{1}{\Gamma(\phi)}\int_0^{\rho_j}(\rho_j-s)^{\phi-1}g(s,x(s),y(s))ds$$
$$-\frac{1}{\Gamma(\sigma)}\int_0^1\int_0^t(t-s)^{\sigma-1}f(s,x(s),y(s))dsdt\Big)\bigg\}. \tag{5.49}$$

For computational convenience, we set the notation:

$$Q_1=\frac{1}{\Gamma(\sigma+1)}+\frac{1}{|\Omega|}\Big[\frac{1}{\phi}\frac{1}{\Gamma(\sigma+2)}+|\Lambda_1|\Big(\sum_{i=2}^{\mu}|\beta'_{i-1}|\frac{\theta_i^{\sigma+1}-\theta_{i-1}^{\sigma+1}}{\Gamma(\sigma+2)}$$
$$+\sum_{j=1}^{\lambda}|\gamma'_j|\frac{\zeta_j^{\sigma}}{\Gamma(\sigma+1)}\Big)\Big], \tag{5.50}$$

$$Q_2=\frac{1}{|\Omega|}\Big[\frac{1}{\phi}\Big(\sum_{i=2}^{p}|\beta_{i-1}|\frac{\eta_i^{\phi+1}-\eta_{i-1}^{\phi+1}}{\Gamma(\phi+2)}+\sum_{j=1}^{q}|\gamma_j|\frac{\rho_j^{\phi}}{\Gamma(\phi+1)}\Big)+|\Lambda_1|\frac{1}{\Gamma(\phi+2)}\Big], \tag{5.51}$$

$$Q_3=\frac{1}{|\Omega|}\Big[\frac{1}{\sigma}\Big(\sum_{i=2}^{\mu}|\beta'_{i-1}|\frac{\theta_i^{\sigma+1}-\theta_{i-1}^{\sigma+1}}{\Gamma(\sigma+2)}+\sum_{j=1}^{\lambda}|\gamma'_j|\frac{\zeta_j^{\sigma}}{\Gamma(\sigma+1)}\Big)+|\Lambda_2|\frac{1}{\Gamma(\sigma+2)}\Big], \tag{5.52}$$

$$Q_4=\frac{1}{\Gamma(\phi+1)}+\frac{1}{|\Omega|}\Big[\frac{1}{\sigma}\frac{1}{\Gamma(\phi+2)}+|\Lambda_2|\Big(\sum_{i=2}^{p}|\beta_{i-1}|\frac{\eta_i^{\phi+1}-\eta_{i-1}^{\phi+1}}{\Gamma(\phi+2)}$$
$$+\sum_{j=1}^{q}|\gamma_j|\frac{\rho_j^{\phi}}{\Gamma(\phi+1)}\Big)\Big]. \tag{5.53}$$

In the following theorem, we prove existence and uniqueness of solutions of the system (5.39)-(5.40) via Banach's contraction principle.

Theorem 5.12. *Assume that:*

(5.12.1) $f,g:[0,1]\times\mathbb{R}\times\mathbb{R}\to\mathbb{R}$ *are continuous functions and there exist positive constants* ℓ_1 *and* ℓ_2 *such that for all* $t\in[0,1]$ *and* $x_i,y_i\in\mathbb{R},\ i=1,2,$

$$|f(t,x_1,x_2)-f(t,y_1,y_2)|\le\ell_1(|x_1-y_1|+|x_2-y_2|),$$
$$|g(t,x_1,x_2)-g(t,y_1,y_2)|\le\ell_2(|x_1-y_1|+|x_2-y_2|).$$

If

$$(Q_1+Q_3)\ell_1+(Q_2+Q_4)\ell_2<1,$$

where $Q_i, i=1,2,3,4$ *are given by (5.50)-(5.53), then the system (5.39)-(5.40) has a unique solution on* $[0,1]$.

Proof. Define $\sup_{t\in[0,1]} f(t,0,0)=N_1<\infty$ and $\sup_{t\in[0,1]} g(t,0,0)=N_2<\infty$ and select $r>0$ such that

$$r>\frac{(Q_1+Q_3)N_1+(Q_2+Q_4)N_2}{1-(Q_1+Q_3)\ell_1-(Q_2+Q_4)\ell_2}.$$

Now we show that $TB_r\subset B_r$, where $B_r=\{(x,y)\in X\times X:\|(x,y)\|\le r\}$. By the assumption (5.12.1), for $(u,v)\in B_r,\ t\in[0,1]$, we have

$$\begin{aligned}|f(t,x(t),y(t))| &\le |f(t,x(t),y(t))-f(t,0,0)|+|f(t,0,0)|\\ &\le \ell_1(|x(t)|+|y(t)|)+N_1\\ &\le \ell_1(\|x\|+\|y\|)+N_1\le \ell_1 r+N_1,\end{aligned}$$

and

$$|g(t,x(t),y(t))|\le \ell_2(\|x\|+\|y\|)+N_2\le \ell_2 r+N_2,$$

which lead to

$$\begin{aligned}|T_1(x,y)(t)| \le{}& \frac{1}{\Gamma(\sigma)}\int_0^t (t-s)^{\sigma-1}(\ell_1 r+N_1)ds\\ &+\frac{1}{|\Omega|}\Bigg\{\frac{1}{\phi}\Bigg(\sum_{i=2}^{p}|\beta_{i-1}|\frac{1}{\Gamma(\phi)}\int_{\eta_{i-1}}^{\eta_i}\int_0^s (s-\tau)^{\phi-1}(\ell_2 r+N_2)d\tau ds\\ &+\sum_{j=1}^{q}|\gamma_j|\frac{1}{\Gamma(\phi)}\int_0^{\rho_j}(\rho_j-s)^{\phi-1}(\ell_2 r+N_2)ds\\ &+\frac{1}{\Gamma(\sigma)}\int_0^1\int_0^t (t-s)^{\sigma-1}(\ell_1 r+N_1)dsdt\Bigg)\\ &+|\Lambda_1|\Bigg(\sum_{i=2}^{\mu}|\beta'_{i-1}|\frac{1}{\Gamma(\sigma)}\int_{\theta_{i-1}}^{\theta_i}\int_0^s (s-\tau)^{\sigma-1}(\ell_1 r+N_1)d\tau ds\\ &+\sum_{j=1}^{\lambda}|\gamma'_j|\frac{1}{\Gamma(\sigma)}\int_0^{\zeta_j}(\zeta_j-s)^{\sigma-1}(\ell_1 r+N_1)ds\\ &+\frac{1}{\Gamma(\phi)}\int_0^1\int_0^t (t-s)^{\phi-1}(\ell_2 r+N_2)dsdt\Bigg)\Bigg\}\\ ={}&(Q_1\ell_1+Q_2\ell_2)r+Q_1N_1+Q_2N_2.\end{aligned}$$

Hence

$$\|T_1(x,y)\|\le (Q_1\ell_1+Q_2\ell_2)r+Q_1N_1+Q_2N_2.$$

In the same way, we can obtain that

$$\|T_2(x,y)\| \leq (Q_3\ell_1 + Q_4\ell_2)r + Q_3N_1 + Q_4N_2.$$

Consequently, we have

$$\|T(x,y)\| \leq [(Q_1+Q_3)\ell_1 + (Q_2+Q_4)\ell_2]r + (Q_1+Q_3)N_1 + (Q_2+Q_4)N_2 \leq r.$$

Now, for $(x_2,y_2),(x_1,y_1) \in X \times X$, and for any $t \in [0,1]$, we get

$$\begin{aligned}
&|T_1(x_2,y_2)(t) - T_1(x_1,y_1)(t)| \\
&\leq \frac{1}{\Gamma(\sigma)}\int_0^t (t-s)^{\sigma-1}\ell_1(\|x_2-x_1\| + \|y_2-y_1\|)ds \\
&+\frac{1}{|\Omega|}\Bigg\{\frac{1}{\phi}\Big(\sum_{i=2}^{p}|\beta_{i-1}|\frac{1}{\Gamma(\phi)}\int_{\eta_{i-1}}^{\eta_i}\int_0^s (s-\tau)^{\phi-1}\ell_2(\|x_2-x_1\| + \|y_2-y_1\|)d\tau ds \\
&+\sum_{j=1}^{q}|\gamma_j|\frac{1}{\Gamma(\phi)}\int_0^{\rho_j}(\rho_j-s)^{\phi-1}\ell_2(\|x_2-x_1\| + \|y_2-y_1\|)ds \\
&+\frac{1}{\Gamma(\sigma)}\int_0^1\int_0^t (t-s)^{\sigma-1}\ell_1(\|x_2-x_1\| + \|y_2-y_1\|)dsdt\Big) \\
&+|\Lambda_1|\Big(\sum_{i=2}^{\mu}|\beta'_{i-1}|\frac{1}{\Gamma(\sigma)}\int_{\theta_{i-1}}^{\theta_i}\int_0^s (s-\tau)^{\sigma-1}\ell_1(\|x_2-x_1\| + \|y_2-y_1\|)d\tau ds \\
&+\sum_{j=1}^{\lambda}|\gamma'_j|\frac{1}{\Gamma(\sigma)}\int_0^{\zeta_j}(\zeta_j-s)^{\sigma-1}\ell_1(\|x_2-x_1\| + \|y_2-y_1\|)ds \\
&+\frac{1}{\Gamma(\phi)}\int_0^1\int_0^t (t-s)^{\phi-1}\ell_2(\|x_2-x_1\| + \|y_2-y_1\|)dsdt\Big)\Bigg\} \\
&\leq (Q_1\ell_1 + Q_2\ell_2)(\|x_2-x_1\| + \|y_2-y_1\|),
\end{aligned}$$

which implies that

$$\|T_1(x_2,y_2) - T_1(x_1,y_1)\| \leq (Q_1\ell_1 + Q_2\ell_2)(\|x_2-x_1\| + \|y_2-y_1\|). \tag{5.54}$$

Similarly, one can obtain

$$\|T_2(x_2,y_2)(t) - T_2(x_1,y_1)\| \leq (Q_3\ell_1 + Q_4\ell_2)(\|x_2-x_1\| + \|y_2-y_1\|). \tag{5.55}$$

It follows from (5.54) and (5.55) that

$$\|T(x_2,y_2) - T(x_1,y_1)\| \leq [(Q_1+Q_3)\ell_1 + (Q_2+Q_4)\ell_2](\|x_2-x_1\| + \|y_2-y_1\|).$$

Since $(Q_1+Q_3)\ell_1 + (Q_2+Q_4)\ell_2 < 1$, therefore, T is a contraction operator. So, by Banach's fixed point theorem, the operator T has a unique fixed point, which is the unique solution of problem (5.39)-(5.40). This completes the proof. □

The second result is based on Leray-Schauder alternative (Theorem 1.7).

Theorem 5.13. *Assume that:*

(5.13.1) $f, g : [0,1] \times \mathbb{R} \times \mathbb{R} \to \mathbb{R}$ *are continuous functions and there exist real constants* $k_i, \gamma_i \geq 0$, $(i = 0,1,2)$ *with* $k_0 > 0, \gamma_0 > 0$ *such that* $\forall x_i \in \mathbb{R}$, $(i = 1,2)$,

$$|f(t,x_1,x_2)| \leq k_0 + k_1|x_1| + k_2|x_2|, \ |g(t,x_1,x_2)| \leq \gamma_0 + \gamma_1|x_1| + \gamma_2|x_2|.$$

If

$$(Q_1 + Q_3)k_1 + (Q_2 + Q_4)\gamma_1 < 1 \quad and \quad (Q_1 + Q_3)k_2 + (Q_2 + Q_4)\gamma_2 < 1,$$

where $M_i, i = 1,2,3,4$, *are given by (5.50)-(5.53), then the system (5.39)-(5.40) has at least one solution on* $[0,1]$.

Proof. First we show that the operator $T : X \times X \to X \times X$ is completely continuous. By continuity of functions f and g, the operator T is continuous.

Let $\Omega \subset X \times X$ be bounded. Then there exist positive constants L_1 and L_2 such that $|f(t,x(t),y(t))| \leq L_1$, $|h(t,x(t),y(t))| \leq L_2$, $\forall (x,y) \in \Omega$. Then, for any $(x,y) \in \Omega$, we have

$$\begin{aligned}
&|T_1(x,y)(t)| \\
&\leq \frac{1}{\Gamma(\sigma)}\int_0^t (t-s)^{\sigma-1}L_1 ds + \frac{1}{|\Omega|}\Bigg\{\frac{1}{\phi}\Bigg(\sum_{i=2}^{p}|\beta_{i-1}|\frac{1}{\Gamma(\phi)}\int_{\eta_{i-1}}^{\eta_i}\int_0^s (s-\tau)^{\phi-1}L_2 d\tau ds \\
&\quad + \sum_{j=1}^{q}|\gamma_j|\frac{1}{\Gamma(\phi)}\int_0^{\rho_j}(\rho_j - s)^{\phi-1}L_2 ds + \frac{1}{\Gamma(\sigma)}\int_0^1\int_0^t (t-s)^{\sigma-1}L_1 dsdt\Bigg) \\
&\quad + |\Lambda_1|\Bigg(\sum_{i=2}^{\mu}|\beta'_{i-1}|\frac{1}{\Gamma(\sigma)}\int_{\theta_{i-1}}^{\theta_i}\int_0^s (s-\tau)^{\sigma-1}L_1 d\tau ds \\
&\quad + \sum_{j=1}^{\lambda}|\gamma'_j|\frac{1}{\Gamma(\sigma)}\int_0^{\zeta_j}(\zeta_j - s)^{\sigma-1}L_1 ds + \frac{1}{\Gamma(\phi)}\int_0^1\int_0^t (t-s)^{\phi-1}L_2 dsdt\Bigg)\Bigg\} \\
&\leq Q_1 L_1 + Q_2 L_2,
\end{aligned}$$

which implies that

$$\|T_1(x,y)\| \leq Q_1 L_1 + Q_2 L_2.$$

Similarly, we get

$$\|T_2(x,y)\| \leq Q_3 L_1 + Q_4 L_2.$$

Thus, it follows from the above inequalities that the operator T is uniformly bounded, since $\|T(x,y)\| \leq (Q_1 + Q_3)L_1 + (Q_2 + Q_4)L_2$.

Next, we show that T is equi-continuous. Let $t_1, t_2 \in [0,1]$ with $t_1 < t_2$. Then we have

$$\begin{aligned}
&|T_1(x(t_2),y(t_2)) - T_1(x(t_1),y(t_1))| \\
&\leq L_1\Bigg\{\frac{1}{\Gamma(\sigma)}\int_0^{t_1}[(t_2-s)^{\sigma-1} - (t_1-s)^{\sigma-1}]ds + \frac{1}{\Gamma(\sigma)}\int_{t_1}^{t_2}(t_2-s)^{\sigma-1}ds\Bigg\}
\end{aligned}$$

$$
\begin{aligned}
&+\frac{|t_2^{\sigma-1}-t_1^{\sigma-1}|}{|\Omega|}\Big\{\Big[\frac{1}{\phi}\frac{1}{\Gamma(\sigma+2)}+|\Lambda_1|\Big(\sum_{i=2}^{\mu}|\beta'_{i-1}|\frac{\theta_i^{\sigma+1}-\theta_{i-1}^{\sigma+1}}{\Gamma(\sigma+2)}\Big)\Big]L_1\\
&+\Big[\frac{1}{\phi}\Big(\sum_{i=2}^{p}|\beta_{i-1}|\frac{\eta_i^{\phi+1}-\eta_{i-1}^{\phi+1}}{\Gamma(\phi+2)}+\sum_{j=1}^{q}|\gamma_j|\frac{\rho_j^{\phi}}{\Gamma(\phi+1)}\Big)+|\Lambda_1|\frac{1}{\Gamma(\phi+2)}\Big]L_2\Big\}\\
\le\ &\frac{L_1}{\Gamma(\sigma+1)}[2(t_2-t_1)^{\sigma}+|t_2^{\sigma}-t_1^{\sigma}|]\\
&+\frac{|t_2^{\sigma-1}-t_1^{\sigma-1}|}{|\Omega|}\Big\{\Big[\frac{1}{\phi}\frac{1}{\Gamma(\sigma+2)}+|\Lambda_1|\Big(\sum_{i=2}^{\mu}|\beta'_{i-1}|\frac{\theta_i^{\sigma+1}-\theta_{i-1}^{\sigma+1}}{\Gamma(\sigma+2)}\Big)\Big]L_1\\
&+\Big[\frac{1}{\phi}\Big(\sum_{i=2}^{p}|\beta_{i-1}|\frac{\eta_i^{\phi+1}-\eta_{i-1}^{\phi+1}}{\Gamma(\phi+2)}+\sum_{j=1}^{q}|\gamma_j|\frac{\rho_j^{\phi}}{\Gamma(\phi+1)}\Big)+|\Lambda_1|\frac{1}{\Gamma(\phi+2)}\Big]L_2\Big\}.
\end{aligned}
$$

Analogously, we can obtain

$$
\begin{aligned}
&|T_2(x(t_2),y(t_2))-T_2(x(t_1),y(t_1))|\\
\le\ &\frac{L_2}{\Gamma(\phi+1)}[2(t_2-t_1)^{\phi}+|t_2-t_1^{\phi}|]\\
&+\frac{|t_2^{\phi-1}-t_1^{\phi-1}|}{|\Omega|}\Big\{\Big[\frac{1}{\sigma}\Big(\sum_{i=2}^{\mu}|\beta'_{i-1}|\frac{\theta_i^{\sigma+1}-\theta_{i-1}^{\sigma+1}}{\Gamma(\sigma+2)}+\sum_{j=1}^{\lambda}|\gamma'_j|\frac{\zeta_j^{\sigma}}{\Gamma(\sigma+1)}\Big)+|\Lambda_2|\frac{1}{\Gamma(\sigma+2)}\Big]L_1\\
&+\Big[\frac{1}{\sigma}\frac{1}{\Gamma(\phi+2)}+|\Lambda_2|\Big(\sum_{i=2}^{p}|\beta_{i-1}|\frac{\eta_i^{\phi+1}-\eta_{i-1}^{\phi+1}}{\Gamma(\phi+2)}+\sum_{j=1}^{q}|\gamma_j|\frac{\rho_j^{\phi}}{\Gamma(\phi+1)}\Big)\Big]L_2\Big\}.
\end{aligned}
$$

Therefore, the operator $T(x,y)$ is equi-continuous, and thus the operator $T(x,y)$ is completely continuous.

Finally, it will be verified that the set $\mathcal{E}=\{(x,y)\in X\times X|(x,y)=\lambda T(x,y),0\le\lambda\le 1\}$ is bounded. Let $(x,y)\in\mathcal{E}$, with $(x,y)=\lambda T(x,y)$. For any $t\in[0,1]$, we have

$$x(t)=\lambda T_1(x,y)(t),\quad y(t)=\lambda T_2(x,y)(t).$$

Then

$$
\begin{aligned}
|x(t)|&\le Q_1(k_0+k_1|x|+k_2|y|)+Q_2(\gamma_0+\gamma_1|x|+\gamma_2|y|)\\
&=Q_1k_0+Q_2\gamma_0+(Q_1k_1+Q_2\gamma_1)|x|+(Q_1k_2+Q_2\gamma_2)|y|,
\end{aligned}
$$

and

$$
\begin{aligned}
|y(t)|&\le Q_3(k_0+k_1|x|+k_2|y|)+Q_4(\gamma_0+\gamma_1|x|+\gamma_2|y|)\\
&=Q_3k_0+Q_4\gamma_0+(Q_3k_1+Q_4\gamma_1)|x|+(Q_3k_2+Q_4\gamma_2)|y|.
\end{aligned}
$$

Hence we have

$$\|x\|\le Q_1k_0+Q_2\gamma_0+(Q_1k_1+Q_2\gamma_1)\|x\|+(Q_1k_2+Q_2\gamma_2)\|y\|$$

and

$$\|y\|\le Q_3k_0+Q_4\gamma_0+(Q_3k_1+Q_4\gamma_1)\|x\|+(Q_3k_2+Q_4\gamma_2)\|y\|,$$

which imply that

$$\|x\| + \|y\| \le (Q_1 + Q_3)k_0 + (Q_2 + Q_4)\gamma_0 + [(Q_1 + Q_3)k_1 + (Q_2 + Q_4)\gamma_1]\|x\| \\ +[(Q_1 + Q_3)k_2 + (Q_2 + Q_4)\gamma_2)]\|y\|.$$

Consequently, we have

$$\|(x, y)\| \le \frac{(Q_1 + Q_3)k_0 + (Q_2 + Q_4)\gamma_0}{M_0},$$

where $M_0 = \min\{1-[(Q_1+Q_3)k_1+(Q_2+Q_4)\gamma_1], 1-[(Q_1+Q_3)k_2+(Q_2+Q_4)\gamma_2)]\}$, which proves that $\mathcal{E}$ is bounded. Thus, by Lemma 1.7, the operator T has at least one fixed point. Hence the system (5.39)-(5.40) has at least one solution. The proof is complete. □

The proof of the next result relies on Krasnosel'skiĭ's fixed point theorem (Theorem 1.4).

Theorem 5.14. *Assume that $f, g : [0, 1] \times \mathbb{R} \to \mathbb{R}$ are continuous functions satisfying assumption (5.13.1) in Theorem 5.12. In addition we suppose that there exist two positive constants L_1, L_2 such that for all $t \in [0, 1]$ and $x_i, y_i \in \mathbb{R}, i = 1, 2,$*

$$|f(t, x_1, x_2)| \le L_1 \quad \text{and} \quad |g(t, x_1, x_2)| \le L_2. \tag{5.56}$$

If

$$\frac{\ell_1}{\Gamma(\sigma + 1)} + \frac{\ell_2}{\Gamma(\phi + 1)} < 1, \tag{5.57}$$

then the problem (5.39)-(5.40) has at least one solution on $[0, 1]$.

Proof. To verify the hypothesis of Theorem 1.4, we decompose the operator T into four operators $T_{1,1}, T_{1,2}, T_{2,1}$ and $T_{2,2}$ on B_δ, where $B_\delta = \{(x, y) \in X \times Y : \|(x, y)\| \le \delta\}$ (δ will be fixed later), as

$$\begin{aligned} T_{1,1}(x, y)(t) = \frac{t^{\sigma-1}}{\Omega}\Bigg\{ & \frac{1}{\phi}\Big(\sum_{i=2}^{p} \beta_{i-1} \frac{1}{\Gamma(\phi)} \int_{\eta_{i-1}}^{\eta_i} \int_0^s (s - \tau)^{\phi-1} g(s, x(s), y(s)) d\tau ds \\ & + \sum_{j=1}^{q} \gamma_j \frac{1}{\Gamma(\phi)} \int_0^{\rho_j} (\rho_j - s)^{\phi-1} g(s, x(s), y(s)) ds \\ & - \frac{1}{\Gamma(\sigma)} \int_0^1 \int_0^t (t - s)^{\sigma-1} f(s, x(s), y(s)) ds dt\Big) \\ & + \Lambda_1\Big(\sum_{i=2}^{\mu} \beta'_{i-1} \frac{1}{\Gamma(\sigma)} \int_{\theta_{i-1}}^{\theta_i} \int_0^s (s - \tau)^{\sigma-1} f(\tau, x(\tau), y(\tau)) d\tau ds \\ & + \sum_{j=1}^{\lambda} \gamma'_j \frac{1}{\Gamma(\sigma)} \int_0^{\zeta_j} (\zeta_j - s)^{\sigma-1} f(s, x(s), y(s)) ds \\ & - \frac{1}{\Gamma(\phi)} \int_0^1 \int_0^t (t - s)^{\phi-1} g(s, x(s), y(s)) ds dt\Big)\Bigg\}, \end{aligned}$$

$$T_{1,2}(x,y)(t) = \frac{1}{\Gamma(\sigma)}\int_0^t (t-s)^{\sigma-1} f(s,x(s),y(s))ds,$$

and

$$\begin{aligned}
T_{2,1}(x,y)(t) = {} & \frac{t^{\phi-1}}{\Omega}\left\{\frac{1}{\sigma}\Big(\sum_{i=2}^{\mu}\beta'_{i-1}\frac{1}{\Gamma(\sigma)}\int_{\theta_{i-1}}^{\theta_i}\int_0^s (s-\tau)^{\sigma-1} f(\tau,x(\tau),y(\tau))d\tau ds\right.\\
& +\sum_{j=1}^{\lambda}\gamma'_j\frac{1}{\Gamma(\sigma)}\int_0^{\zeta_j}(\zeta_j-s)^{\sigma-1} f(s,x(s),y(s))ds\\
& -\frac{1}{\Gamma(\phi)}\int_0^1\int_0^t (t-s)^{\phi-1} g(s,x(s),y(s))dsdt\Big)\\
& +\Lambda_2\Big(\sum_{i=2}^{p}\beta_{i-1}\frac{1}{\Gamma(\phi)}\int_{\eta_{i-1}}^{\eta_i}\int_0^s (s-\tau)^{\phi-1} g(\tau,x(\tau),y(\tau))d\tau ds\\
& +\sum_{j=1}^{q}\gamma_j\frac{1}{\Gamma(\phi)}\int_0^{\rho_j}(\rho_j-s)^{\phi-1} g(s,x(s),y(s))ds\\
& \left.-\frac{1}{\Gamma(\sigma)}\int_0^1\int_0^t (t-s)^{\sigma-1} f(s,x(s),y(s))dsdt\Big)\right\},\\
T_{2,2}(x,y)(t) = {} & \frac{1}{\Gamma(\phi)}\int_0^t (t-s)^{\phi-1} g(s,x(s),y(s))ds.
\end{aligned}$$

Note that $T_1(x,y)(t) = T_{1,1}(x,y)(t) + T_{1,2}(x,y)(t)$ and $T_2(x,y)(t) = T_{2,1}(x,y)(t) + T_{2,2}(x,y)(t)$. Also observe that the ball B_δ is a closed, bounded and convex subset of the Banach space $X \times X$. Choose $\delta \geq \max\{Q_1L_1 + Q_2L_2, Q_3L1, Q_4L_2\}$. Now, we will show that $TB_\delta \subset B_\delta$ for satisfying condition (a) of Theorem 1.4. Setting $x = (x_1, x_2), y = (y_1, y_2) \in B_\delta$, and using condition (5.56), we have

$$\begin{aligned}
& |T_{1,1}(x,y)(t) + T_{1,2}(x,y)(t)|\\
\leq {} & \frac{1}{\Gamma(\sigma)}\int_0^t (t-s)^{\sigma-1}L_1 ds + \frac{1}{|\Omega|}\left\{\frac{1}{\phi}\Big(\sum_{i=2}^{p}|\beta_{i-1}|\frac{1}{\Gamma(\phi)}\int_{\eta_{i-1}}^{\eta_i}\int_0^s (s-\tau)^{\phi-1}L_2 d\tau ds\right.\\
& +\sum_{j=1}^{q}|\gamma_j|\frac{1}{\Gamma(\phi)}\int_0^{\rho_j}(\rho_j-s)^{\phi-1}L_2 ds + \frac{1}{\Gamma(\sigma)}\int_0^1\int_0^t (t-s)^{\sigma-1}L_1 dsdt\Big)\\
& +|\Lambda_1|\Big(\sum_{i=2}^{\mu}|\beta'_{i-1}|\frac{1}{\Gamma(\sigma)}\int_{\theta_{i-1}}^{\theta_i}\int_0^s (s-\tau)^{\sigma-1}L_1 d\tau ds\\
& \left.+\sum_{j=1}^{\lambda}|\gamma'_j|\frac{1}{\Gamma(\sigma)}\int_0^{\zeta_j}(\zeta_j-s)^{\sigma-1}L_1 ds + \frac{1}{\Gamma(\phi)}\int_0^1\int_0^t (t-s)^{\phi-1}L_2 dsdt\Big)\right\}\\
= {} & Q_1L_1 + Q_2L_2 \leq \delta.
\end{aligned}$$

In a similar manner, it can be found that

$$|T_{2,1}(x,y)(t) + T_{2,2}(x,y)(t)| \leq Q_3L_1 + Q_4L_2 \leq \delta.$$

From the foregoing arguments, we deduce that $T_1(x,y) + T_2(x,y) \in B_\delta$.

Next we show that the operator $(T_{1,2}, T_{2,2})$ is a contraction mapping satisfying condition (c) of Theorem 1.4. For $(x_1, y_1), (x_2, y_2) \in B_\delta$, we have

$$\begin{aligned}
&|T_{1,2}(x_1, y_1)(t) - T_{1,2}(x_2, y_2)(t)\\
&= \frac{1}{\Gamma(\sigma)} \int_0^t (t-s)^{\sigma-1} |f(s, x_1(s), y_1(s)) - f(s, x_2(s), y_2(s))| ds\\
&\leq \frac{\ell_1}{\Gamma(\sigma+1)} (\|x_1 - y_1\| + \|x_2 - y_2\|),
\end{aligned} \tag{5.58}$$

and

$$\begin{aligned}
&|T_{2,2}(x_1, y_1)(t) - T_{2,2}(x_2, y_2)(t)\\
&= \frac{1}{\Gamma(\phi)} \int_0^t (t-s)^{\phi-1} |g(s, x_1(s), y_1(s)) - g(s, x_2(s), y_2(s))| ds\\
&\leq \frac{\ell_2}{\Gamma(\phi+1)} (\|x_1 - y_1\| + \|x_2 - y_2\|).
\end{aligned} \tag{5.59}$$

It follows from (5.58) and (5.59) that

$$\begin{aligned}
&\|(T_{1,2}, T_{2,2})(x_1, y_1) - (T_{1,2}, T_{2,2})(x_2, y_2)\|\\
&\leq \left[\frac{\ell_1}{\Gamma(\sigma+1)} + \frac{\ell_2}{\Gamma(\phi+1)}\right] (\|x_1 - y_1\| + \|x_2 - y_2\|),
\end{aligned}$$

which is a contraction by (5.57). Therefore, the condition (c) of Theorem 1.4 is satisfied.

Next we will show that the operator $(T_{1,1}, T_{2,1})$ satisfies the condition (b) of Theorem 1.4. By applying the continuity of the functions f, g on $[0,1] \times \mathbb{R} \times \mathbb{R}$, we can conclude that the operator $(T_{1,1}, T_{2,1})$ is continuous. For each $(x, y) \in B_\delta$, we have

$$\begin{aligned}
&|T_{1,1}(x, y)(t)|\\
&\leq \frac{1}{|\Omega|} \Bigg\{ \frac{1}{\phi} \Bigg(\sum_{i=2}^{p} |\beta_{i-1}| \frac{1}{\Gamma(\phi)} \int_{\eta_{i-1}}^{\eta_i} \int_0^s (s-\tau)^{\phi-1} L_2 d\tau ds\\
&\quad + \sum_{j=1}^{q} |\gamma_j| \frac{1}{\Gamma(\phi)} \int_0^{\rho_j} (\rho_j - s)^{\phi-1} L_2 ds + \frac{1}{\Gamma(\sigma)} \int_0^1 \int_0^t (t-s)^{\sigma-1} L_1 ds dt \Bigg)\\
&\quad + |\Lambda_1| \Bigg(\sum_{i=2}^{\mu} |\beta'_{i-1}| \frac{1}{\Gamma(\sigma)} \int_{\theta_{i-1}}^{\theta_i} \int_0^s (s-\tau)^{\sigma-1} L_1 d\tau ds\\
&\quad + \sum_{j=1}^{\lambda} |\gamma'_j| \frac{1}{\Gamma(\sigma)} \int_0^{\zeta_j} (\zeta_j - s)^{\sigma-1} L_1 ds + \frac{1}{\Gamma(\phi)} \int_0^1 \int_0^t (t-s)^{\phi-1} L_2 ds dt \Bigg) \Bigg\}\\
&= P^*,
\end{aligned}$$

and

$$|T_{2,1}(x, y)(t)|$$

$$
\begin{aligned}
&\le \frac{t^{\phi-1}}{|\Omega|}\bigg\{\frac{1}{\sigma}\Big(\sum_{i=2}^{\mu}\beta'_{i-1}\frac{1}{\Gamma(\sigma)}\int_{\theta_{i-1}}^{\theta_i}\int_0^s(s-\tau)^{\sigma-1}L_1d\tau ds\\
&\quad+\sum_{j=1}^{\lambda}\gamma'_j\frac{1}{\Gamma(\sigma)}\int_0^{\zeta_j}(\zeta_j-s)^{\sigma-1}L_1ds+\frac{1}{\Gamma(\phi)}\int_0^1\int_0^t(t-s)^{\phi-1}L_2dsdt\Big)\\
&\quad+\Lambda_2\Big(\sum_{i=2}^{p}\beta_{i-1}\frac{1}{\Gamma(\phi)}\int_{\eta_{i-1}}^{\eta_i}\int_0^s(s-\tau)^{\phi-1}L_2d\tau ds\\
&\quad+\sum_{j=1}^{q}\gamma_j\frac{1}{\Gamma(\phi)}\int_0^{\rho_j}(\rho_j-s)^{\phi-1}L_2ds+\frac{1}{\Gamma(\sigma)}\int_0^1\int_0^t(t-s)^{\sigma-1}L_1dsdt\Big)\bigg\}\\
&= Q^*.
\end{aligned}
$$

Thus we obtain

$$\|(T_{1,1},T_{2,1})(x,y)\| \le P^*+Q^*,$$

which implies that the set $(T_{1,1},T_{2,1})B_\delta$ is uniformly bounded. In the next step, we will show that the set $(T_{1,1},T_{2,1})B_\delta$ is equi-continuous. For $t_1,t_2\in[0,1]$ such that $t_1<t_2$ and for any $(x,y)\in B_\delta$, we have

$$
\begin{aligned}
&|T_{1,1}(x,y)(t_2)-T_{1,1}(x,y)(t_1)|\\
&\le \frac{|t_2^{\sigma-1}-t_1^{\sigma-1}|}{|\Omega|}\Big\{\Big[\frac{1}{\phi}\frac{1}{\Gamma(\sigma+2)}+|\Lambda_1|\Big(\sum_{i=2}^{\mu}|\beta'_{i-1}|\frac{\theta_i^{\sigma+1}-\theta_{i-1}^{\sigma+1}}{\Gamma(\sigma+2)}\Big)\Big]L_1\\
&\quad+\Big[\frac{1}{\phi}\Big(\sum_{i=2}^{p}|\beta_{i-1}|\frac{\eta_i^{\phi+1}-\eta_{i-1}^{\phi+1}}{\Gamma(\phi+2)}+\sum_{j=1}^{q}|\gamma_j|\frac{\rho_j^{\phi}}{\Gamma(\phi+1)}\Big)+|\Lambda_1|\frac{1}{\Gamma(\phi+2)}\Big]L_2\Big\}\\
&\le +\frac{|t_2^{\sigma-1}-t_1^{\sigma-1}|}{|\Omega|}\Big\{\Big[\frac{1}{\phi}\frac{1}{\Gamma(\sigma+2)}+|\Lambda_1|\Big(\sum_{i=2}^{\mu}|\beta'_{i-1}|\frac{\theta_i^{\sigma+1}-\theta_{i-1}^{\sigma+1}}{\Gamma(\sigma+2)}\Big)\Big]L_1\\
&\quad+\Big[\frac{1}{\phi}\Big(\sum_{i=2}^{p}|\beta_{i-1}|\frac{\eta_i^{\phi+1}-\eta_{i-1}^{\phi+1}}{\Gamma(\phi+2)}+\sum_{j=1}^{q}|\gamma_j|\frac{\rho_j^{\phi}}{\Gamma(\phi+1)}\Big)+|\Lambda_1|\frac{1}{\Gamma(\phi+2)}\Big]L_2\Big\}.
\end{aligned}
$$

Analogously, we can obtain

$$
\begin{aligned}
&|T_{2,1}(x,y)(t_2)-T_{2,1}(x,y)(t_1)|\\
&\le \frac{|t_2^{\phi-1}-t_1^{\phi-1}|}{|\Omega|}\Big\{\Big[\frac{1}{\sigma}\Big(\sum_{i=2}^{\mu}|\beta'_{i-1}|\frac{\theta_i^{\sigma+1}-\theta_{i-1}^{\sigma+1}}{\Gamma(\sigma+2)}+\sum_{j=1}^{\lambda}|\gamma'_j|\frac{\zeta_j^{\sigma}}{\Gamma(\sigma+1)}\Big)+|\Lambda_2|\frac{1}{\Gamma(\sigma+2)}\Big]L_1\\
&+\Big[\frac{1}{\sigma}\frac{1}{\Gamma(\phi+2)}+|\Lambda_2|\Big(\sum_{i=2}^{p}|\beta_{i-1}|\frac{\eta_i^{\phi+1}-\eta_{i-1}^{\phi+1}}{\Gamma(\phi+2)}+\sum_{j=1}^{q}|\gamma_j|\frac{\rho_j^{\phi}}{\Gamma(\phi+1)}\Big)\Big]L_2\Big\}.
\end{aligned}
$$

Thus $|(T_{1,1},T_{2,1})(x,y)(t_2)-(T_{1,1},T_{2,1})(x,y)(t_1)|$ tends to zero as $t_1\to t_2$. Therefore the set $(T_{1,1},T_{2,1})B_\delta$ is equi-continuous. By applying the Arzelá-Ascoli theorem, the operator $(T_{1,1},T_{2,1})$ is compact on B_δ. By application of Theorem 1.4, we have that the problem (5.39)-(5.40) has at least one solution on $[0,1]$. This completes the proof. □

Example 5.10. Consider the following fractional boundary value problem

$$\begin{cases} {}^CD^{3/2}x(t)=\dfrac{1}{4(t+2)^2}\dfrac{|x(t)|}{1+|x(t)|}+1+\dfrac{1}{32}\sin^2 y(t)+\dfrac{1}{\sqrt{t^2+1}}, & t\in[0,1], \\ {}^CD^{5/2}y(t)=\dfrac{1}{32\pi}\sin(2\pi x(t))+\dfrac{|y(t)|}{16(1+|y(t)|)}+\dfrac{1}{2}, \quad t\in[0,1], \\ \displaystyle\int_0^1 x(s)ds=\frac{5}{2}\int_{1/7}^{1/6} y(s)ds+4\int_{1/6}^{1/5} y(s)ds+\frac{3}{2}y\Big(\frac{1}{4}\Big)+2y\Big(\frac{1}{3}\Big), \\ \displaystyle\int_0^1 y(s)ds=2\int_{1/3}^{1/2} x(s)ds+4\int_{1/2}^{2/3} x(s)ds+x\Big(\frac{1}{3}\Big)+\frac{1}{2}x\Big(\frac{2}{3}\Big). \end{cases} \tag{5.60}$$

Here $\sigma = 3/2, \phi = 3/2, \beta_1 = 5/2, \beta_2 = 4, \eta_1 = 1/7, \eta_2 = 1/6, \eta_3 = 1/5, \gamma_1 = 3/2, \gamma_2 = 2, \rho_1 = 1/4, \rho_2 = 1/3, \beta_1' = 2, \beta_2' = 4, \gamma_1' = 1, \gamma_2' = 1/2, \theta_1 = 1/3, \theta_2 = 1/2, \theta_3 = 2/3, \zeta_1 = 1/3, \zeta_2 = 2/3,$ $f(t,x,y) = \dfrac{1}{4(t+2)^2}\dfrac{|x|}{1+|x|}+1+\dfrac{1}{32}\sin^2 y,$ and $g(t,x,y) = \dfrac{1}{32\pi}\sin(2\pi x)+\dfrac{|y|}{16(1+|y|)}+\dfrac{1}{2}$. With the given data, we find that $Q_1 \approx 1.618712$, $Q_2 \approx 0.096199$, $Q_3 \approx 1.276613$, $Q_4 \approx 0.199029$. Note that $|f(t,x_1,x_2)-f(t,y_1,y_2)| \le \dfrac{1}{16}|x_1-x_2|+\dfrac{1}{16}|y_1-y_2|$, $|g(t,x_1,x_2)-g(t,y_1,y_2)| \le \dfrac{1}{16}|x_1-x_2|+\dfrac{1}{16}|y_1-y_2|$, and $(Q_1+Q_3)\ell_1+(Q_2+Q_4)\ell_2 \approx 0.217861 < 1$. Thus all the conditions of Theorem 5.12 are satisfied and consequently, its conclusion applies to the problem (5.60).

5.4 Fractional differential equations involving mixed nonlinearities with nonlocal multi-point and Riemann-Stieltjes integral-multi-strip conditions

In this section, we introduce and study a new boundary value problem of fractional differential equations involving mixed nonlinearities, and nonlocal multi-point and Riemann-Stieltjes integral-multi-strip boundary conditions. Precisely we consider the following problem:

$${}^CD^p[{}^CD^q x(t)+f(t,x(t))] = g(t,x(t)), \quad 0<t<1, \quad 0<p,q\le 1, \tag{5.61}$$

$$x(0)=\sum_{j=1}^m \beta_j x(\sigma_j), \quad bx(1)=a\int_0^1 x(s)dH(s)+\sum_{i=1}^n \alpha_i\int_{\xi_i}^{\eta_i} x(s)ds, \tag{5.62}$$

where ${}^CD^r$ denotes the Caputo fractional derivative operator of order r $(r=p,q)$, f and g are given continuous functions, $0<\sigma_j<\xi_i<\eta_i<1, a,b\in\mathbb{R}, \alpha_i,\beta_j\in\mathbb{R}, i = 1,2,\ldots,n,$ $j=1,2,\ldots,m$ and $H(.)$ is a function of bounded variation. Notice that (5.61) is the neutral fractional differential equation.

Definition 5.1. A function $x \in C^2[0,1]$ satisfying the problem (5.61)-(5.62) is called its solution on $[0,1]$.

Associated with the linear variant of problem (5.61)-(5.62), we consider the following lemma.

Lemma 5.4. *Let $h, k \in C([0,1], \mathbb{R})$, the unique solution of the linear fractional differential equation*

$$^{C}D^{p}(^{C}D^{q}x(t) + h(t)) = k(t), \quad 0 < p, q \leq 1, \tag{5.63}$$

supplemented with the boundary conditions (5.62) is given by

$$\begin{aligned} x(t) = &-\int_0^t \frac{(t-s)^{q-1}}{\Gamma(q)} h(s)ds + \int_0^t \frac{(t-s)^{q+p-1}}{\Gamma(q+p)} k(s)ds \\ &+\lambda_1(t)\Bigg[b\int_0^1 \left(\frac{(1-s)^{q-1}}{\Gamma(q)} h(s) - \frac{(1-s)^{q+p-1}}{\Gamma(q+p)} k(s)\right) ds \\ &+a\int_0^1 \left(\int_0^s \left(\frac{(s-u)^{q+p-1}}{\Gamma(q+p)} k(u) - \frac{(s-u)^{q-1}}{\Gamma(q)} h(u)\right) du\right) dH(s) \\ &+\sum_{i=1}^n \alpha_i \int_{\xi_i}^{\eta_i} \left(\int_0^s \left(\frac{(s-u)^{q+p-1}}{\Gamma(q+p)} k(u) - \frac{(s-u)^{q-1}}{\Gamma(q)} h(u)\right) du\right) ds\Bigg] \\ &-\lambda_2(t)\sum_{j=1}^m \beta_j \left(\int_0^{\sigma_j} \frac{(\sigma_j - s)^{q-1}}{\Gamma(q)} h(s)ds - \int_0^{\sigma_j} \frac{(\sigma_j - s)^{q+p-1}}{\Gamma(q+p)} k(s)ds\right), \end{aligned} \tag{5.64}$$

where

$$\lambda_1(t) = \frac{1}{\kappa}\left(\rho_1 - \frac{\rho_2 t^q}{\Gamma(q+1)}\right), \ \lambda_2(t) = \frac{1}{\kappa}\left(\rho_3 - \frac{\rho_4 t^q}{\Gamma(q+1)}\right),$$

$$\begin{aligned} \rho_1 &= \sum_{j=1}^m \beta_j \frac{\sigma_j^q}{\Gamma(q+1)}, \quad \rho_2 = \sum_{j=1}^m \beta_j - 1, \\ \rho_3 &= \frac{1}{\Gamma(q+1)}\left(b - a\int_0^1 s^q dH(s) - \sum_{i=1}^n \alpha_i \frac{(\eta_i^{q+1} - \xi_i^{q+1})}{q+1}\right), \\ \rho_4 &= b - a\int_0^1 dH(s) - \sum_{i=1}^n \alpha_i(\eta_i - \xi_i), \end{aligned} \tag{5.65}$$

and it is assumed that

$$\kappa = \rho_2\rho_3 - \rho_1\rho_4 \neq 0. \tag{5.66}$$

Proof. Applying the integral operator I^p on (5.63), and then I^q on the resulting equation together with Lemma 2.22 in [108], we get

$$x(t) = -I^q h(t) + I^{q+p}k(t) + c_0 \frac{t^q}{\Gamma(q+1)} + c_1, \tag{5.67}$$

where c_0, c_1 are arbitrary constants. Using the boundary condition (5.62) in (5.67), we obtain

$$\sum_{j=1}^m \beta_j \frac{\sigma_j^q}{\Gamma(q+1)} c_0 + \left(\sum_{j=1}^m \beta_j - 1\right)c_1 = \sum_{j=1}^m \beta_j I^q h(\sigma_j) - \sum_{j=1}^m \beta_j I^{q+p}k(\sigma_j), \tag{5.68}$$

$$\frac{1}{\Gamma(q+1)}\Big(b-a\int_0^1 s^q dHs-\sum_{i=1}^n \alpha_i\frac{(\eta_i^{q+1}-\xi_i^{q+1})}{q+1}\Big)c_0+\Big(b-a\int_0^1 dHs-\sum_{i=1}^n \alpha_i(\eta_i-\xi_i)\Big)c_1$$
$$= bI^q h(1) - bI^{q+p}k(1) + a\int_0^1 \Big(-I^q h(s) + I^{q+p}k(s)\Big)dH(s)$$
$$+\sum_{i=1}^n \alpha_i \int_{\xi_i}^{\eta_i} \Big(I^{q+p}k(s) - I^q h(s)\Big)ds.$$

For the sake of convenience, we use the notation (5.65) in (5.68) and (5.69) to find the following system of equations

$$\begin{cases} \rho_1 c_0 + \rho_2 c_1 = \rho_5, \\ \rho_3 c_0 + \rho_4 c_1 = \rho_6, \end{cases} \tag{5.69}$$

where

$$\rho_5 = \sum_{j=1}^m \beta_j I^q h(\sigma_j) - \sum_{j=1}^m \beta_j I^{q+p}k(\sigma_j),$$
$$\rho_6 = bI^q h(1) - bI^{q+p}k(1) + a\int_0^1 \left(-I^q h(s) + I^{q+p}k(s)\right) dH(s)$$
$$+\sum_{i=1}^n \alpha_i \int_{\xi_i}^{\eta_i} \left(I^{q+p}k(s) - I^q h(s)\right) ds.$$

Solving the system (5.69) for c_0 and c_1, we get

$$c_0 = (\rho_2\rho_6 - \rho_5\rho_4)/\kappa, \ c_1 = (\rho_3\rho_5 - \rho_1\rho_6)/\kappa,$$

where κ is given by (5.66). Substituting the values of c_0, and c_1 in (5.67) together with the notation (5.65), we get the solution (5.64). By direct computation, one can obtain the converse of the lemma. This completes the proof. □

In view of Lemma 5.4, we transform the problem (5.61)-(5.62) into a fixed point problem as $x = \mathcal{G}x$, where the operator $\mathcal{G} : C([0,1],\mathbb{R}) \to C([0,1],\mathbb{R})$ is defined by

$$\begin{aligned}
&\mathcal{G}x(t)\\
=& -\int_0^t \frac{(t-s)^{q-1}}{\Gamma(q)} f(s,x(s))ds + \int_0^t \frac{(t-s)^{q+p-1}}{\Gamma(q+p)} g(s,x(s))ds\\
&+\lambda_1(t)\Bigg[b\int_0^1 \left(\frac{(1-s)^{q-1}}{\Gamma(q)} f(s,x(s)) - \frac{(1-s)^{q+p-1}}{\Gamma(q+p)} g(s,x(s))\right) ds\\
&+a\int_0^1\int_0^s \left(\frac{(s-u)^{q+p-1}}{\Gamma(q+p)} g(u,x(u)) - \frac{(s-u)^{q-1}}{\Gamma(q)} f(u,x(u))\right) du\, dH(s)\\
&+\sum_{i=1}^n \alpha_i \int_{\xi_i}^{\eta_i}\int_0^s \left(\frac{(s-u)^{q+p-1}}{\Gamma(q+p)} g(u,x(u)) - \frac{(s-u)^{q-1}}{\Gamma(q)} f(u,x(u))\right) du\, ds\Bigg]\\
&-\lambda_2(t)\sum_{j=1}^m \beta_j \left(\int_0^{\sigma_j} \frac{(\sigma_j-s)^{q-1}}{\Gamma(q)} f(s,x(s))ds - \int_0^{\sigma_j} \frac{(\sigma_j-s)^{q+p-1}}{\Gamma(q+p)} g(s,x(s))ds\right).
\end{aligned} \tag{5.70}$$

Note that $C([0,1],\mathbb{R})$ denotes the Banach space of all continuous functions $x : [0,1] \to \mathbb{R}$ endowed with the norm defined by $\|x\| = \sup\{|x(t)|, t \in [0,1]\}$.
For the sake of computational convenience, we set

$$\begin{aligned}\Lambda = &\frac{1}{\Gamma(q+1)} + \frac{1}{\Gamma(q+p+1)} + \overline{\lambda}_1 \Bigg[\frac{|b|}{\Gamma(q+1)} + \frac{|b|}{\Gamma(q+p+1)} \\ &+|a| \int_0^1 \left(\frac{s^q}{\Gamma(q+1)} + \frac{s^{q+p}}{\Gamma(q+p+1)} \right) dH(s) \\ &+ \sum_{i=1}^{n-2} |\alpha_i| \left(\frac{(\eta_i^{q+1} - \xi_i^{q+1})}{\Gamma(q+2)} + \frac{(\eta_i^{q+p+1} - \xi_i^{q+p+1})}{\Gamma(q+p+2)} \right) \Bigg] \\ &+ \overline{\lambda}_2 \sum_{j=1}^{m} |\beta_j| \left(\frac{\sigma_j^q}{\Gamma(q+1)} + \frac{\sigma_j^{q+p}}{\Gamma(q+p+1)} \right), \end{aligned} \tag{5.71}$$

and

$$\Lambda_1 = \Lambda - \frac{1}{\Gamma(q+1)} - \frac{1}{\Gamma(q+p+1)}, \tag{5.72}$$

where

$$\overline{\lambda}_1 = \max_{t\in[0,1]} |\lambda_1(t)| = \frac{1}{|\kappa|}(|\rho_2| + |\rho_1|), \ \overline{\lambda}_2 = \max_{t\in[0,1]} |\lambda_2(t)| = \frac{1}{|\kappa|}(|\rho_4| + |\rho_3|).$$

Now we present the existence and uniqueness results for the boundary value problem (5.61)-(5.62).

5.4.1 *Existence result via Leray-Schauder nonlinear alternative*

Theorem 5.15. *Let $f, g : [0,1] \times \mathbb{R} \to \mathbb{R}$ be continuous functions. Assume that:*

(5.15.1) There exist functions $p_1, p_2 \in C([0,1], \mathbb{R}^+)$, with $p = \max\{p_1, p_2\}$ and non-decreasing functions $\psi_1, \psi_2 : \mathbb{R}^+ \to \mathbb{R}^+$, $\psi = \max\{\psi_1, \psi_2\}$ such that $|f(t,x)| \le p_1\psi_1(\|x\|)$ and $|g(t,x)| \le p_2\psi_2(\|x\|)$, for all $(t,x) \in [0,1] \times \mathbb{R}$;

(5.15.2) There exists a constant $M > 0$ such that

$$\frac{M}{\|p\|\psi(M)\Lambda} > 1.$$

Then the boundary value problem (5.61)-(5.62) has at least one solution on $[0,1]$.

Proof. Let us first show that the operator $\mathcal{G} : C([0,1],\mathbb{R}) \to C([0,1],\mathbb{R})$ defined by (5.70) maps bounded sets into bounded sets in $C([0,1],\mathbb{R})$. For a positive number r, let $B_r = \{x \in C([0,1],\mathbb{R}) : \|x\| \le r\}$ be a bounded set in $C([0,1],\mathbb{R})$. Then, in view of the assumption (5.15.1), we have

$$\begin{aligned}&|(\mathcal{G}x)(t)| \\ &\le \|p_1\|\psi_1(\|x\|) \sup_{t\in[0,1]} \Bigg\{ \int_0^t \frac{(t-s)^{q-1}}{\Gamma(q)} ds + |\lambda_1(t)| \Bigg[|b| \int_0^1 \frac{(1-s)^{q-1}}{\Gamma(q)} ds\end{aligned}$$

$$
\begin{aligned}
&+|a|\int_0^1\left(\int_0^s\frac{(s-u)^{q-1}}{\Gamma(q)}du\right)dH(s)+\sum_{i=1}^n|\alpha_i|\int_{\xi_i}^{\eta_i}\Big(\int_0^s\frac{(s-u)^{q-1}}{\Gamma(q)}du\Big)ds\Bigg]\\
&+|\lambda_2(t)|\sum_{j=1}^m|\beta_j|\int_0^{\sigma_j}\frac{(\sigma_j-s)^{q-1}}{\Gamma(q)}ds\Bigg\}\\
&+\|p_2\|\psi_2(\|x\|)\sup_{t\in[0,1]}\Bigg\{\int_0^t\frac{(t-s)^{q+p-1}}{\Gamma(q+p)}ds+|\lambda_1(t)|\Bigg[|b|\int_0^1\frac{(1-s)^{q+p-1}}{\Gamma(q+p)}ds\\
&+|a|\int_0^1\left(\int_0^s\frac{(s-u)^{q+p-1}}{\Gamma(q+p)}du\right)dH(s)+\sum_{i=1}^n|\alpha_i|\int_{\xi_i}^{\eta_i}\Big(\Big(\frac{(s-u)^{q+p-1}}{\Gamma(q+p)}\Big)du\Big)ds\Bigg]\\
&+|\lambda_2(t)|\sum_{j=1}^m|\beta_j|\int_0^{\sigma_j}\frac{(\sigma_j-s)^{q+p-1}}{\Gamma(q+p)}ds\Bigg\}\\
\le\;&\|p_1\|\psi_1(\|x\|)\sup_{t\in[0,1]}\Bigg\{\frac{t^q}{\Gamma(q+1)}+|\lambda_1(t)|\Bigg[\frac{|b|}{\Gamma(q+1)}+|a|\int_0^1\frac{s^q}{\Gamma(q+1)}dH(s)\\
&+\sum_{i=1}^n|\alpha_i|\Big(\frac{\eta_i^{q+1}-\xi_i^{q+1}}{\Gamma(q+2)}\Big)\Bigg]+|\lambda_2(t)|\sum_{j=1}^m|\beta_j|\frac{\sigma_j^q}{\Gamma(q+1)}\Bigg\}\\
&+\|p_2\|\psi_2(\|x\|)\sup_{t\in[0,1]}\Bigg\{\frac{t^{q+p}}{\Gamma(q+p+1)}+|\lambda_1(t)|\Bigg[\frac{b}{\Gamma(q+p+1)}\\
&+|a|\int_0^1\frac{s^{q+p}}{\Gamma(q+p+1)}dH(s)+\sum_{i=1}^n|\alpha_i|\Big(\frac{\eta_i^{q+p+1}-\xi_i^{q+p+1}}{\Gamma(q+p+2)}\Big)\Bigg]\\
&+|\lambda_2(t)|\sum_{j=1}^m|\beta_j|\frac{\sigma_j^{q+p}}{\Gamma(q+p+1)}\Bigg\}\\
\le\;&\|p\|\psi(\|r\|)\Bigg\{\frac{1}{\Gamma(q+1)}+\frac{1}{\Gamma(q+p+1)}+\overline{\lambda}_1\Bigg[|b|\left(\frac{1}{\Gamma(q+1)}+\frac{1}{\Gamma(q+p+1)}\right)\\
&+|a|\int_0^1\left(\frac{s^{q+p}}{\Gamma(q+p+1)}+\frac{s^q}{\Gamma(q+1)}\right)dH(s)\\
&+\sum_{i=1}^n|\alpha_i|\Big(\frac{\eta_i^{q+1}-\xi_i^{q+1}}{\Gamma(q+2)}+\frac{\eta_i^{q+p+1}-\xi_i^{q+p+1}}{\Gamma(q+p+2)}\Big)\Bigg]\\
&+\overline{\lambda}_2\sum_{j=1}^m|\beta_j|\Big(\frac{\sigma_j^q}{\Gamma(q+1)}+\frac{\sigma_j^{q+p}}{\Gamma(q+p+1)}\Big)\Bigg\}.
\end{aligned}
$$

Consequently, using the notation (5.71), we have

$$\|\mathcal{G}x\|\le\|p\|\psi(\|r\|)\Lambda.$$

Next we show that $\mathcal{G}$ maps bounded sets into equi-continuous sets of $C([0,1],\mathbb{R})$. Let $\tau_1, \tau_2 \in [0,1]$ with $\tau_1 < \tau_2$ and $x \in B_r$, where B_r is a bounded set of $C([0,1],\mathbb{R})$. Then we obtain

$$\begin{aligned}
&|\mathcal{G}x(\tau_2) - \mathcal{G}x(\tau_1)| \\
\le& \Big|\int_0^{\tau_1} \frac{(\tau_1 - s)^{q-1} - (\tau_2 - s)^{q-1}}{\Gamma(q)} |f(s,x(s))| ds\Big| \\
&+ \Big|\int_{\tau_1}^{\tau_2} \frac{(\tau_2 - s)^{q-1}}{\Gamma(q)} |f(s,x(s))| ds\Big| + \Big|\int_{\tau_1}^{\tau_2} \frac{(\tau_2 - s)^{q+p-1}}{\Gamma(q+p)} |g(s,x(s))| ds\Big| \\
&+ \Big|\int_0^{\tau_1} \frac{(\tau_2 - s)^{q+p-1} - (\tau_1 - s)^{q+p-1}}{\Gamma(q+p)} |g(s,x(s))| ds\Big| \\
&+ |\lambda_1(\tau_2) - \lambda_1(\tau_1)| \Bigg[|b| \int_0^1 \left(\frac{(1-s)^{q-1}}{\Gamma(q)} |f(s,x(s))| + \frac{(1-s)^{q+p-1}}{\Gamma(q+p)} |g(s,x(s))| \right) ds \\
&+ |a| \int_0^1 \left(\int_0^s \left(\frac{(s-u)^{q+p-1}}{\Gamma(q+p)} |g(u,x(u))| + \frac{(s-u)^{q-1}}{\Gamma(q)} |f(u,x(u))| \right) du \right) dH(s) \\
&+ \sum_{i=1}^n |\alpha_i| \int_{\xi_i}^{\eta_i} \left(\int_0^s \left(\frac{(s-u)^{q-1}}{\Gamma(q)} |f(u,x(u))| + \frac{(s-u)^{q+p-1}}{\Gamma(q+p)} |g(u,x(u))| \right) du \right) ds \Bigg] \\
&+ |\lambda_2(\tau_2) - \lambda_2(\tau_1)| \sum_{j=1}^m |\beta_j| \int_0^{\sigma_j} \Big(\frac{(\sigma_j - s)^{q-1}}{\Gamma(q)} |f(s,x(s))| \\
&+ \frac{(\sigma_j - s)^{q+p-1}}{\Gamma(q+p)} |g(s,x(s))| \Big) ds \\
\le& \|p\| \psi(\|x\|) \Bigg\{ \frac{|\tau_1^q - \tau_2^q| + 2(\tau_2 - \tau_1)^q}{\Gamma(q+1)} + \frac{|\tau_1^{q+p} - \tau_2^{q+p}| + 2(\tau_2 - \tau_1)^{q+p}}{\Gamma(q+p+1)} \\
&+ \Big|\frac{\rho_2(\tau_2^q - \tau_1^q)}{\kappa \Gamma(q+1)}\Big| \Bigg[|b| \left(\frac{1}{\Gamma(q+1)} + \frac{1}{\Gamma(q+p+1)} \right) \\
&+ |a| \int_0^1 \left(\frac{s^{q+p}}{\Gamma(q+p+1)} + \frac{s^q}{\Gamma(q+1)} \right) dH(s) \\
&+ \sum_{i=1}^n |\alpha_i| \Big(\frac{\eta_i^{q+1} - \xi_i^{q+1}}{\Gamma(q+2)} + \frac{\eta_i^{q+p+1} - \xi_i^{q+p+1}}{\Gamma(q+p+2)} \Big) \Bigg] \\
&+ \Big|\frac{\rho_4(\tau_2^q - \tau_1^q)}{\kappa \Gamma(q+1)}\Big| \sum_{j=1}^m |\beta_j| \Big(\frac{\sigma_j^q}{\Gamma(q)} + \frac{\sigma_j^{q+p}}{\Gamma(q+p+1)} \Big) \Bigg\}.
\end{aligned}$$

Obviously the right hand side of the above inequality tends to zero independently of $x \in B_r$ as $\tau_2 - \tau_1 \to 0$. As $\mathcal{G}$ satisfies the above assumptions, therefore it follows by the Arzelá-Ascoli theorem that $\mathcal{G} : C([0,1],\mathbb{R}) \to C([0,1],\mathbb{R})$ is completely continuous.

The conclusion of the Leray-Schauder nonlinear alternative (Theorem 1.8) will apply once we establish the boundedness of the set of all solutions to the equation

$x = \varepsilon \mathcal{G} x$, for $\varepsilon \in (0,1)$. Let x be a solution of (5.61)-(5.62). Then, following the computation used in proving the boundedness of $\mathcal{G}$, we get

$$\frac{\|x\|}{\|p\|\psi(\|x\|)\Lambda} \leq 1.$$

By the condition (5.15.2), we can find a positive number M such that $\|x\| \neq M$. Let us define a set $Y = \{x \in C([0,1],\mathbb{R}) : \|x\| < M\}$ and note that the operator $\mathcal{G} : \overline{Y} \to C([0,1],\mathbb{R})$ is continuous and completely continuous. From the choice of Y, there is no $x \in \partial Y$ such that $x = \varepsilon \mathcal{G}(x)$ for some $\varepsilon \in (0,1)$. In consequence, we deduce by the nonlinear alternative of Leray-Schauder type (Theorem 1.8) that the operator $\mathcal{G}$ has a fixed point $x \in \overline{Y}$ which is a solution of the problem (5.61)-(5.62). The proof is completed. □

5.4.2 *Existence result via Krasnosel'skiĭ's fixed point theorem*

Theorem 5.16. *Let $f, g : [0,1] \times \mathbb{R} \to \mathbb{R}$ be continuous functions satisfying the conditions:*

(5.16.1) $|f(t,x) - f(t,y)| \leq L_1|x-y|$, *and* $|g(t,x) - g(t,y)| \leq L_2|x-y|$ *for all* $t \in [0,1]$, $L > 0$, $x, y \in \mathbb{R}$, *with* $L < 1/\Lambda_1$, *where* Λ_1 *is given by (5.72), and* $L = \max\{L_1, L_2\}$;

(5.16.2) $|f(t,x)| \leq \mu_1(t)$, $|g(t,x)| \leq \mu_2(t)$, *for all* $(t,x) \in [0,1] \times \mathbb{R}$, $\mu_1, \mu_2 \in C([0,1],\mathbb{R}^+)$ *and* $\mu = \max\{\mu_1, \mu_2\}$.

Then the boundary value problem (5.61)-(5.62) has at least one solution on $[0,1]$.

Proof. By the assumption (5.16.2) and (5.71), we fix $\overline{r} \geq \Lambda\|\mu\|$ and consider the closed ball $B_{\overline{r}} = \{x \in C : \|x\| \leq \overline{r}\}$. Next we define operators $\mathcal{G}_1$ and $\mathcal{G}_2$ on $B_{\overline{r}}$ as follows

$$(\mathcal{G}_1 x)(t) = -\int_0^t \frac{(t-s)^{q-1}}{\Gamma(q)} f(s,x(s))ds + \int_0^t \frac{(t-s)^{q+p-1}}{\Gamma(q+p)} g(s,x(s))ds, \quad t \in [0,1],$$

$$\begin{aligned}
&(\mathcal{G}_2 x)(t) \\
&= \lambda_1(t)\Bigg[b\int_0^1 \left(\frac{(1-s)^{q-1}}{\Gamma(q)} f(s,x(s)) - \frac{(1-s)^{q+p-1}}{\Gamma(q+p)} g(s,x(s))\right) ds \\
&\quad + a\int_0^1 \int_0^s \left(\frac{(s-u)^{q+p-1}}{\Gamma(q+p)} g(u,x(u)) - \frac{(s-u)^{q-1}}{\Gamma(q)} f(u,x(u))\right) du\, dH(s) \\
&\quad + \sum_{i=1}^n \alpha_i \int_{\xi_i}^{\eta_i} \int_0^s \left(\frac{(s-u)^{q+p-1}}{\Gamma(q+p)} g(u,x(u)) - \frac{(s-u)^{q-1}}{\Gamma(q)} f(u,x(u))\right) du\, ds\Bigg] \\
&\quad - \lambda_2(t)\sum_{j=1}^m \beta_j \Bigg(\int_0^{\sigma_j} \frac{(\sigma_j-s)^{q-1}}{\Gamma(q)} f(s,x(s))ds \\
&\quad - \int_0^{\sigma_j} \frac{(\sigma_j-s)^{q+p-1}}{\Gamma(q+p)} g(s,x(s))ds\Bigg), \quad t \in [0,1].
\end{aligned}$$

For $x, y \in B_{\overline{r}}$, we find that

$$\begin{aligned}
\|\mathcal{G}_1 x + \mathcal{G}_2 y\| &= \sup_{t\in[0,1]} |\mathcal{G}_1 x + \mathcal{G}_2 y| \\
&\le \|\mu\| \Bigg\{ \frac{1}{\Gamma(q+1)} + \frac{1}{\Gamma(q+p+1)} + \overline{\lambda}_1 \Bigg[\frac{|b|}{\Gamma(q+1)} + \frac{|b|}{\Gamma(q+p+1)} \\
&\quad + |a| \int_0^1 \left(\frac{s^q}{\Gamma(q+1)} + \frac{s^{q+p}}{\Gamma(q+p+1)} \right) dH(s) \\
&\quad + \sum_{i=1}^{n-2} |\alpha_i| \left(\frac{(\eta_i^{q+1} - \xi_i^{q+1})}{\Gamma(q+2)} + \frac{(\eta_i^{q+p+1} - \xi_i^{q+p+1})}{\Gamma(q+p+2)} \right) \Bigg] \\
&\quad + \overline{\lambda}_2 \sum_{j=1}^{m} |\beta_j| \left(\frac{\sigma_j^q}{\Gamma(q+1)} + \frac{\sigma_j^{q+p}}{\Gamma(q+p+1)} \right) \Bigg\} \\
&= \|\mu\| \Lambda \le \overline{r}.
\end{aligned}$$

This shows that $\mathcal{G}_1 x + \mathcal{G}_2 y \in B_{\overline{r}}$. Next we establish that $\mathcal{G}_2$ is a contraction mapping. For $x, y \in C([0,1], \mathbb{R})$ and for each $t \in [0,1]$, we obtain

$$\begin{aligned}
\|\mathcal{G}_2 x - \mathcal{G}_2 y\| &\le \sup_{t\in[0,1]} \Bigg\{ |\lambda_1(t)| \Big[b \int_0^1 \frac{(1-s)^{q-1}}{\Gamma(q)} |f(s,x(s)) - f(s,y(s))| ds \\
&\quad + |b| \int_0^1 \frac{(1-s)^{q+p-1}}{\Gamma(q+p)} |g(s,x(s)) - g(s,y(s))| ds \\
&\quad + |a| \int_0^1 \left(\int_0^s \frac{(s-u)^{q+p-1}}{\Gamma(q+p)} |g(u,x(u)) - g(u,y(u))| du \right) dH(s) \\
&\quad + |a| \int_0^1 \left(\int_0^s \frac{(s-u)^{q-1}}{\Gamma(q)} |f(u,x(u)) - f(u,y(u))| du \right) dH(s) \\
&\quad + \sum_{i=1}^{n} |\alpha_i| \int_{\xi_i}^{\eta_i} \Big(\int_0^s \Big(\frac{(s-u)^{q+p-1}}{\Gamma(q+p)} |g(u,x(u)) - g(u,y(u))| \\
&\quad + \frac{(s-u)^{q-1}}{\Gamma(q)} |f(u,x(u)) - f(u,y(u))| \Big) du \Big) ds \Big] \\
&\quad + |\lambda_2(t)| \sum_{j=1}^{m} |\beta_j| \Big(\int_0^{\sigma_j} \frac{(\sigma_j - s)^{q-1}}{\Gamma(q)} |f(s,x(s)) - f(s,y(s))| ds \\
&\quad + \int_0^{\sigma_j} \frac{(\sigma_j - s)^{q+p-1}}{\Gamma(q+p)} |g(s,x(s)) - g(s,y(s))| \Big) ds \Bigg\} \\
&\le L \Bigg\{ \overline{\lambda}_1 \Big[\frac{|b|}{\Gamma(q+1)} + \frac{|b|}{\Gamma(q+p+1)} \\
&\quad + |a| \int_0^1 \left(\frac{s^{q+p}}{\Gamma(q+p+1)} + \frac{s^q}{\Gamma(q+1)} \right) dH(s) \\
&\quad + \sum_{i=1}^{n} |\alpha_i| \Big(\frac{\eta_i^{q+p+1} - \xi_i^{q+p+1}}{\Gamma(q+p+2)} + \frac{\eta_i^{q+1} - \xi_i^{q+1}}{\Gamma(q+2)} \Big) \Big]
\end{aligned}$$

$$+\overline{\lambda}_2 \sum_{j=1}^{m} |\beta_j| \Big(\frac{\sigma_j^q}{\Gamma(q+1)} + \frac{\sigma_j^{q+p}}{\Gamma(q+p+1)} \Big) \Big\} \|x-y\|$$
$$\leq L\Lambda_1 \|x-y\|,$$

which is a contraction mapping by assumption $L\Lambda_1 < 1$ (Λ_1 is given by (5.72)).

Continuity of f, g implies that the operator $\mathcal{G}_1$ is continuous. Also, $\mathcal{G}_1$ is uniformly bounded on $B_{\overline{r}}$ as

$$\|\mathcal{G}_1 x\| \leq \sup_{t\in[0,1]} \left\{ \int_0^t \frac{(t-s)^{q-1}}{\Gamma(q)} |f(s,x(s))| ds + \int_0^t \frac{(t-s)^{q+p-1}}{\Gamma(q+p)} |g(s,x(s))| ds \right\}$$
$$\leq \|\mu\| \left(\frac{1}{\Gamma(q+1)} + \frac{1}{\Gamma(q+p+1)} \right).$$

Now we prove the compactness of the operator $\mathcal{G}_1$. In view of (5.16.1), we define

$$\sup_{(t,x)\in[0,1]\times B_{\overline{r}}} |f(t,x)| = \overline{f}, \qquad \sup_{(t,x)\in[0,1]\times B_{\overline{r}}} |g(t,x)| = \overline{g}.$$

Consequently, for $0 \leq t_2 < t_1 \leq 1$, we have

$$|\mathcal{G}_1 x(t_1) - \mathcal{G}_1 x(t_2)|$$
$$\leq \frac{\overline{f}}{\Gamma(q+1)} [|t_2^q - t_1^q| + 2(t_1-t_2)^q] + \frac{\overline{g}}{\Gamma(q+p+1)} [|t_2^{q+p} - t_1^{q+p}| + 2(t_1-t_2)^{q+p}],$$

which tends to zero independent of x as $t_1 - t_2 \to 0$. Thus, $\mathcal{G}_1$ is relatively compact on $B_{\overline{r}}$. Hence, by the Arzelá-Ascoli theorem, $\mathcal{G}_1$ is compact on $B_{\overline{r}}$. Thus the hypotheses of Theorem 1.4 are satisfied. Hence we deduce by the conclusion of Theorem 1.4 that the problem (5.61)-(5.62) has at least one solution on $[0,1]$. □

5.4.3 *Existence and uniqueness result via Banach's fixed point theorem*

Theorem 5.17. *Assume that $f, g : [0,1] \times \mathbb{R} \to \mathbb{R}$ are continuous functions satisfying the assumption (5.16.1). Then the problem (5.61)-(5.62) has a unique solution on $[0,1]$ if $L\Lambda < 1$, where Λ is given by (5.71).*

Proof. Define $M = \max\{M_1, M_2\}$, where M_1 and M_2 are positive numbers such that $\sup_{t\in[0,1]} |f(t,0)| = M_1$ and $\sup_{t\in[0,1]} |g(t,0)| = M_2$. Fixing $r \geq \dfrac{M\Lambda}{1-L\Lambda}$, we consider $B_r = \{x \in C : \|x\| \leq r\}$. Then, in view of the assumption (5.16.1), we have $|f(t,x)| = |f(t,x) - f(t,0) + f(t,0)| \leq |f(t,x) - f(t,0)| + |f(t,0)| \leq L_1\|x\| + M_1 \leq L_1 r + M_1$. Similarly one can obtain that $|g(t,x)| \leq L_2 r + M_2$. In the first step, we show that $\mathcal{G}B_r \subset B_r$. For any $x \in B_r$, we have

$$\|\mathcal{G}x\| = \sup_{t\in[0,1]} |\mathcal{G}x(t)|$$
$$\leq (Lr + M) \Bigg\{ \frac{1}{\Gamma(q+1)} + \frac{1}{\Gamma(q+p+1)} + \overline{\lambda}_1 \Bigg[\frac{|b|}{\Gamma(q+1)} + \frac{|b|}{\Gamma(q+p+1)}$$

$$
\begin{aligned}
&+|a|\int_0^1\left(\frac{s^q}{\Gamma(q+1)}+\frac{s^{q+p}}{\Gamma(q+p+1)}\right)dH(s)\\
&+\sum_{i=1}^n|\alpha_i|\left(\frac{(\eta_i^{q+1}-\xi_i^{q+1})}{\Gamma(q+2)}+\frac{(\eta_i^{q+p+1}-\xi_i^{q+p+1})}{\Gamma(q+p+2)}\right)\Bigg]\\
&+\overline{\lambda}_2\sum_{j=1}^m|\beta_j|\left(\frac{\sigma_j^q}{\Gamma(q+1)}+\frac{\sigma_j^{q+p}}{\Gamma(q+p+1)}\right)\Bigg\}\\
&=(Lr+M)\Lambda\le r,
\end{aligned}
$$

which implies that $\mathcal{G}B_r\subset B_r$. Next, for $x,y\in C([0,1],\mathbb{R})$ and for each $t\in[0,1]$, we obtain

$$
\begin{aligned}
\|\mathcal{G}x-\mathcal{G}y\|\le \sup_{t\in[0,1]}&\Bigg\{\int_0^t\frac{(t-s)^{q-1}}{\Gamma(q)}|f(s,x(s))-f(s,y(s))|ds\\
&+\int_0^t\frac{(t-s)^{q+p-1}}{\Gamma(q+p)}|g(s,x(s))-g(s,y(s))|ds\\
&+|\lambda_1(t)|\Bigg[|b|\int_0^1\Big(\frac{(1-s)^{q-1}}{\Gamma(q)}|f(s,x(s))-f(s,y(s))|\\
&+\frac{(1-s)^{q+p-1}}{\Gamma(q+p)}|g(s,x(s))-g(s,y(s))|\Big)ds\\
&+|a|\int_0^1\Big(\int_0^s\frac{(s-u)^{q+p-1}}{\Gamma(q+p)}|g(u,x(u))-g(u,y(u))|\\
&+\frac{(s-u)^{q-1}}{\Gamma(q)}|f(u,x(u))-f(u,y(u))|du\Big)dH(s)\\
&+\sum_{i=1}^n|\alpha_i|\int_{\xi_i}^{\eta_i}\Big(\int_0^s\Big(\frac{(s-u)^{q+p-1}}{\Gamma(q+p)}|g(u,x(u))-g(u,y(u))|\\
&+\frac{(s-u)^{q-1}}{\Gamma(q)}|f(u,x(u))-f(u,y(u))|\Big)du\Big)ds\Bigg]\\
&+|\lambda_2(t)|\sum_{j=1}^m|\beta_j|\Big(\int_0^{\sigma_j}\frac{(\sigma_j-s)^{q+p-1}}{\Gamma(q+p)}|g(s,x(s))-g(s,y(s))|\\
&+\frac{(\sigma_j-s)^{q-1}}{\Gamma(q)}|f(s,x(s))-f(s,y(s))|ds\Big)\Bigg\}\\
&\le L\Lambda\|x-y\|.
\end{aligned}
$$

From the above inequality together with the given condition $L\Lambda<1$ (Λ is given by (5.71)), it follows that the operator $\mathcal{G}$ is a contraction by means of the contraction mapping principle (Banach fixed point theorem). Therefore, there exists a unique solution for the problem (5.61)-(5.62) on $[0,1]$. □

Remark 5.1. In Theorem 5.16, we proved the existence of solutions for the problem (5.61)-(5.62) under the assumption that $L\Lambda_1<1$ by applying Krasnosel'skiĭ's fixed point theorem, which is a hybrid fixed point theorem combining two well known theorems (algebraic (Banach) and one topological (Schauder)). It gives a fixed point for the sum of two operators; one of them is a contraction while the other one is completely continuous. It is well known that the application of Krasnosel'skiĭ fixed point theorem only provides an existence result as only a part of the associated operator is shown to be a contraction. In other words, the entire operator is not a contraction. Indeed, Theorem 5.17 is an existence-uniqueness result obtained by applying Banach contraction mapping principle under the condition $L\Lambda < 1$. Moreover, $L\Lambda < 1$ implies that $L\Lambda_1 < 1$, where $\Lambda = \Lambda_1 + \frac{1}{\Gamma(q+1)} + \frac{1}{\Gamma(q+p+1)}$. This means that an increase of $\frac{1}{\Gamma(q+1)} + \frac{1}{\Gamma(q+p+1)}$ to the value of Λ_1 will lead to the condition $L\Lambda < 1$, ensuring the uniqueness of solutions. This provides the relationship between the contractive conditions imposed in Theorems 5.16 and 5.17. On the other hand, interchanging the roles of the operators $\mathcal{G}_1$ and $\mathcal{G}_2$ in the proof of Theorem 5.16, the difference of the values of Λ and Λ_1 is

$$\Lambda - \Lambda_1 = \overline{\lambda}_1 \Bigg[\frac{|b|}{\Gamma(q+1)} + \frac{|b|}{\Gamma(q+p+1)} + |a| \int_0^1 \left(\frac{s^q}{\Gamma(q+1)} + \frac{s^{q+p}}{\Gamma(q+p+1)} \right) dH(s)$$
$$+ \sum_{i=1}^{n-2} |\alpha_i| \left(\frac{(\eta_i^{q+1} - \xi_i^{q+1})}{\Gamma(q+2)} + \frac{(\eta_i^{q+p+1} - \xi_i^{q+p+1})}{\Gamma(q+p+2)} \right) \Bigg]$$
$$+ \overline{\lambda}_2 \sum_{j=1}^{m} |\beta_j| \left(\frac{\sigma_j^q}{\Gamma(q+1)} + \frac{\sigma_j^{q+p}}{\Gamma(q+p+1)} \right).$$

Thus, Theorem 5.16 provides a precise estimate to extend the contractive condition required for the existence of solutions to the one needed to ensure the uniqueness of solutions in Theorem 5.17 for the problem at hand.

Example 5.11. Consider the following boundary value problem:

$$\begin{aligned} &{}^C D^{1/2}({}^C D^{1/2} x(t) + f(t, x(t))) = g(t, x(t)), \quad t \in [0, 1], \\ &x(0) = \sum_{j=1}^{3} \beta_j x(\sigma_j), \quad x(1) = \int_0^1 x(s) dH(s) + \sum_{i=1}^{3} \alpha_i \int_{\xi_i}^{\eta_i} x(s) ds, \end{aligned} \tag{5.73}$$

where $p = q = 1/2$, $a = 1, b = 1, H(s) = s, \sigma_1 = 1/32, \sigma_2 = 1/26, \sigma_3 = 1/16, \xi_1 = 1/7, \xi_2 = 3/7, \xi_3 = 5/7, \eta_1 = 2/7, \eta_2 = 4/7, \eta_3 = 6/7, \alpha_1 = 1/12, \alpha_2 = 1/6, \alpha_3 = 1/4, \beta_1 = 1/15, \beta_2 = 1/10, \beta_3 = 1/5$, and $f(t, x)$ and $g(t, x)$ will be a fixed later. Using the given values, we find that $\Lambda \approx 15.433601$ and $\Lambda_1 \approx 13.3052219$ (Λ and Λ_1 are respectively given by (5.71) and (5.72)).

In order to illustrate Theorem 5.15, we take

$$f(t, x) = \frac{e^{-t}}{16\pi} \tan^{-1} x + \frac{1}{t^2 + 32}, \; g(t, x) = \frac{e^{-2t}}{\sqrt{289 + t^2}} \left(\frac{|x|}{2(1 + |x|)} + \sin x + \frac{1}{2} \right). \tag{5.74}$$

Clearly $|f(t, x)| \leq [e^{-t}/32 + 1/(t^2 + 32)]$, $|g(t, x)| \leq e^{-2t}(1 + \|x\|)/\sqrt{289 + t^2}$ with $p_1(t) = e^{-t}/32 + 1/(t^2 + 32)$ ($\|p_1\| = 1/16$), $\psi_1(\|x\|) = 1$, $p_2(t) = e^{-2t}/\sqrt{289 + t^2}$

$(\|p_2\| = 1/17)$, $\psi_2(\|x\|) = 1+\|x\|$, $p = \max\{1/16, 1/17\} = 1/16$ and $\psi = \max\{1, 1+\|x\|\} = 1 + \|x\|$. From the assumption (5.15.2), we find that $M > 27.248637$. As all the conditions of Theorem 5.15 are satisfied, there exists at least one solution on $[0, 1]$ for the problem (5.73) with $f(t, x)$ and $g(t, x)$ given by (5.74).

Next we explain Theorem 5.16 by choosing the following functions in the problem (5.73):

$$f(t,x) = \frac{1}{16}\sin x + e^{-t}\cos t, \;\; g(t,x) = \frac{1}{18}\Big(\frac{|x|}{1+|x|}\Big) + 6t. \tag{5.75}$$

Notice that $L_1 = 1/16, L_2 = 1/18$ as

$$|f(t,x) - f(t,y)| \le \frac{1}{16}|x-y|, \; |g(t,x) - g(t,y)| \le \frac{1}{18}|x-y|.$$

Moreover

$$|f(t,x)| \le \frac{1}{16}|\sin x| + e^{-t}|\cos t| \le \frac{1}{16} + e^{-t}\cos t = \mu_1(t), |g(t,x)| \le \frac{1}{18} + 6t = \mu_2(t).$$

Obviously $\|\mu_1\| = 17/16, \|\mu_2\| = 109/18$, $L = \max\{1/16, 1/18\} = 1/16$, $\mu = \max\{17/16,\, 109/18\} = 109/18$, $L\Lambda_1 \approx 0.831576 < 1$ and $L\Lambda \approx 0.9646 < 1$. Clearly all the assumptions of Theorem 5.16 are satisfied. Hence, by the conclusions of Theorem 5.16, we deduce that there exists at least one solution for the problem (5.73) on $[0, 1]$ with $f(t, x(t))$ and $g(t, x(t))$ given by (5.75).

Finally one can notice that the problem (5.73) with $f(t, x(t))$ and $g(t, x(t))$ given by (5.75) has a unique solution on $[0, 1]$ as the hypothesis of Theorem 5.17 holds true.

Remark 5.2. Several new results for the fractional differential equation with mixed nonlinearities (5.61) subject to different boundary conditions follow as special cases by fixing the parameters in (5.62). For example, our results correspond to (i) Dirichlet boundary conditions if we take $\beta_j = 0, \forall j = 1, \ldots, m, a = 0, b \neq 0, \alpha_i = 0, \forall i = 1, \ldots, n$; (ii) multi-point Riemann-Stieltjes integral boundary conditions if we take $\alpha_i = 0, \forall i = 1, \ldots, n$; (iii) multi-point and multi-strip conditions if we take $a = 0$; etc.

5.4.4 *Analogue problems*

In this section, we discuss variants of the problem (5.61)-(5.62). As a first problem we consider

$$\begin{aligned} &{}^C D^{p+q}x(t) + {}^C D^p f(t, x(t))) = g(t, x(t)), \;\; 0 < t < 1, \;\; 0 < p, q \le 1, \\ &x(0) = \sum_{j=1}^{m} \beta_j x(\sigma_j), \; a\int_0^1 x(s)ds - \sum_{i=1}^{n} \alpha_i \int_{\xi_i}^{\eta_i} x(s)ds = \delta, \end{aligned} \tag{5.76}$$

$0 < \sigma_j < \xi_i < \eta_i < 1, i = 1, 2, \ldots, n,\ a, \delta \in \mathbb{R}$.

As argued for the problem (5.61)-(5.62), we can transform the problem (5.76) into a fixed point problem with associated operator $\mathcal{W} : C([0, 1], \mathbb{R}) \to C([0, 1], \mathbb{R})$ defined by

$$
\begin{aligned}
&(\mathcal{W}x)(t)\\
&= -\int_0^t \frac{(t-s)^{q-1}}{\Gamma(q)} f(s,x(s))ds + \int_0^t \frac{(t-s)^{q+p-1}}{\Gamma(q+p)} g(s,x(s))ds\\
&\quad -\omega_1(t)\Bigg[a\int_0^1 \Big(\int_0^s \frac{(s-u)^{q-1}}{\Gamma(q)} f(u,x(u))du\Big)ds\\
&\quad -a\int_0^1 \Big(\int_0^s \frac{(s-u)^{q+p-1}}{\Gamma(q+p)} g(u,x(u))du\Big)ds\\
&\quad -\sum_{i=1}^n \alpha_i \int_{\xi_i}^{\eta_i} \Big(\int_0^s \Big(\frac{(s-u)^{q-1}}{\Gamma(q)} f(u,x(u)) - \frac{(s-u)^{q+p-1}}{\Gamma(q+p)} g(u,x(u))\Big)du\Big)ds + \delta\Bigg]\\
&\quad +\omega_2(t)\sum_{j=1}^m \beta_j\Big(\int_0^{\sigma_j} \frac{(\sigma_j-s)^{q-1}}{\Gamma(q)} f(s,x(s))ds - \int_0^{\sigma_j} \frac{(\sigma_j-s)^{q+p-1}}{\Gamma(q+p)} g(s,x(s))ds\Big),
\end{aligned}
$$

where

$$
\omega_1(t) = \frac{1}{\widehat{\kappa}\Gamma(q+1)}\Big(\sum_{j=1}^m \beta_j\sigma_j^q - \sum_{j=1}^m (\beta_j-1)t^q\Big),
$$

$$
\omega_2(t) = \frac{1}{\widehat{\kappa}\Gamma(q+2)}\Big(\Big(a - \sum_{i=1}^n \alpha_i(\eta_i^{q+1} - \xi_i^{q+1})\Big) - (q+1)(a - \sum_{i=1}^n \alpha_i(\eta_i - \xi_i))t^q\Big),
$$

and

$$
\begin{aligned}
\widehat{\kappa} = \frac{1}{\Gamma(q+2)}\Big[&\Big(\sum_{j=1}^m \beta_j - 1\Big)\Big(a - \sum_{i=1}^n \alpha_i(\eta_i^{q+1} - \xi_i^{q+1})\Big)\\
&-(q+1)\sum_{j=1}^m \beta_j\sigma_j^q\Big(a - \sum_{i=1}^n \alpha_i(\eta_i - \xi_i)\Big)\Big] \neq 0.
\end{aligned}
$$

In relation to the problem (5.76), we define

$$
\begin{aligned}
\widehat{\Lambda} = \frac{1}{\Gamma(q+1)} + \frac{1}{\Gamma(q+p+1)} &+ \widehat{\omega}_1\Big[\frac{|a|}{\Gamma(q+2)} + \frac{|a|}{\Gamma(q+p+2)}\\
&+\sum_{i=1}^{n-2} |\alpha_i|\Big(\frac{(\eta_i^{q+1} - \xi_i^{q+1})}{\Gamma(q+2)} + \frac{(\eta_i^{q+p+1} - \xi_i^{q+p+1})}{\Gamma(q+p+2)}\Big) + |\delta|\Big]\\
&+\widehat{\omega}_2\sum_{j=1}^m |\beta_j|\Big(\frac{\sigma_j^q}{\Gamma(q+1)} + \frac{\sigma_j^{q+p}}{\Gamma(q+p+1)}\Big),
\end{aligned}
$$

$$
\widehat{\Lambda}_1 = \widehat{\Lambda} - \frac{1}{\Gamma(q+1)} - \frac{1}{\Gamma(q+p+1)},
$$

where

$$
\widehat{\omega}_1 = \max_{t\in[0,1]} |\omega_1(t)| = \frac{1}{|\widehat{\kappa}|}\Big[\Big|\sum_{j=1}^m \beta_j - 1\Big| + \Big|\sum_{j=1}^m \beta_j \frac{\sigma_j^q}{\Gamma(q+1)}\Big|\Big],
$$

$$
\widehat{\omega}_2 = \max_{t\in[0,1]} |\omega_2(t)| = \frac{1}{|\widehat{\kappa}|}\Big[\Big|a - \sum_{i=1}^n \alpha_i(\eta_i - \xi_i)\Big|
$$

$$+\Big|\frac{1}{\Gamma(q+2)}\Big(a-\sum_{i=1}^{n}\alpha_i(\eta_i^{q+1}-\xi_i^{q+1})\Big)\Big|\Big].$$

The existence and uniqueness results for the problem (5.76), similar to the ones for the problem (5.61)-(5.62) presented in Section 3, can be obtained in a similar manner.

As a second variant of the problem (5.61)-(5.62), we consider

$$\begin{aligned}&{}^{C}D^{p+q}x(t)+{}^{C}D^{p}f(t,x(t))=g(t,x(t)),\quad 0<t<1,\quad 0<p,q\le 1,\\ &x(0)=\sum_{j=1}^{p}\beta_j x(\sigma_j),\quad x(1)=\sum_{i=1}^{n-2}\alpha_i\int_{\xi_i}^{\eta_i}x(s)ds,\ 0<\sigma_j<\xi_i<\eta_i<1. \end{aligned}\tag{5.77}$$

In relation to the problem (5.77), the fixed point operator $\mathcal{V}: C([0,1],\mathbb{R})\to C([0,1],\mathbb{R})$ is defined by

$$\begin{aligned}&(\mathcal{V}x)(t)\\ &=-\int_0^t\frac{(t-s)^{q-1}}{\Gamma(q)}f(s,x(s))ds+\int_0^t\frac{(t-s)^{q+p-1}}{\Gamma(q+p)}g(s,x(s))ds\\ &\quad-\nu_1(t)\Bigg[\int_0^1\frac{(1-s)^{q-1}}{\Gamma(q)}f(s,x(s))ds-\int_0^1\frac{(1-s)^{q+p-1}}{\Gamma(q+p)}g(s,x(s))ds\\ &\quad-\sum_{i=1}^{n}\alpha_i\int_{\xi_i}^{\eta_i}\Big(\int_0^s\Big(\frac{(s-u)^{q-1}}{\Gamma(q)}f(u,x(u))-\frac{(s-u)^{q+p-1}}{\Gamma(q+p)}g(u,x(u))\Big)du\Big)ds\Bigg]\\ &\quad+\nu_2(t)\Bigg[\sum_{j=1}^{m}\beta_j\Big(\int_0^{\sigma_j}\frac{(\sigma_j-s)^{q-1}}{\Gamma(q)}f(s,x(s))ds-\int_0^{\sigma_j}\frac{(\sigma_j-s)^{q+p-1}}{\Gamma(q+p)}g(s,x(s))ds\Big)\Bigg],\end{aligned}\tag{5.78}$$

where

$$\nu_1(t)=\frac{1}{\Omega\Gamma(q+1)}\Big[\sum_{j=1}^{m}\beta_j\sigma_j^q-\sum_{j=1}^{m}(\beta_j-1)t^q\Big],$$

$$\nu_2(t)=\frac{1}{\Omega\Gamma(q+2)}\Big[\Big(q+1-\sum_{i=1}^{n}\alpha_i(\eta_i^{q+1}-\xi_i^{q+1})\Big)-(q+1)(1-\sum_{i=1}^{n}\alpha_i(\eta_i-\xi_i))t^q\Big],$$

$$\begin{aligned}\Omega=&\frac{1}{\Gamma(q+2)}\Big[\Big(\sum_{j=1}^{m}\beta_j-1\Big)\Big(q+1-\sum_{i=1}^{n}\alpha_i(\eta_i^{q+1}-\xi_i^{q+1})\Big)\\ &-(q+1)\sum_{j=1}^{m}\beta_j\sigma_j^q\Big(1-\sum_{i=1}^{n}\alpha_i(\eta_i-\xi_i)\Big)\Big]\neq 0.\end{aligned}$$

Moreover, we set

$$\varrho=\frac{1}{\Gamma(q+1)}+\frac{1}{\Gamma(q+p+1)}+\overline{\nu}_1\Big(\frac{1}{\Gamma(q+1)}+\frac{1}{\Gamma(q+p+1)}$$

$$+\sum_{i=1}^{n-2}|\alpha_i|\Big(\frac{(\eta_i^{q+1}-\xi_i^{q+1})}{\Gamma(q+2)}+\frac{(\eta_i^{q+p+1}-\xi_i^{q+p+1})}{\Gamma(q+p+2)}\Big)\Big)$$

$$+\overline{\nu}_2\sum_{j=1}^{m}|\beta_j|\Big(\frac{\sigma_j^q}{\Gamma(q+1)}+\frac{\sigma_j^{q+p}}{\Gamma(q+p+1)}\Big),\quad \overline{\nu}_i=\max_{t\in[0,1]}|\nu_i(t)|, i=1,2,$$

$$\overline{\varrho}=\varrho-\frac{1}{\Gamma(q+1)}-\frac{1}{\Gamma(q+p+1)}.$$

As in Section 3, we can obtain the existence and uniqueness results for the problem (5.77) with the aid of the operator $\mathcal{V}$ and the parameters-dependent quantities ϱ and $\overline{\varrho}$ (defined above).

5.5 Existence results for nonlocal multi-point and integral boundary value problems of multi-term fractional differential equations

In this section, we develop the existence theory for multi-term fractional differential equations equipped with nonlocal multi-point and strip boundary conditions. Precisely, we investigate the following boundary value problem:

$$(q_2\,{}^CD^{\sigma+2}+q_1\,{}^CD^{\sigma+1}+q_0\,{}^CD^{\sigma})x(t)=f(t,x(t)),\ 0<\sigma<1,\ 0<t<1,\quad (5.79)$$

$$x(0)=h(x),\ x(\xi)=\sum_{i=1}^{n}j_ix(\eta_i),\ x(1)=\lambda\int_0^{\delta}x(s)ds,$$
$$0<\delta<\xi<\eta_1<\eta_2<\ldots<\eta_n<1,\ \lambda\in\mathbb{R},\quad (5.80)$$

where ${}^CD^{\delta}$ denotes the Caputo fractional derivative operator of order σ, $f:[0,1]\times\mathbb{R}\to\mathbb{R}$, $h:C([0,1],\mathbb{R})\to\mathbb{R}$ are given continuous functions, $q_0, q_1, and\ \ q_2$ are real constants with $q_2\neq 0$.

Definition 5.2. A function $x\in C^3[0,1]$ satisfying the problem (5.79)-(5.80) is called its solution on $[0,1]$.

The following lemma associated with the linear variant of problem (5.79)-(5.80) plays an important role in the sequel.

Lemma 5.5. *For any $y\in C([0,1],\mathbb{R})$ and $q_1^2-4q_0q_2>0$, the solution of linear multi-term fractional differential equation*

$$(q_2\,{}^CD^{\sigma+2}+q_1\,{}^CD^{\sigma+1}+q_0\,{}^CD^{\sigma})x(t)=y(t),\ 0<\sigma<1,\ 0<t<1,\quad (5.81)$$

supplemented with the boundary conditions (5.80) is given by

$$x(t)=\frac{1}{q_2(m_2-m_1)}\Big\{\int_0^t\int_0^s\mathcal{A}(t)\frac{(s-u)^{\sigma-1}}{\Gamma(\sigma)}y(u)du\ ds$$

$$+\rho_1(t)\Big[\int_0^{\xi}\int_0^s\mathcal{A}(\xi)\frac{(s-u)^{\sigma-1}}{\Gamma(\sigma)}y(u)du\ ds$$

$$-\sum_{i=1}^{n} j_i \int_0^{\eta_i}\int_0^s \mathcal{A}(\eta_i)\frac{(s-u)^{\sigma-1}}{\Gamma(\sigma)}y(u)du\, ds\Big]$$
$$+\rho_2(t)\Big[\int_0^1\int_0^s \mathcal{A}(1)\frac{(s-u)^{\sigma-1}}{\Gamma(\sigma)}y(u)du\, ds$$
$$-\lambda\int_0^{\delta}\int_0^s\Big(\frac{(e^{m_2(\delta-s)}-1)}{m_2}-\frac{(e^{m_1(\delta-s)}-1)}{m_1}\Big)\frac{(s-u)^{\sigma-1}}{\Gamma(\sigma)}y(u)du\, ds\Big]\Big\}$$
$$+h(x)\Big[e^{m_2t}+\rho_1(t)\Big(e^{m_2\xi}-\sum_{i=1}^{n}j_i+\sum_{i=1}^{n}j_ie^{m_2\eta_i}\Big)$$
$$+\rho_2(t)\Big(\frac{m_2e^{m_2}-\lambda e^{m_2\delta}-\lambda}{m_2}\Big)\Big], \tag{5.82}$$

where

$$\mathcal{A}(\kappa)=e^{m_2(\kappa-s)}-e^{m_1(\kappa-s)},\quad \kappa=t,\,1,\,\xi\quad\text{and}\quad \eta_i,$$
$$m_1=\frac{-q_1-\sqrt{q_1^2-4q_0q_2}}{2q_2},\quad m_2=\frac{-q_1+\sqrt{q_1^2-4q_0q_2}}{2q_2},$$
$$\rho_1(t)=\frac{\omega_4\varrho_1(t)-\omega_3\varrho_2(t)}{\mu_1},\quad \rho_2(t)=\frac{\omega_1\varrho_2(t)-\omega_2\varrho_1(t)}{\mu_1},$$
$$\varrho_1(t)=\frac{m_1(1-e^{m_2t})-m_2(1-e^{m_1t})}{m_1m_2},\quad \varrho_2(t)=q_2(m_2-m_1)(e^{m_2t}-e^{m_1t}),$$
$$\mu_1=\omega_1\omega_4-\omega_2\omega_3\neq 0, \tag{5.83}$$
$$\omega_1=\frac{1}{m_1m_2}\Big[m_2\Big(1-\sum_{i=1}^{n}j_i-e^{m_1\xi}+\sum_{i=1}^{n}j_ie^{m_1\eta_i}\Big)$$
$$-m_1\Big(1-\sum_{i=1}^{n}j_i-e^{m_2\xi}+\sum_{i=1}^{n}j_ie^{m_2\eta_i}\Big)\Big],$$
$$\omega_2=q_2(m_2-m_1)\Big(e^{m_1\xi}-e^{m_2\xi}-\sum_{i=1}^{n}j_ie^{m_1\eta_i}+\sum_{i=1}^{n}j_ie^{m_2\eta_i}\Big),$$
$$\omega_3=\frac{1}{m_1m_2}\Big[m_2\Big(1-e^{m_1}-\lambda\delta+\lambda/m_1(e^{m_1\delta}-1)\Big)$$
$$-m_1\Big(1-e^{m_2}-\lambda\delta+\lambda/m_2(e^{m_2\delta}-1)\Big)\Big],$$
$$\omega_4=q_2(m_2-m_1)\Big((e^{m_1}+\lambda/m_1(1-e^{m_1\delta}))-(e^{m_2}+\lambda/m_2(1-e^{m_2\delta}))\Big).$$

Proof. Applying the operator I^{σ} on (5.81), we get

$$(q_2D^2+q_1D+q_0)x(t)=\int_0^t\frac{(t-s)^{\sigma-1}}{\Gamma(\sigma)}y(s)ds+c_1, \tag{5.84}$$

where c_1 is an arbitrary constant. By the method of variation of parameters, the solution of (5.84) can be written as

$$x(t)=c_1\left[\frac{m_2(1-e^{m_1t})-m_1(1-e^{m_2t})}{q_2m_1m_2(m_2-m_1)}\right]+c_2e^{m_1t}+c_3e^{m_2t}$$

$$-\frac{1}{q_2(m_2-m_1)}\int_0^t e^{m_1(t-s)}\left(\int_0^s \frac{(s-u)^{\sigma-1}}{\Gamma(\sigma)}y(u)du\right)ds \tag{5.85}$$
$$+\frac{1}{q_2(m_2-m_1)}\int_0^t e^{m_2(t-s)}\left(\int_0^s \frac{(s-u)^{\sigma-1}}{\Gamma(\sigma)}y(u)du\right)ds,$$

where m_1 and m_2 are given in (5.83). Using $x(0) = h(x)$ in (5.85), we get

$$\begin{aligned} x(t) &= c_1\left[\frac{m_2(1-e^{m_1t})-m_1(1-e^{m_2t})}{q_2m_1m_2(m_2-m_1)}\right] \\ &+c_2\left(e^{m_1t}-e^{m_2t}\right)+h(x)e^{m_2t} \\ &+\frac{1}{q_2(m_2-m_1)}\left[\int_0^t\left(e^{m_2(t-s)}-e^{m_1(t-s)}\right)\left(\int_0^s\frac{(s-u)^{\sigma-1}}{\Gamma(\sigma)}y(u)du\right)ds\right], \end{aligned} \tag{5.86}$$

which together with the conditions $x(\xi) = \sum_{i=1}^n j_i x(\eta_i)$ and $x(1) = \lambda\int_0^\delta x(s)ds$ yields the following system of equations in the unknown constants c_1 and c_2:

$$c_1\omega_1 + c_2\omega_2 = V_1, \tag{5.87}$$
$$c_1\omega_3 + c_2\omega_4 = V_2, \tag{5.88}$$

where

$$\begin{aligned} V_1 &= -\int_0^\xi\int_0^s \mathcal{A}(\xi)\frac{(s-u)^{\sigma-1}}{\Gamma(\sigma)}y(u)du\,ds \\ &+\sum_{i=1}^n j_i\int_0^{\eta_i}\int_0^s \mathcal{A}(\eta_i)\frac{(s-u)^{\sigma-1}}{\Gamma(\sigma)}y(u)du\,ds + h(x)\Big(\sum_{i=1}^n j_i e^{m_2\eta_i}-e^{m_2\xi}\Big), \\ V_2 &= -\int_0^1\int_0^s \mathcal{A}(1)\frac{(s-u)^{\sigma-1}}{\Gamma(\sigma)}y(u)du\,ds + h(x)\Big(\frac{\lambda e^{m_2\delta}-\lambda-m_2e^{m_2}}{m_2}\Big) \\ &+\lambda\int_0^\delta\int_0^s\Big[\frac{(e^{m_1(\delta-s)}-1)}{m_1}-\frac{(e^{m_2(\delta-s)}-1)}{m_2}\Big]\frac{(s-u)^{\sigma-1}}{\Gamma(\sigma)}y(u)du\,ds. \end{aligned}$$

Solving the system (5.87)-(5.88) together with the notation (5.83), we find that

$$c_1 = \frac{V_1\omega_4 - V_2\omega_2}{\mu_1}, \quad c_2 = \frac{V_2\omega_1 - V_1\omega_3}{\mu_1}.$$

Substituting the value of c_1 and c_2 in (5.86), we obtain the solution (5.82). The converse follows by direct computation. This completes the proof. □

The proofs of the following lemmas are omitted since they are similar to that of Lemma 5.5.

Lemma 5.6. *For any $y \in C([0,1],\mathbb{R})$ and $q_1^2 - 4q_0q_2 = 0$, the solution of linear multi-term fractional differential equation*

$$(q_2\,{}^CD^{\sigma+2} + q_1\,{}^CD^{\sigma+1} + q_0\,{}^CD^{\sigma})x(t) = y(t),\ 0<\sigma<1,\ 0<t<1, \tag{5.89}$$

supplemented with the boundary conditions (5.80) is given by

$$\begin{aligned}
x(t) &= \frac{1}{q_2}\Bigg\{ \int_0^t \int_0^s \mathcal{B}(t)\frac{(s-u)^{\sigma-1}}{\Gamma(\sigma)} y(u)du\ ds \\
&+\chi_1(t)\Big[\int_0^{\xi} \int_0^s \mathcal{B}(\xi)\frac{(s-u)^{\sigma-1}}{\Gamma(\sigma)} y(u)du\ ds \\
&-\sum_{i=1}^{n} j_i \int_0^{\eta_i} \int_0^s \mathcal{B}(\eta_i)\frac{(s-u)^{\sigma-1}}{\Gamma(\sigma)} y(u)du\ ds\Big] \\
&+\chi_2(t)\Big[\int_0^{1} \int_0^s \mathcal{B}(1)\frac{(s-u)^{\sigma-1}}{\Gamma(\sigma)} y(u)du\ ds \\
&-\lambda \int_0^{\delta} \int_0^s \Big(\frac{m(\delta-s)e^{m(\delta-s)} - e^{m(\delta-s)} + 1}{m^2}\Big)\frac{(s-u)^{\sigma-1}}{\Gamma(\sigma)} y(u)du\ ds\Big]\Bigg\} \\
&+h(x)\Big[e^{mt} + \chi_1(t)\Big(e^{m\xi} - \sum_{i=1}^{n} j_i e^{m\eta_i}\Big) + \chi_2(t)\Big(\frac{me^m - \lambda e^{m\delta} + \lambda}{m}\Big)\Big], \qquad (5.90)
\end{aligned}$$

where

$$\begin{aligned}
&\mathcal{B}(\kappa) = (\kappa - s)e^{m(\kappa-s)}, \quad \kappa = t,\, 1,\, \xi \quad \text{and} \quad \eta_i, \\
&m = \frac{-q_1}{2q_2}, \\
&\chi_1(t) = \frac{\varpi_3 v_2(t) - \varpi_4 v_1(t)}{\mu_2}, \quad \chi_2(t) = \frac{\varpi_2 v_1(t) - \varpi_1 v_2(t)}{\mu_2}, \\
&v_1(t) = \frac{mte^{mt} - e^{mt} + 1}{m^2}, \quad v_2(t) = q_2 te^{mt}, \\
&\varpi_1 = \frac{m\xi e^{m\xi} - e^{m\xi} + 1 - \sum_{i=1}^{n} j_i(m\eta_i e^{m\eta_i} - e^{m\eta_i} + 1)}{m^2}, \qquad (5.91) \\
&\varpi_2 = q_2\Big(\xi e^{m\xi} - \sum_{i=1}^{n} j_i \eta_i e^{m\eta_i}\Big), \\
&\varpi_3 = \frac{m^2 e^m - me^m + m - m\lambda\delta e^{m\delta} + 2\lambda e^{m\delta} - 2\lambda - m\lambda\delta}{m^3}, \\
&\varpi_4 = q_2\Big(\frac{m^2 e^m - \lambda m\delta e^{m\delta} + \lambda e^{m\delta} - \lambda}{m^2}\Big), \\
&\mu_2 = \varpi_1\varpi_4 - \varpi_2\varpi_3 \neq 0.
\end{aligned}$$

Lemma 5.7. *For any $y \in C([0,1],\mathbb{R})$ and $q_1^2 - 4q_0q_2 < 0$, the solution of linear multi-term fractional differential equation*

$$(q_2\ {}^CD^{\sigma+2} + q_1\ {}^CD^{\sigma+1} + q_0\ {}^CD^{\sigma})x(t) = y(t), \ \ 0 < \sigma < 1, \ \ 0 < t < 1, \qquad (5.92)$$

supplemented with the boundary conditions (5.80) is given by

$$x(t) = \frac{1}{q_2 b}\Bigg\{ \int_0^t \int_0^s \mathcal{F}(t)\frac{(s-u)^{\sigma-1}}{\Gamma(\sigma)} y(u)du\ ds$$

$$
\begin{aligned}
&+\tau_1(t)\Big[\int_0^{\xi}\int_0^{s}\mathcal{F}(\xi)\frac{(s-u)^{\sigma-1}}{\Gamma(\sigma)}y(u)du\ ds\\
&-\sum_{i=1}^{n}j_i\int_0^{\eta_i}\int_0^{s}\mathcal{F}(\eta_i)\frac{(s-u)^{\sigma-1}}{\Gamma(\sigma)}y(u)du\ ds\Big]\\
&+\tau_2(t)\Big[\int_0^{1}\int_0^{s}\mathcal{F}(1)\frac{(s-u)^{\sigma-1}}{\Gamma(\sigma)}y(u)du\ ds \qquad (5.93)\\
&-\frac{\lambda}{a^2+b^2}\int_0^{\delta}\int_0^{s}\Big(b-be^{-a(\delta-s)}\cos b(\delta-s)\\
&-ae^{-a(\delta-s)}\sin b(\delta-s)\Big)\frac{(s-u)^{\sigma-1}}{\Gamma(\sigma)}y(u)du\ ds\Big]\Big\}\\
&+h(x)\Big[e^{-at}\cos bt+\tau_1(t)\Big(e^{-a\xi}\cos b\xi-\sum_{i=1}^{n}j_ie^{-a\eta_i}\cos b\eta_i\Big)\\
&+\tau_2(t)\Big(e^{-a}\cos b-\frac{\lambda}{a^2+b^2}(a-ae^{-a\delta}\cos b\delta+be^{-a\delta}\sin b\delta)\Big)\Big],
\end{aligned}
$$

where

$$
\begin{aligned}
&\mathcal{F}(\kappa)=e^{-a(\kappa-s)}\sin b(\kappa-s),\ \ \kappa=t,\,1,\,\xi\ \text{ and }\ \eta_i,\\
&m_{1,2}=-a\pm bi,\ \ a=\frac{q_1}{2q_2},\ \ b=\frac{\sqrt{4q_0q_2-q_1{}^2}}{2q_2},\\
&\tau_1(t)=\frac{p_3\nu_2(t)-p_4\nu_1(t)}{\mu_3},\ \ \tau_2(t)=\frac{p_2\nu_1(t)-p_1\nu_2(t)}{\mu_3},\\
&\nu_1(t)=\frac{b-be^{-at}\cos bt-ae^{-at}\sin bt}{a^2+b^2},\ \ \nu_2(t)=q_2be^{-at}\sin bt\\
&p_1=\frac{b-be^{-a\xi}\cos b\xi-ae^{-a\xi}\sin b\xi-\sum_{i=1}^{n}j_i(b-be^{-a\eta_i}\cos b\eta_i-ae^{-a\eta_i}\sin b\eta_i)}{a^2+b^2},\\
&p_2=q_2b\Big(e^{-a\xi}\sin b\xi-\sum_{i=1}^{n}j_ie^{-a\eta_i}\sin b\eta_i\Big), \qquad (5.94)\\
&p_3=\frac{1}{a^2+b^2}\Big[b-be^{-a}\cos b-ae^{-a}\sin b-b\lambda\delta\\
&\quad+\frac{b\lambda}{a^2+b^2}(a-ae^{-a\delta}\cos b\delta+be^{-a\delta}\sin b\delta)\\
&\quad-\frac{a\lambda}{a^2+b^2}(b-be^{-a\delta}\cos b\delta-ae^{-a\delta}\sin b\delta)\Big],\\
&p_4=q_2b\Big[e^{-a}\sin b-\frac{\lambda}{a^2+b^2}(b-be^{-a\delta}\cos b\delta-ae^{-a\delta}\sin b\delta)\Big],\\
&\mu_3=p_1p_4-p_2p_3\neq 0.
\end{aligned}
$$

5.5.1 *Existence and uniqueness results for the case* $q_1^2-4q_0q_2>0$

Denote by $\mathcal{C}=C([0,1],\mathbb{R})$ the Banach space of all continuous functions from $[0,1]$ to $\mathbb{R}$ endowed with the norm defined by $\|x\|=\sup\{|x(t)|:t\in[0,1]\}$. In relation to the problem (5.79)-(5.80) with $q_1^2 - 4q_0q_2 > 0$, we define an operator $\mathcal{J}: \mathcal{C} \to \mathcal{C}$ by Lemma 5.5 as

$$
\begin{aligned}
&(\mathcal{J}x)(t)\\
&= \frac{1}{q_2(m_2-m_1)}\Bigg\{\int_0^t\int_0^s \mathcal{A}(t)\frac{(s-u)^{\sigma-1}}{\Gamma(\sigma)}f(u,x(u))du\, ds\\
&+\rho_1(t)\Bigg[\int_0^\xi\int_0^s \mathcal{A}(\xi)\frac{(s-u)^{\sigma-1}}{\Gamma(\sigma)}f(u,x(u))du\, ds\\
&-\sum_{i=1}^n j_i\int_0^{\eta_i}\int_0^s \mathcal{A}(\eta_i)\frac{(s-u)^{\sigma-1}}{\Gamma(\sigma)}f(u,x(u))du\, ds\Bigg]\\
&+\rho_2(t)\Bigg[\int_0^1\int_0^s \mathcal{A}(1)\frac{(s-u)^{\sigma-1}}{\Gamma(\sigma)}f(u,x(u))du\, ds\\
&-\lambda\int_0^\delta\int_0^s\Big(\frac{(e^{m_2(\delta-s)}-1)}{m_2}-\frac{(e^{m_1(\delta-s)}-1)}{m_1}\Big)\frac{(s-u)^{\sigma-1}}{\Gamma(\sigma)}f(u,x(u))du\, ds\Bigg]\Bigg\}\\
&+h(x)\Bigg[e^{m_2t}+\rho_1(t)\Big(e^{m_2\xi}-\sum_{i=1}^n j_i+\sum_{i=1}^n j_ie^{m_2\eta_i}\Big)\\
&+\rho_2(t)\Big(\frac{m_2e^{m_2}-\lambda e^{m_2\delta}-\lambda}{m_2}\Big)\Bigg],
\end{aligned}
\tag{5.95}
$$

where $\mathcal{A}(\cdot)$, $\rho_1(t)$ and $\rho_2(t)$ are defined by (5.83).

Observe that the problem (5.79)-(5.80) is equivalent to the operator equation

$$x = \mathcal{J}x, \tag{5.96}$$

In the sequel, for the sake of computational convenience, we set

$$
\begin{aligned}
\widehat{\rho}_1 &= \max_{t\in[0,1]}|\rho_1(t)|, \quad \widehat{\rho}_2 = \max_{t\in[0,1]}|\rho_2(t)|,\\
\varepsilon &= \max_{t\in[0,1]}\Big|m_2(1-e^{m_1t})-m_1(1-e^{m_2t})\Big|,\\
\alpha &= \frac{1}{|q_2m_1m_2(m_2-m_1)|\Gamma(\sigma+1)}\Big\{\varepsilon+\widehat{\rho}_1\Big[\xi^\sigma|m_2(1-e^{m_1\xi})-m_1(1-e^{m_2\xi})|\\
&\quad+\sum_{i=1}^n|j_i|\eta_i^\sigma|m_2(1-e^{m_1\eta_i})-m_1(1-e^{m_2\eta_i})|\Big]\\
&\quad+\widehat{\rho}_2\Big[|m_2(1-e^{m_1})-m_1(1-e^{m_2})|\\
&\quad+\frac{\delta^\sigma|\lambda|}{|m_1m_2|}|m_2^2(m_1\delta-e^{m_1\delta}+1)-m_1^2(m_2\delta-e^{m_2\delta}+1)|\Big]\Big\},
\end{aligned}
\tag{5.97}
$$

$$
\begin{aligned}
\alpha_1 &= \alpha-\frac{\varepsilon}{|q_2m_1m_2(m_2-m_1)|\Gamma(\sigma+1)};\\
\Delta_1 &= \max_{t\in[0,1]}|e^{m_2t}|+\widehat{\rho}_1\Big(|e^{m_2\xi}|+\sum_{i=1}^n|j_i||e^{m_2\eta_i}+1|\Big)
\end{aligned}
$$

$$+\widehat{\rho}_2\Big(\frac{|m_2e^{m_2}|+|\lambda||e^{m_2\delta}+1|}{|m_2|}\Big).$$

Now the platform is set to present our main results. In the first result, we use Krasnosel'skiĭ's fixed point theorem to prove the existence of solutions for the problem (5.79)-(5.80).

In the forthcoming analysis, we need the following assumptions:

(G_1) $|f(t,x)-f(t,y)|\le \ell\|x-y\|$, for all $t\in[0,1]$, $x,y\in\mathbb{R}$, $\ell>0$.
(G_2) $|h(x)-h(y)|\le L\|x-y\|$, for all $t\in[0,1]$, $x,y\in\mathbb{R}$, $L>0$.
(G_3) $|f(t,x)|\le \vartheta(t)$, for all $(t,x)\in[0,1]\times\mathbb{R}$ and $\vartheta\in C([0,1],\mathbb{R}^+)$.

Theorem 5.18. *Let $f:[0,1]\times\mathbb{R}\to\mathbb{R}$ be a continuous function satisfying the conditions (G_1) and (G_3), and $h:C([0,1],\mathbb{R})\to\mathbb{R}$ be a continuous function satisfying the condition (G_2). Then the problem (5.79)-(5.80), with $q_1^2-4q_0q_2>0$, has at least one solution on $[0,1]$ if*

$$\ell\alpha_1+L\Delta_1<1, \tag{5.98}$$

where α_1 and Δ_1 are given in (5.97).

Proof. Setting $\sup_{t\in[0,1]}|\vartheta(t)|=\|\vartheta\|$, we can fix

$$\begin{aligned} r\ge{}& \frac{\|\vartheta\|}{|q_2m_1m_2(m_2-m_1)|\Gamma(\delta+1)}\Big\{\varepsilon+\widehat{\rho}_1\Big[\xi^\sigma|m_2(1-e^{m_1\xi})-m_1(1-e^{m_2\xi})|\\ &+\sum_{i=1}^n|j_i|\eta_i^\sigma|m_2(1-e^{m_1\eta_i})-m_1(1-e^{m_2\eta_i})|\Big]\\ &+\widehat{\rho}_2\Big[|m_2(1-e^{m_1})-m_1(1-e^{m_2})|\\ &+\frac{\delta^\sigma|\lambda|}{|m_1m_2|}|m_2^2(m_1\delta-e^{m_1\delta}+1)-m_1^2(m_2\delta-e^{m_2\delta}+1)|\Big]\Big\}+\Delta_1\|h\|, \end{aligned}$$

and consider $B_r=\{x\in\mathcal{C}:\|x\|\le r\}$. Introduce the operators $\mathcal{J}_1$ and $\mathcal{J}_2$ on B_r as follows:

$$(\mathcal{J}_1x)(t)=\frac{1}{q_2(m_2-m_1)}\int_0^t\int_0^s\mathcal{A}(t)\frac{(s-u)^{\sigma-1}}{\Gamma(\sigma)}f(u,x(u))du\,ds, \tag{5.99}$$

and

$$\begin{aligned} (\mathcal{J}_2x)(t)={}&\frac{1}{q_2(m_2-m_1)}\Big\{\rho_1(t)\Big[\int_0^\xi\int_0^s\mathcal{A}(\xi)\frac{(s-u)^{\sigma-1}}{\Gamma(\sigma)}f(u,x(u))du\,ds\\ &-\sum_{i=1}^n j_i\int_0^{\eta_i}\int_0^s\mathcal{A}(\eta_i)\frac{(s-u)^{\sigma-1}}{\Gamma(\sigma)}y(u)du\,ds\Big]\\ &+\rho_2(t)\Big[\int_0^1\int_0^s\mathcal{A}(1)\frac{(s-u)^{\sigma-1}}{\Gamma(\sigma)}f(u,x(u))du\,ds\\ &-\lambda\int_0^\delta\int_0^s\Big(\frac{(e^{m_2(\delta-s)}-1)}{m_2}-\frac{(e^{m_1(\delta-s)}-1)}{m_1}\Big) \end{aligned}$$

$$\times\frac{(s-u)^{\sigma-1}}{\Gamma(\sigma)}f(u,x(u))du\,ds\Big]\Big\}$$
$$+h(x)\Big[e^{m_2t}+\rho_1(t)\Big(e^{m_2\xi}-\sum_{i=1}^{n}j_i+\sum_{i=1}^{n}j_ie^{m_2\eta_i}\Big)$$
$$+\rho_2(t)\Big(\frac{m_2e^{m_2}-\lambda e^{m_2\delta}-\lambda}{m_2}\Big)\Big]. \tag{5.100}$$

Observe that $\mathcal{J}=\mathcal{J}_1+\mathcal{J}_2$. For $x,y\in B_r$, we have

$$\begin{aligned}
&\|\mathcal{J}_1x+\mathcal{J}_2y\|\\
=&\sup_{t\in[0,1]}|(\mathcal{J}_1x)(t)+(\mathcal{J}_2y)(t)|\\
\le&\frac{1}{|q_2(m_2-m_1)|}\sup_{t\in[0,1]}\Big\{\int_0^t\int_0^s\mathcal{A}(t)\frac{(s-u)^{\sigma-1}}{\Gamma(\sigma)}|f(u,x(u))|du\,ds\\
&+|\rho_1(t)|\Big[\int_0^\xi\int_0^s\mathcal{A}(\xi)\frac{(s-u)^{\sigma-1}}{\Gamma(\sigma)}|f(u,y(u))|du\,ds\\
&+\sum_{i=1}^{n}|j_i|\int_0^{\eta_i}\int_0^s\mathcal{A}(\eta_i)\frac{(s-u)^{\sigma-1}}{\Gamma(\sigma)}y(u)du\,ds\Big]\\
&+|\rho_2(t)|\Big[\int_0^1\int_0^s\mathcal{A}(1)\frac{(s-u)^{\sigma-1}}{\Gamma(\sigma)}|f(u,y(u))|du\,ds\\
&+|\lambda|\int_0^\delta\int_0^s\Big(\frac{(e^{m_2(\delta-s)}-1)}{m_2}-\frac{(e^{m_1(\delta-s)}-1)}{m_1}\Big)\frac{(s-u)^{\sigma-1}}{\Gamma(\sigma)}|f(u,y(u))|du\,ds\Big]\Big\}\\
&+|h(x)|\Big[|e^{m_2t}|+\rho_1(t)\Big(|e^{m_2\xi}|+\sum_{i=1}^{n}|j_i||e^{m_2\eta_i}+1|\Big)\\
&+\rho_2(t)\Big(\frac{|m_2e^{m_2}|+|\lambda||e^{m_2\delta}+1|}{|m_2|}\Big)\Big]\\
\le&\frac{\|\vartheta\|}{|q_2(m_2-m_1)|\Gamma(\sigma+1)}\sup_{t\in[0,1]}\Big\{t^\sigma\int_0^t\Big|e^{m_2(t-s)}-e^{m_1(t-s)}\Big|ds\\
&+|\rho_1(t)|\Big[\xi^\sigma\int_0^\xi\Big|e^{m_2(\xi-s)}-e^{m_1(\xi-s)}\Big|ds\\
&+\sum_{i=1}^{n}|j_i|\eta_i^\sigma\int_0^{\eta_i}\Big|e^{m_2(\eta_i-s)}-e^{m_1(\eta_i-s)}\Big|ds\Big]\\
&+|\rho_2(t)|\Big[\int_0^1\Big|e^{m_2(1-s)}-e^{m_1(1-s)}\Big|ds\\
&+|\lambda|\delta^\sigma\int_0^\delta\Big|\frac{(e^{m_2(\delta-s)}-1)}{m_2}-\frac{(e^{m_1(\delta-s)}-1)}{m_1}\Big|ds\Big]\Big\}+\Delta_1\|h\|\\
\le&\frac{\|\vartheta\|}{|q_2m_1m_2(m_2-m_1)|\Gamma(\sigma+1)}\Big\{\varepsilon+\widehat{\rho}_1[\xi^\sigma|m_2(1-e^{m_1\xi})-m_1(1-e^{m_2\xi})|
\end{aligned}$$

$$+\sum_{i=1}^{n}|j_i|\eta_i^{\sigma}|m_2(1-e^{m_1\eta_i})-m_1(1-e^{m_2\eta_i})|]+\widehat{\rho}_2[|m_2(1-e^{m_1})-m_1(1-e^{m_2})|$$
$$+\frac{\delta^{\sigma}|\lambda|}{|m_1m_2|}|m_2^2(m_1\delta-e^{m_1\delta}+1)-m_1^2(m_2\delta-e^{m_2\delta}+1)|]\Big\}+\Delta_1\|h\|\leq r.$$

Thus $\mathcal{J}_1x+\mathcal{J}_2y\in B_r$. Using the assumptions $(G_1)-(G_3)$ together with (5.98), we get

$$\begin{aligned}
&\|\mathcal{J}_2x-\mathcal{J}_2y\|\\
&=\sup_{t\in[0,1]}|(\mathcal{J}_2x)(t)-(\mathcal{J}_2y)(t)|\\
&\leq\frac{1}{|q_2(m_2-m_1)|}\sup_{t\in[0,1]}\Big\{|\rho_1(t)|\Big[\int_0^{\xi}\int_0^{s}\mathcal{A}(\xi)\frac{(s-u)^{\sigma-1}}{\Gamma(\sigma)}|f(u,x(u))-f(u,y(u))|du\,ds\\
&\quad+\sum_{i=1}^{n}|j_i|\int_0^{\eta_i}\int_0^{s}\mathcal{A}(\eta_i)\frac{(s-u)^{\sigma-1}}{\Gamma(\sigma)}|f(u,x(u))-f(u,y(u))|du\,ds\Big]\\
&\quad+|\rho_2(t)|\Big[\int_0^{1}\int_0^{s}\mathcal{A}(1)\frac{(s-u)^{\sigma-1}}{\Gamma(\sigma)}|f(u,x(u))-f(u,y(u))|du\,ds\\
&\quad+|\lambda|\int_0^{\delta}\int_0^{s}\Big(\frac{(e^{m_2(\delta-s)}-1)}{m_2}-\frac{(e^{m_1(\delta-s)}-1)}{m_1}\Big)\\
&\quad\times\frac{(s-u)^{\sigma-1}}{\Gamma(\sigma)}|f(u,x(u))-f(u,y(u))|du\,ds\Big]\Big\}\\
&\quad+|h(x)-h(y)|\Big[|e^{m_2t}|+\rho_1(t)\Big(|e^{m_2\xi}|+\sum_{i=1}^{n}|j_i||e^{m_2\eta_i}+1|\Big)\\
&\quad+\rho_2(t)\Big(\frac{|m_2e^{m_2}|+|\lambda||e^{m_2\delta}+1|}{|m_2|}\Big)\Big]\\
&\leq\frac{\ell}{|q_2(m_2-m_1)|}\sup_{t\in[0,1]}\Big\{|\rho_1(t)|\Big[\xi^{\sigma}\int_0^{\xi}\Big|e^{m_2(\xi-s)}-e^{m_1(\xi-s)}\Big|ds\\
&\quad+\sum_{i=1}^{n}|j_i|\eta_i^{\sigma}\int_0^{\eta_i}\Big|e^{m_2(\eta_i-s)}-e^{m_1(\eta_i-s)}\Big|ds\Big]+|\rho_2(t)|\Big[\int_0^{1}\int_0^{s}\Big|e^{m_2(1-s)}-e^{m_1(1-s)}\Big|ds\\
&\quad+|\lambda|\delta^{\sigma}\int_0^{\delta}\Big|\frac{(e^{m_2(\delta-s)}-1)}{m_2}-\frac{(e^{m_1(\delta-s)}-1)}{m_1}\Big|ds\Big]\Big\}\|x-y\|+L\Delta_1\|x-y\|\\
&\leq\frac{\ell}{|q_2m_1m_2(m_2-m_1)|\Gamma(\sigma+1)}\Big\{\widehat{\rho}_1[\xi^{\sigma}|m_2(1-e^{m_1\xi})-m_1(1-e^{m_2\xi})|\\
&\quad+\sum_{i=1}^{n}|j_i|\eta_i^{\sigma}|m_2(1-e^{m_1\eta_i})-m_1(1-e^{m_2\eta_i})|]+\widehat{\rho}_2[|m_2(1-e^{m_1})-m_1(1-e^{m_2})|\\
&\quad+\frac{\delta^{\sigma}|\lambda|}{|m_1m_2|}|m_2^2(m_1\delta-e^{m_1\delta}+1)-m_1^2(m_2\delta-e^{m_2\delta}+1))|]\Big\}\|x-y\|+L\Delta_1\|x-y\|\\
&=(\ell\alpha_1+L\Delta_1)\|x-y\|.
\end{aligned}$$

Note that continuity of f implies that the operator $\mathcal{J}_1$ is continuous. Also, $\mathcal{J}_1$ is uniformly bounded on B_r as

$$\|\mathcal{J}_1 x\| = \sup_{t\in[0,1]} |(\mathcal{J}_1 x)(t)| \leq \frac{\|\vartheta\|\varepsilon}{|q_2 m_1 m_2 (m_2 - m_1)|\Gamma(\sigma+1)}.$$

Now we prove the compactness of operator $\mathcal{J}_1$. We define

$$\sup_{(t,x)\in[0,1]\times B_r} |f(t,x)| = \overline{f}.$$

Thus, for $0 < t_1 < t_2 < 1$, we have

$$\begin{aligned}
&|(\mathcal{J}_1 x)(t_2) - (\mathcal{J}_1 x)(t_1)| \\
&= \frac{1}{|q_2(m_2 - m_1)|} \left| \int_0^{t_1}\int_0^s \Big[\mathcal{A}(t_2) - \mathcal{A}(t_1)\Big] \frac{(s-u)^{\sigma-1}}{\Gamma(\sigma)} f(u,x(u))duds \right. \\
&\quad \left. + \int_{t_1}^{t_2}\int_0^s \mathcal{A}(t_2)\frac{(s-u)^{\sigma-1}}{\Gamma(\sigma)} f(u,x(u))duds \right| \\
&\leq \frac{\overline{f}}{|q_2 m_1 m_2 (m_2 - m_1)|\Gamma(\sigma+1)} \Big\{ \Big(t_1^\sigma - t_2^\sigma\Big)\Big| m_1\Big(1 - e^{m_2(t_2 - t_1)}\Big) \\
&\quad - m_2\Big(1 - e^{m_1(t_2-t_1)}\Big)\Big| + t_1^\sigma\Big| m_1\Big(e^{m_2 t_2} - e^{m_2 t_1}\Big) - m_2\Big(e^{m_1 t_2} - e^{m_1 t_1}\Big)\Big|\Big\} \to 0,
\end{aligned}$$

as $t_1 \to t_2$, independent of x. Thus, $\mathcal{J}_1$ is relatively compact on B_r. Hence, by the Arzelá-Ascoli Theorem, $\mathcal{J}_1$ is compact on B_r. Thus all the assumptions of Theorem 1.4 are satisfied. So, by the conclusion of Theorem 1.4, the problem (5.79)-(5.80) has at least one solution [0, 1]. The proof is completed. □

Remark 5.3. In the above theorem we can interchange the roles of the operators $\mathcal{J}_1$ and $\mathcal{J}_2$ to obtain a second result by replacing (5.98) by the following condition:

$$\frac{\ell\varepsilon}{|q_2 m_1 m_2 (m_2 - m_1)|\Gamma(\sigma+1)} < 1.$$

Now we apply Banach's contraction mapping principle to prove uniqueness of solutions for the problem (5.79)-(5.80).

Theorem 5.19. *Assume that $f : [0,1] \times \mathbb{R} \to \mathbb{R}$ is a continuous function such that (G_1) and (G_2) are satisfied, and $h : C([0,1],\mathbb{R}) \to \mathbb{R}$ is a continuous function satisfying the condition (G_2). Then there exists a unique solution for the problem (5.79)-(5.80), with $q_1^2 - 4q_0 q_2 > 0$ on $[0,1]$, if $\ell\alpha + L\Delta_1 < 1$, where α and Δ_1 are given by* (5.97).

Proof. Let us define $\sup_{t\in[0,1]} |f(t,0)| = M$, $|h(0)| = L_0$ and select $\bar{r} \geq \dfrac{\alpha M + L_0 \Delta_1}{1 - (\ell\alpha + L\Delta_1)}$ to show that $\mathcal{J}B_{\bar{r}} \subset B_{\bar{r}}$, where $B_{\bar{r}} = \{x \in \mathcal{C} : \|x\| \leq \bar{r}\}$ and $\mathcal{J}$ is defined by (5.95). Using the condition (G_1) and (G_2), we have

$$\begin{aligned}
|f(t,x)| &= |f(t,x) - f(t,0) + f(t,0)| \leq |f(t,x) - f(t,0)| + |f(x,0)| \\
&\leq \ell\|x\| + M \leq \ell\bar{r} + M, \qquad (5.101)
\end{aligned}$$

and

$$|h(x)| = |h(x) - h(0) + h(0)| \leq |h(x) - h(0)| + |h(0)| \leq L\|x\| + L_0 \leq L\bar{r} + L_0.$$

Then, for $x \in B_{\bar{r}}$, we obtain

$$\begin{aligned}
\|\mathcal{J}(x)\| &= \sup_{t\in[0,1]} |\mathcal{J}(x)(t)| \\
&\leq \frac{1}{|q_2(m_2 - m_1)|} \sup_{t\in[0,1]} \Bigg\{ \int_0^t \int_0^s \mathcal{A}(t) \frac{(s-u)^{\sigma-1}}{\Gamma(\sigma)} |f(u, x(u))| du\, ds \\
&\quad + |\rho_1(t)| \Big[\int_0^\xi \int_0^s \mathcal{A}(\xi) \frac{(s-u)^{\sigma-1}}{\Gamma(\sigma)} |f(u, x(u))| du\, ds \\
&\quad + \sum_{i=1}^n |j_i| \int_0^{\eta_i} \int_0^s \mathcal{A}(\eta_i) \frac{(s-u)^{\sigma-1}}{\Gamma(\sigma)} |f(u, x(u))| du\, ds \Big] \\
&\quad + |\rho_2(t)| \Big[\int_0^1 \int_0^s \mathcal{A}(1) \frac{(s-u)^{\sigma-1}}{\Gamma(\sigma)} |f(u, x(u))| duds \\
&\quad + |\lambda| \int_0^\delta \int_0^s \Big(\frac{(e^{m_2(\delta-s)} - 1)}{m_2} - \frac{(e^{m_1(\delta-s)} - 1)}{m_1} \Big) \\
&\quad \times \frac{(s-u)^{\sigma-1}}{\Gamma(\sigma)} |f(u, y(u))| du\, ds \Big] \Bigg\} \\
&\quad + |h(x)| \Big[|e^{m_2 t}| + \rho_1(t) \Big(|e^{m_2\xi}| + \sum_{i=1}^n |j_i| |e^{m_2\eta_i} + 1| \Big) \\
&\quad + \rho_2(t) \Big(\frac{|m_2 e^{m_2}| + |\lambda| |e^{m_2\delta} + 1|}{|m_2|} \Big) \Big] \\
&\leq \frac{(\ell\bar{r} + M)}{|q_2(m_2 - m_1)|} \sup_{t\in[0,1]} \Bigg\{ \int_0^t \Big| e^{m_2(t-s)} - e^{m_1(t-s)} \Big| \frac{s^\sigma}{\Gamma(\sigma+1)} ds \\
&\quad + |\rho_1(t)| \Big[\int_0^\xi \Big| e^{m_2(\xi-s)} - e^{m_1(\xi-s)} \Big| \frac{s^\sigma}{\Gamma(\sigma+1)} ds \\
&\quad + \sum_{i=1}^n |j_i| \int_0^{\eta_i} \Big| e^{m_2(\eta_i-s)} - e^{m_1(\eta_i-s)} \Big| \frac{s^\sigma}{\Gamma(\sigma+1)} ds \Big] \\
&\quad + |\rho_2(t)| \Big[\int_0^1 \Big| e^{m_2(1-s)} - e^{m_1(1-s)} \Big| \frac{s^\sigma}{\Gamma(\sigma+1)} ds \\
&\quad + |\lambda| \int_0^\delta \Big| \frac{(e^{m_2(\delta-s)} - 1)}{m_2} - \frac{(e^{m_1(\delta-s)} - 1)}{m_1} \Big| \frac{s^\sigma}{\Gamma(\sigma+1)} ds \Big] \Bigg\} + (L\bar{r} + L_0)\Delta_1 \\
&\leq \frac{(\ell\bar{r} + M)}{|q_2 m_1 m_2 (m_2 - m_1)| \Gamma(\sigma+1)} \Big\{ \varepsilon + \widehat{\rho}_1 [\xi^\sigma |m_2(1 - e^{m_1\xi}) - m_1(1 - e^{m_2\xi})| \\
&\quad + \sum_{i=1}^n |j_i| \eta_i^\sigma |m_2(1 - e^{m_1\eta_i}) - m_1(1 - e^{m_2\eta_i})|] \\
&\quad + \widehat{\rho}_2 [|m_2(1 - e^{m_1}) - m_1(1 - e^{m_2})|
\end{aligned}$$

$$
\begin{aligned}
&+\frac{\delta^\sigma|\lambda|}{|m_1m_2|}|m_2^2(m_1\delta-e^{m_1\delta}+1)-m_1^2(m_2\delta-e^{m_2\delta}+1)|]\Big\}+(L\bar{r}+L_0)\Delta_1\\
&=(\ell\bar{r}+M)\alpha+(L\bar{r}+L_0)\Delta_1\le\bar{r},
\end{aligned}
$$

which clearly shows that $\mathcal{J}x\in B_{\bar{r}}$ for any $x\in B_{\bar{r}}$. Thus $\mathcal{J}B_{\bar{r}}\subset B_{\bar{r}}$. Now, for $x,y\in\mathcal{C}$ and for each $t\in[0,1]$, we have

$$
\begin{aligned}
&\|(\mathcal{J}x)-(\mathcal{J}y)\|\\
\le&\frac{1}{|q_2(m_2-m_1)|}\sup_{t\in[0,1]}\bigg\{\int_0^t\int_0^s\mathcal{A}(t)\frac{(s-u)^{\sigma-1}}{\Gamma(\sigma)}|f(u,x(u))-f(u,y(u))|du\,ds\\
&+|\rho_1(t)|\bigg[\int_0^\xi\int_0^s\mathcal{A}(\xi)\frac{(s-u)^{\sigma-1}}{\Gamma(\sigma)}|f(u,x(u))-f(u,y(u))|du\,ds\\
&+\sum_{i=1}^n|j_i|\int_0^{\eta_i}\int_0^s\mathcal{A}(\eta_i)\frac{(s-u)^{\sigma-1}}{\Gamma(\sigma)}|f(u,x(u))-f(u,y(u))|du\,ds\bigg]\\
&+|\rho_2(t)|\bigg[\int_0^1\int_0^s\mathcal{A}(1)\frac{(s-u)^{\sigma-1}}{\Gamma(\sigma)}|f(u,x(u))-f(u,y(u))|du\,ds\\
&+|\lambda|\int_0^\delta\int_0^s\Big(\frac{(e^{m_2(\delta-s)}-1)}{m_2}-\frac{(e^{m_1(\delta-s)}-1)}{m_1}\Big)\\
&\times\frac{(s-u)^{\sigma-1}}{\Gamma(\sigma)}|f(u,x(u))-f(u,y(u))|du\,ds\bigg]\bigg\}\\
&+|h(x)-h(y)|\bigg[|e^{m_2t}|+\rho_1(t)\Big(|e^{m_2\xi}|+\sum_{i=1}^n|j_i||e^{m_2\eta_i}+1|\Big)\\
&+\rho_2(t)\Big(\frac{|m_2e^{m_2}|+|\lambda||e^{m_2\delta}+1|}{|m_2|}\Big)\bigg]\\
\le&\frac{\ell}{|q_2(m_2-m_1)|}\sup_{t\in[0,1]}\bigg\{\int_0^t\Big|e^{m_2(t-s)}-e^{m_1(t-s)}\Big|\frac{s^\sigma}{\Gamma(\sigma+1)}ds\\
&+|\rho_1(t)|\bigg[\int_0^\xi\Big|e^{m_2(\xi-s)}-e^{m_1(\xi-s)}\Big|\frac{s^\sigma}{\Gamma(\sigma+1)}ds\\
&+\sum_{i=1}^n|j_i|\int_0^{\eta_i}\Big|e^{m_2(\eta_i-s)}-e^{m_1(\eta_i-s)}\Big|\frac{s^\sigma}{\Gamma(\sigma+1)}ds\bigg]\\
&+|\rho_2(t)|\bigg[\int_0^1\Big|e^{m_2(1-s)}-e^{m_1(1-s)}\Big|\frac{s^\sigma}{\Gamma(\sigma+1)}ds\\
&+|\lambda|\int_0^\delta\Big|\frac{(e^{m_2(\delta-s)}-1)}{m_2}-\frac{(e^{m_1(\delta-s)}-1)}{m_1}\Big|\frac{s^\sigma}{\Gamma(\sigma+1)}ds\bigg]\bigg\}\|x-y\|+L\Delta_1\|x-y\|\\
\le&\frac{\ell}{|q_2m_1m_2(m_2-m_1)|\Gamma(\sigma+1)}\Big\{\varepsilon+\widehat{\rho}_1[\xi^\sigma|m_2(1-e^{m_1\xi})-m_1(1-e^{m_2\xi})|\\
&+\sum_{i=1}^n|j_i|\eta_i^\sigma|m_2(1-e^{m_1\eta_i})-m_1(1-e^{m_2\eta_i})|]+\widehat{\rho}_2[|m_2(1-e^{m_1})-m_1(1-e^{m_2})|
\end{aligned}
$$

$$+\frac{\delta^\sigma|\lambda|}{|m_1m_2|}|m_2^2(m_1\delta-e^{m_1\delta}+1)-m_1^2(m_2\delta-e^{m_2\delta}+1)|]\Big\}\|x-y\|+L\Delta_1\|x-y\|$$
$$=(\ell\alpha+L\Delta_1)\|x-y\|,$$

where α and Δ_1 are given by (5.97) and depend only on the parameters involved in the problem. In view of the condition $\ell\alpha+L\Delta_1<1$, it follows that $\mathcal{J}$ is a contraction. Thus, by the contraction mapping principle (Banach fixed point theorem), the problem (5.79) and (5.80) has a unique solution on $[0,1]$. This completes the proof. □

The next existence result is based on Leray-Schauder nonlinear alternative. We need the following assumptions:

(H_1) There exist a function $g\in C([0,1],\mathbb{R}^+)$, and a non-decreasing function $\psi:\mathbb{R}^+\to\mathbb{R}^+$ such that $|f(t,y)|\le g(t)\psi(\|y\|),\ \ \forall(t,y)\in[0,1]\times\mathbb{R}$;

(H_2) $h:C([0,1],\mathbb{R})\to\mathbb{R}$ is a continuous function with $h(0)=0$ and there exist constant $L_1>0$, such that
$$|h(x)|\le L_1\|x\|,\quad \forall\ x\in C([0,1],\mathbb{R});$$

(H_3) There exists a constant $K>0$ such that
$$\frac{(1-L_1\Delta_1)K}{\|g\|\psi(K)\alpha}>1,$$
with $L_1\Delta_1<1$.

Theorem 5.20. *Let $f:[0,1]\times\mathbb{R}\to\mathbb{R}$ be a continuous function. Then the problem (5.79)-(5.80), with $q_1^2-4q_0q_2>0$, has at least one solution on $[0,1]$, if (H_1)-(H_3) are satisfied.*

Proof. Consider the operator $\mathcal{J}:\mathcal{C}\to\mathcal{C}$ defined by (5.95). We show that $\mathcal{J}$ maps bounded sets into bounded sets in $\mathcal{C}=C([0,1],\mathbb{R})$. For a positive number ζ, let $\mathcal{E}_\zeta=\{x\in\mathcal{C}:\|x\|\le\zeta\}$ be a bounded set in $\mathcal{C}$. Then we have

$$\|\mathcal{J}(x)\|=\sup_{t\in[0,1]}|\mathcal{J}(x)(t)|$$
$$\le\frac{1}{|q_2(m_2-m_1)|}\sup_{t\in[0,1]}\Bigg\{\int_0^t\int_0^s\mathcal{A}(t)\frac{(s-u)^{\sigma-1}}{\Gamma(\sigma)}|f(u,x(u))|du\,ds$$
$$+|\rho_1(t)|\Big[\int_0^\xi\int_0^s\mathcal{A}(\xi)\frac{(s-u)^{\sigma-1}}{\Gamma(\sigma)}|f(u,x(u))|du\,ds$$
$$+\sum_{i=1}^n|j_i|\int_0^{\eta_i}\int_0^s\mathcal{A}(\eta_i)\frac{(s-u)^{\sigma-1}}{\Gamma(\sigma)}|f(u,x(u))|du\,ds\Big]$$
$$+|\rho_2(t)|\Big[\int_0^1\int_0^s\mathcal{A}(1)\frac{(s-u)^{\sigma-1}}{\Gamma(\sigma)}|f(u,x(u))|du\,ds$$
$$+|\lambda|\int_0^\delta\int_0^s\Big(\frac{(e^{m_2(\delta-s)}-1)}{m_2}-\frac{(e^{m_1(\delta-s)}-1)}{m_1}\Big)$$

$$
\begin{aligned}
&\times\frac{(s-u)^{\sigma-1}}{\Gamma(\sigma)}|f(u,x(u))|du\,ds\Big]\Bigg\}\\
&+|h(x)|\Big[|e^{m_2t}|+\rho_1(t)\Big(|e^{m_2\xi}|+\sum_{i=1}^{n}|j_i||e^{m_2\eta_i}+1|\Big)\\
&+\rho_2(t)\Big(\frac{|m_2e^{m_2}|+|\lambda||e^{m_2\delta}+1|}{|m_2|}\Big)\Big]\\
\le{}&\frac{\|g\|\psi(\zeta)}{|q_2(m_2-m_1)|}\sup_{t\in[0,1]}\Bigg\{\int_0^t\Big|e^{m_2(t-s)}-e^{m_1(t-s)}\Big|\frac{s^\sigma}{\Gamma(\sigma+1)}ds\\
&+|\rho_1(t)|\Big[\int_0^\xi\Big|e^{m_2(\xi-s)}-e^{m_1(\xi-s)}\Big|\frac{s^\sigma}{\Gamma(\sigma+1)}ds\\
&+\sum_{i=1}^{n}|j_i|\int_0^{\eta_i}\Big|e^{m_2(\eta_i-s)}-e^{m_1(\eta_i-s)}\Big|\frac{s^\sigma}{\Gamma(\sigma+1)}ds\Big]\\
&+|\rho_2(t)|\Big[\int_0^1\Big|e^{m_2(1-s)}-e^{m_1(1-s)}\Big|\frac{s^\sigma}{\Gamma(\sigma+1)}ds\\
&+|\lambda|\int_0^\delta\Big|\frac{(e^{m_2(\delta-s)}-1)}{m_2}-\frac{(e^{m_1(\delta-s)}-1)}{m_1}\Big|\frac{s^\sigma}{\Gamma(\sigma+1)}ds\Big]\Bigg\}+L_1\Delta_1\zeta\\
\le{}&\frac{\|g\|\psi(\zeta)}{|q_2m_1m_2(m_2-m_1)|\Gamma(\sigma+1)}\{\varepsilon+\widehat{\rho}_1[\xi^\sigma|m_2(1-e^{m_1\xi})-m_1(1-e^{m_2\xi})|\\
&+\sum_{i=1}^{n}|j_i|\eta_i^\sigma|m_2(1-e^{m_1\eta_i})-m_1(1-e^{m_2\eta_i})|]\\
&+\widehat{\rho}_2[|m_2(1-e^{m_1})-m_1(1-e^{m_2})|\\
&+\frac{\delta^\sigma|\lambda|}{|m_1m_2|}|m_2^2(m_1\delta-e^{m_1\delta}+1)-m_1^2(m_2\delta-e^{m_2\delta}+1)|]\}+L_1\Delta_1\zeta,
\end{aligned}
$$

which yields

$$
\begin{aligned}
\|\mathcal{J}x\|\le{}&\frac{\|g\|\psi(\zeta)}{|q_2m_1m_2(m_2-m_1)|\Gamma(\sigma+1)}\Big\{\varepsilon+\widehat{\rho}_1\xi^\sigma|m_2(1-e^{m_1\xi})-m_1(1-e^{m_2\xi})|\\
&+\sum_{i=1}^{n}|j_i|\eta_i^\sigma|m_2(1-e^{m_1\eta_i})-m_1(1-e^{m_2\eta_i})|]\\
&+\widehat{\rho}_2[|m_2(1-e^{m_1})-m_1(1-e^{m_2})|\\
&+\frac{\delta^\sigma|\lambda|}{|m_1m_2|}|m_2^2(m_1\delta-e^{m_1\delta}+1)-m_1^2(m_2\delta-e^{m_2\delta}+1)|]\}+L_1\Delta_1\zeta.
\end{aligned}
$$

Next we show that $\mathcal{J}$ maps bounded sets into equicontniuous sets of $\mathcal{C}$. Let $t_1,t_2\in[0,1]$ with $t_1<t_2$ and $y\in\mathcal{E}_\zeta$, where $\mathcal{E}_\zeta$ is a bounded set of $\mathcal{C}$. Then we obtain

$$
\begin{aligned}
&|(\mathcal{J}x)(t_2)-(\mathcal{J}x)(t_1)|\\
\le{}&\frac{1}{|q_2(m_2-m_1)|}\Bigg\{\Bigg|\int_0^{t_1}\int_0^s\Big[\mathcal{A}(t_2)-\mathcal{A}(t_1)\Big]\frac{(s-u)^{\sigma-1}}{\Gamma(\sigma)}f(u,x(u))du\,ds
\end{aligned}
$$

$$
\begin{aligned}
&+\int_{t_1}^{t_2}\int_0^s \mathcal{A}(t_2)\frac{(s-u)^{\sigma-1}}{\Gamma(\sigma)}f(u,x(u))du\ ds\Big|\\
&+|\rho_1(t_2)-\rho_1(t_1)|\Big[\int_0^{\xi}\int_0^s \mathcal{A}(\xi)\frac{(s-u)^{\sigma-1}}{\Gamma(\sigma)}|f(u,y(u))|du\ ds\\
&+\sum_{i=1}^{n}|j_i|\int_0^{\eta_i}\int_0^s \mathcal{A}(\eta_i)\frac{(s-u)^{\sigma-1}}{\Gamma(\sigma)}|f(u,y(u))|du\ ds\Big]\\
&+|\rho_2(t_2)-\rho_2(t_1)|\Big[\int_0^{1}\int_0^s \mathcal{A}(1)\frac{(s-u)^{\sigma-1}}{\Gamma(\sigma)}|f(u,y(u))|du\ ds\\
&+|\lambda|\int_0^{\delta}\int_0^s \Big(\frac{(e^{m_2(\delta-s)}-1)}{m_2}-\frac{(e^{m_1(\delta-s)}-1)}{m_1}\Big)\frac{(s-u)^{\sigma-1}}{\Gamma(\sigma)}|f(u,y(u))|du\ ds\Big]\Big\}\\
&+|h(x)|\Big[|e^{m_2t_2}-e^{m_2t_1}|+|\rho_1(t_2)-\rho_1(t_1)|\Big(|e^{m_2\xi}|+\sum_{i=1}^{n}|j_i||e^{m_2\eta_i}+1|\Big)\\
&+|\rho_2(t_2)-\rho_2(t_1)|\Big(\frac{|m_2e^{m_2}|+|\lambda||e^{m_2\delta}+1|}{|m_2|}\Big)\Big]\\
\le\ &\frac{\overline{f}}{|q_2m_1m_2(m_2-m_1)|\Gamma(\sigma+1)}\Big\{\Big(t_1^{\sigma}-t_2^{\sigma}\Big)\Big|m_1(1-e^{m_2(t_2-t_1)})\\
&-m_2(1-e^{m_1(t_2-t_1)})\Big|+t_1^{\sigma}\Big|m_1(e^{m_2t_2}-e^{m_2t_1})-m_2(e^{m_1t_2}-e^{m_1t_1})\Big|\\
&+|\rho_1(t_2)-\rho_1(t_1)|[\xi^{\sigma}|m_2(1-e^{m_1\xi})-m_1(1-e^{m_2\xi})|\\
&+\sum_{i=1}^{n}|j_i|\eta_i^{\sigma}|m_2(1-e^{m_1\eta_i})-m_1(1-e^{m_2\eta_i})|]\\
&+|\rho_2(t_2)-\rho_2(t_1)|[|m_2(1-e^{m_1})-m_1(1-e^{m_2})|\\
&+\frac{\delta^{\sigma}|\lambda|}{|m_1m_2|}|m_1^2(m_2\delta-e^{m_2\delta}+1)-m_2^2(m_1\delta-e^{m_1\delta}+1)|]\Big\}\\
&+|h(x)|\Big[|e^{m_2t_2}-e^{m_2t_1}|+|\rho_1(t_2)-\rho_1(t_1)|\Big(|e^{m_2\xi}|+\sum_{i=1}^{n}|j_i||e^{m_2\eta_i}+1|\Big)\\
&+|\rho_2(t_2)-\rho_2(t_1)|\Big(\frac{|m_2e^{m_2}|+|\lambda||e^{m_2\delta}+1|}{|m_2|}\Big)\Big],
\end{aligned}
$$

which tends to zero independently of $x\in\mathcal{E}_\varsigma$ as $t_2-t_1\to 0$. As $\mathcal{J}$ satisfies the above assumptions, therefore it follows by the Arzelá-Ascoli theorem that $\mathcal{J}:\mathcal{C}\to\mathcal{C}$ is completely continuous.

The result will follow from the Leray-Schauder nonlinear alternative once it is shown that the set of all solutions to the equation $x=\theta\mathcal{J}x$ is bounded for $\theta\in[0,1]$. For that, let x be a solution of $x=\theta\mathcal{J}x$ for $\theta\in[0,1]$. Then, for $t\in[0,1]$, we have

$$
\begin{aligned}
|x(t)|&=|\theta\mathcal{J}x(t)|\\
&\le\frac{1}{|q_2(m_2-m_1)|}\sup_{t\in[0,1]}\Big\{\int_0^t\int_0^s \mathcal{A}(t)\frac{(s-u)^{\sigma-1}}{\Gamma(\sigma)}|f(u,x(u))|du\ ds
\end{aligned}
$$

$$
\begin{aligned}
&+|\rho_1(t)|\Big[\int_0^{\xi}\int_0^{s}\mathcal{A}(\xi)\frac{(s-u)^{\sigma-1}}{\Gamma(\sigma)}|f(u,x(u))|du\,ds \\
&+\sum_{i=1}^{n}|j_i|\int_0^{\eta_i}\int_0^{s}\mathcal{A}(\eta_i)\frac{(s-u)^{\sigma-1}}{\Gamma(\sigma)}|f(u,x(u))|du\,ds\Big] \\
&+|\rho_2(t)|\Big[\int_0^{1}\int_0^{s}\mathcal{A}(1)\frac{(s-u)^{\sigma-1}}{\Gamma(\sigma)}|f(u,x(u))|du\,ds \\
&+|\lambda|\int_0^{\delta}\int_0^{s}\Big(\frac{(e^{m_2(\delta-s)}-1)}{m_2}-\frac{(e^{m_1(\delta-s)}-1)}{m_1}\Big) \\
&\times\frac{(s-u)^{\sigma-1}}{\Gamma(\sigma)}|f(u,x(u))|du\,ds\Big]\Big\} \\
&+|h(x)|\Big[|e^{m_2t}|+\rho_1(t)\Big(|e^{m_2\xi}|+\sum_{i=1}^{n}|j_i||e^{m_2\eta_i}+1|\Big) \\
&+\rho_2(t)\Big(\frac{|m_2e^{m_2}|+|\lambda||e^{m_2\delta}+1|}{|m_2|}\Big)\Big] \\
\le\ &\frac{\|g\|\psi(\|x\|)}{|q_2(m_2-m_1)|}\sup_{t\in[0,1]}\Big\{\int_0^t\Big|e^{m_2(t-s)}-e^{m_1(t-s)}\Big|\frac{s^{\sigma}}{\Gamma(\sigma+1)}ds \\
&+|\rho_1(t)|\Big[\int_0^{\xi}\Big|e^{m_2(\xi-s)}-e^{m_1(\xi-s)}\Big|\frac{s^{\sigma}}{\Gamma(\sigma+1)}ds \\
&+\sum_{i=1}^{n}|j_i|\int_0^{\eta_i}\Big|e^{m_2(\eta_i-s)}-e^{m_1(\eta_i-s)}\Big|\frac{s^{\sigma}}{\Gamma(\sigma+1)}ds\Big] \\
&+|\rho_2(t)|\Big[\int_0^{1}\Big|e^{m_2(1-s)}-e^{m_1(1-s)}\Big|\frac{s^{\sigma}}{\Gamma(\sigma+1)}ds \\
&+|\lambda|\int_0^{\delta}\Big|\frac{(e^{m_2(\delta-s)}-1)}{m_2}-\frac{(e^{m_1(\delta-s)}-1)}{m_1}\Big|\frac{s^{\sigma}}{\Gamma(\sigma+1)}ds\Big]\Big\}+|h(x)|\Delta_1 \\
\le\ &\frac{\|g\|\psi(\|x\|)}{|q_2m_1m_2(m_2-m_1)|\Gamma(\sigma+1)}\Big\{\varepsilon+\widehat{\rho}_1[\xi^{\sigma}|m_2(1-e^{m_1\xi})-m_1(1-e^{m_2\xi})| \\
&+\sum_{i=1}^{n}|j_i|\eta_i^{\sigma}|m_2(1-e^{m_1\eta_i})-m_1(1-e^{m_2\eta_i})|] \\
&+\widehat{\rho}_2[|m_2(1-e^{m_1})-m_1(1-e^{m_2})| \\
&+\frac{\delta^{\sigma}|\lambda|}{|m_1m_2|}|m_1^2(m_2\delta-e^{m_2\delta}+1)-m_2^2(m_1\delta-e^{m_1\delta}+1)|]\Big\}+L_1\Delta_1\|x\| \\
=\ &\|g\|\psi(\|x\|)\alpha+L_1\Delta_1\|x\|,
\end{aligned}
$$

which implies that

$$\frac{(1-L_1\Delta_1)\|x\|}{\|g\|\psi(\|x\|)\alpha}\le 1.$$

In view of (H_3), there is no solution x such that $\|x\|\neq K$. Let us set

$$U=\{x\in\mathcal{C}:\|x\|<K\}.$$

The operator $\mathcal{J}:\overline{U}\to\mathcal{C}$ is continuous and completely continuous. From the choice of U, there is no $u\in\partial U$ such that $u=\theta\mathcal{J}(u)$ for some $\theta\in(0,1)$. Consequently, by the nonlinear alternative of Leray-Schauder type (Theorem 1.8), we deduce that $\mathcal{J}$ has a fixed point $u\in\overline{U}$ which is a solution of the problem (5.79)-(5.80). The proof is completed. □

Example 5.12. Consider the following boundary value problem

$$(2\ {}^{C}D^{12/5}+3\ {}^{C}D^{7/5}+\ {}^{C}D^{2/5})x(t)=\frac{Ae^{-t}}{2\sqrt{4+t^2}}\tan^{-1}x+\cos t,\ \ 0<t<1,\quad(5.102)$$

$$x(0)=\frac{1}{9}\sin x,\ \ x(1/5)=x(1/4)+2x(1/3)+x(1/2),\ \ x(1)=2\int_0^{1/6}x(s)ds.\quad(5.103)$$

Here, $\sigma=2/5$, $\xi=1/5$, $\eta_1=1/4$, $\eta_2=1/3$, $\eta_3=1/2$, $\delta=1/6$, $j_1=1$, $j_2=2$, $j_3=1$, $\lambda=2$, A is positive number and

$$f(t,x)=\frac{Ae^{-t}}{2\sqrt{4+t^2}}\tan^{-1}x+\cos t.$$

Clearly $q_1^2-4q_0q_2>0$, and

$$|f(t,x)-f(t,y)|\le\frac{A}{4}|x-y|,$$

$$|h(x)-h(y)|\le\frac{1}{9}\|x-y\|.$$

where $\ell=A/4$, $L=1/9$. Using the given values, we find that $\alpha\approx0.27045$, $\alpha_1\approx0.095961$ and $\Delta_1\approx6.9171$. It is easy to check that $|f(t,x)|\le\dfrac{A\pi e^{-t}}{4\sqrt{4+t^2}}+\cos t=\vartheta(t)$ and $\ell\alpha_1+L\Delta_1<1$ when $A<9.39715$. As all the conditions of Theorem 5.18 are satisfied, the problem (5.102)-(5.103) has at least one solution on $[0,1]$. On the other hand, $\ell\alpha+L\Delta_1<1$ whenever $A<3.26199$ and thus there exists a unique solution for the problem (5.102)-(5.103) on $[0,1]$ by Theorem 5.19.

Example 5.13. Consider the following boundary value problem

$$(2\ {}^{C}D^{12/5}+3\ {}^{C}D^{7/5}+\ {}^{C}D^{2/5})x(t)=\frac{1}{\pi\sqrt{9+t^2}}\Big(x\tan^{-1}x+\pi/2\Big),\ \ 0<t<1,\quad(5.104)$$

$$x(0)=\frac{1}{9}\sin x,\ \ x(1/5)=x(1/4)+2x(1/3)+x(1/2),\ \ x(1)=2\int_0^{1/6}x(s)ds.\quad(5.105)$$

Here, $\sigma=2/5$, $\xi=1/5$, $\eta_1=1/4$, $\eta_2=1/3$, $\eta_3=1/2$, $\delta=1/6$, $j_1=1$, $j_2=2$, $j_3=1$, $\lambda=2$, $q_1^2-4q_2q_0=1>0$, and

$$f(t,x)=\frac{1}{\pi\sqrt{9+t^2}}(x\tan^{-1}x+\pi/2).$$

Clearly

$$|f(t,x)| \leq \frac{1}{2\sqrt{9+t^2}}\Big(\|x\|+1\Big),$$

where $g(t) = \frac{1}{2\sqrt{9+t^2}}$, $\psi(\|x\|) = \|x\| + 1$. Clearly $L_1\Delta_1 < 1$.

Then, by using the condition (H_3), we find that $K > 0.241877$ (we have used $\alpha = 0.27045$). Thus, the conclusion of Theorem 5.20 applies to problem (5.104)-(5.105).

5.5.2 *Existence results for problem (5.79)-(5.80) with* $q_1^2 - 4q_0q_2 = 0$

By Lemma 5.6, we transform problem (5.79)-(5.80) into an equivalent fixed point problem as

$$x = \mathcal{H}x, \tag{5.106}$$

where the operator $\mathcal{H} : \mathcal{C} \to \mathcal{C}$ is defined by

$$\begin{aligned}
(\mathcal{H}x)(t) = \frac{1}{q_2}\Bigg\{ & \int_0^t \int_0^s \mathcal{B}(t)\frac{(s-u)^{\sigma-1}}{\Gamma(\sigma)} f(u,x(u))du\, ds \\
& +\chi_1(t)\Bigg[\int_0^\xi \int_0^s \mathcal{B}(\xi)\frac{(s-u)^{\sigma-1}}{\Gamma(\sigma)} f(u,x(u))du\, ds \\
& -\sum_{i=1}^n j_i \int_0^{\eta_i}\int_0^s \mathcal{B}(\eta_i)\frac{(s-u)^{\sigma-1}}{\Gamma(\sigma)} f(u,x(u))du\, ds\Bigg] \\
& +\chi_2(t)\Bigg[\int_0^1 \int_0^s \mathcal{B}(1)\frac{(s-u)^{\sigma-1}}{\Gamma(\sigma)} f(u,x(u))du\, ds \\
& -\lambda\int_0^\delta\int_0^s \Big(\frac{m(\delta-s)e^{m(\delta-s)} - e^{m(\delta-s)} + 1}{m^2}\Big) \\
& \times \frac{(s-u)^{\sigma-1}}{\Gamma(\sigma)} f(u,x(u))du\, ds\Bigg]\Bigg\} \\
& +h(x)\Big[e^{mt} + \chi_1(t)\Big(e^{m\xi} - \sum_{i=1}^n j_i e^{m\eta_i}\Big) + \chi_2(t)\Big(\frac{me^m - \lambda e^{m\delta} + \lambda}{m}\Big)\Big],
\end{aligned} \tag{5.107}$$

where $\mathcal{B}(\cdot)$, $\chi_1(t)$ *and* $\chi_2(t)$ are defined by (5.91). We set

$$\widehat{\chi}_1 = \max_{t\in[0,1]} |\chi_1(t)|, \qquad \widehat{\chi}_2 = \max_{t\in[0,1]} |\chi_2(t)|,$$

$$\begin{aligned}
\beta = & \frac{1}{|q_2|m^2\Gamma(\sigma+1)}\Big\{(1+\widehat{\chi}_2)|me^m - e^m + 1| + \widehat{\chi}_1\Big[\xi^\sigma|m\xi e^{m\xi} - e^{m\xi} + 1| \\
& + \sum_{i=1}^n |j_i|\eta_i^\sigma|m\eta_i e^{m\eta_i} - e^{m\eta_i} + 1|\Big] \\
& + \frac{|\lambda|\delta^\sigma\widehat{\chi}_2}{|m|}\Big|m\delta(e^{m\delta}+1) + 2(1-e^{m\delta})\Big|\Big\},
\end{aligned} \tag{5.108}$$

$$\beta_1 = \beta - \frac{|me^m - e^m + 1|}{|q_2|m^2\Gamma(\sigma+1)};$$

$$\Delta_2 = \max_{t\in[0,1]} |e^{mt}| + \widehat{\chi}_1\Big(|e^{m\xi}| + \sum_{i=1}^{n} |j_i||e^{m\eta_i}|\Big) + \widehat{\chi}_2\Big(\frac{|me^m| + |\lambda||e^{m\delta}+1|}{|m|}\Big).$$

In relation to the condition $q_1^2 - 4q_0q_2 = 0$ for the problem (5.79)-(5.80), we formulate the existence results analogue to the ones proved in the previous section as follows.

Theorem 5.21. *Let $f : [0,1]\times\mathbb{R}\to\mathbb{R}$ be a continuous function satisfying the conditions (G_1), (G_3) and $h : C([0,1],\mathbb{R})\to\mathbb{R}$ be a continuous function satisfying the condition (G_2). Then the problem (5.79)-(5.80) with $q_1^2 - 4q_0q_2 = 0$ has at least one solution on $[0,1]$ if*

$$\ell\beta_1 + L\Delta_2 < 1, \tag{5.109}$$

where β_1 and Δ_2 are given by (5.108).

Theorem 5.22. *Assume that $f : [0,1]\times\mathbb{R}\to\mathbb{R}$ is a continuous function such that (G_1) is satisfied and $h : C([0,1],\mathbb{R})\to\mathbb{R}$ is a continuous function satisfying the condition (G_2). Then there exists a unique solution for problem (5.79)-(5.80) with $q_1^2 - 4q_0q_2 = 0$ on $[0,1]$ if $\ell\beta + L\Delta_2 < 1$, where β and Δ_2 are given by* (5.108).

Theorem 5.23. *Let $f : [0,1]\times\mathbb{R}\to\mathbb{R}$ be a continuous function. Then the problem* (5.79)-(5.80) *with $q_1^2 - 4q_0q_2 = 0$ has at least one solution on $[0,1]$, if (H_1), (H_2) and the following condition are satisfied:*

(H_3') *There exists a constant $K_1 > 0$ such that*

$$\frac{(1-L_1\Delta_2)K_1}{\|g\|\psi(K_1)\beta} > 1,$$

with $L_1\Delta_2 < 1$, where β is defined by (5.108).

Example 5.14. Consider the following boundary value problem

$$(2\,{}^CD^{12/5} + 4\,{}^CD^{7/5} + 2\,{}^CD^{2/5})x(t) = \frac{B|x|}{(t+6)(|x|+1)} + e^{-t},\ \ 0<t<1, \tag{5.110}$$

$$x(0) = \frac{1}{9}\sin x,\ \ x(1/5) = x(1/4) + 2x(1/3) + x(1/2),\ \ x(1) = 2\int_0^{1/6} x(s)ds. \tag{5.111}$$

Here, $\sigma = 2/5$, $\xi = 1/5$, $\eta_1 = 1/4$, $\eta_2 = 1/3$, $\eta_3 = 1/2$, $\delta = 1/6$, $j_1 = 1$, $j_2 = 2$, $j_3 = 1$, $\lambda = 2$, B is positive number and

$$f(t,x) = \frac{B|x|}{(t+6)(|x|+1)} + e^{-t}.$$

Clearly $q_1^2 - 4q_0q_2 = 0$, and

$$|f(t,x) - f(t,y)| \le B/6\|x-y\|,$$

$$|h(x) - h(y)| \leq 1/9\|x - y\|,$$

where $\ell = B/6$, $L = 1/9$. Using the given values, we hound $\beta \approx 0.29913$, $\beta_1 \approx 0.15022$ and $\Delta_2 \approx 5.135$.

It is easy to check that $|f(t,x)| \leq \dfrac{B}{t+6} + e^{-t} = \vartheta(t)$ and $\ell\beta_1 + L\Delta_2 < 1$ when $B < 16.49345$. As all the conditions of Theorem 5.21 are satisfied, the problem (5.110)-(5.111) has at least one solution on $[0,1]$. On the other hand, $\ell\beta + L\Delta_2 < 1$ whenever $B < 8.61378$ and thus there exists a unique solution for the problem (5.110)-(5.111) on $[0,1]$ by Theorem 5.22.

5.5.3 *Existence results for problem (5.79)-(5.80) with* $q_1^2 - 4q_0q_2 < 0$

In view of Lemma 5.7, we can transform problem (5.79)-(5.80) into an equivalent fixed point problem as follows:

$$x = \mathcal{K}x, \tag{5.112}$$

where the operator $\mathcal{K} : \mathcal{C} \to \mathcal{C}$ is defined by

$$\begin{aligned}
&(\mathcal{K}x)(t)\\
&= \frac{1}{q_2 b}\Bigg\{\int_0^t \int_0^s \mathcal{F}(t)\frac{(s-u)^{\sigma-1}}{\Gamma(\sigma)} f(u,x(u))du\, ds\\
&+\tau_1(t)\Bigg[\int_0^\xi \int_0^s \mathcal{F}(\xi)\frac{(s-u)^{\sigma-1}}{\Gamma(\sigma)} f(u,x(u))du\, ds\\
&-\sum_{i=1}^n j_i \int_0^{\eta_i} \int_0^s \mathcal{F}(\eta_i)\frac{(s-u)^{\sigma-1}}{\Gamma(\sigma)} f(u,x(u))du\, ds\Bigg]\\
&+\tau_2(t)\Bigg[\int_0^1 \int_0^s \mathcal{F}(1)\frac{(s-u)^{\sigma-1}}{\Gamma(\sigma)} f(u,x(u))du\, ds\\
&-\frac{\lambda}{a^2+b^2}\int_0^\delta \int_0^s \Big(b - be^{-a(\delta-s)}\cos b(\delta-s)\\
&-ae^{-a(\delta-s)}\sin b(\delta-s)\Big)\frac{(s-u)^{\sigma-1}}{\Gamma(\sigma)} f(u,x(u))du\, ds\Bigg]\Bigg\}\\
&+h(x)\Big[e^{-at}\cos bt + \tau_1(t)(e^{-a\xi}\cos b\xi\\
&-\sum_{i=1}^n j_i e^{-a\eta_i}\cos b\eta_i) + \tau_2(t)(e^{-a}\cos b - \frac{\lambda}{a^2+b^2}(a - ae^{-a\delta}\cos b\delta\\
&+be^{-a\delta}\sin b\delta))\Big],
\end{aligned} \tag{5.113}$$

where $\mathcal{F}(\cdot)$, $\tau_1(t)$ *and* $\tau_2(t)$ are defined by (5.94). We set

$$\widehat{\tau}_1 = \max_{t\in[0,1]} |\tau_1(t)|, \quad \widehat{\tau}_2 = \max_{t\in[0,1]} |\tau_2(t)|$$

$$\gamma = \frac{1}{|q_2 b(a^2+b^2)|\Gamma(\sigma+1)}\Big\{(1+\widehat{\tau}_2)\Big[|b - be^{-a}\cos b - ae^{-a}\sin b|\Big]$$

$$+\widehat{\tau}_1\Big[\xi^{\sigma}|b-be^{-a\xi}\cos b\xi-ae^{-a\xi}\sin b\xi|$$
$$+\sum_{i=1}^{n}|j_i|\eta_i^{\sigma}|b-be^{-a\eta_i}\cos b\eta_i-ae^{-a\eta_i}\sin b\eta_i|\Big]$$
$$+|\lambda|\delta^{\sigma}\widehat{\tau}_2\Big[|b\delta-e^{-a\delta}\sin b\delta|\Big]\Big\}, \tag{5.114}$$
$$\gamma_1=\gamma-\frac{|b-be^{-a}\cos b-ae^{-a}\sin b|}{|q_2b(a^2+b^2)|\Gamma(\sigma+1)},$$
$$\Delta_3=\max_{t\in[0,1]}|e^{-at}\cos bt|+\widehat{\tau}_1(|e^{-a\xi}\cos b\xi|+\sum_{i=1}^{n}|j_i||e^{-a\eta_i}\cos b\eta_i|)$$
$$+\widehat{\tau}_2\Big(|e^{-a}\cos b|+\frac{|\lambda|}{a^2+b^2}(|a-ae^{-a\delta}\cos b\delta+be^{-a\delta}\sin b\delta|)\Big).$$

Theorem 5.24. *Let $f:[0,1]\times\mathbb{R}\to\mathbb{R}$ be a continuous function satisfying the conditions (G_1), (G_3) and $h:C([0,1],\mathbb{R})\to\mathbb{R}$ be a continuous function satisfying the condition (G_2). Then the problem (5.79)-(5.80) with $p_1^2-4p_0p_2<0$ has at least one solution on $[0,1]$ provided that*

$$\ell\gamma_1+L\Delta_3<1, \tag{5.115}$$

where γ_1 and Δ_3 are given by (5.114).

Theorem 5.25. *Assume that $f:[0,1]\times\mathbb{R}\to\mathbb{R}$ is a continuous function such that (G_1) and (G_3) are satisfied and $h:C([0,1],\mathbb{R})\to\mathbb{R}$ is a continuous function satisfying the condition (G_2). Then there exists a unique solution for the problem (5.79)-(5.80) with $q_1^2-4q_0q_2<0$ on $[0,1]$ if $\ell\gamma+L\Delta_3<1$, where γ and Δ_3 are given by* (5.114).

Theorem 5.26. *Let $f:[0,1]\times\mathbb{R}\to\mathbb{R}$ be a continuous function. Then the problem* (5.79)-(5.80) *with $q_1^2-4q_0q_2<0$ has at least one solution on $[0,1]$ if (H_1), (H_2) and the following condition hold:*

(H_3'') *There exists a constant $K_2>0$ such that*

$$\frac{(1-L_1\Delta_3)K_2}{\|g\|\psi(K_2)\gamma}>1,$$

with $L_1\Delta_3<1$, where γ and Δ_3 are defined by (5.114).

Example 5.15. Consider the following multi-term nonlocal boundary value problem

$$(2\,{}^CD^{12/5}+3\,{}^CD^{7/5}+2\,{}^CD^{2/5})x(t)=\frac{N}{(t+4)^2}\cos x+\frac{e^{-2t}}{13},\quad 0<t<1, \tag{5.116}$$

$$x(0)=\frac{1}{8}x,\ \ x(1/5)=x(1/4)+2x(1/3)+x(1/2),\ \ x(1)=2\int_0^{1/6}x(s)ds. \tag{5.117}$$

Here, $\sigma = 2/5$, $\xi = 1/5$, $\eta_1 = 1/4$, $\eta_2 = 1/3$, $\eta_3 = 1/2$, $\delta = 1/6$, $j_1 = 1$, $j_2 = 2$, $j_3 = 1$, $\lambda = 2$, N is positive number and

$$f(t,x) = \frac{N}{(t+4)^2}\cos x + \frac{e^{-2t}}{13}.$$

Clearly $q_1^2 - 4q_0q_2 = -7 < 0$, and

$$|f(t,x) - f(t,y)| \leq \frac{N}{16}\|x-y\|, \ |h(x) - h(y)| \leq \frac{1}{8}\|x-y\|,$$

where $\ell = N/16$, $L = 1/8$. Using the given values, it is found that $\gamma \approx 0.34744$, $\gamma_1 \approx 0.17937$ and $\Delta_3 \approx 1.8499$. It is easy to check that $|f(t,x)| \leq \dfrac{N}{(t+4)^2} + \dfrac{e^{-2t}}{13} = \vartheta(t)$ and $\ell\gamma_1 + L\Delta_3 < 1$ when $N < 68.57194$. As all the conditions of Theorem 5.24 are satisfied, the problem (5.116) (5.117) has at least one solution on $[0,1]$. On the other hand, it follows by the condition: $\ell\gamma + L\Delta_3 < 1$ whenever $N < 35.40226$ that there exists a unique solution for the problem (5.110)-(5.111) on $[0,1]$ by Theorem 5.25.

5.6 A study of a fully coupled two-parameter system of sequential fractional integro-differential equations with nonlocal integro-multi-point boundary conditions

In this section, we study the following coupled two-parameter system of sequential fractional integro-differential equations

$$\begin{aligned}
&({}^CD^q + k_1\, {}^CD^{q-1})x(t) = f(t, x(t), y(t), {}^CD^{\alpha}y(t), I^{\alpha_1}y(t)),\\
&\qquad\qquad t \in [0,1], \ \ 2 < q \leq 3, \ 0 < \alpha, \alpha_1 < 1, \ \ k_1 > 0,\\
&({}^CD^p + k_2\, {}^CD^{p-1})y(t) = g(t, x(t), {}^CD^{\delta}x(t), I^{\delta_1}x(t), y(t)),\\
&\qquad\qquad 2 < p \leq 3, \ 0 < \delta, \delta_1 < 1, \ \ k_2 > 0,
\end{aligned} \tag{5.118}$$

associated with coupled nonlocal integro-multi-point boundary conditions:

$$\begin{aligned}
&x(0) = 0, \ x'(0) = 0, \ \sum_{i=1}^{m} a_i x(\zeta_i) = a\int_0^{\eta} \frac{(\eta - s)^{\beta-1}}{\Gamma(\beta)} y(s)ds,\\
&\qquad \beta > 0, \ 0 < \eta < \zeta_1 < \cdots < \zeta_m < 1,\\
&y(0) = 0, \ y'(0) = 0, \ \sum_{j=1}^{n} b_j y(\xi_j) = b\int_0^{\theta} \frac{(\theta - s)^{\gamma-1}}{\Gamma(\gamma)} x(s)ds,\\
&\qquad \gamma > 0, \ 0 < \theta < \xi_1 < \cdots < \xi_n < 1,
\end{aligned} \tag{5.119}$$

where ${}^CD^{(.)}$ denotes the Caputo derivative operator of fractional order (.) , $I^{(.)}$ denotes the Riemann-Liouville integral of fractional order (.), $f, g : [0,1] \times \mathbb{R}^4 \to \mathbb{R}$ are given continuous functions, and $a, \ a_i, \ b, \ b_i$ are real constants.

To define the solution for problem (5.118)-(5.119), we consider the following lemma dealing with its linear variant.

Lemma 5.8. *Let $h_1, h_2 \in C([0,1],\mathbb{R})$ and $x, y \in C^3([0,1],\mathbb{R})$. Then the integral solution for the linear system of fractional differential equations:*

$$\begin{aligned} ({}^CD^q + k_1\, {}^CD^{q-1})x(t) &= h_1(t),\\ ({}^CD^p + k_2\, {}^CD^{p-1})y(t) &= h_2(t), \end{aligned} \tag{5.120}$$

supplemented with the boundary conditions (5.119) is given by

$$\begin{aligned} x(t) = \rho_1(t)\Bigg\{ & A_2\Bigg[b\int_0^\theta \frac{(\theta-s)^{\gamma-1}}{\Gamma(\gamma)}\left(\int_0^s e^{-k_1(s-m)}\left(I^{q-1}h_1(m)\right)dm\right)ds \\ & -\sum_{j=1}^n b_j\int_0^{\xi_j} e^{-k_2(\xi_j-s)}\left(I^{p-1}h_2(s)\right)ds\Bigg] \\ & -B_2\Bigg[a\int_0^\eta \frac{(\eta-s)^{\beta-1}}{\Gamma(\beta)}\left(\int_0^s e^{-k_2(s-m)}\left(I^{p-1}h_2(m)\right)dm\right)ds \\ & -\sum_{i=1}^m a_i\int_0^{\zeta_i} e^{-k_1(\zeta_i-s)}\left(I^{q-1}h_1(s)\right)ds\Bigg]\Bigg\} + \int_0^t e^{-k_1(t-s)}\left(I^{q-1}h_1(s)\right)ds, \end{aligned} \tag{5.121}$$

$$\begin{aligned} y(t) = \rho_2(t)\Bigg\{ & B_1\Bigg[a\int_0^\eta \frac{(\eta-s)^{\beta-1}}{\Gamma(\beta)}\left(\int_0^s e^{-k_2(s-m)}\left(I^{p-1}h_2(m)\right)dm\right)ds \\ & -\sum_{i=1}^m a_i\int_0^{\zeta_i} e^{-k_1(\zeta_i-s)}\left(I^{q-1}h_1(s)\right)ds\Bigg] \\ & -A_1\Bigg[b\int_0^\theta \frac{(\theta-s)^{\gamma-1}}{\Gamma(\gamma)}\left(\int_0^s e^{-k_1(s-m)}\left(I^{q-1}h_1(m)\right)dm\right)ds \\ & -\sum_{j=1}^n b_j\int_0^{\xi_j} e^{-k_2(\xi_j-s)}\left(I^{p-1}h_2(s)\right)ds\Bigg]\Bigg\} + \int_0^t e^{-k_2(t-s)}\left(I^{p-1}h_2(s)\right)ds, \end{aligned} \tag{5.122}$$

where

$$\rho_i(t) = \frac{k_i t - 1 + e^{-k_i t}}{k_i^2\Delta},\quad i=1,2, \tag{5.123}$$

$$\Delta = B_1A_2 - A_1B_2 \neq 0, \tag{5.124}$$

$$A_1 = \sum_{i=1}^m \frac{a_i(k_1\zeta_i - 1 + e^{-k_1\zeta_i})}{k_1^2}, \tag{5.125}$$

$$A_2 = \frac{-a}{k_2^2\Gamma(\beta)}\left[\frac{k_2\eta^{\beta+1}}{\beta(\beta+1)} - \frac{\eta^\beta}{\beta} + \int_0^\eta (\eta-s)^{\beta-1}e^{-k_2 s}ds\right], \tag{5.126}$$

$$B_1 = \frac{-b}{k_1^2\Gamma(\gamma)}\left[\frac{k_1\theta^{\gamma+1}}{\gamma(\gamma+1)} - \frac{\theta^\gamma}{\gamma} + \int_0^\theta (\theta-s)^{\gamma-1}e^{-k_1 s}ds\right], \tag{5.127}$$

$$B_2 = \sum_{j=1}^n \frac{b_j(k_2\xi_j - 1 + e^{-k_2\xi_j})}{k_2^2}. \tag{5.128}$$

Proof. The general solution of the system (5.120) can be written as

$$x(t) = b_0 e^{-k_1 t} + \frac{b_1}{k_1}(1 - e^{-k_1 t}) + \frac{b_2}{k_1^2}(k_1 t - 1 + e^{-k_1 t}) + \int_0^t e^{-k_1(t-s)} \left(\int_0^s \frac{(s-\tau)^{q-2}}{\Gamma(q-1)} h_1(\tau)\, d\tau \right) ds, \tag{5.129}$$

$$y(t) = c_0 e^{-k_2 t} + \frac{c_1}{k_2}(1 - e^{-k_2 t}) + \frac{c_2}{k_2^2}(k_2 t - 1 + e^{-k_2 t}) + \int_0^t e^{-k_2(t-s)} \left(\int_0^s \frac{(s-\tau)^{p-2}}{\Gamma(p-1)} h_2(\tau)\, d\tau \right) ds, \tag{5.130}$$

where $b_i,\ c_i, (i = 0, 1, 2)$ are unknown arbitrary constants. From (5.129) and (5.130), we have

$$x'(t) = -k_1 b_0 e^{-k_1 t} + b_1 e^{-k_1 t} + \frac{b_2}{k_1}(1 - e^{-k_1 t}) - k_1 \int_0^t e^{-k_1(t-s)} \left(\int_0^s \frac{(s-\tau)^{q-2}}{\Gamma(q-1)} h_1(\tau)\, d\tau \right) ds + \int_0^t \frac{(t-\tau)^{q-2}}{\Gamma(q-1)} h_1(\tau)\, d\tau, \tag{5.131}$$

$$y'(t) = -k_2 c_0 e^{-k_2 t} + c_1 e^{-k_2 t} + \frac{c_2}{k_2}(1 - e^{-kt}) - k_2 \int_0^t e^{-k_2(t-s)} \left(\int_0^s \frac{(s-\tau)^{p-2}}{\Gamma(p-1)} h_2(\tau)\, d\tau \right) ds + \int_0^t \frac{(t-\tau)^{p-2}}{\Gamma(p-1)} h_2(\tau)\, d\tau. \tag{5.132}$$

Using the data $x(0) = 0, x'(0) = 0,\ y(0) = 0, y'(0) = 0$ given by (5.119) in (5.129)-(5.132), we find that $b_0 = 0,\ b_1 = 0,\ c_0 = 0,\ c_1 = 0$. Thus (5.129) and (5.130) take the form

$$x(t) = \frac{b_2}{k_1^2}(k_1 t - 1 + e^{-k_1 t}) + \int_0^t e^{-k_1(t-s)} \left(\int_0^s \frac{(s-\tau)^{q-2}}{\Gamma(q-1)} h_1(\tau)\, d\tau \right) ds, \tag{5.133}$$

$$y(t) = \frac{c_2}{k_2^2}(k_2 t - 1 + e^{-k_2 t}) + \int_0^t e^{-k_2(t-s)} \left(\int_0^s \frac{(s-\tau)^{p-2}}{\Gamma(p-1)} h_2(\tau)\, d\tau \right) ds. \tag{5.134}$$

Using the coupled integral boundary conditions:

$$\sum_{i=1}^{m} a_i x(\zeta_i) = a \int_0^\eta \frac{(\eta - s)^{\beta-1}}{\Gamma(\beta)} y(s)\, ds \quad \text{and} \quad \sum_{j=1}^{n} b_j y(\xi_j) = b \int_0^\theta \frac{(\theta - s)^{\gamma-1}}{\Gamma(\gamma)} x(s)\, ds$$

in (5.133) and (5.134), we obtain

$$A_1 b_2 + A_2 c_2 = J_1,\ \ B_1 b_2 + B_2 c_2 = J_2, \tag{5.135}$$

where A_i and B_i $(i = 1, 2)$ are respectively given by (5.126) and (5.128), and

$$J_1 = a \int_0^\eta \frac{(\eta - s)^{\beta-1}}{\Gamma(\beta)} \left(\int_0^s e^{-k_2(s-m)} \left(I^{p-1} h_2(m) \right) dm \right) ds$$

$$-\sum_{i=1}^{m} a_i \int_0^{\zeta_i} e^{-k_1(\zeta_i - s)} \left(I^{q-1}h_1(s)\right) ds, \tag{5.136}$$

$$J_2 = b\int_0^{\theta} \frac{(\theta-s)^{\gamma-1}}{\Gamma(\gamma)} \left(\int_0^s e^{-k_1(s-m)} \left(I^{q-1}h_1(m)\right) dm\right) ds$$
$$-\sum_{j=1}^{n} b_j \int_0^{\xi_j} e^{-k_2(\xi_j - s)} \left(I^{p-1}h_2(s)\right) ds. \tag{5.137}$$

Solving the system (5.135), we find that

$$b_2 = \frac{1}{\Delta}\Bigg\{A_2\Bigg[b\int_0^{\theta} \frac{(\theta-s)^{\gamma-1}}{\Gamma(\gamma)} \left(\int_0^s e^{-k_1(s-m)} \left(I^{q-1}h_1(m)\right) dm\right) ds$$
$$-\sum_{j=1}^{n} b_j \int_0^{\xi_j} e^{-k_2(\xi_j - s)} \left(I^{p-1}h_2(s)\right) ds\Bigg]$$
$$-B_2\Bigg[a\int_0^{\eta} \frac{(\eta-s)^{\beta-1}}{\Gamma(\beta)} \left(\int_0^s e^{-k_2(s-m)} \left(I^{p-1}h_2(m)\right) dm\right) ds$$
$$-\sum_{i=1}^{m} a_i \int_0^{\zeta_i} e^{-k_1(\zeta_i - s)} \left(I^{q-1}h_1(s)\right) ds\Bigg]\Bigg\}, \tag{5.138}$$

$$c_2 = \frac{1}{\Delta}\Bigg\{B_1\Bigg[a\int_0^{\eta} \frac{(\eta-s)^{\beta-1}}{\Gamma(\beta)} \left(\int_0^s e^{-k_2(s-m)} \left(I^{p-1}h_2(m)\right) dm\right) ds$$
$$-\sum_{i=1}^{m} a_i \int_0^{\zeta_i} e^{-k_1(\zeta_i - s)} \left(I^{q-1}h_1(s)\right) ds\Bigg]$$
$$-A_1\Bigg[b\int_0^{\theta} \frac{(\theta-s)^{\gamma-1}}{\Gamma(\gamma)} \left(\int_0^s e^{-k_1(s-m)} \left(I^{q-1}h_1(m)\right) dm\right) ds$$
$$-\sum_{j=1}^{n} b_j \int_0^{\xi_j} e^{-k_2(\xi_j - s)} \left(I^{p-1}h_2(s)\right) ds\Bigg]\Bigg\}, \tag{5.139}$$

where Δ is given by (5.124). Substituting the values of b_2 and c_2 in (5.133) and (5.134), we get the solution (5.121)-(5.122). Note that the converse follows by direct computation. This completes the proof. □

5.6.1 *Existence and uniqueness results*

This section is devoted to the main results concerning the existence and uniqueness of solutions for the system (5.118)-(5.119). First of all, we fix our terminology. Let $X = \{x : x \in C([0,1],\mathbb{R}) \text{ and } {}^{C}D^{\delta}x \in C([0,1],\mathbb{R})\}$ and $Y = \{y : y \in C([0,1],\mathbb{R}) \text{ and } {}^{C}D^{\alpha}y \in C([0,1],\mathbb{R})\}$ denote the spaces equipped respectively with the norms $\|x\|_X = \|x\| + \|{}^{C}D^{\delta}x\| = \sup_{t\in[0,1]}|x(t)| + \sup_{t\in[0,1]}|{}^{C}D^{\delta}x(t)|$ and $\|y\|_Y = \|y\| + \|{}^{C}D^{\alpha}y\| = \sup_{t\in[0,1]}|y(t)| + \sup_{t\in[0,1]}|{}^{C}D^{\alpha}y(t)|$. Observe that $(X, \|.\|_X)$ and $(Y, \|.\|_Y)$ are Banach spaces. In consequence, the product space $(X \times Y, \|.\|_{X\times Y})$ is a Banach space endowed with the norm $\|(x,y)\|_{X\times Y} = \|x\|_X + \|y\|_Y$ for $(x,y) \in X \times Y$.

Using Lemma 5.8, we introduce an operator $F : X \times Y \to X \times Y$ as follows:

$$F(u,v)(t) := (F_1(u,v)(t), F_2(u,v)(t)), \tag{5.140}$$

where

$$\begin{aligned}
&F_1(u,v)(t)\\
&= \rho_1(t)\Bigg\{ A_2\Bigg[b\int_0^\theta \frac{(\theta-s)^{\gamma-1}}{\Gamma(\gamma)}\bigg(\int_0^s e^{-k_1(s-m)}\\
&\quad \times\Big(I^{q-1}f(m,u(m),v(m),{}^CD^\alpha v(m),I^{\alpha_1}v(m))\Big)dm\bigg)ds\\
&\quad -\sum_{j=1}^n b_j\int_0^{\xi_j} e^{-k_2(\xi_j-s)}\left(I^{p-1}g(s,u(s),{}^CD^\delta u(s),I^{\delta_1}u(s),v(s))\right)ds\Bigg]\\
&\quad -B_2\Bigg[a\int_0^\eta \frac{(\eta-s)^{\beta-1}}{\Gamma(\beta)}\bigg(\int_0^s e^{-k_2(s-m)}\\
&\quad \times\Big(I^{p-1}g(m,u(m),{}^CD^\delta u(m),I^{\delta_1}u(m),v(m))\Big)dm\bigg)ds\\
&\quad -\sum_{i=1}^m a_i\int_0^{\zeta_i} e^{-k_1(\zeta_i-s)}\left(I^{q-1}f(s,u(s),v(s),{}^CD^\alpha v(s),I^{\alpha_1}v(s))\right)ds\Bigg]\Bigg\}\\
&\quad +\int_0^t e^{-k_1(t-s)}\left(I^{q-1}f(s,u(s),v(s),{}^CD^\alpha v(s),I^{\alpha_1}v(s))\right)ds,
\end{aligned} \tag{5.141}$$

$$\begin{aligned}
&F_2(u,v)(t)\\
&= \rho_2(t)\Bigg\{ B_1\Bigg[a\int_0^\eta \frac{(\eta-s)^{\beta-1}}{\Gamma(\beta)}\bigg(\int_0^s e^{-k_2(s-m)}\\
&\quad \times\Big(I^{p-1}g(m,u(m),{}^CD^\delta u(m),I^{\delta_1}u(m),v(m))\Big)dm\bigg)ds\\
&\quad -\sum_{i=1}^m a_i\int_0^{\zeta_i} e^{-k_1(\zeta_i-s)}\left(I^{q-1}f(s,u(s),v(s),{}^CD^\alpha v(s),I^{\alpha_1}v(s))\right)ds\Bigg]\\
&\quad -A_1\Bigg[b\int_0^\theta \frac{(\theta-s)^{\gamma-1}}{\Gamma(\gamma)}\bigg(\int_0^s e^{-k_1(s-m)}\\
&\quad \times\Big(I^{q-1}f(m,u(m),v(m),{}^CD^\alpha v(m),I^{\alpha_1}v(m))\Big)dm\bigg)ds\\
&\quad -\sum_{j=1}^n b_j\int_0^{\xi_j} e^{-k_2(\xi_j-s)}\left(I^{p-1}g(s,u(s),{}^CD^\delta u(s),I^{\delta_1}u(s),v(s))\right)ds\Bigg]\Bigg\}
\end{aligned}$$

$$+\int_0^t e^{-k_2(t-s)}\left(I^{p-1}g(s,u(s),{}^CD^{\delta}u(s),I^{\delta_1}u(s),v(s))\right)ds. \tag{5.142}$$

In the forthcoming analysis, we need the following assumptions:

(H_1) Let $f,g:[0,1]\times\mathbb{R}^4\to\mathbb{R}$ be continuous functions and there exist real constants $\mu_j,\lambda_j\geq 0$ $(j=1,2,3,4)$ and $\mu_0>0,\ \lambda_0>0$ such that

$$|f(t,x_1,x_2,x_3,x_4)|\leq \mu_0+\mu_1|x_1|+\mu_2|x_2|+\mu_3|x_3|+\mu_4|x_4|,$$
$$|g(t,x_1,x_2,x_3,x_4)|\leq \lambda_0+\lambda_1|x_1|+\lambda_2|x_2|+\lambda_3|x_3|+\lambda_4|x_4|,$$
$$\forall x_j\in\mathbb{R},\ j=1,2,3,4.$$

(H_2) There exist positive constants $l,\ l_1$ such that

$$\begin{aligned}&|f(t,u_1,u_2,u_3,u_4)-f(t,v_1,v_2,v_3,v_4)|\\ &\leq l\left[|u_1-v_1|+|u_2-v_2|+|u_3-v_3|+|u_4-v_4|\right],\\ &|g(t,u_1,u_2,u_3,u_4)-g(t,v_1,v_2,v_3,v_4)|\\ &\leq l_1\left[|u_1-v_1|+|u_2-v_2|+|u_3-v_3|+|u_4-v_4|\right],\\ &\forall t\in[0,1],\ u_j,\ v_j\in\mathbb{R}.\end{aligned}$$

For computational convenience, we set

$$\sigma_1=1+\frac{1}{\Gamma(1+\alpha_1)},\quad \sigma_2=1+\frac{1}{\Gamma(1+\delta_1)}, \tag{5.143}$$

$$\varrho_i=\sup_{t\in[0,1]}|\rho_i(t)|=\frac{1}{k_i|\Delta|},$$

$$\bar{\varrho}_i=\sup_{t\in[0,1]}|\rho_i'(t)|=\sup_{t\in[0,1]}\left|\frac{1-e^{-k_it}}{k_i\Delta}\right|,\quad i=1,2, \tag{5.144}$$

$$\begin{aligned}M_1=&\frac{\varrho_1}{k_1^2\Gamma(\gamma)\Gamma(q)}\Big\{|b|\ |A_2|\theta^{q+\gamma-2}(\theta k_1+e^{-k_1\theta}-1)\\ &+\sum_{i=1}^m k_1|B_2|\Gamma(\gamma)|a_i|\zeta_i^{q-1}(1-e^{-k_1\zeta_i})\Big\}+\frac{(1-e^{-k_1})}{k_1\Gamma(q)},\end{aligned} \tag{5.145}$$

$$\begin{aligned}M_2=&\frac{\varrho_1}{k_2^2\Gamma(\beta)\Gamma(p)}\Big\{\sum_{j=1}^n k_2|A_2|\Gamma(\beta)|b_j|\xi_j^{p-1}(1-e^{-k_2\xi_j})\\ &+|a||B_2|\eta^{q+\beta-2}(\eta k_2+e^{-k_2\eta}-1)\Big\},\end{aligned} \tag{5.146}$$

$$\begin{aligned}N_1=&\frac{\varrho_2}{k_1^2\Gamma(\gamma)\Gamma(q)}\Big\{\sum_{i=1}^m |a_i|k1|B_1|\zeta_i^{q-1}\Gamma(\gamma)(1-e^{-k_1\zeta_i})\\ &+|b|\ |A_1|\theta^{q+\gamma-2}(\theta k_1+e^{-k_1\theta}-1)\Big\},\end{aligned} \tag{5.147}$$

$$N_2=\frac{\varrho_2}{k_2^2\Gamma(\beta)\Gamma(p)}\Big\{|a|\ |B_1|\eta^{q+\beta-2}(\eta k_2+e^{-k_2\eta}-1) \tag{5.148}$$

$$+\sum_{j=1}^n |b_j|k_2|A_1|\xi_j^{p-1}\Gamma(\beta)(1-e^{-k_2\xi_j})\Big\}+\frac{1}{k_2\Gamma(q)}(1-e^{-k_2}), \tag{5.149}$$

$$M_1^* = \frac{\bar{\varrho}_1}{k_1^2\Gamma(\gamma)\Gamma(q)}\Big\{|b|\ |A_1|\theta^{q+\gamma-2}(\theta k_1 + e^{-k_1\theta} - 1) + \sum_{i=1}^{m} k_1|B_2||a_i|\zeta_i^{q-1}\Gamma(\gamma)(1 - e^{-k_1\zeta_i})\Big\} + \frac{1}{\Gamma(q)}(2 - e^{-k_1}), \tag{5.150}$$

$$M_2^* = \frac{\bar{\varrho}_1}{k_2^2\Gamma(\beta)\Gamma(p)}\Big\{\sum_{j=1}^{n} k_2|b_j||A_2|\xi_j^{p-1}\Gamma(\beta)(1 - e^{-k_2\xi_j}) + |a||B_2|\eta^{q+\beta-2}(\eta k_2 + e^{-k_2\eta} - 1)\Big\}, \tag{5.151}$$

$$N_1^* = \frac{\bar{\varrho}_2}{k_1^2\Gamma(\gamma)\Gamma(q)}\Big\{\sum_{i=1}^{m} |a_i|k_1|B_1|\zeta_i^{q-1}\Gamma(\gamma)(1 - e^{-k_1\zeta_i}) + |b|\ |A_1|\theta^{q+\gamma-2}(\theta k_1 + e^{-k_1\theta} - 1)\Big\}, \tag{5.152}$$

$$N_2^* = \frac{\bar{\varrho}_2}{k_2^2\Gamma(\beta)\Gamma(p)}\Big\{|a|\ |B_1|\eta^{q+\beta-2}(\eta k_2 + e^{-k_2\eta} - 1) + \sum_{j=1}^{n} |b_j|k_2|A_1|\xi_j^{p-1}\Gamma(\beta)(1 - e^{-k_2\xi_j})\Big\} + \frac{(2 - e^{-k_2})}{\Gamma(p)}, \tag{5.153}$$

$$\bar{\omega_1} = \mu_0\left(M_1 + \frac{1}{\Gamma(2-\delta)}M_1^* + N_1 + \frac{1}{\Gamma(2-\alpha)}N_1^*\right) + \lambda\Big(M_2 + \frac{1}{\Gamma(2-\delta)}M_2^* + N_2 + \frac{1}{\Gamma(2-\alpha)}N_2^*\Big), \tag{5.154}$$

$$\bar{\omega_2} = \mu_1 M_1 + (\lambda_1 + \lambda_2 + \sigma_2\lambda_4)M_2 + \frac{\mu_1}{\Gamma(2-\delta)}M_1^* + \frac{1}{\Gamma(2-\delta)}(\lambda_1 + \lambda_2 + \sigma_2\lambda_4)M_2^* + \mu_1 N_1 + (\lambda_1 + \lambda_2 + \sigma_2\lambda_4)N_2 + \frac{\mu_1}{\Gamma(2-\alpha)}N_1^* + \frac{1}{\Gamma(2-\gamma)}(\lambda_1 + \lambda_2 + \sigma_2\lambda_4)N_2^*, \tag{5.155}$$

$$\bar{\omega_3} = (\mu_2 + \mu_3 + \sigma_1\mu_4)M_1 + \lambda_3 M_2 + \frac{1}{\Gamma(2-\delta)}(\mu_2 + \mu_3 + \sigma_1\mu_4)M_1^* + \lambda_3 M_2^* + (\mu_2 + \mu_3 + \sigma_1\mu_4)N_1 + \lambda_3 N_2 + \frac{1}{\Gamma(2-\alpha)}(\mu_2 + \mu_3 + \sigma_1\mu_4)N_1^* + \frac{\lambda_3}{\Gamma(2-\alpha)}N_2^*. \tag{5.156}$$

Our first existence result in this section is based on Leray-Schauder alternative (Theorem 1.7).

Theorem 5.27. *Assume that* (H_1) *holds. In addition it is assumed that* $max\{\bar{\omega_2}, \bar{\omega_3}\} < 1$*, where* $\bar{\omega_2}$ *and* $\bar{\omega_3}$ *are given by (5.155) and (5.156) respectively. Then the system (5.118)-(5.119) has at least one solution on* $[0, 1]$.

Proof. In the first step, we show that the operator $F: X \times Y \to X \times Y$ is completely continuous. By continuity of the functions f and g, it follows that the operators F_1 and F_2 are continuous. In consequence, the operator F is continuous. Next we show that the operator F is uniformly bounded. For that, let $\Omega \subset X \times Y$ be a bounded set. Then there exist positive constants L_1 and L_2 such that $|f(t, u(t), v(t), {}^C D^\alpha v(t), I^{\alpha_1} v(t))| \leq L_1$, $|g(t, u(t), {}^C D^\delta u(t), I^{\delta_1} u(t), v(t))| \leq L_2$, $\forall (u, v) \in \Omega$. Then, for any $(u, v) \in \Omega$, we have

$$
\begin{aligned}
&|F_1(u,v)(t)| \\
&\leq L_1|\rho_1(t)| \Bigg[|b|\,|A_2| \int_0^\theta \frac{(\theta-s)^{\gamma-1}}{\Gamma(\gamma)} \left(\int_0^s e^{-k_1(s-m)} \left(\int_0^m \frac{(m-\tau)^{q-2}}{\Gamma(q-1)} d\tau \right) dm \right) ds \\
&\quad + \sum_{i=1}^m |a_i||B_2| \int_0^{\zeta_i} e^{-k_1(\zeta_i - s)} \left(\int_0^m \frac{(m-\tau)^{q-2}}{\Gamma(q-1)} d\tau \right) ds \Bigg] \\
&\quad + L_2|\rho_1(t)| \Bigg[\sum_{j=1}^n |b_j||A_2| \int_0^{\xi_j} e^{-k_2(\xi_j - s)} \left(\int_0^s \frac{(s-\tau)^{p-2}}{\Gamma(p-1)} d\tau \right) ds \\
&\quad + |a||B_2| \int_0^\eta \frac{(\eta-s)^{\beta-1}}{\Gamma(\beta)} \left(\int_0^s e^{-k_2(s-m)} \left(\int_0^m \frac{(m-\tau)^{p-2}}{\Gamma(p-1)} d\tau \right) dm \right) ds \Bigg] \\
&\quad + L_1 \int_0^t e^{-k_1(t-s)} \left(\int_0^s \frac{(s-\tau)^{q-2}}{\Gamma(q-1)} d\tau \right) ds \\
&\leq L_1 \Bigg\{ \varrho_1 \left[\frac{|b|\,|A_2|\theta^{q+\gamma-2}}{k_1^2 \Gamma(\gamma)\Gamma(q)} (\theta k_1 + e^{-k_1\theta} - 1) + \frac{|B_2| \sum_{i=1}^m |a_i| \zeta_i^{q-1}(1 - e^{-k_1\zeta_i})}{k_1 \Gamma(q)} \right] \\
&\quad + \frac{(1-e^{-k_1})}{k_1\Gamma(q)} \Bigg\} + L_2 \varrho_1 \Bigg[\frac{\sum_{j=1}^n |A_2||b_j| \xi_j^{p-1}(1 - e^{-k_2\xi_j})}{k_2 \Gamma(p)} \\
&\quad + \frac{|a||B_2|\eta^{q+\beta-2}}{k_2^2 \Gamma(\beta)\Gamma(p)} (\eta k_2 + e^{-k_2\eta} - 1) \Bigg],
\end{aligned}
$$

which, on taking the norm for $t \in [0, 1]$ and using (5.145), (5.146) yields

$$\|F_1(u, v)\| \leq L_1 M_1 + L_2 M_2.$$

From (5.141), using the notation given by (5.150) and (5.151), we obtain

$$
\begin{aligned}
&|F_1'(u,v)(t)| \\
&\leq |\rho_1'(t)| \Bigg\{ |b|\,|A_2| \int_0^\theta \frac{(\theta-s)^{\gamma-1}}{\Gamma(\gamma)} \Bigg(\int_0^s e^{-k_1(s-m)} \\
&\quad \times \left(I^{q-1} |f(m, u(m), v(m), {}^C D^\alpha v(m), I^{\alpha_1} v(m))| \right) dm \Bigg) ds \\
&\quad + \sum_{j=1}^n |A_2||b_j| \int_0^{\xi_j} e^{-k_2(\xi_j - s)} \left(I^{p-1} |g(s, u(s), {}^C D^\delta u(s), I^{\delta_1} u(s), v(s))| \right) ds
\end{aligned}
$$

$$
+|a||B_2|\int_0^\eta \frac{(\eta-s)^{\beta-1}}{\Gamma(\beta)}\left(\int_0^s e^{-k_2(s-m)}\right.
$$

$$
\left.\times\left(I^{p-1}|g(m,u(m),{}^CD^{\delta}u(m),I^{\delta_1}u(m),v(m))|\right)dm\right)ds
$$

$$
\left.+\sum_{i=1}^m |B_2||a_i|\int_0^{\zeta_i} e^{-k_1(\zeta_i-s)}\left(I^{q-1}|f(s,u(s),v(s),{}^CD^{\alpha}v(s),I^{\alpha_1}v(s))|\right)ds\right\}
$$

$$
+k_1\int_0^t e^{-k_1(t-s)}\left(I^{q-1}|f(s,u(s),v(s),{}^CD^{\alpha}v(s),I^{\alpha_1}v(s))|\right)ds
$$

$$
+\int_0^t \frac{(t-s)^{q-2}}{\Gamma(q-1)}|f(s,u(s),v(s),{}^CD^{\alpha}v(s),I^{\alpha_1}v(s))|\,ds
$$

$$
\le L_1\left\{\bar{\varrho}_1\left[|b|\ |A_2|\frac{\theta^{q+\gamma-2}}{k_1^2\Gamma(\gamma)\Gamma(q)}(\theta k_1+e^{-k_1\theta}-1)+\sum_{i=1}^m |B_2||a_i|\frac{\zeta_i^{q-1}}{k_1\Gamma(q)}(1-e^{-k_1\zeta_i})\right]\right.
$$

$$
\left.+\frac{1}{\Gamma(q)}(2-e^{-k_1})\right\}+L_2\bar{\varrho}_1\left[\sum_{j=1}^n |A_2||b_j|\frac{\xi_j^{p-1}}{k_2\Gamma(p)}(1-e^{-k_2\xi_j})\right.
$$

$$
\left.+|a||B_2|\frac{\eta^{q+\beta-2}}{k_2^2\Gamma(\beta)\Gamma(p)}(\eta k_2+e^{-k_2\eta}-1)\right]
$$

$$
\le L_1M_1^*+L_2M_2^*.
$$

By definition of Caputo fractional derivative operator with $0<\delta<1$, we get

$$
|{}^CD^{\delta}F_1(u,v)(t)|\le\int_0^t\frac{(t-s)^{-\delta}}{\Gamma(1-\delta)}|F_1'(u,v)(s)|ds\le(L_1M_1^*+L_2M_2^*)\int_0^t\frac{(t-s)^{-\delta}}{\Gamma(1-\delta)}ds
$$

$$
\le\frac{1}{\Gamma(2-\delta)}\left(L_1M_1^*+L_2M_2^*\right).
$$

Hence

$$
\begin{aligned}
\|F_1(u,v)\|_X &= \|F_1(u,v)\|+\|{}^CD^{\delta}F_1(u,v)\| \\
&\le L_1M_1+L_2M_2+\frac{1}{\Gamma(2-\delta)}\left(L_1M_1^*+L_2M_2^*\right). \qquad (5.157)
\end{aligned}
$$

Similarly, we find that

$$
|F_2(u,v)(t)|
$$

$$
\le L_1\varrho_2\left[|b|\ |A_1|\frac{\theta^{q+\gamma-2}}{k_1^2\Gamma(\gamma)\Gamma(q)}(\theta k_1+e^{-k_1\theta}-1)+\sum_{i=1}^m|a_i||B_1|\frac{\zeta_i^{q-1}}{k_1\Gamma(q)}(1-e^{-k_1\zeta_i})\right]
$$

$$
+L_2\left\{\varrho_2\left[\sum_{j=1}^n|b_j||A_1|\frac{\xi_j^{p-1}}{k_2\Gamma(p)}(1-e^{-k_2\xi_j})+|a||B_1|\frac{\eta^{q+\beta-2}}{k_2^2\Gamma(\beta)\Gamma(p)}(\eta k_2+e^{-k_2\eta}-1)\right]\right.
$$

$$
\left.+\frac{1}{k_2\Gamma(q)}(1-e^{-k_2})\right\}
$$

$$\leq L_1N_1 + L_2N_2,$$

$$|F_2'(u,v)(t)|$$
$$\leq L_1\bar{\varrho}_2\left[|b|\ |A_1|\frac{\theta^{q+\gamma-2}}{k_1^2\Gamma(\gamma)\Gamma(q)}(\theta k_1+e^{-k_1\theta}-1)+\sum_{i=1}^{m}|a_i||B_1|\frac{\zeta_i^{q-1}}{k_1\Gamma(q)}(1-e^{-k_1\zeta_i})\right]$$
$$+L_2\left\{\bar{\varrho}_2\left[\sum_{j=1}^{n}|b_j||A_1|\frac{\xi_j^{p-1}}{k_2\Gamma(p)}(1-e^{-k_2\xi_j})+|a||B_1|\frac{\eta^{q+\beta-2}}{k_2^2\Gamma(\beta)\Gamma(p)}(\eta k_2+e^{-k_2\eta}-1)\right]\right.$$
$$\left.+\frac{1}{\Gamma(p)}(2-e^{-k_2})\right\}$$
$$\leq L_1N_1^* + L_2N_2^*,$$

$$\begin{aligned}|^CD^\alpha F_2(u,v)(t)| &\leq \int_0^t \frac{(t-s)^{-\alpha}}{\Gamma(1-\alpha)}|F_2'(u,v)(s)|ds\\ &\leq \frac{1}{\Gamma(2-\alpha)}\left(L_1N_1^*+L_2N_2^*\right),\end{aligned} \tag{5.158}$$

where N_1, N_2, N_1^*, N_2^* are respectively given by (5.147), (5.149), (5.152) and (5.153).

In consequence, we get

$$\begin{aligned}\|F_2(u,v)\|_Y &= \|F_2(u,v)\| + \|^CD^\alpha F_2(u,v)\|\\ &\leq L_1N_1+L_2N_2+\frac{1}{\Gamma(2-\alpha)}\left(L_1N_1^*+L_2N_2^*\right).\end{aligned} \tag{5.159}$$

From the inequalities (5.157) and (5.159), we deduce that F_1 and F_2 are uniformly bounded, which implies that the operator F is uniformly bounded.

Next, we show that F is equi-continuous. Let $t_1,\ t_2 \in [0,1]$ with $t_1 < t_2$. Then we have

$$|F_1(u,v)(t_2)-F_1(u,v)(t_1)|$$
$$\leq L_1\frac{|k_1(t_2-t_1)+e^{-k_1t_1}(e^{-k_1(t_2-t_1)}-1)|}{k_1^2|\Delta|}\left[|b|\ |A_2|\right.$$
$$\times\int_0^\theta\frac{(\theta-s)^{\gamma-1}}{\Gamma(\gamma)}\left(\int_0^s e^{-k_1(s-m)}\left(\int_0^m\frac{(m-\tau)^{q-2}}{\Gamma(q-1)}d\tau\right)dm\right)ds$$
$$\left.+\sum_{i=1}^{m}|a_i||B_2|\int_0^{\zeta_i}e^{-k_1(\zeta_i-s)}\left(\int_0^s\frac{(s-\tau)^{q-2}}{\Gamma(q-1)}d\tau\right)ds\right]$$
$$+L_2\frac{|k_1(t_2-t_1)+e^{-k_1t_1}(e^{-k_1(t_2-t_1)}-1)|}{k_1^2|\Delta|}$$
$$\times\left[\sum_{j=1}^{n}|b_j||A_2|\int_0^{\xi_j}e^{-k_2(\xi_j-s)}\left(\int_0^s\frac{(s-\tau)^{p-2}}{\Gamma(p-1)}d\tau\right)ds\right.$$
$$\left.+\ |a||B_2|\int_0^\eta\frac{(\eta-s)^{\beta-1}}{\Gamma(\beta)}\left(\int_0^s e^{-k_2(s-m)}\left(\int_0^m\frac{(m-\tau)^{p-2}}{\Gamma(p-1)}d\tau\right)dm\right)ds\right]$$

$$+L_1\left[\int_0^{t_1}|e^{-k_1(t_2-s)}-e^{-k_1(t_1-s)}|\left(\int_0^s\frac{(s-\tau)^{q-2}}{\Gamma(q-1)}d\tau\right)ds\right.$$
$$\left.+\int_{t_1}^{t_2}e^{-k_1(t_2-s)}\left(\int_0^s\frac{(s-\tau)^{q-2}}{\Gamma(q-1)}d\tau\right)ds\right].$$

Clearly, $|F_1(u,v)(t_2)-F_1(u,v)(t_1)|\to 0$ as $t_2\to t_1$.

Also $|^CD^\delta F_1(u,v)(t_2)-{}^CD^\delta F_1(u,v)(t_1)|\to 0$ as $t_2\to t_1$ as

$$|^CD^\delta F_1(u,v)(t_2)-{}^CD^\delta F_1(u,v)(t_1)|$$
$$\le\frac{1}{\Gamma(1-\delta)}\int_0^{t_1}\frac{|(t_1-s)^\delta-(t_2-s)^\delta|}{(t_1-s)^\delta(t_2-s)^\delta}|F_1'(u,v)(s)|ds$$
$$+\frac{1}{\Gamma(1-\delta)}\int_{t_1}^{t_2}|(t_2-s)^{-\delta}||F_1'(u,v)(s)|ds$$
$$\le\frac{(L_1M_1^*+L_2M_2^*)}{\Gamma(1-\delta)}\left\{\int_0^{t_1}\frac{|(t_1-s)^\delta-(t_2-s)^\delta|}{(t_1-s)^\delta(t_2-s)^\delta}ds+\int_{t_1}^{t_2}|(t_2-s)^{-\delta}|ds\right\}.$$

In a similar manner, one can obtain that $|F_2(u,v)(t_2)-F_2(u,v)(t_1)|\to 0$ and $|^CD^\alpha F_2(u,v)(t_2)-{}^CD^\alpha F_2(u,v)(t_1)|\to 0$ as $t_2\to t_1$. Thus the operator F is equi-continuous in view of equi-continuity of F_1 and F_2. Therefore, by Arzelá-Ascoli's theorem, we deduce that the operator F is completely continuous.

Finally, it will be established that the set $\varepsilon(F)=\{(u,v)\in X\times Y:(u,v)=\lambda F(u,v)\ ;\ 0\le\lambda\le 1\}$ is bounded. Let $(u,v)\in\varepsilon(F)$. Then $(u,v)=\lambda F(u,v)$. For any $t\in[0,1]$, we have $u(t)=\lambda F_1(u,v)(t),\ v(t)=\lambda F_2(u,v)(t)$. Thus

$$|u(t)|=\lambda|F_1(u,v)(t)|\le|F_1(u,v)(t)|,\quad(|\lambda|\le 1).$$

Using (H_1) in (5.141), we get

$$|u(t)|\le|\rho_1(t)|\left\{|b||A_2|\int_0^\theta\frac{(\theta-s)^{\gamma-1}}{\Gamma(\gamma)}\left(\int_0^s e^{-k_1(s-m)}\left(\int_0^m\frac{(m-s)^{q-2}}{\Gamma(q-1)}\left[\mu_0\right.\right.\right.\right.$$
$$\left.\left.\left.+\mu_1|u(\tau)|+\mu_2|v(\tau)|+\mu_3|^CD^\alpha v(\tau)|+\mu_4|I^{\alpha_1}v(\tau)|\right]d\tau\right)dm\right)ds$$
$$+\sum_{j=1}^n|b_j||A_2|\int_0^{\xi_j}e^{-k_2(\xi_j-s)}\left(\int_0^s\frac{(s-\tau)^{p-2}}{\Gamma(p-1)}\left[\lambda_0\right.\right.$$
$$\left.\left.+\lambda_1|u(\tau)|+\lambda_2|^CD^\delta u(\tau)|+\lambda_3|v(\tau)|+\lambda_4|I^{\delta_1}u(\tau)|\right]d\tau\right)ds$$
$$+|a||B_2|\int_0^\eta\frac{(\eta-s)^{\beta-1}}{\Gamma(\beta)}\left(\int_0^s e^{-k_2(s-m)}\left(\int_0^m\frac{(m-\tau)^{p-2}}{\Gamma(p-1)}\left[\lambda_0\right.\right.\right.$$
$$\left.\left.\left.+\lambda_1|u(\tau)|+\lambda_2|^CD^\delta u(\tau)|+\lambda_3|v(\tau)|+\lambda_4|I^{\delta_1}u(\tau)|\right]d\tau\right)dm\right)ds$$
$$+\sum_{i=1}^m|a_i||B_2|\int_0^{\zeta_i}e^{-k_1(\zeta_i-s)}\left(\int_0^s\frac{(s-\tau)^{q-2}}{\Gamma(q-1)}\left[\mu_0+\mu_1|u(\tau)|\right.\right.$$
$$\left.\left.\left.+\mu_2|v(\tau)|+\mu_3|^CD^\alpha v(\tau)|+\mu_4|I^{\alpha_1}v(\tau)|\right]d\tau\right)ds\right\}$$

$$+\int_0^t e^{-k_1(t-s)}\Big(\int_0^s \frac{(s-\tau)^{q-2}}{\Gamma(q-1)}\Big[\mu_0+\mu_1|u(\tau)|$$
$$+\mu_2|v(\tau)|+\mu_3|^C D^{\alpha}v(\tau)|+\mu_4|I^{\alpha_1}v(\tau)|\Big]d\tau\Big)ds,$$

which, on taking the norm for $t\in[0,1]$, yields

$$\|u\|\le\Big(\mu_0+\mu_1\|u\|_X+\Big(\mu_2+\mu_3+\frac{\mu_4}{\Gamma(1-\alpha_1)}\Big)\|v\|_Y\Big)M_1$$
$$+\Big(\lambda_0+\Big(\lambda_1+\lambda_2+\frac{\lambda_4}{\Gamma(1-\delta_1)}\Big)\|u\|_X+\lambda_3\|v\|_Y\Big)M_2.$$

Similarly, we obtain

$$\|u'\|\le\Big(\mu_0+\mu_1\|u\|_X+\Big(\mu_2+\mu_3+\frac{\mu_4}{\Gamma(1-\alpha_1)}\Big)\|v\|_Y\Big)M_1^*$$
$$+\Big(\lambda_0+\Big(\lambda_1+\lambda_2+\frac{\lambda_4}{\Gamma(1-\delta_1)}\Big)\|u\|_X+\lambda_3\|v\|_Y\Big)M_2^*,$$

which implies that

$$\|^C D^{\delta}u\|\le\frac{1}{\Gamma(2-\delta)}\Big\{\Big(\mu_0+\mu_1\|u\|_X+\Big(\mu_2+\mu_3+\frac{\mu_4}{\Gamma(1-\alpha_1)}\Big)\|v\|_Y\Big)M_1^*$$
$$+\Big(\lambda_0+\Big(\lambda_1+\lambda_2+\frac{\lambda_4}{\Gamma(1-\delta_1)}\Big)\|u\|_X+\lambda_3\|v\|_Y\Big)M_2^*\Big\}.$$

Thus we have

$$\begin{aligned}\|u\|_X&=\|u\|+\|^C D^{\delta}u\|\\&\le\Big(\mu_0+\mu_1\|u\|_X+\Big(\mu_2+\mu_3+\frac{\mu_4}{\Gamma(1-\alpha_1)}\Big)\|v\|_Y\Big)M_1\\&\quad+\Big(\lambda_0+\Big(\lambda_1+\lambda_2+\frac{\lambda_4}{\Gamma(1-\delta_1)}\Big)\|u\|_X+\lambda_3\|v\|_Y\Big)M_2\\&\quad+\frac{1}{\Gamma(2-\delta)}\Big\{\Big(\mu_0+\mu_1\|u\|_X+\Big(\mu_2+\mu_3+\frac{\mu_4}{\Gamma(1-\alpha_1)}\Big)\|v\|_Y\Big)M_1^*\\&\quad+\Big(\lambda_0+\Big(\lambda_1+\lambda_2+\frac{\lambda_4}{\Gamma(1-\delta_1)}\Big)\|u\|_X+\lambda_3\|v\|_Y\Big)M_2^*\Big\}.\end{aligned}\tag{5.160}$$

Likewise, we can have

$$\begin{aligned}\|v\|_Y&\le\Big(\mu_0+\mu_1\|u\|_X+\Big(\mu_2+\mu_3+\frac{\mu_4}{\Gamma(1-\alpha_1)}\Big)\|v\|_Y\Big)N_1\\&\quad+\Big(\lambda_0+\Big(\lambda_1+\lambda_2+\frac{\lambda_4}{\Gamma(1-\delta_1)}\Big)\|u\|_X+\lambda_3\|v\|_Y\Big)N_2\\&\quad+\frac{1}{\Gamma(2-\alpha)}\Big\{\Big(\mu_0+\mu_1\|u\|_X+\Big(\mu_2+\mu_3+\frac{\mu_4}{\Gamma(1-\alpha_1)}\Big)\|v\|_Y\Big)N_1^*\\&\quad+\Big(\lambda_0+\Big(\lambda_1+\lambda_2+\frac{\lambda_4}{\Gamma(1-\delta_1)}\Big)\|u\|_X+\lambda_3\|v\|_Y\Big)N_2^*\Big\}.\end{aligned}\tag{5.161}$$

From (5.160) and (5.161), we find that

$$\|u\|_X+\|v\|_Y$$

$$
\begin{aligned}
&= \mu_0 \left[M_1 + \frac{1}{\Gamma(2-\delta)} M_1^* + N_1 + \frac{1}{\Gamma(2-\alpha)} N_1^* \right] \\
&+\lambda_0 \left[M_2 + \frac{1}{\Gamma(2-\delta)} M_2^* + N_2 + \frac{1}{\Gamma(2-\alpha)} N_2^* \right] \\
&+\|u\|_X \Bigg[\mu_1 M_1 + \Big(\lambda_1 + \lambda_2 + \frac{\lambda_4}{\Gamma(1-\delta_1)}\Big) M_2 + \frac{\mu_1}{\Gamma(2-\delta)} M_1^* \\
&+\frac{1}{\Gamma(2-\delta)} \Big(\lambda_1 + \lambda_2 + \frac{\lambda_4}{\Gamma(1-\delta_1)}\Big) M_2^* \\
&+\Big(\mu_1 N_1 + \lambda_1 + \lambda_2 + \frac{\lambda_4}{\Gamma(1-\delta_1)}\Big) N_2 + \frac{\mu_1}{\Gamma(2-\alpha)} N_1^* \\
&+\frac{1}{\Gamma(2-\gamma)} \Big(\lambda_1 + \lambda_2 + \frac{\lambda_4}{\Gamma(1-\delta_1)}\Big) N_2^* \Bigg] \\
&+\|v\|_Y \Bigg[\Big(\mu_2 + \mu_3 + \frac{\mu_4}{\Gamma(1-\alpha_1)}\Big) M_1 + \lambda_3 M_2 \\
&+\frac{1}{\Gamma(2-\delta)} \Big(\mu_2 + \mu_3 + \frac{\mu_4}{\Gamma(1-\alpha_1)}\Big) M_1^* + \lambda_3 M_2^* \\
&+\Big(\mu_2 + \mu_3 + \frac{\mu_4}{\Gamma(1-\alpha_1)}\Big) N_1 + \lambda_3 N_2 \\
&+\frac{1}{\Gamma(2-\alpha)} \Big(\mu_2 + \mu_3 + \frac{\mu_4}{\Gamma(1-\alpha_1)}\Big) N_1^* + \frac{\lambda_3}{\Gamma(2-\alpha)} N_2^* \Bigg] \\
&\le \bar{\omega_1} + \max\{\bar{\omega_2}, \bar{\omega_3}\} \|(u,v)\|_{X\times Y}, \qquad (5.162)
\end{aligned}
$$

which, together with $\|(u,v)\|_{X\times Y} = \|u\|_X + \|v\|_Y$, yields

$$
\|(u,v)\|_{X\times Y} \le \frac{\bar{\omega_1}}{1 - \max\{\bar{\omega_2}, \bar{\omega_3}\}}.
$$

This shows that $\varepsilon(F)$ is bounded. Thus, Leary-Schauder alternative (Theorem 1.7) applies and therefore F has at least one fixed point. Consequently, the system (5.118)-(5.119) has at least one solution on $[0,1]$. This completes the proof. □

Our next result deals with the uniqueness of solutions for the system (5.118)-(5.119), which relies on Banach's contraction mapping principle. For computational convenience, we introduce the notation:

$$
\begin{aligned}
\Lambda = l\sigma_1 M_1 + l_1\sigma_2 M_2, &\quad \bar{M} = r_1 M_1 + r_2 M_2, \\
\Lambda_1 = l\sigma_1 M_1^* + l_1\sigma_2 M_2^*, &\quad \bar{M}_1 = r_1 M_1^* + r_2 M_2^*, \\
\Lambda' = l\sigma_1 N_1 + l_1\sigma_2 N_2, &\quad \bar{N} = r_1 N_1 + r_2 N_2, \qquad (5.163)\\
\Lambda_1' = l\sigma_1 N_1^* + l_1\sigma_2 N_2^*, &\quad \bar{N}_1 = r_1 N_1^* + r_2 N_2^*, \\
r_1 = \sup_{t\in[0,1]} |f(t,0,0,0,0)| < \infty, &\quad r_2 = \sup_{t\in[0,1]} |g(t,0,0,0,0)| < \infty,
\end{aligned}
$$

where σ_1 and σ_2 are given by (5.143).

Theorem 5.28. *Assume that* (H_2) *holds. Further*

$$\Lambda + \frac{\Lambda_1}{\Gamma(2-\delta)} + \Lambda' + \frac{\Lambda_1'}{\Gamma(2-\alpha)} < 1, \tag{5.164}$$

where $\Lambda,\ \Lambda_1,\ \Lambda'$ *and* Λ_1' *are given by (5.163). Then the system (5.118)-(5.119) has a unique solution on* $[0,1]$.

Proof. Let us fix

$$r \geq \frac{\bar{M} + \dfrac{\bar{M}_1}{\Gamma(2-\delta)} + \bar{N} + \dfrac{\bar{N}_1}{\Gamma(2-\alpha)}}{1 - \left(\Lambda + \dfrac{\Lambda_1}{\Gamma(2-\delta)} + \Lambda' + \dfrac{\Lambda_1'}{\Gamma(2-\alpha)}\right)},$$

where $\Lambda,\ \Lambda_1,\ \Lambda',\ \Lambda_1'$ and $\bar{M},\ \bar{M}_1,\ \bar{N}$ and $\bar{N}_1$ are given by (5.163). Then we show that $FB_r \subset B_r$, where

$$B_r = \{(u,v) \in X \times Y : \|(u,v)\|_{X\times Y} \leq r\}.$$

For $(u,v) \in B_r$, we have

$$\begin{aligned}
&|f(t,u(t),v(t),{}^CD^{\alpha}v(t),I^{\alpha_1}v(t))|\\
&\leq |f(t,u(t),v(t),{}^CD^{\alpha}v(t)I^{\alpha_1}v(t)) - f(t,0,0,0,0)| + |f(t,0,0,0,0)|\\
&\leq l[|u(t)| + |v(t)| + |{}^CD^{\alpha}v(t)| + |I^{\alpha_1}v(t)|] + r_1\\
&\leq l[|u(t)| + \sigma_1|v(t)| + |{}^CD^{\alpha}v(t)|] + r_1\\
&\leq l[\|u\|_X + \sigma_1\|v\|_Y] + r_1\\
&\leqslant l\sigma_1\|(u,v)\|_{X\times Y} + r_1 \leqslant l\sigma_1 r + r_1.
\end{aligned}$$

Similarly, we have

$$|g(t,u(t),{}^CD^{\delta}u(t),I^{\delta_1}u(t),v(t))| \leq l_1\sigma_2 r + r_2.$$

Then

$$\begin{aligned}
|F_1(u,v)(t)| &\leq (l\sigma_1 r + r_1)M_1 + (l_1\sigma_2 r + r_2)M_2\\
&= (l\sigma_1 M_1 + l_1\sigma_2 M_2)r + r_1M_1 + r_2M_2 \leqslant \Lambda r + \bar{M},
\end{aligned}$$

and

$$\begin{aligned}
|F_1'(u,v)(t)| &\leq (l\sigma_1 r + r_1)M_1^* + (l_1\sigma_2 r + r_2)M_2^*\\
&= (l\sigma_1 M_1^* + l_1\sigma_2 M_2^*)r + r_1M_1^* + r_2M_2^* \leqslant \Lambda_1 r + \bar{M}_1,
\end{aligned}$$

which implies that

$$|{}^CD^{\delta}F_1(u,v)(t)| \leqslant \int_0^t \frac{(t-s)^{-\delta}}{\Gamma(1-\delta)}\,|F_1'(u,v)(s)|\,ds \leqslant \frac{1}{\Gamma(2-\delta)}(\Lambda_1 r + \bar{M}).$$

Therefore,

$$\begin{aligned}
\|F_1(u,v)\|_X &= \|F_1(u,v)\| + \|{}^CD^{\delta}F_1(u,v)\|\\
&\leqslant \Lambda r + \bar{M} + \frac{1}{\Gamma(2-\delta)}(\Lambda_1 r + \bar{M}_1)
\end{aligned}$$

$$= \left(\Lambda + \frac{\Lambda_1}{\Gamma(2-\delta)}\right) r + \left(\bar{M} + \frac{\bar{M}_1}{\Gamma(2-\delta)}\right). \tag{5.165}$$

In a similar manner, we obtain

$$|F_2(u,v)(t)| \leq \Lambda' r + \bar{N}, \qquad |F_2'(u,v)(t)| \leq \Lambda_1' r + \bar{N}_1,$$

$$|{}^C D^\alpha F_2(u,v)(t)| \leq \int_0^t \frac{(t-s)^{-\alpha}}{\Gamma(1-\alpha)} |F_2'(u,v)(s)|\, ds \leq \frac{1}{\Gamma(2-\alpha)}(\Lambda_1' r + \bar{N}).$$

In consequence, we get

$$\begin{aligned} \|F_2(u,v)\|_Y &= \|F_2(u,v)\| + \|{}^C D^\alpha F_2(u,v)\| \\ &\leqslant \left(\Lambda' + \frac{\Lambda_1'}{\Gamma(2-\alpha)}\right) r + \left(\bar{N} + \frac{\bar{N}_1}{\Gamma(2-\alpha)}\right). \end{aligned} \tag{5.166}$$

Thus, it follows from (5.165) and (5.166) that

$$\|F(u,v)\|_{X\times Y} = \|F_1(u,v)\|_X + \|F_2(u,v)\|_Y \leqslant r,$$

which implies $FB_r \subset B_r$.

Now we prove that the operator F is a contraction. For $u_i,\ v_i \in B_r;\ i = 1,2$ and for each $t \in [0,1]$ we have

$$\begin{aligned} &|F_1(u_1,v_1)(t) - F_1(u_2,v_2)(t)| \\ &\leq |\rho_1(t)| \Bigg\{ |b||A_2| \int_0^\theta \frac{(\theta-s)^{\gamma-1}}{\Gamma(\gamma)} \left(\int_0^s e^{-k_1(s-m)} \left(\int_0^m \frac{(m-\tau)^{q-2}}{\Gamma(q-1)} \right.\right. \\ &\times \Big| f(\tau, u_1(\tau), v_1(\tau), {}^C D^\alpha v_1(\tau), I^{\alpha_1} v_1(t)) \\ &- f(\tau, u_2(\tau), v_2(\tau), {}^C D^\alpha v_2(\tau), I^{\alpha_1} v_2(t)) \Big| d\tau \Big) dm \Big) ds \\ &+ \sum_{j=1}^n |b_j||A_2| \int_0^{\xi_j} e^{-k_2(\xi_j - s)} \left(\int_0^s \frac{(s-\tau)^{p-2}}{\Gamma(p-1)} \right. \\ &\times \Big| g(\tau, u_1(\tau), {}^C D^\delta u_1(\tau), I^{\delta_1} u_1(t), v_1(\tau)) \\ &- g(\tau, u_2(\tau), {}^C D^\delta u_2(\tau), I^{\delta_1} u_2(t), v_2(\tau)) \Big| d\tau \Big) ds \\ &+ \sum_{i=1}^m |a_i||B_2| \int_0^{\zeta_i} e^{-k_1(\zeta_i - s)} \left(\int_0^s \frac{(s-\tau)^{q-2}}{\Gamma(q-1)} \right. \\ &\times \Big| f(\tau, u_1(\tau), v_1(\tau), {}^C D^\alpha v_1(\tau), I^{\alpha_1} v_1(t)) \\ &- f(\tau, u_2(\tau), v_2(\tau), {}^C D^\alpha v_2(\tau), I^{\alpha_1} v_2(t)) \Big| d\tau \Big) ds \\ &+ |a||B_2| \int_0^\eta \frac{(\eta-s)^{\beta-1}}{\Gamma(\beta)} \left(\int_0^s e^{-k_2(s-m)} \left(\int_0^m \frac{(m-\tau)^{p-2}}{\Gamma(p-1)} \right.\right. \\ &\times \Big| g(\tau, u_1(\tau), {}^C D^\delta u_1(\tau), I^{\delta_1} u_1(t), v_1(\tau)) \\ &- g(\tau, u_2(\tau), {}^C D^\delta u_2(\tau), I^{\delta_1} u_2(t), v_2(\tau)) \Big| d\tau \Big) dm \Big) ds \Bigg\} \end{aligned}$$

$$
\begin{aligned}
&+\int_0^t e^{-k_1(t-s)}\left(\int_0^s \frac{(s-\tau)^{q-2}}{\Gamma(q-1)}\right.\\
&\times\Big|f(\tau,u_1(\tau),v_1(\tau),{}^CD^{\alpha}v_1(\tau),I^{\alpha_1}v_1(t))\\
&\left.-f(\tau,u_2(\tau),v_2(\tau),{}^CD^{\alpha}v_2(\tau),I^{\alpha_1}v_2(t))\Big|d\tau\right)ds\\
\le{}& M_1 l\left[\|u_1-u_2\|+\|v_1-v_2\|+\|{}^CD^{\alpha}v_1-{}^CD^{\alpha}v_2\|+\|I^{\alpha_1}v_1-I^{\alpha_1}v_2\|\right]\\
&+M_2 l_1\left[\|u_1-u_2\|+\|{}^cD^{\delta}u_1-{}^cD^{\delta}u_2\|+\|v_1-v_2\|+\|I^{\delta_1}u_1-I^{\delta_1}u_2\|\right]\\
\le{}& M_1 l\left[\|u_1-u_2\|_X+\sigma_1\|v_1-v_2\|\right]+M_2 l_1\left[\sigma_2\|u_1-u_2\|_X+\|v_1-v_2\|_Y\right]\\
\le{}& \Lambda\left[\|u_1-u_2\|_X+\|v_1-v_2\|_Y\right].
\end{aligned}
$$

Also we have

$$
|F_1'(u_1,v_1)(t)-F_1'(u_2,v_2)(t)|\le\Lambda_1\left[\|u_1-u_2\|_X+\|v_1-v_2\|_Y\right],
$$

which implies that

$$
\begin{aligned}
&|{}^CD^{\delta}F_1(u_1,v_1)(t)-{}^CD^{\delta}F_1(u_2,v_2)(t)|\\
\le{}&\int_0^t\frac{(t-s)^{-\delta}}{\Gamma(1-\delta)}|F_1'(u_1,v_1)(s)-F_1'(u_2,v_2)(s)|ds\\
\le{}&\frac{\Lambda_1}{\Gamma(2-\delta)}[\|u_1-u_2\|_X+\|v_1-v_2\|_Y].
\end{aligned}
$$

From the above inequalities, we get

$$
\begin{aligned}
&\|F_1(u_1,v_1)-F_1(u_2,v_2)\|_X\\
={}&\|F_1(u_1,v_1)-F_1(u_2,v_2)\|+\|{}^CD^{\delta}F_1(u_1,v_1)-{}^CD^{\delta}F_1(u_2,v_2)\|\\
\le{}&\left[\Lambda+\frac{\Lambda_1}{\Gamma(2-\delta)}\right]\left[\|u_1-u_2\|_X+\|v_1-v_2\|_Y\right].
\end{aligned}
\tag{5.167}
$$

Similarly, we can find that

$$
\|F_2(u_1,v_1)-F_2(u_2,v_2)\|_Y\le\left[\Lambda'+\frac{\Lambda_1'}{\Gamma(2-\alpha)}\right]\left[\|u_1-u_2\|_X+\|v_1-v_2\|_Y\right].\tag{5.168}
$$

Consequently, we obtain from (5.167) and (5.168) that

$$
\begin{aligned}
&\|F(u_1,v_1)-F(u_2,v_2)\|_{X\times Y}\\
={}&\|F_1(u_1,v_1)-F_1(u_2,v_2)\|_X+\|F_2(u_1,v_1)-F_2(u_2,v_2)\|_Y\\
\le{}&\left[\Lambda+\Lambda'+\frac{\Lambda_1}{\Gamma(2-\delta)}+\frac{\Lambda_1'}{\Gamma(2-\alpha)}\right]\left[\|u_1-u_2\|_X+\|v_1-v_2\|_Y\right].
\end{aligned}
$$

This shows that F is a contraction in view of (5.164). Hence, by Banach's fixed point theorem, the operator F has a unique fixed point which corresponds to the unique solution of system (5.118)-(5.119). This completes the proof. □

5.6.2 *Example*

Consider the following coupled system of fractional differential equations

$$\begin{aligned}{}^{C}D^{4/3}(D+3)x(t) &= e^{t} + \frac{|x(t)|}{4(t+2)^2(1+|x(t)|)} + \frac{1}{32}\sin^2(y(t)) \\ &\quad + \frac{|{}^{C}D^{1/3}y(t)|}{16+t^2} + \frac{1}{16}I^{1/2}y(t), \quad t\in[0,1], \qquad (5.169)\\ {}^{C}D^{3/2}(D+4)y(t) &= \cos(t) + \frac{1}{32\pi}\sin(2\pi x(t)) + \frac{|{}^{C}D^{1/2}x(t)|}{16(1+|{}^{C}D^{1/2}x(t)|)} \\ &\quad + \frac{|y(t)|}{4\sqrt{16+t^4}} + \frac{1}{16}I^{1/4}x(t),\ t\in[0,1], \qquad (5.170)\end{aligned}$$

supplemented with nonlocal coupled integral boundary conditions:

$$\begin{aligned} &x(0)=0,\ \ x'(0)=0,\ \ x(1/3)+2x(1/2)=\int_0^{1/4} y(s)\,ds, \\ &y(0)=0,\ \ y'(0)=0,\ \ y(1/9)-y(1/8)+3y(1/4)=\int_0^{1/10} x(s)\,ds. \qquad (5.171)\end{aligned}$$

Here, $k_1=3,\ k_2=4,\ q=7/3,\ p=5/2,\ \alpha=1/3,\ \alpha_1=1/2,\ \delta=1/2,\ \delta_1=1/4,\ a=b=1,\ a_1=1,\ a_2=2,\ \zeta_1=1/3,\ \zeta_2=1/2,\ m=2,\ \eta=1/4,\ \gamma=\beta=1, b_1=1,\ b_2=-1,\ b_3=3,\ \xi_1=1/9,\ \xi_2=1/8,\ \xi_3=1/4,\ n=3,\ \theta=1/10,\ f(t,x(t),y(t),{}^{C}D^{\alpha}y(t),I^{\alpha_1}y(t))=e^{t}+\frac{|x(t)|}{4(t+2)^2(1+|x(t)|)}+\frac{1}{32}\sin^2(y(t))+\frac{|{}^{C}D^{1/3}y(t)|}{16+t^2}+\frac{1}{16}I^{1/2}y(t)$ and $g(t,x(t),{}^{C}D^{\delta}x(t),I^{\delta_1}x(t),y(t)) = \cos(t)+\frac{1}{32\pi}\sin(2\pi x(t))+\frac{|{}^{C}D^{1/2}x(t)|}{16(1+|{}^{C}D^{1/2}x(t)|)}+\frac{|y(t)|}{4\sqrt{16+t^4}}+\frac{1}{16}I^{1/4}x(t)$. From the following inequalities, we get $l=1/16$ and $l_1=1/16$:

$$\begin{aligned} &|f(t,x_1(t),y_1(t),{}^{C}D^{\alpha}y_1(t),I^{\alpha_1}y_1(t)) - f(t,x_2(t),y_2(t),{}^{C}D^{\alpha}y_2(t),I^{\alpha_1}y_2(t))| \\ &\le \frac{1}{16}\Big(|x_1(t)-x_2(t)|+|y_1(t)-y_2(t)|+|{}^{C}D^{1/3}y_1(t)-{}^{C}D^{1/3}y_2(t)| \\ &\quad +|I^{1/2}y_1(t)-I^{1/2}y_2(t)|\Big),\end{aligned}$$

and

$$\begin{aligned} &|g(t,x_1(t),{}^{C}D^{\delta}x_1(t),I^{\delta_1}x_1(t),y_1(t)) - g(t,x_2(t),{}^{C}D^{\delta}x_2(t),I^{\delta_1}x_1(t),y_2(t)) \\ &\le \frac{1}{16}\Big(|x_1(t)-x_2(t)|+|{}^{C}D^{1/2}x_1(t)-{}^{C}D^{1/2}x_2(t)|+|y_1(t)-y_2(t)| \\ &\quad +|I^{1/4}x_1(t)-I^{1/4}x_2(t)|\Big).\end{aligned}$$

Using the given data, we find that $A_1\simeq 0.288129,\ |A_2|\simeq 0.002047,\ |B_1|\simeq 0.000129,\ B_2\simeq 0.067878,\ |\Delta|\simeq 0.019557,\ \varrho_1\simeq 17.0439,\ \varrho_2\simeq 12.7829,\ \bar{\varrho}_1\simeq 16.1875,\ \bar{\varrho}_2\simeq 12.5459,\ M_1\simeq 0.432531,\ M_2\simeq 0.005305,\ \Lambda\simeq 0.092137,\ M_1^*\simeq$

1.65183, $M_2^* \simeq 0.005038$, $\Lambda_1 \simeq 0.220395$, $N_1 \simeq 0.000683, N_2 \simeq 0.448889$, $\Lambda' \simeq 0.0591001$, $N_1^* \simeq 0.000669$, $N_2^* \simeq 1.92362$, $\Lambda_1' \simeq 0.252956$. Further

$$\Lambda + \frac{\Lambda_1}{\Gamma(3/2)} + \Lambda' + \frac{\Lambda_1'}{\Gamma(5/3)} \simeq 0.680057 < 1.$$

Thus all the conditions of Theorem 5.28 are satisfied. In consequence, by the conclusion of Theorem 5.28, there exists a unique solution for the problem (5.170)-(5.171) on $[0, 1]$.

Remark 5.4. It has been observed that the work obtained in this section yields some new special results for specific choices of the parameters involved in the given problem. For example, selecting $k_1 = k_2 = k$, our results correspond to the ones for one-parameter coupled system of sequential fractional integro-differential equations. By taking $\beta = 1, \gamma = 1$ in (5.119), our results correspond the multi-point boundary conditions with classical nonlocal strip conditions:

$$\sum_{i=1}^{m} a_i x(\zeta_i) = a \int_0^{\eta} y(s)ds, \ \sum_{j=1}^{n} b_j y(\xi_j) = b \int_0^{\theta} x(s)ds.$$

In case we choose $a_i = 0, i = 1, \ldots, (m-1)$, $a_m = 1$, $\zeta_m \to 1$, and $b_j = 0, j = 1, \ldots, (n-1)$, $b_n = 1$, $\xi_n \to 1$, our results correspond to the conditions

$$x(1) = a \int_0^{\eta} \frac{(\eta - s)^{\beta-1}}{\Gamma(\beta)} y(s)ds, \ y(1) = b \int_0^{\eta} \frac{(\eta - s)^{\gamma-1}}{\Gamma(\gamma)} x(s)ds.$$

Letting $q = 3 = p$, $\alpha, \alpha_1, \delta, \delta_1 \to 1, \gamma = 1 = \beta$ in the results of this paper, we obtain the new ones for the corresponding classical system of coupled third-order integro-differential equations with nonlocal integro-multi-point boundary conditions:

$$(D^3 + k_1 D^2)x(t) = f\left(t, x(t), y(t), Dy(t), \int_0^t y(s)ds\right), D = \frac{d}{dt}, t \in [0, 1], k_1 > 0,$$

$$(D^3 + k_2 D^2)y(t) = g\left(t, x(t), Dx(t), \int_0^t x(s)ds, y(t)\right), \quad k_2 > 0,$$

$$x(0) = 0, \ x'(0) = 0, \ \sum_{i=1}^{m} a_i x(\zeta_i) = a \int_0^{\eta} y(s)ds, \quad 0 < \eta < \zeta_1 < \cdots < \zeta_m < 1,$$

$$y(0) = 0, \ y'(0) = 0, \ \sum_{j=1}^{n} b_j y(\xi_j) = b \int_0^{\theta} x(s)ds, \quad 0 < \theta < \xi_1 < \cdots < \xi_n < 1.$$

5.7 On boundary value problems involving sequential fractional derivative operators and non-separated nonlocal integro-multi-point boundary conditions

In this section, we investigate the existence criteria for the solutions of Caputo type sequential fractional differential equations and inclusions:

$$({}^C D^q + \mu {}^C D^{q-1})x(t) = f(t, x(t), {}^C D^{\kappa} x(t)), \ \mu > 0, \ 0 < \kappa < 1, \ 1 < q \leq 2, \ t \in [0, 1], \tag{5.172}$$

$$({}^C D^q + \mu {}^C D^{q-1})x(t) \in F(t, x(t), {}^C D^\kappa x(t)),\ \mu > 0,\ 0 < \kappa < 1,\ 1 < q \leq 2,\ t \in [0,1], \tag{5.173}$$

supplemented with nonlocal integro-multi-point boundary conditions:

$$\begin{cases} \rho_1 x(0) + \rho_2 x(1) = \displaystyle\sum_{i=1}^{m-2} \alpha_i x(\sigma_i) + \sum_{j=1}^{p-2} r_j \int_{\xi_j}^{\eta_j} x(s)ds, \\ \rho_3 x'(0) + \rho_4 x'(1) = \displaystyle\sum_{i=1}^{m-2} \delta_i x'(\sigma_i) + \sum_{j=1}^{p-2} \gamma_j \int_{\xi_j}^{\eta_j} x'(s)ds, \end{cases} \tag{5.174}$$

where $0 < \sigma_1 < \sigma_2 < \ldots < \sigma_{m-2} < \ldots < \xi_1 < \eta_1 < \xi_2 < \eta_2 < \ldots < \xi_{p-2} < \eta_{p-2} < 1$, ${}^C D^q$, ${}^C D^\kappa$ denote the Caputo fractional derivative operators of order q and κ respectively, f is a given continuous function, $F : [0,1] \times \mathbb{R} \times \mathbb{R} \to \mathcal{P}(\mathbb{R})$ is a multi-valued map ($\mathcal{P}(\mathbb{R})$ is the family of all non-empty subsets of $\mathbb{R}$), $\rho_p (p = 1, 2, 3, 4)$ are real constants and α_i, δ_i $(i = 1, 2, \ldots, m-2), r_j, \gamma_j$ $(j = 1, 2, \ldots, p-2)$, are positive real constants.

To define the solution for problem (5.172) and (5.174), we consider the following lemma dealing with its linear variant.

Lemma 5.9. *Let $h \in C([0,1], \mathbb{R})$ and*

$$\Delta = A_1 B_2 - B_1 A_2 \neq 0. \tag{5.175}$$

Then the integral solution for the sequential fractional differential equation

$$({}^C D^q + \mu {}^C D^{q-1})x(t) = h(t), \tag{5.176}$$

supplemented with the boundary conditions (5.174) is given by

$$\begin{aligned} x(t) = & \sum_{i=1}^{m-2} \phi_i(t) \int_0^{\sigma_i} e^{\mu(s-\sigma_i)} \left(I^{q-1} h(s)\right) ds \\ & + \sum_{j=1}^{p-2} \psi_j(t) \int_{\xi_j}^{\eta_j} \Big(\int_0^s e^{\mu(m-s)} \left(I^{q-1} h(m)\right) dm \Big) ds \\ & + \Omega_1(t) \Big[\sum_{i=1}^{m-2} \delta_i I^{q-1} h(\sigma_i) + \sum_{j=1}^{p-2} \gamma_j \int_{\xi_j}^{\eta_j} \left(I^{q-1} h(s)\right) ds - \rho_4 I^{q-1} h(1) \Big] \\ & + \Omega_2(t) \int_0^1 e^{\mu(s-1)} \Big(I^{q-1} h(s)\Big) ds + \int_0^t e^{\mu(s-t)} \Big(I^{q-1} h(s)\Big) ds, \end{aligned} \tag{5.177}$$

where

$$\begin{aligned} \phi_i(t) &= \frac{1}{\mu\Delta} \Big[\mu e^{-\mu t} \Big(B_2 \alpha_i + \mu B_1 \delta_i \Big) - (1 - e^{-\mu t}) \Big(A_2 \alpha_i + \mu A_1 \delta_i \Big) \Big], \\ \psi_j(t) &= \frac{1}{\mu\Delta} \Big[\mu e^{-\mu t} \Big(B_2 r_j + \mu B_1 \gamma_j \Big) - (1 - e^{-\mu t}) \Big(A_2 r_j + \mu A_1 \gamma_j \Big) \Big], \\ \Omega_1(t) &= \frac{1}{-\mu\Delta} \Big[\mu B_1 e^{-\mu t} - A_1 (1 - e^{-\mu t}) \Big], \end{aligned} \tag{5.178}$$

$$\Omega_2(t) = \frac{1}{-\mu\Delta}\Big[\mu e^{-\mu t}(\rho_4\mu B_1 + \rho_2 B_2) - (1-e^{-\mu t})(\rho_4\mu A_1 + \rho_2 A_2)\Big], \tag{5.179}$$

$$A_1 = \rho_1 + \rho_2 e^{-\mu} - \sum_{i=1}^{m-2}\alpha_i e^{-\mu\sigma_i} + \sum_{j=1}^{p-2}\frac{r_j}{\mu}(e^{-\mu\eta_j} - e^{-\mu\xi_j}),$$

$$A_2 = -\mu\Big[\rho_3 + \rho_4 e^{-\mu} - \sum_{i=1}^{m-2}\delta_i e^{-\mu\sigma_i} + \frac{1}{\mu}\sum_{j=1}^{p-2}\gamma_j(e^{-\mu\eta_j} - e^{-\mu\xi_j})\Big], \tag{5.180}$$

$$B_1 = \frac{1}{\mu}\Big\{\rho_2(1-e^{-\mu}) - \sum_{i=1}^{m-2}\alpha_i(1-e^{-\mu\sigma_i}) - \sum_{j=1}^{p-2}\Big[r_j(\eta_j - \xi_j) + \frac{1}{\mu}(e^{-\mu\eta_j} - e^{-\mu\xi_j})\Big]\Big\},$$

$$B_2 = \rho_3 + \rho_4 e^{-\mu} - \sum_{i=1}^{m-2}\delta_i e^{-\mu\sigma_i} + \frac{1}{\mu}\sum_{j=1}^{p-2}(e^{-\mu\eta_j} - e^{-\mu\xi_j}). \tag{5.181}$$

Proof. Applying the integral operator I^{q-1} on both sides of the equation (5.176) and then solving the resulting equation, we get

$$x(t) = \omega_0 e^{-\mu t} + \frac{\omega_1}{\mu}(1-e^{-\mu t}) + \int_0^t e^{\mu(s-t)}\left(\int_0^s \frac{(s-\tau)^{q-2}}{\Gamma(q-1)}h(\tau)\,d\tau\right)ds, \tag{5.182}$$

where $\omega_i\,(i=0,1)$ are unknown arbitrary constants. From (5.182), we have

$$x'(t) = -\mu\omega_0 e^{-\mu t} + \omega_1 e^{-\mu t} - \mu\int_0^t e^{\mu(s-t)}\left(\int_0^s \frac{(s-\tau)^{q-2}}{\Gamma(q-1)}h(\tau)\,d\tau\right)ds + \int_0^t \frac{(t-\tau)^{q-2}}{\Gamma(q-1)}h(\tau)\,d\tau. \tag{5.183}$$

Using (5.182) and (5.183) in the boundary conditions (5.173), we obtain

$$A_1\omega_0 + B_1\omega_1 = J_1,\ \ A_2\omega_0 + B_2\omega_1 = J_2, \tag{5.184}$$

where A_i and $B_i\,(i=1,2)$ are respectively given by (5.180) and (5.181), and

$$J_1 = \sum_{i=1}^{m-2}\alpha_i\int_0^{\sigma_i} e^{\mu(s-\sigma_i)}\left(I^{q-1}h(s)\right)ds + \sum_{j=1}^{p-2} r_j\int_{\xi_j}^{\eta_j}\Big(\int_0^s e^{\mu(m-s)}\left(I^{q-1}h(m)\right)dm\Big)ds - \rho_2\int_0^1 e^{\mu(s-1)}\Big(I^{q-1}h(s)\Big)ds,$$

$$J_2 = \sum_{i=1}^{m-2}\delta_i\Big[I^{q-1}h(\sigma_i) - \mu\int_0^{\sigma_i} e^{\mu(s-\sigma_i)}\left(I^{q-1}h(s)\right)ds\Big] + \sum_{j=1}^{p-2}\gamma_j\Big[\int_{\xi_j}^{\eta_j}\left(I^{q-1}h(s)\right)ds - \mu\int_{\xi_j}^{\eta_j}\Big(\int_0^s e^{\mu(m-s)}\left(I^{q-1}h(m)\right)dm\Big)ds\Big]$$

$$+\rho_4\Big[\mu\int_0^1 e^{\mu(s-1)}\Big(I^{q-1}h(s)\Big)ds - I^{q-1}h(1)\Big].$$

Solving the system (5.184) for the unknown constants ω_0, ω_1 yields

$$\begin{aligned}\omega_0 = \frac{1}{\Delta}\Big\{&\sum_{i=1}^{m-2}\Big(B_2\alpha_i+\mu B_1\delta_i\Big)\int_0^{\sigma_i} e^{\mu(s-\sigma_i)}\left(I^{q-1}h(s)\right)ds\\ &+\sum_{j=1}^{p-2}\Big(B_2 r_j+\mu B_1\gamma_j\Big)\int_{\xi_j}^{\eta_j}\Big(\int_0^s e^{\mu(m-s)}\left(I^{q-1}h(m)\right)dm\Big)ds\\ &-\Big(\rho_4\mu B_1+\rho_2 B_2\Big)\int_0^1\Big(e^{\mu(s-1)}I^{q-1}h(s)\Big)ds\\ &-B_1\sum_{i=1}^{m-2}\delta_i I^{q-1}h(\sigma_i) - B_1\sum_{j=1}^{p-2}\gamma_j\int_{\xi_j}^{\eta_j}\left(I^{q-1}h(s)\right)ds - \rho_4 B_1 I^{q-1}h(1)\Big\},\end{aligned}$$

$$\begin{aligned}\omega_1 = \frac{1}{\Delta}\Big\{&-\sum_{i=1}^{m-2}\Big(A_2\alpha_i+\mu A_1\delta_i\Big)\int_0^{\sigma_i} e^{\mu(s-\sigma_i)}\left(I^{q-1}h(s)\right)ds\\ &-\sum_{j=1}^{p-2}(A_2 r_j+\mu A_1\gamma_j)\int_{\xi_j}^{\eta_j}\Big(\int_0^s e^{\mu(m-s)}\left(I^{q-1}h(m)\right)dm\Big)ds\\ &+A_1\sum_{i=1}^{m-2}\delta_i I^{q-1}h(\sigma_i) + A_1\sum_{j=1}^{p-2}\gamma_j\int_{\xi_j}^{\eta_j}\left(I^{q-1}h(s)\right)ds - \rho_4 A_1 I^{q-1}h(1)\\ &+\Big(\rho_4\mu A_1+\rho_2 A_2\Big)\int_0^1\Big(e^{\mu(s-1)}I^{q-1}h(s)\Big)ds\Big\},\end{aligned}$$

where Δ is given by (5.175). Substituting the values of ω_0 and ω_1 in (5.182) together with the notation (5.178) and (5.179), we get the solution (5.177). The converse follows by direct computation. This completes the proof. □

5.7.1 *Main results for the problem (5.172) and (5.174)*

Let $X = \{x | x \in C([0,1],\mathbb{R})$ and ${}^CD^\kappa x \in C([0,1],\mathbb{R})\}$ be a space equipped with the norm $\|x\|_X = \sup_{t\in[0,1]}|x(t)| + \sup_{t\in[0,1]}|{}^CD^\kappa x(t)| = \|x\| + \|{}^CD^\kappa x\|$, where ${}^CD^\kappa$ denote the standard Caputo fractional derivative operator of order $0<\kappa\le 1$. As argued in [139], $(X, \|\cdot\|_X)$ is a Banach space.

By means of Lemma 5.9, we transform the problem (5.172) and (5.174) into a fixed point problem as $x = \mathfrak{H}x$, where the operator $\mathfrak{H}: X\to X$ is defined by

$$\begin{aligned}(\mathfrak{H}x)(t) = &\sum_{i=1}^{m-2}\phi_i(t)\int_0^{\sigma_i} e^{\mu(s-\sigma_i)}\left(I^{q-1}f(s,x(s),{}^C D^\kappa x(s))\right)ds\\ &+\sum_{j=1}^{p-2}\psi_j(t)\int_{\xi_j}^{\eta_j}\Big(\int_0^s e^{\mu(m-s)}\left(I^{q-1}f(m,x(m),{}^C D^\kappa x(m))\right)dm\Big)ds\end{aligned}$$

$$
\begin{aligned}
&+\Omega_1(t)\Big[\sum_{i=1}^{m-2}\delta_i\left(I^{q-1}f(\sigma_i,x(\sigma_i),{}^C D^\kappa x(\sigma_i))\right)\\
&+\sum_{j=1}^{p-2}\gamma_j\int_{\xi_j}^{\eta_j}\left(I^{q-1}f(s,x(s),{}^C D^\kappa x(s))\right)ds-\rho_4 I^{q-1}f(1,x(1),{}^C D^\kappa x(1))\Big]\\
&+\Omega_2(t)\int_0^1 e^{\mu(s-1)}\Big(I^{q-1}f(s,x(s),{}^C D^\kappa x(s))\Big)ds\\
&+\int_0^t e^{\mu(s-t)}\Big(I^{q-1}f(s,x(s),{}^C D^\kappa x(s))\Big)ds,\quad t\in[0,1], \qquad (5.185)
\end{aligned}
$$

where ϕ_i, ψ_j and Ω_i $(i = 1, 2)$ are respectively given by (5.178) and (5.179). Furthermore, we have

$$
\begin{aligned}
(\mathfrak{H}x)'(t) =& \sum_{i=1}^{m-2}\phi_i'(t)\int_0^{\sigma_i}e^{\mu(s-\sigma_i)}\left(I^{q-1}f(s,x(s),{}^C D^\kappa x(s))\right)ds\\
&+\sum_{j=1}^{p-2}\psi_j'(t)\int_{\xi_j}^{\eta_j}\Big(\int_0^s e^{\mu(m-s)}\left(I^{q-1}f(m,x(m),{}^C D^\kappa x(m))\right)dm\Big)ds\\
&+\Omega_1'(t)\Big[\sum_{i=1}^{m-2}\delta_i\left(I^{q-1}f(\sigma_i,x(\sigma_i),{}^C D^\kappa x(\sigma_i))\right)\\
&+\sum_{j=1}^{p-2}\gamma_j\int_{\xi_j}^{\eta_j}\left(I^{q-1}f(s,x(s),{}^C D^\kappa x(s))\right)ds-\rho_4 I^{q-1}f(1,x(1),{}^C D^\kappa x(1))\Big]\\
&+\Omega_2'(t)\int_0^1 e^{\mu(s-1)}\Big(I^{q-1}f(s,x(s),{}^C D^\kappa x(s))\Big)ds\\
&-\mu\int_0^t e^{\mu(s-t)}\Big(I^{q-1}f(s,x(s),{}^C D^\kappa x(s))\Big)ds\\
&+I^{q-1}f(t,x(t),{}^C D^\kappa x(t)),\quad t\in[0,1], \qquad (5.186)
\end{aligned}
$$

where

$$
\begin{aligned}
\phi_i'(t) &= \frac{-e^{-\mu t}}{\Delta}[\mu(B_2\alpha_i+\mu B_1\delta_i)+A_2\alpha_i+\mu A_1\delta_i],\\
\psi_i'(t) &= \frac{-e^{-\mu t}}{\Delta}[\mu(B_2 r_j+\mu B_1\gamma_j)+A_2 r_j+\mu A_1\gamma_j],\\
\Omega_1'(t) &= \frac{e^{-\mu t}}{\Delta}[\mu B_1+A_1],\\
\Omega_2'(t) &= \frac{e^{-\mu t}}{\Delta}[\mu(\rho_4\mu B_1+\rho_2 B_1)+\rho_4\mu A_1+\rho_2 A_2].
\end{aligned}
$$

Next we fix some quantities as follows:

$$
\lambda = \frac{1}{\mu\Gamma(q)}\sum_{i=1}^{m-2}\widehat{\phi}_i\sigma_i^{q-1}(1-e^{-\mu\sigma_i})+\frac{1}{\mu\Gamma(q+1)}\sum_{j=1}^{p-2}\widehat{\psi}_j(\eta_j^q-\xi_j^q)(1-e^{-\mu\eta_j})
$$

$$+\frac{\widehat{\Omega}_1}{\Gamma(q+1)}\Big[q\Big(\sum_{i=1}^{m-2}\delta_i\sigma_i^{q-1}+|\rho_4|\Big)+\sum_{j=1}^{p-2}\gamma_j(\eta_j^q-\xi_j^q)\Big]$$
$$+\frac{1}{\mu\Gamma(q)}(\widehat{\Omega}_2+1)(1-e^{-\mu}), \tag{5.187}$$

$$\widehat{\lambda}=\lambda-\frac{1}{\mu\Gamma(q)}(1-e^{-\mu}), \tag{5.188}$$

$$\widehat{\phi}_i=\max_{t\in[0,1]}|\phi_i(t)|=\frac{1}{\mu|\Delta|}\Big[\mu|B_2\alpha_i+\mu B_1\delta_i|+(1-e^{-\mu})|A_2\alpha_i+\mu A_1\delta_i|\Big],$$
$$\widehat{\psi}_j=\max_{t\in[0,1]}|\psi_i(t)|=\frac{1}{\mu|\Delta|}\Big[\mu|B_2r_j+\mu B_1\gamma_j|+|(A_2r_j+\mu A_1\gamma_j)|(1-e^{-\mu})\Big],$$
$$\widehat{\Omega}_1=\max_{t\in[0,1]}|\Omega_1(t)|=\frac{1}{\mu|\Delta|}[\mu|B_1|+|A_1|(1-e^{-\mu})],$$
$$\widehat{\Omega}_2=\max_{t\in[0,1]}|\Omega_2(t)|=\frac{1}{\mu|\Delta|}\Big[\mu|\rho_4\mu B_1+\rho_2B_2|+(1-e^{-\mu})|\rho_4\mu A_1+\rho_2A_2|\Big].$$

$$\lambda_1=\frac{1}{\mu\Gamma(q)}\sum_{i=1}^{m-2}\widehat{\phi}_i'\sigma_i^{q-1}(1-e^{-\mu\sigma_i})+\frac{1}{\mu\Gamma(q+1)}\sum_{j=1}^{p-2}\widehat{\psi}_j'(\eta_j^q-\xi_j^q)(1-e^{-\mu\eta_j})$$
$$+\frac{\widehat{\Omega}_1'}{\Gamma(q+1)}\Big[q\Big(\sum_{i=1}^{m-2}\delta_i\sigma_i^{q-1}+|\rho_4|\Big)+\sum_{j=1}^{p-2}\gamma_j(\eta_j^q-\xi_j^q)\Big]$$
$$+\frac{1}{\mu\Gamma(q)}\widehat{\Omega}_2'(1-e^{-\mu})+\frac{2-e^{-\mu}}{\Gamma(q)}, \tag{5.189}$$

$$\widehat{\lambda}_1=\lambda_1-\frac{2-e^{-\mu}}{\Gamma(q)}. \tag{5.190}$$

$$\widehat{\phi}_i'=\max_{t\in[0,1]}|\phi_i'(t)|=\frac{1}{|\Delta|}\Big[\mu|B_2\alpha_i+\mu B_1\delta_i|+|A_2\alpha_i+\mu A_1\delta_i|\Big],$$
$$\widehat{\psi}_j'=\max_{t\in[0,1]}|\psi_i'(t)|=\frac{1}{|\Delta|}\Big[\mu|B_2r_j+\mu B_1\gamma_j|+|A_2r_j+\mu A_1\gamma_j|\Big],$$
$$\widehat{\Omega}_1'=\max_{t\in[0,1]}|\Omega_1'(t)|=\frac{1}{|\Delta|}[\mu|B_1|+|A_1|],$$
$$\widehat{\Omega}_2'=\max_{t\in[0,1]}|\Omega_2'(t)|=\frac{1}{|\Delta|}\Big[\mu|\rho_4\mu B_1+\rho_2B_2|+|\rho_4\mu A_1+\rho_2A_2|\Big].$$

In the following, for brevity, we use the notation

$$\widehat{f}(x(\theta))=f(\theta,x(\theta),^C D^\kappa x(\theta)). \tag{5.191}$$

5.7.1.1 *Existence result via Krasnosel'skiĭ's fixed point theorem*

Here we present an existence result for the problem (5.172) and (5.174), which relies on Krasnosel'skiĭ's fixed point theorem (Theorem 1.4).

Theorem 5.29. *Let $f : [0,1] \times \mathbb{R} \times \mathbb{R} \to \mathbb{R}$ be a continuous function satisfying the conditions:*

(5.29.1) $|f(t,x_1,y_1) - f(t,x_2,y_2)| \le L(|x_1 - x_2| + |y_1 - y_2|)$, *for* $t \in [0,1]$, $x_i, y_i \in \mathbb{R}, i = 1,2$ *and* $L > 0$ *is a constant.*

(5.29.2) $|f(t,x,y)| \le m(t), \forall (t,x,y) \in [0,1] \times \mathbb{R} \times \mathbb{R}$ *and* $m \in C([0,1], \mathbb{R}^+)$.

Then the problem (5.172) and (5.174) has at least one solution on $[0,1]$ *provided that*

$$L\left(\widehat{\lambda} + \frac{1}{\Gamma(2-\kappa)}\widehat{\lambda}_1\right) < 1, \tag{5.192}$$

where $\widehat{\lambda}, \widehat{\lambda}_1$ *are given by (5.188) and (5.190) respectively.*

Proof. For $\rho > \|m\|\Big(\lambda + \dfrac{\lambda_1}{\Gamma(2-\kappa)}\Big)$, we consider the closed ball $B_\rho = \{x \in X : \|x\|_X \le \rho\}$ and introduce the operators A and B on B_ρ as follows:

$$\begin{aligned}(Ax)(t) = {} & \sum_{i=1}^{m-2} \phi_i(t) \int_0^{\sigma_i} e^{\mu(s-\sigma_i)} \left(I^{q-1}\widehat{f}(x(s))\right) ds \\ & + \sum_{j=1}^{p-2} \psi_j(t) \int_{\xi_j}^{\eta_j} \left(\int_0^s e^{\mu(m-s)} \left(I^{q-1}\widehat{f}(x(m))\right) dm\right) ds \\ & + \Omega_1(t) \Big[\sum_{i=1}^{m-2} \delta_i \left(I^{q-1}\widehat{f}(x(\sigma_i)\right) \\ & + \sum_{j=1}^{p-2} \gamma_j \int_{\xi_j}^{\eta_j} \left(I^{q-1}\widehat{f}(x(s))\right) ds - \rho_4 I^{q-1}\widehat{f}(x(1))\Big] \\ & + \Omega_2(t) \int_0^1 e^{\mu(s-1)} \left(I^{q-1}\widehat{f}(x(s))\right) ds, \quad t \in [0,1],\end{aligned} \tag{5.193}$$

$$(Bx)(t) = \int_0^t e^{\mu(s-t)} \left(I^{q-1}\widehat{f}(x(s))\right) ds, \quad t \in [0,1]. \tag{5.194}$$

For any $x, y \in B_\rho$, it is straightforward to show that

$$\|(\mathfrak{H}x)\|_X = \|Ax + Bx\|_X \le \|m\|\Big(\lambda + \frac{\lambda_1}{\Gamma(2-\kappa)}\Big) < \rho,$$

which implies that $Ax + By \in B_\rho$. Also, the operator B is completely continuous. Indeed, B is uniformly bounded on B_r as

$$\|Bx\|_X \le \frac{1 - e^{-\mu}}{\mu\Gamma(q)} + \frac{2 - e^{-\mu}}{\Gamma(2-\kappa)\Gamma(q)}.$$

Observe that

$$
\begin{aligned}
&|Bx(t_2) - Bx(t_1)| \\
&\leq \left| \int_0^{t_1} [e^{\mu(s-t_2)} - e^{\mu(s-t_1)}](I^{q-1}|\widehat{f}(x(s))|)ds + \int_{t_1}^{t_2} e^{\mu(s-t_2)}(I^{q-1}|\widehat{f}(x(s))|)ds \right| \\
&\leq \|m\| \left| \int_0^{t_1} [e^{\mu(s-t_2)} - e^{\mu(s-t_1)}](I^{q-1}1)ds + \int_{t_1}^{t_2} e^{\mu(s-t_2)}(I^{q-1}1)ds \right| \to 0
\end{aligned}
$$

and

$$
\begin{aligned}
&|Bx'(t_2) - Bx'(t_1)| \\
&\leq \|m\|\mu \left| \int_0^{t_1} [e^{\mu(s-t_2)} - e^{\mu(s-t_1)}](I^{q-1}1)ds + \int_{t_1}^{t_2} e^{\mu(s-t_2)}(I^{q-1}1)ds \right| \\
&\quad + \frac{\|m\|}{\Gamma(q-1)} \left| \int_0^{t_1} [(t_2-s)^{q-2} - (t_1-s)^{q-2}]ds + \int_{t_1}^{t_2} (t_2-s)^{q-2}ds \right| \to 0,
\end{aligned}
$$

as $t_1 \to t_2$ independent of x. So B is equi-continuous. Using the assumption (5.29.1) and the condition (5.192), it is easy to establish that the operator A is a contraction. Thus the hypotheses of Krasnosel'skiĭ's fixed point theorem (Theorem 1.4) hold true and consequently we deduce by its conclusion that the problem (5.172) and (5.173) has at least one solution on $[0, 1]$. This completes the proof. □

5.7.1.2 *Uniqueness of solutions*

Here we establish the uniqueness of solutions for the problem (5.172) and (5.174) by applying a fixed point theorem due to Banach.

Theorem 5.30. *Let the assumption (5.29.1) be satisfied and that*

$$
L\Big(\lambda + \frac{\lambda_1}{\Gamma(2-q)}\Big) < 1, \tag{5.195}
$$

where λ and λ_1 are given by (5.187) and (5.189) respectively. Then there exists a unique solution for the problem (5.172) and (5.173) on $[0, 1]$.

Proof. Let us define $\sup_{t\in[0,1]} |f(t, 0, 0)| = M$ and select

$$
\bar{r} \geq \frac{M\Big(\lambda + \dfrac{\lambda_1}{\Gamma(2-q)}\Big)}{1 - L\Big(\lambda + \dfrac{\lambda_1}{\Gamma(2-q)}\Big)}
$$

to show that $\mathfrak{H}B_{\bar{r}} \subset B_{\bar{r}}$, where $B_{\bar{r}} = \{x \in X : \|x\|_X \leq \bar{r}\}$ and $\mathfrak{H}$ is defined by (5.185).

Using the condition (5.29.1), we have

$$
|\widehat{f}(x(\theta))| = |f(\theta, x(\theta), {}^C D^\kappa x(\theta))| = |f(\theta, x(\theta), {}^C D^\kappa x(\theta)) - f(\theta, 0, 0)| + |f(\theta, 0, 0)|
$$

$$\leq L(\|x\| + \|^C D^\kappa x\|) + M \leq L\|x\|_X + M \leq L\bar{r} + M.$$

Then, for $x \in B_{\bar{r}}$, we obtain

$$\begin{aligned}
\|(\mathfrak{H}x)\| &= \sup_{t\in[0,1]} |(\mathfrak{H}x)(t)| \\
&\leq \sup_{t\in[0,1]} \Big\{ \sum_{i=1}^{m-2} |\phi_i(t)| \int_0^{\sigma_i} e^{\mu(s-\sigma_i)} \Big(I^{q-1}|\widehat{f}(x(s))|\Big)\, ds \\
&\quad + \sum_{j=1}^{p-2} |\psi_j(t)| \int_{\xi_j}^{\eta_j} \Big(\int_0^s e^{\mu(m-s)} \Big(I^{q-1}|\widehat{f}(x(m))|\Big)\, dm \Big) ds \\
&\quad + |\Omega_1(t)| \Big[\sum_{i=1}^{m-2} |\delta_i| \Big(I^{q-1}|\widehat{f}(x(\sigma_i)|\Big) \\
&\quad + \sum_{j=1}^{p-2} |\gamma_j| \int_{\xi_j}^{\eta_j} \Big(I^{q-1}|\widehat{f}(x(s))|\Big)\, ds + |\rho_4| I^{q-1}|\widehat{f}(x(1))| \Big] \\
&\quad + |\Omega_2(t)| \int_0^1 e^{\mu(s-1)} \Big(I^{q-1}|\widehat{f}(x(s))|\Big) ds + \int_0^t e^{\mu(s-t)} \Big(I^{q-1}|\widehat{f}(x(s))|\Big) ds \Big\} \\
&\leq (L\bar{r} + M) \Bigg\{ \sum_{i=1}^{m-2} \widehat{\phi}_i \int_0^{\sigma_i} e^{\mu(s-\sigma_i)} \left(I^{q-1}1\right) ds \\
&\quad + \sum_{j=1}^{p-2} \widehat{\psi}_j \int_{\xi_j}^{\eta_j} \Big(\int_0^s e^{\mu(m-s)} \left(I^{q-1}1\right) dm \Big) ds \\
&\quad + \widehat{\Omega}_1 \Big[\sum_{i=1}^{m-2} \delta_i \left(I^{q-1}1\right) + \sum_{j=1}^{p-2} \gamma_j \int_{\xi_j}^{\eta_j} \left(I^{q-1}1\right) ds - \rho_4 I^{q-1}1 \Big] \\
&\quad + \widehat{\Omega}_2 \int_0^1 e^{\mu(s-1)} \Big(I^{q-1}1\Big) ds + \int_0^t e^{\mu(s-t)} \Big(I^{q-1}1\Big) ds \Bigg\} \\
&\leq (L\bar{r} + M)\lambda.
\end{aligned}$$

Also we have

$$\begin{aligned}
|(\mathfrak{H}x)'(t)| &\leq \sum_{i=1}^{m-2} |\phi_i'(t)| \int_0^{\sigma_i} e^{\mu(s-\sigma_i)} \Big(I^{q-1}|\widehat{f}(x(s))|\Big)\, ds \\
&\quad + \sum_{j=1}^{p-2} |\psi_j'(t)| \int_{\xi_j}^{\eta_j} \Big(\int_0^s e^{\mu(m-s)} \Big(I^{q-1}|\widehat{f}(x(m))|\Big)\, dm \Big) ds \\
&\quad + |\Omega_1'(t)| \Big[\sum_{i=1}^{m-2} |\delta_i| \Big(I^{q-1}|\widehat{f}(x(\sigma_i))|\Big) \\
&\quad + \sum_{j=1}^{p-2} |\gamma_j| \int_{\xi_j}^{\eta_j} \Big(I^{q-1}|\widehat{f}(x(s))|\Big)\, ds + |\rho_4| I^{q-1}|\widehat{f}(x(1))| \Big]
\end{aligned}$$

$$
\begin{aligned}
&+|\Omega_2'(t)|\int_0^1 e^{\mu(s-1)}\Big(I^{q-1}|\widehat{f}(x(s))|\Big)ds\\
&+\mu\int_0^t e^{\mu(s-t)}\Big(I^{q-1}|\widehat{f}(x(s))|\Big)ds+I^{q-1}|\widehat{f}(x(s))|\\
\le\ &(L\bar{r}+M)\lambda_1.
\end{aligned}
$$

Using the above inequality in the definition of Caputo fractional derivative operator with $0<\kappa\le 1$, we get

$$
\begin{aligned}
|{}^C D^\kappa(\mathfrak{H}x)(t)| &\le \int_0^t \frac{(t-s)^{-\kappa}}{\Gamma(1-\kappa)}|(\mathfrak{H}x)'(s)| \le (L\bar{r}+M)\lambda_1\int_0^t \frac{(t-s)^{-\kappa}}{\Gamma(1-\kappa)}ds\\
&\le \frac{1}{\Gamma(2-\kappa)}(L\bar{r}+M)\lambda_1.
\end{aligned}
$$

Hence

$$
\|(\mathfrak{H}x)\|_X=\|(\mathfrak{H}x)\|+\|{}^C D^\kappa(\mathfrak{H}x)\|\le (L\bar{r}+M)\lambda+\frac{1}{\Gamma(2-\kappa)}(L\bar{r}+M)\lambda_1<\bar{r},
$$

which clearly shows that $\mathfrak{H}x\in B_{\bar{r}}$ for any $x\in B_{\bar{r}}$. Thus $\mathfrak{H}B_{\bar{r}}\subset B_{\bar{r}}$.

Now, for $x,y\in X$ and for each $t\in[0,1]$, we have

$$
\begin{aligned}
&|(\mathfrak{H}x)(t)-(\mathfrak{H}y)(t)|\\
\le\ &\sum_{i=1}^{m-2}|\phi_i(t)|\int_0^{\sigma_i} e^{\mu(s-\sigma_i)}\Big(I^{q-1}|\widehat{f}(x(s))-\widehat{f}(y(s))|\Big)\,ds\\
&+\sum_{j=1}^{p-2}|\psi_j(t)|\int_{\xi_j}^{\eta_j}\Big(\int_0^s e^{\mu(m-s)}\Big(I^{q-1}|\widehat{f}(x(m))-\widehat{f}(y(m))|\Big)\,dm\Big)ds\\
&+|\Omega_1(t)|\Big[\sum_{i=1}^{m-2}|\delta_i|\Big(I^{q-1}|\widehat{f}(x(\sigma_i))-\widehat{f}(y(\sigma_i))|\Big)\\
&+\sum_{j=1}^{p-2}|\gamma_j|\int_{\xi_j}^{\eta_j}\Big(I^{q-1}|\widehat{f}(x(s))-\widehat{f}(y(s))|\Big)\,ds\\
&+|\rho_4|I^{q-1}|\widehat{f}(x(1))-\widehat{f}(y(1))|\Big]+|\Omega_2(t)|\int_0^1 e^{\mu(s-1)}\Big(I^{q-1}|\widehat{f}(x(s))-\widehat{f}(y(s))|\Big)ds\\
&+\int_0^t e^{\mu(s-t)}\Big(I^{q-1}|\widehat{f}(x(s))-\widehat{f}(y(s))|\Big)ds\\
\le\ &L\lambda\|x-y\|_X.
\end{aligned}
$$

Also

$$
\begin{aligned}
&|(\mathfrak{H}x)'(t)-(\mathfrak{H}y)'(t)|\\
\le\ &\sum_{i=1}^{m-2}|\phi_i'(t)|\int_0^{\sigma_i} e^{\mu(s-\sigma_i)}\Big(I^{q-1}|\widehat{f}(x(s))-\widehat{f}(y(s))|\Big)\,ds\\
&+\sum_{j=1}^{p-2}|\psi_j'(t)|\int_{\xi_j}^{\eta_j}\Big(\int_0^s e^{\mu(m-s)}\Big(I^{q-1}|\widehat{f}(x(m))-\widehat{f}(y(m))|\Big)\,dm\Big)ds
\end{aligned}
$$

$$
\begin{aligned}
&+|\Omega_1'(t)|\Big[\sum_{i=1}^{m-2}|\delta_i|\Big(I^{q-1}|\widehat{f}(x(\sigma_i))-\widehat{f}(y(\sigma_i))|\Big)\\
&+\sum_{j=1}^{p-2}|\gamma_j|\int_{\xi_j}^{\eta_j}\Big(I^{q-1}|\widehat{f}(x(s))-\widehat{f}(y(s))|\Big)\,ds+|\rho_4|I^{q-1}|\widehat{f}(x(1))-\widehat{f}(y(1))|\Big]\\
&+|\Omega_2'(t)|\int_0^1 e^{\mu(s-1)}\Big(I^{q-1}|\widehat{f}(x(s))-\widehat{f}(y(s))|\Big)ds\\
&+\mu\int_0^t e^{\mu(s-t)}\Big(I^{q-1}|\widehat{f}(x(s))-\widehat{f}(y(s))|\Big)ds+I^{q-1}|\widehat{f}(x(s))-\widehat{f}(y(s))|\\
&\le L\lambda_1\|x-y\|_X,
\end{aligned}
$$

and moreover

$$
|^C D^{\kappa}(\mathfrak{H}x)(t)-{}^C D^{\kappa}(\mathfrak{H}y)(t)|\le\frac{1}{\Gamma(2-\kappa)}L\lambda_1\|x-y\|_X.
$$

Consequently we have

$$
\|(\mathfrak{H}x)-(\mathfrak{H}y)\|_X\le L\Big(\lambda+\frac{\lambda_1}{\Gamma(2-q)}\Big)\|x-y\|_X,
$$

which shows that $\mathfrak{H}$ is a contraction by the condition (5.195). Thus the operator $\mathfrak{H}$ has a unique fixed point by Banach fixed point theorem, which corresponds to a unique solution of the problem (5.172) and (5.174) on $[0,1]$. This completes the proof. □

5.7.2 *Examples*

Consider the following nonlinear sequential fractional differential equation

$$
({}^C D^{7/5}+{}^C D^{2/5})x(t)=\frac{1}{t^2+25}+\frac{e^{-t}}{t^4+16}\Big(\tan^{-1}(x(t))+\frac{|^C D^{1/2}x(t)|}{(1+|^C D^{1/2}x(t)|)}\Big), \tag{5.196}
$$

$t\in[0,1]$, supplemented with the integro-multi-point boundary conditions:

$$
\begin{aligned}
\rho_1 x(0)+\rho_2 x(1)&=\sum_{i=1}^{3}\alpha_i x(\sigma_i)+\sum_{j=1}^{5}r_j\int_{\xi_j}^{\eta_i}x(s)ds,\\
\rho_3 x'(0)+\rho_4 x'(1)&=\sum_{i=1}^{3}\delta_i x'(\sigma_i)+\sum_{j=1}^{5}\gamma_j\int_{\xi_j}^{\eta_j}x'(s)ds.
\end{aligned} \tag{5.197}
$$

Here $q=7/5,\kappa=1/2,\mu=1,m=5,p=7,\sigma_1=1/18,\sigma_2=1/9,\sigma_3=1/6,\xi_1=1/5,\eta_1=1/4,\xi_2=3/10,\eta_2=7/20,\xi_3=2/5,\eta_3=9/20,\xi_4=1/2,\eta_4=11/20,\xi_5=3/5,\eta_5=13/20,\ \alpha_1=1,\alpha_2=1/2,\alpha_3=1,\delta_1=3/4,\delta_2=1,\delta_3=3/2,r_1=1/3,r_2=2/3,r_3=1,r_4=-1,r_5=2,\gamma_1=-1/2,\gamma_2=1/2,\gamma_3=2,\gamma_4=1,\gamma_5=$

$3/2, \rho_1 = 1/2, \rho_2 = 3/4, \rho_3 = -1, \rho_4 = 2$. Using the given data, we find that $\Delta \approx 4.3972, \lambda \approx 2.4121, \widehat{\lambda} \approx 1.6997, \lambda_1 \approx 4.1293, \widehat{\lambda}_1 \approx 2.2898$ ($\Delta, \lambda, \widehat{\lambda}, \lambda_1$ and $\widehat{\lambda}_1$ are respectively given by (5.175), (5.187), (5.188), (5.189) and (5.190)). Further we have

$$m(t) = \frac{1}{t^2+25} + \frac{(\pi+2)e^{-t}}{2(t^4+16)}, \ L = \frac{1}{16}, \ L\Big(\widehat{\lambda} + \frac{1}{\Gamma(2-\kappa)}\widehat{\lambda}_1\Big) \approx 0.2677 < 1.$$

Clearly the hypothesis of Theorem 5.29 holds true. In consequence, there exits a solution to the problem (5.196)-(5.197) on $[0,1]$ by Theorem 5.29. Also

$$L\Big(\lambda + \frac{\lambda_1}{\Gamma(2-\kappa)}\Big) \approx 0.4420 < 1,$$

which implies that the problem (5.196)-(5.197) has a unique solution on $[0,1]$ by Theorem 5.30.

5.7.3 *Existence results for the inclusions problem (5.173) and (5.174)*

In this subsection, we investigate the existence of solutions for the multi-valued (inclusion) boundary value problem (5.173) and (5.174). Our first result deals with the case when the multi-valued map F has convex values, while the second result is concerned with the non-convex multi-valued maps.

For each $x \in C([0,1],\mathbb{R})$, the set

$$S_{F,x} = \{f \in L^1([0,1],\mathbb{R}) : f(t) \in F(t, x(t), {}^C D^\kappa x(t)) \text{ for a.e. } t \in [0,1]\}$$

is known as the set of selections of the multi-valued map F.

5.7.3.1 *The Upper Semi-continuous (convex valued) case*

Here we present an existence result for the problem (5.173) and (5.174) when the multi-valued map F is convex valued. We make use of nonlinear alternative for Kakutani maps (Theorem 1.16) to derive the desired result.

Theorem 5.31. *Assume that:*

(5.31.1) $F : [0,1] \times \mathbb{R}^2 \to \mathcal{P}(\mathbb{R})$ *is* L^1*-Carathéodory and has non-empty compact and convex values;*

(5.31.2) there exist a function $g \in C([0,1],\mathbb{R}^+)$ *with* $\|g\| = \sup_{t\in[0,1]} |g(t)|$ *and non-decreasing function* $Q : \mathbb{R}^+ \to \mathbb{R}^+$ *such that*

$$\|F(t,x,y)\|_{\mathcal{P}} := \sup\{|w| : w \in F(t,x,y)\} \leq g(t)Q(\|x\| + \|y\|),$$

for all $(t,x,y) \in [0,1] \times \mathbb{R} \times \mathbb{R}$*;*

(5.31.3) there exists a constant $K > 0$ such that

$$\frac{K}{\|g\|Q(K)\Big(\lambda + \frac{\lambda_1}{\Gamma(2-\kappa)}\Big)} > 1,$$

where λ, λ_1 are are given by (5.187) and (5.189) respectively.

Then the problem (5.173)-(5.174) *has at least one solution on* $[0,1]$.

Proof. We transform the problem (5.173)-(5.174) into a fixed point problem. Consider the multi-valued map: $N : X \to \mathcal{P}(X)$ defined by

$$N(x) = \left\{ \begin{array}{l} h \in X : \\ h(t) = \left\{ \begin{array}{l} \displaystyle\sum_{i=1}^{m-2} \phi_i(t) \int_0^{\sigma_i} e^{\mu(s-\sigma_i)} \left(I^{q-1} v(s)\right) ds \\ \displaystyle + \sum_{j=1}^{p-2} \psi_j(t) \int_{\xi_j}^{\eta_j} \Big(\int_0^s e^{\mu(m-s)} \left(I^{q-1} v(m)\right) dm \Big) ds \\ \displaystyle + \Omega_1(t) \Big[\sum_{i=1}^{m-2} \delta_i \left(I^{q-1} v(\sigma_i)\right) \\ \displaystyle + \sum_{j=1}^{p-2} \gamma_j \int_{\xi_j}^{\eta_j} \left(I^{q-1} v(s)\right) ds - \rho_4 I^{q-1} v(1) \Big] \\ \displaystyle + \Omega_2(t) \int_0^1 e^{\mu(s-1)} \Big(I^{q-1} v(s)\Big) ds \\ \displaystyle + \int_0^t e^{\mu(s-t)} \Big(I^{q-1} v(s)\Big) ds, \quad t \in [0,1], \quad v \in S_{F,x}. \end{array} \right. \end{array} \right\}$$

It is clear that fixed points of N are solutions of problem (5.173)-(5.174). Now we proceed to show that the operator N satisfies all the conditions of nonlinear alternative for Kakutani maps (Theorem 1.16). The proof is given in several steps.

Step 1. *$N(x)$ is convex for each $x \in X$.*

Indeed, if h_1, h_2 belongs to $N(x)$, then there exist $v_1, v_2 \in S_{F,x}$ such that for each $t \in [0,1]$, we have

$$\begin{aligned} h_k(t) = & \sum_{i=1}^{m-2} \phi_i(t) \int_0^{\sigma_i} e^{\mu(s-\sigma_i)} \left(I^{q-1} v_k(s)\right) ds \\ & + \sum_{j=1}^{p-2} \psi_j(t) \int_{\xi_j}^{\eta_j} \Big(\int_0^s e^{\mu(m-s)} \left(I^{q-1} v_k(m)\right) dm \Big) ds \\ & + \Omega_1(t) \Big[\sum_{i=1}^{m-2} \delta_i \left(I^{q-1} v_k(\sigma_i)\right) + \sum_{j=1}^{p-2} \gamma_j \int_{\xi_j}^{\eta_j} \left(I^{q-1} v_k(s)\right) ds - \rho_4 I^{q-1} v_k(1) \Big] \end{aligned}$$

$$+\Omega_2(t)\int_0^1 e^{\mu(s-1)}\Big(I^{q-1}v_k(s)\Big)ds+\int_0^t e^{\mu(s-t)}\Big(I^{q-1}v_k(s)\Big)ds, \quad k=1,2.$$

Let $0\le\mu\le 1$. Then, for each $t\in[0,1]$, we have

$$\begin{aligned}
&[\mu h_1+(1-\mu)h_2](t)\\
&=\sum_{i=1}^{m-2}\phi_i(t)\int_0^{\sigma_i} e^{\mu(s-\sigma_i)}\left(I^{q-1}[\mu v_1(s)+(1-\mu)v_2(s)]\right)ds\\
&+\sum_{j=1}^{p-2}\psi_j(t)\int_{\xi_j}^{\eta_j}\Big(\int_0^s e^{\mu(m-s)}\left(I^{q-1}[\mu v_1(m)+(1-\mu)v_2(m)]\right)dm\Big)ds\\
&+\Omega_1(t)\Big[\sum_{i=1}^{m-2}\delta_i\left(I^{q-1}[\mu v_1(\sigma_i)+(1-\mu)v_2(\sigma_i)]\right)\\
&+\sum_{j=1}^{p-2}\gamma_j\int_{\xi_j}^{\eta_j}\left(I^{q-1}[\mu v_1(s)+(1-\mu)v_2(s)]\right)ds\\
&-\rho_4 I^{q-1}[\mu v_1(1)+(1-\mu)v_2(1)]\Big]\\
&+\Omega_2(t)\int_0^1 e^{\mu(s-1)}\Big(I^{q-1}[\mu v_1(s)+(1-\mu)v_2(s)]\Big)ds\\
&+\int_0^t e^{\mu(s-t)}\Big(I^{q-1}[\mu v_1(s)+(1-\mu)v_2(s)]\Big)ds.
\end{aligned}$$

Since F has convex values ($S_{F,x}$ is convex), therefore

$$\mu h_1+(1-\mu)h_2\in N(x).$$

Step 2. *$N(x)$ maps bounded sets (balls) into bounded sets in X.*

For a positive number r, let $\mathcal{B}_r=\{x\in X:\|x\|_X\le r\}$ be a bounded set in X. Then, for $h\in N(x)$, $x\in B_r$, there exists $v\in S_{F,x}$ such that

$$\begin{aligned}
h(t)&=\sum_{i=1}^{m-2}\phi_i(t)\int_0^{\sigma_i} e^{\mu(s-\sigma_i)}\left(I^{q-1}v(s)\right)ds\\
&+\sum_{j=1}^{p-2}\psi_j(t)\int_{\xi_j}^{\eta_j}\Big(\int_0^s e^{\mu(m-s)}\left(I^{q-1}v(m)\right)dm\Big)ds+\Omega_1(t)\Big[\sum_{i=1}^{m-2}\delta_i\left(I^{q-1}v(\sigma_i)\right)\\
&+\sum_{j=1}^{p-2}\gamma_j\int_{\xi_j}^{\eta_j}\left(I^{q-1}v(s)\right)ds-\rho_4 I^{q-1}v(1)\Big]\\
&+\Omega_2(t)\int_0^1 e^{\mu(s-1)}\Big(I^{q-1}v(s)\Big)ds+\int_0^t e^{\mu(s-t)}\Big(I^{q-1}v(s)\Big)ds,
\end{aligned}$$

for some $v\in S_{F,x_r}$. Then, for $t\in[0,1]$, we have

$$|h(t)|\le\sum_{i=1}^{m-2}\widehat{\phi}_i\int_0^{\sigma_i} e^{\mu(s-\sigma_i)}\left(I^{q-1}|v(s)|\right)ds$$

$$
\begin{aligned}
&+\sum_{j=1}^{p-2}\widehat{\psi}_j\int_{\xi_j}^{\eta_j}\Big(\int_0^s e^{\mu(m-s)}\left(I^{q-1}|v(m)|\right)dm\Big)ds\\
&+\widehat{\Omega}_1\Big[\sum_{i=1}^{m-2}\delta_i\left(I^{q-1}|v(s)|\right)+\sum_{j=1}^{p-2}\gamma_j\int_{\xi_j}^{\eta_j}\left(I^{q-1}|v(1)|\right)ds-\rho_4 I^{q-1}|v(1)|\Big]\\
&+\widehat{\Omega}_2\int_0^1 e^{\mu(s-1)}\Big(I^{q-1}|v(s)|\Big)ds+\int_0^t e^{\mu(s-t)}\Big(I^{q-1}|v(s)|\Big)ds\\
\le{}& \|g\|Q(\|x\|+\|^C D^\kappa x\|)\Bigg\{\sum_{i=1}^{m-2}\widehat{\phi}_i\int_0^{\sigma_i}e^{\mu(s-\sigma_i)}\left(I^{q-1}1\right)ds\\
&+\sum_{j=1}^{p-2}\widehat{\psi}_j\int_{\xi_j}^{\eta_j}\Big(\int_0^s e^{\mu(m-s)}\left(I^{q-1}1\right)dm\Big)ds\\
&+\widehat{\Omega}_1\Big[\sum_{i=1}^{m-2}\delta_i\left(I^{q-1}1\right)+\sum_{j=1}^{p-2}\gamma_j\int_{\xi_j}^{\eta_j}\left(I^{q-1}1\right)ds-\rho_4 I^{q-1}1\Big]\\
&+\widehat{\Omega}_2\int_0^1 e^{\mu(s-1)}\Big(I^{q-1}1\Big)ds+\int_0^t e^{\mu(s-t)}\Big(I^{q-1}1\Big)ds\Bigg\}\\
\le{}& \|g\|Q(\|x\|_X)\lambda,
\end{aligned}
$$

which, on taking the norm for $t \in [0,1]$, yields $\|h\| \le \|g\|Q(r)\lambda$. Also we have

$$
\begin{aligned}
|h'(t)| \le{}& \sum_{i=1}^{m-2}|\phi_i'(t)|\int_0^{\sigma_i}e^{\mu(s-\sigma_i)}\left(I^{q-1}|v(s)|\right)ds\\
&+\sum_{j=1}^{p-2}|\psi_j'(t)|\int_{\xi_j}^{\eta_j}\Big(\int_0^s e^{\mu(m-s)}\left(I^{q-1}|v(m)|\right)dm\Big)ds\\
&+|\Omega_1'(t)|\Big[\sum_{i=1}^{m-2}|\delta_i|\left(I^{q-1}|v(\sigma_i)|\right)\\
&+\sum_{j=1}^{p-2}|\gamma_j|\int_{\xi_j}^{\eta_j}\left(I^{q-1}|v(s)|\right)ds+|\rho_4|I^{q-1}|v(1)|\Big]\\
&+|\Omega_2'(t)|\int_0^1 e^{\mu(s-1)}\Big(I^{q-1}|v(s)|\Big)ds\\
&+\mu\int_0^t e^{\mu(s-t)}\Big(I^{q-1}|v(s)|\Big)ds+I^{q-1}|v(s)|\\
\le{}& \|g\|Q(\|x\|_X)\lambda_1 \le \|g\|Q(r)\lambda_1.
\end{aligned}
$$

By definition of Caputo fractional derivative operator with $0<\kappa\le 1$, we get

$$
\begin{aligned}
|^C D^\kappa h(t)| \le{}& \int_0^t \frac{(t-s)^{-\kappa}}{\Gamma(1-\kappa)}|h'(t)| \le \|g\|Q(r)\lambda_1\int_0^t\frac{(t-s)^{-\kappa}}{\Gamma(1-\kappa)}ds\\
\le{}& \frac{1}{\Gamma(2-\kappa)}\|g\|Q(r)\lambda_1.
\end{aligned}
$$

Hence

$$\|h\|_X = \|h\| + \|^C D^\kappa h\| \le \|g\| Q(r)\Big(\lambda + \frac{\lambda_1}{\Gamma(2-\kappa)}\Big). \tag{5.198}$$

Step 3. *N maps bounded sets into equi-continuous sets of X.*

Next we show that N maps bounded sets into equi-continuous sets of X. Let $t_1, t_2 \in [0,1]$ with $t_1 < t_2$ and $y \in \mathcal{B}_r$, where $\mathcal{B}_r$ is a bounded set of X. Then we obtain

$$\begin{aligned}
&|h(t_2) - h(t_1)| \\
&\le \sum_{i=1}^{m-2} |\phi_i(t_2) - \phi_i(t_1)| \int_0^{\sigma_i} e^{\mu(s-\sigma_i)} \left(I^{q-1}|v(s)|\right) ds \\
&\quad + \sum_{j=1}^{p-2} |\psi_j(t_2) - \psi_j(t_1)| \int_{\xi_j}^{\eta_j} \Big(\int_0^s e^{\mu(m-s)} \left(I^{q-1}|v(m)|\right) dm \Big) ds \\
&\quad + |\Omega_1(t_2) - \Omega_1(t_1)| \Big[\sum_{i=1}^{m-2} |\delta_i| \left(I^{q-1}|v(\sigma_i)|\right) + \sum_{j=1}^{p-2} |\gamma_j| \int_{\xi_j}^{\eta_j} \left(I^{q-1}|v(s)|\right) ds \\
&\quad + |\rho_4| I^{q-1}|v(1)| \Big] + |\Omega_2(t_2) - \Omega_2(t_1)| \int_0^1 e^{\mu(s-1)} \Big(I^{q-1}|v(s)|\Big) ds \\
&\quad + \Big| \int_0^{t_1} [e^{\mu(s-t_2)} - e^{\mu(s-t_1)}](I^{q-1}|v(s)|) ds + \int_{t_1}^{t_2} e^{\mu(s-t_2)} (I^{q-1}|v(s)|) ds \Big| \\
&\le \|g\| Q(r) \Bigg\{ \sum_{i=1}^{m-2} |\phi_i(t_2) - \phi_i(t_1)| \int_0^{\sigma_i} e^{\mu(s-\sigma_i)} \left(I^{q-1}1\right) ds \\
&\quad + \sum_{j=1}^{p-2} |\psi_j(t_2) - \psi_j(t_1)| \int_{\xi_j}^{\eta_j} \Big(\int_0^s e^{\mu(m-s)} \left(I^{q-1}1\right) dm \Big) ds \\
&\quad + |\Omega_1(t_2) - \Omega_1(t_1)| \Big[\sum_{i=1}^{m-2} |\delta_i| \left(I^{q-1}1\right) + \sum_{j=1}^{p-2} |\gamma_j| \int_{\xi_j}^{\eta_j} \left(I^{q-1}1\right) ds \\
&\quad + |\rho_4| I^{q-1}1 \Big] + |\Omega_2(t_2) - \Omega_2(t_1)| \int_0^1 e^{\mu(s-1)} \Big(I^{q-1}\Big) ds \\
&\quad + \Big| \int_0^{t_1} [e^{\mu(s-t_2)} - e^{\mu(s-t_1)}](I^{q-1}1) ds + \int_{t_1}^{t_2} e^{\mu(s-t_2)} (I^{q-1}1) ds \Big| \Bigg\}
\end{aligned}$$

and

$$\begin{aligned}
&|h'(t_2) - h'(t_1)| \\
&\le \|g\| Q(r) \Bigg\{ \sum_{i=1}^{m-2} |\phi_i'(t_2) - \phi_i'(t_1)| \int_0^{\sigma_i} e^{\mu(s-\sigma_i)} \left(I^{q-1}1\right) ds \\
&\quad + \sum_{j=1}^{p-2} |\psi_j'(t_2) - \psi_j'(t_1)| \int_{\xi_j}^{\eta_j} \Big(\int_0^s e^{\mu(m-s)} \left(I^{q-1}1\right) dm \Big) ds
\end{aligned}$$

$$
\begin{aligned}
&+|\Omega_1'(t_2)-\Omega_1'(t_1)|\Big[\sum_{i=1}^{m-2}|\delta_i|\,(I^{q-1}1)+\sum_{j=1}^{p-2}|\gamma_j|\int_{\xi_j}^{\eta_j}(I^{q-1}1)\,ds\\
&+|\rho_4|I^{q-1}1\Big]+|\Omega_2'(t_2)-\Omega_2'(t_1)|\int_0^1 e^{\mu(s-1)}\Big(I^{q-1}1\Big)ds\\
&+\mu\left|\int_0^{t_1}[e^{\mu(s-t_2)}-e^{\mu(s-t_1)}](I^{q-1}1)ds+\int_{t_1}^{t_2}e^{\mu(s-t_2)}(I^{q-1}1)ds\right|\\
&+\frac{1}{\Gamma(q-1)}\left|\int_0^{t_1}[(t_2-s)^{q-2}-(t_1-s)^{q-2}]ds+\int_{t_1}^{t_2}(t_2-s)^{q-2}ds\right|\Bigg\}.
\end{aligned}
$$

Obviously the right hand sides of the above inequalities tend to zero independently of $x \in \mathcal{B}_r$ as $t_2 - t_1 \to 0$. As N satisfies the above assumptions, it follows by the Arzelá-Ascoli theorem that $N : X \to X$ is completely continuous.

By virtue of the Lemma 1.1, it is enough to establish that the operator $N : X \to X$ has a closed graph, which will imply that N is u.s.c. as it is already shown to be completely continuous. This is done in our next step.

Step 4. *N has a closed graph.*

Let $x_n \to x_*$, $h_n \in N(x_n)$ and $h_n \to h_*$. We need to show that $h_* \in N(x_*)$. Now $h_n \in N(x_n)$ implies that there exists $v_n \in S_{F,x_n}$ such that for each $t \in [0,1]$,

$$
\begin{aligned}
h_n(t)=&\sum_{i=1}^{m-2}\phi_i(t)\int_0^{\sigma_i}e^{\mu(s-\sigma_i)}\left(I^{q-1}v_n(s)\right)ds\\
&+\sum_{j=1}^{p-2}\psi_j(t)\int_{\xi_j}^{\eta_j}\Big(\int_0^s e^{\mu(m-s)}\left(I^{q-1}v_n(m)\right)dm\Big)ds\\
&+\Omega_1(t)\Big[\sum_{i=1}^{m-2}\delta_i\left(I^{q-1}v_n(\sigma_i)\right)+\sum_{j=1}^{p-2}\gamma_j\int_{\xi_j}^{\eta_j}\left(I^{q-1}v_n(s)\right)ds-\rho_4 I^{q-1}v_n(1)\Big]\\
&+\Omega_2(t)\int_0^1 e^{\mu(s-1)}\Big(I^{q-1}v_n(s)\Big)ds+\int_0^t e^{\mu(s-t)}\Big(I^{q-1}v_n(s)\Big)ds.
\end{aligned}
$$

We must show that there exists $v_* \in S_{F,x_*}$ such that for each $t \in [0,1]$,

$$
\begin{aligned}
h_*(t)=&\sum_{i=1}^{m-2}\phi_i(t)\int_0^{\sigma_i}e^{\mu(s-\sigma_i)}\left(I^{q-1}v_*(s)\right)ds\\
&+\sum_{j=1}^{p-2}\psi_j(t)\int_{\xi_j}^{\eta_j}\Big(\int_0^s e^{\mu(m-s)}\left(I^{q-1}v_*(m)\right)dm\Big)ds\\
&+\Omega_1(t)\Big[\sum_{i=1}^{m-2}\delta_i\left(I^{q-1}v_*(\sigma_i)\right)+\sum_{j=1}^{p-2}\gamma_j\int_{\xi_j}^{\eta_j}\left(I^{q-1}v_*(s)\right)ds-\rho_4 I^{q-1}v_*(1)\Big]\\
&+\Omega_2(t)\int_0^1 e^{\mu(s-1)}\Big(I^{q-1}v_*(s)\Big)ds+\int_0^t e^{\mu(s-t)}\Big(I^{q-1}v_*(s)\Big)ds.
\end{aligned}
$$

Consider the continuous linear operator $\Theta : L^1([0,1],\mathbb{R}) \to C([0,1],\mathbb{R})$ given by

$$\begin{aligned}
&v \mapsto \Theta(v)(t) \\
&= \sum_{i=1}^{m-2} \phi_i(t) \int_0^{\sigma_i} e^{\mu(s-\sigma_i)} \left(I^{q-1}v(s)\right) ds \\
&\quad + \sum_{j=1}^{p-2} \psi_j(t) \int_{\xi_j}^{\eta_j} \Big(\int_0^s e^{\mu(m-s)} \left(I^{q-1}v(m)\right) dm \Big) ds \\
&\quad + \Omega_1(t) \Big[\sum_{i=1}^{m-2} \delta_i \left(I^{q-1}v(\sigma_i)\right) + \sum_{j=1}^{p-2} \gamma_j \int_{\xi_j}^{\eta_j} \left(I^{q-1}v(s)\right) ds - \rho_4 I^{q-1}v(1) \Big] \\
&\quad + \Omega_2(t) \int_0^1 e^{\mu(s-1)} \Big(I^{q-1}v(s)\Big) ds + \int_0^t e^{\mu(s-t)} \Big(I^{q-1}v(s)\Big) ds.
\end{aligned}$$

Since $\|h_n(t) - h_*(t)\|_X \to 0$ as $n \to \infty$, it follows from Lemma 1.2 that $\Theta \circ S_{F,x}$ is a closed graph operator. Moreover, we have $h_n \in \Theta(S_{F,x_n})$. Since $x_n \to x_*$, therefore we obtain

$$\begin{aligned}
h_*(t) &= \sum_{i=1}^{m-2} \phi_i(t) \int_0^{\sigma_i} e^{\mu(s-\sigma_i)} \left(I^{q-1}v_*(s)\right) ds \\
&\quad + \sum_{j=1}^{p-2} \psi_j(t) \int_{\xi_j}^{\eta_j} \Big(\int_0^s e^{\mu(m-s)} \left(I^{q-1}v_*(m)\right) dm \Big) ds \\
&\quad + \Omega_1(t) \Big[\sum_{i=1}^{m-2} \delta_i \left(I^{q-1}v_*(\sigma_i)\right) + \sum_{j=1}^{p-2} \gamma_j \int_{\xi_j}^{\eta_j} \left(I^{q-1}v_*(s)\right) ds - \rho_4 I^{q-1}v_*(1) \Big] \\
&\quad + \Omega_2(t) \int_0^1 e^{\mu(s-1)} \Big(I^{q-1}v_*(s)\Big) ds + \int_0^t e^{\mu(s-t)} \Big(I^{q-1}v_*(s)\Big) ds,
\end{aligned}$$

for some $v_* \in S_{F,x_*}$.

Step 5. *We show there exists an open set $U \subseteq C(J,\mathbb{R})$ with $x \notin \theta N(x)$ for any $\theta \in (0,1)$ and all $x \in \partial U$.*

Let $\theta \in (0,1)$ and $x \in \theta N(x)$. Then there exists $v \in L^1([0,1],\mathbb{R})$ with $v \in S_{F,x}$ such that, for $t \in J$, we have

$$\begin{aligned}
x(t) &= \theta \sum_{i=1}^{m-2} \phi_i(t) \int_0^{\sigma_i} e^{\mu(s-\sigma_i)} \left(I^{q-1}v(s)\right) ds \\
&\quad + \theta \sum_{j=1}^{p-2} \psi_j(t) \int_{\xi_j}^{\eta_j} \Big(\int_0^s e^{\mu(m-s)} \left(I^{q-1}v(m)\right) dm \Big) ds \\
&\quad + \theta\Omega_1(t) \Big[\sum_{i=1}^{m-2} \delta_i \left(I^{q-1}v(\sigma_i)\right) + \sum_{j=1}^{p-2} \gamma_j \int_{\xi_j}^{\eta_j} \left(I^{q-1}v(s)\right) ds - \rho_4 I^{q-1}v(1) \Big] \\
&\quad + \theta\Omega_2(t) \int_0^1 e^{\mu(s-1)} \Big(I^{q-1}v(s)\Big) ds + \int_0^t e^{\mu(s-t)} \Big(I^{q-1}v(s)\Big) ds.
\end{aligned}$$

As in the first step, we can find that

$$\frac{\|x\|_X}{\|g\|Q(\|x\|_X)\Big(\lambda + \dfrac{\lambda_1}{\Gamma(2-\kappa)}\Big)} \le 1. \tag{5.199}$$

By the condition (5.31.3), there does not exist any solution x such that $\|x\|_X \neq K$. Let us introduce a set $W = \{x \in X : \|x\|_X < K\}$. The operator $N : \overline{W} \to X$ is continuous and completely continuous. From the choice of W, there is no $w \in \partial W$ such that $w = \theta N(w)$ for some $\theta \in (0,1)$. In cosequence, we deduce by nonlinear alternative for Kakutani maps (Theorem 1.16) that the operator N has a fixed point $w \in \overline{W}$ which is a solution of the problem (5.173) and (5.174). The proof is completed. □

5.7.3.2 *The Lipschitz case*

Now we prove the existence of solutions for the boundary value problem (5.173) and (5.174) with a non-convex valued right hand side by applying a fixed point theorem for multi-valued maps due to Covitz and Nadler (Theorem 1.19).

Theorem 5.32. *Assume that the following conditions hold:*

(5.32.1) $F : [0,1] \times \mathbb{R}^2 \to \mathcal{P}_{cp}(\mathbb{R})$ *is such that* $F(\cdot, x, y) : [0,1] \to \mathcal{P}_{cp}(\mathbb{R})$ *is measurable for each* $x, y \in \mathbb{R}$;

(5.32.2) $H_d(F(t,x,y), F(t,\bar{x},\bar{y})) \le m(t)(|x-\bar{x}| + |y-\bar{y}|)$ *for almost all* $t \in [0,1]$ *and* $x, \bar{x}, y, \bar{y} \in \mathbb{R}$ *with* $m \in C([0,1], \mathbb{R}^+)$ *and* $d(0, F(t,0,0)) \le m(t)$ *for almost all* $t \in [0,1]$.

Then the inclusion problem (5.173) and (5.174) has at least one solution on $[0,1]$ *if*

$$\|m\|\Big(\lambda + \frac{\lambda_1}{\Gamma(2-\kappa)}\Big) < 1, \tag{5.200}$$

where λ, λ_1 *are are given by (5.187) and (5.189) respectively.*

Proof. In order to show that the operator $N : C([0,1], \mathbb{R}) \to \mathcal{P}(C([0,1], \mathbb{R}))$ (introduced in beginning of the proof of Theorem 5.31) satisfies the assumptions of Theorem 1.19, we proceed as follows.

Step 1. *$N(x)$ is non-empty and closed for every* $v \in S_{F,x}$.

Notice that $S_{F,x} \neq \emptyset$ for each x by the assumption (5.32.2) and thus we can find a measurable selection for F (see [78, Theorem III.6]). In order to show that $N(x) \in \mathcal{P}_{cl}(X)$ for each $x \in X$, let $\{u_n\}_{n\ge 0} \in N(x)$ be such that $u_n \to u$ $(n \to \infty)$ in X. Then $u \in C(J, \mathbb{R})$ and there exists $v_n \in S_{F,x_n}$ such that, for each $t \in [0,1]$, we have

$$u_n(t) = \sum_{i=1}^{m-2} \phi_i(t) \int_0^{\sigma_i} e^{\mu(s-\sigma_i)} \left(I^{q-1} v_n(s)\right) ds$$

$$+\sum_{j=1}^{p-2}\psi_j(t)\int_{\xi_j}^{\eta_j}\Big(\int_0^s e^{\mu(m-s)}\left(I^{q-1}v_n(m)\right)dm\Big)ds$$
$$+\Omega_1(t)\Big[\sum_{i=1}^{m-2}\delta_i\left(I^{q-1}v_n(\sigma_i)\right)+\sum_{j=1}^{p-2}\gamma_j\int_{\xi_j}^{\eta_j}\left(I^{q-1}v_n(s)\right)ds-\rho_4 I^{q-1}v_n(1)\Big]$$
$$+\Omega_2(t)\int_0^1 e^{\mu(s-1)}\Big(I^{q-1}v_n(s)\Big)ds+\int_0^t e^{\mu(s-t)}\Big(I^{q-1}v_n(s)\Big)ds.$$

Since F has compact values, one can pass onto a subsequence (if necessary) to find that v_n converges to v in $L^1([0,1],\mathbb{R})$. Thus $v\in S_{F,x}$ and for each $t\in[0,1]$, we have

$$u_n(t)\to u(t)$$
$$=\sum_{i=1}^{m-2}\phi_i(t)\int_0^{\sigma_i}e^{\mu(s-\sigma_i)}\left(I^{q-1}v(s)\right)ds$$
$$+\sum_{j=1}^{p-2}\psi_j(t)\int_{\xi_j}^{\eta_j}\Big(\int_0^s e^{\mu(m-s)}\left(I^{q-1}v(m)\right)dm\Big)ds$$
$$+\Omega_1(t)\Big[\sum_{i=1}^{m-2}\delta_i\left(I^{q-1}v(\sigma_i)\right)+\sum_{j=1}^{p-2}\gamma_j\int_{\xi_j}^{\eta_j}\left(I^{q-1}v(s)\right)ds-\rho_4 I^{q-1}v(1)\Big]$$
$$+\Omega_2(t)\int_0^1 e^{\mu(s-1)}\Big(I^{q-1}v(s)\Big)ds+\int_0^t e^{\mu(s-t)}\Big(I^{q-1}v(s)\Big)ds,$$

which implies that $u\in N(x)$.

Step 2. *There exists $\bar\theta<1$ ($\bar\theta$ is given by (5.200)) such that*

$$H_d(N(x),N(\bar x))\le\bar\theta\|x-\bar x\|_X \quad \textit{for each}\quad x,\bar x\in C([0,1],\mathbb{R}).$$

Let $x,\bar x\in X$ and $h_1\in N(x)$. Then we can find $v_1(t)\in F(t,x(t),y(t))$ such that, for each $t\in[0,1]$,

$$h_1(t)=\sum_{i=1}^{m-2}\phi_i(t)\int_0^{\sigma_i}e^{\mu(s-\sigma_i)}\left(I^{q-1}v_1(s)\right)ds$$
$$+\sum_{j=1}^{p-2}\psi_j(t)\int_{\xi_j}^{\eta_j}\Big(\int_0^s e^{\mu(m-s)}\left(I^{q-1}v_1(m)\right)dm\Big)ds$$
$$+\Omega_1(t)\Big[\sum_{i=1}^{m-2}\delta_i\left(I^{q-1}v_1(\sigma_i)\right)+\sum_{j=1}^{p-2}\gamma_j\int_{\xi_j}^{\eta_j}\left(I^{q-1}v_1(s)\right)ds-\rho_4 I^{q-1}v_1(1)\Big]$$
$$+\Omega_2(t)\int_0^1 e^{\mu(s-1)}\Big(I^{q-1}v_1(s)\Big)ds+\int_0^t e^{\mu(s-t)}\Big(I^{q-1}v_1(s)\Big)ds.$$

By (5.32.2), we have

$$H_d(F(t,x,y),F(t,\bar x,\bar y)\le m(t)(|x(t)-\bar x(t)|+|y(t)-\bar y(t)|).$$

So there exists $w(t) \in F(t, \bar{x}(t), \bar{y}(t))$ such that

$$|v_1(t) - w| \leq m(t)(|x(t) - \bar{x}(t)| + |y(t) - \bar{y}(t)|), \quad t \in [0,1].$$

Define $\mathcal{V} : [0,1] \to \mathcal{P}(\mathbb{R})$ by

$$\mathcal{V}(t) = \{w \in \mathbb{R} : |v_1(t) - w| \leq m(t)(|x(t) - \bar{x}(t)| + |y(t) - \bar{y}(t)|)\}.$$

In view of the fact that the multi-valued operator $\mathcal{V}(t) \cap F(t, \bar{x}(t), \bar{y}(t))$ is measurable (see Proposition III.4 [78]), we can find a function $v_2(t)$ which is a measurable selection for $\mathcal{V}$ and that $v_2(t) \in F(t, \bar{x}(t), \bar{y}(t))$. Then, for each $t \in [0,1]$, we have $|v_1(t) - v_2(t)| \leq m(t)(|x(t) - \bar{x}(t)| + |y(t) - \bar{y}(t)|)$. For each $t \in [0,1]$, we define

$$\begin{aligned}
h_2(t) = & \sum_{i=1}^{m-2} \phi_i(t) \int_0^{\sigma_i} e^{\mu(s-\sigma_i)} \left(I^{q-1} v_2(s)\right) ds \\
& + \sum_{j=1}^{p-2} \psi_j(t) \int_{\xi_j}^{\eta_j} \Big(\int_0^s e^{\mu(m-s)} \left(I^{q-1} v_2(m)\right) dm \Big) ds \\
& + \Omega_1(t) \Big[\sum_{i=1}^{m-2} \delta_i \left(I^{q-1} v_2(\sigma_i)\right) + \sum_{j=1}^{p-2} \gamma_j \int_{\xi_j}^{\eta_j} \left(I^{q-1} v_2(s)\right) ds - \rho_4 I^{q-1} v_2(1) \Big] \\
& + \Omega_2(t) \int_0^1 e^{\mu(s-1)} \Big(I^{q-1} v_2(s)\Big) ds + \int_0^t e^{\mu(s-t)} \Big(I^{q-1} v_2(s)\Big) ds.
\end{aligned}$$

Thus,

$$\begin{aligned}
|h_1(t) - h_2(t)| \leq & \sum_{i=1}^{m-2} |\phi_i(t)| \int_0^{\sigma_i} e^{\mu(s-\sigma_i)} \left(I^{q-1} |v_2(s) - v_1(s)|\right) ds \\
& + \sum_{j=1}^{p-2} |\psi_j(t)| \int_{\xi_j}^{\eta_j} \Big(\int_0^s e^{\mu(m-s)} \left(I^{q-1} |v_2(m) - v_1(m)|\right) dm \Big) ds \\
& + |\Omega_1(t)| \Big[\sum_{i=1}^{m-2} |\delta_i| \left(I^{q-1} |v_2(\sigma) - v_1(\sigma)|\right) \\
& + \sum_{j=1}^{p-2} |\gamma_j| \int_{\xi_j}^{\eta_j} \left(I^{q-1} |v_2(s) - v_1(s)|\right) ds - \rho_4 I^{q-1} |v_2(1) - v_1(1)| \Big] \\
& + |\Omega_2(t)| \int_0^1 e^{\mu(s-1)} \Big(I^{q-1} |v_2(s) - v_1(s)|\Big) ds \\
& + \int_0^t e^{\mu(s-t)} \Big(I^{q-1} |v_2(s) - v_1(s)|\Big) ds \\
\leq & \Bigg\{ \sum_{i=1}^{m-2} \widehat{\phi}_i \int_0^{\sigma_i} e^{\mu(s-\sigma_i)} \left(I^{q-1} 1\right) ds \\
& + \sum_{j=1}^{p-2} \widehat{\psi}_j \int_{\xi_j}^{\eta_j} \Big(\int_0^s e^{\mu(m-s)} \left(I^{q-1} 1\right) dm \Big) ds
\end{aligned}$$

$$+\widehat{\Omega}_1\Big[\sum_{i=1}^{m-2}\delta_i\left(I^{q-1}1\right)+\sum_{j=1}^{p-2}\gamma_j\int_{\xi_j}^{\eta_j}\left(I^{q-1}1\right)ds-\rho_4 I^{q-1}1\Big]$$
$$+\widehat{\Omega}_2\int_0^1 e^{\mu(s-1)}\Big(I^{q-1}1\Big)ds+\int_0^t e^{\mu(s-t)}\Big(I^{q-1}1\Big)ds\Bigg\}\|m\|\|x-\overline{x}\|$$
$$\leq \lambda\|m\|\|x-\overline{x}\|_X.$$

Hence $\|h_1-h_2\|\leq \lambda\|m\|\|x-\overline{x}\|_X$. In a similar manner, we obtain

$$|h_1'(t)-h_2'(t)|$$
$$\leq\Bigg\{\sum_{i=1}^{m-2}|\phi_i'(t)|\int_0^{\sigma_i}e^{\mu(s-\sigma_i)}\left(I^{q-1}1\right)ds$$
$$+\sum_{j=1}^{p-2}|\psi_j'(t)|\int_{\xi_j}^{\eta_j}\Big(\int_0^{s}e^{\mu(m-s)}\left(I^{q-1}1\right)dm\Big)ds+|\Omega_1'(t)|\Big[\sum_{i=1}^{m-2}|\delta_i|\left(I^{q-1}1\right)$$
$$+\sum_{j=1}^{p-2}|\gamma_j|\int_{\xi_j}^{\eta_j}\left(I^{q-1}1\right)ds+|\rho_4|I^{q-1}1\Big]+|\Omega_2'(t)|\int_0^1 e^{\mu(s-1)}\Big(I^{q-1}1\Big)ds$$
$$+\mu\int_0^t e^{\mu(s-t)}\Big(I^{q-1}1\Big)ds+I^{q-1}1\Bigg\}\|m\|\|x-\overline{x}\|_X$$
$$\leq\lambda_1\|m\|\|x-\overline{x}\|_X,$$

and

$$|{}^CD^\kappa h_1(t)-{}^CD^\kappa h_2(t)|\leq\int_0^t\frac{(t-s)^{-\kappa}}{\Gamma(1-\kappa)}|h_1'(t)-h_2'(t)|ds$$
$$\leq\frac{1}{\Gamma(2-\kappa)}\lambda_1\|m\|\|x-\overline{x}\|_X.$$

Thus

$$\|h_1-h_2\|_X\leq\|m\|\left(\lambda+\frac{\lambda_1}{\Gamma(2-\kappa)}\right)\|x-\bar{x}\|_X.$$

Analogously, interchanging the roles of x and $\overline{x}$, we obtain

$$H_d(N(x),N(\bar{x}))\leq\|m\|\left(\lambda+\frac{\lambda_1}{\Gamma(2-\kappa)}\right)\|x-\bar{x}\|_X.$$

From the foregoing arguments, we deduce that N is a contraction. Thus it follows by Theorem 1.19 that N has a fixed point x which is a solution of (5.173) and (5.174). This completes the proof. □

5.7.4 *Examples*

Consider the sequential fractional differential inclusion

$$({}^CD^{7/5}+{}^CD^{2/5})x(t)\in F(t,x(t),{}^CD^{1/2}x(t)),\ t\in[0,1],\tag{5.201}$$

equipped with the conditions (5.197).

In order to illustrate Theorem 5.31, we take

$$F(t,x(t),{}^CD^{1/2}x(t)) = \Big[\frac{1}{\sqrt{t^2+144}}\left(\frac{1}{3}\sin(x(t)) + \frac{1}{2}\frac{|{}^CD^{1/2}x(t)|}{(1+|{}^CD^{1/2}x(t)|)} + \frac{1}{2}\right),$$
$$\frac{1}{(t+15)}\left(\frac{1}{16}e^{-x^4(t)} + \frac{1}{5}\sin({}^CD^{1/2}x(t)) + \frac{1}{2}\right)\Big]. \quad (5.202)$$

It is easy to find that $g(t) = \frac{1}{\sqrt{t^2+144}}$ with $\|g\| = 1/12, \Omega(K) = (K+3)/3$. By the condition (5.31.3), we find that $K > 0.7333$. Thus all the assumptions of Theorem 5.31 hold and consequently the inclusion (5.201) equipped with boundary conditions (5.197) and F given by (5.202), has a solution on $[0,1]$.

For illustrating Theorem 5.32, we consider the following multi-valued map

$$F(t,x(t),{}^CD^{1/2}x(t)) = \Big[\frac{1}{t+10}x(t) + \frac{1}{t^2+15}\cos(D^{1/2}x(t)),$$
$$\frac{1}{t^2+25}\tan^{-1}(x(t)) + \frac{1}{t^2+49}\,{}^CD^{1/2}x(t) + \frac{1}{98}\Big]. \quad (5.203)$$

By the condition (5.32.2), we get $m(t) = 1/(t+10)$ with $\|m\| = 1/10$. Moreover,

$$\|m\|\left(\lambda + \frac{\lambda_1}{\Gamma(2-\kappa)}\right) \approx 0.70715 < 1$$

(the values of λ and λ_1 are taken from subsection 3.1). Clearly all the hypotheses of Theorem 5.32 are satisfied. Therefore there exists at least one solution on $[0,1]$ for the inclusion (5.201) with F given by (5.203), equipped with boundary conditions (5.197).

5.8 Fractional order coupled systems for mixed fractional derivatives with nonlocal multi-point and Riemann-Stieltjes integral-multi-strip conditions

This section deals with the existence and uniqueness of solutions for neutral fractional order coupled systems containing mixed Caputo and Riemann-Liouville sequential fractional derivative operators, complemented with nonlocal multi-point and Riemann-Stieltjes integral multi-strip conditions. In precise terms, we study the following coupled system

$$\begin{cases} {}^CD^q({}^{RL}D^px(t) + f(t,x(t)) = g(t,x(t),y(t)), & t \in J := [0,1], \\ {}^CD^{q_1}({}^{RL}D^{p_1}y(t) + f_1(t,y(t)) = g_1(t,x(t),y(t)), & t \in J := [0,1], \\ x(0) = 0, \quad bx(1) = a\displaystyle\int_0^1 y(s)dH(s) + \sum_{i=1}^{n}\alpha_i\int_{\xi_i}^{\eta_i} y(s)ds, \\ y(0) = 0, \quad b_1y(1) = a_1\displaystyle\int_0^1 x(s)dH(s) + \sum_{j=1}^{m}\beta_j\int_{\theta_j}^{\zeta_j} x(s)ds, \end{cases} \quad (5.204)$$

where ${}^{RL}D^p, {}^{RL}D^{p_1}$ and ${}^CD^q, {}^CD^{q_1}$ denote the Riemann-Liouville and Caputo fractional derivative operators of order p, p_1 and q, q_1 respectively, $0 < p, p_1, q, q_1 \le 1$, with $1 < p+q \leq 2, 1 < p_1 + q_1 \leq 2$, f, f_1 and g, g_1 are given continuous functions, $0 < \xi_i < \eta_i < 1$, $0 < \theta_j < \zeta_j < 1$, $\alpha_i, \beta_j \in \mathbb{R}, i = 1, 2, \dots, n$, $j = 1, 2, \dots, m, a, a_1, b, b_1 \in \mathbb{R}$, and $H(\cdot)$ is a function of bounded variation.

Now we prove an auxiliary lemma dealing with the linear variant of the problem (5.204), which plays a pivotal role in the forthcoming analysis.

Lemma 5.10. *Let $h, h_1, k, k_1 \in C([0,1], \mathbb{R})$. Then the unique solution of the linear fractional differential system*

$$\begin{cases} {}^{C}D^{q}({}^{RL}D^{p}x(t) + h(t)) = k(t), \quad t \in J := [0,1], \\ {}^{C}D^{q_1}({}^{RL}D^{p_1}y(t) + h_1(t)) = k_1(t), \quad t \in J := [0,1], \\ x(0) = 0, \quad bx(1) = a\displaystyle\int_0^1 y(s)dH(s) + \sum_{i=1}^{n}\alpha_i \int_{\xi_i}^{\eta_i} y(s)ds, \\ y(0) = 0, \quad b_1 y(1) = a_1 \displaystyle\int_0^1 x(s)dH(s) + \sum_{j=1}^{m}\beta_j \int_{\theta_j}^{\zeta_j} x(s)ds, \end{cases} \tag{5.205}$$

is given by

$$\begin{aligned} x(t) = {} & -I^{p}h(t) + I^{q+p}k(t) + \frac{t^{p}}{\Lambda\Gamma(1+p)}\Bigg[\Delta\Big(\sum_{i=1}^{n}\alpha_i \int_{\xi_i}^{\eta_i}\Big[-I^{p_1}h_1(s) + I^{q_1+p_1}k_1(s)\Big]ds \\ & + a\int_0^1 \Big[-I^{p_1}h_1(s) + I^{q_1+p_1}k_1(s)\Big]dH(s) + bI^{p}h(1) - bI^{q+p}k(1)\Big) \\ & + B\Big(\sum_{j=1}^{m}\beta_j \int_{\theta_j}^{\zeta_j}\Big[-I^{p}h(s) + I^{q+p}k(s)\Big]ds + a_1\int_0^1\Big[-I^{p}h(s) + I^{q+p}k(s)\Big]dH(s) \\ & + b_1 I^{p_1}h_1(1) - b_1 I^{q_1+p_1}k_1(1)\Big)\Bigg], \end{aligned} \tag{5.206}$$

and

$$\begin{aligned} y(t) = {} & -I^{p_1}h_1(t) + I^{q_1+p_1}k_1(t) + \frac{t^{p_1}}{\Lambda\Gamma(1+p_1)}\Bigg[A\Big(\sum_{j=1}^{m}\beta_j \int_{\theta_j}^{\zeta_j}\Big[-I^{p}h(s) + I^{q+p}k(s)\Big]ds, \\ & + a_1\int_0^1 \Big[-I^{p}h(s) + I^{q+p}k(s)\Big]dH(s) + b_1 I^{p}h_1(1) - b_1 I^{q_1+p_1}k_1(1)\Big) \\ & + \Gamma\Big(\sum_{i=1}^{n}\alpha_i \int_{\xi_i}^{\eta_i}\Big[-I^{p_1}h_1(s) + I^{q_1+p_1}k_1(s)\Big]ds \\ & + a\int_0^1 \Big[-I^{p_1}h_1(s) + I^{q_1+p_1}k_1(s)\Big]dH(s) + bI^{p}h(1) - bI^{q+p}k(1)\Big)\Bigg], \end{aligned} \tag{5.207}$$

where

$$A = \frac{b}{\Gamma(1+p)}, \quad B = a\int_0^1 \frac{s^{p_1}}{\Gamma(1+p_1)}dH(s) + \sum_{i=1}^{n}\alpha_i \frac{\eta_i^{p_1+1} - \xi_i^{p_1+1}}{\Gamma(2+p_1)}, \tag{5.208}$$

$$\Gamma = a_1 \int_0^1 \frac{s^p}{\Gamma(1+p)} dH(s) + \sum_{j=1}^{m} \beta_j \frac{\zeta_j^{p+1} - \theta_j^{p+1}}{\Gamma(2+p)}, \quad \Delta = \frac{b_1}{\Gamma(1+p_1)}, \tag{5.209}$$

and it is assumed that

$$\Lambda := A\Delta - B\Gamma \neq 0. \tag{5.210}$$

Proof. Applying the integral operators I^q, I^{q_1} on the fractional differential equations in (5.205), and then I^p, I^{p_1} on the resulting equations, we get

$$x(t) = -I^p h(t) + I^{q+p} k(t) + c_0 \frac{t^p}{\Gamma(1+p)} + c_1 t^{p-1}, \tag{5.211}$$

and

$$y(t) = -I^{p_1} h_1(t) + I^{q_1+p_1} k_1(t) + d_0 \frac{t^{p_1}}{\Gamma(1+p_1)} + d_1 t^{p_1-1}, \tag{5.212}$$

where c_0, c_1, d_0, d_1 are arbitrary constants. Using the boundary conditions of (5.205) in (5.211) and (5.212), we get $c_1 = 0, d_1 = 0$, and

$$Ac_0 - Bd_0 = P, \ -\Gamma c_0 + \Delta d_0 = Q, \tag{5.213}$$

where

$$P = \sum_{i=1}^{n} \alpha_i \int_{\xi_i}^{\eta_i} \Big[- I^{p_1} h_1(s) + I^{q_1+p_1} k_1(s) \Big] ds + a \int_0^1 \Big[- I^{p_1} h_1(s) + I^{q_1+p_1} k_1(s) \Big] dH(s)$$
$$+bI^p h(1) - bI^{q+p} k(1),$$
$$Q = \sum_{j=1}^{m} \beta_j \int_{\theta_j}^{\zeta_j} \Big[- I^p h(s) + I^{q+p} k(s) \Big] ds + a_1 \int_0^1 \Big[- I^p h(s) + I^{q+p} k(s) \Big] dH(s)$$
$$+b_1 I^{p_1} h_1(1) - b_1 I^{q_1+p_1} k_1(1).$$

Solving the system (5.213) for c_0 and $d0$, we find that

$$c_0 = \frac{1}{\Lambda}(\Delta P + BQ), \ d_0 = \frac{1}{\Lambda}(AQ + \Gamma P).$$

Substituting the values of c_0, c_1, d_0, and d_1 in (5.211) and (5.212) yields the solution (5.206) and (5.207). By direct computation, one can obtain the converse of the lemma. This completes the proof. □

5.8.1 *Main results*

Let us introduce the space $X = \{x(t)|x(t) \in C([0,1], \mathbb{R})\}$ endowed with the norm $\|x\| = \sup\{|x(t)|, t \in [0,1]\}$. Obviously $(X, \|\cdot\|)$ is a Banach space. Then the product space $(X \times X, \|(x,y)\|)$ is also a Banach space equipped with norm $\|(x,y)\| = \|x\| + \|y\|$.

In view of Lemma 5.10, we define an operator $T : X \times X \to X \times X$ associated with the system (5.204) by

$$T(x,y)(t) = \begin{pmatrix} T_1(x,y)(t) \\ T_2(x,y)(t) \end{pmatrix},$$

where

$$
\begin{aligned}
&T_1(x,y)(t)\\
&= -I^p f(t,x(t)) + I^{q+p} g(t,x(t),y(t)) + \frac{t^p}{\Lambda\Gamma(1+p)}\Big[\Delta P + BQ\Big]\\
&= -I^p f(t,x(t)) + I^{q+p} g(t,x(t),y(t))\\
&+\frac{t^p}{\Lambda\Gamma(1+p)}\Bigg[\Delta\Big(\sum_{i=1}^{n}\alpha_i\int_{\xi_i}^{\eta_i}\Big[-I^{p_1}f_1(s,y(s)) + I^{q_1+p_1}g_1(s,x(s),y(s))\Big]ds\\
&+a\int_0^1\Big[-I^{p_1}f_1(s,y(s)) + I^{q_1+p_1}g_1(s,y(s),y(s))\Big]dH(s)\\
&+bI^p f(s,x(s))(1) - bI^{q+p}g(s,x(s),y(s))(1)\Big)\\
&+B\Big(\sum_{j=1}^{m}\beta_j\int_{\theta_j}^{\zeta_j}\Big[-I^p f(s,x(s)) + I^{q+p}g(s,x(s),y(s))\Big]ds\\
&+a_1\int_0^1\Big[-I^p f(s,x(s)) + I^{q+p}g(s,x(s),y(s))\Big]dH(s)\\
&+b_1 I^{p_1}f_1(s,y(s))(1) - b_1 I^{q_1+p_1}g_1(s,x(s),y(s))(1)\Big)\Bigg],
\end{aligned}
\tag{5.214}
$$

$$
\begin{aligned}
&T_2(x,y)(t)\\
&= -I^{p_1} f_1(t,y(t)) + I^{q_1+p_1} g_1(t,x(t),y(t)) + \frac{t_1^p}{\Lambda\Gamma(1+p_1)}\Big[AQ + \Gamma P\Big]\\
&= -I^{p_1} f_1(t,y(t)) + I^{q_1+p_1} g_1(t,x(t),y(t))\\
&+\frac{t^{p_1}}{\Lambda\Gamma(1+p_1)}\Bigg[A\Big(\sum_{j=1}^{m}\beta_j\int_{\theta_j}^{\zeta_j}\Big[-I^p f(s,x(s)) + I^{q+p}g(s,x(s),y(s))\Big]ds\\
&+a_1\int_0^1\Big[-I^p f(s,x(s)) + I^{q+p}g(s,x(s),y(s))\Big]dH(s)\\
&+b_1 I^p f_1(s,y(s))(1) - b_1 I^{q_1+p_1}g_1(s,x(s),y(s))(1)\Big)\\
&+\Gamma\Big(\sum_{i=1}^{n}\alpha_i\int_{\xi_i}^{\eta_i}\Big[-I^{p_1}f_1(s,y(s)) + I^{q_1+p_1}g_1(s,x(s),y(s))\Big]ds\\
&+a\int_0^1\Big[-I^{p_1}f_1(s,y(s)) + I^{q_1+p_1}g_1(s,x(s),y(s))\Big]dH(s)\\
&+bI^p f(s,y(s))(1) - bI^{q+p}g(s,x(s),y(s))(1)\Big)\Bigg].
\end{aligned}
\tag{5.215}
$$

For the sake of computational convenience, we put

$$
Q_1 = \frac{1}{\Gamma(p+1)} + \frac{1}{|\Lambda|\Gamma(1+p)}\Big[|\Delta||b|\frac{1}{\Gamma(p+1)}
$$

$$+|B|\Big(\sum_{i=1}^{m}|\beta_j|\frac{\zeta_j^p-\theta_j^p}{\Gamma(p+1)}+|a_1|\int_0^1\frac{s^p}{\Gamma(p+1)}dH(s)\Big)\Big], \tag{5.216}$$

$$Q_2=\frac{1}{\Gamma(p+q+1)}+\frac{1}{|\Lambda|\Gamma(1+p)}\Big[|\Delta||b|\frac{1}{\Gamma(p+q+1)}$$
$$+|B|\Big(\sum_{i=1}^{m}|\beta_j|\frac{\zeta_j^{p+q}-\theta_j^{p+q}}{\Gamma(p+q+1)}+|a_1|\int_0^1\frac{s^{p+q}}{\Gamma(p+q+1)}dH(s)\Big)\Big], \tag{5.217}$$

$$Q_3=\frac{1}{|\Lambda|\Gamma(1+p)}\Big[|B||b_1|\frac{1}{\Gamma(p_1+1)}$$
$$+|\Delta|\Big(\sum_{i=1}^{n}|\alpha_i|\frac{\eta_i^{p_1}-\xi_i^{p_1}}{\Gamma(p_1+1)}+|a|\int_0^1\frac{s^{p_1}}{\Gamma(p_1+1)}dH(s)\Big)\Big], \tag{5.218}$$

$$Q_4=\frac{1}{|\Lambda|\Gamma(1+p)}\Big[|B||b_1|\frac{1}{\Gamma(p_1+q_1+1)}$$
$$+|\Delta|\Big(\sum_{i=1}^{n}|\alpha_i|\frac{\eta_i^{p_1+q_1}-\xi_i^{p_1+q_1}}{\Gamma(p_1+q_1+1)}+|a|\int_0^1\frac{s^{p_1+q_1}}{\Gamma(p_1+q_1+1)}dH(s)\Big)\Big], \tag{5.219}$$

$$\overline{Q}_1=\frac{1}{|\Lambda|\Gamma(1+p_1)}\Big[|\Gamma||b|\frac{1}{\Gamma(p+1)}$$
$$+|A|\Big(\sum_{i=1}^{m}|\beta_j|\frac{\zeta_j^p-\theta_j^p}{\Gamma(p+1)}+|a_1|\int_0^1\frac{s^p}{\Gamma(p+1)}dH(s)\Big)\Big], \tag{5.220}$$

$$\overline{Q}_2=\frac{1}{|\Lambda|\Gamma(1+p_1)}\Big[|\Gamma||b|\frac{1}{\Gamma(p+q+1)}$$
$$+|A|\Big(\sum_{i=1}^{m}|\beta_j|\frac{\zeta_j^{p+q}-\theta_j^{p+q}}{\Gamma(p+q+1)}+|a_1|\int_0^1\frac{s^{p+q}}{\Gamma(p+q+1)}dH(s)\Big)\Big], \tag{5.221}$$

$$\overline{Q}_3=\frac{1}{\Gamma(p_1+1)}+\frac{1}{|\Lambda|\Gamma(1+p_1)}\Big[|A||b_1|\frac{1}{\Gamma(p_1+1)}$$
$$+|\Gamma|\Big(\sum_{i=1}^{n}|\alpha_i|\frac{\eta_i^{p_1}-\xi_i^{p_1}}{\Gamma(p_1+1)}+|a|\int_0^1\frac{s^{p_1}}{\Gamma(p_1+1)}dH(s)\Big)\Big], \tag{5.222}$$

$$\overline{Q}_4=\frac{1}{\Gamma(p_1+q_1+1)}+\frac{1}{|\Lambda|\Gamma(1+p_1)}\Big[|A||b_1|\frac{1}{\Gamma(p_1+q_1+1)}$$
$$+|\Gamma|\Big(\sum_{i=1}^{n}|\alpha_i|\frac{\eta_i^{p_1+q_1}-\xi_i^{p_1+q_1}}{\Gamma(p_1+q_1+1)}+|a|\int_0^1\frac{s^{p_1+q_1}}{\Gamma(p_1+q_1+1)}dH(s)\Big)\Big]. \tag{5.223}$$

In the proof of the following existence result, we make use of Leray-Schauder alternative (Theorem 1.7).

Theorem 5.33. *Assume that:*

(5.33.1) $f, f_1 : [0,1]\times\mathbb{R}\to\mathbb{R}$, $g, g_1 : [0,1]\times\mathbb{R}\times\mathbb{R}\to\mathbb{R}$ *are continuous functions and there exist real constants* $\lambda_1,\mu_1,k_i,\gamma_i\geq 0$, $(i=1,2)$ *and* $\lambda_0,k_0,\mu_0,\gamma_0>0$ *such that,* $\forall x,y,x_i\in\mathbb{R}$, $(i=1,2)$,

$$|f(t,x)|\leq\lambda_0+\lambda_1|x|,\quad |g(t,x_1,x_2)|\leq k_0+k_1|x_1|+k_2|x_2|,$$

$$|f_1(t,y)| \le \mu_0 + \mu_1|y|, \quad |g_1(t,x_1,x_2)| \le \gamma_0 + \gamma_1|x_1| + \gamma_2|x_2|.$$

If

$$M_1 = [Q_1 + \overline{Q}_1]\lambda_1 + [Q_2 + \overline{Q}_2]k_1 + [Q_4 + \overline{Q}_4]\gamma_1 < 1,$$

$$M_2 = [Q_2 + \overline{Q}_2]k_2 + [Q_3 + \overline{Q}_3]\mu_1 + [Q_4 + \overline{Q}_4]\gamma_2 < 1,$$

where $Q_i, \overline{Q}_i$, $i = 1,2,3,4$ *are given by (5.216)-(5.223), then the system (5.204) has at least one solution on* $[0,1]$.

Proof. First we show that the operator $T : X \times X \to X \times X$ is completely continuous. By continuity of functions f, f_1, g and g_1, the operator T is continuous.

Let $\Omega \subset X \times X$ be bounded. Then there exist positive constants L_i, $i = 1,2,3,4$ such that $|f(t,x(t))| \le L_1$, $|g(t,x(t),y(t))| \le L_2$, $|f_1(t,x(t)(| \le L_3$, $|g_1(t,x(t),y(t))| \le L_4$, $\forall (x,y) \in \Omega$. Then, for any $(x,y) \in \Omega$, we have

$$\begin{aligned}
&|T_1(x,y)(t)| \\
&\le \frac{1}{\Gamma(p+1)}L_1 + \frac{1}{\Gamma(p+q+1)}L_2 \\
&\quad + \frac{1}{|\Lambda|\Gamma(1+p)}\Bigg\{\Big[|\Delta||b|\frac{1}{\Gamma(p+1)} + |B|\Big(\sum_{i=1}^{m}|\beta_j|\frac{\zeta_j^p - \theta_j^p}{\Gamma(p+1)} + |a_1|\int_0^1 \frac{s^p}{\Gamma(p+1)}dH(s)\Big)\Big]L_1 \\
&\quad + \Big[|\Delta||b|\frac{1}{\Gamma(p+q+1)} + |B|\Big(\sum_{i=1}^{m}|\beta_j|\frac{\zeta_j^{p+q} - \theta_j^{p+q}}{\Gamma(p+q+1)} + |a_1|\int_0^1 \frac{s^{p+q}}{\Gamma(p+q+1)}dH(s)\Big)\Big]L_2 \\
&\quad + \Big[|B||b_1|\frac{1}{\Gamma(p_1+1)} + |\Delta|\Big(\sum_{i=1}^{n}|\alpha_i|\frac{\eta_i^{p_1} - \xi_i^{p_1}}{\Gamma(p_1+1)} + |a|\int_0^1 \frac{s^{p_1}}{\Gamma(p_1+1)}dH(s)\Big)\Big]L_3 \\
&\quad + \Big[|\Delta|\Big(\sum_{i=1}^{n}|\alpha_i|\frac{\eta_i^{p_1+q_1} - \xi_i^{p_1+q_1}}{\Gamma(p_1+q_1+1)} + |a|\int_0^1 \frac{s^{p_1+q_1}}{\Gamma(p_1+q_1+1)}dH(s)\Big) \\
&\quad + |B||b_1|\frac{1}{\Gamma(p_1+q_1+1)}\Big]L_4\Bigg\} \\
&\le Q_1L_1 + Q_2L_2 + Q_3L_3 + Q_4L_4,
\end{aligned}$$

which implies that

$$\|T_1(x,y)\| \le Q_1L_1 + Q_2L_2 + Q_3L_3 + Q_4L_4.$$

Similarly, it can be shown that

$$\|T_2(x,y)\| \le \overline{Q}_1L_1 + \overline{Q}_2L_2 + \overline{Q}_3L_3 + \overline{Q}_4L_4.$$

Thus, it follows from the above inequalities that the operator T is uniformly bounded, since

$$\|T(x,y)\| \le [Q_1 + \overline{Q}_1]L_1 + [Q_2 + \overline{Q}_2]L_2 + [Q_3 + \overline{Q}_3]L_3 + [Q_4 + \overline{Q}_4]L_4.$$

Next, we show that T is equi-continuous. Let $t_1, t_2 \in [0,1]$ with $t_1 < t_2$. Then we have

$$|T_1(x(t_2),y(t_2)) - T_1(x(t_1),y(t_1))|$$

$$\begin{aligned}
&\leq L_1\left|\int_0^{t_1}\frac{[(t_2-s)^{p-1}-(t_1-s)^{p-1}]}{\Gamma(p)}ds+\int_{t_1}^{t_2}\frac{(t_2-s)^{p-1}}{\Gamma(p)}ds\right|\\
&+L_2\left|\int_0^{t_1}\frac{[(t_2-s)^{p+q-1}-(t_1-s)^{p+q-1}]}{\Gamma(p+q)}ds+\int_{t_1}^{t_2}\frac{(t_2-s)^{p+q-1}}{\Gamma(p+q)}ds\right|\\
&+\frac{|t_2^p-t_1^p|}{|\Lambda|\Gamma(1+p)}|\Delta P+BQ|\\
&\leq L_1\frac{2(t_2-t_1)^p+|t_2^p-t_1^p|}{\Gamma(p+1)}+L_2\frac{2(t_2-t_1)^{p+q}+|t_2^{p+q}-t_1^{p+q}|}{\Gamma(p+q+1)}+\frac{|t_2^p-t_1^p|}{|\Lambda|\Gamma(1+p)}|\Delta P+BQ|.
\end{aligned}$$

Analogously, we can obtain

$$\begin{aligned}
&|T_2(x(t_2),y(t_2))-T_2(x(t_1),y(t_1))|\\
&\leq L_3\frac{2(t_2-t_1)^{p_1}+|t_2^{p_1}-t_1^{p_1}|}{\Gamma(p_1+1)}+L_4\frac{2(t_2-t_1)^{p_1+q_1}+|t_2^{p_1+q_1}-t_1^{p_1+q_1}|}{\Gamma(p_1+q_1+1)}\\
&+\frac{|t_2^{p_1}-t_1^{p_1}|}{|\Lambda|\Gamma(1+p_1)}|AQ+\Gamma P|.
\end{aligned}$$

Therefore, the operator $T(x,y)$ is equi-continuous, and thus the operator $T(x,y)$ is completely continuous.

Finally, it will be verified that the set $\mathcal{E}=\{(x,y)\in X\times X|(x,y)=\lambda T(x,y),0\leq\lambda\leq 1\}$ is bounded. Let $(x,y)\in\mathcal{E}$ with $(x,y)=\lambda T(x,y)$. For any $t\in[0,1]$, we have

$$x(t)=\lambda T_1(x,y)(t),\quad y(t)=\lambda T_2(x,y)(t).$$

Then

$$\begin{aligned}
&|x(t)|\\
&\leq Q_1(\lambda_0+\lambda_1|x|)+Q_2(k_0+k_1|x|+k_2|y|)+Q_3(\mu_0+\mu_1|y|)+Q_4(\gamma_0+\gamma_1|x|+\gamma_2|y|)\\
&=Q_1\lambda_0+Q_2k_0+Q_3\mu_0+Q_4\gamma_0+[Q_1\lambda_1+Q_2k_1+Q_4\gamma_1]|x|+[Q_2k_2+Q_3\mu_1+Q_4\gamma_2]|y|,
\end{aligned}$$

and

$$\begin{aligned}
&|y(t)|\\
&\leq \overline{Q}_1(\lambda_0+\lambda_1|x|)+\overline{Q}_2(k_0+k_1|x|+k_2|y|)+\overline{Q}_3(\mu_0+\mu_1|y|)+\overline{Q}_4(\gamma_0+\gamma_1|x|+\gamma_2|y|)\\
&=\overline{Q}_1\lambda_0+\overline{Q}_2k_0+\overline{Q}_3\mu_0+\overline{Q}_4\gamma_0+[\overline{Q}_1\lambda_1+\overline{Q}_2k_1+\overline{Q}_4\gamma_1]|x|+[\overline{Q}_2k_2+\overline{Q}_3\mu_1+\overline{Q}_4\gamma_2]|y|.
\end{aligned}$$

Hence we have

$$\begin{aligned}
\|x\|\leq\ &Q_1\lambda_0+Q_2k_0+Q_3\mu_0+Q_4\gamma_0+[Q_1\lambda_1+Q_2k_1+Q_4\gamma_1]\|x\|\\
&+[Q_2k_2+Q_3\mu_1+Q_4\gamma_2]\|y\|,
\end{aligned}$$

and

$$\begin{aligned}
\|y\|\leq\ &\overline{Q}_1\lambda_0+\overline{Q}_2k_0+\overline{Q}_3\mu_0+\overline{Q}_4\gamma_0+[\overline{Q}_1\lambda_1+\overline{Q}_2k_1+\overline{Q}_4\gamma_1]\|x\|\\
&+[\overline{Q}_2k_2+\overline{Q}_3\mu_1+\overline{Q}_4\gamma_2]\|y\|,
\end{aligned}$$

which imply that

$$\begin{aligned}\|x\| + \|y\| \le & [Q_1 + \overline{Q}_1]\lambda_0 + [Q_2 + \overline{Q}_2]k_0 + [Q_3 + \overline{Q}_3]\mu_0 + [Q_4 + \overline{Q}_4]\gamma_0 \\ & + \{[Q_1 + \overline{Q}_1]\lambda_1 + [Q_2 + \overline{Q}_2]k_1 + [Q_4 + \overline{Q}_4]\gamma_1\}\|x\| \\ & + \{[Q_2 + \overline{Q}_2]k_2 + [Q_3 + \overline{Q}_3]\mu_1 + [Q_4 + \overline{Q}_4]\gamma_2\}\|y\|.\end{aligned}$$

Consequently,

$$\|(x, y)\| \le \frac{[Q_1 + \overline{Q}_1]\lambda_0 + [Q_2 + \overline{Q}_2]k_0 + [Q_3 + \overline{Q}_3]\mu_0 + [Q_4 + \overline{Q}_4]\gamma_0}{\min\{1 - M_1, 1 - M_2\}},$$

which proves that $\mathcal{E}$ is bounded. Thus, by Theorem 1.7, the operator T has at least one fixed point. Hence the boundary value problem (5.204) has at least one solution on $[0, 1]$. The proof is complete. □

Now we prove the uniqueness of solutions for the system (5.204) via Banach's contraction mapping principle.

Theorem 5.34. *Assume that:*

(5.34.1) $f, f_1 : [0, 1] \times \mathbb{R} \to \mathbb{R}$, $g, g_1 : [0, 1] \times \mathbb{R} \times \mathbb{R} \to \mathbb{R}$ *are continuous functions and there exist positive constants* ℓ_i, $i = 1, 2, 3, 4$, *such that for all* $t \in [0, 1]$ *and* $x, y, x_i, y_i \in \mathbb{R}$, $i = 1, 2$,

$$|f(t, x) - f(t, y)| \le \ell_1|x - y|, \tag{5.224}$$
$$|g(t, x_1, x_2) - g(t, y_1, y_2)| \le \ell_2(|x_1 - y_1| + |x_2 - y_2|), \tag{5.225}$$
$$|f_1(t, x) - f_1(t, y)| \le \ell_3|x - y|, \tag{5.226}$$
$$|g_1(t, x_1, x_2) - g_1(t, y_1, y_2)| \le \ell_4(|x_1 - y_1| + |x_2 - y_2|). \tag{5.227}$$

Then there exists a unique solution for the system (5.204) on $[0, 1]$ *provided that*

$$[Q_1 + \overline{Q}_1]\ell_1 + [Q_2 + \overline{Q}_2]\ell_2 + [Q_3 + \overline{Q}_3]\ell_3 + [Q_4 + \overline{Q}_4]\ell_4 < 1, \tag{5.228}$$

where $Q_i, \overline{Q}_i, i = 1, 2, 3, 4$ *are given by (5.216)-(5.223).*

Proof. Define $\sup_{t\in[0,1]} f(t, 0) = N_1 < \infty$, $\sup_{t\in[0,1]} g(t, 0, 0) = N_2 < \infty$, $\sup_{t\in[0,1]} f_1(t, 0) = N_3 < \infty$, $\sup_{t\in[0,1]} g_1(t, 0, 0) = N_4 < \infty$ and $r > 0$ such that

$$r > \frac{[Q_1 + \overline{Q}_1]N_1 + [Q_2 + \overline{Q}_2]N_2 + [Q_3 + \overline{Q}_3]N_3 + [Q_4 + \overline{Q}_4]N_4}{1 - \{[Q_1 + \overline{Q}_1]\ell_1 + [Q_2 + \overline{Q}_2]\ell_2 + [Q_3 + \overline{Q}_3]\ell_3 + [Q_4 + \overline{Q}_4]\ell_4\}}.$$

In the first step, we show that $TB_r \subset B_r$, where $B_r = \{(x, y) \in X \times X : \|(x, y)\| \le r\}$. By the assumption (H_2), for $(x, y) \in B_r$, $t \in [0, 1]$, we have

$$\begin{aligned}|f(t, x(t)| &\le |f(t, x(t)) - f(t, 0)| + |f(t, 0)| \le \ell_1|x(t)| + N_1 \\ &\le \ell_1(\|x\| + \|y\|) + N_1 \le \ell_1 r + N_1,\end{aligned}$$

$$|g(t, x(t), y(t))| \le |g(t, x(t), y(t)) - g(t, 0, 0)| + |g(t, 0, 0)| \le \ell_2(|x(t)| + |y(t)|) + N_2$$

$$\leq \ell_2(\|x\| + \|y\|) + N_2 \leq \ell_2 r + N_2,$$

and

$$|f_1(t, x(t))| \leq \ell_3 r + N_3, \quad |g_1(t, x(t), y(t))| \leq \ell_4 r + N_4.$$

Using the above estimates, we obtain

$$\begin{aligned}
&|T_1(x,y)(t)| \\
&\leq I^p|f(t,x(t))| + I^{q+p}|g(t,x(t),y(t))| + \frac{t^p}{|\Lambda|\Gamma(1+p)}\Big[|\Delta P + BQ|\Big] \\
&\leq \frac{1}{\Gamma(p+1)}(\ell_1 r + N_1) + \frac{1}{\Gamma(p+q+1)}(\ell_2 r + N_2) \\
&\quad + \frac{1}{|\Lambda|\Gamma(1+p)}\Bigg\{\Big[|\Delta||b|\frac{1}{\Gamma(p+1)} + |B|\Big(\sum_{i=1}^{m}|\beta_j|\frac{\zeta_j^p - \theta_j^p}{\Gamma(p+1)} + |a_1|\int_0^1 \frac{s^p}{\Gamma(p+1)}dH(s)\Big)\Big](\ell_1 r + N_1) \\
&\quad + \Big[|\Delta||b|\frac{1}{\Gamma(p+q+1)} + |B|\Big(\sum_{i=1}^{m}|\beta_j|\frac{\zeta_j^{p+q} - \theta_j^{p+q}}{\Gamma(p+q+1)} + |a_1|\int_0^1 \frac{s^{p+q}}{\Gamma(p+q+1)}dH(s)\Big)\Big](\ell_2 r + N_2) \\
&\quad + \Big[|B||b_1|\frac{1}{\Gamma(p_1+1)} + |\Delta|\Big(\sum_{i=1}^{n}|\alpha_i|\frac{\eta_i^{p_1} - \xi_i^{p_1}}{\Gamma(p_1+1)} + |a|\int_0^1 \frac{s^{p_1}}{\Gamma(p_1+1)}dH(s)\Big)\Big](\ell_3 r + N_3) \\
&\quad + \Big[|\Delta|\Big(\sum_{i=1}^{n}|\alpha_i|\frac{\eta_i^{p_1+q_1} - \xi_i^{p_1+q_1}}{\Gamma(p_1+q_1+1)} + |a|\int_0^1 \frac{s^{p_1+q_1}}{\Gamma(p_1+q_1+1)}dH(s)\Big) \\
&\quad + |B||b_1|\frac{1}{\Gamma(p_1+q_1+1)}\Big](\ell_4 r + N_4)\Bigg\} \\
&= [Q_1\ell_1 + Q_2\ell_2 + Q_3\ell_3 + Q_4\ell_4]r + Q_1N_1 + Q_2N_2 + Q_3N_3 + Q_4N_4.
\end{aligned}$$

Hence

$$\|T_1(x,y)\| \leq [Q_1\ell_1 + Q_2\ell_2 + Q_3\ell_3 + Q_4\ell_4]r + Q_1N_1 + Q_2N_2 + Q_3N_3 + Q_4N_4.$$

In the same way, we can obtain that

$$\|T_2(x,y)\| \leq [\overline{Q}_1\ell_1 + \overline{Q}_2\ell_2 + \overline{Q}_3\ell_3 + \overline{Q}_4\ell_4]r + \overline{Q}_1N_1 + \overline{Q}_2N_2 + \overline{Q}_3N_3 + \overline{Q}_4N_4.$$

In consequence, it follows that

$$\begin{aligned}
\|T(x,y)\| \leq{}& \{[Q_1 + \overline{Q}_1]\ell_1 + [Q_2 + \overline{Q}_2]\ell_2 + [Q_3 + \overline{Q}_3]\ell_3 + [Q_4 + \overline{Q}_4]\ell_4\}r \\
&+ [Q_1 + \overline{Q}_1]N_1 + [Q_2 + \overline{Q}_2]N_2 + [Q_3 + \overline{Q}_3]N_3 + [Q_4 + \overline{Q}_4]N_4 \leq r.
\end{aligned}$$

Now, for $(x_2, y_2), (x_1, y_1) \in X \times X$, and for any $t \in [0,1]$, we get

$$\begin{aligned}
&|T_1(x_2,y_2)(t) - T_1(x_1,y_1)(t)| \\
&\leq \frac{1}{\Gamma(p+1)}\ell_1(\|x_2 - x_1\| + \|y_2 - y_1\|) + \frac{1}{\Gamma(p+q+1)}\ell_2(\|x_2 - x_1\| + \|y_2 - y_1\|) \\
&\quad + \frac{1}{|\Lambda|\Gamma(1+p)}\Bigg\{\Big[|\Delta||b|\frac{1}{\Gamma(p+1)} + |B|\Big(\sum_{i=1}^{m}|\beta_j|\frac{\zeta_j^{p+1} - \theta_j^{p+1}}{\Gamma(p+2)} \\
&\quad + |a_1|\int_0^1 \frac{s^p}{\Gamma(p+1)}dH(s)\Big)\Big]\ell_1(\|x_2 - x_1\| + \|y_2 - y_1\|) \\
&\quad + \Big[|\Delta||b|\frac{1}{\Gamma(p+q+1)} + |B|\Big(\sum_{i=1}^{m}|\beta_j|\frac{\zeta_j^{p+q+1} - \xi_j^{p+q+1}}{\Gamma(p+q+2)}
\end{aligned}$$

$$
\begin{aligned}
&+|a_1|\int_0^1 \frac{s^{p+q}}{\Gamma(p+q+1)}dH(s)\Big)\Big]\ell_2(\|x_2-x_1\|+\|y_2-y_1\|)\\
&+\Big[|B||b_1|\frac{1}{\Gamma(p_1+1)}+|\Delta|\Big(\sum_{i=1}^{n}|\alpha_i|\frac{\eta_i^{p_1+1}-\xi_i^{p_1+1}}{\Gamma(p_1+2)}\\
&+|a|\int_0^1 \frac{s^{p_1}}{\Gamma(p_1+1)}dH(s)\Big)\Big]\ell_3(\|x_2-x_1\|+\|y_2-y_1\|)\\
&+\Big[|\Delta|\Big(\sum_{i=1}^{n}|\alpha_i|\frac{\eta_i^{p_1+q_1+1}-\xi_i^{p_1+q_1+1}}{\Gamma(p_1+q_1+2)}+|a|\int_0^1 \frac{s^{p_1+q_1}}{\Gamma(p_1+q_1+1)}dH(s)\Big)\\
&+|B||b_1|\frac{1}{\Gamma(p_1+q_1+1)}\Big]\ell_4(\|x_2-x_1\|+\|y_2-y_1\|)\Big\}\\
&\le [Q_1\ell_1+Q_2\ell_2+Q_3\ell_3+Q_4\ell_4](\|x_2-x_1\|+\|y_2-y_1\|),
\end{aligned}
$$

and consequently we obtain

$$
\begin{aligned}
&\|T_1(x_2,y_2)-T_1(x_1,y_1)\|\\
&\le [Q_1\ell_1+Q_2\ell_2+Q_3\ell_3+Q_4\ell_4](\|x_2-x_1\|+\|y_2-y_1\|).
\end{aligned}
\tag{5.229}
$$

Similarly, one can find that

$$
\begin{aligned}
&\|T_2(x_2,y_2)(t)-T_2(x_1,y_1)\|\\
&\le [\overline{Q}_1\ell_1+\overline{Q}_2\ell_2+\overline{Q}_3\ell_3+\overline{Q}_4\ell_4](\|x_2-x_1\|+\|y_2-y_1\|).
\end{aligned}
\tag{5.230}
$$

It follows from (5.229) and (5.230) that

$$
\begin{aligned}
\|T(x_2,y_2)-T(x_1,y_1)\| \le \{&[Q_1+\overline{Q}_1]\ell_1+[Q_2+\overline{Q}_2]\ell_2+[Q_3+\overline{Q}_3]\ell_3\\
&+[Q_4+\overline{Q}_4]\ell_4\}(\|x_2-x_1\|+\|y_2-y_1\|).
\end{aligned}
$$

By assumption (5.228), the foregoing inequality implies that the operator T is a contraction. So, by Banach's fixed point theorem, the operator T has a unique fixed point, which is the unique solution of problem (5.204). This completes the proof. □

Example 5.16. Consider the following coupled system:

$$
\begin{aligned}
&{}^CD^{1/2}({}^CD^{1/2}x(t)+f(t,x(t)))=g(t,x(t),y(t)),\quad t\in[0,1],\\
&{}^CD^{1/3}({}^CD^{1/4}y(t)+f_1(t,y(t)))=g_1(t,x(t),y(t)),\quad t\in[0,1],\\
&x(0)=0,\quad x(1)=\int_0^1 y(s)dH(s)+\sum_{i=1}^{3}\alpha_i\int_{\xi_i}^{\eta_i}y(s)ds,\\
&y(0)=0,\quad y(1)=\int_0^1 x(s)dH(s)+\sum_{j=1}^{3}\beta_j\int_{\theta_j}^{\zeta_j}x(s)ds.
\end{aligned}
\tag{5.231}
$$

Here $p=q=1/2$, $q_1=1/3, p_1=1/4$, $a=1, b=1, a_1=1, b_1=1$, $H(s)=s, \xi_1=1/7, \xi_2=3/7, \xi_3=5/7, \eta_1=2/7, \eta_2=4/7, \eta_3=6/7, \alpha_1=1/12, \alpha_2=$

$1/6, \alpha_3 = 1/4, \beta_1 = 1/5, \beta_2 = 1/10, \beta_3 = 1/5, \ \theta_1 = 1/8, \theta_2 = 3/8, \theta_3 = 5/8,$ $\zeta_1 = 1/4, \zeta_2 = 1/2, \zeta_3 = 3/4,$

$$f(t,x) = \frac{1}{160}\sin x + e^{-t}\cos t, \ f_1(t,y) = \frac{1}{180}\Big(\frac{|y|}{1+|y|}\Big) + 6t.$$

$$g(t,x,y) = \frac{1}{8(t+2)^2}\frac{|x|}{1+|x|} + 1 + \frac{1}{64}\sin^2 y,$$

$$g_1(t,x,y) = \frac{1}{64\pi}\sin(2\pi x) + \frac{|y|}{32(1+|y|)} + \frac{1}{2}.$$

Using the given values, we find that $A \approx 1.12838, B \approx 0.950675, \Gamma \approx 0.609051, \Delta \approx 1.10326, \Lambda \approx 0.665890, Q_1 \approx 4.26043, Q_2 \approx 3.77571, Q_3 \approx 3.57472, Q_4 \approx 3.24588, \bar{Q}_1 \approx 2.32525, \bar{Q}_2 \approx 2.06069, \bar{Q}_3 \approx 4.13600, \ \bar{Q}_4 \approx 3.99478$. Notice that $\ell_1 = 1/160, \ell_3 = 1/180$ as $|f(t,x) - f(t,y)| \leq \frac{1}{160}|x-y|, |f_1(t,x) - f_1(t,y)| \leq \frac{1}{180}|x-y|$, and $\ell_2 = \ell_4 = \frac{1}{32}$ as $|g(t,x_1,x_2) - g(t,y_1,y_2)| \leq \frac{1}{32}(|x_1 - x_2| + |y_1 - y_2|), |g_1(t,x_1,x_2) - g_1(t,y_1,y_2)| \leq \frac{1}{32}(|x_1-x_2|+|y_1-y_2|)$. Moreover, it is found that $(Q_1+\overline{Q}_1)\ell_1 + (Q_2+\overline{Q}_2)\ell_2 + (Q_3+\overline{Q}_3)\ell_3 + (Q_4+\overline{Q}_4)\ell_4 \approx 0.492656 < 1$. Thus all the conditions of Theorem 5.34 are satisfied and consequently, its conclusion applies to the problem (5.231).

5.9 Nonlinear multi-term fractional differential equations with Riemann-Stieltjes integro-multi-point boundary conditions

In the present section, we investigate the existence of solutions for an abstract nonlinear multi-term Caputo fractional differential equation with nonlinearity depending on the unknown function together with its lower-order Caputo fractional derivatives given by

$$\mu\ {}^C D_a^q x(t) + \xi\ {}^C D_a^r x(t) = f(t, x(t), {}^C D_a^p x(t), {}^C D_a^{p+1} x(t)), \ t \in [a,b], \qquad (5.232)$$

supplemented with Riemann-Stieltjes integro-multi-point boundary conditions

$$x(a) = \sum_{i=1}^{n-2} \alpha_i x(\eta_i) + \int_a^b x(s)dA(s), \ x'(a) = 0, \ x(b) = 0, \ x'(b) = 0, \qquad (5.233)$$

where $3 < q \leq 4, 0 < p, r \leq 1$, ${}^C D_a^\theta$ denotes the Caputo fractional derivative operator of order θ with $\theta = q, r, p$, $a < \eta_1 < \eta_2 < \cdots < \eta_{n-2} < b$, $f : [a,b] \times \mathbb{R}^3 \longrightarrow \mathbb{R}$ is a given continuous function, A is a function of bounded variation, and $\mu\ (\mu \neq 0), \xi, \alpha_i \in \mathbb{R}, \ i = 1, 2, \ldots, n-2$.

Lemma 5.11. *Let $\psi \in C([a,b])$. Then the unique solution of the linear multi-term fractional differential equation*

$$\mu\ {}^C D_a^q x(t) + \xi\ {}^C D_a^r x(t) = \psi(t), \ 3 < q \leq 4, \ 0 < r < 1, \ t \in [a,b], \qquad (5.234)$$

subject to the boundary conditions (5.233) is given by

$$x(t) = \frac{-\xi}{\mu} {}_aI^{q-r} x(t) + \frac{1}{\mu} {}_aI^{q} \psi(t) + \frac{1}{\mu}\Big[\phi_1(t)\Big(\xi\, {}_aI^{q-r} x(b) - {}_aI^{q}\psi(b)\Big)$$

$$+\phi_2(t)\Big(\xi\, {}_aI^{q-r-1} x(b) - {}_aI^{q-1}\psi(b)\Big) \tag{5.235}$$

$$+\phi_3(t)\Big(\xi \sum_{i=1}^{n-2} \alpha_i\, {}_aI^{q-r} x(\eta_i) - \sum_{i=1}^{n-2} \alpha_i\, {}_aI^{q}\psi(\eta_i) \tag{5.236}$$

$$+\xi \int_a^b {}_aI^{q-r} x(s) dA(s) - \int_a^b {}_aI^{q}\psi(s) dA(s)\Big)\Big],$$

where

$$\phi_i(t) = (t-a)^3\sigma_i + (t-a)^2\delta_i + \lambda_i, \;\; i = 1, 2, 3, \tag{5.237}$$

$$\lambda_1 = 1-(b-a)^3\sigma_1-(b-a)^2\delta_1, \lambda_j = -(b-a)^3\sigma_j-(b-a)^2\delta_j, j=2,3, \tag{5.238}$$

$$\delta_1 = \frac{-3(b-a)\sigma_1}{2}, \quad \delta_2 = \frac{1-3(b-a)^2\sigma_2}{2(b-a)}, \quad \delta_3 = \frac{-3(b-a)\sigma_3}{2}, \tag{5.239}$$

$$\sigma_1 = \frac{-2A_1}{\gamma_1}, \quad \sigma_2 = \frac{2\gamma_2}{\gamma_1}, \quad \sigma_3 = \frac{2}{\gamma_1}, \tag{5.240}$$

$$\gamma_1 = (b-a)^3 A_1 - 3(b-a)A_2 + 2A_3, \;\; \gamma_2 = \frac{(b-a)^2A_1 - A_2}{2(b-a)}, \tag{5.241}$$

$$A_1 = \sum_{i=1}^{n-2} \alpha_i + \int_a^b dA(s) - 1, \quad A_2 = \sum_{i=1}^{n-2} \alpha_i(\eta_i - a)^2 + \int_a^b (s-a)^2 dA(s), \tag{5.242}$$

$$A_3 = \sum_{i=1}^{n-2} \alpha_i(\eta_i - a)^3 + \int_a^b (s-a)^3 dA(s), \tag{5.243}$$

and it is assumed that $\gamma_1 \neq 0$.

Proof. Applying the integral operator ${}_aI^q$ on both sides of fractional differential equation (5.234), we get

$$x(t) = \frac{-\xi}{\mu} {}_aI^{q-r} x(t) + \frac{1}{\mu} {}_aI^{q}\psi(t) + c_0 + c_1(t-a) + c_2(t-a)^2 + c_3(t-a)^3, \tag{5.244}$$

$$x'(t) = \frac{-\xi}{\mu} {}_aI^{q-r-1} x(t) + \frac{1}{\mu} {}_aI^{q-1}\psi(s) + c_1 + 2c_2(t-a) + 3c_3(t-a)^2, \tag{5.245}$$

where $c_i \in \mathbb{R}$, $i = 0, 1, 2, 3$ are unknown arbitrary constants. Using the boundary conditions (5.233) in (5.244) and (5.245), we obtain $c_1 = 0$ and

$$c_0 + (b-a)^2 c_2 + (b-a)^3 c_3 = J_1, \tag{5.246}$$

$$2(b-a)c_2 + 3(b-a)^2 c_3 = J_2, \tag{5.247}$$

$$A_1 c_0 + A_2 c_2 + A_3 c_3 = J_3, \tag{5.248}$$

where A_i $(i = 1, 2, 3)$ are given by (5.242), (5.243) and

$$J_1 = \frac{1}{\mu}\Big(\xi I_a^{q-r}x(b) - I_a^q\psi(b)\Big), \quad J_2 = \frac{1}{\mu}\Big(\xi I_a^{q-r-1}x(b) - I_a^{q-1}\psi(b)\Big),$$

$$J_3 = \frac{1}{\mu}\Big(\xi\sum_{i=1}^{n-2}\alpha_i I_a^{q-r}x(\eta_i) - \sum_{i=1}^{n-2}\alpha_i I_a^q\psi(\eta_i) + \xi\int_a^b I_a^{q-r}x(s)dA(s) - \int_a^b I_a^q\psi(s)dA(s)\Big). \tag{5.249}$$

Eliminating c_0 from (5.246) and (5.248), we get

$$(A_2 - (b-a)^2A_1)c_2 + (A_3 - (b-a)^3A_1)c_3 = J_3 - A_1J_1. \tag{5.250}$$

Solving (5.247) and (5.250), we find that

$$c_2 = \delta_1J_1 + \delta_2J_2 + \delta_3J_3, \tag{5.251}$$

$$c_3 = \sigma_1J_1 + \sigma_2J_2 + \sigma_3J_3, \tag{5.252}$$

where δ_i and σ_i $(i = 1, 2, 3)$ are defined by (5.239) and (5.240) respectively. Using (5.251) and (5.252) in (5.246), we get

$$c_0 = \lambda_1J_1 + \lambda_2J_2 + \lambda_3J_3, \tag{5.253}$$

where λ_i $(i = 1, 2, 3)$ are given by (5.238). Inserting the values of c_0, c_1, c_2 and c_3 in (5.244) together with notation (5.237), we obtain the solution (5.235). The converse of the lemma follows by direct computation. □

5.9.1 *Existence of solutions*

For $0 < p \leq 1$, let $\mathcal{A} = \{x : x, {}^CD_a^px(t), {}^CD_a^{p+1}x(t) \in C([a,b],\mathbb{R})\}$ denote the Banach space of all continuous functions from $[a,b] \longrightarrow \mathbb{R}$ endowed with the norm defined by

$$\|x\|^* = \sup_{t\in[a,b]}\{|x(t)| + |{}^CD_a^px(t)| + |{}^CD_a^{p+1}x(t)|\}. \tag{5.254}$$

In view of Lemma 5.11, we transform the problem (5.232)-(5.233) into an equivalent fixed point problem as

$$x = \mathcal{F}x, \tag{5.255}$$

where $\mathcal{F} : \mathcal{A} \longrightarrow \mathcal{A}$ is defined by

$$(\mathcal{F}x)(t)$$
$$= \frac{-\xi}{\mu}I_a^{q-r}x(t) + \frac{1}{\mu}I_a^q\widehat{f}(x(t)) + \frac{\phi_1(t)}{\mu}\Big[\xi\int_a^b\frac{(b-s)^{q-r-1}}{\Gamma(q-r)}x(s)ds - \int_a^b\frac{(b-s)^{q-1}}{\Gamma(q)}\widehat{f}(x(s))ds\Big] + \frac{\phi_2(t)}{\mu}\Big[\xi\int_a^b\frac{(b-s)^{q-r-2}}{\Gamma(q-r-1)}x(s)ds$$

$$-\int_a^b \frac{(b-s)^{q-2}}{\Gamma(q-1)}\widehat{f}(x(s))ds\Big] + \frac{\phi_3(t)}{\mu}\Big[\xi\sum_{i=1}^{n-2}\alpha_i\int_a^{\eta_i}\frac{(\eta_i-s)^{q-r-1}}{\Gamma(q-r)}x(s)ds$$

$$-\sum_{i=1}^{n-2}\alpha_i\int_a^{\eta_i}\frac{(\eta_i-s)^{q-1}}{\Gamma(q)}\widehat{f}(x(s))ds + \xi\int_a^b\Big(\int_a^s\frac{(s-u)^{q-r-1}}{\Gamma(q-r)}x(u)du\Big)dA(s)$$

$$-\int_a^b\Big(\int_a^s\frac{(s-u)^{q-1}}{\Gamma(q)}\widehat{f}(x(u))du\Big)dA(s)\Big], \tag{5.256}$$

where $\phi_i(t),\ i=1,2,3$ are defined by (5.237) and

$$\widehat{f}(x(t)) = f(t, x(t), {}^CD_a^p x(t), {}^CD_a^{p+1}x(t)).$$

From (5.256), we have

$$({}^CD_a^p\mathcal{F}x)(t)$$

$$= \frac{-\xi}{\mu}I_a^{q-p-r}x(t) + \frac{1}{\mu}I_a^{q-p}\widehat{f}(x(t)) + \frac{\omega_1(t)}{\mu}\Big[\xi\int_a^b\frac{(b-s)^{q-r-1}}{\Gamma(q-r)}x(s)ds$$

$$-\int_a^b\frac{(b-s)^{q-1}}{\Gamma(q)}\widehat{f}(x(s))ds\Big] + \frac{\omega_2(t)}{\mu}\Big[\xi\int_a^b\frac{(b-s)^{q-r-2}}{\Gamma(q-r-1)}x(s)ds$$

$$-\int_a^b \frac{(b-s)^{q-2}}{\Gamma(q-1)}\widehat{f}(x(s))ds\Big] + \frac{\omega_3(t)}{\mu}\Big[\xi\sum_{i=1}^{n-2}\alpha_i\int_a^{\eta_i}\frac{(\eta_i-s)^{q-r-1}}{\Gamma(q-r)}x(s)ds$$

$$-\sum_{i=1}^{n-2}\alpha_i\int_a^{\eta_i}\frac{(\eta_i-s)^{q-1}}{\Gamma(q)}\widehat{f}(x(s))ds + \xi\int_a^b\Big(\int_a^s\frac{(s-u)^{q-r-1}}{\Gamma(q-r)}x(u)du\Big)dA(s)$$

$$-\int_a^b\Big(\int_a^s\frac{(s-u)^{q-1}}{\Gamma(q)}\widehat{f}(x(u))du\Big)dA(s)\Big], \tag{5.257}$$

where

$$\begin{aligned}
\omega_1(t) &= {}^CD_a^p\phi_1(t) = {}^CD_a^p((t-a)^3\sigma_1 + (t-a)^2\delta_1 + \lambda_1)\\
&= 6\sigma_1\frac{(t-a)^{3-p}}{\Gamma(4-p)} + 2\delta_1\frac{(t-a)^{2-p}}{\Gamma(3-p)},\\
\omega_2(t) &= {}^CD_a^p\phi_2(t) = {}^CD_a^p((t-a)^3\sigma_2 + (t-a)^2\delta_2 + \lambda_2)\\
&= 6\sigma_2\frac{(t-a)^{3-p}}{\Gamma(4-p)} + 2\delta_2\frac{(t-a)^{2-p}}{\Gamma(3-p)},\\
\omega_3(t) &= {}^CD_a^p\phi_3(t) = {}^CD_a^p((t-a)^3\sigma_3 + (t-a)^2\delta_3 + \lambda_3)\\
&= 6\sigma_3\frac{(t-a)^{3-p}}{\Gamma(4-p)} + 2\delta_3\frac{(t-a)^{2-p}}{\Gamma(3-p)},
\end{aligned} \tag{5.258}$$

and

$$({}^CD^{p+1}\mathcal{F}x)(t)$$

$$= \frac{-\xi}{\mu}I_a^{q-p-r-1}x(t) + \frac{1}{\mu}I_a^{q-p-1}\widehat{f}(x(t)) + \frac{\nu_1(t)}{\mu}\Big[\xi\int_a^b\frac{(b-s)^{q-r-1}}{\Gamma(q-r)}x(s)ds$$

$$-\int_a^b \frac{(b-s)^{q-1}}{\Gamma(q)}\widehat{f}(x(s))ds\Big] + \frac{\nu_2(t)}{\mu}\Big[\xi\int_a^b \frac{(b-s)^{q-r-2}}{\Gamma(q-r-1)}x(s)ds$$
$$-\int_a^b \frac{(b-s)^{q-2}}{\Gamma(q-1)}\widehat{f}(x(s))ds\Big] + \frac{\nu_3(t)}{\mu}\Big[\xi\sum_{i=1}^{n-2}\alpha_i\int_a^{\eta_i}\frac{(\eta_i-s)^{q-r-1}}{\Gamma(q-r)}x(s)ds -$$
$$\sum_{i=1}^{n-2}\alpha_i\int_a^{\eta_i}\frac{(\eta_i-s)^{q-1}}{\Gamma(q)}\widehat{f}(x(s))ds + \xi\int_a^b\Big(\int_a^s\frac{(s-u)^{q-r-1}}{\Gamma(q-r)}x(u)du\Big)dA(s)$$
$$-\int_a^b\Big(\int_a^s\frac{(s-u)^{q-1}}{\Gamma(q)}\widehat{f}(x(u))du\Big)dA(s)\Big], \tag{5.259}$$

where

$$\begin{aligned}
\nu_1(t) &= {}^C D_a^{p+1}\phi_1(t) = {}^C D_a^{p+1}((t-a)^3\sigma_1 + (t-a)^2\delta_1 + \lambda_1)\\
&= 6\sigma_1\frac{(t-a)^{2-p}}{\Gamma(3-p)} + 2\delta_1\frac{(t-a)^{1-p}}{\Gamma(2-p)},\\
\nu_2(t) &= {}^C D_a^{p+1}\phi_2(t) = {}^C D_a^{p+1}((t-a)^3\sigma_2 + (t-a)^2\delta_2 + \lambda_2)\\
&= 6\sigma_2\frac{(t-a)^{2-p}}{\Gamma(3-p)} + 2\delta_2\frac{(t-a)^{1-p}}{\Gamma(2-p)},\\
\nu_3(t) &= {}^C D_a^{p+1}\phi_3(t) = {}^C D_a^{p+1}((t-a)^3\sigma_3 + (t-a)^2\delta_3 + \lambda_3)\\
&= 6\sigma_3\frac{(t-a)^{2-p}}{\Gamma(3-p)} + 2\delta_3\frac{(t-a)^{1-p}}{\Gamma(2-p)}.
\end{aligned} \tag{5.260}$$

For the sake of computational convenience, we introduce

$$\begin{aligned}
\Lambda_1 &= \frac{|\xi|}{|\mu|}\Big[\frac{(b-a)^{q-r}}{\Gamma(q-r+1)} + \bar{\phi}_1\frac{(b-a)^{q-r}}{\Gamma(q-r+1)} + \bar{\phi}_2\frac{(b-a)^{q-r-1}}{\Gamma(q-r)}\\
&\quad + \bar{\phi}_3\Big(\sum_{i=1}^{n-2}|\alpha_i|\frac{(\eta_i-a)^{q-r}}{\Gamma(q-r+1)} + \int_a^b\frac{(s-a)^{q-r}}{\Gamma(q-r+1)}dA(s)\Big)\Big],\\
\bar{\Lambda}_1 &= \frac{1}{|\mu|}\Big[\frac{(b-a)^{q}}{\Gamma(q+1)} + \bar{\phi}_1\frac{(b-a)^{q}}{\Gamma(q+1)} + \bar{\phi}_2\frac{(b-a)^{q-1}}{\Gamma(q)} + \bar{\phi}_3\Big(\sum_{i=1}^{n-2}|\alpha_i|\frac{(\eta_i-a)^{q}}{\Gamma(q+1)}\\
&\quad + \int_a^b\frac{(s-a)^{q}}{\Gamma(q+1)}dA(s)\Big)\Big],
\end{aligned} \tag{5.261}$$

$$\begin{aligned}
\Lambda_2 &= \frac{|\xi|}{|\mu|}\Big[\frac{(b-a)^{q-p-r}}{\Gamma(q-p-r+1)} + \bar{\omega}_1\frac{(b-a)^{q-r}}{\Gamma(q-r+1)} + \bar{\omega}_2\frac{(b-a)^{q-r-1}}{\Gamma(q-r)}\\
&\quad + \bar{\omega}_3\Big(\sum_{i=1}^{n-2}|\alpha_i|\frac{(\eta_i-a)^{q-r}}{\Gamma(q-r+1)} + \int_a^b\frac{(s-a)^{q-r}}{\Gamma(q-r+1)}dA(s)\Big)\Big],\\
\bar{\Lambda}_2 &= \frac{1}{|\mu|}\Big[\frac{(b-a)^{q-p}}{\Gamma(q-p+1)} + \bar{\omega}_1\frac{(b-a)^{q}}{\Gamma(q+1)} + \bar{\omega}_2\frac{(b-a)^{q-1}}{\Gamma(q)}\\
&\quad + \bar{\omega}_3\Big(\sum_{i=1}^{n-2}|\alpha_i|\frac{(\eta_i-a)^{q}}{\Gamma(q+1)} + \int_a^b\frac{(s-a)^{q}}{\Gamma(q+1)}dA(s)\Big)\Big],
\end{aligned} \tag{5.262}$$

$$\Lambda_3 = \frac{|\xi|}{|\mu|}\Big[\frac{(b-a)^{q-p-r-1}}{\Gamma(q-p-r)} + \bar{\nu}_1\frac{(b-a)^{q-r}}{\Gamma(q-r+1)} + \bar{\nu}_2\frac{(b-a)^{q-r-1}}{\Gamma(q-r)}$$
$$+\bar{\nu}_3\Big(\sum_{i=1}^{n-2}|\alpha_i|\frac{(\eta_i-a)^{q-r}}{\Gamma(q-r+1)} + \int_a^b \frac{(s-a)^{q-r}}{\Gamma(q-r+1)}dA(s)\Big)\Big],$$
$$\bar{\Lambda}_3 = \frac{1}{|\mu|}\Big[\frac{(b-a)^{q-p-1}}{\Gamma(q-p)} + \bar{\nu}_1\frac{(b-a)^{q}}{\Gamma(q+1)} + \bar{\nu}_2\frac{(b-a)^{q-1}}{\Gamma(q)}$$
$$+\bar{\nu}_3\Big(\sum_{i=1}^{n-2}|\alpha_i|\frac{(\eta_i-a)^{q}}{\Gamma(q+1)} + \int_a^b \frac{(s-a)^{q}}{\Gamma(q+1)}dA(s)\Big)\Big], \tag{5.263}$$

where $\bar{\phi}_i = \sup_{t\in[a,b]}|\phi_i(t)|$, $\bar{\omega}_i = \sup_{t\in[a,b]}|\omega_i(t)|$, $\bar{\nu}_i = \sup_{t\in[a,b]}|\nu_i(t)|$, $i=1,2,3$,

$$\Delta = \max\{\Lambda_1,\ \Lambda_2,\ \Lambda_3\}, \tag{5.264}$$
$$\bar{\Delta} = max\{\bar{\Lambda}_1,\ \bar{\Lambda}_2,\ \bar{\Lambda}_3\}. \tag{5.265}$$

In the following result, we prove the existence of solutions for the problem (5.232)-(5.233) by applying Theorem 1.14.

Theorem 5.35. *Let $f:[a,b]\times\mathbb{R}^3\longrightarrow\mathbb{R}$ be a continuous function. Assume that:*

(5.35.1) there exists a function $\rho\in C([a,b],\mathbb{R}_+)$ such that

$$|f(t,x_1,x_2,x_3)|\le\rho(t),\ \textit{for } t\in[a,b],\ \textit{and each } x_i\in\mathbb{R},\ i=1,2,3;$$

(5.35.2) $\kappa<1$, where $\kappa=3\Delta$ and Δ is defined by (5.264).

Then the problem (5.232)-(5.233) has at least one solution on $[a,b]$.

Proof. Consider a closed bounded and convex ball $B_\tau=\{x\in\mathcal{A}:\|x\|^*\le\tau\}\subseteq\mathcal{A}$, where τ is a fixed constant. Let us define $\mathcal{F}_1,\mathcal{F}_2:B_\tau\longrightarrow B_\tau$ by

$$(\mathcal{F}_1x)(t) = \frac{-\xi}{\mu}I_a^{q-r}x(t) + \frac{\xi\,\phi_1(t)}{\mu}\int_a^b\frac{(b-s)^{q-r-1}}{\Gamma(q-r)}x(s)ds$$
$$+\frac{\xi\,\phi_2(t)}{\mu}\int_a^b\frac{(b-s)^{q-r-2}}{\Gamma(q-r-1)}x(s)ds$$
$$+\frac{\xi\,\phi_3(t)}{\mu}\Big[\sum_{i=1}^{n-2}\alpha_i\int_a^{\eta_i}\frac{(\eta_i-s)^{q-r-1}}{\Gamma(q-r)}x(s)ds$$
$$+\int_a^b\Big(\int_a^s\frac{(s-u)^{q-r-1}}{\Gamma(q-r)}x(u)du\Big)dA(s)\Big],\ t\in[a,b]$$

$$(\mathcal{F}_2x)(t) = \frac{1}{\mu}I_a^q\widehat{f}(x(t)) - \frac{\phi_1(t)}{\mu}\int_a^b\frac{(b-s)^{q-1}}{\Gamma(q)}\widehat{f}(x(s))ds$$
$$-\frac{\phi_2(t)}{\mu}\int_a^b\frac{(b-s)^{q-2}}{\Gamma(q-1)}\widehat{f}(x(s))ds$$

$$-\frac{\phi_3(t)}{\mu}\Big[\sum_{i=1}^{n-2}\alpha_i\int_a^{\eta_i}\frac{(\eta_i-s)^{q-1}}{\Gamma(q)}\widehat{f}(x(s))ds$$
$$+\int_a^b\Big(\int_a^s\frac{(s-u)^{q-1}}{\Gamma(q)}\widehat{f}(x(u))du\Big)dA(s)\Big],\ t\in[a,b].$$

Observe that,

$$(\mathcal{F}x)(t)=(\mathcal{F}_1x)(t)+(\mathcal{F}_2x)(t),\ t\in[a,b].$$

Now we show that $\mathcal{F}_1$ and $\mathcal{F}_2$ satisfy all the conditions of Theorem 1.14. The proof will be given in several steps.

Step 1. $\mathcal{F}B_\tau\subset B_\tau$.

Let us choose $\tau\geq\dfrac{3\|\rho\|\bar{\Delta}}{1-\kappa}$, where κ is defined in (5.35.2) and $\bar{\Delta}$ is given by (5.265). For $x\in B_\tau$, we have

$$|\mathcal{F}x(t)|\leq\|x\|^*\frac{|\xi|}{|\mu|}\Bigg[\frac{(b-a)^{q-r}}{\Gamma(q-r+1)}+\bar{\phi}_1\frac{(b-a)^{q-r}}{\Gamma(q-r+1)}+\bar{\phi}_2\frac{(b-a)^{q-r-1}}{\Gamma(q-r)}$$
$$+\bar{\phi}_3\Big(\sum_{i=1}^{n-2}|\alpha_i|\frac{(\eta_i-a)^{q-r}}{\Gamma(q-r+1)}+\int_a^b\frac{(s-a)^{q-r}}{\Gamma(q-r+1)}dA(s)\Big)\Bigg]$$
$$+\frac{\|\rho\|}{|\mu|}\Bigg[\frac{(b-a)^q}{\Gamma(q+1)}+\bar{\phi}_1\frac{(b-a)^q}{\Gamma(q+1)}+\bar{\phi}_2\frac{(b-a)^{q-1}}{\Gamma(q)}$$
$$+\bar{\phi}_3\Big(\sum_{i=1}^{n-2}|\alpha_i|\frac{(\eta_i-a)^q}{\Gamma(q+1)}+\int_a^b\frac{(s-a)^q}{\Gamma(q+1)}dA(s)\Big)\Bigg]$$
$$\leq\tau\Lambda_1+\|\rho\|\bar{\Lambda}_1\leq\tau(\kappa/3)+\|\rho\|\bar{\Delta}.$$

Similarly, it can be shown that

$$|{}^CD_a^p\mathcal{F}x(t)|\leq\tau\Lambda_2+\|\rho\|\bar{\Lambda}_2\leq\tau(\kappa/3)+\|\rho\|\bar{\Delta},$$
$$|{}^CD_a^{p+1}\mathcal{F}x(t)|\leq\tau\Lambda_3+\|\rho\|\bar{\Lambda}_3\leq\tau(\kappa/3)+\|\rho\|\bar{\Delta}.$$

Hence

$$\|\mathcal{F}x\|^*=\sup_{x\in[a,b]}\{|\mathcal{F}x(t)|+|{}^CD_a^p\mathcal{F}x(t)|+|{}^CD_a^{p+1}\mathcal{F}x(t)|\}$$
$$\leq\tau\kappa+3\|\rho\|\bar{\Delta}<\tau.$$

Thus we get $\mathcal{F}B_\tau\subset B_\tau$.

Step 2. $\mathcal{F}_1$ is a κ-contractive.

For $x,y\in B_\tau$ and using the condition (5.35.2), we have

$$\Big|(\mathcal{F}_1x)(t)-(\mathcal{F}_1y)(t)\Big|$$
$$\leq\frac{|\xi|}{|\mu|}I_a^{q-r}|x(t)-y(t)|+\frac{|\xi||\phi_1(t)|}{|\mu|}I_a^{q-r}|x(b)-y(b)|+\frac{|\xi||\phi_2(t)|}{|\mu|}I_a^{q-r-1}|x(b)-y(b)|$$

$$+\frac{|\xi||\phi_3(t)|}{|\mu|}\Big(\sum_{i=1}^{n-2}|\alpha_i|I_a^{q-r}|x(\eta_i)-y(\eta_i)|+\int_a^b I_a^{q-r}|x(s)-y(s)|dA(s)\Big)$$
$$\le \frac{|\xi|}{|\mu|}\Big[\frac{(b-a)^{q-r}}{\Gamma(q-r+1)}+\bar{\phi}_1\frac{(b-a)^{q-r}}{\Gamma(q-r+1)}+\bar{\phi}_2\frac{(b-a)^{q-r-1}}{\Gamma(q-r)}$$
$$+\bar{\phi}_3\Big(\sum_{i=1}^{n-2}|\alpha_i|\frac{(\eta_i-a)^{q-r}}{\Gamma(q-r+1)}+\int_a^b\frac{(s-a)^{q-r}}{\Gamma(q-r+1)}dA(s)\Big)\Big]\|x-y\|$$

$$=\Lambda_1\|x-y\|\le(\kappa/3)\|x-y\|.$$

Similarly, we can obtain

$$|{}^CD_a^p\mathcal{F}_1x(t)-{}^CD_a^p\mathcal{F}_1y(t)|\le\Lambda_2\|x-y\|\le(\kappa/3)\|x-y\|,$$
$$|{}^CD_a^{p+1}\mathcal{F}_1x(t)-{}^CD_a^{p+1}\mathcal{F}_1y(t)|\le\Lambda_3\|x-y\|\le(\kappa/3)\|x-y\|.$$

Hence

$$\|\mathcal{F}_1x-\mathcal{F}_1y\|^*\le\kappa\|x-y\|,$$

which proves that $\mathcal{F}_1$ is κ-contractive.

Step 3. $\mathcal{F}_2$ is compact.

Continuity of f implies that the operator $\mathcal{F}_2$ is continuous. Also, $\mathcal{F}_2$ is uniformly bounded on B_τ as

$$|\mathcal{F}_2x(t)|\le\|\rho\|\bar{\Lambda}_1,\ |{}^CD_a^p\mathcal{F}_2x(t)|\le\|\rho\|\bar{\Lambda}_2,\quad\text{and}\quad|{}^CD_a^{p+1}\mathcal{F}_2x(t)|\le\|\rho\|\bar{\Lambda}_3,$$

which imply that $\|\mathcal{F}_2x\|^*\le 3\|\rho\|\bar{\Delta}$.

Let $t_1,\ t_2\in[a,b]$ with $t_1<t_2$ and $x\in B_\tau$. Then we have

$$\begin{aligned}&|\mathcal{F}_2x(t_2)-\mathcal{F}_2x(t_1)|\\&\le\frac{1}{|\mu|\Gamma(q)}\Big[\int_a^{t_1}|(t_2-s)^{q-1}-(t_1-s)^{q-1}||\rho(s)|ds+\int_{t_1}^{t_2}|(t_2-s)^{q-1}||\rho(s)|ds\Big]\\&\quad+\frac{|\phi_1(t_2)-\phi_1(t_1)|}{|\mu|}\int_a^b\frac{(b-s)^{q-1}}{\Gamma(q)}|\rho(s)|ds\\&\quad+\frac{|\phi_2(t_2)-\phi_2(t_1)|}{|\mu|}\int_a^b\frac{(b-s)^{q-2}}{\Gamma(q-1)}|\rho(s)|ds\\&\quad+\frac{|\phi_3(t_2)-\phi_3(t_1)|}{|\mu|}\Big[\sum_{i=1}^{n-2}|\alpha_i|\int_a^{\eta_i}\frac{(\eta_i-s)^{q-1}}{\Gamma(q)}|\rho(s)|ds\\&\quad+\int_a^b\Big(\int_a^s\frac{(s-u)^{q-1}}{\Gamma(q)}|\rho(u)|du\Big)dA(s)\Big]\\&\le\frac{\|\rho\|}{|\mu|\Gamma(q+1)}\Big[|(t_2-a)^q-(t_1-a)^q|+2(t_2-t_1)^q\Big]\\&\quad+\frac{\|\rho\|(b-a)^q|\phi_1(t_2)-\phi_1(t_1)|}{|\mu|\Gamma(q+1)}+\frac{\|\rho\|(b-a)^{q-1}|\phi_2(t_2)-\phi_2(t_1)|}{|\mu|\Gamma(q)}\end{aligned}\tag{5.266}$$

$$+\frac{\|\rho\||\phi_3(t_2)-\phi_3(t_1)|}{|\mu|\Gamma(q+1)}\Big[\sum_{i=1}^{n-2}|\alpha_i|(\eta_i-a)^q+\int_a^b(s-a)^q dA(s)\Big], \tag{5.267}$$

$$\begin{aligned}&|{}^CD_a^p\mathcal{F}_2x(t_2)-{}^CD_a^p\mathcal{F}_2x(t_1)|\\ &\le \frac{\|\rho\|}{|\mu|\Gamma(q-p+1)}\Big[|(t_2-a)^{q-p}-(t_1-a)^{q-p}|+2(t_2-t_1)^{q-p}\Big]\\ &\quad+\frac{\|\rho\|(b-a)^q|\omega_1(t_2)-\omega_1(t_1)|}{|\mu|\Gamma(q+1)}+\frac{\|\rho\|(b-a)^{q-1}|\omega_2(t_2)-\omega_2(t_1)|}{|\mu|\Gamma(q)}\\ &\quad+\frac{\|\rho\||\omega_3(t_2)-\omega_3(t_1)|}{|\mu|\Gamma(q+1)}\Big[\sum_{i=1}^{n-2}|\alpha_i|(\eta_i-a)^q+\int_a^b(s-a)^q dA(s)\Big],\end{aligned} \tag{5.268}$$

and

$$\begin{aligned}&|{}^CD_a^{p+1}\mathcal{F}_2x(t_2)-{}^CD_a^{p+1}\mathcal{F}_2x(t_1)|\\ &\le \frac{\|\rho\|}{|\mu|\Gamma(q-p)}\Big[|(t_2-a)^{q-p-1}-(t_1-a)^{q-p-1}|+2(t_2-t_1)^{q-p-1}\Big]\\ &\quad+\frac{\|\rho\|(b-a)^q|\nu_1(t_2)-\nu_1(t_1)|}{|\mu|\Gamma(q+1)}+\frac{\|\rho\|(b-a)^{q-1}|\nu_2(t_2)-\nu_2(t_1)|}{|\mu|\Gamma(q)}\\ &\quad+\frac{\|\rho\||\nu_3(t_2)-\nu_3(t_1)|}{|\mu|\Gamma(q+1)}\Big[\sum_{i=1}^{n-2}|\alpha_i|(\eta_i-a)^q+\int_a^b(s-a)^q dA(s)\Big].\end{aligned} \tag{5.269}$$

The right hand sides of the inequalities (5.266)-(5.269) tend to zero as $t_2-t_1\longrightarrow 0$ independent of x. Thus, $\mathcal{F}_2$ is equi-continuous on B_τ. Therefore, by Arzelá-Ascoli theorem, $\mathcal{F}_2$ is relatively compact on B_τ.

Step 4. $\mathcal{F}$ is condensing. Since $\mathcal{F}_1$ is continuous, κ-contractive and $\mathcal{F}_2$ is compact, therefore, by Theorem 1.13, the operator $\mathcal{F}: B_\tau \longrightarrow B_\tau$, with $\mathcal{F}=\mathcal{F}_1+\mathcal{F}_2$ is a condensing map on B_τ.

Hence, by Theorem 1.14, the operator $\mathcal{F}$ has a fixed point. Therefore, the problem (5.232)-(5.233) has at least one solution on $[a,b]$. □

Example 5.17. Consider the fractional boundary value problem.

$$\begin{cases} 45\ {}^CD^{\frac{35}{9}}x(t)+9\ {}^CD^{\frac{1}{8}}x(t)=f(t,x(t),{}^CD^{\frac{1}{4}}x(t),{}^CD^{\frac{5}{4}}x(t)),\ t\in(0,1),\\ x(0)=\sum_{i=1}^{4}\alpha_i x(\eta_i)+\int_0^1 x(s)dA(s),\ x'(0)=0,\ x(1)=0,\ x'(1)=0, \end{cases} \tag{5.270}$$

where $q=\frac{35}{9}$, $r=\frac{1}{8}$, $p=\frac{1}{4}$, $a=0$, $b=1$, $\mu=45$, $\xi=9$, $\alpha_1=\frac{-1}{300}$, $\alpha_2=\frac{-1}{22}$, $\alpha_3=\frac{1}{240}$, $\alpha_4=\frac{2}{39}$, $\eta_1=\frac{1}{17}$, $\eta_2=\frac{2}{17}$, $\eta_3=\frac{3}{17}$, $\eta_4=\frac{4}{17}$, and

$$f(t,x(t),{}^CD^{\frac{1}{4}}x(t),{}^CD^{\frac{5}{4}}x(t))=\frac{1}{\sqrt{e^{2t}+80}}\Big(\frac{|x(t)|}{1+|x(t)|}+\sin^2({}^CD^{\frac{1}{4}}x(t))+\cos({}^CD^{\frac{5}{4}}x(t))\Big).$$

Let us take $A(s) = \frac{s^2}{2}$. Using the given data, we have that $A_1 \simeq -0.493339$, $A_2 \simeq 0.252328$, $A_3 \simeq 0.200616$, $\gamma_1 \simeq -0.849088$, $\gamma_2 \simeq -0.372834$, $\sigma_1 \simeq -1.16204$, $\sigma_2 \simeq 0.878198$, $\sigma_3 \simeq -2.35546$, $\delta_1 \simeq 1.74306$, $\delta_2 \simeq -0.817295$, $\delta_3 \simeq 3.53319$, $\lambda_1 \simeq 0.41898$, $\lambda_2 \simeq -0.060903$, $\lambda_3 \simeq -1.17773$, $\bar{\phi}_1 \simeq 1.00000$, $\bar{\phi}_2 \simeq 0.113338$, $\bar{\phi}_3 \simeq 1.17773$, $\bar{\omega}_1 \simeq 0.591136$, $\bar{\omega}_2 \simeq 0.175009$, $\bar{\omega}_3 \simeq 1.19823$, $\bar{\nu}_1 \simeq 0.541872$, $\bar{\nu}_2 \simeq 1.49758$, $\bar{\nu}_3 \simeq 1.09837$, $\Lambda_1 \simeq 0.031094$, $\Lambda_2 \simeq 0.034096$, $\Lambda_3 \simeq 0.134535$, $\bar{\Lambda}_1 \simeq 0.000700$, $\bar{\Lambda}_2 \simeq 0.002538$, $\bar{\Lambda}_3 \simeq 0.012289$.

Also, the conditions (5.35.1) and (5.35.2) are satisfied as

$$|f(t, x(t), {}^{C}D^{\frac{1}{4}}x(t), {}^{C}D^{\frac{5}{4}}x(t))| \leq \frac{3}{\sqrt{e^{2t}+80}} = \rho(t),$$

and $\kappa \approx 0.403605 < 1$, where κ is defined in (5.35.2). Hence, the conditions of Theorem (5.35) hold. Therefore, by the conclusion of Theorem 5.35, the problem (5.270) has at least one solution on $[0, 1]$.

5.9.2 *Uniqueness of solutions*

Next, we prove the uniqueness of solutions for the problem (5.232)-(5.233) via Banach fixed point theorem.

Theorem 5.36. *Assume that $f : [a, b] \times \mathbb{R}^3 \longrightarrow \mathbb{R}$ is a continuous function such that,*

$$|f(t, x_1, x_2, x_3) - f(t, y_1, y_2, y_3)| \leq L\Big(|x_1 - y_1| + |x_2 - y_2| + |x_3 - y_3|\Big),$$

$L > 0$, for all $t \in [a, b]$, $x_i, y_i \in \mathbb{R}$, $i = 1, 2, 3$.

Then the problem (5.232)-(5.233) has a unique solution on $[a, b]$ if

$$\kappa + 3L\bar{\Delta} < 1, \tag{5.271}$$

where κ is defined in (5.35.2) and $\bar{\Delta}$ is given by (5.265).

Proof. Setting $\sup_{t\in[a,b]} |f(t, 0, 0, 0)| = \mathcal{N} < \infty$, and selecting

$$r^* \geq \frac{3\mathcal{N}\bar{\Delta}}{1 - \kappa - 3L\bar{\Delta}},$$

we define $B_{r^*} = \{x \in \mathcal{A} : \|x\|^* \leq r^*\}$, and show that $\mathcal{F}B_{r^*} \subset B_{r^*}$, where the operator $\mathcal{F}$ is defined by (5.256). For $x \in B_{r^*}$, we use (5.271) to find that

$$\begin{aligned}
&|f(t, x(t), {}^{C}D_a^{p}x(t), {}^{C}D_a^{p+1}x(t))| \\
&= |f(t, x(t), {}^{C}D_a^{p}x(t), {}^{C}D_a^{p+1}x(t)) - f(t, 0, 0, 0) + f(t, 0, 0, 0)| \\
&\leq |f(t, x(t), {}^{C}D_a^{p}x(t), {}^{C}D_a^{p+1}x(t)) - f(t, 0, 0, 0)| + |f(t, 0, 0, 0)| \\
&\leq L(|x(t)| + |{}^{C}D_a^{p}x(t)| + |{}^{C}D_a^{p+1}x(t)|) + \mathcal{N} \\
&\leq L\|x\|^* + \mathcal{N} \leq Lr^* + \mathcal{N},
\end{aligned}$$

where we used the norm given by (5.254).

Then, we have

$$
\begin{aligned}
|\mathcal{F}x(t)| \leq\ & r^* \frac{|\xi|}{|\mu|}\bigg[\frac{(b-a)^{q-r}}{\Gamma(q-r+1)} + \bar{\phi}_1\frac{(b-a)^{q-r}}{\Gamma(q-r+1)} + \bar{\phi}_2\frac{(b-a)^{q-r-1}}{\Gamma(q-r)} \\
& + \bar{\phi}_3\Big(\sum_{i=1}^{n-2}|\alpha_i|\frac{(\eta_i-a)^{q-r}}{\Gamma(q-r+1)} + \int_a^b \frac{(s-a)^{q-r}}{\Gamma(q-r+1)}dA(s)\Big)\bigg] \\
& + \frac{(Lr^*+\mathcal{N})}{|\mu|}\bigg[\frac{(b-a)^{q}}{\Gamma(q+1)} + \bar{\phi}_1\frac{(b-a)^{q}}{\Gamma(q+1)} + \bar{\phi}_2\frac{(b-a)^{q-1}}{\Gamma(q)} \\
& + \bar{\phi}_3\Big(\sum_{i=1}^{n-2}|\alpha_i|\frac{(\eta_i-a)^{q}}{\Gamma(q+1)} + \int_a^b \frac{(s-a)^{q}}{\Gamma(q+1)}dA(s)\Big)\bigg] \\
\leq\ & r^*\Lambda_1 + (Lr^*+\mathcal{N})\bar{\Lambda}_1 \leq (\kappa/3)r^* + (Lr^*+\mathcal{N})\bar{\Delta}.
\end{aligned}
$$

Similarly, we have

$$
\begin{aligned}
|{}^C D_a^p \mathcal{F}x(t)| &\leq r^*\Lambda_2 + (Lr^*+\mathcal{N})\bar{\Lambda}_2 \leq (\kappa/3)r^* + (Lr^*+\mathcal{N})\bar{\Delta}, \\
|{}^C D_a^{p+1} \mathcal{F}x(t)| &\leq r^*\Lambda_3 + (Lr^*+\mathcal{N})\bar{\Lambda}_3 \leq (\kappa/3)r^* + (Lr^*+\mathcal{N})\bar{\Delta}.
\end{aligned}
$$

Hence we have

$$
\|\mathcal{F}x\|^* \leq \kappa r^* + 3(Lr^*+\mathcal{N})\bar{\Delta} < r^*.
$$

Thus, $\mathcal{F}x \in B_{r^*}$ for any $x \in B_{r^*}$. Therefore, $\mathcal{F}B_{r^*} \subset B_{r^*}$. Now, we show that $\mathcal{F}$ is a contraction. For $x, y \in \mathcal{A}$ and $t \in [a,b]$, we obtain

$$
\begin{aligned}
& \Big|(\mathcal{F}x)(t) - (\mathcal{F}y)(t)\Big| \\
\leq\ & \frac{|\xi|}{|\mu|} I_a^{q-r}|x(t)-y(t)| + \frac{1}{|\mu|} I_a^q |\widehat{f}(x(t)) - \widehat{f}(y(t))| \\
& + \frac{1}{|\mu|}\bigg[|\phi_1(t)|\Big(|\xi|\int_a^b \frac{(b-s)^{q-r-1}}{\Gamma(q-r)}|x(s)-y(s)|ds \\
& + \int_a^b \frac{(b-s)^{q-1}}{\Gamma(q)}|\widehat{f}(x(s)) - \widehat{f}(y(s))|ds\Big) \\
& + |\phi_2(t)|\Big(|\xi|\int_a^b \frac{(b-s)^{q-r-2}}{\Gamma(q-r-1)}|x(s)-y(s)|ds \\
& + \int_a^b \frac{(b-s)^{q-2}}{\Gamma(q-1)}|\widehat{f}(x(s)) - \widehat{f}(y(s))|ds\Big) \\
& + |\phi_3(t)|\bigg(|\xi|\sum_{i=1}^{n-2}|\alpha_i|\int_a^{\eta_i} \frac{(\eta_i-s)^{q-r-1}}{\Gamma(q-r)}|x(s)-y(s)|ds \\
& + \sum_{i=1}^{n-2}|\alpha_i|\int_a^{\eta_i} \frac{(\eta_i-s)^{q-1}}{\Gamma(q)}|\widehat{f}(x(s)) - \widehat{f}(y(s))|ds
\end{aligned}
$$

$$
\begin{aligned}
&+|\xi|\int_a^b\Big(\int_a^s\frac{(s-u)^{q-r-1}}{\Gamma(q-r)}|x(u)-y(u)|du\Big)dA(s)\\
&+\int_a^b\Big(\int_a^s\frac{(s-u)^{q-1}}{\Gamma(q)}|\widehat{f}(x(u))-\widehat{f}(y(u))|du\Big)dA(s)\Big)\Big)\\
\le\ &\frac{|\xi|}{|\mu|}\Big[\frac{(b-a)^{q-r}}{\Gamma(q-r+1)}+\bar{\phi}_1\frac{(b-a)^{q-r}}{\Gamma(q-r+1)}+\bar{\phi}_2\frac{(b-a)^{q-r-1}}{\Gamma(q-r)}\\
&+\bar{\phi}_3\Big(\sum_{i=1}^{n-2}|\alpha_i|\frac{(\eta_i-a)^{q-r}}{\Gamma(q-r+1)}+\int_a^b\frac{(s-a)^{q-r}}{\Gamma(q-r+1)}dA(s)\Big)\Big]\|x-y\|\\
&+\frac{1}{|\mu|}\Big[\frac{(b-a)^{q}}{\Gamma(q+1)}+\bar{\phi}_1\frac{(b-a)^{q}}{\Gamma(q+1)}+\bar{\phi}_2\frac{(b-a)^{q-1}}{\Gamma(q)}\\
&+\bar{\phi}_3\Big(\sum_{i=1}^{n-2}|\alpha_i|\frac{(\eta_i-a)^{q}}{\Gamma(q+1)}+\int_a^b\frac{(s-a)^{q}}{\Gamma(q+1)}dA(s)\Big)\Big]L\|x-y\|\\
=\ &(\Lambda_1+L\bar{\Lambda}_1)\|x-y\|\le(\kappa/3+L\bar{\Delta})\|x-y\|.
\end{aligned}
$$

In a similar mannar, we have

$$
\begin{aligned}
|{}^CD_a^p\mathcal{F}x(t)-{}^CD_a^p\mathcal{F}y(t)|&\le(\Lambda_2+L\bar{\Lambda}_2)\|x-y\|\le(\kappa/3+L\bar{\Delta})\|x-y\|,\\
|{}^CD_a^{p+1}\mathcal{F}x(t)-{}^CD_a^{p+1}\mathcal{F}y(t)|&\le(\Lambda_3+L\bar{\Lambda}_3)\|x-y\|\le(\kappa/3+L\bar{\Delta})\|x-y\|.
\end{aligned}
$$

Consequently, we obtain $\|(\mathcal{F}x)-(\mathcal{F}y)\|^*\le(\kappa+3L\bar{\Delta})\|x-y\|$, which in view of (5.271), implies that the operator $\mathcal{F}$ is a contraction. Therefore, $\mathcal{F}$ has a unique fixed point, which corresponds to a unique solution of the problem (5.232)-(5.233) on $[a,b]$. This completes the proof. □

Example 5.18. Consider the following fractional differential equation

$$
\begin{aligned}
&45\ {}^CD^{\frac{35}{9}}x(t)+9\ {}^CD^{\frac{1}{8}}x(t)\\
&=\frac{1}{6(t^2+2)}\Big(\tan^{-1}x(t)+\cos({}^CD^{\frac{1}{4}}x(t))+\frac{|{}^CD^{\frac{5}{4}}x(t)|}{1+|{}^CD^{\frac{5}{4}}x(t)|}\Big),
\end{aligned}
\tag{5.272}
$$

$t\in[0,1]$, supplemented with the boundary conditions of Example 5.17.
Obviously

$$
f(t,x(t),{}^CD^{\frac{1}{4}}x(t),{}^CD^{\frac{5}{4}}x(t))=\frac{1}{6(t^2+2)}\Big(\tan^{-1}x(t)+\cos({}^CD^{\frac{1}{4}}x(t))+\frac{|{}^CD^{\frac{5}{4}}x(t)|}{1+|{}^CD^{\frac{5}{4}}x(t)|}\Big).
$$

Using the given data, we find that $\Delta\simeq0.134535$ and $\bar{\Delta}\simeq0.012289$, where Δ and $\bar{\Delta}$ are respectively given by (5.264) and (5.265). By the following inequality

$$
\begin{aligned}
&|f(t,x(t),{}^CD^{\frac{1}{4}}x(t),{}^CD^{\frac{5}{4}}x(t))-f(t,y(t),{}^CD^{\frac{1}{4}}y(t),{}^CD^{\frac{5}{4}}y(t))|\\
\le\ &\frac{1}{12}\Big(|x-y|+|{}^CD^{\frac{1}{4}}x-{}^CD^{\frac{1}{4}}y|+|{}^CD^{\frac{5}{4}}x-{}^CD^{\frac{5}{4}}y|\Big)\le\frac{1}{12}\|x-y\|,
\end{aligned}
$$

we have $L = \frac{1}{12}$. Clearly $(\kappa + 3L\bar{\Delta}) \simeq 0.406677 < 1$. Therefore, the hypothesis of Theorem 5.36 is satisfied and consequently the problem (5.272) has a unique solution on $[0,1]$.

Remark 5.5. Letting $\xi = 0$ and $\mu = 1$ in the results of this section, we obtain the ones for the fractional differential equation of the form:

$$^{C}D^{q}x(t) = f(t, x(t), {}^{C}D^{p}x(t), {}^{C}D^{p+1}x(t)), \ 3 < q \leq 4, \ 0 < p \leq 1, \ t \in [a,b],$$

supplemented with Riemann-Stieltjes integro-multi-point boundary conditions (5.233). In this case the fixed point operator takes the form:

$$\begin{aligned}(\mathcal{F}x)(t) &= I^{q}\widehat{f}(x(t)) - \phi_1(t)\int_a^b \frac{(b-s)^{q-1}}{\Gamma(q)}\widehat{f}(x(s))ds \\ &\quad -\phi_2(t)\int_a^b \frac{(b-s)^{q-2}}{\Gamma(q-1)}\widehat{f}(x(s))ds - \phi_3(t)\Big[\sum_{i=1}^{n-2}\alpha_i\int_a^{\eta_i}\frac{(\eta_i - s)^{q-1}}{\Gamma(q)}\widehat{f}(x(s))ds \\ &\quad +\int_a^b\Big(\int_a^s \frac{(s-u)^{q-1}}{\Gamma(q)}\widehat{f}(x(u))du\Big)dA(s)\Big].\end{aligned}$$

Remark 5.6. In Theorem 5.35, the existence of solutions for the given problem is established by means of Sadovskii fixed point theorem. The proof of this result is based on the idea of splitting the operator $\mathcal{F}$ into the sum of two operators $\mathcal{F}_1$ and $\mathcal{F}_2$ such that $\mathcal{F}_1$ is κ-contractive and $\mathcal{F}_2$ is compact. One can notice that the entire operator $\mathcal{F}$ is not required to be contractive. On the other hand, Theorem 5.36 deals with the existence of a unique solution of the given problem via Banach contraction mapping principle, in which the entire operator $\mathcal{F}$ is shown to be contractive. Thus, the linkage between contractive conditions imposed in Theorems 5.35 and 5.36 provides a precise estimate to pass onto a unique solution from the existence of a solution for the problem at hand.

Remark 5.7. In case we take $\alpha_i = 0$ for all $i = 1, \ldots, n-2$, then our results correspond to the the ones associated with integral boundary conditions:

$$x(a) = \int_a^b x(s)dA(s), \ x'(a) = 0, \ x(b) = 0, \ x'(b) = 0.$$

5.10 Notes and remarks

The results in this chapter are based on the papers [133], [1], [38], [50], [39], [13], [9], [55], [51], [7], [14], [5], [10] and [63].

Chapter 6

Positive solutions for nonlocal fractional-order boundary value problems

By applying monotone iterative techniques, we discuss the existence of positive solutions for nonlocal fractional-order boundary value problems involving Riemann-Liouville fractional differential and integrodifferential equations on an unbounded domain, Hadamard fractional differential equations, the right-handed Riemann-Liouville fractional differential system, and generalized fractional differential equations. Our study includes nonlocal fractional integro-differential, multi-point discrete and Hadamard integral, and Steiltjes type integral boundary conditions.

6.1 Successive iterations for positive extremal solutions of nonlinear fractional differential equations on a half line

In this section, we consider a nonlinear fractional boundary value problem on a half line given by

$$\begin{cases} D^{\alpha}u(t)+f(t,u(t),D^{\alpha-1}u(t))=0, & 1<\alpha\leq 2, \\ u(0)=0, \quad D^{\alpha-1}u(\infty)=\beta u(\xi), \end{cases} \tag{6.1}$$

where $t\in J=[0,+\infty)$, $f\in C[J\times\mathbb{R}\times\mathbb{R},J]$ and D^{α} is the standard Riemann-Liouville fractional derivative operator of order α.

For the forthcoming analysis, we define two Banach spaces:

$$X=\Big\{u\in C(J,\mathbb{R}):\sup_{t\in J}\frac{|u(t)|}{1+t^{\alpha-1}}<+\infty\Big\},$$

$$Y=\{u\in X:D^{\alpha-1}u(t)\in C(J,\mathbb{R}),\sup_{t\in J}|D^{\alpha-1}u(t)|<+\infty\},$$

equipped with the respective norms $\|u\|_X=\sup_{t\in J}\frac{|u(t)|}{1+t^{\alpha-1}}$ and $\|u\|_Y=\max\{\|u\|_X,\sup_{t\in J}|D^{\alpha-1}u(t)|\}$. Define a cone $P\subset Y$ by $P=\{u\in Y:u(t)\geq 0,t\in J\}$.

We need the following known lemma in the forthcoming analysis.

Lemma 6.1. *[118] Let $U\subset X$ be a bounded set. Then U is relatively compact in X if the following conditions hold:*

(i) for any $u \in U$, $\dfrac{u(t)}{1+t^{\rho-1}}$ *is equi-continuous on any compact interval of* J*;*
(ii) for any $\varepsilon > 0$, *there exists a constant* $T = T(\varepsilon) > 0$ *such that*

$$\left|\frac{u(t_1)}{1+t_1^{\rho-1}} - \frac{u(t_2)}{1+t_2^{\rho-1}}\right| < \varepsilon,$$

for any $t_1, t_2 \geq T$ *and* $u \in U$.

Lemma 6.2. *[153] Let* $h \in C([0,+\infty))$ *with* $\int_0^\infty h(s)ds < \infty$. *If* $\Gamma(\alpha) \neq \beta\xi^{\alpha-1}$, *then the boundary value problem*

$$\begin{cases} D^\alpha u(t) + h(t) = 0, \\ u(0) = 0, \qquad D^{\alpha-1}u(\infty) = \beta u(\xi), \quad \beta, \xi > 0, \end{cases} \tag{6.2}$$

has a unique solution given by

$$u(t) = \int_0^{+\infty} G(t,s)h(s)ds, \tag{6.3}$$

where

$$G(t,s) = \frac{1}{\Delta}\begin{cases} [\Gamma(\alpha) - \beta(\xi-s)^{\alpha-1}]t^{\alpha-1} \\ \quad - [\Gamma(\alpha) - \beta\xi^{\alpha-1}](t-s)^{\alpha-1}, & s \leq t,\ s \leq \xi, \\ [\Gamma(\alpha) - \beta(\xi-s)^{\alpha-1}]t^{\alpha-1}, & 0 \leq t \leq s \leq \xi, \\ \Gamma(\alpha)t^{\alpha-1} - [\Gamma(\alpha) - \beta\xi^{\alpha-1}](t-s)^{\alpha-1}, & 0 \leq \xi \leq s \leq t, \\ \Gamma(\alpha)t^{\alpha-1}, & s \geq t,\ s \geq \xi, \end{cases} \tag{6.4}$$

and $\Delta = \Gamma(\alpha)[\Gamma(\alpha) - \beta\xi^{\alpha-1}]$.

From (6.3), we have

$$D^{\alpha-1}u(t) = \int_0^{+\infty} G^*(t,s)h(s)ds,$$

where

$$G^*(t,s) = \frac{1}{\Gamma(\alpha) - \beta\xi^{\alpha-1}}\begin{cases} \beta\xi^{\alpha-1} - \beta(\xi-s)^{\alpha-1}, & s \leq t,\ s \leq \xi, \\ \Gamma(\alpha) - \beta(\xi-s)^{\alpha-1}, & 0 \leq t \leq s \leq \xi, \\ \beta\xi^{\alpha-1}, & 0 \leq \xi \leq s \leq t, \\ \Gamma(\alpha), & s \geq t,\ s \geq \xi. \end{cases} \tag{6.5}$$

Lemma 6.3. *For* $(s,t) \in J \times J$, *let the following condition holds:*

(6.3.1) $\beta, \xi > 0$, $\Gamma(\alpha) > \beta\xi^{\alpha-1}$.

Then

$$0 \leq \frac{G(t,s)}{1+t^{\alpha-1}} \leq \frac{1}{\Gamma(\alpha) - \beta\xi^{\alpha-1}}, \qquad 0 \leq G(t,s) \leq \frac{t^{\alpha-1}}{\Gamma(\alpha) - \beta\xi^{\alpha-1}} \tag{6.6}$$

and

$$0 \leq G^*(t,s) \leq \frac{1}{\Gamma(\alpha) - \beta\xi^{\alpha-1}}. \tag{6.7}$$

Proof. It is obvious from (6.4) and (6.5). □

Lemma 6.4. *Suppose that*

(6.4.1) there exist non-negative functions $a(t), b(t), c(t)$ defined on $[0,\infty)$ and constant $p, q \geq 0$, such that $f(t,u,v) \leq a(t) + b(t)|u|^p + c(t)|v|^q$ with $\int_0^{+\infty} a(t)dt = a^* < +\infty$, $\int_0^{+\infty} b(t)(1+t^{\alpha-1})^p dt = b^* < +\infty$, $\int_0^{+\infty} c(t)dt = c^* < +\infty$.

Then

$$\int_0^{+\infty} f(s,u(s),D^{\alpha-1}u(s))ds \leq a^* + b^*\|u\|_Y^p + c^*\|u\|_Y^q, \quad \forall u \in Y. \tag{6.8}$$

Proof. For $\forall u \in Y$, by condition (6.4.1), we have

$$\begin{aligned}
&\int_0^{+\infty} f(s,u(s),D^{\alpha-1}u(s))ds \\
&\leq \int_0^{+\infty} \left[a(s) + b(s)|u(s)|^p + c(s)|D^{\alpha-1}u(s)|^q\right]ds \\
&\leq a^* + \int_0^{+\infty} b(s)(1+s^{\alpha-1})^p \frac{|u(s)|^p}{(1+s^{\alpha-1})^p}ds + \int_0^{+\infty} c(s)|D^{\alpha-1}u(s)|^q ds \\
&\leq a^* + b^*\|u\|_Y^p + c^*\|u\|_Y^q.
\end{aligned} \tag{6.9}$$

□

6.1.1 *Main results*

Using Lemma 6.2 with $h(t) = f(t,u(t),D^{\alpha-1}u(t))$, we define an integral operator $Q: Y \to Y$ associated with the problem (6.1) as

$$Qu(t) = \int_0^{+\infty} G(t,s)f(s,u(s),D^{\alpha-1}u(s))ds. \tag{6.10}$$

Notice that the problem (6.1) has a solution if and only if the operator equation $u = Qu$ has a fixed point, where Q is given by (6.10).

Lemma 6.5. *Assume that (6.3.1) and (6.4.1) hold. Then the operator $Q: Y \to Y$ is completely continuous.*

Proof. The proof consists of two steps.

(a) The operator $Q: Y \to Y$ is relatively compact.

Let Ω be any bounded subset of Y, then for any $u \in \Omega$, there exists a constant $M > 0$ such that $\|u\|_Y \leq M$. By Lemmas 6.3 and 6.4, we have

$$\|Qu\|_X = \sup_{t\in J} \int_0^{\infty} \frac{G(t,s)}{1+t^{\alpha-1}}|f(s,u(s),D^{\alpha-1}u(s))|ds$$

$$\begin{aligned}
&\le \frac{1}{\Gamma(\alpha)-\beta\xi^{\alpha-1}}\int_0^{+\infty}|f(s,u(s),D^{\alpha-1}u(s))|ds\\
&\le \frac{1}{\Gamma(\alpha)-\beta\xi^{\alpha-1}}[a^*+b^*\|u\|_Y^p+c^*\|u\|_Y^q]\\
&\le \frac{1}{\Gamma(\alpha)-\beta\xi^{\alpha-1}}[a^*+M^pb^*+M^qc^*], \qquad (6.11)
\end{aligned}$$

which implies that $Q\Omega$ is uniformly bounded.

Next we show that the operator $Q: Y \to Y$ is equi-continuous. Let $I \subset J$ be any compact interval and let Ω be any bounded subset of Y. Then, $\forall t_1, t_2 \in I, t_2 > t_1$ and $u \in \Omega$, we have

$$\begin{aligned}
\left|\frac{Qu(t_2)}{1+t_2^{\alpha-1}}-\frac{Qu(t_1)}{1+t_1^{\alpha-1}}\right| &= \left|\int_0^\infty\left(\frac{G(t_2,s)}{1+t_2^{\alpha-1}}-\frac{G(t_1,s)}{1+t_1^{\alpha-1}}\right)f(s,u(s),D^{\alpha-1}u(s))ds\right|\\
&\le \int_0^\infty\left|\frac{G(t_2,s)}{1+t_2^{\alpha-1}}-\frac{G(t_1,s)}{1+t_1^{\alpha-1}}\right|\left|f(s,u(s),D^{\alpha-1}u(s))\right|ds.
\end{aligned} \qquad (6.12)$$

Since $G(t,s) \in C(J\times J)$, for any compact set $I \times I$, $\frac{G(t,s)}{1+t^{\alpha-1}}$ is uniformly continuous. Note that this function only depends on t for $s \ge t$. So it is uniformly continuous on $I \times (J\backslash I)$. Thus, for all $s \in J$ and $t_1,\ t_2 \in I$, we have

$$\forall\varepsilon>0,\ \exists\,\delta(\varepsilon)>0 \text{ such that if } |t_1-t_2|<\delta, \text{ then } \left|\frac{G(t_2,s)}{1+t_2^{\alpha-1}}-\frac{G(t_1,s)}{1+t_1^{\alpha-1}}\right|<\varepsilon. \qquad (6.13)$$

By Lemma 6.4, $\forall u \in \Omega$, we get

$$\int_0^\infty\left|f(s,u(s),D^{\alpha-1}u(s))\right|ds<\infty, \quad \forall u\in\Omega. \qquad (6.14)$$

This, together with (6.12) and (6.13), implies that $\dfrac{Qu(t)}{1+t^{\alpha-1}}$ is equi-continuous on I.

Observe that

$$D^{\alpha-1}Qu(t)=\int_0^{+\infty}G^*(t,s)f(s,u(s),D^{\alpha-1}u(s))ds \qquad (6.15)$$

and the function $G^*(t,s) \in C(J\times J)$ does not depend on t. Thus it is obvious that $D^{\alpha-1}Qu(t)$ is equi-continuous on I. Furthermore,

$$\lim_{t\to\infty}\frac{G(t,s)}{1+t^{\alpha-1}}=\frac{1}{\Gamma(\alpha)[\Gamma(\alpha)-\beta\xi^{\alpha-1}]}\begin{cases}\beta\xi^{\alpha-1}-\beta(\xi-s)^{\alpha-1}, & 0\le s\le\xi,\\ \beta\xi^{\alpha-1}, & \xi\le s.\end{cases} \qquad (6.16)$$

In view of the above argument, for any given $\varepsilon > 0$, it is easy to verify that there exists a constant $T' = T'(\varepsilon) > 0$ such that

$$\left|\frac{G(t_2,s)}{1+t_2^{\alpha-1}}-\frac{G(t_1,s)}{1+t_1^{\alpha-1}}\right|<\varepsilon$$

for any $t_1, t_2 \geq T'$ and $s \in J$. Thus, by Lemma 6.4 and (6.12), we infer that the same property holds for $Q\Omega$, uniformly on Ω. Hence, the operator Q is equi-convergent at ∞. As the function $G^*(t,s)$ does not depend on t, we easily obtain that $D^{\alpha-1}Qu(t)$ is equi-convergent at ∞. Therefore, it follows by Lemma 6.1 that $Q\Omega$ is relatively compact on J.

(b) The operator $Q : Y \to Y$ is continuous.

Let $u_n, u \in X$ such that $u_n \to u(n \to \infty)$, then, $\|u_n\|_Y < \infty$, $\|u\|_Y < \infty$. By Lemmas 6.3 and 6.4, we have

$$\begin{aligned} Qu_n(t) &= \int_0^\infty \frac{G(t,s)}{1+t^{\alpha-1}} f(s, u_n(s), D^{\alpha-1}u_n(s))ds \\ &\leq \frac{1}{\Gamma(\alpha)-\beta\xi^{\alpha-1}} \int_0^\infty |f(s, u_n(s), D^{\alpha-1}u_n(s))|ds \\ &\leq \frac{1}{\Gamma(\alpha)-\beta\xi^{\alpha-1}} [a^* + b^*\|u_n\|_Y^p + c^*\|u_n\|_Y^q] < \infty, \end{aligned} \tag{6.17}$$

and

$$\begin{aligned} D^{\alpha-1}Qu_n(t) &= \int_0^{+\infty} G^*(t,s) f(s, u_n(s), D^{\alpha-1}u_n(s))ds \\ &\leq \frac{1}{\Gamma(\alpha)-\beta\xi^{\alpha-1}} \int_0^\infty |f(s, u_n(s), D^{\alpha-1}u_n(s))|ds \\ &\leq \frac{1}{\Gamma(\alpha)-\beta\xi^{\alpha-1}} [a^* + b^*\|u_n\|_Y^p + c^*\|u_n\|_Y^q] < \infty. \end{aligned} \tag{6.18}$$

By the Lebesgue dominated convergence theorem and continuity of f, we obtain

$$\lim_{n\to\infty} \int_0^\infty \frac{G(t,s)}{1+t^{\alpha-1}} f(s, u_n(s), D^{\alpha-1}u_n(s))ds = \int_0^\infty \frac{G(t,s)}{1+t^{\alpha-1}} f(s, u(s), D^{\alpha-1}u(s))ds,$$

and

$$\lim_{n\to\infty} \int_0^\infty G^*(t,s) f(s, u_n(s), D^{\alpha-1}u_n(s))ds = \int_0^\infty G^*(t,s) f(s, u(s), D^{\alpha-1}u(s))ds.$$

In consequence, we have

$$\begin{aligned} &\|Qu_n - Qu\|_X \\ &= \sup_{t\in J} \int_0^\infty \frac{G(t,s)}{1+t^{\alpha-1}} |f(s, u_n(s), D^{\alpha-1}u_n(s)) - f(s, u(s), D^{\alpha-1}u(s))|ds \to 0, \end{aligned}$$

as $n \to \infty$ and $\sup_{t\in J} |D^{\alpha-1}Qu_n - D^{\alpha-1}Qu| \to 0(n \to \infty)$. This shows that the operator Q is continuous.

From the above steps, we conclude that the operator $Q : Y \to Y$ is completely continuous. This completes the proof. □

Theorem 6.1. *Let the conditions (6.3.1), (6.4.1) hold. In addition we assume that:*

(6.1.1) f is non-decreasing with respect to the second and last variables.

Then there exists a positive constant R such that the problem (6.1) has minimal and maximal positive solutions v^*, u^* respectively in $(0, Rt^{\alpha-1}]$, which can be obtained by means of the following two explicit monotone iterative sequences:

$$v_{n+1}=\int_0^{+\infty} G(t,s)a(s)f(s,v_n(s),D^{\alpha-1}v_n(s))ds \text{ with initial value } v_0(t)=0,$$

$$u_{n+1}=\int_0^{+\infty} G(t,s)a(s)f(s,u_n(s),D^{\alpha-1}u_n(s))ds \text{ with initial value } u_0(t)=Rt^{\alpha-1}. \tag{6.19}$$

Moreover,

$$v_0 \le v_1 \le \cdots \le v_n \le \cdots \le v^* \le \cdots \le u^* \le \cdots \le u_n \le \cdots \le u_1 \le u_0. \tag{6.20}$$

Proof. Let us set $L = \dfrac{1}{\Gamma(\alpha) - \beta\xi^{\alpha-1}}$. In view of $\Gamma(\alpha) > \beta\xi^{\alpha-1}$, Lemma 6.3 leads to the fact that $(Qu)(t) \ge 0$ for any $u \in P, t \in J$. Thus, $Q(P) \subset P$. For $0 \le p, q < 1$, choose

$$R \ge \max\left\{3La^*,\ (3Lb^*)^{\frac{1}{1-p}},\ (3Lc^*)^{\frac{1}{1-q}}\right\}$$

and define $B = \{u \in Y, \|u\|_Y \le R\}$. In what follows, we first show that $Q(B) \subset B$.

For any $u \in B$, by Lemmas 6.3 and 6.4, we have

$$\begin{aligned}
\|Qu\|_X &= \sup_{t\in J}\int_0^{\infty} \frac{G(t,s)}{1+t^{\alpha-1}} f(s,u(s),D^{\alpha-1}u(s))ds \\
&\le L\int_0^{+\infty} f(s,u(s),D^{\alpha-1}u(s))ds \\
&\le L[a^* + b^*\|u\|_Y^p + c^*\|u\|_Y^q] \\
&\le L[a^* + R^p b^* + R^q c^*] \\
&\le R,
\end{aligned} \tag{6.21}$$

and

$$\begin{aligned}
\sup_{t\in J}|D^{\alpha-1}Qu(t)| &= \sup_{t\in J}\int_0^{\infty} G^*(t,s)f(s,u(s),D^{\alpha-1}u(s))ds \\
&\le L\int_0^{+\infty} f(s,u(s),D^{\alpha-1}u(s))ds \\
&\le R.
\end{aligned} \tag{6.22}$$

So it holds that $\|Qu\|_Y \le R, \forall u \in B$. Thus $Q(B) \subset B$.

The definition of the operator Q and condition (6.1.1) imply that the operator Q is non-decreasing.

Denote $v_0(t) = 0,\ v_1 = Q0 = Qv_0,\ v_2 = Q^2 0 = Qv_1,\ \forall t \in J$. Since $v_0(t) = 0 \in B$ and $Q : B \to B$, therefore $v_1 \in Q(B) \subset B$ and $v_2 \in Q(B) \subset B$. So, we have

$$v_1(t) = (Q0)(t) \ge 0 = v_0(t), \quad \forall t \in J.$$

By the non-decreasing nature of the operator Q, we get

$$v_2(t) = (Qv_1)(t) \ge (Qv_0)(t) = v_1(t), \quad \forall t \in J.$$

By induction, we can now define a sequence $v_{n+1} = Qv_n$, $n = 0, 1, 2, \ldots$. Clearly the sequence $\{v_n\}_{n=1}^{\infty} \subset Q(B) \subset B$ and satisfies

$$v_{n+1}(t) \geq v_n(t), \quad \forall t \in J, \quad n = 0, 1, 2, \ldots. \tag{6.23}$$

By the complete continuity of the operator Q, we have that $\{v_n\}_{n=1}^{\infty}$ has a convergent subsequence $\{v_{n_k}\}_{k=1}^{\infty}$ and there exists a $v^* \in B$ such that $v_{n_k} \to v^*$ as $k \to \infty$. This together with (6.23) implies that $\lim_{n\to\infty} v_n = v^*$. Since Q is continuous and $v_{n+1} = Qv_n$, we have $Qv^* = v^*$, that is, v^* is a fixed point of the operator Q.

Denote $u_0(t) = Rt^{\alpha-1}$, $u_1 = Qu_0$, $u_2 = Q^2u_0 = Qu_1$, $\forall t \in J$. Since $u_0(t) \in B$ and $Q : B \to B$, we get $u_1 \in Q(B) \subset B$ and $u_2 \in Q(B) \subset B$. By Lemmas 6.3 and Lemmas 6.4, we obtain

$$\begin{aligned} u_1(t) &= \int_0^{+\infty} G(t,s) f(s, u_0(s), D^{\alpha-1}u_0(s))ds \\ &\leq \int_0^{+\infty} Lt^{\alpha-1} f(s, u_0(s), D^{\alpha-1}u_0(s))ds \\ &\leq Lt^{\alpha-1}(a^* + b^*\|u_0\|_Y^p + c^*\|u_0\|_Y^q) \\ &\leq Rt^{\alpha-1} = u_0(t), \qquad \forall t \in J. \end{aligned}$$

Noting that Q is non-decreasing, we get

$$u_2(t) = (Qu_1)(t) \leq (Qu_0)(t) = u_1(t), \quad \forall t \in J.$$

As before, by induction, we define $u_{n+1} = Qu_n$, $n = 0, 1, 2, \ldots$. Then the sequence $\{u_n\}_{n=1}^{\infty} \subset Q(B) \subset B$ and satisfies the relation

$$u_{n+1}(t) \leq u_n(t), \quad \forall t \in J, \quad n = 0, 1, 2, \ldots. \tag{6.24}$$

Similar to the earlier arguments, it can be shown that there exists a $u^* \in B$ such that $\lim_{n\to\infty} u_n = u^*$. Since Q is continuous and $u_{n+1} = Qu_n$, we have $Qu^* = u^*$ which, in turn, implies that u^* is a fixed point of the operator Q.

Now, we are in a position to show that u^* and v^* are the maximal and minimal positive solutions of (6.1) respectively in $(0, Rt^{\alpha-1}]$. We first establish if $w \in [0, bt^{\alpha-1}]$ is any solution of (6.1), then $v_0(t) = 0 \leq w(t) \leq Rt^{\alpha-1} = u_0(t)$ and $Qw = w$. Using the monotone nature of Q, we have that $v_1(t) = Qv_0(t) \leq w(t) \leq Qu_0(t) = u_1(t), \forall t \in J$.

Repeating the above process several times, we obtain

$$v_n(t) \leq w(t) \leq u_n(t), \qquad \forall t \in J, \quad n = 0, 1, 2, \ldots. \tag{6.25}$$

In view of $u^* = \lim_{n\to\infty} u_n$ and $v^* = \lim_{n\to\infty} v_n$, it follows from (6.23) – (6.25) that

$$v_0 \leq v_1 \leq \cdots \leq v_n \cdots \leq v^* \leq w \leq u^* \cdots \leq u_n \leq \cdots \leq u_1 \leq u_0. \tag{6.26}$$

Since $f(t, 0) \not\equiv 0, \forall t \in J$, therefore 0 is not a solution of problem (6.1). It follows from (6.26) that u^* and v^* are the maximal and minimal positive solutions of (6.1) respectively in $(0, Rt^{\alpha-1}]$, which can be obtained by the corresponding iterative sequences in (6.19). For the other details concerning the range of p and q, the method is similar; so we omit the details. This completes the proof. □

Example 6.1. Consider the following nonlocal fractional boundary value problem for nonlinear fractional differential equation on a half line:

$$\begin{cases} D^{1.25}u(t) + \dfrac{2}{(5+t)^2} + \dfrac{e^{-3t}|u(t)|^p}{(1+\sqrt[4]{t})^p} + \dfrac{2\ln(1+t)|D^{0.25}u(t)|^q}{(2+t^2)^2} = 0, \ t \in [0, +\infty), \\ u(0) = 0, \qquad D^{0.25}u(\infty) = 0.5u(1), \end{cases} \tag{6.27}$$

where $\alpha = 1.25, \beta = 0.5, \xi = 1$, and

$$f(t,u,v) = \frac{2}{(5+t)^2} + \frac{e^{-3t}|u|^p}{(1+\sqrt[4]{t})^p} + \frac{2\ln(1+t)|v|^q}{(2+t^2)^2}, \ 0 \le p, q < 1.$$

Obviously $\Gamma(1.25) \approx 0.91315$, $\beta\xi^{\alpha-1} = 0.5$. Thus (6.5.1) holds.

Taking $a(t) = \dfrac{2}{(5+t)^2}$, $b(t) = \dfrac{e^{-3t}}{(1+\sqrt[4]{t})^p}$, $c(t) = \dfrac{2t}{(2+t^2)^2}$, we have

$$\begin{aligned} |f(t,u,v)| &= \frac{2}{(5+t)^2} + \frac{e^{-3t}|u|^p}{(1+\sqrt[4]{t})^p} + \frac{2\ln(1+t)|v|^q}{(2+t^2)^2} \\ &\le \frac{2}{(5+t)^2} + \frac{e^{-3t}|u|^p}{(1+\sqrt[4]{t})^p} + \frac{2t|v|^q}{(2+t^2)^2} \\ &\triangleq a(t) + b(t)|u|^p + c(t)|v|^q, \end{aligned}$$

and

$$\begin{aligned} a^* &= \int_0^{+\infty} a(t)dt = \int_0^{+\infty} \frac{2}{(5+t)^2}dt = \frac{2}{5} < +\infty, \\ b^* &= \int_0^{+\infty} b(t)(1+t^{\alpha-1})^p dt = \int_0^{+\infty} \frac{e^{-3t}}{(1+\sqrt[4]{t})^p}(1+\sqrt[4]{t})^p dt = \frac{1}{3} < +\infty, \\ c^* &= \int_0^{+\infty} c(t)dt = \int_0^{+\infty} \frac{2t}{(2+t^2)^2}dt = \frac{1}{2} < +\infty, \end{aligned}$$

which imply that (6.5.2) holds.

From the expression of f, it is easy to see that f is non-decreasing with respect to the second and last variables. This means that (6.1.1) holds.

Hence, by Theorem 6.1, it follows that there exists a positive constant R such that the fractional boundary value problem (6.27) has the minimal and maximal positive solutions v^*, u^* respectively in $(0, Rt^{\alpha-1}]$, which can be approximated by the following iterative sequences:

$$v_{n+1} = \int_0^{+\infty} G(t,s)a(s)f(s, v_n(s), D^{\alpha-1}v_n(s))ds \text{ with initial value } v_0(t) = 0,$$

$$u_{n+1} = \int_0^{+\infty} G(t,s)a(s)f(s, u_n(s), D^{\alpha-1}u_n(s))ds \text{ with initial value } u_0(t) = Rt^{\alpha-1}.$$

6.2 Existence and approximation of positive solutions for nonlinear fractional integro-differential boundary value problems on an unbounded domain

In this section, we study the existence, uniqueness, and approximation of positive solutions for a Riemann-Liouville type boundary value problem of nonlinear fractional differential equations with nonlocal fractional integro-differential boundary conditions on an unbounded domain by using a monotone iterative procedure. Precisely, we investigate the following problem:

$$\begin{cases} D^\rho u(t) + q(t)f(t,u(t)) = 0, \quad 2 < \rho \leq 3, \quad t \in J = (0,+\infty), \\ u(0) = u'(0) = 0, \quad D^{\rho-1}u(+\infty) = \mu I^{\rho-1}u(\xi), \quad 0 < \mu, \xi < +\infty, \end{cases} \tag{6.28}$$

where D^ρ denote Riemann-Liouville fractional derivative operator of order ρ, $f : J \times J \to J$ and $q : J \to J$.

6.2.1 *Auxiliary lemmas*

In this subsection, we develop some auxiliary material for the main results.

Lemma 6.6. *Let $\sigma(t) \in C([0,+\infty))$ with $\int_0^{+\infty} \sigma(s)ds < +\infty$. For $\Gamma(2\rho-1) \neq \mu\xi^{2\rho-2}$, the linear fractional boundary value problem*

$$\begin{cases} D^\rho u(t) + \sigma(t) = 0, \quad 2 < \rho \leq 3, \\ u(0) = u'(0) = 0, \quad D^{\rho-1}u(+\infty) = \mu I^{\rho-1}u(\xi), \quad 0 < \mu, \xi < +\infty, \end{cases} \tag{6.29}$$

has a unique solution given by

$$u(t) = \int_0^{+\infty} G(t,s)\sigma(s)ds,$$

where

$$G(t,s) = \frac{1}{\Delta} \begin{cases} -(\Gamma(2\rho-1) - \mu\xi^{2\rho-2})(t-s)^{\rho-1} \\ \qquad + (\Gamma(2\rho-1) - \mu(\xi-s)^{2\rho-2})t^{\rho-1}, s \leq t, \ s \leq \xi, \\ (\Gamma(2\rho-1) - \mu(\xi-s)^{2\rho-2})t^{\rho-1}, 0 \leq t \leq s \leq \xi, \\ -(\Gamma(2\rho-1) - \mu\xi^{2\rho-2})(t-s)^{\rho-1} \\ \qquad + \Gamma(2\rho-1)t^{\rho-1}, 0 \leq \xi \leq s \leq t, \\ \Gamma(2\rho-1)t^{\rho-1}, s \geq t, \ s \geq \xi, \end{cases} \tag{6.30}$$

and

$$\Delta = \Gamma(\rho)(\Gamma(2\rho-1) - \mu\xi^{2\rho-2}).$$

Proof. It is well known that the fractional differential equation $D^\rho u(t) + \sigma(t) = 0$ in (6.29) is equivalent to the integral equation

$$u(t) = -I^\rho\sigma(t) + c_1 t^{\rho-1} + c_2 t^{\rho-2} + c_3 t^{\rho-3}, \tag{6.31}$$

where $c_i \in \mathbb{R}(i = 1,2,3)$ are arbitrary constants. By the condition $u(0) = 0$, we find that $c_3 = 0$. From (6.31), we have

$$\begin{aligned} u'(t) &= -\int_0^t \frac{(t-s)^{\rho-2}}{\Gamma(\rho-1)}\sigma(s)ds + (\rho-1)c_1 t^{\rho-2} + (\rho-2)c_2 t^{\rho-3}, \\ D^{\rho-1}u(t) &= -I^1\sigma(t) + c_1\Gamma(\rho) = c_1\Gamma(\rho) - \textstyle\int_0^t \sigma(s)ds. \end{aligned} \tag{6.32}$$

Using the given boundary conditions $u'(0) = 0$ and $D^{\rho-1}u(+\infty) = \mu I^{\rho-1}u(\xi)$ in (6.32), we find that $c_2 = 0$ and

$$\begin{aligned} c_1 &= \frac{1}{\Gamma(\rho) - A}\int_0^{+\infty}\sigma(s)ds - \frac{\mu}{\Gamma(\rho) - A}\int_0^\xi \frac{(\xi-s)^{\rho-2}}{\Gamma(\rho-1)}\Big[\int_0^s \frac{(s-x)^{\rho-1}}{\Gamma(\rho)}\sigma(x)dx\Big]ds \\ &= \frac{1}{\Gamma(\rho) - A}\int_0^{+\infty}\sigma(s)ds - \frac{\mu}{\Gamma(\rho) - A}I^{2\rho-1}\sigma(\xi), \end{aligned}$$

where

$$A = \mu \int_0^\xi \frac{s^{\rho-1}(\xi - s)^{\rho-2}}{\Gamma(\rho - 1)} ds = \frac{\mu\Gamma(\rho)\xi^{2\rho-2}}{\Gamma(2\rho - 1)}.$$

Substituting the values of c_1, c_2, c_3 in (6.31) gives

$$\begin{aligned}
&u(t)\\
&= -\int_0^t \frac{(t-s)^{\rho-1}}{\Gamma(\rho)}\sigma(s)ds - \frac{\mu t^{\rho-1}}{\Gamma(\rho) - A} I^{2\rho-1}\sigma(\xi) + \frac{t^{\rho-1}}{\Gamma(\rho) - A}\int_0^{+\infty} \sigma(s)ds\\
&= -\int_0^t \frac{(t-s)^{\rho-1}}{\Gamma(\rho)}\sigma(s)ds - \frac{\Gamma(2\rho-1)\mu t^{\rho-1}}{\Gamma(\rho)(\Gamma(2\rho-1) - \mu\xi^{2\rho-2})}\int_0^\xi \frac{(\xi - s)^{2\rho-2}}{\Gamma(2\rho - 1)}\sigma(s)ds\\
&\quad + \frac{\Gamma(2\rho-1)t^{\rho-1}}{\Gamma(\rho)(\Gamma(2\rho-1) - \mu\xi^{2\rho-2})}\int_0^{+\infty} \sigma(s)ds\\
&= \int_0^{+\infty} G(t,s)\sigma(s)ds,
\end{aligned}$$

where $G(t, s)$ is defined by (6.30). This completes the proof. □

Remark 6.1. If $\Gamma(2\rho - 1) > \mu\xi^{2\rho-2}$, then the Green's function $G(t, s)$ satisfies the following properties:

(1) $G(t, s) \geq 0$,

(2)

$$\frac{G(t,s)}{1 + t^{\rho-1}} \leq \frac{\Gamma(2\rho - 1)}{\Gamma(\rho)[\Gamma(2\rho - 1) - \mu\xi^{2\rho-2}]} \triangleq L. \tag{6.33}$$

(3)

$$G(t,s) \leq \frac{\Gamma(2\rho - 1)t^{\rho-1}}{\Gamma(\rho)[\Gamma(2\rho - 1) - \mu\xi^{2\rho-2}]} = Lt^{\rho-1}.$$

For the forthcoming analysis, we introduce a space

$$X = \Big\{u \in C(J, \mathbb{R}) : \sup_{t\in J} \frac{|u(t)|}{1 + t^{\rho-1}} < +\infty\Big\},$$

equipped with the norm

$$\|u\|_X = \sup_{t\in J} \frac{|u(t)|}{1 + t^{\rho-1}}.$$

Notice that X is a Banach space. Define a cone $P \subset X$ by

$$P = \{u \in X : u(t) \geq 0, t \in J\},$$

and an operator $T : X \to X$ as

$$Tu(t) = \int_0^{+\infty} G(t,s)q(s)f(s,u(s))ds. \tag{6.34}$$

Observe that problem (6.28) has a solution only if the operator T has a fixed point.

Lemma 6.7. *Assume that:*

(6.7.1) $f \in C(J\times J, J)$, $f(t,0) \not\equiv 0$ *on any sub-interval of* J*, and when* u *is bounded,* $f(t,(1+t^{\rho-1})u)$ *is bounded on* J.
(6.7.2) $q : J \to J$ *does not identically vanish on any sub-interval of* J *and* $0 < \int_0^{+\infty} q(t)dt < +\infty$.
(6.7.3) $\Gamma(2\rho-1) > \mu\xi^{2\rho-2}$.

Then the operator $T : P \to P$ *defined by (6.34) is completely continuous.*

Proof. We divide the proof into several steps.

(i) $T : P \to P$ is continuous.

Obviously, $T(P) \subset P$. Let u_n, $u \in P$ such that $u_n \to u$ as $n \to +\infty$, then there exists a number $r_0 > 0$ such that $\sup\limits_{n\in N\setminus\{0\}} \|u_n\|_X < r_0$. Take $B_{r_0} = \sup\{f(t,(1+t^{\rho-1})u)|(t,u) \in J\times[0,r_0]\}$. By (6.7.2), we have

$$\int_0^{+\infty} \frac{G(t,s)}{1+t^{\rho-1}} q(s)f(s,u_n(s))ds \le LB_{r_0}\int_0^{+\infty} q(s)ds < \infty,$$

where L is defined by (6.33).

By the Lebesgue dominated convergence theorem and continuity of f, we can obtain

$$\lim_{n\to\infty}\int_0^{+\infty} \frac{G(t,s)}{1+t^{\rho-1}} q(s)f(s,u_n(s))ds = \int_0^{+\infty} \frac{G(t,s)}{1+t^{\rho-1}} q(s)f(s,u(s))ds.$$

Taking limit as $n \to \infty$, we get

$$\|Tu_n - Tu\|_X = \sup_{t\in J}\int_0^{+\infty} \frac{G(t,s)}{1+t^{\rho-1}} |q(s)f(s,u_n(s)) - q(s)f(s,u(s))|ds \to 0.$$

This shows that $T : P \to P$ is continuous.

(ii) T maps bounded set into bounded set. Let Ω be any bounded subset of cone P, then there exists a number $r_1 > 0$ such that $\|u\|_X < r_1, \forall u \in \Omega$. Let $B_{r_1} = \sup\{f(t,(1+t^{\rho-1})u)|(t,u) \in J\times[0,r_1]\}$. By (6.7.2) and (6.33), we have

$$\|Tu\|_X = \sup_{t\in J}\int_0^{+\infty} \frac{G(t,s)}{1+t^{\rho-1}} |q(s)f(s,u(s))|ds \le LB_{r_1}\int_0^{+\infty} q(s)ds < \infty,$$

which implies that $T\Omega$ is uniformly bounded.

(iii) T is relatively compact.

(a) For any Ω taken in step (ii), $T\Omega$ is equi-continuous on any compact interval.

Let $I \subset J$ be any compact interval and $t_1, t_2 \in I$. Without loss of generality, we assume that $t_1 < t_2$. Then, for any $u \in \Omega$, we have

$$\begin{aligned}\left|\frac{Tu(t_2)}{1+t_2^{\rho-1}} - \frac{Tu(t_1)}{1+t_1^{\rho-1}}\right| &= \left|\int_0^{+\infty}\left(\frac{G(t_2,s)}{1+t_2^{\rho-1}} - \frac{G(t_1,s)}{1+t_1^{\rho-1}}\right) q(s)f(s,u(s))ds\right| \\ &\le B_{r_1}\int_0^{+\infty}\left|\frac{G(t_2,s)}{1+t_2^{\rho-1}} - \frac{G(t_1,s)}{1+t_1^{\rho-1}}\right| q(s)ds. \qquad (6.35)\end{aligned}$$

Since $G(t,s) \in C(J \times J)$, we have that $\dfrac{G(t,s)}{1+t^{\rho-1}}$ is uniformly continuous on the compact set $I \times I$. For $s \geq t$, $\dfrac{G(t,s)}{1+t^{\rho-1}}$ only depends on t. Obviously, it is uniformly continuous on $I \times (J \backslash I)$. In consequence, for all $t_1,\, t_2 \in I$ and $s \in J$, we have

$$\forall \varepsilon > 0,\ \exists \delta(\varepsilon) > 0 \text{ such that if } |t_1 - t_2| < \delta \text{ then } \left| \frac{G(t_2,s)}{1+t_2^{\rho-1}} - \frac{G(t_1,s)}{1+t_1^{\rho-1}} \right| < \varepsilon. \tag{6.36}$$

Since $\int_0^{+\infty} q(s)ds < \infty$, it follows from (6.35) and (6.36) that $T\Omega$ is equi-continuous on I.

(b) T is equi-convergent at ∞. Observe that

$$\lim_{t\to\infty} \frac{G(t,s)}{1+t^{\rho-1}} = \frac{1}{\triangle} \begin{cases} \mu\xi^{2\rho-2} - \mu(\xi - s)^{2\rho-2}, s \leq \xi, \\ \mu\xi^{2\rho-2}, 0 \leq \xi \leq s. \end{cases}$$

It is easy to show that for any given $\varepsilon > 0$, there exists a constant $T' = T'(\varepsilon) > 0$ such that

$$\left| \frac{G(t_1,s)}{1+t_1^{\rho-1}} - \frac{G(t_2,s)}{1+t_2^{\rho-1}} \right| < \varepsilon$$

for any $t_1, t_2 \geq T'$ and $s \in J$. Hence, T is equi-convergent at ∞. Thus, Theorem 6.1 applies and the operator T is relatively compact on J. Hence $T : P \to P$ is completely continuous. This completes the proof. □

6.2.2 *Main results*

Algorithm 1. For chosen initial values $v_0(t) = 0$ and $u_0(t) = rt^{\rho-1}, t \in J$, we consider the sequences $\{v_n\}$, $\{u_n\}$ such that

$$u_{n+1} = \int_0^{+\infty} G(t,s)q(s)f(s,u_n(s))ds,$$

$$v_{n+1} = \int_0^{+\infty} G(t,s)q(s)f(s,v_n(s))ds.$$

Theorem 6.2. *Assume that the conditions (6.7.1)-(6.7.3) hold. In addition we suppose that:*

(6.2.1) $f(t,\cdot)$ is non-decreasing for each $t \in J$, and there exists a positive constant r, such that

$$f(t,(1+t^{\rho-1})u) \leq \frac{r}{L\int_0^{+\infty} q(s)ds}, \quad \text{for } (t,u) \in J \times [0,r].$$

Then the fractional boundary value problem (6.28) has two positive solutions v^*, u^* such that $\|v^*\|_X, \|u^*\|_X \in (0,r]$. Moreover, the iterative sequences $\{v_n\}$, $\{u_n\}$ generated by the Algorithm 1 converge to v^*, u^*, respectively.

Proof. We define $P_r = \{u \in P, \|u\|_X < r\}$ and show that $T(P_r) \subset P_r$. Let $u \in P_r$, then $\|u\|_X \le r$. From (6.2.1) and (6.33), we have

$$\|Tu\|_X = \sup_{t\in J} \int_0^{+\infty} \frac{G(t,s)}{1+t^{\rho-1}} |q(s)f(s,u(s))|ds \\ \le L\int_0^{+\infty} q(s)ds \cdot \frac{r}{L\int_0^{+\infty} q(s)ds} = r.$$

This implies that $T(P_r) \subset P_r$.

Next we construct two monotone iterative sequences which converge to two positive solutions of the problem (6.28). It follows from (6.2.1) that the operator T is non-decreasing.

Let $v_0(t) = 0, \forall t \in J$ and define $v_{n+1} = Tv_n,\ n = 0,1,2,\ldots$. It is clear that $v_0 \subset P_r$ and $v_{n+1} = T^n v_0,\ n = 0,1,2,\ldots$. This, together with $T(P_r) \subset P_r$, implies that $v_n \in P_r,\ n = 1,2,\ldots$. By the complete continuity of T, we infer that $\{v_n\}_{n=1}^{\infty}$ possesses a convergent subsequence $\{v_{n_k}\}_{k=1}^{\infty}$ and there exists a $v^* \in P_r$ such that $v_{n_k} \to v^*$ as $k \to \infty$.

For $v_0 \in P_r$, we can obtain that $v_1 = Tv_0 = T0 \ge 0 = v_0$. Since T is non-decreasing, therefore, $v_2 = Tv_1 \ge Tv_0 = v_1$. By induction, we have

$$v_{n+1} \ge v_n,\ n = 0,1,2,\ldots. \tag{6.37}$$

Hence, $v_n \to v^*(n \to \infty)$. By the complete continuity of T and $v_{n+1} = Tv_n$, we have that $Tv^* = v^*$, which means that v^* is a fixed point of the operator T.

Let $u_0(t) = rt^{\rho-1}, \forall t \in J$ and define $u_{n+1} = Tu_n,\ n = 0,1,2,\ldots$. It is clear that $u_0 \in P_r$ and $u_{n+1} = T^n u_0,\ n = 0,1,2,\ldots$. This, together with $T(P_r) \subset P_r$, gives $u_n \in P_r,\ n = 1,2,\ldots$. Further, complete continuity of T implies that $\{u_n\}_{n=1}^{\infty}$ possesses a convergent subsequence $\{u_{n_k}\}_{k=1}^{\infty}$ and there exists a $u^* \in P_r$ such that $u_{n_k} \to u^*$ as $k \to \infty$.

By (6.2.1), we have

$$u_1(t) = \int_0^{+\infty} G(t,s)q(s)f(s,u_0(s))ds \\ \le \int_0^{+\infty} Lt^{\rho-1}q(s)ds \cdot \frac{r}{\int_0^{+\infty} Lq(s)ds} \\ = rt^{\rho-1} = u_0(t).$$

Since T is non-decreasing, we have $u_2 = Tu_1 \le Tu_0 = u_1$. By induction, we have

$$u_{n+1} \le u_n,\ n = 0,1,2,\ldots. \tag{6.38}$$

Hence, $u_n \to u^*(n \to \infty)$. By the complete continuity of T and $u_{n+1} = Tu_n$, we have $Tu^* = u^*$, which means that u^* is also a fixed point of the operator T.

Since $f(t,0) \not\equiv 0, \forall t \in J$, therefore $u = 0$ is not a solution of problem (6.28). Thus, u^*, v^* are two positive solutions of problem (6.28), which can be obtained by using the corresponding iterative sequences $\{u_n\}, \{v_n\}$ defined in Algorithm 1. □

Algorithm 2. For chosen initial value $w_0(t) = 0$ or $w_0(t) = rt^{\rho-1}, t \in J$, define the sequence $\{w_n\}$ by

$$w_{n+1} = \int_0^{+\infty} G(t,s)q(s)f(s,w_n(s))ds.$$

Theorem 6.3. *Assume that the condition (6.7.1)-(6.7.3) and (6.2.1) hold. In addition suppose that:*

(6.3.1) There exists a constant $M \geq 0$, such that $f(t,u)-f(t,v) \leq M(u-v)$, where $0 \leq v \leq u \leq rt^{\rho-1}$.

Then there exists $w^* \in [0, rt^{\rho-1}]$, which is a unique positive solution of problem (6.28) in $[0, rt^{\rho-1}]$. Moreover, the iterative sequence $\{w_n\}$ generated by the Algorithm 2 converges to w^*. Furthermore, the following error estimate formula holds:

$$\|w_n - w^*\|_X \leq H^n r, \tag{6.39}$$

where $H = ML\int_0^{+\infty} q(s)(1+s^{\rho-1})ds < 1$.

Proof. By Theorem 6.2, we know that the problem (6.28) has two positive solutions u^* and v^*, which can be obtained by $u_{n+1} = T^n u_0$, $v_{n+1} = T^n v_0$ with initial values $u_0(t) = rt^{\rho-1}$, $v_0(t) = 0, \forall t \in J$, respectively.

Since T is non-decreasing and $v_0(t) \leq u_0(t), \forall t \in J$, therefore, we have

$$v_1 = Tv_0 \leq Tu_0 = u_1$$

By induction, we have that $v_n \leq u_n$, $\forall t \in J, n = 0,1,2,\ldots$. This, together with (6.37) and (6.38), yields

$$v_0 \leq v_1 \leq \cdots \leq v_n \leq \cdots \leq u_n \leq \cdots \leq u_1 \leq u_0. \tag{6.40}$$

By (6.3.1), we have

$$\begin{aligned} 0 \leq (u_n - v_n)(t) &= Tu_{n-1}(t) - Tv_{n-1}(t) \\ &= \int_0^{+\infty} G(t,s)q(s)[f(s,u_{n-1}(s)) - f(s,v_{n-1}(s))]ds \\ &\leq M\int_0^{+\infty} G(t,s)q(s)[u_{n-1} - v_{n-1}](s)ds \\ &\leq M\int_0^{+\infty} G(t,s)q(s)(1+s^{\rho-1})ds\|u_{n-1} - v_{n-1}\|_X, \end{aligned}$$

which gives

$$\|u_n - v_n\|_X \leq ML\int_0^{+\infty} q(s)(1+s^{\rho-1})ds\|u_{n-1} - v_{n-1}\|_X = H\|u_{n-1} - v_{n-1}\|_X.$$

Thus, we can obtain

$$\|u_n - v_n\|_X \leq H^n\|u_0 - v_0\|_X = H^n r. \tag{6.41}$$

So, for any $m,\ n \in N$, we have

$$\begin{cases} \|v_{n+m} - v_n\|_X \leq \|u_n - v_n\|_X \leq H^n r, \\ \|u_n - u_{n+m}\|_X \leq \|u_n - v_n\|_X \leq H^n r. \end{cases} \tag{6.42}$$

Next, we show that u^*, v^* are the maximal and minimal solutions of (6.28) in $[v_0, u_0]$. Let $w \in [v_0, u_0]$ be any solution of (6.28), that is, $Tw = w$. By Theorem

6.2, we know that $u^* = \lim_{n\to\infty} u_n$ and $v^* = \lim_{n\to\infty} v_n$ are two positive solutions of problem (6.28). This, together with (6.40) and the fact that T is non-decreasing, it follows

$$v^* \leq w \leq u^*, \ (n \to \infty).$$

Hence, u^*, v^* are the maximal and minimal solutions of (6.28) in $[v_0, u_0]$, respectively. Finally, we prove that the problem (6.28) has a unique solution $w^* \in [v_0, u_0]$ and the error estimate formula (6.39) holds.

By (6.41), we have

$$\|u^* - v^*\|_X \leq H^n r \to 0 \ (n \to +\infty),$$

which shows $u^* = v^* \triangleq w^*$. Hence the problem (6.28) has a unique solution $w^* \in [v_0, u_0]$. In fact, the sequences $\{u_n\}, \{v_n\}$ correspond to $\{w_n\}$ with $w_0(t) = rt^{\rho-1}$ and $w_0(t) = 0$ respectively. Furthermore, from (6.42), we can obtain the error estimate formula (6.39). This completes the proof. □

Example 6.2. Consider the boundary value problem:

$$\begin{cases} D^{\frac{5}{2}}u(t) + \dfrac{1}{1+t^2} f(t, u(t)) = 0, \ t \in (0, +\infty), \\ u(0) = u'(0) = 0, \quad D^{\frac{3}{2}}u(+\infty) = 2I^{\frac{3}{2}}u\left(\dfrac{1}{2}\right), \end{cases} \tag{6.43}$$

where $\rho = \frac{5}{2}$, $\mu = 2$, $\xi = \frac{1}{2}$, $q(t) = \frac{1}{1+t^2}$,

$$f(t,u) = \begin{cases} 5e^{-t} + \sqrt{\frac{u}{1+t^{\frac{3}{2}}}}, & 0 \leq u \leq 10, \\ 5e^{-t} + \dfrac{10}{\sqrt{1+t^{\frac{3}{2}}}}, & u \geq 10. \end{cases}$$

With the given values, $\Gamma(\rho) = \Gamma\left(\frac{5}{2}\right) = \frac{3\sqrt{\pi}}{4}$, $\Gamma(2\rho-1) = \Gamma(4) = 6$, $\mu\xi^{2\rho-2} = \frac{1}{4}$, $\int_0^{+\infty} q(t)dt = \frac{\pi}{2}$, $L = \frac{\Gamma(2\rho-1)}{\Gamma(\rho)[\Gamma(2\rho-1) - \mu\xi^{2\rho-2}]} = \frac{32}{23\sqrt{\pi}}$, we find that the condition (6.7.1)-(6.7.1) hold. Take $r = 16\sqrt{\pi}$. By a simple computation, we have that

$$f(t, (1+t^{\frac{3}{2}})u) \leq 15 < \frac{r}{L\int_0^{+\infty} q(s)ds} = 23, \quad \text{for } (t,u) \in J \times [0, 16\sqrt{\pi}].$$

So, the condition (6.2.1) holds. Thus all the conditions of Theorem 6.2 are satisfied. Hence, by the conclusion of Theorem 6.2, the fractional boundary value problem (6.43) has two positive solutions v^*, u^* such that $\|v^*\|_X, \|u^*\|_X \in (0, 16\sqrt{\pi}]$, which can be obtained with the aid of monotone iterative sequences.

6.3 A nonlocal Hadamard fractional boundary value problem with Hadamard integral and discrete boundary conditions on a half line

By employing the monotone iterative method, we investigate the iterative positive solutions for a nonlinear Hadamard type fractional differential equation

$$^{H}D^{q}x(t)+\sigma(t)f(t,x(t))=0,\quad 2<q\leq 3,\quad t\in(1,+\infty),\tag{6.44}$$

supplemented with nonlocal Hadamard integral and discrete boundary conditions

$$\begin{cases} x(1)=x'(1)=0,\\ ^{H}D^{q-1}x(\infty)=a\ ^{H}I^{\beta}x(\xi)+b\sum\limits_{i=1}^{m-2}\alpha_i x(\eta_i), 1<\xi<\eta_1<\eta_2<\ldots<\eta_{m-2}<+\infty,\end{cases}\tag{6.45}$$

where $^{H}D^{q}$ denotes Hadamard fractional derivative operator of order q and is defined by

$$^{H}D^{q}x(t)=\frac{1}{\Gamma(n-q)}\left(t\frac{d}{dt}\right)^{n}\int_1^t\left(\log\frac{t}{s}\right)^{n-q-1}\frac{x(s)}{s}ds,\quad n-1<q<n, n=[q]+1,$$

and $^{H}I^{\beta}$ is the Hadamard fractional integral of order β defined by

$$^{H}I^{\beta}x(t)=\frac{1}{\Gamma(\beta)}\int_1^t\left(\log\frac{t}{s}\right)^{\beta-1}\frac{x(s)}{s}ds,\quad \beta>0,$$

a and b are real constants, and α_i are positive real constants. We obtain the twin positive solutions and a unique positive solution of the nonlocal boundary value problem (6.44)-(6.45) by constructing convergent monotone iterative schemes and derive the error estimate formula for the unique positive solution.

Lemma 6.8. *[108] If $a,\gamma,\beta>0$, then*

$$\left(^{H}I_a^{\gamma}\log\left(\frac{t}{a}\right)^{\beta-1}\right)(x)=\frac{\Gamma(\beta)}{\Gamma(\beta+\gamma)}\left(\log\frac{x}{a}\right)^{\beta+\gamma-1},$$

$$\left(^{H}D_a^{\gamma}\log\left(\frac{t}{a}\right)^{\beta-1}\right)(x)=\frac{\Gamma(\beta)}{\Gamma(\beta-\gamma)}\left(\log\frac{x}{a}\right)^{\beta-\gamma-1}.$$

Lemma 6.9. *[108] For $n-1<\gamma<n$ and $x\in C[1,\infty)\cap L^1[1,\infty)$, the solution of Hadamard fractional differential equation $^{H}D^{\gamma}x(t)=0$ is*

$$x(t)=\sum_{i=1}^{n}c_i(\log t)^{\gamma-i},$$

where $c_i\in\mathbb{R}, i=1,2,\ldots,n$. Moreover, the following formula holds:

$$^{H}I^{\gamma}\,{}^{H}D^{\gamma}x(t)=x(t)+\sum_{i=1}^{n}c_i(\log t)^{\gamma-i}.$$

Lemma 6.10. *Let* $h \in C[1,\infty)$ *with* $0 < \int_1^\infty h(s)\frac{ds}{s} < \infty$, *and*

$$\Delta = \Gamma(q) - \frac{a\Gamma(q)}{\Gamma(q+\beta)}(\log\xi)^{q+\beta-1} - b\sum_{i=1}^{m-2}\alpha_i(\log\eta_i)^{q-1} > 0, \tag{6.46}$$

then the Hadamard type fractional differential equation

$${}^H D^q x(t) + h(t) = 0, \quad 2 < q \le 3, \quad t \in J = (1,+\infty), \tag{6.47}$$

subject to the Hadamard integral and discrete boundary conditions (6.45) has a unique solution

$$x(t) = \int_1^\infty G(t,s)h(s)\frac{ds}{s}, \tag{6.48}$$

where

$$G(t,s) = m(t,s,q) + \frac{a(\log t)^{q-1}}{\Delta}m(\xi,s,q+\beta) + \sum_{i=1}^{m-2}\frac{b\alpha_i(\log t)^{q-1}}{\Delta}m(\eta_i,s,q), \tag{6.49}$$

and

$$m(t,s,q) = \frac{1}{\Gamma(q)}\begin{cases}(\log t)^{q-1} - \left(\log\frac{t}{s}\right)^{q-1}, & 1 \le s \le t < \infty,\\ (\log t)^{q-1}, & 1 \le t \le s < \infty.\end{cases} \tag{6.50}$$

Proof. The solution of Hadamard fractional differential equation (6.47) can be written as

$$x(t) = -{}^H I^q h(t) + c_1(\log t)^{q-1} + c_2(\log t)^{q-2} + c_3(\log t)^{q-3}, \tag{6.51}$$

where $c_i \in \mathbb{R}$ $(i = 1,2,3)$ are arbitrary constants. By the condition $x(1) = 0$, we find that $c_3 = 0$. From (6.51), we have

$$x'(t) = -\tfrac{1}{t}\, {}^H I^{q-1}h(t) + \tfrac{(q-1)}{t}c_1(\log t)^{q-2} + \tfrac{(q-2)}{t}c_2(\log t)^{q-3}. \tag{6.52}$$

In view of the condition $x'(1) = 0$, we find that $c_2 = 0$. Therefore, (6.51) takes the form:

$$x(t) = c_1(\log t)^{q-1} - \frac{1}{\Gamma(q)}\int_1^t \left(\log\frac{t}{s}\right)^{q-1} h(s)\frac{ds}{s}. \tag{6.53}$$

In accordance with Lemma 6.8, we have

$${}^H D^{q-1}x(t) = c_1\Gamma(q) - \int_1^t h(s)\frac{ds}{s}. \tag{6.54}$$

Applying the condition ${}^H D^{q-1}x(\infty) = a\,{}^H I^\beta x(\xi) + b\sum_{i=1}^{m-2}\alpha_i x(\eta_i)$ to (6.53) yields

$$\begin{aligned} c_1 = \frac{1}{\Delta}\Big\{ &\int_1^\infty h(s)\frac{ds}{s} - \frac{a}{\Gamma(q+\beta)}\int_1^\xi \left(\log\frac{\xi}{s}\right)^{q+\beta-1} h(s)\frac{ds}{s} \\ &- \sum_{i=1}^{m-2}\frac{b\alpha_i}{\Gamma(q)}\int_1^{\eta_i}\left(\log\frac{\eta_i}{s}\right)^{q-1} h(s)\frac{ds}{s}\Big\}, \end{aligned}$$

where Δ is defined by (6.46). Substituting the value of c_1 in (6.53), we get

$$
\begin{aligned}
&x(t)\\
&= (\log t)^{q-1}\Big[1+\frac{\big[\frac{a\Gamma(q)}{\Gamma(q+\beta)}(\log\xi)^{q+\beta-1}+b\sum_{i=1}^{m-2}\alpha_i(\log\eta_i)^{q-1}\big]}{\Delta}\Big]\int_1^\infty\frac{h(s)}{\Gamma(q)}\frac{ds}{s}\\
&\quad-\frac{(\log t)^{q-1}}{\Delta}\Big\{\frac{a}{\Gamma(q+\beta)}\int_1^\xi\Big(\log\frac{\xi}{s}\Big)^{q+\beta-1}h(s)\frac{ds}{s}\\
&\quad+\sum_{i=1}^{m-2}\frac{b\alpha_i}{\Gamma(q)}\int_1^{\eta_i}\Big(\log\frac{\eta_i}{s}\Big)^{q-1}h(s)\frac{ds}{s}\Big\}-\frac{1}{\Gamma(q)}\int_1^t\Big(\log\frac{t}{s}\Big)^{q-1}h(s)\frac{ds}{s}\\
&=\frac{1}{\Gamma(q)}\int_1^t\Big[(\log t)^{q-1}-\Big(\log\frac{t}{s}\Big)^{q-1}\Big]h(s)\frac{ds}{s}+\frac{1}{\Gamma(q)}\int_t^\infty(\log t)^{q-1}h(s)\frac{ds}{s}\\
&\quad+\frac{a(\log t)^{q-1}}{\Delta\Gamma(q+\beta)}\int_1^\xi\Big[(\log\xi)^{q+\beta-1}-\Big(\log\frac{\xi}{s}\Big)^{q+\beta-1}\Big]h(s)\frac{ds}{s}\\
&\quad+\frac{a(\log t)^{q-1}}{\Delta\Gamma(q+\beta)}\int_\xi^\infty(\log\xi)^{q+\beta-1}h(s)\frac{ds}{s}\\
&\quad+\sum_{i=1}^{m-2}\frac{b\alpha_i(\log t)^{q-1}}{\Delta\Gamma(q)}\int_1^{\eta_i}\Big[(\log\eta_i)^{q-1}-\Big(\log\frac{\eta_i}{s}\Big)^{q-1}\Big]h(s)\frac{ds}{s}\\
&\quad+\sum_{i=1}^{m-2}\frac{b\alpha_i(\log t)^{q-1}}{\Delta\Gamma(q)}\int_{\eta_i}^\infty(\log\eta_i)^{q-1}h(s)\frac{ds}{s}\\
&=\int_1^\infty m(t,s,q)h(s)\frac{ds}{s}+\frac{a(\log t)^{q-1}}{\Delta}\int_1^\infty m(\xi,s,q+\beta)h(s)\frac{ds}{s}\\
&\quad+\sum_{i=1}^{m-2}\frac{b\alpha_i}{\Delta}(\log t)^{q-1}\int_1^\infty m(\eta_i,s,q)h(s)\frac{ds}{s}\\
&=\int_1^\infty G(t,s)h(s)\frac{ds}{s}.
\end{aligned}
$$

This completes the proof. □

Lemma 6.11. *For $(t,s)\in[1,\infty)\times[1,\infty)$, the Green's function $G(t,s)$ defined by (6.49) satisfies the following properties:*

(C_1) : *$G(t,s)$ is continuous and $G(t,s)\geq 0$;*

$$(C_2) : G(t,s)\leq\Big(\frac{1}{\Gamma(q)}+\frac{a(\log\xi)^{q+\beta-1}}{\Delta\Gamma(q+\beta)}+\sum_{i=1}^{m-2}\frac{b\alpha_i(\log\eta_i)^{q-1}}{\Delta\Gamma(q)}\Big)(\log t)^{q-1}.$$

Proof. It is easy to see that (C_1) holds.

In order to establish (C_2), for $(t,s)\in[1,\infty)\times[1,\infty)$, we have

$$G(t,s)=m(t,s,q)+\frac{a(\log t)^{q-1}}{\Delta}m(\xi,s,q+\beta)+\sum_{i=1}^{m-2}\frac{b\alpha_i(\log t)^{q-1}}{\Delta}m(\eta_i,s,q)$$

$$\leq \frac{(\log t)^{q-1}}{\Gamma(q)} + \frac{a(\log t)^{q-1}(\log \xi)^{q+\beta-1}}{\Delta\Gamma(q+\beta)} + \sum_{i=1}^{m-2} \frac{b\alpha_i(\log t)^{q-1}(\log \eta_i)^{q-1}}{\Delta\Gamma(q)}$$

$$= \Big(\frac{1}{\Gamma(q)} + \frac{a(\log \xi)^{q+\beta-1}}{\Delta\Gamma(q+\beta)} + \sum_{i=1}^{m-2} \frac{b\alpha_i(\log \eta_i)^{q-1}}{\Delta\Gamma(q)}\Big)(\log t)^{q-1}.$$

The proof is completed. □

Remark 6.2. From (C_2) in Lemma 6.11, we deduce that

$$\frac{G(t,s)}{1+(\log t)^{q-1}} \leq \frac{1}{\Gamma(q)} + \frac{a(\log \xi)^{q+\beta-1}}{\Delta\Gamma(q+\beta)} + \sum_{i=1}^{m-2} \frac{b\alpha_i(\log \eta_i)^{q-1}}{\Delta\Gamma(q)} \triangleq A$$

for $(t,s) \in [1,\infty) \times [1,\infty)$.

Let us introduce the Banach space E defined by

$$E = \Big\{x \in C([1,\infty)) : \sup_{t\in[1,\infty)} \frac{|x(t)|}{1+(\log t)^{q-1}} < \infty\Big\},$$

and equipped with norm $\|x\|_E = \sup_{t\in[1,\infty)} \frac{|x(t)|}{1+(\log t)^{q-1}}$.

In view of Lemma 6.10, we define an operator $T : P \to P$ by

$$Tx(t) = \int_1^\infty G(t,s)\sigma(s)f(s,x(s))\frac{ds}{s}, \tag{6.55}$$

where $P \subset E$ is a cone given by $P = \{x \in E : x(t) \geq 0,\ \forall t \in [1,\infty)\}$.

Observe that the nonlocal Hadamard fractional boundary value problem (6.44)-(6.45) has a solution if and only if the operator equation $Tx = x$ has a fixed point.

6.3.1 *Twin positive solutions*

In this subsection, we construct two monotone iterative sequences which converge to twin positive solutions of the nonlocal Hadamard fractional boundary value problem (6.44)-(6.45). Let us first enlist the assumptions that we need in constructing the desired results.

(H_1) $f \in C([1,\infty) \times [0,\infty), [0,\infty))$ be such that $f(t,0) \not\equiv 0$ on any sub-interval of $[0,\infty)$, and $f(t,(1+(\log t)^{q-1})x)$ is bounded on $[0,\infty)$ when x is bounded.

(H_2) $\sigma : [1,\infty) \to [0,\infty)$ does not identically vanish on any sub-interval of $[0,\infty)$ and

$$0 < \int_1^{+\infty} \sigma(s)\frac{ds}{s} < +\infty.$$

(H_3) $f(t,\cdot)$ is non-decreasing for each $t \in [1,\infty)$, and there exists a positive constant ζ, such that

$$f(t,(1+(\log t)^{q-1})x) \leq \frac{\zeta}{A\int_1^{+\infty}\sigma(s)\frac{ds}{s}},\ \ \textit{for}\ (t,x) \in [1,\infty) \times [0,\zeta].$$

(H_4) There exists a constant $L \geq 0$, such that

$$f(t,u) - f(t,v) \leq L(u-v),$$

where $0 \leq v \leq u \leq \zeta(\log t)^{q-1}$.

Theorem 6.4. *Suppose the conditions $(H_1) - (H_3)$ are satisfied. Then the nonlocal Hadamard fractional boundary value problem (6.44)-(6.45) has twin positive solutions v^*, u^* such that $\|v^*\|_E$, $\|u^*\|_E \in (0,\zeta]$. Furthermore, the twin positive solutions can be achieved by the sequences $\{v_n\}_{n=1}^{\infty}$ and $\{u_n\}_{n=1}^{\infty}$ such that*

$$v_0 \leq v_1 \leq \cdots \leq v_n \leq \cdots \leq v^* \leq u^* \leq \cdots \leq u_n \leq \cdots \leq u_1 \leq u_0, \tag{6.56}$$

where

$$\begin{aligned} v_{n+1}(t) &= \int_1^{+\infty} G(t,s)\sigma(s)f(s,v_n(s))\frac{ds}{s}, \quad n = 0,1,2,\ldots, \\ u_{n+1}(t) &= \int_1^{+\infty} G(t,s)\sigma(s)f(s,u_n(s))\frac{ds}{s}, \quad n = 0,1,2,\ldots, \end{aligned} \tag{6.57}$$

with the initial values $v_0(t) = 0$ and $u_0(t) = \zeta(\log t)^{q-1}, \forall t \in (1,\infty)$, which converge to v^*, u^*, respectively.

Proof. In the first step, we show that the operator $T : P \to P$ defined by (6.55) is completely continuous. Let $x_n \to x$ as $n \to +\infty$ in P. Then there exists a constant $\zeta_0 > 0$ such that $\sup_{n\in N\setminus\{0\}} \|x_n\|_E < \zeta_0$. Let $B_{\zeta_0} = \sup\{f(t,(1+(\log t)^{q-1})x)|(t,x) \in [1,\infty) \times [0,\zeta_0]\}$. Following the conditions (H_2), we have

$$\begin{aligned} \int_1^{+\infty} \frac{G(t,s)}{1+(\log t)^{q-1}}\sigma(s)f(s,x_n(s))\frac{ds}{s} &\leq A\int_1^{+\infty} \sigma(s)f(s,x_n(s))\frac{ds}{s} \\ &\leq AB_{\zeta_0}\int_1^{+\infty}\sigma(s)\frac{ds}{s} < \infty, \end{aligned}$$

where A is defined in Remark 6.2. The Lebesgue dominated convergence theorem and continuity of f guarantee that

$$\lim_{n\to\infty}\int_1^{+\infty} \frac{G(t,s)}{1+(\log t)^{q-1}}\sigma(s)f(s,x_n(s))\frac{ds}{s} = \int_1^{+\infty} \frac{G(t,s)}{1+(\log t)^{q-1}}\sigma(s)f(s,x(s))\frac{ds}{s}.$$

So, we have

$$\|Tx_n - Tx\|_E = \sup_{t\in[1,\infty)}\int_1^{+\infty} \frac{G(t,s)}{1+(\log t)^{q-1}}\Big|\sigma(s)f(s,x_n(s)) - \sigma(s)f(s,x(s))\Big|\frac{ds}{s} \to 0,$$

as $n \to \infty$. Thus, $T : P \to P$ is continuous.

Let Φ be any bounded subset of cone P. It is clear that there exists a constant $\zeta_1 > 0$ such that $\|x\|_E < \zeta_1, \forall x \in \Phi$. Let $B_{\zeta_1} = \sup\{f(t,(1+(\log t)^{q-1})x)|(t,x) \in [1,\infty) \times [0,\zeta_1]\}$. By the conditions (H_2), we have

$$\|Tx\|_E = \sup_{t\in[1,\infty)}\int_1^{+\infty} \frac{G(t,s)}{1+(\log t)^{q-1}}\Big|\sigma(s)f(s,x(s))\Big|\frac{ds}{s} \leq AB_{\zeta_1}\int_1^{+\infty}\sigma(s)\frac{ds}{s} < \infty,$$

for $x \in P$, which implies that $T\Phi$ is uniformly bounded.

Next we show that the operator $T\Phi$ is equi-continuous on any compact interval of $[1,\infty)$ for any Φ mentioned above. Let $I \subset [1,\infty)$ be any compact interval. Then, $\forall\ t_1, t_2 \in I$, $t_1 < t_2$, we have

$$\begin{aligned}
&\left|\frac{Tx(t_2)}{1+(\log t_2)^{q-1}} - \frac{Tx(t_1)}{1+(\log t_1)^{q-1}}\right| \\
&= \left|\int_1^{+\infty}\left(\frac{G(t_2,s)}{1+(\log t_2)^{q-1}} - \frac{G(t_1,s)}{1+(\log t_1)^{q-1}}\right)\sigma(s)f(s,x(s))\frac{ds}{s}\right| \\
&\le B_{\zeta_1}\int_1^{+\infty}\left|\frac{G(t_2,s)}{1+(\log t_2)^{q-1}} - \frac{G(t_1,s)}{1+(\log t_1)^{q-1}}\right|\sigma(s)\frac{ds}{s}. \qquad (6.58)
\end{aligned}$$

For any compact set $I \times I$, $\dfrac{G(t,s)}{1+(\log t)^{q-1}}$ is uniformly continuous as $G(t,s) \in C([1,\infty)\times[1,\infty))$. Considering this function to be only related to t when $s \ge t$, it follows that it is uniformly continuous on $I \times ([1,\infty)\backslash I)$. In consequence, for all t_1, $t_2 \in I$ and $s \in [1,\infty)$, we obtain
$\forall\varepsilon > 0$, $\exists\delta(\varepsilon) > 0$ such that if $|t_1 - t_2| < \delta$, then

$$\left|\frac{G(t_2,s)}{1+(\log t_2)^{q-1}} - \frac{G(t_1,s)}{1+(\log t_1)^{q-1}}\right| < \varepsilon. \qquad (6.59)$$

By (H_2), (6.58) and (6.59), we deduce that $T\Phi$ is equi-continuous on I.

Next we show that the operator T is equi-convergent at ∞. For any $x \in \Phi$, we have

$$\begin{aligned}
\lim_{t\to\infty}\left|\frac{Tx(t)}{1+(\log t)^{q-1}}\right| &= \lim_{t\to\infty}\int_1^{+\infty}\frac{G(t,s)}{1+(\log t)^{q-1}}\left|\sigma(s)f(s,x(s))\right|\frac{ds}{s} \\
&\le AB_{\zeta_1}\int_1^{+\infty}\sigma(s)\frac{ds}{s} < \infty.
\end{aligned}$$

From the above steps, we infer that the operator $T : P \to P$ is completely continuous.

Next we define $P_\zeta = \{x \in P, \|x\|_E \le \zeta\}$. If $x \in P_\zeta$, then $\|x\|_E \le \zeta$. In view of (H_3) and Remark 6.2, we have

$$\begin{aligned}
\|Tx\|_E &= \sup_{t\in[1,\infty)}\int_1^{+\infty}\frac{G(t,s)}{1+(\log t)^{q-1}}\left|\sigma(s)f(s,x(s))\right|\frac{ds}{s} \\
&\le A\int_1^{+\infty}\sigma(s)\frac{ds}{s}\cdot\frac{\zeta}{A\int_1^{+\infty}\sigma(s)\frac{ds}{s}} = \zeta.
\end{aligned}$$

Hence, we have $T(P_\zeta) \subset P_\zeta$.

Define $v_0(t) = 0$ and a sequence $v_{n+1} = Tv_n$, $n = 0,1,2,\ldots$. By the non-decreasing property of T, which follows from the condition (H_3), it is clear that the sequence $\{v_n\}_{n=1}^{\infty} \subset T(P_\zeta) \subset P_\zeta$ and satisfies

$$v_{n+1}(t) \ge v_n(t), \qquad \forall t \in [1,\infty), \quad n = 0,1,2,\ldots. \qquad (6.60)$$

The complete continuity of the operator T implies that there exists a $v^* \in P_\zeta$ such that $v_{n_k} \to v^*$ as $k \to \infty$, which follows from the fact that $\{v_n\}_{n=1}^{\infty}$ possesses a convergent subsequence $\{v_{n_k}\}_{k=1}^{\infty}$. This, together with (6.60), implies that $\lim_{n\to\infty} v_n = v^*$.

By the complete continuity of T and $v_{n+1} = Tv_n$, we have that $Tv^* = v^*$, which means that v^* is a fixed point of the operator T.

Let $u_0(t) = \zeta(\log t)^{q-1}$, $u_1 = Tu_0$, $u_2 = T^2u_0 = Tu_1$. For $u_0(t) \in P_\zeta$ and $T : P_\zeta \to P_\zeta$, it is clear that $u_1 \in T(P_\zeta) \subset P_\zeta$ and $u_2 \in T(P_\zeta) \subset P_\zeta$. By (H_3), we obtain

$$\begin{aligned} u_1(t) &= \int_1^{+\infty} G(t,s)\sigma(s)f(s,u_0(s))\frac{ds}{s} \\ &\le \int_1^{+\infty} A(\log t)^{q-1}\sigma(s)\frac{ds}{s} \cdot \frac{\zeta}{A\int_1^{+\infty}\sigma(s)\frac{ds}{s}} \\ &= \zeta(\log t)^{q-1} = u_0(t). \end{aligned} \tag{6.61}$$

Since T is non-decreasing, we can obtain that $u_2 = Tu_1 \le Tu_0 = u_1$.

We define $u_{n+1} = Tu_n, (n = 0, 1, 2, \ldots)$ by induction. Clearly the sequence $\{u_n\}_{n=1}^{\infty} \subset T(P_\zeta) \subset P_\zeta$ and satisfies

$$u_{n+1}(t) \le u_n(t), \quad n = 0, 1, 2, \ldots. \tag{6.62}$$

Therefore, it follows that $\lim_{n\to\infty} u_n = u^*$. By the complete continuity of T and $u_{n+1} = Tu_n$, we have $Tu^* = u^*$, which means that u^* is also a fixed point of the operator T. In view of the non-decreasing property of the operator T and $v_0(t) \le u_0(t), \forall t \in [1,\infty)$, we have

$$v_1 = Tv_0 \le Tu_0 = u_1.$$

By induction, we obtain

$$v_n \le u_n, \ \forall t \in [1,\infty), n = 0, 1, 2, \ldots. \tag{6.63}$$

According to (6.60), (6.62) and (6.63), we get

$$v_0 \le v_1 \le \cdots \le v_n \le \cdots \le u_n \le \cdots \le u_1 \le u_0.$$

This, together with $u^* = \lim_{n\to\infty} u_n$ and $v^* = \lim_{n\to\infty} v_n$, implies that

$$v_0 \le v_1 \le \cdots \le v_n \le \cdots \le v^* \le u^* \le \cdots \le u_n \le \cdots \le u_1 \le u_0.$$

Observe that it follows from $f(t,0) \not\equiv 0$ that $x = 0$ is not a solution of the nonlocal Hadamard fractional boundary value problem (6.44)-(6.45). Thus we deduce that the nonlocal Hadamard fractional boundary value problem (6.44)-(6.45) has twin positive solutions u^*, v^*, which can be obtained by means of the two monotone iterative sequences $\{u_n\}$ and $\{v_n\}$ defined in (6.57). The proof is completed. □

6.3.2 *Unique positive solution*

Here we construct a monotone iterative sequence which converges to a unique positive solution to the nonlocal Hadamard fractional boundary value problem (6.44)-(6.45).

Theorem 6.5. *If the conditions* $(H_1) - (H_4)$ *hold, then the nonlocal Hadamard fractional boundary value problem (6.44)-(6.45) has a unique positive solution* $w^* \in (0, \zeta(\log t)^{q-1}]$. *Moreover, the unique solution* w^* *can be obtained by the iterative sequence*

$$w_{n+1}(t) = \int_1^{+\infty} G(t,s)\sigma(s)f(s, w_n(s))\frac{ds}{s} \tag{6.64}$$

with initial value $w_0(t) = 0$ *or* $w_0(t) = \zeta(\log t)^{q-1}$. *Furthermore, the following error estimate formula holds:*

$$\|w_n - w^*\|_E \leq Z^n \zeta, \tag{6.65}$$

where $Z = AL\int_1^{+\infty} \sigma(s)(1 + (\log s)^{q-1})\frac{ds}{s} < 1$.

Proof. By Theorem 6.4, we know that the nonlocal Hadamard fractional boundary value problem (6.44)-(6.45) has twin positive solutions u^* and v^*, which can be achieved by $u_{n+1} = Tu_n$ and $v_{n+1} = Tv_n$ $(n = 0, 1, 2, \ldots)$ with initial values $u_0(t) = \zeta(\log t)^{q-1}$, $v_0(t) = 0$, respectively. In view of (H_4) and (6.56), we have

$$\begin{aligned} 0 \leq (u_n - v_n)(t) &= Tu_{n-1}(t) - Tv_{n-1}(t) \\ &= \int_1^{+\infty} G(t,s)\sigma(s)[f(s, u_{n-1}(s)) - f(s, v_{n-1}(s))]\frac{ds}{s} \\ &\leq L\int_1^{+\infty} G(t,s)\sigma(s)[u_{n-1} - v_{n-1}](s)\frac{ds}{s} \\ &\leq L\int_1^{+\infty} G(t,s)\sigma(s)(1 + (\log s)^{q-1})\frac{ds}{s}\|u_{n-1} - v_{n-1}\|_E. \end{aligned} \tag{6.66}$$

Hence, we have

$$\|u_n - v_n\|_E \leq AL\int_1^{+\infty} \sigma(s)(1 + (\log s)^{q-1})\frac{ds}{s}\|u_{n-1} - v_{n-1}\|_E = Z\|u_{n-1} - v_{n-1}\|_E.$$

By induction, we have

$$\|u_n - v_n\|_E \leq Z^n\|u_0 - v_0\|_E = Z^n\zeta. \tag{6.67}$$

In consequence, for any $m,\ n \in N$, it guarantees that

$$\|v_{n+m} - v_n\|_E \leq \|u_n - v_n\|_E \leq Z^n\zeta,\ \|u_n - u_{n+m}\|_E \leq \|u_n - v_n\|_E \leq Z^n\zeta. \tag{6.68}$$

Let $w \in [v_0, u_0]$ be any solution of the nonlocal Hadamard fractional boundary value problem (6.44)-(6.45). It is clear that $v_0 \leq w(t) \leq u_0$ and $Tw = w$. Owing to the non-decreasing property of the operator T and (6.56), it follows that

$$v^* \leq w \leq u^*,\ (n \to \infty).$$

From (6.67), we deduce that

$$\|u^* - v^*\|_E \le Z^n \zeta \to 0 \ (n \to +\infty).$$

This shows that $u^* = v^* \triangleq w^*$. So the nonlocal Hadamard fractional boundary value problem (6.44)-(6.45) has a unique positive solution $w^* \in [v_0, u_0]$. In fact, the sequences $\{u_n\}, \{v_n\}$ correspond to $\{w_n\}$ with $w_0(t) = \zeta(\log t)^{q-1}$ and $w_0(t) = 0$, respectively. In addition, according to (6.68), the error estimate formula (6.65) holds. This completes the proof. □

Example 6.3. Consider the nonlocal Hadamard fractional boundary value problem on unbounded domain:

$$\begin{cases} {}^H D^{\frac{5}{2}} x(t) + \dfrac{5t}{(5+t)^2} f(t, x(t)) = 0, \qquad t \in (1, +\infty), \\ u(1) = u'(1) = 0, \quad {}^H D^{\frac{3}{2}} u(\infty) = \dfrac{1}{2}\, {}^H I^{\frac{1}{4}} x\Big(\dfrac{9}{5}\Big) + \dfrac{1}{3} \sum\limits_{i=1}^{4} \alpha_i x(\eta_i), \end{cases} \tag{6.69}$$

where $q = \frac{5}{2}$, $\beta = \frac{1}{4}$, $\xi = \frac{9}{5}$, $a = \frac{1}{2}$, $b = \frac{1}{3}$, $m = 6$, $\alpha_1 = \frac{1}{5}$, $\alpha_2 = \frac{2}{5}$, $\alpha_3 = \frac{3}{5}$, $\alpha_4 = \frac{4}{5}$, $\eta_1 = \frac{8}{7}$, $\eta_2 = \frac{7}{6}$, $\eta_3 = \frac{6}{5}$, $\eta_4 = \frac{5}{4}$, $\sigma(t) = \dfrac{5t}{(5+t)^2}$ and

$$f(t,x) = \begin{cases} 10e^{-t} + \sqrt{\frac{x}{1+(\log t)^{\frac{3}{2}}}}, & 0 \le x \le e^2, \\ 10e^{-t} + \dfrac{e^2}{\sqrt{1+(\log t)^{\frac{3}{2}}}}, & x \ge e^2. \end{cases}$$

With the given data, we find that

$$\int_1^{+\infty} \sigma(s) \frac{ds}{s} = \frac{5}{6}, \quad \Delta \approx 1.1115 > 0, \quad A \approx 0.8329,$$

which imply that the conditions $(H_1) - (H_2)$ hold. Take $\zeta = 15$. By a simple computation, we have that

$$f(t, (1+(\log t)^{\frac{3}{2}})x) \le 15 < \frac{\zeta}{A \int_1^{\infty} \sigma(s) \frac{ds}{s}} \approx 21.6112, \quad \text{for } (t,x) \in [1,\infty) \times [0,15].$$

Therefore, the condition (H_3) holds. So, by Theorem 6.4, the nonlocal Hadamard fractional boundary value problem (6.69) has twin positive solutions v^*, u^* such that $\|v^*\|_E, \|u^*\|_E \in (0, 15]$, which can be achieved with the aid of monotone iterative sequences.

6.4 Existence of extremal solutions to a nonlinear system with the right-handed Riemann-Liouville fractional derivative operator

In this section, we apply a monotone iterative technique to show the existence of an extremal solution for a nonlinear system of fractional differential equations involving right-handed Riemann-Liouville fractional derivative operators and complemented with nonlocal coupled integral boundary conditions. In precise terms, we consider a nonlinear system of the form:

$$\begin{cases} D_T^q u(t) = f(t, u(t), v(t)), \ t \in [0, T), \ 0 < T < \infty, \\ D_T^q v(t) = g(t, v(t), u(t)), \end{cases} \tag{6.70}$$

subject to nonlocal coupled integral boundary conditions:

$$\begin{cases} (T-t)^{1-q} u(t)|_{t=T} = \displaystyle\int_0^{\tau} W(s)v(s)ds + k_1, 0 < \tau < T, \\ (T-t)^{1-q} v(t)|_{t=T} = \displaystyle\int_0^{\tau} W(s)u(s)ds + k_2, \ k_1 \le k_2, k_1, k_2 \in R, \end{cases} \tag{6.71}$$

where $f, g \in C(J \times R \times R, R)$, $J = [0, T]$, $W(s) \in C(J, (-\infty, 0])$, D_T^q is the right-handed Riemann-Liouville fractional derivative operator of order $0 < q < 1$ defined by

$$D_T^q u(t) = -\frac{1}{\Gamma(1-q)} \frac{d}{dt} \int_t^T (s-t)^{-q} u(s) ds, \quad t \in [0, T),$$

and the right-sided fractional integral I_T^q of order $q > 0$ is given by

$$I_T^q u(t) = \frac{1}{\Gamma(q)} \int_t^T (s-t)^{q-1} u(s) ds, \quad t \in [0, T).$$

In passing, we remark that the nonlocal integral boundary conditions of the type (6.71) may be regarded as a continuous distribution of arbitrary finite length in practical situations.

6.4.1 *Comparison results*

In this subsection, we establish two comparison results which play a key role in the proof of the main result. Let us introduce the following set

$$C_{1-q}([0, T]) = \{u \in C([0, T)); \ (T-t)^{1-q} u \in C([0, T])\}.$$

Lemma 6.12. *(Comparison result) Let $M(t)$ be a non-negative bounded integrable function on J. If $w \in C_{1-q}(J, \mathbb{R})$ satisfies*

$$\begin{cases} D_T^q w(t) \le -M(t) w(t), \\ (T-t)^{1-q} w(t)|_{t=T} \le 0, \end{cases} \tag{6.72}$$

then $w(t) \le 0, \ \forall t \in J$.

Proof. For the sake of contradiction, we assume that $w(t) \le 0$ is not true. Then there exist $t^*, t_* \in [0, T)$ such that $w(t^*) > 0, \quad w(t_*) = 0$ and

$$w(t) \le 0, \quad t \in (t_*, T), \quad w(t) > 0, \quad t \in (t^*, t_*).$$

Since $M(t) \ge 0, \forall t \in J$, therefore, we have $D_T^q w(t) \le 0, \forall t \in (t^*, t_*)$. This, together with $D_T^q w(t) = -\frac{d}{dt} I_T^{1-q} w(t)$, implies that $I_T^{1-q} w(t)$ is non-decreasing on (t^*, t_*). Hence, for any $t \in (t^*, t_*)$, we get

$$I_T^{1-q} w(t) - I_T^{1-q} w(t_*) \le 0.$$

On the other hand,

$$\begin{aligned} I_T^{1-q}w(t) - I_T^{1-q}w(t_*) &= \int_t^T \frac{(s-t)^{-q}}{\Gamma(1-q)} w(s)ds - \int_{t_*}^T \frac{(s-t_*)^{-q}}{\Gamma(1-q)} w(s)ds \\ &= \int_{t_*}^T \frac{(s-t)^{-q} - (s-t_*)^{-q}}{\Gamma(1-q)} w(s)ds + \int_t^{t_*} \frac{(s-t)^{-q}}{\Gamma(1-q)} w(s)ds \\ &> 0, \ \forall t \in (t^*, t_*), \end{aligned}$$

which is a contradiction. Thus, the conclusion of Lemma 6.12 holds. □

Lemma 6.13. *(Comparison theorem) Let $M(t), N(t)$ be non-negative bounded integrable functions on J with $M(t) \geq N(t), \forall t \in J$. Assume that $u, v \in C_{1-q}([0,T])$ satisfy the inequalities*

$$\begin{cases} D_T^q u(t) \leq -M(t)u(t) + N(t)v(t), & t \in [0,T), \\ D_T^q v(t) \leq -M(t)v(t) + N(t)u(t), & t \in [0,T), \\ (T-t)^{1-q}u(t)|_{t=T} \leq 0, \\ (T-t)^{1-q}v(t)|_{t=T} \leq 0. \end{cases} \tag{6.73}$$

Then $u(t) \leq 0, \ v(t) \leq 0, \forall t \in J$.

Proof. Let us set $p(t) = u(t) + v(t), \forall t \in [0,T)$. Then, by (6.73), we have

$$\begin{cases} D_T^q p(t) \leq -(M-N)(t)p(t), & t \in [0,T), \\ (T-t)^{1-q}p(t)|_{t=T} \leq 0. \end{cases} \tag{6.74}$$

Further, it follows by (6.74) and Lemma 6.12 that

$$p(t) \leq 0, \quad \forall t \in J, \tag{6.75}$$

that is, $u(t) + v(t) \leq 0, \ \forall t \in J$.

Next, we show $u(t) \leq 0, \ v(t) \leq 0, \ \forall \ t \in J$. Using (6.73) and (6.75), we find that

$$\begin{cases} D_T^q u(t) \leq -(M+N)(t)u(t), & t \in [0,T), \\ (T-t)^{1-q}u(t)|_{t=T} \leq 0, \end{cases} \tag{6.76}$$

which, in view of (6.76) and Lemma 6.12, yields $u(t) \leq 0, \ \forall \ t \in J$. In a similar manner, it can be shown that $v(t) \leq 0, \ \forall \ t \in J$. □

6.4.2 *Linear fractional differential system*

In the sequel, we need the following lemma which deals with the following linear fractional problem:

$$\begin{cases} D_T^q u(t) = \sigma_1(t) - M(t)u(t) - N(t)v(t), & t \in [0,T), \\ D_T^q v(t) = \sigma_2(t) - M(t)v(t) - N(t)u(t), & t \in [0,T), \\ (T-t)^{1-q}u(t)|_{t=T} = r_1, \quad (T-t)^{1-q}v(t)|_{t=T} = r_2, \end{cases} \tag{6.77}$$

where $M(t), N(t)$ are non-negative bounded integrable functions on J, $\sigma_1, \sigma_2 \in C(J, \mathbb{R})$, $r_1, r_2 \in \mathbb{R}$.

Theorem 6.6. *[102] Let $q \in (0,1), f \in C(J \times \mathbb{R}, \mathbb{R})$. In addition, we assume that there exists a non-negative constant K such that*

$$|f(t,u_1) - f(t,v_1)| \leq K|v_1 - u_1|.$$

Then the problem

$$(D_T^q x)(t) = f(t, x(t)), \quad \widetilde{x}(t) = k \tag{6.78}$$

has a unique solution.

Lemma 6.14. *The linear fractional differential system (6.77) has a unique solution in $C_{1-q}([0,T]) \times C_{1-q}([0,T])$.*

Proof. Let

$$u(t) = \frac{p(t)+q(t)}{2} \quad \text{and} \quad v(t) = \frac{p(t)-q(t)}{2} \quad \forall t \in [0,T). \tag{6.79}$$

Using (6.77), we have

$$\begin{cases} D_T^q p(t) = (\sigma_1 + \sigma_2)(t) - (M+N)(t)p(t), & t \in [0,T), \\ (T-t)^{1-q} p(t)|_{t=T} = r_1 + r_2 \end{cases} \tag{6.80}$$

and

$$\begin{cases} D_T^q q(t) = (\sigma_1 - \sigma_2)(t) - (M-N)q(t), & t \in [0,T), \\ (T-t)^{1-q} q(t)|_{t=T} = r_1 - r_2. \end{cases} \tag{6.81}$$

According to Theorem 6.6, the problem (6.80) and (6.81) have a unique solution in $C_{1-q}([0,T])$. Hence, the problem (6.77) has a unique solution in $C_{1-q}([0,T]) \times C_{1-q}([0,T])$. □

6.4.3 *Nonlinear fractional differential system*

In this subsection we present our main result which ensures the existence of an extremal solution for the nonlinear problem (6.70)-(6.71).

Theorem 6.7. *Assume that*

(6.7.1) There exist $u_0, v_0 \in C_{1-q}([0,T])$ such that $u_0(t) \le v_0(t)$ and

$$\begin{cases} D_T^q u_0(t) \le f(t, u_0(t), v_0(t)), & t \in [0,T), \\ (T-t)^{1-q} u_0(t)|_{t=T} \le \int_0^T W(s) v_0(s) ds + k_1, \\ D_T^q v_0(t) \ge g(t, v_0(t), u_0(t)), & t \in [0,T), \\ (T-t)^{1-q} v_0(t)|_{t=T} \ge \int_0^T W(s) u_0(s) ds + k_2. \end{cases}$$

(6.7.2) There exist non-negative bounded integrable functions $M(t), N(t)$ with $M(t) \ge N(t)$ such that

$$\begin{cases} f(t, \overline{u}(t), \overline{v}(t)) - f(t, u(t), v(t)) \le M(t)(u - \overline{u})(t) + N(t)(v - \overline{v})(t), \\ g(t, \overline{u}(t), \overline{v}(t)) - g(t, u(t), v(t)) \le M(t)(u - \overline{u})(t) + N(t)(v - \overline{v})(t), \end{cases}$$

where $u_0(t) \le \overline{u} \le u \le v_0(t)$, $u_0(t) \le v \le \overline{v} \le v_0(t)$. and

$$f(t, u(t), v(t)) - g(t, v(t), u(t)) \le M(t)(v-u)(t) + N(t)(u-v)(t)$$

for $u_0(t) \le u(t) \le v(t) \le v_0(t)$.

Then the problem (6.70)-(6.71) has an extremal solution $(u^*, v^*) \in [u_0, v_0] \times [u_0, v_0]$. *Moreover, there exist monotone iterative sequences* $\{u_n\}, \{v_n\} \subset [u_0, v_0]$ *such that* $u_n \to u^*$, $v_n \to v^*$ *uniformly on* $t \in [0, T)$ *as* $n \to \infty$ *and*

$$u_0 \le u_1 \le \cdots \le u_n \le \cdots \le u^* \le v^* \le \cdots \le v_n \le \cdots \le v_1 \le v_0. \tag{6.82}$$

Proof. For any $u_{n-1}, v_{n-1} \in C_{1-q}([0,T])$, $n \ge 1$, we define

$$\begin{cases} \sigma_n^1(t) = f(t, u_{n-1}(t), v_{n-1}(t)) + M(t)u_{n-1}(t) + N(t)v_{n-1}(t), \\ \sigma_n^2(t) = g(t, v_{n-1}(t), u_{n-1}(t)) + M(t)v_{n-1}(t) + N(t)u_{n-1}(t), \\ r_n^1 = (T-t)^{1-q} \displaystyle\int_0^T W(s)v_{n-1}(s)ds + k_1, \\ r_n^2 = (T-t)^{1-q} \int_0^T W(s)u_{n-1}(s)ds + k_2, \end{cases}$$

and consider (6.77) as follows

$$\begin{cases} D_T^q u_n(t) = \sigma_n^1(t) - M(t)u_n(t) - N(t)v_n(t), \quad t \in [0,T), \\ D_T^q v_n(t) = \sigma_n^2(t) - M(t)v_n(t) - N(t)u_n(t), \quad t \in [0,T), \\ (T-t)^{1-q}u_n(t)|_{t=T} = r_n^1, \quad (T-t)^{1-q}v_n(t)|_{t=T} = r_n^2. \end{cases} \tag{6.83}$$

In view of Lemma 6.14, the problem (6.83) has a unique solution in $C_{1-q}([0,T]) \times C_{1-q}([0,T])$.

Next we show that $\{u_n(t)\}, \{v_n(t)\}$ satisfy the relation:

$$u_{n-1} \le u_n \le v_n \le v_{n-1}, \quad n = 1, 2, \ldots. \tag{6.84}$$

Let $x = u_0 - u_1, y = v_1 - v_0$. Then, by the condition (6.7.1), it follows from (6.83) that

$$\begin{cases} D_T^q x(t) \le -M(t)x(t) + N(t)y(t), & D_T^q y(t) \le -M(t)y(t) + N(t)x(t), \\ (T-t)^{1-q}x(t)|_{t=T} \le 0, & (T-t)^{1-q}y(t)|_{t=T} \le 0. \end{cases}$$

Thus, in view of Lemma 6.13, we have that $x(t) \le 0$, $y(t) \le 0$, $\forall t \in J$.

Letting $z = u_1 - v_1$, and using the condition (6.7.2) and (6.83), we obtain

$$\begin{aligned} &D_T^q z(t) \\ &= D_T^q u_1(t) - D_T^q v_1(t) \\ &= f(t, u_0(t), v_0(t)) + M(t)u_0(t) + N(t)v_0(t) - M(t)u_1(t) - N(t)v_1(t) \\ &\quad - g(t, v_0(t), u_0(t)) - M(t)v_0(t) - N(t)u_0(t) + M(t)v_1(t) + N(t)u_1(t) \\ &\le M(t)(v_0 - u_0)(t) + N(t)(u_0 - v_0)(t) + M(t)u_0(t) + N(t)v_0(t) - M(t)u_1(t) \\ &\quad - N(t)v_1(t) - M(t)v_0(t) - N(t)u_0(t) + M(t)v_1(t) + N(t)u_1(t) \\ &= -(M-N)(t)z(t). \end{aligned}$$

Also, $(T-t)^{1-q}z(t)|_{t=T} = \displaystyle\int_0^T W(s)(u_0 - v_0)(s)ds + k_1 - k_2 \le 0$. Thus, Lemma 6.12 applies and yields $z(t) \le 0$, $\forall t \in J$. Therefore, $u_0 \le u_1 \le v_1 \le v_0$ holds.

Next, we suppose that $u_{k-1} \le u_k \le v_k \le v_{k-1}$ holds for some $k \ge 1$. Then, using the earlier arguments, one can show that $u_k \le u_{k+1} \le v_{k+1} \le v_k$. Also our supposition $(u_k \le v_k)$ implies that $r_{k+1}^1 - r_{k+1}^2 \le 0$. Thus, by induction, it follows that

$$u_0 \le u_1 \le \cdots \le u_n \le \cdots \le v_n \le \cdots \le v_1 \le v_0, \quad r_n^1 - r_n^2 \le 0. \tag{6.85}$$

We note that $\{u_n\}$ is monotone non-decreasing and bounded from above while $\{v_n\}$ is monotone non-increasing and bounded from below. Thus we have

$$\lim_{n\to\infty} u_n(t) = u^*(t), \qquad \lim_{n\to\infty} v_n(t) = v^*(t)$$

uniformly on compact subsets of $[0, T)$, and the limit functions u^*, v^* respectively satisfy (6.70) and (6.71). Moreover, $u^*, v^* \in [u_0, v_0]$. Taking the limits $n \to \infty$ in (6.83), we find that (u^*, v^*) is a solution of the problem (6.70)-(6.71) in $[u_0, v_0] \times [u_0, v_0]$ satisfying (6.82).

Finally, we show that (u^*, v^*) is an extremal solution of the system (6.70) and (6.71). Assume that $(u, v) \in [u_0, v_0] \times [u_0, v_0]$ is any solution for the problem (6.70)-(6.71), that is,

$$\begin{cases} D_T^q u(t) = f(t, u(t), v(t)), & t \in [0, T), \\ D_T^q v(t) = g(t, v(t), u(t)), & t \in [0, T), \\ (T-t)^{1-q} u(t)|_{t=T} = \displaystyle\int_0^\tau W(s)v(s)ds + k_1, \\ (T-t)^{1-q} v(t)|_{t=T} = \displaystyle\int_0^\tau W(s)u(s)ds + k_2. \end{cases} \tag{6.86}$$

Applying (6.83), (6.86), (6.7.2) and Lemma 6.13, we obtain

$$u_n \le u, \; v \le v_n, \qquad n = 1, 2, \ldots. \tag{6.87}$$

Then, taking the limit $n \to \infty$ in (6.87), we obtain $u^* \le u, v \le v^*$. This shows that (u^*, v^*) is an extremal solution of the problem (6.70) and (6.71) in $[u_0, v_0] \times [u_0, v_0]$. □

Example 6.4. Consider the following nonlinear problem

$$\begin{cases} D_T^q u(t) = \frac{1}{2}\sin^2 t[1 + \frac{1}{\sqrt{1-t}} - u(t)] - te^t v^2(t), & t \in [0, 1), \\ D_T^q v(t) = \frac{1}{2}t^2[1 + \frac{1}{\sqrt{1-t}} - v(t)] - \frac{1}{3}t^2 u^2(t), & t \in [0, 1), \\ (T-t)^{1-q} u(t)|_{t=T} = \int_0^{\frac{1}{2}} \left(-\frac{1}{2\sqrt{1-s}}\right) v(s)ds + \frac{2}{3}, \\ (T-t)^{1-q} v(t)|_{t=T} = \int_0^{\frac{1}{2}} \left(-\frac{1}{2\sqrt{1-s}}\right) u(s)ds + 1, \end{cases} \tag{6.88}$$

where $q = \frac{1}{2}$, $T = 1$, $k_1 = \frac{2}{3}$, $k_2 = 1$, $W(t) = -\frac{1}{2\sqrt{1-t}}$, $\tau = 1/2$ and

$$\begin{cases} f(t, u(t), v(t)) = \frac{1}{2}\sin^2 t[1 + \frac{1}{\sqrt{1-t}} - u(t)] - te^t v^2(t), \\ g(t, v(t), u(t)) = \frac{1}{2}t^2[1 + \frac{1}{\sqrt{1-t}} - v(t)] - \frac{1}{3}t^2 u^2(t). \end{cases}$$

Taking $u_0(t) = 0, v_0(t) = 1 + \frac{1}{\sqrt{1-t}}$, it is easy to show that condition (H_1) of Theorem 6.7 holds. On the other hand, for $M(t) = \frac{1}{2}t^2, N(t) = 0$, we notice that (H_2) holds. In consequence, the nonlinear fractional problem (6.88) has an extremal solution $(u^*, v^*) \in [u_0, v_0] \times [u_0, v_0]$.

6.5 Nonlinear fractional integro-differential equations on unbounded domains in a Banach space

This section is concerned with the existence of a unique solution of a fractional nonlinear integro-differential equation on a semi-infinite interval in a Banach space E. Precisely, we investigate the following problem:

$$\begin{cases} D^{\alpha}u(t) + f(t, u(t), Tu(t), Su(t)) = \theta, & n-1 < \alpha \leq n, \ n \in \mathbb{N}, \ n \geq 2, \\ u(0) = u'(0) = u''(0) = \cdots = u^{(n-2)}(0) = \theta, & D^{\alpha-1}u(\infty) = u_{\infty}, \end{cases} \tag{6.89}$$

where $t \in J = [0, +\infty)$, $f \in C[J \times E \times E \times E, E], u_{\infty} \in E$, D^{α} is the Riemann-Liouville fractional derivative operator, θ refers to the zero in the space E and

$$(Tu)(t) = \int_0^t k(t,s)u(s)ds, \quad (Su)(t) = \int_0^{\infty} h(t,s)u(s)ds,$$

with $k(t,s) \in C[D, \mathbb{R}]$, $h(t,s) \in C[D_0, \mathbb{R}]$, $D = \{(t,s) \in \mathbb{R}^2 \mid 0 \leq s \leq t\}$, $D_0 = \{(t,s) \in J \times J\}$.

Throughout this section, we assume that the following condition holds:

(H) there exists constants k^* and h^* such that

$$k^* = \sup_{t \in J} \int_0^t |k(t,s)|ds < \infty, \qquad h^* = \sup_{t \in J} \frac{1}{(1+t^{\alpha-1})} \int_0^{\infty} |h(t,s)|(1+s^{\alpha-1})ds < \infty$$

and

$$\lim_{t' \to t} \int_0^{\infty} |h(t',s) - h(t,s)|(1+s^{\alpha-1})ds = 0, \quad t, t' \in J.$$

Define

$$FC(J,E) = \left\{ u \in C(J,E) : \sup_{t \in J} \frac{\|u(t)\|}{1+t^{\alpha-1}} < \infty \right\}$$

equipped with norm

$$\|u\|_F = \sup_{t \in J} \frac{\|u(t)\|}{1+t^{\alpha-1}}.$$

Notice that $FC(J,E)$ is a Banach space.

Lemma 6.15. *Assume that*

(6.15.1) there exist non-negative constants a, b, c and a positive function $p(t)$ with $p^* = \int_0^{\infty}(1+t^{\alpha-1})p(t)dt < \infty$ such that

$$\|f(t,u,v,w) - f(t,\overline{u},\overline{v},\overline{w})\| \leq p(t)(a\|u-\overline{u}\| + b\|v-\overline{v}\| + c\|w-\overline{w}\|),$$
$$t \in J, \ u, v, w, \overline{u}, \overline{v}, \overline{w} \in E.$$

Then, for any $u \in FC(J,E)$,

$$\int_0^{\infty} \|f(s, u(s), (Tu)(s), (Su)(s))\|ds \leq p^*(a + bk^* + ch^*)\|u\|_F + \lambda, \tag{6.90}$$

where

$$\lambda = \int_0^{\infty} \|f(t, \theta, \theta, \theta)\|dt < \infty. \tag{6.91}$$

Proof. In view of the condition (6.15.1), we find that

$$\|f(t,u,v,w)\| \leq p(t)(a\|u\| + b\|v\| + c\|w\|) + \|f(t,\theta,\theta,\theta)\|,$$

where $\overline{u} = \overline{v} = \overline{w} = \theta$ (θ refers to the zero in the space E). Then, for any $u \in FC(J,E)$, we have

$$\begin{aligned}
&\|f(t,u(t),(Tu)(t),(Su)(t))\| \\
&\leq p(t)(a\|u(t)\| + b\|Tu(t)\| + c\|Su(t)\|) + \|f(t,\theta,\theta,\theta)\| \\
&\leq p(t)(1+t^{\alpha-1})(a\frac{\|u(t)\|}{1+t^{\alpha-1}} + b\frac{\|(Tu)(t)\|}{1+t^{\alpha-1}} + c\frac{\|(Su)(t)\|}{1+t^{\alpha-1}}) + \|f(t,\theta,\theta,\theta)\| \\
&\leq p(t)(1+t^{\alpha-1})(a\|u\|_F + bk^*\|u\|_F + ch^*\|u\|_F) + \|f(t,\theta,\theta,\theta)\|. \qquad (6.92)
\end{aligned}$$

Integrating both sides of (6.92) and using (H), we get (6.90). □

Lemma 6.16. *Suppose that (6.15.1) holds. Then $u(t) \in FC(J,E)$ with its Riemann-Liouville derivative of order α existing on J is called a solution of problem (6.89) if and only if $u(t) \in FC(J,E)$ is a solution of the integral equation*

$$u(t) = \int_0^\infty G(t,s)f(s,u(s),(Tu)(s),(Su)(s))ds + \frac{u_\infty t^{\alpha-1}}{\Gamma(\alpha)}, \qquad (6.93)$$

where

$$G(t,s) = \frac{1}{\Gamma(\alpha)}\begin{cases} t^{\alpha-1} - (t-s)^{\alpha-1}, & 0 \leq s \leq t, \\ t^{\alpha-1}, & 0 \leq t \leq s. \end{cases} \qquad (6.94)$$

Proof. We only transform (6.89) to the integral equation (6.93) as the converse follows by direct computation. We know that the general solution of the fractional differential equation in (6.89) can be written as

$$\begin{aligned}
u(t) = &-\frac{1}{\Gamma(\alpha)}\int_0^t (t-s)^{\alpha-1}f(s,u(s),(Tu)(s),(Su)(s))ds + c_1t^{\alpha-1} + c_2t^{\alpha-2} \\
&+ \cdots + c_nt^{\alpha-n}, \qquad (6.95)
\end{aligned}$$

where $c_i (i = 1,2,\ldots,n) \in \mathbb{R}$ are arbitrary constants. Clearly, the condition $u(0) = \theta$ implies that $c_n = 0$. By (6.95), we have

$$\begin{aligned}
u'(t) = &-\frac{1}{\Gamma(\alpha-1)}\int_0^t (t-s)^{\alpha-2}f(s,u(s),(Tu)(s),(Su)(s))ds \\
&+(\alpha-1)c_1t^{\alpha-2} + (\alpha-2)c_2t^{\alpha-3} + \cdots + (\alpha-n+1)c_{n-1}t^{\alpha-n}, \qquad (6.96)
\end{aligned}$$

substituting $u'(0) = \theta$ in (6.96) gives $c_{n-1} = 0$. Repeating the above steps with $u''(0) = \cdots = u^{(n-2)}(0) = \theta$, we obtain

$$c_{n-2} = c_{n-3} = \cdots = c_2 = 0.$$

Using the condition $D^{\alpha-1}u(\infty) = u_\infty$ in (6.95), we get

$$D^{\alpha-1}u(\infty) = -\int_0^\infty f(s,u(s),(Tu)(s),(Su)(s))ds + c_1\Gamma(\alpha) = u_\infty, \qquad (6.97)$$

which yields

$$c_1 = \frac{1}{\Gamma(\alpha)}\Big(\int_0^\infty f(s,u(s),(Tu)(s),(Su)(s))ds + u_\infty\Big).$$

Substituting the values of $c_i (i = 1, 2, \ldots, n)$ in (6.95), we obtain

$$\begin{aligned} u(t) &= \frac{u_\infty t^{\alpha-1}}{\Gamma(\alpha)} + \frac{t^{\alpha-1}}{\Gamma(\alpha)}\int_0^\infty f(s,u(s),(Tu)(s),(Su)(s))ds \\ &\quad - \frac{1}{\Gamma(\alpha)}\int_0^t (t-s)^{\alpha-1} f(s,u(s),(Tu)(s),(Su)(s))ds \\ &= \frac{u_\infty t^{\alpha-1}}{\Gamma(\alpha)} + \int_0^\infty G(t,s) f(s,u(s),(Tu)(s),(Su)(s))ds, \end{aligned} \tag{6.98}$$

where $G(t,s)$ is defined by (6.94). □

Remark 6.3. It can easily be shown that the Green's function $G(t,s)$ satisfies the conditions: $G(t,s) \geq 0$ and $\dfrac{G(t,s)}{1+t^{\alpha-1}} < \dfrac{1}{\Gamma(\alpha)}$.

6.5.1 *Main result*

Theorem 6.8. *Let (6.15.1) and the following condition hold:*

$$m = \frac{p^*(a + bk^* + ch^*)}{\Gamma(\alpha)} < 1. \tag{6.99}$$

Then the problem (6.89) has a unique solution $u(t)$ in $FC(J,E)$. Furthermore, there exists a monotone iterative sequence $\{u_i(t)\}$ such that $u_i(t) \to \overline{u}(t)$ as $i \to \infty$ uniformly on any finite sub-interval of J, where

$$u_i(t) = \int_0^\infty G(t,s) f(s,u_{i-1}(s),(Tu_{i-1})(s),(Su_{i-1})(s))ds + \frac{u_\infty t^{\alpha-1}}{\Gamma(\alpha)}. \tag{6.100}$$

In addition, there exists an error estimate for the sequence $\{u_i(t)\}$ of the form:

$$\|u_i - \overline{u}\|_F \leq \frac{m^i}{1-m}\|u_1 - u_0\|_F, \quad i = 1, 2, \ldots. \tag{6.101}$$

Proof. Define an operator $A : FC(J,E) \to FC(J,E)$ by

$$(Au)(t) = \int_0^\infty G(t,s) f(s,u(s),(Tu)(s),(Su)(s))ds + \frac{u_\infty t^{\alpha-1}}{\Gamma(\alpha)}, \tag{6.102}$$

and observe that

$$\begin{aligned} \frac{\|(Au)(t)\|}{1+t^{\alpha-1}} &\leq \int_0^\infty \frac{G(t,s)}{1+t^{\alpha-1}}\|f(s,u(s),(Tu)(s),(Su)(s))\|ds + \frac{\|u_\infty\| t^{\alpha-1}}{\Gamma(\alpha)(1+t^{\alpha-1})} \\ &\leq \frac{1}{\Gamma(\alpha)}[\|u_\infty\| + \lambda + p^*(a + bk^* + ch^*)\|u\|_F] \\ &= l + m\|u\|_F, \end{aligned}$$

that is, $\|Au\|_F \le l + m\|u\|_F$, $\forall u \in FC(J, E)$, where $l = (\|u_\infty\| + \lambda)/\Gamma(\alpha)$ and m is given by (6.99).

On the other hand, for any $u, v \in FC(J, E)$, we have

$$\begin{aligned}
&\frac{\|(Au)(t) - (Av)(t)\|}{1 + t^{\alpha-1}} \\
&\le \int_0^\infty \frac{G(t,s)}{1 + t^{\alpha-1}} \|f(s, u(s), (Tu)(s), (Su)(s)) - f(s, v(s), (Tv)(s), (Sv)(s))\| ds \\
&\le \frac{1}{\Gamma(\alpha)} \int_0^\infty p(s)(1 + s^{\alpha-1})(a\|u - v\|_F + bk^*\|u - v\|_F + ch^*\|u - v\|_F) ds \\
&\le \frac{1}{\Gamma(\alpha)} p^*(a + bk^* + ch^*)\|u - v\|_F.
\end{aligned}$$

In consequence, we get

$$\|Au - Av\|_F \le m\|u - v\|_F, \ \forall u, v \in FC(J, E). \tag{6.103}$$

Since $m < 1$ by (6.99), therefore, it follows by Banach's fixed point theorem that the operator A has a unique fixed point $\overline{u}$ in $FC(J, E)$. Moreover, for any $u \in FC(J, E)$, $\|u_i - \overline{u}\|_F \to 0$ as $i \to \infty$, where $u_i = Au_{i-1}$. Hence, by Lemma 6.16, the unique fixed point $\overline{u}$ is the unique solution of the problem (6.89).

From (6.103), we have

$$\|u_i - u_{i-1}\|_F \le m^{i-1}\|u_1 - u_0\|_F, \tag{6.104}$$

and

$$\begin{aligned}
\|u_i - u_j\|_F &\le \|u_i - u_{i-1}\|_F + \|u_{i-1} - u_{i-2}\|_F + \cdots + \|u_{j+1} - u_j\|_F \\
&\le \frac{m^i(1 - m^{i-j})}{1 - m}\|u_1 - u_0\|_F.
\end{aligned} \tag{6.105}$$

Taking the limit $i \to \infty$, we get

$$\|u_i - \overline{u}\|_F \le \frac{m^i}{1 - m}\|u_1 - u_0\|_F.$$

This completes the proof. □

Example 6.5. Consider the problem

$$\begin{cases}
D^{\frac{7}{2}} u_n(t) + \dfrac{e^{-2t}}{n^2(1 + t^{\frac{5}{2}})^2} \sin(n^2 t + u_n(t)) \\
+ \dfrac{e^{-t}}{(n+1)(1 + t^{\frac{5}{2}})^3} \ln\left[1 + \left(\int_0^t \dfrac{u_{2n}(s)}{(1 + t + s)^2} ds\right)^2\right] \\
+ \dfrac{e^{-3t}}{n(3 + t^{\frac{5}{2}})} \left[\int_0^\infty \dfrac{\cos(t - s)}{(1 + s^2)(1 + s^{\frac{5}{2}})} u_{n+1}(s) ds\right] = \theta, \ t \in J = [0, \infty), \\
u_n(0) = u_n'(0) = u_n''(0) = \theta, \quad D^{\frac{5}{2}} u_n(\infty) = \dfrac{1}{n^3}, \qquad n = 1, 2, 3, \ldots.
\end{cases} \tag{6.106}$$

Let $E = \{u = (u_1, \ldots, u_n, \ldots) : u_n \to 0\}$ with norm $\|u\| = \sup\limits_n |u_n|$. Then (6.106) can be considered as a boundary value problem of form (6.89) in E, where $\theta = \{0, \ldots, 0, \ldots\} \in E$, $u_\infty = \{1, \ldots, \frac{1}{n^3}, \ldots\} \in E$, $k(t,s) = \dfrac{1}{(1+t+s)^2}$, $h(t,s) = \dfrac{\cos(t-s)}{(1+s^2)(1+s^{\frac{5}{2}})}$, $u = (u_1, \ldots, u_n, \ldots)$, $v = (v_1, \ldots, v_n, \ldots)$, $w = (w_1, \ldots, w_n, \ldots)$, $f = (f_1, \ldots, f_n, \ldots)$ with

$$f_n(t,u,v,w) = \frac{e^{-2t}}{n^2(1+t^{\frac{5}{2}})^2}\sin(n^2 t + u_n) + \frac{e^{-t}}{(n+1)(1+t^{\frac{5}{2}})^3}\ln(1+v_n^2) + \frac{e^{-3t}}{n(3+t^{\frac{5}{2}})}w_n.$$

It is clear that $f \in C(J \times E \times E \times E, E)$, and

$$\begin{aligned}
&|f_n(t,u,v,w) - f_n(t,\overline{u},\overline{v},\overline{w})| \\
&\le \frac{e^{-2t}}{n^2(1+t^{\frac{5}{2}})^2}|\sin(n^2t+u_n) - \sin(n^2t+\overline{u}_n)| \\
&\quad + \frac{e^{-t}}{(n+1)(1+t^{\frac{5}{2}})^3}|\ln(1+v_n^2) - \ln(1+\overline{v}_n^2)| + \frac{e^{-3t}}{n(3+t^{\frac{5}{2}})}|w_n - \overline{w}_n| \\
&\le \frac{e^{-2t}}{(1+t^{\frac{5}{2}})^2}|u_n - \overline{u}_n| + \frac{e^{-t}}{2(1+t^{\frac{5}{2}})^3}|v_n - \overline{v}_n| \\
&\quad + \frac{e^{-3t}}{(3+t^{\frac{5}{2}})}|w_n - \overline{w}_n| \\
&\le \frac{e^{-t}}{1+t^{\frac{5}{2}}}\Big[|u_n - \overline{u}_n| + \frac{1}{2}|v_n - \overline{v}_n| + |w_n - \overline{w}_n|\Big],
\end{aligned}$$

which implies that

$$\|f(t,u,v,w) - f(t,\overline{u},\overline{v},\overline{w})\| \le \frac{e^{-t}}{1+t^{\frac{5}{2}}}\Big[\|u-\overline{u}\| + \frac{1}{2}\|v-\overline{v}\| + \|w-\overline{w}\|\Big].$$

Here, $a = c = 1$, $b = \frac{1}{2}$, $p(t) = \dfrac{e^{-t}}{1+t^{\frac{5}{2}}}$.

Obviously the conditions (6.15.1) is satisfied as

$$p^* = \int_0^\infty (1+t^{\frac{5}{2}})p(t)dt = \int_0^\infty e^{-t}dt = 1 < \infty$$

and

$$\lambda = \int_0^\infty \|f(t,\theta,\theta,\theta)\|dt \le \int_0^\infty e^{-2t}dt = 1 < \infty.$$

In addition,

$$k^* = \sup_{t\in J}\int_0^t \frac{1}{(1+t+s)^2}ds = \sup_{t\in J}\Big(\frac{1}{t+1} - \frac{1}{2t+1}\Big)ds \le 1,$$

$$h^* = \sup_{t\in J}\frac{1}{1+t^{\frac{5}{2}}}\int_0^\infty \frac{|\cos(t-s)|}{(1+s^2)(1+s^{\frac{5}{2}})}(1+s^{\frac{5}{2}})ds \le \sup_{t\in J}\frac{1}{1+t^{\frac{5}{2}}}\int_0^\infty \frac{1}{1+s^2}ds = \frac{\pi}{2},$$

and

$$\int_0^\infty |h(t',s) - h(t,s)|(1+s^{\frac{5}{2}})ds \le \int_0^\infty \frac{|t'-t|}{1+s^2}ds = \frac{\pi}{2}|t'-t| \to 0, \text{ as } t' \to t.$$

Finally, by a simple computation, we get

$$m = \frac{p^*(a + bk^* + ch^*)}{\Gamma(\alpha)} = \frac{4(3 + \pi)}{15\sqrt{\pi}} < 0.924006 < 1.$$

Thus all the conditions of Theorem 6.8 are satisfied. Therefore, the conclusion of Theorem 6.8 applies and the problem (6.106) has a unique solution, which can be obtained by constructing iterative sequences.

6.6 Extremal solutions for generalized Caputo fractional differential equations with Steiltjes-type fractional integro-initial conditions

In this section, we introduce a new class of initial value problems consisting of Caputo type generalized fractional differential equations and Steiltjes-type fractional integro-initial conditions. Precisely, we investigate the following problem:

$$\begin{cases} {}^{\rho}_{c}D^{\alpha}_{0+}y(t) = f(t, y(t)), \quad t \in [0, T], \\ y(0) = \lambda I^{\gamma}_{0+;g}y(T) + \kappa := \dfrac{\lambda}{\Gamma(\gamma)} \displaystyle\int_0^T \frac{g'(s)y(s)}{[g(T) - g(s)]^{1-\gamma}} ds + \kappa, \ t^{1-\rho}\frac{d}{dt}y(t)|_{t=0} = 0, \end{cases} \tag{6.107}$$

where ${}^{\rho}_{c}D^{\alpha}_{0+}$ is the Caputo-type generalized fractional derivative operator of order $\alpha \in]1, 2]$, $\rho > 0$, $I^{\gamma}_{0+;g}y(T)$ is the Stieltjes-type fractional integral with respect to a function $g : [0, T] \to \mathbb{R}$, which is monotone increasing on $(0, T]$ with continuous derivative $g'(t)$ existing on $(0, T)$, $f : [0, T] \times \mathbb{R} \to \mathbb{R}$ is a continuous function and $\lambda, \kappa \in \mathbb{R}$.

We apply a monotone iterative technique to obtain the existence of extremal solutions to the problem (6.107). In order to implement the idea of monotone iterates, we need the following result [94].

Lemma 6.17. *Let $[a, b]$ be a non-empty order interval of a subset Y of an ordered Banach space X and let $P : [a, b] \to [a, b]$ be a non-decreasing mapping. If each sequence $\{Py_n\} \subset P([a, b])$ converges whenever $\{y_n\}$ is a monotone sequence in $[a,b]$, then the sequence of P-iteration of a converges to the least fixed point y_* of P and the sequence of P-iteration of b converges to the greatest fixed point y^* of P. Moreover,*

$$y_* = \min\{x \in [a, b] : x \geq Px\}, \quad y^* = \max\{x \in [a, b] : x \leq Px\}.$$

For $c \in \mathbb{R}, 1 \leq p \leq \infty$, let $X^p_c(0, T)$ denote the space of all complex-valued Lebesgue measurable functions ϕ on $(0, T)$ equipped with the norm:

$$\|\phi\|_{X^p_c} = \left(\int_0^T |x^c\phi(x)|^p \frac{dx}{x}\right)^{1/p} < \infty.$$

Let $\mathcal{C} = C([0, T], \mathbb{R})$ denote the Banach space of all continuous functions from $[0, T]$ to $\mathbb{R}$ endowed with the norm defined by $\|y\| = \sup_{t\in[0,T]} |y(t)|$. Further, we introduce a class of functions from $[0, T]$ to $\mathbb{R}$ as follows:

$$\mathcal{E}^n([0, T], \mathbb{R}) = \left\{f : [0, T] \to \mathbb{R} : (t^{1-\rho}\frac{d}{dt})^{n-1} f \in AC([0, T], \mathbb{R})\right\},$$

which is equipped with the norm $\|f\|_{\mathcal{E}^n} = \sum_{k=0}^{n-1} \|(t^{1-\rho}\frac{d}{dt})^k f(t)\|_C$, where

$$(t^{1-\rho}\frac{d}{dt})^k = \underbrace{(t^{1-\rho}\frac{d}{dt})(t^{1-\rho}\frac{d}{dt})\dots\left(t^{1-\rho}\frac{d}{dt}\right)}_{k\ times}.$$

Definition 6.1. [108] The fractional integral of order $\alpha > 0$ of the function $f \in X_c^p(0,T)$ for $0 < t < T < \infty$, with respect to another function g on $[0,T]$, is defined by

$$(I_{0+;g}^{\alpha} f)(t) = \frac{1}{\Gamma(\alpha)} \int_0^t \frac{g'(s)f(s)}{[g(t)-g(s)]^{1-\alpha}} ds. \tag{6.108}$$

where g is a monotone increasing function on $(0,T]$, possessing a continuous derivative on (a,b).

Definition 6.2. A function $y \in C([0,T],\mathbb{R})$ is said to be a solution of (6.107) if y satisfies the equation ${}_c^{\rho}D_{0+}^{\alpha} y(t) = f(t,y(t))$ on $[0,T]$ and the conditions $y(0) = \lambda I_{0+;g}^{\gamma} y(T) + \kappa,\ \ t^{1-\rho}\frac{d}{dt}y(t)|_{t=0} = 0$ (that is, y satisfies (6.107)).

In the following lemma, we solve the linear variant of the problem (6.107).

Lemma 6.18. *Let $h \in C(0,T)$, $y \in \mathcal{E}^2([0,T],\mathbb{R})$ and*

$$\Omega = 1 - \frac{\lambda}{\Gamma(\gamma+1)}[g(T)-g(0)]^{\gamma} \neq 0. \tag{6.109}$$

Then the unique solution of the linear integro-initial value problem:

$$\begin{cases} {}_c^{\rho}D_{0+}^{\alpha} y(t) = h(t), & t \in [0,T], \\ y(0) = \lambda I_{0+;g}^{\gamma} y(T) + \kappa, & t^{1-\rho}\frac{d}{dt}y(t)|_{t=0} = 0, \end{cases} \tag{6.110}$$

is given by

$$y(t) = {}^{\rho}I_{0+}^{\alpha} h(t) + \frac{1}{\Omega}\Big\{\lambda I_{0+;g}^{\gamma}[{}^{\rho}I_{0+}^{\alpha} h(s)](T) + \kappa\Big\}. \tag{6.111}$$

Proof. Applying the operator ${}^{\rho}I_{0+}^{\alpha}$ on both sides of the generalized Caputo fractional differential equation in (6.110), we obtain

$$y(t) = {}^{\rho}I_{0+}^{\alpha} h(t) + c_1 + c_2\frac{t^{\rho}}{\rho} = \frac{\rho^{1-\alpha}}{\Gamma(\alpha)} \int_0^t s^{\rho-1}(t^{\rho}-s^{\rho})^{\alpha-1} h(s) ds + c_1 + c_2\frac{t^{\rho}}{\rho}, \tag{6.112}$$

where $c_1, c_2 \in \mathbb{R}$ are arbitrary constants. Differentiating (6.112) with respect to t and rearranging the terms, we get

$$t^{1-\rho}\frac{d}{dt}y(t) = \frac{\rho^{2-\alpha}}{\Gamma(\alpha-1)} \int_0^t s^{\rho-1}(t^{\rho}-s^{\rho})^{\alpha-2} h(s) ds + c_2. \tag{6.113}$$

Using the boundary condition $t^{1-\rho}\frac{d}{dt}y(t)|_{t=0} = 0$ in (6.113), we get $c_2 = 0$. Applying the operator $I_{0+;g}^{\gamma}$ to (6.112) after inserting the value $c_2 = 0$, we get

$$I_{0+;g}^{\gamma} y(t) = I_{0+;g}^{\gamma}[{}^{\rho}I_{0+}^{\alpha} h(s)](t) + c_1\frac{1}{\Gamma(\gamma+1)}[g(t)-g(0)]^{\gamma}, \tag{6.114}$$

which, together with the condition $y(0) = \lambda I_{0+;g}^{\gamma} y(T) + \kappa$, yields

$$c_1 = \frac{1}{\Omega}\Big\{\lambda I_{0+;g}^{\gamma}[{}^{\rho}I_{0+}^{\alpha} h(s)](T) + \kappa\Big\}.$$

Substituting the values of c_1 and c_2 in (6.112), we get the solution (6.111). The converse follows by direct computation. The proof is completed. □

Next, we establish a comparison result.

Lemma 6.19. *(Comparison result) Let* $\lambda > 0$, $\frac{\lambda}{\Gamma(\gamma+1)}[g(T) - g(0)]^\gamma < 1$ *and* $y \in \mathcal{E}^2 = \mathcal{E}^2([0,T], \mathbb{R})$ *satisfies*

$$\begin{cases} {}^\rho_c D^\alpha_{0+} y(t) \geq 0, \quad t \in [0,T], \\ y(0) \geq \lambda I^\gamma_{0+;g} y(T), \quad t^{1-\rho}\frac{d}{dt}y(t)|_{t=0} = 0. \end{cases} \tag{6.115}$$

Then $y(t) \geq 0$ *for all* $t \in [0,T]$.

Proof. Consider a modified form of problem (6.110) given by

$$\begin{cases} {}^\rho_c D^\alpha_{0+} y(t) = p(t), \quad t \in [0,T], \\ y(0) = \lambda I^\gamma_{0+;g} y(T) + \kappa, \quad t^{1-\rho}\frac{d}{dt}y(t)|_{t=0} = 0, \end{cases} \tag{6.116}$$

where $p \in C([0,T], \mathbb{R}^+)$ and κ, λ are non-negative constants. Then the solution of the problem (6.116) is

$$y(t) = {}^\rho I^\alpha_{0+} p(t) + \frac{1}{\Omega}\Big\{\lambda I^\gamma_{0+;g}[{}^\rho I^\alpha_{0+} p(s)](T) + \kappa\Big\}. \tag{6.117}$$

In view of the non-negative nature of the function p and the constants λ, κ, Ω ($\frac{\lambda}{\Gamma(\gamma+1)}[g(T) - g(0)]^\gamma < 1$ ensures that $\Omega > 0$), the conclusion of Lemma 6.19 follows from (6.117). □

6.6.1 *Main result*

Before proceeding to the main result concerning the existence of extremal solutions for the problem (6.107), we define lower and upper solutions for the problem (6.107).

Definition 6.3. A function $y \in \mathcal{E}^2$ is said to be a lower solution of the problem (6.107) if

$$\begin{cases} {}^\rho_c D^\alpha_{0+} y(t) \leq f(t, y(t)), \quad t \in [0,T], \\ y(0) \leq \lambda I^\gamma_{0+;g} y(T) + \kappa, \quad t^{1-\rho}\frac{d}{dt}y(t)|_{t=0} = 0. \end{cases} \tag{6.118}$$

By reversing the inequalities in the above definition, it corresponds to the one for an upper solution of (6.107).

By Lemma 6.18, we define an operator $F : \mathcal{E}^2 \to \mathcal{E}^2$ associated with the problem (6.107) by

$$F(y)(t) = {}^\rho I^\alpha_{0+} f(t, y(t)) + \frac{1}{\Omega}\Big\{\lambda I^\gamma_{0+;g}[{}^\rho I^\alpha_{0+} f(s, y(s))](T) + \kappa\Big\}. \tag{6.119}$$

Theorem 6.9. *Assume that*

(6.9.1) the functions $f(t,y)$ *is continuous and non-decreasing in* y*;*
(6.9.2) there exist lower and upper solutions y_0 *and* $x_0 \in \mathcal{E}^2$ *for the problem (6.107) respectively such that* $y_0 \leq x_0$*;*
(6.9.3) $\lambda > 0$ *and* $\frac{\lambda}{\Gamma(\gamma+1)}[g(T) - g(0)]^\gamma < 1$.

Then the problem (6.107) has extremal solutions in the sector $[y_0, x_0]$.

Proof. Let us consider the operator F from $[y_0, x_0]$ to $\mathcal{E}^2$, defined by (6.119), such that $y(t) = Fv(t)$. First it will be shown that F maps $[y_0, x_0]$ into $[y_0, x_0]$.

Let $y_1 = Fy_0$, $x_1 = Fx_0$. Then y_1, x_1 are well defined and respectively satisfy the problems

$$\begin{cases} {}^{\rho}_{c}D^{\alpha}_{0+}y_1(t) = f(t, y_0(t)), \quad t \in [0, T], \\ y_1(0) = \lambda I^{\gamma}_{0+;g}y_1(T) + \kappa, \quad t^{1-\rho}\frac{d}{dt}y_1(t)|_{t=0} = 0, \end{cases} \tag{6.120}$$

and

$$\begin{cases} {}^{\rho}_{c}D^{\alpha}_{0+}x_1(t) = f(t, x_0(t)), \quad t \in [0, T], \\ x_1(0) = \lambda I^{\gamma}_{0+;g}x_1(T) + \kappa, \quad t^{1-\rho}\frac{d}{dt}x_1(t)|_{t=0} = 0. \end{cases} \tag{6.121}$$

Setting $u = y_1 - y_0$ and using the definition of lower solution, we get

$$\begin{cases} {}^{\rho}_{c}D^{\alpha}_{0+}u(t) \geq 0, \quad t \in [0, T], \\ u(0) \geq \lambda I^{\gamma}_{0+;g}u(T), \quad t^{1-\rho}\frac{d}{dt}u(t)|_{t=0} = 0, \end{cases} \tag{6.122}$$

which, by Lemma 6.19, implies that $u(t) \geq 0$, $\forall t \in [0, T]$. Thus $Fy_0 \geq y_0$. Similarly, using the definition of upper solution, one can show that $Fx_0 \leq x_0$.

Now we define $\omega = x_1 - y_1$ and use (6.120) and (6.121) together with the assumption (6.9.1) to obtain

$$\begin{cases} {}^{\rho}_{c}D^{\alpha}_{0+}\omega(t) = f(t, x_0(t)) - f(t, y_0(t)) \geq 0, \quad t \in [0, T], \\ \omega(0) = \lambda I^{\gamma}_{0+;g}\omega(T), \quad t^{1-\rho}\frac{d}{dt}\omega(t)|_{t=0} = 0. \end{cases} \tag{6.123}$$

Applying Lemma 6.19, we deduce that $\omega(t) \geq 0$, that is, $Fx_0 \geq Fy_0$. Thus F is non-decreasing and $y_0 \leq Fy \leq x_0$ for any $y \in [y_0, x_0]$. In consequence, $F[y_0, x_0] \subset [y_0, x_0]$ and $\|Fy\| \leq \max\{\|y_0\|, \|x_0\|\} := A$.

Let $\{y_n\}$ be a monotone sequence in $[y_0, x_0]$. Then $y_0 \leq Fy_n \leq x_0$ and $\|Fy_n\| \leq A$. Next we show that the sequence $\{Fy_n\}$ is equi-continuous. For any $(t, y) \in [0, T] \times [-A, A]$, there exists a positive constant K such that $|f(t, y)| \leq K$. Then, for any $t_1, t_2 \in [0, T]$ with $t_1 \leq t_2$, we obtain

$$\begin{aligned} &\|F(y)(t_2) - F(y)(t_1)\| \\ &= \Bigg\| \frac{\rho^{1-\alpha}}{\Gamma(\alpha)} \Bigg[\int_0^{t_1} s^{\rho-1}[(t_2^{\rho} - s^{\rho})^{\alpha-1} - (t_1^{\rho} - s^{\rho})^{\alpha-1}] f(s, y(s))ds \\ &\quad + \int_{t_1}^{t_2} s^{\rho-1}(t_2^{\rho} - s^{\rho})^{\alpha-1} f(s, y(s))ds \Bigg] \Bigg\| \\ &\leq \frac{K}{\rho^{\alpha}\Gamma(\alpha+1)} \Big\{ |t_2^{\rho\alpha} - t_1^{\rho\alpha}| + 2(t_2^{\rho} - t_1^{\rho})^{\alpha} \Big\}, \end{aligned}$$

which tends to zero as $t_2 - t_1 \to 0$ independent of y. Thus, $\{Fy_n\}$ is equi-continuous on all $[0, T]$. So F is relatively compact on $[y_0, x_0]$. Hence, by the Arzelá-Ascoli theorem, F is compact on $[y_0, x_0]$ and consequently $\{Fy_n\}$ converges in $F([y_0, x_0])$. Thus all the hypotheses of Lemma 6.17 hold true and the conclusion of Lemma 6.17 implies that F has least and greatest fixed points in $[y_0, x_0]$. This shows that the problem (6.107) has extremal solutions on $[y_0, x_0]$. □

Remark 6.4. Letting $\gamma \to 1$ in the problem (6.107), the integro-initial conditions become

$$y(0) = \int_0^T g'(s)y(s)ds + \kappa, \quad t^{1-\rho}\frac{d}{dt}y(t)|_{t=0} = 0, \tag{6.124}$$

which is a special case of the following problem when g is monotone increasing function on $[0, T]$:

$$\begin{cases} {}^\rho_c D^\alpha_{0+} y(t) = f(t, y(t)), \quad t \in [0, T], \\ y(0) = \lambda \int_0^T y(s)dg(s) + \kappa, \quad t^{1-\rho}\frac{d}{dt}y(t)|_{t=0} = 0. \end{cases} \tag{6.125}$$

In this case, the operator (6.119) takes the form:

$$F(y)(t) = {}^\rho I^\alpha_{0+} f(t, y(t)) + \frac{1}{\Omega}\Big\{\lambda \int_0^T [{}^\rho I^\alpha_{0+} f(s, y(s))]dg(s) + \kappa\Big\},$$

where

$$\Omega = 1 - \lambda(g(T) - g(0))$$

and the condition (6.9.3) in Theorem 6.9 reduces to

$$\lambda > 0 \quad \text{and} \quad \lambda[g(T) - g(0)] < 1.$$

Remark 6.5. As a special case, we obtain the results for a Liouville-Caputo type fractional differential equation with Steiltjes-type fractional integro-initial conditions by taking $\rho = 1$.

Example 6.6. Consider the following integro-initial value problem

$$\begin{cases} {}^{1/2}_{c} D^{7/5}_{0+} y(t) = \dfrac{t^{1/2}}{20}\Big(\tan^{-1} y + 1/8\Big), \ t \in [0, 1], \\ y(0) = 1/25\ I^{2/7}_{0+;g} y(1) + 1/4, \quad t^{1/2}\frac{d}{dt}y(t)|_{t=0} = 0, \end{cases} \tag{6.126}$$

where $\rho = 1/2$, $\alpha = 7/5$, $\gamma = 2/7$, $\lambda = 1/25$, $\kappa = 1/4$, $T = 1$ and $f(t, y) = \frac{t^{1/2}}{20}(\tan^{-1} y + 1/8)$.

Let us take $g(t) = t^2 + 1$. Clearly g is monotone increasing on $[0, 1]$ and $g'(t) = 2t$ is a continuous function on $(0, 1)$. We take $y_0(t) = 0$ as the lower solution and $x_0(t) = 1 + 4t$ as the upper solution of the problem (6.126). With the given data, it is found that

$$\frac{\lambda}{\Gamma(\gamma + 1)}[g(T) - g(0)]^\gamma \approx 0.044457 < 1.$$

Also, the assumption (6.9.1) is clearly satisfied. Thus, by Theorem 6.9, the problem (6.126) has extremal solutions on $[y_0, x_0]$.

Remark 6.6. The above example can be reproduced for a nonlinear function $f(t, y) = \frac{t^{1/2}}{150}\left(e^y + \frac{1}{8}\right)$, which is not bounded on $[0, 1] \times \mathbb{R}$ and satisfies the assumption (H).

6.7 Notes and remarks

The content of the chapter is based on the papers [64], [151], [152], [145], [149] and [150].

Chapter 7

Nonlocal boundary value problems with generalized fractional integral and derivative

This chapter is devoted to the study of different kinds of boundary value problems involving generalized fractional derivative operators and integrals. We prove several existence results for fractional differential equations and inclusions, and their generalized counterparts, fractional differential equation involving multiple Caputo type fractional derivative operators and mixed nonlinearities, subject to nonlocal generalized Riemann-Liouville integral, Stieltjes, and nonlocal multi-point boundary conditions.

7.1 Fractional differential equations and inclusions with nonlocal generalized Riemann-Liouville integral boundary conditions

In this section, we introduce a new class of boundary value problems of fractional differential equations and inclusions supplemented with nonlocal and generalized Riemann-Liouville fractional integral boundary conditions. In precise terms, we consider the following nonlocal single-valued problem:

$$\begin{cases} {}^cD^{\alpha}x(t) = f(t, x(t)), \quad t \in [0, T], \\ x(0) = g(x), \\ x(T) = \beta \dfrac{\rho^{1-q}}{\Gamma(q)} \displaystyle\int_0^{\xi} \dfrac{s^{\rho-1}}{(\xi^{\rho} - s^{\rho})^{1-q}} x(s)ds := \beta\ {}^{\rho}I^{q}x(\xi), \quad 0 < \xi < T, \end{cases} \tag{7.1}$$

and nonlocal multi-valued problem:

$$\begin{cases} {}^cD^{\alpha}x(t) \in F(t, x(t)), \quad t \in [0, T], \\ x(0) = g(x), \quad x(T) = \beta\ {}^{\rho}I^{q}x(\xi), \quad 0 < \xi < T, \end{cases} \tag{7.2}$$

where ${}^cD^{\alpha}$ is the Caputo fractional derivative operator of order $1 < \alpha \leq 2$, $f : [0, T] \times \mathbb{R} \to \mathbb{R}$ is a continuous function, $g : C([0, T], \mathbb{R}) \to \mathbb{R}$ is a continuous function, ${}^{\rho}I^{q}$ is the generalized Riemann-Liouville fractional integral of order $q > 0$, $\rho > 0$, $\xi \in (0, T)$, and $F : [0, T] \times \mathbb{R} \to \mathcal{P}(\mathbb{R})$ is a multi-valued function, ($\mathcal{P}(\mathbb{R})$ is the family of all non-empty subsets of $\mathbb{R}$).

Lemma 7.1. *For any $y \in C([0,T],\mathbb{R})$, the following linear fractional boundary value problem*

$$\begin{cases} {}^cD^{\alpha}x(t) = y(t), & 1 < \alpha \le 2, \\ x(0) = g(x), & x(T) = \beta\ {}^{\rho}I^{q}x(\xi), \quad 0 < \xi < T, \end{cases} \tag{7.3}$$

is equivalent to fractional integral equation:

$$x(t) = J^{\alpha}y(t) + \frac{t}{\Lambda}\Big\{\beta\ {}^{\rho}I^{q}J^{\alpha}y(\xi) - J^{\alpha}y(T)\Big\} + \Big[1 + \beta\frac{t}{\Lambda}\frac{\xi^{\rho q}}{\rho^{q}}\frac{1}{\Gamma(q+1)}\Big]g(x), \tag{7.4}$$

where

$$\Lambda = T - \beta\frac{\xi^{\rho q+1}}{\rho^{q}}\frac{\Gamma(\frac{1+\rho}{\rho})}{\Gamma(\frac{1+\rho q+\rho}{\rho})} \neq 0. \tag{7.5}$$

Proof. It is well known that the general solution of the fractional differential equation in (7.3) can be written as

$$x(t) = c_0 + c_1 t + J^{\alpha}y(t), \tag{7.6}$$

where $c_0, c_1 \in \mathbb{R}$ are arbitrary constants and J^{α} denotes Riemann-Liouville fractional integral of order α. Using the first condition of the problem (7.3) in (7.6), we get $c_0 = g(x)$. Applying the generalized Riemann-Liouville fractional integral operator ${}^{\rho}I^{q}$ to both sides of (7.6), we obtain

$${}^{\rho}I^{q}x(t) = {}^{\rho}I^{q}J^{\alpha}y(t) + c_0\frac{t^{\rho q}}{\rho^{q}}\frac{1}{\Gamma(q+1)} + c_1\frac{t^{\rho q+1}}{\rho^{q}}\frac{\Gamma(\frac{1+\rho}{\rho})}{\Gamma(\frac{1+\rho q+\rho}{\rho})}, \tag{7.7}$$

which together with the second condition of (7.3) yields

$$J^{\alpha}y(T) + c_1T + c_0 = \beta\ {}^{\rho}I^{q}J^{\alpha}y(\xi) + \beta c_0\frac{\xi^{\rho q}}{\rho^{q}}\frac{1}{\Gamma(q+1)} + \beta c_1\frac{\xi^{\rho q+1}}{\rho^{q}}\frac{\Gamma(\frac{1+\rho}{\rho})}{\Gamma(\frac{1+\rho q+\rho}{\rho})}. \tag{7.8}$$

Using $c_0 = g(x)$ in (7.8), we find that

$$c_1 = \frac{1}{\Lambda}\left\{\beta\ {}^{\rho}I^{q}J^{\alpha}y(\xi) - J^{\alpha}y(T) + \beta g(x)\frac{\xi^{\rho q}}{\rho^{q}}\frac{1}{\Gamma(q+1)}\right\}.$$

Substituting the values of c_0, c_1 in (7.6), we get (7.4). Conversely, it follows by direct computation that the integral equation (7.4) satisfies the problem (7.3). This completes the proof. □

For computational convenience, we introduce the notations:

$$J^{z}f(s,x(s))(t) = \frac{1}{\Gamma(z)}\int_0^t (t-s)^{z-1}f(s,x(s))ds,$$

$${}^{\rho}I^{z}f(s,x(s))(t) = \frac{\rho^{1-z}}{\Gamma(z)}\int_0^t \frac{s^{\rho-1}f(s,x(s))}{(t^{\rho}-s^{\rho})^{1-z}}ds,$$

where $z > 0$ and $t \in [0,T]$.

7.1.1 *Existence and uniqueness results for problem (7.1)*

We denote by $\mathcal{C} = C([0,T],\mathbb{R})$ the Banach space of all continuous functions from $[0,T] \to \mathbb{R}$ endowed with a topology of uniform convergence with the norm defined by $\|x\| = \sup\{|x(t)| : t \in [0,T]\}$. Also by $L^1([0,T],\mathbb{R})$ we denote the Banach space of measurable functions $x : [0,T] \to \mathbb{R}$ which are Lebesgue integrable and normed by $\|x\|_{L^1} = \int_0^T |x(t)|dt$.

In view of Lemma 7.1, we define an operator $\mathcal{P} : \mathcal{C} \to \mathcal{C}$ associated with problem (7.1) as

$$(\mathcal{P}x)(t) = J^\alpha f(s,x(s))(t) + \frac{t}{\Lambda}\Big\{\beta\ {}^\rho I^q J^\alpha f(s,x(s))(\xi) - J^\alpha f(s,x(s))(T)\Big\}$$
$$+ \Big[1 + \beta\frac{t}{\Lambda}\frac{\xi^{\rho q}}{\rho^q}\frac{1}{\Gamma(q+1)}\Big]g(x), \quad t \in [0,T]. \tag{7.9}$$

Let us define $\mathcal{P}_{1,2} : \mathcal{C} \to \mathcal{C}$ by

$$(\mathcal{P}_1 x)(t) = J^\alpha f(s,x(s))(t) + \frac{t}{\Lambda}\Big\{\beta\ {}^\rho I^q J^\alpha f(s,x(s))(\xi) - J^\alpha f(s,x(s))(T)\Big\}, \tag{7.10}$$

and

$$(\mathcal{P}_2 x)(t) = \Big[1 + \beta\frac{t}{\Lambda}\frac{\xi^{\rho q}}{\rho^q}\frac{1}{\Gamma(q+1)}\Big]g(x). \tag{7.11}$$

Clearly

$$(\mathcal{P}x)(t) = (\mathcal{P}_1 x)(t) + (\mathcal{P}_2 x)(t), \quad t \in [0,T]. \tag{7.12}$$

In the sequel, we use the notation:

$$p_0 := \frac{T^\alpha}{\Gamma(\alpha+1)}\left(1 + \frac{T}{|\Lambda|}\right) + \frac{T}{|\Lambda|}\frac{|\beta|}{\Gamma(\alpha+1)}\frac{\xi^{\alpha+\rho q}}{\rho^q}\frac{\Gamma(\frac{\alpha+\rho}{\rho})}{\Gamma(\frac{\alpha+\rho q+\rho}{\rho})}, \tag{7.13}$$

and

$$k_0 := 1 + \frac{T}{|\Lambda|}|\beta|\frac{\xi^{\rho q}}{\rho^q}\frac{1}{\Gamma(q+1)}. \tag{7.14}$$

Theorem 7.1. *Let $f : [0,T] \times \mathbb{R} \to \mathbb{R}$ be a continuous function. Assume that*

(7.1.1) $|f(t,x) - f(t,y)| \le L\|x - y\|, \forall t \in [0,T],\ L > 0,\ x,y \in \mathbb{R}$;
(7.1.2) $g : C([0,1],\mathbb{R}) \to \mathbb{R}$ *is a continuous function satisfying the condition:*

$$|g(u) - g(v)| \le \ell\|u - v\|,\ \ell < k_0^{-1},\ \ \forall\ u,v \in C([0,1],\mathbb{R}),\ \ell > 0;$$

(7.1.3) $\gamma := Lp_0 + k_0\ell < 1$, *where* p_0 *and* k_0 *are respectively given by (7.13) and (7.14).*

Then the boundary value problem (7.1) has a unique solution on $[0,T]$.

Proof. For $x, y \in \mathcal{C}$ and for each $t \in [0,T]$, from the definition of $\mathcal{P}$ and assumptions (7.1.1) and (7.1.2), we obtain

$$
\begin{aligned}
&|(\mathcal{P}x)(t) - (\mathcal{P}y)(t)| \\
&\le \sup_{t\in[0,T]} \Bigg\{ J^{\alpha}|f(s,x(s)) - f(s,y(s))|(t) + \frac{t}{|\Lambda|} J^{\alpha}|f(s,x(s)) - f(s,y(s))|(T) \\
&\quad + \frac{|\beta| t}{|\Lambda|}\, {}^{\rho}I^{q} J^{\alpha}|f(s,x(s)) - f(s,y(s))|(\xi) \\
&\quad + \left|1 + \beta \frac{t}{\Lambda} \frac{\xi^{\rho q}}{\rho^{q}} \frac{1}{\Gamma(q+1)}\right| |g(x) - g(y)| \Bigg\} \\
&\le L\|x-y\| J^{\alpha}(1)(T) + L\|x-y\| \frac{T}{|\Lambda|} J^{\alpha}(1)(T) + L\|x-y\| \frac{|\beta| T}{|\Lambda|}\, {}^{\rho}I^{q} J^{\alpha}(1)(\xi) \\
&\quad + \left(1 + |\beta| \frac{T}{|\Lambda|} \frac{\xi^{\rho q}}{\rho^{q}} \frac{1}{\Gamma(q+1)}\right) |g(x) - g(y)| \\
&\le L \left\{ \frac{T^{\alpha}}{\Gamma(\alpha+1)} + \frac{T}{|\Lambda|} \frac{T^{\alpha}}{\Gamma(\alpha+1)} + \frac{T}{|\Lambda|} \frac{|\beta|}{\Gamma(\alpha+1)} \frac{\xi^{\alpha+\rho q}}{\rho^{q}} \frac{\Gamma(\frac{\alpha+\rho}{\rho})}{\Gamma(\frac{\alpha+\rho q+\rho}{\rho})} \right\} \|x-y\| \\
&\quad + \left(1 + |\beta| \frac{T}{|\Lambda|} \frac{\xi^{\rho q}}{\rho^{q}} \frac{1}{\Gamma(q+1)}\right) \ell \|x-y\| \\
&= (Lp_0 + k_0 \ell)\|x-y\|.
\end{aligned}
$$

Hence

$$
\|(\mathcal{P}x) - (\mathcal{P}y)\| \le \gamma \|x-y\|.
$$

As $\gamma < 1$ by (7.1.1), the operator $\mathcal{P}$ is a contraction map from the Banach space $\mathcal{C}$ into itself. Hence the conclusion of the theorem follows by the contraction mapping principle (Banach fixed point theorem). □

Example 7.1. Consider the following fractional boundary value problem

$$
\begin{cases}
{}^{c}D^{3/2}x(t) = \dfrac{\sin^2(\pi t)}{2(e^t+9)}\left(\dfrac{|x(t)|}{|x(t)|+1} + 1\right)|x(t)| + \dfrac{\sqrt{3}}{4}, \quad t \in [0,1], \\[2ex]
x(0) = \dfrac{1}{2} + \dfrac{1}{12}\tan^{-1}(x(1/8)), \quad x(1) = \dfrac{1}{2}\,{}^{2/3}I^{3/2}x(3/4).
\end{cases}
\tag{7.15}
$$

Here, $\alpha = 3/2$, $T = 1$, $\beta = 1/2$, $\xi = 3/4$, $\rho = 2/3$, $q = 3/2$,

$$
f(t,x) = \frac{\sin^2(\pi t)}{2(e^t+9)}\Big(\frac{|x|}{|x|+1} + 1\Big)|x| + \frac{\sqrt{3}}{4}, \quad g(x) = (1/12)\tan^{-1}(x(1/8)).
$$

Since $|f(t,x) - f(t,y)| \le \frac{1}{10}\|x-y\|$, $|g(x) - g(y)| \le \frac{1}{12}\|x-y\|$, therefore, $L = 1/10$ and $\ell = 1/12$. Using the given values, it is found that $\Lambda \approx 0.8851733$, $p_0 \approx 1.6599468$, $k_0 \approx 1.563258$ and $\gamma = Lp_0 + k_0\ell \approx 0.2962661 < 1$. Clearly all the conditions of Theorem 7.1 hold true. Therefore, by the conclusion of Theorem 7.1, the boundary value problem (7.15) has a solution on $[0, 1]$.

Our second existence result is based on Sadovskii's fixed point theorem (Theorem 1.14).

Theorem 7.2. *Let $f : [0, T] \times \mathbb{R} \to \mathbb{R}$ be a continuous function and condition (7.1.2) holds. In addition we assume that:*

(7.2.1) $g(0) = 0$;

(7.2.2) there exists a non-negative function $p \in C([0, T], \mathbb{R})$ and a non-decreasing function $\psi : [0, \infty) \to (0, \infty)$ such that

$$|f(t, u)| \leq p(t)\psi(\|u\|) \quad \textit{for any} \quad (t, u) \in [0, T] \times \mathbb{R}.$$

Then the problem (7.2) has at least one solution on $[0, T]$.

Proof. Let $B_r = \{x \in \mathcal{C} : \|x\| \leq r\}$ be a closed bounded and convex subset of $\mathcal{C}$, where r will be fixed later. We define a map $\mathcal{P} : B_r \to \mathcal{C}$ as

$$(\mathcal{P}x)(t) = (\mathcal{P}_1 x)(t) + (\mathcal{P}_2 x)(t), \quad t \in [0, T],$$

where $\mathcal{P}_1$ and $\mathcal{P}_2$ are defined by (7.10) and (7.11) respectively. Notice that the problem (7.2) is equivalent to a fixed point problem $\mathcal{P}(x) = x$. Now we verify the hypothesis of Sadovskii's fixed point (Theorem 1.14) in several steps.

Step 1. $(\mathcal{P}x)B_r \subset B_r$.

Let us select $r \geq \dfrac{\psi(r)\|p\|p_0}{1 - \ell k_0}$, where p_0 and k_0 are defined by (7.13) and (7.14) respectively. For any $x \in B_r$, we have

$$\begin{aligned}
\|\mathcal{P}x\| &\leq J^\alpha |f(s, x(s))|(T) + \frac{T}{|\Lambda|}\Big\{|\beta| \ {}^\rho I^q J^\alpha |f(s, x(s))|(\xi) + J^\alpha |f(s, x(s))|(T)\Big\} \\
&\quad + \left(1 + |\beta| \frac{T}{|\Lambda|} \frac{\xi^{\rho q}}{\rho^q} \frac{1}{\Gamma(q+1)}\right)|g(x) - g(y)| \\
&\leq \psi(\|x\|) J^\alpha p(s)(T) + \psi(\|x\|) \frac{|\beta| T}{|\Lambda|} \ {}^\rho I^q J^\alpha p(s)(\xi) + \psi(\|x\|) \frac{T}{|\Lambda|} J^\alpha p(s)(T) \\
&\quad + \left(1 + |\beta| \frac{T}{|\Lambda|} \frac{\xi^{\rho q}}{\rho^q} \frac{1}{\Gamma(q+1)}\right)\ell \|x\| \\
&\leq \|p\|\psi(r)\left\{\frac{T^\alpha}{\Gamma(\alpha+1)} + \frac{T}{|\Lambda|}\frac{T^\alpha}{\Gamma(\alpha+1)} + \frac{T}{|\Lambda|}\frac{|\beta|}{\Gamma(\alpha+1)}\frac{\xi^{\alpha+\rho q}}{\rho^q}\frac{\Gamma(\frac{\alpha+\rho}{\rho})}{\Gamma(\frac{\alpha+\rho q+\rho}{\rho})}\right\} \\
&\quad + \left(1 + |\beta| \frac{T}{|\Lambda|} \frac{\xi^{\rho q}}{\rho^q} \frac{1}{\Gamma(q+1)}\right)\ell r \\
&= \psi(r)\|p\|p_0 + k_0 \ell r < r,
\end{aligned}$$

which implies that $(\mathcal{P}x)(B_r) \subset B_r$.

Step 2. $\mathcal{P}_1$ is compact.

Observe that the operator $\mathcal{P}_1$ is uniformly bounded in view of Step 1. Let $\tau_1, \tau_2 \in [0, T]$ with $\tau_1 < \tau_2$ and $x \in B_r$. Then we obtain

$$\begin{aligned}
&|(\mathcal{P}_1 x)(\tau_2) - (\mathcal{P}_1 x)(\tau_1)| \\
&\le |J^\alpha f(s, x(s))(\tau_2) - J^\alpha f(s, x(s))(\tau_1)| + \frac{|\tau_2 - \tau_1|}{|\Lambda|} J^\alpha |f(s, x(s))|(T) \\
&\quad + \frac{|\beta||\tau_2 - \tau_1|}{|\Lambda|}\ {}^\rho I^q J^\alpha |f(s, x(s))|(\xi) \\
&\le \frac{\psi(r)}{\Gamma(\alpha)} \left| \int_0^{\tau_1} [(\tau_2 - s)^{\alpha-1} - (\tau_1 - s)^{\alpha-1}] p(s) ds + \int_{\tau_1}^{\tau_2} (\tau_2 - s)^{\alpha-1} p(s) ds \right| \\
&\quad + \frac{\psi(r)|\tau_2 - \tau_1|}{|\Lambda|} \left(J^\alpha p(s)(T) + |\beta|\ {}^\rho I^q J^\alpha p(s)(\xi) \right),
\end{aligned}$$

which is independent of x and tends to zero as $\tau_2 - \tau_1 \to 0$. Thus, $\mathcal{P}_1$ is equicontinuous. Hence, by the Arzelá-Ascoli Theorem, $\mathcal{P}_1(B_r)$ is a relatively compact set.

Step 3. $\mathcal{P}_2$ is continuous and γ-contractive.

To show the continuity of $\mathcal{P}_2$ for $t \in [0, T]$, let us consider a sequence x_n converging to x. Then, we have

$$\begin{aligned}
\|\mathcal{P}_2 x_n - \mathcal{P}_2 x\| &\le \left(1 + |\beta| \frac{T}{|\Lambda|} \frac{\xi^{\rho q}}{\rho^q} \frac{1}{\Gamma(q+1)}\right) |g(x_n) - g(x)| \\
&\le \left(1 + |\beta| \frac{T}{|\Lambda|} \frac{\xi^{\rho q}}{\rho^q} \frac{1}{\Gamma(q+1)}\right) \ell \|x_n - x\|,
\end{aligned}$$

which, in view of (7.1.2), implies that $\mathcal{P}_2$ is continuous. Also $\mathcal{P}_2$ is γ-contractive, since

$$\gamma = \left(1 + |\beta| \frac{T}{|\Lambda|} \frac{\xi^{\rho q}}{\rho^q} \frac{1}{\Gamma(q+1)}\right) \ell = k_0 \ell < 1.$$

Step 4. $\mathcal{P}$ is condensing.

Since $\mathcal{P}_2$ is continuous, γ-contraction and $\mathcal{P}_1$ is compact, therefore, by Theorem 1.13, $\mathcal{P} : B_r \to B_r$ with $\mathcal{P} = \mathcal{P}_1 + \mathcal{P}_2$ is a condensing map on B_r.

From the above four steps, we conclude by Theorem 1.14 that the map $\mathcal{P}$ has a fixed point which, in turn, implies that the problem (7.2) has a solution on $[0, T]$. □

Example 7.2. Consider the following boundary value problem

$$\begin{cases} {}^c D^{3/2} x(t) = \dfrac{e^{-2t}}{\pi\sqrt{9 + t^2}} \left(x \tan^{-1} x + \pi/2\right),\ 0 < t < 1, \\ x(0) = \dfrac{1}{4}\left(1 - \cos x\right), \quad x(1) = \dfrac{1}{2}\, {}^{2/3} I^{3/2} x(3/4). \end{cases} \tag{7.16}$$

Observe that $|f(t,x)| \leq p(t)\psi(|x|)$ with $p(t) = \dfrac{e^{-2t}}{2\sqrt{9+t^2}}$, $\psi(|x|) = 1 + |x|$, and $g(0) = 0$, $\ell = 1/4$ as $|g(u) - g(v)| \leq (1/4)|u - v|$. Thus, all the conditions of Theorem 7.2 are satisfied and hence by its conclusion, the problem (7.16) has at least one solution on $[0, 1]$.

Our next result relies on the following fixed point theorem due to Burton and Kirk (Theorem 1.6).

Theorem 7.3. *Assume that $f, g : [0,T] \times \mathbb{R} \to \mathbb{R}$ are continuous functions and conditions (7.1.2) and (7.2.1) hold. In addition we suppose that:*

(7.3.1) there exists a function $p \in L^1(J, \mathbb{R}_+)$ such that

$$|f(t,u)| \leq p(t), \quad \textit{for a.e. } t \in J, \ \textit{and each } u \in \mathbb{R}.$$

Then the boundary value problem (7.1) has at least one solution on $[0,T]$.

Proof. To transform the problem (7.1) into a fixed point problem, we consider the map $\mathcal{P} : \mathcal{C} \to \mathcal{C}$ given by

$$(\mathcal{P}x)(t) = (\mathcal{P}_1 x)(t) + (\mathcal{P}_2 x)(t), \quad t \in [0,T],$$

where $\mathcal{P}_1$ and $\mathcal{P}_2$ are defined by (7.10) and (7.11) respectively.

We shall show that the operators $\mathcal{P}_1$ and $\mathcal{P}_2$ satisfy all the conditions of Theorem 1.6.

Step 1. The operator $\mathcal{P}_1$ defined by (7.10) is continuous.

Let $\{x_n\} \subset B_r = \{x \in \mathcal{C} : \|x\| \leq r\}$ with $\|x_n - x\| \to 0$. Then the limit $\|x_n(t) - x(t)\| \to 0$ is uniformly valid on $[0,T]$. From the uniform continuity of $f(t,x)$ on the compact set $[0,T] \times [-r,r]$, it follows that $\|f(t,x_n(t)) - f(t,x(t))\| \to 0$ uniformly on $[0,T]$. Hence $\|\mathcal{P}_1 x_n - \mathcal{P}_1 x\| \to 0$ as $n \to \infty$ which implies that the operator $\mathcal{P}_1$ is continuous.

Step 2. The operator $\mathcal{P}_1$ maps bounded sets into bounded sets in $\mathcal{C}$.

It is indeed enough to show that for any $r > 0$ there exists a positive constant L such that for each $x \in B_r = \{x \in \mathcal{C} : \|x\| \leq r\}$, we have $\|\mathcal{P}_1 x\| \leq L$. Let $x \in B_r$. Then

$$\begin{aligned}
\|\mathcal{P}_1 x\| &\leq J^\alpha |f(s,x(s))|(T) + \frac{T}{|\Lambda|}\Big\{|\beta| \ {}^\rho I^q J^\alpha |f(s,x(s))|(\xi) + J^\alpha |f(s,x(s))|(T)\Big\} \\
&\leq J^\alpha p(s)(T) + \frac{|\beta| T}{|\Lambda|} \ {}^\rho I^q J^\alpha p(s)(\xi) + \frac{T}{|\Lambda|} J^\alpha p(s)(T) := L.
\end{aligned}$$

Step 3. The operator $\mathcal{P}_1$ maps bounded sets into equi-continuous sets in $\mathcal{C}$.

Let $\tau_1, \tau_2 \in [0,T]$ with $\tau_1 < \tau_2$ and $x \in B_r$. Then, for each $x \in B_r$, we obtain

$$|(\mathcal{P}_1 x)(\tau_2) - (\mathcal{P}_1 x)(\tau_1)|$$

$$\begin{aligned}
&\le |J^\alpha f(s,x(s))(\tau_2) - J^\alpha f(s,x(s))(\tau_1)| + \frac{|\tau_2-\tau_1|}{|\Lambda|} J^\alpha |f(s,x(s))|(T) \\
&\quad + \frac{|\beta||\tau_2-\tau_1|}{|\Lambda|} |{}^\rho I^q J^\alpha |f(s,x(s))|(\xi) \\
&\le \frac{1}{\Gamma(\alpha)} \left| \int_0^{\tau_1} [(\tau_2-s)^{\alpha-1} - (\tau_1-s)^{\alpha-1}] p(s)ds + \int_{\tau_1}^{\tau_2} (\tau_2-s)^{\alpha-1} p(s)ds \right| \\
&\quad + \frac{|\tau_2-\tau_1|}{|\Lambda|} \left(J^\alpha p(s)(T) + |\beta| \, {}^\rho I^q J^\alpha p(s)(\xi) \right),
\end{aligned}$$

which is independent of x and tends to zero as $\tau_2 - \tau_1 \to 0$. Thus, $\mathcal{P}_1$ is equicontinuous.

Step 4. The operator $\mathcal{P}_2$ defined by (7.11) is a contraction.

This was established in Step 3 of Theorem 7.2.

Step 5. A priori bounds on solutions.

Now it remains to show that the set $\mathcal{E} = \left\{ u \in \mathcal{C} : \lambda \mathcal{P}_2\left(\frac{u}{\lambda}\right) + \lambda \mathcal{P}_1(u) = u \right\}$ is unbounded for some $\lambda \in (0,1)$.

Let $\lambda \in (0,1)$ and $x \in \mathcal{E}$ be a solution of the integral equation

$$\begin{aligned}
x(t) &= \lambda J^\alpha f(s,x(s))(t) + \lambda \frac{t}{\Lambda} \left\{ \beta \, {}^\rho I^q J^\alpha f(s,x(s))(\xi) - J^\alpha f(s,x(s))(T) \right\} \\
&\quad + \lambda \left[1 + \beta \frac{t}{\Lambda} \frac{\xi^{\rho q}}{\rho^q} \frac{1}{\Gamma(q+1)} \right] g(x), \quad t \in [0,T].
\end{aligned}$$

Then, for each $t \in [0,T]$, we have

$$\begin{aligned}
|x(t)| &\le J^\alpha p(s)(T) + \frac{|\beta| T}{|\Lambda|} \, {}^\rho I^q J^\alpha p(s)(\xi) + \frac{T}{|\Lambda|} J^\alpha p(s)(T) \\
&\quad + \lambda \left[1 + |\beta| \frac{T}{|\Lambda|} \frac{\xi^{\rho q}}{\rho^q} \frac{1}{\Gamma(q+1)} \right] \left[\left| g\left(\frac{x(s)}{\lambda}\right) - g(0) \right| + |g(0)| \right] \\
&\le J^\alpha p(s)(T) + \frac{|\beta| T}{|\Lambda|} \, {}^\rho I^q J^\alpha p(s)(\xi) + \frac{T}{|\Lambda|} J^\alpha p(s)(T) \\
&\quad + \left[1 + |\beta| \frac{T}{|\Lambda|} \frac{\xi^{\rho q}}{\rho^q} \frac{1}{\Gamma(q+1)} \right] \ell \|x\|,
\end{aligned}$$

which, on taking the norm for $t \in [0,T]$ and solving for $\|x\|$, yields

$$\|x\| \le M := \frac{1}{(1 - k_0 \ell)} \left[J^\alpha p(s)(T) + \frac{|\beta| T}{|\Lambda|} \, {}^\rho I^q J^\alpha p(s)(\xi) + \frac{T}{|\Lambda|} J^\alpha p(s)(T) \right].$$

This shows that the set $\mathcal{E}$ is bounded, since $k_0 \ell < 1$. Hence, $\mathcal{P}$ has a fixed point in $[0,T]$ by Theorem 1.6, and consequently the problem (7.1) has a solution. This completes the proof. □

In the last result of this section, we show the existence of solutions for the boundary value problem (7.1) by means of a fixed point theorem due to O'Regan (Theorem 1.9).

In relation to this result, we use the terminology:

$$\Omega_r = \{x \in \mathcal{C} : \|x\| < r\}, \ M_r = \max\{|f(t,x)| : (t,x) \in [0,T] \times [-r,r]\}.$$

Theorem 7.4. *Let $f : [0,T] \times \mathbb{R} \to \mathbb{R}$ be a continuous function and conditions (7.1.1), (7.1.2), (7.2.1) and (7.2.2) hold. In addition we assume that:*

(7.4.1) $\displaystyle\sup_{r\in(0,\infty)} \frac{r}{p_0\psi(r)\|p\|} > \frac{1}{1-k_0\ell}$, *where p_0 and k_0 are defined in (7.13) and (7.14) respectively.*

Then the boundary value problem (7.1) has at least one solution on $[0,T]$.

Proof. By the assumption (7.4.1), there exists a number $r_0 > 0$ such that

$$\frac{r_0}{p_0\psi(r_0)\|p\|} > \frac{1}{1-k_0\ell}. \tag{7.17}$$

We shall show that the operators $\mathcal{P}_1$ and $\mathcal{P}_2$ defined by (7.10) and (7.11) respectively, satisfy all the conditions of Lemma 1.9.

Step 1. *The operator $\mathcal{P}_1$ is continuous and completely continuous.* We first show that $\mathcal{P}_1(\bar{\Omega}_{r_0})$ is bounded. For any $x \in \bar{\Omega}_{r_0}$, we have

$$\begin{aligned}
&\|\mathcal{P}_1 x\| \\
&\le J^\alpha|f(s,x(s))|(T) + \frac{T}{|\Lambda|}\Big\{|\beta| \ {}^\rho I^q J^\alpha|f(s,x(s))|(\xi) + J^\alpha|f(s,x(s))|(T)\Big\} \\
&\le M_r J^\alpha p(s)(T) + M_r\frac{|\beta|T}{|\Lambda|} \ {}^\rho I^q J^\alpha p(s)(\xi) + M_r\frac{T}{|\Lambda|}J^\alpha p(s)(T) \\
&\le \|p\|M_r\left\{\frac{T^\alpha}{\Gamma(\alpha+1)} + \frac{T}{|\Lambda|}\frac{T^\alpha}{\Gamma(\alpha+1)} + \frac{T}{|\Lambda|}\frac{|\beta|}{\Gamma(\alpha+1)}\frac{\xi^{\alpha+\rho q}}{\rho^q}\frac{\Gamma(\frac{\alpha+\rho}{\rho})}{\Gamma(\frac{\alpha+\rho q+\rho}{\rho})}\right\} \\
&= M_r\|p\|p_0.
\end{aligned}$$

Thus the operator $\mathcal{P}_1(\bar{\Omega}_{r_0})$ is uniformly bounded. Let $\tau_1, \tau_2 \in [0,T]$ with $\tau_1 < \tau_2$ and $x \in B_r$. Then

$$\begin{aligned}
&|(\mathcal{P}_1x)(\tau_2) - (\mathcal{P}_1x)(\tau_1)| \\
&\le |J^\alpha f(s,x(s))(\tau_2) - J^\alpha f(s,x(s))(\tau_1)| + \frac{|\tau_2-\tau_1|}{|\Lambda|}J^\alpha|f(s,x(s))|(T) \\
&\quad + \frac{|\beta||\tau_2-\tau_1|}{|\Lambda|}|{}^\rho I^q J^\alpha|f(s,x(s))|(\xi) \\
&\le \frac{M_r}{\Gamma(\alpha)}\left|\int_0^{\tau_1}[(\tau_2-s)^{\alpha-1} - (\tau_1-s)^{\alpha-1}]p(s)ds + \int_{\tau_1}^{\tau_2}(\tau_2-s)^{\alpha-1}p(s)ds\right| \\
&\quad + \frac{M_r|\tau_2-\tau_1|}{|\Lambda|}\left(\frac{T^\alpha}{\Gamma(\alpha+1)} + \frac{|\beta|}{\Gamma(\alpha+1)}\frac{\xi^{\alpha+\rho q}}{\rho^q}\frac{\Gamma(\frac{\alpha+\rho}{\rho})}{\Gamma(\frac{\alpha+\rho q+\rho}{\rho})}\right),
\end{aligned}$$

which is independent of x and tends to zero as $\tau_2 - \tau_1 \to 0$. Thus, $\mathcal{P}_1$ is equicontinuous. Hence, by the Arzelá-Ascoli Theorem, $\mathcal{P}_1(\bar{\Omega}_{r_0})$ is a relatively compact set. Now let $\{x_n\} \subset \bar{\Omega}_{r_0}$ with $\|x_n - x\| \to 0$. Then the limit $\|x_n(t) - x(t)\| \to 0$ is uniformly valid on $[0,T]$. From the uniform continuity of $f(t,x)$ on the compact set $[0,T] \times [-r_0, r_0]$, it follows that $\|f(t,x_n(t)) - f(t,x(t))\| \to 0$ uniformly on $[0,T]$. Hence $\|\mathcal{P}_1 x_n - \mathcal{P}_1 x\| \to 0$ as $n \to \infty$ which proves the continuity of $\mathcal{P}_1$. This completes the proof of Step 1.

Step 2. *The operator $\mathcal{P}_2 : \bar{\Omega}_{r_0} \to C([0,T],\mathbb{R})$ is contractive.* This is a consequence of (7.1.2)

Step 3. *The set $\mathcal{P}(\bar{\Omega}_{r_0})$ is bounded.* The assumptions (7.1.2) and (7.2.1) imply that

$$\|\mathcal{P}_2(x)\| \leq k_0 \ell r_0,$$

for any $x \in \bar{\Omega}_{r_0}$. This together with the boundedness of the set $\mathcal{P}_1(\bar{\Omega}_{r_0})$ implies that the set $\mathcal{P}(\bar{\Omega}_{r_0})$ is bounded.

Step 4. *Finally, it will be shown that the case (C2) in Theorem 1.9 does not hold.* On the contrary, we suppose that (C2) holds. Then, we have that there exist $\lambda \in (0,1)$ and $x \in \partial\Omega_{r_0}$ such that $x = \lambda \mathcal{P} x$. So, we have $\|x\| = r_0$ and

$$\begin{aligned} x(t) &= \lambda J^{\alpha} f(s,x(s))(t) + \lambda \frac{t}{\Lambda}\Big\{\beta\, {}^{\rho}I^{q} J^{\alpha} f(s,x(s))(\xi) - J^{\alpha} f(s,x(s))(T)\Big\} \\ &\quad + \lambda\Big[1 + \beta \frac{t}{\Lambda}\frac{\xi^{\rho q}}{\rho^q}\frac{1}{\Gamma(q+1)}\Big] g(x), \quad t \in [0,T]. \end{aligned}$$

Using our assumptions, we get

$$\begin{aligned} r_0 &\leq \|p\|\psi(r_0)\left\{\frac{T^{\alpha}}{\Gamma(\alpha+1)} + \frac{T}{|\Lambda|}\frac{T^{\alpha}}{\Gamma(\alpha+1)} + \frac{T}{|\Lambda|}\frac{|\beta|}{\Gamma(\alpha+1)}\frac{\xi^{\alpha+\rho q}}{\rho^q}\frac{\Gamma(\frac{\alpha+\rho}{\rho})}{\Gamma(\frac{\alpha+\rho q+\rho}{\rho})}\right\} \\ &\quad + \left(1 + |\beta|\frac{T}{|\Lambda|}\frac{\xi^{\rho q}}{\rho^q}\frac{1}{\Gamma(q+1)}\right)\ell r_0, \end{aligned}$$

which yields

$$r_0 \leq p_0 \psi(r_0)\|p\| + k_0 \ell r_0.$$

Thus we get a contradiction:

$$\frac{r_0}{p_0\psi(r_0)\|p\|} \leq \frac{1}{1 - k_0\ell}.$$

Clearly the operators $\mathcal{P}_1$ and $\mathcal{P}_2$ satisfy all the conditions of Theorem 1.9. Hence, the operator $\mathcal{P}$ has at least one fixed point $x \in \bar{\Omega}_{r_0}$, which is a solution of the problem (7.2). This completes the proof. □

Example 7.3. Consider the following fractional order boundary value problem

$$\begin{cases} {}^{c}D^{3/2}x(t) = \dfrac{e^{-2t}}{2\pi\sqrt{9+t^2}}\Big(x\tan^{-1}x + \pi/2\Big), \ 0 < t < 1, \\ x(0) = \dfrac{1}{4}(1 - \cos x), \quad x(1) = \dfrac{1}{2}{}^{2/3}I^{3/2}x(3/4). \end{cases} \tag{7.18}$$

Observe that $|f(t,x)| \leq p(t)\psi(|x|)$ with $p(t) = \dfrac{e^{-2t}}{4\sqrt{9+t^2}}$, $\psi(|x|) = 1 + |x|$, and $g(0) = 0$, $\ell = 1/4$ as $|g(u) - g(v)| \leq (1/4)|u - v|$. With $\psi(r) = 1 + r$, $\|p\| = 1/12$, $\Lambda \approx 0.8851733$, $p_0 \approx 1.6599468$, $k_0 \approx 1.563258$ (as found in Example 7.1), we have that (7.4.1) holds, since $\sup_{r\in(0,\infty)} \dfrac{r}{p_0\psi(r)\|p\|} \approx 7.2291473 > 1.6415361 \approx \dfrac{1}{1-k_0\ell}$. Thus, all the conditions of Theorem 7.4 are satisfied and hence, by its conclusion, the problem (7.18) has at least one solution on $[0,1]$.

7.1.2 *Existence results for the problem (7.2)*

Definition 7.1. A function $x \in C^2([0,T],\mathbb{R})$ is a solution of the problem (7.2) if $x(0) = g(x)$, $x(T) = \beta\ {}^\rho I^q x(\xi)$, and there exists a function $f \in L^1([0,T],\mathbb{R})$ such that $f(t) \in F(t,x(t))$ a.e. on $[0,T]$ and

$$x(t) = J^\alpha f(s)(t) + \frac{t}{\Lambda}\Big\{\beta\ {}^\rho I^q J^\alpha f(s)(\xi) - J^\alpha f(s)(T)\Big\}$$
$$+ \left[1 + \beta\frac{t}{\Lambda}\frac{\xi^{\rho q}}{\rho^q}\frac{1}{\Gamma(q+1)}\right]g(x).$$

Theorem 7.5. *Assume that (7.1.2) holds. In addition we suppose that:*

(7.5.1) $F : [0,T] \times \mathbb{R} \to \mathcal{P}_{cp,c}(\mathbb{R})$ *is* L^1-*Carathéodory multi-valued map;*
(7.5.2) there exists a function $p \in C([0,T],\mathbb{R}^+)$ *such that*

$$\|F(t,x)\|_{\mathcal{P}} := \sup\{|y| : y \in F(t,x)\} \leq p(t), \quad \textit{for each } (t,x) \in [0,T] \times \mathbb{R}.$$

Then the boundary value problem (7.2) has at least one solution on $[0,T]$.

Proof. To transform the problem (7.2) into a fixed point problem, we define an operator $\mathcal{N} : \mathcal{C} \longrightarrow \mathcal{P}(\mathcal{C})$ by

$$\mathcal{N}(x) = \left\{\begin{array}{l} h \in \mathcal{C} : \\ h(t) = \left\{\begin{array}{l} J^\alpha f(s)(t) + \dfrac{t}{\Lambda}\Big\{\alpha^\rho I^q J^\alpha f(s)(\xi) - J^\alpha f(s)(T)\Big\} \\ + \left[1 + \beta\dfrac{t}{\Lambda}\dfrac{\xi^{\rho q}}{\rho^q}\dfrac{1}{\Gamma(q+1)}\right]g(x), \end{array}\right. \end{array}\right\}$$

for $f \in S_{F,x}$.

Next we introduce operators $\mathcal{A} : \mathcal{C} \longrightarrow \mathcal{C}$ and $\mathcal{B} : \mathcal{C} \longrightarrow \mathcal{P}(\mathcal{C})$ by

$$\mathcal{A}x(t) = \left[1 + \beta\frac{t}{\Lambda}\frac{\xi^{\rho q}}{\rho^q}\frac{1}{\Gamma(q+1)}\right]g(x), \tag{7.19}$$

$$\mathcal{B}(x) = \left\{\begin{array}{l} h \in \mathcal{C} : \\ h(t) = J^\alpha f(s)(t) + \dfrac{t}{\Lambda}\Big\{\beta\ {}^\rho I^q J^\alpha f(s)(\xi) - J^\alpha f(s)(T)\Big\} \end{array}\right\}. \tag{7.20}$$

Observe that $\mathcal{N} = \mathcal{A} + \mathcal{B}$. We shall show that the operators $\mathcal{A}$ and $\mathcal{B}$ satisfy all the conditions of Theorem 1.17 on $[0,T]$. First, we show that the operators $\mathcal{A}$ and

$\mathcal{B}$ define the multi-valued operators $\mathcal{A}, \mathcal{B} : B_r \to \mathcal{P}_{cp,c}(\mathcal{C})$, where $B_r = \{x \in \mathcal{C} : \|x\| \leq r\}$ is a bounded set in $\mathcal{C}$. Firstly, we prove that $\mathcal{B}$ is compact-valued on B_r. Note that the operator $\mathcal{B}$ is equivalent to the composition $\mathcal{L} \circ S_F$, where $\mathcal{L}$ is the continuous linear operator on $L^1([0,T],\mathbb{R})$ into $\mathcal{C}$, defined by

$$\mathcal{L}(v)(t) = J^\alpha v(s)(t) + \frac{t}{\Lambda}\Big\{\beta\ {}^\rho I^q J^\alpha v(s)(\xi) - J^\alpha v(s)(T)\Big\}.$$

Suppose that $x \in B_r$ is arbitrary and let $\{v_n\}$ be a sequence in $S_{F,x}$. Then, by definition of $S_{F,x}$, we have $v_n(t) \in F(t,x(t))$ for almost all $t \in [0,T]$. Since $F(t,x(t))$ is compact for all $t \in J$, there is a convergent subsequence of $\{v_n(t)\}$ (we denote it by $\{v_n(t)\}$ again) that converges in measure to some $v(t) \in S_{F,x}$ for almost all $t \in J$. On the other hand, $\mathcal{L}$ is continuous, so $\mathcal{L}(v_n)(t) \to \mathcal{L}(v)(t)$ pointwise on $[0,T]$.

In order to show that the convergence is uniform, we have to show that $\{\mathcal{L}(v_n)\}$ is an equi-continuous sequence. Let $t_1, t_2 \in [0,T]$ with $t_1 < t_2$. Then, we have

$$\begin{aligned}
&|\mathcal{L}(v_n)(t_2) - \mathcal{L}(v_n)(t_1)| \\
&\leq |J^\alpha v_n(s)(t_2) - J^\alpha v_n(s)(t_1)| + \frac{|t_2-t_1|}{|\Lambda|} J^\alpha |v_n(s)|(T) + \frac{|\beta||t_2-t_1|}{|\Lambda|} |{}^\rho I^q J^\alpha |v_n(s)|(\xi) \\
&\leq \frac{1}{\Gamma(\alpha)} \left| \int_0^{t_1} [(t_2-s)^{\alpha-1} - (t_1-s)^{\alpha-1}] p(s)ds + \int_{t_1}^{t_2} (t_2-s)^{\alpha-1} p(s)ds \right| \\
&\quad + \frac{|t_2-t_1|}{|\Lambda|} \left(J^q p(s)(T) + |\beta|\ {}^\rho I^q J^q p(s)(\xi) \right).
\end{aligned}$$

We see that the right hand of the above inequality tends to zero as $t_2 \to t_1$. Thus, the sequence $\{\mathcal{L}(v_n)\}$ is equi-continuous and hence, by the Arzelá-Ascoli theorem, we deduce that there exits a uniformly convergent subsequence. So, there is a subsequence of $\{v_n\}$ (we denote it again by $\{v_n\}$) such that $\mathcal{L}(v_n) \to \mathcal{L}(v)$. Note that $\mathcal{L}(v) \in \mathcal{L}(S_{F,x})$. Hence, $\mathcal{B}(x) = \mathcal{L}(S_{F,x})$ is compact for all $x \in B_r$. So $\mathcal{B}(x)$ is compact.

Now, we show that $\mathcal{B}(x)$ is convex for all $x \in \mathcal{C}$. Let $z_1, z_2 \in \mathcal{B}(x)$. We select $f_1, f_2 \in S_{F,x}$ such that

$$z_i(t) = J^\alpha f_i(s)(t) + \frac{t}{\Lambda}\Big\{\beta\ {}^\rho I^q J^\alpha f_i(s)(\xi) - J^\alpha f_i(s)(T)\Big\}, \quad i = 1,2,$$

for almost all $t \in [0,T]$. Let $0 \leq \lambda \leq 1$. Then, we have

$$\begin{aligned}
&[\lambda z_1 + (1-\lambda) z_2](t) = J^\alpha[\lambda f_1(s) + (1-\lambda) f_2(s)](t) \\
&+\frac{t}{\Lambda}\Big\{\beta\ {}^\rho I^q J^\alpha [\lambda f_1(s) + (1-\lambda) f_2(s)](\xi) - J^\alpha[\lambda f_1(s) + (1-\lambda) f_2(s)](s)(T)\Big\}.
\end{aligned}$$

Since F has convex values, so $S_{F,u}$ is convex and $\lambda f_1(s) + (1-\lambda) f_2(s) \in S_{F,x}$. Thus

$$\lambda z_1 + (1-\lambda) z_2 \in \mathcal{B}(x),$$

which shows that $\mathcal{B}$ is convex-valued. Obviously, $\mathcal{A}$ is compact and convex-valued.

The rest of the proof consists of several steps and claims.

Step 1. We show that $\mathcal{A}$ *is a contraction on* $\mathcal{C}$. For $x, y \in \mathcal{C}$, we have

$$\begin{aligned}|\mathcal{A}x(t) - \mathcal{A}y(t)| &= \left|1 + \beta \frac{t}{\Lambda} \frac{\xi^{\rho q}}{\rho^q} \frac{1}{\Gamma(q+1)}\right| |g(x) - g(y)| \\ &\leq \left(1 + |\beta| \frac{T}{|\Lambda|} \frac{\xi^{\rho q}}{\rho^q} \frac{1}{\Gamma(q+1)}\right) |g(x) - g(y)|, \\ &\leq k_0 \ell \|x - y\|,\end{aligned}$$

which, on taking supremum over $t \in [0, T]$, yields

$$\|\mathcal{A}x - \mathcal{A}y\| \leq L_0 \|x - y\|, \quad L_0 = k_0 \ell.$$

This shows that $\mathcal{A}$ is a contraction as $L_0 < 1$.

Step 2. $\mathcal{B}$ *is compact and upper semi-continuous.* This will be established in several claims.

CLAIM I. *$\mathcal{B}$ maps bounded sets into bounded sets in $\mathcal{C}$.*

Let $B_r = \{x \in \mathcal{C} : \|x\| \leq r\}$ be a bounded set in $\mathcal{C}$. Then, for each $h \in \mathcal{B}(x), x \in B_r$, there exists $f \in S_{F,x}$ such that

$$h(t) = J^\alpha f(s)(t) + \frac{t}{\Lambda}\Big\{\beta\ {}^\rho I^q J^\alpha f(s)(\xi) - J^\alpha f(s)(T)\Big\}.$$

Then, for $t \in [0, T]$, we have

$$\|h\| \leq \|p\| \left\{\frac{T^\alpha}{\Gamma(\alpha+1)} + \frac{T}{|\Lambda|} \frac{T^\alpha}{\Gamma(\alpha+1)} + \frac{T}{|\Lambda|} \frac{|\beta|}{\Gamma(\alpha+1)} \frac{\xi^{\alpha+\rho q}}{\rho^q} \frac{\Gamma(\frac{\alpha+\rho}{\rho})}{\Gamma(\frac{\alpha+\rho q+\rho}{\rho})}\right\}.$$

CLAIM II. *$\mathcal{B}$ maps bounded sets into equi-continuous sets.*

Let $\tau_1, \tau_2 \in [0, T]$ with $\tau_1 < \tau_2$ and $x \in B_r$. Then, for each $h \in \mathcal{B}(x)$, we obtain

$$\begin{aligned}&|h(\tau_2) - h(\tau_1)| \\ &\leq |J^\alpha f(s)(\tau_2) - J^\alpha f(s)(\tau_1)| + \frac{|\tau_2 - \tau_1|}{|\Lambda|} J^\alpha |f(s)|(T) + \frac{|\beta||\tau_2 - \tau_1|}{|\Lambda|}\ {}^\rho I^q J^\alpha |f(s)|(\xi) \\ &\leq \frac{1}{\Gamma(\alpha)} \left| \int_0^{\tau_1} [(\tau_2 - s)^{\alpha-1} - (\tau_1 - s)^{\alpha-1}] p(s) ds + \int_{\tau_1}^{\tau_2} (\tau_2 - s)^{\alpha-1} p(s) ds \right| \\ &\quad + \|p\| \frac{|\tau_2 - \tau_1|}{|\Lambda|} \left(\frac{T^\alpha}{\Gamma(\alpha+1)} + \frac{|\beta|}{\Gamma(\alpha+1)} \frac{\xi^{\alpha+\rho q}}{\rho^q} \frac{\Gamma(\frac{\alpha+\rho}{\rho})}{\Gamma(\frac{\alpha+\rho q+\rho}{\rho})}\right) \\ &\leq \frac{\|p\|}{\Gamma(\alpha+1)} [\tau_2^\alpha - \tau_1^\alpha] + \|p\| \frac{|\tau_2 - \tau_1|}{|\Lambda|} \left(\frac{T^\alpha}{\Gamma(\alpha+1)} + \frac{|\beta|}{\Gamma(\alpha+1)} \frac{\xi^{\alpha+\rho q}}{\rho^q} \frac{\Gamma(\frac{\alpha+\rho}{\rho})}{\Gamma(\frac{\alpha+\rho q+\rho}{\rho})}\right).\end{aligned}$$

Obviously the right hand side of the above inequality tends to zero independently of $x \in B_r$ as $\tau_2 - \tau_1 \to 0$. Therefore it follows by the Arzelá-Ascoli theorem that $\mathcal{B} : \mathcal{C} \to \mathcal{P}(\mathcal{C})$ is completely continuous.

By Claims I and II, $\mathcal{B}$ is completely continuous. By Lemma 1.1, $\mathcal{B}$ will be upper semi-continuous (since it is completely continuous) if we prove that it has a closed graph.

CLAIM III. *$\mathcal{B}$ has a closed graph.*

Let $x_n \to x_*, h_n \in \mathcal{B}(x_n)$ and $h_n \to h_*$. Then we need to show that $h_* \in \mathcal{B}(x_*)$. Associated with $h_n \in \mathcal{B}(x_n)$, there exists $f_n \in S_{F,x_n}$ such that, for each $t \in [0,T]$,

$$h(t) = J^\alpha f_n(s)(t) + \frac{t}{\Lambda}\Big\{\beta\ {}^\rho I^q J^\alpha f_n(s)(\xi) - J^\alpha f_n(s)(T)\Big\}.$$

Thus it suffices to show that there exists $f_* \in S_{F,x_*}$ such that for each $t \in [0,T]$,

$$h_*(t) = J^\alpha f_*(s)(t) + \frac{t}{\Lambda}\Big\{\beta\ {}^\rho I^q J^\alpha f_*(s)(\xi) - J^\alpha f_*(s)(T)\Big\}.$$

Let us consider the linear operator $\Theta : L^1([0,T],\mathbb{R}) \to \mathcal{C}$ given by

$$f \mapsto \Theta(f)(t) = J^\alpha f(s)(t) + \frac{t}{\Lambda}\Big\{\beta\ {}^\rho I^q J^\alpha f(s)(\xi) - J^\alpha f(s)(T)\Big\}.$$

Observe that

$$\begin{aligned}\|h_n(t) - h_*(t)\| = \bigg\| & J^\alpha(f_n(s) - f_*(s))(t) \\ & + \frac{t}{\Lambda}\Big\{\beta\ {}^\rho I^q J^\alpha(f_n(s) - f_*(s))(\xi) - J^\alpha(f_n(s) - f_*(s))(T)\Big\}\bigg\| \to 0,\end{aligned}$$

as $n \to \infty$.

Thus, it follows by Lemma 1.2 that $\Theta \circ S_F$ is a closed graph operator. Further, we have $h_n(t) \in \Theta(S_{F,x_n})$. Since $x_n \to x_*$, we have that

$$h_*(t) = J^\alpha f_*(s)(t) + \frac{t}{\Lambda}\Big\{\beta\ {}^\rho I^q J^\alpha f_*(s)(\xi) - J^\alpha f_*(s)(T)\Big\},$$

for some $f_* \in S_{F,x_*}$. Hence $\mathcal{B}$ has a closed graph (and therefore has closed values). In consequence, the operator $\mathcal{B}$ is compact and upper semi-continuous.

Step 3. Here, we show that $\mathcal{A}(x) + \mathcal{B}(x) \subset B_r$ for all $x \in B_r$. Suppose $x \in B_r$, with $r > \dfrac{p_0\|p\|}{1 - k_0\ell}$ and $h \in \mathcal{B}$ are arbitrary elements. Choose $f \in S_{F,x}$ such that

$$\begin{aligned}h(t) = {} & J^\alpha f(s)(t) + \frac{t}{\Lambda}\Big\{\beta\ {}^\rho I^q J^\alpha f(s)(\xi) - J^\alpha f(s)(T)\Big\} \\ & + \left[1 + \beta\frac{t}{\Lambda}\frac{\xi^{\rho q}}{\rho^q}\frac{1}{\Gamma(q+1)}\right] g(x), \quad t \in [0,T].\end{aligned}$$

Following the method for proof for Claim I, we can obtain

$$\begin{aligned}|h(t)| \le {} & \|p\|\left\{\frac{T^\alpha}{\Gamma(\alpha+1)} + \frac{T}{|\Lambda|}\frac{T^\alpha}{\Gamma(\alpha+1)} + \frac{T}{|\Lambda|}\frac{|\beta|}{\Gamma(\alpha+1)}\frac{\xi^{\alpha+\rho q}}{\rho^q}\frac{\Gamma(\frac{\alpha+\rho}{\rho})}{\Gamma(\frac{\alpha+\rho q+\rho}{\rho})}\right\} \\ & + \left(1 + |\beta|\frac{T}{|\Lambda|}\frac{\xi^{\rho q}}{\rho^q}\frac{1}{\Gamma(q+1)}\right)\ell\|x\|.\end{aligned}$$

Thus

$$\|h\| \leq p_0\|p\| + k_0\ell r < r. \tag{7.21}$$

Hence $\|h\| \leq r$, which means that $\mathcal{A}(x) + \mathcal{B}(x) \subset B_r$ for all $x \in B_r$.

Thus, the operators $\mathcal{A}$ and $\mathcal{B}$ satisfy all the conditions of Theorem 1.17 and hence its conclusion implies that $x \in \mathcal{A}(x) + \mathcal{B}(x)$ in B_r. Therefore the boundary value problem (7.2) has a solution in B_r and the proof is completed. □

To prove our next result, we make use of the nonlinear alternative for contractive maps (Theorem 1.18).

Theorem 7.6. *Assume that (7.1.2) and (7.5.1) are satisfied. In addition we suppose that:*

(7.6.1) there exists a continuous non-decreasing function $\psi : [0, \infty) \to (0, \infty)$ and a function $p \in L^1([0, T], \mathbb{R}^+)$ such that

$$\|F(t, x)\|_{\mathcal{P}} := \sup\{|y| : y \in F(t, x)\} \leq p(t)\psi(\|x\|) \text{ for each } (t, x) \in [0, T]\times\mathbb{R};$$

(7.6.2) there exists a number $M > 0$ such that

$$\frac{M}{\psi(M)} > \frac{1}{(1 - k_0\ell)}\left[J^\alpha p(s)(T) + \frac{T}{|\Lambda|}\left\{|\beta|\ {}^\rho I^q J^\alpha p(s)(\xi) + J^\alpha p(s)(T)\right\}\right], \tag{7.22}$$

where k_0 is defined in (7.14).

Then the boundary value problem (7.2) has at least one solution on $[0, T]$.

Proof. To transform the problem (7.2) into a fixed point, we define an operator $\mathcal{N} : C([0, T], \mathbb{R}) \longrightarrow \mathcal{P}(\mathcal{C})$ and consider the operators $\mathcal{A}$ and $\mathcal{B}$ defined in the beginning of the proof of Theorem 7.5. As in Theorem 7.5, one can show that the operators $\mathcal{A}$ and $\mathcal{B}$ are indeed the multi-valued operators $\mathcal{A}, \mathcal{B} : B_r \to \mathcal{P}_{cp,c}(\mathcal{C})$, where $B_r = \{x \in \mathcal{C} : \|x\| \leq r\}$ is a bounded set in $\mathcal{C}$, $\mathcal{A}$ is a contraction on $\mathcal{C}$ and $\mathcal{B}$ is u.s.c. and compact.

Thus the operators $\mathcal{A}$ and $\mathcal{B}$ satisfy all the conditions of Theorem 1.18 and hence its conclusion implies either condition (i) or condition (ii) holds. We show that the conclusion (ii) is not possible. If $x \in \lambda\mathcal{A}(x) + \lambda\mathcal{B}(x)$ for $\lambda \in (0, 1)$, then there exists $f \in S_{F,x}$ such that

$$\begin{aligned} x(t) &= J^\alpha f(s)(t) + \frac{t}{\Lambda}\left\{\beta\ {}^\rho I^q J^\alpha f(s)(\xi) - J^\alpha f(s)(T)\right\} \\ &\quad + \left[1 + \beta\frac{t}{\Lambda}\frac{\xi^{\rho q}}{\rho^q}\frac{1}{\Gamma(q+1)}\right]g(x), \quad t \in [0, T], \end{aligned}$$

and

$$|x(t)| \leq J^\alpha p(s)|\psi(\|x\|)(T) + \frac{T}{|\Lambda|}\left\{|\beta|\ {}^\rho I^q J^\alpha p(s)\psi(\|x\|)(\xi) + J^\alpha p(s)\psi(\|x\|)(T)\right\}$$

$$+\left(1+|\beta|\frac{T}{|\Lambda|}\frac{\xi^{\rho q}}{\rho^q}\frac{1}{\Gamma(q+1)}\right)\ell\|x\|$$
$$\le \psi(\|x\|)\left[J^\alpha p(s)(T)+\frac{T}{|\Lambda|}\Big\{|\beta|\ {}^\rho I^q J^\alpha p(s)(\xi)+J^\alpha p(s)(T)\Big\}\right]$$
$$+\left(1+|\beta|\frac{T}{|\Lambda|}\frac{\xi^{\rho q}}{\rho^q}\frac{1}{\Gamma(q+1)}\right)\ell\|x\|.$$

Thus

$$(1-k_0\ell)\|x\|\le \psi(\|x\|)\left[J^\alpha p(s)(T)+\frac{T}{|\Lambda|}\Big\{|\beta|\ {}^\rho I^q J^\alpha p(s)(\xi)+J^\alpha p(s)(T)\Big\}\right]. \quad (7.23)$$

If condition (ii) of Theorem 1.18 holds, then there exists $\lambda \in (0,1)$ and $x \in \partial B_r$ with $x = \lambda\mathcal{N}(x)$. Thus x is a solution of (7.2) with $\|x\| = r$. Now, by the inequality (7.23), we get

$$\frac{M}{\psi(M)}\le\frac{1}{(1-k_0\ell)}\left[J^\alpha p(s)(T)+\frac{T}{|\Lambda|}\Big\{|\beta|\ {}^\rho I^q J^\alpha p(s)(\xi)+J^\alpha p(s)(T)\Big\}\right],$$

which contradicts (7.22). Hence, $\mathcal{N}$ has a fixed point in $[0,T]$ by Theorem 1.18, and consequently the problem (7.2) has a solution. This completes the proof. □

Example 7.4. Consider the following boundary value problem of fractional differential inclusions

$$\begin{cases} D^{3/2}x(t)\in F(t,x(t)),\ t\in[0,1],\ 0<t<1,\\ x(0)=\dfrac{1}{8}x(1/4),\quad x(1)=\dfrac{1}{2}{}^{2/3}I^{3/2}x(3/4),\end{cases} \quad (7.24)$$

where

$$F(t,x(t))=\left[\frac{2}{\sqrt{t^2+64}}\left(\frac{|x(t)|}{2}\left(\frac{|x(t)|}{|x(t)|+1}+1\right)+\frac{1}{5}\right),\frac{e^{-t}}{(10+t)}\left(\sin x(t)+\frac{1}{15}\right)\right].$$

Clearly $|F(t,x)| \le p(t)\psi(|x|)$, where $p(t) = 2/\sqrt{t^2+64}$, $\psi_1(|x|) = |x| + 1/5$ and $\ell = 1/8$. Using the values: $\Lambda \approx 0.8851733$, $p_0 \approx 1.6599468$, $k_0 \approx 1.563258$ (see Example 7.1) and the condition (7.6.2), we find that $M > M_1 \simeq 0.2130287$. Since the hypotheses of Theorem 7.6 are satisfied, the problem (7.24) has a solution on $[0,1]$.

7.2 On perturbed fractional differential equations and inclusions with generalized Riemann-Liouville fractional integral boundary conditions

In this section, we study perturbed boundary value problems of fractional differential equations and inclusions supplemented with generalized Riemann-Liouville fractional integral conditions. In precise terms, we consider the single-valued and multi-valued nonlocal problems respectively given by

$$\begin{cases} {}^cD^\alpha x(t) = f(t, x(t)) + g(t, x(t)), \quad t \in [0, T], \\ x(0) = 0, \\ x(T) = \beta \dfrac{\rho^{1-q}}{\Gamma(q)} \displaystyle\int_0^\xi \frac{s^{\rho-1}}{(\xi^\rho - s^\rho)^{1-q}} x(s) ds := \beta\ {}^\rho I^q x(\xi), \quad 0 < \xi < T, \end{cases} \tag{7.25}$$

and

$$\begin{cases} {}^cD^\alpha x(t) \in F(t, x(t)) + G(t, x(t)), \quad t \in [0, T], \\ x(0) = 0, \quad x(T) = \beta\ {}^\rho I^q x(\xi), \quad 0 < \xi < T, \end{cases} \tag{7.26}$$

where ${}^cD^\alpha$ is the Caputo fractional derivative operator of order $1 < \alpha \leq 2$, $f, g : [0, T] \times \mathbb{R} \to \mathbb{R}$ are continuous functions, ${}^\rho I^q$ is the generalized Riemann-Liouville fractional integral of orders $q > 0$, $\rho > 0$, and $F, G : [0, T] \times \mathbb{R} \to \mathcal{P}(\mathbb{R})$ are multi-valued functions, ($\mathcal{P}(\mathbb{R})$ is the family of all non-empty subjects of $\mathbb{R}$).

The following lemma plays a fundamental role in the study of the problem (7.25). We omit the proof of this lemma as it is similar to that of Lemma 7.1.

Lemma 7.2. *For any $y \in C([0, T], \mathbb{R})$, a function $x \in C^2([0, T], \mathbb{R})$ is a solution of the linear fractional boundary value problem*

$$\begin{cases} {}^cD^\alpha x(t) = y(t), \quad 1 < \alpha \leq 2, \\ x(0) = 0, \quad x(T) = \beta\ {}^\rho I^q x(\xi), \quad 0 < \xi < T, \end{cases} \tag{7.27}$$

if and only if

$$x(t) = J^\alpha y(t) + \frac{t}{\Lambda}\Big\{\beta\ {}^\rho I^q J^\alpha y(\xi) - J^\alpha y(T)\Big\}, \tag{7.28}$$

where

$$\Lambda = T - \beta \frac{\xi^{\rho q + 1}}{\rho^q} \frac{\Gamma(\frac{1+\rho}{\rho})}{\Gamma(\frac{1+\rho q+\rho}{\rho})} \neq 0, \tag{7.29}$$

where J^α denotes Riemann-Liouville fractional integral of order α.

Next we set the notation:

$$J^z f(s, x(s))(t) = \frac{1}{\Gamma(z)} \int_0^t (t-s)^{z-1} f(s, x(s)) ds,$$

$${}^\rho I^z f(s, x(s))(t) = \frac{\rho^{1-z}}{\Gamma(z)} \int_0^t \frac{s^{\rho-1} f(s, x(s))}{(t^\rho - s^\rho)^{1-z}} ds,$$

where $z > 0$ and $t \in [0, T]$.

7.2.1 *Existence result for the problem (7.25)*

We denote by $\mathcal{C} = C([0,T],\mathbb{R})$ the Banach space of all continuous functions from $[0,T] \to \mathbb{R}$ endowed with a topology of uniform convergence with the norm defined by $\|x\| = \sup\{|x(t)| : t \in [0,T]\}$. Also by $L^1([0,T],\mathbb{R})$ we denote the Banach space of measurable functions $x : [0,T] \to \mathbb{R}$ which are Lebesgue integrable and normed by $\|x\|_{L^1} = \int_0^T |x(t)|dt$.

Our main result in this section is based upon the following fixed point theorem due to Burton and Kirk (Theorem 1.6).

Theorem 7.7. *Assume that $f, g : [0,T] \times \mathbb{R} \to \mathbb{R}$ are continuous functions. In addition we suppose that:*

(7.7.1) there exist a function $k(t) \in L^1(J,\mathbb{R}_+)$ such that

$$|g(t,u) - g(t,\overline{u})| \le k(t)|u - \overline{u}|, \quad \text{for a.e. } t \in J, \ u, \ \overline{u} \in \mathbb{R},$$

with

$$\gamma := J^\alpha k(s)(T) + \frac{T}{|\Lambda|} J^\alpha k(s)(T) + \frac{|\beta| T}{|\Lambda|} \, {}^\rho I^q J^\alpha k(s)(\xi) < \frac{1}{2}; \tag{7.30}$$

(7.7.2) there exists a function $p \in L^1(J,\mathbb{R}_+)$ such that

$$|f(t,u)| \le p(t), \quad \text{for a.e. } t \in J, \text{ and each } u \in \mathbb{R}.$$

Then the boundary value problem (7.25) has at least one solution on $[0,T]$.

Proof. Let us transform the problem (7.25) into a fixed point problem by using Lemma (7.2). For that, we introduce an operator $\mathcal{P} : \mathcal{C} \to \mathcal{C}$ associated with the problem (7.25) as

$$\begin{aligned}(\mathcal{P}x)(t) &= J^\alpha[f(s,x(s)) + g(s,x(s))](t) \\ &\quad + \frac{t}{\Lambda}\Big\{\beta \, {}^\rho I^q J^\alpha [f(s,x(s)) + g(s,x(s))](\xi) \\ &\quad - J^\alpha[f(s,x(s)) + g(s,x(s))](T)\Big\}, \quad t \in [0,T]. \end{aligned} \tag{7.31}$$

Then we split the operator $\mathcal{P} : \mathcal{C} \to \mathcal{C}$ defined by (7.31) as

$$(\mathcal{P}x)(t) = (\mathcal{P}_1 x)(t) + (\mathcal{P}_2 x)(t), \quad t \in [0,T], \tag{7.32}$$

where $\mathcal{P}_{1,2} : \mathcal{C} \to \mathcal{C}$ are given by

$$(\mathcal{P}_1 x)(t) = J^\alpha f(s,x(s))(t) + \frac{t}{\Lambda}\Big\{\beta \, {}^\rho I^q J^\alpha f(s,x(s))(\xi) - J^\alpha f(s,x(s))(T)\Big\}, \tag{7.33}$$

$$(\mathcal{P}_2 x)(t) = J^\alpha g(s,x(s))(t) + \frac{t}{\Lambda}\Big\{\beta \, {}^\rho I^q J^\alpha g(s,x(s))(\xi) - J^\alpha g(s,x(s))(T)\Big\}. \tag{7.34}$$

Obviously the existence of a fixed point for the operator $\mathcal{P}$ is equivalent to the existence of one for $\mathcal{P}_1 + \mathcal{P}_2$. So we shall show that the operators $\mathcal{P}_1$ and $\mathcal{P}_2$ satisfies all conditions of Theorem 1.6. The proof will be given in several steps.

Step 1. The operator $\mathcal{P}_1$ defined by (7.33) is continuous.

Let $\{x_n\} \subset B_r = \{x \in \mathcal{C} : \|x\| \leq r\}$ with $\|x_n - x\| \to 0$. Then the limit $\|x_n(t)-x(t)\| \to 0$ is uniformly valid on $[0,T]$. From the uniform continuity of $f(t,x)$ on the compact set $[0,T] \times [-r,r]$, it follows that $\|f(t,x_n(t)) - f(t,x(t))\| \to 0$ is uniformly valid on $[0,T]$. Hence $\|\mathcal{P}_1 x_n - \mathcal{P}_1 x\| \to 0$ as $n \to \infty$, which proves the continuity of $\mathcal{P}_1$.

Step 2. The operator $\mathcal{P}_1$ maps bounded sets into bounded sets in $\mathcal{C}$.

It is enough to show that for any $r > 0$, there exists a positive constant L such that, for each $x \in B_r$, we have $\|\mathcal{P}_1 x\| \leq L$. Let $x \in B_r$. Then

$$\begin{aligned}\|\mathcal{P}_1 x\| &\leq J^\alpha |f(s,x(s))|(T) + \frac{T}{|\Lambda|}\Big\{|\beta|\ {}^\rho I^q J^\alpha |f(s,x(s))|(\xi) + J^\alpha|f(s,x(s))|(T)\Big\}\\ &\leq J^\alpha p(s)(T) + \frac{|\beta| T}{|\Lambda|}\ {}^\rho I^q J^\alpha p(s)(\xi) + \frac{T}{|\Lambda|} J^\alpha p(s)(T) := L.\end{aligned}$$

Step 3. The operator $\mathcal{P}_1$ maps bounded sets into equi-continuous sets in $\mathcal{C}$.

Let $\tau_1, \tau_2 \in [0,T]$ with $\tau_1 < \tau_2$ and $x \in B_r$. Then, for each $x \in B_r$, we obtain

$$\begin{aligned}&|(\mathcal{P}_1 x)(\tau_2) - (\mathcal{P}_1 x)(\tau_1)|\\ &\leq |J^\alpha f(s,x(s))(\tau_2) - J^\alpha f(s,x(s))(\tau_1)| + \frac{|\tau_2-\tau_1|}{|\Lambda|} J^\alpha |f(s,x(s))|(T)\\ &\quad+\frac{|\beta||\tau_2-\tau_1|}{|\Lambda|} |{}^\rho I^q J^\alpha |f(s,x(s))|(\xi)\\ &\leq \frac{1}{\Gamma(\alpha)}\left|\int_0^{\tau_1} [(\tau_2-s)^{\alpha-1} - (\tau_1-s)^{\alpha-1}]p(s)ds + \int_{\tau_1}^{\tau_2} (\tau_2-s)^{\alpha-1}p(s)ds\right|\\ &\quad+\frac{|\tau_2-\tau_1|}{|\Lambda|}\Big(J^\alpha p(s)(T) + |\beta|\ {}^\rho I^q J^\alpha p(s)(\xi)\Big),\end{aligned}$$

which is independent of x and tends to zero as $\tau_2 - \tau_1 \to 0$. Thus, $\mathcal{P}_1$ is equi-continuous. Thus, by the above three steps, the operator $\mathcal{P}_1$ is completely continuous.

Step 4. The operator $\mathcal{P}_2$ defined by (7.34) is a contraction.

For $x, y \in \mathcal{C}$ and for each $t \in [0,T]$, from the definition of $\mathcal{P}_2$ and assumption (7.7.1), we obtain

$$\begin{aligned}&|(\mathcal{P}_2 x)(t) - (\mathcal{P}_2 y)(t)|\\ &\leq J^\alpha |g(s,x(s)) - g(s,y(s))|(t) + \frac{T}{|\Lambda|} J^\alpha |g(s,x(s)) - g(s,y(s))|(T)\\ &\quad+\frac{|\beta| T}{|\Lambda|}\ {}^\rho I^q J^\alpha |g(s,x(s)) - g(s,y(s))|(\xi)\\ &\leq \Big\{J^\alpha k(s)(T) + \frac{T}{|\Lambda|} J^\alpha k(s)(T) + \frac{|\beta| T}{|\Lambda|}\ {}^\rho I^q J^\alpha k(s)(\xi)\Big\}\|x-y\|.\end{aligned}$$

Hence

$$\|(\mathcal{P}_2x) - (\mathcal{P}_2y)\| \le \gamma\|x - y\|.$$

As $\gamma < 1$ by (7.30), the operator $\mathcal{P}_2$ is a contraction map from the Banach space $\mathcal{C}$ into itself.

Step 5. A priori bounds on solutions.

Now it remains to show that the set $\mathcal{E} = \left\{u \in \mathcal{C} : \lambda\mathcal{P}_1\left(\frac{u}{\lambda}\right) + \lambda\mathcal{P}_2(u) = u\right\}$ is unbounded for some $\lambda \in (0,1)$.

Let $\lambda \in (0,1)$ and $x \in \mathcal{E}$ be a solution of the integral equation

$$\begin{aligned} x(t) &= \lambda J^\alpha f(s,x(s))(t) + \lambda\frac{t}{\Lambda}\left\{\beta\, {}^\rho I^q J^\alpha f(s,x(s))(\xi) - J^\alpha f(s,x(s))(T)\right\} \\ &\quad + \lambda J^\alpha g\Big(s,\frac{x(s)}{\lambda}\Big)(t) + \lambda\frac{t}{\Lambda}\left\{\beta\, {}^\rho I^q J^\alpha g\Big(s,\frac{x(s)}{\lambda}\Big)(\xi) - J^\alpha g\Big(s,\frac{x(s)}{\lambda}\Big)(T)\right\},\ t \in [0,T]. \end{aligned}$$

Then, for each $t \in [0,T]$, we have

$$\begin{aligned} |x(t)| &\le J^\alpha p(s)(T) + \frac{|\beta|T}{|\Lambda|}\, {}^\rho I^q J^\alpha p(s)(\xi) + \frac{T}{|\Lambda|}J^\alpha p(s)(T) \\ &\quad + \lambda J^\alpha\left[\left|g\Big(s,\frac{x(s)}{\lambda}\Big) - g(s,0)\right| + |g(s,0)|\right](T) \\ &\quad + \frac{|\beta|T}{|\Lambda|}\, {}^\rho I^q J^\alpha\left[\left|g\Big(s,\frac{x(s)}{\lambda}\Big) - g(s,0)\right| + |g(s,0)|\right](\xi) \\ &\quad + \frac{T}{|\Lambda|}J^\alpha\left[\left|g\Big(s,\frac{x(s)}{\lambda}\Big) - g(s,0)\right| + |g(s,0)|\right](T) \\ &\le J^\alpha p(s)(T) + \frac{|\beta|T}{|\Lambda|}\, {}^\rho I^q J^\alpha p(s)(\xi) + \frac{T}{|\Lambda|}J^\alpha p(s)(T) \\ &\quad + \|x\|\left\{J^\alpha k(s)(T) + \frac{|\beta|T}{|\Lambda|}\, {}^\rho I^q J^\alpha k(s)(\xi) + \frac{T}{|\Lambda|}J^\alpha k(s)(T)\right\} \\ &\quad + g^*\left\{\frac{T^\alpha}{\Gamma(\alpha+1)} + \frac{T}{|\Lambda|}\frac{T^\alpha}{\Gamma(\alpha+1)} + \frac{T}{|\Lambda|}\frac{|\beta|}{\Gamma(\alpha+1)}\frac{\xi^{\alpha+\rho q}}{\rho^q}\frac{\Gamma(\frac{\alpha+\rho}{\rho})}{\Gamma(\frac{\alpha+\rho q+\rho}{\rho})}\right\}, \end{aligned}$$

which implies that

$$\begin{aligned} (1-\gamma)\|x\| &\le J^\alpha p(s)(T) + \frac{|\beta|T}{|\Lambda|}\, {}^\rho I^q J^\alpha p(s)(\xi) + \frac{T}{|\Lambda|}J^\alpha p(s)(T) \\ &\quad + g^*\left\{\frac{T^\alpha}{\Gamma(\alpha+1)} + \frac{T}{|\Lambda|}\frac{T^\alpha}{\Gamma(\alpha+1)} + \frac{T}{|\Lambda|}\frac{|\beta|}{\Gamma(\alpha+1)}\frac{\xi^{\alpha+\rho q}}{\rho^q}\frac{\Gamma(\frac{\alpha+\rho}{\rho})}{\Gamma(\frac{\alpha+\rho q+\rho}{\rho})}\right\}, \end{aligned}$$

where $g^* = \sup\{|g(t,0)|, t \in [0,T]\}$. Consequently we get

$$\|x\| \le M := \frac{1}{(1-\gamma)}\left[J^\alpha p(s)(T) + \frac{|\beta|T}{|\Lambda|}\, {}^\rho I^q J^\alpha p(s)(\xi) + \frac{T}{|\Lambda|}J^\alpha p(s)(T)\right.$$

$$+g^*\left\{\frac{T^\alpha}{\Gamma(\alpha+1)}+\frac{T}{|\Lambda|}\frac{T^\alpha}{\Gamma(\alpha+1)}+\frac{T}{|\Lambda|}\frac{|\beta|}{\Gamma(\alpha+1)}\frac{\xi^{\alpha+\rho q}}{\rho^q}\frac{\Gamma(\frac{\alpha+\rho}{\rho})}{\Gamma(\frac{\alpha+\rho q+\rho}{\rho})}\right\}\Bigg],$$

which means that the set $\mathcal{E}$ is bounded, since $\gamma < 1/2$. Hence, (b) of Theorem 1.6 does not hold. Thus (a) holds, that is, $\mathcal{P}$ has a fixed point in $[0,T]$, and consequently the problem (7.25) has a solution. This completes the proof. □

7.2.2 *Existence result for the problem (7.26)*

Definition 7.2. A function $x \in C([0,T],\mathbb{R})$ possessing a fractional derivative of order $\alpha \in (1,2]$ is a solution of the problem (7.26) if $x(0)=0$, $x(T)=\beta\ {}^\rho I^q x(\xi)$, and there exist functions $f,g \in L^1([0,T],\mathbb{R})$ such that $f(t)\in F(t,x(t))$ and $g(t)\in G(t,x(t))$ a.e. on $[0,T]$ and

$$x(t)=J^\alpha[f(s)+g(s)](t)+\frac{t}{\Lambda}\Big\{\beta\ {}^\rho I^q J^\alpha[f(s)+g(s)](\xi)-J^\alpha[f(s)+g(s)](T)\Big\}. \quad (7.35)$$

Theorem 7.8. *Assume that:*

(7.8.1) $F:[0,T]\times\mathbb{R}\to\mathcal{P}_{cp,c}(\mathbb{R})$ *is* L^1*-Carathéodory;*

(7.8.2) there exists a continuous non-decreasing function $\Phi:[0,\infty)\to(0,\infty)$ *and a function* $p\in L^1([0,T],\mathbb{R}^+)$ *such that*

$$\|F(t,x)\|_{\mathcal{P}}:=\sup\{|y|:y\in F(t,x)\}\le p(t)\Phi(\|x\|)\ \textit{for each}\ (t,x)\in[0,T]\times\mathbb{R};$$

(7.8.3) the multi-valued map $t\to G(t,x)$ *is measurable for each* $x\in\mathbb{R}$ *and integrably bounded, i.e., there exists a function* $M\in L^1([0,T],\mathbb{R})$ *such that*

$$\|G(t,x)\|_{\mathcal{P}}:=\sup\{|g|:g(t)\in G(t,x)\}\le M(t),\ \textit{for a.e.}\ t\in[0,T]\ \textit{and}\ x\in\mathbb{R};$$

(7.8.4) $G:[0,T]\times\mathbb{R}\to\mathcal{P}_{cp,c}(\mathbb{R})$ *and there exists a function* $\ell\in L^1([0,T],\mathbb{R})$ *such that*

$$H_d(G(t,x)-G(t,y))\le\ell(t)|x-y|,\ t\in[0,T]$$

for all $x,y\in\mathbb{R}$ *with*

$$J^q\ell(s)(T)+\frac{T}{|\Lambda|}\left(|\beta|\ {}^\rho I^q J^q\ell(s)(\xi_i)+J^q\ell(s)(T)(T)\right)<1; \quad (7.36)$$

(7.8.5) there exists a constant $M>0$ *such that*

$$\frac{M}{\Phi(M)\Psi_1+\Psi_2}>1, \quad (7.37)$$

where

$$\Psi_1=J^\alpha p(s)(T)+\frac{|\beta|T}{|\Lambda|}\ {}^\rho I^q J^\alpha p(s)(\xi)+\frac{T}{|\Lambda|}J^\alpha p(s)(T)$$

and

$$\Psi_2=J^\alpha M(s)(T)+\frac{|\beta|T}{|\Lambda|}\ {}^\rho I^q J^\alpha M(s)(\xi)+\frac{T}{|\Lambda|}J^\alpha M(s)(T).$$

Then the boundary value problem (7.26) has at least one solution on $[0, T]$.

Proof. To transform the problem (7.26) into a fixed point, we define an operator $\mathcal{N} : \mathcal{C} \longrightarrow \mathcal{P}(\mathcal{C})$ by

$$\mathcal{N}(x) = \left\{ \begin{array}{l} h \in \mathcal{C} : \\ h(t) = \left\{ \begin{array}{l} J^{\alpha}[f(s) + g(s)](t) + \dfrac{t}{\Lambda}\Big\{\alpha^{\rho} I^{q} J^{\alpha}[f(s) + g(s)](\xi) \\ -J^{\alpha}[f(s) + g(s)](T)\Big\}, \end{array} \right. \end{array} \right\}$$

for $f \in S_{F,x}$ and $g \in S_{G,x}$.

Next we introduce the multi-valued operators $\mathcal{A}, \mathcal{B} : C([0,T], \mathbb{R}) \longrightarrow \mathcal{P}(C([0,T], \mathbb{R}))$ by

$$\mathcal{A}(x) = \left\{ h \in \mathcal{C} : h(t) = J^{\alpha} f(s)(t) + \frac{t}{\Lambda}\Big\{\beta\ {}^{\rho}I^{q} J^{\alpha} f(s)(\xi) - J^{\alpha} f(s)(T)\Big\} \right\} \quad (7.38)$$

for $f \in S_{F,x}$ and

$$\mathcal{B}(x) = \left\{ z \in \mathcal{C} : z(t) = J^{\alpha} g(s)(t) + \frac{t}{\Lambda}\Big\{\beta\ {}^{\rho}I^{q} J^{\alpha} g(s)(\xi) - J^{\alpha} g(s)(T)\Big\} \right\} \quad (7.39)$$

for $g \in S_{G,x}$. Observe that $\mathcal{N} = \mathcal{A} + \mathcal{B}$. We shall show that the operators $\mathcal{A}$ and $\mathcal{B}$ satisfy all the conditions of Theorem 1.18 on $[0, T]$. First, we show that the operators $\mathcal{A}$ and $\mathcal{B}$ define the multi-valued operators $\mathcal{A}, \mathcal{B} : B_r \to \mathcal{P}_{cp,c}(\mathcal{C})$, where $B_r = \{x \in \mathcal{C} : \|x\| \leq r\}$ is a bounded set in $\mathcal{C}$. First we prove that $\mathcal{A}$ is compact-valued on B_r. Note that the operator $\mathcal{A}$ is equivalent to the composition $\mathcal{L} \circ S_F$, where $\mathcal{L}$ is the continuous linear operator on $L^1([0,T], \mathbb{R})$ into $\mathcal{C}$, defined by

$$\mathcal{L}(v)(t) = J^{\alpha} v(s)(t) + \frac{t}{\Lambda}\Big\{\beta\ {}^{\rho}I^{q} J^{\alpha} v(s)(\xi) - J^{\alpha} v(s)(T)\Big\}.$$

Suppose that $x \in B_r$ is arbitrary and let $\{v_n\}$ be a sequence in $S_{F,x}$. Then, by definition of $S_{F,x}$, we have $v_n(t) \in F(t, x(t))$ for almost all $t \in [0, T]$. Since $F(t, x(t))$ is compact for all $t \in J$, there is a convergent subsequence of $\{v_n(t)\}$ (we denote it by $\{v_n(t)\}$ again) that converges in measure to some $v(t) \in S_{F,x}$ for almost all $t \in J$. On the other hand, $\mathcal{L}$ is continuous, so $\mathcal{L}(v_n)(t) \to \mathcal{L}(v)(t)$ pointwise on $[0, T]$.

In order to show that the convergence is uniform, we have to show that $\{\mathcal{L}(v_n)\}$ is an equi-continuous sequence. Let $t_1, t_2 \in [0, T]$ with $t_1 < t_2$. Then, we have

$$\begin{aligned}
&|\mathcal{L}(v_n)(t_2) - \mathcal{L}(v_n)(t_1)| \\
&\leq |J^{\alpha} v_n(s)(t_2) - J^{\alpha} v_n(s)(t_1)| + \frac{|t_2 - t_1|}{|\Lambda|} J^{q}|v_n(s)|(T) \\
&\quad + \frac{|\beta||t_2 - t_1|}{|\Lambda|} |{}^{\rho}I^{q} J^{q}|v_n(s)|(\xi) \\
&\leq \frac{\psi(r)}{\Gamma(q)} \left| \int_0^{t_1} [(t_2 - s)^{q-1} - (t_1 - s)^{q-1}] p(s) ds + \int_{t_1}^{t_2} (t_2 - s)^{q-1} p(s) ds \right|
\end{aligned}$$

$$+\psi(r)\frac{|t_2-t_1|}{|\Lambda|}\left(J^q p(s)(T)+|\beta||^\rho I^q J^q p(s)(\xi)\right).$$

We see that the right hand of the above inequality tends to zero, independently of x, as $t_2 \to t_1$. Thus, the sequence $\{\mathcal{L}(v_n)\}$ is equi-continuous and by using the Arzelá-Ascoli theorem, we get that there is a uniformly convergent subsequence. So, there is a subsequence of $\{v_n\}$ (we denote it again by $\{v_n\}$) such that $\mathcal{L}(v_n) \to \mathcal{L}(v)$. Note that, $\mathcal{L}(v) \in \mathcal{L}(S_{F,x})$. Hence $\mathcal{A}(x) = \mathcal{L}(S_{F,x})$ is compact for all $x \in B_r$. So $\mathcal{A}(x)$ is compact.

Now, we show that $\mathcal{A}(x)$ is convex for all $x \in \mathcal{C}$. Let $h_1, h_2 \in \mathcal{A}(x)$. We select $f_1, f_2 \in S_{F,x}$ such that

$$h_i(t) = J^\alpha f_i(s)(t) + \frac{t}{\Lambda}\Big\{\beta\ {}^\rho I^q J^\alpha f_i(s)(\xi) - J^\alpha f_i(s)(T)\Big\},\quad i=1,2,$$

for almost all $t \in [0,T]$. Let $0 \le \lambda \le 1$. Then, we have

$$[\lambda h_1 + (1-\lambda)h_2](t) = J^\alpha[\lambda f_1(s) + (1-\lambda)f_2(s)](t)$$
$$+\frac{t}{\Lambda}\Big\{\beta\ {}^\rho I^q J^\alpha[\lambda f_1(s)+(1-\lambda)f_2(s)](\xi) - J^\alpha[\lambda f_1(s)+(1-\lambda)f_2(s)](s)(T)\Big\}.$$

Since F has convex values, so $S_{F,u}$ is convex and $\lambda f_1(s) + (1-\lambda)f_2(s) \in S_{F,x}$. Thus

$$\lambda h_1 + (1-\lambda)h_2 \in \mathcal{A}(x).$$

Consequently, $\mathcal{A}$ is convex-valued. For the operator $\mathcal{B}$ we work in a similar way.

The rest of the proof consists of several steps and claims.

Step 1. We show that $\mathcal{B}$ *is a contraction on* $\mathcal{C}$. Let $x, \bar{x} \in C([0,T],\mathbb{R})$ and $z_1 \in \mathcal{G}(x)$. Then there exists $g_1(t) \in G(t,x(t))$ such that, for each $t \in [0,T]$,

$$z_1(t) = J^q g_1(s)(t) + \frac{t^{q-1}}{\Lambda}\Big(\beta\ {}^\rho I^q J^\alpha g_1(s)(\xi_i) - J^q g_1(s)(T)\Big).$$

By (7.8.4), we have

$$H_d(G(t,x), G(t,\bar{x})) \le \ell(t)|x(t) - \bar{x}(t)|.$$

So, there exists $w \in G(t,\bar{x}(t))$ such that

$$|v_1(t) - w| \le \ell(t)|x(t) - \bar{x}(t)|,\quad t \in [0,T].$$

Define $U : [0,T] \to \mathcal{P}(\mathbb{R})$ by

$$U(t) = \{w \in \mathbb{R} : |v_1(t) - w| \le \ell(t)|x(t) - \bar{x}(t)|\}.$$

Since the multi-valued operator $U(t) \cap G(t,\bar{x}(t))$ is measurable (Proposition III.4 [78]), there exists a function $v_2(t)$ which is a measurable selection for U. So $v_2(t) \in G(t,\bar{x}(t))$ and for each $t \in [0,T]$, we have $|v_1(t) - v_2(t)| \le \ell(t)|x(t) - \bar{x}(t)|$.

For each $t \in [0,T]$, let us define

$$z_2(t) = J^q g_2(s)(t) + \frac{t}{\Lambda}\Big(\beta\ {}^\rho I^q J^\alpha g_2(s)(\xi_i) - J^q g_2(s)(T)\Big).$$

Then it is easy to find that

$$\|z_1 - z_2\| \leq \left\{ J^q\ell(s)(T) + \frac{T}{|\Lambda|}\Big(|\beta|\ {}^\rho I^q J^q \ell(s)(\xi_i) + J^q\ell(s)(T)(T)\Big) \right\} \|x - \bar{x}\|.$$

Analogously, interchanging the roles of x and $\overline{x}$, we obtain

$$H_d(\mathcal{A}(x), \mathcal{A}(\bar{x})) \leq \delta \|x - \bar{x}\|$$

where

$$\delta = J^q\ell(s)(T) + \frac{T}{|\Lambda|}\Big(|\beta|\ {}^\rho I^q J^q \ell(s)(\xi_i) + J^q\ell(s)(T)(T)\Big).$$

So $\mathcal{A}$ is a contraction, since $\delta < 1$ by (7.36).

Step 2. *$\mathcal{A}$ is compact and upper semi-continuous.* This will be established in several claims.

CLAIM I. *$\mathcal{A}$ maps bounded sets into bounded sets in $\mathcal{C}$.* Let $B_r = \{x \in \mathcal{C} : \|x\| \leq r\}$ be a bounded set in $\mathcal{C}$. Then, for each $h \in \mathcal{A}(x), x \in B_r$, there exists $f \in S_{F,x}$ such that

$$h(t) = J^\alpha f(s)(t) + \frac{t}{\Lambda}\Big\{\beta\ {}^\rho I^q J^\alpha f(s)(\xi) - J^\alpha f(s)(T)\Big\}.$$

Then, for $t \in [0,T]$, we have

$$\|h\| \leq \psi(r)\left\{ J^\alpha p(s)(T) + \frac{|\beta| T}{|\Lambda|}\ {}^\rho I^q J^\alpha p(s)(\xi) + \frac{T}{|\Lambda|} J^\alpha p(s)(T) \right\}.$$

CLAIM II. *$\mathcal{B}$ maps bounded sets into equi-continuous sets.* Let $\tau_1, \tau_2 \in [0,T]$ with $\tau_1 < \tau_2$ and $x \in B_r$. Then, for each $h \in \mathcal{A}(x)$, we obtain

$$\begin{aligned}
&|h(\tau_2) - h(\tau_1)| \\
&\leq \frac{\psi(r)}{\Gamma(q)} \left| \int_0^{\tau_1} [(\tau_2 - s)^{q-1} - (\tau_1 - s)^{q-1}] p(s) ds + \int_{\tau_1}^{\tau_2} (\tau_2 - s)^{q-1} p(s) ds \right| \\
&\quad + \psi(r) \frac{|\tau_2 - \tau_1|}{|\Lambda|}\Big(J^q p(s)(T) + |\beta|\ {}^\rho I^q J^q p(s)(\xi)\Big).
\end{aligned}$$

Obviously the right hand side of the above inequality tends to zero independently of $x \in B_r$ as $\tau_2 - \tau_1 \to 0$. Therefore it follows by the Ascoli-Arzelá theorem that $\mathcal{A} : \mathcal{C} \to \mathcal{P}(\mathcal{C})$ is completely continuous.

Next we show that $\mathcal{B}$ is an upper semi-continuous multi-valued mapping. It is known by Lemma 1.1 that $\mathcal{B}$ will be upper semi-continuous if we establish that it has a closed graph, since it is already shown to be completely continuous. Thus we will prove that $\mathcal{A}$ *has a closed graph.*

CLAIM III. *$\mathcal{A}$ has a closed graph.* Let $x_n \to x_*, h_n \in \mathcal{A}(x_n)$ and $h_n \to h_*$. Then we need to show that $h_* \in \mathcal{A}(x_*)$. Associated with $h_n \in \mathcal{A}(x_n)$, there exists $f_n \in S_{F,x_n}$ such that, for each $t \in [0,T]$,

$$h(t) = J^\alpha f_n(s)(t) + \frac{t}{\Lambda}\Big\{\beta\ {}^\rho I^q J^\alpha f_n(s)(\xi) - J^\alpha f_n(s)(T)\Big\}.$$

Thus it suffices to show that there exists $f_* \in S_{F,x_*}$ such that, for each $t \in [0,T]$,

$$h_*(t) = J^\alpha f_*(s)(t) + \frac{t}{\Lambda}\Big\{\beta\ {}^\rho I^q J^\alpha f_*(s)(\xi) - J^\alpha f_*(s)(T)\Big\}.$$

Let us consider the linear operator $\Theta : L^1([0,T],\mathbb{R}) \to \mathcal{C}$ given by

$$f \mapsto \Theta(f)(t) = J^\alpha f(s)(t) + \frac{t}{\Lambda}\Big\{\beta\ {}^\rho I^q J^\alpha f(s)(\xi) - J^\alpha f(s)(T)\Big\}.$$

Observe that

$$\begin{aligned}\|h_n(t) - h_*(t)\| = \Big\| & J^\alpha(f_n(s) - f_*(s))(t) \\ & + \frac{t}{\Lambda}\Big\{\beta\ {}^\rho I^q J^\alpha(f_n(s) - f_*(s))(\xi) - J^\alpha(f_n(s) - f_*(s))(T)\Big\}\Big\| \to 0,\end{aligned}$$

as $n \to \infty$. Thus, it follows by Lemma 1.2 that $\Theta \circ S_F$ is a closed graph operator. Further, we have $h_n(t) \in \Theta(S_{F,x_n})$. Since $x_n \to x_*$, we have that

$$h_*(t) = J^\alpha f_*(s)(t) + \frac{t}{\Lambda}\Big\{\beta\ {}^\rho I^q J^\alpha f_*(s)(\xi) - J^\alpha f_*(s)(T)\Big\},$$

for some $f_* \in S_{F,x_*}$. Hence $\mathcal{A}$ has a closed graph (and therefore has closed values). In consequence, the operator $\mathcal{A}$ is compact valued and upper semi-continuous.

Thus the operators $\mathcal{A}$ and $\mathcal{B}$ satisfy all the conditions of Theorem 1.18 and hence its conclusion implies either condition (i) or condition (ii) holds. We show that the conclusion (ii) is not possible. If $x \in \lambda\mathcal{A}(x) + \lambda\mathcal{B}(x)$ for $\lambda \in (0,1)$, then there exist $f \in S_{F,x}$ and $g \in S_{G,x}$ such that

$$\begin{aligned}x(t) = {} & J^\alpha f(s)(t) + \frac{t}{\Lambda}\Big\{\beta\ {}^\rho I^q J^\alpha f(s)(\xi) - J^\alpha f(s)(T)\Big\} \\ & + J^\alpha g(s)(t) + \frac{t}{\Lambda}\Big\{\beta\ {}^\rho I^q J^\alpha g(s)(\xi) - J^\alpha g(s)(T)\Big\}, \quad t \in [0,T].\end{aligned}$$

By our assumptions, we can obtain

$$\begin{aligned}\|x\| \le \psi(\|x\|)\Bigg\{ & J^\alpha p(s)(T) + \frac{|\beta| T}{|\Lambda|}\ {}^\rho I^q J^\alpha p(s)(\xi) + \frac{T}{|\Lambda|} J^\alpha p(s)(T)\Bigg\} \\ & + J^\alpha M(s)(T) + \frac{|\beta| T}{|\Lambda|}\ {}^\rho I^q J^\alpha M(s)(\xi) + \frac{T}{|\Lambda|} J^\alpha M(s)(T). \end{aligned} \tag{7.40}$$

If condition (ii) of Theorem 1.18 holds, then there exists $\lambda \in (0,1)$ and $x \in \partial B_M$ with $x = \lambda\mathcal{N}(x)$. So, x is a solution of (7.26) with $\|x\| = M$. Now, by the inequality (7.40), we get

$$\frac{M}{\psi(M)\Psi_1 + \Psi_2} \le 1,$$

which contradicts (7.37). Hence, $\mathcal{N}$ has a fixed point in $[0,T]$ by Theorem 1.18, and consequently the problem (7.26) has a solution. This completes the proof. □

7.3 Fractional differential equations involving generalized derivative with Stieltjes and fractional integral boundary conditions

In this section, we initiate the study of boundary value problems of generalized fractional differential equations supplemented with Stieltjes and generalized fractional integral boundary conditions. In precise terms, we investigate the uniqueness of solutions for the following generalized fractional differential equation:

$$ {}^{\rho}D^{\alpha}y(t) = f(t, y(t)), \quad t \in [0, T], \tag{7.41}$$

supplemented with Stieltjes and generalized fractional integral boundary conditions of the form

$$y(0) = 0, \quad \int_0^T y(s)dH(s) = \lambda \frac{\rho^{1-\beta}}{\Gamma(\beta)} \int_0^{\xi} \frac{s^{\rho-1}}{(\xi^{\rho} - s^{\rho})^{1-\beta}} y(s)ds := \lambda\, {}^{\rho}I^{\beta}y(\xi), \quad \xi \in (0, T). \tag{7.42}$$

Here ${}^{\rho}D^{\alpha}$ is the generalized fractional derivative operator of order $1 < \alpha \le 2, \rho > 0$, ${}^{\rho}I^{\beta}$ is the generalized fractional integral of order $\beta > 0, \rho > 0$, $\int_0^T y(s)dH(s)$ is the Stieltjes integral with respect the function $H : [0, T] \to \mathbb{R}$, $f : [0, T] \times \mathbb{R} \to \mathbb{R}$ is a continuous function and $\lambda \in \mathbb{R}$. Here we use the concept of generalized fractional derivative operator and integral due to Katugampola.

Let $\mathcal{P} = C([0, T], \mathbb{R})$ denote the Banach space of all continuous functions from $[0, T]$ to $\mathbb{R}$ endowed with the norm defined by $\|y\| = \sup_{t\in[0,T]} |y(t)|$. Let $AC[0, T]$ denote the space of all absolutely continuous real valued functions on $[0, T]$. Moreover, we introduce the space AC^2_{ρ} by

$$AC^2_{\rho}[0, T] = \Big\{ f : [0, T] \to \mathbb{R} : (t^{1-\rho}\frac{d}{dt}) f(t) \in AC[0, T] \Big\}.$$

Lemma 7.3. *[119] Let $1 < \alpha \le 2, \rho > 0$, $u \in X^p_c(0, T)$ and ${}^{\rho}I^{2-\alpha}u \in AC^2_{\rho}[0, T]$. Then the general solution of the fractional differential equation ${}^{\rho}D^{\alpha}u(t) = 0$ is*

$$u(t) = c_1 t^{\rho(\alpha-1)} + c_2 t^{\rho(\alpha-2)},$$

where $c_i \in \mathbb{R}, i = 1, 2$. Moreover,

$${}^{\rho}I^{\alpha}\, {}^{\rho}D^{\alpha}u(t) = u(t) + c_1 t^{\rho(\alpha-1)} + c_2 t^{\rho(\alpha-2)}.$$

Definition 7.3. The variation of a function $f : [a, b] \to \mathbb{R}$ on $[a, b]$, denoted by $V_a^b f$, is

$$V_a^b f = \sup \sum_{i=0}^{n} |f(t_i) - f(t_{i-1})|,$$

where the supremum is taken over all possible partitions $P : a = t_0 < t_1 < \ldots < t_n = b$ of $[a, b]$. If $V_a^b f < \infty$, then f is called a function of bounded variation on $[a, b]$.

Lemma 7.4. *If $h \in C([a, b], \mathbb{R})$ and $f : [a, b] \to \mathbb{R}$ is a function of bounded variation on $[a, b]$, then the integral $\int_a^b h(s)df(s)$ exists.*

Lemma 7.5. *If $h \in C([a,b],\mathbb{R})$, $f : [a,b] \to \mathbb{R}$ is a function of bounded variation on $[a,b]$ and $M = \max_{t\in[a,b]} |h(t)|$, then*

$$\left|\int_a^b h(s)df(s)\right| \leq MV_a^b f.$$

Corollary 7.1. *Let $h \in C([a,b],\mathbb{R})$ and $f : [a,b] \to \mathbb{R}$ be monotonically increasing on $[a,b]$. Then the integral $\int_a^b h(s)df(s)$ exists and*

$$\left|\int_a^b h(s)df(s)\right| \leq M(f(b) - f(a)),$$

where $M = \max_{t\in[a,b]} |h(t)|$.

Lemma 7.6. *If h is Riemann integrable and $f' \in C[a,b]$. Then the integral $\int_a^b h(s)df(s)$ exists and*

$$\int_a^b h(s)df(s) = \int_a^b h(s)f'(s)ds.$$

Definition 7.4. A function $y \in C([0,T],\mathbb{R})$ with ${}^\rho I^{2-\alpha}y \in AC_\rho^2[0,T]$ is said to be a solution of (7.41) and (7.42) if y satisfies the equation ${}^\rho D^\alpha y(t) = f(t,y(t))$ on $[0,T]$ and the conditions $y(0) = 0,\ \int_0^T y(s)dH(s) = \lambda\ {}^\rho I^\beta y(\xi)$.

Lemma 7.7. *For any $h \in C(0,T)$ and $y \in C([0,T],\mathbb{R})$ with ${}^\rho I^{2-\alpha}y \in AC_\rho^2[0,T]$, the function y is the solution of the problem*

$$\begin{cases} {}^\rho D^\alpha y(t) = h(t), \quad t \in [0,T], \\ y(0) = 0, \quad \int_0^T y(s)dH(s) = \lambda\ {}^\rho I^\beta y(\xi), \quad \xi \in (0,T), \end{cases} \tag{7.43}$$

if and only if

$$y(t) = {}^\rho I^\alpha h(t) + \frac{t^{\rho(\alpha-1)}}{\Lambda}\Big\{\lambda\ {}^\rho I^{\alpha+\beta}h(\xi) - \int_0^T {}^\rho I^\alpha h(s)dH(s)\Big\}, \tag{7.44}$$

where it is assumed that

$$\Lambda = \int_0^T s^{\rho(\alpha-1)}dH(s) - \lambda\frac{\Gamma(\alpha)}{\rho^\beta\Gamma(\alpha+\beta)}\xi^{\rho(\alpha+\beta-1)} \neq 0. \tag{7.45}$$

Proof. Applying the operator ${}^\rho I^\alpha$ to the generalized linear fractional differential equation in (7.43) and using Lemma 7.3, we obtain

$$y(t) = {}^\rho I^\alpha h(t) + c_1 t^{\rho(\alpha-1)} + c_2 t^{\rho(\alpha-2)}, \tag{7.46}$$

where c_1 and c_2 are arbitrary unknown constants. Making use of the condition $y(0) = 0$ in (7.46), we get $c_2 = 0$ and applying the generalized integral ${}^\rho I^\beta$ to (7.46) after inserting $c_2 = 0$ in it, we get

$${}^\rho I^\beta y(t) = {}^\rho I^{\alpha+\beta}h(t) + c_1\frac{\Gamma(\alpha)}{\rho^\beta\Gamma(\alpha+\beta)}t^{\rho(\alpha+\beta-1)},$$

which, together with the condition $\int_0^T y(s)dH(s) = \lambda\ {}^\rho I^\beta y(\xi)$, yields

$$\int_0^T {}^\rho I^\alpha h(s)dH(s) + c_1 \int_0^T s^{\rho(\alpha-1)} dH(s) \\ = \lambda\ {}^\rho I^{\alpha+\beta} h(\xi) + c_1 \lambda \frac{\Gamma(\alpha)}{\rho^\beta \Gamma(\alpha+\beta)} \xi^{\rho(\alpha+\beta-1)}. \quad (7.47)$$

Solving (7.47) for c_1 and using (7.45), we get

$$c_1 = \frac{1}{\Lambda}\Big\{\lambda\ {}^\rho I^{\alpha+\beta} h(\xi) - \int_0^T {}^\rho I^\alpha h(s)dH(s)\Big\}.$$

Substituting the value of c_1 in (7.46) completes the solution (7.44). The converse follows by direct computation. The proof is completed. □

7.3.1 *Uniqueness of solutions*

By Lemma 7.7, we can define an operator $N : \mathcal{P} \to \mathcal{P}$ by

$$N(y)(t) = {}^\rho I^\alpha f(t, y(t)) \\ + \frac{t^{\rho(\alpha-1)}}{\Lambda}\Big\{\lambda\ {}^\rho I^{\alpha+\beta} f(\xi, y(\xi)) - \int_0^T {}^\rho I^\alpha f(s, y(s))dH(s)\Big\}. \quad (7.48)$$

For computational convenience, we set

$$A = \frac{T^{\rho\alpha}}{\rho^\alpha \Gamma(\alpha+1)} + \frac{T^{\rho(\alpha-1)}}{|\Lambda|}\Big\{|\lambda| \frac{\xi^{\rho(\alpha+\beta)}}{\rho^{\alpha+\beta}\Gamma(\alpha+\beta+1)} + \frac{T^{\rho\alpha}}{\rho^\alpha \Gamma(\alpha+1)} V_0^T H\Big\}. \quad (7.49)$$

Now we are in a position to present our main result, which deals with the uniqueness of solutions for the problem (7.41) and (7.42) and relies on Banach's fixed point theorem.

Theorem 7.9. *Assume the following conditions hold:*

(7.9.1) $H : [0, T] \times \mathbb{R} \to \mathbb{R}$ *is s function of bounded variation on* $[0, T]$*;*
(7.9.2) $f : [0, T] \times \mathbb{R} \to \mathbb{R}$ *is a continuous function satisfying the Lipschitz condition:*

$$|f(t, x) - f(t, y)| \leq L\|x - y\|, \quad \forall t \in [0, T],\ L > 0,\ x, y \in \mathbb{R}.$$

Then the problem (7.41) and (7.42) has a unique solution on $[0, T]$ *if*

$$LA < 1, \quad (7.50)$$

where A *is given by (7.49).*

Proof. Consider the operator $N : \mathcal{P} \to \mathcal{P}$ defined by (7.48) and set $\sup_{t\in[0,T]} |f(t, 0)| = f_0$. Choosing $r \geq \frac{AM}{1-LA}$, we show that $NB_r \subset B_r$, where $B_r = \{y \in C([0, T], \mathbb{R}) : \|y\| \leq r\}$. For $y \in B_r$, using (7.9.1) and (7.9.2), we get

$$|N(y)(t)| \leq {}^\rho I^\alpha [|f(t, y(t)) - f(t, 0)| + |f(t, 0)|] \\ + \frac{t^{\rho(\alpha-1)}}{|\Lambda|}\Big\{|\lambda|\ {}^\rho I^{\alpha+\beta}[|f(\xi, y(\xi)) - f(\xi, 0)| + |f(\xi, 0)|]$$

$$
\begin{aligned}
&+\int_0^T {}^{\rho}I^{\alpha}[|f(s,y(s))-f(s,0)|+|f(s,0)|]dV_0^s H(s)\Big\}\\
&\le (Lr+f_0)\Big[\frac{T^{\rho\alpha}}{\rho^{\alpha}\Gamma(\alpha+1)}+\frac{T^{\rho(\alpha-1)}}{|\Lambda|}\Big\{|\lambda|\frac{\xi^{\rho(\alpha+\beta)}}{\rho^{\alpha+\beta}\Gamma(\alpha+\beta+1)}\\
&+\frac{T^{\rho\alpha}}{\rho^{\alpha}\Gamma(\alpha+1)}V_0^T H\Big\}\Big]\\
&\le (Lr+f_0)A\le r,
\end{aligned}
$$

which, on taking the norm for $t\in[0,T]$, yields $\|N(y)\|\le r$. This shows that N maps B_r into itself. In order to show that the operator N is a contraction, let $y,z\in C([0,T],\mathbb{R})$. Then, for each $t\in[0,T]$, we obtain

$$
\begin{aligned}
&|N(y)(t)-N(z)(t)|\\
&\le {}^{\rho}I^{\alpha}|f(t,y(t))-f(t,z(t))|+\frac{t^{\rho(\alpha-1)}}{|\Lambda|}\Big\{|\lambda|\ {}^{\rho}I^{\alpha+\beta}|f(\xi,y(\xi))-f(\xi,z(\xi)|\\
&+\int_0^T {}^{\rho}I^{\alpha}|f(s,y(s))-f(s,z(s))|dV_0^s H(s)\Big\}\\
&\le L\|y-z\|\Big[\frac{T^{\rho\alpha}}{\rho^{\alpha}\Gamma(\alpha+1)}+\frac{T^{\rho(\alpha-1)}}{|\Lambda|}\Big\{|\lambda|\frac{\xi^{\rho(\alpha+\beta)}}{\rho^{\alpha+\beta}\Gamma(\alpha+\beta+1)}\\
&+\frac{T^{\rho\alpha}}{\rho^{\alpha}\Gamma(\alpha+1)}V_0^T H\Big\}\Big]\\
&\le LA\|y-z\|.
\end{aligned}
$$

Taking the norm of the above inequality for $t\in[0,T]$, we obtain $\|N(y)-N(z)\|\le LA\|y-z\|$. Thus, in view of the condition (7.50), it follows that the operator N is a contraction. Hence the operator N has a unique fixed point by Banach's contraction principle, which corresponds to a unique solution of the problem (7.41) and (7.42). The proof is completed. □

Corollary 7.2. *Let $H:[0,T]\to\mathbb{R}$ be monotonically increasing on $[0,T]$ and the assumption (7.9.2) holds. Then the problem (7.41) and (7.42) has a unique solution on $[0,T]$, provided that*

$$LA<1, \tag{7.51}$$

where A is given by

$$
\begin{aligned}
A=&\frac{T^{\rho\alpha}}{\rho^{\alpha}\Gamma(\alpha+1)}+\frac{T^{\rho(\alpha-1)}}{|\Lambda|}\Big\{|\lambda|\frac{\xi^{\rho(\alpha+\beta)}}{\rho^{\alpha+\beta}\Gamma(\alpha+\beta+1)}\\
&+\frac{T^{\rho\alpha}}{\rho^{\alpha}\Gamma(\alpha+1)}(H(T)-H(0))\Big\}.
\end{aligned}
\tag{7.52}
$$

Remark 7.1. A monotonically increasing character of the function H on $[0,T]$ implies that $H'\in C[0,T]$. So, by Lemma 7.6, Λ given by (7.45) takes the form:

$$\Lambda=\int_0^T s^{\rho(\alpha-1)}H'(s)ds-\lambda\frac{\Gamma(\alpha)}{\rho^{\beta}\Gamma(\alpha+\beta)}\xi^{\rho(\alpha+\beta-1)}.$$

Remark 7.2. Setting $H(s) = s$ in the problem (7.41) and (7.42), the integral boundary condition reduces to

$$\int_0^T y(s)ds = \lambda \frac{\rho^{1-\beta}}{\Gamma(\beta)} \int_0^\xi \frac{s^{\rho-1}}{(\xi^\rho - s^\rho)^{1-\beta}} y(s)ds := \lambda\ {}^\rho I^\beta y(\xi), \quad \xi \in (0,T). \tag{7.53}$$

In this case, the value of A is found to be

$$A = \frac{T^{\rho\alpha}}{\rho^\alpha \Gamma(\alpha+1)} + \frac{T^{\rho(\alpha-1)}}{|\Lambda|}\Big\{|\lambda| \frac{\xi^{\rho(\alpha+\beta)}}{\rho^{\alpha+\beta}\Gamma(\alpha+\beta+1)} + \frac{T^{\rho\alpha+1}}{\rho^\alpha\Gamma(\alpha+1)}\Big\}$$

and the operator (7.48) modifies to the form:

$$N(y)(t) = {}^\rho I^\alpha f(t,y(t)) + \frac{t^{\rho(\alpha-1)}}{\Lambda}\Big\{\lambda\ {}^\rho I^{\alpha+\beta} f(\xi, y(\xi)) - \int_0^T {}^\rho I^\alpha f(s,y(s))ds\Big\},$$

where

$$\Lambda = \frac{T^{\rho(\alpha-1)+1}}{\rho(\alpha-1)+1} - \lambda \frac{\Gamma(\alpha)}{\rho^\beta \Gamma(\alpha+\beta)} \xi^{\rho(\alpha+\beta-1)}.$$

Example 7.5. Let us consider the following boundary value problem

$$\begin{cases} {}^{1/3}D^{7/5} y(t) = \dfrac{1}{20(t+1)}\Big(\dfrac{|y(t)|+2}{|y(t)|+1} + \cos t\Big), & t \in [0,2], \\ y(0) = 0, \quad \displaystyle\int_0^2 y(s)dH(s) = 1/2\ {}^{1/3}I^{2/5} y(3/2), \end{cases} \tag{7.54}$$

where $\rho = 1/3$, $\alpha = 7/5$, $\lambda = 1/2$, $\beta = 2/5$, $\xi = 3/2$, $T = 2$ and $f(t,y(t)) = \frac{1}{20(t+1)}\Big(\frac{|y(t)|+2}{|y(t)|+1} + \cos t\Big)$. Let us take $H(t) = 3t^2 + 1$. Clearly $H(t)$ is monotonically increasing function on $[0,2]$. By using the given data, we find that $|\Lambda| = 11.51571147$, $A = 11.36070263$, where Λ and A are respectively given by (7.45) and (7.52). It is easy to check that $f(t,y)$ is continuous and satisfies the condition (7.9.2) of Theorem 7.9 with $L = 1/20$. Also $LA \approx 0.5680351315 < 1$. Thus all the assumptions of Corollary 7.2 are satisfied. So the conclusion of Corollary 7.2 implies that the problem (7.54) with $H(t) = 3t^2 + 1$ has a unique solution on $[0,2]$.

7.4 Generalized fractional differential equations and inclusions equipped with nonlocal generalized fractional integral boundary conditions

In this section, we study a new class of boundary value problems involving generalized fractional derivative operators and integrals. Precisely, we investigate the existence and uniqueness of solutions for the following generalized fractional differential equation and inclusions:

$${}^\rho D^\alpha y(t) = f(t,y(t)), \quad t \in [0,T], \tag{7.55}$$

$${}^\rho D^\alpha y(t) \in F(t,y(t)), \quad t \in [0,T], \tag{7.56}$$

supplemented with the following nonlocal boundary conditions:

$$y(0) = 0, \ y(T) = \lambda \frac{\rho^{1-\beta}}{\Gamma(\beta)} \int_0^{\xi} \frac{s^{\rho-1}}{(\xi^\rho - s^\rho)^{1-\beta}} y(s)ds := \lambda \ ^\rho I^\beta y(\xi), \ \xi \in (0,T), \quad (7.57)$$

$$y(0) = 0, \ y(T) = \sum_{i=1}^{n} \lambda_i y(\xi_i), \ \xi_i \in (0,T), \quad (7.58)$$

and

$$y(0) = 0, \ y(T) = \sum_{i=1}^{n} \lambda_i \ ^\rho I^\beta y(\xi_i), \ \xi_i \in (0,T), \quad (7.59)$$

where $^\rho D^\alpha$ is the generalized (Katugampola) fractional derivative operator of order $1 < \alpha \leq 2, \rho > 0$, $^\rho I^\beta$ is the generalized (Katugampola) type fractional integral of order $\beta > 0, \rho > 0$, $f : [0,T] \times \mathbb{R} \to \mathbb{R}$ is a continuous function, $\lambda, \lambda_i \in \mathbb{R}, i = 1,2,\ldots,n$, and $F : [0,T] \times \mathbb{R} \to \mathcal{P}(\mathbb{R})$ is a multi-valued function ($\mathcal{P}(\mathbb{R})$ is the family of all non-empty subjects of $\mathbb{R}$). Note that the boundary conditions (7.57) and (7.59) involve generalized fractional integrals.

Definition 7.5. A function $y \in C^2([0,T],\mathbb{R})$ is said to be a solution of (7.55) and (7.57) if y satisfies the equation $^\rho D^\alpha y(t) = f(t,y(t))$ on $[0,T]$, and the condition $y(0) = 0, \ y(T) = \lambda \ ^\rho I^\beta y(\xi)$.

Relative to the problem (7.55) and (7.57), we consider the following lemma.

Lemma 7.8. *Let $h \in C(0,T) \cap L(0,T)$ and*

$$\Lambda_1 = T^{\rho(\alpha-1)} - \lambda \frac{\Gamma(\alpha)}{\rho^\beta \Gamma(\alpha+\beta)} \xi^{\rho(\alpha+\beta-1)} \neq 0. \quad (7.60)$$

Then the solution of the linear equation

$$^\rho D^\alpha y(t) = h(t), \quad t \in [0,T], \quad (7.61)$$

supplemented with the conditions (7.57) is equivalent to the integral equation:

$$y(t) = {}^\rho I^\alpha h(t) + \frac{t^{\rho(\alpha-1)}}{\Lambda_1} \Big\{ \lambda \ ^\rho I^{\alpha+\beta} h(\xi) - {}^\rho I^\alpha h(T) \Big\}. \quad (7.62)$$

Proof. Applying the operator $^\rho I^\alpha$ to the generalized linear fractional differential equation (7.61), we obtain

$$y(t) = {}^\rho I^\alpha h(t) + c_1 t^{\rho(\alpha-1)} + c_2 t^{\rho(\alpha-2)}, \quad (7.63)$$

where c_1 and c_2 are arbitrary unknown constants. Using the first boundary condition $y(0) = 0$ in (7.63), we get $c_2 = 0$. Applying the generalized integral to (7.63) after inserting the value of c_2, we get

$$^\rho I^\beta y(t) = {}^\rho I^{\alpha+\beta} h(t) + c_1 \frac{\Gamma(\alpha)}{\rho^\beta \Gamma(\alpha+\beta)} t^{\rho(\alpha+\beta-1)}. \quad (7.64)$$

Using the second boundary condition: $y(T) = \lambda\ {}^{\rho}I^{\beta} y(\xi)$ in (7.64), we have

$$ {}^{\rho}I^{\alpha}h(T) + c_1 T^{\rho(\alpha-1)} = \lambda\ {}^{\rho}I^{\alpha+\beta}h(\xi) + c_1\lambda\frac{\Gamma(\alpha)}{\rho^{\beta}\Gamma(\alpha+\beta)}\xi^{\rho(\alpha+\beta-1)}, $$

which yields

$$ c_1 = \frac{1}{\Lambda_1}\Big\{\lambda\ {}^{\rho}I^{\alpha+\beta}h(\xi) - {}^{\rho}I^{\alpha}h(T)\Big\}. $$

Substituting the value of c_1 and $c_2 = 0$ in (7.63), we get the solution (7.62). The converse follows by direct computation. The proof is completed. □

Now we give two lemmas associated with the nonlinear boundary value problem (7.55) and (7.58), and (7.55) and (7.59). One can obtain the proof of these results by using the arguments employed in Lemma 7.8.

Lemma 7.9. *For any $h \in C(0,T) \cap L(0,T)$ and*

$$ \Lambda_2 = T^{\rho(\alpha-1)} - \sum_{i=1}^{n}\lambda_i \xi_i^{\rho(\alpha-1)} \neq 0, \tag{7.65} $$

the solution of the equation (7.61) with the boundary conditions (7.58) is given by

$$ y(t) = {}^{\rho}I^{\alpha}h(t) + \frac{t^{\rho(\alpha-1)}}{\Lambda_2}\Big\{\sum_{i=1}^{n}\lambda_i\ {}^{\rho}I^{\alpha}h(\xi_i) - {}^{\rho}I^{\alpha}h(T)\Big\}. \tag{7.66} $$

Lemma 7.10. *For any $h \in C(0,T) \cap L(0,T)$ and*

$$ \Lambda_3 = T^{\rho(\alpha-1)} - \sum_{i=1}^{n}\lambda_i \frac{\Gamma(\alpha)}{\rho^{\beta}\Gamma(\alpha+\beta)}\xi_i^{\rho(\alpha+\beta-1)} \neq 0, \tag{7.67} $$

the solution of the equation (7.61) with the boundary conditions (7.59) is given by

$$ y(t) = {}^{\rho}I^{\alpha}h(t) + \frac{t^{\rho(\alpha-1)}}{\Lambda_3}\Big\{\sum_{i=1}^{n}\lambda_i\ {}^{\rho}I^{\alpha+\beta}h(\xi_i) - {}^{\rho}I^{\alpha}h(T)\Big\}. \tag{7.68} $$

Main results for the single valued problems

This section is devoted to the existence of solutions for the fractional differential equation (7.55) supplemented with the boundary conditions (7.57), (7.58) and (7.59).

7.4.1 *Existence results for the problem (7.55) and (7.57)*

Here we discuss the existence and uniqueness of solutions for the problem (7.55) and (7.57). In relation to the problem (7.55) and (7.57), we define an operator $\mathbf{F}_1 : C([0,T],\mathbb{R}) \to C([0,T],\mathbb{R})$ by

$$ \mathbf{F}_1(y)(t) = {}^{\rho}I^{\alpha}f(t,y(t)) + \frac{t^{\rho(\alpha-1)}}{\Lambda_1}\Big\{\lambda\ {}^{\rho}I^{\alpha+\beta}f(\xi,y(\xi)) - {}^{\rho}I^{\alpha}f(T,y(T))\Big\}, \tag{7.69} $$

where $C([0,T],\mathbb{R})$ denotes the Banach space of all continuous functions from $[0,T]$ to $\mathbb{R}$ endowed with the norm defined by $\|y\| = \sup_{t\in[0,T]} |y(t)|$. For computational convenience, we set

$$A_1 = \frac{T^{\rho\alpha}}{\rho^{\alpha}\Gamma(\alpha+1)} + \frac{T^{\rho(\alpha-1)}}{|\Lambda_1|}\Big\{|\lambda|\frac{\xi^{\rho(\alpha+\beta)}}{\rho^{\alpha+\beta}\Gamma(\alpha+\beta+1)} + \frac{T^{\rho\alpha}}{\rho^{\alpha}\Gamma(\alpha+1)}\Big\}. \tag{7.70}$$

In the first result we establish the uniqueness of solutions for the problem (7.55) and (7.57) via Banach's fixed point theorem.

Theorem 7.10. *Let $f : [0,T]\times\mathbb{R}\to\mathbb{R}$ be a continuous function satisfying the assumption:*

(7.10.1) $|f(t,x) - f(t,y)| \le \vartheta\|x-y\|, \quad \forall t\in[0,T],\ \vartheta>0,\ x,y\in\mathbb{R}.$

Then the problem (7.55) and (7.57) has a unique solution on $[0,T]$ if

$$\vartheta A_1 < 1, \tag{7.71}$$

where A_1 is given by (7.70).

Proof. In view of the condition (7.70), consider the set $B_r = \{y\in C([0,T],\mathbb{R}) : \|y\|\le r\}$ with $r > \frac{f_0 A_1}{1-\vartheta A_1}$, $\sup_{t\in[0,T]}|f(t,0)| = f_0$, and show that $\mathbf{F_1}B_r\subset B_r$, where $\mathbf{F_1}$ is defined by (7.69). For $y\in B_r$, we have

$$\begin{aligned}
|\mathbf{F_1}(y)(t)| &= \Big|{}^{\rho}I^{\alpha}f(t,y(t)) + \frac{t^{\rho(\alpha-1)}}{\Lambda_1}\Big\{\lambda\ {}^{\rho}I^{\alpha+\beta}f(\xi,y(\xi)) - {}^{\rho}I^{\alpha}f(T,y(T))\Big\}\Big| \\
&\le {}^{\rho}I^{\alpha}(|f(t,y(t)) - f(t,0)| + |f(t,0)|) \\
&\quad + \frac{t^{\rho(\alpha-1)}}{|\Lambda_1|}\Bigg\{|\lambda|\ {}^{\rho}I^{\alpha+\beta}(|f(\xi,y(\xi)) - f(\xi,0)| + |f(\xi,0)|) \\
&\quad + {}^{\rho}I^{\alpha}(|f(T,y(T)) - f(T,0)| + |f(T,0)|)\Bigg\} \\
&\le (\vartheta r + f_0)\Big[\frac{T^{\rho\alpha}}{\rho^{\alpha}\Gamma(\alpha+1)} + \frac{T^{\rho(\alpha-1)}}{|\Lambda_1|}\Big\{|\lambda|\frac{\xi^{\rho(\alpha+\beta)}}{\rho^{\alpha+\beta}\Gamma(\alpha+\beta+1)} \\
&\quad + \frac{T^{\rho\alpha}}{\rho^{\alpha}\Gamma(\alpha+1)}\Big\}\Big] \\
&= (\vartheta r + f_0)A_1 < r,
\end{aligned}$$

which implies that $\|\mathbf{F_1}y\| < r$, for any $y\in B_r$. Thus $\mathbf{F_1}B_r\subset B_r$. Now, for $y,z\in C([0,T],\mathbb{R})$ and for each $t\in[0,T]$, we obtain

$$\begin{aligned}
&|\mathbf{F_1}(y)(t) - \mathbf{F_1}(z)(t)| \\
&\le {}^{\rho}I^{\alpha}|f(t,y(t)) - f(t,z(t))| + \frac{t^{\rho(\alpha-1)}}{\Lambda_1}\Bigg\{|\lambda|\ {}^{\rho}I^{\alpha+\beta}|f(\xi,y(\xi)) - f(\xi,z(\xi))| \\
&\quad + {}^{\rho}I^{\alpha}|f(T,y(T)) - f(T,z(T))|\Bigg\}
\end{aligned}$$

$$\leq \vartheta\Big[\frac{T^{\rho\alpha}}{\rho^{\alpha}\Gamma(\alpha+1)} + \frac{T^{\rho(\alpha-1)}}{|\Lambda_1|}\Big\{|\lambda|\frac{\xi^{\rho(\alpha+\beta)}}{\rho^{\alpha+\beta}\Gamma(\alpha+\beta+1)} + \frac{T^{\rho\alpha}}{\rho^{\alpha}\Gamma(\alpha+1)}\Big\}\Big]\|y-z\|$$
$$= \vartheta A_1\|y-z\|,$$

which, on taking the norm for $t \in [0,T]$, yields

$$\|\mathbf{F_1}(y) - \mathbf{F_1}(z)\| \leq \vartheta A_1\|y-z\|. \tag{7.72}$$

By the given condition $\vartheta A_1 < 1$, it follows from (7.72) that the operator $\mathbf{F_1}$ is a contraction. Thus, the conclusion of the theorem follows by the contraction mapping principle (Banach fixed point theorem). The proof is completed. □

In the next result, we prove the existence of solutions for the problem (7.55) and (7.57) by applying Leray-Schauder nonlinear alternative (Theorem 1.8).

Theorem 7.11. *Assume that:*

(7.11.1) there exist a function $p \in L^1([0,T],\mathbb{R}^+)$, and a non-decreasing function $\Omega : \mathbb{R}^+ \to \mathbb{R}^+$ such that $|f(t,y)| \leq p(t)\Omega(\|y\|)$, $\forall (t,y) \in [0,T]\times\mathbb{R}$;

(7.11.2) there exists a constant $K > 0$ such that

$$\frac{K}{\Omega(K)\Big(\,{}^{\rho}I^{\alpha}p(T) + \frac{T^{\rho(\alpha-1)}}{|\Lambda_1|}\{|\lambda|{}^{\rho}I^{\alpha+\beta}p(\xi) + {}^{\rho}I^{\alpha}p(T)\}\Big)} > 1.$$

Then the problem (7.55) and (7.57) has at least one solution on $[0,T]$.

Proof. First of all, we show that the operator $\mathbf{F_1} : C([0,T],\mathbb{R}) \to C([0,T],\mathbb{R})$ defined by (7.69) is continuous and completely continuous. This will be established in several steps.

Step 1. $\mathbf{F_1}$ *is continuous.*

Let $\{y_n\}$ be a sequence such that $y_n \to y$ in $C([0,T],\mathbb{R})$. Then

$$|\mathbf{F_1}(y_n)(t) - \mathbf{F_1}(y)(t)|$$
$$\leq {}^{\rho}I^{\alpha}|f(t,y_n(t)) - f(t,y(t))| + \frac{t^{\rho(\alpha-1)}}{\Lambda_1}\Bigg\{|\lambda|\ {}^{\rho}I^{\alpha+\beta}|f(\xi,y_n(\xi)) - f(\xi,y(\xi))|$$
$$+{}^{\rho}I^{\alpha}|f(T,y_n(T)) - f(T,y(T))|\Bigg\}$$
$$\leq A_1\|f(\cdot,y_n) - f(\cdot,y)\|.$$

Since f is a continuous function, therefore, we have

$$\|\mathbf{F_1}(y_n) - \mathbf{F_1}(y)\| \leq A_1\|f(\cdot,y_n) - f(\cdot,y)\| \to 0, \ as\, n \to \infty.$$

Step 2. $\mathbf{F_1}$ *maps bounded sets into bounded sets in $C([0,T],\mathbb{R})$.*

Indeed, it is enough to show that for any $\mu > 0$ there exists a positive constant ℓ such that, for $y \in B_\mu = \{y \in C([0,T],\mathbb{R}) : \|y\| \le \mu\}$, we have $\|\mathbf{F_1}(y)\| \le \ell$. By (7.11.1), for each $t \in [0,T]$, we have

$$\begin{aligned}
|\mathbf{F_1}(y)(t)| &\le {}^\rho I^\alpha |f(t,y(t))| + \frac{t^{\rho(\alpha-1)}}{|\Lambda_1|}\left\{|\lambda|\ {}^\rho I^{\alpha+\beta}|f(\xi,y(\xi))| + {}^\rho I^\alpha|f(T,y(T))|\right\}\\
&\le {}^\rho I^\alpha p(T)\Omega(\|y\|) + \frac{t^{\rho(\alpha-1)}}{|\Lambda_1|}\left\{|\lambda|\ {}^\rho I^{\alpha+\beta}p(\xi)\Omega(\|y\|) + {}^\rho I^\alpha p(T)\Omega(\|y\|)\right\}\\
&\le \Omega(\|y\|)\left({}^\rho I^\alpha p(T) + \frac{T^{\rho(\alpha-1)}}{|\Lambda_1|}\left\{|\lambda|\ {}^\rho I^{\alpha+\beta}p(\xi) + {}^\rho I^\alpha p(T)\right\}\right).
\end{aligned}$$

Thus

$$\|\mathbf{F_1}(y)\| \le \Omega(\|\mu\|)\left({}^\rho I^\alpha p(T) + \frac{T^{\rho(\alpha-1)}}{|\Lambda_1|}\left\{|\lambda|\ {}^\rho I^{\alpha+\beta}p(\xi) + {}^\rho I^\alpha p(T)\right\}\right) := \ell.$$

Step 3. *F maps bounded sets into equi-continuous sets of $C([0,T],\mathbb{R})$.*

Let $0 < t_1 < t_2 < T$, B_μ be a bounded set of $C([0,T],\mathbb{R})$ as in Step 2, and let $y \in B_\mu$. Then

$$\begin{aligned}
&|\mathbf{F}_1(y)(t_2) - \mathbf{F}_1(y)(t_1)|\\
&\le \left|{}^\rho I^\alpha |f(t_2,y(t_2))| - {}^\rho I^\alpha|f(t_1,y(t_1))|\right|\\
&\quad + \frac{\Omega(\mu)|t_2^{\rho(\alpha-1)} - t_1^{\rho(\alpha-1)}|}{|\Lambda_1|}\left\{|\lambda|\ {}^\rho I^{\alpha+\beta}p(\xi) + {}^\rho I^\alpha p(T)\right\}\\
&\le \frac{\rho^{1-\alpha}\Omega(\mu)}{\Gamma(\alpha)}\left|\int_0^{t_1}\left[\frac{s^{\rho-1}}{(t_2^\rho - s^\rho)^{1-\alpha}} - \frac{s^{\rho-1}}{(t_1^\rho - s^\rho)^{1-\alpha}}\right]p(s)ds + \int_{t_1}^{t_2}\frac{s^{\rho-1}}{(t_2^\rho - s^\rho)^{1-\alpha}}p(s)ds\right|\\
&\quad + \frac{\Omega(\mu)|t_2^{\rho(\alpha-1)} - t_1^{\rho(\alpha-1)}|}{|\Lambda_1|}\left\{|\lambda|\ {}^\rho I^{\alpha+\beta}p(\xi) + {}^\rho I^\alpha p(T)\right\}.
\end{aligned}$$

The right-hand side of the above inequality is independent of y and tends to zero as $t_1 \longrightarrow t_2$. In view of the foregoing three steps, the Arzelá-Ascoli theorem applies. In consequence, we conclude that $\mathbf{F}_1 : C([0,T],\mathbb{R}) \longrightarrow C([0,T],\mathbb{R})$ is completely continuous.

Step 4. *We show that there exists an open set $U \subseteq C([0,T],\mathbb{R})$ with $y \ne \nu\mathbf{F}_1(y)$ for $\nu \in (0,1)$ and $y \in \partial U$.*

Let $y \in C([0,T],\mathbb{R})$ be a solution of $y = \nu\mathbf{F_1}y$ for $\nu \in [0,1]$. Then, for $t \in [0,T]$, we have

$$\begin{aligned}
|y(t)| &= |\nu(\mathbf{F_1}y)(t)|\\
&\le {}^\rho I^\alpha |f(t,y(t))| + \frac{t^{\rho(\alpha-1)}}{|\Lambda_1|}\left\{|\lambda|\ {}^\rho I^{\alpha+\beta}|f(\xi,y(\xi))| + {}^\rho I^\alpha|f(T,y(T))|\right\}\\
&\le {}^\rho I^\alpha p(T)\Omega(\|y\|) + \frac{t^{\rho(\alpha-1)}}{|\Lambda_1|}\left\{|\lambda|\ {}^\rho I^{\alpha+\beta}p(\xi)\Omega(\|y\|) + {}^\rho I^\alpha p(T)\Omega(\|y\|)\right\}
\end{aligned}$$

$$\leq \Omega(\|y\|)\Big({}^{\rho}I^{\alpha}p(T) + \frac{T^{\rho(\alpha-1)}}{|\Lambda_1|}\Big\{|\lambda|\ {}^{\rho}I^{\alpha+\beta}p(\xi) + {}^{\rho}I^{\alpha}p(T)\Big\}\Big),$$

which, on taking the norm for $t \in [0,T]$, implies that

$$\frac{\|y\|}{\Omega(\|y\|)\Big({}^{\rho}I^{\alpha}p(T) + \frac{T^{\rho(\alpha-1)}}{|\Lambda_1|}\Big\{|\lambda|\ {}^{\rho}I^{\alpha+\beta}p(\xi) + {}^{\rho}I^{\alpha}p(T)\Big\}\Big)} \leq 1.$$

In view of (7.11.3), there exists K such that $\|y\| \neq \mathrm{K}$. Let us set

$$U = \{y \in C([0,T],\mathbb{R}) : \|y\| < \mathrm{K}\}.$$

Note that the operator $\mathbf{F_1} : \overline{U} \to C([0,T],\mathbb{R})$ is continuous and completely continuous.

From the choice of U, there is no $y \in \partial U$ such that $y = \nu\mathbf{F_1}(y)$ for some $\nu \in (0,1)$. Consequently, by the nonlinear alternative of Leray-Schauder type (Theorem 1.8), we deduce that $\mathbf{F_1}$ has a fixed point $y \in \overline{U}$ which is a solution of the problem (7.55) and (7.57). This completes the proof. □

Remark 7.3. The condition (7.11.1) can be modified by assuming $p \in C([0,T],\mathbb{R}^+)$ instead of $p \in L^1([0,T],\mathbb{R}^+)$. In this case, the constant K in condition (7.11.2) will take the form:

$$\frac{K}{\Omega(K)\|p\|A_1} > 1.$$

As a special case, when $p(t) = 1$ and $\Omega(\|y\|) = d_1\|y\| + d_2$, we have the following corollary.

Corollary 7.3. *Let $f : [0,T] \times \mathbb{R} \to \mathbb{R}$ be a continuous function. Assume that*

(7.3.1) there exist constants $0 \leq d_1 < \dfrac{1}{A_1}$, and $d_2 > 0$ such that

$$|f(t,y)| \leq d_1|y| + d_2 \quad \textit{for all } t \in [0,T],\ y \in \mathbb{R}.$$

Then the problem (7.55) and (7.57) has at least one solution on $[0,T]$.

In the last result, we prove the existence of solutions for the problem (7.55) and (7.57) by applying Krasnosel'skiĭ's fixed point theorem.

Theorem 7.12. *Let $f : [0,T]\times\mathbb{R} \to \mathbb{R}$ be a continuous function, and the assumption (7.10.1) holds. In addition, we assume that*

(7.12.1) $|f(t,y)| \leq \theta(t)$, $\forall (t,y) \in [0,T] \times \mathbb{R}$, and $\theta \in C([0,T],\mathbb{R}^+)$.

Then the problem (7.55) and (7.57) has at least one solution on $[0,T]$, provided that

$$\vartheta\frac{T^{\rho(\alpha-1)}}{|\Lambda_1|}\Big\{|\lambda|\frac{\xi^{\rho(\alpha+\beta)}}{\rho^{\alpha+\beta}\Gamma(\alpha+\beta+1)} + \frac{T^{\rho\alpha}}{\rho^{\alpha}\Gamma(\alpha+1)}\Big\} < 1. \tag{7.73}$$

Proof. Letting $\sup_{t\in[0,T]}|\theta(t)| = \|\theta\|$ and $\overline{r} \geq A_1\|\theta\|$, we define $B_{\overline{r}} = \{y \in C([0,T],\mathbb{R}) : \|y\| \leq \overline{r}\}$. Introduce operators $\mathcal{P}$ and $\mathcal{Q}$ on $B_{\overline{r}}$ as

$$\mathcal{P}(y)(t) = {}^{\rho}I^{\alpha} f(s, y(s))(t),$$

$$\mathcal{Q}(y)(t) = \frac{t^{\rho(\alpha-1)}}{\Lambda_1}\Big\{\lambda\ {}^{\rho}I^{\alpha+\beta} f(s, y(s))(\xi) - {}^{\rho}I^{\alpha} f(s, y(s))(T)\Big\}.$$

Note that $\mathbf{F_1} = \mathcal{P} + \mathcal{Q}$.

For $y, z \in B_{\overline{r}}$, it easy to find that

$$\begin{aligned}
&\|\mathcal{P}y + \mathcal{Q}z\| \\
&\leq \|\theta\|\Big[\frac{T^{\rho\alpha}}{\rho^{\alpha}\Gamma(\alpha+1)} + \frac{T^{\rho(\alpha-1)}}{|\Lambda_1|}\Big\{|\lambda|\frac{\xi^{\rho(\alpha+\beta)}}{\rho^{\alpha+\beta}\Gamma(\alpha+\beta+1)} + \frac{T^{\rho\alpha}}{\rho^{\alpha}\Gamma(\alpha+1)}\Big\}\Big] \leq \overline{r}.
\end{aligned}$$

Thus, $\mathcal{P}y + \mathcal{Q}z \in B_{\overline{r}}$.

It follows from the assumptions (7.10.1) and (7.73) that $\mathcal{Q}$ is a contraction mapping, that is, for $y, z \in C([0,T],\mathbb{R})$ and for each $t \in [0,T]$, we have

$$\begin{aligned}
&|\mathcal{Q}(y)(t) - \mathcal{Q}(z)(t)| \\
&\leq \vartheta\frac{T^{\rho(\alpha-1)}}{|\Lambda_1|}\Big\{|\lambda|\frac{\xi^{\rho(\alpha+\beta)}}{\rho^{\alpha+\beta}\Gamma(\alpha+\beta+1)} + \frac{T^{\rho\alpha}}{\rho^{\alpha}\Gamma(\alpha+1)}\Big\}\|y - z\|.
\end{aligned}$$

Continuity of f implies that the operator $\mathcal{P}$ is continuous. Also, $\mathcal{P}$ is uniformly bounded on $B_{\overline{r}}$ as

$$\|\mathcal{P}x\| \leq \|\theta\|\frac{T^{\rho\alpha}}{\rho^{\alpha}\Gamma(\alpha+1)}.$$

Finally we prove the compactness of the operator $\mathcal{P}$. For that we define $\sup_{(t,x)\in[0,T]\times B_{\overline{r}}}|f(t,y)| = \bar{f} < \infty$. Then, for $0 < t_1 < t_2 < T$, we have

$$\begin{aligned}
&|\mathcal{P}y(t_2) - \mathcal{P}y(t_1)| \\
&\leq \frac{\rho^{1-\alpha}\bar{f}}{\Gamma(\alpha)}\left|\int_0^{t_1}\Big[\frac{s^{\rho-1}}{(t_2^{\rho} - s^{\rho})^{1-\alpha}} - \frac{s^{\rho-1}}{(t_1^{\rho} - s^{\rho})^{1-\alpha}}\Big]ds + \int_{t_1}^{t_2}\frac{s^{\rho-1}}{(t_2^{\rho} - s^{\rho})^{1-\alpha}}ds\right| \\
&\leq \frac{\bar{f}}{\rho^{\alpha}\Gamma(\alpha+1)}\left|2(t_2^{\rho} - t_1^{\rho})^{\alpha} + t_1^{\rho\alpha} - t_2^{\rho\alpha}\right|,
\end{aligned}$$

which is independent of y and tends to zero as $t_2 - t_1 \to 0$. Thus, $\mathcal{P}$ is equicontinuous. So $\mathcal{P}$ is relatively compact on $B_{\overline{r}}$. Hence, by the Arzelá-Ascoli theorem, $\mathcal{P}$ is compact on $B_{\overline{r}}$. Thus all the assumptions of Theorem 7.12 are satisfied. Hence, by Theorem 1.4, we deduce that the problem (7.55) and (7.57) has at least one solution on $[0,T]$. □

Example 7.6. Consider the following boundary value problem

$$\begin{cases} {}^{1/2}D^{7/5}y(t) = \dfrac{1}{20(t+1)}\Big(\dfrac{|y|+2}{|y|+1} + \cos t\Big), \quad t \in [0,2], \\ y(0) = 0, \quad y(2) = 1/3\ {}^{1/2}I^{4/5}y(3/2), \end{cases} \tag{7.74}$$

where $\rho = 1/2$, $\alpha = 7/5$, $\lambda = 1/3$, $\beta = 4/5$, $\xi = 3/2$ and $T = 2$. Using the given data, we find that $|\Lambda_1| \approx 0.5526156533$, $A_1 \approx 12.67699740$, where Λ_1 and A_1 are respectively given by (7.60) and (7.70). One can easily check that $f(t, y)$ is continuous and the condition (7.10.1) of Theorem 7.10 holds true with $\vartheta = 1/20$. Also $\vartheta A_1 \approx 0.63384987 < 1$. Thus all the conditions of Theorem 7.10 are satisfied. So, by the conclusion of Theorem 7.10, the problem (7.74) has a unique solution on $[0, 2]$.

Further the hypothesis of Theorem 7.12 is satisfied with $\theta(t) = \dfrac{2 + \cos t}{20(1 + t)}$. Also

$$\vartheta \frac{T^{\rho(\alpha-1)}}{|\Lambda_1|} \Big\{ |\lambda| \frac{\xi^{\rho(\alpha+\beta)}}{\rho^{\alpha+\beta}\Gamma(\alpha+\beta+1)} + \frac{T^{\rho\alpha}}{\rho^\alpha \Gamma(\alpha+1)} \Big\} \approx 0.4612850793 < 1.$$

Thus the conclusion of Theorem 7.12 applies to the problem (7.74).

7.4.2 *Existence results for the problem (7.55) with (7.58) and (7.59)*

Associated with the problem (7.55) and (7.58), we introduce an operator $\mathbf{F_2}$: $C([0, 1], \mathbb{R}) \to C([0, T], \mathbb{R})$ as

$$\mathbf{F_2}(y)(t) = {}^\rho I^\alpha f(t, y(t)) + \frac{t^{\rho(\alpha-1)}}{\Lambda_2} \Big\{ \sum_{i=1}^n \lambda_i \, {}^\rho I^\alpha f(\xi_i, y(\xi_i)) - {}^\rho I^\alpha f(T, y(T)) \Big\}, \quad (7.75)$$

where Λ_2 is defined by (7.65).

Furthermore, we set

$$A_2 = \frac{T^{\rho\alpha}}{\rho^\alpha \Gamma(\alpha+1)} + \frac{T^{\rho(\alpha-1)}}{|\Lambda_2|} \Big\{ \sum_{i=1}^n |\lambda_i| \frac{\xi_i^{\rho\alpha}}{\rho^\alpha \Gamma(\alpha+1)} + \frac{T^{\rho\alpha}}{\rho^\alpha \Gamma(\alpha+1)} \Big\}. \quad (7.76)$$

The existence results for the problem (7.55) and (7.58), similar to ones for the problem (7.55) and (7.57) obtained in Section 3, can be proved with the aid of the operator $\mathbf{F_2}$ and the constant A_2 given by (7.75) and (7.76) respectively. So, we formulate the results without proof.

Theorem 7.13. *Suppose that the condition (7.10.1) holds. Then the problem (7.55) and (7.58) has a unique solution on $[0, T]$ if $\vartheta A_2 < 1$, where A_2 is given by (7.76).*

Theorem 7.14. *Assume that (7.11.1) and the following condition hold:*

(7.14.1) there exists a constant $K > 0$ such that

$$\frac{K}{\Omega(K)\Big({}^\rho I^\alpha p(T) + \dfrac{T^{\rho(\alpha-1)}}{|\Lambda_2|} \{ \sum_{i=1}^n |\lambda_i| {}^\rho I^\alpha p(\xi_i) + {}^\rho I^\alpha p(T) \} \Big)} > 1.$$

Then the problem (7.55) and (7.58) has at least one solution on $[0, T]$.

Theorem 7.15. *Assume that the condition (7.10.1) and (7.12.1) hold. Then the problem (7.55) and (7.58) has at least one solution on* $[0, T]$, *provided that*

$$\vartheta \frac{T^{\rho(\alpha-1)}}{|\Lambda_1|} \Big\{ \sum_{i=1}^{n} |\lambda_i| \frac{\xi_i^{\rho(\alpha)}}{\rho^\alpha \Gamma(\alpha+1)} + \frac{T^{\rho\alpha}}{\rho^\alpha \Gamma(\alpha+1)} \Big\} < 1. \tag{7.77}$$

Next, for the problem (7.55) and (7.59), we define an operator $\mathbf{F_3} : C([0,1], \mathbb{R}) \to C([0,T], \mathbb{R})$ by

$$\mathbf{F_3}(y)(t) = {}^{\rho}I^{\alpha} f(t, y(t)) + \frac{t^{\rho(\alpha-1)}}{\Lambda_3} \Big\{ \sum_{i=1}^{n} \lambda_i \, {}^{\rho}I^{\alpha+\beta} f(\xi_i, y(\xi_i)) - {}^{\rho}I^{\alpha} f(T, y(T)) \Big\}, \tag{7.78}$$

where Λ_3 is defined by (7.67).

Let us set

$$A_3 = \frac{T^{\rho\alpha}}{\rho^\alpha \Gamma(\alpha+1)} + \frac{T^{\rho(\alpha-1)}}{|\Lambda_3|} \Big\{ \sum_{i=1}^{n} |\lambda_i| \frac{\xi_i^{\rho(\alpha+\beta)}}{\rho^{\alpha+\beta} \Gamma(\alpha+\beta+1)} + \frac{T^{\rho\alpha}}{\rho^\alpha \Gamma(\alpha+1)} \Big\}. \tag{7.79}$$

As argued before, the existence results for the problem (7.55) and (7.59) can be obtained with the help of the operator $\mathbf{F_3}$ and the constant A_3. However, we state these results below.

Theorem 7.16. *Suppose that the condition (7.10.1) holds. Then the problem (7.55) and (7.59) has a unique solution on* $[0, T]$ *if* $\vartheta A_3 < 1$, *where* A_3 *is given by (7.79).*

Theorem 7.17. *Assume that (7.11.1) and the following condition hold:*

(7.16.1) there exists a constant $K > 0$ *such that*

$$\frac{K}{\Omega(K)\Big({}^{\rho}I^{\alpha} p(T) + \frac{T^{\rho(\alpha-1)}}{|\Lambda_3|} \Big\{ \sum_{i=1}^{n} |\lambda_i| {}^{\rho}I^{\alpha+\beta} p(\xi_i) + {}^{\rho} I^{\alpha} p(T) \Big\} \Big)} > 1.$$

Then the problem (7.55) and (7.59) has at least one solution on $[0, T]$.

Theorem 7.18. *Assume that the conditions (7.10.1) and (7.12.1) hold. Then the problem (7.55) and (7.59) has at least one solution on* $[0, T]$, *provided that*

$$\vartheta \frac{T^{\rho(\alpha-1)}}{|\Lambda_3|} \Big\{ \sum_{i=1}^{n} |\lambda_i| \frac{\xi_i^{\rho(\alpha+\beta)}}{\rho^{\alpha+\beta} \Gamma(\alpha+\beta+1)} + \frac{T^{\rho\alpha}}{\rho^\alpha \Gamma(\alpha+1)} \Big\} < 1. \tag{7.80}$$

Example 7.7. Let us consider the following boundary value problem

$$\begin{cases} {}^{1/2}D^{7/5} y(t) = f(t, y(t)), \ t \in [0, 2], \\ y(0) = 0, \quad y(2) = y(3/4) + 1/3 \; y(3/2), \end{cases} \tag{7.81}$$

where $\rho = 1/2$, $\alpha = 7/5$, $\lambda_1 = 1$, $\lambda_2 = 1/3$, $\xi_1 = 3/4$, $\xi_2 = 3/2$ and $T = 2$ and $f(t, y(t))$ will be fixed later.

Using the given values, we find that $|\Lambda_2| \approx 0.1568797472$, $A_2 \approx 48.32814774$, where Λ_2 and A_2 are respectively given by (7.65) and (7.76). For illustrating Theorem 7.13, we take

$$f(t,y) = \frac{1}{3\sqrt{900+t^2}} e^{-t}\Big(\tan^{-1} y + \cos t\Big). \tag{7.82}$$

Notice that $f(t,x)$ is continuous and satisfies the condition (7.10.1) with $\vartheta = 1/90$. Also $\vartheta A_2 \approx 0.5369794193 < 1$. Thus all the conditions of Theorem 7.13 are satisfied. Hence the conclusion of Theorem 7.13 applies to the problem (7.81) with $f(t,y)$ given by (7.82).

In order to illustrate Theorem 7.14, we take

$$f(t,y(t)) = \frac{(1+t)}{90}\Big(\tan^{-1} y + y + \frac{1}{8}\Big), \tag{7.83}$$

where $p(t) = \dfrac{(1+t)}{90}$ and $\psi(||y||) = ||y|| + \dfrac{4\pi+1}{8}$. By condition (7.11.2), we have $K > 5.021036057$. Thus all the conditions of Theorem 7.14 are satisfied and consequently the problem 7.81 with $f(t,y)$ given by (7.83) has at least one solution on $[0,2]$.

Example 7.8. Consider the following boundary value problem

$$\begin{cases} {}^{1/2}D^{7/5}y(t) = f(t,y(t)),\ t \in [0,2], \\ y(0) = 0, \quad y(2) = {}^{1/2}I^{4/5}y(3/4) + 1/3\ {}^{1/2}I^{4/5}y(3/2), \end{cases} \tag{7.84}$$

where $\rho = 1/2$, $\alpha = 7/5$, $\lambda_1 = 1$, $\lambda_2 = 1/3$, $\beta = 4/5$, $\xi_1 = 3/4$, $\xi_2 = 3/2$ and $T = 2$ and $f(t,y(t))$ will be defined later.

Using the given data, we find that $|\Lambda_3| \approx 0.6271881077, A_3 \approx 14.11003320$, where Λ_3 and A_3 are respectively given by (7.67) and (7.79). For illustrating Theorem 7.16, we take

$$f(t,y) = \frac{1}{20(t+1)}\Big(\frac{|y|+2}{|y|+1} + \cos t\Big). \tag{7.85}$$

Observe that $f(t,x)$ is continuous and satisfies the condition (7.10.1) with $\vartheta = 1/20$. Also $\vartheta A_3 \approx 0.70550166 < 1$. Thus all the conditions of Theorem 7.16 are satisfied and hence its conclusion applies to the problem (7.84) with $f(t,y)$ given by (7.85).

Next we illustrate Theorem 7.18 by taking $f(t,y)$ given by (7.85). Clearly the hypothesis of Theorem 7.18 holds true with $\theta(t) = \dfrac{2+\cos t}{20(1+t)}$. Also

$$\vartheta\frac{T^{\rho(\alpha-1)}}{|\Lambda_3|}\Big\{\sum_{i=1}^{n}|\lambda_i|\frac{\xi_i^{\rho(\alpha+\beta)}}{\rho^{\alpha+\beta}\Gamma(\alpha+\beta+1)} + \frac{T^{\rho\alpha}}{\rho^{\alpha}\Gamma(\alpha+1)}\Big\} \approx 0.5329368695 < 1.$$

Hence we deduce by Theorem 7.18 that there exists at least one solution for the problem (7.84) with $f(t,y)$ given by (7.85).

Main results for the multi-valued problem (7.56) and (7.57)

This section is devoted to the existence of solutions for the problem (7.56) and (7.57). The existence of solutions for the problems (7.56) and (7.58), and (7.56) and (7.59) can be studied by employing the strategy used for the problem (7.56) and (7.57).

7.4.3 *The Carathéodory case*

In this subsection, we consider the case when F has convex values and is of Carathéodory type, and prove an existence result for the problem (7.56) and (7.57) by applying nonlinear alternative of Leray-Schauder type (Theorem 1.16).

Definition 7.6. A function $y \in C^2([0,T],\mathbb{R})$ is said to be a solution of the boundary value problem (7.56)-(7.57) if $y(0)=0, y(T)=\lambda\ {}^{\rho}I^{\beta}y(\xi)$ and there exists function $v \in L^1([0,1],\mathbb{R})$ such that $v(t)\in F(t,y(t))$ a.e. on $[0,T]$ and

$$y(t) = {}^{\rho}I^{\alpha}v(t) + \frac{t^{\rho(\alpha-1)}}{\Lambda_1}\Big\{\lambda\ {}^{\rho}I^{\alpha+\beta}v(\xi) - {}^{\rho}I^{\alpha}v(T)\Big\}.$$

Theorem 7.19. *Assume that*

(7.19.1) $F:[0,T]\times\mathbb{R}\to\mathcal{P}_{cp,c}(\mathbb{R})$ *is* L^1*-Carathéodory;*

(7.19.2) there exists a continuous non-decreasing function $\Phi:[0,\infty)\to(0,\infty)$ *and a function* $p\in L^1([0,T],\mathbb{R}^+)$ *such that*

$$\|F(t,y)\|_{\mathcal{P}} := \sup\{|x| : x\in F(t,y)\} \le p(t)\Phi(\|y\|),$$

for each $(t,y)\in[0,T]\times\mathbb{R}$;

(7.19.3) there exists a constant $\widehat{K}>0$ *such that*

$$\frac{\widehat{K}}{\Phi(\widehat{K})\Big({}^{\rho}I^{\alpha}p(T) + \dfrac{T^{\rho(\alpha-1)}}{\Lambda_1}\Big(|\lambda|\ {}^{\rho}I^{\alpha+\beta}p(\xi) + {}^{\rho}I^{\alpha}p(T)\Big)\Big)} > 1.$$

Then the problem (7.56) and (7.57) has at least one solution on $[0,T]$.

Proof. To transform the problem (7.56) and (7.57) into a fixed point problem, we define an operator $\mathcal{F}: C([0,T],\mathbb{R}) \longrightarrow \mathcal{P}(C([0,T],\mathbb{R}))$ by

$$\mathcal{F}(y) = \{h\in C([0,T],\mathbb{R}) : h(t) = N(y)(t)\}, \tag{7.86}$$

where

$$N(y)(t) = {}^{\rho}I^{\alpha}v(t) + \frac{t^{\rho(\alpha-1)}}{\Lambda_1}\Big\{\lambda\ {}^{\rho}I^{\alpha+\beta}v(\xi) + {}^{\rho}I^{\alpha}v(T)\Big\},$$

for $v\in S_{F,x}$. It is obvious that the fixed points of $\mathcal{F}$ are solutions of the problem (7.56) and (7.57).

We will show that $\mathcal{F}$ satisfies the assumptions of Leray-Schauder nonlinear alternative (Theorem 1.16) in several steps.

Step 1. *$\mathcal{F}(y)$ is convex for each $y\in C([0,T],\mathbb{R})$.*

This step is obvious since $S_{F,y}$ is convex (F has convex values).

Step 2. *$\mathcal{F}$ maps bounded sets (balls) into bounded sets in $C([0,T],\mathbb{R})$.*

For a positive number r, let $B_r = \{y \in C([0,T],\mathbb{R}) : \|y\| \leq r\}$ be a bounded ball in $C([0,T],\mathbb{R})$. Then, for each $h \in \mathcal{F}(y)$, $y \in B_r$, there exists $v \in S_{F,y}$ such that

$$h(t) = {}^\rho I^\alpha v(t) + \frac{t^{\rho(\alpha-1)}}{\Lambda_1}\Big\{\lambda\ {}^\rho I^{\alpha+\beta} v(\xi) + {}^\rho I^\alpha v(T)\Big\}.$$

So, for $t \in [0,T]$, we have

$$\begin{aligned}|h(t)| &\leq {}^\rho I^\alpha |v(T)| + \frac{t^{\rho(\alpha-1)}}{|\Lambda_1|}\Big\{|\lambda|\ {}^\rho I^{\alpha+\beta}|v(\xi)| + {}^\rho I^\alpha|v(T)|\Big\} \\ &\leq \Phi(\|y\|)\Big({}^\rho I^\alpha p(T) + \frac{T^{\rho(\alpha-1)}}{|\Lambda_1|}\Big\{|\lambda|\ {}^\rho I^{\alpha+\beta}p(\xi) + {}^\rho I^\alpha p(T)\Big\}\Big).\end{aligned}$$

Thus

$$\|h\| \leq \Phi(r)\Big({}^\rho I^\alpha p(T) + \frac{T^{\rho(\alpha-1)}}{|\Lambda_1|}\Big\{|\lambda|\ {}^\rho I^{\alpha+\beta}p(\xi) + {}^\rho I^\alpha p(T)\Big\}\Big) := \ell.$$

Step 3. *$\mathcal{F}$ maps bounded sets into equi-continuous sets of $C([0,T],\mathbb{R})$.*

Let $t_1, t_2 \in (0,T]$, $t_1 < t_2$, and let $y \in B_r$. Then

$$\begin{aligned}&|h(t_2) - h(t_1)| \\ \leq\ & \Big|{}^\rho I^\alpha|v(t_2)| - {}^\rho I^\alpha|v(t_1)|\Big| \\ &+ \frac{\Phi(r)|t_2^{\rho(\alpha-1)} - t_1^{\rho(\alpha-1)}|}{\Lambda_1}\Big\{|\lambda|{}^\rho I^{\alpha+\beta}p(\xi) + {}^\rho I^\alpha p(T)\Big\} \\ \leq\ & \frac{\rho^{1-\alpha}\Phi(r)}{\Gamma(\alpha)}\Big|\int_0^{t_1}\Big[\frac{s^{\rho-1}}{(t_2^\rho - s^\rho)^{1-\alpha}} - \frac{s^{\rho-1}}{(t_1^\rho - s^\rho)^{1-\alpha}}\Big]p(s)ds + \int_{t_1}^{t_2}\frac{s^{\rho-1}}{(t^\rho - s^\rho)^{1-\alpha}}p(s)ds\Big| \\ &+ \frac{\Phi(r)|t_2^{\rho(\alpha-1)} - t_1^{\rho(\alpha-1)}|}{\Lambda_1}\Big\{|\lambda|{}^\rho I^{\alpha+\beta}p(\xi) + {}^\rho I^\alpha p(T)\Big\}.\end{aligned}$$

Obviously the right hand side of the above inequality tends to zero independently of $y \in B_r$ as $t_2 - t_1 \to 0$. Therefore it follows by the Arzelá-Ascoli theorem that $\mathcal{F} : C([0,T],\mathbb{R}) \to \mathcal{P}(C([0,T],\mathbb{R}))$ is completely continuous.

In our next step, we show that $\mathcal{F}$ is u.s.c. Since $\mathcal{F}$ is completely continuous, it is enough to establish that it has a closed graph.

Step 4. *$\mathcal{F}$ has a closed graph.*

Let $y_n \to y_*, h_n \in \mathcal{F}(y_n)$ and $h_n \to h_*$. Then we need to show that $h_* \in \mathcal{F}(y_*)$. Associated with $h_n \in \mathcal{F}(y_n)$, there exists $v_n \in S_{F,y_n}$ such that, for each $t \in [0,T]$,

$$h_n(t) = {}^\rho I^\alpha v_n(t) + \frac{t^{\rho(\alpha-1)}}{\Lambda_1}\Big\{\lambda\ {}^\rho I^{\alpha+\beta} v_n(\xi) - {}^\rho I^\alpha v_n(s) v_n(T)\Big\}.$$

Thus it suffices to show that there exists $v_* \in S_{F,y_*}$ such that, for each $t \in [0,T]$,

$$h_*(t) = {}^\rho I^\alpha v_*(t) + \frac{t^{\rho(\alpha-1)}}{\Lambda_1}\Big\{\lambda\ {}^\rho I^{\alpha+\beta} v_*(\xi) - {}^\rho I^\alpha v_*(T)\Big\}.$$

Let us consider the linear operator $\Theta : L^1([0,T],\mathbb{R}) \to C([0,T],\mathbb{R})$ given by

$$v \mapsto \Theta v(t) = {}^\rho I^\alpha v(t) + \frac{t^{\rho(\alpha-1)}}{\Lambda_1}\Big\{\lambda\ {}^\rho I^{\alpha+\beta} v(\xi) - {}^\rho I^\alpha v(T)\Big\}.$$

Observe that $\|h_n(t) - h_*(t)\| \to 0$ as $n \to \infty$, so it follows by Lemma 1.2 that $\Theta \circ S_F$ is a closed graph operator. Further, we have $h_n(t) \in \Theta(S_{F,y_n})$. Since $y_n \to y_*$, therefore, we have

$$h_*(t) = {}^\rho I^\alpha v_*(t) + \frac{t^{\rho(\alpha-1)}}{\Lambda_1}\Big\{\lambda\ {}^\rho I^{\alpha+\beta} v_*(\xi) - {}^\rho I^\alpha v_*(T)\Big\},$$

for some $v_* \in S_{F,y_*}$.

Step 5. *We show that there exists an open set $U \subseteq C([0,T],\mathbb{R})$ with $y \notin \nu\mathcal{F}(y)$ for any $\nu \in (0,1)$ and all $y \in \partial U$.*

Let $\nu \in (0,1)$ and $y \in \nu\mathcal{F}(y)$. Then there exists $v \in L^1([0,T],\mathbb{R})$ with $v \in S_{F,y}$ such that, for $t \in [0,T]$, we have

$$y(t) = {}^\rho I^\alpha v(t) + \frac{t^{\rho(\alpha-1)}}{\Lambda_1}\Big\{\lambda\ {}^\rho I^{\alpha+\beta} v(\zeta) - {}^\rho I^\alpha v(T)\Big\}.$$

As in the second step, one can obtain

$$\begin{aligned}|y(t)| &\le {}^\rho I^\alpha |v(T)| + \frac{t^{\rho(\alpha-1)}}{|\Lambda_1|}\Big\{|\lambda|\ {}^\rho I^\alpha |v(\xi)| + {}^\rho I^\alpha |v(T)|\Big\}\\ &\le \Phi(\|y\|)\Big({}^\rho I^\alpha p(T) + \frac{T^{\rho(\alpha-1)}}{|\Lambda_1|}\Big\{|\lambda|\ {}^\rho I^{\alpha+\beta} p(\xi) + {}^\rho I^\alpha p(T)\Big\}\Big),\end{aligned}$$

which implies that

$$\frac{\|y\|}{\Phi(\|y\|)\Big({}^\rho I^\alpha p(T) + \frac{T^{\rho(\alpha-1)}}{|\Lambda_1|}\Big\{|\lambda|\ {}^\rho I^{\alpha+\beta} p(\xi) + {}^\rho I^\alpha p(T)\Big\}\Big)} \le 1.$$

In view of (7.19.3), there exists $\widehat{K}$ such that $\|y\| \ne \widehat{K}$. Let us set

$$U = \{y \in C(I,\mathbb{R}) : \|y\| < \widehat{K}\}.$$

Note that the operator $\mathcal{F} : \overline{U} \to \mathcal{P}(C(I,\mathbb{R}))$ is a compact multi-valued map and u.s.c. with convex closed values. From the choice of U, there is no $y \in \partial U$ such that $y \in \nu\mathcal{F}(y)$ for some $\nu \in (0,1)$. Consequently, by the nonlinear alternative of Leray-Schauder type (Theorem 1.16), we deduce that $\mathcal{F}$ has a fixed point $y \in \overline{U}$ which is a solution of the problem (7.56) and (7.57). This completes the proof. □

7.4.4 *The Lipschitz case*

In this subsection, we prove the existence of solutions for the problem (7.56) and (7.57) with non-convex valued right hand side by applying a fixed point theorem for multi-valued maps due to Covitz and Nadler (Theorem 1.19).

Theorem 7.20. *Assume that*

(7.20.1) $F : [0,T] \times \mathbb{R} \to \mathcal{P}_{cp}(\mathbb{R})$ *is such that* $F(\cdot, y) : [0,T] \to \mathcal{P}_{cp}(\mathbb{R})$ *is measurable for each* $y \in \mathbb{R}$*;*

(7.20.2) $H_d(F(t,y),F(t,\bar{y})) \le m(t)|y-\bar{y}|$ *for almost all* $t \in [0,T]$ *and* $y,\bar{y} \in \mathbb{R}$ *with* $m \in C([0,T],\mathbb{R}^+)$ *and* $d(0,F(t,0)) \le m(t)$ *for almost all* $t \in [0,T]$.

Then the problem (7.56) and (7.57) has at least one solution on $[0,T]$ *if* $\delta = \|m\|A_1 < 1$, *i.e.,*

$$\delta := \|m\|\Big[\frac{T^{\rho\alpha}}{\rho^{\alpha}\Gamma(\alpha+1)} + \frac{T^{\rho(\alpha-1)}}{|\Lambda_1|}\Big\{|\lambda|\frac{\xi^{\rho(\alpha+\beta)}}{\rho^{\alpha+\beta}\Gamma(\alpha+\beta+1)} + \frac{T^{\rho\alpha}}{\rho^{\alpha}\Gamma(\alpha+1)}\Big\}\Big] < 1. \tag{7.87}$$

Proof. Notice that the set $S_{F,y}$ is non-empty for each $y \in C([0,T],\mathbb{R})$ by the assumption (7.20.1). So F has a measurable selection (see Theorem III.6 [78]). Now we show that the operator $\mathcal{F}$ defined by (7.86) satisfies the assumptions of Theorem 1.19. Firstly we show that $\mathcal{F}(y) \in \mathcal{P}_{cl}((C[0,T],\mathbb{R}))$ for each $y \in C([0,T],\mathbb{R})$. Let $\{u_n\}_{n\ge 0} \in \mathcal{F}(y)$ be such that $u_n \to u$ $(n \to \infty)$ in $C([0,T],\mathbb{R})$. Then $u \in C([0,T],\mathbb{R})$ and there exists $v_n \in S_{F,y_n}$ such that, for each $t \in [0,T]$,

$$u_n(t) = {}^{\rho}I^{\alpha}v_n(t) + \frac{t^{\rho(\alpha-1)}}{\Lambda_1}\Big\{\lambda\ {}^{\rho}I^{\alpha+\beta}v_n(\xi) - {}^{\rho}I^{\alpha}v_n(T)\Big\}.$$

As F has compact values, we pass onto a subsequence (if necessary) to obtain that v_n converges to v in $L^1([0,T],\mathbb{R})$. Thus, $v \in S_{F,y}$ and for each $t \in [0,T]$, we have

$$u_n(t) \to u(t) = {}^{\rho}I^{\alpha}v(t) + \frac{t^{\rho(\alpha-1)}}{\Lambda_1}\Big\{\lambda\ {}^{\rho}I^{\alpha+\beta}v(\xi) - {}^{\rho}I^{\alpha}v(T)\Big\}.$$

Hence, $u \in \mathcal{F}(y)$.

Next we show that there exists $\delta < 1$ (defined by (7.87)) such that

$$H_d(\mathcal{F}(y),\mathcal{F}(\bar{y})) \le \delta\|y-\bar{y}\| \quad \text{for each} \quad y,\bar{y} \in C^2([0,T],\mathbb{R}).$$

Let $y,\bar{y} \in C^2([0,T],\mathbb{R})$ and $h_1 \in \mathcal{F}(y)$. Then there exists $v_1(t) \in F(t,y(t))$ such that, for each $t \in [0,T]$,

$$h_1(t) = {}^{\rho}I^{\alpha}v_1(t) + \frac{t^{\rho(\alpha-1)}}{\Lambda_1}\Big\{\lambda\ {}^{\rho}I^{\alpha+\beta}v_1(\xi) - {}^{\rho}I^{\alpha}v_1(T)\Big\}.$$

By (7.20.2), we have

$$H_d(F(t,y),F(t,\bar{y})) \le m(t)|y(t)-\bar{y}(t)|.$$

So, there exists $w \in F(t,\bar{y}(t))$ such that

$$|v_1(t)-w| \le m(t)|y(t)-\bar{y}(t)|, \quad t \in [0,T].$$

Define $U : [0,T] \to \mathcal{P}(\mathbb{R})$ by

$$U(t) = \{w \in \mathbb{R} : |v_1(t)-w| \le m(t)|y(t)-\bar{y}(t)|\}.$$

Since the multi-valued operator $U(t) \cap F(t,\bar{y}(t))$ is measurable (Proposition III.4 [78]), there exists a function $v_2(t)$ which is a measurable selection for U. So $v_2(t) \in F(t,\bar{y}(t))$ and for each $t \in [0,T]$, we have $|v_1(t)-v_2(t)| \le m(t)|y(t)-\bar{y}(t)|$.

For each $t \in [0, T]$, let us define

$$h_2(t) = {}^\rho I^\alpha v_2(t) + \frac{t^{\rho(\alpha-1)}}{\Lambda_1}\Big\{\lambda\ {}^\rho I^{\alpha+\beta} v_2(\xi) - {}^\rho I^\alpha v_2(T)\Big\}.$$

Then

$$\begin{aligned}
&|h_1(t) - h_2(t)| \\
&\le {}^\rho I^\alpha |v_1(t) - v_2(t)| + \frac{T^{\rho(\alpha-1)}}{|\Lambda_1|}\Big\{|\lambda|\ {}^\rho I^{\alpha+\beta}|v_1(\xi) - v_2(\xi)| + {}^\rho I^\alpha |v_1(T) - v_2(T)|\Big\} \\
&\le \|m\|\Big[\frac{T^{\rho\alpha}}{\rho^\alpha \Gamma(\alpha+1)} + \frac{T^{\rho(\alpha-1)}}{|\Lambda_1|}\Big\{|\lambda|\frac{\xi^{\rho(\alpha+\beta)}}{\rho^{\alpha+\beta}\Gamma(\alpha+\beta+1)} + \frac{T^{\rho\alpha}}{\rho^\alpha\Gamma(\alpha+1)}\Big\}\Big]\|y - \bar{y}\|.
\end{aligned}$$

Hence

$$\|h_1 - h_2\| \le \|m\|\Big[\frac{T^{\rho\alpha}}{\rho^\alpha \Gamma(\alpha+1)} + \frac{T^{\rho(\alpha-1)}}{|\Lambda_1|}\Big\{|\lambda|\frac{\xi^{\rho(\alpha+\beta)}}{\rho^{\alpha+\beta}\Gamma(\alpha+\beta+1)} + \frac{T^{\rho\alpha}}{\rho^\alpha\Gamma(\alpha+1)}\Big\}\Big]\|y - \bar{y}\|.$$

Analogously, interchanging the roles of y and $\overline{y}$, we can obtain

$$\begin{aligned}
&H_d(\mathcal{F}(y), \mathcal{F}(\bar{y})) \\
&\le \|m\|\Big[\frac{T^{\rho\alpha}}{\rho^\alpha \Gamma(\alpha+1)} + \frac{T^{\rho(\alpha-1)}}{|\Lambda_1|}\Big\{|\lambda|\frac{\xi^{\rho(\alpha+\beta)}}{\rho^{\alpha+\beta}\Gamma(\alpha+\beta+1)} + \frac{T^{\rho\alpha}}{\rho^\alpha\Gamma(\alpha+1)}\Big\}\Big]\|y - \bar{y}\|.
\end{aligned}$$

So $\mathcal{F}$ is a contraction. Therefore, it follows by Theorem 1.19, that $\mathcal{F}$ has a fixed point y which is a solution of (7.56) and (7.57). This completes the proof. □

Example 7.9. Consider the following boundary value problem

$$\begin{cases} {}^{1/2}D^{7/5} y(t) \in F(t, y(t)), \ t \in [0, 2], \\ y(0) = 0, \quad y(2) = 1/3\ {}^{1/2}I^{4/5} y(3/2), \end{cases} \tag{7.88}$$

where $\rho = 1/2$, $\alpha = 7/5$, $\lambda = 1/3$, $\beta = 4/5$, $\xi = 3/2$ and $T = 2$, and $F(t, y(t))$ will be fixed later.

Using the given data, we find that $|\Lambda_1| \approx 0.5526156533$, $A_1 \approx 12.67699740$, where Λ_1 and A_1 are respectively given by (7.60) and (7.70). For illustrating Theorem 7.19, we take

$$F(t, y(t)) = \Big[\frac{1}{250 + e^t}\Big(\frac{|y|}{|y|+1} + \frac{1}{2}\Big), \frac{(1+t)}{30}\Big(\tan^{-1} y + y + \frac{1}{8}\Big)\Big], \tag{7.89}$$

which, in view of (7.19.2), implies that $p(t) = \dfrac{(1+t)}{30}$ and $\Phi(\|y\|) = \|y\| + \dfrac{4\pi + 1}{8}$. By condition (7.19.3), it is found that $\widehat{K} > 2.668611421$. Thus all conditions of Theorem 7.19 are satisfied and consequently, there exists at least one solution for the problem (7.88) with $F(t, y(t))$ given by (7.89) on $[0, 2]$.

In order to demonstrate the application of Theorem 7.20, let us consider

$$F(t, y(t)) = \left[\frac{\sin t}{250}\Big(\frac{|y|}{|y|+1} + 1\Big), \frac{(t+1)\tan^{-1} y + t}{50}\right]. \tag{7.90}$$

Clearly

$$H_d(F(t, y), F(t, \bar{y})) \le \frac{(t+1)}{50}\|y - \bar{y}\|.$$

Letting $m(t) = \dfrac{(t+1)}{50}$, it is easy to check that $d(0, F(t,0)) \leq m(t)$ holds for almost all $t \in [0,2]$ and that $\delta \approx 0.760619844 < 1$ (δ is given by (7.87)). As the hypotheses of Theorem 7.20 are satisfied, we conclude that the problem (7.88) with $F(t, y(t))$ given by (7.90) has at least one solution on $[0,2]$.

7.5 Multi-term fractional differential equations and inclusions with generalized nonlocal fractional integro-differential boundary conditions

In this section, we introduce a new boundary value problem of fractional differential equations involving multiple Caputo type fractional derivative operators and generalized nonlocal fractional integro-differential boundary conditions. In precise terms, we consider the following problem:

$$\begin{cases} {}^cD^{\alpha}[{}^cD^{\beta}y(t) - g(t, y(t))] = f(t, y(t)), \ t \in [0, T], \\ y(0) = 0, \ \ {}^cD^{p}y(\eta) = \displaystyle\sum_{i=1}^{n} \lambda_i \ {}^{\rho_i}I^{\gamma_i,\delta_i}_{\theta_i,\kappa_i} y(\xi_i), \ \ \eta, \xi_i \in (0, T), \end{cases} \tag{7.91}$$

where ${}^cD^{\chi}$ is the Caputo fractional derivative operator of order $\chi \in (0,1)$, $\chi = \alpha, \beta, p$, ${}^{\rho_i}I^{\gamma_i,\delta_i}_{\theta_i,\kappa_i}$ is the generalized (Katugampola) fractional integral with $\gamma_i > 0, \rho_i, \eta_i, \kappa_i, \lambda_i \in \mathbb{R}$, $i = 1, 2, \ldots, n$, and $f, g : [0, T] \times \mathbb{R} \to \mathbb{R}$ are given continuous functions.

Lemma 7.11. *Let*

$$\Omega := \frac{\eta^{\beta-p}}{\Gamma(1+\beta-p)} - \sum_{i=1}^{n} \lambda_i \frac{1}{\Gamma(\delta_i+1)} \frac{\xi_i^{\kappa_i+\rho_i(\theta_i+\gamma_i)+\beta}}{\rho_i^{\delta_i}} \frac{\Gamma\Big(\theta_i + \frac{\beta}{\rho_i} + 1\Big)}{\Gamma\Big(\theta_i + \frac{\beta}{\rho_i} + \gamma_i + 1\Big)} \neq 0. \tag{7.92}$$

The function y is a solution of the problem (7.91) if and only if

$$\begin{aligned} y(t) &= J^{\beta}g(s, y(s))(t) + J^{\alpha+\beta}f(s, y(s))(t) \\ &\quad + \frac{t^{\beta}}{\Gamma(\beta+1)} \frac{1}{\Omega} \Big\{ \sum_{i=1}^{n} \lambda_i \ {}^{\rho_i}I^{\gamma_i,\delta_i}_{\theta_i,\kappa_i} (J^{\beta}g(s, y(s)) + J^{\alpha+\beta}f(s, y(s)))(\xi_i) \\ &\quad - J^{\beta-p}g(s, y(s))(\eta) - J^{\alpha+\beta-p}f(s, y(s))(\eta) \Big\}, \end{aligned} \tag{7.93}$$

where $J^{(.)}$ denotes Riemann-Liouville fractional integral of order (.).

Proof. The solution of the differential equation in (7.91) can be written as

$$y(t) = J^{\beta}g(s, y(s))(t) + J^{\alpha+\beta}f(s, y(s))(t) + \frac{t^{\beta}}{\Gamma(\beta+1)} c_1 + c_2, \tag{7.94}$$

where c_1 and $c_2 \in \mathbb{R}$ are arbitrary constants. From the first boundary condition in (7.91), we find that $c_2 = 0$. Taking Caputo fractional derivative operator of both sides of (7.94), after inserting $c_2 = 0$, we get

$${}^cD^{p}y(t) = J^{\beta-p}g(s, y(s))(t) + J^{\alpha+\beta-p}f(s, y(s))(t) + \frac{t^{\beta-p}}{\Gamma(\beta-p+1)} c_1, \tag{7.95}$$

while applying the generalized fractional integral of order $\delta_i, \gamma_i > 0$ to (7.94) yields

$$
\begin{aligned}
{}^{\rho_i}I^{\gamma_i,\delta_i}_{\theta_i,\kappa_i} y(t) &= {}^{\rho_i}I^{\gamma_i,\delta_i}_{\theta_i,\kappa_i}(I^{\beta} g(s,y(s)) + J^{\alpha+\beta} f(s,y(s)))(t) \\
&\quad + c_1 \frac{1}{\Gamma(1+\beta)} \frac{t^{\kappa_i+\rho_i(\theta_i+\gamma_i)+\beta}}{\rho_i^{\delta_i}} \frac{\Gamma\Big(\theta_i + \frac{\beta}{\rho_i} + 1\Big)}{\Gamma\Big(\theta_i + \frac{\beta}{\rho_i} + \gamma_i + 1\Big)}. \qquad (7.96)
\end{aligned}
$$

Using (7.95) and (7.96) in the second boundary condition of (7.91), we obtain

$$
\begin{aligned}
c_1 &= \frac{1}{\Omega}\Big\{ \sum_{i=1}^{n} \lambda_i \, {}^{\rho_i}I^{\gamma_i,\delta_i}_{\theta_i,\kappa_i}(I^{\beta} g(s,y(s)) + J^{\alpha+\beta} f(s,y(s)))(\xi_i) \\
&\quad - J^{\beta-p} g(s,y(s))(\eta) - J^{\alpha+\beta-p} f(s,y(s))(\eta) \Big\}.
\end{aligned}
$$

Inserting the values of c_1 and c_2 in (7.94), we get the solution (7.91). The converse follows by direct computation. This completes the proof. □

7.5.1 *Existence and uniqueness results for the problem (7.91)*

Let $\mathcal{C} = C([0,T],\mathbb{R})$ denote the Banach space of all continuous functions from $[0,T]$ to $\mathbb{R}$ endowed with the norm defined by $\|y\| = \sup_{t\in[0,T]} |x(t)|$. Using Lemma 7.11, we define an operator $\mathcal{N} : \mathcal{C} \to \mathcal{C}$ as

$$
\begin{aligned}
Ny(t) &= J^{\beta} g(s,y(s))(t) + J^{\alpha+\beta} f(s,y(s))(t) \\
&\quad + \frac{t^{\beta}}{\Gamma(\beta+1)} \frac{1}{\Omega}\Big\{ \sum_{i=1}^{n} \lambda_i \, {}^{\rho_i}I^{\gamma_i,\delta_i}_{\theta_i,\kappa_i}(J^{\beta} g(s,y(s)) + J^{\alpha+\beta} f(s,y(s)))(\xi_i) \\
&\quad - J^{\beta-p} g(s,y(s))(\eta) - J^{\alpha+\beta-p} f(s,y(s))(\eta) \Big\}. \qquad (7.97)
\end{aligned}
$$

Notice that the problem (7.91) has solutions if and only if the operator $\mathcal{N}$ has fixed points.

In the following, for brevity, we use the notation:

$$
\Lambda_1 = \frac{T^{\beta}}{\Gamma(\beta+1)} \left(1 + \frac{1}{|\Omega|} \left\{ \sum_{i=1}^{n} \lambda_i \frac{\xi_i^{\kappa_i+\rho_i(\theta_i+\gamma_i)+\beta}}{\rho_i^{\delta_i}} \frac{\Gamma(\theta_i + \frac{\beta}{\rho_i} + 1)}{\Gamma(\gamma_i + \theta_i + \frac{\beta}{\rho_i} + 1)} + \frac{\eta^{\beta-p}}{\Gamma(\beta-p+1)} \right\} \right), \qquad (7.98)
$$

and

$$
\begin{aligned}
\Lambda_2 &= \frac{T^{\alpha+\beta}}{\Gamma(\alpha+\beta+1)} + \frac{T^{\beta}}{|\Omega|\Gamma(\beta+1)} \left(\sum_{i=1}^{n} \lambda_i \frac{\xi_i^{\kappa_i+\rho_i(\theta_i+\gamma_i)+\alpha+\beta}}{\rho_i^{\delta_i}} \frac{\Gamma(\theta_i + \frac{\alpha+\beta}{\rho_i} + 1)}{\Gamma(\gamma_i + \theta_i + \frac{\alpha+\beta}{\rho_i} + 1)} \right. \\
&\quad \left. + \frac{\eta^{\alpha+\beta-p}}{\Gamma(\alpha+\beta-p+1)} \right). \qquad (7.99)
\end{aligned}
$$

Now we are in a position to present our main results. Our first existence result is based on Leray-Schauder nonlinear alternative (Theorem 1.8).

Theorem 7.21. *Assume that the following hypotheses hold:*

(7.21.1) $f, g : [0,T] \times \mathbb{R} \to \mathbb{R}$ *are continuous functions;*
(7.21.2) *there exist constants* $d_1 < 1/\Lambda_1$ *and* $d_2 \geq 0$ *such that*

$$|g(t,u)| \leq d_1\|u\| + d_2, \quad t \in [0,T], \ u \in \mathbb{R};$$

(7.21.3) *there exist a continuous non-decreasing function* $\psi : [0,\infty) \to (0,\infty)$ *and a function* $p \in C([0,T], \mathbb{R}^+)$ *such that*

$$|f(t,u)| \leq p(t)\psi(\|u\|) \ \textit{for each} \ (t,u) \in [0,T] \times \mathbb{R};$$

(7.21.4) *there exists a constant* $M > 0$ *such that*

$$\frac{(1 - d_1\Lambda_1)M}{d_2\Lambda_1 + \Lambda_2\|p\|\psi(M)} > 1,$$

where Λ_1 *and* Λ_2 *are respectively given by* (7.98) *and* (7.99).

Then the problem (7.91) has at least one solution on $[0,T]$.

Proof. We complete the proof in several steps. First of all, it will be shown that the operator $\mathcal{N} : C([0,T], \mathbb{R}) \to C([0,T], \mathbb{R})$ defined by (7.97) is continuous and completely continuous.

Step 1. *N is continuous.*

Let $\{y_n\}$ be a sequence such that $y_n \to y$ in $C([0,T], \mathbb{R})$. Then

$$\begin{aligned}
&|\mathcal{N}(y_n)(t) - \mathcal{N}(y)(t)| \\
&\leq J^{\beta}|g(s,y_n(s)) - g(s,y(s))|(t) + J^{\alpha+\beta}|f(s,y_n(s)) - f(s,y(s))|(t) \\
&\quad + \frac{T^{\beta}}{|\Omega|\Gamma(\beta+1)}\Big\{\sum_{i=1}^{n} \lambda_i \, {}^{\rho_i}I_{\theta_i,\kappa_i}^{\gamma_i,\delta_i}(J^{\beta}|g(s,y_n(s)) - g(s,y(s))|)(\xi_i) \\
&\quad + \sum_{i=1}^{n} \lambda_i \, {}^{\rho_i}I_{\theta_i,\kappa_i}^{\gamma_i,\delta_i}(J^{\alpha+\beta}|f(s,y_n(s)) - f(s,y(s))|)(\xi_i) \\
&\quad + J^{\beta-p}|g(s,y_n(s)) - g(s,y(s))|(\eta) + J^{\alpha+\beta-p}|f(s,y_n(s)) - f(s,y(s))|(\eta)\Big\} \\
&\leq \Lambda_1\|g(\cdot,y_n) - g(\cdot,y)\| + \Lambda_2\|f(\cdot,y_n) - f(\cdot,y)\|.
\end{aligned}$$

Since f, g are continuous functions, therefore, we have

$$\|\mathcal{N}(y_n) - \mathcal{N}(y)\| \leq \Lambda_1\|g(\cdot,y_n) - g(\cdot,y)\| + \Lambda_2\|f(\cdot,y_n) - f(\cdot,y)\| \to 0,$$

as $n \to \infty$.

Step 2. *$\mathcal{N}$ maps bounded sets into bounded sets in $C([0,T], \mathbb{R})$.*

Indeed, it is enough to show that for any $\theta > 0$, there exists a positive constant $\tilde{\ell}$ such that for each $y \in B_\theta = \{y \in C([0,T], \mathbb{R}) : \|y\| \leq \theta\}$, we have $\|N(y)\| \leq \tilde{\ell}$. By (7.21.2) and (7.21.3), for each $t \in [0,T]$, we have

$$|\mathcal{N}(y)(t)|$$

$$
\begin{aligned}
&\leq J^{\beta}|g(s,y(s))|(t) + J^{\alpha+\beta}|f(s,y(s))|(t) \\
&\quad + \frac{t^{\beta}}{|\Omega|\Gamma(\beta+1)}\Big\{\sum_{i=1}^{n}\lambda_i\, {}^{\rho_i}I^{\gamma_i,\delta_i}_{\theta_i,\kappa_i}(J^{\beta}|g(s,y(s))|)(\xi_i) \\
&\quad + \sum_{i=1}^{n}\lambda_i\, {}^{\rho_i}I^{\gamma_i,\delta_i}_{\theta_i,\kappa_i}(J^{\alpha+\beta}|f(s,y(s))|)(\xi_i) \\
&\quad + J^{\beta-p}|g(s,y(s))|(\eta) + J^{\alpha+\beta-p}|f(s,y(s))|(\eta)\Big\} \\
&\leq (d_1\|y\|+d_2)\frac{T^{\beta}}{\Gamma(\beta+1)} + \|p\|\psi(\|y\|)\frac{T^{\alpha+\beta}}{\Gamma(\alpha+\beta+1)} \\
&\quad + \frac{T^{\beta}}{|\Omega|\Gamma(\beta+1)}\Bigg\{\sum_{i=1}^{n}\lambda_i(d_1\|y\|+d_2)\frac{\xi_i^{\kappa_i+\rho_i(\theta_i+\gamma_i)+\beta}}{\rho_i^{\delta_i}}\frac{\Gamma(\theta_i+\frac{\beta}{\rho_i}+1)}{\Gamma(\gamma_i+\theta_i+\frac{\beta}{\rho_i}+1)} \\
&\quad + \sum_{i=1}^{n}\lambda_i\|p\|\psi(\|y\|)\frac{\xi_i^{\kappa_i+\rho_i(\theta_i+\gamma_i)+\alpha+\beta}}{\rho_i^{\delta_i}}\frac{\Gamma(\theta_i+\frac{\alpha+\beta}{\rho_i}+1)}{\Gamma(\gamma_i+\theta_i+\frac{\alpha+\beta}{\rho_i}+1)} \\
&\quad + (d_1\|y\|+d_2)\frac{\eta^{\beta-p}}{\Gamma(\beta-p+1)} + \|p\|\psi(\|y\|)\frac{\eta^{\alpha+\beta-p}}{\Gamma(\alpha+\beta-p+1)}\Bigg\},
\end{aligned}
$$

which, on taking the norm for $t \in [0,T]$, and using the notation (7.98) and (7.99), yields

$$
\|\mathcal{N}(y)\| \leq \Lambda_1(d_1\theta + d_2) + \Lambda_2\|p\|\psi(\theta) := \tilde{\ell}.
$$

Step 3. *$\mathcal{N}$ maps bounded sets into equi-continuous sets of $C([0,T],\mathbb{R})$.*

Let $t_1, t_2 \in [0,T]$, $t_1 < t_2$, B_θ be a bounded set of $C([0,T],\mathbb{R})$ as in Step 2, and let $y \in B_\theta$. Then

$$
\begin{aligned}
&|\mathcal{N}(y)(t_2) - \mathcal{N}(y)(t_1)| \\
&\leq \Big|\frac{1}{\Gamma(\beta)}\int_0^{t_1}\Big[(t_2-s)^{\beta-1} - (t_1-s)^{\beta-1}\Big]g(s,y(s))ds + \frac{1}{\Gamma(\beta)}\int_{t_1}^{t_2}(t_2-s)^{\beta-1}g(s,y(s))ds\Big| \\
&\quad + \Big|\frac{1}{\Gamma(\alpha+\beta)}\int_0^{t_1}\Big[(t_2-s)^{\alpha+\beta-1} - (t_1-s)^{\alpha+\beta-1}\Big]f(s,y(s))ds \\
&\quad + \frac{1}{\Gamma(\alpha+\beta)}\int_{t_1}^{t_2}(t_2-s)^{\alpha+\beta-1}f(s,y(s))ds\Big| \\
&\quad + \frac{|t_2^{\beta}-t_1^{\beta}|}{|\Omega|\Gamma(\beta+1)}\Big\{\sum_{i=1}^{n}\lambda_i\, {}^{\rho_i}I^{\gamma_i,\delta_i}_{\theta_i,\kappa_i}(J^{\beta}|g(s,y(s))|)(\xi_i) + \sum_{i=1}^{n}\lambda_i\, {}^{\rho_i}I^{\gamma_i,\delta_i}_{\theta_i,\kappa_i}(J^{\alpha+\beta}|f(s,y(s))|)(\xi_i) \\
&\quad + J^{\beta-p}|g(s,y(s))|(\eta) + J^{\alpha+\beta-p}|f(s,y(s))|(\eta)\Big\} \\
&\leq \frac{d_1\theta+d_2}{\Gamma(\beta+1)}[t_2^{\beta} - t_1^{\beta} + 2(t_2-t_1)^{\beta}] + \frac{\|p\|\psi(\theta)}{\Gamma(\alpha+\beta+1)}[t_2^{\alpha+\beta} - t_1^{\alpha+\beta} + 2(t_2-t_1)^{\alpha+\beta}] \\
&\quad + \frac{|t_2^{\beta}-t_1^{\beta}|}{|\Omega|\Gamma(\beta+1)}\Bigg\{\Bigg(\sum_{i=1}^{n}\lambda_i\frac{\xi_i^{\kappa_i+\rho_i(\theta_i+\gamma_i)+\beta}}{\rho_i^{\delta_i}}\frac{\Gamma(\theta_i+\frac{\beta}{\rho_i}+1)}{\Gamma(\gamma_i+\theta_i+\frac{\beta}{\rho_i}+1)} + \frac{\eta^{\beta-p}}{\Gamma(\beta-p+1)}\Bigg)(d_1\theta+d_2) \\
&\quad + \Bigg(\sum_{i=1}^{n}\lambda_i\frac{\xi_i^{\kappa_i+\rho_i(\theta_i+\gamma_i)+\alpha+\beta}}{\rho_i^{\delta_i}}\frac{\Gamma(\theta_i+\frac{\alpha+\beta}{\rho_i}+1)}{\Gamma(\gamma_i+\theta_i+\frac{\alpha+\beta}{\rho_i}+1)} + \frac{\eta^{\alpha+\beta-p}}{\Gamma(\alpha+\beta-p+1)}\Bigg)\|p\|\psi(\theta)\Bigg\}.
\end{aligned}
$$

Clearly the right-hand side of the above inequality tends to zero independent of y as $t_1 \longrightarrow t_2$. In consequence of Steps 1 to 3, it follows by the Arzelá-Ascoli theorem that the operator $\mathcal{N} : C([0,T],\mathbb{R}) \longrightarrow C([0,T],\mathbb{R})$ is continuous and completely continuous.

Step 4. *We show that there exists an open set $U \subseteq C([0,T],\mathbb{R})$ with $y \neq \mu\mathcal{N}(y)$ for $\mu \in (0,1)$ and $y \in \partial U$.*

Let $y \in C([0,T],\mathbb{R})$ and $y = \mu N(y)$ for some $0 < \mu < 1$. Then, for each $t \in [0,T]$, we have

$$\begin{aligned} y(t) &= \mu J^{\beta} g(s,y(s))(t) + \mu J^{\alpha+\beta} f(s,y(s))(t) \\ &+\mu\frac{t^{\beta}}{\Gamma(\beta+1)}\frac{1}{\Omega}\Big\{\sum_{i=1}^{n}\lambda_i\, {}^{\rho_i}I_{\theta_i,\kappa_i}^{\gamma_i,\delta_i}(J^{\beta}g(s,y(s)) + J^{\alpha+\beta}f(s,y(s)))(\xi_i) \\ &-J^{\beta-p}g(s,y(s))(\eta) - J^{\alpha+\beta-p}f(s,y(s))(\eta)\Big\}. \end{aligned}$$

As in Step 2, for each $t \in [0,T]$, it can be established that

$$|y(t)| \leq \Lambda_1(d_1\|y\| + d_2) + \Lambda_2\|p\|\psi(\|y\|),$$

which can be expressed as

$$\frac{(1 - d_1\Lambda_1)\|y\|}{d_2\Lambda_1 + \Lambda_2\|p\|\psi(\|y\|)} \leq 1.$$

In view of (7.21.4), there exists M such that $\|y\| \neq$ M. Let us set

$$U = \{y \in C([0,T],\mathbb{R}) : \|y\| < \mathrm{M}\}.$$

Note that the operator $\mathcal{N} : \overline{U} \to C([0,T],\mathbb{R})$ is continuous and completely continuous. From the choice of U, there is no $y \in \partial U$ such that $y = \mu\mathcal{N}y$ for some $\mu \in (0,1)$. Consequently, by the nonlinear alternative of Leray-Schauder type (Theorem 1.8), we deduce that $\mathcal{N}$ has a fixed point $y \in \overline{U}$ which is a solution of the problem (7.91). This completes the proof. □

In the following result, we prove the uniqueness of solutions for the problem (7.91) by using Banach's contraction mapping principle.

Theorem 7.22. *Assume that the following conditions hold:*

(7.22.1) there exists a non-negative constant k such that

$$|g(t,u) - g(t,v)| \leq k|u - v|, \quad \textit{for } t \in [0,T] \textit{ and every } u,v \in \mathbb{R};$$

(7.22.2) there exists $\ell > 0$ such that

$$|f(t,u) - f(t,v)| \leq \ell|u - v|, \quad \textit{for } t \in [0,T] \textit{ and every } u,v \in \mathbb{R}.$$

If

$$k\Lambda_1 + \ell\Lambda_2 < 1, \tag{7.100}$$

where Λ_1, Λ_2 are defined by (7.98) and (7.99) respectively, then there exists a unique solution for the problem (7.91) on $[0,T]$.

Proof. With Λ_1 and Λ_2 respectively given by (7.98) and (7.99), we fix

$$r \geq \frac{\Lambda_1 g_0 + \Lambda_2 f_0}{1 - k\Lambda_1 - \ell\Lambda_2}, \ g_0 = \sup_{t\in[0,T]} |g(t,0)|, \ f_0 = \sup_{t\in[0,T]} |f(t,0)|,$$

and show that $\mathcal{N}B_r \subset B_r$, where the operator $\mathcal{N} : C([0,T],\mathbb{R}) \to C([0,T],\mathbb{R})$ is given by (7.97) and $B_r = \{y \in C([0,T],\mathbb{R}) : \|y\| \leq r\}$. For $x \in B_r$, using (7.22.1) and (7.22.2), we have

$$\begin{aligned}
|\mathcal{N}(y)(t)| &\leq J^{\beta}[|g(s,y(s)) - g(s,0)| + |g(s,0)|](t) \\
&\quad + J^{\alpha+\beta}[|f(s,y(s)) - f(s,0)| + |f(s,0)|](t) \\
&\quad + \frac{t^{\beta}}{|\Omega|\Gamma(\beta+1)}\Bigg\{\sum_{i=1}^{n}\lambda_i \ {}^{\rho_i}I_{\theta_i,\kappa_i}^{\gamma_i,\delta_i}(J^{\beta}[|g(s,y(s)) - g(s,0)| + |g(s,0)|])(\xi_i) \\
&\quad + \sum_{i=1}^{n}\lambda_i \ {}^{\rho_i}I_{\theta_i,\kappa_i}^{\gamma_i,\delta_i}(J^{\alpha+\beta}[|f(s,y(s)) - f(s,0)| + |f(s,0)|])(\xi_i) \\
&\quad + J^{\beta-p}[|g(s,y(s)) - g(s,0)| + |g(s,0)|](\eta) \\
&\quad + J^{\alpha+\beta-p}[|f(s,y(s)) - f(s,0)| + |f(s,0)|](\eta)\Bigg\} \\
&\leq (k\|y\| + g_0)\frac{T^{\beta}}{\Gamma(\beta+1)} + (\ell\|y\| + f_0)\frac{T^{\alpha+\beta}}{\Gamma(\alpha+\beta+1)} \\
&\quad + \frac{T^{\beta}}{|\Omega|\Gamma(\beta+1)}\Bigg\{\sum_{i=1}^{n}\lambda_i(k\|y\| + g_0)\frac{\xi_i^{\kappa_i+\rho_i(\theta_i+\gamma_i)+\beta}}{\rho_i^{\delta_i}}\frac{\Gamma(\theta_i + \frac{\beta}{\rho_i} + 1)}{\Gamma(\gamma_i + \theta_i + \frac{\beta}{\rho_i} + 1)} \\
&\quad + \sum_{i=1}^{n}\lambda_i(\ell\|y\| + f_0)\frac{\xi_i^{\kappa_i+\rho_i(\theta_i+\gamma_i)+\alpha+\beta}}{\rho_i^{\delta_i}}\frac{\Gamma(\theta_i + \frac{\alpha+\beta}{\rho_i} + 1)}{\Gamma(\gamma_i + \theta_i + \frac{\alpha+\beta}{\rho_i} + 1)} \\
&\quad + (k\|y\| + g_0)\frac{\eta^{\beta-p}}{\Gamma(\beta-p+1)} + (\ell\|y\| + f_0)\frac{\eta^{\alpha+\beta-p}}{\Gamma(\alpha+\beta-p+1)}\Bigg\} \\
&\leq \Lambda_1(kr + g_0) + \Lambda_2(\ell r + f_0) \leq r,
\end{aligned}$$

which, on taking the norm for $t \in [0,T]$, implies that $\|\mathcal{N}(y)\| \leq r$. Thus $\mathcal{N}$ maps B_r into itself.

In order to show that the operator $\mathcal{N}$ is a contraction, let $y, z \in C([0,T],\mathbb{R})$. Then, using (7.98) and (7.99), we get

$$\begin{aligned}
&|\mathcal{N}(y)(t) - \mathcal{N}(z)(t)| \\
&\leq J^{\beta}|g(s,y(s)) - g(s,z(s))|(t) + J^{\alpha+\beta}|f(s,y(s)) - f(s,z(s))|(t) \\
&\quad + \frac{T^{\beta}}{|\Omega|\Gamma(\beta+1)}\Bigg\{\sum_{i=1}^{n}\lambda_i \ {}^{\rho_i}I_{\theta_i,\kappa_i}^{\gamma_i,\delta_i}(J^{\beta}|g(s,y(s)) - g(s,z(s))|)(\xi_i) \\
&\quad + \sum_{i=1}^{n}\lambda_i \ {}^{\rho_i}I_{\theta_i,\kappa_i}^{\gamma_i,\delta_i}(J^{\alpha+\beta}|f(s,y(s)) - f(s,z(s))|)(\xi_i) \\
&\quad + J^{\beta-p}|g(s,y(s)) - g(s,z(s))|(\eta) + J^{\alpha+\beta-p}|f(s,y(s)) - f(s,z(s))|(\eta)\Bigg\}
\end{aligned}$$

$$
\begin{aligned}
&\leq k\|y-z\|\frac{T^{\beta}}{\Gamma(\beta+1)}+\ell\|y-z\|\frac{T^{\alpha+\beta}}{\Gamma(\alpha+\beta+1)}\\
&\quad+\frac{T^{\beta}}{|\Omega|\Gamma(\beta+1)}\Bigg\{\sum_{i=1}^{n}\lambda_i\frac{\xi_i^{\kappa_i+\rho_i(\theta_i+\gamma_i)+\beta}}{\rho_i^{\delta_i}}\frac{\Gamma(\theta_i+\frac{\beta}{\rho_i}+1)}{\Gamma(\gamma_i+\theta_i+\frac{\beta}{\rho_i}+1)}k\|y-z\|\\
&\quad+\sum_{i=1}^{n}\lambda_i\frac{\xi_i^{\kappa_i+\rho_i(\theta_i+\gamma_i)+\alpha+\beta}}{\rho_i^{\delta_i}}\frac{\Gamma(\theta_i+\frac{\alpha+\beta}{\rho_i}+1)}{\Gamma(\gamma_i+\theta_i+\frac{\alpha+\beta}{\rho_i}+1)}\ell\|y-z\|\\
&\quad+k\|y-z\|\frac{\eta^{\beta-p}}{\Gamma(\beta-p+1)}+\ell\|y-z\|\frac{\eta^{\alpha+\beta-p}}{\Gamma(\alpha+\beta-p+1)}\Bigg\}\\
&=(k\Lambda_1+\ell\Lambda_2)\|y-z\|.
\end{aligned}
$$

Consequently we obtain

$$\|\mathcal{N}(y)-\mathcal{N}(z)\|\leq(k\Lambda_1+\ell\Lambda_2)\|y-z\|,$$

which, in view of (7.100), implies that $\mathcal{N}$ is a contraction. Hence $\mathcal{N}$ has a unique fixed point by Banach's contraction principle. This, in turn, shows that the problem (7.91) has a unique solution on $[0,T]$. The proof is completed. □

Our final existence result is based on Krasnosel'skiĭ's fixed point theorem.

Theorem 7.23. *Assume that (7.22.1) and (7.21.1) hold. In addition we assume that:*

(7.23.1) $|g(t,x)|\leq\phi(t)$, $|f(t,x)|\leq q(t)$, $\forall(t,x)\in[0,T]\times\mathbb{R}$, *and* $q,\phi\in C([0,T],\mathbb{R}^+)$.

Then the problem (7.91) has at least one solution on $[0,T]$, *provided that*

$$k\Lambda_1<1, \tag{7.101}$$

where Λ_1 *is given by (7.98).*

Proof. Let us split the operator $\mathcal{N}:C([0,T],\mathbb{R})\to C([0,T],\mathbb{R})$ defined by (7.97) as $\mathcal{N}=\mathcal{A}_1+\mathcal{A}_2$, where $\mathcal{A}_1$ and $\mathcal{A}_2$ are given by

$$
\begin{aligned}
\mathcal{A}_1(x)(t)=J^{\beta}g(s,y(s))(t)+\frac{t^{\beta}}{\Gamma(\beta+1)}\frac{1}{\Omega}\Bigg\{&\sum_{i=1}^{n}\lambda_i\,{}^{\rho_i}I_{\theta_i,\kappa_i}^{\gamma_i,\delta_i}(J^{\beta}g(s,y(s)))(\xi_i)\\
&-J^{\beta-p}g(s,y(s))(\eta)\Bigg\},
\end{aligned}
$$

and

$$
\begin{aligned}
\mathcal{A}_2(x)(t)=J^{\alpha+\beta}f(s,y(s))(t)+\frac{t^{\beta}}{\Gamma(\beta+1)}\frac{1}{\Omega}\Bigg\{&\sum_{i=1}^{n}\lambda_i\,{}^{\rho_i}I_{\theta_i,\kappa_i}^{\gamma_i,\delta_i}(J^{\alpha+\beta}f(s,y(s)))(\xi_i)\\
&-J^{\alpha+\beta-p}f(s,y(s))(\eta)\Bigg\}.
\end{aligned}
$$

Setting $\sup_{t\in[0,T]}|\phi(t)| = \|\phi\|, \sup_{t\in[0,T]}|q(t)| = \|q\|$ and choosing $\rho \geq \|\phi\|\Lambda_1 + \|q\|\Lambda_2$, we consider $B_\rho = \{y \in C([0,T],\mathbb{R}) : \|y\| \leq \rho\}$. For any $y, z \in B_\rho$, we have

$$\begin{aligned}|\mathcal{A}_1 y(t) + \mathcal{A}_2 z(t)| &\leq J^\beta|g(s,y(s))|(t) + J^{\alpha+\beta}|f(s,z(s))|(t)\\ &\quad+\frac{t^\beta}{|\Omega|\Gamma(\beta+1)}\Big\{\sum_{i=1}^n \lambda_i\ {}^{\rho_i}I^{\gamma_i,\delta_i}_{\theta_i,\kappa_i}(J^\beta|g(s,y(s))|)(\xi_i)\\ &\quad+\sum_{i=1}^n \lambda_i\ {}^{\rho_i}I^{\gamma_i,\delta_i}_{\theta_i,\kappa_i}(J^{\alpha+\beta}|f(s,z(s))|)(\xi_i)\\ &\quad+J^{\beta-p}|g(s,y(s))|(\eta) + J^{\alpha+\beta-p}|f(s,z(s))|(\eta)\Big\}\\ &\leq \|\phi\|\Lambda_1 + \|q\|\Lambda_2 \leq \rho.\end{aligned}$$

Hence $\|\mathcal{A}_1 y + \mathcal{A}_2 z\| \leq \rho$, which shows that $\mathcal{A}_1 y + \mathcal{A}_2 z \in B_\rho$. In view of (7.101), it is easy to show that $\mathcal{A}_1$ is a contraction. Continuity of f implies that the operator $\mathcal{A}_2$ is continuous. Also, $\mathcal{A}_2$ is uniformly bounded on B_ρ as $\|\mathcal{A}_2 y\| \leq \|q\|\Lambda_2$. Finally we prove the compactness of the operator $\mathcal{A}_2$. For that, we define $\sup_{(t,y)\in[0,T]\times B_\rho}|f(t,y)| = \bar{f} < \infty$. Then, for $t_1, t_2 \in [0,T]$, $t_1 < t_2$, we have

$$\begin{aligned}&|\mathcal{A}_2 y(t_2) - \mathcal{A}_2 y(t_1)|\\ &\leq \Big|\frac{1}{\Gamma(\alpha+\beta)}\int_0^{t_1}\Big[(t_2-s)^{\alpha+\beta-1} - (t_1-s)^{\alpha+\beta-1}\Big]f(s,y(s))ds\\ &\quad+\frac{1}{\Gamma(\alpha+\beta)}\int_{t_1}^{t_2}(t_2-s)^{\alpha+\beta-1}f(s,y(s))ds\Big|\\ &\quad+\frac{|t_2^\beta - t_1^\beta|}{|\Omega|\Gamma(\beta+1)}\Big\{\sum_{i=1}^n \lambda_i\ {}^{\rho_i}I^{\gamma_i,\delta_i}_{\theta_i,\kappa_i}(J^{\alpha+\beta}|f(s,y(s))|)(\xi_i) + J^{\alpha+\beta-p}|f(s,y(s))|(\eta)\Big\}\\ &\leq \frac{\bar{f}}{\Gamma(\alpha+\beta+1)}[t_2^{\alpha+\beta} - t_1^{\alpha+\beta} + 2(t_2-t_1)^{\alpha+\beta}]\\ &\quad+\frac{|t_2^\beta - t_1^\beta|\bar{f}}{|\Omega|\Gamma(\beta+1)}\left\{\sum_{i=1}^n \lambda_i\frac{\xi_i^{\kappa_i+\rho_i(\theta_i+\gamma_i)+\alpha+\beta}}{\rho_i^{\delta_i}}\frac{\Gamma(\theta_i+\frac{\alpha+\beta}{\rho_i}+1)}{\Gamma(\gamma_i+\theta_i+\frac{\alpha+\beta}{\rho_i}+1)} + \frac{\eta^{\alpha+\beta-p}}{\Gamma(\alpha+\beta-p+1)}\right\},\end{aligned}$$

which is independent of y and tends to zero as $t_2 - t_1 \to 0$. Thus $\mathcal{A}_2$ is equicontinuous. So $\mathcal{A}_2$ is relatively compact on B_ρ. Hence, by the Arzelá-Ascoli theorem, $\mathcal{A}_2$ is compact on B_ρ. Thus all the assumptions of Theorem 1.4 are satisfied. So the conclusion of Theorem 1.4 implies that the problem (7.91) has at least one solution on $[0,T]$. □

7.5.2 *Examples*

In this section, we give examples to illustrate the usefulness of our main results. For that, let us consider the problem

$$\begin{cases} {}^cD^{1/2}[{}^cD^{3/4}y(t) - g(t,x(t))] = f(t,x(t)), \ t \in [0,2], \\ y(0) = 0, \quad D^{1/4}y(3/4) = I^{1/3,1}y(1) + 1/2 \ {}^{2/5}I^{1/2,1/2}y(3/2), \end{cases} \tag{7.102}$$

where $\alpha = 1/2, \beta = 3/4, p = 1/4, \gamma_1 = 1/3, \delta_1 = 1, \rho_1 = 1, \kappa_1 = \theta_1 = 0, \lambda_1 = 1, \xi_1 = 1, \gamma_2 = \delta_2 = 1/2, \rho_2 = 2/5, \kappa_2 = \theta_2 = 0, \lambda_2 = 1/2, \xi_2 = 3/2$ and

$$f(t,y) = \frac{1}{2\sqrt{900+t^2}} e^{-t}\Big(\tan^{-1} y + y + \frac{1}{8}\Big) \ \text{and} \ g(t,y) = \frac{1}{30(1+t^2)} \sin y + \frac{e^{-t}}{25}.$$

Using the given data, it is found that $|\Omega| \approx 0.7156097036, \Lambda_1 \approx 8.422530006, \Lambda_2 \approx 7.915897126$, where Ω, Λ_1 and Λ_2 are respectively given by (7.92), (7.98) and (7.99).

(i) In order to illustrate Theorem 7.21, note that $d_1 = 1/30$, $d_2 = 1/25$, $\psi(\|y\|) = \|y\| + \dfrac{4\pi+1}{8}$, $p(t) = \dfrac{1}{2\sqrt{900+t^2}} e^{-t}$ and by the condition:

$$\frac{(1-d_1\Lambda_1)M}{d_2\Lambda_1 + \Lambda_2\|p\|\psi(M)} > 1,$$

it follows that $M > 0.95456176$. Clearly the hypothesis of Theorem 7.21 holds true and consequently there exists at least one solution for the problem (7.102).

(ii) It is easy to check that $f(t,x)$ and $g(t,x)$ satisfy the condition (7.22.1) and (7.22.2) respectively with $\ell = 1/30$ and $k = 1/30$. Also

$$k\Lambda_1 + \ell\Lambda_2 \approx 0.5446142377 < 1.$$

Thus all the conditions of Theorem 7.22 are satisfied. So, by the conclusion of Theorem 7.22, the problem (7.102) has a unique solution on $[0,2]$.

(iii) For illustrating Theorem 7.23, we take

$$f(t,y) = \frac{1}{2\sqrt{900+t^2}}\Big(\tan^{-1} y + e^{-t}\Big) \ \text{and} \ g(t,y) = \frac{1}{30(1+t^2)} \sin y + \frac{e^{-t}}{25}. \tag{7.103}$$

With $\phi(t) = \dfrac{1}{30(1+t^2)} + \dfrac{e^{-t}}{25}$, $q(t) = \dfrac{(\pi + 2e^{-t})}{4\sqrt{900+t^2}}$ and $k\Lambda_1 \approx 0.2807510002 < 1$, the hypothesis of Theorem 7.23 is satisfied, which implies that there exists at least one solution for the problem (7.102) with $f(t,x)$ and $g(t,x)$ given by (7.103).

7.6 Multi-term fractional differential equations involving mixed nonlinearities with generalized integral boundary conditions

A new class of boundary value problems of fractional differential equations consisting of multiple Caputo type fractional derivative operators and nonlocal generalized integral boundary conditions, is studied in this section. In precise terms, we consider the following problem:

$$\begin{cases} {}^cD^{\alpha}[{}^cD^{\beta}y(t) - g(t, y(t))] = f(t, y(t)), \ t \in [0, T], \\ a_1 y(\eta) + b_1 y(T) = c, \\ a_2^c D^p y(\eta) + b_2^c D^p y(T) = \displaystyle\sum_{i=1}^{n} \lambda_i \ {}^{\rho_i}I_{\theta_i,\kappa_i}^{\gamma_i,\delta_i} y(\xi_i), \quad \eta, \xi_i \in (0, T), \end{cases} \tag{7.104}$$

where ${}^cD^{\chi}$ is the Caputo fractional derivative operator of order $\chi \in \{\alpha, \beta, p\}$ with $0 < \chi < 1$, ${}^{\rho_i}I_{\theta_i,\kappa_i}^{\gamma_i,\delta_i}$ is the generalized fractional integral, $\gamma_i > 0, \rho_i, \eta_i, \kappa_i, \lambda_i \in \mathbb{R}$, $i = 1, 2, \dots, n$, $a_i, b_i \in \mathbb{R}, i = 1, 2$ and $f, g : [0, T] \times \mathbb{R} \to \mathbb{R}$ are continuous functions.

Here we emphasize that the consideration of generalized integral boundary conditions in problem (7.104) makes it of quite a general nature as such integral boundary conditions reduce to Riemann-Liouville, Hadamard, Erdélyi-Kober and Katugampola type integral boundary conditions for appropriate choice of the parameters involved in the conditions.

Lemma 7.12. *Let $\alpha, \beta > 0$ and $\eta, \kappa, \rho \in \mathbb{R}$. Then*

$${}^{\rho}I_{\eta,\kappa}^{\alpha,\beta} t^q = \frac{t^{\kappa+\rho(\eta+\alpha)+q}}{\rho^{\beta}} \frac{\Gamma\left(\eta + \frac{q}{\rho} + 1\right)}{\Gamma\left(\eta + \frac{q}{\rho} + \alpha + 1\right)}.$$

In the sequel, we set the following notation for the sake of computational convenience:

$$A_1 = \frac{a_1 \eta^{\beta} + b_1 T^{\beta}}{\Gamma(\beta + 1)}, \tag{7.105}$$

$$A_2 = \frac{(a_2 \eta^{\beta-p} + b_2 T^{\beta-p})}{\Gamma(1 + \beta - p)} - \sum_{i=1}^{n} \lambda_i \frac{\Gamma(\theta_i + \frac{\beta}{\rho_i} + 1)\xi_i^{\kappa_i+\rho_i(\theta_i+\gamma_i)+\beta}}{\Gamma(\beta + 1)\Gamma(\theta_i + \frac{\beta}{\rho_i} + \gamma_i + 1)\rho_i^{\delta_i}}, \tag{7.106}$$

$$B_1 = a_1 + b_1, \tag{7.107}$$

$$B_2 = \sum_{i=1}^{n} \lambda_i \frac{\Gamma(\theta_i + 1)\xi_i^{\kappa_i+\rho_i(\theta_i+\gamma_i)}}{\Gamma(\theta_i + \gamma_i + 1)\rho_i^{\delta_i}}, \tag{7.108}$$

$$\Omega = -(A_1 B_2 + B_1 A_2). \tag{7.109}$$

Lemma 7.13. *Let $\Omega \neq 0$. Then the function y is a solution of the problem*

$$\begin{cases} {}^cD^{\alpha}[{}^cD^{\beta}y(t) - g(t, y(t))] = f(t, y(t)), \ t \in [0, T], \\ a_1 y(\eta) + b_1 y(T) = c, \\ a_2^c D^p y(\eta) + b_2^c D^p y(T) = \displaystyle\sum_{i=1}^{n} \lambda_i \ {}^{\rho_i}I_{\theta_i,\kappa_i}^{\gamma_i,\delta_i} y(\xi_i), \quad \eta, \xi_i \in (0, T), \end{cases} \tag{7.110}$$

if and only if

$$y(t) = J^{\beta} g(t, y(t)) + J^{\alpha+\beta} f(t, y(t)) + \frac{t^{\beta}}{\Omega\Gamma(\beta + 1)} \Big\{ - B_2 \Big[c - a_1 J^{\beta} g(\eta, y(\eta))$$
$$-b_1 J^{\beta} g(T, y(T)) - a_1 J^{\alpha+\beta} f(\eta, y(\eta)) - b_1 J^{\alpha+\beta} f(T, y(T))\Big]$$

$$-B_1\Big[\sum_{i=1}^{n}\lambda_i\, {}^{\rho_i}I^{\gamma_i,\delta_i}_{\theta_i,\kappa_i}(J^{\beta}g(\xi_i,y(\xi_i))+J^{\alpha+\beta}f(\xi_i,y(\xi_i)))-a_2J^{\beta-p}g(\eta,y(\eta))$$
$$-b_2J^{\beta-p}g(T,y(T))-a_2J^{\alpha+\beta-p}f(\eta,y(\eta))-b_2J^{\alpha+\beta-p}f(T,y(T))\Big]\Big\}$$
$$+\frac{1}{\Omega}\Big\{A_1\Big[\sum_{i=1}^{n}\lambda_i\, {}^{\rho_i}I^{\gamma_i,\delta_i}_{\theta_i,\kappa_i}(J^{\beta}g(\xi_i,y(\xi_i))+J^{\alpha+\beta}f(\xi_i,y(\xi_i)))-a_2J^{\beta-p}g(\eta,y(\eta))$$
$$-b_2J^{\beta-p}g(T,y(T))-a_2J^{\alpha+\beta-p}f(\eta,y(\eta))-b_2J^{\alpha+\beta-p}f(T,y(T))\Big]$$
$$-A_2\Big[c-a_1J^{\beta}g(\eta,y(\eta))-b_1J^{\beta}g(T,y(T))-a_1J^{\alpha+\beta}f(\eta,y(\eta))$$
$$-b_1J^{\alpha+\beta}f(T,y(T))\Big]\Big\}, \tag{7.111}$$

where $J^{(.)}$ denotes the classical Riemann-Liouville fractional integral of order (.).

Proof. Applying the Riemann-Liouville fractional integral operator J^{β} to the differential equation in (7.110) and followed by J^{α} on the resulting equation, we get

$$y(t)=J^{\beta}g(t,y(t))+J^{\alpha+\beta}f(t,y(t))+\frac{t^{\beta}}{\Gamma(\beta+1)}c_1+c_2, \tag{7.112}$$

where $c_1, c_2 \in \mathbb{R}$ are arbitrary constants. Taking Riemann-Liouville fractional derivative and the generalized fractional integral of order $\delta_i, \gamma_i > 0$ of (7.112) respectively yields

$$D^{p}y(t)=J^{\beta-p}g(t,y(t))+J^{\alpha+\beta-p}f(t,y(t))+\frac{t^{\beta-p}}{\Gamma(\beta-p+1)}c_1, \tag{7.113}$$

$$\begin{aligned}{}^{\rho_i}I^{\gamma_i,\delta_i}_{\theta_i,\kappa_i}y(t)&={}^{\rho_i}I^{\gamma_i,\delta_i}_{\theta_i,\kappa_i}(J^{\beta}g(t,y(t))+J^{\alpha+\beta}f(t,y(t)))\\&\quad+c_1\frac{1}{\Gamma(1+\beta)}\frac{t^{\kappa_i+\rho_i(\theta_i+\gamma_i)+\beta}}{\rho_i^{\delta_i}}\frac{\Gamma\Big(\theta_i+\frac{\beta}{\rho_i}+1\Big)}{\Gamma\Big(\theta_i+\frac{\beta}{\rho_i}+\gamma_i+1\Big)}.\end{aligned} \tag{7.114}$$

Combining the boundary conditions given by (7.110) with (7.112)-(7.114), we obtain

$$\begin{aligned}A_1c_1+B_1c_2&=c-a_1J^{\beta}g(\eta,y(\eta))-b_1J^{\beta}g(T,y(T))\\&\quad-a_1J^{\alpha+\beta}f(\eta,y(\eta))-b_1J^{\alpha+\beta}f(T,y(T)),\end{aligned} \tag{7.115}$$

$$\begin{aligned}A_1c_1+B_1c_2&=c-a_1J^{\beta}g(\eta,y(\eta))-b_1J^{\beta}g(T,y(T))\\&\quad-a_1J^{\alpha+\beta}f(\eta,y(\eta))-b_1J^{\alpha+\beta}f(T,y(T)),\end{aligned} \tag{7.116}$$

where $A_i, B_i, i = 1, 2$ are given by (7.105)-(7.108). Solving the system of equations (7.115) and (7.116) for c_1 and c_2, we find that

$$\begin{aligned}c_1=\frac{1}{\Omega}\Big[&-B_2\Big\{c-a_1J^{\beta}g(\eta,y(\eta))-b_1J^{\beta}g(T,y(T))\\&-a_1J^{\alpha+\beta}f(\eta,y(\eta))-b_1J^{\alpha+\beta}f(T,y(T))\Big\}\end{aligned}$$

$$
\begin{aligned}
&-B_1\Big\{\sum_{i=1}^{n}\lambda_i\ {}^{\rho_i}I^{\gamma_i,\delta_i}_{\theta_i,\kappa_i}(J^{\beta}g(\xi_i,y(\xi_i))+J^{\alpha+\beta}f(\xi_i,y(\xi_i)))\\
&-a_2J^{\beta-p}g(\eta,y(\eta))-b_2J^{\beta-p}g(T,y(T))\\
&-a_2J^{\alpha+\beta-p}f(\eta,y(\eta))-b_2J^{\alpha+\beta-p}f(T,y(T))\Big\}\Big]
\end{aligned}
$$

and

$$
\begin{aligned}
c_2=\frac{1}{\Omega}\Big[&A_1\Big\{\sum_{i=1}^{n}\lambda_i\ {}^{\rho_i}I^{\gamma_i,\delta_i}_{\theta_i,\kappa_i}(J^{\beta}g(\xi_i,y(\xi_i))+J^{\alpha+\beta}f(\xi_i,y(\xi_i)))\\
&-a_2J^{\beta-p}g(\eta,y(\eta))-b_2J^{\beta-p}g(T,y(T))-a_2J^{\alpha+\beta-p}f(\eta,y(\eta))\\
&-b_2J^{\alpha+\beta-p}f(T,y(T))\Big\}\\
&-A_2\Big\{c-a_1J^{\beta}g(\eta,y(\eta))-b_1J^{\beta}g(T,y(T))-a_1J^{\alpha+\beta}f(\eta,y(\eta))\\
&-b_1J^{\alpha+\beta}f(T,y(T))\Big\}\Big].
\end{aligned}
$$

Inserting the values of c_1 and c_2 in (7.112) completes the solution (7.111). The converse follows by direct computation. This completes the proof. □

7.6.1 *Existence and uniqueness results*

Let $\mathcal{C}=C([0,T],\mathbb{R})$ denote the Banach space of all continuous functions from $[0,T]$ to $\mathbb{R}$ endowed with the norm defined by $\|y\|=\sup_{t\in[0,T]}|y(t)|$. Using Lemma 7.13, we can define an operator $F:\mathcal{C}\to\mathcal{C}$ by

$$
\begin{aligned}
Fy(t)=\ &J^{\beta}g(t,y(t))+J^{\alpha+\beta}f(t,y(t))+\frac{t^{\beta}}{\Omega\Gamma(\beta+1)}\Big\{-B_2\Big[c-a_1J^{\beta}g(\eta,y(\eta))\\
&-b_1J^{\beta}g(T,y(T))-a_1J^{\alpha+\beta}f(\eta,y(\eta))-b_1J^{\alpha+\beta}f(T,y(T))\Big]\\
&-B_1\Big[\sum_{i=1}^{n}\lambda_i\ {}^{\rho_i}I^{\gamma_i,\delta_i}_{\theta_i,\kappa_i}(J^{\beta}g(\xi_i,y(\xi_i))+J^{\alpha+\beta}f(\xi_i,y(\xi_i)))-a_2J^{\beta-p}g(\eta,y(\eta))\\
&-b_2J^{\beta-p}g(T,y(T))-a_2J^{\alpha+\beta-p}f(\eta,y(\eta))-b_2J^{\alpha+\beta-p}f(T,y(T))\Big]\Big\} \qquad (7.117)\\
&+\frac{1}{\Omega}\Big\{A_1\Big[\sum_{i=1}^{n}\lambda_i\ {}^{\rho_i}I^{\gamma_i,\delta_i}_{\theta_i,\kappa_i}(J^{\beta}g(\xi_i,y(\xi_i))+J^{\alpha+\beta}f(\xi_i,y(\xi_i)))-a_2I^{\beta-p}g(\eta,y(\eta))\\
&-b_2J^{\beta-p}g(T,y(T))-a_2J^{\alpha+\beta-p}f(\eta,y(\eta))-b_2J^{\alpha+\beta-p}f(T,y(T))\Big]\\
&-A_2\Big[c-a_1J^{\beta}g(\eta,y(\eta))-b_1J^{\beta}g(T,y(T))-a_1J^{\alpha+\beta}f(\eta,y(\eta))\\
&-b_1J^{\alpha+\beta}f(T,y(T))\Big]\Big\}.
\end{aligned}
$$

In the following, for brevity, we use the notation:

$$\begin{aligned}\Lambda_1 = {} & \frac{T^\beta}{\Gamma(\beta+1)} + \frac{T^\beta}{|\Omega|\Gamma(\beta+1)}\Bigg\{|B_2|\frac{|a_1|\eta^\beta + |b_1|T^\beta}{\Gamma(\beta+1)} \\ & + |B_1|\Bigg[\sum_{i=1}^n |\lambda_i| \frac{\xi_i^{\kappa_i+\rho_i(\theta_i+\gamma_i)+\beta}\Gamma(\theta_i + \frac{\beta}{\rho_i}+1)}{\rho_i^{\delta_i}\Gamma(\beta+1)\Gamma(\gamma_i+\theta_i+\frac{\beta}{\rho_i}+1)} \\ & + \frac{|a_2|\eta^{\beta-p} + |b_2|T^{\beta-p}}{\Gamma(\beta-p+1)}\Bigg]\Bigg\} \\ & + \frac{1}{|\Omega|}\Bigg\{|A_1|\Bigg[\sum_{i=1}^n |\lambda_i| \frac{\xi_i^{\kappa_i+\rho_i(\theta_i+\gamma_i)+\beta}\Gamma(\theta_i + \frac{\beta}{\rho_i}+1)}{\rho_i^{\delta_i}\Gamma(\beta+1)\Gamma(\gamma_i+\theta_i+\frac{\beta}{\rho_i}+1)} \\ & + \frac{|a_2|\eta^{\beta-p} + |b_2|T^{\beta-p}}{\Gamma(\beta-p+1)}\Bigg] + |A_2|\frac{|a_1|\eta^\beta + |b_1|T^\beta}{\Gamma(\beta+1)}\Bigg\}, \end{aligned} \tag{7.118}$$

and

$$\begin{aligned}\Lambda_2 = {} & \frac{T^{\alpha+\beta}}{\Gamma(\alpha+\beta+1)} + \frac{T^\beta}{|\Omega|\Gamma(\beta+1)}\Bigg\{|B_2|\frac{|a_1|\eta^{\alpha+\beta} + |b_1|T^{\alpha+\beta}}{\Gamma(\alpha+\beta+1)} \\ & + |B_1|\Bigg[\sum_{i=1}^n |\lambda_i| \frac{\xi_i^{\kappa_i+\rho_i(\theta_i+\gamma_i)+\alpha+\beta}\Gamma(\theta_i + \frac{\alpha+\beta}{\rho_i}+1)}{\rho_i^{\delta_i}\Gamma(\alpha+\beta+1)\Gamma(\gamma_i+\theta_i+\frac{\alpha+\beta}{\rho_i}+1)} \\ & + \frac{|a_2|\eta^{\alpha+\beta-p} + |b_2|T^{\alpha+\beta-p}}{\Gamma(\alpha+\beta-p+1)}\Bigg]\Bigg\} \\ & + \frac{1}{|\Omega|}\Bigg\{|A_1|\Bigg[\sum_{i=1}^n |\lambda_i| \frac{\xi_i^{\kappa_i+\rho_i(\theta_i+\gamma_i)+\alpha+\beta}\Gamma(\theta_i + \frac{\alpha+\beta}{\rho_i}+1)}{\rho_i^{\delta_i}\Gamma(\alpha+\beta+1)\Gamma(\gamma_i+\theta_i+\frac{\alpha+\beta}{\rho_i}+1)} \\ & + \frac{|a_2|\eta^{\alpha+\beta-p} + |b_2|T^{\alpha+\beta-p}}{\Gamma(\alpha+\beta-p+1)}\Bigg] + |A_2|\frac{|a_1|\eta^{\alpha+\beta} + |b_1|T^{\alpha+\beta}}{\Gamma(\alpha+\beta+1)}\Bigg\}. \end{aligned} \tag{7.119}$$

7.6.2 *Existence result via Kransnosel'skiǐ's fixed point theorem*

Theorem 7.24. *Assume that the following conditions hold:*

(7.24.1) $f, g : [0,T] \times \mathbb{R} \to \mathbb{R}$ *are continuous functions;*

(7.24.2) there exists a non-negative constant k *such that*

$$|g(t,y) - g(t,x)| \le k\|y - x\|, \quad \text{for } t \in [0,T] \text{ and every } y, x \in \mathbb{R};$$

(7.24.3) there exist continuous non-negative functions $\psi, \phi \in C([0,T], \mathbb{R}^+)$ *such that*

$$|g(t,y)| \le \psi(t), \quad |f(t,y)| \le \phi(t), \quad \forall (t,y) \in [0,T] \times \mathbb{R}.$$

Then the BVP (7.104) has at least one solution on $[0,T]$*, provided that*

$$k\Lambda_1 < 1, \tag{7.120}$$

where Λ_1 *is given by (7.118).*

Proof. By assumption (7.24.3), we can fix

$$\rho \geq ||\psi||\Lambda_1 + ||\phi||\Lambda_2 + \frac{|c|}{|\Omega|}\Big[|B_2|\frac{T^\beta}{\Gamma(\beta+1)} + |A_2|\Big],$$

where $\sup_{t\in[0,T]}|\psi(t)| = \|\psi\|, \sup_{t\in[0,T]}|\phi(t)| = \|\phi\|$ and consider $B_\rho = \{y \in C([0,T],\mathbb{R}) : ||y|| \leq \rho\}$. Let us split the operator $F : C([0,T],\mathbb{R}) \to C([0,T],\mathbb{R})$ defined by (7.117) as $F = F_1 + F_2$, where F_1 and F_2 are given by

$$\begin{aligned}
&F_1(y)(t)\\
&= J^\beta g(t,y(t)) + \frac{t^\beta}{\Omega\Gamma(\beta+1)}\Big\{ - B_2\Big[c - a_1 J^\beta g(\eta,y(\eta)) - b_1 I^\beta g(T,y(T))\Big]\\
&-B_1\Big[\sum_{i=1}^n \lambda_i\ {}^{\rho_i}I^{\gamma_i,\delta_i}_{\theta_i,\kappa_i}(J^\beta g(\xi_i,y(\xi_i))) - a_2 J^{\beta-p} g(\eta,y(\eta)) - b_2 J^{\beta-p} g(T,y(T))\Big]\Big\}\\
&+\frac{1}{\Omega}\Big\{A_1\Big[\sum_{i=1}^n \lambda_i\ {}^{\rho_i}I^{\gamma_i,\delta_i}_{\theta_i,\kappa_i}(J^\beta g(\xi_i,y(\xi_i))) - a_2 J^{\beta-p} g(\eta,y(\eta)) - b_2 J^{\beta-p} g(T,y(T))\Big]\\
&-A_2\Big[c - a_1 J^\beta g(\eta,y(\eta)) - b_1 J^\beta g(T,y(T))\Big]\Big\},
\end{aligned}$$

and

$$\begin{aligned}
&F_2(y)(t)\\
&= J^{\alpha+\beta} f(t,y(t)) + \frac{t^\beta}{\Omega\Gamma(\beta+1)}\Big\{ - B_2\Big[- a_1 J^{\alpha+\beta} f(\eta,y(\eta)) - b_1 J^{\alpha+\beta} f(T,y(T))\Big]\\
&-B_1\Big[\sum_{i=1}^n \lambda_i\ {}^{\rho_i}I^{\gamma_i,\delta_i}_{\theta_i,\kappa_i}(J^{\alpha+\beta} f(\xi_i,y(\xi_i)))\\
&-a_2 J^{\alpha+\beta-p} f(\eta,y(\eta)) - b_2 J^{\alpha+\beta-p} f(T,y(T))\Big]\Big\}\\
&+\frac{1}{\Omega}\Big\{A_1\Big[\sum_{i=1}^n \lambda_i\ {}^{\rho_i}I^{\gamma_i,\delta_i}_{\theta_i,\kappa_i}(J^{\alpha+\beta} f(\xi_i,y(\xi_i))) - a_2 J^{\alpha+\beta-p} f(\eta,y(\eta))\\
&-b_2 J^{\alpha+\beta-p} f(T,y(T))\Big] - A_2\Big[- a_1 J^{\alpha+\beta} f(\eta,y(\eta)) - b_1 J^{\alpha+\beta} f(T,y(T))\Big]\Big\}.
\end{aligned}$$

For any $x, y \in B_\rho$, we have

$$\begin{aligned}
&|F_1x(t) + F_2y(t)|\\
&\leq J^\beta|g(t,x(t))| + J^{\alpha+\beta}|f(t,y(t))| + \frac{T^\beta}{\Omega\Gamma(\beta+1)}\Big\{|B_2|\Big[|c| + |a_1|J^\beta|g(\eta,x(\eta))|\\
&+|b_1|J^\beta|g(T,x(T))| + |a_1|J^{\alpha+\beta}|f(\eta,y(\eta))| + |b_1|J^{\alpha+\beta}|f(T,y(T))|\Big]\\
&+|B_1|\Big[\sum_{i=1}^n |\lambda_i|\ {}^{\rho_i}I^{\gamma_i,\delta_i}_{\theta_i,\kappa_i}(J^\beta|g(\xi_i,x(\xi_i))| + J^{\alpha+\beta}|f(\xi_i,y(\xi_i))|) + |a_2|J^{\beta-p}|g(\eta,x(\eta))|\\
&+|b_2|J^{\beta-p}|g(T,x(T))| + |a_2|J^{\alpha+\beta-p}|f(\eta,y(\eta))| + |b_2|J^{\alpha+\beta-p}|f(T,y(T))|\Big]\Big\}\\
&+\frac{1}{|\Omega|}\Big\{|A_1|\Big[\sum_{i=1}^n |\lambda_i|\ {}^{\rho_i}I^{\gamma_i,\delta_i}_{\theta_i,\kappa_i}(J^\beta|g(\xi_i,x(\xi_i))| + J^{\alpha+\beta}|f(\xi_i,y(\xi_i))|)
\end{aligned}$$

$$+|a_2|J^{\beta-p}|g(\eta,x(\eta))| + |b_2|J^{\beta-p}|g(T,x(T))| + |a_2|J^{\alpha+\beta-p}|f(\eta,y(\eta))|$$
$$+|b_2|J^{\alpha+\beta-p}|f(T,y(T))|\Big] + |A_2|\Big[|c| + |a_1|J^{\beta}|g(\eta,x(\eta))| + |b_1|J^{\beta}|g(T,x(T))|$$
$$+|a_1|J^{\alpha+\beta}|f(\eta,y(\eta))| + |b_1|J^{\alpha+\beta}|f(T,y(T))|\Big]\Big\}$$
$$\leq \|\psi\|\Lambda_1 + \|\phi\|\Lambda_2 + \frac{|c|}{|\Omega|}\Big[|B_2|\frac{T^{\beta}}{\Gamma(\beta+1)} + |A_2|\Big] \leq \rho.$$

Hence $\|F_1x + F_2y\| \leq \rho$, which shows that $F_1x + F_2y \in B_\rho$. In view of (7.120), it is easy to show that F_1 is a contraction mapping. Continuity of f implies that the operator F_2 is continuous. Also, F_2 is uniformly bounded on B_ρ as $\|F_2y\| \leq \|\phi\|\Lambda_2$. Finally we prove the compactness of the operator F_2. For that we define $\sup_{(t,x)\in[0,T]\times B_\rho}|f(t,y)| = \bar{f} < \infty$. Then, for $t_1, t_2 \in [0,T]$, $t_1 < t_2$, we have

$$|F_2y(t_2) - F_2y(t_1)|$$
$$\leq \Big|\frac{1}{\Gamma(\alpha+\beta)}\int_0^{t_1}\Big[(t_2-s)^{\alpha+\beta-1} - (t_1-s)^{\alpha+\beta-1}\Big]f(s,y(s))ds$$
$$+\frac{1}{\Gamma(\alpha+\beta)}\int_{t_1}^{t_2}(t_2-s)^{\alpha+\beta-1}f(s,y(s))ds\Big|$$
$$+\frac{|t_2^{\beta}-t_1^{\beta}|}{|\Omega|\Gamma(\beta+1)}\Big\{|B_2|\Big[|a_1|J^{\alpha+\beta}|f(\eta,y(\eta))| + |b_1|J^{\alpha+\beta}|f(T,y(T))|\Big]$$
$$+|B_1|\Big[\sum_{i=1}^{n}|\lambda_i|\ {}^{\rho_i}I_{\theta_i,\kappa_i}^{\gamma_i,\delta_i}(J^{\alpha+\beta}|f(\xi_i,y(\xi_i))|) + |a_2|J^{\alpha+\beta-p}|f(\eta,y(\eta))|$$
$$+|b_2|J^{\alpha+\beta-p}|f(T,y(T))|\Big]\Big\}$$
$$\leq \frac{\bar{f}}{\Gamma(\alpha+\beta+1)}[t_2^{\alpha+\beta} - t_1^{\alpha+\beta} + 2(t_2-t_1)^{\alpha+\beta}]$$
$$+\frac{|t_2^{\beta}-t_1^{\beta}|\bar{f}}{|\Omega|\Gamma(\beta+1)}\Big\{|B_2|\frac{|a_1|\eta^{\alpha+\beta} + |b_1|T^{\alpha+\beta}}{\Gamma(\alpha+\beta+1)}$$
$$+|B_1|\Big[\sum_{i=1}^{n}|\lambda_i|\frac{\xi_i^{\kappa_i+\rho_i(\theta_i+\gamma_i)+\alpha+\beta}\Gamma(\theta_i + \frac{\alpha+\beta}{\rho_i}+1)}{\rho_i^{\delta_i}\Gamma(\alpha+\beta+1)\Gamma(\gamma_i+\theta_i+\frac{\alpha+\beta}{\rho_i}+1)}$$
$$+\frac{|a_2|\eta^{\alpha+\beta-p} + |b_2|T^{\alpha+\beta-p}}{\Gamma(\alpha+\beta-p+1)}\Big]\Big\},$$

which is independent of y and tends to zero as $t_2 - t_1 \to 0$. Thus, F_2 is equicontinuous. So F_2 is relatively compact on B_ρ. Hence, by the Arzelá-Ascoli theorem, F_2 is compact on B_ρ. Thus all the assumptions of Theorem 1.4 are satisfied. So the conclusion of Theorem 1.4 implies that the problem (7.104) has at least one solution on $[0,T]$. □

7.6.3 *Existence result via Leray-Schauder nonlinear alternative.*

Theorem 7.25. *Assume that the following hypotheses hold:*

(7.25.1) $f, g : [0,T] \times \mathbb{R} \to \mathbb{R}$ *are continuous functions;*
(7.25.2) there exist constants $d_1 < 1/\Lambda_1$ *and* $d_2 \geq 0$ *such that*

$$|g(t,y)| \leq d_1\|y\| + d_2, \quad t \in [0,T], \ y \in \mathbb{R};$$

(7.25.3) there exist a continuous non-decreasing function $\psi : [0,\infty) \to (0,\infty)$ *and a function* $p \in C([0,T], \mathbb{R}^+)$ *such that*

$$|f(t,y)| \leq p(t)\psi(\|y\|) \ \text{ for each } \ (t,y) \in [0,T] \times \mathbb{R};$$

(7.25.4) there exists a constant $M > 0$ *such that*

$$\frac{(1 - d_1\Lambda_1)M}{d_2\Lambda_1 + \Lambda_2\|p\|\psi(M) + \frac{|c|}{|\Omega|}\left[|B_2|\frac{T^\beta}{\Gamma(\beta+1)} + |A_2|\right]} > 1.$$

Then the BVP (7.104) has at least one solution on $[0,T]$.

Proof. Consider the operator $F : C([0,T],\mathbb{R}) \to C([0,T],\mathbb{R})$ defined by (7.117). We show that the operator F is continuous and completely continuous.

Step 1. *F is continuous.*

Let $\{y_n\}$ be a sequence such that $y_n \to y$ in $C([0,T],\mathbb{R})$. Then

$$\begin{aligned}
&|F(y_n)(t) - F(y)(t)| \\
\leq\ & J^\beta|g(t,y_n(t)) - g(t,y(t))| + J^{\alpha+\beta}|f(t,y_n(t)) - f(t,y(t))| \\
&+\frac{T^\beta}{\Omega\Gamma(\beta+1)}\Big\{|B_2|\Big[|a_1|J^\beta|g(\eta,y_n(\eta)) - g(\eta,y(\eta))| \\
&+|b_1|J^\beta|g(T,y_n(T)) - g(T,y(T))| + |a_1|J^{\alpha+\beta}|f(\eta,y_n(\eta)) - f(\eta,y(\eta))| \\
&+|b_1|J^{\alpha+\beta}|f(T,y_n(T)) - f(T,y(T))|\Big] \\
&+|B_1|\Big[\sum_{i=1}^n|\lambda_i|\ {}^{\rho_i}I_{\theta_i,\kappa_i}^{\gamma_i,\delta_i}(J^\beta|g(\xi_i,y_n(\xi_i)) - g(\xi_i,y(\xi_i))| \\
&+J^{\alpha+\beta}|f(\xi_i,y_n(\xi_i)) - f(\xi_i,y(\xi_i))|) \\
&+|a_2|J^{\beta-p}|g(\eta,y_n(\eta)) - g(\eta,y(\eta))| + |b_2|J^{\beta-p}|g(T,y_n(T)) - g(T,y(T))| \\
&+|a_2|J^{\alpha+\beta-p}|f(\eta,y_n(\eta)) - f(\eta,y(\eta))| \\
&+|b_2|J^{\alpha+\beta-p}|f(T,y_n(T)) - f(T,y(T))|\Big]\Big\} \\
&+\frac{1}{|\Omega|}\Big\{|A_1|\Big[\sum_{i=1}^n|\lambda_i|\ {}^{\rho_i}I_{\theta_i,\kappa_i}^{\gamma_i,\delta_i}(J^\beta|g(\xi_i,y_n(\xi_i)) - g(\xi_i,y(\xi_i))| \\
&+J^{\alpha+\beta}|f(\xi_i,y_n(\xi_i)) - f(\xi_i,y(\xi_i))|) \\
&+|a_2|J^{\beta-p}|g(\eta,y_n(\eta)) - g(\eta,y(\eta))| + |b_2|J^{\beta-p}|g(T,y_n(T)) - g(T,y(T))|
\end{aligned}$$

$$
\begin{aligned}
&+|a_2|J^{\alpha+\beta-p}|f(\eta,y_n(\eta))-f(\eta,y(\eta)|+|b_2|J^{\alpha+\beta-p}|f(T,y_n(T))-f(T,y(T))|\Big]\\
&+|A_2|\Big[|a_1|J^{\beta}|g(\eta,y_n(\eta))-g(\eta,y(\eta))|+|b_1|J^{\beta}|g(T,y_n(T))-g(T,y(T))|\\
&+|a_1|J^{\alpha+\beta}|f(\eta,y_n(\eta))-f(\eta,y(\eta)|+|b_1|J^{\alpha+\beta}|f(T,y_n(T))-f(T,y(T))|\Big]\Big\}\\
&\leq \Lambda_1\|g(\cdot,y_n)-g(\cdot,y)\|+\Lambda_2\|f(\cdot,y_n)-f(\cdot,y)\|.
\end{aligned}
$$

Since f, g are continuous functions, therefore, we have

$$
\|F(y_n)-F(y)\|\leq \Lambda_1\|g(\cdot,y_n)-g(\cdot,y)\|+\Lambda_2\|f(\cdot,y_n)-f(\cdot,y)\|\to 0
$$

as $n\to\infty$.

Step 2. *F maps bounded sets into bounded sets in $C([0,T],\mathbb{R})$.*

Indeed it is enough to show that for any $\theta>0$, there exists a positive constant $\tilde{\ell}$ such that, for each $y\in B_\theta=\{y\in C([0,T],\mathbb{R}):\|y\|\leq\theta\}$, we have $\|F(y)\|\leq\tilde{\ell}$. By (7.25.2) and (7.25.3), as in the proof of the last theorem, we can obtain for each $t\in[0,T]$ that

$$
\|F(y)\|\leq \Lambda_1(d_1\theta+d_2)+\Lambda_2\|p\|\psi(\theta)+\frac{|c|}{|\Omega|}\Big[|B_2|\frac{T^{\beta}}{\Gamma(\beta+1)}+|A_2|\Big]:=\tilde{\ell}.
$$

Step 3. *F maps bounded sets into equi-continuous sets of $C([0,T],\mathbb{R})$.*

Let $t_1,t_2\in[0,T]$, $t_1<t_2$, B_θ be a bounded set of $C([0,T],\mathbb{R})$ as in Step 2, and let $y\in B_\theta$. Then

$$
\begin{aligned}
&|F(y)(t_2)-F(y)(t_1)|\\
&\leq\Big|\frac{1}{\Gamma(\beta)}\int_0^{t_1}\Big[(t_2-s)^{\beta-1}-(t_1-s)^{\beta-1}\Big]g(s,y(s))ds+\frac{1}{\Gamma(\beta)}\int_{t_1}^{t_2}(t_2-s)^{\beta-1}g(s,y(s))ds\Big|\\
&+\Big|\frac{1}{\Gamma(\alpha+\beta)}\int_0^{t_1}\Big[(t_2-s)^{\alpha+\beta-1}-(t_1-s)^{\alpha+\beta-1}\Big]f(s,y(s))ds\\
&+\frac{1}{\Gamma(\alpha+\beta)}\int_{t_1}^{t_2}(t_2-s)^{\alpha+\beta-1}f(s,y(s))ds\Big|\\
&+\frac{|t_2^{\beta}-t_1^{\beta}|(d_1\theta+d_2))}{|\Omega|\Gamma(\beta+1)}\Big\{|B_2|\frac{|a_1|\eta^{\beta}+|b_1|T^{\beta}}{\Gamma(\beta+1)}\\
&+|B_1|\Big[\sum_{i=1}^{n}|\lambda_i|\frac{\xi_i^{\kappa_i+\rho_i(\theta_i+\gamma_i)+\beta}\Gamma(\theta_i+\frac{\beta}{\rho_i}+1)}{\rho_i^{\delta_i}\Gamma(\beta+1)\Gamma(\gamma_i+\theta_i+\frac{\beta}{\rho_i}+1)}\\
&+\frac{|a_2|\eta^{\beta-p}+|b_1|T^{\beta-p}}{\Gamma(\alpha+\beta-p+1)}\Big]\Big\}+\frac{|t_2^{\beta}-t_1^{\beta}|\|p\|\psi(\theta)}{|\Omega|\Gamma(\beta+1)}\Big\{|B_2|\frac{|a_1|\eta^{\alpha+\beta}+|b_1|T^{\alpha+\beta}}{\Gamma(\alpha+\beta+1)}\\
&+|B_1|\Big[\sum_{i=1}^{n}|\lambda_i|\frac{\xi_i^{\kappa_i+\rho_i(\theta_i+\gamma_i)+\alpha+\beta}\Gamma(\theta_i+\frac{\alpha+\beta}{\rho_i}+1)}{\rho_i^{\delta_i}\Gamma(\alpha+\beta+1)\Gamma(\gamma_i+\theta_i+\frac{\alpha+\beta}{\rho_i}+1)}+\frac{|a_2|\eta^{\alpha+\beta-p}+|b_2|T^{\alpha+\beta-p}}{\Gamma(\alpha+\beta-p+1)}\Big]\Big\}\\
&\leq\frac{d_1\theta+d_2}{\Gamma(\beta+1)}[t_2^{\beta}-t_1^{\beta}+2(t_2-t_1)^{\beta}]+\frac{\|p\|\psi(\theta)}{\Gamma(\alpha+\beta+1)}[t_2^{\alpha+\beta}-t_1^{\alpha+\beta}+2(t_2-t_1)^{\alpha+\beta}]
\end{aligned}
$$

$$+\frac{|t_2^\beta - t_1^\beta|(d_1\theta + d_2))}{|\Omega|\Gamma(\beta+1)}\bigg\{|B_2|\frac{|a_1|\eta^\beta + |b_1|T^\beta}{\Gamma(\beta+1)}$$

$$+|B_1|\Big[\sum_{i=1}^{n}|\lambda_i|\frac{\xi_i^{\kappa_i+\rho_i(\theta_i+\gamma_i)+\beta}\Gamma(\theta_i+\frac{\beta}{\rho_i}+1)}{\rho_i^{\delta_i}\Gamma(\beta+1)\Gamma(\gamma_i+\theta_i+\frac{\beta}{\rho_i}+1)}$$

$$+\frac{|a_2|\eta^{\beta-p} + |b_1|T^{\beta-p}}{\Gamma(\alpha+\beta-p+1)}\Big]\bigg\} + \frac{|t_2^\beta - t_1^\beta|\|p\|\psi(\theta)}{|\Omega|\Gamma(\beta+1)}\bigg\{|B_2|\frac{|a_1|\eta^{\alpha+\beta} + |b_1|T^{\alpha+\beta}}{\Gamma(\alpha+\beta+1)}$$

$$+|B_1|\Big[\sum_{i=1}^{n}|\lambda_i|\frac{\xi_i^{\kappa_i+\rho_i(\theta_i+\gamma_i)+\alpha+\beta}\Gamma(\theta_i+\frac{\alpha+\beta}{\rho_i}+1)}{\rho_i^{\delta_i}\Gamma(\alpha+\beta+1)\Gamma(\gamma_i+\theta_i+\frac{\alpha+\beta}{\rho_i}+1)} + \frac{|a_2|\eta^{\alpha+\beta-p} + |b_2|T^{\alpha+\beta-p}}{\Gamma(\alpha+\beta-p+1)}\Big]\bigg\}.$$

Clearly the right-hand side of the above inequality tends to zero independently of y as $t_1 \longrightarrow t_2$.

In consequence of Steps 1 to 3, it follows by the Arzelá-Ascoli theorem that the operator $F : C([0,T],\mathbb{R}) \longrightarrow C([0,T],\mathbb{R})$ is continuous and completely continuous.

Step 4. *We show that there exists an open set $U \subseteq C([0,T],\mathbb{R})$ with $y \neq \mu F(y)$ for $\mu \in (0,1)$ and $y \in \partial U$.*

Let $y \in C([0,T],\mathbb{R})$ and $y = \mu F(y)$ for some $0 < \mu < 1$. Thus, for each $t \in [0,T]$, we have

$$y(t)$$

$$= \mu J^\beta g(t,y(t)) + \mu J^{\alpha+\beta} f(t,y(t)) + \mu\frac{t^\beta}{\Omega\Gamma(\beta+1)}\bigg\{-B_2\Big[c - a_1 J^\beta g(\eta,y(\eta))$$

$$-b_1 J^\beta g(T,y(T)) - a_1 J^{\alpha+\beta} f(\eta,y(\eta)) - b_1 J^{\alpha+\beta} f(T,y(T))\Big]$$

$$-B_1\Big[\sum_{i=1}^{n}\lambda_i\ {}^{\rho_i}I_{\theta_i,\kappa_i}^{\gamma_i,\delta_i}(J^\beta g(\xi_i,y(\xi_i)) + J^{\alpha+\beta} f(\xi_i,y(\xi_i))) - a_2 J^{\beta-p} g(\eta,y(\eta))$$

$$-b_2 J^{\beta-p} g(T,y(T)) - a_2 J^{\alpha+\beta-p} f(\eta,y(\eta)) - b_2 J^{\alpha+\beta-p} f(T,y(T))\Big]\bigg\}$$

$$+\mu\frac{1}{\Omega}\bigg\{A_1\Big[\sum_{i=1}^{n}\lambda_i\ {}^{\rho_i}I_{\theta_i,\kappa_i}^{\gamma_i,\delta_i}(J^\beta g(\xi_i,y(\xi_i)) + J^{\alpha+\beta} f(\xi_i,y(\xi_i))) - a_2 J^{\beta-p} g(\eta,y(\eta))$$

$$-b_2 J^{\beta-p} g(T,y(T)) - a_2 J^{\alpha+\beta-p} f(\eta,y(\eta)) - b_2 J^{\alpha+\beta-p} f(T,y(T))\Big]$$

$$-A_2\Big[c - a_1 J^\beta g(\eta,y(\eta)) - b_1 J^\beta g(T,y(T)) - a_1 J^{\alpha+\beta} f(\eta,y(\eta))$$

$$-b_1 J^{\alpha+\beta} f(T,y(T))\Big]\bigg\}.$$

As in Step 2, for each $t \in [0,T]$, it can be established that

$$|y(t)| = |\mu(Fy)(t)| \le \Lambda_1(d_1\|y\| + d_2) + \Lambda_2\|p\|\psi(\|y\|) + \frac{|c|}{|\Omega}\Big[|B_2|\frac{T^\beta}{\Gamma(\beta+1)} + |A_2|\Big],$$

which can be expressed as

$$\frac{(1-d_1\Lambda_1)\|y\|}{d_2\Lambda_1+\Lambda_2\|p\|\psi(\|y\|)+\dfrac{|c|}{|\Omega|}\Big[|B_2|\dfrac{T^\beta}{\Gamma(\beta+1)}+|A_2|\Big]}\leq 1.$$

In view of (7.25.4), there exists M such that $\|y\|\neq \mathrm{M}$. Let us set

$$U=\{y\in C([0,T],\mathbb{R}):\|y\|<\mathrm{M}\}.$$

Note that the operator $F:\overline{U}\to C([0,T],\mathbb{R})$ is continuous and completely continuous. From the choice of U, there is no $y\in\partial U$ such that $y=\mu F y$ for some $\mu\in(0,1)$. Consequently, by the nonlinear alternative of Leray-Schauder type (Theorem 1.8), we deduce that F has a fixed point $y\in\overline{U}$ which is a solution of the problem (7.104). This completes the proof. □

7.6.4 *Existence and uniqueness result via Banach fixed point theorem*

Theorem 7.26. *Assume that the following conditions hold:*

(7.26.1) there exists a non-negative constant k such that

$$|g(t,x)-g(t,y)|\leq k\|x-y\|,\quad \textit{for } t\in[0,T]\ \textit{ and every } x,y\in\mathbb{R};$$

(7.26.2) there exists $\ell>0$ such that

$$|f(t,x)-f(t,y)|\leq \ell\|y-x\|,\quad \textit{for } t\in[0,T]\ \textit{ and every } x,y\in\mathbb{R}.$$

If

$$k\Lambda_1+\ell\Lambda_2<1, \tag{7.121}$$

where Λ_1,Λ_2 are defined by (7.118) and (7.119) respectively, then there exists a unique solution for the BVP (7.104) on $[0,T]$.

Proof. Consider the operator $F:C([0,T],\mathbb{R})\to C([0,T],\mathbb{R})$ associated with the problem (7.104) defined by (7.117). With Λ_1 and Λ_2 respectively given by (7.118) and (7.119), we fix

$$r\geq\frac{\Lambda_1 g_0+\Lambda_2 f_0+\dfrac{|c|}{|\Omega|}\Big[|B_2|\dfrac{T^\beta}{\Gamma(\beta+1)}+|A_2|\Big]}{1-k\Lambda_1-\ell\Lambda_2},$$

$g_0=\sup_{t\in[0,T]}|g(t,0)|$, $f_0=\sup_{t\in[0,T]}|f(t,0)|$, and show that $FB_r\subset B_r$, where $B_r=\{y\in C([0,T],\mathbb{R}):\|y\|\leq r\}$. For $y\in B_r$, using (7.26.1) and (7.26.2), we can obtain that $|g(t,y(t))|\leq|g(t,y(t))-g(t,0)|+|g(t,0)|\leq L\|y\|+g_0\leq k\tilde{r}+g_0$ and $|f(t,y(t))|\leq|f(t,y(t))-f(t,0)|+|f(t,0)|\leq L\|y\|+f_0\leq\ell\tilde{r}+f_0$ for all $t\in[0,T]$. In consequence, we get

$$|F(y)(t)|$$

$$
\begin{aligned}
&\leq (k\|y\|+g_0)\Bigg[\frac{T^\beta}{\Gamma(\beta+1)}+\frac{T^\beta}{|\Omega|\Gamma(\beta+1)}\Bigg\{|B_2|\frac{|a_1|\eta^\beta+|b_1|T^\beta}{\Gamma(\beta+1)}\\
&+|B_1|\Bigg[\sum_{i=1}^n|\lambda_i|\frac{\xi_i^{\kappa_i+\rho_i(\theta_i+\gamma_i)+\beta}\Gamma(\theta_i+\frac{\beta}{\rho_i}+1)}{\rho_i^{\delta_i}\Gamma(\beta+1)\Gamma(\gamma_i+\theta_i+\frac{\beta}{\rho_i}+1)}+\frac{|a_2|\eta^{\beta-p}+|b_2|T^{\beta-p}}{\Gamma(\beta-p+1)}\Bigg]\Bigg\}\\
&+\frac{1}{|\Omega|}\Bigg\{|A_1|\Bigg[\sum_{i=1}^n|\lambda_i|\frac{\xi_i^{\kappa_i+\rho_i(\theta_i+\gamma_i)+\beta}\Gamma(\theta_i+\frac{\beta}{\rho_i}+1)}{\rho_i^{\delta_i}\Gamma(\beta+1)\Gamma(\gamma_i+\theta_i+\frac{\beta}{\rho_i}+1)}+\frac{|a_2|\eta^{\beta-p}+|b_2|T^{\beta-p}}{\Gamma(\beta-p+1)}\Bigg]\\
&+|A_2|\frac{|a_1|\eta^\beta+|b_1|T^\beta}{\Gamma(\beta+1)}\Bigg\}\Bigg]\\
&+(\ell\|y\|+f_0)\Bigg[\frac{T^{\alpha+\beta}}{\Gamma(\alpha+\beta+1)}+\frac{T^\beta}{|\Omega|\Gamma(\beta+1)}\Bigg\{|B_2|\frac{|a_1|\eta^{\alpha+\beta}+|b_1|T^{\alpha+\beta}}{\Gamma(\alpha+\beta+1)}\\
&+|B_1|\Bigg[\sum_{i=1}^n|\lambda_i|\frac{\xi_i^{\kappa_i+\rho_i(\theta_i+\gamma_i)+\alpha+\beta}\Gamma(\theta_i+\frac{\alpha+\beta}{\rho_i}+1)}{\rho_i^{\delta_i}\Gamma(\alpha+\beta+1)\Gamma(\gamma_i+\theta_i+\frac{\alpha+\beta}{\rho_i}+1)}\\
&+\frac{|a_2|\eta^{\alpha+\beta-p}+|b_2|T^{\alpha+\beta-p}}{\Gamma(\alpha+\beta-p+1)}\Bigg]\Bigg\}\\
&+\frac{1}{|\Omega|}\Bigg\{|A_1|\Bigg[\sum_{i=1}^n|\lambda_i|\frac{\xi_i^{\kappa_i+\rho_i(\theta_i+\gamma_i)+\alpha+\beta}\Gamma(\theta_i+\frac{\alpha+\beta}{\rho_i}+1)}{\rho_i^{\delta_i}\Gamma(\alpha+\beta+1)\Gamma(\gamma_i+\theta_i+\frac{\alpha+\beta}{\rho_i}+1)}\\
&+\frac{|a_2|\eta^{\alpha+\beta-p}+|b_2|T^{\alpha+\beta-p}}{\Gamma(\alpha+\beta-p+1)}\Bigg]\\
&+|A_2|\frac{|a_1|\eta^{\alpha+\beta}+|b_1|T^{\alpha+\beta}}{\Gamma(\alpha+\beta+1)}\Bigg\}\Bigg]+\frac{|c|}{|\Omega|}\Bigg[|B_2|\frac{T^\beta}{\Gamma(\beta+1)}+|A_2|\Bigg]\\
&\leq \Lambda_1(kr+g_0)+\Lambda_2(\ell r+f_0)+\frac{|c|}{|\Omega|}\Bigg[|B_2|\frac{T^\beta}{\Gamma(\beta+1)}+|A_2|\Bigg]\leq r,
\end{aligned}
$$

which, on taking the norm for $t\in[0,T]$, yields $\|F(y)\|\leq r$. This shows that F maps B_r into itself. In order to show that the operator F is a contraction, let $y,z\in C([0,T],\mathbb{R})$. Then, using (7.118) and (7.119), we get

$$
\begin{aligned}
&|F(y)(t)-F(z)(t)|\\
&\leq J^\beta|g(t,y(t))-g(t,z(t))|+J^{\alpha+\beta}|f(t,y(t))-f(t,z(t))|\\
&+\frac{T^\beta}{|\Omega|\Gamma(\beta+1)}\Bigg\{|B_2|\Big[|a_1|J^\beta|g(\eta,y(\eta))-g(\eta,z(\eta))|\\
&+|b_1|J^\beta|g(T,y(T))-g(T,z(T))|+|a_1|J^{\alpha+\beta}|f(\eta,y(\eta))-f(\eta,z(\eta))|\\
&+|b_1|J^{\alpha+\beta}|f(T,y(T))-f(T,z(T))|\Big]\\
&+|B_1|\Bigg[\sum_{i=1}^n|\lambda_i|\;{}^{\rho_i}I_{\theta_i,\kappa_i}^{\gamma_i,\delta_i}\Big\{J^\beta|g(\xi_i,y(\xi_i))-g(\xi_i,z(\xi_i))|
\end{aligned}
$$

$$
\begin{aligned}
&+J^{\alpha+\beta}|f(\xi_i,y(\xi_i))-f(\xi_i,z(\xi_i))|\Big\}\\
&+|a_2|J^{\beta-p}|g(\eta,y(\eta))-g(\eta,z(\eta))|+|b_2|J^{\beta-p}|g(T,y(T))-g(T,z(T))|\\
&+|a_2|J^{\alpha+\beta-p}|f(\eta,y(\eta))-f(\eta,z(\eta))|\\
&+|b_2|J^{\alpha+\beta-p}|f(T,y(T))-f(T,z(T))|\Big]\Bigg\}\\
&+\frac{1}{|\Omega|}\Bigg\{|A_1|\Big[\sum_{i=1}^{n}|\lambda_i|\ {}^{\rho_i}I_{\theta_i,\kappa_i}^{\gamma_i,\delta_i}\Big\{J^{\beta}|g(\xi_i,y(\xi_i))-g(\xi_i,z(\xi_i))|\\
&+J^{\alpha+\beta}|f(\xi_i,y(\xi_i))-f(\xi_i,z(\xi_i))|\Big\}\\
&+|a_2|J^{\beta-p}|g(\eta,y(\eta))-g(\eta,z(\eta))|+|b_2|J^{\beta-p}|g(T,y(T))-g(T,z(T))|\\
&+|a_2|J^{\alpha+\beta-p}|f(\eta,y(\eta))-f(\eta,z(\eta))|+|b_2|J^{\alpha+\beta-p}|f(T,y(T))-f(T,z(T))|\Big]\\
&+|A_2|\Big[|a_1|J^{\beta}|g(\eta,y(\eta))-g(\eta,z(\eta))|+|b_1|J^{\beta}|g(T,y(T))-g(T,z(T))|\\
&+|a_1|J^{\alpha+\beta}|f(\eta,y(\eta))-f(\eta,z(\eta))|+|b_1|J^{\alpha+\beta}|f(T,y(T))-f(T,z(T))|\Big]\Bigg\}\\
\le\ & k\|y-z\|\Bigg[\frac{T^{\beta}}{\Gamma(\beta+1)}+\frac{T^{\beta}}{|\Omega|\Gamma(\beta+1)}\Bigg\{|B_2|\frac{|a_1|\eta^{\beta}+|b_1|T^{\beta}}{\Gamma(\beta+1)}\\
&+|B_1|\Big[\sum_{i=1}^{n}|\lambda_i|\frac{\xi_i^{\kappa_i+\rho_i(\theta_i+\gamma_i)+\beta}\Gamma(\theta_i+\frac{\beta}{\rho_i}+1)}{\rho_i^{\delta_i}\Gamma(\beta+1)\Gamma(\gamma_i+\theta_i+\frac{\beta}{\rho_i}+1)}+\frac{|a_2|\eta^{\beta-p}+|b_2|T^{\beta-p}}{\Gamma(\beta-p+1)}\Big]\Bigg\}\\
&+\frac{1}{|\Omega|}\Bigg\{|A_1|\Big[\sum_{i=1}^{n}|\lambda_i|\frac{\xi_i^{\kappa_i+\rho_i(\theta_i+\gamma_i)+\beta}\Gamma(\theta_i+\frac{\beta}{\rho_i}+1)}{\rho_i^{\delta_i}\Gamma(\beta+1)\Gamma(\gamma_i+\theta_i+\frac{\beta}{\rho_i}+1)}+\frac{|a_2|\eta^{\beta-p}+|b_2|T^{\beta-p}}{\Gamma(\beta-p+1)}\Big]\\
&+|A_2|\frac{|a_1|\eta^{\beta}+|b_1|T^{\beta}}{\Gamma(\beta+1)}\Bigg\}\Bigg]\\
&+\ell\|y-z\|\Bigg[\frac{T^{\alpha+\beta}}{\Gamma(\alpha+\beta+1)}+\frac{T^{\beta}}{|\Omega|\Gamma(\beta+1)}\Bigg\{|B_2|\frac{|a_1|\eta^{\alpha+\beta}+|b_1|T^{\alpha+\beta}}{\Gamma(\alpha+\beta+1)}\\
&+|B_1|\Big[\sum_{i=1}^{n}|\lambda_i|\frac{\xi_i^{\kappa_i+\rho_i(\theta_i+\gamma_i)+\alpha+\beta}\Gamma(\theta_i+\frac{\alpha+\beta}{\rho_i}+1)}{\rho_i^{\delta_i}\Gamma(\alpha+\beta+1)\Gamma(\gamma_i+\theta_i+\frac{\alpha+\beta}{\rho_i}+1)}\\
&+\frac{|a_2|\eta^{\alpha+\beta-p}+|b_2|T^{\alpha+\beta-p}}{\Gamma(\alpha+\beta-p+1)}\Big]\Bigg\}\\
&+\frac{1}{|\Omega|}\Bigg\{|A_1|\Big[\sum_{i=1}^{n}|\lambda_i|\frac{\xi_i^{\kappa_i+\rho_i(\theta_i+\gamma_i)+\alpha+\beta}\Gamma(\theta_i+\frac{\alpha+\beta}{\rho_i}+1)}{\rho_i^{\delta_i}\Gamma(\alpha+\beta+1)\Gamma(\gamma_i+\theta_i+\frac{\alpha+\beta}{\rho_i}+1)}\\
&+\frac{|a_2|\eta^{\alpha+\beta-p}+|b_2|T^{\alpha+\beta-p}}{\Gamma(\alpha+\beta-p+1)}\Big]+|A_2|\frac{|a_1|\eta^{\alpha+\beta}+|b_1|T^{\alpha+\beta}}{\Gamma(\alpha+\beta+1)}\Bigg\}\Bigg]
\end{aligned}
$$

$$= (k\Lambda_1 + \ell\Lambda_2)\|y - z\|.$$

Consequently, we obtain

$$\|F(y) - F(z)\| \leq (k\Lambda_1 + \ell\Lambda_2)\|y - z\|,$$

which, in view of (7.121), implies that F is a contraction. Hence F has a unique fixed point by Banach's contraction principle. This, in turn, shows that the problem (7.104) has a unique solution on $[0, T]$. The proof is completed. □

7.6.4.1 *Examples*

In this subsection, we give examples to illustrate the usefulness of our main results.

Let us consider the following boundary value problem

$$\begin{cases} D^{2/5}[D^{4/5}y(t) - g(t, x(t))] = f(t, x(t)), \ t \in [0, 2], \\ y(1/2) + 2y(2) = 1, \\ D^{1/5}y(1/2) + D^{1/5}y(2) \\ \quad = {}^{1}I_{0,0}^{1/4,1}y(1/3) - 3\ {}^{1/2}I_{1/3,1/7}^{3/4,1/2}y(1) + 2\ {}^{3/5}I_{0,0}^{1/2,1/2}y(3/2), \end{cases} \tag{7.122}$$

where $\alpha = 2/5$, $\beta = 4/5$, $p = 1/5$, $\eta = 1/2$, $a_1 = a_2 = b_2 = c = 1$, $b_1 = 2$, $\gamma_1 = 1/4$, $\delta_1 = 1$, $\rho_1 \to 1$, $\kappa_1 = \theta_1 = 0$, $\lambda_1 = 1$, $\xi_1 = 1/3$, $\gamma_2 = 3/4$, $\delta_2 = 1/2$, $\rho_2 = 1/2$, $\kappa_2 = 1/7$, $\theta_2 = 1/3$, $\lambda_2 = -3$, $\xi_2 = 1$, $\gamma_3 = \delta_3 = 1/2$, $\rho_3 = 3/5$, $\kappa_3 = \theta_3 = 0$, $\lambda_3 = 2$, $\xi_3 = 3/2$, and $g(t, y)$, $f(t, y)$ will be fixed later.

Using the given data, we find that $|A_1| \approx 4.355402744$, $|A_2| \approx 1.232504437$, $|B_1| = 3$, $|B_2| \approx 0.479215198$, $|\Omega| \approx 5.784688499$, $\Lambda_1 \approx 16.95703626$, $\Lambda_2 \approx 15.33660937$, where A_1, A_2, B_1, B_2, Ω, Λ_1 and Λ_2 are respectively given by (7.105), (7.106), (7.107), (7.108), (7.109), (7.118) and (7.119). For illustrating Theorem 7.24, we take

$$f(t, y) = \frac{1}{6\sqrt{900 + t^2}}\Big(\tan^{-1} y + e^{-t}\Big) \ \text{ and } \ g(t, y) = \frac{1}{90(1 + t^2)}\sin y + \frac{\cos t}{25}. \tag{7.123}$$

It is easy to check that $f(t, y)$ and $g(t, y)$ satisfy the hypothesis of Theorem 7.24 with $\psi(t) = \dfrac{1}{90(1 + t^2)} + \dfrac{1}{25}$, $\phi(t) = \dfrac{(\pi + 2e^{-t})}{12\sqrt{900 + t^2}}$ and $k = 1/90$. Also

$$k\Lambda_1 \approx 0.1884115140 < 1.$$

Thus, by the conclusion of Theorem 7.24, there exists at least one solution for the problem (7.122) with $f(t, y)$ and $g(t, y)$ given by (7.123).

In order to explain,Theorem 7.25, we choose

$$f(t, y) = \frac{1}{6\sqrt{900 + t^2}}e^{-t}\Big(\tan^{-1} y + y + \frac{1}{8}\Big) \ \text{ and } \ g(t, y) = \frac{1}{90(1 + t^2)}\sin y + \frac{\cos t}{25}, \tag{7.124}$$

where $d_1 = 1/90$, $d_2 = 1/25$, $\psi(||y||) = ||y|| + \dfrac{4\pi + 1}{8}$ and $p(t) = \dfrac{1}{6\sqrt{900 + t^2}}e^{-t}$. Furthermore, by the condition (7.25.4), we find that $M > 1.639205778$. Clearly the hypothesis of Theorem 7.25 holds true and consequently there exists at least one solution for the problem (7.122) with $f(t, y)$ and $g(t, y)$ given by (7.124).

Next we illustrate Theorem 7.26 by taking $f(t, y)$ and $g(t, y)$ given by (7.124). Clearly the condition (7.26.1) and (7.26.2) hold true with $\ell = 1/90$ and $k = 1/90$. Also

$$k\Lambda_1 + \ell\Lambda_2 \approx 0.3588182848 < 1.$$

Thus all the conditions of Theorem 7.26 are satisfied. So, the conclusion of Theorem 7.26 implies that the problem (7.122) with $f(t, y)$ and $g(t, y)$ given by (7.124) has a unique solution on $[0, 2]$.

Remark 7.4. Several new results for multi-term Caputo type fractional differential equations equations supplemented with different kinds of boundary conditions follow by fixing the parameters in the problem (7.104). Here are some examples.

- Letting $a_1 = a_2 = 0$ and $b_1 = b_2 = 1$ in the boundary conditions of the problem (7.104), the results presented in this section correspond to the ones with the boundary conditions of the form:

$$y(T) = c,\ D^p y(T) = \sum_{i=1}^{n} \lambda_i\ {}^{\rho_i}I_{\theta_i,\kappa_i}^{\gamma_i,\delta_i} y(\xi_i), \quad \eta, \xi_i \in (0, T).$$

- If we choose $a_1 = a_2 = 1$ and $b_1 = b_2 = 0$ in the problem (7.104), then the results presented in this section become the ones associated with the nonlocal multi-point conditions:

$$y(\eta) = c,\ D^p y(\eta) = \sum_{i=1}^{n} \lambda_i\ {}^{\rho_i}I_{\theta_i,\kappa_i}^{\gamma_i,\delta_i} y(\xi_i), \quad \eta, \xi_i \in (0, T).$$

- Fixing $\lambda_i = 0$ $(i = 1, 2, 3, \ldots, n), c = 0$ and $a_1 = a_2 = b_1 = b_2 = 1$ in the present results, we obtain the ones for multi-term fractional differential equations supplemented with parametric type anti-periodic fractional boundary conditions:

$$y(\eta) = -y(T),\ D^p y(\eta) = -D^p y(T), \quad \eta \in (0, T).$$

7.7 Generalized Caputo fractional differential equations and inclusions with nonlocal generalized fractional integral and multi-point boundary conditions

In this section, we study a new class of boundary value problems of Caputo-type generalized fractional differential equations and inclusions (instead of usual Caputo fractional order derivative) supplemented with nonlocal generalized fractional integral and multi-point boundary conditions. Precisely, we consider the problems:

$$\begin{cases} {}^{\rho}_{c}D^{\alpha}_{0+} y(t) = f(t, y(t)), \quad t \in [0, T], \\ y(T) = \sum_{i=1}^{m} \sigma_i {}^{\rho}I^{\beta}_{0+} y(\eta_i) + \kappa, \quad \delta y(0) = \sum_{j=1}^{k} \mu_j y(\xi_j), \\ 0 < \eta_1 < \cdots < \eta_i < \cdots < \eta_m < \xi_1 < \cdots < \xi_j < \cdots < \xi_k < T, \end{cases} \tag{7.125}$$

and

$$\begin{cases} {}^{\rho}_{c}D^{\alpha}_{0+}y(t) \in F(t, y(t)), \quad t \in [0, T], \\ y(T) = \sum_{i=1}^{m} \sigma_i {}^{\rho}I^{\beta}_{0+}y(\eta_i) + \kappa, \quad \delta y(0) = \sum_{j=1}^{k} \mu_j y(\xi_j), \\ 0 < \eta_1 < \cdots < \eta_i < \cdots < \eta_m < \xi_1 < \cdots < \xi_j < \cdots < \xi_k < T, \end{cases} \tag{7.126}$$

where ${}^{\rho}_{c}D^{\alpha}_{0+}$ is the Caputo-type generalized fractional derivative operator of order $1 < \alpha \leq 2$, ${}^{\rho}I^{\beta}_{0+}$ is the generalized (Katugampola type) fractional integral of order $\beta > 0, \rho > 0$, $f : [0, T] \times \mathbb{R} \to \mathbb{R}$ is a continuous function, $\sigma_i, \mu_j, \kappa \in \mathbb{R}, i = 1, 2, \ldots, m, j = 1, 2, \ldots, k$, $\delta = t^{1-\rho}\frac{d}{dt}$, and $F : [0, T] \times \mathbb{R} \to \mathcal{P}(\mathbb{R})$ is a multi-valued function ($\mathcal{P}(\mathbb{R})$ is the family of all non-empty subsets of $\mathbb{R}$).

For computational convenience, we introduce the notation:

$$A_1 = 1 - \sum_{j=1}^{k} \mu_j \frac{\xi_j^{\rho}}{\rho}, \qquad A_2 = \sum_{j=1}^{k} \mu_j, \tag{7.127}$$

$$B_1 = \frac{T^{\rho}}{\rho} - \sum_{i=1}^{m} \sigma_i \frac{\eta_i^{\rho(\beta+1)}}{\rho^{\beta+1}\Gamma(\beta+2)}, \qquad B_2 = 1 - \sum_{i=1}^{m} \sigma_i \frac{\eta_i^{\rho\beta}}{\rho^{\beta}\Gamma(\beta+1)}, \tag{7.128}$$

$$\Omega = A_1 B_2 + B_1 A_2. \tag{7.129}$$

The following lemma, related to the linear variant of the problem (7.125), plays a key role in converting the given problem into a fixed point problem.

Lemma 7.14. *Let $h \in C(0, T) \cap L(0, T)$, $y \in AC^2_{\delta}(J)$ and $\Omega \neq 0$, where Ω is given by (7.129). Then the solution of the linear boundary value problem (BVP):*

$$\begin{cases} {}^{\rho}_{c}D^{\alpha}_{0+}y(t) = h(t), \quad t \in [0, T], \\ y(T) = \sum_{i=1}^{m} \sigma_i {}^{\rho}I^{\beta}_{0+}y(\eta_i) + \kappa, \quad \delta y(0) = \sum_{j=1}^{k} \mu_j y(\xi_j), \\ 0 < \eta_1 < \cdots < \eta_i < \cdots < \eta_m < \xi_1 < \cdots < \xi_j < \cdots < \xi_k < T, \end{cases} \tag{7.130}$$

is given by

$$\begin{aligned} y(t) = {}^{\rho}I^{\alpha}_{0+}h(t) + \frac{1}{\Omega}\Big\{ - B_1 \sum_{j=1}^{k} \mu_j {}^{\rho}I^{\alpha}_{0+}h(\xi_j) + A_1 \Big[\sum_{i=1}^{m} \sigma_i \, {}^{\rho}I^{\alpha+\beta}_{0+}h(\eta_i) - {}^{\rho}I^{\alpha}_{0+}h(T) + \kappa \Big] \Big\} \\ + \frac{t^{\rho}}{\rho\Omega}\Big\{ B_2 \sum_{j=1}^{k} \mu_j {}^{\rho}I^{\alpha}_{0+}h(\xi_j) + A_2 \Big[\sum_{i=1}^{m} \sigma_i \, {}^{\rho}I^{\alpha+\beta}_{0+}h(\eta_i) - {}^{\rho}I^{\alpha}_{0+}h(T) + \kappa \Big] \Big\}. \end{aligned} \tag{7.131}$$

Proof. Applying ${}^{\rho}I^{\alpha}_{0+}$ to the fractional differential equation in (7.130) and using Lemma 1.8, the solution of fractional differential equation in (7.130) for $t \in [0, T]$ is

$$y(t) = {}^{\rho}I^{\alpha}_{0+}h(t) + c_1 + c_2\frac{t^{\rho}}{\rho} = \frac{\rho^{1-\alpha}}{\Gamma(\alpha)} \int_0^t s^{\rho-1}(t^{\rho} - s^{\rho})^{\alpha-1} h(s) ds + c_1 + c_2 \frac{t^{\rho}}{\rho}, \tag{7.132}$$

for some $c_1, c_2 \in \mathbb{R}$. Taking δ-derivative of (7.132), we get

$$\delta y(t) = {}^{\rho}I_{0+}^{\alpha-1}h(t) + c_2 = \frac{\rho^{2-\alpha}}{\Gamma(\alpha-1)}\int_0^t s^{\rho-1}(t^{\rho}-s^{\rho})^{\alpha-2}h(s)ds + c_2. \tag{7.133}$$

Using the boundary condition $\delta y(0) = \sum_{j=1}^{k}\mu_j y(\xi_j)$ in (7.133), together with the notation (7.127), we get

$$A_1c_2 - A_2c_1 = \sum_{j=1}^{k}\mu_j {}^{\rho}I_{0+}^{\alpha}h(\xi_j). \tag{7.134}$$

Applying the generalized integral operator ${}^{\rho}I_{0+}^{\beta}$ to (7.132), we get

$${}^{\rho}I_{0+}^{\beta}y(t) = {}^{\rho}I_{0+}^{\alpha+\beta}h(t) + c_1\frac{t^{\rho\beta}}{\rho^{\beta}\Gamma(\beta+1)} + c_2\frac{t^{\rho(\beta+1)}}{\rho^{\beta+1}\Gamma(\beta+2)}, \tag{7.135}$$

which, together with the boundary condition $y(T) = \sum_{i=1}^{m}\sigma_i\ {}^{\rho}I_{0+}^{\beta}y(\eta_i) + \kappa$ and notation (7.128), yields

$$B_1c_2 + B_2c_1 = \sum_{i=1}^{m}\sigma_i\ {}^{\rho}I_{0+}^{\alpha+\beta}h(\eta_i) - {}^{\rho}I_{0+}^{\alpha}h(T) + \kappa. \tag{7.136}$$

Solving the system of equations (7.134) and (7.136) for c_1 and c_2, we find that

$$c_1 = \frac{1}{\Omega}\Big\{-B_1\sum_{j=1}^{k}\mu_j {}^{\rho}I_{0+}^{\alpha}h(\xi_j) + A_1\Big[\sum_{i=1}^{m}\sigma_i\ {}^{\rho}I_{0+}^{\alpha+\beta}h(\eta_i) - {}^{\rho}I_{0+}^{\alpha}h(T) + \kappa\Big]\Big\}, \tag{7.137}$$

and

$$c_2 = \frac{1}{\Omega}\Big\{B_2\sum_{j=1}^{k}\mu_j {}^{\rho}I_{0+}^{\alpha}h(\xi_j) + A_2\Big[\sum_{i=1}^{m}\sigma_i\ {}^{\rho}I_{0+}^{\alpha+\beta}h(\eta_i) - {}^{\rho}I^{\alpha}h(T) + \kappa\Big]\Big\}. \tag{7.138}$$

Substituting the values of c_1 and c_2 in (7.132), we get the equation (7.131). The converse follows by direct computation. The proof is completed. □

7.7.1 *Main results for the problem (7.125)*

Let $\mathcal{C} = C(J, \mathbb{R})$ denote the Banach space of all continuous functions from $[0, T]$ to $\mathbb{R}$ endowed with the norm defined by $\|y\| = \sup_{t\in[0,T]}|y(t)|$.

By Lemma 7.14, we define an operator $\mathcal{G} : \mathcal{C} \to \mathcal{C}$ associated with the problem (7.125) as

$$\begin{aligned}\mathcal{G}y(t) = {}&{}^{\rho}I_{0+}^{\alpha}f(t, y(t)) + \frac{1}{\Omega}\Big\{-B_1\sum_{j=1}^{k}\mu_j {}^{\rho}I_{0+}^{\alpha}f(\xi_j, y(\xi_j)) + A_1\Big[\sum_{i=1}^{m}\sigma_i\ {}^{\rho}I_{0+}^{\alpha+\beta}f(\eta_i, y(\eta_i))\\ &-{}^{\rho}I_{0+}^{\alpha}f(T, y(T)) + \kappa\Big]\Big\} + \frac{t^{\rho}}{\rho\Omega}\Big\{B_2\sum_{j=1}^{k}\mu_j {}^{\rho}I_{0+}^{\alpha}f(\xi_j, y(\xi_j))\\ &+A_2\Big[\sum_{i=1}^{m}\sigma_i\ {}^{\rho}I_{0+}^{\alpha+\beta}f(\eta_i, y(\eta_i)) - {}^{\rho}I_{0+}^{\alpha}f(T, y(t)) + \kappa\Big]\Big\}.\end{aligned} \tag{7.139}$$

In the following, for brevity, we use the notation:

$$\Lambda = \frac{T^{\rho\alpha}}{\rho^{\alpha}\Gamma(\alpha+1)} + \frac{1}{|\Omega|}\Big\{|B_1|\sum_{j=1}^{k}|\mu_j|\frac{\xi_j^{\rho\alpha}}{\rho^{\alpha}\Gamma(\alpha+1)}$$
$$+|A_1|\Big[\sum_{i=1}^{m}|\sigma_i|\frac{\eta_i^{\rho(\alpha+\beta)}}{\rho^{\alpha+\beta}\Gamma(\alpha+\beta+1)} + \frac{T^{\rho\alpha}}{\rho^{\alpha}\Gamma(\alpha+1)}\Big]\Big\}$$
$$+\frac{T^{\rho}}{\rho|\Omega|}\Big\{|B_2|\sum_{j=1}^{k}|\mu_j|\frac{\xi_j^{\rho\alpha}}{\rho^{\alpha}\Gamma(\alpha+1)}$$
$$+|A_2|\Big[\sum_{i=1}^{m}|\sigma_i|\frac{\eta_i^{\rho(\alpha+\beta)}}{\rho^{\alpha+\beta}\Gamma(\alpha+\beta+1)} + \frac{T^{\rho\alpha}}{\rho^{\alpha}\Gamma(\alpha+1)}\Big]\Big\}. \tag{7.140}$$

In the first result, we establish the existence of solutions for the problem (7.125) via Leray-Schauder nonlinear alternative (theorem 1.8).

Theorem 7.27. *Assume that:*

(7.27.1) there exist a function $\phi \in L^1([0,T],\mathbb{R}^+)$, and a non-decreasing function $\psi : \mathbb{R}^+ \to \mathbb{R}^+$ such that $|f(t,y)| \le \phi(t)\psi(\|y\|)$, $\forall (t,y) \in [0,T]\times\mathbb{R}$;

(7.27.2) there exists a constant $M > 0$ such that

$$\frac{M}{\psi(M)\Lambda_1 + \dfrac{|\kappa|(\rho|A_1| + T^{\rho}|A_2|)}{\rho|\Omega|}} > 1,$$

where

$$\Lambda_1 = {}^{\rho}I_{0+}^{\alpha}\phi(T) + \frac{1}{|\Omega|}\Big\{|B_1|\sum_{j=1}^{k}|\mu_j|{}^{\rho}I_{0+}^{\alpha}\phi(\xi_j)$$
$$+|A_1|\Big[\sum_{i=1}^{m}|\sigma_i|\ {}^{\rho}I_{0+}^{\alpha+\beta}\phi(\eta_i) + {}^{\rho}I_{0+}^{\alpha}\phi(T)\Big]\Big\}$$
$$+\frac{T^{\rho}}{\rho|\Omega|}\Big\{|B_2|\sum_{j=1}^{k}|\mu_j|{}^{\rho}I_{0+}^{\alpha}\phi(\xi_j)$$
$$+|A_2|\Big[\sum_{i=1}^{m}|\sigma_i|\ {}^{\rho}I_{0+}^{\alpha+\beta}\phi(\eta_i) + {}^{\rho}I_{0+}^{\alpha}\phi(T)\Big]\Big\}. \tag{7.141}$$

Then the problem (7.125) has at least one solution on $[0,T]$.

Proof. Firstly, we show that the operator $\mathcal{G} : \mathcal{C} \to \mathcal{C}$ defined by (7.139) is continuous and completely continuous.

Step 1. *$\mathcal{G}$ is continuous.*

Let $\{y_n\}$ be a sequence such that $y_n \to y$ in $\mathcal{C}$. Then

$$\begin{aligned}
&|\mathcal{G}(y_n)(t) - \mathcal{G}(y)(t)| \\
&\le {}^\rho I^\alpha_{0+}|f(t,y_n(t)) - f(t,y(t))| + \frac{1}{|\Omega|}\Bigg\{|B_1|\sum_{j=1}^{k}|\mu_j|{}^\rho I^\alpha_{0+}|f(\xi_j,y_n(\xi_j)) - f(\xi_j,y(\xi_j))| \\
&\quad +|A_1|\Big[\sum_{i=1}^{m}|\sigma_i|\ {}^\rho I^{\alpha+\beta}_{0+}|f(\eta_i,y_n(\eta_i)) - f(\eta_i,y(\eta_i))| + {}^\rho I^\alpha_{0+}|f(T,y_n(T)) - f(T,y(T))|\Big]\Bigg\} \\
&\quad +\frac{t^\rho}{\rho|\Omega|}\Big\{|B_2|\sum_{j=1}^{k}|\mu_j|{}^\rho I^\alpha_{0+}|f(\xi_j,y_n(\xi_j)) - f(\xi_j,y(\xi_j))| \\
&\quad +|A_2|\Big[\sum_{i=1}^{m}|\sigma_i|\ {}^\rho I^{\alpha+\beta}_{0+}|f(\eta_i,y_n(\eta_i)) - f(\eta_i,y(\eta))| + {}^\rho I^\alpha_{0+}|f(T,y_n(T)) - f(T,y(T))|\Big]\Big\} \\
&\le \Lambda\|f(\cdot,y_n) - f(\cdot,y)\|.
\end{aligned}$$

Since f is continuous function, therefore, we have

$$\|\mathcal{G}(y_n) - \mathcal{G}(y)\| \le \Lambda\|f(\cdot,y_n) - f(\cdot,y)\| \to 0, \text{ as } n \to \infty.$$

Step 2. *$\mathcal{G}$ maps bounded sets into bounded sets in $\mathcal{C}$.*

Indeed, it is enough to show that there exists a positive constant ℓ for any number $r > 0$ such that, for $y \in B_r = \{y \in \mathcal{C} : \|y\| \le r\}$, we have $\|\mathcal{G}(y)\| \le \ell$. By (7.27.1), for each $t \in J$, we have

$$\begin{aligned}
&|\mathcal{G}(y)(t)| \\
&\le {}^\rho I^\alpha_{0+}|f(t,y(t))| + \frac{1}{|\Omega|}\Big\{|B_1|\sum_{j=1}^{k}|\mu_j|{}^\rho I^\alpha_{0+}|f(\xi_j,y(\xi_j))| + |A_1|\Big[\sum_{i=1}^{m}|\sigma_i|\ {}^\rho I^{\alpha+\beta}_{0+}|f(\eta_i,y(\eta_i))| \\
&\quad +{}^\rho I^\alpha_{0+}|f(T,y(T))| + |\kappa|\Big]\Big\} + \frac{t^\rho}{\rho|\Omega|}\Big\{|B_2|\sum_{j=1}^{k}|\mu_j|{}^\rho I^\alpha_{0+}|f(\xi_j,y(\xi_j))| \\
&\quad +|A_2|\Big[\sum_{i=1}^{m}|\sigma_i|\ {}^\rho I^{\alpha+\beta}_{0+}|f(\eta_i,y(\eta_i))| + {}^\rho I^\alpha_{0+}|f(T,y(T))| + |\kappa|\Big]\Big\} \\
&\le \psi(\|r\|)\Big({}^\rho I^\alpha_{0+}\phi(T) + \frac{1}{|\Omega|}\Big\{|B_1|\sum_{j=1}^{k}|\mu_j|{}^\rho I^\alpha_{0+}\phi(\xi_j) + |A_1|\Big[\sum_{i=1}^{m}|\sigma_i|\ {}^\rho I^{\alpha+\beta}_{0+}\phi(\eta_i) \\
&\quad +{}^\rho I^\alpha_{0+}\phi(T)\Big]\Big\} + \frac{T^\rho}{\rho|\Omega|}\Big\{|B_2|\sum_{j=1}^{k}|\mu_j|{}^\rho I^\alpha_{0+}\phi(\xi_j) \\
&\quad +|A_2|\Big[\sum_{i=1}^{m}|\sigma_i|\ {}^\rho I^{\alpha+\beta}_{0+}\phi(\eta_i) + {}^\rho I^\alpha_{0+}\phi(T)\Big]\Big\}\Big) + \frac{|\kappa|(\rho|A_1| + T^\rho|A_2|)}{\rho|\Omega|} := \ell.
\end{aligned}$$

Step 3. *$\mathcal{G}$ maps bounded sets into equi-continuous sets of $\mathcal{C}$.*

Let $t_1, t_2 \in [0,T]$, $t_1 < t_2$, B_r be a bounded set of $\mathcal{C}$ as in Step 2, and let $y \in B_r$.

Then

$$|\mathcal{G}(y)(t_2) - \mathcal{G}(y)(t_1)|$$

$$\leq \frac{\rho^{1-\alpha}\psi(r)}{\Gamma(\alpha)}\Big|\int_0^{t_1}\Big[\frac{s^{\rho-1}}{(t_2^\rho - s^\rho)^{1-\alpha}} - \frac{s^{\rho-1}}{(t_1^\rho - s^\rho)^{1-\alpha}}\Big]\phi(s)ds + \int_{t_1}^{t_2}\frac{s^{\rho-1}}{(t_2^\rho - s^\rho)^{1-\alpha}}\phi(s)ds\Big|$$

$$+\frac{|t_2^\rho - t_1^\rho|}{\rho|\Omega|}\Big\{\psi(r)\Big(|B_2|\sum_{j=1}^{k}|\mu_j|{}^\rho I_{0+}^\alpha\phi(\xi_j)$$

$$+|A_2|\Big[\sum_{i=1}^{m}|\sigma_i|\ {}^\rho I_{0+}^{\alpha+\beta}\phi(\eta_i) + {}^\rho I_{0+}^\alpha\phi(T)\Big]\Big) + |A_2||\kappa|\Big\}.$$

Clearly the right-hand side of the above inequality tends to zero as $t_2 \to t_1$, independently of y. In view of Steps 1-3, it follows from the Arzelá-Ascoli theorem that the operator $\mathcal{G} : \mathcal{C} \longrightarrow \mathcal{C}$ is completely continuous.

Step 4. *We show that there exists an open set $U \subseteq \mathcal{C}$ with $y \neq \lambda\mathcal{G}(y)$ for $\lambda \in (0,1)$ and $y \in \partial U$.*

Let $y \in \mathcal{C}$ be a solution of $y = \lambda\mathcal{G}y$ for $\lambda \in [0,1]$. Then, for $t \in [0,T]$, we have

$$|y(t)| = |\lambda(\mathcal{G}y)(t)|$$

$$\leq {}^\rho I_{0+}^\alpha|f(t,y(t))| + \frac{1}{|\Omega|}\Big\{|B_1|\sum_{j=1}^{k}|\mu_j|{}^\rho I_{0+}^\alpha|f(\xi_j,y(\xi_j))| + |A_1|\Big[\sum_{i=1}^{m}|\sigma_i|\ {}^\rho I_{0+}^{\alpha+\beta}|f(\eta_i,y(\eta_i))|$$

$$+{}^\rho I_{0+}^\alpha|f(T,y(T))| + |\kappa|\Big]\Big\} + \frac{t^\rho}{\rho|\Omega|}\Big\{|B_2|\sum_{j=1}^{k}|\mu_j|{}^\rho I_{0+}^\alpha|f(\xi_j,y(\xi_j))|$$

$$+|A_2|\Big[\sum_{i=1}^{m}|\sigma_i|\ {}^\rho I_{0+}^{\alpha+\beta}|f(\eta_i,y(\eta_i))| + {}^\rho I_{0+}^\alpha|f(T,y(T))| + |\kappa|\Big]\Big\}$$

$$\leq \psi(\|y\|)\Big({}^\rho I_{0+}^\alpha|\phi(T) + \frac{1}{|\Omega|}\Big\{|B_1|\sum_{j=1}^{k}|\mu_j|{}^\rho I_{0+}^\alpha\phi(\xi_j) + |A_1|\Big[\sum_{i=1}^{m}|\sigma_i|\ {}^\rho I_{0+}^{\alpha+\beta}\phi(\eta_i)$$

$$+{}^\rho I_{0+}^\alpha\phi(T)\Big]\Big\} + \frac{T^\rho}{\rho|\Omega|}\Big\{|B_2|\sum_{j=1}^{k}|\mu_j|{}^\rho I_{0+}^\alpha\phi(\xi_j) + |A_2|\Big[\sum_{i=1}^{m}|\sigma_i|\ {}^\rho I_{0+}^{\alpha+\beta}\phi(\eta_i) + {}^\rho I_{0+}^\alpha\phi(T)\Big]\Big\}\Big)$$

$$+\frac{|\kappa|(\rho|A_1| + T^\rho|A_2|)}{\rho|\Omega|},$$

which, on taking the norm for $t \in [0,T]$, implies that

$$\frac{\|y\|}{\psi(\|y\|)\Lambda_1 + \dfrac{|\kappa|(\rho|A_1| + T^\rho|A_2|)}{\rho|\Omega|}} \leq 1.$$

In view of (7.27.2), there exists M such that $\|y\| \neq M$. Let us set

$$U = \{y \in \mathcal{C} : \|y\| < M\}.$$

Note that the operator $\mathcal{G} : \overline{U} \to \mathcal{C}$ is continuous and completely continuous. From the choice of U, there is no $y \in \partial U$ such that $y = \lambda \mathcal{G}(y)$ for some $\lambda \in (0,1)$. Consequently, by the nonlinear alternative of Leray-Schauder type (Theorem 1.8), we deduce that $\mathcal{G}$ has a fixed point $y \in \overline{U}$ which is a solution of the problem (7.125). This completes the proof. □

In the next result, we prove the existence of solutions for the problem (7.125) by applying Krasnosel'skiĭ's fixed point theorem (Theorem 1.4).

Theorem 7.28. *Let $f : [0,T] \times \mathbb{R} \to \mathbb{R}$ be a continuous function satisfying the assumptions:*

(7.28.1) $|f(t,x) - f(t,y)| \le L\|x-y\|, \quad \forall t \in [0,T],\ L > 0,\ x, y \in \mathbb{R}$;
(7.28.2) $|f(t,y)| \le \Phi(t), \ \forall (t,y) \in [0,T] \times \mathbb{R},$ *and* $\Phi \in C([0,T], \mathbb{R}^+)$.

Then the problem (7.125) has at least one solution on $[0,T]$, provided that

$$L\left(\frac{T^\rho}{\rho|\Omega|}\left\{|B_2|\sum_{j=1}^{k}|\mu_j|\frac{\xi_j^{\rho\alpha}}{\rho^\alpha\Gamma(\alpha+1)} + |A_2|\left[\sum_{i=1}^{m}|\sigma_i|\frac{\eta_i^{\rho(\alpha+\beta)}}{\rho^{\alpha+\beta}\Gamma(\alpha+\beta+1)} + \frac{T^{\rho\alpha}}{\rho^\alpha\Gamma(\alpha+1)}\right]\right\}\right) < 1. \quad (7.142)$$

Proof. Let us fix $\bar{r} \ge \|\Phi\|\Lambda + \frac{|\kappa|(\rho|A_1|+T^\rho|A_2|)}{\rho|\Omega|}$, where $\|\Phi\| = \sup_{t\in[0,T]}|\Phi(t)|$ and consider $B_{\bar{r}} = \{y \in \mathcal{C} : \|y\| \le \bar{r}\}$. Let us split the operator $\mathcal{G} : \mathcal{C} \to \mathcal{C}$ defined by (7.139) on $B_{\bar{r}}$ as $\mathcal{G} = \mathcal{A} + \mathcal{B}$, where $\mathcal{A}$ and $\mathcal{B}$ are given by

$$\begin{aligned}&\mathcal{A}(t)\\ &= {}^\rho I_{0+}^\alpha f(t,y(t)) + \frac{1}{\Omega}\Big\{ - B_1\sum_{j=1}^{k}\mu_j {}^\rho I_{0+}^\alpha f(\xi_j, y(\xi_j)) + A_1\Big[\sum_{i=1}^{m}\sigma_i\ {}^\rho I_{0+}^{\alpha+\beta} f(\eta_i, y(\eta_i))\\ &\quad - {}^\rho I_{0+}^\alpha f(T, y(T)) + \kappa\Big]\Big\},\end{aligned}$$

and

$$\begin{aligned}\mathcal{B}(t) = \frac{t^\rho}{\rho\Omega}\Big\{B_2\sum_{j=1}^{k}\mu_j {}^\rho I_{0+}^\alpha f(\xi_j, y(\xi_j)) + A_2\Big[\sum_{i=1}^{m}\sigma_i\ {}^\rho I_{0+}^{\alpha+\beta} f(\eta_i, y(\eta_j))\\ - {}^\rho I_{0+}^\alpha f(T, y(t)) + \kappa\Big]\Big\}.\end{aligned}$$

For $x, y \in B_{\bar{r}}$, we find that

$$\begin{aligned}&\|\mathcal{A}x + \mathcal{B}y\|\\ &\le \sup_{t\in J}\Bigg\{{}^\rho I^\alpha|f(t,x(t))| + \frac{1}{\Omega}\Big\{|B_1|\sum_{j=1}^{k}|\mu_j|{}^\rho I_{0+}^\alpha f(\xi_j, x(\xi_j)) + A_1\Big[\sum_{i=1}^{m}|\sigma_i|\ {}^\rho I_{0+}^{\alpha+\beta}|f(\eta_i, x(\eta_i))|\\ &\quad + {}^\rho I_{0+}^\alpha|f(T, x(T))| + |\kappa|\Big]\Big\} + \frac{t^\rho}{\rho|\Omega|}\Big\{|B_2|\sum_{j=1}^{k}|\mu_j|{}^\rho I_{0+}^\alpha|f(\xi_j, y(\xi_j))|\end{aligned}$$

$$+|A_2|\Big[\sum_{i=1}^{m}|\sigma_i|\ {}^{\rho}I_{0+}^{\alpha+\beta}|f(\eta_i,y(\eta_j))|+{}^{\rho}I_{0+}^{\alpha}|f(T,y(T))|+|\kappa|\Big]\Big\}$$

$$\le \|\Phi\|\Bigg\{\frac{T^{\rho\alpha}}{\rho^{\alpha}\Gamma(\alpha+1)}+\frac{1}{|\Omega|}\Big\{|B_1|\sum_{j=1}^{k}|\mu_j|\frac{\xi_j^{\rho\alpha}}{\rho^{\alpha}\Gamma(\alpha+1)}$$

$$+|A_1|\Big[\sum_{i=1}^{m}|\sigma_i|\frac{\eta_i^{\rho(\alpha+\beta)}}{\rho^{\alpha+\beta}\Gamma(\alpha+\beta+1)}+\frac{T^{\rho\alpha}}{\rho^{\alpha}\Gamma(\alpha+1)}\Big]\Big\}+\frac{T^{\rho}}{\rho|\Omega|}\Big\{|B_2|\sum_{j=1}^{k}|\mu_j|\frac{\xi_j^{\rho\alpha}}{\rho^{\alpha}\Gamma(\alpha+1)}$$

$$+|A_2|\Big[\sum_{i=1}^{m}|\sigma_i|\frac{\eta_i^{\rho(\alpha+\beta)}}{\rho^{\alpha+\beta}\Gamma(\alpha+\beta+1)}+\frac{T^{\rho\alpha}}{\rho^{\alpha}\Gamma(\alpha+1)}\Big]\Big\}\Bigg\}+\frac{|\kappa|(\rho|A_1|+T^{\rho}|A_2|)}{\rho|\Omega|}$$

$$\le \|\Phi\|\Lambda+\frac{|\kappa|(\rho|A_1|+T^{\rho}|A_2|)}{\rho|\Omega|}<\bar{r}.$$

Thus $\mathcal{A}x+\mathcal{B}y\in B_{\bar{r}}$. Now, for $x,y\in B_{\bar{r}}$ and for each $t\in[0,T]$, we obtain

$$\|\mathcal{B}x-\mathcal{B}y\|$$

$$\le \sup_{t\in J}\Big\{\frac{t^{\rho}}{\rho|\Omega|}\Big\{|B_2|\sum_{j=1}^{k}|\mu_j|{}^{\rho}I_{0+}^{\alpha}|f(\xi_j,x(\xi_j))-f(\xi_j,y(\xi_j))|$$

$$+|A_2|\Big[\sum_{i=1}^{m}|\sigma_i|\ {}^{\rho}I_{0+}^{\alpha+\beta}|f(\eta_i,x(\eta_i))-f(\eta_i,y(\eta_i))|+{}^{\rho}I_{0+}^{\alpha}|f(T,x(T))-f(T,y(T))|\Big]\Big\}\Big\}$$

$$\le L\Big(\frac{T^{\rho}}{\rho|\Omega|}\Big\{|B_2|\sum_{j=1}^{k}|\mu_j|\frac{\xi_j^{\rho\alpha}}{\rho^{\alpha}\Gamma(\alpha+1)}$$

$$+|A_2|\Big[\sum_{i=1}^{m}|\sigma_i|\frac{\eta_i^{\rho(\alpha+\beta)}}{\rho^{\alpha+\beta}\Gamma(\alpha+\beta+1)}+\frac{T^{\rho\alpha}}{\rho^{\alpha}\Gamma(\alpha+1)}\Big]\Big\}\Big)\|x-y\|,$$

which, together with the condition (7.142), implies that $\mathcal{B}$ is a contraction. Continuity of f implies that the operator $\mathcal{A}$ is continuous. Also, $\mathcal{A}$ is uniformly bounded on $B_{\bar{r}}$ as

$$\|\mathcal{A}y\|\le\|\Phi\|\Big(\frac{T^{\rho\alpha}}{\rho^{\alpha}\Gamma(\alpha+1)}+\frac{1}{|\Omega|}\Big\{|B_1|\sum_{j=1}^{k}|\mu_j|\frac{\xi_j^{\rho\alpha}}{\rho^{\alpha}\Gamma(\alpha+1)}$$

$$+|A_1|\Big[\sum_{i=1}^{m}|\sigma_i|\frac{\eta_i^{\rho(\alpha+\beta)}}{\rho^{\alpha+\beta}\Gamma(\alpha+\beta+1)}+\frac{T^{\rho\alpha}}{\rho^{\alpha}\Gamma(\alpha+1)}\Big]\Big\}\Big)+\frac{|A_1||\kappa|}{|\Omega|}.$$

In order to show the compactness of the operator $\mathcal{A}$, let $\sup_{(t,y)\in J\times B_{\bar{r}}}|f(t,y)|=\bar{f}<\infty$. Consequently, for $t_1,t_2\in[0,T]$, $t_1<t_2$, we have

$$|(\mathcal{A}y)(t_2)-(\mathcal{A}y)(t_1)|$$

$$\le\Big|\frac{\rho^{1-\alpha}}{\Gamma(\alpha)}\Big[\int_0^{t_1}s^{\rho-1}[(t_2^{\rho}-s^{\rho})^{\alpha-1}-(t_1^{\rho}-s^{\rho})^{\alpha-1}]f(s,y(s))ds$$

$$+\int_{t_1}^{t_2}s^{\rho-1}(t_2^{\rho}-s^{\rho})^{\alpha-1}f(s,y(s))ds\Big]\Big|$$

$$\leq \frac{\bar{f}}{\rho^\alpha\Gamma(\alpha+1)}\Big\{2(t_2^\rho - t_1^\rho)^\alpha + |t_2^{\rho\alpha} - t_1^{\rho\alpha}|\Big\},$$

which is independent of $y \in B_{\bar{r}}$ and tends to zero as $t_2 \to t_1$. Thus $\mathcal{A}$ is equicontinuous. So $\mathcal{A}$ is relatively compact on $B_{\bar{r}}$. Hence, by the Arzelá-Ascoli theorem, we deduce that $\mathcal{A}$ is compact on $B_{\bar{r}}$. Thus all the assumptions of Theorem 1.4 are satisfied. So the conclusion of Theorem 1.4 implies that the problem (7.125) has at least one solution on J. □

Our final result in this section deals with the uniqueness of solutions for the problem (7.125) and is based on Banach contraction mapping principle.

Theorem 7.29. *Let $f : [0, T] \times \mathbb{R} \to \mathbb{R}$ be a continuous function satisfying the assumption (7.28.1). Then there exists a unique solution for the problem (7.125) on $[0, T]$, provided that*

$$L\Lambda < 1, \tag{7.143}$$

where Λ is defined by (7.140).

Proof. In view of the condition (7.143), consider a set $B_{\tilde{r}} = \{y \in \mathcal{C} : \|y\| \leq \tilde{r}\}$, where

$$\tilde{r} > \frac{f_0\Lambda + \frac{|\kappa|(\rho|A_1|+T^\rho|A_2|)}{\rho|\Omega|}}{1 - L\Lambda}, \quad \sup_{t\in[0,T]} |f(t, 0)| = f_0$$

and show that $\mathcal{G}B_{\tilde{r}} \subset B_{\tilde{r}}$ ($\mathcal{G}$ is defined by (7.139)). For $y \in B_{\tilde{r}}$ and for all $t \in [0, T]$, it follows by (7.28.1) that $|f(t, y(t))| \leq |f(t, y(t)) - f(t, 0)| + |f(t, 0)| \leq L\|y\| + f_0 \leq L\tilde{r} + f_0$. In consequence, we have

$$\begin{aligned}
|\mathcal{G}(y)(t)| &\leq {}^\rho I_{0+}^\alpha |f(t, y(t))| + \frac{1}{|\Omega|}\Big\{|B_1| \sum_{j=1}^{k} |\mu_j| {}^\rho I_{0+}^\alpha |f(\xi_j, y(\xi_j))| \\
&\quad + A_1\Big[\sum_{i=1}^{m} |\sigma_i| \, {}^\rho I_{0+}^{\alpha+\beta} |f(\eta_i, y(\eta_i))| + {}^\rho I_{0+}^\alpha |f(T, y(T))|\Big] + |\kappa|\Big\} \\
&\quad + \frac{t^\rho}{\rho|\Omega|}\Big\{|B_2| \sum_{j=1}^{k} \mu_j {}^\rho I_{0+}^\alpha |f(\xi_j, y(\xi_j))) \\
&\quad + |A_2|\Big[\sum_{i=1}^{m} \sigma_i \, {}^\rho I_{0+}^{\alpha+\beta} |f(\eta_i, y(\eta_i))| + {}^\rho I_{0+}^\alpha |f(T, y(T))| + |\kappa|\Big]\Big\} \\
&\leq (L\tilde{r} + f_0)\Lambda + \frac{|\kappa|(\rho|A_1| + T^\rho|A_2|)}{\rho|\Omega|} \leq \tilde{r},
\end{aligned}$$

which, on taking the norm for $t \in J$, yields $\|\mathcal{G}(y)\| \leq \tilde{r}$. This shows that $\mathcal{G}$ maps $B_{\tilde{r}}$ into itself. In order to show that the operator $\mathcal{G}$ is a contraction, let $y, z \in \mathcal{C}$. Then, we get

$$
\begin{aligned}
&|\mathcal{G}(y)(t) - \mathcal{G}(z)(t)| \\
&\le {}^{\rho}I_{0+}^{\alpha}|f(t,y(t)) - f(t,z(t))| + \frac{1}{|\Omega|}\Big\{|B_1|\sum_{j=1}^{k}|\mu_j|^{\rho}I_{0+}^{\alpha}|f(\xi_j,y(\xi_j)) - f(\xi_j,z(\xi_j))| \\
&\quad + A_1\Big[\sum_{i=1}^{m}|\sigma_i|\ {}^{\rho}I_{0+}^{\alpha+\beta}|f(\eta_i,y(\eta_i)) - f(\eta_i,z(\eta_i))| + {}^{\rho}I_{0+}^{\alpha}|f(T,y(T)) - f(T,z(T))|\Big]\Big\} \\
&\quad + \frac{t^{\rho}}{\rho|\Omega|}\Big\{|B_2|\sum_{j=1}^{k}|\mu_j|^{\rho}I_{0+}^{\alpha}|f(\xi_j,y(\xi_j)) - f(\xi_j,z(\xi_j))| \\
&\quad + |A_2|\Big[\sum_{i=1}^{m}|\sigma_i|\ {}^{\rho}I_{0+}^{\alpha+\beta}|f(\eta_i,y(\eta_i)) - f(\eta_i,z(\eta_i))| + {}^{\rho}I_{0+}^{\alpha}|f(T,y(T)) - f(T,z(T))|\Big]\Big\} \\
&\le L\Lambda\|y - z\|.
\end{aligned}
$$

Consequently we obtain

$$\|\mathcal{G}(y) - \mathcal{G}(z)\| \le L\Lambda\|y - z\|,$$

which, in view of (7.143), implies that $\mathcal{G}$ is a contraction. Hence $\mathcal{G}$ has a unique fixed point by Banach's contraction mapping principle. This, in turn, shows that the problem (7.125) has a unique solution on $[0,T]$. The proof is completed. □

Example 7.10. Consider the following boundary value problem

$$
\begin{cases}
{}^{1/3}{}_{c}D^{7/5}y(t) = f(t,y(t)),\ t \in [0,2], \\
y(2) = 1/2\ {}^{1/3}I^{3/5}y(1/4) + 2/3\ {}^{1/3}I^{3/5}y(3/4) + 2/9, \\
\delta y(0) = 2/5\ y(1) + 4/5\ y(3/2),
\end{cases}
\tag{7.144}
$$

where $\rho = 1/3$, $\alpha = 7/5$, $\sigma_1 = 1/2$, $\sigma_2 = 2/3$, $\beta = 3/5$, $\eta_1 = 1/4$, $\eta_2 = 3/4$, $\mu_1 = 2/5$, $\mu_2 = 4/5$, $\kappa = 2/9$, $\xi_1 = 1$, $\xi_2 = 3/2$, $T = 2$ and $f(t,y(t))$ will be fixed later.

Using the given data, we find that $|A_1| \approx 2.947314182$, $|A_2| \approx 1.2$, $|B_1| \approx 0.491608875$, $|B_2| \approx 1.181571585$, $|\Omega| \approx 4.072393340$, and $\Lambda \approx 27.12293267$ (A_i, B_i $(i = 1,2), \Omega$ and Λ are respectively given by (7.127), (7.128), (7.129) and (7.140)).

For illustrating Theorem 7.27, we take

$$f(t,y) = \frac{(1+t)}{60}\Big(\frac{|y|}{|y|+1} + y + \frac{1}{8}\Big). \tag{7.145}$$

Clearly $f(t,x)$ is continuous and satisfies the condition (7.27.1) with $\phi(t) = \frac{(1+t)}{60}, \psi(\|y\|) = \|y\| + \frac{9}{8}$. By the condition (7.27.2), we find that $M > 2.390158$. Thus all conditions of Theorem 7.27 are satisfied and consequently, there exists at least one solution for the problem (7.144) with $f(t,y(t))$ given by (7.145) on $[0,2]$.

In order to illustrate Theorem 7.28, we choose

$$f(t,y) = \frac{\tan^{-1}y + e^{-t}}{4\sqrt{81 + \sin t}}. \tag{7.146}$$

It is easy to check that $f(t,x)$ is continuous and satisfies the conditions (7.28.1) and (7.28.2) with $L = 1/36$ and $\Phi(t) = \dfrac{\pi + 2e^{-t}}{8\sqrt{81+\sin t}}$. Also

$$L\Big(\frac{T^\rho}{\rho|\Omega|}\Big\{|B_2|\sum_{j=1}^{k}|\mu_j|\frac{\xi_j^{\rho\alpha}}{\rho^\alpha\Gamma(\alpha+1)}+|A_2|\Big[\sum_{i=1}^{m}|\sigma_i|\frac{\eta_i^{\rho(\alpha+\beta)}}{\rho^{\alpha+\beta}\Gamma(\alpha+\beta+1)}+\frac{T^{\rho\alpha}}{\rho^\alpha\Gamma(\alpha+1)}\Big]\Big\}\Big)$$
$$\approx 0.420512 < 1.$$

Thus all the conditions of Theorem 7.28 hold true. So, by the conclusion of Theorem 7.28, the problem (7.144) has at least one solution on $[0, 2]$.

With $L\Lambda \approx 0.753415 < 1$, one can note that the assumptions of Theorem 7.29 are also satisfied. Hence the conclusion of Theorem 7.29 applies and that the problem (7.144) with $f(t, y)$ given (7.146) has a unique solution on $[0, 2]$.

7.7.2 *Existence results for the problem (7.126)*

This section is devoted to the existence of solutions for the problem (7.126).

Definition 7.7. A function $y \in C([0,T],\mathbb{R})$ with its Caputo-type generalized deravative of order α is said to be a solution of the boundary value problem (7.126) if $y(T) = \sum_{i=1}^{m}\sigma_i{}^\rho I_{0+}^{\beta}y(\eta_i) + \kappa$, $\delta y(0) = \sum_{j=1}^{k}\mu_j y(\xi_j)$ and there exists function $v \in L^1([0,1],\mathbb{R})$ such that $v(t) \in F(t, y(t))$ a.e. on $[0, T]$ and

$$y(t) = {}^\rho I_{0+}^{\alpha}v(t) + \frac{1}{\Omega}\Big\{-B_1\sum_{j=1}^{k}\mu_j{}^\rho I_{0+}^{\alpha}v(\xi_j) + A_1\Big[\sum_{i=1}^{m}\sigma_i\,{}^\rho I_{0+}^{\alpha+\beta}v(\eta_i) - {}^\rho I_{0+}^{\alpha}v(T) + \kappa\Big]\Big\}$$
$$+\frac{t^\rho}{\rho\Omega}\Big\{B_2\sum_{j=1}^{k}\mu_j{}^\rho I_{0+}^{\alpha}v(\xi_j) + A_2\Big[\sum_{i=1}^{m}\sigma_i\,{}^\rho I_{0+}^{\alpha+\beta}v(\eta_i) - {}^\rho I_{0+}^{\alpha}v(T) + \kappa\Big]\Big\}. \quad (7.147)$$

7.7.2.1 *The Carathéodory case*

In this subsection, we consider the case when F has convex values and is of Carathéodory type, and prove an existence result for the problem (7.126) by applying nonlinear alternative of Leray-Schauder type (Theoren 1.16).

Theorem 7.30. *Assume that:*

(7.30.1) $F : [0,T]\times\mathbb{R} \to \mathcal{P}_{cp,c}(\mathbb{R})$ *is* L^1*-Carathéodory;*

(7.30.2) there exists a continuous non-decreasing function $\Psi : [0,\infty) \to (0,\infty)$ *and a function* $\Phi \in L^1([0,T],\mathbb{R}^+)$ *such that*

$$\|F(t,y)\|_{\mathcal{P}} := \sup\{|x| : x \in F(t,y)\} \le \Phi(t)\Psi(\|y\|),$$

for each $(t,y) \in [0,T]\times\mathbb{R}$*;*

(7.30.3) there exists a constant $W > 0$ such that

$$\frac{\|W\|}{\Psi(\|W\|)\Lambda_2 + \dfrac{|\kappa|(\rho|A_1| + T^\rho|A_2|)}{\rho|\Omega|}} > 1,$$

where

$$\begin{aligned}\Lambda_2 = {}& {}^\rho I^\alpha_{0+}\Phi(T) + \frac{1}{|\Omega|}\Big\{|B_1|\sum_{j=1}^{k}|\mu_j|{}^\rho I^\alpha_{0+}\Phi(\xi_j) \\ & +|A_1|\Big[\sum_{i=1}^{m}|\sigma_i|\ {}^\rho I^{\alpha+\beta}_{0+}\Phi(\eta_i) + {}^\rho I^\alpha_{0+}\Phi(T)\Big]\Big\} \\ & +\frac{T^\rho}{\rho|\Omega|}\Big\{|B_2|\sum_{j=1}^{k}|\mu_j|{}^\rho I^\alpha_{0+}\Phi(\xi_j) \\ & +|A_2|\Big[\sum_{i=1}^{m}|\sigma_i|\ {}^\rho I^{\alpha+\beta}_{0+}\Phi(\eta_i) + {}^\rho I^\alpha_{0+}\Phi(T)\Big]\Big\}. \end{aligned} \tag{7.148}$$

Then the problem (7.126) has at least one solution on $[0, T]$.

Proof. To transform the problem (7.126) into a fixed point problem, we define an operator $\mathcal{M} : \mathcal{C} \longrightarrow \mathcal{P}(\mathcal{C})$ by

$$\mathcal{M}(y) = \{h \in \mathcal{C} : h(t) = \mathcal{F}(y)(t)\}, \tag{7.149}$$

where

$$\begin{aligned}&\mathcal{F}(y)(t) \\ &= {}^\rho I^\alpha_{0+}v(t) + \frac{1}{\Omega}\Big\{-B_1\sum_{j=1}^{k}\mu_j{}^\rho I^\alpha_{0+}v(\xi_j) + A_1\Big[\sum_{i=1}^{m}\sigma_i\ {}^\rho I^{\alpha+\beta}_{0+}v(\eta_i) - {}^\rho I^\alpha_{0+}v(T) + \kappa\Big]\Big\} \\ &\quad +\frac{t^\rho}{\rho\Omega}\Big\{B_2\sum_{j=1}^{k}\mu_j{}^\rho I^\alpha_{0+}v(\xi_j) + A_2\Big[\sum_{i=1}^{m}\sigma_i\ {}^\rho I^{\alpha+\beta}_{0+}v(\eta_i) - {}^\rho I^\alpha_{0+}v(T) + \kappa\Big]\Big\},\end{aligned}$$

for $v \in S_{F,y}$. Obviously the fixed points of the operator $\mathcal{M}$, given by (7.149), are solutions of the problem (7.126).

We will show that $\mathcal{M}$ satisfies the assumptions of Leray-Schauder nonlinear alternative (Theorem 1.16) in several steps.

Step 1. *$\mathcal{M}(y)$ is convex for each $y \in \mathcal{C}$.*

This step is obvious since $S_{F,y}$ is convex (F has convex values).

Step 2. *$\mathcal{M}$ maps bounded sets (balls) into bounded sets in $\mathcal{C}$.*

For a positive number $\bar{r}$, let $B_R = \{y \in \mathcal{C} : \|y\| \leq R\}$ be a bounded ball in $\mathcal{C}$. Then, for each $h \in \mathcal{M}(y)$, $y \in B_R$, there exists $v \in S_{F,y}$ such that

$$\begin{aligned}h(t) = {}& {}^\rho I^\alpha_{0+}v(t) + \frac{1}{\Omega}\Big\{-B_1\sum_{j=1}^{k}\mu_j{}^\rho I^\alpha_{0+}v(\xi_j) + A_1\Big[\sum_{i=1}^{m}\sigma_i\ {}^\rho I^{\alpha+\beta}_{0+}v(\eta_i) - {}^\rho I^\alpha_{0+}v(T) + \kappa\Big]\Big\} \\ & +\frac{t^\rho}{\rho\Omega}\Big\{B_2\sum_{j=1}^{k}\mu_j{}^\rho I^\alpha_{0+}v(\xi_j) + A_2\Big[\sum_{i=1}^{m}\sigma_i\ {}^\rho I^{\alpha+\beta}_{0+}v(\eta_i) - {}^\rho I^\alpha_{0+}v(T) + \kappa\Big]\Big\}.\end{aligned}$$

Then, for $t \in [0,T]$, we have

$$\begin{aligned}
&|h(t)|\\
&\leq {}^\rho I_{0+}^\alpha |v(t)| + \frac{1}{|\Omega|}\Big\{|B_1|\sum_{j=1}^{k}\mu_j {}^\rho I_{0+}^\alpha |v(\xi_j)|\\
&\quad + |A_1|\Big[\sum_{i=1}^{m}\sigma_i\ {}^\rho I_{0+}^{\alpha+\beta}|v(\eta_i)| + {}^\rho I_{0+}^\alpha |v(T)| + |\kappa|\Big]\Big\}\\
&\quad + \frac{t^\rho}{\rho|\Omega|}\Big\{|B_2|\sum_{j=1}^{k}\mu_j {}^\rho I^\alpha |v(\xi_j)| + |A_2|\Big[\sum_{i=1}^{m}\sigma_i\ {}^\rho I_{0+}^{\alpha+\beta}|v(\eta_i)| + {}^\rho I_{0+}^\alpha |v(T)| + |\kappa|\Big]\Big\}\\
&\leq \Psi(\|y\|)\Big({}^\rho I_{0+}^\alpha \Phi(T) + \frac{1}{|\Omega|}\Big\{|B_1|\sum_{j=1}^{k}\mu_j {}^\rho I_{0+}^\alpha \Phi(\xi_j)\\
&\quad + |A_1|\Big[\sum_{i=1}^{m}\sigma_i\ {}^\rho I_{0+}^{\alpha+\beta}|p(\eta_i)| + {}^\rho I_{0+}^\alpha \Phi(T)\Big]\Big\}\\
&\quad + \frac{T^\rho}{\rho|\Omega|}\Big\{|B_2|\sum_{j=1}^{k}\mu_j {}^\rho I^\alpha \Phi(\xi_j) + |A_2|\Big[\sum_{i=1}^{m}\sigma_i\ {}^\rho I_{0+}^{\alpha+\beta}\Phi(\eta_i) + {}^\rho I_{0+}^\alpha \Phi(T)\Big]\Big\}\Big)\\
&\quad + \frac{|\kappa|(\rho|A_1| + T^\rho|A_2|)}{\rho|\Omega|}.
\end{aligned}$$

Thus

$$\begin{aligned}
\|h\| \leq \Psi(R)\Big(&{}^\rho I_{0+}^\alpha \Phi(T) + \frac{1}{|\Omega|}\Big\{|B_1|\sum_{j=1}^{k}|\mu_j| {}^\rho I_{0+}^\alpha \Phi(\xi_j)\\
&+ |A_1|\Big[\sum_{i=1}^{m}|\sigma_i|\ {}^\rho I_{0+}^{\alpha+\beta}\Phi(\eta_i) + {}^\rho I_{0+}^\alpha \Phi(T)\Big]\Big\}\\
&+ \frac{T^\rho}{\rho|\Omega|}\Big\{|B_2|\sum_{j=1}^{k}|\mu_j| {}^\rho I_{0+}^\alpha \Phi(\xi_j)| + |A_2|\Big[\sum_{i=1}^{m}|\sigma_i|\ {}^\rho I_{0+}^{\alpha+\beta}\Phi(\eta_i) + {}^\rho I_{0+}^\alpha \Phi(T)\Big]\Big\}\Big)\\
&+ \frac{|\kappa|(\rho|A_1| + T^\rho|A_2|)}{\rho|\Omega|} := \tilde{\ell}.
\end{aligned}$$

Step 3. *$\mathcal{M}$ maps bounded sets into equi-continuous sets of $\mathcal{C}$.*

Let $t_1, t_2 \in (0,T]$, $t_1 < t_2$, and let $y \in B_R$. Then

$$\begin{aligned}
&|h(t_2) - h(t_1)|\\
&\leq \Big|{}^\rho I_{0+}^\alpha v(t_2) - {}^\rho I_{0+}^\alpha v(t_1)\Big| + \frac{|t_2^\rho - t_1^\rho|}{\rho|\Omega|}\Big\{|B_2|\sum_{j=1}^{k}|\mu_j| {}^\rho I_{0+}^\alpha |v(\xi_j|\\
&\quad + |A_2|\Big[\sum_{i=1}^{m}|\sigma_i|\ {}^\rho I_{0+}^{\alpha+\beta}|v(\eta_i)| + {}^\rho I_{0+}^\alpha |v(T)| + |\kappa|\Big]\Big\}\\
&\leq \frac{\rho^{1-\alpha}\Psi(R)}{\Gamma(\alpha)}\Big|\int_0^{t_1}\Big[\frac{s^{\rho-1}}{(t_2^\rho - s^\rho)^{1-\alpha}} - \frac{s^{\rho-1}}{(t_1^\rho - s^\rho)^{1-\alpha}}\Big]\Phi(s)ds + \int_{t_1}^{t_2}\frac{s^{\rho-1}}{(t_2^\rho - s^\rho)^{1-\alpha}}\Phi(s)ds\Big|
\end{aligned}$$

$$+\frac{|t_2^\rho - t_1^\rho|}{\rho|\Omega|}\Big\{\Psi(R)\Big(|B_2|\sum_{j=1}^{k}|\mu_j|{}^\rho I_{0+}^{\alpha}\Phi(\xi_j)$$

$$+|A_2|\Big[\sum_{i=1}^{m}|\sigma_i|\ {}^\rho I_{0+}^{\alpha+\beta}\Phi(\eta_i) + {}^\rho I_{0+}^{\alpha}\Phi(T)\Big]\Big) + |A_2||\kappa|\Big\}.$$

Obviously the right hand side of the above inequality tends to zero independently of $y \in B_R$ as $t_2 - t_1 \to 0$. Therefore it follows by the Arzelá-Ascoli theorem that $\mathcal{M} : \mathcal{C} \to \mathcal{P}(\mathcal{C})$ is completely continuous.

In our next step, we show that $\mathcal{M}$ is u.s.c. Since $\mathcal{M}$ is completely continuous, it is enough to establish that it has a closed graph.

Step 4. *$\mathcal{M}$ has a closed graph.*

Let $y_n \to y_*, h_n \in \mathcal{M}(y_n)$ and $h_n \to h_*$. Then we need to show that $h_* \in \mathcal{M}(y_*)$. Associated with $h_n \in \mathcal{M}(y_n)$, there exists $v_n \in S_{F,y_n}$ such that, for each $t \in [0,T]$,

$$h_n(t)$$

$$= {}^\rho I^\alpha v_n(t) + \frac{1}{\Omega}\Big\{ - B_1\sum_{j=1}^{k}\mu_j{}^\rho I^\alpha v_n(\xi_j) + A_1\Big[\sum_{i=1}^{m}\sigma_i\ {}^\rho I_{0+}^{\alpha+\beta}v_n(\eta_i) - {}^\rho I^\alpha v_n(T) + \kappa\Big]\Big\}$$

$$+\frac{t^\rho}{\rho\Omega}\Big\{B_2\sum_{j=1}^{k}\mu_j{}^\rho I_{0+}^{\alpha}v_n(\xi_j) + A_2\Big[\sum_{i=1}^{m}\sigma_i\ {}^\rho I_{0+}^{\alpha+\beta}v_n(\eta_i) - {}^\rho I_{0+}^{\alpha}v_n(T) + \kappa\Big]\Big\}.$$

Thus it suffices to show that there exists $v_* \in S_{F,y_*}$ such that, for each $t \in [0,T]$,

$$h_*(t)$$

$$= {}^\rho I_{0+}^\alpha v_*(t) + \frac{1}{\Omega}\Big\{ - B_1\sum_{j=1}^{k}\mu_j{}^\rho I^\alpha v_*(\xi_j) + A_1\Big[\sum_{i=1}^{m}\sigma_i\ {}^\rho I_{0+}^{\alpha+\beta}v_*(\eta_i) - {}^\rho I^\alpha v_*(T) + \kappa\Big]\Big\}$$

$$+\frac{t^\rho}{\rho\Omega}\Big\{B_2\sum_{j=1}^{k}\mu_j{}^\rho I_{0+}^{\alpha}v_*(\xi_j) + A_2\Big[\sum_{i=1}^{m}\sigma_i\ {}^\rho I_{0+}^{\alpha+\beta}v_*(\eta_i) - {}^\rho I_{0+}^{\alpha}v_*(T) + \kappa\Big]\Big\}.$$

Let us consider the linear operator $\Theta : L^1([0,T],\mathbb{R}) \to \mathcal{C}$ given by

$$v \mapsto \Theta v(t)$$

$$= {}^\rho I_{0+}^\alpha v(t) + \frac{1}{\Omega}\Big\{ - B_1\sum_{j=1}^{k}\mu_j{}^\rho I^\alpha v(\xi_j) + A_1\Big[\sum_{i=1}^{m}\sigma_i\ {}^\rho I_{0+}^{\alpha+\beta}v(\eta_i) - {}^\rho I_{0+}^\alpha v(T) + \kappa\Big]\Big\}$$

$$+\frac{t^\rho}{\rho\Omega}\Big\{B_2\sum_{j=1}^{k}\mu_j{}^\rho I^{\alpha}v_*(\xi_j) + A_2\Big[\sum_{i=1}^{m}\sigma_i\ {}^\rho I_{0+}^{\alpha+\beta}v_*(\eta_i) - {}^\rho I_{0+}^{\alpha}v(T) + \kappa\Big]\Big\}.$$

Observe that $\|h_n - h_*\| \to 0$ as $n \to \infty$, so it follows by Lemma 1.2 that $\Theta \circ S_F$ is a closed graph operator. Further, we have $h_n(t) \in \Theta(S_{F,y_n})$. Since $y_n \to y_*$, therefore, we have

$$h_*(t) = {}^\rho I_{0+}^\alpha v_*(t) + \frac{1}{\Omega}\Big\{ - B_1\sum_{j=1}^{k}\mu_j{}^\rho I^\alpha v_*(\xi_j)$$

$$+A_1\Big[\sum_{i=1}^{m}\sigma_i\ {}^{\rho}I_{0^+}^{\alpha+\beta}v_*(\eta_i)-{}^{\rho}I_{0^+}^{\alpha}v_*(T)+\kappa\Big]\Big\}$$

$$+\frac{t^\rho}{\rho\Omega}\Big\{B_2\sum_{j=1}^{k}\mu_j{}^{\rho}I^{\alpha}v_*(\xi_j)+A_2\Big[\sum_{i=1}^{m}\sigma_i\ {}^{\rho}I_{0^+}^{\alpha+\beta}v_*(\eta_i)-{}^{\rho}I_{0^+}^{\alpha}v_*(T)+\kappa\Big]\Big\}.$$

for some $v_* \in S_{F,y_*}$.

Step 5. *We show there exists an open set $U \subseteq C([0,T],\mathbb{R})$ with $y \notin \lambda\mathcal{M}(y)$ for any $\lambda \in (0,1)$ and all $y \in \partial U$.*

Let $\lambda \in (0,1)$ and $y \in \lambda\mathcal{M}(y)$. Then there exists $v \in L^1([0,T],\mathbb{R})$ with $v \in S_{F,y}$ such that, for $t \in [0,T]$, we have

$$y(t)=\lambda^{\rho}I_{0^+}^{\alpha}v(t)+\frac{\lambda}{\Omega}\Big\{-B_1\sum_{j=1}^{k}\mu_j{}^{\rho}I_{0^+}^{\alpha}v(\xi_j)+A_1\Big[\sum_{i=1}^{m}\sigma_i\ {}^{\rho}I_{0^+}^{\alpha+\beta}v(\eta_i)-{}^{\rho}I_{0^+}^{\alpha}v(T)+\kappa\Big]\Big\}$$

$$+\lambda\frac{t^\rho}{\rho\Omega}\Big\{B_2\sum_{j=1}^{k}\mu_j{}^{\rho}I_{0^+}^{\alpha}v(\xi_j)+A_2\Big[\sum_{i=1}^{m}\sigma_i\ {}^{\rho}I_{0^+}^{\alpha+\beta}v(\eta_i)-{}^{\rho}I_{0^+}^{\alpha}v(T)+\kappa\Big]\Big\}.$$

As in the second step, one can obtain

$$|y(t)|\le\Psi(y)\Big({}^{\rho}I^{\alpha}\Phi(T)+\frac{1}{|\Omega|}\Big\{|B_1|\sum_{j=1}^{k}|\mu_j|{}^{\rho}I^{\alpha}\Phi(\xi_j)$$

$$+|A_1|\Big[\sum_{i=1}^{m}|\sigma_i|\ {}^{\rho}I_{0^+}^{\alpha+\beta}|\Phi(\eta_i)|+{}^{\rho}I^{\alpha}|\Phi(T)|\Big]\Big\}$$

$$+\frac{T^\rho}{\rho|\Omega|}\Big\{|B_2|\sum_{j=1}^{k}|\mu_j|{}^{\rho}I^{\alpha}\Phi(\xi_j)+|A_2|\Big[\sum_{i=1}^{m}|\sigma_i|\ {}^{\rho}I_{0^+}^{\alpha+\beta}\Phi(\eta_i)+{}^{\rho}I^{\alpha}\Phi(T)\Big]\Big\}\Big)$$

$$+\frac{|\kappa|(\rho|A_1|+T^\rho|A_2|)}{\rho|\Omega|},$$

which implies that

$$\frac{\|y\|}{\Psi(\|y\|)\Lambda_2+\dfrac{|\kappa|(\rho|A_1|+T^\rho|A_2|)}{\rho|\Omega|}}\le 1.$$

In view of (7.30.3), there exists W such that $\|y\| \neq W$. Let us set

$$U=\{y\in C(J,\mathbb{R}):\|y\|<W\}.$$

Note that the operator $\mathcal{M} : \overline{U} \to \mathcal{P}(C(J,\mathbb{R}))$ is a compact multi-valued map, and u.s.c. with convex closed values. From the choice of U, there is no $y \in \partial U$ such that $y \in \lambda\mathcal{M}(y)$ for some $\lambda \in (0,1)$. Consequently, by the nonlinear alternative of Leray-Schauder type (Lemma 1.16), we deduce that $\mathcal{M}$ has a fixed point $y \in \overline{U}$ which is a solution of the problem (7.126). This completes the proof. □

7.7.2.2 *The Lipschitz case*

In this subsection, we prove the existence of solutions for the problem (7.126) with non-convex valued right hand side by applying a fixed point theorem for multi-valued maps due to Covitz and Nadler (Theorem 1.19).

Theorem 7.31. *Assume that:*

(7.31.1) $F : [0,T] \times \mathbb{R} \to \mathcal{P}_{cp}(\mathbb{R})$ *is such that* $F(\cdot, y) : [0,T] \to \mathcal{P}_{cp}(\mathbb{R})$ *is measurable for each* $y \in \mathbb{R}$;

(7.31.2) $H_d(F(t,y), F(t,\bar{y})) \leq \theta(t)|y - \bar{y}|$ *for almost all* $t \in [0,T]$ *and* $y, \bar{y} \in \mathbb{R}$ *with* $\theta \in C([0,T], \mathbb{R}^+)$ *and* $d(0, F(t,0)) \leq \theta(t)$ *for almost all* $t \in [0,T]$.

Then the problem (7.126) has at least one solution on $[0,T]$ *if* $\|\theta\|\Lambda < 1$, *i.e.,*

$$\begin{aligned} K := \|\theta\| \Big[& \frac{T^{\rho\alpha}}{\rho^\alpha \Gamma(\alpha+1)} + \frac{1}{|\Omega|}\Big\{ |B_1| \sum_{j=1}^{k} |\mu_j| \frac{\xi_j^{\rho\alpha}}{\rho^\alpha \Gamma(\alpha+1)} \\ & + |A_1| \Big[\sum_{i=1}^{m} |\sigma_i| \frac{\eta_i^{\rho(\alpha+\beta)}}{\rho^{\alpha+\beta}\Gamma(\alpha+\beta+1)} + \frac{T^{\rho\alpha}}{\rho^\alpha \Gamma(\alpha+1)} \Big] \Big\} \\ & + \frac{T^\rho}{\rho|\Omega|} \Big\{ |B_2| \sum_{j=1}^{k} |\mu_j| \frac{\xi_j^{\rho\alpha}}{\rho^\alpha \Gamma(\alpha+1)} \\ & + |A_2| \Big[\sum_{i=1}^{m} |\sigma_i| \frac{\eta_i^{\rho(\alpha+\beta)}}{\rho^{\alpha+\beta}\Gamma(\alpha+\beta+1)} + \frac{T^{\rho\alpha}}{\rho^\alpha \Gamma(\alpha+1)} \Big] \Big\} \Big] < 1. \end{aligned} \tag{7.150}$$

Proof. Notice that the set $S_{F,y}$ is non-empty for each $y \in \mathcal{C}$ by the assumption (7.31.1). So F has a measurable selection (see Theorem III.6 [78]). Now we show that the operator $\mathcal{M}$ defined by (7.149) satisfies the assumptions of Theorem 1.19. Firstly we show that $\mathcal{M}(y) \in \mathcal{P}_{cl}(\mathcal{C})$ for each $y \in \mathcal{C}$. Let $\{u_n\}_{n\geq 0} \in \mathcal{F}(y)$ be such that $u_n \to u$ $(n \to \infty)$ in $\mathcal{C}$. Then $u \in \mathcal{C}$ and there exists $v_n \in S_{F,y_n}$ such that, for each $t \in [0,T]$,

$$\begin{aligned} u_n(t) = {}& {}^\rho I_{0+}^{\alpha} v_n(t) + \frac{1}{\Omega}\Big\{ - B_1 \sum_{j=1}^{k} \mu_j {}^\rho I^\alpha v_n(\xi_j) \\ & + A_1 \Big[\sum_{i=1}^{m} \sigma_i \, {}^\rho I_{\eta_0}^{\alpha+\beta} v_n(\eta_i) - {}^\rho I^\alpha v_n(T) + \kappa \Big] \Big\} \\ & + \frac{t^\rho}{\rho\Omega}\Big\{ B_2 \sum_{j=1}^{k} \mu_j {}^\rho I_{0+}^{\alpha} v_n(\xi_j) + A_2 \Big[\sum_{i=1}^{m} \sigma_i \, {}^\rho I_{0+}^{\alpha+\beta} v_n(\eta_i) - {}^\rho I_{0+}^{\alpha} v_n(T) + \kappa \Big] \Big\}. \end{aligned}$$

As F has compact values, we pass onto a subsequence (if necessary) to obtain that v_n converges to v in $L^1([0,T], \mathbb{R})$. Thus, $v \in S_{F,y}$ and for each $t \in [0,T]$, we have

$$u_n(t) \to u(t)$$
$$= {}^\rho I_{0+}^\alpha v(t) + \frac{1}{\Omega}\Big\{ - B_1 \sum_{j=1}^{k} \mu_j {}^\rho I_{0+}^\alpha v_n(\xi_j) + A_1 \Big[\sum_{i=1}^{m} \sigma_i \; {}^\rho I_{0+}^{\alpha+\beta} v(\eta_i) - {}^\rho I_{0+}^\alpha v(T) + \kappa \Big] \Big\}$$
$$+ \frac{t^\rho}{\rho\Omega}\Big\{ B_2 \sum_{j=1}^{k} \mu_j {}^\rho I_{0+}^\alpha v(\xi_j) + A_2 \Big[\sum_{i=1}^{m} \sigma_i \; {}^\rho I_{0+}^{\alpha+\beta} v(\eta_i) - {}^\rho I_{0+}^\alpha v(T) + \kappa \Big] \Big\}.$$

Hence, $u \in \mathcal{M}(y)$.

Next we show that there exists $K < 1$ (defined by (7.150)) such that
$$H_d(\mathcal{M}(y), \mathcal{M}(\bar{y})) \le K\|y - \bar{y}\| \quad \text{for each} \quad y, \bar{y} \in \mathcal{C}.$$
Let $y, \bar{y} \in \mathcal{C}$ and $h_1 \in \mathcal{F}(y)$. Then there exists $v_1(t) \in F(t, y(t))$ such that, for each $t \in [0, T]$,

$$h_1(t) = {}^\rho I_{0+}^\alpha v_1(t) + \frac{1}{\Omega}\Big\{ - B_1 \sum_{j=1}^{k} \mu_j {}^\rho I_{0+}^\alpha v_1(\xi_j)$$
$$+ A_1 \Big[\sum_{i=1}^{m} \sigma_i \; {}^\rho I_{0+}^{\alpha+\beta} v_1(\eta_i) - {}^\rho I_{0+}^\alpha v_1(T) + \kappa \Big] \Big\}$$
$$+ \frac{t^\rho}{\rho\Omega}\Big\{ B_2 \sum_{j=1}^{k} \mu_j {}^\rho I_{0+}^\alpha v_1(\xi_j) + A_2 \Big[\sum_{i=1}^{m} \sigma_i \; {}^\rho I_{0+}^{\alpha+\beta} v_1(\eta_i) - {}^\rho I_{0+}^\alpha v_1(T) + \kappa \Big] \Big\}.$$

By (7.31.1), we have
$$H_d(F(t, y), F(t, \bar{y})) \le \theta(t)|y(t) - \bar{y}(t)|.$$
So, there exists $w \in F(t, \bar{y}(t))$ such that
$$|v_1(t) - w| \le \theta(t)|y(t) - \bar{y}(t)|, \quad t \in [0, T].$$
Define $U : [0, T] \to \mathcal{P}(\mathbb{R})$ by
$$U(t) = \{w \in \mathbb{R} : |v_1(t) - w| \le \theta(t)|y(t) - \bar{y}(t)|\}.$$
Since the multi-valued operator $U(t) \cap F(t, \bar{y}(t))$ is measurable (Proposition III.4 [78]), there exists a function $v_2(t)$ which is a measurable selection for U. So $v_2(t) \in F(t, \bar{y}(t))$ and for each $t \in [0, T]$, we have $|v_1(t) - v_2(t)| \le \theta(t)|y(t) - \bar{y}(t)|$. For each $t \in [0, T]$, let us define

$$h_2(t) = {}^\rho I_{0+}^\alpha v_2(t) + \frac{1}{\Omega}\Big\{ - B_1 \sum_{j=1}^{k} \mu_j {}^\rho I_{0+}^\alpha v_2(\xi_j)$$
$$+ A_1 \Big[\sum_{i=1}^{m} \sigma_i \; {}^\rho I_{0+}^{\alpha+\beta} v_n(\eta_i) - {}^\rho I_{0+}^\alpha v_2(T) + \kappa \Big] \Big\}$$
$$+ \frac{t^\rho}{\rho\Omega}\Big\{ B_2 \sum_{j=1}^{k} \mu_j {}^\rho I_{0+}^\alpha v_2(\xi_j) + A_2 \Big[\sum_{i=1}^{m} \sigma_i \; {}^\rho I_{0+}^{\alpha+\beta} v_2(\eta_i) - {}^\rho I_{0+}^\alpha v_2(T) + \kappa \Big] \Big\}.$$

Then we have

$$\begin{aligned}
&|h_1(t) - h_2(t)| \\
&\leq {}^\rho I_{0+}^\alpha |v_1(t) - v_2(t)| + \frac{1}{|\Omega|}\Big\{|B_1|\sum_{j=1}^{k}|\mu_j|{}^\rho I_{0+}^\alpha |v_1(\xi_j) - v_2(\xi_j)| \\
&\quad + |A_1|\Big[\sum_{i=1}^{m}|\sigma_i|\ {}^\rho I_{0+}^{\alpha+\beta}|v_1(\eta_i) - v_2(\eta_i)| \\
&\quad + {}^\rho I_{0+}^\alpha |v_1(T) - v_2(T)|\Big]\Big\} + \frac{t^\rho}{\rho|\Omega|}\Big\{|B_2|\sum_{j=1}^{k}\mu_j{}^\rho I_{0+}^\alpha |v_2(\xi_j) - v_1(\xi_j)| \\
&\quad + |A_2|\Big[\sum_{i=1}^{m}\sigma_i\ {}^\rho I_{0+}^{\alpha+\beta}|v_2(\eta_i) - v_1(\eta_i)| + {}^\rho I_{0+}^\alpha |v_2(T) - v_1(T)|\Big]\Big\} \\
&\leq \|\theta\|\Big[\frac{T^{\rho\alpha}}{\rho^\alpha\Gamma(\alpha+1)} + \frac{1}{|\Omega|}\Big\{|B_1|\sum_{j=1}^{k}|\mu_j|\frac{\xi_j^{\rho\alpha}}{\rho^\alpha\Gamma(\alpha+1)} \\
&\quad + |A_1|\Big[\sum_{i=1}^{m}|\sigma_i|\frac{\eta_i^{\rho(\alpha+\beta)}}{\rho^{\alpha+\beta}\Gamma(\alpha+\beta+1)} + \frac{T^{\rho\alpha}}{\rho^\alpha\Gamma(\alpha+1)}\Big]\Big\} \\
&\quad + \frac{T^\rho}{\rho|\Omega|}\Big\{|B_2|\sum_{j=1}^{k}|\mu_j|\frac{\xi_j^{\rho\alpha}}{\rho^\alpha\Gamma(\alpha+1)} \\
&\quad + |A_2|\Big[\sum_{i=1}^{m}|\sigma_i|\frac{\eta_i^{\rho(\alpha+\beta)}}{\rho^{\alpha+\beta}\Gamma(\alpha+\beta+1)} + \frac{T^{\rho\alpha}}{\rho^\alpha\Gamma(\alpha+1)}\Big]\Big\}\Big]\|y - \bar{y}\|.
\end{aligned}$$

Hence

$$\begin{aligned}
\|h_1 - h_2\| \leq \|\theta\|\Big[&\frac{T^{\rho\alpha}}{\rho^\alpha\Gamma(\alpha+1)} + \frac{1}{|\Omega|}\Big\{|B_1|\sum_{j=1}^{k}|\mu_j|\frac{\xi_j^{\rho\alpha}}{\rho^\alpha\Gamma(\alpha+1)} \\
&+ |A_1|\Big[\sum_{i=1}^{m}|\sigma_i|\frac{\eta_i^{\rho(\alpha+\beta)}}{\rho^{\alpha+\beta}\Gamma(\alpha+\beta+1)} \\
&+ \frac{T^{\rho\alpha}}{\rho^\alpha\Gamma(\alpha+1)}\Big]\Big\} + \frac{T^\rho}{|\rho\Omega|}\Big\{|B_2|\sum_{j=1}^{k}|\mu_j|\frac{\xi_j^{\rho\alpha}}{\rho^\alpha\Gamma(\alpha+1)} \\
&+ |A_2|\Big[\sum_{i=1}^{m}|\sigma_i|\frac{\eta_i^{\rho(\alpha+\beta)}}{\rho^{\alpha+\beta}\Gamma(\alpha+\beta+1)} + \frac{T^{\rho\alpha}}{\rho^\alpha\Gamma(\alpha+1)}\Big]\Big\}\Big]\|y - \bar{y}\|.
\end{aligned}$$

Analogously, interchanging the roles of y and $\overline{y}$, we can obtain

$$\begin{aligned}
&H_d(\mathcal{M}(y), \mathcal{M}(\bar{y})) \\
&\leq \|\theta\|\Big[\frac{T^{\rho\alpha}}{\rho^\alpha\Gamma(\alpha+1)} + \frac{1}{|\Omega|}\Big\{|B_1|\sum_{j=1}^{k}|\mu_j|\frac{\xi_j^{\rho\alpha}}{\rho^\alpha\Gamma(\alpha+1)} + |A_1|\Big[\sum_{i=1}^{m}|\sigma_i|\frac{\eta_i^{\rho(\alpha+\beta)}}{\rho^{\alpha+\beta}\Gamma(\alpha+\beta+1)} \\
&+ \frac{T^{\rho\alpha}}{\rho^\alpha\Gamma(\alpha+1)}\Big]\Big\} + \frac{T^\rho}{|\rho\Omega|}\Big\{|B_2|\sum_{j=1}^{k}|\mu_j|\frac{\xi_j^{\rho\alpha}}{\rho^\alpha\Gamma(\alpha+1)} \\
&+ |A_2|\Big[\sum_{i=1}^{m}|\sigma_i|\frac{\eta_i^{\rho(\alpha+\beta)}}{\rho^{\alpha+\beta}\Gamma(\alpha+\beta+1)} + \frac{T^{\rho\alpha}}{\rho^\alpha\Gamma(\alpha+1)}\Big]\Big\}\Big]\|y - \bar{y}\|.
\end{aligned}$$

So $\mathcal{M}$ is a contraction. Therefore, it follows by Theorem 1.19, that $\mathcal{M}$ has a fixed point y which is a solution of (7.126). This completes the proof. □

Example 7.11. Consider the following boundary value problem of fractional differential inclusions:

$$\begin{cases} {}^{1/3}{}_cD^{7/5}y(t) \in F(t,y(t)), \ t \in [0,2], \\ y(2) = 1/2 \ {}^{1/3}I^{3/5}y(1/4) + 2/3 \ {}^{1/3}I^{3/5}y(3/4) + 2/9, \\ \delta y(0) = 2/5 \ y(1) + 4/5 \ y(3/2), \end{cases} \tag{7.151}$$

where $F(t,y(t))$ will be defined later.

The values of $|A_1|$, $|A_2|$, $|B_1|$, $|B_2|$, $|\Omega|$ and Λ are the same as obtained in Example 7.10. For illustrating Theorem 7.30, we take

$$F(t,y(t)) = \Big[\frac{e^{-t}}{\sqrt{4000+t}}\Big(\sin y + \frac{1}{2}\Big), \frac{(1+t)}{60}\Big(\tan^{-1} y + y + \frac{1}{4}\Big)\Big]. \tag{7.152}$$

It is easy to check that $F(t,y(t))$ is L^1-Carathéodory. In view of (7.30.2), we find that $\Phi(t) = \dfrac{(1+t)}{60}$, $\Psi(\|y\|) = \|y\| + \dfrac{2\pi+1}{4}$ and the condition (7.30.3) implies that $W > 3.289470$. Thus all conditions of Theorem 7.30 are satisfied and consequently, there exists at least one solution for the problem (7.151) with $F(t,y(t))$ given by (7.152) on $[0,2]$.

In order to demonstrate the application of Theorem 7.31, let us choose

$$F(t,y(t)) = \Big[\frac{(t+1)}{4\sqrt{900+t}}\Big(\tan^{-1} y + \sin t\Big), \frac{e^{-t}\cos t}{250}\Big(\frac{|y|}{|y|+1} + \frac{1}{8}\Big)\Big]. \tag{7.153}$$

Clearly

$$H_d(F(t,y), F(t,\bar{y})) \leq \frac{(t+1)}{120}\|y-\bar{y}\|.$$

Letting $\theta(t) = \dfrac{(t+1)}{120}$, it is easy to check that $d(0, F(t,0)) \leq \theta(t)$ holds for almost all $t \in [0,2]$ and that $K \approx 0.6780733168 < 1$ (K is given by (7.150)). As the hypotheses of Theorem 7.31 are satisfied, we conclude that the problem (7.151) with $F(t,y(t))$ given by (7.153) has at least one solution on $[0,2]$.

7.8 Notes and remarks

The content of the chapter is based on the papers [48], [34], [8], [125], [52], [6] and [65].

Bibliography

[1] R.P. Agarwal, A. Alsaedi, N. Alghamdi, S.K. Ntouyas, B. Ahmad, Existence results for multi-term fractional differential equations with nonlocal multi-point and multi-strip boundary conditions, *Adv. Difference Equ.* (2018) **2018:342**.

[2] R. Agarwal, S.K. Ntouyas, B. Ahmad, M.S. Alhothuali, Existence of solutions for integro-differential equations of fractional order with nonlocal three-point fractional boundary conditions, *Adv. Difference Equ.* 2013, **2013:128**.

[3] B. Ahmad, T. Hayat, Diffraction of a plane wave by an elastic knife-edge adjacent to a rigid strip, *Canad. Appl. Math. Quart.* **9** (2001) 303-316.

[4] B. Ahmad, On nonlocal boundary value problems for nonlinear integro-differential equations of arbitrary fractional order, *Result. Math.* **63** (2013), 183-194.

[5] B. Ahmad, R.P. Agarwal, A. Alsaedi, S.K. Ntouyas, Y. Alruwaily, Fractional order coupled systems for mixed fractional derivative operators with nonlocal multi-point and Riemann-Stieltjes integral-multi-strip conditions, *Dynam. Systems Appl.* **29** (2020), 71-86.

[6] B. Ahmad, M. Alghanmi, A. Alsaedi, S.K. Ntouyas, Multi-term fractional differential equations with generalized integral boundary conditions, *Fractal Fract.* 2019, 3, 44.

[7] B. Ahmad, Y. Alruwaily, A. Alsaedi, S.K. Ntouyas, Existence and stability results for a fractional order differential equation with non-conjugate Riemann-Stieltjes integro-multipoint boundary conditions, *Mathematics 2019, 7, 249, Special Issue "Fractional Differential Equations, Inclusions and Inequalities with Appllications"*.

[8] B. Ahmad, M. Alghanmi, S.K. Ntouyas, A. Alsaedi, Fractional differential equations involving generalized derivative with Stieltjes and fractional integral boundary conditions, *Appl. Math. Lett.* **84** (2018), 111-117.

[9] B. Ahmad, A. Alsaedi, Y. Alruwaily, S.K. Ntouyas, On a coupled system of higher order nonlinear Caputo fractional differential equations with coupled Riemann-Stieltjes type integro-multipoint boundary conditions, *Adv. Difference Equ.* (2019) **2019:474**.

[10] B. Ahmad, A. Alsaedi, Y. Alruwaily, S.K. Ntouyas, Nonlinear multi-term fractional differential equations with Riemann-Stieltjes integro-multipoint boundary conditions, *AIMS Mathematics*, **5** (2) (2020), 1446-1461.

[11] B. Ahmad, A. Alsaedi, S.K. Ntouyas, Nonlocal fractional order systems with mixed Caputo and Riemann-Liouville fractional derivative operators, *Bound. Value Probl.* (2019), **2019:109**.

[12] B. Ahmad, A. Alsaedi, S.K. Ntouyas, Fractional order systems for mixed Riemann-Liouville and Caputo fractional derivative operators with integro-differential boundary conditions, *J. Appl. Anal. Comput.* **10** (3), (2020), 892-903.

[13] B. Ahmad, A. Alsaedi, S.K. Ntouyas, Nonlinear coupled fractional order systems with integro-multistrip-multipoint boundary conditions, *Intern. J. Anal. Appl.* **17** (6) (2019), 940-957.

[14] B. Ahmad, A. Alsaedi, S.K. Ntouyas, On more general boundary value problems involving sequential fractional derivative operators, *Adv. Difference Equ.* (2019) **2019:290**.

[15] B. Ahmad, S. Al-Sulami, S.K. Ntouyas, Existence results for fractional differential inclusions involving non-convex valued maps with four-point nonlocal boundary conditions, *Commun. Math. Anal.* **14** (2013), pp. 15-27.

[16] B. Ahmad, A. Broom, A. Alsaedi, S.K. Ntouyas, Nonlinear integro-differential equations involving mixed right and left fractional derivative operators and integrals with nonlocal boundary data, *Mathematics 2020, 8, 336*; *Special Issue "Fractional Differential Equations, Inclusions and Inequalities with Applications"*.

[17] B. Ahmad, J.R. Graef, Coupled systems of nonlinear fractional differential equations with nonlocal boundary conditions, *Panam. Math. J.* **19** (2009), 29-39.

[18] B. Ahmad, J.J. Nieto, Existence of solutions for nonlocal boundary value problems of higher-order nonlinear fractional differential equations, *Abstr. Appl. Anal.* **2009**, Article ID 494720, 9 pp.

[19] B. Ahmad, J.J. Nieto, Existence results for a coupled system of nonlinear fractional differential equations with three-point boundary conditions, *Comput. Math. Appl.* **58** (2009), 1838-1843.

[20] B. Ahmad, J.J. Nieto, Riemann-Liouville fractional integro-differential equations with fractional nonlocal integral boundary conditions, *Bound. Value Probl.* 2011:36, 9 pp.

[21] B. Ahmad, S.K. Ntouyas, Existence results for fractional differential inclusions with four-point boundary conditions, *Commun. Appl. Anal.* **15** (2011), 141-152.

[22] B. Ahmad, S.K. Ntouyas, Existence results for nonlinear fractional differential equations with four-point nonlocal type integral boundary conditions, *African Diaspora J. Math.* **11** (2011), 29-39.

[23] B. Ahmad, S.K. Ntouyas, Boundary value problems for fractional differential inclusions with four point integral boundary conditions, *Surv. Math. Appl.* **6** (2011), 175-194.

[24] B. Ahmad, S.K. Ntouyas, A four-point integral boundary value problem for fractional differential equations of arbitrary order, *Electron. J. Qual. Theory Differ. Equ.* **2011**, No. 22, 15 pp.

[25] B. Ahmad, S.K. Ntouyas, Existence of solutions for fractional differential inclusions with four-point nonlocal Riemann-Liouville type integral boundary conditions, *Filomat* 27:6 (2013), 1027-1036.

[26] B. Ahmad, S.K. Ntouyas, Integro-differential equations of fractional order with nonlocal fractional boundary conditions associated with financial asset model, *Electron. J. Diff. Eqns. Vol. 2013 (2013), No. 60, pp. 1-10.*

[27] B. Ahmad, S.K. Ntouyas, A new kind of nonlocal-integral fractional boundary value problems, *Bull. Malays. Math. Sci. Soc. (2)* 39 (2016) 1343-1361.

[28] B. Ahmad, S.K. Ntouyas, Existence of solutions for fractional differential inclusions with nonlocal strip conditions, *Arab J. Math. Sci.* **18** (2012), 121-134.

[29] B. Ahmad, S.K. Ntouyas, Existence results for higher-order fractional differential inclusions with Riemann-Stieltjes type integral boundary conditions, *Commun. Appl. Anal.* **17** (2013), 87-98.

[30] B. Ahmad, S.K. Ntouyas, A coupled system of nonlocal fractional differential equations with coupled, uncoupled slit-strips integral boundary conditions, *Nonlinear Oscil.* **19** (2016), 291-310.

[31] B. Ahmad, S.K. Ntouyas, Some fractional-order one-dimensional semi-linear problems under nonlocal integral boundary conditions, *Rev. R. Acad. Cienc. Exactas Fysıs. Nat. Ser. A-Mat. RACSAM*, (2016) 110:159-172.

[32] B. Ahmad, S.K. Ntouyas, Existence results for fractional differential inclusions with Erdélyi-Kober fractional integral conditions, *An. Ştiinţ. Univ. "Ovidius" Constanţa Scr. Mat.* **25** (2), (2017), 5-24.

[33] B. Ahmad, S.K. Ntouyas, On perturbed fractional differential inclusions with nonlocal multi-point Erdélyi-Kober fractional integral boundary conditions, *Meditter. J. Math.* (2017) 14: 27.

[34] B. Ahmad, S.K. Ntouyas, On perturbed fractional differential equations, inclusions with generalized Riemann-Liouville fractional integral boundary conditions, *Intern. J. Pure Appl. Math.* **109** (2016), 29-48.

[35] B. Ahmad, S.K. Ntouyas, A boundary value problem of fractional differential inclusions with three-point nonlocal fractional boundary conditions, *Electr. J. Math. Anal. Appl.* **1** (2013), 1-15.

[36] B. Ahmad, S.K. Ntouyas, Boundary value problems for fractional differential equations with integral, ordinary-fractional flux boundary conditions, *J. Nonlinear Sci. Appl.* **9** (2016), 3622-3637.

[37] B. Ahmad, S.K. Ntouyas, Fractional-order multivalued problems with non-separated integral-flux boundary conditions, *Electron. J. Qual. Theory Differ. Equ.* **2015**, No. 26, 17 pp.

[38] B. Ahmad, S.K. Ntouyas, Nonlinear fractional differential equations, inclusions of arbitrary order, multi-strip boundary conditions, *Electron. J. Differential Equations* **2012**, No. 98, 22 pp.

[39] B. Ahmad, S.K. Ntouyas, Existence results for higher order fractional differential inclusions with multi-strip fractional integral boundary conditions, *Electron. J. Qual. Theory Differ. Equ.* **2013**, No. 20, 19 pp.

[40] B. Ahmad, S.K. Ntouyas, R.P. Agarwal, A. Alsaedi, On fractional differential equations, inclusions with nonlocal, average valued (integral) boundary conditions, *Adv. Difference Equ.* **2016**, Paper No. 80, 18 pp.

[41] B. Ahmad, S.K. Ntouyas, A. Alsaedi, Existence theory for nonlocal boundary value problems involving mixed fractional derivatives, *Nonlinear Anal. Model. Control* **24** (2019), 1-21.

[42] B. Ahmad, S.K. Ntouyas, A. Alsaedi, On fully coupled nonlocal multi-point boundary value problems of nonlinear mixed-order fractional differential equations on an arbitrary domain, *Filomat* **32** (2018), 4503-4511.

[43] B. Ahmad, S.K. Ntouyas, A. Alsaedi, New existence results for nonlinear fractional differential equations with three-point integral boundary conditions, *Adv. Difference Equ.* **2011**, Art. ID 107384, 11 pp.

[44] B. Ahmad, S.K. Ntouyas, A. Alsaedi, On a coupled system of fractional differential equations with coupled nonlocal, integral boundary conditions, *Chaos Solitons & Fractals* **83** (2016), 234-241.

[45] B. Ahmad, S.K. Ntouyas, A. Alsaedi, On nonlinear neutral Liouville-Caputo-type fractional differential equations with Riemann-Liouville integral boundary conditions, *J. Appl. Anal.* **25** (2019), 119-130.

[46] B. Ahmad, S.K. Ntouyas, A. Alsaedi, On solvability of a coupled system of fractional differential equations supplemented with a new kind of flux type integral boundary conditions, *J. Comput. Appl. Anal.* **24** (2018), 1304-1312.

[47] B. Ahmad, S.K. Ntouyas, A. Alsaedi, Existence of solutions for fractional differential equations with nonlocal , average type integral boundary conditions, *J. Appl. Math. Comput.* **53** (2017), 129-145.

[48] B. Ahmad, S.K. Ntouyas, A. Alsaedi, Fractional differential equations, inclusions with nonlocal generalized Riemann-Liouville integral boundary conditions, *Int. J. Anal. Appl.* **13** (2017), 231-247.

[49] B. Ahmad, S.K. Ntouyas, A. Alsaedi, Differential equations and inclusions involving mixed fractional derivative operators with four-point nonlocal fractional boundary conditions, *J. Comput. Anal. Appl.* **24** (2019), 1-21.

[50] B. Ahmad, S.K. Ntouyas, A. Alsaedi, A study of nonlinear fractional differential equations of arbitrary order with Riemann-Liouville type multistrip boundary conditions, *Math. Probl. Eng.* **2013**, Art. ID 320415, 9 pp.

[51] B. Ahmad, S.K. Ntouyas, A. Alsaedi, A study of a more general class of nonlocal integro-multipoint boundary value problems of fractional integro-differential inclusions, *Ukrainian Math. J.* (accepted).

[52] B. Ahmad, S.K. Ntouyas, A. Alsaedi, M. Alghanmi, Multi-term fractional differential equations and inclusions with generalized nonlocal fractional integro-differential boundary conditions, *J. Nonlinear Func. Anal.* 2018 (2018), Article ID 36.

[53] B. Ahmed, S.K. Ntouyas, A. Alsaedi, F. Alzahrani, New fractional-order multivalued problem with nonlocal nonlinear flux type integral boundary conditions, *Bound. Value Probl.* **2015**, 2015:83, 16 pp.

[54] B. Ahmad, S.K. Ntouyas, A. Alsaedi, A. Hobiny, Solvability of a coupled system of fractional differential equations with nonlocal, integral boundary conditions, *Fund. Inform.* **150** (2017) 1-18. Special Issue on Theory & Applications of Fractional Fourier Transforms & its Variants.

[55] B. Ahmad, S.K. Ntouyas, A. Alsaedi, W. Shammakh, R. Agarwal, Existence theory for fractional differential equations with non-separated type nonlocal multi-point and multi-strip boundary conditions, *Adv. Difference Equ.* **2018**, Paper No. 89, 20 pp.

[56] B. Ahmad, S.K. Ntouyas, A. Assolami, Caputo type fractional differential equations with nonlocal Riemann-Liouville integral boundary conditions, *J. Appl. Math. Comput.* **41** (2013), 339-350.

[57] B. Ahmad, S.K. Ntouyas, J. Tariboon, A nonlocal hybrid boundary value problem of Caputo fractional integro-differential equations, *Acta Math. Sci. Ser. B (Engl. Ed.)* **36** (2016), 1631-1640.

[58] B. Ahmad, S.K. Ntouyas, J. Tariboon, On hybrid Caputo fractional integro-differential inclusions with nonlocal conditions, *J. Nonlinear Sci. Appl.* **9** (2016), 4235-4246.

[59] B. Ahmad, S.K. Ntouyas, J. Tariboon, A. Alsaedi, Caputo type fractional differential equations with nonlocal Riemann-Liouville, Erdélyi-Kober type integral boundary conditions, *Filomat* **31** (2017), 4515-4529.

[60] B. Ahmad, S.K. Ntouyas, Y. Zhou, A. Alsaedi, A study of fractional differential equations, inclusions with nonlocal Erdélyi-Kober type integral boundary conditions, *Bull. Iranian Math. Soc.* **44** (2018), 1315-1328.

[61] B. Ahmad, S. Sivasundaram, On four-point nonlocal boundary value problems of nonlinear integro-differential equations of fractional order, *Appl. Math. Comput.* **217** (2010), 480-487.

[62] A. Alsaedi, B. Ahmad, Existence of solutions for nonlinear fractional integro-differential equations with three-point nonlocal fractional boundary conditions, *Adv. Difference Equ.* **2010**, Art. ID 691721, 10 pp.

[63] A. Alsaedi, B. Ahmad, Sh. Aljoudi, S.K. Ntouyas, A study of a fully coupled two-parameter system of sequential fractional integro-differential equations with nonlocal integro-multipoint boundary conditions, *Acta Math. Sci. Ser. B (Engl. Ed.)* **39** (2019), 927-944.

[64] A. Alsaedi, B. Ahmad, M. Alghanmi, Extremal solutions for generalized Caputo fractional differential equations with Steiltjes-type fractional integro-initial conditions., *Appl. Math. Lett.* **91** (2019), 113-120.

[65] A. Alsaedi, M. Alghanmi, B. Ahmad, S.K. Ntouyas, Generalized Liouville-Caputo fractional differential equations and inclusions with nonlocal generalized fractional integral and multipoint boundary conditions, *Symmetry* **2018**, 10, 667.

[66] A. Alsaedi, S.K. Ntouyas, B. Ahmad, H.H. Alsulami, Existence results for fractional integral inclusions via Nonlinear Alternative for contractive maps, *Bound. Value Probl.* **2014**, 2014:25, 14 pp.

[67] A. Alsaedi, S.K. Ntouyas, B. Ahmad, New existence results for fractional integro-differential equations with nonlocal integral boundary conditions, *Abstr. Appl. Anal.* **2015**, Art. ID 205452, 10 pp.

[68] A. Alsaedi, S.K. Ntouyas, D. Garout, B. Ahmad, Coupled fractional-order systems with nonlocal coupled integral, discrete boundary conditions, *Bull. Malays. Math. Sci. Soc.* **42** (2019), 241-266.

[69] A. Amini-Harandi, Endpoints of set-valued contractions in metric spaces, *Nonlinear Anal.* **72** (2010), 132-134.

[70] T.M. Atanackovic, B. Stankovic, On a differential equation with left and right fractional derivative operators, *Fract. Calc. Appl. Anal.* **10** (2007), 139-150.

[71] A.V. Bicadze, A.A. Samarskii, Some elementary generalizations of linear elliptic boundary value problems (Russian), *Anal. Dokl. Akad. Nauk SSSR*, **185** (1969), 739-740.

[72] D.W. Boyd, J.S.W. Wong, On nonlinear contractions, *Proc. Amer. Math. Soc.* **20** (1969), 458-464.

[73] A. Bressan, G. Colombo, Extensions and selections of maps with decomposable values, *Studia Math.* **90** (1988), 69-86.

[74] T. Burton, C. Kirk, A fixed point theorem of Krasnosel'skiĭ-Schaefer type, *Math. Nachr.* **189** (1998), 23-31.

[75] L. Byszewski, Theorems about existence and uniqueness of solutions of a semilinear evolution nonlocal Cauchy problem, *J. Math. Anal. Appl.* **162** (1991), 494-505.

[76] J.R. Cannon, The solution of the heat equation subject to the specification of energy, *Q. Appl. Math.* **21** (1963), 155-160.

[77] A. Carvalho, C.M.A. Pinto, A delay fractional order model for the co-infection of malaria and HIV/AIDS, *Int. J. Dyn. Control* **5** (2017), 168-186.

[78] C. Castaing, M. Valadier, *Convex Analysis and Measurable Multifunctions*, Lecture Notes in Mathematics 580, Springer-Verlag, Berlin-Heidelberg-New York, 1977.

[79] R. Yu. Chegis, Numerical solution of a heat conduction problem with an integral condition, *Liet. Mat. Rink.* **24** (1984), 209-215.

[80] H. Covitz, S.B. Nadler Jr., Multivalued contraction mappings in generalized metric spaces, *Israel J. Math.* **8** (1970), 5-11.

[81] K. Deimling, *Multivalued Differential Equations*, Walter De Gruyter, Berlin-New York, 1992.

[82] K. Deimling, *Nonlinear Functional Analysis*, Springer-Verlag, New York, 1985.

[83] A.S. Deshpande, V. Daftardar-Gejji, On disappearance of chaos in fractional systems, *Chaos Solitons Fractals* **102** (2017), 119-126.

[84] Y. Ding, Z. Wang, H. Ye, Optimal control of a fractional-order HIV-immune system with memory, *IEEE Trans. Contr. Sys. Techn.* **20** (2012), 763-769.

[85] B.C. Dhage, A fixed point theorem in Banach algebras with applications to functional integral equations, *Kyungpook Math. J.* **44** (2004), 145-155.

[86] B.C. Dhage, On solvability of operator inclusions $x \in AxBx+Cx$ in Banach algebras and differential inclusions, *Commun. Appl. Anal.* **14** (2010), 567-596.

[87] M. Faieghi, S. Kuntanapreeda, H. Delavari, D. Baleanu, LMI-based stabilization of a class of fractional-order chaotic systems, *Nonlinear Dynam.* **72** (2013), 301-309.

[88] M. Frigon, Théorèmes d'existence de solutions d'inclusions différentielles, *Topological Methods in Differential Equations and Inclusions* (edited by A. Granas and M. Frigon), NATO ASI Series C, Vol. 472, Kluwer Acad. Publ., Dordrecht, (1995), 51-87.

[89] Z.M. Ge, C.Y. Ou, Chaos synchronization of fractional order modified Duffing systems with parameters excited by a chaotic signal, *Chaos Solitons Fractals* **35** (2008), 705-717.

[90] W. Glockle, T. Nonnenmacher, A fractional calculus approach to self-similar protein dynamics, *Biophys. J.* **68** (1995), 46-53.

[91] L. Gorniewicz,*Topological Fixed Point Theory of Multivalued Mappings, Mathematics and its Applications,* 495, Kluwer Academic Publishers, Dordrecht, 1999.

[92] A. Granas, J. Dugundji, *Fixed Point Theory*, Springer-Verlag, New York, 2003.

[93] D. Guo, V. Lakshmikantham, *Nonlinear Problems in Abstract Cones*, Academic Press, New York, 1988.

[94] S. Heikkila, V. Lakshmikantham, *Monotone Iterative Techniques for Discontinuous Nonlinear Differential Equations*, Marcel Dekker, Inc., New York, 1994.

[95] B. Henry, S. Wearne, Existence of Turing instabilities in a two-species fractional reaction-diffusion system, *SIAM J. Appl. Math.* **62** (2002) 870-887.

[96] N. Heymans, J.C. Bauwens, Fractal rheological models and fractional differential equations for viscoelastic behavior, *Rheol. Acta* **33** (1994), 210-219.

[97] R. Hilfer, *Anomalous transport: Foundations and applications, Anomalous Transport: Foundations and Applications* (Eds. R. Klages, G Radons and I.M. Sokolov), Wiley-VCH, 2008.

[98] Sh. Hu, N. Papageorgiou, *Handbook of Multivalued Analysis, Theory I*, Kluwer, Dordrecht, 1997.

[99] V.A. Il'in, E.I. Moiseev, A nonlocal boundary value problem of the first kind for the Sturm-Liouville operator in differential and difference interpretations (Russian), *Differentsial'nye Uravneniya* **23** (1987), 1198-1207.

[100] N.I. Ionkin, The solution of a certain boundary value problem of the theory of heat conduction with a nonclassical boundary condition (Russian), *Differentsial'nye Uravneniya* **13** (1977), 294-304.

[101] A. Jakubowska, J. Walczak, Analysis of the transient state in a series circuit of the class $RL_{\beta}C_{\alpha}$, *Circ. Syst. Signal Pr.* **35** (2016), 1831-1853.

[102] T. Jankowski, Boundary problems for fractional differential equations, *Appl. Math. Lett.* **28** (2014), 14-19.

[103] F. Jarad, T. Abdeljawad, D. Baleanu, On the generalized fractional derivative operators and their caputo modification, *J. Nonlinear Sci. Appl.* **10** (2017), 2607-2619.

[104] M. Javidi, B. Ahmad, Dynamic analysis of time fractional order phytoplankton-toxic phytoplankton–zooplankton system, *Ecol. Model.* **318** (2015), 8-18.

[105] U.N. Katugampola, New Approach to a generalized fractional integral, *Appl. Math. Comput.* **218** (2011), 860-865.

[106] U.N. Katugampola, A new approach to generalized fractional derivative operators, *Bull. Math. Anal. Appl.* **6** (2014), 1-15.

[107] U.N. Katugampola, New fractional integral unifying six existing fractional integrals, arXiv:1612.08596v1 [math.CA] 22 Dec 2016.

[108] A.A. Kilbas, H.M. Srivastava, J.J. Trujillo, *Theory and Applications of Fractional Differential Equations*, North-Holland Mathematics Studies, 204. Elsevier Science B.V., Amsterdam, 2006.

[109] M. Kisielewicz, *Differential Inclusions and Optimal Control*, Kluwer, Dordrecht, The Netherlands, 1991.

[110] J. Klafter, S.C. Lim, R. Metzler (Eds.), *Fractional Dynamics in Physics*, World Scientific, Singapore, 2011.

[111] H. Kober, On fractional integrals and derivatives, *Quart. J. Math. Oxford*, (1940), Ser. ll, 193-211.

[112] M.A. Krasnoselskii, Two remarks on the method of successive approximations, *Uspekhi Mat. Nauk* **10** (1955), 123-127.

[113] M.A. Krasnoselskii, P.P. Zabreiko, *Geometrical methods of nonlinear analysis*, Springer-verlag, New York, 1984.

[114] A. Lasota, Z. Opial, An application of the Kakutani-Ky Fan theorem in the theory of ordinary differential equations, *Bull. Acad. Polon. Sci. Ser. Sci. Math. Astronom. Phys.* **13** (1965), 781-786.

[115] R.W. Leggett and L.R. Williams, Multiple positive fixed points of nonlinear operators on ordered Banach spaces, *Indiana Univ. Math. J.* **28** (1979), 673-688.

[116] S. Liang and J. Zhang, Existence of three positive solutions of m-point boundary value problems for some nonlinear fractional differential equations on an infinite interval, *Comput. Math. Appl.* **61** (2011) 3343-3354.

[117] S.C. Lim, C.H. Eab, *Fractional quantum fields. Handbook of fractional calculus with applications*, Vol. 5, 237-256, De Gruyter, Berlin, 2019.

[118] Y.S. Liu, Existence and unboundedness of positive solutions for singular boundary value problems on half-line, *Appl. Math. Comput.* **144** (2003), 543-556.

[119] B. Lupinska, T. Odzijewicz, A Lyapunov-type inequality with the Katugampola fractional derivative operator, *Math. Methods Appl. Sci.* **41** (2018), 8985-8996.

[120] F. Mainardi, *Some basic problems in continuum and statistical mechanics, in: A. Carpinteri, F. Mainardi (Eds.), Fractals and Fractional Calculus in Continuum Mechanics*, Springer, Berlin, 1997.

[121] T. Matsuzaki, M. Nakagawa, A chaos neuron model with fractional differential equation, *J. Phys. Soc. Jpn.* **72** (2003), 2678-2684.

[122] F. Nicoud, T. Schfonfeld, Integral boundary conditions for unsteady biomedical CFD applications, *Int. J. Numer. Meth. Fluids* **40** (2002), 457-465.

[123] S.K. Ntouyas, A. Alsaedi, B. Ahmad, Existence theorems for mixed Riemann-Liouville and Caputo fractional differential equations and inclusions with nonlocal fractional integro-differential boundary conditions, *Fractal Fract.* **2019**,3, 21.

[124] S.K. Ntouyas, H.H. Al-Sulami, A study of coupled systems of mixed order fractional differential equations and inclusions with coupled integral fractional boundary conditions, *Adv. Difference Equ.* (2020) **2020:73**.

[125] S.K. Ntouyas, B. Ahmad, M. Alghanmi, A. Alsaedi, Generalized fractional differential equations and inclusions equipped with nonlocal generalized fractional integral boundary conditions, *Topol. Methods Nonlinear Anal.* **54**, (2019), 1051-1073.

[126] N. Nyamoradi, M. Javidi, B. Ahmad, Dynamics of SVEIS epidemic model with distinct incidence, *Int. J. Biomath.* **8** (2015), no. 6, 1550076, 19 pp.

[127] D. O'Regan, Fixed-point theory for the sum of two operators, *Appl. Math. Lett.* **9** (1996), 1-8.

[128] I. Petras, "Fractional-Order Systems" in Fractional-Order Nonlinear Systems, Nonlinear Physical Science, pp. 43-54, Springer Berlin Heidelberg, Berlin, Heidelberg, 2011.

[129] I. Petras, R.L. Magin, Simulation of drug uptake in a two compartmental fractional model for a biological system, *Commun. Nonlinear Sci. Numer.* **16** (2011), 4588-4595.

[130] A. Petrusel, Fixed points and selections for multivalued operators, *Seminar on Fixed Point Theory Cluj-Napoca* **2** (2001), 3-22.

[131] W.V. Petryshyn, P.M. Fitzpatric, A degree theory, fixed point theorems, and mapping theorems for multivalued noncompact maps, *Trans. Amer. Math. Soc.*, **194** (1974), 1-25.

[132] S. Picozzi, B.J. West, Fractional Langevin model of memory in financial markets, *Phys. Rev. E* **66** (2002), 46-118.

[133] D. Qarout, B. Ahmad, A. Alsaedi, Existence theorems for semi-linear Caputo fractional differential equations with nonlocal discrete and integral boundary conditions, *Fract. Calc. Appl. Anal.* **19** (2016), 463-479.

[134] T.V. Renterghem, D. Botteldooren, K. Verheyen, Road traffic noise shielding by vegetation belts of limited depth. *J. Sound Vib.* **331** (2012), 2404-2425.

[135] B.N. Sadovskii, On a fixed point principle, *Funct. Anal. Appl.* **1** (1967), 74-76.

[136] R. Schumer, D. Benson, M.M. Meerschaert, S.W. Wheatcraft, Eulerian derivative of the fractional advection-dispersion equation, *J. Contam. Hydrol.* **48** (2001), 69-88.

[137] D.R. Smart, *Fixed Point Theorems*, Cambridge University Press, 1980.

[138] I.M. Sokolov, J. Klafter, A. Blumen, Fractional kinetics, *Phys. Today* **55** (2002), 48-54.

[139] X. Su, Boundary value problem for a coupled system of nonlinear fractional differential equations, *Appl. Math. Lett.* **22** (2009), 64-69.

[140] P.J. Torvik, R.L. Bagley, On the appearance of the fractional derivative operator in the behavior of real materials, *J. Appl. Mech.* **51** (1984), 294-298.

[141] V.E. Tarasov, Fractional integro-differential equations for electromagnetic waves in dielectric media, *Theor. Math. Phys.* **158** (2009) , 355-359.

[142] G.S. Wang, A.F. Blom, A strip model for fatigue crack growth predictions under general load conditions, *Eng. Fract. Mech.* **40** (1991), 507-533.

[143] G. Wang, B. Ahmad, L. Zhang, On nonlocal integral boundary value problems for impulsive nonlinear differential equations of fractional order, *Fixed Point Theory* **15** (2014), 265-284.

[144] G. Wang, B. Ahmad, L. Zhang, A coupled system of nonlinear fractional differential equations with multipoint fractional boundary conditions on an unbounded domain, *Abstr. Appl. Anal. 2012, Article ID 248709, 11 p. (2012).*

[145] G. Wang, K. Pei, R.P. Agarwal, L. Zhang, B. Ahmad, Nonlocal Hadamard fractional boundary value problem with Hadamard integral and discrete boundary conditions on a half-line, *J. Comput. Appl. Math.* **343** (2018), 230-239.

[146] R. Wegrzyk, Fixed point theorems for multifunctions and their applications to functional equations, *Dissertationes Math. (Rozprawy Mat.)* **201** (1982), 28p.

[147] A.M. Yang, Y. Han, Y.Z. Zhang, L.T. Wang, D. Zhang, X.J. Yang, On nonlocal fractional Volterra integro-differential equations in fractional steady heat transfer, *Therm. Sci.* **20** (2016), S789-S793.

[148] E. Yusufoglu, I. Turhan, A mixed boundary value problem in orthotropic strip containing a crack, *J. Franklin Inst.* **349** (2012), 2750-2769.

[149] L. Zhang, B. Ahmad, G. Wang, Successive iterations for positive extremal solutions of nonlinear fractional differential equations on a half-line, *Bull. Aust. Math. Soc.* **91** (2015), 116-128.

[150] L. Zhang, B. Ahmad, G. Wang, Existence and approximation of positive solutions for nonlinear fractional integro-differential boundary value problems on an unbounded domain, *Appl. Comput. Math.* **15** (2016), 149-158.

[151] L. Zhang, B. Ahmad, G. Wang, The existence of an extremal solution to a nonlinear system with the right-handed Riemann-Liouville fractional derivative operator, *Appl. Math. Lett.* **31** (2014), 1-6.

[152] L. Zhang, B. Ahmad, G. Wang, R.P. Agarwal, Nonlinear fractional integro-differential equations on unbounded domains in a Banach space, *J. Comput. Appl. Math.* **249** (2013), 51-56.

[153] X.K. Zhao, W.G. Ge, Unbounded solutions for a fractional boundary value problem on the infinite interval, *Acta Appl. Math.* **109** (2010), 495-505.

[154] E. Zeidler, *Nonlinear Functional Analysis and Its Application: Fixed Point-Theorems, vol.1* Springer, New York, 1986.

[155] L. Zheng, X. Zhang, Modeling and Analysis of Modern Fluid Problems, *Mathematics in Science and Engineering*, Elsevier/Academic Press: London, UK, 2017.

Index

Lightning Source UK Ltd.
Milton Keynes UK
UKHW051343280321
381105UK00002B/5

9 789811 230400